Theory of Machines and Mechanisms

Uniquely comprehensive and precise, this thoroughly updated sixth edition of the well-established and respected textbook is ideal for the complete study of the kinematics and dynamics of machines. With a strong emphasis on intuitive graphical methods, and accessible approaches to vector analysis, students are given all the essential background, notation, and nomenclature needed to understand the various independent technical approaches that exist in the field of mechanisms, kinematics, and dynamics, which are presented with clarity and coherence. This revised edition features updated coverage, and new worked examples alongside over 840 figures, over 620 end-of-chapter problems, and a solutions manual for instructors.

The late **John J. Uicker, Jr.** was Professor Emeritus of Mechanical Engineering at the University of Wisconsin-Madison. A pioneering researcher on matrix methods of linkage analysis, he was the first person to derive the general dynamic equations of motion for rigid-body articulated mechanical systems. He served on several national committees of ASME and SAE. John was a founding member of the US Council for the Theory of Machines and Mechanisms, and served as Editor-in-Chief of the federation journal *Mechanism and Machine Theory*.

Gordon R. Pennock is Associate Professor of Mechanical Engineering at Purdue University. He is a member of the Commission on Standards and Terminology, the International Federation for the Theory of Machines and Mechanisms. He has also served as the Technical Committee Chairman of Mechanical Design, Internal Combustion Engine Division, and Chairman of the Mechanisms and Robotics Committee, ASME. Gordon is a fellow of ASME, a fellow of SAE, and a fellow and chartered engineer of the Institution of Mechanical Engineers, United Kingdom.

The late **Joseph E. Shigley** was Professor Emeritus of Mechanical Engineering at the University of Michigan and a fellow of ASME. He received the Mechanisms Committee Award in 1974, the Worcester Reed Warner medal in 1977, and the Machine Design Award in 1985. He was author of eight books, including *Mechanical Engineering Design* (with Charles R. Mischke) and *Applied Mechanics of Materials*, and was Co-editor-in-Chief of the *Standard Handbook of Machine Design*.

"In the sixth edition of this classic and comprehensive machine design text, the authors have produced an update that stays true to the strengths of the previous editions (e.g., kinematic coefficients) and incorporates state of the art advances to both content and presentation."

Pierre Larochelle, *South Dakota School of Mines & Technology*

"The analytical approaches implemented in this book paired with the graphical methods facilitate learning for students. The order of sections in both kinematics and kinetics chapters are well thought out. The chapters on cams and gears are very inclusive. The example problems are very practical, ranging from easy to complicated."

Ahmad Ghasemloonia, *University of Calgary*

"This is the most comprehensive undergraduate textbook available on the theory of mechanisms and their kinematics. It covers linkages, cams, gears, engine dynamics, and more with rigorous mathematics, yet it is sufficiently thorough to serve as an introduction to the material for students seeing it for the first time."

Christopher Barrett, *Mississippi State University*

Theory of Machines and Mechanisms

SIXTH EDITION

John J. Uicker, Jr.
University of Wisconsin, Madison

Gordon R. Pennock
Purdue University, Indiana

Joseph E. Shigley

CAMBRIDGE
UNIVERSITY PRESS

Shaftesbury Road, Cambridge CB2 8EA, United Kingdom

One Liberty Plaza, 20th Floor, New York, NY 10006, USA

477 Williamstown Road, Port Melbourne, VIC 3207, Australia

314–321, 3rd Floor, Plot 3, Splendor Forum, Jasola District Centre, New Delhi – 110025, India

103 Penang Road, #05–06/07, Visioncrest Commercial, Singapore 238467

Cambridge University Press is part of Cambridge University Press & Assessment,
a department of the University of Cambridge.

We share the University's mission to contribute to society through the pursuit of
education, learning and research at the highest international levels of excellence.

www.cambridge.org
Information on this title: www.cambridge.org/highereducation/isbn/9781009303675

DOI: 10.1017/9781009303644

First and Second editions © McGraw-Hill 1980, 1995
Third, Fourth and Fifth editions © Oxford University Press 2003, 2010, 2018
Sixth edition © Cambridge University Press & Assessment 2024

Printed in the United Kingdom by CPI Group Ltd, Croydon, CR0 4YY

A catalogue record for this publication is available from the British Library.

Library of Congress Cataloging-in-Publication Data
Names: Uicker, John Joseph, author. | Pennock, G. R., author. | Shigley, Joseph Edward, author.
Title: Theory of machines and mechanisms / John J. Uicker Jr., University of Wisconsin, Madison, Gordon R. Pennock,
    Purdue University, Indiana, Joseph E. Shigley.
Description: Sixth edition. | Cambridge, United Kingdom ; New York, NY : Cambridge University Press, 2023. |
    Includes bibliographical references and index.
Identifiers: LCCN 2022050177 (print) | LCCN 2022050178 (ebook) | ISBN 9781009303675 (hardback) |
    ISBN 9781009303644 (epub)
Subjects: LCSH: Mechanical engineering.
Classification: LCC TJ145 .U33 2023 (print) | LCC TJ145 (ebook) | DDC 621.8/1–dc23/eng/20221101
LC record available at https://lccn.loc.gov/2022050177
LC ebook record available at https://lccn.loc.gov/2022050178

ISBN 978-1-009-30367-5 Hardback

Additional resources for this publication at www.cambridge.org/MaM6.

This textbook is dedicated to the memory of my late father, John J. Uicker, Emeritus Dean of Engineering, University of Detroit; to my late mother, Elizabeth F. Uicker; and to my six children, Theresa A. Zenchenko, John J. Uicker III, Joseph M. Uicker, Dorothy J. Winger, Barbara A. Peterson, and Joan E. Horne.

John J. Uicker, Jr.

This work is also dedicated first and foremost to my wife, Mollie B., and my son, Callum R. Pennock. The work is also dedicated to my friend and mentor, the late Dr. An (Andy) Tzu Yang and my colleagues in the School of Mechanical Engineering, Purdue University, West Lafayette, Indiana.

Gordon R. Pennock

Finally, this text is dedicated to the memory of the late **Joseph E. Shigley**, Professor Emeritus, Mechanical Engineering Department, University of Michigan, Ann Arbor. Although this sixth edition contains significant changes from earlier editions, much of the text remains consistent with his previous writings.

John J. Uicker, Jr., July 11, 1938 – April 25, 2023

My professional association with Professor Uicker commenced during the summer of 2000 while attending a conference at Trinity College, Cambridge University, England. John asked me to join him as coauthor of the third edition of *Theory of Machines and Mechanisms*. I was honored to accept this invitation from such a highly respected educator and researcher.

Professor Uicker received his BME degree from the University of Detroit and his MS and PhD degrees in mechanical engineering from Northwestern University. He joined the University of Wisconsin-Madison faculty in 1967, where he served until his retirement in 2007. He then received Professor Emeritus of Mechanical Engineering status from the university. His teaching and research specialties were in solid geometric modeling and the modeling of mechanical motion and their application to computer-aided design and manufacture. He was a registered Mechanical Engineer in the state of Wisconsin and served for many years as an active consultant to industry. As an ASEE Resident Fellow he spent 1972–1973 at the Ford Motor Company. He was awarded a Fulbright–Hayes Senior Lectureship and in 1978–1979 was a visiting professor at the Cranfield Institute of Technology, Cranfield, England. He was one of the founding members of the US Council for the Theory of Machines and Mechanisms and of IFToMM (the International Federation for the Promotion of Mechanism and Machine Science) and served for several years as Editor-in-Chief of the federation journal *Mechanism and Machine Theory*.

Professor Uicker was the pioneering researcher on matrix methods of linkage analysis and was the first to derive the general dynamic equations of motion for rigid-body articulated mechanical systems. He generated a number of design automation programs and made them available to the engineering profession. The solid modeling software system known as Geometric Modeling of Solids, SWIFT, the casting solidification software, and IMP, the mechanism simulation software, are all supported and distributed by the Wisconsin company JML Research, Inc. In addition to advising graduate students, John made important contributions through formal classroom instruction. In 1979 he started the first course on Computer Graphics at the University of Wisconsin and he was instrumental in initiating a new era of computing for education on the university campus. The Computer-Aided Engineering Center, used by students and staff of the College of Engineering for education and research, was formed as a result of his activities. He served as the Director for the first several years of its existence.

Professor Uicker made numerous contributions to the American Society of Mechanical Engineers (ASME) Design Automation Division through the research that he and his graduate students conducted over the years. A large number of these past students are currently faculty members who have made significant contributions to, and continue to advance, the field of design automation. John was actively involved with ASME for a period of some 40 years and served on several national committees of both ASME and SAE (the Society of Automotive Engineers). He made many important contributions to the Design Automation Division of ASME since his first journal paper (with J. Denavit and R.S. Hartenberg) in 1964, entitled 'An iterative method for the displacement analysis of spatial mechanisms' (*ASME Journal of Applied Mechanics, Series E*, Vol. 31, No. 2, pp. 309–314, June 1964). The authors showed, for the first time, a numerical method for the position solution of spatial linkages. Another of John's papers, entitled 'Dynamic force analysis of spatial linkages' (*ASME Journal of Applied Mechanics, Series E*, Vol. 34, No. 2, pp. 418–424, June 1967) was awarded a Certificate of Commendation by the US Army Munitions Command and judged a *historic* paper by the Mechanisms Committee, ASME Design Division.

Professor Uicker received the SAE Ralph R. Teetor Educational Award in 1969, the ASME Design Engineering Division Mechanisms and Robotics Committee Award in 2004, the ASME Fellow Award in 2007, and the ASME Machine Design Award in 2018. He also received the outstanding teaching award on two occasions, outstanding research publications on three occasions, and historically significant publications on two separate occasions.

Professor Uicker made significant contributions to the kinematics community and he will be greatly missed by his colleagues in the machine design family. I am proud to have known him as a friend and colleague for more than two decades. He was in the final stages of proofreading the sixth edition of *Theory of Machines and Mechanisms* at the time of his death. This book is dedicated to his memory.

Gordon R. Pennock
May 3, 2023

Contents

Preface

The tremendous growth of scientific knowledge over the past 50 years has resulted in an intense pressure on the engineering curricula of many universities to substitute "modern" subjects in place of subjects perceived as weaker or outdated. The result is that, for some, the kinematics and dynamics of machines has remained a critical component of the curriculum and a requirement for all mechanical engineering students, while at others, a course on these subjects is only made available as an elective topic for specialized study by a small number of engineering students. Some schools, depending largely on the faculty, require a greater emphasis on mechanical design at the expense of depth of knowledge in analysis techniques. Rapid advances in technology, however, have produced a need for a textbook that satisfies the requirement of new and changing course structures.

Much of the new knowledge in the theory of machines and mechanisms currently exists in a large variety of technical journals and manuscripts, each couched in its own singular language and nomenclature and each requiring additional background for clear comprehension. It is possible that the individual published contributions could be used to strengthen engineering courses if the necessary foundation was provided and a common notation and nomenclature was established. These new developments could then be integrated into existing courses to provide a logical, modern, and comprehensive whole. The purpose of this book is to provide the background that will allow such an integration.

This book is intended to cover that field of engineering theory, analysis, design, and practice that is generally described as mechanisms or as kinematics and dynamics of machines. Although this text is written primarily for students of mechanical engineering, the content can also be of considerable value to practicing engineers throughout their professional careers.

To develop a broad and basic comprehension the text presents numerous methods of analysis and synthesis that are common to the literature of the field. The authors have included graphic methods of analysis and synthesis extensively throughout the book because they are firmly of the opinion that graphic methods provide visual feedback that enhances the student's understanding of the basic nature of, and interplay between, the underlying equations. Therefore, graphic methods are presented as one possible solution technique, but are always accompanied by vector equations defined by the fundamental laws of mechanics, rather than as graphical "tricks" to be learned by rote and applied blindly. In addition, although graphic techniques, performed by hand, may lack accuracy, they can be performed quickly and even inaccurate sketches can often provide reasonable estimates of a solution and can be used to check the results of analytic or numeric solution techniques.

The authors also use conventional methods of vector analysis throughout the book, both in deriving and presenting the governing equations and in their solution. Raven's methods using complex algebra for the solution of two-dimensional vector equations are included because of their compactness, because of the ease of taking derivatives, because they are employed so frequently in the literature, and because they are so easy to program for computer evaluation. In the chapter dealing with three-dimensional kinematics and robotics, the authors present a brief introduction to Denavit and Hartenberg's methods using transformation matrices.

Another feature of this text is its focus on the method of kinematic coefficients, which are derivatives of motion variables with respect to the input position variable rather than with respect to time. The authors believe that this analytic technique provides several important advantages, namely: (1) kinematic coefficients clarify for the student those parts of a motion problem that are kinematic (geometric) in their nature, and clearly separate these from the parts that are dynamic or speed-dependent; and (2) kinematic coefficients help to integrate the analysis of different types of mechanical systems, such as gears, cams, and linkages, which might not otherwise seem similar.

One dilemma that all writers on the subject of this book have faced is how to distinguish between the motions of different points of a single moving body and the motions of coincident points of different moving bodies. In other texts it has been customary to describe both of these as "relative motion"; however, because they are two distinctly different situations and are described by different equations, this causes the student confusion in distinguishing between them. We believe that we have greatly relieved this problem by the introduction of the terms *motion difference* and *apparent motion* and by using different terminology and different notation for the two cases. Thus, for example, this book uses the two terms *velocity difference* and *apparent velocity*, instead of the term "relative velocity," which will not be found when speaking rigorously. This approach is introduced beginning with position and displacement, used extensively in the chapter on velocity, and brought to fulfillment in the chapter on accelerations, where the Coriolis component *always* arises in, and *only* arises in, the apparent acceleration equation.

Access to personal computers, programmable calculators, and laptop computers is commonplace and is of considerable importance to the material of this book. Yet engineering educators have told us very forcibly that they do not want computer programs included in the text. They prefer to write their own programs and they expect their students to do so as well. Having programmed almost all the material in the book many times, we also understand that the book should not include such programs and thus become obsolete with changes in computers or programming languages.

The authors have endeavored to use US customary units and SI units in about equal proportions throughout the book. However, there are certain exceptions. For example, in Chapter 14 (Dynamics of Reciprocating Engines) only SI units are presented because engines are designed for an international marketplace, even by US companies. Therefore, they are always rated in kilowatts rather than horsepower; they have displacements in liters rather than cubic inches; and their cylinder pressures are measured in kilopascals rather than pounds per square inch. In addition, appropriate units are shown with all numeric values throughout the solution of all example problems.

Part I of this book deals mostly with theory, nomenclature, notation, and methods of analysis. Serving as an introduction, Chapter 1 tells what a mechanism is, what a mechanism can do, how mechanisms can be classified, and what some of their limitations are. Chapters 2, 3, and 4 are concerned totally with analysis, specifically with kinematic analysis, because they cover position, velocity, and acceleration analyses, respectively, of single-degree-of-freedom planar mechanisms. Chapter 5 expands this background to include multi-degree-of-freedom planar mechanisms.

Part II of the book goes on to demonstrate engineering applications involving the selection, the specification, the design, and the sizing of mechanisms to accomplish specific motion objectives. This part includes chapters on cam systems, gears, gear trains, synthesis of linkages, spatial mechanisms, and an introduction to robotics. Chapter 6 is a study of the geometry, kinematics, proper design of high-speed cam systems, and now includes material on the dynamics of elastic cam systems. Chapter 7 studies the geometry and kinematics of spur gears, particularly of involute tooth profiles, their manufacture, and proper tooth meshing, and then studies gear trains, with an emphasis on epicyclic and differential gear trains. Chapter 8 expands this background to include helical gears, bevel gears, worms, and worm gears. Chapter 9 is an introduction to the kinematic synthesis of planar linkages. Chapter 10 is a brief introduction to the kinematic analysis of spatial mechanisms and robotics, including the forward and inverse kinematics problems.

Part III of the book adds the dynamics of machines. In a sense this part is concerned with the consequences of the mechanism design specifications. In other words, having designed a machine by selecting, specifying, and sizing the various components, what happens during the operation of the machine? What forces are produced? Are there any unexpected operating results? Will the proposed design be satisfactory in all respects? Chapter 11 presents the static force analysis of machines. This chapter also includes sections focusing on the buckling of two-force members subjected to axial loads. Chapter 12 studies the planar and spatial aspects of the dynamic force analysis of machines. Chapter 13 then presents the vibration analysis of mechanical systems. Chapter 14 is a more detailed study of one particular type of mechanical system, namely the dynamics of both single- and multi-cylinder reciprocating engines. Chapter 15 next addresses the static and dynamic balancing of rotating and reciprocating systems. Finally, Chapter 16 is on the study of the dynamics of flywheels, governors, and gyroscopes.

As with all texts, the subject matter of this book also has limitations. Probably the clearest boundary on the coverage in this text is that it is limited to the study of rigid-body mechanical systems. It does study planar multibody systems with movable connections or constraints between them. However, all motion effects are assumed to come within the connections; the shapes of the individual bodies are assumed constant, except for the dynamics of elastic cam systems. This assumption is necessary to allow the separate study of kinematic effects from those of dynamics. Because each individual body is assumed rigid, it can have no strain; therefore, except for buckling of axially loaded members, the study of stress is also outside the scope of this text. It is hoped, however, that courses using this text can provide background for the later study of stress, strength, fatigue life, modes of failure, lubrication, and other aspects important to the proper design of mechanical systems.

Despite the limitations on the scope of this book, it is still clear that it is not reasonable to expect that all of the material presented here can be covered in a single-semester course. As stated above, a variety of methods and applications have been included to allow the instructor to choose those topics which best fit the course objectives and to still provide a reference for follow-on courses and to help build the student's library. Yet many instructors have asked for suggestions regarding a choice of topics that might fit a 3 hours per week, 15 week course. Two such outlines follow, as used by the authors to teach such courses at their institutions. It is hoped that these might be used as helpful guidelines to assist others in making their own parallel choices.

	Tentative Schedule I	
	Kinematics and Dynamics of Machine Systems	
Week	Topics	Sections
1	Introduction to Mechanisms	1.1–1.10
	Kutzbach and Grashof Criteria	1.6, 1.9
	Advance-to-Return Time Ratio	1.7
	Overlay Method of Synthesis	9.8
2	Vector Loop-Closure Equation	2.6, 2.7
	Velocity Difference Equation	3.1–3.3
	Velocity Polygons; Velocity Images	3.4
3	Apparent Velocity Equation	3.5–3.6, 3.8
	Direct and Rolling Contact Velocity	3.7
4	Instantaneous Centers of Velocity	3.12
	Aronhold–Kennedy Theorem of Three Centers	3.13, 3.14
	Use of Instant Centers to Find Velocities	3.15, 3.16
5	Exam #1	
	Acceleration Difference Equation	4.1–4.3
	Acceleration Polygons; Acceleration Images	4.4
6	Apparent Acceleration Equation	4.5, 4.6
	Coriolis Component of Acceleration	
7	Direct and Rolling Contact Acceleration	4.7, 4.8
	Review of Velocity and Acceleration Analyses	
8	Raven's Method of Kinematic Analysis	2.10, 3.10, 4.10
	Kinematic Coefficients	3.11, 4.11
	Computer Methods in Kinematics	10.9
9	Exam #2	
	Static Forces	11.1–11.6
	Two-, Three-, and Four-Force Members	11.7, 11.8
	Force Polygons	
10	Coulomb Friction Forces in Machines	11.9, 11.10
11	D'Alembert's Principle	12.1–12.4
	Dynamic Forces in Machine Members	12.4, 12.5
12	Introduction to Cam Design	6.1–6.4
	Choice of Cam Profiles; Matching Displacement Curves	6.5–6.8

(cont.)

<table>
<tr><td colspan="3" align="center">Tentative Schedule I</td></tr>
<tr><td colspan="3" align="center">Kinematics and Dynamics of Machine Systems</td></tr>
<tr><td>Week</td><td>Topics</td><td>Sections</td></tr>
<tr><td>13</td><td>First-Order Kinematic Coefficients; Face-Width; Pressure Angle
Second-Order Kinematic Coefficients; Pointing and Undercutting</td><td>6.9
6.10</td></tr>
<tr><td>14</td><td>Exam #3
Introduction to Gearing
Involute Tooth Geometry; Contact Ratio; Undercutting</td><td>
7.1–7.6
7.7–7.9, 7.11</td></tr>
<tr><td>15</td><td>Epicyclic and Differential Gear Trains
Review
Final Exam</td><td>7.15–7.17</td></tr>
</table>

<table>
<tr><td colspan="3" align="center">Tentative Schedule II</td></tr>
<tr><td colspan="3" align="center">Machine Design I</td></tr>
<tr><td>Week</td><td>Topics</td><td>Sections</td></tr>
<tr><td>1</td><td>The World of Mechanisms
Measures of Performance (Indices of Merit)
Quick Return Mechanisms</td><td>1.1–1.6
1.10, 3.19
1.7</td></tr>
<tr><td>2</td><td>Position Analysis. Vector Loops
Newton–Raphson Technique</td><td>2.1–2.7
2.8, 2.11</td></tr>
<tr><td>3</td><td>Velocity Analysis
First-Order Kinematic Coefficients
Instant Centers of Zero Velocity</td><td>3.1–3.9
3.11
3.12–3.17</td></tr>
<tr><td>4</td><td>Rolling Contact, Rack and Pinion, Two Gears
Acceleration Analysis
Second-Order Kinematic Coefficients</td><td>3.10
4.1–4.4
4.5–4.11</td></tr>
<tr><td>5</td><td>Geometry of a Point Path
Kinematic Coefficients for Point Path
Radius and Center of Curvature</td><td>4.15
4.15
4.16</td></tr>
<tr><td>6</td><td>Cam Design
Lift Curve
Exam #1</td><td>6.1–6.4
6.1–6.4</td></tr>
<tr><td>7</td><td>Kinematic Coefficients of the Follower
Roller Follower
Flat-Face Follower</td><td>6.5
6.9, 6.10
6.9, 6.10</td></tr>
<tr><td>8</td><td>Graphic Approach
Two-, Three-, and Four-Force Members
Friction–Force Models</td><td>11.5, 11.6
11.7, 11.8
11.9, 11.10</td></tr>
</table>

(cont.)

A solutions manual is available for instructors only from the resources page of the textbook website.

The authors wish to thank the reviewers for their very helpful criticisms and recommendations.

The many instructors and students who have tolerated previous versions of this book and made their suggestions for its improvement also deserve our continuing gratitude.

The authors would also like to offer our sincere thanks to the production team at Cambridge University Press for their continuing cooperation and assistance in bringing this edition to completion.

John J. Uicker, Jr.
Gordon R. Pennock

Part I

Kinematics and Mechanisms

1   The World of Mechanisms

1.1   Introduction

The theory of machines and mechanisms is an applied science that allows us to understand the relationships between the geometry and motions of the parts of a machine, or mechanism, and the forces that produce these motions. The subject, and therefore this book, divides itself naturally into three parts. Part I, which includes Chapters 1–5, is concerned with mechanisms and the kinematics of mechanisms, which is the analysis of their motions. Part I lays the groundwork for Part II, comprising Chapters 6–10, in which we study methods of designing mechanisms. Finally, in Part III, which includes Chapters 11–16, we take up the study of kinetics, the time-varying forces in machines and the resulting dynamic phenomena that must be considered in their design.

The design of a modern machine is often very complex. In the design of a new engine, for example, the automotive engineer must deal with many interrelated questions. What is the relationship between the motion of the piston and the motion of the crankshaft? What are the sliding velocities and the loads at the lubricated surfaces, and what lubricants are available for this purpose? How much heat is generated, and how is the engine cooled? What are the synchronization and control requirements, and how are they satisfied? What is the cost to the consumer, both for initial purchase and for continued operation and maintenance? What materials and manufacturing methods are used? What are the fuel economy, noise, and exhaust emissions, and do they meet legal requirements? Although all these and many other important questions must be answered before the design is completed, obviously not all can be addressed in a book of this size. Just as people with diverse skills must be brought together to produce an adequate design, so too must many branches of science be brought together. This book assembles material that falls into the science of mechanics as it relates to the design of mechanisms and machines.

1.2   Analysis and Synthesis

There are two completely different aspects of the study of mechanical systems: *design* and *analysis*. The concept embodied in the word "design" is more properly termed *synthesis*, the process of contriving a scheme or a method of accomplishing a given purpose. Design is the process of prescribing the sizes, shapes, material compositions, and arrangements of parts so that the resulting machine will perform the prescribed task.

Although there are many phases in the design process that can be approached in a well-ordered, scientific manner, the overall process is by its very nature as much an art as a science. It calls for imagination, intuition, creativity, judgment, and experience. The role of science in the design process is merely to provide tools to be used by designers as they practice their art.

In the process of evaluating the various interacting alternatives, designers find a need for a large collection of mathematical and scientific tools. These tools, when applied properly, provide more accurate and more reliable information for judging a design than one achieves through intuition or estimation. Thus, the tools are of tremendous help in deciding among alternatives. However, scientific tools cannot make decisions for designers; designers have every right to exert their imagination and creative abilities, even to the extent of overruling the mathematical recommendations.

Probably the largest collection of scientific methods at the designer's disposal fall into the category called *analysis*. These are techniques that allow the designer to critically examine an already existing or proposed design to judge its suitability for a task. Thus, analysis in itself is not a creative science but one of evaluation and rating things already conceived.

We should bear in mind that, although most of our effort may be spent on analysis, the real goal is synthesis: the design of a machine or system. Analysis is simply a tool; however, it is a vital tool and will inevitably be used as one step in the design process.

1.3   Science of Mechanics

The branch of scientific analysis that deals with motion, time, and force is called *mechanics*, and is made up of two parts: statics and dynamics. *Statics* deals with the analysis of stationary systems – that is, those in which time is not a factor – and *dynamics* deals with systems that change with time.

As shown in Fig. 1.1, dynamics is also made up of two major disciplines, first recognized as separate entities by Euler[1] in 1765 [2]:[2]

The investigation of the motion of a rigid body may be conveniently separated into two parts, the one, geometric and the other mechanical. In the first part, the transference of the body from a given posture to any other posture must be investigated without respect to the causes of the motion, and must be represented by analytic formulae, which define the position of each point of the body. This investigation will therefore be referable solely to geometry, or rather to stereotomy [the art of stonecutting, now referred to as descriptive geometry].

It is clear that by the separation of this part of the question from the other, which belongs properly to Mechanics, the determination of the motion from dynamic principles will be made much easier than if the two parts are undertaken conjointly.

---

[1] Leonhard Euler (1707–1783).   [2] Numbers in square brackets refer to references at the end of each chapter.

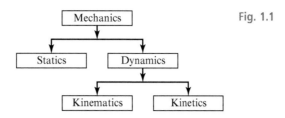

Fig. 1.1

These two aspects of dynamics were later recognized as the distinct sciences of *kinematics* (*cinématique* was a term coined by Ampère[3] and derived from the Greek word *kinema*, meaning motion) and kinetics, and deal with motion and the forces producing the motion, respectively.

The initial problem in the design of a mechanical system, therefore, is understanding the kinematics. *Kinematics* is the study of motion, quite apart from the forces that produce the motion. In particular, kinematics is the study of position, displacement, rotation, speed, velocity, acceleration, and jerk. The study, say, of planetary or orbital motion is also a problem in kinematics, but in this book we shall concentrate our attention on kinematic problems that arise in the design and operation of mechanical systems. Thus, the kinematics of machines and mechanisms is the focus of the next several chapters of this book. In addition, statics and kinetics are also vital parts of a complete design analysis, and they are also covered in detail in later chapters.

It should be carefully noted in the previous quotation that Euler based his separation of dynamics into kinematics and kinetics on the assumption that they deal with *rigid* bodies. It is this very important assumption that allows the two to be treated separately. For flexible bodies, the shapes of the bodies themselves, and therefore their motions, depend on the forces exerted on them. In this situation, the study of force and motion must take place simultaneously, thus significantly increasing the complexity of the analysis.

Fortunately, although all real machine parts are flexible to some degree, machines are usually designed from relatively rigid materials, keeping part deflections to a minimum. Therefore, it is common practice to assume that deflections are negligible and parts are rigid while analyzing a machine's kinematic performance and then, during dynamic analysis when loads are sought, to design the parts so that the assumption is justified. A more detailed discussion of a rigid body compared to a deformable, or flexible, body is presented in the introduction to static force analysis in Section 11.1.

1.4 Terminology, Definitions, and Assumptions

Reuleaux[4] defines a *machine*[5] as a "combination of resistant bodies so arranged that by their means the mechanical forces of nature can be compelled to do work accompanied by certain

---

[3] André-Marie Ampère (1775–1836).

[4] Much of the material of this section is based on definitions originally set down by Franz Reuleaux (1829–1905), a German kinematician whose work marked the beginning of a systematic treatment of kinematics [7].

[5] There appears to be no agreement at all on the proper definition of a machine. In a footnote Reuleaux gives 17 definitions, and his translator gives 7 more and discusses the whole problem in detail [7].

determinate motions." He also defines a *mechanism* as an "assemblage of resistant bodies, connected by movable joints, to form a closed kinematic chain with one link fixed and having the purpose of transforming motion."

Some light can be shed on these definitions by contrasting them with the term *structure*. A structure is also a combination of resistant (rigid) bodies connected by joints, but the purpose of a structure (such as a truss) is not to do work or to transform motion, but to be rigid. A truss can perhaps be moved from place to place and is movable in this sense of the word; however, it has no *internal* mobility. A structure has no *relative motions* between its various links, whereas both machines and mechanisms do. Indeed, the whole purpose of a machine or mechanism is to utilize these relative internal motions in transmitting or transforming motion.

A machine is an arrangement of parts for doing work, a device for applying power or changing the direction of motion. It differs from a mechanism in its purpose. In a machine, terms such as force, torque, work, and power describe the predominant concepts. In a mechanism, though it may transmit power or force, the predominant idea in the mind of the designer is one of achieving a desired motion. There is a direct analogy between the terms *structure*, *mechanism*, and *machine* and the branches of mechanics illustrated in Fig. 1.1. The term *structure* is to statics as the term *mechanism* is to kinematics and as the term *machine* is to kinetics.

We use the word *link* to designate a machine part or a component of a mechanism. As discussed in the previous section, a link is assumed to be completely rigid. Machine components that do not fit this assumption of rigidity, such as springs, usually have no effect on the kinematics of a device, but do play a role in supplying forces. Such parts or components are not called links; they are usually ignored during kinematic analysis, and their force effects are introduced during force analysis (see the analysis of buckling in Sections 11.14–11.18). Sometimes, as with a belt or chain, a machine part may possess one-way rigidity; such a body can be considered a link when in tension but not when under compression.

The links of a mechanism must be connected in some manner in order to transmit motion from the *driver*, or input link, to the *driven*, or *follower* or output link. The connections, the joints between the links, are called *kinematic pairs* (or simply *pairs*), because each joint consists of a pair of mating surfaces, two elements, one mating surface or element being a part of each of the joined links. Thus, we can also define a link as *the rigid connection between two or more joint elements*.

Stated explicitly, the assumption of rigidity is that there can be no relative motion (no change in distance) between two arbitrarily chosen points on the same link. In particular, the relative postures of joint elements on any given link do not change no matter what loads are applied. In other words, the purpose of a link is to hold a constant spatial relationship between its joint elements.

As a result of the assumption of rigidity, many of the intricate details of the actual part shapes are unimportant when studying the kinematics of a machine or mechanism. For this reason, it is common practice to draw highly simplified schematic diagrams that contain important features of the shape of each link, such as the relative locations of joint elements, but that completely subdue the real geometry of the manufactured part. The slider-crank linkage of the internal combustion engine, for example, can be simplified for the purposes of analysis to the schematic diagram shown later in Fig. 1.3b. Such simplified schematics are a great help since they eliminate confusing factors that do not affect the analysis; such diagrams are used extensively throughout this text. However, these schematics also have the drawback of bearing little resemblance to physical hardware. As a result they may give the impression that they represent only academic constructs rather than real machinery. We should continually bear in mind that these simplified diagrams are intended to carry only the minimum necessary information so as not to confuse the issue with unimportant detail (for kinematic purposes) or complexity of the true machine parts.

When several links are connected together by joints, they are said to form a *kinematic chain*. Links containing only two joint elements are called *binary* links, those having three joint elements are called *ternary* links, those having four joint elements are called *quaternary* links, and so on. If every link in a chain is connected to at least two other links, the chain forms one or more closed loops and is called a *closed* kinematic chain; if not, the chain is referred to as *open*. If a chain consists entirely of binary links, it is a *simple closed* chain. *Compound closed* chains include other than binary links and thus form more than a single closed loop.

Recalling Reuleaux's definition of a mechanism, we see that it is necessary to have a closed kinematic chain *with one link fixed*. When we say that one link is fixed, we mean that it is chosen as the frame of reference for the postures and motions of all other links; that is, the motions of all points on the links of the mechanism are measured with respect to the fixed link. This link, in a practical machine, usually takes the form of a stationary platform or base (or a housing rigidly attached to such a base) and is commonly referred to as the *ground, frame*, or *base link*.[6] The question of whether this reference frame is truly stationary (in the sense of being an inertial reference frame) is immaterial in the study of kinematics, but becomes important in the investigation of kinetics, where forces are considered. In either case, once a frame link is designated (and other conditions are met), the kinematic chain becomes a mechanism and, as the driver is moved through various postures, all other links have well-defined motions with respect to the chosen frame of reference. We use the term *kinematic chain* to specify a particular arrangement of links and joints when it is not clear which link is to be treated as the frame. When the frame link is specified, the kinematic chain is called a *mechanism*.

For a mechanism to be useful, the motions between links cannot be completely arbitrary; they too must be constrained to produce the proper relative motions – those chosen by the designer for the particular task to be performed. These desired relative motions are achieved by proper choice of the number of links and the kinds of joints used to connect them. Thus we are led to the concept that, in addition to the distances between successive joints, the nature of the joints themselves and the relative motions they permit are essential in determining the kinematics of a mechanism. For this reason, it is important to look more closely at the nature of joints in general terms, and in particular at several of the more common types.

The controlling factors that determine the relative motions allowed by a given joint are the shapes of the mating surfaces or elements. Each type of joint has its own characteristic shapes for the elements, and each allows a given type of motion, which is determined by the possible

---

[6] In this text, the ground, frame, or base of the mechanism is commonly numbered 1.

ways in which these elemental surfaces can move with respect to each other. For example, the pin joint in Fig. 1.2a has cylindric elements, and, assuming that the links cannot slide axially, these surfaces permit only relative rotational motion. Thus a pin joint allows the two connected links to experience relative rotation about the pin axis. So, too, other joints each have their own characteristic element shapes and relative motions. These shapes restrict the totally arbitrary motion of two unconnected links to some prescribed type of relative motion and form constraining conditions (constraints) on the mechanism's motion.

It should be pointed out that the element shapes may often be subtly disguised and difficult to recognize. For example, a pin joint might include a needle bearing, so that two mating surfaces, as such, are not distinguishable. Nevertheless, if the motions of the individual rollers are not of interest, the motions allowed by the joints are equivalent, and the joints are of the same generic type. Thus the criterion for distinguishing different joint types is the relative motions they permit and not necessarily the shapes of the elements, though these may provide vital clues. The diameter of the pin used (or other dimensional data) is also of no more importance than the exact sizes and shapes of the connected links. As stated previously, the kinematic function of a link is to hold a fixed geometric relationship between its joint elements. Similarly, the only kinematic function of a joint, or pair, is to determine the relative motion between the connected links. All other features are determined for other reasons and are unimportant in the study of kinematics.

When a kinematic problem is formulated, it is necessary to recognize the type of relative motion permitted in each of the joints and to assign to it some variable parameter(s) for measuring or calculating the motion. There will be as many of these parameters as there are degrees of freedom of the joint in question, and they are referred to as *joint variables*. Thus, the joint variable of a pinned joint will be a single angle measured between reference lines fixed in the adjacent links, while a spheric joint will have three joint variables (all angles) to specify its three-dimensional rotation.

Reuleaux separated kinematic pairs into two categories, namely, *higher pairs* and *lower pairs*, with the latter category consisting of the six prescribed types to be discussed next. He distinguished between the categories by noting that lower pairs, such as the pin joint, have surface contact between the joint elements, while higher pairs, such as the connection between a cam and its follower, have line or point contact between the elemental surfaces. This criterion, however, can be misleading (as noted in the case of a needle bearing). We should rather look for distinguishing features in the relative motion(s) that the joint allows between the connected links.

Lower pairs consist of the six prescribed types shown in Fig. 1.2.

The names and the symbols [4] that are commonly employed for the six lower pairs are presented in Table 1.1. The table also includes the number of degrees of freedom and the joint variables that are associated with each lower pair.

The *revolute* or *turning pair*, R (Fig. 1.2a), permits only relative rotation and is often referred to as a pin joint. This joint has one degree of freedom.

The *prism* or *prismatic pair*, P (Fig. 1.2b), permits only relative sliding motion and therefore is often called a sliding joint. This joint also has one degree of freedom.

The *screw* or *helical pair*, H (Fig. 1.2c), permits both rotation and sliding motion. However, it only has one degree of freedom, since the rotation and sliding motions are related by the helix

Table 1.1 **Lower pairs**

Pair	Symbol	Pair variable	Degrees of freedom	Relative motion
Revolute	R	$\Delta\theta$	1	Circular
Prism	P	Δs	1	Rectilinear
Screw	H	$\Delta\theta$ or Δs	1	Helical
Cylinder	C	$\Delta\theta$ and Δs	2	Cylindric
Sphere	S	$\Delta\theta, \Delta\phi, \Delta\psi$	3	Spheric
Flat	F	$\Delta x, \Delta y, \Delta\theta$	3	Planar

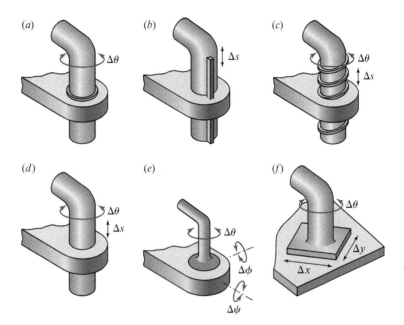

Fig. 1.2 Lower pairs: (*a*) Revolute; (*b*) prism; (*c*) screw; (*d*) cylinder; (*e*) sphere; (*f*) flat pairs.

angle of the thread. Thus, the joint variable may be chosen as either Δs or $\Delta\theta$, but not both. Note that the helical pair reduces to a revolute if the helix angle is made zero, and to a prism if the helix angle is made 90°.

The *cylinder* or *cylindric pair*, C (Fig. 1.2*d*), permits both a rotation and an independent sliding motion. Thus, the cylindric pair has two degrees of freedom.

The *sphere* or *globular pair*, S (Fig. 1.2*e*), is a ball-and-socket joint. It has three degrees of freedom, sometimes taken as rotations about each of the coordinate axes.

The *flat* or *planar pair*, sometimes called an *ebene pair* (German), F (Fig. 1.2*f*), is seldom found in mechanisms in its undisguised form, except at a support point. It has three degrees of freedom, that is, two translations and a rotation.

All other joint types are called higher pairs. Examples include mating gear teeth, a wheel rolling and/or sliding on a rail, a ball rolling on a flat surface, and a cam contacting its follower.

Since an unlimited variety of higher pairs exist, a systematic accounting of them is not a realistic objective. We shall treat each separately as it arises.

Among the higher pairs is a subcategory known as *wrapping pairs*. Examples are the connections between a belt and a pulley, a chain and a sprocket, or a rope and a drum. In each case, one of the links has only one-way rigidity.

The treatment of various joint types, whether lower or higher pairs, includes another important limiting assumption. Throughout the book, we assume that the actual joint, as manufactured, can be reasonably represented by a mathematical abstraction having perfect geometry. That is, when a real machine joint is assumed to be a spheric joint, for example, it is also assumed that there is no "play" or clearance between the joint elements and that any deviation from spheric geometry of the elements is negligible. When a pin joint is treated as a revolute, it is assumed that no axial motion takes place; if it is necessary to study the small axial motions resulting from clearances between real elements, the joint must be treated as cylindrical, thus allowing the axial motion.

The term "mechanism," as defined earlier, can refer to a wide variety of devices, including both higher and lower pairs. A more limited term, however, refers to those mechanisms having only lower pairs; such a mechanism is commonly called a *linkage*. A linkage, then, is connected only by the lower pairs shown in Fig. 1.2.

1.5   Planar, Spheric, and Spatial Mechanisms

Mechanisms may be categorized in several different ways to emphasize their similarities and differences. One such grouping divides mechanisms into planar, spheric, and spatial categories. All three groups have many things in common; the criterion that distinguishes the groups, however, is to be found in the characteristics of the motions of the links.

A *planar mechanism* is one in which all particles describe planar curves in space, and all these curves lie in parallel planes; that is, the loci of all points are planar curves parallel to a single common plane. This characteristic makes it possible to represent the locus of any chosen point of a planar mechanism in its true size and shape in a single drawing or figure. The motion transformation of any such mechanism is called *coplanar*. The planar four-bar linkage, the slider-crank linkage, the plate cam-and-follower mechanism, and meshing gears are familiar examples of planar mechanisms.

Planar mechanisms utilizing only lower pairs are called *planar linkages*; they include only revolute and prismatic joints. Although the planar pair might theoretically be included in a planar linkage, this would impose no constraint on the motion. Planar motion also requires that all revolute axes be normal to the plane of motion, and that all prismatic joint axes be parallel to the plane.

As already pointed out, it is possible to observe the motions of all particles of a planar mechanism in true size and shape from a single direction. In other words, all motions can be represented graphically in a single view. Thus, graphic techniques are well suited to their analysis, and this background is beneficial to the student once mastered. Since spheric and spatial mechanisms do not have this special geometry, visualization becomes more difficult and more powerful techniques must be used for their study.

A *spheric mechanism* is one in which each moving link has a point that remains stationary as the mechanism moves. Also, arbitrary points fixed in each moving link travel on spheric surfaces; the spheric surfaces must all be *concentric*. Therefore, the motions of all these points can be completely described by their radial projections (or shadows) on the surface of a sphere with a properly chosen center. Note that the only lower pairs (Table 1.1) that allow spheric motion are the revolute pair and the spheric pair. In a spheric linkage, the axes of all revolute pairs must intersect at a single point. In addition, a spheric pair center must be concentric with this point and, then, would not produce any constraint on the motions of the other links. Therefore, a *spheric linkage* must be comprised entirely of revolute pairs, and the axes of all such pairs must intersect at a single point. A familiar example of a spheric mechanism is the Hooke universal joint (also referred to as the Cardan joint) shown in Fig. 1.21*b*.

Spatial mechanisms include *no* restrictions on the relative motions of the links. For example, a mechanism that contains a screw joint (Fig. 1.2*c*) must be a spatial mechanism, since the relative motion within a screw joint is helical. An example of a spatial mechanism is the differential screw shown later in Fig. 1.11. Because of the more complex motion characteristics of spatial mechanisms, and since these motions cannot be analyzed graphically from a single viewing direction, more powerful techniques are required for their analysis. Such techniques are introduced in Chapter 10 for a detailed study of spatial mechanisms and robotics.

Since the majority of mechanisms in modern machinery are planar, one might question the need to study these complex mathematical techniques. However, even though the simpler graphic techniques for planar mechanisms may have been mastered, an understanding of the more complex techniques is of value for the following reasons:

1. They provide new, alternative methods that can solve problems in a different way. Thus, they provide a means for checking results. Certain problems by their nature may also be more amenable to one method than another.
2. Methods that are analytic in nature are better suited to solution by a calculator or a digital computer than by graphic techniques.
3. One reason why planar mechanisms are so common is that good methods for the analysis of spatial mechanisms have not been available until relatively recently. Without these methods, the design and application of spatial mechanisms has been hindered, even though they may be inherently better suited to certain applications.
4. We will discover that spatial mechanisms are, in fact, much more common in practice than their formal description indicates.

Consider the planar four-bar linkage (Fig. 1.3*c*), which has four links connected by four revolute pairs whose axes are parallel. This "parallelism" is a mathematical hypothesis; it is not a reality. The axes, as produced in a machine shop – in any machine shop, no matter how precise the machining – are only approximately parallel. If the axes are far out of parallel, there is binding in no uncertain terms, and the linkage moves only because the "rigid" links flex and twist, producing loads in the bearings. If the axes are nearly parallel, the linkage operates because of looseness of the running fits of the bearings or flexibility of the links. A common way of compensating for small nonparallelism is to connect the links with self-aligning bearings, actually spheric joints allowing three-dimensional rotation. Such a "planar" linkage is thus a low-grade spatial linkage.

Thus, the overwhelmingly large category of planar mechanisms and the category of spheric mechanisms are special cases, or subsets, of the all-inclusive category of spatial mechanisms. They occur as a consequence of the special orientations of their joint axes.

1.6   Mobility

One of the first concerns in either the design or the analysis of a mechanism is the number of degrees of freedom, also called the *mobility* of the device. The mobility[7] of a mechanism is the number of input parameters (usually joint variables) that must be controlled independently to bring the device into a particular posture. Ignoring, for the moment, certain exceptions to be mentioned later, it is possible to determine the mobility of a mechanism directly from a count of the number of links and the number and types of joints comprising the mechanism.

To develop this relationship, consider that – before they are connected together – each link of a planar mechanism has three degrees of freedom when moving with planar motion relative to the fixed link. Not counting the fixed link, therefore, an n-link planar mechanism has $3(n-1)$ degrees of freedom before any of the joints are connected. Connecting two of the links by a joint that has one degree of freedom, such as a revolute, has the effect of providing two constraints between the connected links. If the two links are connected by a two-degree-of-freedom joint, it provides one constraint. When the constraints for all joints are subtracted from the total freedoms of the unconnected links, we find the resulting mobility of the assembled mechanism.

If we denote the number of single-degree-of-freedom joints as j_1 and the number of two-degree-of-freedom joints as j_2, then the resulting mobility, m, of a planar n-link mechanism is given by

$$m = 3(n-1) - 2j_1 - j_2. \tag{1.1}$$

Written in this form, Eq. (1.1) is called the *Kutzbach criterion* for the mobility of a planar mechanism [8]. Its application is shown for several simple examples in Fig. 1.3.

If the Kutzbach criterion yields $m > 0$, the mechanism has m degrees of freedom. If $m = 1$, the mechanism can be driven by a single input motion to produce constrained (uniquely defined) motion. Two examples are the slider-crank linkage and the four-bar linkage, shown in Figs. 1.3b and 1.3c, respectively. If $m = 2$, then two separate input motions are necessary to produce constrained motion for the mechanism; such a case is the five-bar linkage shown in Fig. 1.3d.

If the Kutzbach criterion yields $m = 0$, as in Figs. 1.3a and 1.4a, motion is impossible and the mechanism forms a structure.

---

[7] The German literature distinguishes between *movability* and *mobility*. Movability includes the six degrees of freedom of the device as a whole, as though the ground link were not fixed, and thus applies to a kinematic chain. Mobility neglects these degrees of freedom and considers only the internal relative motions, thus applying to a mechanism. The English literature seldom recognizes this distinction, and the terms are used somewhat interchangeably.

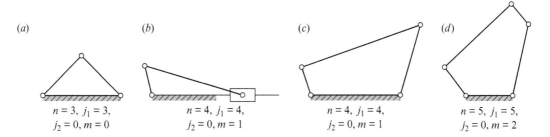

Fig. 1.3 Applications of the Kutzbach criterion.

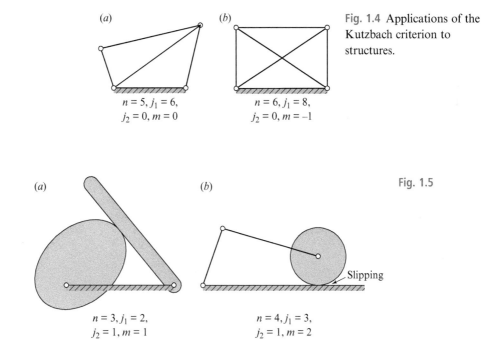

Fig. 1.4 Applications of the Kutzbach criterion to structures.

Fig. 1.5

If the criterion yields $m < 0$, then there are redundant constraints in the chain, and it forms a statically indeterminate structure. An example is shown in Fig. 1.4b. Note in the examples of Fig. 1.4 that when three links are joined by a single pin, such a connection is treated as two separate but concentric joints; two j_1 joints must be counted.

Figure 1.5 shows two examples of the Kutzbach criterion applied to mechanisms with two-degree-of-freedom joints – that is, j_2 joints. Particular attention should be paid to the contact (joint) between the wheel and the fixed link in Fig. 1.5b. Here it is assumed that slipping is possible between the two links. If this contact prevents slipping, the joint would be counted as a one-degree-of-freedom joint – that is, a j_1 joint – because only one relative motion would then be possible between the links. Recall that, in this case, the mechanism is generally referred to as a "linkage."

Fig. 1.6   (a)   (b)

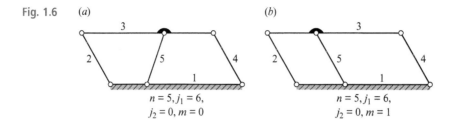

$$n = 5, j_1 = 6,\qquad\qquad n = 5, j_1 = 6,$$
$$j_2 = 0, m = 0\qquad\qquad j_2 = 0, m = 1$$

It is important to realize that the Kutzbach criterion can give an incorrect result. For example, note that Fig. 1.6a represents a structure and that the criterion properly predicts $m = 0$. However, if link 5 is arranged as in Fig. 1.6b, the result is a double-parallelogram linkage with a mobility of $m = 1$, even though Eq. (1.1) indicates that it is a structure. The actual mobility of $m = 1$ results only if the parallelogram geometry is achieved. In the development of the Kutzbach criterion, no consideration was given to the lengths of the links or other dimensional properties. Therefore, it should not be surprising that exceptions to the criterion are found for particular cases with equal link lengths, parallel links, or other special geometric features.

Although there are exceptions, the Kutzbach criterion remains useful, because it is so easily applied during mechanism design. To avoid exceptions, it would be necessary to include all the dimensional properties of the mechanism. The resulting criterion would be very complex and would be useless at the early stages of design when dimensions may not be known.

An earlier mobility criterion, named after Grübler [3], applies to a planar linkage where the overall mobility is $m = 1$. Substituting $j_2 = 0$ and $m = 1$ into Eq. (1.1) and rearranging, we find that Grübler's criterion for planar linkages can be written as

$$3n - 2j_1 - 4 = 0. \tag{1.2}$$

Rearranging this equation, the number of links is

$$n = \frac{2j_1 + 4}{3}. \tag{1.3}$$

From this equation, we see that a planar linkage, with a mobility of $m = 1$, cannot have an odd number of links. Also, the simplest possible linkage with all binary links has $n = j_1 = 4$. This explains one reason why the slider-crank linkage (Fig. 1.3b) and the four-bar linkage (Fig. 1.3c) appear so commonly in machines.

Both the Kutzbach criterion, Eq. (1.1), and the Grübler criterion, Eq. (1.2), were derived for the case of planar mechanisms and linkages, respectively. If similar criteria are developed for spatial mechanisms and linkages, which is the subject of Chapter 10, we must recall that each unconnected link has six degrees of freedom and each single-degree-of-freedom joint provides five constraints, each two-degree-of-freedom joint provides four constraints, and so on. Similar arguments then lead to the Kutzbach criterion for spatial mechanisms,

$$m = 6(n - 1) - 5j_1 - 4j_2 - 3j_3 - 2j_4 - j_5,$$

and the Grübler criterion for spatial linkages,

$$6n - 5j_1 - 7 = 0. \tag{1.4}$$

Therefore, the simplest form of a spatial linkage,[8] with mobility of $m = 1$, is $n = j_1 = 7$.

Example 1.1

Determine the mobility of the planar mechanism shown in Fig. 1.7a.

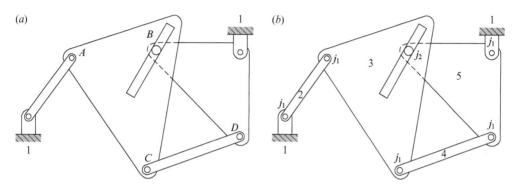

Fig. 1.7 Planar mechanism.

Solution

The link numbers and the joint types for the mechanism are shown in Fig. 1.7b. The number of links is $n = 5$, the number of lower pairs is $j_1 = 5$, and the number of higher pairs is $j_2 = 1$. Substituting these values into the Kutzbach criterion, Eq. (1.1), the mobility of the mechanism is

$$m = 3(5 - 1) - 2(5) - 1(1) = 1. \qquad\qquad Ans.$$

Note that the Kutzbach criterion gives the correct answer for the mobility of this mechanism; that is, a single input motion is required to give a unique output motion. For example, rotation of link 2 could be used as the input and rotation of link 5 could be used as the output.

Example 1.2

For the mechanism shown in Fig. 1.8a, determine: (a) the number of lower pairs (j_1 joints) and the number of higher pairs (j_2 joints); and (b) the mobility of the mechanism using the Kutzbach criterion. Treating rolling contact to mean rolling with no slipping, does this criterion provide the correct answer for the mobility of this mechanism? Briefly explain why or why not.

---

[8] Note that all planar linkages are exceptions to the spatial mobility criterion. They have the special geometric characteristics that all revolute axes are parallel and perpendicular to the plane of motion and that all prismatic axes lie in the plane of motion.

Example 1.2 (cont.)

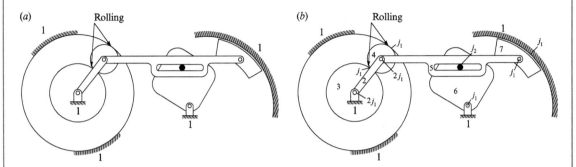

Fig. 1.8 Planar mechanism.

Solution

a. The links and the joint types of the mechanism are labeled in Fig. 1.8*b*. The number of links is $n = 7$, the number of lower pairs is $j_1 = 9$, and the number of higher pairs is $j_2 = 1$.

b. Substituting these values into the Kutzbach criterion, Eq. (1.1), the mobility of the mechanism is

$$m = 3(7 - 1) - 2(9) - 1(1) = -1.$$

However, this answer is not correct; that is, the Kutzbach criterion does not give the correct mobility for this mechanism. The mobility of this mechanism is, in fact, $m = 1$; that is, a single input motion gives a unique output motion.

Reasoning: Links 3 and 4 are superfluous to the constraints of the mechanism. If links 3 and 4 were removed, the motion of the remaining links would be unaffected. With links 3 and 4 removed, the mobility of the mechanism using the Kutzbach criterion is $m = 1$. Note that if links 3 and 4 were attached with no special conditions – that is, not pinned at their centers, for example – then the mechanism would indeed be locked and the answer $m = -1$ would be correct.

Example 1.3

For the mechanism shown in Fig. 1.9*a*, determine: (a) the number of lower pairs and the number of higher pairs; and (b) the mobility of the mechanism predicted by the Kutzbach criterion. Does this criteria provide the correct answer for this mechanism? Briefly explain why or why not.

Example 1.3 (cont.)

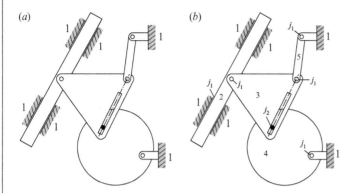

(a)   (b)   Fig. 1.9 Planar mechanism.

Solution

a. The links and the joints of the mechanism are labeled as shown in Fig. 1.9*b*. The number of links is $n = 5$, the number of lower pairs is $j_1 = 5$, and the number of higher pairs is

$$j_2 = 1. \qquad \qquad Ans.$$

b. Substituting these values into the Kutzbach criterion, Eq. (1.1), the mobility of the mechanism is

$$m = 3(5 - 1) - 2(5) - 1(1) = 1. \qquad \qquad Ans.$$

For this mechanism, the mobility is indeed 1, which indicates that the Kutzbach criterion gives the correct answer for this mechanism.

For a mechanism, or a linkage, with a mobility of $m = 1$, the input or driving link is, in general, numbered 2 in this text.

1.7   Characteristics of Mechanisms

An ideal system for the classification of mechanisms would be a system that allows a designer to enter the system with a set of specifications and leave with one or more mechanisms that satisfy those specifications. Although history[9] records that many attempts have been made, few have been particularly successful in devising a satisfactory classification system. In view of the fact that the purpose of a mechanism is the transformation of motion, we will follow Torfason's lead [9] and classify mechanisms according to the type of motion transformation. In total, Torfason displays 262 mechanisms, each of which can have a variety of dimensions. His categories are as follows:

[9] For an excellent short history of the kinematics of mechanisms, see [4].

(a)

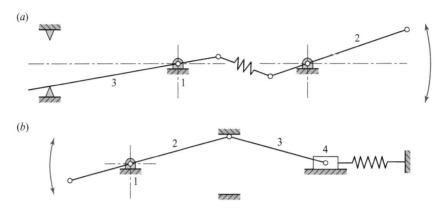

(b)

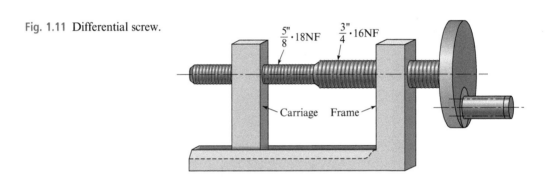

Fig. 1.10 (a) Bistable mechanism; (b) true toggle mechanism.

Fig. 1.11 Differential screw.

Snap-Action Mechanisms.   Snap-action, toggle, or flip-flop mechanisms are used for switches, clamps, or fasteners. Torfason also includes spring clips and circuit breakers. Figure 1.10 shows examples of bistable and true toggle mechanisms.

Linear Actuators.   Linear actuators include stationary screws with traveling nuts, stationary nuts with traveling screws, and single-acting and double-acting hydraulic and pneumatic cylinders.

Fine Adjustments.   Fine adjustments may be obtained with screws, including differential screws, worm gearing, wedges, levers, levers in series, and various motion-adjusting mechanisms. For the differential screw shown in Fig. 1.11, you should be able to determine that the translation of the carriage resulting from one turn of the handle is 0.0069 in to the left (see Prob. 1.17).

Clamping Mechanisms.   Typical clamping mechanisms are the C-clamp, the woodworker's screw clamp, cam-actuated and lever-actuated clamps, vises, presses (such as the toggle press shown in Fig. 1.10b), collets, and stamp mills.

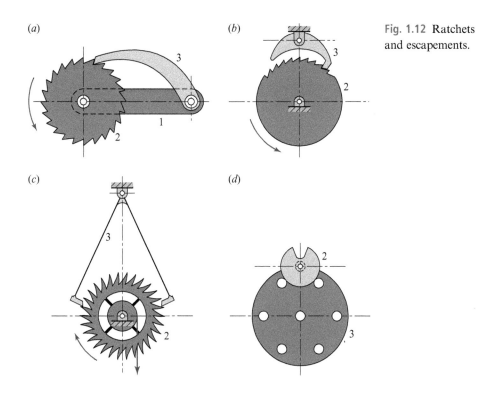

(a)　(b)　(c)　(d)

Fig. 1.12 Ratchets and escapements.

Locational Devices.   Torfason [9] shows 15 locational mechanisms. These are usually self-centering and locate either axially or angularly using springs and detents.

Ratchets and Escapements.   There are many different forms of ratchets and escapements, some quite clever. They are used in locks, jacks, clockworks, and other applications requiring some form of intermittent motion. Figure 1.12 shows four typical applications.

The ratchet in Fig. 1.12a allows only one direction of rotation of wheel 2. Pawl 3 is held against the wheel by gravity or by a spring. A similar arrangement is used for lifting jacks, which then employ a toothed rack for rectilinear motion.

Figure 1.12b is an escapement used for rotary adjustments.

Graham's escapement, shown in Fig. 1.12c, is used to regulate the movement of clockwork. Anchor 3 drives a pendulum whose oscillating motion is caused by the two clicks on engaging wheel 2. One is a push click, the other is a pull click. The lifting and engaging of each click caused by oscillation of the pendulum results in a wheel motion that, at the same time, presses each respective click and adds a gentle force to the motion of the pendulum.

The escapement shown in Fig. 1.12d has a control wheel, 2, that may rotate continuously to allow wheel 3 to be driven (by another source) in either direction.

Indexing Mechanisms.   The indexer shown in Fig. 1.13a uses standard gear teeth; for light loads, pins can be used in the input wheel 2 with corresponding slots in wheel 3, but neither form should be used if the shaft inertias are large.

Fig. **1.13** Indexing mechanisms.   (*a*)

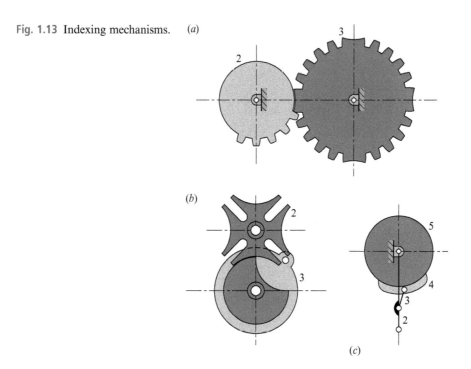

(*b*)

(*c*)

Figure 1.13*b* shows a Geneva-wheel, sometimes called a "Maltese-cross," indexer. Three or more slots may be used in the driven link, 2, which can be attached to, or geared to, the output to be indexed. High speeds and large inertias may cause problems with this indexer.

Toothless ratchet 5 in Fig. 1.13*c* is driven by the oscillating crank, 2, of variable throw. Note the similarity of this indexing mechanism to the ratchet of Fig. 1.12*a*.

Torfason [9] lists nine different indexing mechanisms, and many variations are possible.

Swinging or Rocking Mechanisms.

The category of swinging, or rocking mechanisms is often termed *oscillators*; in each case, the output rocks, or swings, through an angle that is generally less than 360°. However, the output shaft can be geared to a successor shaft to produce a larger angle of oscillation.

Figure 1.14*a* is a mechanism consisting of a rotating crank 2 and a coupler 3 containing a rack, which meshes with output gear 4 to produce the oscillating motion.

In Fig. 1.14*b*, crank 2 drives link 3, which slides on output link 4, producing a rocking motion. This mechanism is described as a *quick-return mechanism*, because crank 2 rotates through a larger angle on the forward stroke of link 4 than on the return stroke.

Figure 1.14*c* is a *four-bar linkage* called the *crank-rocker linkage* (Section 1.9). Crank 2 drives rocker 4 through coupler 3. Of course, link 1 is the frame. The characteristics of the rocking motion depend on the dimensions of the links and the placement of the frame pivots.

Figure 1.14*d* shows a *cam-and-follower mechanism*, in which the rotating link 2, called the *cam*, drives link 3, called the *follower*, in a rocking motion. An endless variety of cam-and-follower mechanisms is possible, many of which are discussed in Chapter 6. In each case, the cam can be designed to produce an output motion with the desired characteristics.

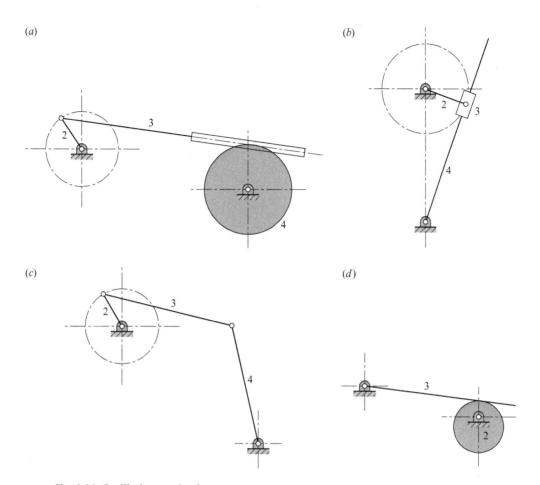

Fig. 1.14 Oscillating mechanisms.

Reciprocating Mechanisms.   Repeating straight-line motion is commonly obtained using either a pneumatic or hydraulic cylinder, a stationary screw with a traveling nut, a rectilinear drive using a reversible motor or reversing gears, or a cam-and-follower mechanism. A variety of typical linkages for obtaining reciprocating motion are shown in Figs. 1.15 and 1.16 [5].

The *offset slider-crank linkage* shown in Fig. 1.15a has kinematic characteristics that differ from the in-line (or on-center) slider-crank mechanism, shown in Fig. 1.3b. If the length of connecting rod 3 is long compared to the length of crank 2, then the resulting motion is nearly harmonic. Exact harmonic motion can be obtained from link 4 of the *Scotch-yoke linkage* shown in Fig. 1.15b.

The six-bar linkage shown in Fig. 1.15c is often called the *shaper linkage*, after the name of the machine tool in which it is used. Note that it is obtained from Fig. 1.14b by adding coupler 5 and slider 6. The stroke of the slider has a quick-return characteristic.

Figure 1.15d shows another version of the shaper linkage, which is termed the *Whitworth quick-return linkage*. The linkage is presented in an upside-down posture to illustrate its similarity to Fig. 1.15c.

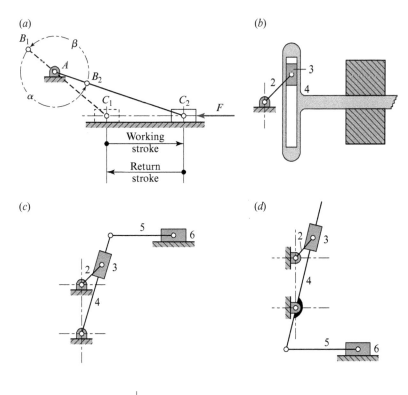

Fig. 1.15 Reciprocating linkages.

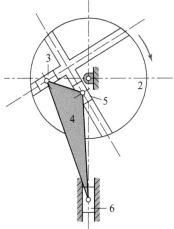

Fig. 1.16 Wanzer needle-bar linkage (Richard Mott Wanzer, 1812–1900).

Another example of a six-bar linkage is the Wanzer needle-bar linkage [5] shown in Fig. 1.16.

Figure 1.17a shows a six-bar linkage derived from the crank-rocker linkage of Fig. 1.14c by expanding coupler 3 and adding coupler 5 and slider 6. Coupler point C should be located to produce the desired motion characteristic for slider 6.

A crank-driven toggle linkage is shown in Fig. 1.17b. With this linkage, a high mechanical advantage is obtained at one end of the stroke of slider 6. (For a detailed discussion of the mechanical advantage of a mechanism, see Sections 1.10 and 3.19.)

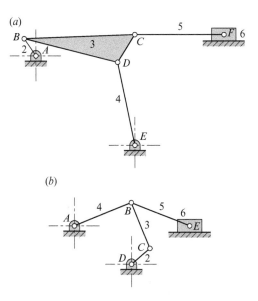

Fig. 1.17 Additional six-bar reciprocating linkages.

In many applications, mechanisms are used to perform repetitive operations, such as pushing parts along an assembly line, clamping parts together while they are welded, or folding cardboard boxes in an automated packaging machine. In such applications it is often desirable to use a constant-speed motor; this leads us to a discussion of Grashof's law in Section 1.9. In addition, however, we should give some consideration to the power and timing requirements.

In such repetitive operations, there is usually a part of the cycle when the mechanism is under load, called the *advance* or *working stroke*, and a part of the cycle, called the *return stroke*, when the mechanism is not working but simply returning to repeat the operation. For example, consider the offset slider-crank linkage shown in Fig. 1.15a. Work may be required to overcome the load, F, while the piston moves to the right from position C_1 to position C_2 but not during its return to position C_1, since the load may have been removed. In such situations, in order to keep the power requirements of the motor to a minimum and to avoid wasting valuable time, it is desirable to design a mechanism so that the piston moves much faster through the return stroke than it does during the advance (or working) stroke – that is, to use a higher portion of the cycle time for doing work than for returning.

A measure of the suitability of a mechanism from this viewpoint, called the *advance-to-return ratio* is defined as

$$Q = \frac{\text{cycle fraction for advance stroke}}{\text{cycle fraction for return stroke}}. \qquad (a)$$

A mechanism for which the value of Q is high is more desirable for such repetitive operations than one in which the value of Q is lower. Certainly, any such operation would call for a mechanism for which Q is greater than unity. Because of this, mechanisms with Q greater than unity are called *quick-return mechanisms*.

As shown in Fig. 1.15a, the first step is to determine the two crank postures, AB_1 and AB_2, that mark the beginning and the end of the working stroke. Next, noting the direction of rotation of the crank, we determine the crank angle α traveled through during the advance stroke and the remaining crank angle β of the return stroke. Then,

$$\text{cycle fraction for advance stroke} = \frac{\alpha}{2\pi}, \qquad (b)$$

and

$$\text{cycle fraction for return stroke} = \frac{\beta}{2\pi}. \qquad (c)$$

Finally, substituting Eqs. (b) and (c) into Eq. (a), the advance-to-return ratio can be written as

$$Q = \frac{\alpha}{\beta}. \qquad (1.5)$$

Note that the advance-to-return ratio depends only on geometry (that is, on changes in the crank posture); this ratio does not depend on the amount of work being done or on the speed of the driving motor. It is a kinematic property of the mechanism itself. Therefore, this ratio can be used for either design or analysis totally by graphic constructions. The following two examples illustrate applications in design.

Example 1.4

The rocker of a crank-rocker four-bar linkage is required to have a length of 4 in and swing through a total angle of 45°. Also, the advance-to-return ratio of the linkage is required to be 2.0. Determine a suitable set of link lengths for the remaining three links.

Solution

Equation (1.5) requires

$$Q = \alpha/\beta = 2.0, \qquad (1)$$

where

$$\alpha = 180° + \phi \qquad (2)$$

and

$$\beta = 180° - \phi. \qquad (3)$$

Substituting Eqs. (2) and (3) into Eq. (1) allows us to solve for

$$\phi = 60°, \quad \alpha = 240°, \quad \text{and} \quad \beta = 120°.$$

Example 1.4 (cont.)

Now, referring to Fig. 1.18, we apply the following graphic procedure:

a. Draw the rocker ($r_4 = 4.0$ in) to a suitable scale in its two extreme postures; that is, show the swing angle of the rocker of $45°$. Label the ground pivot O_4 and label the pin B in the two positions B_1 and B_2.

b. Through point B_1, draw an arbitrary line (labeled the X-line). Through B_2, draw a line parallel to the X-line.

c. Measure the angle $\phi = 60°$, counterclockwise from the X-line through point B_1. The intersection of this line with the line parallel to the X-line is the required position of the input crank pivot O_2.

d. The length O_2O_4 of the ground link can be measured from the drawing – that is

$$r_1 = 1.50 \text{ in.} \qquad \textit{Ans.}$$

e. The lengths of the crank r_2 and the coupler r_3 can be determined from the measurements,

$$O_2B_1 = r_3 + r_2 = 3.50 \text{ in} \quad \text{and} \quad O_2B_2 = r_3 - r_2 = 2.50 \text{ in.}$$

That is,

$$r_2 = 0.5(O_2B_1 - O_2B_2) = 0.50 \text{ in} \quad \text{and} \quad r_3 = 0.5(O_2B_2 + O_2B_1) = 3.00 \text{ in.} \qquad \textit{Ans.}$$

The solution for the synthesized four-bar linkage is shown in Fig. 1.18.

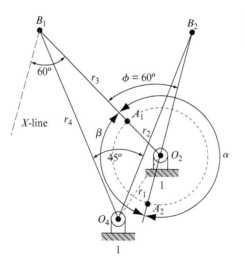

Fig. 1.18 Synthesized four-bar linkage.

In the two limiting postures, O_2B_1 and O_2B_2, the output link is momentarily stopped and, for this reason, these two postures of the linkage are referred to as limit postures (see Section 1.10). Also, note that this problem is an example of two-posture synthesis. For a detailed discussion of the general problems of two-, three-, and four-posture synthesis, see Chapter 9.

Example 1.5

Determine a suitable set of link lengths for a slider-crank linkage such that the stroke is 2.50 in and the advance-to-return ratio is 1.4.

Solution

Equation (1.5) requires

$$Q = \alpha/\beta = 1.40, \tag{1}$$

where

$$\alpha = 180° + \phi \tag{2}$$

and

$$\beta = 180° - \phi. \tag{3}$$

Substituting Eqs. (2) and (3) into Eq. (1) allows us to solve for

$$\phi = 30°, \quad \alpha = 210°, \quad \text{and} \quad \beta = 150°.$$

Now, referring to Fig. 1.19, we apply the following graphic procedure:

a. Draw the stroke (shown horizontal) of 2.50 in of the slider-crank linkage to a suitable scale. Label the pin B in its two extreme positions B_1 and B_2.
b. Through point B_2, draw an arbitrary line (labeled the X-line). Through point B_1, draw a line parallel to the X-line.
c. Measure the angle $\phi = 30°$ clockwise from the X-line. The intersection of this line with the line parallel to the X-line is the ground pivot O_2.

Fig. 1.19 Synthesized slider-crank linkage.

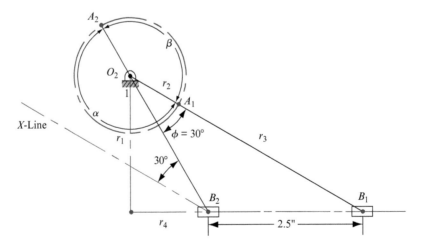

Example 1.5 (cont.)

d. The length of the ground link – that is, the offset or eccentricity (the perpendicular distance from the ground pivot O_2 to the line of travel of the slider) – can be measured from the drawing. That is,

$$r_1 = 2.17 \text{ in.} \qquad\qquad Ans.$$

e. The lengths of the crank r_2 and coupler r_3 can be determined from the measurements,

$$O_2B_1 = r_3 + r_2 = 4.33 \text{ in} \quad \text{and} \quad O_2B_2 = r_3 - r_2 = 2.50 \text{ in.}$$

That is,

$$r_2 = 0.5(O_2B_1 - O_2B_2) = 0.92 \text{ in} \quad \text{and} \quad r_3 = 0.5(O_2B_2 + O_2B_1) = 3.42 \text{ in.} \qquad Ans.$$

The solution of the synthesized slider-crank linkage is shown in Fig. 1.19.

Note that there is a proper and an improper direction of rotation for the input of such a device. If the direction of crank rotation were reversed in the example of Fig. 1.19, the roles of α and β would also be reversed, and the advance-to-return ratio would be less than 1.0. Thus, the motor must rotate clockwise for this mechanism to have the quick-return property.

Other mechanisms with quick-return characteristics are the shaper linkage, shown in Fig. 1.15c, and the Whitworth linkage, shown in Fig. 1.15d. The synthesis of a quick-return mechanism, as well as mechanisms with other properties, is discussed in detail in Chapter 9.

Reversing Mechanisms. When a mechanism capable of delivering output rotation in either direction is desired, some form of reversing mechanism is required. Many such devices make use of a two-way clutch that connects the output shaft to either of two drive shafts turning in opposite directions. This method is used in both gear and belt drives and does not require that the drive be stopped to change direction. Gear-shift devices, as in automotive transmissions, are also in common use.

Couplings and Connectors. Couplings and connectors are used to transmit motion between coaxial, parallel, intersecting, or skewed shafts. Gears of one kind or another can be used for any of these situations. These are discussed in Chapters 7 and 8.

Flat belts can be used to transmit motion between parallel shafts. They can also be used between intersecting or skewed shafts if guide pulleys are used, as shown in Fig. 1.20a. When parallel shafts are involved, the belts can be open or crossed, depending on the direction of rotation desired.

Figure 1.20b shows a *drag-link* (also referred to as a *double-crank*) four-bar linkage used to transmit rotary motion between parallel shafts. Here, crank 2 is the driver and crank 4 is the output. This is a very interesting linkage; you should try to construct one using cardboard strips

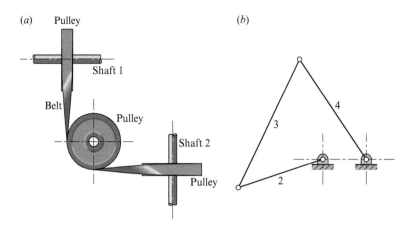

Fig. 1.20 Two-shaft coupling mechanisms.

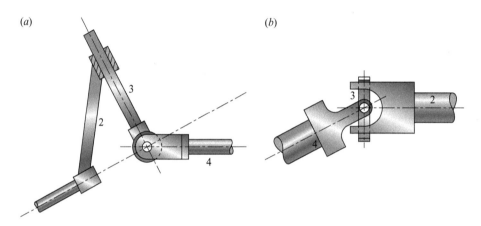

Fig. 1.21 Coupling mechanisms for intersecting shafts.

and thumbtacks for joints to observe its motion. Can you devise a working model that allows complete rotation of both links 2 and 4 (see Prob. 1.14)?

The *Reuleaux coupling*, shown in Fig. 1.21a, for intersecting shafts is recommended only for light loads. The *Hooke joint*, shown in Fig. 1.21b, is also used for intersecting shafts. However, this joint can withstand heavy loads and is commonly used with a drive-shaft in rear-wheel-drive vehicles. It is customary to use two of these joints in series for connecting parallel shafts.

Sliding Connectors.   Sliding connectors are used when one slider (the input) is to drive another slider (the output). The usual problem is that the two sliders operate in the same plane but in different directions. The possible solutions are:

1. a rigid link pivoted at each end to a slider;
2. a belt or chain connecting the two sliders with the use of a guide pulley or sprocket;
3. rack gear teeth cut on each slider and the connection completed using one or more gears; and
4. a flexible cable connector.

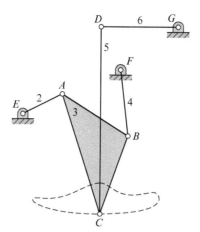

Fig. 1.22 Six-bar stop-and-dwell linkage.

Stop, Pause, and Hesitation Mechanisms.   In an automotive engine a valve must open, remain open for a short period of time, and then close. A conveyor line may need to halt for an interval of time while an operation is being performed and then continue its motion. Many similar requirements occur in the design of machines. Torfason [9] classifies these as *stop-and-dwell*, *stop-and-return*, *stop-and-advance*, and so on. Such requirements can often be met using cam-and-follower mechanisms (Chapter 6), indexing mechanisms, including those of Fig. 1.13, ratchets, linkages at the limits of their motion, and gear-and-clutch mechanisms.

The six-bar linkage of Fig. 1.22 is a clever method to obtain a rocking motion (of link 6) containing a dwell. This linkage, an extension of the four-bar linkage, consisting of frame 1, crank 2, coupler 3, and rocker 4, can be designed such that point C on the coupler generates the curve shown by dashed lines. A portion of this curve will then fit closely to a circular arc whose radius is equal to the length of link 5 – that is, distance DC. Thus, when point C traverses this portion of the coupler curve, link 6, the output rocker, is stationary.

Curve Generators.   The connecting rod, or coupler, of a planar four-bar linkage may be imagined as an infinite plane extending in all directions, but pin-connected to the input and output cranks. Then, during motion of the linkage, any point attached to the plane of the coupler generates a path with respect to the fixed link; this path is called a *coupler curve*. Two of these paths, namely those generated by the pin connectors of the coupler, are true circles with centers at the two fixed pivots. However, other points can be found that trace much more complex curves.

One of the best sources of coupler curves for the four-bar linkage is the Hrones and Nelson atlas [6]. This book consists of a set of 11×17 in drawings containing over 7 000 coupler curves of crank-rocker linkages. Figure 1.23 is a reproduction of a typical page of this atlas (by permission of the publishers). In each case, the crank has unit length, and the lengths of the remaining links vary from page to page to produce the different combinations. On each page a number of different coupler points are chosen and their coupler curves are shown. This atlas of coupler curves has proven to be invaluable to the designer who needs a linkage to generate a curve with specified characteristics.

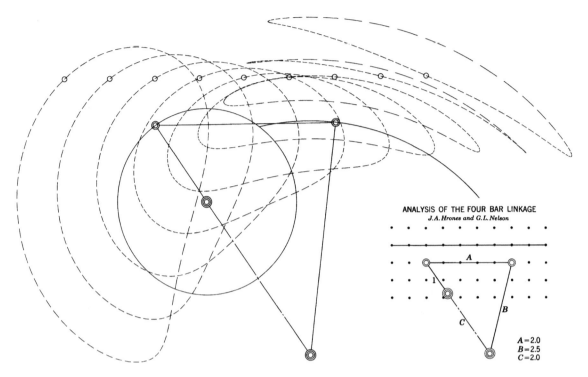

Fig. 1.23 A set of coupler curves [6].

The algebraic equation of a four-bar linkage coupler curve is, in general, a sixth-order polynomial [1]; thus, it is possible to find coupler curves with a wide variety of shapes and many interesting features. Some coupler curves have sections that are nearly straight-line segments; others have almost exact circular arc segments; still others have one or more cusps or cross over themselves as in a figure-eight. Therefore, it is often not necessary to use a mechanism with a large number of links to obtain a complex motion for a coupler point.

Yet the complexity of the coupler-curve equation is also a hinderance; it means that hand-calculation methods can become very cumbersome. Thus, over the years, many mechanisms have been designed by strictly intuitive procedures and proven with cardboard models, without the use of kinematic principles or procedures. Until quite recently, those techniques that did offer a rational approach have been graphic, avoiding tedious computations. Finally, with the availability of digital computers, and particularly with computer graphics, useful design methods are now emerging that can deal directly with the complex calculations required without burdening the designer with the computational drudgery (Section 10.9 has details on some of these).

One of the more curious and interesting facts about the coupler-curve equation is that the same curve can always be generated by three different four-bar linkages. These are called *cognate linkages*, and the theory is developed in Section 9.10.

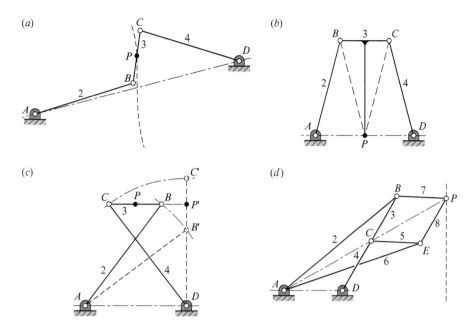

Fig. 1.24 (*a*) Watt's linkage; (*b*) Roberts' linkage; (*c*) Chebyshev linkage; (*d*) Peaucillier inversor.

Straight-Line Generators.   In the late seventeenth century, before the development of the milling machine, it was extremely difficult to machine straight, flat surfaces. For this reason, good prismatic joints with close clearances were not available. During that era, much thought was given to the problem of attaining a straight line as a part of the coupler curve of a linkage having only revolute connections. Probably the best-known result of this search is the straight-line mechanism developed by Watt for guiding the piston of early steam engines. Figure 1.24*a* shows a four-bar linkage, known as Watt's linkage, which generates an approximate straight line as a part of its coupler curve. Although the coupler point (tracing point *P*) does not generate an exact straight line, a good approximation is achieved over a considerable distance of travel.

Another four-bar linkage in which the tracing point *P* generates an approximate straight-line coupler-curve segment is Roberts' linkage (Fig. 1.24*b*). The dashed lines *BP* and *CP* in the figure indicate that the linkage is defined by forming three congruent isosceles triangles; thus $BC = AP = PD = AD/2$.

The tracing point *P* (the midpoint of coupler link *BC*) of the Chebyshev linkage shown in Fig. 1.24*c* also generates an approximate straight line. The linkage forms a 3-4-5 triangle when link 4 is in the vertical posture, as shown by the dashed lines; thus, $DB' = 3$ units, $AD = 4$ units, and $AB' = 5$ units. Note that $AB = DC, DC' = 5$ units, and the tracing point P' is the midpoint of the dashed line $B'C'$. Also, note that $DP'C$ forms another 3-4-5 triangle and hence the line containing *P* and P' is parallel to the ground link *AD*.

A linkage that generates an exact straight line is the Peaucillier inversor shown in Fig. 1.24*d*. The conditions describing its geometry are that $BC = BP = EC = EP$ and $AB = AE$ such that,

Fig. 1.25 $AB = AP = O_2A$.

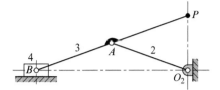

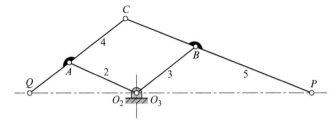

Fig. 1.26 Pantograph linkage $CA = O_2B$ and $BC = O_2A$.

by symmetry, points $A, C,$ and P always lie on a straight line passing through A. Under these conditions $AC = AP = k$, a constant, and the curves generated by C and P are said to be *inverses* of each other. If we place the other fixed pivot D such that $AD = CD$, then point C must trace a circular arc while point P follows an exact straight line. Another interesting property is that if AD is not equal to CD, then point P traces a true circular arc of very large radius. This was the first straight-line generator, and was important in the development of the steam engine.

Figure 1.25 shows another linkage which generates exact straight-line motion, the Scott–Russell linkage. However, note that it employs a slider.

The *pantograph* shown in Fig. 1.26 is used to trace figures at a larger or smaller size. If, for example, point P traces a map, then a pen at Q will draw a similar map at a smaller scale. The dimensions $O_2A, AC, CB,$ and BO_2 must conform to an equal-sided parallelogram.

Torfason [9] also includes robots, speed-changing devices, computing mechanisms, function generators, loading mechanisms, and transportation devices in his classification. Many of these utilize arrangements of mechanisms already presented. Others appear in some of the chapters to follow.

1.8   Kinematic Inversion

In Section 1.4 we noted that every mechanism has a fixed link called the frame. When different links are chosen as the frame, the *relative* motions between the various links are not altered, but their *absolute* motions (those measured with respect to the frame) may change significantly. The process of choosing different links for the frame is known as *kinematic inversion*. In an *n*-link mechanism, choosing each link in turn as the frame yields *n* distinct kinematic inversions – that is, *n* different mechanisms.

For illustration, consider the slider-crank linkage. Figure 1.27a shows the first inversion, which is found in most internal combustion engines, where the frame is the cylinder block,

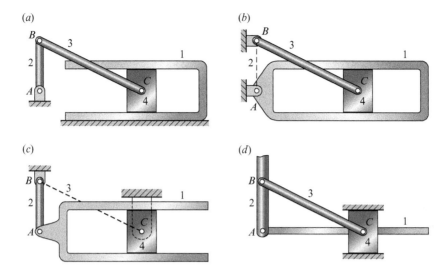

Fig. 1.27 Four inversions of the slider-crank linkage.

link 1. Link 4, the piston, is driven by the expanding gases, and this movement provides the input. Link 3, the connecting rod, connects link 4 to link 2, the crank, which rotates as the driven output. By reversing the roles of the input and output, this linkage can be used as a compressor.

Figure 1.27b shows another kinematic inversion, where link 2 is stationary. Link 1, formerly the frame, now rotates about revolute A. This inversion of the slider-crank linkage was the basis of the rotary engine found in many early aircraft.

Yet another inversion is shown in Fig. 1.27c; it has link 3, formerly the connecting rod, as the fixed link. This linkage was used to drive the wheels of early steam locomotives, link 2 being attached to a wheel.

The final inversion, shown in Fig. 1.27d, has the piston, link 4, stationary. Although it is not found in engines, by rotating the figure 90° clockwise this linkage can be recognized as part of a garden water pump. Note in this inversion that the prismatic joint connecting links 1 and 4 is also inverted – that is, the "inside" and "outside" elements of the joint are reversed.

1.9   Grashof's Law

A very important consideration when designing a mechanism to be driven by a motor, obviously, is to ensure that the input crank can make a complete revolution. Mechanisms in which no link can make a complete revolution would not be useful in such applications. For the four-bar linkage, there is a very simple test of whether this is the case.

Grashof's law states that, *for a planar four-bar linkage, the sum of the shortest and longest link lengths cannot be greater than the sum of the remaining two link lengths if there is to be continuous relative rotation between two links.* This is shown in Fig. 1.28, where the longest link has length l, the shortest link has length s, and the other two links have lengths p and q.

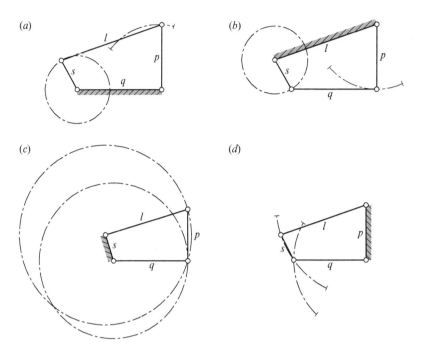

Fig. 1.28 (*a, b*) Crank-rocker linkages; (*c*) drag-link linkage; (*d*) double-rocker linkage.

In this notation, Grashof's law states that one of the links, in particular the shortest link, will rotate continuously relative to the other three links if and only if

$$s + l \leq p + q. \tag{1.6}$$

If this inequality is not satisfied, no link will make a complete revolution relative to another.

Attention is called to the fact that nothing in Grashof's law specifies the order in which the links are connected or which link of the four-bar linkage is fixed. We are free, therefore, to fix any of the four links. When we do so, we create variations of the four-bar linkage, four of which are shown in Fig. 1.28. All of these fit Grashof's law and, in each, the link s makes a complete revolution relative to the other links. The different variations are distinguished by the location of the link s relative to the fixed link.

If the shortest link, s, is adjacent to the fixed link, as shown in Figs. 1.28*a* and 1.28*b*, we obtain what is called a *crank-rocker linkage*. Link s is, of course, the crank, since it is able to rotate continuously, and link p, which can only oscillate between limits, is the rocker.

The drag-link (or double-crank linkage) is obtained by fixing the shortest link, s, as the frame. In this variation, shown in Fig. 1.28*c*, both links adjacent to s can rotate continuously, and both are properly described as cranks; the shorter of the two is generally used as the input. Although this is a very common linkage, you will find it an interesting challenge to devise a practical working model that can operate through the full cycle.

By fixing the link opposite to link s, we obtain the final variation, the double-rocker linkage of Fig. 1.28*d*. Note that although link s is able to make a complete revolution, neither link adjacent to the frame can do so; both must oscillate between limits and, therefore, are rockers.

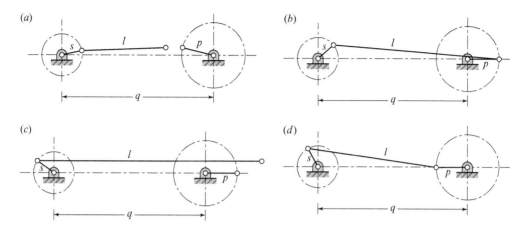

Fig. 1.29 (a) Eq. (1.7a); $s + l + p < q$, and the links cannot be connected; (b) Eq. (1.7b); $s + l - p > q$ and s is incapable of rotation; (c) Eq. (1.7c); $s + q + p < l$ and the links cannot be connected; (d) Eq. (1.7d); $s + q - p > l$ and s is incapable of rotation.

In each of the variations shown, the shortest link, s, is adjacent to the longest link, l. However, exactly the same types of linkages will occur if the longest link, l, is opposite the shortest link, s; you should demonstrate this for your own satisfaction.

Reuleaux approaches the problem somewhat differently but, of course, obtains the same results. In his approach, and using Fig. 1.29, the links are labeled

$$s \text{ is the crank}, \qquad p \text{ is the lever},$$

$$l \text{ is the coupler}, \qquad q \text{ is the frame},$$

where l need not be the longest link. Then, the following conditions apply:

$$s + l + p \geq q, \tag{1.7a}$$

$$s + l - p \leq q, \tag{1.7b}$$

$$s + q + p \geq l, \tag{1.7c}$$

$$s + q - p \leq l. \tag{1.7d}$$

If these four conditions are not satisfied, then the results are as shown in Fig. 1.29.

Example 1.6

Is the linkage shown in Fig. 1.30 a crank-rocker, a double-rocker, or a drag-link four-bar linkage?

Solution

Substituting the link lengths into Eq. (1.6) gives

$$3 \text{ in} + 6 \text{ in} \leq 4 \text{ in} + 5 \text{ in}$$

or

$$9 \text{ in} \leq 9 \text{ in}.$$

Example 1.6 (cont.)

Fig. 1.30 $p = 4$ in,
$l = 6$ in, $q = 5$ in, and $s = 3$ in.

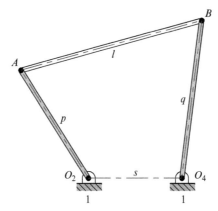

Therefore, the linkage satisfies Grashof's law; that is, the linkage is a Grashof four-bar linkage. Since the shortest link is the frame, the two links adjacent to the shortest link can both rotate continuously (Fig. 1.28c) and both are properly described as cranks. Therefore, Fig. 1.30 shows a drag-link linkage.   *Ans.*

1.10 Mechanical Advantage

In general, the *mechanical advantage* of a mechanism is defined as the ratio of the force or torque exerted by the driven link to the necessary force or torque required at the driver. With the widespread use of the four-bar linkage, a few remarks are in order here that will help us judge the quality of such a linkage for its intended application.

Consider the crank-rocker four-bar linkage shown in Fig. 1.31, where link 2 is the driver and link 4 is the follower.

In Section 3.19, we will show that the mechanical advantage of the four-bar linkage can be written as

$$MA = \frac{R_{CD} \sin \gamma}{R_{BA} \sin \beta}. \tag{1.8}$$

Note that this is directly proportional to the sine of the angle γ between the coupler and the follower, and is inversely proportional to the sine of the angle β between the coupler and the driver. Of course, both of these angles, and therefore the mechanical advantage, are continuously changing as the linkage moves.

When the sine of the angle β becomes zero, the mechanical advantage becomes infinite; thus, at such a posture, only a small input torque is necessary to produce a very large output torque.

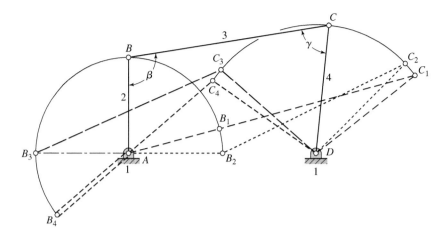

Fig. 1.31 Crank-rocker four-bar linkage.

This is the case when the driver AB is directly in line with the coupler BC, as shown in Fig. 1.31; it occurs when the crank is in posture AB_1 and again when the crank is in posture AB_4. Note that these also define the extreme postures of travel of the rocker DC_1 and DC_4. When the four-bar linkage is in either of these postures, the mechanical advantage is infinite – that is, $\beta = 0°$ or $\beta = 180°$ – and the linkage is said to be in a *toggle* (or *limit*) posture.

The angle γ between the coupler and the follower is called the *transmission angle*. As this angle becomes smaller, the mechanical advantage decreases and even a small amount of friction might cause the mechanism to lock or jam. A common rule of thumb is that a four-bar linkage should not be used to overcome a load in a region where the transmission angle is less than, say, $45°$ or $50°$. The extreme values of the transmission angle occur when crank AB lies along the line of the frame, AD. The transmission angle is minimal when the crank is in the posture AB_2 and is maximal when the crank is in the posture AB_3 (see Fig. 1.31). Because of the ease with which it can be visually inspected, the transmission angle has become a commonly accepted measure of the quality of the design of the four-bar linkage. A double-rocker four-bar linkage has a dead-center posture when links 3 and 4 lie along a straight line. In a dead-center posture, the transmission angle is $\gamma = 0°$ or $\gamma = 180°$, and the linkage is locked. The designer must either avoid such a posture or provide an external force, such as a spring, to unlock the linkage.

Example 1.7

Determine the mechanical advantage of the four-bar linkage in the posture shown in Fig. 1.32.

Solution

Angles γ and β for Eq. (1.8) are as shown in Fig. 1.33 and can be obtained from trigonometry. Using the law of cosines,

$$R_{DB} = \sqrt{(60 \text{ mm})^2 + (180 \text{ mm})^2 - 2(60 \text{ mm})(180 \text{ mm}) \cos 60°} = 158.74 \text{ mm}. \quad (1)$$

Example 1.7 (cont.)

Fig. 1.32 $R_{DA} = 180$ mm,
$R_{BA} = 60$ mm, $R_{CB} = 210$ mm,
and $R_{CD} = 120$ mm.

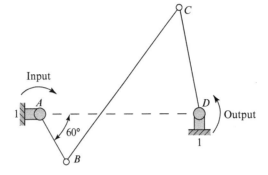

Fig. 1.33 Angles γ and β.

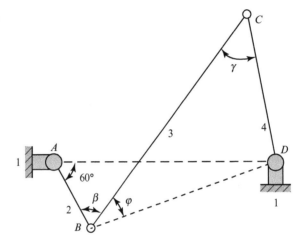

Also, the angles

$$\gamma = \cos^{-1}\left[\frac{(158.74 \text{ mm})^2 - (210 \text{ mm})^2 - (120 \text{ mm})^2}{-2(210 \text{ mm})(120 \text{ mm})}\right] = 48.64°, \tag{2}$$

and

$$\varphi = \cos^{-1}\left[\frac{(120 \text{ mm})^2 - (210 \text{ mm})^2 - (158.74 \text{ mm})^2}{-2(210 \text{ mm})(158.74 \text{ mm})}\right] = 34.57°. \tag{3}$$

Finally, the sum of the angles

$$\beta + \varphi = \cos^{-1}\left[\frac{(180 \text{ mm})^2 - (60 \text{ mm})^2 - (158.74 \text{ mm})^2}{-2(60 \text{ mm})(158.74 \text{ mm})}\right] = 100.90°. \tag{4}$$

Example 1.7 (cont.)

Subtracting Eq. (3) from Eq. (4) gives

$$\beta = 100.90° - 34.57° = 66.33°. \tag{5}$$

Then, substituting Eqs. (2) and (5) into Eq. (1.8), the mechanical advantage of the four-bar linkage in the given posture is

$$MA = \frac{(120 \text{ mm}) \sin 48.64°}{(60 \text{ mm}) \sin 66.33°} = 1.64. \qquad\qquad Ans.$$

Note that mechanical advantage, toggle posture, transmission angle, and dead-center posture depend on the choice of the driver and driven links. For example, in Fig. 1.31, if link 4 is used as the driver and link 2 as the driven link, then the roles of γ and β are reversed. In this case, the linkage has no toggle posture, and its mechanical advantage becomes zero when link 2 is in posture AB_1 or AB_4, because the transmission angle is zero. These and other methods of rating the suitability of the four-bar linkage or other mechanisms are discussed more thoroughly in Section 3.19.

PROBLEMS

1.1 Sketch at least six different examples of the use of a planar four-bar linkage in practice. These can be found in the workshop, in domestic appliances, on vehicles, on agricultural machines, and so on.

1.2 The link lengths of a planar four-bar linkage are 25 mm, 75 mm, 125 mm, and 125 mm. Assemble the links in all possible combinations and sketch the four inversions of each. Do these linkages satisfy Grashof's law? Describe each inversion by name (for example, a crank-rocker linkage or a drag-link linkage).

1.3 A crank-rocker linkage has a 4 in frame, a 1 in crank, a 3.6 in coupler, and a 3 in rocker. Draw the linkage and find the maximum and minimum values of the transmission angle. Locate both toggle postures and record the corresponding crank angles and transmission angles.

1.4 Plot the complete path of coupler point C.

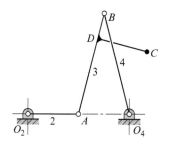

Fig. P1.4 $O_2A = 38$ mm,
$AB = 75$ mm, $O_2O_4 = 75$ mm,
$O_4B = 75$ mm, $AD = 56$ mm,
and $DC = 38$ mm.

1.5 Find the mobility of each mechanism.

Fig. P1.5   (a)   (b)

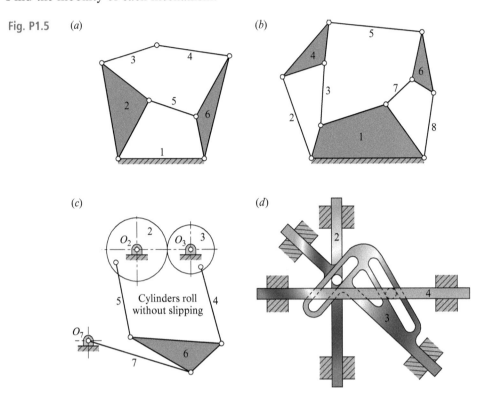

(c)   (d)

1.6 Use the Kutzbach criterion to determine the mobility of the mechanism.

Fig. P1.6

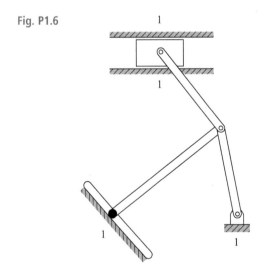

1.7 Sketch a planar linkage with only revolute joints and a mobility of $m = 1$ that contains a quaternary link. How many distinct variations of this linkage can you find?

1.8 Use the Kutzbach criterion to determine the mobility of the mechanism. Clearly number each link and label the lower pairs (j_1 joints) and higher pairs (j_2 joints).

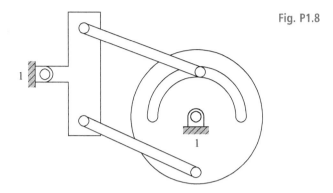

Fig. P1.8

1.9 Determine the number of links, the number of lower pairs, and the number of higher pairs. Use the Kutzbach criterion to determine the mobility of the mechanism. Is the answer correct? Briefly explain.

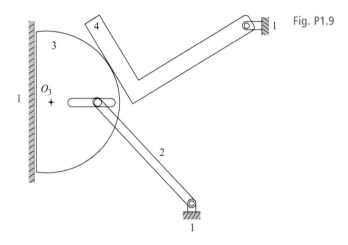

Fig. P1.9

1.10 Use the Kutzbach criterion to determine the mobility of the mechanism. Clearly number each link, and label the lower pairs and higher pairs.

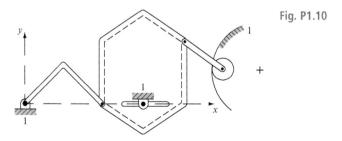

Fig. P1.10

1.11 Determine the number of links, the number of lower pairs, and the number of higher pairs. Treat rolling contact to mean rolling with no slipping. Using the Kutzbach criterion, determine the mobility. Is the answer correct? Briefly explain.

Fig. P1.11

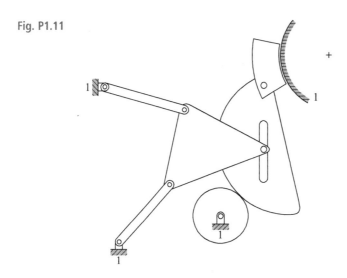

1.12 Does the Kutzbach criterion provide the correct result for this mechanism? Briefly explain why or why not.

Fig. P1.12

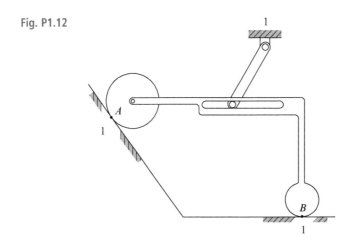

1.13 The mobility of the mechanism is $m = 1$. Use the Kutzbach criterion to determine the number of lower pairs and the number of higher pairs. Is the wheel rolling without slipping, or rolling and slipping, at point A on the wall?

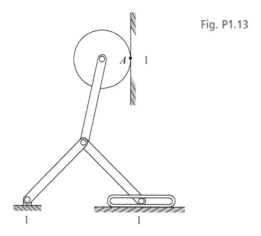

Fig. P1.13

1.14 Devise a practical working model of the drag-link linkage.

1.15 Find the advance-to-return ratio of the linkage of Prob. 1.3.

1.16 Plot the complete coupler curve of Roberts' linkage shown in Fig. 1.24*b*. Use $AB = CD = AD = 62$ mm and $BC = 31$ mm.

1.17 If the handle of the differential screw in Fig. 1.11 is turned 30 revolutions clockwise shown viewed from the right, how far and in what direction does the carriage move?

1.18 Show how the linkage of Fig. 1.15*b* can be used to generate a sine wave.

1.19 Devise a crank-rocker four-bar linkage, as in Fig. 1.14*c*, having a rocker angle of 60°. The rocker length is to be 20 in.

1.20 A crank-rocker four-bar linkage is required to have an advance-to-return ratio $Q = 1.2$. The rocker is to have a length of 62 mm and oscillate through a total angle of 60°. Determine a suitable set of link lengths for the remaining three links of the four-bar linkage.

1.21 Determine the mobility of the mechanism. Number each link and label the lower pairs and the higher pairs. Identify a suitable input, or inputs, for the mechanism.

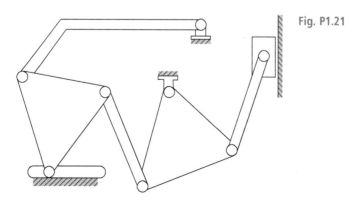

Fig. P1.21

1.22 Determine the mobility of the mechanism. Number each link and label the lower pairs and the higher pairs. Identify a suitable input, or inputs, for the mechanism.

Fig. P1.22

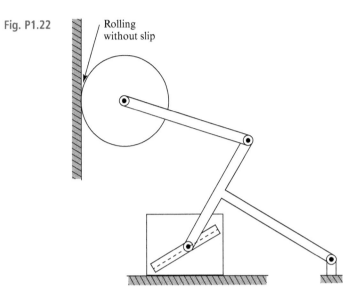

1.23 Determine the mobility of the mechanism. Number each link and label the lower pairs and the higher pairs. Identify a suitable input, or inputs, for the mechanism.

Fig. P1.23

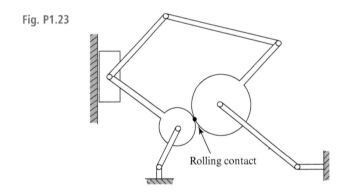

1.24 Determine the mobility of the mechanism. Number each link and label the lower pairs and the higher pairs. Identify a suitable input, or inputs, for the mechanism.

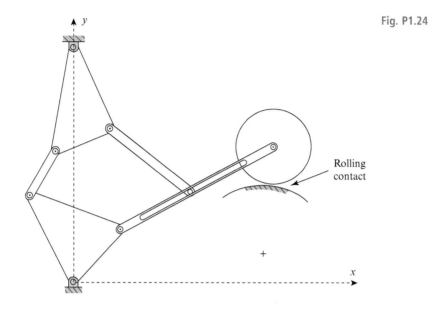

Fig. P1.24

1.25 Determine the mobility of the mechanism. Number each link and label the lower pairs and the higher pairs. Identify a suitable input, or inputs, for the mechanism.

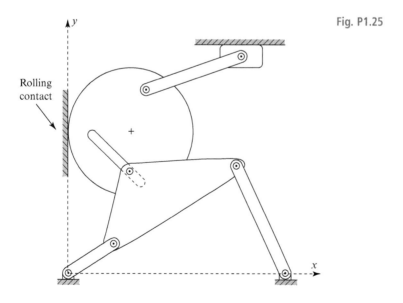

Fig. P1.25

1.26 Determine the mobility of the mechanism. Number each link and label the lower pairs and the higher pairs. Identify a suitable input, or inputs, for the mechanism.

Fig. P1.26

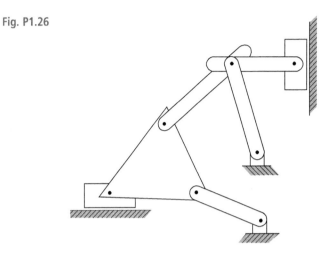

1.27 Determine the transmission angle and the mechanical advantage of the four-bar linkage in the posture shown.

Fig. P1.27 $O_2O_4 = 150$ mm, $O_2A = 75$ mm, $AB = 125$ mm, $O_4B = 162$ mm, and $\theta_2 = 30°$.

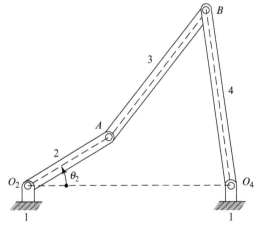

1.28 Determine the mobility of the mechanism. Number each link and label the lower pairs and the higher pairs. Identify a suitable input, or inputs, for the mechanism.

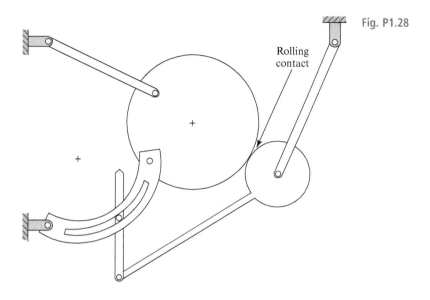

Fig. P1.28

Rolling contact

1.29 Determine the mobility of the mechanism. Number each link and label the lower pairs and the higher pairs. Identify a suitable input, or inputs, for the mechanism.

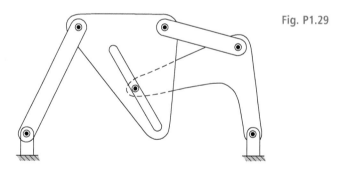

Fig. P1.29

1.30 The rocker of a crank-rocker four-bar linkage is required to have a length of 150 mm and swing through a total angle of 30°. Also, the advance-to-return ratio of the linkage is required to be 1.75. Determine a suitable set of link lengths for the remaining three links.

1.31 Determine a suitable set of link lengths for a slider-crank linkage such that the stroke will be 20 in and the advance-to-return ratio will be 1.8.

1.32 Determine the transmission angle and the mechanical advantage of the four-bar linkage in the posture shown. What type of four-bar linkage is this?

Fig. P1.32 $R_{AO_2} = 25$ mm,
$R_{BA} = 88$ mm, $R_{BO_4} = 112$ mm,
$R_{O_4O_2} = 75$ mm, and $\theta_2 = 30°$.

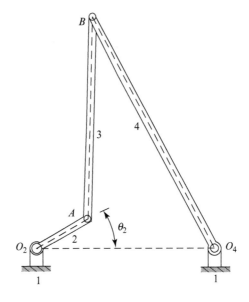

1.33 Determine the mobility of the mechanism. Number each link and label the lower pairs and the higher pairs. Identify a suitable input, or inputs, for the mechanism.

Fig. P1.33

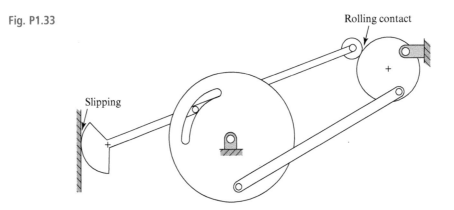

1.34 A crank-rocker four-bar linkage is shown in one of its two toggle postures. Find θ_2 and θ_4 corresponding to each toggle posture. What is the total rocking angle of link 4? What are the transmission angles at the extremes?

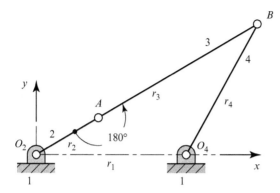

Fig. P1.34 $R_{AO_2} = 8$ in, $R_{BA} = 20$ in, $R_{BO_4} = 16$ in, and $R_{O_4O_2} = 16$ in.

1.35 Find the values of θ_2 and θ_4 which correspond to the two toggle postures of the non-Grashof four-bar linkage. Also find the values of θ_2 and θ_4, which correspond to the two dead-center postures of this linkage.

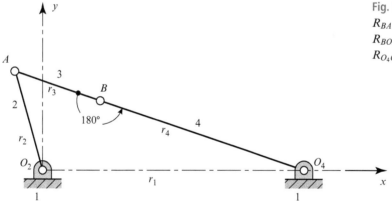

Fig. P1.35 $R_{AO_2} = 138$ mm, $R_{BA} = 125$ mm, $R_{BO_4} = 300$ mm, and $R_{O_4O_2} = 350$ mm.

1.36 Determine the advance-to-return ratio for the slider-crank linkage with the offset e. Also, determine in which direction the crank should rotate to provide quick return.

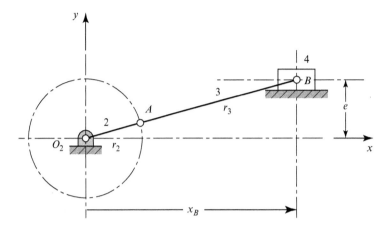

Fig. P1.36 Offset slider-crank linkage in a dead-center posture.

1.37 If the mobility of the mechanism is $m = 1$, use the Kutzbach mobility criterion to determine if the point of contact C is rolling with or without slipping.

Fig. P1.37

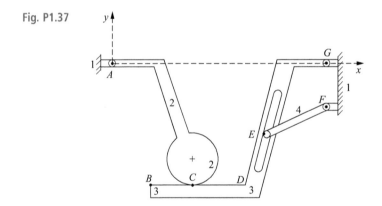

1.38 Determine the mobility of the mechanism. Number each link and identify a suitable input, or inputs, for the mechanism.

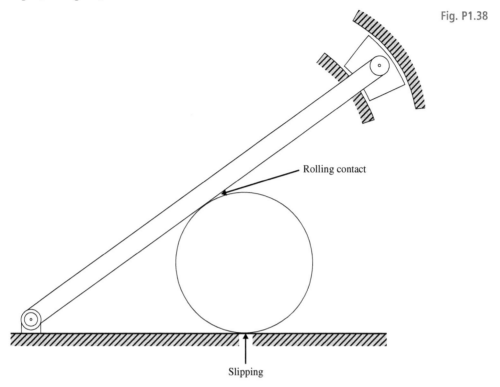

Fig. P1.38

Rolling contact

Slipping

1.39 Determine the mobility of the mechanism. Number each link and identify a suitable input, or inputs, for the mechanism.

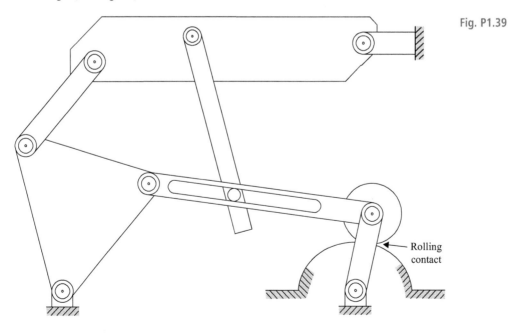

Fig. P1.39

Rolling contact

1.40 Determine the mobility of the mechanism. Number each link and identify a suitable input, or inputs, for the mechanism.

Fig. P1.40

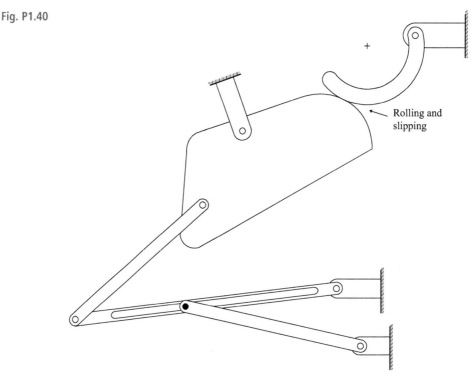

Rolling and slipping

1.41 Determine the mobility of the mechanism. Number each link and identify a suitable input, or inputs, for the mechanism.

Fig. P1.41

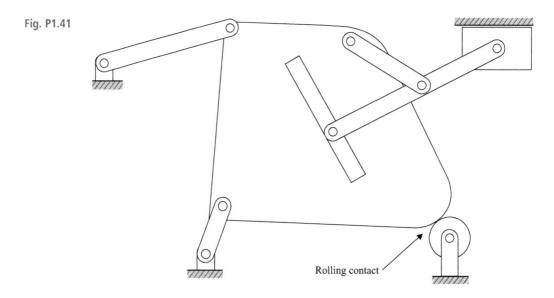

Rolling contact

1.42 Determine the mobility of the mechanism. Number each link and identify a suitable input, or inputs, for the mechanism.

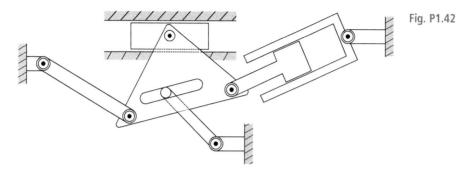

Fig. P1.42

1.43 Determine the mobility of the mechanism. Number each link and identify a suitable input, or inputs, for the mechanism.

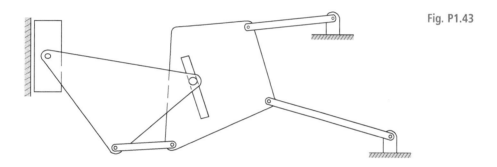

Fig. P1.43

1.44 Determine the input angles and the corresponding transmission angles of the four-bar linkage when the mechanical advantage is infinite.

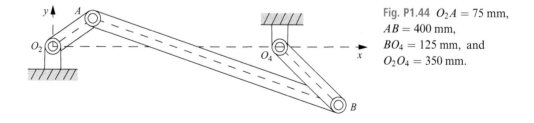

Fig. P1.44 $O_2A = 75$ mm,
$AB = 400$ mm,
$BO_4 = 125$ mm, and
$O_2O_4 = 350$ mm.

1.45 Determine the mechanical advantage of the four-bar linkage in the given posture.

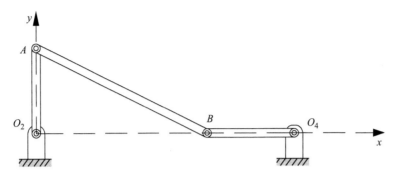

Fig. P1.45 $O_2A = 6$ in, $AB = 10$ in, $O_2O_4 = 12$ in, and $BO_4 = 4$ in.

REFERENCES

[1] Beyer, R., 1963. *The Kinematic Synthesis of Mechanisms*, London: Chapman and Hall.

[2] Euler, L., 1765. Theoria motus corporum solidorum seu rigidorum, *Commentarii Academiae Scientiarum Imperialis Petropolitanae*. Translated by R. W. Willis, 1841. *Principles of Mechanisms*, John W. Parker, London: Cambridge University Press.

[3] Grübler, M. F., 1883. Allgemeine Eigenschaften der Zwangläufigen ebenen kinematische Kette: 1, *Civilingenieur* **29**:167–200.

[4] Hartenberg, R. S., and J. Denavit, 1964. *Kinematic Synthesis of Linkages*, New York: McGraw-Hill.

[5] Holowenko, A. R., 1955. *Dynamics of Machinery*, New York: Wiley.

[6] Hrones, J. A., and G. L. Nelson, 1951. *Analysis of the Four-Bar Linkagenalysis of the Four-Bar Linkage*, Cambridge, MA: Technology Press; New York: Wiley.

[7] Kennedy, A. B. W., 1876. *Reuleaux's Kinematics of Machinery*, London: Macmillan; republished New York: Dover, 1963.

[8] Kutzbach, K., 1929. Mechanische Leitungsverzweigung, ihre Gesetze und Anwendungen, *Maschinenbau*, **8**:710–716.

[9] Torfason, L. E., 1990. A thesaurus of mechanisms, in *Mechanical Designer's Notebooks*, **5**, *Mechanisms*, edited by J. E. Shigley and C. R. Mischke, New York: McGraw-Hill, chapter 1. (Alternatively, Torfason, L. E. 1986. A thesaurus of mechanisms, in *Standard Handbook of Machine Design*, edited by J. E. Shigley and C. R. Mischke, New York: McGraw-Hill, chapter 39.)

2   Position, Posture, and Displacement

In analyzing motion, the first and most basic problem encountered is that of defining and dealing with the concepts of position, posture, and displacement. Since motion can be thought of as a series of displacements between successive positions of a point or postures of a body, it is important to understand exactly the meaning of the terms position and posture. Rules or conventions are established here to make the definitions precise.

2.1   Locus of a Moving Point

In speaking of the position of a point, or particle, we are really answering the question: Where is the point, or what is its location? We are speaking of something that exists in nature, and we are considering the question of how to express this (in words or symbols or numbers) in such a way that the meaning is clear. We soon discover that position cannot be defined on a truly absolute basis. We must define the position of a point in terms of some agreed-upon frame of reference, that is, some reference frame or reference coordinate system.

Once we have agreed upon the right-handed three-dimensional coordinate system xyz as the reference, as shown in Fig. 2.1, we can say that point P is located x units along the x axis, y units along the y axis, and z units along the z axis *from the origin O*. In this statement, we see that three vitally important parts of the definition depend on the existence of the coordinate system:

1. The *origin O* provides an agreed-upon location from which to measure the location of point P.
2. The *axes* provide agreed-upon *directions* along which the measurements are to be made; they also provide known lines and planes for the definition and measurement of angles.
3. The unit distance along any of the axes provides a scale for quantifying distances.

These observations are not restricted to the Cartesian coordinates (x, y, z) of point P. All three properties are also necessary in defining cylindric coordinates (r, θ, z), spheric coordinates (R, θ, ϕ), or any other coordinates of point P. The same properties are also required if point P is restricted to remain in a single plane and a two-dimensional coordinate system is used. No matter how it is defined, the position of a point cannot be documented without the definition of a reference frame.

The direction cosines for locating point P in Fig. 2.1 are defined as

Fig. 2.1 Right-handed three-dimensional coordinate system showing how point P is located algebraically.

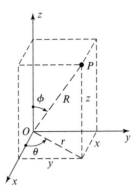

$$\cos\alpha = \frac{x}{R}, \quad \cos\beta = \frac{y}{R}, \quad \text{and} \quad \cos\gamma = \frac{z}{R}, \tag{2.1}$$

where α, β, and γ are the angles measured from the positive x-, y-, and z-coordinate axes, respectively, to the directed line OP.

One means of expressing the motion of a point, or particle, is to define its components along the reference axes as functions of some parameter, such as time:

$$x = x(t), \quad y = y(t), \quad z = z(t). \tag{2.2}$$

If these relations are known, then the position of P can be found for any time t. This is the general case for the motion of a point, or particle, and is illustrated in the following example.

Example 2.1

Describe the motion of a particle, P, whose position changes with time according to the equations $x = a\cos 2\pi t$, $y = a\sin 2\pi t$, and $z = bt$.

Solution

Substituting values for t from 0 to 2 s gives the coordinates listed in Table 2.1. As shown in Fig. 2.2, the particle moves with *helical motion* with radius a around the z axis and with a lead of b. Note that if $b = 0$, then $z(t) = 0$, the moving particle is confined to the xy plane, and the motion is a circle with its center at the origin.

Table 2.1 **Helical motion of a particle**

t	x	y	z
0	a	0	0
1/4	0	a	$b/4$
1/2	$-a$	0	$b/2$
3/4	0	$-a$	$3b/4$
1	a	0	b
5/4	0	a	$5b/4$
3/2	$-a$	0	$3b/2$
7/4	0	$-a$	$7b/4$
2	a	0	$2b$

Example 2.1 (cont.)

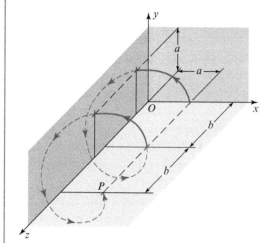

Fig. 2.2 *Helical motion of a particle, P.*

So far, we have been using the words *point* and *particle* interchangeably. Henceforth, when the word *point* is used, we have in mind something that has no dimensions – that is, something with zero length, zero width, and zero thickness. When the word *particle* is used, we have in mind something whose dimensions are unimportant – that is, a material body whose dimensions are negligible or, in other words, a body whose dimensions have no effect on the analysis to be performed.

The successive positions of a moving point define a line or curve. This curve has no thickness, since the point has no dimensions. However, the curve does have length, since the point occupies different positions as time changes. This curve, representing the successive positions of the point, is called the *path* or *locus* of the moving point in the reference coordinate system.

If three coordinates are necessary to describe the path of a moving point, the point is said to have *spatial motion*. If the path can be described by only two coordinates – that is, if the coordinate axes can be chosen such that one coordinate is always zero or constant – the path lies in a single plane, and the point is said to have *planar motion*. Sometimes it happens that the path of a point can be described by a single coordinate – that is, two of the spatial position coordinates are zero or constant. In this case, the path of the point is a straight line and the point is said to have *rectilinear motion*.

In each of these three cases, it is assumed that the coordinate system is chosen so as to obtain the least number of coordinates necessary to describe the motion of the point. Thus, a point whose locus is a *spatial curve*, sometimes called a *skew curve*, requires three position coordinates; a point whose locus is a *planar curve* requires two coordinates; and a point whose locus is a straight line (rectilinear motion) requires only one coordinate.

2.2   Position of a Point

The physical process involved in observing the position of a point, as shown in Fig. 2.3, implies that the observer is actually keeping track of the relative location of two points, P and O. The observer looks at both points, performs a mental comparison, and recognizes that point P has a certain location *with respect to point O*. In this determination, two properties are noted: the distance from O to P (based on the unit distance or grid size of the reference coordinate system), and the *relative* angular orientation of the line OP with respect to the coordinate system. Note that these properties, magnitude and direction, are precisely those required for a vector. Therefore, we define the *position of a point* as the *vector from the origin of a specified reference coordinate system to the point*. We choose the symbol $\mathbf{R}_{PO}$ to denote the vector position of point P relative to point O, which points from O to P and is read as *the position of P with respect to O*.[1]

The coordinate system is related in a very special way to what is seen by a specific observer. This raises the important question: What properties must this coordinate system have to ensure that position measurements made in this system are actually those of the observer? The key to this question is that the observer *must* be stationary with respect to the coordinate system. This means that, if the observer moves, the coordinate system also moves – through a rotation, a distance, or both. If points or bodies are fixed in this coordinate system, then they always appear stationary to the observer, regardless of what movements the observer (and the coordinate system) may execute. Their positions with respect to the observer do not change, and hence their position vectors remain unchanged. The actual location of the observer within the frame of reference has no significance because the positions of observed points are always defined with respect to the origin of the coordinate system.

Fig. 2.3 Position of a point
defined by a vector.

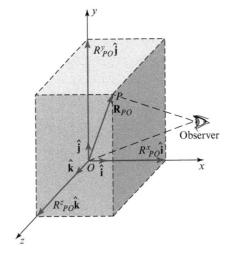

Often it is convenient to express the position vector in terms of its components along the coordinate axes:

$$\mathbf{R}_{PO} = R_{PO}^{x}\hat{\mathbf{i}} + R_{PO}^{y}\hat{\mathbf{j}} + R_{PO}^{z}\hat{\mathbf{k}}, \tag{2.3}$$

where superscripts are used to denote the direction of each component. As in the remainder of this text, $\hat{\mathbf{i}}, \hat{\mathbf{j}},$ and $\hat{\mathbf{k}}$ are used to designate unit vectors in the directions of the x axis, y axis, and z axis, respectively.

> Whereas vectors are denoted throughout this text by boldface symbols, the scalar magnitude of a vector is signified by the same symbol in italics, without boldface.

For example, the magnitude of the position vector is

$$R_{PO} = |\mathbf{R}_{PO}| = \sqrt{\mathbf{R}_{PO} \cdot \mathbf{R}_{PO}} = \sqrt{\left(R_{PO}^{x}\right)^{2} + \left(R_{PO}^{y}\right)^{2} + \left(R_{PO}^{z}\right)^{2}}. \tag{2.4}$$

The unit vector in the direction of $\mathbf{R}_{PO}$ is denoted by the same boldface symbol with a caret:

$$\hat{\mathbf{R}}_{PO} = \frac{\mathbf{R}_{PO}}{R_{PO}}. \tag{2.5}$$

A distinction can be made between the direction of a line and the orientation of a directed line – that is, a line that is assigned a positive or a negative sense. In general, a vector has a magnitude, a direction, and a sense. The sense defines the positive or negative attribute of the vector and can be used to distinguish between the direction of a vector and its orientation.

2.3   Position Difference between Two Points

We now investigate the relationship between the position vectors of two different points. For the purposes of illustration, consider points P and Q in Fig. 2.4. An observer fixed in the xyz coordinate system would observe the positions of P and Q by comparing each with the position of the origin O – see Eq. (2.3). The positions of the two points are defined by the vectors $\mathbf{R}_{PO}$ and $\mathbf{R}_{QO}$. These two vectors are related by a third vector, $\mathbf{R}_{PQ}$, the *position difference* to point P from point Q; that is,

$$\mathbf{R}_{PQ} = \mathbf{R}_{PO} - \mathbf{R}_{QO}. \tag{2.6}$$

The physical interpretation of this equation is different from that of the position vector itself. The observer is no longer defining the position of P with respect to O but is now defining the position of P with respect to Q. Put another way, in $\mathbf{R}_{PQ}$ the position of P is defined as if it were in the $x'y'z'$ coordinate system with the origin at Q and axes directed parallel to those of the xyz coordinate system. Either point of view can be used for the interpretation, but we should understand both of them since both are used in future developments.

Fig. 2.4 Definition of the
position-difference vector $\mathbf{R}_{PQ}$.

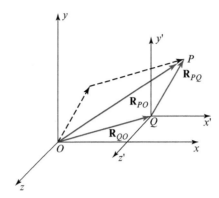

Finally, it is worth remarking that having the $x'y'z'$ axes parallel to the xyz axes is only a matter of convenience and not a necessary condition; it is only necessary that $x'y'z'$ does not rotate with respect to the xyz coordinate system. This parallelism causes no loss of generality and it simplifies the visualization when the coordinate systems are in motion.

Having now generalized our concept of position to include the position difference between any two points, we reflect again on the previous discussion of the position vector itself. We note that it is merely the special case where we agree to use the origin of coordinates as the second point. Thus, to be consistent in notation, we have denoted the position vector of a single point P by the dual subscripted symbol $\mathbf{R}_{PO}$. However, in the interest of brevity, we will henceforth agree that when the second subscript is not given explicitly, it is understood to be the origin of the observer's coordinate system; that is,

$$\mathbf{R}_P = \mathbf{R}_{PO}. \tag{2.7}$$

2.4   Apparent Position of a Point

In discussing the position vector, our point of view, up to now, has been entirely that of a single observer in a single coordinate system. However, it is sometimes desirable to make observations in a secondary coordinate system – that is, as seen by a second observer in a different coordinate system – and then to convert this information into the primary coordinate system. Such a situation is shown in Fig. 2.5.

If observer 1, using the primary coordinate system $x_1 y_1 z_1$ and observer 2, using the secondary coordinate system $x_2 y_2 z_2$ were asked to give the location of a point at P, they would report different observations. Observer 1 would observe position vector $\mathbf{R}_{PO_1}$, whereas observer 2 would report position vector $\mathbf{R}_{PO_2}$. These position vectors are related by

$$\mathbf{R}_{PO_1} = \mathbf{R}_{O_2 O_1} + \mathbf{R}_{PO_2}. \tag{2.8}$$

The difference in the positions of the two origins is not the only discrepancy between the two observations of the position of point P. Since the two coordinate systems may not be

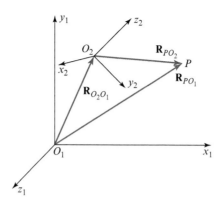

Fig. 2.5 Definition of the apparent-position vector $\mathbf{R}_{PO_2}$ of point P.

aligned, the two observers may be using different reference lines for their measurements of direction; observer 1 measures along the x_1, y_1, z_1 axes while observer 2 measures along the x_2, y_2, z_2 axes.

Another very important distinction between these two observations arises when we consider that the two coordinate systems may be moving with respect to each other. Whereas point P may appear stationary with respect to one observer, it may be in motion with respect to the other observer; that is, the position vector $\mathbf{R}_{PO_1}$ may appear constant to observer 1 while $\mathbf{R}_{PO_2}$ appears to vary as seen by observer 2.

When any of these conditions exists, it is convenient to add an additional subscript to our notation that will distinguish which observer is being considered. When we are considering the position of P as seen by observer 1 using coordinate system $x_1y_1z_1$ we denote this by the symbol $\mathbf{R}_{PO_1/1}$, or, since O_1 is the origin for this observer,[2] by $\mathbf{R}_{P/1}$. The observations made by observer 2, in coordinate system $x_2y_2z_2$ are denoted as $\mathbf{R}_{PO_2/2}$ or $\mathbf{R}_{P/2}$. With this extension of the notation, Eq. (2.8) can be written as

$$\mathbf{R}_{P/1} = \mathbf{R}_{O_2/1} + \mathbf{R}_{P/2}. \tag{2.9}$$

We refer to $\mathbf{R}_{P/2}$ as *the apparent position* of point P to an *observer in coordinate system* 2, and we note that it is by no means equal to the apparent-position vector $\mathbf{R}_{P/1}$ seen by observer 1.

We have now made note of certain intrinsic differences between $\mathbf{R}_{P/1}$ and $\mathbf{R}_{P/2}$ and found Eq. (2.9) to relate them. However, there is no reason why components of either vector must be taken along the natural axes of the observer's coordinate system. As with all vectors, components can be found along any desired set of axes; in applying Eq. (2.9), however, we must use a single consistent set of axes during the numeric evaluation. Although the observer in coordinate system 2 may find it natural to measure the components of $\mathbf{R}_{P/2}$ along the $x_2y_2z_2$ axes, these components must be transformed into the equivalent components along the $x_1y_1z_1$ axes before the addition is actually performed; that is,

---

[2] Note that $\mathbf{R}_{PO_2/1}$ cannot be abbreviated as $\mathbf{R}_{P/1}$, since O_2 is not the origin used by observer 1.

$$\begin{aligned}
\mathbf{R}_{P/1} &= \mathbf{R}_{O_2/1} + \mathbf{R}_{P/2} \\
&= R_{O_2/1}^{x_1}\hat{\mathbf{i}}_1 + R_{O_2/1}^{y_1}\hat{\mathbf{j}}_1 + R_{O_2/1}^{z_1}\hat{\mathbf{k}}_1 + R_{P/2}^{x_1}\hat{\mathbf{i}}_1 + R_{P/2}^{y_1}\hat{\mathbf{j}}_1 + R_{P/2}^{z_1}\hat{\mathbf{k}}_1 \\
&= \left(R_{O_2/1}^{x_1} + R_{P/2}^{x_1}\right)\hat{\mathbf{i}}_1 + \left(R_{O_2/1}^{y_1} + R_{P/2}^{y_1}\right)\hat{\mathbf{j}}_1 + \left(R_{O_2/1}^{z_1} + R_{P/2}^{z_1}\right)\hat{\mathbf{k}}_1 \\
&= R_{P/1}^{x_1}\hat{\mathbf{i}}_1 + R_{P/1}^{y_1}\hat{\mathbf{j}}_1 + R_{P/1}^{z_1}\hat{\mathbf{k}}_1 .
\end{aligned}$$

The addition can be performed equally well if all vector components are transformed into the $x_2 y_2 z_2$ system or, for that matter, into any other consistent coordinate system. However, they cannot be added algebraically when they have been evaluated along inconsistent coordinate systems. The additional subscript in the apparent-position vector, therefore, does not specify a set of directions to be used in the evaluation of components; it merely states the coordinate system in which the vector is defined, the coordinate system in which the observer is stationary.

2.5   Absolute Position of a Point

In Section 2.2 we learned that every position vector is defined relative to a second point, the origin of the observer's reference frame. It is one particular case of the position-difference vector studied in Section 2.3, where the reference point is the origin of the coordinate system. Then, in Section 2.4, we noted that, for certain problems, it may be convenient to consider the apparent positions of a single point as viewed by more than one observer using different coordinate systems. When a particular problem leads us to consider multiple coordinate systems, the application will lead us to choose one of the coordinate systems as primary. This primary coordinate system is commonly referred to as the *absolute coordinate system*. Most often this is the coordinate system in which the final result is to be expressed, and this coordinate system is usually stationary. The absolute position of a point is defined as the apparent position of the point as seen by an observer in the *absolute coordinate system*.

Which coordinate system is designated absolute is an arbitrary decision and unimportant in the study of kinematics. Whether the absolute coordinate system is truly stationary is also a moot point, since, as we have seen, all position (and motion) information is measured relative to something else; nothing is truly absolute in the strict sense. When analyzing the kinematics of an automobile suspension, for example, it may be convenient to choose an "absolute" coordinate system attached to the frame of the car and to study the motion of the suspension relative to this system. It is unimportant whether the car is moving; motions of the suspension relative to the frame are then defined as absolute.

It is a common convention to number the absolute coordinate system 1 and to use other numbers for other coordinate systems. Since we adopt this convention throughout this text, absolute-position vectors are those apparent-position vectors viewed by an observer in coordinate system 1 and carry symbols of the form $\mathbf{R}_{P/1}$. In the interest of brevity and to reduce complexity, we will agree that when the coordinate system number is not given explicitly, it is assumed to be 1; thus, $\mathbf{R}_{P/1}$ can be abbreviated as $\mathbf{R}_P$. Similarly, the apparent-position equation, Eq. (2.9), can be written[3] as

$$\mathbf{R}_P = \mathbf{R}_{O_2} + \mathbf{R}_{P/2}. \tag{2.10}$$

Example 2.2

The path of a moving point is defined by the equation $y = 2x^2 - 28$. Find the position difference from point P to point Q on the path where $R_P^x = 4$ and $R_Q^x = -3$.

Solution

The y components of the two vectors can be written as

$$R_P^y = 2(4)^2 - 28 = 4 \quad \text{and} \quad R_Q^y = 2(-3)^2 - 28 = -10.$$

Therefore, the two vectors can be written as

$$\mathbf{R}_P = 4\hat{\mathbf{i}} + 4\hat{\mathbf{j}} \quad \text{and} \quad \mathbf{R}_Q = -3\hat{\mathbf{i}} - 10\hat{\mathbf{j}}$$

The position difference from point P to point Q is

$$\mathbf{R}_{QP} = \mathbf{R}_Q - \mathbf{R}_P = -7\hat{\mathbf{i}} - 14\hat{\mathbf{j}} = 15.65\angle{-116.6°}. \qquad \textit{Ans.}$$

2.6   Posture of a Rigid Body

Now let us consider the term *position* when applied to something other than a point. In order to specify the location of a rigid body, for example, it is necessary to specify more than just three coordinates. It is necessary to specify enough coordinates that the location of every point of the item being located is uniquely determined. If all of these coordinates are grouped into a single quantity according to some agreed-upon set of conventions, then the result describes the location of the item.

In the case of a single rigid body, the location and orientation of a coordinate system fixed to the body with respect to a stationary reference, or world coordinate system, describes the *posture* of that body. Posture is described in terms of the location of the origin of the coordinate system fixed to the body as well as a description of the orientation of this coordinate system, both specified with respect to the world coordinate system.

---

[3] Reviewing Sections 2.1–2.3 will verify that the position-difference vector $\mathbf{R}_{PQ}$ was treated entirely from the absolute coordinate system and is an abbreviation of the notation $\mathbf{R}_{PQ/1}$. We have no need to treat the completely general case $\mathbf{R}_{PQ/2}$, the apparent-position-difference vector.

In the robotics literature, for example, a matrix is often used to describe the location and orientation of a coordinate system attached to the end-effector with respect to a base or world coordinate system (Chapter 10). In some robotics literature, the word "position" is used loosely to describe the location of only a single point (such as the origin) of a coordinate system attached to the end-effector. In such literature, the end-effector is said to have a certain position; the orientation may then be added and the term *pose* is sometimes used for the combination of the two. It should be pointed out, however, that the term *posture* is more appropriate[4] and is utilized throughout this text.

The term posture becomes even more suitable when dealing with a mechanism or multibody mechanical system, because we are not concerned with the position of only a single point or the position and orientation of only a single rigid body, but we wish to describe the positions and orientations of an assembly of rigid bodies. We use the term posture to describe the location and orientation of a rigid body or for the configuration of a mechanism, including both the locations and orientations of every link, all at a particular instant in time.

The problem of posture analysis is to determine the values of all position variables (the positions of all points and joints) and the postures of all links, given the dimensions of each link and the value(s) of the independent variable(s) – that is, the variables chosen to represent the degree(s) of freedom of the system.

2.7   Loop-Closure Equations

Our discussion of the position-difference and apparent-position vectors has been somewhat abstract so far, the intent being to develop a rigorous foundation for the analysis of motion in mechanical systems. Certainly, precision is not without merit, since it is rigor that permits science to predict a correct result despite the personal prejudices and emotions of the analyst. However, tedious developments are not interesting unless they lead to real-life applications. Although many fundamental principles are yet to be discovered, it may be worthwhile at this point to show the relationship between the relative-position vectors discussed so far and some typical situations encountered in real machines.

As pointed out in Chapter 1, one of the most common and most useful of all mechanisms is the planar four-bar linkage. A practical example of this linkage is the clamping device shown in Figs. 2.6 and 2.7. A brief study of the assembly drawing indicates that, as the handle of the clamp is lifted, the clamping bar swings away from the clamping surface, thereby opening the clamp. As the handle is pressed down, the clamping bar swings down and the clamp closes again. If we wish to design such a clamp, however, things are not quite this simple. It may be desirable, for example, for the clamp to open at a given rate for a specified rate of lift of the handle. Such relationships are not obvious; they depend on the exact dimensions of the various parts and the relationships, or interactions, between the parts. To discover these relationships, a

---

[4] The *Webster Comprehensive Dictionary: International Edition*, states under its definition of attitude, "Synonyms: pose, position, posture. A posture is assumed without any special reference to expression of feeling; ... A pose is a position studied for artistic effect or considered with reference to such an effect."

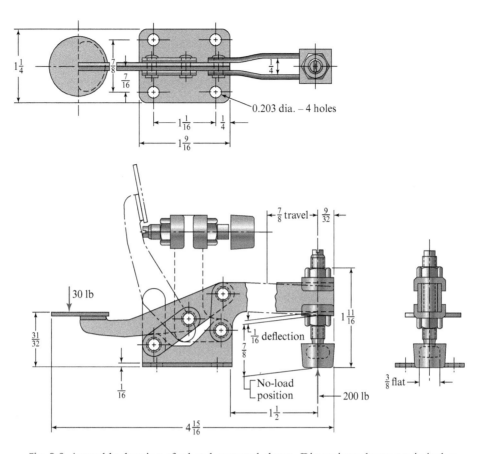

Fig. 2.6 Assembly drawing of a hand-operated clamp. Dimensions shown are in inches.

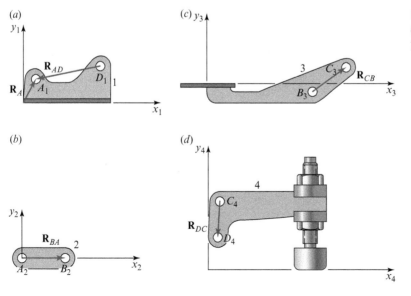

Fig. 2.7 (*a*) Frame link; (*b*) connecting link; (*c*) handle; (*d*) clamping bar.

rigorous description of the essential features of the device is required. The position-difference and apparent-position vectors can be used to provide a description of any posture of the clamp.

Figure 2.7 shows the detail drawings of the individual links of the disassembled clamp. Although not shown, the detail drawings would be completely dimensioned, thus fixing the complete geometry of each link. The assumption that each link is rigid ensures that the position of any point is determined precisely relative to any other point on the same link. However, the detail drawings do not provide the interrelationships between the individual parts – that is, the constraints that ensure each link moves relative to its neighboring links in a prescribed fashion. These constraints are, of course, provided by the four pinned joints. Anticipating that they will be of importance in any description of the linkage, we label these pin centers as $A, B, C,$ and $D,$ and we identify the appropriate points on link 1 as A_1 and D_1, those on link 2 as A_2 and B_2, and so on. We also choose a coordinate system fixed to each link, as shown in Fig. 2.7.

Since it is necessary to relate the positions of the successive pin centers, we define the position-difference vectors $\mathbf{R}_{AD}$ on link 1, $\mathbf{R}_{BA}$ on link 2, $\mathbf{R}_{CB}$ on link 3, and $\mathbf{R}_{DC}$ on link 4. We note that each vector appears constant to an observer fixed in the coordinate system of that link; the magnitude and direction of each vector can be obtained from the constant dimensions of that link.

A vector equation can also be written to describe the constraints provided by each of the pinned joints. Note that no matter which position or which observer is chosen, the two points describing each pin center (for example, A_1 and A_2) remain coincident. Thus,

$$\mathbf{R}_{A_2 A_1} = \mathbf{R}_{B_3 B_2} = \mathbf{R}_{C_4 C_3} = \mathbf{R}_{D_1 D_4} = \mathbf{0}. \tag{2.11}$$

Let us now develop vector equations for the absolute position of each pin center. Since link 1 is the frame, absolute positions are those defined relative to an observer in coordinate system 1. Point A_1 is, of course, at the position described by $\mathbf{R}_A$. Next, we connect link 2 to link 1 (mathematically) by writing

$$\mathbf{R}_{A_2} = \mathbf{R}_{A_1} + \overset{0}{\mathbf{R}_{A_2 A_1}} = \mathbf{R}_A. \tag{a}$$

Transferring to the other end of link 2, we attach link 3. Therefore,

$$\mathbf{R}_B = \mathbf{R}_A + \mathbf{R}_{BA}. \tag{b}$$

Connecting joints C and D in the same manner, we obtain

$$\mathbf{R}_C = \mathbf{R}_B + \mathbf{R}_{CB} = \mathbf{R}_A + \mathbf{R}_{BA} + \mathbf{R}_{CB}, \tag{c}$$

$$\mathbf{R}_D = \mathbf{R}_C + \mathbf{R}_{DC} = \mathbf{R}_A + \mathbf{R}_{BA} + \mathbf{R}_{CB} + \mathbf{R}_{DC}. \tag{d}$$

Then, transferring back across link 1 to point A, we have

$$\mathbf{R}_A = \mathbf{R}_D + \mathbf{R}_{AD} = \mathbf{R}_A + \mathbf{R}_{BA} + \mathbf{R}_{CB} + \mathbf{R}_{DC} + \mathbf{R}_{AD}. \tag{e}$$

Finally, rearranging Eq. (e), we obtain

$$\mathbf{R}_{BA} + \mathbf{R}_{CB} + \mathbf{R}_{DC} + \mathbf{R}_{AD} = \mathbf{0}. \tag{2.12}$$

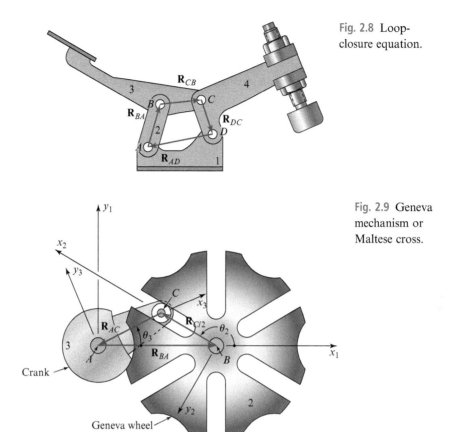

Fig. 2.8 Loop-closure equation.

Fig. 2.9 Geneva mechanism or Maltese cross.

This important equation is called the *loop-closure* equation, or the *vector loop* equation, for the clamp. As shown in Fig. 2.8, it expresses the fact that this linkage forms a closed loop. Therefore, the polygon formed by the position-difference vectors through successive links and joints must remain closed as the linkage moves. The constant lengths of these vectors ensure that the joint centers remain separated by constant distances, a requirement for rigid links. The rotations between successive vectors indicate the motions within the pinned joints. Note that the rotation of a position-difference vector shows the rotation of a particular link. Thus, the loop-closure equation holds within it all the important constraints that illustrate the operation of the clamp. This equation forms a mathematical description, or *model*, of the linkage, and many later developments are based on this loop-closure model as a starting point.

Of course, the form of the loop-closure equation depends on the type of mechanism. This is demonstrated by another example, the *Geneva* mechanism or *Maltese* cross, shown in Fig. 2.9. One early application of this mechanism was to prevent over-winding a watch. Today the mechanism finds wide use as an indexing device, for example, in a milling machine with an automatic tool changer.

First, we define vectors for the absolute posture of each link. Although the frame of the mechanism, link 1, is not shown in Fig. 2.9, it is an important part of the mechanism, since it holds the two shafts, with centers A and B, a constant distance apart. Thus we define the vector

R_{BA} to show this dimension. The left crank, link 3, is attached to a shaft, usually rotating at constant speed, and carries a roller at C, running in a slot of the Geneva wheel. The vector $\mathbf{R}_{AC}$ has constant magnitude equal to the crank length, the distance from the center of roller C to shaft center A. The rotation of this vector relative to link 1 is used later to describe the angular speed of the crank. The x_2 axis is aligned along one slot of the wheel; thus the roller is constrained to ride along this slot. The vector $\mathbf{R}_{C/2}$ has the same rotation as the wheel, link 2; also, the change in length, $\Delta\mathbf{R}_{C/2}$, demonstrates the relative sliding motion taking place between the roller C and the slot in link 2. Note that $\mathbf{R}_{C/2}$ is equivalent to $\mathbf{R}_{CB/2}$ since point B is the origin of coordinate system 2. Therefore, the loop-closure equation for this mechanism can be written as

$$\mathbf{R}_{BA} + \mathbf{R}_{C/2} + \mathbf{R}_{AC} = \mathbf{0}. \tag{2.13}$$

This form of the loop-closure equation is a valid mathematical model only while roller C travels in the slot along the x_2 axis. However, this condition does not hold throughout the entire cycle of motion. Once the roller leaves the slot, the motion is controlled by the two mating circular arcs on links 2 and 3. A new form of the loop-closure equation rules that portion of the cycle.

A mechanism can, of course, be based on a multiple-loop kinematic chain. In such a case, more than one loop-closure equation is required to completely model the mechanism. The procedures for obtaining the loop-closure equations, however, are similar to those shown in the previous two examples.

2.8   Graphic Posture Analysis

When the paths of the points in the moving links of a mechanism lie in a single plane or in parallel planes, the mechanism is called a *planar* mechanism. Since a substantial portion of the mechanisms in use today are planar mechanisms, the development of methods specific to such problems is justified. As we will see in the following section, the nature of the loop-closure equation approached analytically often leads to the solution of simultaneous nonlinear equations and can become quite cumbersome. Yet, particularly for planar mechanisms, the solution is usually straightforward when approached graphically.

First, let us briefly review the process of graphic vector addition. Any two known vectors $\mathbf{A}$ and $\mathbf{B}$ can be added graphically, as shown in Fig. 2.10a. After a scale is chosen, the vectors are drawn tip to tail in either order and their sum $\mathbf{C}$ is identified:

$$\mathbf{C} = \mathbf{A} + \mathbf{B} = \mathbf{B} + \mathbf{A}. \tag{2.14}$$

Note that the magnitudes and orientations of both vectors $\mathbf{A}$ and $\mathbf{B}$ are used in performing the addition, and that both the magnitude and orientation of the sum $\mathbf{C}$ are found as a result.

The operation of graphic vector subtraction is shown in Fig. 2.10b, where the vectors are drawn tip to tip (or tail to tail) in solving the equation

$$\mathbf{A} = \mathbf{C} - \mathbf{B}. \tag{2.15}$$

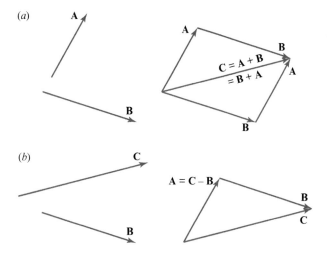

Fig. 2.10 (a) Vector addition; (b) vector subtraction.

These graphic vector operations should be studied carefully and understood, since they are used extensively throughout the book.

A spatial (three-dimensional) vector equation, such as

$$\mathbf{C} = \mathbf{D} + \mathbf{E} + \mathbf{B}, \tag{a}$$

can be divided into components along any convenient axes, leading to the three scalar equations

$$C^x = D^x + E^x + B^x, \quad C^y = D^y + E^y + B^y, \quad \text{and} \quad C^z = D^z + E^z + B^z. \tag{b}$$

Since they are components of the same vector equation, these three scalar equations must be consistent. If the three equations are also linearly independent, they can be solved simultaneously for three unknowns, which may be three magnitudes, three directions, or any combination of three magnitudes and directions. For some combinations, however, the problem is highly nonlinear and quite difficult to solve analytically. Therefore, we shall delay consideration of the three-dimensional case until it is needed in Chapter 10.

A planar (two-dimensional) vector equation can be solved for two unknowns – namely, two magnitudes, two directions, or one magnitude and one direction. In general, it is desirable to indicate the known ($\surd$) and unknown (?) quantities above each vector in an equation, as shown here:

$$\overset{?\surd}{\mathbf{C}} = \overset{\surd\surd}{\mathbf{D}} + \overset{\surd\surd}{\mathbf{E}} + \overset{?\surd}{\mathbf{B}}, \tag{c}$$

where the first symbol ($\surd$ or ?) above each vector indicates the state of its magnitude and the second symbol indicates the state of its direction. Another, equivalent, form is

$$\overset{?\ \surd}{C\hat{\mathbf{C}}} = \overset{\surd\ \surd}{D\hat{\mathbf{D}}} + \overset{\surd\ \surd}{E\hat{\mathbf{E}}} + \overset{?\ \surd}{B\hat{\mathbf{B}}}. \tag{d}$$

Either of Eqs. (c) or (d) clearly identifies the unknowns and indicates whether a solution is possible. In Eq. (c) or (d), the vectors $\mathbf{D}$ and $\mathbf{E}$ are completely defined and can be replaced by their sum,

$$\overset{\surd\surd}{\mathbf{A}} = \overset{\surd\surd}{\mathbf{D}} + \overset{\surd\surd}{\mathbf{E}}, \tag{e}$$

Table 2.2 **Unknowns in planar vector equations**

Case	Unknowns		
1	C, θ_C	Fig. 2.10	Eq. (2.14)
2	A, θ_B	Fig. 2.11	Eq. (2.17)
3	A, B	Fig. 2.12	Eq. (2.18)
4	θ_A, θ_B	Fig. 2.13	Eq. (2.19)

giving

$$\overset{?\surd}{\mathbf{C}} = \overset{\surd\surd}{\mathbf{A}} + \overset{?\surd}{\mathbf{B}}. \tag{2.16}$$

Note that any planar vector equation, if it is solvable, can be reduced to a three-term equation with two unknowns. Depending on the forms of the two unknowns, four distinct cases occur. These cases and the corresponding unknowns are presented in Table 2.2.

We will now show the graphic solutions of these four cases.

Case 1: The two unknowns are the magnitude and the direction of the same vector (say, C and θ_C). This case can be solved by graphic addition or subtraction of the remaining two vectors, which are completely defined. This case was shown in Fig. 2.10.

Case 2: A magnitude and a direction from different vectors (say, A and θ_B) are to be found. The vector equation is

$$\overset{\surd\surd}{\mathbf{C}} = \overset{?\surd}{\mathbf{A}} + \overset{\surd?}{\mathbf{B}}. \tag{2.17}$$

The solution, shown in Fig. 2.11, is obtained as follows:
1. Choose a convenient scale factor and draw vector $\mathbf{C}$.
2. Construct a line through the origin of $\mathbf{C}$ in the direction θ_A.
3. Adjust a compass to the scaled magnitude B and construct a circular arc with center at the terminus of $\mathbf{C}$.
4. The two intersections of the line and the arc define the two sets of solutions A, θ_B and A', θ'_B.

Case 3: The two unknowns are two magnitudes (say, A and B). The vector equation is

$$\overset{\surd\surd}{\mathbf{C}} = \overset{?\surd}{\mathbf{A}} + \overset{?\surd}{\mathbf{B}}. \tag{2.18}$$

The solution, shown in Fig. 2.12, is obtained as follows:
1. Choose a convenient scale factor and draw vector $\mathbf{C}$.
2. Construct a line through the origin of $\mathbf{C}$ in the direction θ_A.
3. Construct another line through the terminus of $\mathbf{C}$ in the direction θ_B.
4. The intersection of these two lines defines both magnitudes, A and B, each of which may be either positive or negative.

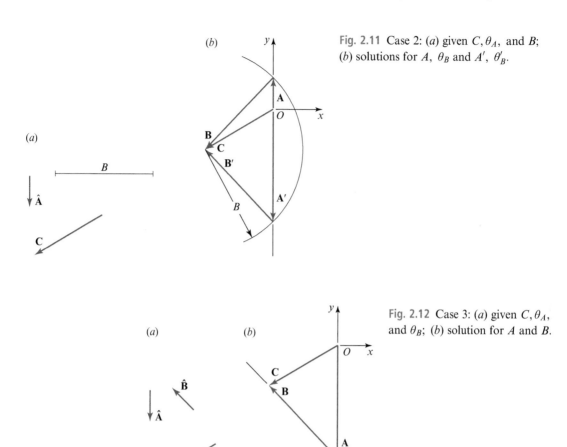

Fig. 2.11 Case 2: (a) given C, θ_A, and B; (b) solutions for A, θ_B and A', θ'_B.

Fig. 2.12 Case 3: (a) given C, θ_A, and θ_B; (b) solution for A and B.

Note that case 3 has a unique solution unless the lines are collinear or parallel. If the lines are collinear, the magnitudes A and B are both indeterminate. If the lines are parallel but distinct, the magnitudes A and B are both infinite.

Case 4: The two unknowns are the directions of two vectors, θ_A and θ_B. The vector equation is

$$\overset{\sqrt{\sqrt{}}}{\mathbf{C}} = \overset{\sqrt{?}}{\mathbf{A}} + \overset{\sqrt{?}}{\mathbf{B}}. \tag{2.19}$$

The solution, shown in Fig. 2.13, is obtained as follows:
1. Choose a convenient scale factor and draw vector $\mathbf{C}$.
2. Construct a circular arc of radius A centered at the origin of $\mathbf{C}$.
3. Construct a circular arc of radius B centered at the terminus of $\mathbf{C}$.
4. The two intersections of these arcs define the two sets of solutions θ_A, θ_B and θ'_A, θ'_B.

Note that a real solution is possible only if $A + B \geq C$.

Fig. 2.13 Case 4: (*a*) given **C**, **A**, and **B**; (*b*) solution for θ_A, θ_B and θ'_A, θ'_B.

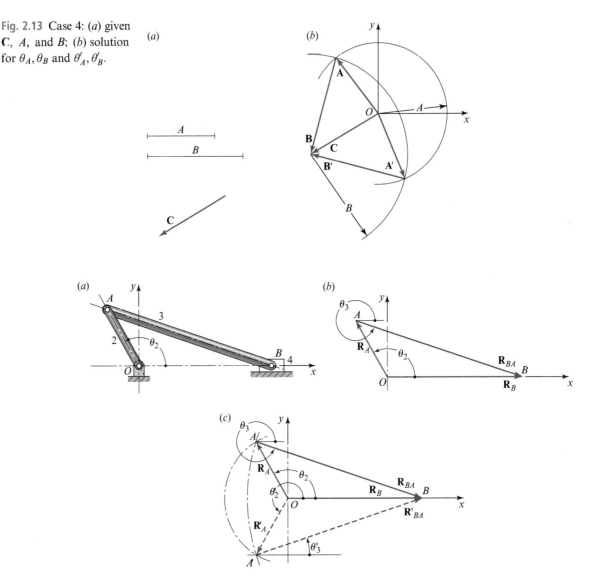

Fig. 2.14 (*a*) Slider-crank linkage; (*b*) vectors replace the link centerlines; (*c*) graphic posture analysis.

The graphic procedures of this section are now applied to obtain solutions of the loop-closure equations for the slider-crank linkage and the four-bar linkage.

Slider-Crank Linkage.

Consider the slider-crank linkage shown in Fig. 2.14*a*. Given the position of link 4, the problem is to determine the postures of links 2 and 3.

Replace the link centerlines with vectors $\mathbf{R}_A$ for the crank, $\mathbf{R}_B$ for the slider, and $\mathbf{R}_{BA}$ for the connecting rod, as shown in Fig. 2.14*b*. With a known location of the slider (that is, the distance R_B) and the angles θ_2 and θ_3 unknown, the loop-closure equation for the linkage is

$$\overset{\sqrt{?}}{\mathbf{R}_A} + \overset{\sqrt{?}}{\mathbf{R}_{BA}} - \overset{I\sqrt{}}{\mathbf{R}_B} = \mathbf{0},$$

which can be written as

$$\overset{I\sqrt{}}{\mathbf{R}_B} = \overset{\sqrt{?}}{\mathbf{R}_A} + \overset{\sqrt{?}}{\mathbf{R}_{BA}}, \tag{a}$$

where the "I" symbol, for the magnitude of the vector $\mathbf{R}_B$, denotes that the location of the slider is the given input.

We recognize Eq. (a) as case 4 in Table 2.2.

The graphic solution procedure explained in Fig. 2.13 is now carried out in Fig. 2.14c. We note that there are two possible solutions (θ_2, θ_3 and θ'_2, θ'_3), which correspond to two different postures of the linkage; that is, two configurations of the links, both of which are consistent with the given position of the slider. We know in advance which of the two solutions is sought; both sets of results are equally valid roots to the loop-closure equation, and the choice is based on the application.

Four-Bar Linkage. Consider the four-bar linkage shown in Fig. 2.15. Given the posture of link 2, the problem is to define the postures of links 3 and 4 and the position of coupler point P.

Replace the link centerlines with vectors $\mathbf{R}_A$ for the crank, $\mathbf{R}_{BA}$ for the connecting rod, $\mathbf{R}_{BC}$ for the output link, and $\mathbf{R}_C$ for the frame, as shown in Fig. 2.16a. With a known posture of

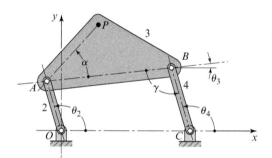

Fig. 2.15 Four-bar linkage and coupler point P.

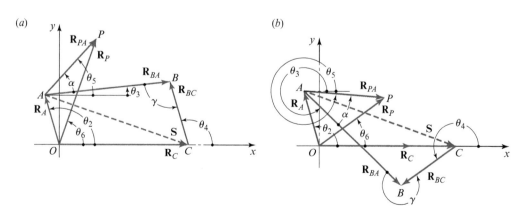

Fig. 2.16 Vector diagrams showing the graphic solutions for: (a) the open posture; (b) the crossed posture.

link 2, that is, the angle θ_2, and with angles θ_3 and θ_4 unknown, the loop-closure equation for the linkage is

$$\overset{\sqrt{I}}{\mathbf{R}_A} + \overset{\sqrt{?}}{\mathbf{R}_{BA}} - \overset{\sqrt{?}}{\mathbf{R}_{BC}} - \overset{\sqrt{\sqrt{}}}{\mathbf{R}_C} = \mathbf{0}, \qquad (a)$$

which can be rearranged as

$$\overset{\sqrt{I}}{\mathbf{R}_A} + \overset{\sqrt{?}}{\mathbf{R}_{BA}} = \overset{\sqrt{\sqrt{}}}{\mathbf{R}_C} + \overset{\sqrt{?}}{\mathbf{R}_{BC}}, \qquad (b)$$

where the "I" symbol, for the direction of vector $\mathbf{R}_A$, denotes that the orientation of the crank is the given input.

Also, the position of coupler point P is given by the position-difference equation,

$$\overset{??}{\mathbf{R}_P} = \overset{\sqrt{I}}{\mathbf{R}_A} + \overset{\sqrt{?}}{\mathbf{R}_{PA}}. \qquad (c)$$

Note that this equation contains three unknowns, but this can be reduced to two after Eq. (b) is solved by noting the constant angular relationship between $\hat{\mathbf{R}}_{PA}$ and $\hat{\mathbf{R}}_{BA}$ – namely,

$$\theta_5 = \theta_3 + \alpha, \qquad (d)$$

where the constant coupler angle $\alpha = \angle BAP$. Therefore, Eq. (c) reduces to

$$\overset{??}{\mathbf{R}_P} = \overset{\sqrt{I}}{\mathbf{R}_A} + \overset{\sqrt{C}}{\mathbf{R}_{PA}}, \qquad (e)$$

where the "C" symbol, for the direction of $\mathbf{R}_{PA}$, denotes that this direction is known from the constraint equation – that is, Eq. (d) – and, therefore, is not a third unknown.

We begin the graphic solution by combining the two unknown terms in Eq. (b), thus locating the positions of points A and C as shown in Figs. 2.16a and 2.16b:

$$\overset{\sqrt{\sqrt{}}}{\mathbf{S}} = \overset{\sqrt{\sqrt{}}}{\mathbf{R}_C} - \overset{\sqrt{I}}{\mathbf{R}_A} = \overset{\sqrt{?}}{\mathbf{R}_{BA}} - \overset{\sqrt{?}}{\mathbf{R}_{BC}}. \qquad (f)$$

We recognize this equation as case 4 in Table 2.2. The solution procedure (that is, two unknown orientations) is used to locate point B. Note that two solutions are possible, which indicates that the four-bar linkage can be assembled in two different postures. The two postures are referred to here as: (a) the open posture shown in Fig. 2.16a; and (b) the crossed posture shown in Fig. 2.16b.

To determine the position of coupler point P, first we obtain the two orientations of $\hat{\mathbf{R}}_{PA}$ from Eq. (d). Then we solve Eq. (c) by the procedure for case 1. Two solutions are obtained for the position of point P, as shown in Figs. 2.16a and 2.16b. Both are valid solutions to Eqs. (b) through (d), although, in this instance, the crossed posture shown could not be reached from the open posture without first disassembling the linkage.

From the two examples presented here it is clear that graphic posture analysis usually requires precisely the same constructions that would be chosen naturally in drafting scale drawings of a mechanism at the posture under consideration. For this reason, the procedure may seem trivial and not truly worthy of the title "analysis." Yet this is highly misleading. As we shall see in later sections of this chapter, the posture analysis of a mechanism is a nonlinear algebraic problem when approached by analytic or numeric methods. It is, in fact, the most difficult problem in kinematic analysis, and this is one primary reason why graphic solution techniques have retained their attraction.

2.9 Algebraic Posture Analysis

To illustrate the classic algebraic approach for the posture analysis of planar mechanisms, this section presents the same two examples of the previous section – namely, the slider-crank linkage and the four-bar linkage.

Slider-Crank Linkage. For generality, the offset version of this linkage, shown in Fig. 2.17, is chosen for analysis. By making the offset or eccentricity $e = R_{O_2 O} = 0$, the resulting equations can be used for the in-line (on-center or symmetric) version. The notation in Fig. 2.17 shows that the input crank angle θ_2 is measured around the origin of vector $\mathbf{r}_2 = \mathbf{R}_{AO_2}$, and the connecting rod angle θ_3 is measured around the origin of vector $\mathbf{r}_3 = \mathbf{R}_{BA}$. Both are taken as positive (counterclockwise) angles.

There are two problems that occur in the posture analysis of the slider-crank linkage – namely:

Problem 1: Given the input crank angle θ_2, find the connecting rod angle θ_3 and the slider position x_B.

Problem 2: Given the slider position x_B, find the crank angle θ_2 and the connecting rod angle θ_3. This is the problem that we addressed in the previous section.

Solution to Problem 1: First, we define the position of point A by the equations

$$x_A = r_2 \cos \theta_2 \quad \text{and} \quad y_A = e + r_2 \sin \theta_2. \qquad (2.20)$$

Next, we note that

$$e + r_2 \sin \theta_2 = r_3 \sin \theta_3, \qquad (a)$$

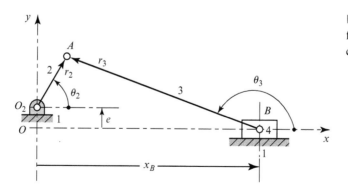

Fig. 2.17 Notation for the offset slider-crank linkage.

Fig. 2.18 Newton–Raphson method.

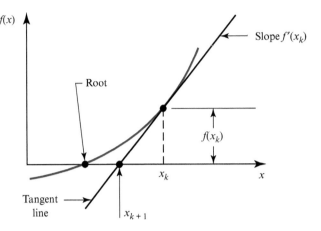

so that

$$\sin \theta_3 = \frac{1}{r_3}(e + r_2 \sin \theta_2). \tag{2.21}$$

From the geometry of Fig. 2.17, we see that

$$x_B = r_2 \cos \theta_2 - r_3 \cos \theta_3. \tag{b}$$

Substituting Eq. (2.21) into the trigonometric identity $\cos \theta_3 = \pm\sqrt{1 - \sin^2 \theta_3}$ gives

$$\cos \theta_3 = -\frac{1}{r_3}\sqrt{r_3^2 - (e + r_2 \sin \theta_2)^2}. \tag{c}$$

Here the negative sign corresponds to an obtuse angle for θ_3 when measured around point B, as shown in Fig. 2.17. (A positive sign would imply an acute angle for θ_3.)

Finally, substituting Eq. (c) into Eq. (b) gives

$$x_B = r_2 \cos \theta_2 + \sqrt{r_3^2 - (e + r_2 \sin \theta_2)^2}. \tag{2.22}$$

Thus, with the angle θ_2 given, the unknowns θ_3 and x_B can be obtained by solving Eqs. (2.21), (c), and (2.22).

Solution to Problem 2: Given x_B, we must solve Eq. (2.22) for the angle θ_2. This requires the use of a calculator or a computer together with a root-finding technique. Here we select the well-known Newton–Raphson method.[5] This method can be explained by reference to Fig. 2.18, which is a graph of some function $f(x)$ versus x.

Suppose we wish to find a root of the homogeneous equation $f(x) = 0$, and we choose x_k as an approximation (a rough estimate) of the root which we wish to find. A tangent to the curve at $x = x_k$ intersects the x axis at x_{k+1}, which is a better approximation to the root. The slope of the tangent is equal to the derivative of the function at $x = x_k$ and is

---

[5] See, for example, [1].

$$f'(x_k) = \frac{f(x_k)}{x_k - x_{k+1}}. \tag{d}$$

Solving this for x_{k+1} gives

$$x_{k+1} = x_k - \frac{f(x_k)}{f'(x_k)}. \tag{2.23}$$

Using a calculator, for example, to start a solution, we enter an estimate, x_k, solve Eq. (2.23) for x_{k+1}, use this as the next estimate, and repeat this process as many times as needed to obtain the result with satisfactory accuracy. The accuracy is evaluated by comparing $(x_{k+1} - x_k)$ with a small number ε after each repetition and continuing until $(x_{k+1} - x_k) < \varepsilon$.

We note that the root-finding programs built into many calculators utilize an approximation to obtain $f'(x_k)$. These are sometimes of little value in solving Eq. (2.22), so we proceed as follows. In Eq. (2.22), we replace the angle θ_2 with the symbol z, and let r_2, r_3, e, and x_B be given constants. Then

$$f(z) = r_2 \cos z + \sqrt{r_3^2 - (e + r_2 \sin z)^2} - x_B, \tag{e}$$

and

$$f'(z) = -r_2 \sin z - \frac{(e + r_2 \sin z)r_2 \cos z}{\sqrt{r_3^2 - (e + r_2 \sin z)^2}}. \tag{f}$$

These two equations can now be programmed with Eq. (2.23) to solve for the unknown value of the angle $\theta_2 = z$ when x_B is given. We note that the angle θ_2 has two possible values. These may be found separately by using different initial estimates.

A closed-form algebraic solution is possible if the eccentricity e is zero – that is, if the linkage is centered. For this case, we take Eqs. (a) and (b), square them, and add them together. Noting a trigonometric identity, the result is

$$x_B^2 - 2x_B r_2 \cos \theta_2 + r_2^2 = r_3^2. \tag{g}$$

Solving for θ_2 gives

$$\theta_2 = \cos^{-1} \frac{x_B^2 + r_2^2 - r_3^2}{2x_B r_2}. \tag{2.24}$$

The solution of Problem 1 for the centered version is, of course, obtained directly from Eq. (2.22) by setting $e = 0$.

Four-Bar Linkage.

To obtain the analytic solution, the distance R_{AO_4} is designated as S in Fig. 2.19 (as it was in Fig. 2.16a and 2.16b). The cosine law for each of the two triangles AO_2O_4 and ABO_4 can now be written in terms of the angles and link lengths; that is,

$$S = \sqrt{r_1^2 + r_2^2 - 2r_1 r_2 \cos \theta_2}, \tag{2.25}$$

Fig. 2.19 Four-bar linkage.

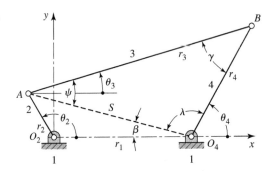

$$\beta = \cos^{-1} \frac{r_1^2 + S^2 - r_2^2}{2r_1 S}, \qquad (2.26)$$

$$\psi = \cos^{-1} \frac{r_3^2 + S^2 - r_4^2}{2r_3 S}, \qquad (2.27)$$

$$\lambda = \cos^{-1} \frac{r_4^2 + S^2 - r_3^2}{2r_4 S}. \qquad (2.28)$$

Note that Eqs. (2.26)–(2.28) can each yield double values, since they involve inverse cosines. Here we take β, ψ, and λ as positive numeric values. Then a careful study of the figure shows that, when θ_2 is in the range $0 \le \theta_2 \le \pi$, the unknown angles are

$$\theta_3 = -\beta \pm \psi, \qquad (2.29)$$

$$\theta_4 = \pi - \beta \mp \lambda. \qquad (2.30)$$

However, when θ_2 is in the range $\pi \le \theta_2 \le 2\pi$, then

$$\theta_3 = \beta \pm \psi, \qquad (2.31)$$

$$\theta_4 = \pi + \beta \mp \lambda. \qquad (2.32)$$

In these results, the upper signs correspond to the open posture whereas the lower signs correspond to the crossed posture.

In Section 1.10, the concept of transmission angle was discussed in connection with the subject of mechanical advantage. Now, with Fig. 2.19 and using the cosine law, the transmission angle can be written as

$$\gamma = \pm \cos^{-1} \frac{r_3^2 + r_4^2 - S^2}{2r_3 r_4}. \qquad (2.33)$$

2.10 Complex Polar Algebra

In planar problems, it is often desirable to express a vector by specifying its magnitude and orientation in *polar notation*:

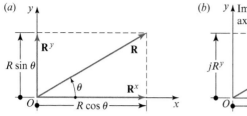
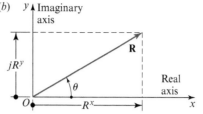

Fig. 2.20 Correlation of planar vectors and complex values.

$$\mathbf{R} = R \angle \theta. \tag{a}$$

In Fig. 2.20a, the two-dimensional vector

$$\mathbf{R} = R^x \hat{\mathbf{i}} + R^y \hat{\mathbf{j}} \tag{2.34}$$

has two rectangular components with magnitudes

$$R^x = R \cos \theta \quad \text{and} \quad R^y = R \sin \theta, \tag{2.35}$$

where

$$R = \sqrt{(R^x)^2 + (R^y)^2} \quad \text{and} \quad \theta = \tan^{-1} \frac{R^y}{R^x}. \tag{2.36}$$

Note that we have made the arbitrary choice here of accepting the positive square root for magnitude R when calculating from the components of $\mathbf{R}$. Therefore, we must be careful to interpret the signs of R^x and R^y individually when deciding upon the quadrant of the angle θ. Note that θ is defined as the angle from the positive x axis to the positive end of vector $\mathbf{R}$, measured about the origin of the vector, and this angle is positive when measured counterclockwise.

If we had chosen to accept the negative square root for the magnitude R, then we would get $\theta \pm 180°$ for the orientation, and the two components in Eq. (2.35) would remain the same. However, in the interest of clarity and consistency, we will use the positive square root so that a vector always has a positive magnitude throughout this text.

Example 2.3

Express vectors $\mathbf{A} = 10 \angle 30°$ and $\mathbf{B} = 8 \angle -15°$ in rectangular notation[6] and find their sum.

Solution

The vectors are shown in Fig. 2.21 and can be written as

$$\mathbf{A} = 10 \cos 30° \hat{\mathbf{i}} + 10 \sin 30° \hat{\mathbf{j}} = 8.660 \hat{\mathbf{i}} + 5.000 \hat{\mathbf{j}} \qquad \textit{Ans.}$$

$$\mathbf{B} = 8 \cos(-15°) \hat{\mathbf{i}} + 8 \sin(-15°) \hat{\mathbf{j}} = 7.727 \hat{\mathbf{i}} - 2.071 \hat{\mathbf{j}} \qquad \textit{Ans.}$$

$$\mathbf{C} = \mathbf{A} + \mathbf{B} = (8.660 + 7.727) \hat{\mathbf{i}} + (5.000 - 2.071) \hat{\mathbf{j}} = 16.387 \hat{\mathbf{i}} + 2.929 \hat{\mathbf{j}}.$$

[6] Many calculators are equipped to perform polar–rectangular and rectangular–polar conversions directly.

Example 2.3 (cont.)

Fig. 2.21 Addition of vectors.

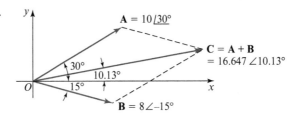

The magnitude of the resultant is determined from Eq. (2.36),

$$C = \sqrt{16.387^2 + 2.929^2} = 16.647,$$

and the angle is

$$\theta = \tan^{-1}\frac{2.929}{16.387} = 10.13°.$$

The final result in polar notation is

$$\mathbf{C} = 16.647\angle 10.13°. \qquad\qquad Ans.$$

Another method of treating two-dimensional vector problems analytically makes use of complex algebra. Although complex numbers are not vectors, they can be used to represent vectors in a plane by choosing an origin and real and imaginary axes. In two-dimensional kinematic problems, these axes can conveniently be chosen coincident with the $x_1 y_1$ axes of the absolute coordinate system.

As shown in Fig. 2.20b, the location of any point in the plane can be specified either by its absolute-position vector or by its corresponding real and imaginary coordinates,

$$\mathbf{R} = R^x + jR^y, \qquad\qquad (2.37)$$

where the operator j is defined as the unit imaginary number,

$$j = \sqrt{-1}. \qquad\qquad (2.38)$$

Complex numbers are commonly used in planar analysis because of the ease with which they can be changed into polar form. Employing complex rectangular notation for vector $\mathbf{R}$, we can write

$$\mathbf{R} = R\angle\theta = R\cos\theta + jR\sin\theta. \qquad\qquad (2.39)$$

But using the well-known Euler equation from trigonometry,

$$e^{j\theta} = \cos\theta + j\sin\theta, \qquad\qquad (2.40)$$

we can also write **R** in complex polar form as

$$\mathbf{R} = Re^{j\theta}, \tag{2.41}$$

where the magnitude and direction of the vector appear explicitly. As we will see in Chapters 3 and 4, expressing a vector in this form is also especially useful when differentiation is required.

2.11 Complex-Algebraic Solutions of Planar Vector Equations

Some familiarity with useful manipulation techniques for vectors written in complex polar forms can be gained by again solving the four cases of the loop-closure equation of Table 2.2. Writing Eq. (2.16) in complex polar form, we have

$$Ce^{j\theta_C} = Ae^{j\theta_A} + Be^{j\theta_B}. \tag{2.42}$$

Case 1: The two unknowns for case 1 are C and θ_C. We begin the solution by substituting Euler's equation, Eq. (2.40), into Eq. (2.42), which gives

$$C(\cos\theta_C + j\sin\theta_C) = A(\cos\theta_A + j\sin\theta_A) + B(\cos\theta_B + j\sin\theta_B). \tag{a}$$

Then, equating the real terms and the imaginary terms separately, we obtain two real equations corresponding to the horizontal and vertical components of the two-dimensional vector equation:

$$C\cos\theta_C = A\cos\theta_A + B\cos\theta_B, \tag{b}$$

$$C\sin\theta_C = A\sin\theta_A + B\sin\theta_B. \tag{c}$$

The unknown angle θ_C can be eliminated by squaring and adding these two equations. The result is

$$C = \sqrt{A^2 + B^2 + 2AB\cos(\theta_B - \theta_A)}. \tag{2.43}$$

The positive square root is chosen arbitrarily; the negative square root would yield a negative solution for C, with θ_C differing by 180°. Dividing Eq. (c) by Eq. (b), and rearranging, the angle θ_C can be written as

$$\theta_C = \tan^{-1}\frac{A\sin\theta_A + B\sin\theta_B}{A\cos\theta_A + B\cos\theta_B}, \tag{2.44}$$

where the signs of the numerator and the denominator must be considered separately in determining the proper quadrant of θ_C.[7] Only a single solution is found for case 1, as previously shown in Fig. 2.10.

---

[7] When writing computer programs it will be noted that most programming languages, including both ANSI/ISO standard FORTRAN and C, provide a library function named ATAN2(y, x) that accepts the numerator and denominator separately and provides the solution (in radians) in the proper quadrant. If such a function is not available, then a solution (in radians) in either the first or fourth quadrant is usually provided, and π radians (180°) should be added to the angle if the denominator is negative.

Fig. 2.22 (a) Original axes; (b) rotated axes.

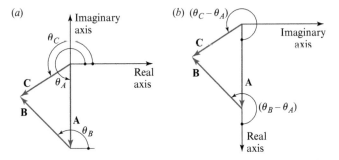

Case 2: The two unknowns for case 2 are A and θ_B. One convenient way of solving this case, in complex polar form, is first to divide Eq. (2.42) by $e^{j\theta_A}$. The resulting equation can be written as

$$Ce^{j(\theta_C - \theta_A)} = A + Be^{j(\theta_B - \theta_A)}. \tag{d}$$

Comparing this equation with Fig. 2.22, we see that division by the complex polar form of a unit vector $e^{j\theta_A}$ has the effect of rotating the real and imaginary axes by the angle θ_A *clockwise* such that the real axis lies along the vector **A**.

We can now use Euler's equation, Eq. (2.40), to separate the real and imaginary components; that is,

$$C\cos(\theta_C - \theta_A) = A + B\cos(\theta_B - \theta_A), \tag{e}$$

$$C\sin(\theta_C - \theta_A) = B\sin(\theta_B - \theta_A). \tag{f}$$

The solutions are then obtained directly from Eqs. (f) and (e), respectively, as

$$\theta_B = \theta_A + \sin^{-1}\frac{C\sin(\theta_C - \theta_A)}{B}, \tag{2.45}$$

$$A = C\cos(\theta_C - \theta_A) - B\cos(\theta_B - \theta_A). \tag{2.46}$$

The solutions given by Eqs. (2.45) and (2.46) are intentionally presented in this order since Eq. (2.46) cannot be evaluated numerically until after θ_B is found. We also note that the inverse sine term in Eq. (2.45) is double-valued. Therefore, case 2 has two distinct solutions, A, θ_B and A', θ'_B; these are shown in Fig. 2.11b.

Case 3: The two unknowns are the magnitudes A and B. Proceeding as in case 2, we obtain Eqs. (e) and (f). Then, from Eq. (f), the solution for B is

$$B = C\frac{\sin(\theta_C - \theta_A)}{\sin(\theta_B - \theta_A)}. \tag{2.47}$$

The solution for the unknown magnitude, A, is obtained by dividing Eq. (2.42) by $e^{j\theta_B}$. The resulting equation is then separated into real and imaginary parts and yields

$$A = C \frac{\sin(\theta_C - \theta_B)}{\sin(\theta_A - \theta_B)}. \tag{2.48}$$

Note that this case yields a unique solution, as shown in Fig. 2.12.

Case 4: The two unknowns are θ_A and θ_B. We begin by dividing Eq. (2.42) by $e^{j\theta_C}$ to align the real axis along vector **C** :

$$C = Ae^{j(\theta_A - \theta_C)} + Be^{j(\theta_B - \theta_C)}. \tag{g}$$

Using Euler's equation to separate components and then rearranging terms, we obtain

$$A\cos(\theta_A - \theta_C) = C - B\cos(\theta_B - \theta_C), \tag{h}$$

$$A\sin(\theta_A - \theta_C) = -B\sin(\theta_B - \theta_C). \tag{i}$$

Squaring both equations and adding the results gives

$$A^2 = C^2 + B^2 - 2BC\cos(\theta_B - \theta_C). \tag{j}$$

We recognize this as the law of cosines for the vector triangle. Rearranging this equation, the angle θ_B can be written as

$$\theta_B = \theta_C \pm \cos^{-1}\frac{C^2 + B^2 - A^2}{2BC}. \tag{2.49}$$

Moving C to the left-hand side of Eq. (h) before the squaring and adding operations results in another form of the law of cosines, from which

$$\theta_A = \theta_C \pm \cos^{-1}\frac{C^2 + A^2 - B^2}{2AC}. \tag{2.50}$$

The plus-or-minus signs in these two equations are a reminder that the inverse cosines are each double-valued and, therefore, θ_B and θ_A each have two solutions. These two pairs of angles can be paired naturally together as θ_A, θ_B and θ'_A, θ'_B under the restriction of Eq. (i). Therefore, case 4 has two distinct solutions, as shown in Fig. 2.13.

2.12 Posture Analysis Techniques

There are many different approaches to the posture analysis of mechanisms. Here we identify the following five approaches:

• graphic approach
• analytic approach
• complex-algebraic approach
• vector-algebraic approach
• numeric approach.

These approaches are illustrated in the following two examples.

Example 2.4

Consider the inverted slider-crank linkage in the posture shown in Fig. 2.23. Perform the posture analysis; that is, find θ_4 and distance R_{AO_4}.

Fig. 2.23 $R_{O_2O_4} = 9.0$ in, $R_{AO_2} = 4.5$ in, and $\theta_2 = 135°$.

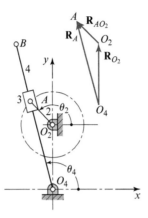

Solution

Using the vector diagram shown in Fig. 2.23 we recognize this as a case 1 problem, and the vector loop equation can be written as

$$\overset{??}{\mathbf{R}_{AO_4}} = \overset{\sqrt\surd}{\mathbf{R}_{O_2O_4}} + \overset{\sqrt{I}}{\mathbf{R}_{AO_2}}, \tag{1}$$

where the "I" symbol, for the direction of vector $\mathbf{R}_{AO_2}$, denotes that the posture of link 2 is the given input.

(a) **Graphic approach.** When Fig. 2.23 is drawn to a suitable scale (for example, 6 in/in), then, by direct measurement of this figure, we find that

$$\theta_4 = 105.3° \qquad\qquad Ans.$$

$$R_{AO_4} = 2.09 \text{ in}(6 \text{ in/in}) = 12.48 \text{ in}. \qquad\qquad Ans.$$

(b) **Analytic approach.** For case 1, we employ Eqs. (2.43) and (2.44); therefore,

$$R_{AO_4} = \sqrt{R_{O_2O_4}^2 + R_{AO_2}^2 + 2R_{O_2O_4}R_{AO_2}\cos(\theta_2 - 90°)}$$

$$= \sqrt{(9 \text{ in})^2 + (4.5 \text{ in})^2 + 2(9 \text{ in})(4.5 \text{ in})\cos(135° - 90°)} \qquad Ans.$$

$$= 12.59 \text{ in},$$

$$\begin{aligned}\theta_4 &= \tan^{-1}\frac{R_{O_2O_4}\sin 90° + R_{AO_2}\sin\theta_2}{R_{O_2O_4}\cos 90° + R_{AO_2}\cos\theta_2}\\[2mm] &= \tan^{-1}\frac{(9 \text{ in})\sin 90° + (4.5 \text{ in})\sin 135°}{(9 \text{ in})\cos 90° + (4.5 \text{ in})\cos 135°} = \tan^{-1}\left(\frac{12.182 \text{ in}}{-3.182 \text{ in}}\right) = -75.36°.\end{aligned}$$

Example 2.4 (cont.)

However, note that the calculator used here did not recognize for the inverse tangent function that the negative sign in the denominator indicates an angle in the second or third quadrant. Therefore, the proper value is

$$\theta_4 = -75.36° \pm 180° = 104.64°. \qquad \textit{Ans.}$$

(c) **Complex-algebraic approach.** The terms for Eq. (1) are

$$\mathbf{R}_{AO_4} = R_{AO_4}\angle\theta_4 = R_{AO_4}\cos\theta_4 + jR_{AO_4}\sin\theta_4,$$
$$\mathbf{R}_{O_2O_4} = 9 \text{ in} \angle 90° = (9 \text{ in})\cos 90° + j(9 \text{ in})\sin 90° = 0 + j9 \text{ in},$$
$$\mathbf{R}_{AO_2} = 4.5 \text{ in} \angle 135° = (4.5 \text{ in})\cos 135° + j(4.5 \text{ in})\sin 135°$$
$$= -3.182 \text{ in} + j3.182 \text{ in}.$$

Substituting these into Eq. (1) yields

$$\mathbf{R}_{AO_4} = (0 + j9 \text{ in}) + (-3.182 \text{ in} + j3.182 \text{ in}) = -3.182 \text{ in} + j12.182 \text{ in}.$$

Thus,

$$R_{AO_4} = \sqrt{(-3.182 \text{ in})^2 + (12.182 \text{ in})^2} = 12.59 \text{ in} \qquad \textit{Ans.}$$

and

$$\theta_4 = \tan^{-1}\frac{R^y_{AO_4}}{R^x_{AO_4}} = \tan^{-1}\frac{12.182 \text{ in}}{-3.182 \text{ in}} = 104.64°. \qquad \textit{Ans.}$$

Again, note the comment in (b) regarding inverse tangent results from the calculator.

(d) **Vector-algebraic approach.** Here we employ a scientific hand calculator which will add and subtract complex numbers in rectangular notation and will convert from rectangular to polar notation or vice-versa. Thus,

$$\mathbf{R}_{O_2O_4} = 9 \text{ in} \angle 90° = 0 + j9 \text{ in},$$
$$\mathbf{R}_{AO_2} = 4.5 \text{ in} \angle 135° = -3.182 \text{ in} + j3.182 \text{ in},$$
$$\mathbf{R}_{AO_4} = \mathbf{R}_{O_2O_4} + \mathbf{R}_{AO_2} = (0 + j9 \text{ in}) + (-3.182 \text{ in} + j3.182)$$
$$= -3.182 \text{ in} + j12.182 \text{ in} = 12.59 \text{ in} \angle 104.64°.$$

Ans.

(e) **Numeric approach.** Some scientific hand calculators permit equations containing complex numbers to be solved in either rectangular or polar form or both, with the results displayed in either mode. Using such a calculator, we enter

$$\mathbf{R}_{AO_4} = 9 \text{ in} \angle 90° + 4.5 \text{ in} \angle 135° = 12.59 \text{ in} \angle 104.64°. \qquad \textit{Ans.}$$

Note that all algebraic methods give results which are in exact agreement and are more accurate than graphic results obtained manually.

Example 2.5

A crank-rocker four-bar linkage with a given input angle is shown in both the open and crossed postures in Fig. 2.24. Perform a posture analysis of the linkage; that is, find θ_3 and θ_4 for both postures.

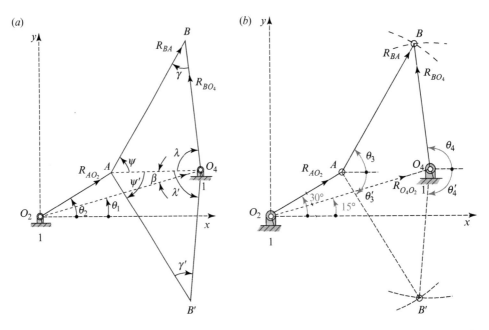

Fig. 2.24 $\mathbf{R}_{AO_2} = 200$ mm$\angle 30°$, $R_{BA} = 350$ mm, $R_{BO_4} = 300$ mm, and $\mathbf{R}_{O_4O_2} = 400$ mm$\angle 15°$.

Solution

Note that the line connecting the ground pins O_2 and O_4 (that is, the ground link) is not horizontal. In the previous examples, the ground link was chosen to be coincident with the x axis; that is, $\theta_1 = 0$. In this example, however, the angle $\theta_1 = 15°$ and must be accounted for when using Eqs. (2.25)–(2.30).

For the input angle $\theta_2 = 30°$, we observe that

$$R_{AO_2} = 0.200 \text{ m} \angle 30°, \quad \text{and} \quad \mathbf{R}_{O_4O_2} = 0.400 \text{ m} \angle 15°.$$

Therefore,

$$\mathbf{S} = \mathbf{R}_{O_4O_2} - \mathbf{R}_{AO_2} = 0.400 \text{ m} \angle 15° - 0.200 \text{ m} \angle 30° = 0.213 \text{ m} \angle 0.95°.$$

Note that the vectors in triangle ABO_4 are related by the equation

$$\mathbf{S} = \mathbf{R}_{BA} - \mathbf{R}_{BO_4}. \tag{1}$$

Example 2.5 (cont.)

There are two unknown postures in this equation, and so we identify this as case 4 of Table 2.2.

(a) **Graphic approach.** When Fig. 2.24 is drawn at a chosen scale, then, by direct measurement of the figure in the open posture, we find

$$\theta_3 = 60° \quad \text{and} \quad \theta_4 = 97°, \qquad \qquad Ans.$$

and in the crossed posture, we find

$$\theta_3' = -57° \quad \text{and} \quad \theta_4' = -95°. \qquad \qquad Ans.$$

(b) **Analytic approach.** Comparing the notation in Fig. 2.24 with Fig. 2.19, we set $r_1 = R_{O_4 O_2} = 0.400\,\text{m}$, $r_2 = R_{AO_2} = 0.200\,\text{m}$, $r_3 = R_{BA} = 0.350\,\text{m}$, and $r_4 = R_{BO_4} = 0.300\,\text{m}$. Also, note that the ground link of Fig. 2.19 must be rotated by $\theta_1 = 15°\text{ccw}$, and $\theta_2 = 30° - 15° = 15°$; therefore, Eqs. (2.25)–(2.30) give

$$S = \sqrt{(0.400\,\text{m})^2 + (0.200\,\text{m})^2 - 2(0.400\,\text{m})(0.200\,\text{m})\cos(30° - 15°)}$$
$$= 0.213\,\text{m},$$

$$\beta = \cos^{-1} \frac{(0.400\,\text{m})^2 + (0.213\,\text{m})^2 - (0.200\,\text{m})^2}{2(0.400\,\text{m})(0.213\,\text{m})} = 14.05°,$$

$$\psi = \cos^{-1} \frac{(0.350\,\text{m})^2 + (0.213\,\text{m})^2 - (0.300\,\text{m})^2}{2(0.350\,\text{m})(0.213\,\text{m})} = 58.51°,$$

$$\lambda = \cos^{-1} \frac{(0.300\,\text{m})^2 + (0.213\,\text{m})^2 - (0.350\,\text{m})^2}{2(0.300\,\text{m})(0.213\,\text{m})} = 84.19°.$$

Then, in the open posture, remembering that $\theta_1 = 15°$,

$$\theta_3 = 15° - 14.05° + 58.51° = 59.46°, \qquad \qquad Ans.$$

$$\theta_4 = 15° + 180° - 14.05° - 84.19° = 96.76°, \qquad \qquad Ans.$$

and, in the crossed posture,

$$\theta_3' = 15° - 14.05° - 58.51° = -57.56°, \qquad \qquad Ans.$$

$$\theta_4' = 15° + 180° - 14.05° + 84.19° = -94.86°. \qquad \qquad Ans.$$

Example 2.5 (cont.)

(c) **Complex-algebraic approach.** Substituting S for C, R_{BA} for A, $-R_{BO_4}$ for B, θ_S for θ_C, and θ_4 for θ_B, into Eq. (2.49) gives

$$\theta_4 = \theta_S \pm \cos^{-1}\frac{S^2 + R_{BO_4}^2 - R_{BA}^2}{2(-R_{BO_4})S} = (\theta_1 - \beta) \pm \cos^{-1}\frac{S^2 + R_{BO_4}^2 - R_{BA}^2}{2(-R_{BO_4})S}$$

$$= 0.95° \pm \cos^{-1}\frac{(0.213\ \text{m})^2 + (-0.300\ \text{m})^2 - (0.350\ \text{m})^2}{2(-0.300\ \text{m})(0.213\ \text{m})}$$

$$= 0.95° \pm 95.81° = 96.76° \quad \text{or} \quad -94.86°. \qquad\qquad Ans.$$

In addition, substituting θ_3 for θ_A into Eq. (2.50) gives

$$\theta_3 = \theta_S \pm \cos^{-1}\frac{S^2 + R_{BA}^2 - R_{BO_4}^2}{2R_{BA}S}$$

$$= 0.95° \pm \cos^{-1}\frac{(0.213\ \text{m})^2 + (0.350\ \text{m})^2 - (-0.300\ \text{m})^2}{2(0.350\ \text{m})(0.213\ \text{m})}$$

$$= 0.95° \pm 58.51° = 59.46° \quad \text{or} \quad -57.56°, \qquad\qquad Ans.$$

where the first result corresponds to the open posture and the second result is for the crossed posture. We note that the answers in parts (b) and (c) are in good agreement with those of part (a), but are of higher accuracy than measurements taken from the scale drawing.

(d) **Vector-algebraic approach.** The vector loop equation for the four-bar linkage can be written as

$$\overset{\sqrt{?}}{\mathbf{R}_{BA}} = \overset{\sqrt{\surd}}{\mathbf{R}_{O_4O_2}} + \overset{\sqrt{?}}{\mathbf{R}_{BO_4}} - \overset{\sqrt{I}}{\mathbf{R}_{AO_2}}. \qquad (2)$$

Separating this equation into horizontal and vertical components,

$$R_{BA}\cos\theta_3 = R_{O_4O_2}\cos\theta_1 + R_{BO_4}\cos\theta_4 - R_{AO_2}\cos\theta_2, \qquad (3a)$$

$$R_{BA}\sin\theta_3 = R_{O_4O_2}\sin\theta_1 + R_{BO_4}\sin\theta_4 - R_{AO_2}\sin\theta_2. \qquad (3b)$$

By squaring and adding these two equations, we can eliminate the variable θ_3:

$$R_{BA}^2 = R_{O_4O_2}^2 + R_{BO_4}^2 + R_{AO_2}^2 + 2R_{O_4O_2}R_{BO_4}[\cos\theta_1\cos\theta_4 + \sin\theta_1\sin\theta_4]$$
$$- 2R_{O_4O_2}R_{AO_2}[\cos\theta_1\cos\theta_2 + \sin\theta_1\sin\theta_2] - 2R_{BO_4}R_{AO_2}[\cos\theta_4\cos\theta_2 + \sin\theta_4\sin\theta_2].$$

Then we can write this equation as

$$A\cos\theta_4 + B\sin\theta_4 = C, \qquad (4)$$

where

$$A = 2R_{O_4O_2}R_{BO_4}\cos\theta_1 - 2R_{BO_4}R_{AO_2}\cos\theta_2, \qquad (5a)$$

$$B = 2R_{O_4O_2}R_{BO_4}\sin\theta_1 - 2R_{BO_4}R_{AO_2}\sin\theta_2, \qquad (5b)$$

$$C = R_{BA}^2 - R_{O_4O_2}^2 - R_{BO_4}^2 - R_{AO_2}^2 + 2R_{O_4O_2}R_{AO_2}\cos(\theta_2 - \theta_1). \qquad (5c)$$

Example 2.5 (cont.)

Equation (4) is one form of Freudenstein's equation (Section 9.11 shows another variation that is more suitable for kinematic synthesis).

Substituting the known data into Eqs. (5) gives

$$A = 2(0.400 \text{ m})(0.300 \text{ m})\cos 15° - 2(0.300 \text{ m})(0.200 \text{ m})\cos 30° = 0.127\,90 \text{ m}^2,$$
$$B = 2(0.400 \text{ m})(0.300 \text{ m})\sin 15° - 2(0.300 \text{ m})(0.200 \text{ m})\sin 30° = 0.002\,12 \text{ m}^2,$$
$$C = (0.350 \text{ m})^2 - (0.400 \text{ m})^2 - (0.300 \text{ m})^2 - (0.200 \text{ m})^2$$
$$+ 2(0.400 \text{ m})(0.200 \text{ m})\cos(30° - 15°)$$
$$= -0.012\,95 \text{ m}^2.$$

Therefore, Eq. (4) can be written as

$$(0.127\,90 \text{ m}^2)\cos\theta_4 + (0.002\,12 \text{ m}^2)\sin\theta_4 = -0.012\,95 \text{ m}^2. \tag{6}$$

To determine the output angle θ_4, this transcendental equation can be written as a quadratic equation. The procedure is to use the half-angle relationships; that is, to define

$$Z = \tan(\theta_4/2), \tag{7a}$$

which gives

$$\sin\theta_4 = \frac{2Z}{1 + Z^2} \quad \text{and} \quad \cos\theta_4 = \frac{1 - Z^2}{1 + Z^2}. \tag{7b}$$

Substituting Eqs. (7b) into Eq. (6), multiplying through by $(1 + Z^2)$, and rearranging, gives the quadratic equation

$$(0.114\,95 \text{ m}^2)Z^2 + (-0.004\,23 \text{ m}^2)Z + (-0.140\,85 \text{ m}^2) = 0.$$

The solutions to this equation can be written as

$$Z = \frac{-(-0.004\,23 \text{ m}) \pm \sqrt{(-0.004\,23 \text{ m})^2 - 4(0.114\,95 \text{ m})(-0.140\,85 \text{ m})}}{2(0.114\,95 \text{ m})} \tag{8}$$
$$= 0.018\,41 \pm 1.106\,80 = 1.125\,21 \quad \text{or} \quad -1.088\,39.$$

Substituting Eq. (8) into Eq. (7a) gives the two output angles corresponding to the open and crossed postures, namely

$$\theta_4 = 96.76° \quad \text{and} \quad \theta_4' = -94.86°. \qquad\qquad Ans.$$

Substituting these results into Eqs. (3a) and (3b) gives the corresponding coupler angles

$$\theta_3 = 59.46° \quad \text{and} \quad \theta_3' = -57.56°. \qquad\qquad Ans.$$

(e) **Numeric approach.** Rearrange the vector loop equation, Eq. (2), into the form

$$\mathbf{f} = \mathbf{R}_{AO_2} + \mathbf{R}_{BA} - \mathbf{R}_{O_4O_2} - \mathbf{R}_{BO_4} = \mathbf{0}, \tag{9}$$

Example 2.5 (cont.)

and separate the horizontal and vertical components:

$$f^x = R_{AO_2} \cos\theta_2 + R_{BA}\cos\theta_3 - R_{O_4O_2}\cos\theta_1 - R_{BO_4}\cos\theta_4 = 0, \qquad (10a)$$

$$f^y = R_{AO_2} \sin\theta_2 + R_{BA}\sin\theta_3 - R_{O_4O_2}\sin\theta_1 - R_{BO_4}\sin\theta_4 = 0. \qquad (10b)$$

We recognize these as two equations in two unknowns, θ_3 and θ_4, and we solve them to an accuracy of $0.01°$ by the Newton–Raphson method introduced in Section 2.9. To do this, we expand Eqs. (10) to first-order in Taylor series in the two unknowns as follows:

$$R_{AO_2}\cos\theta_2 + R_{BA}\cos\theta_3 - R_{BA}\sin\theta_3\Delta\theta_3 - R_{O_4O_2}\cos\theta_1 - R_{BO_4}\cos\theta_4 + R_{BO_4}\sin\theta_4\Delta\theta_4 = 0,$$

$$R_{AO_2}\sin\theta_2 + R_{BA}\sin\theta_3 + R_{BA}\cos\theta_3\Delta\theta_3 - R_{O_4O_2}\sin\theta_1 - R_{BO_4}\sin\theta_4 - R_{BO_4}\cos\theta_4\Delta\theta_4 = 0,$$

or, after rearranging,

$$R_{BA}\sin\theta_3\Delta\theta_3 - R_{BO_4}\sin\theta_4\Delta\theta_4 = R_{AO_2}\cos\theta_2 + R_{BA}\cos\theta_3 - R_{O_4O_2}\cos\theta_1 - R_{BO_4}\cos\theta_4,$$

$$-R_{BA}\cos\theta_3\Delta\theta_3 + R_{BO_4}\cos\theta_4\Delta\theta_4 = R_{AO_2}\sin\theta_2 + R_{BA}\sin\theta_3 - R_{O_4O_2}\sin\theta_1 - R_{BO_4}\sin\theta_4.$$

Then, writing these equations in matrix form,

$$J\begin{bmatrix}\Delta\theta_3\\\Delta\theta_4\end{bmatrix} = \begin{bmatrix}f^x\\f^y\end{bmatrix} = \begin{bmatrix}R_{AO_2}\cos\theta_2 + R_{BA}\cos\theta_3 - R_{O_4O_2}\cos\theta_1 - R_{BO_4}\cos\theta_4\\R_{AO_2}\sin\theta_2 + R_{BA}\sin\theta_3 - R_{O_4O_2}\sin\theta_1 - R_{BO_4}\sin\theta_4\end{bmatrix}, \qquad (11a)$$

where J, referred to as the Jacobian, is

$$J = \begin{bmatrix}R_{BA}\sin\theta_3 & -R_{BO_4}\sin\theta_4\\-R_{BA}\cos\theta_3 & R_{BO_4}\cos\theta_4\end{bmatrix}, \qquad (11b)$$

and the determinant of the Jacobian is

$$\Delta = (R_{BA}\sin\theta_3)(R_{BO_4}\cos\theta_4) - (-R_{BA}\cos\theta_3)(-R_{BO_4}\sin\theta_4) = R_{BA}R_{BO_4}\sin(\theta_3-\theta_4). \qquad (12)$$

Note that $\Delta = 0$ when $\theta_3 = \theta_4$ or $\theta_3 = \theta_4 + 180°$, that is, when the four-bar linkage is in a toggle posture (see Section 1.10).

By Cramer's rule, the corrections can be written from Eq. (11a) as

$$\Delta\theta_3 = \frac{R_{BO_4}(f^x\cos\theta_4 + f^y\sin\theta_4)}{\Delta} \qquad (13a)$$

and

$$\Delta\theta_4 = \frac{R_{BA}(f^x\cos\theta_3 + f^y\sin\theta_3)}{\Delta}. \qquad (13b)$$

Example 2.5 (cont.)

Note that Eqs. (13) give dimensionless values, meaning values in radians, for $\Delta\theta_3$ and $\Delta\theta_4$; therefore, we must be careful when combining these with θ_3 and θ_4 if the two angles have units of degrees. The new values for the two angles are

$$(\theta_3)_{\text{new}} = (\theta_3)_{\text{old}} + \frac{180°}{\pi \text{ rad}}\Delta\theta_3 \tag{14a}$$

and

$$(\theta_4)_{\text{new}} = (\theta_4)_{\text{old}} + \frac{180°}{\pi \text{ rad}}\Delta\theta_4. \tag{14b}$$

To use the iteration procedure, we begin with initial estimates of the unknowns θ_3 and θ_4. Let us suppose that we visually estimate from a scale drawing that, for the open posture, our starting estimates are

$$\theta_3 = 60° \quad \text{and} \quad \theta_4 = 100°. \tag{15}$$

Inserting these estimates into Eq. (12) gives us the initial estimate of the determinant of the Jacobian as

$$\Delta = (0.350 \text{ m})(0.300 \text{ m}) \sin(-40°) = -0.067\ 492\ 7 \text{ m}^2. \tag{16}$$

Substituting Eqs. (15) into Eqs. (11) gives

$$\begin{bmatrix} 0.303\ 11 \text{ m} & -0.295\ 44 \text{ m} \\ -0.175\ 00 \text{ m} & -0.052\ 09 \text{ m} \end{bmatrix}\begin{bmatrix} \Delta\theta_3 \\ \Delta\theta_4 \end{bmatrix} = \begin{bmatrix} 0.013\ 93 \text{ m} \\ 0.004\ 14 \text{ m} \end{bmatrix}.$$

Using Cramer's rule, these equations give a solution of

$$\Delta\theta_3 = -0.007\ 371\ 5 \text{ rad} = -0.422° \quad \text{and} \quad \Delta\theta_4 = -0.054\ 711\ 4 \text{ rad} = -3.135°.$$

Combining these with Eqs. (15) gives updated values of

$$\theta_3 = 59.58° \quad \text{and} \quad \theta_4 = 96.87°. \tag{17}$$

Inserting these improved estimates into Eqs. (11) gives

$$\begin{bmatrix} 0.301\ 82 \text{ m} & -0.297\ 85 \text{ m} \\ -0.177\ 22 \text{ m} & -0.035\ 89 \text{ m} \end{bmatrix}\begin{bmatrix} \Delta\theta_3 \\ \Delta\theta_4 \end{bmatrix} = \begin{bmatrix} -0.577\ 13(10)^{-4} \text{ m} \\ 0.441\ 91(10)^{-3} \text{ m} \end{bmatrix}.$$

Solving again by Cramer's rule, these equations give a solution of

$$\Delta\theta_3 = -0.002\ 101\ 6 \text{ rad} = -0.120° \quad \text{and} \quad \Delta\theta_4 = -0.001\ 94 \text{ rad} = -0.111°,$$

and, combining these with Eqs. (17), gives updated values of

$$\theta_3 = 59.46° \quad \text{and} \quad \theta_4 = 96.76°. \tag{18}$$

Example 2.5 (cont.)

Inserting these improved estimates into Eq. (11) gives us

$$\begin{bmatrix} 0.301\,45 \text{ m} & -0.297\,91 \text{ m} \\ -0.177\,85 \text{ m} & -0.035\,31 \text{ m} \end{bmatrix} \begin{bmatrix} \Delta\theta_3 \\ \Delta\theta_4 \end{bmatrix} = \begin{bmatrix} 0.213\,51(10)^{-5} \text{ m} \\ 0.173\,94(10)^{-5} \text{ m} \end{bmatrix}.$$

Solving again by Cramer's rule, these equations give

$$\Delta\theta_3 = -0.000\,069\,59 \text{ rad} = -0.004\,0° \quad \text{and} \quad \Delta\theta_4 = -0.000\,014\,2 \text{ rad} = -0.000\,81°$$

and, combining these with Eqs. (18), gives updated values of

$$\theta_3 = 59.46° \quad \text{and} \quad \theta_4 = 96.76°. \qquad\qquad Ans.$$

Since the corrections are now both smaller than the desired accuracy of 0.01°, we accept these values as final results for the open posture.

For the crossed posture, we might begin with initial estimates from our drawing of $\theta_3 = -60°$ and $\theta_4 = -100°$. Then, using these as starting estimates with Eq. (11), we repeat the same process, and iterate until we converge to a result of

$$\theta_3 = -57.56° \quad \text{and} \quad \theta_4 = -94.86°. \qquad\qquad Ans.$$

A computer code could be written using a programming language such as MATLAB or Java.

2.13 Coupler-Curve Generation

In Chapter 1, Fig. 1.23, we learned of the vast variety of useful coupler curves that can be generated by a planar four-bar linkage. These curves are quite easy to obtain graphically, but computer-generated curves can be obtained more quickly and are easier to vary to obtain the desired curve characteristics. Here we present the basic equations but omit the programming details required for presentation of a curve on an electronic display.

Considering the four-bar linkage of Fig. 2.16, we first write the vector equation

$$\mathbf{S} = \mathbf{R}_C - \mathbf{R}_A.$$

This is case 1, where S and θ_S are the two unknowns. The solutions are given by Eqs. (2.43) and (2.44). Substituting gives

$$S = \sqrt{R_C^2 + R_A^2 - 2R_C R_A \cos(\theta_A - \theta_C)}, \qquad\qquad (2.51)$$

$$\theta_S = \tan^{-1} \frac{R_C \sin\theta_C - R_A \sin\theta_A}{R_C \cos\theta_C - R_A \cos\theta_A}. \qquad\qquad (2.52)$$

Then we note that

$$\mathbf{S} = \mathbf{R}_{BA} - \mathbf{R}_{BC},$$

where the two orientations θ_3 and θ_4 are the two unknowns. The solutions are given by Eqs. (2.49) and (2.50). Substituting gives

$$\theta_4 = \theta_S \pm \cos^{-1} \frac{S^2 + R_{BC}^2 - R_{BA}^2}{2R_{BC}S}, \qquad (2.53)$$

$$\theta_3 = \theta_S \mp \cos^{-1} \frac{S^2 + R_{BA}^2 - R_{BC}^2}{2R_{BA}S}. \qquad (2.54)$$

The solution for the open posture of the linkage, shown in Fig. 2.16a, corresponds to the lower set of signs; the crossed posture of the linkage, shown in Fig. 2.16b, corresponds to the upper set.

Coupler point P generates a coupler curve when crank 2 is rotated. From Fig. 2.16, we see that

$$\mathbf{R}_P = R_P e^{j\theta_6} = R_A e^{j\theta_2} + R_{PA} e^{j(\theta_3 + \alpha)}. \qquad (2.55)$$

We recognize this as case 1, where R_p and θ_6 are the two unknowns. The solutions are given by Eqs. (2.43) and (2.44); that is,

$$R_P = \sqrt{R_A^2 + R_{PA}^2 + 2R_A R_{PA} \cos(\theta_3 + \alpha - \theta_2)} \qquad (2.56)$$

and

$$\theta_6 = \tan^{-1} \frac{R_A \sin\theta_2 + R_{PA}\sin(\theta_3 + \alpha)}{R_A \cos\theta_2 + R_{PA}\cos(\theta_3 + \alpha)}. \qquad (2.57)$$

Note that each of these equations gives two solutions because of the double values for θ_3, corresponding to the two postures of the linkage. The following example uses Eqs. (2.56) and (2.57) to plot the coupler curve of a four-bar linkage.

Example 2.6

Consider the crank-rocker four-bar linkage shown in Fig. 2.25 with $R_{BA} = 100$ mm, $R_{CB} = 250$ mm, $R_{CD} = 300$ mm, and $R_{DA} = 200$ mm. The paths of the coupler pins B and C are shown by the circle and the circular arc, respectively. The location of the coupler point P is given by $R_{PB} = 150$ mm and $\alpha = \angle CBP = -45°$. Calculate the coordinates of coupler point P and plot the path of this point (the coupler curve) for a complete rotation of the crank.

Solution

For each value of crank angle θ_2, the angles β, ψ, and γ are calculated from Eqs. (2.26), (2.27), and (2.33), respectively. This notation corresponds to that of Figs. 2.15 and 2.16. Next, Eqs.

Example 2.6 (cont.)

(2.29) and (2.31) are applied to give the posture of link 3, θ_3, as displayed in Table 2.3. Note that the same result can be obtained from Eq. (2.50).

Finally, the polar coordinates of coupler point P are calculated from Eqs. (2.56) and (2.57). From these, we obtain the Cartesian coordinates of point P. The solutions for crank angles from $0°$ to $90°$ for the open posture are displayed in Table 2.3. The coupler curve is shown in Fig. 2.25. The interested reader can repeat the solution to this problem for the crossed posture.

Table 2.3 Posture of link 3 and coordinates of coupler point P

θ_2, deg	θ_3, deg	R_P, mm	θ_6, deg	R_P^x, mm	R_P^y, mm
0.0	110.5	212.0	40.1	162.2	136.5
10.0	99.4	232.2	36.9	185.8	139.3
20.0	87.8	245.3	33.7	204.0	136.1
30.0	77.5	249.9	31.5	213.1	130.6
40.0	69.2	247.7	30.5	213.4	125.8
50.0	62.9	240.7	30.7	207.0	122.7
60.0	58.3	230.4	31.7	196.0	121.1
70.0	55.1	218.0	33.5	181.9	120.3
80.0	53.0	204.4	35.7	165.9	119.4
90.0	51.8	189.9	38.3	148.9	117.8

Fig. 2.25 The coupler curve traced by point P.

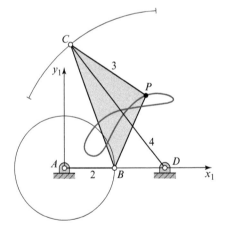

2.14 Displacement of a Moving Point

In the study of motion, we must be concerned with the relationship between successive positions of a point and postures of a link. In Fig. 2.26, a point, originally at position P, is

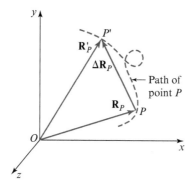

Fig. 2.26 Displacement of a moving point.

moving along the path shown and, sometime later, arrives at position P'. The *displacement* of the point during the time interval is defined as the *net change in position*; that is,

$$\Delta \mathbf{R}_P = \mathbf{R}'_P - \mathbf{R}_P. \tag{2.58}$$

Note that this displacement is a vector quantity having the magnitude and direction of the vector from point P to point P'.

It is important to note that the displacement $\Delta \mathbf{R}_P$ is the net change in position and does not depend on the particular path taken from P to P'. The magnitude of this vector is *not* necessarily equal to the length of the path (the distance traveled), and its direction is *not* necessarily along the tangent to the path, although both are true when the displacement is infinitesimally small. Knowledge of the path actually traveled from P to P' is not even necessary to find the displacement vector, providing the initial and displaced positions of the point can be found.

2.15 Displacement Difference between Two Points

In this section we consider the difference in the displacements of two points. In particular, we are concerned with the case where two moving points are both fixed in the same rigid body. The situation is shown in Fig. 2.27, where a rigid body moves from an initial posture defined by $x_2 y_2 z_2$ to a later posture defined by $x'_2 y'_2 z'_2$.

From Eq. (2.6), the position difference between the two points P and Q at the initial instant is

$$\mathbf{R}_{PQ} = \mathbf{R}_P - \mathbf{R}_Q. \tag{a}$$

After the displacement of this body, the two points are located at P' and Q'. At that time, the position difference is

$$\mathbf{R}'_{PQ} = \mathbf{R}'_P - \mathbf{R}'_Q. \tag{b}$$

During the time interval of the movement, the two points have undergone individual displacements of $\Delta \mathbf{R}_P$ and $\Delta \mathbf{R}_Q$, respectively.

Fig. 2.27 Displacement difference between two points fixed in the same rigid body.

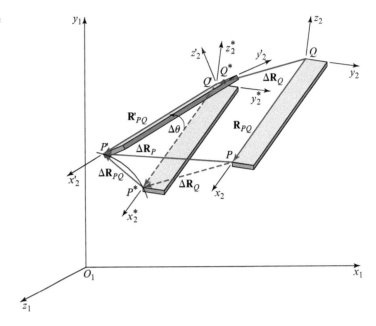

As the name implies, the *displacement difference* between the two points is defined as the net difference between their respective displacements and can be written as

$$\Delta \mathbf{R}_{PQ} = \Delta \mathbf{R}_P - \Delta \mathbf{R}_Q. \tag{2.59}$$

Note that this equation corresponds to the vector triangle PP^*P' in Fig. 2.27. As stated in the previous section, the displacement depends only on the net change in position and not on the particular path taken. Therefore, no matter how the body (containing points P and Q) was *actually* displaced, we are free to visualize the path as we choose. Equation (2.59) leads us to visualize the displacement as taking place in two stages. First, the body translates (slides without rotation) from $x_2 y_2 z_2$ to $x_2^* y_2^* z_2^*$; during this movement, all points, including P and Q, have the same displacement $\Delta \mathbf{R}_Q$. Second, the body rotates about point Q' to its final posture $x_2' y_2' z_2'$.

A different interpretation can be obtained by manipulating Eq. (2.59) as follows:

$$\begin{aligned} \Delta \mathbf{R}_{PQ} &= \left(\mathbf{R}_P' - \mathbf{R}_P \right) - \left(\mathbf{R}_Q' - \mathbf{R}_Q \right) \\ &= \left(\mathbf{R}_P' - \mathbf{R}_Q' \right) - \left(\mathbf{R}_P - \mathbf{R}_Q \right), \end{aligned} \tag{c}$$

and then, from Eqs. (*a*) and (*b*),

$$\Delta \mathbf{R}_{PQ} = \mathbf{R}_{PQ}' - \mathbf{R}_{PQ}. \tag{2.60}$$

This equation corresponds to the vector triangle $Q'P^*P'$ in Fig. 2.27 and demonstrates that the displacement difference, defined as the difference between two displacements, is equal to the net change between the two position-difference vectors.

For either interpretation, we are illustrating that *any displacement of a rigid body is equivalent to the sum of a net translation of one point and a net rotation of the body about that point.* We also see that only the rotation contributes to the displacement difference between two points fixed in the same rigid body; that is, *there is no difference between the displacements of any two points fixed in the same rigid body as the result of a translation.* (The next section gives the definition of the term *translation*.)

In view of the previous discussion, we can visualize the displacement difference $\Delta \mathbf{R}_{PQ}$ as the displacement that would be seen for point P by a moving observer, who always stays coincident with point Q *but does not rotate* with the moving body – that is, always using the absolute coordinate axes $x_1 y_1 z_1$ for the measurement of direction. It is important to understand the difference between this interpretation of an observer moving with point Q but not rotating and the case of an observer *on* the moving body. To an observer *on* the moving body, both points P and Q would appear stationary; neither would be seen to have any displacement because they do not move relative to the observer, and the displacement difference seen by such an observer would be zero.

2.16 Translation and Rotation

Using the concept of displacement difference between two points fixed in the same rigid body, we are now able to define the terms *translation* and *rotation*.

Translation is defined as *a motion of a body for which the displacement difference between any two points P and Q in the body is zero*; that is, from the displacement-difference equation, Eq. (2.60),

$$\Delta \mathbf{R}_{PQ} = \Delta \mathbf{R}_P - \Delta \mathbf{R}_Q = \mathbf{0}.$$

That is,

$$\Delta \mathbf{R}_P = \Delta \mathbf{R}_Q, \tag{2.61}$$

which states that *the displacements of any two points in the body are equal.*

Rotation is a motion of a body for which different points of the body exhibit different displacements.

Figure 2.28a shows a situation where the body has moved along a curved path from posture $x_2 y_2$ to posture $x_2' y_2'$. Despite the fact that the point paths are curved,[8] $\Delta \mathbf{R}_P$ is still equal to $\Delta \mathbf{R}_Q$, and the body has undergone a translation. Note that in translation, the point paths described by any two points on the body are identical, and there is no change of angular orientation between the moving coordinate system and the coordinate system of the observer; that is, $\Delta \theta_2 = \theta_2' - \theta_2 = 0$.

Now consider the midpoint of a body that is *constrained* to move along a straight-line path, as shown in Fig. 2.28b. As it does so, the body rotates so that $\Delta \theta_2 = \theta_2' - \theta_2 \neq 0$ and the displacements $\Delta \mathbf{R}_P$ and $\Delta \mathbf{R}_Q$ are not equal. Although there is no obvious point on the body about which it has rotated, the coordinate system $x_2' y_2'$ has changed angular orientation relative to $x_1 y_1$, and the body is said to have undergone a rotation. Note that the paths described by points P and Q are not equal.

---

[8] Translation in which the point paths are straight lines is called *rectilinear* translation; translation in which the point paths are identical curves, but not straight lines, is referred to as *curvilinear* translation.

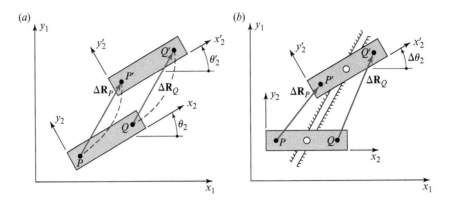

Fig. 2.28 (a) Translation: $\Delta \mathbf{R}_P = \Delta \mathbf{R}_Q, \Delta\theta_2 = 0$; (b) rotation: $\Delta \mathbf{R}_P \neq \Delta \mathbf{R}_Q, \Delta\theta_2 \neq 0$.

In conclusion, we note that translation and rotation of a body cannot be defined from the motion of a single point. These are characteristics of a body or a coordinate system. It is improper to speak of "rotation of a point" since there is no meaning for angular orientation of a point. It is also improper to associate the terms "translation" and "rotation" with the path of a single point of a moving body. Although it does not matter which points of the body are chosen, the motions of two or more points must be compared before meaningful definitions exist for these terms.

2.17 Apparent Displacement

We have stated in previous sections that the displacement of a point does not depend on the particular path along which it travels. However, since displacement is computed from the position vectors of the endpoints of the path, knowledge of the coordinate system of the observer is essential.

In Fig. 2.29 we identify three bodies: body 1 is a fixed or stationary body containing the absolute reference system $x_1 y_1 z_1$; body 2 is a moving body containing the reference system $x_2 y_2 z_2$; and body 3 moves with respect to body 2. We also consider two observers identified as HE and SHE. We designate that HE is an observer fixed to body 1, the stationary system, and we designate that SHE is another observer on body 2, fixed to the moving system $x_2 y_2 z_2$. Thus, we might say that HE is on the ground observing, whereas SHE is going for a ride on body 2.

Now consider a point P_3, fixed in body 3 and moving along a known path on body 2. HE and SHE are both observing the motion of point P_3, and we want to compare what they see. We must define another point P_2, which is fixed in (or rigidly attached to) body 2 and initially is instantaneously coincident with P_3.

Now let body 2 and the axes $x_2 y_2 z_2$ be displaced to a new posture $x_2' y_2' z_2'$. While this motion is taking place, let P_3 move to another position on body 2, which we identify as P_3'. But P_2, being fixed in body 2, is now in the new position identified as P_2'. SHE, on body 2, reports the motion of P_3 as the vector $\Delta \mathbf{R}_{P_3/2}$, which is read as the displacement of P_3 as it appears to an

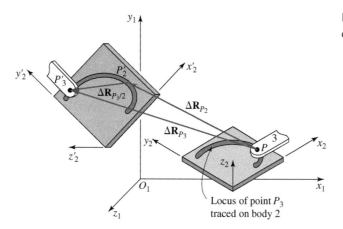

Fig. 2.29 Apparent
displacement of a point.

observer on body 2. This is called the apparent-displacement vector. Note that SHE sees no
motion of P_2, since it appears to HER to be stationary on body 2. Therefore, $\Delta\mathbf{R}_{P_2/2} = \mathbf{0}$.

However, HE, on the stationary body, reports the displacement of P_3 as vector $\Delta\mathbf{R}_{P_3}$. Note
that HE also reports the displacement of P_2 as the vector $\Delta\mathbf{R}_{P_2}$. From the vector triangle in
Fig. 2.29, we see that the observations of the two observers are related by the *apparent-
displacement equation*,

$$\Delta\mathbf{R}_{P_3} = \Delta\mathbf{R}_{P_2} + \Delta\mathbf{R}_{P_3/2}. \tag{2.62}$$

We can take this equation as the definition of the apparent-displacement vector, although it is
important also to understand the physical concepts involved. Note that the apparent-
displacement vector relates the absolute displacements of two *coincident points* that are points
of *different* moving bodies. Note also that there is no restriction on the actual location of the
observer moving with coordinate system 2, only that SHE be fixed in that coordinate system so
that SHE senses no displacement for point P_2.

One primary use of the apparent displacement is to determine an absolute displacement. It is
not uncommon in machines to find a point, such as P_3, which is constrained to move along a
known slot or path or guideway defined by the shape of another moving link 2. In such cases, it
may be much more convenient to measure or calculate $\Delta\mathbf{R}_{P_2}$ and $\Delta\mathbf{R}_{P_3/2}$ and to use Eq. (2.62)
than to measure the absolute displacement $\Delta\mathbf{R}_{P_3}$ directly.

2.18 Absolute Displacement

In reflecting on the definition and concept of the apparent-displacement vector, we conclude
that the absolute displacement of a moving point, say $\Delta\mathbf{R}_{P_3/1}$, is the special case of an apparent
displacement where the observer is fixed in the absolute coordinate system. As explained for the
position vector, the notation is often abbreviated to read $\Delta\mathbf{R}_{P_3}$ or just $\Delta\mathbf{R}_P$, and an absolute
observer is implied whenever this is not noted explicitly.

Perhaps a better physical understanding of apparent displacement can be achieved by relating it to absolute displacement. Imagine a car, P_3, traveling along a roadway and under observation by an absolute observer some distance to one side. Consider how this observer visually senses the motion of the car. Although HE may not be conscious of all of the following steps, the idea here is that the observer first imagines a fixed point P_1 *coincident* with P_3, which HE defines in his mind as stationary; HE may relate to a fixed point of the roadway or a nearby tree or road sign, for example. HE then compares his later observation of the car P_3 with that of P_1 to sense displacement. Note that HE does not compare with his own location but with the initially coincident point P_1. In this instance, the apparent-displacement equation becomes an identity:

$$\Delta\mathbf{R}_{P_3} = \Delta\overset{0}{\mathbf{R}}_{P_1} + \Delta\mathbf{R}_{P_3/1}.$$

2.19 Apparent Angular Displacement

In general, rotations cannot be treated as vectors (this is explained in some detail in Section 3.2). However, for planar motion, the concept of apparent displacement also extends to include rotation. For example, suppose we consider the rotations of the two gears 2 and 3 connected by an arm, body 4, as shown in Fig. 2.30. In Fig. 2.30a, we see the rotations as they would appear in HER coordinate system attached to the arm. In Fig. 2.30b we see the same situation, but as it might appear to HIM in the absolute frame of reference.

Although it may not be easy to write equations relating the absolute rotations of the gears shown in Fig. 2.30, it is quite easy to relate their apparent rotations if we take HER point of view as an observer riding on the moving coordinate system attached to the arm. From HER vantage point it is obvious that, if there is no slipping between bodies 2 and 3, then the arc lengths of the two gear sectors that pass the arm at the point of contact must be equal. That is, if ρ_2 is the radius of gear 2 and ρ_3 is the radius of gear 3, then

$$\rho_2 \Delta\theta_{2/4} = -\rho_3 \Delta\theta_{3/4}, \tag{a}$$

where $\Delta\theta_{2/4}$ and $\Delta\theta_{3/4}$ are the angular displacements of gears 2 and 3 as they appear to HER in coordinate system 4, and the negative sign accounts for the difference in the senses of the two apparent rotations.

When these angular displacements are replaced by those seen by HIM from the absolute coordinate system, Eq. (a) becomes

$$\rho_2(\Delta\theta_2 - \Delta\theta_4) = -\rho_3(\Delta\theta_3 - \Delta\theta_4). \tag{b}$$

Using these ideas, the loop-closure equations for mechanisms having rolling contact without slipping can be solved. The following two examples make the procedure clear.

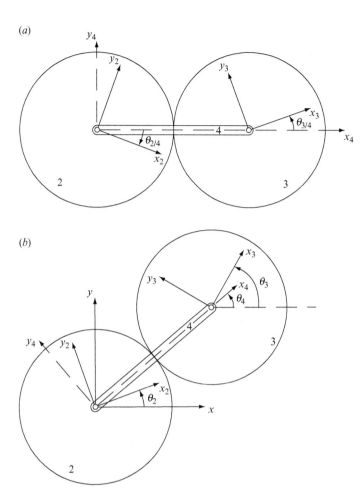

(a)

(b)

Fig. 2.30 (a) Apparent rotations observed from the arm; (b) absolute rotations.

Example 2.7

For the cam-and-follower mechanism in the posture shown in Fig. 2.31, define a set of vectors that is suitable for a complete kinematic analysis. Label and show the sense and orientation of each vector. Write the vector loop equation(s) and identify suitable input(s) for the mechanism. Identify the known quantities, the unknown variables, and any constraints. If you identify constraints, then write the constraint equation(s).

Solution

Since the circular cam, link 2, and the roller follower, link 3, are in rolling contact, let us consider them linked together by a (fictitious) arm, link 23. This leads us to consider the set of vectors shown in Fig. 2.32.

The vector loop equation for the mechanism can be written as

$$\overset{\sqrt{I}}{\mathbf{R}_2} + \overset{\sqrt{?}}{\mathbf{R}_{23}} - \overset{\sqrt{?}}{\mathbf{R}_4} - \overset{\sqrt{\sqrt{}}}{\mathbf{R}_{12}} - \overset{\sqrt{\sqrt{}}}{\mathbf{R}_{11}} = \mathbf{0}. \qquad\qquad Ans. \ (1)$$

Example 2.7 (cont.)

Fig. 2.31 Cam-and-follower
mechanism.

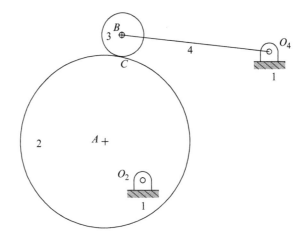

Fig. 2.32 Vectors chosen for
analysis of the mechanism.

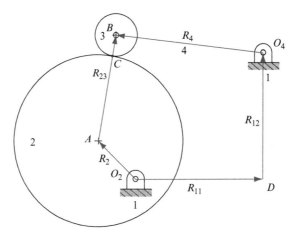

This is a valid loop-closure equation and, given the angle θ_2 as input, it can be solved for the orientations of link 4 and of the fictitious link 23. However, this does not yet allow a solution for the rotation of the roller, link 3.

Since there is rolling contact between cam 2 and roller 3 at point C, let us consider the apparent angular displacements of links 2 and 3 as they would appear to an observer riding on the fictitious arm, link 23. Such an observer would observe

$$\rho_3 \Delta\theta_{3/23} = -\rho_2 \Delta\theta_{2/23}, \tag{2}$$

which gives the following relationship between the absolute angular displacements:

$$\rho_3(\Delta\theta_3 - \Delta\theta_{23}) = -\rho_2(\Delta\theta_2 - \Delta\theta_{23}). \tag{3}$$

Rearranging this equation gives

$$\rho_3 \Delta\theta_3 = (\rho_2 + \rho_3)\Delta\theta_{23} - \rho_2 \Delta\theta_2. \qquad Ans. \ (4)$$

Example 2.7 (cont.)

Since the angular displacement of link 2 is known (the input) and $\Delta\theta_{23}$ is determined from Eq. (1), the angular displacement, $\Delta\theta_3$, can be determined from Eq. (4). It is worth noting that we can obtain a solution for the posture of link 3 even though no vector is attached to or rotates with link 3.

Example 2.8

For the rack-and-pinion mechanism shown in Fig. 2.33, define a set of vectors that is suitable for a complete kinematic analysis. Label and show the sense and orientation of each vector. Write the vector loop equation(s) and identify suitable input(s) for the mechanism. Identify the known quantities, the unknown variables, and any constraints. If you identify constraints, then write the constraint equation(s).

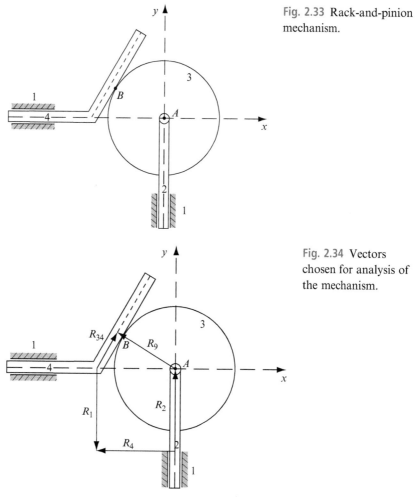

Fig. 2.33 Rack-and-pinion mechanism.

Fig. 2.34 Vectors chosen for analysis of the mechanism.

Example 2.8 (cont.)

Solution

Let us consider the set of vectors shown in Fig. 2.34.

The length of vector $\mathbf{R}_2$ represents the input, and the length of vector $\mathbf{R}_4$ indicates the output. The loop-closure equation is

$$\overset{I\surd}{\mathbf{R}_2} + \overset{\surd C}{\mathbf{R}_9} - \overset{?\surd}{\mathbf{R}_{34}} + \overset{\surd\surd}{\mathbf{R}_1} - \overset{?\surd}{\mathbf{R}_4} = \mathbf{0}. \qquad\qquad Ans.$$

The constraint that gives the orientation of vector $\mathbf{R}_9$ is

$$\theta_9 = \theta_{34} + 90°. \qquad\qquad Ans.$$

The angular displacement of pinion 3 can be determined by considering the apparent rotation of pinion 3 to an observer riding on the rack, link 4. As such an observer observes the changing position of the point of contact B, SHE must see the same increment of surface on both links 3 and 4. The constraint relationship can be written as

$$\Delta R_{34} = \rho_3 \Delta\theta_{3/4} = \rho_3\left(\Delta\theta_3 - \overset{0}{\Delta\theta}_4\right) = \rho_3 \Delta\theta_3.$$

Therefore, the angular displacement of the pinion is

$$\Delta\theta_3 = \Delta R_{34}/\rho_3. \qquad\qquad Ans.$$

PROBLEMS[9]

2.1 Describe and sketch the locus of a point A, which moves according to the equations $R_A^x = at\cos(2\pi t)$, $R_A^y = at\sin(2\pi t)$, and $R_A^z = 0$.

2.2 Find the position difference from point P to point Q on the curve $y = x^2 + x - 16$, where $R_P^x = 2$ and $R_Q^x = 4$.

2.3 The path of a moving point is defined by the equation $y = 2x^2 - 28$. Find the position difference from point P to point Q if $R_P^x = 4$ and $R_Q^x = -3$.

2.4 The path of a moving point P is defined by the equation $y = 60 - x^3/3$. What is the displacement of the point if its motion begins at $R_P^x = 0$ and ends at $R_P^x = 3$?

2.5 If point A moves on the locus of Prob. 2.1, find its displacement from $t = 2$ to $t = 2.5$.

2.6 The position of a point is given by the equation $\mathbf{R} = 100e^{j2\pi t}$. What is the path of the point? Determine the displacement of the point from $t = 0.10$ to $t = 0.40$.

---

[9] When assigning problems, the instructor may wish to specify the method of solution to be used, since a variety of approaches are presented in the text.

2.7 The equation $\mathbf{R} = (t^2 + 4)e^{-j\pi t/10}$ defines the position of a point. In which direction is the position vector rotating? Where is the point located when $t = 0$? What is the next value t can have if the orientation of the position vector is to be the same as it is when $t = 0$? What is the displacement from the first position of the point to the second?

2.8 The location of a point is defined by the equation $\mathbf{R} = (4t + 2)e^{j\pi t^2/30}$, where t is time in seconds. Motion of the point is initiated when $t = 0$. What is the displacement during the first 3 s? Find the change in angular orientation of the position vector during the same time interval.

2.9 Link 2 rotates according to the equation $\theta = \pi t/4$. Block 3 slides outward on link 2 according to the equation $r = t^2 + 2$. What is the absolute displacement $\Delta \mathbf{R}_{P_3}$ from $t = 1$ to $t = 2$? What is the apparent displacement $\Delta \mathbf{R}_{P_{3/2}}$?

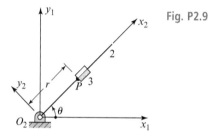

Fig. P2.9

2.10 A wheel with center at O and with 150 mm radius rolls without slipping on the ground at point P. If point O is displaced 250 mm to the right, determine the displacement of point P during this interval.

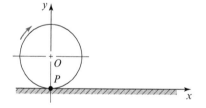

Fig. P2.10 Rolling wheel.

2.11 A point Q moves from A to B along link 3 while link 2 rotates from $\theta_2 = 30°$ to $\theta_2' = 120°$. Find the absolute displacement of Q.

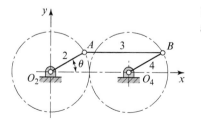

Fig. P2.11 $R_{AO_2} = R_{BO_4} = 75$ mm and $R_{BA} = R_{O_4O_2} = 150$ mm.

2.12 The double-slider linkage is driven by moving sliding block 2. Write the loop-closure equation. Solve analytically for the position of sliding block 4. Check the result graphically for the posture where $\phi = -45^\circ$.

Fig. P2.12 $R_{AB} = 8$ in and $\psi = 15^\circ$.

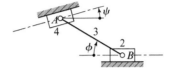

2.13 The offset slider-crank linkage is driven by crank 2. Write the loop-closure equation. Solve for the position of slider 4 as a function of θ_2.

Fig. P2.13 $R_{AO} = 20$ mm, $R_{BA} = 50$ mm, and $R_{CB} = 140$ mm.

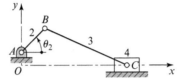

2.14 Define a set of vectors that is suitable for a complete kinematic analysis of the mechanism. Label and show the sense and orientation of each vector. Write the vector loop equation(s) for the mechanism. Identify suitable input(s), known quantities, unknown variables, and any constraints. If you identify constraints, then write the constraint equation(s).

Fig. P2.14

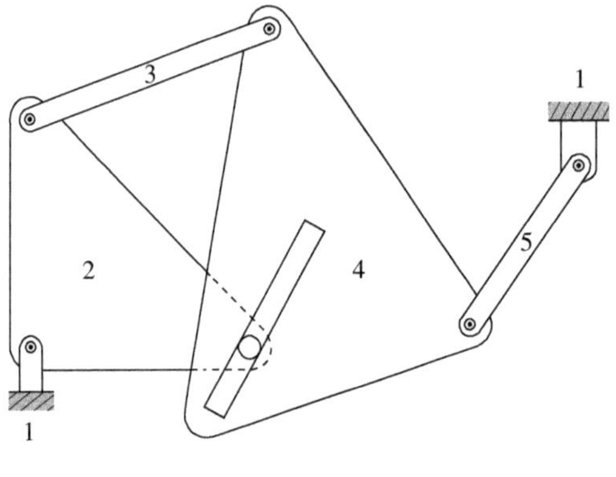

2.15 Define a set of vectors that is suitable for a complete kinematic analysis of the rack-and-pinion mechanism. Label and show the sense and orientation of each vector. Assuming rolling with no slip between rack 4 and pinion 5, write the vector loop equation(s) for the mechanism. Identify suitable input(s), known quantities, unknown variables, and any constraints. If you identify constraints, then write the constraint equation(s).

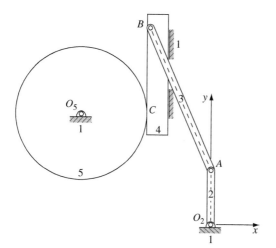

Fig. P2.15 Rack-and-pinion mechanism.

2.16 Define a set of vectors that is suitable for a complete kinematic analysis of the mechanism. Label and show the sense and orientation of each vector. Assuming rolling with no slip between gears 2 and 5, write the vector loop equation(s) for the mechanism. Identify suitable input(s), known quantities, unknown variables, and any constraints. If you identify constraints, then write the constraint equation(s).

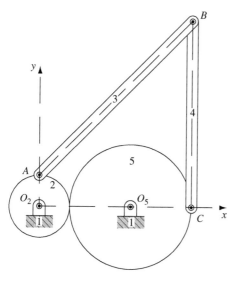

Fig. P2.16 Geared five-bar mechanism.

2.17 Gear 3, which is pinned to link 4 at point B, is rolling without slip on semicircular ground link 1. The radius of gear 3 is ρ_3, and the radius of the ground link is ρ_1. Define a set of vectors that is suitable for a complete kinematic analysis of the mechanism. Label and show the sense and orientation of each vector. Write the vector loop equation(s) for the mechanism. Identify suitable input(s), known quantities, unknown variables, and any constraints. If you identify constraints, then write the constraint equation(s).

Fig. P2.17

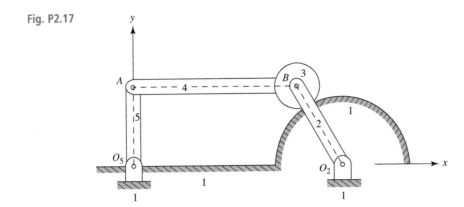

2.18 For the mechanism in Prob. 1.6, define a set of vectors that is suitable for a complete kinematic analysis of the mechanism. Label and show the sense and orientation of each vector. Write the vector loop equation(s) for the mechanism. Identify suitable input(s), known quantities, unknown variables, and any constraints. If you identify constraints, then write the constraint equation(s).

2.19 For the mechanism in Prob. 1.8, define a set of vectors that is suitable for a complete kinematic analysis of the mechanism. Label and show the sense and orientation of each vector. Write the vector loop equation(s) for the mechanism. Identify suitable input(s), known quantities, unknown variables, and any constraints. If you identify constraints, then write the constraint equation(s).

2.20 For the mechanism in Prob. 1.9, define a set of vectors that is suitable for a complete kinematic analysis of the mechanism. Label and show the sense and orientation of each vector. Write the vector loop equation(s) for the mechanism. Identify suitable input(s), known quantities, unknown variables, and any constraints. If you identify constraints, then write the constraint equation(s).

2.21 For the mechanism in Prob. 1.10, define a set of vectors that is suitable for a complete kinematic analysis of the mechanism. Label and show the sense and orientation of each vector. Write the vector loop equation(s) for the mechanism. Identify suitable input(s), known quantities, unknown variables, and any constraints. If you identify constraints, then write the constraint equation(s).

2.22 Write a calculator program to find the sum of any number of two-dimensional vectors expressed in mixed rectangular or polar forms. The result should be obtainable in either form with the magnitude and angle of the polar form having only positive values.

2.23 Write a computer program to plot the coupler curve of any crank-rocker or double-crank form of the four-bar linkage. The program should accept four link lengths and either rectangular or polar coordinates of the coupler point with respect to the coupler.

2.24 Plot the path of point P for: (a) an inverted slider-crank linkage; (b) second inversion of the slider-crank linkage; (c) a Scott–Russell straight-line linkage; and (d) a drag-link linkage.

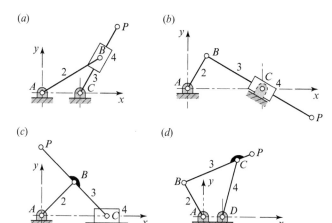

Fig. P2.24 (a) $R_{CA} = 50$ mm,
$R_{BA} = 88$ mm, and
$R_{PC} = 100$ mm; (b) $R_{CA} = 1.60$ in,
$R_{BA} = 0.80$ in, and $R_{PB} = 2.60$ in;
(c) $R_{BA} = R_{CB} = R_{PB} = 1$ in; (d)
$R_{DA} = 25$ mm, $R_{BA} = 50$ mm,
$R_{CB} = R_{CD} = 75$ mm, and
$R_{PB} = 100$ mm.

2.25 Using the offset slider-crank linkage in Fig. P2.13, find the crank angles corresponding to the extreme values of the transmission angle.

2.26 Section 1.10 states that the transmission angle, denoted as the angle γ in Fig. 1.31, reaches an extreme value for the four-bar linkage when the crank lies on the line between the fixed pivots. This means that γ reaches a maximum or minimum when crank 2 is collinear with the ground link – that is, the line O_2O_4. Show, analytically, that this statement is true.

2.27 Define a set of vectors that is suitable for a complete kinematic analysis of the mechanism. Label and show the sense and orientation of each vector. Write the vector loop equation (s) for the mechanism. Identify suitable input(s), known quantities, unknown variables, and any constraints. If you identify constraints, then write the constraint equation(s).

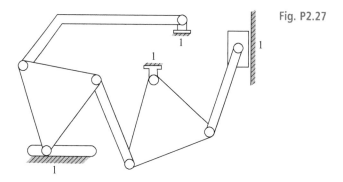

Fig. P2.27

2.28 Define a set of vectors that is suitable for a complete kinematic analysis of the mechanism. Label and show the sense and orientation of each vector. Write the vector loop equation (s) for the mechanism. Identify suitable input(s), known quantities, unknown variables, and any constraints. If you identify constraints, then write the constraint equation(s).

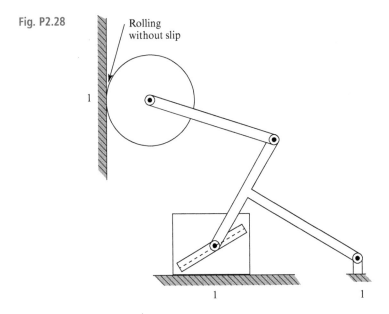

Fig. P2.28

2.29 Define a set of vectors that is suitable for a complete kinematic analysis of the mechanism. Label and show the sense and orientation of each vector. Write the vector loop equation (s) for the mechanism. Identify suitable input(s), known quantities, unknown variables, and any constraints. If you identify constraint(s) then write the constraint equation(s).

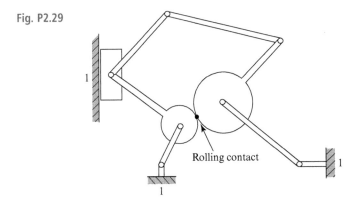

Fig. P2.29

2.30 Define a set of vectors that is suitable for a complete kinematic analysis of the mechanism. Label and show the sense and orientation of each vector. Write the vector loop equation (s) for the mechanism. Identify suitable input(s), known quantities, unknown variables, and any constraints. If you identify constraint(s) then write the constraint equation(s).

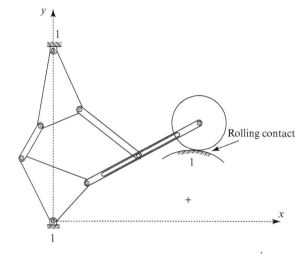

Rolling contact

2.31 For the input angle $\theta_2 = 300°$, measured counterclockwise from the x axis, determine the two postures of link 4.

Fig. P2.31 $r_2 = 2.4$ in, $r_3 = 5.6$ in,
$r_4 = 5.6$ in, and $r_1 = 6.4$ in.

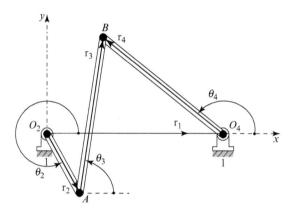

2.32 For the input angle $\theta_2 = 60°$, measured counterclockwise from the x axis, determine the two postures of link 4.

Fig. P2.32 $r_2 = 3.2$ in, $r_3 = 2$ in,
$r_4 = 4$ in, and $r_1 = 2.8$ in.

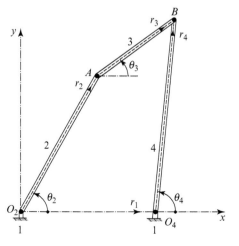

2.33 Consider a four-bar linkage for which ground link 1 is 350 mm, input link 2 is 175 mm, coupler link 3 is 250 mm, and output link 4 is 200 mm. The fixed x and y axes are specified as horizontal and vertical, respectively. The origin of this reference frame is coincident with the ground pivot of link 2 and the ground link is aligned with the x axis. For input angle $\theta_2 = 60°$ (counterclockwise from the x axis): (a) Using a suitable scale, draw the linkage in the open and crossed postures and measure the values of the variables θ_3 and θ_4 for each posture. (b) Use trigonometry (that is, the laws of sines and cosines) to determine θ_3 and θ_4 for the open posture. (c) Use Freudenstein's equation to determine θ_3 and θ_4 for both postures. (d) Use the Newton–Raphson iteration procedure to determine θ_3 and θ_4 for the open posture. Using the measurements in (a) as initial estimates for θ_3 and θ_4, iterate until the two variables converge to within $0.01°$.

2.34 A crank-rocker four-bar linkage is shown in two different postures for which $\theta_2 = 150°$ and $\theta_2' = 240°$. Determine θ_3 and θ_4 for the open posture and θ_3' and θ_4' for the crossed posture.

Fig. P2.34 $R_{O_4 O_2} = 24$ in, $R_{AO_2} = 5.6$ in, $R_{BA} = 27.6$ in, and $R_{BO_4} = 16$ in.

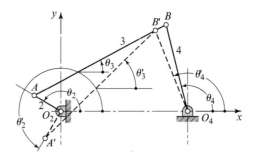

2.35 Define a set of vectors that is suitable for a complete kinematic analysis of the mechanism of Prob. 1.37. Write the vector loop equation(s) for the mechanism, clearly identifying the sense and orientation of each vector. Identify suitable input(s), known quantities, unknown variables, and any constraints. If you identify constraints, then write the constraint equation(s).

2.36 Define a set of vectors that is suitable for a complete kinematic analysis of the mechanism of Prob. 1.38. Write the vector loop equation(s) for the mechanism, clearly identifying the sense and orientation of each vector. Identify suitable input(s), known quantities, unknown variables, and any constraints. If you identify constraints, then write the constraint equation(s).

2.37 Define a set of vectors that is suitable for a complete kinematic analysis of the mechanism of Prob. 1.43. Write the vector loop equation(s) for the mechanism, clearly identifying the sense and orientation of each vector. Identify suitable input(s), known quantities, unknown variables, and any constraints. If you identify constraints, then write the constraint equation(s).

2.38 Define a set of vectors that is suitable for a complete kinematic analysis of the mechanism. Write the vector loop equation(s) for the mechanism, clearly identifying the sense and orientation of each vector. Identify suitable input(s), known quantities, unknown variables, and any constraints. If you identify constraints, then write the constraint equation(s).

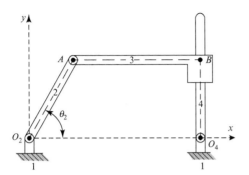

Fig. P2.38 $R_{O_4O_2} = 12$ in,
$R_{AO_2} = 6$ in, $R_{BA} = 9$ in, and
$\theta_2 = 60°$.

2.39 Define a set of vectors that is suitable for a complete kinematic analysis of the mechanism. Write the vector loop equation(s) for the mechanism, clearly identifying the sense and orientation of each vector. Identify suitable input(s), known quantities, unknown variables, and any constraints. If you identify constraints, then write the constraint equation(s).

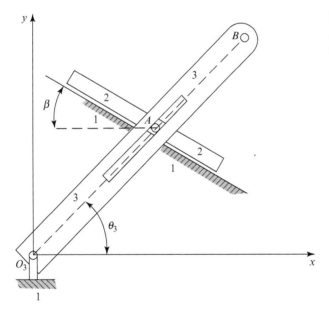

Fig. P2.39 $R_{A_2O_3} = 750$ mm,
$R_{BO_3} = 1\,250$ mm, $\beta = -30°$, and
$\theta_3 = 45°$.

2.40 Write the rolling contact equations for: (a) gear 2 rolling without slip on gear 3 at point C; (b) gear 3 rolling on the ground link 1 at point E; and (c) gear 2 rolling on the rack (link 5) at point at point F.

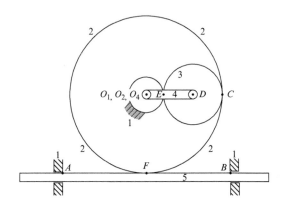

Fig. P2.40 $\rho_1 = R_{EO_1} = 4$ in,
$\rho_2 = R_{CO_2} = 18$ in, $\rho_3 = R_{CD} = 7$ in, and
$R_{FA} = R_{BF} = 20$ in.

2.41 For the mechanism in the posture shown, the input arm, link 2, is pinned to the ground at O_1 and is pinned to the center of gear 3 at point A. The center of gear 4 is also pinned to the ground at O_1 and gear 5 is pinned to the ground at O_5. Gears 3, 4, and 5 are all in rolling contact at point B. Write the rolling contact equation for gear 3 which is rolling without slip on the fixed gear 1.

Fig. P2.41 $\rho_1 = 100$ mm, $\rho_3 = 100$ mm,
$\rho_4 = 300$ mm, and $\rho_5 = 500$ mm.

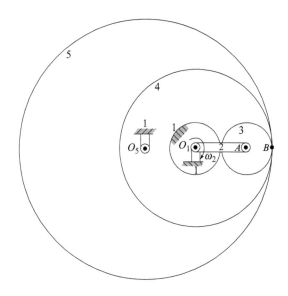

2.42 Define a set of vectors that is suitable for a complete kinematic analysis of the mechanism. The wheel (link 5) is rolling without slipping on the circular ground link 1 at point D. Label and show the sense and orientation of each vector. Write the vector loop equation(s) for the mechanism. Identify suitable input(s), known quantities, unknown variables, and any constraints. If you identify constraints, then write the constraint equation(s).

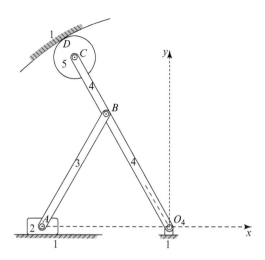

Fig. P2.42 $R_{BO_4} = 150$ mm, $R_{CO_4} = 225$ mm,
$R_{BA} = 150$ mm, $\rho_5 = 25$ mm, and $\theta_4 = 120°$.

2.43 For the mechanism in the posture shown, pinion 3 rolls without slip on rack 4 at point B. Define a set of vectors that is suitable for a complete kinematic analysis of the mechanism. Label and show the sense and orientation of each vector. Write the vector loop equation(s) for the mechanism. Identify suitable input(s), known quantities, unknown variables, and any constraints. If you identify constraints, then write the constraint equation(s).

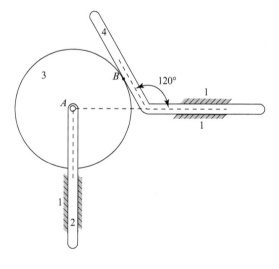

Fig. P2.43 $\rho_3 = 2$ in.

2.44 Define a set of vectors that is suitable for a complete kinematic analysis of the mechanism. Label and show the sense and orientation of each vector. Write the vector loop equation(s) for the mechanism. Identify suitable input(s), known quantities, unknown variables, and any constraints. If you identify constraints, then write the constraint equation(s).

Fig. P2.44 $R_{AO_2} = 4$ in, $R_{BA} = 8$ in,
$R_{DO_4} = 4$ in, and $R_{O_2O_4} = 2$ in.

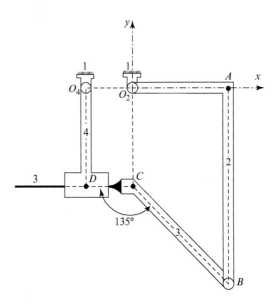

2.45 For the mechanism shown, gear 3 is rolling without slip on ground link 1 at point C. Define a set of vectors that is suitable for a complete kinematic analysis of the mechanism. Label and show the sense and orientation of each vector. Write the vector loop equation(s) for the mechanism. Identify suitable input(s), known quantities, unknown variables, and any constraints. If you identify constraints, then write the constraint equation(s).

Fig. P2.45

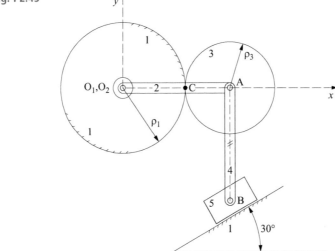

2.46 Define a set of vectors that is suitable for a complete kinematic analysis of the mechanism. Label and show the sense and orientation of each vector. Write the vector loop equation(s) for the mechanism. Identify suitable input(s), known quantities, unknown variables, and any constraints. If you identify constraints, then write the constraint equation(s).

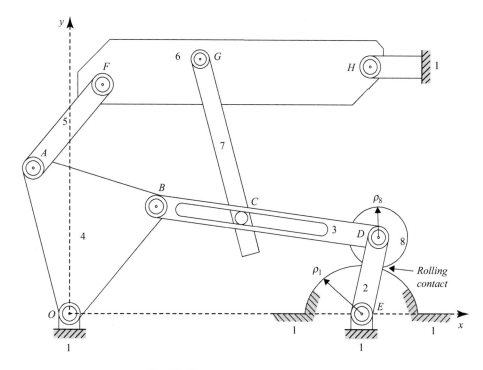

Fig. P2.46

REFERENCES

[1] Mischke, C. R., 1980. *Mathematical Model Building*, Ames, IA: Iowa State University Press.

3 Velocity

3.1 Definition of Velocity

In Fig. 3.1, a moving point is first observed at location P, defined by the absolute position vector $\mathbf{R}_P$. After a short time interval Δt, its location is observed to have changed to P', defined by $\mathbf{R}'_P$. Recall from Eq. (2.58) that the displacement during this time interval is defined as

$$\Delta \mathbf{R}_P = \mathbf{R}'_P - \mathbf{R}_P.$$

The *average velocity* of the point during the time increment Δt is defined by the ratio $\Delta \mathbf{R}_P / \Delta t$. The *instantaneous velocity* (hereafter simply called *velocity*) is defined by the limit of this ratio as the time increment goes to zero and is given by

$$\mathbf{V}_P = \lim_{\Delta t \to 0} \frac{\Delta \mathbf{R}_P}{\Delta t} = \frac{d \mathbf{R}_P}{dt}. \tag{3.1}$$

Since $\Delta \mathbf{R}_P$ is a vector, there are two convergences in taking this limit, the magnitude and the direction. Therefore, the velocity of a point is a vector quantity equal to the time rate of change of its position. Like the position and displacement vectors, the velocity vector is defined for a specific point. The term "velocity" should not be applied to a line, coordinate system, volume, or other collection of points, since the velocity of each point may be different.

We recall from Chapter 2 that, for their definitions, the position vectors $\mathbf{R}_P$ and $\mathbf{R}'_P$ depend on the posture of the coordinate system of the observer. The displacement vector $\Delta \mathbf{R}_P$ and the velocity vector $\mathbf{V}_P$, on the other hand, are independent of the location of the coordinate system or the location of the observer within the coordinate system. However, the velocity vector $\mathbf{V}_P$ does depend on the motion, if any, of the observer or the coordinate system during the time interval; it is for this reason that the observer is assumed to remain stationary within the coordinate system. If the coordinate system involved is the absolute coordinate system, the velocity is referred to as an *absolute velocity* and is denoted by $\mathbf{V}_{P/1}$ or simply $\mathbf{V}_P$. This is consistent with the notation used for absolute position and absolute displacement.

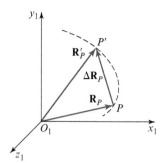

Fig. 3.1 Displacement of a moving point.

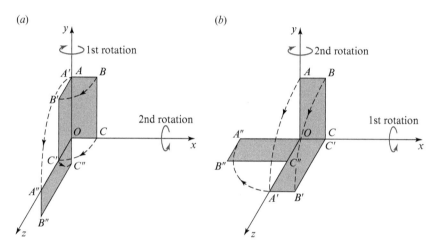

Fig. 3.2 Three-dimensional angular displacements.

3.2 Rotation of a Rigid Body

When a rigid body translates, the motion of any particular point is equal to the motion of every other point of the body. When a rigid body rotates, however, two arbitrarily chosen points, P and Q, do not undergo the same motion, and a coordinate system attached to the body does not remain parallel to its initial orientation; that is, the body undergoes some angular displacement $\Delta\theta$. This was discussed in Section 2.16.

Angular displacements were not treated in detail in Chapter 2, since, in general, they cannot be treated as vectors. The reason is that they do not obey the laws of vector addition. If a rigid body undergoes multiple finite angular displacements in succession, in three dimensions, the result depends on the order in which the displacements take place. To show this, consider the rectangular body $ABCO$ in Fig. 3.2a. If the body is first rotated by $-90°$ (that is, clockwise) about the y axis and then rotated by $+90°$ (that is, counterclockwise) about the x axis, the final posture of the body is in the yz plane. In Fig. 3.2b, the body occupies the same starting posture and is again rotated about the same axes, through the same angles, and in the same directions; however, in this case, the first rotation is about the x axis and the second is about the y axis. The order of the rotations is reversed, and the final posture of the body is now in the zx plane rather

Fig. 3.3 Displacement difference
between points P and Q.

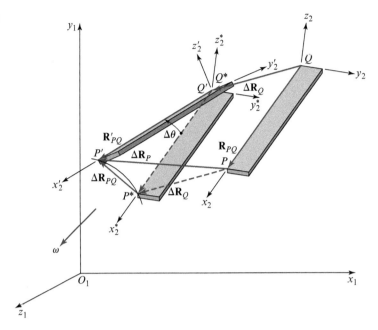

Fig. 3.3 Displacement difference between points P and Q.

than in the yz plane, as it was before. Since this characteristic does not correspond to the commutative law of vector addition, three-dimensional angular displacements cannot be treated as vectors.

Angular displacements that occur about the same axis or parallel axes, on the other hand, do follow the commutative law. Also, infinitesimally small angular displacements are commutative. To avoid confusion, we will treat all finite angular displacements as scalar quantities. However, we will have occasion to treat infinitesimal angular displacements as vectors.

Recall from Section 2.15 the definition of the displacement difference between two points, P and Q, both attached to the same rigid body, as shown again in Fig. 3.3. The displacement-difference vector is entirely attributable to the rotation of the body; there is no displacement difference between points in a rigid body undergoing a translation. We reach this conclusion by picturing the displacement as occurring in two steps. First, the body is assumed to translate through the displacement $\Delta \mathbf{R}_Q$ to the posture $x_2^* y_2^* z_2^*$. Next, the body is assumed to rotate about point Q^* to the posture $x_2' y_2' z_2'$.

Another way to interpret the displacement difference $\Delta \mathbf{R}_{PQ}$ is to imagine a moving coordinate system whose origin travels with point Q but whose axes remain parallel to the absolute axes $x_1 y_1 z_1$. Note that this coordinate system does not rotate. An observer in this moving coordinate system observes no motion for point Q since it remains fixed in that observer's coordinate system. For the displacement of point P, such a translating observer only sees the displacement-difference vector $\Delta \mathbf{R}_{PQ}$. It seems to such an observer that point Q remains fixed and that the body rotates about this seemingly fixed point, as shown in Fig. 3.4.

No matter whether the observer is in the fixed coordinate system $x_1 y_1 z_1$ or in the moving coordinate system, the body appears to rotate through some total angle $\Delta \theta$ in its displacement from $x_2 y_2 z_2$ to $x_2' y_2' z_2'$. If we take the point of view of the fixed observer, the location of the axis

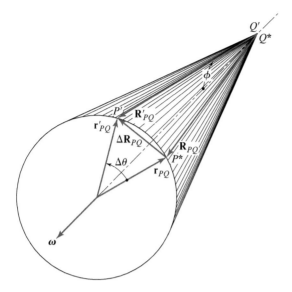

Fig. 3.4 Displacement difference $\Delta\mathbf{R}_{PQ}$ as seen by a translating observer.

of rotation is not obvious. As seen by the translating observer, the axis passes through the apparently stationary point Q; all points in the body appear to travel in circular paths about this axis, and any line in the body whose direction is normal to this axis appears to undergo an identical angular displacement $\Delta\theta$. The rotating position-difference vector $\mathbf{R}_{PQ}$ generates a cone.

The *angular velocity* of a rotating body is now defined as a vector quantity $\boldsymbol{\omega}$ having a direction parallel to the instantaneous axis of rotation. The magnitude of the angular velocity is defined as the time rate of change of the angular orientation of any line in the body whose direction is normal to the axis of rotation. If we designate the angular displacement of any of these lines as $\Delta\theta$ and the time interval as Δt, the magnitude of the angular-velocity vector is

$$\omega = \lim_{\Delta t \to 0} \frac{\Delta\theta}{\Delta t} = \frac{d\theta}{dt}. \tag{3.2}$$

Since we agree to treat counterclockwise rotations as positive, the sense of the angular-velocity vector along the axis of rotation is in accordance with the right-hand rule.

3.3   Velocity Difference between Points of a Rigid Body

Figure 3.5*a* shows another view of the same rigid-body displacement that was presented in Fig. 3.3. This is the view seen by an observer in the absolute coordinate system looking directly along the axis of rotation of the moving body, from the tip of the angular-velocity vector. Therefore, the angular displacement $\Delta\theta$ is observed in true size, and the projections of *all* lines in the body rotate through this same angle during the displacement. The displacement vectors and the position-difference vectors, however, are not necessarily observed in true size; their projections may appear foreshortened under the current viewing angle.

Fig. 3.5 (*a*) True view of angular
displacements; (*b*) vector subtraction to
form displacement difference $\Delta\mathbf{R}_{PQ}$.

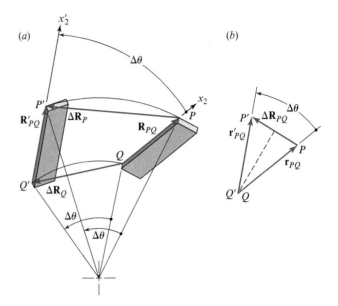

Fig. 3.5 (*a*) True view of angular
displacements; (*b*) vector subtraction to
form displacement difference $\Delta\mathbf{R}_{PQ}$.

Figure 3.5*b* shows the same rigid-body displacement from the same viewing angle, but this time from the point of view of a translating observer. Thus, Fig. 3.5*a* corresponds to the base of the cone of Fig. 3.4. We note that the two vectors labeled $\mathbf{r}_{PQ}$ and $\mathbf{r}'_{PQ}$ are the foreshortened projections of $\mathbf{R}_{PQ}$ and $\mathbf{R}'_{PQ}$, and from Fig. 3.4 we observe that their magnitudes are

$$r_{PQ} = r'_{PQ} = R_{PQ}\sin\phi, \tag{a}$$

where ϕ is the angle from the angular-velocity vector $\boldsymbol{\omega}$ to the rotating position-difference vector $\mathbf{R}_{PQ}$ as it traverses the cone. Figure 3.5*b* can also be interpreted as a scale drawing corresponding to Eq. (2.60); that is,

$$\Delta\mathbf{R}_{PQ} = \mathbf{R}'_{PQ} - \mathbf{R}_{PQ}. \tag{b}$$

We can calculate the magnitude of the displacement-difference vector $\Delta\mathbf{R}_{PQ}$ by drawing the perpendicular bisector of the vector shown in Fig. 3.5*b*. From this construction we have

$$\Delta R_{PQ} = 2r_{PQ}\sin\frac{\Delta\theta}{2}. \tag{c}$$

Substituting Eq. (*a*) into Eq. (*c*), we have

$$\Delta R_{PQ} = 2\left(R_{PQ}\sin\phi\right)\sin\frac{\Delta\theta}{2}. \tag{d}$$

If we limit ourselves to infinitesimal displacements, the sine of the angular displacement can be approximated by the angle itself; that is,

$$\Delta R_{PQ} = 2\left(R_{PQ}\sin\phi\right)\frac{\Delta\theta}{2} = \Delta\theta R_{PQ}\sin\phi. \tag{e}$$

Dividing by the time increment Δt, noting that the magnitude R_{PQ} and the angle ϕ are constant during the interval, and taking the limit gives

$$\lim_{\Delta t \to 0}\left(\frac{\Delta R_{PQ}}{\Delta t}\right) = \lim_{\Delta t \to 0}\left(\frac{\Delta \theta}{\Delta t}\right) R_{PQ}\sin\phi = \omega R_{PQ}\sin\phi. \qquad (f)$$

Recalling the definition of ϕ as the angle from vector $\boldsymbol{\omega}$ to vector $\mathbf{R}_{PQ}$, we can restore the vector attributes of Eq. (f) by recognizing this result as the magnitude of a vector cross-product. Therefore,

$$\lim_{\Delta t \to 0}\left(\frac{\Delta \mathbf{R}_{PQ}}{\Delta t}\right) = \frac{d\mathbf{R}_{PQ}}{dt} = \boldsymbol{\omega} \times \mathbf{R}_{PQ}. \qquad (g)$$

This form is so important and so useful that it is given its own name and notation; it is called the *velocity-difference* vector and is denoted $\mathbf{V}_{PQ}$; that is,

$$\mathbf{V}_{PQ} = \frac{d\mathbf{R}_{PQ}}{dt} = \boldsymbol{\omega} \times \mathbf{R}_{PQ}. \qquad (3.3)$$

Let us now recall the displacement-difference equation – see Eq. (2.59):

$$\Delta \mathbf{R}_P = \Delta \mathbf{R}_Q + \Delta \mathbf{R}_{PQ}. \qquad (h)$$

Dividing this equation by Δt and taking the limit as the time increment goes to zero, gives

$$\lim_{\Delta t \to 0}\left(\frac{\Delta \mathbf{R}_P}{\Delta t}\right) = \lim_{\Delta t \to 0}\left(\frac{\Delta \mathbf{R}_Q}{\Delta t}\right) + \lim_{\Delta t \to 0}\left(\frac{\Delta \mathbf{R}_{PQ}}{\Delta t}\right), \qquad (i)$$

which, from Eqs. (3.1) and (3.3), can be written as

$$\mathbf{V}_P = \mathbf{V}_Q + \mathbf{V}_{PQ}. \qquad (3.4)$$

This important equation is called the *velocity-difference equation* and, together with Eq. (3.3), it forms one of the basic equations of all velocity analysis techniques.

Equation (3.4) can be written for any two points with no restriction. However, reviewing the previous derivation, we recognize that Eq. (3.3) cannot be applied to any arbitrary pair of points. *This equation is only valid if the two points are fixed in the same rigid body.*[1] This restriction can, perhaps, be better remembered if all subscripts are written explicitly; that is,

$$\mathbf{V}_{P_2 Q_2} = \boldsymbol{\omega}_2 \times \mathbf{R}_{P_2 Q_2}. \qquad (j)$$

Note that the link-number subscripts are the same throughout this equation, and so, in the interest of brevity, are often suppressed. If a mistaken attempt is made to apply Eq. (3.3) when points P and Q are not in the same rigid body, the error should be discovered, since it will not be clear which angular-velocity vector should be used.

---

[1] More precisely, the restriction is the requirement that the distance R_{PQ} remains constant. However, in applications, the above wording fits most real situations.

3.4   Velocity Polygons; Velocity Images

One important approach to velocity analysis is graphic. As observed in graphic posture analysis (Section 2.8), it is primarily of use in two-dimensional mechanism problems when the solution of only a single posture is required. The major advantages are that a solution can be achieved quickly and that visualization of, and insight into, the problem are enhanced by the graphic approach [7].

Consider the planar motion of the unconstrained link ABC shown in Fig. 3.6a. Given the velocities of points A and B, the problem is to determine: (a) the angular velocity of the link, and (b) the velocity of point C. Let us agree that a scale drawing of the link has been prepared for the given posture and that any required position-difference vectors can be scaled from the drawing.

For the solution of part (a), we first consider the velocity-difference equation, Eq. (3.4), relating points A and B; that is,

$$\overset{\sqrt{\sqrt{}}}{\mathbf{V}_B} = \overset{\sqrt{\sqrt{}}}{\mathbf{V}_A} + \overset{??}{\mathbf{V}_{BA}}, \tag{a}$$

where the two unknowns are the magnitude and the direction of the velocity-difference vector, $\mathbf{V}_{BA}$. Figure 3.6b shows the graphic solution to this equation. After an appropriate scale is chosen to represent velocity vectors, the vectors $\mathbf{V}_A$ and $\mathbf{V}_B$ are both drawn to scale, starting from a common origin and in the two known directions. The vector spanning the termini of $\mathbf{V}_A$ and $\mathbf{V}_B$ is the velocity-difference vector $\mathbf{V}_{BA}$ and is correct, within graphic accuracy, in both magnitude and direction.

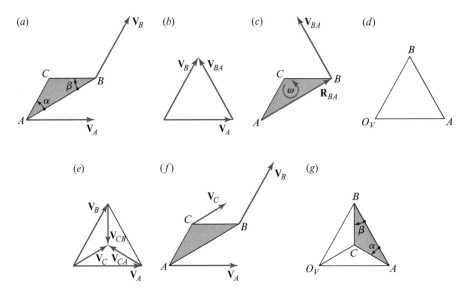

Fig. 3.6 Graphic velocity analysis of link ABC.

The velocity-difference vector can be written from Eq. (3.3) as

$$\mathbf{V}_{BA} = \boldsymbol{\omega} \times \mathbf{R}_{BA}. \qquad (b)$$

Since the link is in planar motion, the angular-velocity vector, $\boldsymbol{\omega}$, lies perpendicular to the plane of motion – that is, perpendicular to vectors $\mathbf{V}_{BA}$ and $\mathbf{R}_{BA}$. Considering only the magnitudes in Eq. (b) gives

$$V_{BA} = \omega R_{BA} \sin 90° = \omega R_{BA}.$$

Rearranging this equation, the angular velocity of the link is

$$\omega = \frac{V_{BA}}{R_{BA}}. \qquad (c)$$

Therefore, the magnitude of the angular velocity can be obtained by scaling V_{BA} from Fig. 3.6b and R_{BA} from Fig. 3.6a, being careful to properly apply the scale factors for units; it is common practice to evaluate ω in units of radians per second.

The magnitude, ω, is not a complete solution for the angular-velocity vector; the direction must also be determined. As observed earlier, the $\boldsymbol{\omega}$ vector is perpendicular to the plane of the link itself, since the motion is planar. However, this does not indicate whether $\boldsymbol{\omega}$ is directed into or out of the plane of the motion. Taking the point of view of a translating observer, moving with point A but not rotating, we can visualize the link as rotating about point A; this is shown in Fig. 3.6c. The velocity difference $\mathbf{V}_{BA}$ is the only velocity seen by such an observer. Therefore, interpreting $\mathbf{V}_{BA}$ to indicate the direction of rotation of point B about point A, we discover the direction of $\boldsymbol{\omega}$, counterclockwise in this example. Although it is not strict vector notation, it is common practice in two-dimensional problems to indicate the final solution in the form $\boldsymbol{\omega} = xxx$ rad/s ccw, where xxx is a numeric value indicating the magnitude and "ccw" or "cw" indicates the direction.

The practice of constructing vector diagrams using thick dark lines, such as in Fig. 3.6b, makes them easy to read, but when the diagram is the graphic solution of an equation, it is not very accurate. For this reason, it is preferable to construct the graphic solution with thin sharp lines, made with a hard drawing pencil, as shown in Fig. 3.6d. The solution begins by choosing a scale and a point labeled O_V to represent zero velocity. Absolute velocities such as $\mathbf{V}_A$ and $\mathbf{V}_B$ are constructed with their origins at O_V, and their termini are labeled as points A and B. The line *from A to B* then represents the velocity difference $\mathbf{V}_{BA}$. As we continue, we observe that these labels at the vertices are sufficient to determine the precise notation of all velocity differences represented by lines in the diagram. Note, for example, that $\mathbf{V}_{BA}$ is represented by the vector *from point A to point B*. With this labeling convention, no arrowheads or additional notation are necessary and do not clutter the diagram. Such a diagram is called a *velocity polygon* and adds considerable convenience to the graphic solution technique.

A danger of this convention, however, is that the reader may begin to think of the technique as a series of graphic "tricks" and lose sight of the fact that each line that is drawn can be, and should be, fully justified by a corresponding vector equation. The graphics are merely a convenient solution technique and not a substitute for a sound theoretical basis.

Returning to Fig. 3.6c, it may have appeared coincidental that the velocity difference $\mathbf{V}_{BA}$ was perpendicular to $\mathbf{R}_{BA}$. Recalling Eq. (b), however, we see that it is a necessary outcome, resulting from the cross-product with the $\boldsymbol{\omega}$ vector. We will take advantage of this relationship in the next step.

For the solution of part (b), we can relate the absolute velocity of point C to the absolute velocities of both points A and B, by two velocity-difference equations; that is,

$$\mathbf{V}_C = \overset{\sqrt{\sqrt{}}}{\mathbf{V}_A} + \overset{?\sqrt{}}{\mathbf{V}_{CA}} = \overset{\sqrt{\sqrt{}}}{\mathbf{V}_B} + \overset{?\sqrt{}}{\mathbf{V}_{CB}}. \qquad (d)$$

Since points A, B, and C are all fixed in the same link, each of the velocity-difference vectors, $\mathbf{V}_{CA}$ and $\mathbf{V}_{CB}$, is of the form $\boldsymbol{\omega} \times \mathbf{R}$, using $\mathbf{R}_{CA}$ and $\mathbf{R}_{CB}$, respectively. As a result, $\mathbf{V}_{CA}$ is perpendicular to $\mathbf{R}_{CA}$, and $\mathbf{V}_{CB}$ is perpendicular to $\mathbf{R}_{CB}$. The directions of these two vectors are, therefore, indicated as known in Eq. (d).

Since $\boldsymbol{\omega}$ has already been determined, it is easy to calculate the magnitudes of $\mathbf{V}_{CA}$ and $\mathbf{V}_{CB}$ using equations similar to Eq. (c); however, this is avoided here. Instead, we form the graphic solution to Eq. (d). Equation (d) states that a vector that is perpendicular to $\mathbf{R}_{CA}$ must be added to $\mathbf{V}_A$ and that the result is equal to the sum of $\mathbf{V}_B$ and a vector perpendicular to $\mathbf{R}_{CB}$. The solution is shown in Fig. 3.6e. In practice the solution is commonly continued on the same diagram as Fig. 3.6d and results in Fig. 3.6g. A line perpendicular to $\mathbf{R}_{CA}$ (representing $\mathbf{V}_{CA}$) is drawn starting at point A (representing addition to $\mathbf{V}_A$); similarly, a line is drawn perpendicular to $\mathbf{R}_{CB}$ starting at point B. The point of intersection of these two lines is labeled C and represents the solution to Eq. (d). The line from O_V to point C now represents the absolute velocity $\mathbf{V}_C$. This velocity can be transferred back to the link and interpreted as $\mathbf{V}_C$ in both magnitude and direction, as shown in Fig. 3.6f.

In seeing the shading and the labeled angles α and β in Figs. 3.6g and 3.6a, we are led to investigate whether the two triangles labeled ABC in each of these figures are similar in shape, as they appear to be. In reviewing the construction steps we see that indeed they are similar, since the velocity-difference vectors $\mathbf{V}_{BA}, \mathbf{V}_{CA}$, and $\mathbf{V}_{CB}$ are each perpendicular to the respective position-difference vectors $\mathbf{R}_{BA}$, $\mathbf{R}_{CA}$, and $\mathbf{R}_{CB}$. This property is always true, regardless of the shape of the moving link; a similarly shaped figure appears in the velocity polygon. The sides of the polygon are always scaled up or down by a factor equal to the angular velocity of the link, and they are always rotated by 90° in the directions of their angular velocities. These properties result from the fact that each velocity-difference vector between two points on the link results from a cross-product of the same $\boldsymbol{\omega}$ vector with the corresponding position-difference vector. This similarly shaped figure in the velocity polygon is commonly referred to as the *velocity image* of the link, and every link has a corresponding velocity image in the velocity polygon.

The graphic procedures in this section were first published in 1883 by Mehmke as the following theorem [10]:

The end points of the velocity vectors of the points of a rigid plane body, when plotted from a common origin, produce a figure that is geometrically similar to the original figure [image diagram].

It is this theorem that allows clarity in the velocity polygon despite the minimal labeling that is required. This becomes evident in the upcoming two examples and continues throughout the text.

The concept of the velocity image allows the graphic solution to be obtained more efficiently. For example, there is no need to explicitly write Eq. (d). Once the solution has progressed to the state of Fig. 3.6*d*, the velocity-image points *A* and *B* are known. Therefore, one can use these two points as the base of a triangle similar to the link shape and construct the image point *C* directly. Care must be taken not to allow the triangle to be flipped over between the position diagram and the velocity image, but the solution can proceed quickly, accurately, and naturally, resulting in Fig. 3.6*g*. Again, we note that all steps in the solution are based on strictly derived vector equations and are not graphic tricks. It is wise to continue to write the corresponding vector equations until one is thoroughly familiar with the procedure and its vector basis.

The following are important properties of velocity images:

1. The velocity image of each link in the velocity polygon is a scale reproduction of the shape of the real link.
2. The velocity image of each link is rotated 90° from the real link in the direction of the angular velocity of that link.
3. The labels identifying the vertices of each link in the velocity polygon are the same as, and progress around the velocity image in the same order and in the same angular sense as, around the real link.
4. The ratio of the size of the velocity image of a link to the size of the real link is equal to the magnitude of the angular velocity of the link. In general, this is not the same for different links in a mechanism.
5. The velocities of all points on a translating link are equal, and the angular velocity of the link is zero. Therefore, the velocity image of a link that is translating shrinks to a single point in the velocity polygon.
6. Point O_V in the velocity polygon is the image of all points with zero absolute velocity; it is the velocity image of the stationary link.
7. The absolute velocity of any point on any link is represented in the velocity polygon by the line from O_V to the image of the point. The velocity-difference vector between any two points, say *P* and *Q*, is represented by the line to image point *P* from image point *Q*.

To illustrate the graphic velocity analysis of mechanisms and the role of velocity images, we consider two typical linkage problems.

Example 3.1

The four-bar linkage in the posture shown in Fig. 3.7*a* is driven by crank 2 at a constant angular velocity $\omega_2 = 900$ rev/min ccw. For this posture, determine the angular velocities of coupler link 3 and output link 4 and the velocities of point *E* in link 3 and point *F* in link 4.

Graphic Solution

First, we draw the linkage to a suitable scale. Then, we calculate the angular velocity of link 2 in radians per second; that is,

$$\omega_2 = \left(900\,\frac{\text{rev}}{\text{min}}\right)\left(2\pi\,\frac{\text{rad}}{\text{rev}}\right)\left(\frac{1\,\text{min}}{60\,\text{s}}\right) = 94.25\,\text{rad/s ccw.} \tag{1}$$

Example 3.1 (cont.)

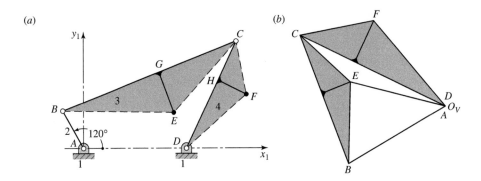

Fig. 3.7 (a) $R_{BA} = 4$ in, $R_{CB} = 18$ in, $R_{CD} = 11$ in, $R_{DA} = 10$ in, $R_{GB} = 10$ in, $R_{EG} = 4$ in, $R_{HD} = 7$ in, and $R_{FH} = 3$ in; (b) velocity polygon.

Next we calculate the velocity of point B from Eqs. (3.3) and (3.4), noting that point A is fixed:

$$\mathbf{V}_B = \mathbf{V}_A^0 + \mathbf{V}_{BA} = \boldsymbol{\omega}_2 \times \mathbf{R}_{BA},$$
$$V_B = (94.25 \text{ rad/s})\left(\frac{4}{12}\text{ft}\right) = 31.42 \text{ ft/s.} \tag{2}$$

It is important to note that the form "$\boldsymbol{\omega} \times \mathbf{R}$" was used for the velocity difference and not for the absolute velocity $\mathbf{V}_B$ directly. Next, we choose a suitable velocity scale and an arbitrary location for an origin, denoted as O_V, shown in Fig. 3.7b. Since $\mathbf{V}_A = \mathbf{0}$, the image point A is coincident with O_V. We construct the line AB perpendicular to $\mathbf{R}_{BA}$ and toward the lower left because of the counterclockwise sense of $\boldsymbol{\omega}_2$; this line represents $\mathbf{V}_{BA}$.

Now we write two equations for the velocity of point C. Since the velocities of points C_3 and C_4 must be equal (links 3 and 4 are pinned together at C), we have

$$\mathbf{V}_C = \overset{\sqrt{\sqrt{}}}{\mathbf{V}_B} + \overset{?\sqrt{}}{\mathbf{V}_{CB}} = \overset{0}{\mathbf{V}_D} + \overset{?\sqrt{}}{\mathbf{V}_{CD}}. \tag{3}$$

We construct two lines in the velocity polygon; line BC is drawn from image point B perpendicular to $\mathbf{R}_{CB}$, and line DC is drawn from image point D (coincident with O_V since $\mathbf{V}_D = \mathbf{0}$) perpendicular to $\mathbf{R}_{CD}$. We label the point of intersection of these two lines as image point C. When the lengths of these two lines are measured, we find that $V_{CB} = 38.4 \text{ ft/s}$ and $V_C = V_{CD} = 45.5 \text{ ft/s}$. The angular velocities of links 3 and 4 can now be obtained:

$$\omega_3 = \frac{V_{CB}}{R_{CB}} = \frac{38.4 \text{ ft/s}}{(18/12) \text{ ft}} = 25.6 \text{ rad/s ccw,} \qquad Ans. \text{ (4)}$$

$$\omega_4 = \frac{V_{CD}}{R_{CD}} = \frac{45.5 \text{ ft/s}}{(11/12) \text{ ft}} = 49.64 \text{ rad/s ccw,} \qquad Ans. \text{ (5)}$$

where the directions of ω_3 and ω_4 are obtained using the technique shown in Fig. 3.6c.

Example 3.1 (cont.)

There are several graphic methods for finding the velocity of point E. One method is to measure R_{EB} from the scale drawing of the linkage. Then, since points B and E are both attached to link 3, we calculate[2]

$$V_{EB} = \omega_3 R_{EB} = (25.6 \text{ rad/s})\left(\frac{10.8}{12}\text{ ft}\right) = 23.04 \text{ ft/s}. \qquad (6)$$

We next construct the line BE in the velocity polygon, drawn to the chosen scale and perpendicular to $\mathbf{R}_{EB}$. Then we solve[3] the velocity-difference equation

$$\overset{??}{\mathbf{V}_E} = \overset{\sqrt{}\sqrt{}}{\mathbf{V}_B} + \overset{\sqrt{}\sqrt{}}{\mathbf{V}_{EB}}. \qquad (7)$$

From the velocity polygon, the velocity of point E is measured as

$$V_E = 27.6 \text{ ft/s} \angle 165.3°. \qquad \textit{Ans.}$$

Alternatively, the velocity of point E can be obtained from the velocity-difference equation

$$\overset{??}{\mathbf{V}_E} = \overset{\sqrt{}\sqrt{}}{\mathbf{V}_C} + \overset{\sqrt{}\sqrt{}}{\mathbf{V}_{EC}} \qquad (8)$$

by a procedure identical to that used for Eq. (7). This yields the triangle $O_V EC$ in the velocity polygon.

Suppose we wish to find $\mathbf{V}_E$ without the intermediate step of calculating ω_3. In this case, we write Eqs. (7) and (8) simultaneously; that is,

$$\mathbf{V}_E = \overset{\sqrt{}\sqrt{}}{\mathbf{V}_B} + \overset{?\sqrt{}}{\mathbf{V}_{EB}} = \overset{\sqrt{}\sqrt{}}{\mathbf{V}_C} + \overset{?\sqrt{}}{\mathbf{V}_{EC}}. \qquad (9)$$

Drawing lines EB (perpendicular to $\mathbf{R}_{EB}$) and EC (perpendicular to $\mathbf{R}_{EC}$) in the velocity polygon, we find their point of intersection and so solve Eq. (9).

Perhaps the easiest method of solving for $\mathbf{V}_E$, however, is to take advantage of the concept of the velocity image of link 3. Recognizing that the velocity-image points B and C have already been found, we construct the triangle BEC in the velocity polygon, similar in shape to the triangle BEC in Fig. 3.7a. This locates point E in the velocity polygon and, therefore, gives a solution for the velocity of point E.

The velocity of point F can also be found by any of the above methods using points C, D, and F of link 4. The result is

$$V_F = 31.8 \text{ ft/s} \angle 130.9°. \qquad \textit{Ans.}$$

For the purpose of comparison, an analytic solution is also presented here.

---

[2] There is no restriction in our derivation that requires that $\mathbf{R}_{EB}$ lie along the material portion of link 3 in order to use Eq. (6), only that points E and B remain a constant distance apart.

[3] Note that numeric values should not be substituted into Eq. (7) directly: This is a vector equation and requires vector addition, not scalar; performing these vector operations is precisely the purpose of constructing the velocity polygon.

Example 3.1 (cont.)

Analytic Solution

The first step is to perform a posture analysis of the linkage. Since this step was presented in Chapter 2, only the results are shown here. For the given link dimensions and the specified input variable $\theta_2 = 120°$, the postures of links 3 and 4 are $\theta_3 = 20.92°$ and $\theta_4 = 64.05°$, respectively.

In vector form, the position-difference vectors corresponding to the links are:

$$\mathbf{R}_{BA} = 4\text{ in}\angle 120° = -2\hat{\mathbf{i}} + 3.464\hat{\mathbf{j}}\text{ in},$$
$$\mathbf{R}_{CB} = 18\text{ in}\angle 20.92° = 16.813\hat{\mathbf{i}} + 6.427\hat{\mathbf{j}}\text{ in},$$
$$\mathbf{R}_{CD} = 11\text{ in}\angle 64.05° = 4.813\hat{\mathbf{i}} + 9.891\hat{\mathbf{j}}\text{ in},$$
$$\mathbf{R}_{DA} = 10\text{ in}\angle 0° = 10\hat{\mathbf{i}}\text{ in},$$
$$\mathbf{R}_{EB} = 10.77\text{ in}\angle -0.88° = 10.769\hat{\mathbf{i}} - 0.165\hat{\mathbf{j}}\text{ in},$$
$$\mathbf{R}_{FD} = 7.616\text{ in}\angle 40.85° = 5.761\hat{\mathbf{i}} + 4.981\hat{\mathbf{j}}\text{ in}.$$

We proceed as we did in the graphic solution – namely, the given input angular velocity is

$$\omega_2 = \left(900\,\frac{\text{rev}}{\text{min}}\right)\left(2\pi\,\frac{\text{rad}}{\text{rev}}\right)\left(\frac{1\text{ min}}{60\text{ s}}\right) = 94.25\text{ rad/s ccw}. \tag{10}$$

Then the velocity of point B is written as

$$\mathbf{V}_B = \mathbf{V}_A^0 + \mathbf{V}_{BA} = \omega_2 \times \mathbf{R}_{BA}$$
$$= -326.482\hat{\mathbf{i}} - 188.500\hat{\mathbf{j}}\text{ in/s} = 377\text{ in/s}\angle -150°. \tag{11}$$

The velocity of point C can be written as

$$\mathbf{V}_C = \mathbf{V}_B + \mathbf{V}_{CB} = \mathbf{V}_D^0 + \mathbf{V}_{CD}$$
$$= \mathbf{V}_B + \omega_3 \times \mathbf{R}_{CB} = \omega_4 \times \mathbf{R}_{CD}. \tag{12}$$

Substituting the position-difference vectors $\mathbf{R}_{CB}$ and $\mathbf{R}_{CD}$ and Eq. (11) into Eq. (12) and writing the result in matrix form gives

$$\begin{bmatrix} -326.482\text{ in/s} \\ -188.500\text{ in/s} \end{bmatrix} + \begin{bmatrix} -6.427\text{ in} \\ 16.813\text{ in} \end{bmatrix}\omega_3 = \begin{bmatrix} -9.891\text{ in} \\ 4.813\text{ in} \end{bmatrix}\omega_4. \tag{13}$$

Solving Eq. (13), the angular velocities of links 3 and 4, respectively, are

$$\omega_3 = 25.382\text{ rad/s (ccw)} \quad\text{and}\quad \omega_4 = 49.501\text{ rad/s (ccw)}. \qquad Ans. \tag{14}$$

The velocities of points E and F can be obtained from the velocity-difference equation using the known velocities of points B and D; that is,

$$\mathbf{V}_E = \mathbf{V}_B + \mathbf{V}_{EB} = \mathbf{V}_B + \omega_3 \times \mathbf{R}_{EB} \quad\text{and}\quad \mathbf{V}_F = \mathbf{V}_D + \mathbf{V}_{FD} = \mathbf{V}_D + \omega_4 \times \mathbf{R}_{FD}. \tag{15}$$

Example 3.1 (cont.)

Substituting the known data into Eq. (15) gives

$$\mathbf{V}_E = -322.294\hat{\mathbf{i}} + 84.839\hat{\mathbf{j}} \text{ in/s}$$
$$= 27.8 \text{ ft/s} \angle 165.25°$$

and

$$\mathbf{V}_F = -246.564\hat{\mathbf{i}} + 285.175\hat{\mathbf{j}} \text{ in/s}$$
$$= 31.4 \text{ ft/s} \angle 130.85°.$$

Ans.

Note that these answers are in good agreement with the results obtained by the graphic method.

Example 3.2

The offset slider-crank linkage in the posture shown in Fig. 3.8*a* is driven by slider 4 at a constant velocity $\mathbf{V}_C = -10\hat{\mathbf{i}}$ m/s. Determine the velocity of point D and the angular velocities of links 2 and 3.

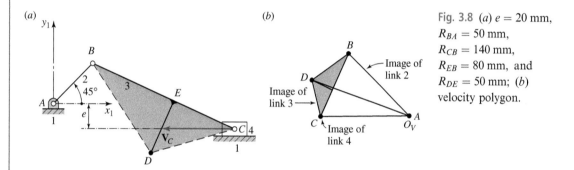

Fig. 3.8 (*a*) $e = 20$ mm, $R_{BA} = 50$ mm, $R_{CB} = 140$ mm, $R_{EB} = 80$ mm, and $R_{DE} = 50$ mm; (*b*) velocity polygon.

Solution

Choosing a suitable velocity scale and origin O_V, we draw the velocity vector $\mathbf{V}_C$. This locates the velocity image point C as shown in Fig. 3.8*b*. Solving the simultaneous velocity-difference equations for the velocity of point B, namely,

$$\mathbf{V}_B = \overset{\sqrt{}}{\mathbf{V}_C} + \overset{?\sqrt{}}{\mathbf{V}_{BC}} = \overset{0}{\mathbf{X}_A} + \overset{?\sqrt{}}{\mathbf{V}_{BA}},$$

provides the location of the velocity image point B in the velocity polygon.

Having found the velocity image points B and C, we construct the image of link 3 as shown to locate image point D. We then measure

$$V_D = 12.0 \text{ m/s,}$$

Ans.

with the direction shown in the velocity polygon.

Example 3.2 (cont.)

The angular velocities of links 2 and 3, respectively, are

$$\omega_2 = \frac{V_{BA}}{R_{BA}} = \frac{10.0 \text{ m/s}}{0.050 \text{ m}} = 200 \text{ rad/s ccw,}$$ *Ans.*

$$\omega_3 = \frac{V_{BC}}{R_{BC}} = \frac{7.5 \text{ m/s}}{0.140 \text{ m}} = 53.6 \text{ rad/s cw.}$$ *Ans.*

Note that the velocity image of each link is shown in Fig. 3.8*b*.

3.5   Apparent Velocity of a Point in a Moving Coordinate System

In analyzing the velocities of various machine components, we frequently encounter situations in which it is convenient to describe how a point fixed in a moving link moves with respect to another point of a different moving link. However, it is not at all convenient to describe the absolute motion. This situation occurs when a rotating link contains a slot along which a point of another link is constrained to slide. With the motion of the link containing the slot and the relative sliding motion taking place within the slot as known quantities, we may wish to find the absolute motion of the sliding link. It is for this type of problem that the apparent-displacement vector is defined in Section 2.16 (see Fig. 2.29). We now wish to extend this concept to velocity analysis.

Consider Fig. 3.9, in which a rigid link with arbitrary motion has a coordinate system $x_2 y_2 z_2$ attached to it. After a short time increment Δt, the coordinate system lies at $x_2' y_2' z_2'$. All points of link 2 – for example, point P_2 – move with the coordinate system. During the same time interval, the initially coincident point P of link 3, denoted P_3, is constrained to move along a known path with respect to link 2. This constraint is depicted in the figure as a slot carrying a pin from link 3; the center of the pin is point P_3. Although it is pictured in this way, the constraint may occur in a variety of different forms. The only assumption here is that we know the path that point P_3 traces in coordinate system $x_2 y_2 z_2$ – that is, the locus of the tip of the apparent-position vector, $\mathbf{R}_{P_3/2}$.

Fig. 3.9 Apparent displacement.

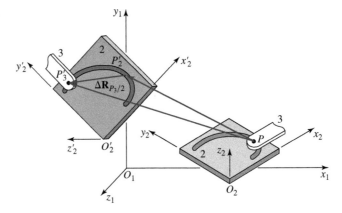

Recall the apparent-displacement equation, Eq. (2.62); that is,

$$\Delta \mathbf{R}_{P_3} = \Delta \mathbf{R}_{P_2} + \Delta \mathbf{R}_{P_3/2}.$$

Dividing this equation by the small time increment Δt and taking the limit gives

$$\lim_{\Delta t \to 0}\left(\frac{\Delta \mathbf{R}_{P_3}}{\Delta t}\right) = \lim_{\Delta t \to 0}\left(\frac{\Delta \mathbf{R}_{P_2}}{\Delta t}\right) + \lim_{\Delta t \to 0}\left(\frac{\Delta \mathbf{R}_{P_3/2}}{\Delta t}\right). \tag{a}$$

Defining the *apparent-velocity* vector as

$$\mathbf{V}_{P_3/2} = \lim_{\Delta t \to 0}\left(\frac{\Delta \mathbf{R}_{P_3/2}}{\Delta t}\right) = \frac{d\mathbf{R}_{P_3/2}}{dt}, \tag{3.5}$$

then, in the limit, Eq. (*a*) becomes

$$\mathbf{V}_{P_3} = \mathbf{V}_{P_2} + \mathbf{V}_{P_3/2}, \tag{3.6}$$

which is called the *apparent-velocity equation*.

We note from Eq. (3.5) that the apparent velocity resembles the absolute velocity except that it comes from the apparent displacement rather than the absolute displacement. Thus, $\mathbf{V}_{P_3/2}$ is the velocity of point P_3 *as it would appear to an observer attached to moving link* 2 and making observations in coordinate system $x_2 y_2 z_2$. This concept accounts for its name: the apparent-velocity vector. We also note that the absolute velocity, $\mathbf{V}_{P/1}$, is a special case of the apparent velocity where the observer happens to be fixed to the $x_1 y_1 z_1$ coordinate system.

Insight into the use of the apparent-velocity equation, Eq. (3.6), can be obtained from the following examples.

Example 3.3

An inversion of the slider-crank linkage is shown in Fig. 3.10*a*. For the given posture, with $\theta_2 = 30°$, the angular velocity of link 2 is a constant 36 rad/s cw. Link 3 is pinned to the crank at A and slides on link 4 at point B. Find the angular velocity of link 4.

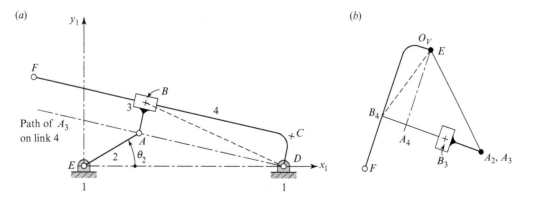

Fig. 3.10 (*a*) $R_{AE} = 3$ in, $R_{BA} = R_{CD} = 2$ in, and $R_{DE} = 14$ in; (*b*) velocity polygon.

Example 3.3 (cont.)

Solution

From Eq. (3.4), the velocity of point A is

$$\mathbf{V}_A = \mathbf{V}_E^0 + \mathbf{V}_{AE} = \omega_2 \times \mathbf{R}_{AE},$$

$$V_A = (36\ \text{rad/s})\left(\frac{3}{12}\text{ft}\right) = 9\ \text{ft/s}. \tag{1}$$

Choosing a suitable velocity scale and origin O_V, we draw velocity vector $\mathbf{V}_A$ to locate image point A in the velocity polygon (Fig. 3.10b).

Next, we distinguish two different points, B_3 and B_4, at the location of sliding. Point B_3 is fixed in link 3, and point B_4 is fixed in link 4, but at the instant shown, the two points are coincident. Note that, as seen by an observer on link 4, point B_3 slides along link 4, thus defining a straight path along line CF. Thus, we can write the apparent-velocity equation, Eq. (3.6), as

$$\mathbf{V}_{B_3} = \mathbf{V}_{B_4} + \mathbf{V}_{B_3/4}. \tag{2}$$

When image point B_3 is related to image point A, and image point B_4 is related to image point D by velocity differences, expansion of Eq. (2) gives

$$\overset{\surd\surd_A}{\mathbf{V}} + \overset{?\surd}{\mathbf{V}_{B_3A}} = \overset{\mathbf{0}}{\mathbf{V}_D} + \overset{?\surd}{\mathbf{V}_{B_4D}} + \overset{?\surd}{\mathbf{V}_{B_3/4}}, \tag{3}$$

where $\mathbf{V}_{B_3A}$ is perpendicular to $\mathbf{R}_{BA}$, $\mathbf{V}_{B_4D}$ (dashed line in Fig. 3.10b) is perpendicular to $\mathbf{R}_{BD}$, and $\mathbf{V}_{B_3/4}$ is directed along the tangent to the path of sliding at B.

Although Eq. (3) appears to have three unknowns, if we note that $\mathbf{V}_{B_3A}$ and $\mathbf{V}_{B_3/4}$ have identical orientations, the equation can be rearranged as

$$\overset{\surd\surd}{\mathbf{V}}_A + \overbrace{\left(\mathbf{V}_{B_3A} - \mathbf{V}_{B_3/4}\right)}^{?\surd} = \overset{?\surd}{\mathbf{V}_{B_4D}}, \tag{4}$$

and the vector difference indicated in parentheses can be treated as a single vector with known orientation. The equation is thereby reduced to two unknowns and can be solved graphically to locate image point B_4 in the velocity polygon.

The magnitude of $\mathbf{R}_{BD}$ can be computed or measured from the scale drawing of the linkage, and the velocity-difference $\mathbf{V}_{B_4D}$ can be determined from the velocity polygon (the dashed line from O_V to B_4). Therefore, the angular velocity of link 4 is

$$\omega_4 = \frac{V_{B_4D}}{R_{BD}} = \frac{7.3\ \text{ft/s}}{(11.6/12)\ \text{ft}} = 7.55\ \text{rad/s ccw}. \qquad Ans.\ (5)$$

Another approach to this problem, which avoids the need to combine terms as in Eq. (4), is to extend the velocity polygon to include the images of links 2, 3, and 4. In so doing, it is necessary to note that, since links 3 and 4 always remain perpendicular to each other ($\theta_3 = \theta_4 - 90°$), they

Example 3.3 (cont.)

must rotate at the same rate. Therefore, the angular velocities of the two links are the same: $\omega_3 = \omega_4$. This allows the calculation of $\mathbf{V}_{BA} = \boldsymbol{\omega}_3 \times \mathbf{R}_{BA}$ and the plotting of the velocity image point, B_3. We also note that the velocity images of links 3 and 4 are of comparable size, since $\omega_3 = \omega_4$. However, they have quite different scales from the velocity image of link 2, line $O_V A$, since link 2 has a larger angular velocity.

Now, consider an observer riding on link 4 and ask what SHE would see for the path of point A in HER coordinate system. The answer to this is that this path is a straight line parallel to the line CF, as indicated in Fig. 3.10a. Let us now define one point of this path as A_4. Note that this point *does not move together with pins A_2 and A_3; it is a point in link 4 and it rotates along with the path around the fixed point D.* Since we can identify the path traced by A_2 and A_3 on the extension of link 4, we can write the apparent-velocity equation,[4]

$$\mathbf{V}_{A_2} = \mathbf{V}_{A_4} + \mathbf{V}_{A_2/4}. \tag{6}$$

Since A_4 is a point in link 4, we can write

$$\mathbf{V}_{A_4} = \mathbf{V}_D^0 + \mathbf{V}_{A_4 D}. \tag{7}$$

Substituting Eq. (7) into Eq. (6) gives

$$\mathbf{V}_{A_2}^{\sqrt{\sqrt{}}} = \mathbf{V}_{A_4 D}^{?\sqrt{}} + \mathbf{V}_{A_2/4}^{?\sqrt{}}, \tag{8}$$

where $\mathbf{V}_{A_4 D}$ is perpendicular to $\mathbf{R}_{AD}$, and $\mathbf{V}_{A_2/4}$ is tangent to the path that A_2 traces on extended link 4. Solving Eq. (8) locates velocity-image point A_4 in the velocity polygon and gives $V_{A_4 D} = 7.17$ ft/s and $V_{A_2/4} = 5.48$ ft/s. Also, measuring the diagram gives $R_{AD} = 11.4$ in. Substituting these values into the following equation, we obtain

$$\omega_4 = \frac{V_{A_4 D}}{R_{AD}} = \frac{7.17 \text{ ft/s}}{(11.4/12) \text{ ft}} = 7.55 \text{ rad/s ccw}. \tag{9}$$

Note that this is in complete agreement with the result of Eq. (5).

The velocity of other points in link 4 – for example, points C and F – can be obtained using the velocity polygon and are left as an exercise for the reader.

Further insight into the nature and use of the apparent-velocity equation is provided by the following example.

---

[4] It is wrong to use the equation $\mathbf{V}_{A_4} = \mathbf{V}_{A_2} + \mathbf{V}_{A_4/2}$, since the *path* traced by point A_4 in a coordinate system attached to link 2 is *not known*. Although this equation shows poor understanding, it still yields a correct solution. If the corresponding path (of point A_4 as observed from link 2) were found, it would be tangent to the path used for point A_2 as observed from link 4. Since the *tangents* to the two paths are the same, although the paths are not, the two solutions both yield the same numeric result. However, this is *not* true in acceleration analysis in Chapter 4. Therefore, the concept should be studied and this "backward" use should be avoided.

Example 3.4

An airplane is traveling at a speed of $V_{P_2} = 300$ km/h on a circular path of radius $R_{P_2C} = 5$ km with a center at C, as shown in Fig. 3.11. A rocket at a distance of $R_{R_3P_2} = 30$ km away from the airplane is traveling on a straight path at a speed of $V_{R_3} = 2\,000$ km/h. Determine the velocity of the rocket as observed by the pilot of the airplane.

Fig. 3.11 Airplane traveling
on a circular path.

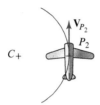

Solution

Since the airplane is on a circular course, the point C_2, attached to the coordinate system of the airplane but coincident with C_1, has no motion. Therefore, the angular velocity of the airplane is

$$\omega_2 = \frac{V_{PC}}{R_{PC}} = \frac{(V_{P_2} - V_{C_2})}{R_{PC}} = \frac{(300\ \text{km/h} - 0)}{5\ \text{km}} = 60\ \text{rad/h ccw}.$$

The question posed explicitly requires the calculation of the apparent velocity $\mathbf{V}_{R_3/2}$, but the apparent-velocity equation applies *only between coincident points*. Therefore, we define (fictitious) point, R_2, attached to the rotating coordinate system of the airplane but located coincident with the rocket R_3 at the instant depicted. As part of the airplane, the velocity of this point can be written from Eq. (3.6) as

$$\mathbf{V}_{R_2} = \mathbf{V}_{P_2} + \boldsymbol{\omega}_2 \times \mathbf{R}_{RP}$$
$$= \left(300\hat{\mathbf{j}}\ \text{km/h}\right) + \left(60\hat{\mathbf{k}}\ \text{rad/h}\right) \times \left(30\hat{\mathbf{i}}\ \text{km}\right) = 2\,100\hat{\mathbf{j}}\ \text{km/h}.$$

The apparent velocity of the rocket as observed by the pilot of the airplane can be written as

$$\mathbf{V}_{R_3/2} = \mathbf{V}_{R_3} - \mathbf{V}_{R_2}$$
$$= 2\,000\hat{\mathbf{j}}\ \text{km/h} - 2\,100\hat{\mathbf{j}}\ \text{km/h} = -100\hat{\mathbf{j}}\ \text{km/h}.$$

Ans.

Thus, as seen by the pilot of the airplane, the rocket appears to be *backing up* at a speed of 100 km/h. This result becomes better understood as we consider the motion of point R_2. This imagined point is *attached* to the airplane and, therefore, seems stationary to the pilot. Yet, in the absolute coordinate system, this point is traveling faster than the rocket; the rocket is not keeping up with this point and, therefore, appears to the pilot to be backing up.

We can gain further insight into the nature of the apparent-velocity vector by studying Fig. 3.12. This figure shows the view of a moving point P_3 as it would appear to a moving observer. To HER, the path traced on (moving) link 2 appears stationary, and the point moves along this path from P_3 to P'_3. Working in this coordinate system on link 2, suppose we locate

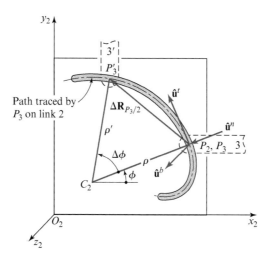

Fig. 3.12 Apparent displacement of point P_3 as seen by an observer on link 2.

point C_2 as the center of curvature of the path of point P_3. For small distances from P_2, the path follows the circular arc $P_3 P_3'$ with center C_2 and radius of curvature ρ. We now define the unit vector $\hat{\mathbf{u}}^t$ tangent to the path with positive sense in the direction of positive movement. The plane defined by this unit tangent vector $\hat{\mathbf{u}}^t$ and the center of curvature C_2 is called the *osculating plane*. If we choose a preferred side of this plane as the positive side and denote it by the positive unit vector $\hat{\mathbf{u}}^b$ (commonly referred to as the *binormal* vector), we can complete a right-hand Cartesian coordinate system by defining the unit vector normal to the path

$$\hat{\mathbf{u}}^n = \hat{\mathbf{u}}^b \times \hat{\mathbf{u}}^t. \tag{3.7}$$

Therefore, looking from the positive binormal, the unit normal vector $\hat{\mathbf{u}}^n$ is always 90° counterclockwise from the unit tangent vector $\hat{\mathbf{u}}^t$. This implies that the radius of curvature ρ has a positive value when $\hat{\mathbf{u}}^n$ points from point P_3 toward the center of curvature of the path C_2, and a negative value when $\hat{\mathbf{u}}^n$ points from point P_3 away from the center of curvature.

This coordinate system, $\hat{\mathbf{u}}^t \hat{\mathbf{u}}^n \hat{\mathbf{u}}^b$ (commonly referred to as *path coordinates*), moves with its origin, tracking the motion of point P_3. However, it rotates with the radius of curvature (through angle $\Delta\phi$) as the motion progresses, but *not* the same rotation as either link 2 or link 3. Note that, if positive movement were chosen in the opposite direction along this curve, the sense of both the unit tangent vector $\hat{\mathbf{u}}^t$ and the unit normal vector $\hat{\mathbf{u}}^n$ would be reversed. This would mean that the radius of curvature ρ would have a negative value; however, the unit normal vector $\hat{\mathbf{u}}^n$ would still be obtained from Eq. (3.7) rather than from the sense of the radius of curvature. Since the actual movement would then be in the negative direction, angle $\Delta\phi$ would still be counterclockwise (as seen from the positive $\hat{\mathbf{u}}^b$ side of the plane) and would still have a positive value.

We now define the scalar distance, Δs, as the arc length along the curve from P_3 to P_3' and note that $\Delta\mathbf{R}_{P_3/2}$ is a chord of the same arc. However, for a very short time interval, Δt, the magnitude of the chord and the arc distance approach equality. Therefore,

$$\lim_{\Delta t \to 0} \left(\frac{\Delta\mathbf{R}_{P_3/2}}{\Delta s} \right) = \frac{d\mathbf{R}_{P_3/2}}{ds} = \hat{\mathbf{u}}^t. \tag{3.8}$$

Here, both $\Delta\mathbf{R}_{P_3/2}$ and Δs are considered functions of time. Therefore, from Eq. (3.5), the *apparent-velocity* vector can be written as

$$\mathbf{V}_{P_3/2} = \lim_{\Delta t \to 0} \left(\frac{\Delta\mathbf{R}_{P_3/2}}{\Delta s} \frac{\Delta s}{\Delta t} \right) = \frac{d\mathbf{R}_{P_3/2}}{ds} \frac{ds}{dt} = \frac{ds}{dt}\hat{\mathbf{u}}^t$$

or as

$$\mathbf{V}_{P_3/2} = \dot{s}\hat{\mathbf{u}}^t, \tag{3.9}$$

where $\dot{s}$ is the instantaneous speed of P_3 along the path. There are two important conclusions from this result: (a) the magnitude of the apparent velocity is equal to the speed with which the point P_3 progresses along the path, and (b) the apparent-velocity vector is always tangent to the path traced by the point in the coordinate system of the observer.

The first of these two conclusions is seldom useful in the solution of problems, although it is an important concept. The second conclusion is extremely useful, since the apparent path traced by a moving point can often be visualized from the nature of the constraints, and, thus, the direction of the apparent-velocity vector becomes known. Note that only the tangent to the path $\hat{\mathbf{u}}^t$ is used in this chapter; the radius of curvature ρ of the path is not needed yet but will become important when we analyze acceleration in Chapter 4.

3.6   Apparent Angular Velocity

Completeness suggests that we should define the term *apparent angular velocity*. When two rigid bodies rotate with different angular velocities, the vector difference between the two angular velocities is defined as the *apparent angular velocity*. For example, the apparent angular velocity of rotating link 3 with respect to rotating link 2 can be written as

$$\boldsymbol{\omega}_{3/2} = \boldsymbol{\omega}_3 - \boldsymbol{\omega}_2. \tag{3.10}$$

Therefore, the angular velocity of link 3 is

$$\boldsymbol{\omega}_3 = \boldsymbol{\omega}_2 + \boldsymbol{\omega}_{3/2}. \tag{3.11}$$

Note that $\boldsymbol{\omega}_{3/2}$ is the angular velocity of body 3 as it would appear to an observer attached to, and rotating with, link 2.

3.7   Direct Contact and Rolling Contact

Two links of a mechanism that are in direct contact have relative motion that may or may not involve sliding[5] at the point of contact. For example, consider the cam-and-follower mechanism shown in Fig. 3.13a; the cam, link 2, drives the follower, link 3, by direct contact. We see

---

[5] The words *sliding* and *slipping* are used interchangeably in this text.

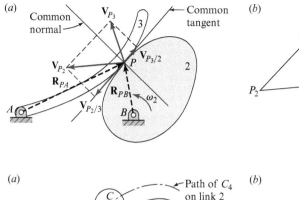

Fig. 3.13 (a) Cam-and-follower mechanism; (b) velocity polygon.

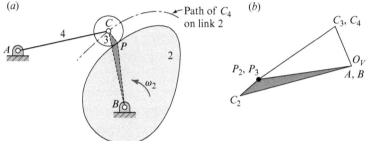

Fig. 3.14 (a) Cam-and-roller-follower mechanism; (b) velocity polygon.

that if slip were not possible between links 2 and 3 at the point of contact P, then the triangle PAB would form a truss; that is, the mechanism would have a mobility of $m = 0$. Therefore, for the cam to drive the follower, sliding as well as rotation must take place at point P. We distinguish between two coincident points, both located at P at the instant shown: point P_2, attached to link 2, and point P_3, attached to link 3. Therefore, we can rearrange the apparent-velocity equation, Eq. (3.6), as

$$\mathbf{V}_{P_3/2} = \mathbf{V}_{P_3} - \mathbf{V}_{P_2}. \qquad (3.12)$$

The graphic solution to this equation is as follows. First, we note that the normal component of the apparent velocity must be zero; otherwise, the two links would either separate or interfere, both contrary to our basic assumption that contact continues. Therefore, the apparent velocity must act along the common tangent and is the velocity of the relative sliding motion along the direct-contact interface.

For a given angular velocity of the driver (the cam) ω_2, the velocity of point P_2 can be obtained from

$$\mathbf{V}_{P_2} = \omega_2 \times \mathbf{R}_{P_2 B},$$

thus locating image point P_2 in the velocity polygon; see Fig. 3.13b. The component of $\mathbf{V}_{P_3}$ along the common normal must be equal to the component of $\mathbf{V}_{P_2}$ along the common normal, and this locates image point P_3. Therefore, the apparent velocity, Eq. (3.12), can be measured from the velocity polygon.

It is also possible for there to be direct contact between two links of a mechanism without sliding between the links. For example, the cam-and-roller-follower mechanism of Fig. 3.14a could have sufficient friction between the cam surface, link 2, and the roller, link 3, to restrain the roller to roll on the cam without slip. Henceforth, we will restrict our use of the term *rolling contact* to situations where *no slip* takes place. By "no slip" we imply that the apparent "slipping" velocity of Eq. (3.12) is zero:

$$\mathbf{V}_{P_3/2} = \mathbf{0}. \tag{3.13}$$

This equation is commonly referred to as the *rolling contact condition for velocity*. Substituting Eq. (3.13) into Eq. (3.12), the rolling contact condition can also be written as

$$\mathbf{V}_{P_3} = \mathbf{V}_{P_2}. \tag{3.14}$$

This says that *the absolute velocities of two points in rolling contact are equal.*

The graphic velocity analysis of this mechanism, assuming rolling contact at point *P*, is as follows. Given ω_2, the velocity-difference $\mathbf{V}_{P_2B}$ can be calculated and drawn to scale, thus locating point P_2 in the velocity polygon; see Fig. 3.14*b*. Using Eq. (3.13), the rolling contact condition, we also label this point P_3. Next, writing simultaneous equations for the velocity of pin *C*, using $\mathbf{V}_{CP_3}$ and $\mathbf{V}_{CA_4}$, we locate the velocity-image points C_3 and C_4. Finally, the angular velocities of links 3 and 4 are obtained from $\omega_3 = V_{CP}/R_{CP}$ and $\omega_4 = V_{CA}/R_{CA}$, respectively.

Another approach involves defining a fictitious point C_2, which is located instantaneously coincident with points C_3 and C_4, but which is attached to and moving with link 2, as shown by the shaded triangle BPC_2. When the velocity-image concept is used for link 2, the velocity-image point C_2 can be located (see Fig. 3.14*b*). Noting that point C_4 (and C_3) traces a known path on link 2, we can write and solve the apparent-velocity equation involving $\mathbf{V}_{C_4/2}$. Then the velocity $\mathbf{V}_{C_4}$ (and angular velocity ω_4, if desired) is obtained without dealing with the point of direct contact. This approach would be necessary if we had not assumed rolling contact (no slip) at *P*.

3.8 Systematic Strategy for Velocity Analysis

A careful review of the preceding sections and example problems indicates that we have developed sufficient tools for dealing with situations that normally arise in the velocity analysis of planar mechanisms. It will also be noted that the word "relative" velocity has been carefully avoided. Instead, we note that whenever the desire for using "relative" velocity arises, there are always two points whose velocities are to be "related." These two points are fixed either in the same rigid body or in two different rigid bodies. When the two points are in the same body, the velocity-difference equation, Eq. (3.4), is appropriate. However, when it is desirable to switch to a point in a different body, then coincident points should be chosen and the apparent-velocity equation, Eq. (3.6), should be used. We can organize all possible situations into four cases, as shown in Table 3.1.

The notation has intentionally been made different to remind us that these are totally different situations, and the formulae are not interchangeable between them. We should not try to use the "$\omega \times \mathbf{R}$" formula when using the apparent velocity; if we do try, then we will not

Table 3.1 "Relative" velocity equations

Points are	Coincident	Separated
In same body	*Trivial case:* $\mathbf{V_P} = \mathbf{V_Q}$.	*Velocity difference:* $V_P = V_Q + V_{PQ}$ $V_{PQ} = \omega_j \times R_{PQ}$.
In different bodies	*Apparent velocity:* $\mathbf{V_{P_i}} = \mathbf{V_{P_j}} + \mathbf{V_{P_{i/j}}}$ where path $P_{i/j}$ is known. *Rolling contact velocity:* $\mathbf{V_{P_i}} = \mathbf{V_{P_j}}$ and $\mathbf{V_{P_{i/j}}} = 0$.	*Too general;* *use two steps.*

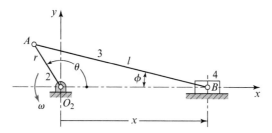

Fig. 3.15 Central slider-crank linkage.

find a single **ω** or a useful **R**. Similarly, when using the velocity difference, there is no question of which **ω** to use since only one link pertains. Further advantages will become clear when we study accelerations in Chapter 4.

A detailed study of the graphic solutions of Example 3.1 through Example 3.4 clearly indicates how the suggested strategy in Table 3.1, and the labeling of the velocity polygons, is applied to the velocity analysis of planar mechanisms.

3.9 Algebraic Velocity Analysis

For some types of mechanisms, a numeric solution by calculator or computer is often the most convenient. Also, when solutions for multiple postures are required, graphic methods become cumbersome. In this section we present an algebraic method for velocity analysis.

As an illustration, consider the central (or in-line) slider-crank linkage shown in Fig. 3.15. This linkage is the mechanism found in most internal combustion engines, and, for this reason, we use the standard engine notation r and l to designate the length of the crank and the length of the connecting rod, respectively.

From the geometry of the linkage we can write

$$r \sin \theta = l \, \sin \phi, \tag{a}$$

$$x = r \cos \theta + l \, \cos \phi. \tag{b}$$

The angle ϕ can be eliminated as follows. From trigonometry,

$$l \cos \phi = l\sqrt{1 - \sin^2\phi} = \sqrt{l^2 - l^2 \sin^2\phi}. \qquad (c)$$

Substituting Eq. (a) into Eq. (c) gives

$$l \cos \phi = \sqrt{l^2 - r^2 \sin^2 \theta}. \qquad (d)$$

Then, substituting Eq. (d) into Eq. (b), the position of the slider (or piston) can be written as

$$x = r \cos \theta + \sqrt{l^2 - r^2 \sin^2 \theta}$$

or as

$$x = r \cos \theta + l\sqrt{1 - \left(\frac{r}{l}\sin\theta\right)^2}. \qquad (3.15a)$$

Differentiating this equation with respect to time gives the exact equation for the velocity of the slider:

$$\dot{x} = -r\omega\left[\sin\theta + \frac{r\sin 2\theta}{2l\sqrt{1 - (r/l)^2\sin^2\theta}}\right]. \qquad (3.15b)$$

We note that for most internal combustion engines, the ratio r/l is in the range $1/10 \le r/l \le 1/4$. This implies that the maximum value of the second term under the radical is about $1/16$ or less. Therefore, it is common practice to use approximations for the position and the velocity of the slider. These approximations are obtained by using the first two terms of the binomial expansion to write the radical in Eq. (3.15a) as

$$\sqrt{1 - \left(\frac{r}{l}\sin\theta\right)^2} \approx 1 - \frac{r^2}{2l^2}\sin^2\theta = 1 - \frac{r^2}{4l^2}(1 - \cos 2\theta). \qquad (e)$$

Substituting this into Eq. (3.15a) gives

$$x \approx r\cos\theta + l\left[1 - \frac{r^2}{4l^2}(1 - \cos 2\theta)\right]. \qquad (3.15c)$$

Rearranging this equation, the approximate position of the slider can be written as

$$x \approx l - \frac{r^2}{4l} + r\left(\cos\theta + \frac{r}{4l}\cos 2\theta\right). \qquad (3.15d)$$

Differentiating Eq. (3.15a) with respect to time and rearranging, the approximate velocity of the slider can be written as

$$\dot{x} \approx -r\omega\left[\sin\theta + \frac{r}{2l}\sin 2\theta\right]. \qquad (3.15e)$$

3.10 Complex-Algebraic Velocity Analysis

We recall from Section 2.9 that complex algebra provides an alternative algebraic formulation for problems in two-dimensional kinematics. As we noted there, the complex-algebraic formulation provides the advantage of increased accuracy over graphic methods, and, once a program is written, it is convenient for solution by computer at a large number of postures. On the other hand, the solution of the loop-closure equation for the unknown posture variables is a nonlinear problem and can lead to tedious algebraic manipulations. Fortunately, the extension of the complex-algebraic approach to velocity analysis leads to a set of *linear* equations, and the solution is quite straightforward.

Recalling the complex polar form of a two-dimensional vector from Eq. (2.41), that is,

$$\mathbf{R} = Re^{j\theta},$$

we find the general form of its time derivative,

$$\dot{\mathbf{R}} = \frac{d\mathbf{R}}{dt} = \dot{R}e^{j\theta} + j\dot{\theta}Re^{j\theta}, \tag{3.16}$$

where $\dot{R}$ and $\dot{\theta}$ denote, respectively, the time rates of change of the magnitude and angle of vector $\mathbf{R}$.

We will see in the following two examples that the first term on the right-hand side of Eq. (3.16) commonly represents an apparent velocity, and the second term usually represents a velocity difference. The method shown in these examples was developed by Raven [12]. Although the original work provides methods applicable to both planar and spatial mechanisms, only the planar aspects are shown here.

Example 3.5

For the inverted slider-crank linkage in the posture shown in Fig. 3.16a, link 2, the driver, has a known posture θ_2 and a known angular velocity ω_2. Derive expressions for the angular velocity of link 4 and the absolute velocity of point P in terms of the angular velocity of the driver.

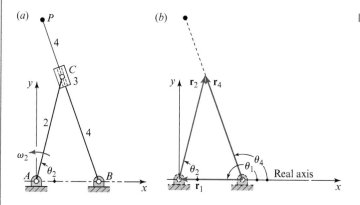

Fig. 3.16 Inverted slider-crank linkage.

Example 3.5 (cont.)

Solution

To simplify the notation, we will use the symbols[6] shown in Fig. 3.16b for the position-difference vectors; thus, $\mathbf{R}_{AB}$ is denoted $\mathbf{r}_1$, $\mathbf{R}_{C_2A}$ is denoted $\mathbf{r}_2$, and $\mathbf{R}_{C_4B}$ is denoted $\mathbf{r}_4$. In terms of these symbols, the loop-closure equation can be written as

$$\overset{??}{\mathbf{r}_4} = \overset{\sqrt{}\sqrt{}}{\mathbf{r}_1} + \overset{\sqrt{}I}{\mathbf{r}_2}, \tag{1}$$

where vector $\mathbf{r}_4$ has unknown magnitude and direction, vector $\mathbf{r}_1$ has constant magnitude and direction,[7] and vector $\mathbf{r}_2$ has constant magnitude (its direction, θ_2, varies). The variable angle θ_2 is the given input angle or, more precisely, the unknown variables will be solved as functions of θ_2.

Recognizing this as case 1 (Section 2.8), we obtain the posture solution from Eqs. (2.43) and (2.44) as

$$r_4 = \sqrt{r_1^2 + r_2^2 - 2r_1r_2 \cos\theta_2} \tag{2}$$

and

$$\theta_4 = \tan^{-1}\left(\frac{r_2 \sin\theta_2}{r_2 \cos\theta_2 - r_1}\right). \tag{3}$$

The velocity solution is initiated by differentiating the loop-closure equation, Eq. (1), with respect to time. Applying the general formula, Eq. (3.16), to each term of Eq. (1) in turn, and keeping in mind that r_1, θ_1, and r_2 are constants, we obtain

$$\dot{r}_4 e^{j\theta_4} + j\dot{\theta}_4 r_4 e^{j\theta_4} = j\dot{\theta}_2 r_2 e^{j\theta_2}. \tag{4}$$

Since $\dot{\theta}_2$ and $\dot{\theta}_4$ are the same as ω_2 and ω_4, respectively, and since we recognize that

$$\dot{\theta}_2 r_2 = V_{C_2}, \qquad \dot{r}_4 = V_{C_2/4}, \qquad \dot{\theta}_4 r_4 = V_{C_4},$$

we see that Eq. (4) is, in fact, the complex polar form of the apparent-velocity equation,

$$\mathbf{V}_{C_4} + \mathbf{V}_{C_2/4} = \mathbf{V}_{C_2}.$$

(This is pointed out for comparison only and is not a necessary step in the solution process.)

The velocity solution is performed using Euler's formula, Eq. (2.40), to separate Eq. (4) into its real and imaginary components. This gives

$$\dot{r}_4 \cos\theta_4 - \omega_4 r_4 \sin\theta_4 = -\omega_2 r_2 \sin\theta_2, \tag{5}$$

$$\dot{r}_4 \sin\theta_4 + \omega_4 r_4 \cos\theta_4 = \omega_2 r_2 \cos\theta_2. \tag{6}$$

---

[6] The symbols $\mathbf{R}$ and $\mathbf{r}$ are used interchangeably throughout this chapter.

[7] Note particularly that the angle of $\mathbf{r}_1$ *is* $\theta_1 = 180°$, not $\theta_1 = 0°$.

Example 3.5 (cont.)

Using Cramer's rule, the expressions for the absolute velocity of point P and the angular velocity of link 4, in terms of the angular velocity of the driver, are

$$\dot{r}_4 = \omega_2 r_2 \sin(\theta_4 - \theta_2), \tag{7}$$

$$\omega_4 = \omega_2 \frac{r_2}{r_4} \cos(\theta_4 - \theta_2). \tag{Ans. (8)}$$

Although the variables r_4 and θ_4 given by Eqs. (2) and (3) could be substituted into Eqs. (7) and (8) to reduce the results to functions of θ_2 and ω_2 alone, the previous forms are considered sufficient. The reason is that, in writing a computer program, numeric values are normally found first for r_4 and θ_4 while performing the posture analysis, and then these numeric values are used to obtain numeric values for $\dot{r}_4$ and ω_4 at each input angle θ_2.

To find the velocity of point P, we first write the position of point P as

$$\mathbf{R}_P = R_{PB} e^{j\theta_4}. \tag{9}$$

Then, differentiating this equation with respect to time, and remembering that R_{PB} is a constant length, gives

$$\mathbf{V}_P = j\omega_4 R_{PB} e^{j\theta_4}. \tag{10}$$

Next, substituting Eq. (8) into this equation gives

$$\mathbf{V}_P = j\omega_2 R_{PB} \frac{r_2}{r_4} \cos(\theta_4 - \theta_2) e^{j\theta_4}. \tag{11}$$

Therefore, the horizontal and vertical components of the velocity of point P in terms of the angular velocity of the driver are

$$V_P^x = -\omega_2 R_{PB} \frac{r_2}{r_4} \cos(\theta_4 - \theta_2) \sin\theta_4 \tag{Ans. (12)}$$

and

$$V_P^y = \omega_2 R_{PB} \frac{r_2}{r_4} \cos(\theta_4 - \theta_2) \cos\theta_4. \tag{Ans. (13)}$$

Example 3.6

For the four-bar linkage in the arbitrary posture shown in Fig. 3.17a, derive expressions for the relationships between the angular velocity of the input link 2 and the angular velocities of the coupler link 3 and the output link 4.

Solution

First, replace each link of Fig. 3.17a by a vector as shown in Fig. 3.17b. Then, the loop-closure equation can be written as

Example 3.6 (cont.)

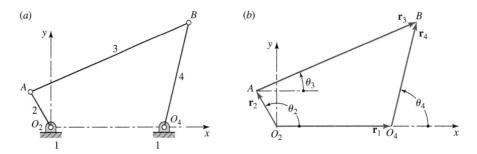

Fig. 3.17 (a) Four-bar linkage; (b) the links are replaced by vectors.

$$\overset{\sqrt{I}}{\mathbf{r}_2} + \overset{\sqrt{?}}{\mathbf{r}_3} - \overset{\sqrt{?}}{\mathbf{r}_4} - \overset{\sqrt{\sqrt{}}}{\mathbf{r}_1} = \mathbf{0}. \tag{1}$$

Next, write each vector in Eq. (1) in complex polar form; that is,

$$r_2 e^{j\theta_2} + r_3 e^{j\theta_3} - r_4 e^{j\theta_4} - r_1 e^{j\theta_1} = 0. \tag{2}$$

Taking the time derivative of this equation and noting that link 1 is fixed gives

$$j r_2 \dot{\theta}_2 e^{j\theta_2} + j r_3 \dot{\theta}_3 e^{j\theta_3} - j r_4 \dot{\theta}_4 e^{j\theta_4} = 0. \tag{3}$$

Now, transform Eq. (3) into rectangular form and separate the real and the imaginary terms. Then, noting that $\dot{\theta}_2 = \omega_2, \dot{\theta}_3 = \omega_3$, and $\dot{\theta}_4 = \omega_4$, gives

$$\begin{aligned} r_2\omega_2 \cos\theta_2 + r_3\omega_3 \cos\theta_3 - r_4\omega_4 \cos\theta_4 = 0, \\ r_2\omega_2 \sin\theta_2 + r_3\omega_3 \sin\theta_3 - r_4\omega_4 \sin\theta_4 = 0, \end{aligned} \tag{4}$$

where the input angular velocity ω_2 is known and the angular velocities ω_3 and ω_4 are the two unknown variables.

Finally, using Cramer's rule, the expressions for the angular velocities of the coupler link 3 and the output link 4 in terms of the input angular velocity, respectively, are

$$\omega_3 = \frac{r_2 \sin(\theta_2 - \theta_4)}{r_3 \sin(\theta_4 - \theta_3)}\omega_2 \quad \text{and} \quad \omega_4 = \frac{r_2 \sin(\theta_2 - \theta_3)}{r_4 \sin(\theta_4 - \theta_3)}\omega_2. \qquad \textit{Ans. (5)}$$

Note that the factor $\sin(\theta_4 - \theta_3)$ appears in the denominators of both results. In general, the velocity analysis of a planar mechanism will have similar denominators in the solutions for each of the velocity unknowns; these denominators are the determinant of the matrix of coefficients of the unknowns, as is recognized by recalling Cramer's rule. The angle $(\theta_4 - \theta_3)$ is, in fact, the transmission angle γ (see Fig. 1.31, Section 1.10). Recall that when the transmission angle becomes small, the ratio of the output speed, ω_4, to the input speed, ω_2, becomes very large and difficulties will result (Section 1.10).

From Examples 3.5 and 3.6, we note that the simultaneous equations to be solved are linear in the unknown variables. This is not a coincidence but is true for all velocity problems. It results from the fact that the general form of each velocity term, Eq. (3.16), is linear in the velocity variables. When real and imaginary components are taken, the *coefficients* may become complicated, but the equations remain linear with respect to the velocity unknowns. Therefore, their solution is straightforward.

Recall that in Sections 3.4–3.8, the graphic velocity methods, it was possible to choose an arbitrary scale factor for a velocity polygon. Therefore, if the input speed of a mechanism is doubled, the scale factor of the velocity polygon can also be doubled, and the same polygon is valid. This is another indication of the linearity of the velocity equations.

3.11 Method of Kinematic Coefficients

A method that provides substantial geometric insight into the kinematic analysis and synthesis of a linkage is to differentiate the loop-closure equation with respect to the input variable(s), rather than differentiating with respect to time. This analytic approach is referred to as the *method of kinematic coefficients* [8]. The numeric values of the first-order kinematic coefficients can also be checked with the graphic approach of finding the locations of the instantaneous centers of velocity (Section 3.12).

This method is shown here by again solving the four-bar linkage problem that was presented in Example 3.1 and the offset slider-crank linkage problem that was presented in Example 3.2.

Example 3.7

The four-bar linkage in the posture shown in Fig. 3.7a is driven by crank 2 at a constant angular velocity $\omega_2 = 900$ rev/min ccw. Determine the angular velocities of the coupler link and the output link, and the velocities of points E and F.

Solution

The loop-closure equation is given by Eq. (1) in Example 3.6 (Fig. 3.17b), and is reproduced here:

$$\mathbf{r}_2^{\sqrt{I}} + \mathbf{r}_3^{\sqrt{?}} - \mathbf{r}_4^{\sqrt{?}} - \mathbf{r}_1^{\sqrt{\sqrt{}}} = \mathbf{0}. \tag{1}$$

The two scalar component equations are

$$r_2 \cos\theta_2 + r_3 \cos\theta_3 - r_4 \cos\theta_4 - r_1 \cos\theta_1 = 0,$$

$$r_2 \sin\theta_2 + r_3 \sin\theta_3 - r_4 \sin\theta_4 - r_1 \sin\theta_1 = 0. \tag{2}$$

Since the input is the angular displacement of link 2, the method of kinematic coefficients calls for differentiating Eqs. (2) with respect to the input variable θ_2. Rearranging the result gives

Example 3.7 (cont.)

$$-r_3 \sin\theta_3\theta_3' + r_4 \sin\theta_4\theta_4' = r_2 \sin\theta_2$$

and

$$r_3 \cos\theta_3\theta_3' - r_4 \cos\theta_4\theta_4' = -r_2 \cos\theta_2, \qquad (3)$$

where

$$\theta_3' = \frac{d\theta_3}{d\theta_2} \quad \text{and} \quad \theta_4' = \frac{d\theta_4}{d\theta_2} \qquad (4)$$

are referred to as the *first-order kinematic coefficients* of links 3 and 4, respectively.

Algebraic forms for the kinematic coefficients can be obtained using Cramer's rule. Writing Eqs. (3) in matrix form gives

$$\begin{bmatrix} -r_3 \sin\theta_3 & r_4 \sin\theta_4 \\ r_3 \cos\theta_3 & -r_4 \cos\theta_4 \end{bmatrix} \begin{bmatrix} \theta_3' \\ \theta_4' \end{bmatrix} = \begin{bmatrix} r_2 \sin\theta_2 \\ -r_2 \cos\theta_2 \end{bmatrix}. \qquad (5)$$

The determinant of the (2×2) coefficient matrix can be written as

$$\Delta = -r_3 r_4 \sin(\theta_4 - \theta_3), \qquad (6)$$

and provides geometric insight into special postures of the linkage (Example 3.6). For example, when this determinant tends toward zero, the kinematic coefficients tend toward infinity. Note that the determinant is zero (that is, the transmission angle $\theta_4 - \theta_3$ is zero) either when (a) $\theta_3 = \theta_4$ or when (b) $\theta_3 = \theta_4 \pm 180°$ – that is, when links 3 and 4 are fully extended or aligned on top of each other.

From Eq. (5), the first-order kinematic coefficients of links 3 and 4 are

$$\theta_3' = \frac{r_2 \sin(\theta_2 - \theta_4)}{r_3 \sin(\theta_4 - \theta_3)} \quad \text{and} \quad \theta_4' = \frac{r_2 \sin(\theta_2 - \theta_3)}{r_4 \sin(\theta_4 - \theta_3)}. \qquad (7)$$

Recall from Example 3.1 that the posture values are $\theta_3 = 20.92°$ and $\theta_4 = 64.05°$. Therefore, the first-order kinematic coefficients of links 3 and 4 are

$$\theta_3' = +0.271\,8 \text{ rad/rad} \quad \text{and} \quad \theta_4' = +0.526\,5 \text{ rad/rad}, \qquad (8)$$

where the positive signs indicate that links 3 and 4 are rotating in the same direction as the input link 2.

The angular velocities of links 3 and 4, obtained from the chain rule, are

$$\omega_3 = \theta_3'\omega_2 \quad \text{and} \quad \omega_4 = \theta_4'\omega_2. \qquad (9)$$

Substituting Eqs. (7) into Eqs. (9) gives the same symbolic results as Eqs. (5) in Example 3.6. Also, substituting Eqs. (8) and the input angular velocity into Eqs. (9), the angular velocities of links 3 and 4 are

Example 3.7 (cont.)

$$\omega_3 = 25.62 \text{ rad/s ccw} \quad \text{and} \quad \omega_4 = 49.62 \text{ rad/s ccw.} \qquad \textit{Ans.} \text{ (10)}$$

These answers are in good agreement with the results obtained in Example 3.1 from the graphic method [Eqs. (4) and (5)] and the analytic method [Eqs. (14)].

The position of coupler point E with respect to the ground pivot A (Fig. 3.7a) can be written as

$$\overset{??}{\mathbf{r}_E} = \overset{\sqrt{I}}{\mathbf{r}_2} + \overset{\sqrt{V}}{\mathbf{r}_{EB}}. \qquad (11)$$

The x and y components of this vector equation are

$$x_E = r_2 \cos\theta_2 + r_{EB}\cos(\theta_3 - \phi),$$
$$y_E = r_2 \sin\theta_2 + r_{EB}\sin(\theta_3 - \phi), \qquad (12)$$

where $\phi = \tan^{-1}(R_{EG}/R_{GB}) = \tan^{-1}(4\text{ in}/10\text{ in}) = 21.80°$.

Differentiating Eqs. (12) with respect to the input variable θ_2, the first-order kinematic coefficients for point E are

$$x'_E = \frac{dx_E}{d\theta_2} = -r_2\sin\theta_2 - r_{EB}\sin(\theta_3 - \phi)\theta'_3,$$
$$y'_E = \frac{dy_E}{d\theta_2} = r_2\cos\theta_2 + r_{EB}\cos(\theta_3 - \phi)\theta'_3. \qquad (13)$$

Substituting the known values into these equations, the first-order kinematic coefficients for point E are

$$x'_E = -3.419\,1 \text{ in/rad} \quad \text{and} \quad y'_E = +0.926\,9 \text{ in/rad.} \qquad (14)$$

The velocity of point E can be written from the chain rule as

$$\mathbf{V}_E = (x'_E\hat{\mathbf{i}} + y'_E\hat{\mathbf{j}})\omega_2. \qquad (15)$$

Substituting Eqs. (14) and the input angular velocity into this equation, the velocity of point E is

$$\mathbf{V}_E = -26.85\hat{\mathbf{i}} + 7.28\hat{\mathbf{j}} \text{ ft/s.} \qquad \textit{Ans.}$$

Therefore, the speed of point E is $V_E = 27.8$ ft/s, which agrees closely with the result obtained from the graphic method in Example 3.1 ($V_E = 27.6$ ft/s).

The velocity of point F can be obtained in a similar manner, and the result is

$$\mathbf{V}_F = -20.60\hat{\mathbf{i}} + 23.82\hat{\mathbf{j}} \text{ ft/s.} \qquad \textit{Ans.}$$

Therefore, the speed of point F is $V_F = 31.5$ ft/s, which agrees closely with the result obtained from the graphic method in Example 3.1 ($V_F = 31.8$ ft/s).

Example 3.8

The offset slider-crank linkage in the posture shown in Fig. 3.8 is driven by slider 4 at a constant speed of $V_C = 10\,\text{m/s}$ to the left. Determine the angular velocities of links 2 and 3 and the velocity of point D.

Solution

The loop-closure equation for the offset slider-crank linkage, in complex polar form, is

$$jr_1 + r_2 e^{j\theta_2} + r_3 e^{j\theta_3} - r_4 = 0. \tag{1}$$

The two scalar equations are

$$\begin{aligned} r_2 \cos\theta_2 + r_3 \cos\theta_3 - r_4 &= 0, \\ r_1 + r_2 \sin\theta_2 + r_3 \sin\theta_3 &= 0. \end{aligned} \tag{2}$$

Since the input is the linear displacement of link 4, differentiating Eqs. (2) with respect to r_4 gives

$$\begin{aligned} -r_2 \sin\theta_2\theta_2' - r_3 \sin\theta_3\theta_3' &= 1, \\ r_2 \cos\theta_2\theta_2' + r_3 \cos\theta_3\theta_3' &= 0, \end{aligned} \tag{3}$$

where

$$\theta_2' = \frac{d\theta_2}{dr_4} \quad \text{and} \quad \theta_3' = \frac{d\theta_3}{dr_4} \tag{4}$$

are the first-order kinematic coefficients of links 2 and 3. Symbolic equations for the first-order kinematic coefficients are obtained here using Cramer's rule. The first step is to write Eqs. (3) in matrix form; that is,

$$\begin{bmatrix} -r_2 \sin\theta_2 & -r_3 \sin\theta_3 \\ r_2 \cos\theta_2 & r_3 \cos\theta_3 \end{bmatrix} \begin{bmatrix} \theta_2' \\ \theta_3' \end{bmatrix} = \begin{bmatrix} 1 \\ 0 \end{bmatrix}. \tag{5}$$

The determinant of the (2×2) coefficient matrix in Eq. (5) is written as

$$\Delta = r_2 r_3 \sin(\theta_3 - \theta_2). \tag{6}$$

Note that this determinant is zero when (a) $\theta_2 = \theta_3$ or when (b) $\theta_2 = \theta_3 \pm 180°$ – that is, when links 2 and 3 are fully extended or aligned on top of each other.

From Eq. (5), the first-order kinematic coefficients of links 2 and 3 can be written as

$$\theta_2' = \frac{\cos\theta_3}{r_2 \sin(\theta_3 - \theta_2)} \quad \text{and} \quad \theta_3' = \frac{-\cos\theta_2}{r_3 \sin(\theta_3 - \theta_2)}. \tag{7}$$

For the given link dimensions and the given input position $r_4 = 164\,\text{mm}$, the postures of links 2 and 3 are $\theta_2 = 45°$ and $\theta_3 = 337°$, respectively. Substituting these values into Eqs. (7), the first-order kinematic coefficients of links 2 and 3 are

$$\theta_2' = -19.856\,\text{rad/m} \quad \text{and} \quad \theta_3' = +5.447\,\text{rad/m}, \tag{8}$$

Example 3.8 (cont.)

where the negative sign shows that link 2 rotates clockwise while the positive sign shows that link 3 rotates counterclockwise for positive input motion of link 4 (to the right).

The angular velocities of links 2 and 3 can be written by the chain rule as

$$\omega_2 = \theta_2' \dot{r}_4 \quad \text{and} \quad \omega_3 = \theta_3' \dot{r}_4, \tag{9}$$

Substituting Eqs. (8) and the time rate of change of the input vector, $\dot{r}_4 = V_C = -10$ m/s, into Eqs. (9), the angular velocities of the two links are

$$\omega_2 = 198.56 \text{ rad/s ccw} \quad \text{and} \quad \omega_3 = -54.47 \text{ rad/s (cw)}. \qquad \textit{Ans.}$$

These answers agree closely with the results obtained from the velocity polygon method in Example 3.2 ($\omega_2 = 200$ rad/s ccw and $\omega_3 = 53.6$ rad/s cw), and are more accurate.

Note that the signs of the first-order kinematic coefficients depend on the direction of the position vector that is chosen to represent the input. In this example, the input vector is chosen from the y axis on the ground link to point C on link 4 (see Fig. 3.8a). However, if the input position vector were chosen to point C from a point on the ground link to the right of link 4, then the first-order kinematic coefficients would have opposite signs to those of Eq. (8); that is,

$$\theta_2' = 19.856 \text{ rad/m} \quad \text{and} \quad \theta_3' = -5.447 \text{ rad/m}.$$

This agrees with our definition that a first-order kinematic coefficient is the differential change in the unknown variable with respect to the differential change in the input. Also, for this choice of the input vector, the time rate of change of the input vector has changed sign; that is, $\dot{r}_4 = V_C = 10$ m/s. Substituting these values into Eqs. (9), the angular velocities of links 2 and 3 are

$$\omega_2 = (19.856)(10) = 198.56 \text{ rad/s ccw} \quad \text{and} \quad \omega_3 = (-5.447)(10) = -54.47 \text{ rad/s (cw)},$$
$$\textit{Ans.}$$

which agree with the previous results.

Note that the direction of the input vector can change the signs of the kinematic coefficients and the sign of the change of the input vector. However, it has no effect on the angular velocities of links 2 and 3. Therefore, the vector loop equation that represents the mechanism need not be unique.

The position of point D with respect to ground pivot A (Fig. 3.8a) can be written as

$$\overset{??}{\mathbf{r}_D} = \overset{\surd I}{\mathbf{r}_2} + \overset{\surd\surd}{\mathbf{r}_{DB}}. \tag{10}$$

The x and y components of this vector equation are

$$x_D = r_2 \cos\theta_2 + r_{DB} \cos(\theta_3 - \beta), \tag{11a}$$

$$y_D = r_2 \sin\theta_2 + r_{DB} \sin(\theta_3 - \beta), \tag{11b}$$

where $\beta = (R_{DE}/R_{EB}) = \tan^{-1}(50 \text{ mm}/80 \text{ mm}) = 32.01°$.

Example 3.8 (cont.)

Differentiating Eqs. (11) with respect to input variable r_4, the first-order kinematic coefficients for point D are

$$x'_D = -r_2 \sin\theta_2\theta'_2 - r_{DB}\sin(\theta_3 - \beta)\theta'_3, \tag{12a}$$

$$y'_D = r_2 \cos\theta_2\theta'_2 + r_{DB}\cos(\theta_3 - \beta)\theta'_3. \tag{12b}$$

Substituting the known values into these equations, the first-order kinematic coefficients are

$$x'_D = 1.123 \text{ m/m} \quad \text{and} \quad y'_D = -0.407 \text{ m/m}. \tag{13}$$

The velocity of the point D can be written as

$$\mathbf{V}_D = \left(x'_D\hat{\mathbf{i}} + y'_D\hat{\mathbf{j}}\right)\dot{r}_4. \tag{14}$$

Substituting Eqs. (13) and the input velocity $\dot{r}_4 = -10$ m/s into this equation, the velocity of point D is

$$\mathbf{V}_D = -11.23\hat{\mathbf{i}} + 4.07\hat{\mathbf{j}} \text{ m/s} = 11.94 \text{ m/s}\angle160.08°. \qquad Ans. \ (15)$$

Therefore, the instantaneous speed of point D agrees closely with the result obtained from the velocity polygon method in Example 3.2 ($V_D = 12$ m/s), but is more accurate.

The method of kinematic coefficients also provides geometric insight into the motion of mechanisms that have links that are in rolling contact. The following two examples help to illustrate this important concept.

Example 3.9

For the mechanism in the posture shown in Fig. 3.18, the wheel is rolling without slipping on the ground. Determine the first-order kinematic coefficients of links 3 and 4. If input link 2 is rotating with a constant angular velocity of 10 rad/s ccw, then determine: (a) the angular velocities of links 3 and 4; and (b) the velocity of the center of the wheel, point G.

Fig. 3.18 $R_{AO_2} = r_2 = 150$ mm,
$R_{BA} = r_3 = 200$ mm,
$R_{GB} = r_4 = 50$ mm, and
$\rho = 100$ mm.

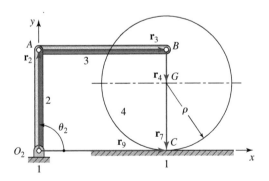

Example 3.9 (cont.)

Solution

The loop-closure equation for the mechanism is

$$\overset{\sqrt{I}}{\mathbf{r}_2} + \overset{\sqrt{?}}{\mathbf{r}_3} + \overset{\sqrt{?}}{\mathbf{r}_4} + \overset{\sqrt{C1}}{\mathbf{r}_7} - \overset{C2\sqrt{}}{\mathbf{r}_9} = \mathbf{0}, \tag{1}$$

where the first constraint, $C1$, is $\theta_7 = \theta_9 + 270°$ and the second constraint, $C2$, designates the rolling contact constraint defined later by Eq. (5).

The horizontal and vertical components of Eq. (1) give the scalar equations

$$r_2 \cos\theta_2 + r_3 \cos\theta_3 + r_4 \cos\theta_4 + r_7 \cos\theta_7 - r_9 \cos\theta_9 = 0, \tag{2a}$$

$$r_2 \sin\theta_2 + r_3 \sin\theta_3 + r_4 \sin\theta_4 + r_7 \sin\theta_7 - r_9 \sin\theta_9 = 0. \tag{2b}$$

Since the input is the angular displacement of link 2, we differentiate Eqs. (2) with respect to θ_2 and, by setting $\theta_9 = 0°$, we obtain

$$-r_2 \sin\theta_2 - r_3 \sin\theta_3\theta_3' - r_4 \sin\theta_4\theta_4' - r_9' = 0 \tag{3a}$$

and

$$r_2 \cos\theta_2 + r_3 \cos\theta_3\theta_3' + r_4 \cos\theta_4\theta_4' = 0, \tag{3b}$$

where the first-order kinematic coefficients of links 3 and 4 and the change in position of the contact point C, respectively, are

$$\theta_3' = \frac{d\theta_3}{d\theta_2}, \quad \theta_4' = \frac{d\theta_4}{d\theta_2}, \quad \text{and} \quad r_9' = \frac{dr_9}{d\theta_2}. \tag{4}$$

Also, we note that the rotation of the wheel, θ_4, and the change in the distance, r_9, are not independent, since the wheel is rolling on the ground without slipping. This rolling contact (no slip) condition is constraint $C2$ and can be written as shown in Section 2.19.

$$-\Delta r_9 = \rho(\Delta\theta_4 - \Delta\theta_7),$$

where ρ is the radius of the wheel and, from the constraint $C1$, we note that $\Delta\theta_7 = 0$. Therefore, the rolling contact condition is

$$-\Delta r_9 = \rho\Delta\theta_4. \tag{5}$$

The negative sign on the left-hand side of this constraint equation is the result of the decreasing magnitude of r_9 for a positive change in the input variable – that is, $\Delta\theta_2$.

In the limit, for infinitesimal displacements, constraint Eq. (5) can be written in terms of first-order kinematic coefficients as

$$-r_9' = \rho\theta_4'. \tag{6}$$

Example 3.9 (cont.)

Substituting Eq. (6) into Eqs. (3), and writing the resulting equations in matrix form, gives

$$\begin{bmatrix} -r_3\sin\theta_3 & \rho - r_4\sin\theta_4 \\ r_3\cos\theta_3 & r_4\cos\theta_4 \end{bmatrix}\begin{bmatrix} \theta_3' \\ \theta_4' \end{bmatrix} = \begin{bmatrix} +r_2\sin\theta_2 \\ -r_2\cos\theta_2 \end{bmatrix}. \tag{7}$$

The determinant of the (2×2) coefficient matrix in Eq. (7) can be written as

$$\Delta = r_3[r_4\sin(\theta_4 - \theta_3) - \rho\cos\theta_3]. \tag{8}$$

From Eq. (7), using Cramer's rule, the symbolic equations for the first-order kinematic coefficients of links 3 and 4 are

$$\theta_3' = \frac{r_2[\rho\cos\theta_2 - r_4\sin(\theta_4 - \theta_2)]}{\Delta}, \tag{9a}$$

$$\theta_4' = \frac{r_2 r_3 \sin(\theta_3 - \theta_2)}{\Delta}. \tag{9b}$$

For the posture shown in Fig. 3.18, we have $\theta_2 = 90°$, $\theta_3 = 0°$, and $\theta_4 = -90°$. Therefore, the determinant in Eq. (8) can be written as

$$\Delta = -r_3(r_4 + \rho). \tag{10}$$

Substituting the known data and Eq. (10) into Eqs. (9), and simplifying, gives

$$\theta_3' = 0 \quad \text{and} \quad \theta_4' = \frac{r_2}{r_4 + \rho} = 1 \text{ rad/rad}. \qquad \textit{Ans.} \tag{11}$$

Therefore, the angular velocity of link 3 is

$$\omega_3 = \theta_3'\omega_2 = 0, \qquad \textit{Ans.}$$

and the angular velocity of the wheel, link 4, is

$$\omega_4 = \theta_4'\omega_2 = 10 \text{ rad/s ccw}. \qquad \textit{Ans.}$$

Note that these results are in agreement with our intuition; in this posture, link 3 is not rotating, and link 4 is rotating at the same angular velocity as the input.

The velocity of the center of the wheel, point G, can be written as

$$\mathbf{V}_G = r_9'\omega_2\angle\theta_9. \tag{12}$$

Substituting Eq. (11) into Eq. (6) gives $r_9' = -\rho = -100$ mm, and substituting this into Eq. (12) gives

$$\mathbf{V}_G = r_9'\omega_2\angle\theta_9 = (-100 \text{ mm/rad})(10 \text{ rad/s})\angle 0° = -1.0 \text{ m/s}\angle 0°. \qquad \textit{Ans.}$$

This result can be verified by direct calculation by noting that

$$\mathbf{V}_G = \mathbf{V}_{C_4}^0 + \boldsymbol{\omega}_4 \times (\rho\hat{\mathbf{j}}) = (10 \text{ rad/s})\hat{\mathbf{k}} \times (100 \text{ mm})\hat{\mathbf{j}} = -1.0\hat{\mathbf{i}} \text{ m/s}. \qquad \textit{Ans.}$$

Example 3.10

For the mechanism in the posture shown in Fig. 3.19, gear 3 is rolling without slip on input gear 2. Determine: (a) the first-order kinematic coefficients of links 3, 4, and 5; and (b) the angular velocities of links 3, 4, and 5 if the input gear is rotating with a constant angular velocity of 10 rad/s ccw.

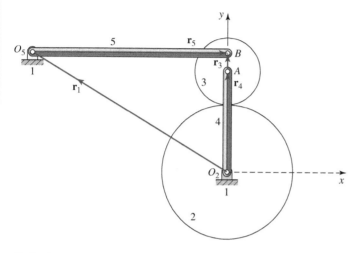

Fig. 3.19 $\rho_2 = 20$ mm, $\rho_3 = 10$ mm, $R_{BA} = r_3 = 5$ mm, $R_{AO_2} = r_4 = 30$ mm, and $R_{BO_3} = r_5 = 60$ mm.

Solution

The loop-closure equation for the mechanism is

$$\overset{\sqrt{?}}{\mathbf{r}_4} + \overset{\sqrt{C}}{\mathbf{r}_3} - \overset{\sqrt{?}}{\mathbf{r}_5} - \overset{\sqrt{\sqrt{}}}{\mathbf{r}_1} = \mathbf{0}, \tag{1}$$

where the C refers to the rolling contact constraint that is defined later by Eq. (5a).

The two scalar component equations are

$$r_4 \cos \theta_4 + r_3 \cos \theta_3 - r_5 \cos \theta_5 - r_1 \cos \theta_1 = 0, \tag{2a}$$

$$r_4 \sin \theta_4 + r_3 \sin \theta_3 - r_5 \sin \theta_5 - r_1 \sin \theta_1 = 0. \tag{2b}$$

Since the input is the angular displacement of gear 2, differentiating Eqs. (2) with respect to θ_2 gives

$$-r_4 \sin \theta_4 \theta_4' - r_3 \sin \theta_3 \theta_3' + r_5 \sin \theta_5 \theta_5' = 0, \tag{3a}$$

$$r_4 \cos \theta_4 \theta_4' + r_3 \cos \theta_3 \theta_3' - r_5 \cos \theta_5 \theta_5' = 0, \tag{3b}$$

where

$$\theta_3' = \frac{d\theta_3}{d\theta_2}, \quad \theta_4' = \frac{d\theta_4}{d\theta_2}, \quad \text{and} \quad \theta_5' = \frac{d\theta_5}{d\theta_2}, \tag{4}$$

are the first-order kinematic coefficients of links 3, 4, and 5.

Example 3.10 (cont.)

Note that the variables θ_2, θ_3, and θ_4 are not independent, since gear 3 is constrained to roll on gear 2. The rolling contact condition can be written to ensure that the arc lengths passed during a small movement are equal on the two surfaces. This rolling contact (no slip) condition can be written as shown in Section 2.19:

$$\rho_2(\Delta\theta_2 - \Delta\theta_4) = -\rho_3(\Delta\theta_3 - \Delta\theta_4), \tag{5a}$$

where $\Delta\theta_2, \Delta\theta_3$, and $\Delta\theta_4$ are the angular displacements of links 2, 3, and 4 from the posture shown in Fig. 3.19. The negative sign for the right-hand term arises from the reversal in the direction of the angular displacement differences.

Equation (5a) can be divided by a change in the input angular displacement, $\Delta\theta_2$, and the limit taken for small increments. This yields the equivalent constraint equation written in terms of first-order kinematic coefficients; that is,

$$\rho_2(\theta_2' - \theta_4') = -\rho_3(\theta_3' - \theta_4'). \tag{5b}$$

Since the input is the angular displacement of gear 2, then, by definition, $\theta_2' = 1$, and by substituting $\rho_2 = 2\rho_3$, Eq. (5b) can be written as

$$\theta_3' = 3\theta_4' - 2. \tag{6}$$

Substituting Eq. (6) into Eqs. (3) and writing the resulting equations in matrix form gives

$$\begin{bmatrix} -r_4 \sin\theta_4 - 3r_3 \sin\theta_3 & r_5 \sin\theta_5 \\ r_4 \cos\theta_4 + 3r_3 \cos\theta_3 & -r_5 \cos\theta_5 \end{bmatrix} \begin{bmatrix} \theta_4' \\ \theta_5' \end{bmatrix} = \begin{bmatrix} -2r_3 \sin\theta_3 \\ 2r_3 \cos\theta_3 \end{bmatrix}. \tag{7}$$

The determinant of the (2×2) coefficient matrix in Eq. (7) can be written as

$$\Delta = r_5[r_4\sin(\theta_4 - \theta_5) + 3r_3 \sin(\theta_3 - \theta_5)]. \tag{8}$$

From Eq. (7), using Cramer's rule, the symbolic equations for the first-order kinematic coefficients of links 4 and 5 are

$$\theta_4' = \frac{2r_3r_5 \sin(\theta_3 - \theta_5)}{\Delta}, \tag{9a}$$

$$\theta_5' = \frac{2r_3r_4 \sin(\theta_3 - \theta_4)}{\Delta}. \tag{9b}$$

For the posture shown in Fig. 3.19, we have $\theta_3 = 90°, \theta_4 = 90°$, and $\theta_5 = 0°$; therefore, Eq. (8) can be written as

$$\Delta = r_5(r_4 + 3r_3). \tag{10}$$

Also, Eqs. (9) reduce to

$$\theta_4' = \frac{2r_3}{r_4 + 3r_3} = \frac{2}{9} \text{ rad/rad} \quad \text{and} \quad \theta_5' = 0. \tag{11}$$

Example 3.10 (cont.)

It is interesting to note that link 5 is not rotating in this posture; that is, the angular velocity $\omega_5 = 0$ at this instant. Substituting Eq. (11) into Eq. (6), the first-order kinematic coefficient of gear 3 is

$$\theta_3' = -\frac{4}{3} \text{ rad/rad}. \tag{12}$$

The angular velocity of link j can now be written as

$$\omega_j = \theta_j' \omega_2. \tag{13}$$

Therefore, substituting Eqs. (11) and (12) into Eq. (13), and setting $\omega_2 = 10$ rad/s ccw, the angular velocities of gear 3 and links 4 and 5 are

$$\omega_3 = -13.33 \text{ rad/s}, \quad \omega_4 = +2.22 \text{ rad/s}, \quad \text{and} \quad \omega_5 = 0, \qquad \textit{Ans.} \tag{14}$$

where the negative sign indicates clockwise, and the positive sign indicates counterclockwise rotation.

Note that kinematic coefficients are functions of *posture only*; that is, they are not directly functions of time. Also, note that the units of the first-order kinematic coefficients depend on the specified input variable and the unknown variable under consideration. The units may be nondimensional (rad/rad or length/length), length (length/rad), or the reciprocal of length (rad/length). Table 3.2 summarizes the first-order kinematic coefficients that are related to link j of a mechanism (that is, vector $\mathbf{r}_j$), having (a) unknown angle θ_j and/or (b) unknown magnitude r_j.

3.12 Instantaneous Centers of Velocity

One of the more interesting concepts in planar kinematics is that of an instantaneous center of velocity for a pair of rigid bodies that move with respect to one another. For spatial motion, an

Table 3.2 **Summary of first-order kinematic coefficients**

	Variable of interest Angle θ_j (use symbol θ_j' for kinematic coefficient regardless of input)	**Variable of interest** Magnitude r_j (use symbol r_j' for kinematic coefficient regardless of input)
Input $\psi = $ angle θ_i	$\omega_j = \theta_j' \dot{\psi}$ $\theta_j' = \dfrac{d\theta_j}{d\psi}$ (dimensionless, rad/rad)	$\dot{r}_j = r_j' \dot{\psi}$ $r_j' = \dfrac{dr_j}{d\psi}$ (length, length/rad)
Input $\psi = $ magnitude r_i	$\omega_j = \theta_j' \dot{\psi}$ $\theta_j' = \dfrac{d\theta_j}{d\psi}$ (1/length, rad/length)	$\dot{r}_j = r_j' \dot{\psi}$ $r_j' = \dfrac{dr_j}{d\psi}$ (dimensionless, length/length)

Fig. 3.20 Rigid body 2 in planar motion.

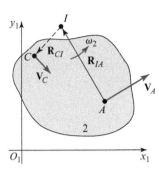

axis exists that is common to both bodies and about which each body is rotating and translating with respect to the other.[8] The study here is restricted to planar motion; therefore, each axis is perpendicular to the plane of the motion and reduces to a pair of coincident points in the plane. We shall refer to these points as *instantaneous centers of velocity*, henceforth referred to as *instant centers*. (Some texts prefer the name *velocity poles*; see the footnote in Section 3.20.)

The instantaneous center of velocity is defined as *the instantaneous location of a pair of coincident points of two different rigid bodies in planar motion for which the absolute velocities of the two points are equal*. The instantaneous center of velocity may also be defined as the location of a pair of coincident points of two different rigid bodies in planar motion for which the apparent velocity of one of the points is zero, as seen by an observer on the other body.

This chapter only considers instant centers for mechanisms with mobility $m = 1$. However, instant centers can also be used for velocity analysis of mechanisms with mobility greater than 1 (Chapter 5).

For planar motion, an instant center between two rigid bodies can be regarded as a pair of coincident points, one attached to each body, about which one body may have a rotational velocity, but no translational velocity, with respect to the other. This property is true only instantaneously, and the location of a new pair of coincident points becomes the instant center at the next instant. It is not correct, therefore, to speak of an instant center as the center of rotation, since it is generally not located at the center of curvature of the apparent locus that a point of one body generates with respect to the coordinate system of the other. Even with this restriction, however, we will find that instant centers contribute substantially to our understanding of the kinematics of planar motion.

Let us consider the rigid body 2, shown in Fig. 3.20, having an arbitrary planar motion with respect to the x_1y_1 plane (body 1); the motion might be translation, rotation, or a combination of both. Suppose that a chosen point A of the body has a known velocity $\mathbf{V}_A$ and that the body has a known angular velocity ω_2. With these two quantities known, the velocity of any other point of the body, such as point C, can be found from the velocity-difference equation, Eq. (3.4):

$$\mathbf{V}_C = \mathbf{V}_A + \mathbf{V}_{CA} = \mathbf{V}_A + \omega_2 \times \mathbf{R}_{CA}. \qquad (a)$$

---

[8] For three-dimensional motion, this axis is referred to as the *instantaneous screw axis*. The classic work covering the theory of screws is [1].

Suppose we now choose a point I of body 2 whose position-difference from point A is

$$\mathbf{R}_{IA} = \frac{\boldsymbol{\omega}_2 \times \mathbf{V}_A}{\omega_2^2}. \tag{3.17}$$

As shown by the vector cross-product, point I is located on the perpendicular to $\mathbf{V}_A$, and vector $\mathbf{R}_{IA}$ is rotated from the direction of $\mathbf{V}_A$ in the direction of $\boldsymbol{\omega}_2$, as shown in Fig. 3.20. The magnitude of $\mathbf{R}_{IA}$ can be calculated from Eq. (3.17), and point I can be located. The velocity of point I can then be written from Eq. (a) as

$$\mathbf{V}_I = \mathbf{V}_A + \mathbf{V}_{IA} = \mathbf{V}_A + \boldsymbol{\omega}_2 \times \mathbf{R}_{IA} = \mathbf{V}_A + \boldsymbol{\omega}_2 \times \frac{\boldsymbol{\omega}_2 \times \mathbf{V}_A}{\omega_2^2}. \tag{b}$$

Recalling the vector triple product identity $\boldsymbol{a} \times (\boldsymbol{b} \times \boldsymbol{c}) = \boldsymbol{b}(\boldsymbol{c} \cdot \boldsymbol{a}) - \boldsymbol{c}(\boldsymbol{a} \cdot \boldsymbol{b})$ and substituting this identity into Eq. (b) gives

$$\mathbf{V}_I = \mathbf{V}_A + \frac{\boldsymbol{\omega}_2 \overbrace{(\mathbf{V}_A \cdot \boldsymbol{\omega}_2)}^{0} - \mathbf{V}_A \overbrace{(\boldsymbol{\omega}_2 \cdot \boldsymbol{\omega}_2)}^{\omega_2^2}}{\omega_2^2} = \mathbf{V}_A - \mathbf{V}_A = \mathbf{0}. \tag{c}$$

Since the absolute velocity of point I is zero, the same as the velocity of the coincident point of the fixed link, then point I is the instant center of velocity between body 2 and body 1.

The velocity of any other point of the moving body can now be determined using the instant center, point I. Note that the choice of the instant center simplifies the velocity-difference equation. For example, the velocity of point C in Eq. (a) can now be written as

$$\mathbf{V}_C = \cancel{\mathbf{V}_I}^{0} + \mathbf{V}_{CI} = \boldsymbol{\omega}_2 \times \mathbf{R}_{CI}. \tag{3.18}$$

The direction of the velocity of this point is as shown in Fig. 3.20.

An instant center can be located more easily when the absolute velocities of two points are known. In Fig. 3.21a, suppose that the points A and C have known velocities $\mathbf{V}_A$ and $\mathbf{V}_C$, respectively. The perpendiculars to $\mathbf{V}_A$ and $\mathbf{V}_C$ intersect at the instant center, point I. Figure 3.21b shows how to locate instant center I when points A, C, and I happen to fall on the same straight line.

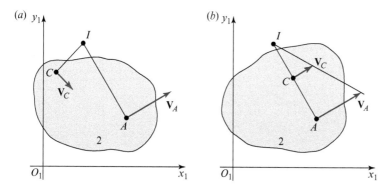

Fig. 3.21 Locating an instant center from two known velocities.

The instant center between two bodies, in general, is not a stationary point. Its location changes with respect to both bodies as the motion progresses and describes a path or locus on each. These paths of the instant centers, called *centrodes*, are discussed in Section 3.20.

Since we have adopted the convention of numbering the links of a mechanism, it is convenient to designate an instant center by using the numbers of the two links associated with it. Thus, I_{32}, for example, identifies the instant center between links 3 and 2. This same instant center can also be identified as I_{23}, since the order of the numbers has no significance. A mechanism has as many instant centers as there are ways of pairing the link numbers. Thus, the number of instant centers in an *n*-link mechanism is

$$N = \frac{n(n-1)}{2}. \tag{3.19}$$

3.13 Aronhold–Kennedy Theorem of Three Centers

Consider the four-bar linkage shown in Fig. 3.22*a*. According to Eq. (3.19), the number of instant centers for this linkage is six. We can identify four of the instant centers (henceforth called *primary instant centers*) by inspection. The four pins are primary instant centers I_{12}, I_{23}, I_{34}, *and* I_{14}, since each satisfies the definition. For example, instant center I_{23} is a point of link 2 about which link 3 appears to rotate; it is a point of link 3 that has no apparent velocity as seen from link 2. This instant center is a pair of coincident points of links 2 and 3 that have the same absolute velocities.

A good method of keeping track of which instant centers have been identified is to arrange the link numbers around the perimeter of a circle (called the *Kennedy circle*), as shown in Fig. 3.22*b*. Then, as each instant center is identified, a solid line is drawn connecting the corresponding pair of link numbers. Figure 3.22*b* shows that I_{12}, I_{23}, I_{34}, and I_{14} (primary instant centers) have been identified, as indicated by the solid lines. The figure also includes two dashed lines indicating that I_{13} and I_{24} (henceforth called *secondary instant centers*) have not yet been identified. The locations of the secondary instant centers cannot be determined simply by applying the definition visually. After locating the primary instant centers by inspection, the secondary instant centers are located by applying the Aronhold–Kennedy theorem of three centers (often just called *Kennedy's theorem*). This theorem states: *Three instant centers shared*

Fig. 3.22 Locations of the primary instant centers.

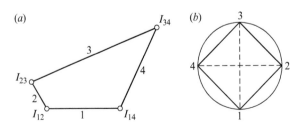

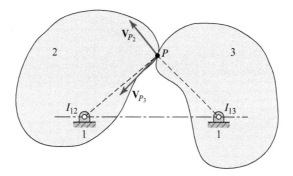

Fig. 3.23 Proof by contradiction of the Aronhold–Kennedy theorem.

by three rigid bodies in motion with respect to one another (whether or not they are connected) all lie on the same straight line.[9]

The theorem can be proven by contradiction, as shown in Fig. 3.23. Link 1 is a stationary frame, and instant center I_{12} is located where link 2 is pin-connected to link 1. Similarly, I_{13} is located at the pin connecting links 1 and 3. The shapes of links 2 and 3 are arbitrary (they can be regarded as infinite planes). The Aronhold–Kennedy theorem states that the three instant centers I_{12}, I_{13}, and I_{23} must all lie on the same straight line (the line connecting the two pins), commonly referred to as the *line of centers*. Let us suppose that this were not true; in fact, let us suppose that I_{23} were located at the point labeled P in Fig. 3.23. Then the velocity of P as a point of link 2 would have the orientation $\mathbf{V}_{P_2}$, perpendicular to $\mathbf{R}_{PI_{12}}$. However, the velocity of P as a point of link 3 would have the orientation $\mathbf{V}_{P_3}$, perpendicular to $\mathbf{R}_{PI_{13}}$. The orientations of these two velocities are inconsistent with the definition that an instant center must have equal absolute velocities as a part of either link. The point P chosen, therefore, cannot be the instant center I_{23}. This same contradiction in the orientations of $\mathbf{V}_{P_2}$ and $\mathbf{V}_{P_3}$ occurs for any location chosen for point P unless the point is chosen on the straight line through the instant centers I_{12} and I_{13}.

3.14 Locating Instantaneous Centers of Velocity

In the previous two sections, we have considered several methods of locating instant centers. They can often be located by inspecting the figure of a mechanism and visually seeking out coincident point pairs that fit the definition, such as pin-joint centers. Also, after these primary instant centers are identified, the secondary instant centers can be determined using the theorem of three centers. Section 3.13 demonstrated that an instant center between a moving body and the fixed link can be identified if the directions of the absolute velocities of two points of the body are known or if the absolute velocity of one point and the angular velocity of the body are

---

[9] The Aronhold–Kennedy theorem is named after its two independent discoverers, S. H. Aronhold, 1872, and A. B. W. Kennedy, 1886. It is usually known as the *Aronhold theorem* in German-speaking countries, and is usually called *Kennedy's theorem* in English-speaking countries.

Fig. 3.24 Instant centers of a disk cam with a flat-faced follower.

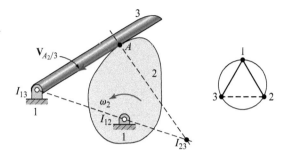

Fig. 3.25 Instant center at a point of rolling contact.

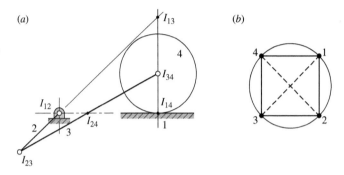

known. The purpose of this section is to expand the list of techniques to include instant centers of mechanisms involving direct contact and instant centers at infinity.

For example, consider the cam-and-follower system of Fig. 3.24. The instant centers I_{12} and I_{13} are primary instant centers, since, by inspection, they are located at the two pin centers. However, the location of the secondary instant center I_{23} (indicated by the dashed line in the Kennedy circle) is not immediately obvious. According to the Aronhold–Kennedy theorem, it must lie on the straight line connecting I_{12} and I_{13}, but where on this line? After some reflection, we see that the orientation of the apparent velocity, $\mathbf{V}_{A_2/3}$, must be along the common tangent to the two moving links at their point of contact. As seen by an observer on link 3, this velocity must result from the apparent relative rotation of body 2 about the instant center I_{23}. Therefore, I_{23} must lie on the line that is perpendicular to $\mathbf{V}_{A_2/3}$. This line now locates I_{23}, as shown in Fig. 3.24. The concept shown in this example should be remembered, since it is often useful in locating the instant centers of mechanisms involving direct contact.

A special case of direct contact, as we have previously noted, is rolling contact with no slip. Considering the mechanism of Fig. 3.25, we can immediately locate the primary instant centers I_{12}, I_{23}, and I_{34}. As demonstrated by the previous example, if the contact between links 1 and 4 involves any slippage, we can only say that instant center I_{14} is located on the vertical line through the point of contact. However, if we also know that there is no slip – that is, *if there is rolling contact – then the instant center is located at the point of contact*. This is also a general principle, as can be seen by comparing the definition of rolling contact, Eq. (3.14), and the

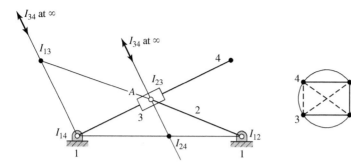

Fig. 3.26 Inverted slider-crank linkage and locations of instant centers.

definition of an instant center; they are equivalent. Therefore, I_{14} is regarded as a primary instant center, as shown in Fig. 3.25b.

Another special case of direct contact is evident between links 3 and 4 of the inverted slider-crank linkage in Fig. 3.26. In this case, there *is* an apparent (slip) velocity, $V_{A_3/4}$, between points A of links 3 and 4, but there is *no apparent rotation between the links*. Here, as in Fig. 3.24, the instant center I_{34} lies along a common perpendicular to the known line of slipping, but now, since there is no apparent rotation, it is located infinitely far away in the direction defined by this perpendicular line. This infinite distance can be demonstrated by considering the kinematic inversion of the mechanism in which link 4 becomes stationary. In this case, Eq. (3.17) for the inverted mechanism can be written as

$$|\mathbf{R}_{I_{34}A}| = \left| \frac{\boldsymbol{\omega}_{3/4} \times \mathbf{V}_{A_3/4}}{\omega_{3/4}^2} \right| = \left| \hat{\boldsymbol{\omega}}_{3/4} \times \left(\frac{\mathbf{V}_{A_3/4}}{\omega_{3/4}} \right) \right| = \infty. \tag{3.20}$$

Since there is no relative rotation between links 3 and 4, the denominator in this equation is zero, and the distance from I_{23} to I_{34} is infinite. The direction of I_{34}, stated earlier, is confirmed by the cross-product in Eq. (3.20).

Note that I_{34} can be regarded as either a primary instant center or a secondary instant center. The remaining secondary instant centers, I_{13} and I_{24}, can be obtained from the Aronhold–Kennedy theorem. To locate I_{13}, a line must be drawn through I_{14} parallel to the line through the instant centers I_{23} and I_{34}. The point of intersection of this line with the line through I_{12} and I_{23} is I_{13}. The point of intersection of the line through I_{23} and I_{34} with the line through I_{14} and I_{12} is I_{24}.

An interesting example that has instant centers at infinity is the *Scotch-yoke* linkage shown in Fig. 3.27, which is a variation of the slider-crank linkage. The four primary instant centers are I_{12}, I_{23}, I_{34}, and I_{14}. The secondary instant center I_{24} can be found directly from the Aronhold–Kennedy theorem. Also, from the same theorem, the secondary instant center I_{13} must lie on the line $I_{12}I_{23}$, but the intersecting line $I_{14}I_{43}$ cannot be drawn in finite space. However, since link 3 is in curvilinear translation (that is, link 3 does not change angle with respect to the fixed link), instant center I_{13} is located at infinity along the known line.

One more example is presented here to reinforce the above concepts.

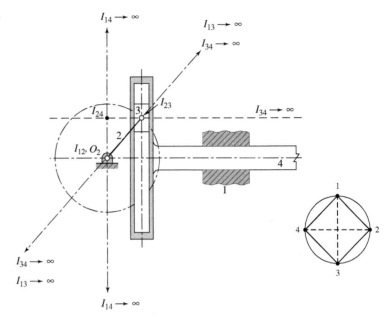

Fig. **3.27** Scotch-yoke linkage and locations of instant centers.

Example 3.11

Locate all instant centers of the five-bar mechanism shown in Fig. 3.28. There is rolling contact between links 1 and 2, and slipping between links 2 and 4 and between links 2 and 5.

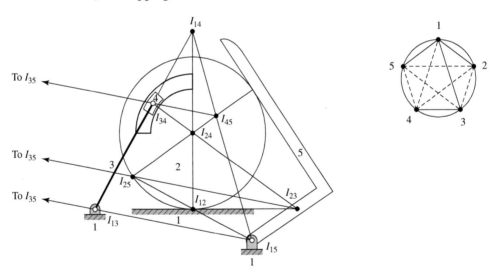

Fig. **3.28** Five-bar mechanism and locations of instant centers.

Solution

According to Eq. (3.19), the number of instant centers for this mechanism is 10. As indicated by the Kennedy circle, there are 4 primary instant centers and 6 secondary instant centers. The three

Example 3.11 (cont.)

pin joints are the primary instant centers I_{13}, I_{34}, and I_{15}, and the point of rolling contact between links 1 and 2 is the primary instant center I_{12}. The instant center I_{24} is regarded as a secondary instant center, since it can be located by drawing perpendicular lines to the directions of the apparent velocities at two of the corners of link 4. However, this instant center could also be regarded as a primary instant center, since it can be obtained from observation, as the center of the apparent rotation between links 2 and 4. One line for the secondary instant center I_{25} is the perpendicular to the direction of slipping between links 2 and 5, and the other line is the line through the instant centers $I_{12}I_{15}$. The locations of the remaining four secondary instant centers I_{23}, I_{35}, I_{45}, and I_{14}, can be found by repeated applications of the theorem of three centers. This is left as an exercise for the reader.

It should be noted before closing this section that in the above examples the locations of instant centers were all found without requiring knowledge of the actual operating speed of the mechanism. This is another indication of the linearity of the equations relating velocities, as indicated in Section 3.9. For a single-degree-of-freedom mechanism, the locations of all instant centers are uniquely determined by geometry alone and do not depend on the operating speed [5].

3.15 Velocity Analysis Using Instant Centers

In this section we demonstrate how the properties of instant centers provide a simple graphic approach to the velocity analysis of planar mechanisms. For example, consider the four-bar linkage in the posture shown in Fig. 3.29a. For a known angular velocity of input link 2, the objective is to determine the velocities of pin B, coupler point D, and point E of link 4.

The procedure is to denote the four pins as the four primary instant centers and then locate the two secondary instant centers by Kennedy's theorem (Section 3.13). Now consider the instant center I_{24}. This instant center, by definition, is a pair of coincident points common to both links 2 and 4 and has the same absolute velocities in both links. Therefore, the velocity of this instant center can be written from Eq. (3.18) as

$$\mathbf{V}_{I_{24}} = \omega_2 \times \mathbf{R}_{I_{24}I_{12}} = \omega_4 \times \mathbf{R}_{I_{24}I_{14}},$$

and is shown on Fig. 3.29b.

Now we can obtain the velocity of any other point of link 4. For example, consider points B' and E', which are on the line of centers $I_{24}I_{12}I_{14}$ and are chosen to have the same radii from I_{14} as pin B and point E. Therefore, their velocities have magnitudes equal to those of $\mathbf{V}_B$ and $\mathbf{V}_E$, respectively, and these can be laid out in their proper directions, as shown in Fig. 3.29b.

To obtain the velocity of the coupler point D, consider the line of centers $I_{23}I_{12}I_{13}$, which is shown in Fig. 3.30. Using the given input angular velocity, ω_2, and the primary instant center I_{12}, we find the absolute velocity of the instant center I_{23}. Here, this step is straightforward since $\mathbf{V}_{I_{23}} = \mathbf{V}_A$. Locating point D' on this line of centers (in the same manner as B' and E',

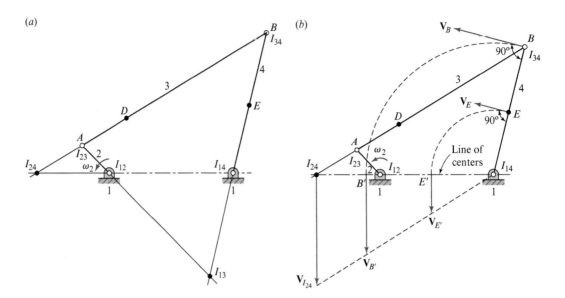

Fig. 3.29 Graphic velocity analysis using instant centers.

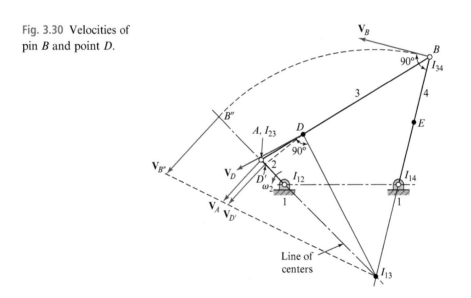

Fig. 3.30 Velocities of pin B and point D.

above), we find $\mathbf{V}_{D'}$ as shown and use its magnitude to draw the desired velocity $\mathbf{V}_D$. Note that, according to its definition, the instant center I_{13}, as a point of link 3, has zero velocity at this instant. Also, note that pin B can be considered a point of link 3; its velocity can be determined in a similar manner by determining $\mathbf{V}_{B''}$, as shown in Fig. 3.30.

The *line-of-centers* method of velocity analysis using instant centers is summarized as follows:

1. Identify the three link numbers associated with the given velocity, the velocity to be determined, and the reference link. The reference link, in general, is link 1, since typically absolute velocity information is given and requested.
2. Locate the three instant centers defined by the links of step 1 and draw the line of centers.
3. Find the velocity of the common instant center by treating it as a point of the link whose velocity is given.
4. With the velocity of the common instant center known, consider it as a point of the link whose velocity is to be determined. The velocity of any other point in that link can now be determined.

The following example illustrates the line-of-centers method and also demonstrates how to treat instant centers at infinity.

Example 3.12

For the device shown in Fig. 3.31, some of the links are visible, whereas other links are enclosed in a housing. However, the location of instant center I_{25} is known, as shown in the figure. If the velocity of point C of link 6 is $\mathbf{V}_C = 10\,\text{m/s}$ to the right, determine the angular velocity of link 2.

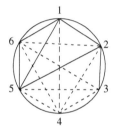

Fig. 3.31 Some links enclosed in a housing.

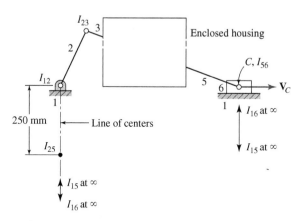

Solution

Since $\mathbf{V}_{C_5/1} = \mathbf{V}_C$, and we are required to find $\omega_{2/1}$, we need to use instant centers I_{15}, I_{12}, and I_{25}. Note that $\omega_{2/1}$ is the same as ω_2, but the additional subscript has been written to emphasize the presence of the third link (the frame). The primary instant center I_{16} is at infinity, as shown. The instant center I_{15} lies on line $I_{12}I_{25}$, and on line $I_{16}I_{56}$, and since these two lines are parallel, then I_{15} is also at infinity.

Example 3.12 (cont.)

Considering I_{25} to be a point of link 5, we wish to find the velocity of this point from the given $\mathbf{V}_C$. We have difficulty in locating point C' on the line of centers at the same radius from I_{15} as point C, since I_{15} is at infinity. How can we proceed?

Recalling the discussion of Section 3.14 and Eq. (3.20), since I_{15} is at infinity, the relative motion between links 5 and 1 is translation, and $\omega_{5/1} = 0$. Therefore, *every* point of link 5 has the same absolute velocity, including $\mathbf{V}_{I_{25}} = \mathbf{V}_C$.

Next, we turn our attention to link 2. We treat I_{25} as a point of link 2 which is rotating about I_{12}, and solve for the angular velocity of link 2:

$$\omega_2 = \frac{V_{I_{25}}}{R_{I_{25}I_{12}}} = \frac{10 \text{ m/s}}{0.25 \text{ m}} = 40 \text{ rad/s ccw.} \qquad\qquad Ans.$$

Noticing the apparent paradox between the directions of $\mathbf{V}_C$ and ω_2, we may speculate on the validity of our solution. This is resolved, however, by opening the enclosed housing and discovering that the linkage is as shown in Fig. 3.32.

Fig. 3.32 Explanation of the velocity paradox.

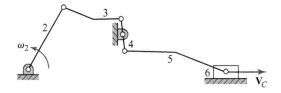

3.16 Angular-Velocity-Ratio Theorem

The instant center I_{24} of the four-bar linkage shown in Fig. 3.33 is a pair of coincident points of links 2 and 4. The absolute velocity $\mathbf{V}_{I_{24}}$ is the same whether I_{24} is considered as a point of link 2 or a point of link 4. Considering it each way, we can write

Fig. 3.33 Angular-velocity-ratio theorem.

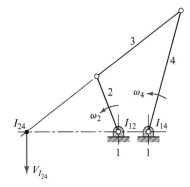

$$\mathbf{V}_{I_{24}} = \mathbf{V}_{I_{12}}^{\mathbf{0}} + \boldsymbol{\omega}_{2/1} \times \mathbf{R}_{I_{24}I_{12}} = \mathbf{V}_{I_{14}}^{\mathbf{0}} + \boldsymbol{\omega}_{4/1} \times \mathbf{R}_{I_{24}I_{14}}. \tag{a}$$

Again, note that $\boldsymbol{\omega}_{2/1}$ and $\boldsymbol{\omega}_{4/1}$ are the same as $\boldsymbol{\omega}_2$ and $\boldsymbol{\omega}_4$, respectively, but the additional subscript is used to emphasize the presence of the third link (the frame).

Considering only the magnitudes of Eq. (a), the equation can be rewritten as

$$\frac{\omega_{4/1}}{\omega_{2/1}} = \frac{R_{I_{24}I_{12}}}{R_{I_{24}I_{14}}}. \tag{b}$$

This shows the *angular-velocity-ratio theorem*, which states that *the angular-velocity ratio of any two bodies in planar motion with respect to a third body is inversely proportional to the segments into which the common instant center cuts the line of centers.* Written in general notation for the motion of two bodies j and k with respect to a third body i, the equation is

$$\frac{\omega_{k/i}}{\omega_{j/i}} = \frac{R_{I_{jk}I_{ji}}}{R_{I_{jk}I_{ki}}}. \tag{3.21}$$

Choosing an arbitrary positive direction along the line of centers, you should prove for yourself that *the angular-velocity ratio is positive when the common instant center falls outside the other two instant centers and negative when it falls between them.*

3.17 Relationships between First-Order Kinematic Coefficients and Instant Centers

First-order kinematic coefficients (Section 3.11) can be expressed in terms of the locations of instant centers. For a planar linkage, the angular velocity of link j can be written from the chain rule as

$$\omega_j = \theta_j' \omega_i, \tag{3.22}$$

where ω_i is the input angular velocity. Therefore, the first-order kinematic coefficient of link j can be written as

$$\theta_j' = \frac{\omega_j}{\omega_i}.$$

Consistent with the angular-velocity-ratio theorem, Eq. (3.21), the first-order kinematic coefficient of link j can be written as

$$\theta_j' = \frac{R_{I_{ij}I_{i1}}}{R_{I_{ij}I_{j1}}}, \tag{3.23}$$

where I_{i1} and I_{j1} are the absolute instant centers of links i and j, respectively, and I_{ij} is the common instant center to links i and j.

Example 3.13

For the four-bar linkage in Example 3.1, the lengths of the input link and the frame can be written as $R_{I_{12}I_{23}} = 4$ in and $R_{I_{12}I_{14}} = 10$ in, respectively. Determine the angular velocities of the coupler link and the output link using instant centers and first-order kinematic coefficients.

Solution

The first-order kinematic coefficients of links 3 and 4 can be written from Eq. (3.23) as

$$\theta_3' = \frac{R_{I_{23}I_{12}}}{R_{I_{23}I_{13}}} \quad \text{and} \quad \theta_4' = \frac{R_{I_{24}I_{12}}}{R_{I_{24}I_{14}}}.$$

From Fig. 3.29 – that is, the Aronhold–Kennedy theorem – we find $R_{I_{23}I_{13}} = 15$ in and $R_{I_{24}I_{12}} = 11.2$ in. Therefore, the first-order kinematic coefficients of the two links are

$$\theta_3' = \frac{4 \text{ in}}{15 \text{ in}} = 0.267 \text{ rad/rad} \quad \text{and} \quad \theta_4' = \frac{11.2 \text{ in}}{21.2 \text{ in}} = 0.528 \text{ rad/rad}.$$

Substituting these values and the input angular velocity $\omega_2 = 94.25$ rad/s ccw into Eq. (3.22), the angular velocity of the coupler link and the output link, respectively, are

$$\omega_3 = 25.16 \text{ rad/s ccw} \quad \text{and} \quad \omega_4 = 49.77 \text{ rad/s ccw}.$$

These answers agree closely with the results obtained from the velocity polygon method of Example 3.1 ($\omega_3 = 25.6$ rad/s ccw and $\omega_4 = 49.64$ rad/s ccw).

Note that Eqs. (3.22) and (3.23), as now written, are not valid for a linkage where link i is a slider. For example, consider a slider-crank linkage with the slider (denoted as link i) regarded as the input. The first-order kinematic coefficient of link j can be written from the chain rule as

$$\theta_j' = \frac{\omega_j}{\dot{r}_i}, \tag{3.24}$$

where $\dot{r}_i$ is the velocity of the slider. When $R_{I_{ij}I_{1i}}$ becomes infinite, Eq. (3.23) for the first-order kinematic coefficient of link j becomes

$$\theta_j' = \frac{\pm 1}{R_{I_{ij}I_{1j}}}, \tag{3.25}$$

where the plus or minus sign depends on the positive direction chosen along the line of centers. Consistent with the statement after Eq. (3.21), the angular-velocity ratio is positive when the relative instant center falls outside the two absolute centers and negative when it falls between them. The following example illustrates Eqs. (3.24) and (3.25).

Example 3.14

For the offset slider-crank linkage in the posture shown in Fig. 3.34, the input velocity of the slider 4 is 10 m/s to the right. Determine the angular velocities of the crank and coupler link using instant centers and first-order kinematic coefficients.

Example 3.14 (cont.)

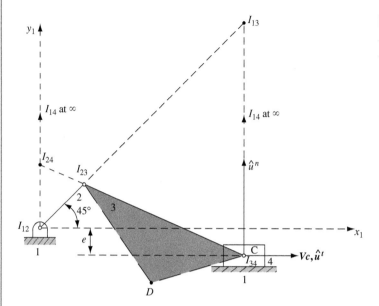

Fig. 3.34 $e = 20$ mm,
$R_{BA} = 50$ mm, $R_{CB} = 140$ mm,
$R_{EB} = 80$ mm, and $R_{DE} = 50$ mm.

Solution

The first-order kinematic coefficients of links 2 and 3 can be written from Eq. (3.25) as

$$\theta_2' = \frac{\pm 1}{R_{I_{24}I_{12}}} \quad \text{and} \quad \theta_3' = \frac{\pm 1}{R_{I_{34}I_{13}}}.$$

We choose the velocity of slider 4 as positive to the right; therefore we have $\hat{u}^t = \hat{i}$, and then, by Eq. (3.7), we have $\hat{u}^n = \hat{u}^b \times \hat{u}^t = \hat{k} \times \hat{i} = \hat{j}$ (upward), as the positive direction along the line of centers.[10] From a scale drawing of the linkage, we measure $R_{I_{24}I_{12}} = +51$ mm and $R_{I_{34}I_{13}} = -185$ mm, and find that the kinematic coefficients are

$$\theta_2' = \frac{\pm 1}{0.051 \text{ m}} = -19.6 \text{ rad/m} \quad \text{and} \quad \theta_3' = \frac{\pm 1}{-0.185 \text{ m}} = +5.41 \text{ rad/m},$$

where θ_2' is negative since I_{24} lies between I_{12} and I_{14} (which is at infinity upward, in the positive $\hat{u}^n$ direction along the line of centers). Similarly, θ_3' is positive since I_{34} is outside of I_{13} and I_{14}.

Rearranging Eq. (3.24), and substituting the kinematic coefficients and the input velocity, $\dot{r}_4 = 10$ m/s, the angular velocities of links 2 and 3, respectively, are

$$\omega_2 = 196 \text{ rad/s cw} \quad \text{and} \quad \omega_3 = 54.1 \text{ rad/s ccw}.$$

---

[10] Note that this sign convention is the same as saying that positive input motion of slider 4 is the same as a positive (counterclockwise) rotation of link 4 about instant center I_{14} now located at infinity in the positive $\hat{u}^n$ direction.

Example 3.14 (cont.)

Allowing for the reversal of the input velocity, these answers agree closely with the results obtained from the graphic method in Example 3.2 ($\omega_2 = 200$ rad/s ccw and $\omega_3 = 53.6$ rad/s cw). Also, since measurements are used for finding the locations of the instant centers, these results are probably still not as accurate as those of Example 3.8 ($\omega_2 = 198.56$ rad/s ccw and $\omega_3 = 54.47$ rad/s cw).

The velocity of a point, say P, fixed in a link of a mechanism can be written as

$$\mathbf{V}_P = V_P \hat{\mathbf{u}}_P^t \tag{3.26}$$

or as

$$\mathbf{V}_P = \left(x_P' \hat{\mathbf{i}} + y_P' \hat{\mathbf{j}} \right) \dot{\psi}, \tag{3.27}$$

where $\dot{\psi}$ is the generalized input velocity to the mechanism. The magnitude of the velocity, commonly referred to as the speed, can be written from the chain rule as

$$V_P = r_P' \dot{\psi}, \tag{3.28}$$

where the first-order kinematic coefficient is defined as

$$r_P' = \pm\sqrt{x_P'^2 + y_P'^2}. \tag{3.29}$$

Here, the sign convention is as follows: We use the positive sign if the instantaneous change in the input position is positive, and we use the negative sign if the instantaneous change in the input position is negative.

Rearranging Eq. (3.26), the unit tangent vector to the point trajectory can be written as

$$\hat{\mathbf{u}}_P^t = \frac{\mathbf{V}_P}{V_P}. \tag{3.30}$$

Then, substituting Eqs. (3.27) and (3.28) into this relation gives

$$\hat{\mathbf{u}}_P^t = \left(\frac{x_P'}{r_P'} \right)\hat{\mathbf{i}} + \left(\frac{y_P'}{r_P'} \right)\hat{\mathbf{j}}. \tag{3.31}$$

Consistent with Eq. (3.7), the unit normal vector to the point trajectory can now be written as $\hat{\mathbf{u}}_P^n = \hat{\mathbf{u}}^b \times \hat{\mathbf{u}}_P^t$. Substituting Eq. (3.31) into this relation, the unit normal vector can be written as

$$\hat{\mathbf{u}}_P^n = \left(\frac{-y_P'}{r_P'} \right)\hat{\mathbf{i}} + \left(\frac{x_P'}{r_P'} \right)\hat{\mathbf{j}}. \tag{3.32}$$

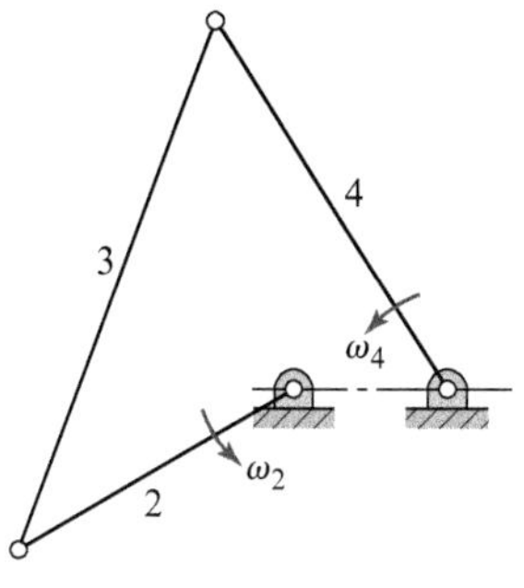

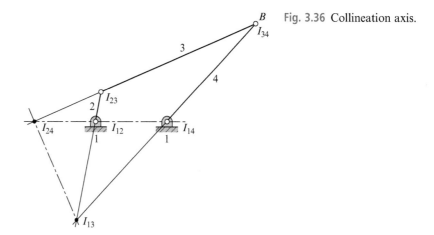

Fig. 3.35 Drag-link four-bar linkage.

Fig. 3.36 Collineation axis.

3.18 Freudenstein's Theorem

In the analysis or design of a linkage, it is sometimes important to know the postures of the linkage at which the extreme values of the output velocity occur, or, more precisely, the input position values at which the ratio of the output and input velocities reaches its extremes.

The earliest work in determining extreme values is apparently that of Krause [9], who stated that the velocity ratio ω_2/ω_4 of the drag-link four-bar linkage (Fig. 3.35) reaches an extreme value when the connecting rod and follower, links 3 and 4, become perpendicular to each other. Rosenauer, however, demonstrated that this is not strictly true [13]. Following Krause, Freudenstein developed a simple graphic method for determining the postures of the four-bar linkage at which the extreme values of the output velocity do occur [6].

Freudenstein's theorem makes use of the line connecting instant centers I_{13} *and* I_{24} (Fig. 3.36), called the *collineation axis*. The theorem states: *At an extreme of the output to input angular-velocity ratio of a four-bar linkage, the collineation axis is perpendicular to the coupler* link.[11]

---

[11] A. S. Hall, Jr. contributed a rigorous proof of this theorem in an appendix to Freudenstein's paper [8].

Using the angular-velocity-ratio theorem, Eq. (3.21), we write

$$\frac{\omega_4}{\omega_2} = \frac{R_{I_{24}I_{12}}}{R_{I_{24}I_{12}} + R_{I_{12}I_{14}}}.$$

Since $R_{I_{12}I_{14}}$ is the fixed length of the frame, the extremes of the velocity ratio occur when $R_{I_{24}I_{12}}$ is either a maximum or a minimum. Such positions may occur on either or both sides of the instant center I_{12}. Thus the problem reduces to finding the geometry of the linkage for which $R_{I_{24}I_{12}}$ is an extremum.

During motion of the linkage, I_{24} travels along the line $I_{12}I_{14}$ as shown by the theorem of three centers, but, at an extreme value of the velocity ratio I_{24} must instantaneously be at rest (its direction of travel on this line must be reversing). This occurs when the velocity of I_{24}, considered a point of link 3, is directed along the coupler link. This is true only when the coupler link is perpendicular to the collineation axis, since I_{13} is the instant center of link 3.

An inversion of the theorem (treating link 2 as fixed) states that an *extreme value of the velocity ratio ω_3/ω_2 of a four-bar linkage occurs when the collineation axis is perpendicular to the follower (link 4).*

3.19   Indices of Merit; Mechanical Advantage

In this section we will study various ratios, and other parameters of mechanisms that tell whether a mechanism is acceptable for a particular application (Section 1.10). Many such parameters have been defined by various authors over the years, and there is no common agreement on a single "index of merit" for all mechanisms. Yet the many used have a number of features in common, including the fact that most can be related to velocity ratios of the mechanism, and, therefore, can be determined solely from its geometry. In addition, most depend on some knowledge of the application of the mechanism, especially the input and output variables. It is often desirable in the analysis or synthesis of a mechanism to plot an index of merit for a revolution of the input crank and to note in particular the minimum and maximum values when evaluating a design or its suitability for a given application.

In Section 1.10 we defined the mechanical advantage of a mechanism as the instantaneous ratio of the output force (or torque) to the input force (or torque). Note that this definition may result in a force divided by a torque ($MA = F_{out}/T_{in}$) or a torque divided by a force ($MA = T_{out}/F_{in}$), depending on the particular application.

In Section 3.16 we learned that the ratio of the angular velocity of the output link to the input velocity of a mechanism is inversely proportional to the segments into which the common instant center cuts the line of centers. Also, Section 3.17 showed that first-order kinematic coefficients can be expressed in terms of the locations of instant centers. Thus, if links 2 and 4 are the input and output links, respectively, of the four-bar linkage shown in Fig. 3.37, then from Eqs. (3.21) and (3.23), the output-to-input angular-velocity ratio is

$$\frac{\omega_4}{\omega_2} = \theta_4' = \frac{R_{I_{12}I_{24}}}{R_{I_{14}I_{24}}}. \qquad (a)$$

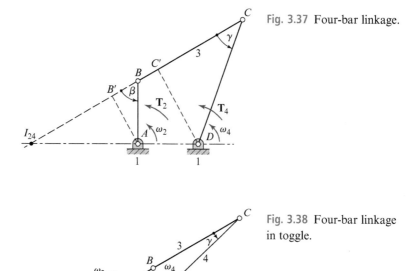

Fig. 3.37 Four-bar linkage.

Fig. 3.38 Four-bar linkage in toggle.

We also learned in Section 3.18 that the extremes of this ratio occur when the collineation axis is perpendicular to the coupler, link 3.

If we now assume that the four-bar linkage of Fig. 3.37 has no friction or inertia forces during its operation or that these are negligible compared with the input torque, T_2, applied to link 2, and the output torque, T_4, the resisting load torque on link 4, then we can derive a relation between T_2 and T_4. Since friction and inertia forces are negligible, the input power applied to link 2 is equal to the power applied to link 4 by the load; hence,

$$T_2\omega_2 = T_4\omega_4. \tag{b}$$

Rearranging this equation, and using Eq. (a), the mechanical advantage is defined as

$$MA = \frac{T_4}{T_2} = \frac{\omega_2}{\omega_4} = \frac{1}{\theta_4'}. \tag{c}$$

This equation shows that the mechanical advantage is the reciprocal of the velocity ratio (first-order kinematic coefficient). Either can be used as an index of merit in judging the ability of a mechanism to transmit force or power.

Substituting Eq. (a) into Eq. (c), the mechanical advantage is

$$MA = \frac{R_{I_{14}I_{24}}}{R_{I_{12}I_{24}}}. \tag{3.33}$$

The four-bar linkage shown in Fig. 3.37 is redrawn in Fig. 3.38 at a posture where links 2 and 3 are aligned. At such a posture, R_{IA} and ω_4 are passing through zero; hence an extreme (infinite) value of the mechanical advantage is obtained.

Recall from Section 1.10 that a mechanism in such a posture is said to be *in toggle*. Such toggle postures are often used to produce a high mechanical advantage; an example is the clamping mechanism of Fig. 2.8. Another practical application is illustrated in the following example of a vise-grip wrench.

Example 3.15

Determine the mechanical advantage of the vise-grip wrench shown in Fig. 3.39.

Fig. 3.39 Vise-grip wrench.

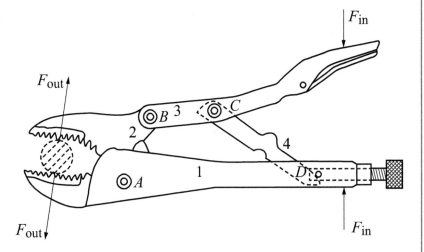

Solution

The vise-grip wrench can be modeled as an adjustable four-bar linkage $ABCD$, as shown in Fig. 3.39. Note that the input is the coupler, link 3, and the output is the jaw, link 2. The lower handle is regarded as fixed in the hand and is denoted as link 1.

If we assume that the four-bar linkage has no friction or inertia forces during its operation or that these are negligible compared with the input torque T_3 applied to link 3, and the output torque T_2 applied to link 2, then we can define the mechanical advantage as

$$MA = \frac{F_{out}}{F_{in}}. \tag{1}$$

The input torque and the output torque can be written, respectively, as

$$T_3 = d_{in}F_{in} \quad \text{and} \quad T_2 = d_{out}F_{out};$$

that is, the input force and the output force are

$$F_{in} = \frac{T_3}{d_{in}} \quad \text{and} \quad F_{out} = \frac{T_2}{d_{out}}. \tag{2}$$

Substituting Eqs. (2) into Eq. (1), the mechanical advantage for the vise-grip wrench can be written as

Example 3.15 (cont.)

$$MA = \left(\frac{d_{\text{in}}}{d_{\text{out}}}\right)\left(\frac{T_2}{T_3}\right). \tag{3}$$

The torque ratio can be written as

$$\frac{T_2}{T_3} = \frac{\omega_3}{\omega_2}. \tag{4}$$

From Eq. (3.22), the angular-velocity ratio can be written as

$$\frac{\omega_3}{\omega_2} = \frac{R_{I_{12}I_{23}}}{R_{I_{13}I_{23}}}. \tag{5}$$

The locations of the instant centers I_{12}, I_{23}, I_{34}, I_{14}, and I_{13} are shown in Fig. 3.40.

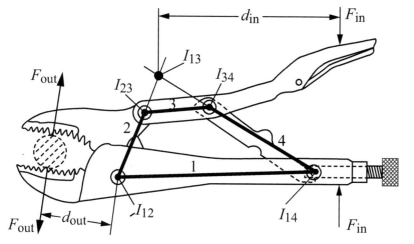

Fig. 3.40 Locations of five of the six instant centers.

Substituting Eq. (5) into Eq. (4), the torque ratio is

$$\left(\frac{T_2}{T_3}\right) = \frac{R_{I_{12}I_{23}}}{R_{I_{13}I_{23}}}. \tag{6}$$

Therefore, substituting Eq. (6) into Eq. (3), the mechanical advantage of the vise-grip wrench is

$$MA = \left(\frac{d_{\text{in}}}{d_{\text{out}}}\right)\left(\frac{R_{I_{12}I_{23}}}{R_{I_{13}I_{23}}}\right). \qquad\qquad Ans.$$

Note that, as the jaws close on an object, the instant center I_{24} approaches the instant center I_{14}. When the three instant centers I_{14}, I_{34}, and I_{23} lie nearly on a straight line, then I_{24} is almost coincident with I_{14} and instant center I_{13} approaches instant center I_{23}. Therefore, the mechanical advantage approaches infinity.

Example 3.15 (cont.)

The screw can be adjusted so that the maximum mechanical advantage occurs at the required distance between the jaws of the vise-grip wrench. Some vise-grips have a stop located close to the dead center posture. This gives both a very high mechanical advantage and a stable grip for the linkage, since it requires an almost infinite force at the jaws to move the vise-grip wrench back through the toggle posture.

Example 3.16

Determine the mechanical advantage for the crimping tool shown in Fig. 3.41.

Fig. 3.41 Crimping tool.

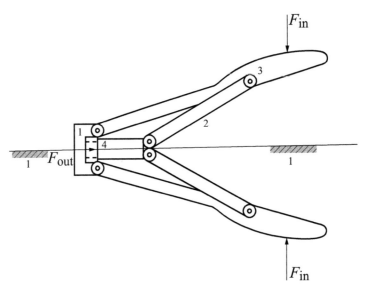

Solution

For this application, mechanical advantage is defined as

$$MA = \frac{F_{out}}{F_{in}},\tag{1}$$

where the input is the coupler, link 3, and the output is the slider, link 4. From conservation of power we can write

$$T_{in}\,\omega_3 = F_{out}\,V_4.\tag{2}$$

The input torque can be written as

$$T_{in} = d_{in}F_{in}.\tag{3}$$

Example 3.16 (cont.)

Substituting Eqs. (2) and (3) into Eq. (1), and simplifying, the mechanical advantage of the crimping tool is

$$MA = d_{in}\left(\frac{\omega_3}{V_{in}}\right). \tag{4}$$

The velocity of the output link, slider 4, can be written as

$$V_4 = R_{I_{13}I_{34}}\omega_3. \tag{5}$$

The locations of the instant centers of the crimping tool are shown in Fig. 3.42.

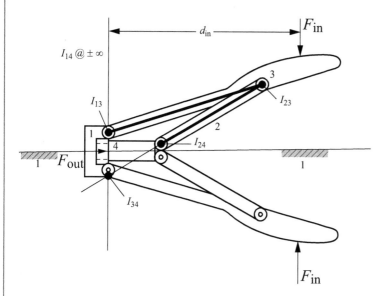

Fig. 3.42 Locations of five of the instant centers.

Substituting Eq. (5) into Eq. (4), and simplifying, the mechanical advantage is

$$MA = \frac{d_{in}}{R_{I_{13}I_{34}}}. \qquad\qquad Ans.$$

From this result we observe that the mechanical advantage of the crimping tool can be improved in the following two ways: (1) increase the distance d_{in}, and/or (2) decrease the distance between instant center I_{34} and instant center I_{13}. The mechanical advantage will become infinite when instant center I_{34} is coincident with instant center I_{13}, that is, when $R_{I_{13}I_{34}} = 0$. In this configuration, links 2 and 3 are aligned.

As further examples, consider any of the reciprocating linkages of Figs. 1.15, 1.16, or 1.17. For each of these, the input is link 2 and requires an input torque, while the output is a sliding link, numbered either 4 or 6, and is intended to produce an output force. Note that, for any of these the mechanical advantage is given by $MA = F_{\text{out}}/T_{\text{in}}$, and has units of reciprocal distance.

Another index of merit that has been proposed [4] is the determinant of the matrix of coefficients of the simultaneous equations relating the dependent velocities of a mechanism. To illustrate this method, consider the four-bar linkage of Example 2.5; in Eq. (11) of that example we saw that the displacements of the dependent posture variables are related by

$$R_{AO_2} \cos\theta_2 + R_{BA} \cos\theta_3 - R_{BA} \sin\theta_3 \Delta\theta_3 - R_{O_4O_2} \cos\theta_1 - R_{BO_4} \cos\theta_4 + R_{BO_4} \sin\theta_4 \Delta\theta_4 = 0,$$

$$R_{AO_2} \sin\theta_2 + R_{BA} \sin\theta_3 + R_{BA} \cos\theta_3 \Delta\theta_3 - R_{O_4O_2} \sin\theta_1 - R_{BO_4} \sin\theta_4 - R_{BO_4} \cos\theta_4 \Delta\theta_4 = 0.$$

The matrix of the coefficients of the dependent variables is called the *Jacobian*. In this example, the determinant of the Jacobian is

$$\Delta = \begin{vmatrix} R_{BA} \sin\theta_3 & -R_{BO_4} \sin\theta_4 \\ R_{BA} \cos\theta_3 & -R_{BO_4} \cos\theta_4 \end{vmatrix} = R_{BA} R_{BO_4} \sin(\theta_4 - \theta_3). \qquad (a)$$

As is clear from Cramer's rule, the solutions for the dependent variables, in this case the corrections $\Delta\theta_3$ and $\Delta\theta_4$, must include this determinant in the denominator.

This same determinant also appears in the denominators of the solutions for the dependent velocities (Section 3.10), the dependent accelerations (Section 4.11), and all other quantities that require derivatives of the loop-closure equation. For example, we saw that the dependent velocities of the four-bar linkage in Example 3.6 are related by

$$(r_3\sin\theta_3)\omega_3 - (r_4\sin\theta_4)\omega_4 = -(r_2\sin\theta_2)\omega_2,$$
$$(r_3\cos\theta_3)\omega_3 - (r_4\cos\theta_4)\omega_4 = -(r_2\cos\theta_2)\omega_2.$$

Here, the determinant of the Jacobian is

$$\Delta = \begin{vmatrix} r_3 \sin\theta_3 & -r_4 \sin\theta_4 \\ r_3 \cos\theta_3 & -r_4 \cos\theta_4 \end{vmatrix} = r_3 r_4 \sin(\theta_4 - \theta_3). \qquad (b)$$

Note that, except for notation changes, the determinants in Eqs. (a) and (b) are identical. Again, the solutions for the dependent velocities, in this case, ω_3 and ω_4, must include this determinant in the denominator. This is borne out in the velocity solution of the four-bar linkage, Eqs. (5) of Example 3.6. If this determinant becomes zero, then the Jacobian is singular and the mechanism is said to be in a singular (toggle) configuration, and the angular velocities of links 3 and 4 tend to infinity. When the determinant of the Jacobian becomes small, the mechanical advantage becomes small and the usefulness of the mechanism is reduced in such a region. The mechanism functions poorly in all respects – force transmission, motion transformation, sensitivity to manufacturing errors, and so on.

Although the form of the Jacobian determinant changes for different mechanisms, such a determinant can always be defined and always appears in the denominators of the derivatives of all dependent variable solutions. For a multi-loop mechanism, there is more than one sub-determinant, as shown by the following example.

Example 3.17

Plot the sub-determinants of the Jacobian of the Watt II six-bar linkage shown in Fig. 3.43 for a complete rotation of input crank 2.

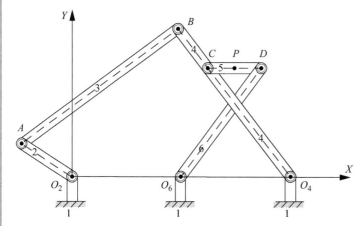

Fig. 3.43 $R_1 = R_{O_4O_2} = 203.20$ mm, $R_2 = R_{AO_2} = 57.15$ mm, $R_3 = R_{BA} = 184.15$ mm, $R_4 = R_{BO_4} = 177.80$ mm, $R_{44} = R_{CO_4} = 127$ mm, $R_5 = R_{DC} = 50.80$ mm, $R_6 = R_{DO_6} = 127$ mm, and $R_{11} = R_{O_6O_4} = 101.60$ mm.

Solution

There are two independent vector loops for the six-bar linkage; therefore, we form two sub-determinants. The vectors for links 1, 2, 3, and 4 are as shown in Fig. 3.44.

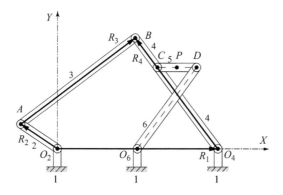

Fig. 3.44 First vector loop.

The determinant of this first coefficient matrix is

$$\Delta_1 = R_3 R_4 \sin(\theta_3 - \theta_4).$$

The plot of the sub-determinant for this first vector loop versus the posture of input link 2 is shown in Fig. 3.45.

We observe that this determinant is never zero (that is, this matrix cannot be singular); therefore, this vector loop can never be in a singular (or toggle) posture, and the angular velocities of links 3 and 4 are always finite.

The second vector loop, including links 1, 4, 5, and 6 is shown in Fig. 3.46.

Example 3.17 (cont.)

Fig. 3.45 Sub-determinant 1 versus the posture of input link 2.

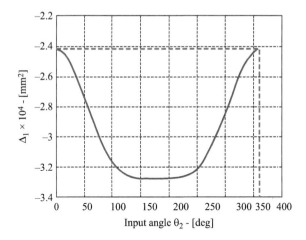

Fig. 3.46 Second vector loop.

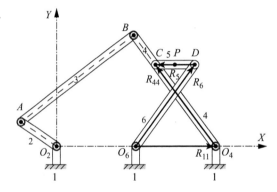

Fig. 3.47 Sub-determinant 2 versus the posture of the input link.

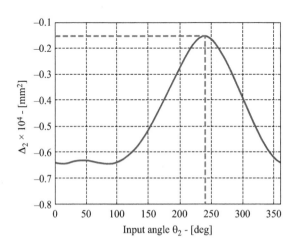

Example 3.17 (cont.)

The sub-determinant of the coefficient matrix for the second vector loop can be written as

$$\Delta_2 = R_5 R_6 \sin(\theta_5 - \theta_6).$$

The plot of this sub-determinant versus the posture of the input link 2 is shown in Fig. 3.47.

Again, we observe that this sub-determinant is never zero (that is, this matrix cannot be singular); therefore, the angular velocities of links 5 and 6 are always finite. The dashed lines in Figs. 3.45 and 3.47 indicate the maximum values of the two sub-determinants and the corresponding postures of the input link.

3.20 Centrodes

We noted in Section 3.12 that the location of an instant center is only defined instantaneously and changes as the mechanism moves. When the changing locations of an instant center are found for all possible postures of a single-degree-of-freedom mechanism, a curve or locus, called a *centrode*,[12] is defined. For example, consider the four-bar linkage shown in Fig. 3.48. The instant center I_{13} (located at the intersection of the extensions of links 2 and 4) traces out a curve, called the *fixed centrode* on link 1, as the linkage moves through all possible postures.

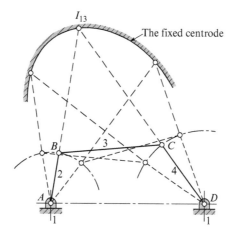

Fig. 3.48 Fixed centrode.

[12] Opinion seems divided on whether these loci should be called *centrodes* or *polodes*. Generally, those preferring the name *instant center* call them *centrodes*, and those preferring the name *velocity pole* call them *polodes*. The French name *roulettes* has also been applied. The three-dimensional equivalents are ruled surfaces and are referred to as *axodes*. Note that there is a subtle distinction between the terms *instant center* and *velocity pole*, which is explained in Section 4.12.

Fig. 3.49 Moving centrode.

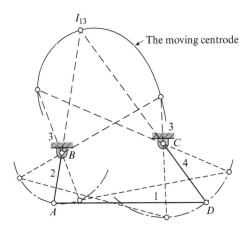

Fig. 3.50 Rolling contact between centrodes.

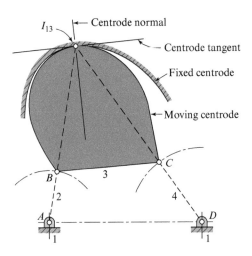

The literature [11] indicates that the fixed centrode of a four-bar linkage is a planar algebraic curve of degree eight.

Figure 3.49 shows the inversion of the same four-bar linkage in which link 3 is fixed and link 1 is movable. When this inversion moves through all possible input positions, I_{13} traces a *different* curve on link 3. For the original linkage, with link 1 fixed, this is the curve traced by I_{13} on the coordinate system of the moving link 3; it is called the *moving centrode*.

Figure 3.50 shows the moving centrode, attached to link 3, and the fixed centrode, attached to link 1. It is imagined here that links 1 and 3 have been machined to the actual shapes of the respective centrodes and that links 2 and 4 have been removed entirely. If the moving centrode is now permitted to roll on the fixed centrode without slip, link 3 has exactly the same motion as it had in the original linkage. This remarkable property, which stems from the fact that a point of rolling contact is an instant center, turns out to be quite useful in the synthesis of linkages.

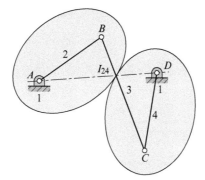

Fig. 3.51 Elliptic gears.

We can restate this property as follows: *The plane motion of one rigid body with respect to another is completely equivalent to the rolling motion of one centrode on the other.* Beyer [2] provides a detailed discussion of centrodes and their fundamental significance in the synthesis of planar mechanisms. The instantaneous point of rolling contact of the centrodes is the instant center, as shown in Fig. 3.50. Also shown are the common tangent to the two centrodes and the common normal, called the *centrode tangent* and the *centrode normal*. These coordinates are sometimes used as the axes of a coordinate system, commonly referred to as the *canonical coordinate system*. This system is used for developing equations of a coupler curve and other geometric properties of the motion, called *instantaneous invariants* [3].

The centrodes of Fig. 3.50 were generated by the instant center I_{13} on links 1 and 3. Another pair of centrodes, both moving, is generated on links 2 and 4 when instant center I_{24} is considered. These two centrodes roll upon each other and describe the identical motion between links 2 and 4 that would result from the operation of the original four-bar linkage. Figure 3.51 shows these centrodes as two ellipses for the case of a crossed double-crank linkage with equal cranks. This construction can be used as the basis for the development of a pair of elliptic gears.

PROBLEMS[13]

3.1 The position vector of a point is given by the equation $\mathbf{R} = 2e^{j\pi t}$, where R is in meters. Find the velocity of the point at $t = 0.40$ s.

3.2 The path of a point is defined by the equation $\mathbf{R} = (t^2 + 160)e^{-j\pi t/10}$, where R is in inches. Find the velocity of the point at $t = 20$ s.

3.3 Automobile A is traveling south at 55 km/h and automobile B is traveling north 60° east at 40 km/h. Find the velocity difference between B and A and the apparent velocity of B to the driver of A.

---

[13] When assigning problems, the instructor may wish to specify the method of solution to be used, because a variety of approaches are presented in the text.

3.4 Wheel 2 rotates at 600 rev/min cw and drives wheel 3 without slipping. Find the velocity difference between points B and A.

Fig. P3.4 $R_{AO_2} = 100$ mm, $R_{AO_3} = 225$ mm, and $R_{BO_3} = 200$ mm.

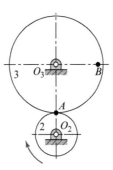

3.5 The distance between points A and B, located along the radius of a wheel, is $R_{BA} = 15$ in. The speeds of points A and B are $V_A = 4\,000$ in/s and $V_B = 7\,000$ in/s, respectively. Find the diameter of the wheel, the velocities $\mathbf{V}_{AB}$ and $\mathbf{V}_{BA}$, and the angular velocity of the wheel.

Fig. P3.5

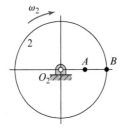

3.6 An airplane takes off from point B and flies east at 350 km/h. Simultaneously, another airplane at point A, 200 km southeast, takes off and flies northeast at 390 km/h. (a) How close do the airplanes come to each other if they fly at the same altitude? (b) If both airplanes leave at 6:00 p.m., at what time does this occur?

Fig. P3.6 $R_{AB} = 200$ km.

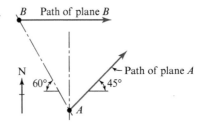

3.7 Include a wind of 30 km/h from the west with the data of Prob. 3.6. (a) If airplane A flies the same heading, what is its new path? (b) What change does the wind make in the results of Prob. 3.6?

3.8 For the double-slider linkage in the posture shown, the velocity of point B is 1 600 in/s. Find the velocity of point A and the angular velocity of link 3.

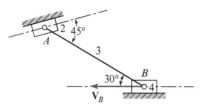

Fig. P3.8 $R_{AB} = 16$ in.

3.9 The four-bar linkage in the posture shown is driven by crank 2 at $\omega_2 = 45$ rad/s ccw. Find the angular velocities of links 3 and 4.

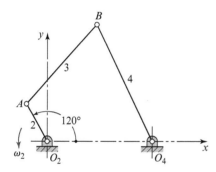

Fig. P3.9 $R_{AO_2} = 100$ mm,
$R_{BA} = 250$ mm,
$R_{O_4O_2} = 250$ mm, and
$R_{BO_4} = 300$ mm.

3.10 The four-bar linkage in the posture shown is driven by crank 2 at $\omega_2 = 60$ rad/s cw. Find the angular velocities of links 3 and 4 and the velocity of pin B and point C on link 3.

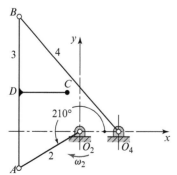

Fig. P3.10 $R_{AO_2} = 6$ in,
$R_{BA} = 12$ in, $R_{O_4O_2} = 3$ in,
$R_{BO_4} = 12$ in, $R_{DA} = 6$ in, and
$R_{CD} = 4$ in.

3.11 The four-bar linkage in the posture shown is driven by crank 2 at $\omega_2 = 48$ rad/s ccw. Find the angular velocity of link 3 and the velocity of point C on link 4.

Fig. P3.11 $R_{AO_2} = 200$ mm,
$R_{BA} = 800$ mm,
$R_{O_4O_2} = 400$ mm,
$R_{BO_4} = 400$ mm, and
$R_{CO_4} = 300$ mm.

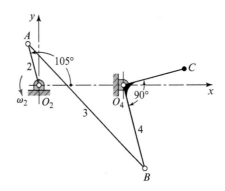

3.12 For the parallelogram four-bar linkage, demonstrate that ω_3 is always zero and that $\omega_4 = \omega_2$. How would you describe the motion of link 4 with respect to link 2?

Fig. P3.12

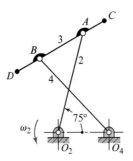

3.13 The antiparallel ogram, or crossed, four-bar linkage in the posture shown is driven by link 2 at $\omega_2 = 1$ rad/s ccw. Find the velocities of points C and D.

Fig. P3.13 $R_{AO_2} = R_{BO_4} = 12$ in,
$R_{BA} = R_{O_4O_2} = 6$ in, and
$R_{CA} = R_{DB} = 3$ in.

3.14 For the four-bar linkage in the posture shown, link 2 has an angular velocity of 60 rad/s ccw. Find the angular velocities of links 3 and 4 and the velocity of point C.

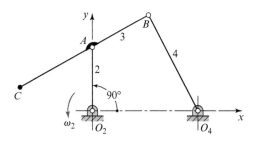

Fig. P3.14 $R_{AO_2} = R_{BA} = 150$ mm, $R_{O_4O_2} = R_{BO_4} = 250$ mm, and $R_{CA} = 200$ mm.

3.15 Crank 2 of the inverted slider-crank linkage, in the posture shown, is driven at $\omega_2 = 60$ rad/s ccw. Find the angular velocities of links 3 and 4 and the velocity of point B.

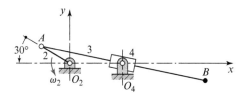

Fig. P3.15 $R_{AO_2} = 3$ in, $R_{BA} = 16$ in, and $R_{O_4O_2} = 5$ in.

3.16 For the four-bar linkage in the posture shown, crank 2 has an angular velocity of 30 rad/s cw. Find the velocity of the coupler point C and the angular velocities of links 3 and 4.

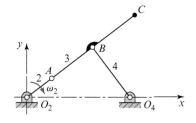

Fig. P3.16 $R_{AO_2} = 75$ mm, $R_{BA} = R_{CB} = 125$ mm, $R_{O_4O_2} = 250$ mm, and $R_{BO_4} = 150$ mm.

3.17 For the modified slider-crank linkage in the posture shown, crank 2 has an angular velocity of 10 rad/s ccw. Find the angular velocity of link 6 and the velocities of points B, C, and D.

Fig. P3.17 $R_{AO_2} = 50$ mm,
$R_{BA} = 200$ mm, $e = 30$ mm,
$R_{CB} = 160$ mm,
$R_{CA} = R_{DC} = 80$ mm,
$R_{O_6O_2} = 160$ mm,
$R_{DO_6} = 120$ mm, and $\theta_2 = 135°$.

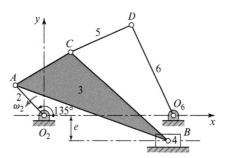

3.18 For the four-bar linkage shown, the angular velocity of crank 2 is a constant 16 rad/s cw. Plot a polar velocity diagram for the velocity of point B for all crank positions. Check the positions of maximum and minimum velocities by using Freudenstein's theorem.

Fig. P3.18 $R_{AO_2} = 14$ in,
$R_{BA} = 17$ in, $R_{O_4O_2} = 4$ in, and
$R_{BO_4} = 16$ in.

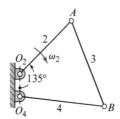

3.19 For the four-bar linkage in the posture shown, link 2 is driven at $\omega_2 = 36$ rad/s cw. Find the angular velocity of link 3 and the velocity of point B.

Fig. P3.19 $R_{AO_2} = 125$ mm,
$R_{BA} = R_{BO_4} = 200$ mm, and
$R_{O_4O_2} = 175$ mm.

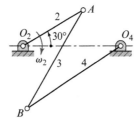

3.20 For the four-bar linkage in the posture shown, the angular velocity of the input link 2 is 8 rad/s ccw. Find the velocity of point C and the angular velocity of link 3.

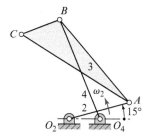

Fig. P3.20 $R_{AO_2} = 6$ in,
$R_{BA} = R_{BO_4} = 10$ in,
$R_{O_4O_2} = 3$ in, $R_{CA} = 12$ in, and
$R_{CB} = 4$ in.

3.21 For the four-bar linkage in the posture shown, link 2 has an angular velocity of 56 rad/s ccw. Find the velocity of point C.

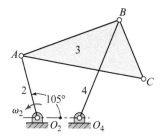

Fig. P3.21 $R_{AO_2} = 6$ in,
$R_{BA} = R_{BO_4} = 10$ in,
$R_{O_4O_2} = 4$ in, and $R_{CA} = 12$ in.

3.22 For the double-slider linkage in the posture shown, the angular velocity of the input crank 2 is 42 rad/s cw. Find the velocities of points B, C, and D.

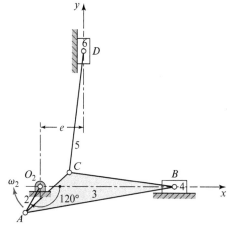

Fig. P3.22 $R_{AO_2} = 50$ mm,
$R_{BA} = 250$ mm, $e = 75$ mm,
$R_{CA} = 100$ mm, $R_{CB} = 175$ mm,
$R_{DC} = 200$ mm, and
$\theta_2 = -120°$.

3.23 For the linkage used in a two-cylinder 60° V-engine consisting, in part, of an articulated connecting rod, crank 2 rotates at 2 000 rev/min cw. Find the velocities of points B, C, and D.

Fig. P3.23 $R_{AO_2} = 50$ mm,
$R_{BA} = R_{CB} = 150$ mm,
$R_{CA} = 50$ mm, and $R_{DC} = 125$ mm.

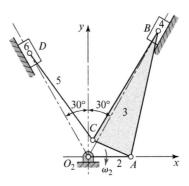

3.24 For the inverted slider-crank linkage in the posture shown, the angular velocity of the crank is $\omega_2 = 24$ rad/s cw. Make a complete velocity analysis of the linkage. What is the absolute velocity of point B? What is its apparent velocity to an observer moving with link 4?

Fig. P3.24 $R_{AO_2} = 200$ mm,
$R_{CA} = R_{DO_4} = 150$ mm,
$R_{BC} = 950$ mm, $R_{O_4O_2} = 500$ mm, and
$\theta_2 = 135°$.

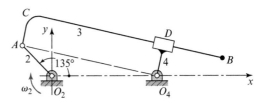

3.25 For the linkage in the posture shown, the velocity of point A is $-0.3\hat{\mathbf{i}}$ m/s. Find the velocity of coupler point B.

Fig. P3.25 $R_{CA} = 50$ mm,
$R_{BC} = 400$ mm, and
$R_A^x = 225$ mm.

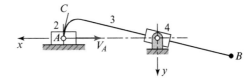

3.26 A variation of the Scotch-yoke linkage in the posture shown is driven by crank 2 at $\omega_2 = 36$ rad/s ccw. Find the velocity of the crosshead, link 4.

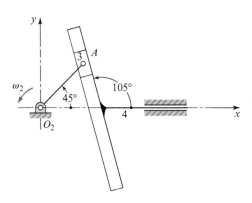

Fig. P3.26 $R_{AO_2} = 10$ in.

3.27 Perform a complete velocity analysis of the modified four-bar linkage for $\omega_2 = 72$ rad/s ccw.

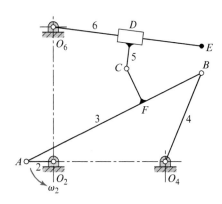

Fig. P3.27 $R_{AO_2} = 30$ mm, $R_{BA} = 210$ mm, $R_{O_4O_2} = 120$ mm, $R_{BO_4} = 100$ mm, $R_{FA} = 140$ mm, $R_{CF} = 40$ mm, $R_{DC} = 30$ mm, $R_{O_6O_2} = 140$ mm, and $R_{EO_6} = 160$ mm.

3.28 For the mechanism in the posture shown, the velocity of point C is $V_C = 0.2$ m/s to the left. There is rolling contact between links 1 and 2, but slip is possible between links 2 and 3. Determine the angular velocity of link 3.

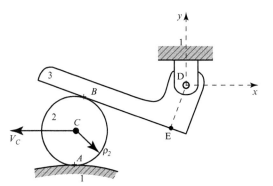

Fig. P3.28 $\rho_1 = 120$ mm, $\rho_2 = 15$ mm, $\mathbf{R}_{AD} = -50\hat{\mathbf{i}} - 35\hat{\mathbf{j}}$ mm, and $R_{ED} = 20$ mm.

3.29 For the circular cam in the posture shown, the angular velocity of the cam is $\omega_2 = 15$ rad/s ccw. There is rolling contact between the cam and the roller, link 3. Find the angular velocity of the oscillating follower, link 4.

Fig. P3.29 $\rho_2 = 40$ mm,
$R_{BA} = 25$ mm, $\rho_3 = 10$ mm,
$R_{DE} = 70$ mm,
$\mathbf{R}_{EA} = 60\hat{\mathbf{i}} + 60\hat{\mathbf{j}}$ mm, and
$\theta_2 = 135°$.

3.30 The mechanism in the posture shown is driven by link 2 at 10 rad/s ccw. There is rolling contact at point F. Determine the velocities of points E and G and the angular velocities of links 3, 4, 5, and 6.

Fig. P3.30 $R_{BA} = 20$ mm,
$R_{CB} = 80$ mm,
$R_{DA} = R_{CD} = 40$ mm,
$R_{EB} = R_{EC} = 50$ mm, $\rho_5 = 10$ mm,
$\mathbf{R}_{HA} = -20\hat{\mathbf{i}} + 30\hat{\mathbf{j}}$ mm, and
$\mathbf{R}_{GH/6} = 60\hat{\mathbf{i}}_6 + 20\hat{\mathbf{j}}_6$ mm,

3.31 The two-piston pump, in the posture shown, is driven by a circular eccentric, link 2, at $\omega_2 = 25$ rad/s ccw. Find the velocities of the two pistons, links 6 and 7.

Fig. P3.31 $\rho_2 = 40$ mm,
$R_{FE} = 20$ mm, $R_{EG} = 120$ mm,
$\mathbf{R}_{CG/3} = 220\hat{\mathbf{i}}_3 - 50\hat{\mathbf{j}}_3$ mm,
$\mathbf{R}_{DG/3} = 220\hat{\mathbf{i}}_3 + 50\hat{\mathbf{j}}_3$ mm,
$R_{AC} = R_{BD} = 120$ mm,
$\mathbf{R}_A = x_A\hat{\mathbf{i}} - 40\hat{\mathbf{j}}$ mm, and
$\mathbf{R}_B = x_B\hat{\mathbf{i}} + 40\hat{\mathbf{j}}$ mm.

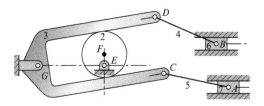

3.32 The epicyclic gear train is driven by the arm, link 2, at $\omega_2 = 10$ rad/s cw. Determine the angular velocity of the output shaft, attached to gear 3.

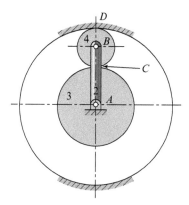

Fig. P3.32 $\rho_1 = R_{DA} = 100$ mm, $R_{BA} = 75$ mm, $\rho_3 = 50$ mm, and $\rho_4 = 25$ mm.

3.33 The diagram shows a planar schematic approximation of an automotive front suspension. The *roll center* is the term used by the industry to describe the point about which the auto body seems to rotate (roll) with respect to the ground. The assumption can be made that there is pivoting but no slip between the tires and the road. After making a sketch, use the concepts of instant centers to find a technique to locate the roll center.

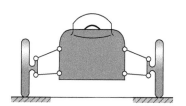

Fig. P3.33

3.34 Locate all instant centers for the linkage of Prob. 3.22.
3.35 Locate all instant centers for the mechanism of Prob. 3.25.
3.36 Locate all instant centers for the mechanism of Prob. 3.26.
3.37 Locate all instant centers for the mechanism of Prob. 3.27.
3.38 Locate all instant centers for the mechanism of Prob. 3.28.
3.39 Locate all instant centers for the mechanism of Prob. 3.29.

3.40 The posture of the input link 2 is $\mathbf{R}_{AO_4} = -6\hat{\mathbf{i}}$ in, and the velocity of point A is $\mathbf{V}_A = 750\hat{\mathbf{i}}$ in/s. Determine the first-order kinematic coefficients for the mechanism. Find the angular velocities of links 3 and 4.

Fig. P3.40 $R_{BO_4} = R_{BA} = 6$ in.

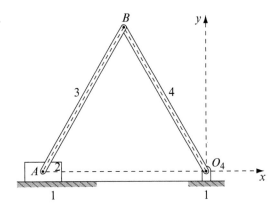

3.41 For the rack-and-pinion mechanism in the posture shown, link 2 is the input, and pinion 3 is rolling without slipping on rack 4 at point D. Determine the first-order kinematic coefficients of links 3 and 4. If the constant input velocity is $\mathbf{V}_G = 75\hat{\mathbf{i}}$ mm/s, determine the angular velocities of rack 4 and pinion 3.

Fig. P3.41 $R_{GO_4} = 250$ mm and $R_{DG} = r_7 = 125$ mm.

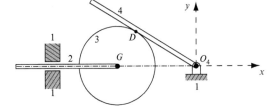

3.42 For the rack-and-pinion mechanism of Example 2.8 (Figs. 2.33 and 2.34), the dimensions are $R_1 = 32$ in, $R_9 = 22$ in, $\theta_{34} = 60°$, and $\rho_3 = 20$ in. In the posture where $R_2 = 30$ in, input link 2 has a velocity of $\mathbf{V}_A = 6\hat{\mathbf{j}}$ in/s. Determine the first-order kinematic coefficients to obtain the velocity of rack 4 and the angular velocity of pinion 3.

3.43 For the mechanism in the posture shown, $R_{AO_4} = 250$ mm, and the input velocity is $\mathbf{V}_A = -125\hat{\mathbf{i}}$ mm/s. Determine the first-order kinematic coefficients to obtain the angular velocity of link 3 and the slipping velocity between links 3 and 4.

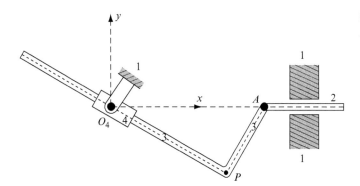

Fig. P3.43 $R_{PA} = 125$ mm and $\angle APO_4 = 90°$.

3.44 For the mechanism in the posture shown in Prob. 3.30, input link 2 has an angular velocity $\omega_2 = 10$ rad/s ccw, and there is rolling contact between links 5 and 6 at point F. Determine the first-order kinematic coefficients of links 3, 4, 5, and 6. Find the angular velocities of links 3, 4, 5, and 6, and the velocities of points E and G.

3.45 For the mechanism in the posture shown, where $R_{AO_4} = 2$ in, pinion 4 is rolling without slipping on rack 3 at point B. Determine the first-order kinematic coefficients of rack 3 and pinion 4. If $\mathbf{V}_A = 7.5\hat{\mathbf{j}}$ in/s, determine the angular velocities of rack 3 and pinion 4, and the velocity of point E. Also, determine the velocity along rack 3 of the point of contact between links 3 and 4 (that is, point B).

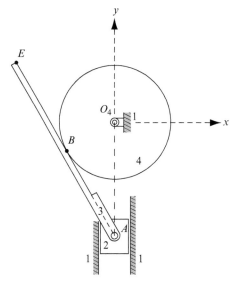

Fig. P3.45 $\rho_4 = 1$ in and $R_{EB} = R_{BA}$.

3.46 For the mechanism in the posture shown, where $\theta_2 = 150°$, $R_{PA} = R_{AO_4}$, and $R_{PB} = R_{BA}$, determine the first-order kinematic coefficients of links 3, 4, and 5. If the angular velocity of the input link 2 is $\omega_2 = 5$ rad/s cw, determine: (a) the angular velocities of links 3 and 4, (b) the velocity of link 5, and (c) the velocity of point P fixed in link 4.

Fig. P3.46 $R_{AO_2} = 250$ mm and $R_{PO_4} = 500$ mm.

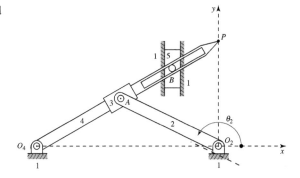

3.47 For the inverted slider-crank linkage in the posture shown, where $\theta_4 = 60°$, the input link 2 is moving parallel to the x axis. Determine the first-order kinematic coefficients of links 3 and 4. Also, determine the conditions for the determinant of the coefficient matrix to become zero. If $V_B = 0.3$ m/s constant to the right, determine the angular velocities of links 3 and 4.

Fig. P3.47 $R_{AO_4} = R_{BA} = 80$ mm.

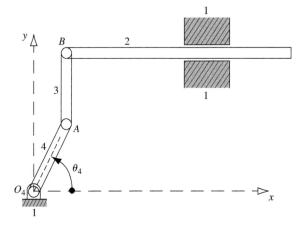

3.48 For the rack-and-pinion mechanism in the posture shown in Prob. 2.15, the input link 2 is vertical and $\angle BAO_2$ is $150°$. The dimensions are $\rho_5 = 50$ mm, $\mathbf{R}_{O_5O_2} = -160\hat{\mathbf{i}} + 80\hat{\mathbf{j}}$ mm, $R_{AO_2} = 40$ mm, and $R_{BA} = 120$ mm. (a) Show the locations of all instant centers. (b) Using instant centers, determine the first-order kinematic coefficients of link 3, rack 4, and pinion 5. (c) If $\omega_2 = 10$ rad/s cw, determine the angular velocity of link 3, the velocity of rack 4, and the angular velocity of pinion 5.

3.49 For the mechanism in the posture shown in Prob. 2.16, $\rho_2 = 25$ mm, $\rho_5 = 50$ mm, $R_{BA} = 176.8$ mm, and $R_{BC} = 150$ mm. Determine the first-order kinematic coefficients of links 3, 4, and 5. If link 2 is driven at $\omega_2 = 5$ rad/s ccw, determine the angular velocities of links 3, 4, and 5.

3.50 For the mechanism in the posture shown in Prob. 2.17, link 4 is parallel to the x axis and link 5 is coincident with the y axis. The radius of wheel 3 is $\rho_3 = 15$ mm, $R_{O_2 O_5} = 140$ mm, $R_{BA} = 110$ mm, and $R_{AO_5} = 52$ mm. Determine the first-order kinematic coefficients of links 3, 4, and 5. If the input link 2 has an angular velocity of $\omega_2 = 15$ rad/s cw, determine the angular velocities of links 3, 4, and 5.

3.51 For the mechanism in the posture shown, wheel 3 is rolling without slipping on the ground link at point C while sliding in the slot in link 2. Write the vector loop equation and determine the first-order kinematic coefficients of the mechanism. If the angular velocity of the input is $\omega_2 = 30$ rad/s ccw, determine the angular velocity of the wheel and the apparent velocity of the center of the wheel, point A, with respect to the slot in link 2.

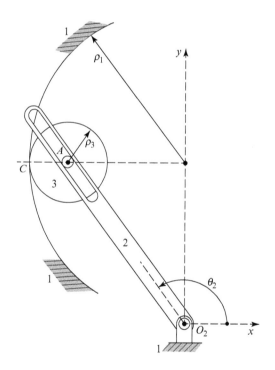

Fig. P3.51 $\rho_1 = 3$ in, $\rho_3 = 0.75$ in,
$\mathbf{R}_{CO_2} = -3\hat{\mathbf{i}} + 3\hat{\mathbf{j}}$ in, and $R_{AO_2} = 3.75$ in.

3.52 Determine the first-order kinematic coefficients for the linkage in the posture shown in Prob. 2.38. If the constant angular velocity of link 2 is $\omega_2 = 30$ rad/s ccw, determine the angular velocities of links 3 and 4.

3.53 Determine the first-order kinematic coefficients for the mechanism in the posture shown in Prob. 2.39. If the velocity of link 2 is $V_{A_2} = 12$ in/s down the inclined plane, determine (a) the angular velocity of link 3, (b) the apparent velocity of pin A_2 with respect to the slot in link 3, and (c) the velocity of point B.

3.54 For the mechanism in the posture shown in Prob. 2.40, determine the first-order kinematic coefficients of gear 3 and the rack. Determine the angular velocities of gear 3 and link 4, and the velocity of the rack if the angular velocity of gear 2 is $\omega_2 = 77$ rad/s ccw.

3.55 For the mechanism in the posture shown in Prob. 2.41, determine the first-order kinematic coefficients of gears 3, 4, and 5. If the angular velocity $\omega_2 = 15$ rad/s cw, determine the angular velocities of gears 3, 4, and 5.

3.56 For the mechanism in the posture shown in Prob. 2.42, the wheel (link 5) is rolling without slipping on the circular ground link. Determine the first-order kinematic coefficients of the mechanism. If the input link has a constant velocity of $\mathbf{V}_{A_2} = 250\mathbf{i}$ in/s, determine the angular velocities of links 3, 4, and 5.

3.57 Determine the first-order kinematic coefficients of the mechanism in the posture shown in Prob. 2.43 assuming no slip at point B. If the velocity of input link 2 is $V_{A_2} = 120$ mm/s constant upward, determine the angular velocity of the pinion and the velocity of the rack.

3.58 For the mechanism in the posture shown, determine the first-order kinematic coefficients of links 3, 4, 5, and 6. If the constant input velocity is $\mathbf{V}_{A_2} = 4.5\hat{\mathbf{j}}$ in/s, determine the angular velocities of links 3 and 5, and the velocity of point P.

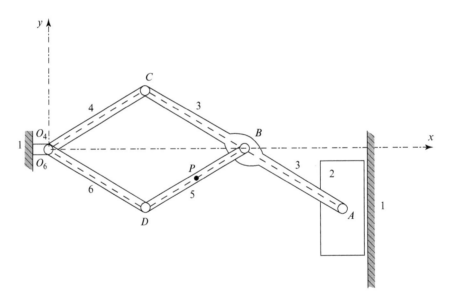

Fig. P3.58 $R_{CO_4} = R_{DO_6} = R_{BC} = R_{BD} = R_{AB} = 2$ in, $R_{PD} = 1$ in, and $\mathbf{R}_A = (5.196 \text{ in})\hat{\mathbf{i}} + y_A\hat{\mathbf{j}}$.

3.59 For the mechanism in the posture shown in Prob. 2.44, link AB is vertical and link CD is horizontal. Determine the first-order kinematic coefficients of the mechanism. If the angular velocity of the input link 2 is $\omega_2 = 10$ rad/s cw, determine the angular velocities of links 3 and 4, and the velocity of point D fixed in link 4 with respect to point C fixed in link 3.

3.60 For the mechanism in the posture shown, gear 3 rolls without slipping on link 4 at point C. Determine the first-order kinematic coefficients of the mechanism. If the velocity of input link 2 is a constant $\mathbf{V}_B = -20\hat{\mathbf{j}}$ in/s, determine the angular velocity of gear 3 and the velocity of link 4.

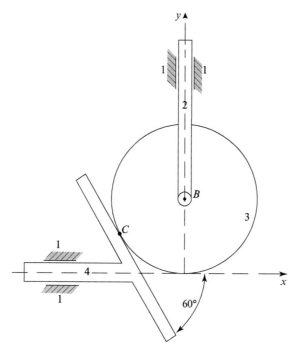

Fig. P3.60 $\rho_3 = 1$ in.

3.61 Determine the first-order kinematic coefficients of the mechanism and the coupler point C. If the input link 2 has a constant angular velocity $\omega_2 = 15$ rad/s ccw, then determine the velocity of point C.

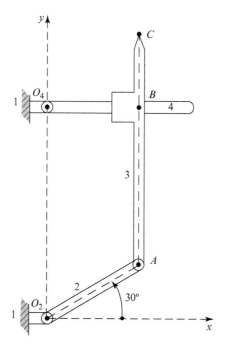

Fig. P3.61 $R_{O_4O_2} = 300$ mm, $R_{AO_2} = 150$ mm, and $R_{CA} = 325$ mm.

REFERENCES

[1] Ball, R. S., 1900. *A Treatise on the Theory of Screws*, Cambridge: Cambridge University Press; reprinted in paperback, 1998.

[2] Beyer, R., 1963. *The Kinematic Synthesis of Mechanisms*, translated by H. Kuenzel, New York: McGraw-Hill.

[3] Bottema, O., and B. Roth, 1979. *Theoretical Kinematics*, Amsterdam: North-Holland Publishing Co.

[4] Denavit, J., R. S. Hartenberg, R. Razi, and J. J. Uicker, Jr., 1965. Velocity, acceleration, and static force analysis of spatial linkages, *J. Appl. Mech. ASME Trans. E* **87**:903–910.

[5] Foster, D. E., and G. R. Pennock, 2005. Graphical methods to locate the secondary instant centers of single-degree-of-freedom indeterminate linkages, *J. Mech. Des. ASME Trans.* **127**(2):249–256.

[6] Freudenstein, F., 1956. On the maximum and minimum velocities and accelerations in four-link mechanisms. *J. Appl. Mech. ASME Trans.* **78**:779–787.

[7] Hain, K., 1967. *Applied Kinematics*, 2nd ed., New York: McGraw-Hill.

[8] Hall, A. S., Jr., 1981. *Notes on Mechanism Analysis*, Chicago: Waveland Press.

[9] Krause, R., 1939. Die Doppelkurbel und Ihre Geschwindigkeitssgrenzen, *Maschinenbau/Getriebetechnik* **18**:37–41; Zur Synthese der Doppelkurbel, *Maschinenbau/Getriebetechnik* **18**:93–94.

[10] Mehmke, R., 1883. Über die Geschwindigkeiten beliebiger Ordnung eines in seiner Ebene bewegten ähnlich veränderlichen Systems, *Civilingenieur*, **487**.

[11] Müller, R., 1903. Über einige Kurven, die mit der Theorie des ebenen Gelenkvierecks im Zusammenhang stehen. *Zeitschrift für Mathematik und Physik*. **48**:224–248. Translated: 1962. *Kansas State University, Bulletin* **46**, No. 6. Special Report No. 21, 217–247.

[12] Raven, F. H., 1958. Velocity and acceleration analysis of plane and space mechanisms by means of independent-position equations, *J. Appl. Mech. ASME Trans. E* **80**:1–6.

[13] Rosenauer, N., 1957. Synthesis of drag-link mechanisms for producing nonuniform rotational motion with prescribed reduction ratio limits, *Aus. J. Appl. Sci.*, **8**:1–6.

4   Acceleration

4.1   Definition of Acceleration

Consider a moving point, first observed at location P, where it has a velocity, $\mathbf{V}_P$, as shown in Fig. 4.1a. After a short time interval, Δt, the point is observed to have moved along a path to a new location, P', where it has a velocity $\mathbf{V}'_P$, which may be different from $\mathbf{V}_P$ in both magnitude and direction. The change in the velocity of the point between the two positions can be evaluated as shown in the velocity polygon, Fig. 4.1b; namely,

$$\Delta\mathbf{V}_P = \mathbf{V}'_P - \mathbf{V}_P.$$

The *average acceleration* of point P during the given time interval is defined as $\Delta\mathbf{V}_P/\Delta t$. The *instantaneous acceleration* (henceforth, referred to simply as the *acceleration*) of point P is defined as the time rate of change of its velocity – that is, the limit of the average acceleration for an infinitesimal time interval:

$$\mathbf{A}_P = \lim_{\Delta t \to 0}\left(\frac{\Delta\mathbf{V}_P}{\Delta t}\right) = \frac{d\mathbf{V}_P}{dt} = \frac{d^2\mathbf{R}_P}{dt^2}. \tag{4.1}$$

Since velocity is a vector quantity, the change in velocity, $\Delta\mathbf{V}_P$, and the acceleration, $\mathbf{A}_P$, are also vector quantities – that is, they have both magnitude and direction. Also, like velocity, the acceleration vector is properly defined only for a point; the term should not be applied to a line, a coordinate system, a volume, or any other collection of points, since the accelerations of different points may be different.

As in the case of velocity (Chapter 3), the acceleration of a moving point may appear differently to different observers. Acceleration does not depend on the actual location of the observer but does depend on the motion of the observer or, more precisely, on the motion of the observer's coordinate system. If acceleration is sensed by an observer using the absolute coordinate system, it is referred to as *absolute acceleration* and is denoted by the symbol $\mathbf{A}_{P/1}$ or simply $\mathbf{A}_P$, which is consistent with the notation used for position, displacement, and velocity.

Recall that the velocity of point P, moving along the path AB (Fig. 4.2a) can be written from Eq. (3.9) as

$$\mathbf{V}_P = \dot{s}\hat{\mathbf{u}}^t, \tag{a}$$

203

Fig. 4.1 (*a*) Change in velocity of a moving point; (*b*) velocity polygon.

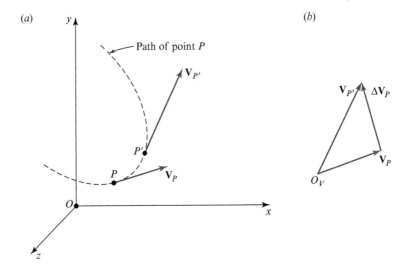

Fig. 4.2 (*a*) Motion of point *P* generates path *AB*; (*b*) change in unit tangent vector.

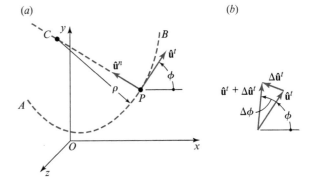

where $\dot{s}$ is the instantaneous speed of *P* along the path, and $\hat{\mathbf{u}}^t$ is the unit vector tangent to the path and with positive sense in the direction of positive displacement Δs (Section 3.5) of point *P*. We identify the *osculating plane*, defined by $\hat{\mathbf{u}}^t$ and the instantaneous *center of curvature* of the path of *P* (denoted as *C*). Designating the preferred positive side of the osculating plane by the *binormal* unit vector, $\hat{\mathbf{u}}^b$, we complete the right-hand vector triad $\hat{\mathbf{u}}^b\hat{\mathbf{u}}^t\hat{\mathbf{u}}^n$ by defining the *unit normal* vector, $\hat{\mathbf{u}}^n = \hat{\mathbf{u}}^b \times \hat{\mathbf{u}}^t$, consistent with Eq. (3.7). Thus, $\hat{\mathbf{u}}^t$ and $\hat{\mathbf{u}}^n$ are tangent and normal, respectively, to the path at the instantaneous position of point *P*.

The acceleration of point *P* is obtained by differentiating Eq. (*a*) with respect to time; that is

$$\mathbf{A}_P = \dot{s}\dot{\hat{\mathbf{u}}}^t + \ddot{s}\,\hat{\mathbf{u}}^t, \tag{b}$$

where $\ddot{s}$ is the instantaneous time rate of change in the speed of *P* along the path. The first term on the right-hand side of this equation can be written in the form

$$\dot{s}\dot{\hat{\mathbf{u}}}^t = \frac{ds}{dt}\left(\frac{d\hat{\mathbf{u}}^t}{d\phi}\frac{d\phi}{ds}\frac{ds}{dt}\right), \tag{c}$$

where ϕ represents the inclination angle of the unit tangent vector, $\hat{\mathbf{u}}^t$, with respect to an arbitrary axis selected in the osculating plane (Fig. 4.2a shows an axis parallel to the x axis). As point P moves along the path AB, $\hat{\mathbf{u}}^t$ is a function of ϕ, which in turn is a function of the distance, s, along the path.

In Fig. 4.2b, the unit tangent vector $\hat{\mathbf{u}}^t$ changes; the vectors $\hat{\mathbf{u}}^t$ and $\hat{\mathbf{u}}^t + \Delta \hat{\mathbf{u}}^t$ have been transferred to a common origin, and we observe that

$$\lim_{\Delta\phi\to 0}\left(\frac{\Delta\hat{\mathbf{u}}^t}{\Delta\phi}\right) = \frac{d\hat{\mathbf{u}}^t}{d\phi}. \qquad (d)$$

From trigonometry, the left-hand side of this equation can be written as

$$\lim_{\Delta\phi\to 0}\left(\frac{2\sin(\Delta\phi/2)\,\hat{\mathbf{u}}^n}{\Delta\phi}\right) = \hat{\mathbf{u}}^n. \qquad (e)$$

Therefore, equating Eq. (d) with Eq. (e), we have the relation

$$\frac{d\hat{\mathbf{u}}^t}{d\phi} = \hat{\mathbf{u}}^n. \qquad (f)$$

The term $d\phi/ds$ in Eq. (c) is the rate of change of the inclination angle with respect to the change in distance, s, along the path and can be written as

$$\frac{d\phi}{ds} = \kappa = \frac{1}{\rho}, \qquad (g)$$

where κ is called the *curvature* of the path and, as indicated in Eq. (g), is the reciprocal of the *radius of curvature* of the path.

Substituting Eqs. (f) and (g) into Eq. (c), and rearranging, gives

$$\dot{s}\hat{\mathbf{u}}^t = \frac{\dot{s}^2}{\rho}\hat{\mathbf{u}}^n. \qquad (h)$$

Finally, substituting Eq. (h) into Eq. (b), the acceleration of point P can now be written as

$$\mathbf{A}_P = \frac{\dot{s}^2}{\rho}\hat{\mathbf{u}}^n + \ddot{s}\hat{\mathbf{u}}^t. \qquad (4.2)$$

The important observation from Eq. (4.2) is that, in general, the acceleration vector of a point has two perpendicular components: a normal component of magnitude $\dot{s}^2/\rho$ (called normal since it is oriented along the $\hat{\mathbf{u}}^n$ axis),[1] and a tangential component of magnitude $\ddot{s}$ along the $\hat{\mathbf{u}}^t$ axis. Hence, the acceleration of point P can be written as

$$\mathbf{A}_P = \mathbf{A}_P^n + \mathbf{A}_P^t, \qquad (4.3)$$

---

[1] The reader should verify that the normal component of acceleration is always directed toward the center of curvature of the path, no matter which orientations are positive for $\hat{\mathbf{u}}^t$ or $\hat{\mathbf{u}}^n$.

where the superscript ^{n} designates the normal direction and the superscript ^{t} designates the tangential direction. Note that the normal component of acceleration is commonly referred to as centripetal acceleration.

4.2   Angular Acceleration

The previous section demonstrated that the acceleration of a point is a vector quantity having a magnitude and a direction. But a point has no dimensions (Section 2.1), so we cannot speak of the angular acceleration of a point. Rather, *rectilinear acceleration*, or simply "acceleration," *deals with the motion of a point*, whereas *angular acceleration deals with the motion of a rigid body*.

Suppose a rigid body has an angular velocity ω at one instant in time, and an instant later has an angular velocity ω'. The difference,

$$\Delta\omega = \omega' - \omega, \qquad (a)$$

is also a vector quantity. The angular velocities ω and ω' may have different magnitudes as well as different directions. Thus, we define *angular acceleration* as *the time rate of change of the angular velocity of a rigid body* and designate it by the symbol $\boldsymbol{\alpha}$; that is,

$$\boldsymbol{\alpha} = \lim_{\Delta t \to 0} \left(\frac{\Delta\omega}{\Delta t}\right) = \frac{d\omega}{dt} = \dot{\omega}. \qquad (4.4)$$

As is the case with $\Delta\omega$, there is no reason to believe that $\boldsymbol{\alpha}$ has the same direction as either ω or ω'; it may have an entirely new direction.

We further note that the angular acceleration vector, $\boldsymbol{\alpha}$, applies to the absolute rotation of the entire rigid body and hence may be subscripted by the number of the coordinate system of the rigid body (for example, $\boldsymbol{\alpha}_2$ or $\boldsymbol{\alpha}_{2/1}$).

4.3   Acceleration Difference between Points of a Rigid Body

In Section 3.3, we determined the velocity difference between two points of a rigid body moving with both translation and rotation. Also, we showed that the velocity of a point in a rigid body can be obtained as the sum of the velocity of a reference point of the body and a term, called the *velocity difference*, caused by the angular velocity of the body. Thus, the velocity of any point, P, in a rigid body can be obtained from the velocity-difference equation, Eq. (3.4); that is,

$$\mathbf{V}_P = \mathbf{V}_Q + \mathbf{V}_{PQ}, \qquad (a)$$

where $\mathbf{V}_Q$ is the velocity of the reference point, Q, and $\mathbf{V}_{PQ}$ is the velocity difference and is given by

$$\mathbf{V}_{PQ} = \omega \times \mathbf{R}_{PQ}. \qquad (b)$$

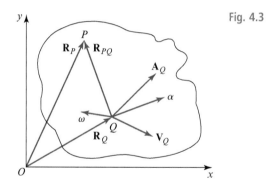

Fig. 4.3

Here, $\boldsymbol{\omega}$ is the angular velocity of the body and $\mathbf{R}_{PQ}$ is the position-difference vector that defines the position of point P with respect to point Q. Substituting Eq. (b) into Eq. (a) gives

$$\mathbf{V}_P = \mathbf{V}_Q + \boldsymbol{\omega} \times \mathbf{R}_{PQ}. \tag{c}$$

We employ a similar nomenclature in Fig. 4.3 and specify that a reference point, Q, of the rigid body has an acceleration $\mathbf{A}_Q$ and that the body may have an angular acceleration $\boldsymbol{\alpha}$ in addition to its angular velocity $\boldsymbol{\omega}$. Note that, in general, $\boldsymbol{\alpha}$ need not have the same orientation as $\boldsymbol{\omega}$.

The acceleration of point P is obtained by differentiating Eq. (c) with respect to time; that is

$$\dot{\mathbf{V}}_P = \dot{\mathbf{V}}_Q + \boldsymbol{\omega} \times \dot{\mathbf{R}}_{PQ} + \dot{\boldsymbol{\omega}} \times \mathbf{R}_{PQ}. \tag{d}$$

Since we know that $\dot{\mathbf{V}}_P = \mathbf{A}_P$, $\dot{\mathbf{V}}_Q = \mathbf{A}_Q$, $\dot{\mathbf{R}}_{PQ} = \boldsymbol{\omega} \times \mathbf{R}_{PQ}$, and $\dot{\boldsymbol{\omega}} = \boldsymbol{\alpha}$, then the acceleration of point P can be written from Eq. (d) as

$$\mathbf{A}_P = \mathbf{A}_Q + \boldsymbol{\omega} \times \left(\boldsymbol{\omega} \times \mathbf{R}_{PQ} \right) + \boldsymbol{\alpha} \times \mathbf{R}_{PQ}, \tag{4.5}$$

where the first term, $\mathbf{A}_Q$, is the acceleration of the reference point, Q, and the remaining two terms are caused by the rotation of the body. To visualize the directions of these two terms, let us first study them for a two-dimensional application.

In Fig. 4.4, let P and Q be two points of a rigid body that has a combination of translation and rotation with respect to the ground reference plane, $x_1 y_1$. Let us also define a moving system, $x_2 y_2$, with origin at Q, but restrict this system to only translation; thus, x_2 must remain parallel to x_1. As given quantities, we specify the velocity and acceleration of reference point Q; we also specify the angular velocity, $\boldsymbol{\omega}$, and the angular acceleration, $\boldsymbol{\alpha}$, of the rigid body. These angular rates can be treated as scalars, since the corresponding vectors always have axes perpendicular to the plane of the motion. They may have different senses, however, and the scalar quantities may be either positive or negative.

The location of point P can be specified by the position-difference equation,

$$\mathbf{R}_P = \mathbf{R}_Q + \mathbf{R}_{PQ}. \tag{e}$$

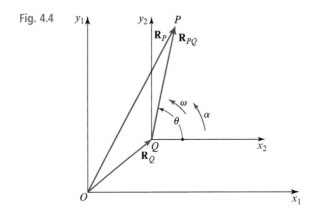

Fig. 4.4

Equation (*a*) can also be written in two-dimensional form as

$$\mathbf{R}_p = \mathbf{R}_Q + R_{PQ}\angle\theta, \tag{f}$$

or in complex polar form as

$$\mathbf{R}_P = \mathbf{R}_Q + R_{PQ}e^{j\theta}. \tag{g}$$

Differentiating Eq. (*g*) with respect to time gives the velocity of point P; that is,

$$\mathbf{V}_P = \mathbf{V}_Q + j\omega R_{PQ}e^{j\theta}. \tag{h}$$

Note that this is the complex polar form of Eq. (*c*) for planar motion, where the second term on the right-hand side corresponds to the velocity-difference vector, $\mathbf{V}_{PQ}$. The magnitude of this term is ωR_{PQ} and the direction is $je^{j\theta}$, which is perpendicular to $\mathbf{R}_{PQ}$, rotated in the sense of $\boldsymbol{\omega}$, as shown in Fig. 4.5.

Differentiating Eq. (*h*) with respect to time gives the acceleration of point P; that is,

$$\mathbf{A}_P = \mathbf{A}_Q - \omega^2 R_{PQ}e^{j\theta} + j\alpha R_{PQ}e^{j\theta}. \tag{i}$$

The second and third terms on the right-hand side of this equation correspond exactly with the second and third terms of Eq. (4.5). The second term is called the *normal* or *centripetal* component of the acceleration difference. For planar motion, the magnitude of this component is

$$\omega^2 R_{PQ} = \frac{V_{PQ}^2}{R_{PQ}},$$

and the direction $-e^{j\theta}$ is opposite to the position-difference vector, $\mathbf{R}_{PQ}$. This is the normal component of the acceleration difference and we designate its magnitude

$$A_{PQ}^n = \omega^2 R_{PQ} = \frac{V_{PQ}^2}{R_{PQ}}. \tag{4.6}$$

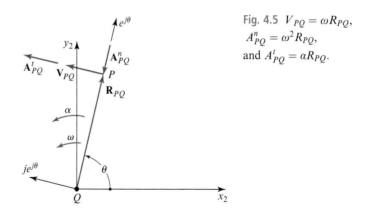

Fig. 4.5 $V_{PQ} = \omega R_{PQ}$,
$A^n_{PQ} = \omega^2 R_{PQ}$,
and $A^t_{PQ} = \alpha R_{PQ}$.

The third term of Eq. (*i*) is associated with the angular acceleration of the body. The magnitude is αR_{PQ} and the direction is $je^{j\theta}$, which is along the same line as the velocity difference, $\mathbf{V}_{PQ}$. Note that P traces out a circular arc in its motion about the translating reference point Q. Since the third term is perpendicular to the position-difference vector $\mathbf{R}_{PQ}$ and hence tangent to the circular arc, it is convenient to call it the *tangential* component of the acceleration difference and designate its magnitude

$$A^t_{PQ} = \alpha R_{PQ}. \tag{4.7}$$

The normal and tangential components of the acceleration difference – that is, Eqs. (4.6) and (4.7) – are shown in Fig. 4.5 for two-dimensional motion.

Let us now examine the final two terms of Eq. (4.5) again, but this time for three-dimensional applications. The direction of the normal component of the acceleration difference, that is

$$\mathbf{A}^n_{PQ} = \boldsymbol{\omega} \times \left(\boldsymbol{\omega} \times \mathbf{R}_{PQ} \right), \tag{4.8}$$

is shown in Fig. 4.6. This component is in the plane containing $\boldsymbol{\omega}$ and $\mathbf{R}_{PQ}$ and is perpendicular to $\boldsymbol{\omega}$. The magnitude is

$$\left| \boldsymbol{\omega} \times \left(\boldsymbol{\omega} \times \mathbf{R}_{PQ} \right) \right| = \omega^2 R_{PQ} \sin \phi = \frac{V^2_{PQ}}{R_{PQ} \sin \phi},$$

where $R_{PQ} \sin \phi$ is the radius of the circle in Fig. 4.6.

According to the definition of the vector cross-product, the tangential component of the acceleration difference,

$$\mathbf{A}^t_{PQ} = \boldsymbol{\alpha} \times \mathbf{R}_{PQ}, \tag{4.9}$$

is perpendicular to the plane containing $\boldsymbol{\alpha}$ and $\mathbf{R}_{PQ}$ with a sense indicated by the right-hand rule. Because of the angular acceleration, $\boldsymbol{\alpha}$, we visualize point P accelerating around a circle,

Fig. 4.6 $\mathbf{V}_{PQ} = \mathbf{\omega} \times \mathbf{R}_{PQ}$ and
$\mathbf{A}_{PQ}^n = \mathbf{\omega} \times (\mathbf{\omega} \times \mathbf{R}_{PQ}).$

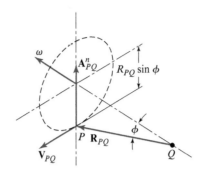

Fig. 4.7 $\mathbf{A}_{PQ}^t = \mathbf{\alpha} \times \mathbf{R}_{PQ}.$

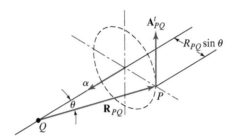

as shown in Fig. 4.7; the plane of this circle is normal to the plane containing $\mathbf{\alpha}$ and $\mathbf{R}_{PQ}$. Using the definition of the vector cross-product, the magnitude of this tangential component of the acceleration difference can be written as

$$\left| \mathbf{\alpha} \times \mathbf{R}_{PQ} \right| = \alpha R_{PQ} \sin \theta,$$

where $R_{PQ} \sin \theta$ is the radius of the circle in Fig. 4.7.

Again we emphasize that, in general, $\mathbf{\alpha}$ and $\mathbf{\omega}$ may have different orientations in three-dimensional applications.

Let us now summarize the results of this section. The acceleration of a point fixed in a rigid body can be obtained from the sum of three terms. The first term is the acceleration of a reference point (Q in Fig. 4.3), which, of course, depends on the motion of the particular point selected. The remaining two terms are due to the rotation of the body. One of these is the normal component and depends on the angular velocity of the body; the other is the tangential component and depends on the time rate of change of the angular velocity.

Equation (4.5) can also be written as

$$\mathbf{A}_P = \mathbf{A}_Q + \mathbf{A}_{PQ}, \tag{4.10}$$

which is called the *acceleration-difference equation*. It may be convenient to designate the components of the acceleration-difference term as

$$\mathbf{A}_{PQ} = \mathbf{A}_{PQ}^n + \mathbf{A}_{PQ}^t.$$

Then, the acceleration-difference equation can be written as

$$\mathbf{A}_P = \mathbf{A}_Q + \mathbf{A}^n_{PQ} + \mathbf{A}^t_{PQ}. \tag{4.11}$$

Many acceleration problems can be solved using the acceleration-difference equation in a manner similar to our use of the velocity-difference equation, Eq. (3.4).

The following example shows both graphic and analytic methods of acceleration analysis.

Example 4.1

For the four-bar linkage in the posture shown in Fig. 4.8a, the constant input angular velocity is $\omega_2 = 200$ rad/s ccw. Determine the accelerations of points A and B, and the angular accelerations of links 3 and 4.

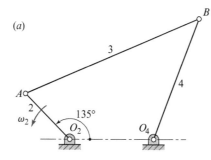

(a)

Fig. 4.8 (a) $R_{O_4 O_2} = 8$ in, $R_{AO_2} = 6$ in, $R_{BA} = 18$ in, and $R_{BO_4} = 12$ in.

Graphic Solution

The velocity polygon is drawn first, since the velocities of points A and B and the angular velocities of links 3 and 4 are required for the acceleration analysis.

The magnitude of the velocity of point A is

$$V_A = \omega_2 R_{AO_2} = (200 \text{ rad/s})(6/12\text{ft}) = 100 \text{ ft/s}.$$

Using this value, the velocity polygon can be drawn as shown in Fig. 4.8b. Then, from the polygon, we measure the magnitudes of the velocity difference of points B and A, and the velocity of point B, respectively, as

$$V_{BA} = 128 \text{ ft/s} \quad \text{and} \quad V_B = 129 \text{ ft/s}.$$

Therefore, the angular velocities of links 3 and 4, respectively, are

$$\omega_3 = \frac{V_{BA}}{R_{BA}} = \frac{128 \text{ ft/s}}{(18/12) \text{ ft}} = 85.3 \text{ rad/s ccw},$$

$$\omega_4 = \frac{V_{BO_4}}{R_{BO_4}} = \frac{129 \text{ ft/s}}{(12/12) \text{ ft}} = 129 \text{ rad/s ccw},$$

where the directions are obtained from an examination of the velocity polygon.

Example 4.1 (cont.)

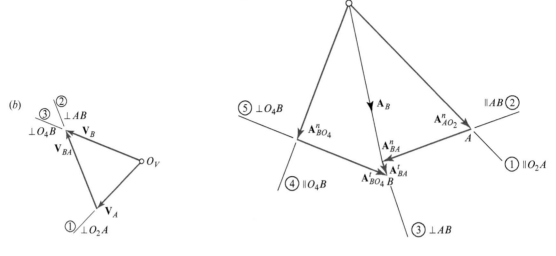

Fig. 4.8 (cont.) (b) velocity polygon; (c) acceleration polygon.

Now, the acceleration of point B can be obtained as follows. First, the acceleration-difference equation, Eq. (4.10), is written as

$$\mathbf{A}_B = \mathbf{A}_A + \mathbf{A}_{BA}$$

or as

$$\cancel{\mathbf{A}_{O_4}}^{\mathbf{0}} + \mathbf{A}_{BO_4} = \cancel{\mathbf{A}_{O_2}}^{\mathbf{0}} + \mathbf{A}_{AO_2} + \mathbf{A}_{BA}.$$

Next, this equation is written in terms of components as

$$\mathbf{A}_{BO_4}^n + \mathbf{A}_{BO_4}^t = \mathbf{A}_{AO_2}^n + \cancel{\mathbf{A}_{AO_2}^t}^{\mathbf{0}} + \mathbf{A}_{BA}^n + \mathbf{A}_{BA}^t, \qquad (1)$$

where

$$A_{AO_2}^t = \alpha_2 R_{AO_2} = 0,$$

$$A_A = A_{AO_2}^n = \omega_2^2 R_{AO_2} = (200 \text{ rad/s})^2 (6/12)\text{ft} = 20\,000 \text{ ft/s}^2,$$

$$A_{BA}^n = \frac{V_{BA}^2}{R_{BA}} = \frac{(128 \text{ ft/s})^2}{(18/12) \text{ ft}} = 10\,923 \text{ ft/s}^2,$$

$$A_{BO_4}^n = \frac{V_{BO_4}^2}{R_{BO_4}} = \frac{(129 \text{ ft/s})^2}{(12/12) \text{ ft}} = 16\,641 \text{ ft/s}^2. \qquad \textit{Ans.}$$

Example 4.1 (cont.)

The two unknowns – that is, the magnitudes of the two tangential components of the acceleration-difference vectors – can now be obtained from the acceleration polygon. Choosing a convenient scale and an acceleration origin, O_A, we construct $\mathbf{A}^n_{AO_2}$ (the terminus is denoted as point A), $\mathbf{A}^n_{BA}$, and then $\mathbf{A}^t_{BA}$, which is temporarily of indefinite length since the magnitude of this vector is not yet known (see Fig. 4.8c).

Beginning again at the acceleration origin, O_A, and following the left-hand side of Eq. (1), we now construct $\mathbf{A}^n_{BO_4}$ and then $\mathbf{A}^t_{BO_4}$ (which is temporarily of indefinite length, since the magnitude of this vector is also not yet known). The intersection of the two vectors, $\mathbf{A}^t_{BA}$ and $\mathbf{A}^t_{BO_4}$, completes the acceleration polygon, as shown in Fig. 4.8c. The point of intersection is labeled the acceleration image point B. The circled numbers indicate the order of the construction steps; the methods of finding the directions of the vectors are also indicated, using the symbol $\parallel$ to indicate parallelism and the symbol $\perp$ to indicate perpendicularity.

From the acceleration polygon, the two unknown magnitudes are measured as

$$A^t_{BO_4} = 11\,900 \text{ ft/s}^2 \quad \text{and} \quad A^t_{BA} = 2\,000 \text{ ft/s}^2.$$

The line from point O_A to point B in the polygon is the magnitude of the acceleration of point B, and is measured as

$$A_B = 20\,500 \text{ ft/s}^2. \qquad \textit{Ans.}$$

The angular accelerations of links 3 and 4 are then found as follows:

$$\alpha_3 = \frac{A^t_{BA}}{R_{BA}} = \frac{2\,000 \text{ ft/s}^2}{(18/12) \text{ ft}} = 1\,333 \text{ rad/s}^2 \text{ cw}, \qquad \textit{Ans.}$$

$$\alpha_4 = \frac{A^t_{BO_4}}{R_{BO_4}} = \frac{11\,900 \text{ ft/s}^2}{(12/12) \text{ ft}} = 11\,900 \text{ rad/s}^2 \text{ cw}, \qquad \textit{Ans.}$$

where the directions are obtained from an examination of the tangential components in the acceleration polygon.

Analytic Solution

The first two steps are to perform a posture analysis and a velocity analysis of the linkage. Since these steps were presented in Chapters 2 and 3, only the results are presented here.

Example 4.1 (cont.)

The results of the posture analysis are shown in Fig. 4.9. In vector form, the position-difference vectors corresponding to the links are:

$$\mathbf{R}_{AO_2} = \left(\frac{6}{12}\right) \text{ ft } \angle 135° = -0.353\,55\hat{\mathbf{i}} + 0.353\,55\hat{\mathbf{j}} \text{ ft,}$$

$$\mathbf{R}_{BA} = \left(\frac{18}{12}\right) \text{ ft } \angle 22.4° = 1.386\,82\hat{\mathbf{i}} + 0.571\,61\hat{\mathbf{j}} \text{ ft,}$$

$$\mathbf{R}_{BO_4} = \left(\frac{12}{12}\right) \text{ ft } \angle 68.4° = 0.368\,12\hat{\mathbf{i}} + 0.929\,78\hat{\mathbf{j}} \text{ ft.}$$

Fig. 4.9

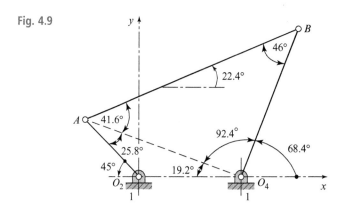

For the given input angular velocity $\omega_2 = 200\hat{\mathbf{k}}$ rad/s, the results of the velocity analysis are:

$$\omega_3 = 84.253\hat{\mathbf{k}} \text{ rad/s,} \quad \text{and} \quad \omega_4 = 129.39\hat{\mathbf{k}} \text{ rad/s.}$$

Next, the known acceleration components are

$$\mathbf{A}^n_{AO_2} = \omega_2 \times (\omega_2 \times \mathbf{R}_{AO_2}) = 14\,142\hat{\mathbf{i}} - 14\,142\hat{\mathbf{j}} \text{ ft/s}^2 = \mathbf{A}_A, \qquad Ans. \text{ (2)}$$

$$\mathbf{A}^n_{BA} = \omega_3 \times (\omega_3 \times \mathbf{R}_{BA}) = -9\,844\hat{\mathbf{i}} - 4\,058\hat{\mathbf{j}} \text{ ft/s}^2, \qquad (3)$$

$$\mathbf{A}^n_{BO_4} = \omega_4 \times (\omega_4 \times \mathbf{R}_{BO_4}) = -6\,163\hat{\mathbf{i}} - 15\,567\hat{\mathbf{j}} \text{ ft/s}^2. \qquad (4)$$

Although the angular accelerations α_3 and α_4 have unknown magnitudes, we can incorporate them into the solution in the following manner:

$$\mathbf{A}^t_{BA} = \alpha_3 \times \mathbf{R}_{BA} = \begin{vmatrix} \hat{\mathbf{i}} & \hat{\mathbf{j}} & \hat{\mathbf{k}} \\ 0 & 0 & \alpha_3 \\ 1.386\,82 \text{ ft} & 0.571\,61 \text{ ft} & 0 \end{vmatrix} = -0.571\,61\alpha_3\hat{\mathbf{i}} + 1.386\,82\alpha_3\hat{\mathbf{j}} \text{ ft,} \qquad (5)$$

Example 4.1 (cont.)

$$\mathbf{A}^t_{BO_4} = \boldsymbol{\alpha}_4 \times \mathbf{R}_{BO_4} = \begin{vmatrix} \hat{\mathbf{i}} & \hat{\mathbf{j}} & \hat{\mathbf{k}} \\ 0 & 0 & \alpha_4 \\ 0.368\,12 \text{ ft} & 0.929\,78 \text{ ft} & 0 \end{vmatrix} = -0.929\,78\alpha_4\hat{\mathbf{i}} + 0.368\,12\alpha_4\hat{\mathbf{j}} \text{ ft.} \quad (6)$$

Writing the acceleration-difference equation for point B and noting that $\mathbf{A}^t_{AO_2} = \mathbf{0}$ gives

$$\mathbf{A}^n_{BO_4} + \mathbf{A}^t_{BO_4} = \mathbf{A}^n_{AO_2} + \mathbf{A}^n_{BA} + \mathbf{A}^t_{BA}. \quad (7)$$

Then, substituting Eqs. (2) through (6) into Eq. (7) and separating the $\hat{\mathbf{i}}$ *and* $\hat{\mathbf{j}}$ components, we obtain the following pair of simultaneous equations:

$$(0.571\,61 \text{ ft})\alpha_3 - (0.929\,78 \text{ ft})\alpha_4 = 10\,461 \text{ ft/s}^2, \quad (8)$$

$$-(1.386\,82 \text{ ft})\alpha_3 + (0.368\,12 \text{ ft})\alpha_4 = -2\,633 \text{ ft/s}^2. \quad (9)$$

Solving these equations, the angular accelerations of links 3 and 4, respectively, are

$$\alpha_3 = -1\,300\hat{\mathbf{k}} \text{ rad/s}^2 \quad \text{and} \quad \alpha_4 = -12\,050\hat{\mathbf{k}} \text{ rad/s}^2, \qquad \textit{Ans.} \quad (10)$$

where the two negative signs indicate that both angular accelerations are clockwise.

The acceleration of point B can now be written from Eq. (4.11) as

$$\mathbf{A}_B = \mathbf{A}^n_{BO_4} + \mathbf{A}^t_{BO_4}. \quad (11)$$

The procedure is to substitute Eq. (10) into Eq. (6), which gives

$$\mathbf{A}^t_{BO_4} = 11\,204\hat{\mathbf{i}} - 4\,436\hat{\mathbf{j}} \text{ ft/s}^2. \quad (12)$$

Then, substituting Eqs. (4) and (12) into Eq. (11), the acceleration of point B is

$$\mathbf{A}_B = 5\,041\hat{\mathbf{i}} - 20\,003\hat{\mathbf{j}} \text{ ft/s}^2 = 20\,628 \text{ ft/s}^2\angle{-75.86°}. \qquad \textit{Ans.}$$

Note that the four answers from the analytic method are in good agreement with, but more accurate than, the results of the graphic method.

4.4 Acceleration Polygons; Acceleration Images

The acceleration image of a link in an acceleration polygon can be obtained in much the same manner as the velocity image of a link in a velocity polygon (Section 3.4).

According to Rosenauer and Willis [7] the theorem of Mehmke can be written as:

The end points of the acceleration vectors of the points of a plane rigid body, when plotted from a common origin, produce a figure which is geometrically similar to the original figure [image diagram].

It is this theorem that results in the acceleration image presented in this section. Also, it is this theorem that allows clarity in the acceleration polygon despite the very minimal labeling that is required. This becomes evident in the following examples, and continues throughout the text.

Example 4.2

For the four-bar linkage of Example 3.1, in the posture shown in Fig. 4.10, the angular velocity of crank 2 is a constant 900 rev/ min $= 94.25$ rad/s ccw. Determine the angular accelerations of links 3 and 4 and the accelerations of points E and F.

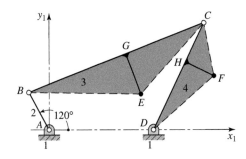

Fig. 4.10 $R_{BA} = 4$ in,
$R_{CB} = 18$ in, $R_{CD} = 11$ in,
$R_{DA} = 10$ in, $R_{GB} = 10$ in,
$R_{EG} = 4$ in, $R_{HD} = 7$ in, and
$R_{FH} = 3$ in.

Solution

First, we consider the accelerations of points B and C. Since the angular acceleration of link 2 is zero, point B has only a normal component of acceleration, namely,

$$A_B = A_{BA}^n = \omega_2^2 R_{BA} = (94.25 \text{ rad/s})^2 (4/12 \text{ ft}) = 2961 \text{ ft/s}^2.$$

Now, we choose a convenient scale and an acceleration origin, O_A. Note that points A and D have zero acceleration, and therefore their images are coincident with the acceleration origin, O_A. Then, we draw the acceleration vector, $\mathbf{A}_B$ (opposite in direction to the vector $\mathbf{R}_{BA}$), to locate the acceleration image point B, as shown in Fig. 4.11. Next, we write Eq. (4.11) to relate the acceleration of point C to those of points B and D in the acceleration polygon; that is,

$$\mathbf{A}_C = \overset{\sqrt{}}{\mathbf{A}_B} + \overset{\sqrt{}}{\mathbf{A}_{CB}^n} + \overset{?\sqrt{}}{\mathbf{A}_{CB}^t} = \overset{\sqrt{}}{\mathbf{A}_{CD}^n} + \overset{?\sqrt{}}{\mathbf{A}_{CD}^t}. \tag{1}$$

Using the posture and velocity results obtained from Example 3.1 (see Fig. 3.7b), we calculate the magnitudes of the two normal components of Eq. (1); that is,

$$A_{CB}^n = \frac{V_{CB}^2}{R_{CB}} = \frac{(38.4 \text{ ft/s})^2}{(18/12) \text{ ft}} = 983 \text{ ft/s}^2,$$

$$A_{CD}^n = \frac{V_{CD}^2}{R_{CD}} = \frac{(45.5 \text{ ft/s})^2}{(11/12) \text{ ft}} = 2258 \text{ ft/s}^2.$$

Example 4.2 (cont.)

These two normal components are constructed with directions opposite to the vectors $\mathbf{R}_{CB}$ and $\mathbf{R}_{CD}$, respectively. As required by Eq. (1), they are added to the acceleration polygon originating from acceleration image points B and D, respectively, as indicated by two dashed lines in Fig. 4.11. Perpendicular dashed lines are then drawn through the termini of these two normal components, representing the addition of the two tangential components, A^t_{CB} and A^t_{CD}. The point of intersection of the two tangential components is labeled the acceleration image point C.

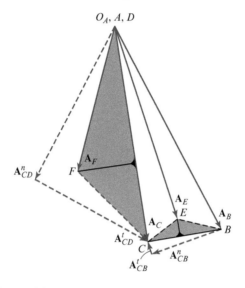

O_A, A, D

Fig. 4.11 Acceleration polygon.

The angular accelerations of links 3 and 4 are now obtained from measurements of the two tangential components:

$$\alpha_3 = \frac{A^t_{CB}}{R_{CB}} = \frac{170 \ \text{ft/s}^2}{(18/12)\text{ft}} = 113 \ \text{rad/s}^2 \ \text{ccw}, \qquad \qquad Ans.$$

$$\alpha_4 = \frac{A^t_{CD}}{R_{CD}} = \frac{1\,670 \ \text{ft/s}^2}{(11/12)\text{ft}} = 1\,822 \ \text{rad/s}^2 \ \text{cw}. \qquad \qquad Ans.$$

There are several methods for determining the acceleration of point E of link 3. One method is to write two acceleration-difference equations, starting from points B and C, which are also in link 3; that is

$$\mathbf{A}_E = \mathbf{A}_B + \mathbf{A}^n_{EB} + \mathbf{A}^t_{EB} = \mathbf{A}_C + \mathbf{A}^n_{EC} + \mathbf{A}^t_{EC}. \qquad (2)$$

The solution of these two equations can be obtained by the same methods used for Eq. (1), if desired. A second method to determine the acceleration of point E is to use the now known angular acceleration, α_3, to calculate one or both of the tangential components in Eq. (2).

A third method and probably the easiest (and the one used in Fig. 4.11), is to form an acceleration image triangle, BCE, for link 3. This image triangle is formed by using the line of

Example 4.2 (cont.)

the acceleration-difference vector, $\mathbf{A}_{CB}$, as a base line, then setting a scale for triangle BCE of link 3.[2] Any of the three methods provides the location of the acceleration image point E. The magnitude of the acceleration of point E is measured as

$$A_E = 2\,580 \text{ ft/s}^2. \qquad\qquad Ans.$$

Similarly, the magnitude of the acceleration of point F is found by using the acceleration image triangle DCF and is measured as

$$A_F = 1\,960 \text{ ft/s}^2. \qquad\qquad Ans.$$

Example 4.3

Find the acceleration images corresponding to the links of the slider-crank linkage in the posture shown in Fig. 4.12a. The crank (link 2) is rotating counterclockwise with constant angular velocity $\omega_2 = 1$ rad/s.

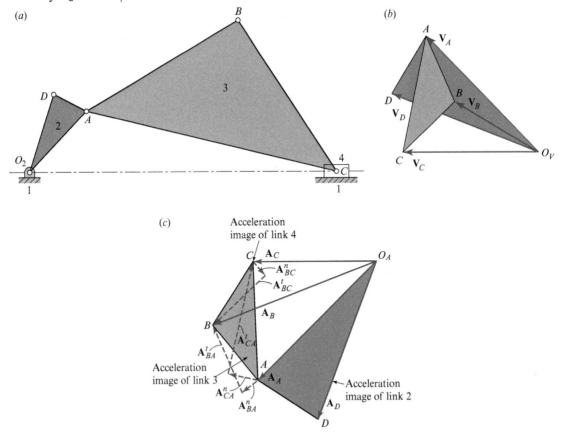

Fig. 4.12 (a) Slider-crank linkage; (b) velocity polygon; (c) acceleration polygon.

---

[2] We must be careful that the shape of the acceleration image is not "flipped over" with respect to the original shape. A convenient test is to notice that, for link 3 of this example, because the labels BCE appear in clockwise order for the original link shape, they also appear in clockwise order in the velocity and the acceleration images.

Example 4.3 (cont.)

Solution

In Fig. 4.12a, links 2 and 3 are represented by the triangles O_2DA and ABC, respectively. Note that the velocity and acceleration polygons are found in exactly the same manner as in the solution of the previous example, and the results are shown in Figs. 4.12b and 4.12c, respectively. Note also that each acceleration image is formed from the *total acceleration-difference* vectors, not from the component vectors.

Since the angular acceleration of the crank is zero ($\alpha_2 = 0$), the corresponding acceleration image is rotated 180° from the orientation of the crank. Note that link 3 has counterclockwise angular acceleration and that its acceleration image is oriented much less than 180° from the orientation of the link. Thus, the orientation of each acceleration image depends on both the angular velocity and the angular acceleration of the link under consideration.

Summary

The following are important properties of acceleration images:

1. The acceleration image of a link in the acceleration polygon is a scale reproduction of the shape of the link itself.
2. The letters identifying the vertices in the acceleration polygon are the same as those of the corresponding link, and they progress around the acceleration image in the same order and in the same angular sense as around the link itself.
3. The acceleration origin, O_A, in the acceleration polygon is the image of all points with zero absolute acceleration. It is the acceleration image of the fixed link.
4. The absolute acceleration of any point on any link is represented by the line from O_A to the image of the point in the acceleration polygon. The acceleration difference between two points in the same link, say B and C, is represented by the line to acceleration image point B from acceleration image point C.
5. For a nonrotating (translating) link, the accelerations of all points on the link are equal, and the angular velocity and angular acceleration of the link are both zero. Therefore, the acceleration image of a link that is translating shrinks to a single point in the acceleration polygon.
6. The orientation of the acceleration image of link j is given by

$$\delta_j = 180° - \tan^{-1}\frac{\alpha_j}{\omega_j^2}, \qquad (4.12)$$

where δ_j is the angle in degrees, measured in the (positive) counterclockwise direction from the orientation of link j to its acceleration image.

4.5   Apparent Acceleration of a Point in a Moving Coordinate System

In Section 3.5 we found it helpful to develop the apparent-velocity equation for situations where it is convenient to describe the path along which a point moves relative to another moving link, but where it is not convenient to describe the absolute motion of the same point. Let us now investigate the acceleration of such a point.

Figure 4.13 shows a point, P, of link 3 (denoted as P_3) that moves along a known path (sometimes referred to as the "slot") relative to the moving reference frame $x_2y_2z_2$. Recall from Chapters 2 and 3 that point P_2 is a point fixed in link 2 that is instantaneously coincident with point P_3. The problem now is to find an equation relating the accelerations of points P_3 and P_2 in terms of parameters that can either be calculated or measured in a typical mechanism.

In Fig. 4.14, we recall how the same situation would be perceived by a moving observer attached to link 2. To HER, the path of P_3 (the "slot") would appear stationary, and point P_3 would appear to move along the tangent to the path with the apparent velocity $\mathbf{V}_{P_3/2}$.

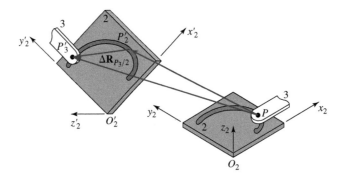

Fig. 4.13 Apparent displacement.

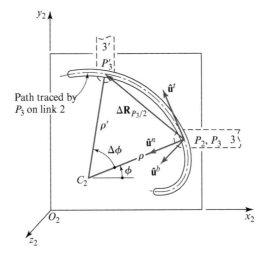

Fig. 4.14 Apparent displacement of point P_3 as seen by an observer on link 2.

In Section 3.5 (Fig. 3.12), we define the moving coordinate system $\hat{\mathbf{u}}^t\hat{\mathbf{u}}^n\hat{\mathbf{u}}^b$, where $\hat{\mathbf{u}}^t$ is the unit tangent vector to the path of point P in the direction of positive movement, and $\hat{\mathbf{u}}^b$ is normal to the plane containing $\hat{\mathbf{u}}^t$ and the center of curvature C, and is directed positive to the preferred side of the plane. A third unit vector, $\hat{\mathbf{u}}^n$, is obtained from Eq. (3.7), $\hat{\mathbf{u}}^n = \hat{\mathbf{u}}^b \times \hat{\mathbf{u}}^t$, thus completing a right-handed Cartesian coordinate system. Also, we derive the apparent-velocity equation, Eq. (3.9), as

$$\mathbf{V}_{P_3/2} = \frac{ds}{dt}\hat{\mathbf{u}}^t, \qquad (a)$$

where the scalar, s, is the arc distance along the path measuring the travel of P_3.

Now, the radius of curvature, ρ, sweeps through some small angle, $\Delta\phi$, as P_3 travels the small arc distance Δs (P_3P_3', see Fig. 4.14), as seen by the moving observer during a short time interval, Δt. The changes of the angle and the arc distance are related by

$$\Delta\phi = \frac{\Delta s}{\rho}. \qquad (b)$$

Note here that the center of curvature C can lie along either the positive or the negative extension of $\hat{\mathbf{u}}^n$. Therefore, the radius of curvature ρ, measured *from P to C*, can have either a positive value or a negative value according to the sense of $\hat{\mathbf{u}}^n$. Also, this implies that the angle $\Delta\phi$ is positive when counterclockwise as seen from positive $\hat{\mathbf{u}}^b$.

Dividing Eq. (b) by Δt and taking the limit for infinitesimally small Δt, we find with the aid of Eq. (a),

$$\frac{d\phi}{dt} = \frac{1}{\rho}\frac{ds}{dt} = \frac{V_{P_3/2}}{\rho}. \qquad (c)$$

This is the angular rate at which the radius of curvature ρ (and also the unit vectors $\hat{\mathbf{u}}^t$ and $\hat{\mathbf{u}}^n$) appears to rotate as seen by an observer in coordinate system 2 as point P_3 moves along its path. We can give this rotation rate its proper vector properties as an apparent angular velocity by noting that the axis of this rotation is parallel to $\hat{\mathbf{u}}^b$. Thus, we define the apparent angular velocity vector as

$$\dot{\boldsymbol{\phi}} = \frac{d\phi}{dt}\hat{\mathbf{u}}^b = \frac{V_{P_3/2}}{\rho}\hat{\mathbf{u}}^b. \qquad (d)$$

Next, we seek the time derivative of unit tangent vector $\hat{\mathbf{u}}^t$ so that we can differentiate Eq. (a) to obtain the apparent acceleration. Since $\hat{\mathbf{u}}^t$ is a unit vector, its length does not change; however, it does have a derivative because of its change in direction: its rotation. In the absolute coordinate system, $\hat{\mathbf{u}}^t$ is subject to the apparent angular velocity, $\dot{\boldsymbol{\phi}}$, and also to the angular velocity, $\boldsymbol{\omega}_2$, with which the moving coordinate system 2 is rotating. Therefore, the time derivative of $\hat{\mathbf{u}}^t$ can be written as

$$\frac{d\hat{\mathbf{u}}^t}{dt} = \left(\boldsymbol{\omega}_2 + \dot{\boldsymbol{\phi}}\right) \times \hat{\mathbf{u}}^t = \boldsymbol{\omega}_2 \times \hat{\mathbf{u}}^t + \dot{\boldsymbol{\phi}} \times \hat{\mathbf{u}}^t. \qquad (e)$$

Substituting Eq. (*d*) into this equation gives

$$\frac{d\hat{\mathbf{u}}^t}{dt} = \boldsymbol{\omega}_2 \times \hat{\mathbf{u}}^t + \frac{V_{P_3/2}}{\rho}\hat{\mathbf{u}}^b \times \hat{\mathbf{u}}^t = \boldsymbol{\omega}_2 \times \hat{\mathbf{u}}^t + \frac{V_{P_3/2}}{\rho}\hat{\mathbf{u}}^n. \qquad (f)$$

Similarly, the time derivative of $\hat{\mathbf{u}}^n$ can be written as

$$\begin{aligned}\frac{d\hat{\mathbf{u}}^n}{dt} &= \left(\boldsymbol{\omega}_2 + \dot{\boldsymbol{\phi}}\right) \times \hat{\mathbf{u}}^n = \boldsymbol{\omega}_2 \times \hat{\mathbf{u}}^n + \dot{\boldsymbol{\phi}} \times \hat{\mathbf{u}}^n \\ &= \boldsymbol{\omega}_2 \times \hat{\mathbf{u}}^n + \frac{V_{P_3/2}}{\rho}\hat{\mathbf{u}}^b \times \hat{\mathbf{u}}^n = \boldsymbol{\omega}_2 \times \hat{\mathbf{u}}^n - \frac{V_{P_3/2}}{\rho}\hat{\mathbf{u}}^t.\end{aligned} \qquad (g)$$

Now, taking the time derivative of Eq. (*a*), and using Eq. (*f*), we find that

$$\frac{d\mathbf{V}_{P_3/2}}{dt} = \frac{ds}{dt}\frac{d\hat{\mathbf{u}}^t}{dt} + \frac{d^2s}{dt^2}\hat{\mathbf{u}}^t = \frac{ds}{dt}\,\hat{\boldsymbol{\omega}}_2 \times \hat{\mathbf{u}}^t + \frac{ds}{dt}\frac{V_{P_3/2}}{\rho}\hat{\mathbf{u}}^n + \frac{d^2s}{dt^2}\hat{\mathbf{u}}^t.$$

Finally, using Eqs. (*a*) and (*c*), this equation reduces to

$$\frac{d\mathbf{V}_{P_3/2}}{dt} = \boldsymbol{\omega}_2 \times \mathbf{V}_{P_3/2} + \frac{V_{P_3/2}^2}{\rho}\hat{\mathbf{u}}^n + \frac{d^2s}{dt^2}\hat{\mathbf{u}}^t. \qquad (h)$$

Note that the three components on the right-hand side of Eq. (*h*) are *not* all defined as *apparent-acceleration* components. To be consistent, the apparent acceleration includes only those components *that are seen by an observer attached to the moving coordinate system.* Equation (*h*) is derived in the absolute coordinate system and includes the rotation effect of $\boldsymbol{\omega}_2$ that is not sensed by the moving observer. The apparent acceleration can easily be determined, however, by setting $\boldsymbol{\omega}_2$ to zero in Eq. (*h*). The two remaining components define the apparent acceleration and can be written as

$$\mathbf{A}_{P_3/2} = \mathbf{A}_{P_3/2}^n + \mathbf{A}_{P_3/2}^t, \qquad (4.13)$$

where

$$\mathbf{A}_{P_3/2}^n = \frac{V_{P_3/2}^2}{\rho}\hat{\mathbf{u}}^n \qquad (4.14)$$

is called the *normal component* of the apparent acceleration, indicating that it is always normal to the path and is always directed from point P toward the center of curvature (that is, in the $\hat{\mathbf{u}}^n$ direction when ρ is positive, or in the $-\hat{\mathbf{u}}^n$ direction when ρ is negative), and

$$\mathbf{A}_{P_3/2}^t = \frac{d^2s}{dt^2}\hat{\mathbf{u}}^t \qquad (4.15)$$

is called the *tangential component*, indicating that it is always tangent to the path (along the $\hat{\mathbf{u}}^t$ axis, but it may be either positive or negative).

Next, from Fig. 4.14, we write the position of point P_3 as

$$\mathbf{R}_{P_3} = \mathbf{R}_{C_2} - \rho\hat{\mathbf{u}}^n.$$

With the help of Eq. (g), the time derivative of this equation can be written as[3]

$$\mathbf{V}_{P_3} = \mathbf{V}_{C_2} - \boldsymbol{\omega}_2 \times (\rho\hat{\mathbf{u}}^n) + \rho\frac{V_{P_3/2}}{\rho}\hat{\mathbf{u}}^t = \mathbf{V}_{C_2} - \boldsymbol{\omega}_2 \times (\rho\hat{\mathbf{u}}^n) + \mathbf{V}_{P_3/2}. \qquad (i)$$

Then, differentiating this equation, again with respect to time, the acceleration of point P_3 can be written as

$$\mathbf{A}_{P_3} = \mathbf{A}_{C_2} - \boldsymbol{\alpha}_2 \times (\rho\hat{\mathbf{u}}^n) - \boldsymbol{\omega}_2 \times \frac{d(\rho\hat{\mathbf{u}}^n)}{dt} + \frac{d\mathbf{V}_{P_3/2}}{dt},$$

and, with the help of Eqs. (g) and (h), this becomes

$$\mathbf{A}_{P_3} = \mathbf{A}_{C_2} + \boldsymbol{\alpha}_2 \times (-\rho\hat{\mathbf{u}}^n) + \boldsymbol{\omega}_2 \times [\boldsymbol{\omega}_2 \times (-\rho\hat{\mathbf{u}}^n)] + 2\boldsymbol{\omega}_2 \times \mathbf{V}_{P_3/2} + \frac{V_{P_3/2}^2}{\rho}\hat{\mathbf{u}}^n + \frac{d^2s}{dt^2}\hat{\mathbf{u}}^t. \qquad (j)$$

The first three terms on the right-hand side of Eq. (j) are the components of the acceleration, $\mathbf{A}_{P_2}$ [Eq. (4.5)], and the final two terms are the normal and tangential components of the apparent acceleration, $\mathbf{A}_{P_3/2}$ [Eqs. (4.13)–(4.15)]. For the fourth term, we define the following new symbol:

$$\mathbf{A}_{P_3P_2}^c = 2\boldsymbol{\omega}_2 \times \mathbf{V}_{P_3/2}. \qquad (4.16)$$

This vector is called the *Coriolis component of acceleration*. Unlike the components of the apparent acceleration, it is not sensed by an observer attached to the moving coordinate system, 2. Still, it is a necessary term in Eq. (j), and it is a part of the difference between $\mathbf{A}_{P_3}$ and $\mathbf{A}_{P_2}$ sensed by an absolute observer.

Finally, Eq. (j) can now be written in the following form, called the *apparent-acceleration equation*,

$$\mathbf{A}_{P_3} = \mathbf{A}_{P_2} + \mathbf{A}_{P_3P_2}^c + \mathbf{A}_{P_3/2}^n + \mathbf{A}_{P_3/2}^t, \qquad (4.17)$$

where the definitions of the individual components are given by Eqs. (4.16), (4.14), and (4.15), respectively.

It is extremely important to recognize the following features of the apparent-acceleration equation:

1. It serves the objectives of this section, since it relates the accelerations of two *coincident points on different links* in a meaningful way.
2. There is only *one unknown* among the three new components defined. The normal component and the Coriolis component can be calculated from Eqs. (4.14) and (4.16) from velocity information; they do not contribute any new unknowns. The tangential component given by

---

[3] The first two terms on the right-hand side of Eq. (i) are equal to $\mathbf{V}_{P_2}$; thus, Eq. (i) is equivalent to the apparent-velocity equation. Note, however, that although $\rho\hat{\mathbf{u}}^n = \mathbf{R}_{C_2P_2}$ is true instantaneously, the derivatives of the two terms are not equal – that is, the vectors do not rotate at the same rate. Thus, several terms could be missed if the apparent-velocity equation were differentiated instead.

Eq. (4.15), however, almost always has an unknown magnitude in application, since d^2s/dt^2 is usually not known.

3. It is important in each application to note the dependence of Eq. (4.17) on the ability to recognize the point path that P_3 traces on coordinate system 2. This path is the basis for the axes of the normal and tangential components and is also necessary for determining the radius of curvature ρ for Eq. (4.14).

A word of warning: *The path described by P_3 on link 2 is not necessarily the same as the path described by P_2 on link 3.* In Fig. 4.14, the path of P_3 on link 2 is clear; it is the curved slot. However, the path of P_2 on link 3 is not at all clear. As a result, there is a natural "right" and "wrong" way to write the apparent-acceleration equation for that situation. The equation

$$\mathbf{A}_{P_2} = \mathbf{A}_{P_3} + \mathbf{A}^c_{P_2 P_3} + \mathbf{A}^n_{P_2/3} + \mathbf{A}^t_{P_2/3}$$

is a perfectly valid equation, but is *useless*, since that path, and hence the radius of curvature of that path, are not known for the normal component. Note, also, that $\mathbf{A}^c_{P_3 P_2}$ makes use of ω_2, while $\mathbf{A}^c_{P_2 P_3}$ would make use of ω_3. *We must be extremely careful to write the appropriate equation for each application, recognizing which path is known.*

The following three examples demonstrate the importance of the apparent-acceleration equation, Eq. (4.17), in the graphic approach to the acceleration analysis of mechanisms.

Example 4.4

For the sliding-block linkage in the posture shown in Fig. 4.15a, the block (link 3) is sliding outward on link 2 at a uniform rate of 30 m/s, while link 2 is rotating at a constant angular velocity of 50 rad/s ccw. Determine the acceleration of point A of the block.

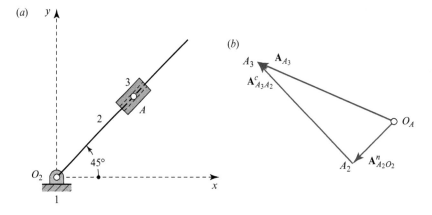

Fig. 4.15 (a) $R_{AO_2} = 500$ mm; (b) acceleration polygon.

Solution

The acceleration of A_2 – that is, the point coincident with point A of the block (point A_3) but attached to link 2 – can be written from Eq. (4.11) as

$$\mathbf{A}_{A_2} = \overset{0}{\cancel{\mathbf{A}_{O_2}}} + \mathbf{A}^n_{A_2 O_2} + \overset{0}{\cancel{\mathbf{A}^t_{A_2 O_2}}},$$

Example 4.4 (cont.)

where

$$A^n_{A_2 O_2} = \omega_2^2 R_{A_2 O_2} = (50 \text{ rad/s})^2 (0.500 \text{ m}) = 1\,250 \text{ m/s}^2.$$

Therefore, we can draw this normal component of acceleration to scale, locating the acceleration image point A_2 in Fig. 4.15b. Next, we recognize that point A_3 is constrained to travel along link 2. This provides a path for which we can write the apparent-acceleration equation:

$$\mathbf{A}_{A_3} = \mathbf{A}_{A_2} + \mathbf{A}^c_{A_3 A_2} + \mathbf{A}^n_{A_3/2} + \mathbf{A}^t_{A_3/2}.$$

The final three components for this equation can be computed; that is

$$A^c_{A_3 A_2} = 2\omega_2 V_{A_3/2} = 2(50 \text{ rad/s})(30 \text{ m/s}) = 3\,000 \text{ m/s}^2, \tag{1}$$

$$A^n_{A_3/2} = \frac{V^2_{A_3/2}}{\rho} = \frac{(30 \text{ m/s})^2}{\infty} = 0, \tag{2}$$

and

$$A^t_{A_3/2} = \frac{d^2 s}{dt^2} = 0 \quad \text{(because of uniform rate along the path).} \tag{3}$$

The component given by Eq. (1) can now be drawn on the acceleration polygon. Noting that the sense of this component comes from the sense of the vector cross-product of Eq. (4.16), this locates the acceleration image point A_3. Therefore, the acceleration of point A_3 of the block is measured from the polygon as

$$A_{A_3} = 3\,250 \text{ m/s}^2. \qquad \textit{Ans.}$$

Example 4.5

Perform an acceleration analysis of the inverted slider-crank linkage in the posture shown in Fig. 4.16. The input link 2 is rotating at a constant input angular velocity $\omega_2 = 18$ rad/s cw.

Solution

First, a complete velocity analysis is performed, as shown by the velocity polygon in Fig. 4.16b, and the results are

$$V_A = 12.0 \text{ ft/s}, \quad V_{B_3 A} = 10.0 \text{ ft/s}, \quad \text{and} \quad V_{B_3/4} = 6.7 \text{ ft/s},$$

$$\omega_3 = \omega_4 = 7.67 \text{ rad/s cw}.$$

Example 4.5 (cont.)

Fig. 4.16 (a) $R_{AO_2} = 8$ in and $R_{BO_2} = 10$ in; (b) velocity polygon; (c) acceleration polygon.

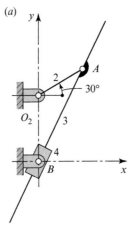

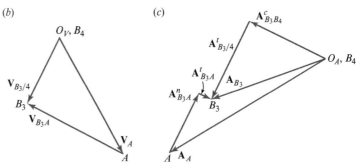

To solve for accelerations, we first calculate the acceleration of pin A:

$$\mathbf{A}_A = \cancel{\mathbf{A}_{O_2}}^{\mathbf{0}} + \mathbf{A}^n_{AO_2} + \cancel{\mathbf{A}^t_{AO_2}}^{\mathbf{0}},$$

where,

$$A^n_{AO_2} = \omega_2^2 R_{AO_2} = (18 \text{ rad/s})^2 (8/12)\text{ft} = 216.0 \text{ ft/s}^2. \qquad Ans.$$

Then, we draw this acceleration to scale, locating the acceleration image point A in Fig. 4.16c. Next we write the acceleration-difference equation

$$\overset{??}{\mathbf{A}_{B_3}} = \overset{\sqrt{\sqrt{}}}{\mathbf{A}_A} + \overset{\sqrt{\sqrt{}}}{\mathbf{A}^n_{B_3A}} + \overset{?\sqrt{}}{\mathbf{A}^t_{B_3A}}, \qquad (1)$$

where

$$A^n_{B_3A} = \frac{V^2_{B_3A}}{R_{BA}} = \frac{(10.0 \text{ ft/s})^2}{(15.6/12)\text{ft}} = 76.9 \text{ ft/s}^2,$$

and is directed from B toward A. This component is added to the acceleration polygon in Fig. 4.16c.

Example 4.5 (cont.)

The final component, $A^t_{B_3A}$, has an unknown magnitude but is perpendicular to $\mathbf{R}_{BA}$. Since Eq. (1) has three unknowns, it cannot be solved by itself. Therefore, a second equation is required to solve for $\mathbf{A}_{B_3}$. Consider the view of an observer located on link 4; SHE would see point B_3 moving on a straight-line path along the centerline of block 4. Using this path, we can write the apparent-acceleration equation, Eq. (4.17), as

$$\mathbf{A}_{B_3} = \overset{0}{\cancel{\mathbf{A}_{B_4}}} + \overset{\sqrt{}\sqrt{}}{\mathbf{A}^c_{B_3B_4}} + \overset{\sqrt{}\sqrt{}}{\mathbf{A}^n_{B_3/4}} + \overset{?\sqrt{}}{\mathbf{A}^t_{B_3/4}}. \tag{2}$$

Note that point B_4 has zero acceleration, since it is pinned to the ground link. The two known component magnitudes of Eq. (2) are

$$A^c_{B_3B_4} = 2\omega_4 V_{B_3/4} = 2(7.67 \text{ rad/s})(6.7 \text{ ft/s}) = 103 \text{ ft/s}^2$$

and

$$A^n_{B_3/4} = \frac{V^2_{B_3/4}}{\rho} = \frac{(6.5 \text{ ft/s})^2}{\infty} = 0.$$

The Coriolis component is added to the acceleration polygon, originating at point B_4, which is coincident with O_A. Finally, $\mathbf{A}^t_{B_3/4}$, with unknown magnitude and sense, is graphically added along a line defined by the path tangent. It crosses the unknown-length line of $\mathbf{A}^t_{B_3A}$, Eq. (1), locating the acceleration image point B_3. From the acceleration polygon, the results are measured as

$$A^t_{B_3/4} = 103 \text{ ft/s}^2, \quad A^t_{B_3A} = 17 \text{ ft/s}^2, \quad \text{and} \quad A_{B_3} = 145 \text{ ft/s}^2. \qquad Ans.$$

The angular accelerations of links 3 and 4 are

$$\alpha_3 = \alpha_4 = \frac{A^t_{B_3A}}{R_{BA}} = \frac{17 \text{ ft/s}^2}{(15.6/12)\text{ft}} = 13.1 \text{ rad/s}^2 \text{ ccw}. \qquad Ans.$$

We note that, in this example, the path of B_3 on link 4 and the path of B_4 on link 3 can both be visualized, and either can be used in deciding the approach. However, even though B_4 is pinned to ground (link 1), the path of point B_3 on link 1 is not known. Therefore, the term $\mathbf{A}^n_{B_3/1}$ cannot be calculated directly.

Example 4.6

For the inverted slider-crank linkage of Example 3.3 in the posture shown in Fig. 4.17a, the constant angular velocity of link 2 is $\omega_2 = 36$ rad/s cw. Determine the angular acceleration of link 4.

Example 4.6 (cont.)

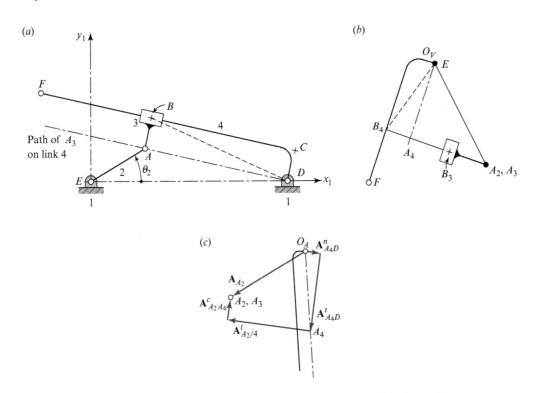

Fig. 4.17 (a) Inverted slider-crank linkage; (b) velocity polygon; (c) acceleration polygon.

Solution

The velocity analysis of this linkage in the given posture was performed in Example 3.3. The results are

$$V_{A_2} = 9 \text{ ft/s}, \quad V_{A_4D} = 7.17 \text{ ft/s}, \quad V_{A_2/4} = 5.48 \text{ ft/s}, \quad \text{and} \quad \omega_3 = \omega_4 = 7.55 \text{ rad/s ccw}.$$

The velocity polygon is shown in Fig. 3.10b, and is repeated here as Fig. 4.17b.
The acceleration of point A on link 2 can be written as

$$\mathbf{A}_{A_2} = \overset{\mathbf{0}}{\cancel{\mathbf{A}_E}} + \mathbf{A}^n_{A_2E} + \overset{\mathbf{0}}{\cancel{\mathbf{A}^t_{A_2E}}}, \tag{1}$$

where

$$A^n_{A_2E} = \omega_2^2 R_{A_2E} = (36 \text{ rad/s})^2 \left(\frac{3}{12} \text{ ft}\right) = 324 \text{ ft/s}^2. \tag{2}$$

We plot this to scale, locating the acceleration image point, A_2, as shown in Fig. 4.17c.

Example 4.6 (cont.)

Note that point A_2 travels along the straight-line path indicated, as seen by an observer on link 4. Knowing this path, we write

$$\mathbf{A}_{A_2} \overset{\sqrt{\sqrt{}}}{=} \mathbf{A}_{A_4} \overset{?\sqrt{}}{+} \mathbf{A}^c_{A_2 A_4} \overset{\sqrt{\sqrt{}}}{+} \mathbf{A}^n_{A_2/4} \overset{0}{+} \mathbf{A}^t_{A_2/4}, \tag{3}$$

where

$$A^c_{A_2 A_4} = 2\omega_4 V_{A_2/4} = 2(7.55 \text{ rad/s})(5.48 \text{ ft/s}) = 82.7 \text{ ft/s}^2, \tag{4}$$

and $A^n_{A_2/4} = 0$, since $\rho = \infty$. The acceleration, $\mathbf{A}_{A_4}$, in Eq. (3) was marked as having only one unknown, since

$$\mathbf{A}_{A_4} \overset{0}{=} \mathbf{A}_D \overset{\sqrt{\sqrt{}}}{+} \mathbf{A}^n_{A_4 D} \overset{?\sqrt{}}{+} \mathbf{A}^t_{A_4 D}, \tag{5}$$

where

$$A^n_{A_4 D} = \frac{V^2_{A_4 D}}{R_{A_4 D}} = \frac{(7.17 \text{ ft/s})^2}{(11.5/12) \text{ ft}} = 53.6 \text{ ft/s}^2. \tag{6}$$

This component can now be added from O_A, followed by a line of unknown length for the $\mathbf{A}^t_{A_4 D}$ component. Since image point A_4 is not yet known, components $\mathbf{A}^c_{A_2 A_4}$ and $\mathbf{A}^t_{A_2/4}$ cannot be added as indicated by Eq. (3). However, these two components can be transferred to the other side of the equation and graphically subtracted from image point A_2, thus completing the acceleration polygon. The angular acceleration of link 4 can then be found:

$$\alpha_4 = \frac{A^t_{A_4 D}}{R_{AD}} = \frac{279.9 \text{ ft/s}^2}{(11.5/12)\text{ft}} = 292 \text{ rad/s}^2 \text{ ccw.} \qquad\qquad \textit{Ans.}$$

This need to subtract vectors is common in acceleration problems involving the Coriolis component and should be studied carefully. Note that the equation involving $\mathbf{A}_{A_4/2}$ cannot be used, since ρ and, therefore, $A^n_{A_4/2}$ would become an additional (third) unknown.

Since the angular acceleration of link 3 must be equal to the angular acceleration of link 4, then the acceleration of point B_3 can also be obtained.

4.6 Apparent Angular Acceleration

Completeness suggests that we should define the term *apparent angular acceleration*. When two rigid bodies rotate with different angular accelerations, the vector difference between them is defined as the apparent angular acceleration,

$$\boldsymbol{\alpha}_{3/2} = \boldsymbol{\alpha}_3 - \boldsymbol{\alpha}_2.$$

The apparent-angular-acceleration equation can also be written as

$$\alpha_3 = \alpha_2 + \alpha_{3/2}. \tag{4.18}$$

We recognize that $\alpha_{3/2}$ is the angular acceleration of body 3 as it would appear to an observer attached to, and rotating with, body 2.

4.7   Direct Contact and Rolling Contact

We recall from Section 3.7 that the relative motion between two bodies in direct contact at a point can occur in two different ways: there may be an apparent slipping velocity between the bodies or there may be no such slip. The purpose of this section is to extend these concepts to include acceleration. The following two examples illustrate these two cases; the first example is for direct contact with slipping, and the second example is for rolling contact.

Example 4.7

Consider the circular cam, link 2, in direct contact with the oscillating flat-faced follower, link 3, in the posture shown in Fig. 4.18a. The angular velocity and angular acceleration of link 2 are $\omega_2 = 10$ rad/s cw and $\alpha_2 = 25$ rad/s^2 cw, respectively. Determine the angular acceleration of link 3.

Fig. 4.18 (a) $R_{BA} = 3$ in, $R_{CB} = 1.5$ in, $R_{CD}^x = -1.4$ in, $R_{CD}^y = 2.14$ in, and $R_{DA} = 5$ in; (b) velocity polygon.

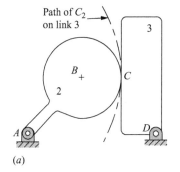

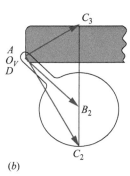

(a)   (b)

Solution

The velocity polygon must be drawn first, since the velocities of points B and C and the angular velocity of link 3 are required for acceleration analysis.

The magnitude of the velocity of point B is

$$V_B = \omega_2 R_{BA} = (10 \text{ rad/s})(3/12\text{ft}) = 2.5 \text{ ft/s}.$$

Using this value, the velocity polygon is now drawn as shown in Fig. 4.18b. Then, from the polygon, we measure the magnitude of the velocity difference of points C_3 and D as

$$V_{C_3D} = 2.13 \text{ ft/s}.$$

Example 4.7 (cont.)

Therefore, the angular velocity of link 3 is

$$\omega_3 = \frac{V_{C_3D}}{R_{CD}} = \frac{2.13\ \text{ft/s}}{(2.56/12)\ \text{ft}} = 9.98\ \text{rad/s cw},$$

where the direction is obtained from an examination of the velocity polygon.

Now, to write an acceleration equation relating points of link 2 and link 3, we look for a pair of coincident points where the curvature of the path is known. We note that the curvature is not known for the path of point C_2 on link 3 (Fig. 4.18a). However, considering the path traced by point B_2 on (extended) link 3, we note that it remains a constant distance from the surface of (the actual) link 3. This path is a straight line, as shown by the "slot" in the equivalent mechanism of Fig. 4.18c. Note that link 3 of this equivalent mechanism has motion equivalent to the original mechanism.

Since this "slot" can be used for a path, it becomes clear how to proceed; the appropriate equation is

$$\mathbf{A}_{B_2} \overset{\sqrt{\sqrt{}}}{=} \mathbf{A}_{B_3} \overset{??}{+} \mathbf{A}^c_{B_2B_3} \overset{\sqrt{\sqrt{}}}{+} \mathbf{A}^n_{B_2/3} \overset{\sqrt{\sqrt{}}}{+} \mathbf{A}^t_{B_2/3}, \overset{?\sqrt{}}{} \tag{1}$$

where B_3 is a point coincident with B_2 but attached to an extension of link 3.

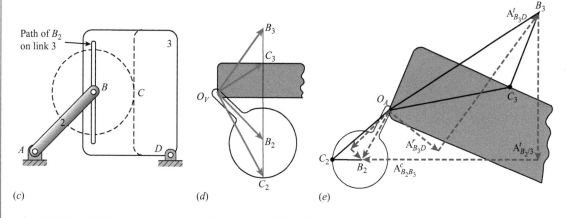

Fig. 4.18 (cont.) (c) equivalent mechanism; (d) extended velocity polygon; (e) acceleration polygon.

Since the path (the slot) of Eq. (1) is a straight line, we can extend the velocity polygon (Fig. 4.18b) as shown in Fig. 4.18d to find the velocities,

$$V_{B_2/3} = 4.17\ \text{ft/s} \quad \text{and} \quad V_{B_3} = 2.99\ \text{ft/s}.$$

Next, the term on the left-hand side of Eq. (1) can be written as

$$\mathbf{A}_{B_2} = \overset{0}{\cancel{\mathbf{A}_A}} + \mathbf{A}^n_{B_2A} + \mathbf{A}^t_{B_2A},$$

Example 4.7 (cont.)

where

$$A^n_{B_2A} = \omega_2^2 R_{B_2A} = (10 \text{ rad/s})^2(3/12 \text{ ft}) = 25 \text{ ft/s}^2,$$

$$A^t_{B_2A} = \alpha_2 R_{B_2A} = (25 \text{ rad/s}^2)(3/12 \text{ ft}) = 6.25 \text{ ft/s}^2.$$

These are plotted on the acceleration polygon as shown in Fig. 4.18e.

Then, we calculate the known magnitudes of the terms on the right-hand side of Eq. (1):

$$A^n_{B_2/3} = \frac{V^2_{B_2/3}}{\rho} = \frac{(4.17 \text{ ft/s})^2}{\infty} = 0 \tag{2}$$

and

$$A^c_{B_2B_3} = 2\omega_3 V_{B_2/3} = 2(10 \text{ rad/s})(4.17 \text{ ft/s}) = 83.4 \text{ ft/s}^2. \tag{3}$$

The acceleration of B_3 is determined from the acceleration-difference equation,

$$\overset{??}{\mathbf{A}_{B_3}} = \overset{\mathbf{0}}{\cancel{\mathbf{A}_D}} + \overset{\sqrt{\sqrt{}}}{\mathbf{A}^n_{B_3D}} + \overset{?\sqrt{}}{\mathbf{A}^t_{B_3D}}, \tag{4}$$

where

$$A^n_{B_3D} = \frac{V^2_{B_3D}}{R_{B_3D}} = \frac{(2.99 \text{ ft/s})^2}{(3.6/12 \text{ ft})} = 29.8 \text{ ft/s}^2. \tag{5}$$

Substituting Eqs. (2) through (5) into Eq. (1) and rearranging terms, we arrive at an equation with only two unknowns:

$$\overset{\sqrt{\sqrt{}}}{\mathbf{A}_{B_2}} - \overset{\sqrt{\sqrt{}}}{\mathbf{A}^c_{B_2B_3}} - \overset{?\sqrt{}}{\mathbf{A}^t_{B_2/3}} = \overset{\sqrt{\sqrt{}}}{\mathbf{A}^n_{B_3D}} + \overset{?\sqrt{}}{\mathbf{A}^t_{B_3D}}. \tag{6}$$

This equation is solved graphically as shown in Fig. 4.18e, and the results are

$$A^t_{B_2/3} = 67.2 \text{ ft/s}^2 \quad \text{and} \quad A^t_{B_3D} = 78.7 \text{ ft/s}^2.$$

Once image point B_3 has been determined, image point C_3 is found by constructing the acceleration image of triangle DB_3C_3, all on link 3. Figure 4.18e has been extended to show the acceleration images of links 2 and 3 to aid in visualization and to show once again that there is no obvious relation between the final locations of image points C_2 and C_3.

Finally, the angular acceleration of link 3 is determined:

$$\alpha_3 = \frac{A^t_{B_3D}}{R_{B_3D}} = \frac{78.7 \text{ ft/s}^2}{(3.6/12 \text{ ft})} = 262 \text{ rad/s}^2 \text{ cw.} \qquad\qquad \textit{Ans.}$$

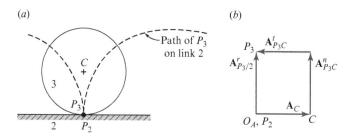

Fig. 4.19 (*a*) A rolling wheel; (*b*) Apparent acceleration at the point of rolling contact.

Recall from Section 3.7 that we defined the term *rolling contact* to imply that no slip is in progress, and developed the rolling contact condition, Eq. (3.13), to indicate that the apparent velocity at such a point is zero. Here we will investigate the apparent acceleration at a point of rolling contact.

Consider the case of a circular wheel 3, in rolling contact with the fixed straight link 2, as shown in Fig. 4.19. Although this is admittedly a very simplified case, the arguments made and the conclusions reached are completely general and apply to any rolling contact situation, no matter what the shapes of the two bodies or whether either is the ground link. To keep this clear in our minds, the ground link has been numbered here as link 2.

Once the acceleration, $\mathbf{A}_C$, of the center point of the wheel is known, an origin, O_A, can be chosen, and an acceleration polygon can be started by plotting $\mathbf{A}_C$, see Fig. 4.19*b*. In relating the accelerations of points. P_3 and P_2 at the point of rolling contact, however, we are dealing with two coincident points of different bodies. Therefore, it is appropriate to use the apparent-acceleration equation, Eq. (4.17) [and not the acceleration-difference equation, Eq. (4.10)]. To do this, we must identify a path that one of these points traces on the other body. The path that point P_3 traces on link 2 is sketched in the figure.[4] Although the precise shape of this path depends on the shapes of the two contacting links, provided that there is no slip, there is always a cusp at the point of rolling contact, and the tangent to this cusp-shaped path is always normal to the surfaces that are in contact.

Since this path is known, we are free to write the apparent-acceleration equation:

$$\mathbf{A}_{P_3} = \mathbf{A}_{P_2} + \mathbf{A}^c_{P_3 P_2} + \mathbf{A}^n_{P_3/2} + \mathbf{A}^t_{P_3/2}.$$

In evaluating the components, we keep in mind the rolling contact velocity condition, Eq. (3.13), that is, $\mathbf{V}_{P_3/2} = \mathbf{0}$. Therefore,

$$\mathbf{A}^c_{P_3 P_2} = 2\boldsymbol{\omega}_2 \times \mathbf{V}_{P_3/2} = \mathbf{0} \quad \text{and} \quad A^n_{P_3/2} = \frac{V^2_{P_3/2}}{\rho} = 0.$$

Therefore, only one component of the apparent-acceleration equation – namely, $\mathbf{A}^t_{P_3/2}$ – can be nonzero. To avoid possible confusion in calling this term a tangential component (tangent to the cusp-shaped path) while its direction is *normal* to the rolling surfaces, we adopt a new superscript and refer to it as the *rolling contact acceleration*, $\mathbf{A}^r_{P_3/2}$.

---

[4] This particular curve is called a *cycloid*.

For rolling contact with no slip, the apparent-acceleration equation (henceforth referred to as the *rolling contact condition for acceleration*) becomes

$$\mathbf{A}_{P_3} = \mathbf{A}_{P_2} + \mathbf{A}^r_{P_3/2}, \tag{4.19}$$

where the component $\mathbf{A}^r_{P_3/2}$ is always normal to the surfaces at the point of rolling contact.

Example 4.8

Consider the circular roller, 4, rolling without slip on the oscillating flat-faced follower, 3, in the posture shown in Fig. 4.20a. The angular velocity and angular acceleration of input link 2 are $\omega_2 = 10$ rad/s cw and $\alpha_2 = 25$ rad/s² cw, respectively. Determine the angular accelerations of the roller and the follower.

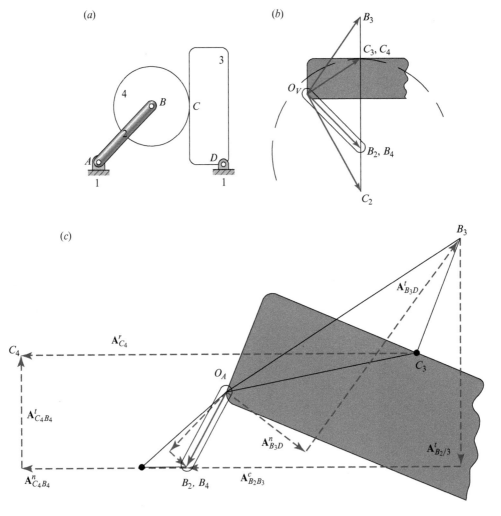

Fig. 4.20 (*a*) Rolling contact mechanism; (*b*) velocity polygon; (*c*) acceleration polygon.

Example 4.8 (cont.)

Solution

Since this problem is an extension of Example 4.7, in which a roller (link 4) has been included, then the velocity polygon can be easily completed as in Fig. 4.20b. The rolling contact condition for velocity implies that $\mathbf{V}_{C_4} = \mathbf{V}_{C_3}$, which allows drawing the velocity image of link 4 as indicated.

Next, we might be tempted to write the apparent-acceleration equation for point C_4 :

$$\mathbf{A}_{C_4} = \overset{\sqrt{\sqrt{}}}{\mathbf{A}^n_{B_2A_2}} + \overset{\sqrt{\sqrt{}}}{\mathbf{A}^t_{B_2A_2}} + \overset{\sqrt{\sqrt{}}}{\mathbf{A}^n_{C_4B_4}} + \overset{?\sqrt{}}{\mathbf{A}^t_{C_4B_4}} = \overset{\sqrt{\sqrt{}}}{\mathbf{A}^n_{C_3D_3}} + \overset{?\sqrt{}}{\mathbf{A}^t_{C_3D_3}} + \overset{?\sqrt{}}{\mathbf{A}^r_{C_4/3}}. \tag{1}$$

Unfortunately, this equation cannot be solved, since it contains three unknowns – namely, the two angular accelerations, α_3 and α_4, and the rolling contact acceleration, $A^r_{C_4/3}$. A different solution strategy must be sought.

We proceed as in Example 4.7, Fig. 4.18e, all the way to the solution for $\boldsymbol{\alpha}_3$ and to finding $\mathbf{A}_{C_3}$. From this, we will have constructed a good portion of the acceleration polygon of Fig. 4.20c. Now we can relate the acceleration of point C_4 to that of B_4; that is,

$$\overset{??}{\mathbf{A}_{C_4}} = \overset{\sqrt{\sqrt{}}}{\mathbf{A}_{B_4}} + \overset{\sqrt{\sqrt{}}}{\mathbf{A}^n_{C_4B_4}} + \overset{?\sqrt{}}{\mathbf{A}^t_{C_4B_4}}, \tag{2}$$

where

$$A^n_{C_4B_4} = \frac{V^2_{C_4B_4}}{R_{CB}} = \frac{(2.87 \text{ ft/s})^2}{(1.5/12 \text{ ft})} = 65.9 \text{ ft/s}^2. \tag{3}$$

We can also write the rolling contact acceleration condition [Eq. (4.19)] as

$$\overset{??}{\mathbf{A}_{C_4}} = \overset{\sqrt{\sqrt{}}}{\mathbf{A}_{C_3}} + \overset{?\sqrt{}}{\mathbf{A}^r_{C_4/3}}. \tag{4}$$

Remembering that $\mathbf{A}^r_{C_4/3}$ is perpendicular to the surfaces at C, we graphically construct the simultaneous solution to Eqs. (2) and (4) as shown in Fig. 4.20c. Finally, the angular acceleration of the roller 4 can be obtained as follows:

$$\alpha_4 = \frac{A^t_{C_4B_4}}{R_{CB}} = \frac{65.9 \text{ ft/s}^2}{(1.5/12 \text{ ft})} = 527 \text{ rad/s}^2 \text{ ccw.} \qquad \textit{Ans.}$$

The angular acceleration of the follower is identical to that found in Example 4.7, namely

$$\alpha_3 = 262 \text{ rad/s}^2 \text{ cw.} \qquad \textit{Ans.}$$

In general, it is almost always necessary to determine the motions of the two links adjacent on either side of a point of rolling contact (links 3 and 4 in Example 4.8) and then to perform the rolling contact computation working from both sides. It is almost never possible to work straight through a point of direct contact, as tried in Eq. (1) of Example 4.8. This almost always requires the visualization of an equivalent mechanism and the solution of a Coriolis equation. However, with a little practice, it is usually quite straightforward.

4.8   Systematic Strategy for Acceleration Analysis

Review of the preceding sections and example problems will demonstrate that we have now developed sufficient tools for dealing with those situations that arise repeatedly in the acceleration analysis of planar rigid-body mechanical systems. It will also be noted that the word "relative" acceleration has been carefully avoided. Instead, as with velocity analysis, whenever the desire for using "relative" acceleration arises, there are always two points whose accelerations are to be "related"; also, these two points are fixed either in the same rigid body or in two different rigid bodies. Therefore, as with velocity analysis (Table 3.1), we can organize all situations into the four cases shown in Table 4.1.

In Table 4.1 we see that, when the two points are separated by a distance, only the acceleration-difference equation is appropriate for use, and two points in the *same link* should be used. When it is desirable to switch to another link, then *coincident points* should be chosen and the apparent-acceleration equation should be used. The path of one of these points in a coordinate system on the other link is then required.

Even the notation has been made different to continually remind us that these are two totally different situations and the formulae are not interchangeable between the two. We should not

Table 4.1 "Relative" acceleration equations

Points are	Coincident	Separated
In same body	*Trivial case:* $\mathbf{A}_P = \mathbf{A}_Q.$	*Acceleration difference:* $\mathbf{A}_P = \mathbf{A}_Q + \mathbf{A}_{PQ}^n + \mathbf{A}_{PQ}^t$ $\mathbf{A}_{PQ}^n = \boldsymbol{\omega} \times (\boldsymbol{\omega} \times \mathbf{R}_{PQ})$ $\mathbf{A}_{PQ}^t = \boldsymbol{\alpha} \times \mathbf{R}_{PQ}.$
In different bodies	*Apparent acceleration:* $\mathbf{A}_{P_i} = \mathbf{A}_{P_j} + \mathbf{A}_{P_iP_j}^c + \mathbf{A}_{P_i/j}^n + \mathbf{A}_{P_i/j}^t$ *where path $P_{i/j}$ is known, and* $\mathbf{A}_{P_iP_j}^c = 2\boldsymbol{\omega}_j \times \mathbf{V}_{P_{i/j}}$ $\mathbf{A}_{P_i/j}^n = \dfrac{V_{P_{i/j}}^2}{\rho}\,\hat{\mathbf{u}}^n$ $\mathbf{A}_{P_i/j}^t = \dfrac{d^2s}{dt^2}\,\hat{\mathbf{u}}^t.$ *Rolling contact acceleration:* $\mathbf{A}_{P_i} = \mathbf{A}_{P_j} + \mathbf{A}_{P_i/j}^r$ *where path* $\mathbf{A}_{P_i/j}^r$ *is normal to surfaces at point* *of contact.*	*Too general; use two steps.*

try to use an "$\boldsymbol{\alpha} \times \mathbf{R}$" formula when the apparent acceleration is required; however, if we do try, then we will not find an appropriate $\boldsymbol{\alpha}$ or $\mathbf{R}$. Similarly, when using the acceleration difference, there is no question of which $\boldsymbol{\alpha}$ to use since only one link pertains. The two questions that always arise with respect to apparent acceleration are: (a) When should we include the Coriolis term and when should we only use normal and tangential components? (b) Which $\boldsymbol{\omega}$ should we use in the Coriolis term? The answer to the first question is relatively straightforward: Whenever we use the apparent-acceleration equation, the Coriolis term should always be included; if it should not be there, such as when the "path" is not rotating, it should be included anyway, and the calculation will give it a magnitude of zero. The answer to the second question is also straightforward: Whenever we use the apparent-acceleration equation, we must always visualize a path that a point P_i makes on another link j; then $\boldsymbol{\omega}_j$ for the link that contains the path is used. As indicated in Section 4.7 and Table 4.1, rolling contact acceleration is a special case of apparent acceleration.

Careful review of Examples 4.2–4.8 demonstrates how the strategy suggested in Table 4.1 and the labeling strategy is applied in a variety of problems of acceleration analysis.

4.9 Algebraic Acceleration Analysis

In this section we will continue the analytic approach that we began in Section 3.9. Again, we focus on the central (in-line) slider-crank linkage of Fig. 3.15, reproduced here as Fig. 4.21.

The equation for the exact acceleration of the slider might be obtained by differentiating Eq. (3.15b) with respect to time. However, this is a difficult differentiation and the resulting expression is very unwieldy. Alternatively, we differentiate Eq. (a) of Section 3.9 twice with respect to time, and after considerable manipulation, the exact acceleration of the slider becomes

$$\ddot{x} = -r\omega^2 \left(\cos\theta + \frac{r\cos 2\theta}{l\cos\phi} + \frac{r^3 \sin^2 2\theta}{4l^3 \cos^3 \phi} \right) - r\alpha \left(\sin\theta + \frac{r\sin 2\theta}{2l\cos\phi} \right). \qquad (4.20)$$

This is still a rather complex expression, and it becomes even more complex when ϕ is eliminated by using Eq. (d) of Section 3.9; that is

$$\cos\phi = \sqrt{1 - \left(\frac{r}{l}\sin\theta \right)^2}. \qquad (4.21)$$

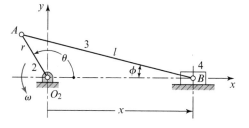

Fig. 4.21 Central slider-crank linkage.

In some mechanism applications, approximate expressions are used for velocity and acceleration analyses. In Section 3.9, the binomial expansion was used to approximate the velocity of the slider [Eq. (3.15e)]. Here we will obtain an approximate expression for the acceleration of the slider. For small values of (r/l), the last term in the first bracket of Eq. (4.20) can be neglected. Also, in this case, $\cos\phi$ is near unity; therefore, the acceleration of the slider can be written as

$$\ddot{x} = -r\omega^2\left(\cos\theta + \frac{r}{l}\cos 2\theta\right) - r\alpha\left(\sin\theta + \frac{r}{2l}\sin 2\theta\right). \tag{4.22a}$$

Note that this result could have been obtained directly by differentiating Eq. (3.16b) with respect to time. Also, in the special case that the input angular velocity is constant ($\alpha = 0$), then the acceleration of the slider is

$$\ddot{x} = -r\omega^2\left(\cos\theta + \frac{r}{l}\cos 2\theta\right). \tag{4.22b}$$

4.10 Complex-Algebraic Acceleration Analysis

Let us now see how Raven's method (Section 3.10) is extended to the analysis of accelerations. The general approach is outlined here for the offset slider-crank linkage shown in Fig. 4.22.

In complex polar form, the loop-closure equation is

$$r_2 e^{j\theta_2} + r_3 e^{j\theta_3} - r_1 e^{-j(\pi/2)} - r_4 e^{j0} = 0. \tag{a}$$

If we separate this equation into its real and imaginary components, we obtain the two posture equations:

$$r_2 \cos\theta_2 + r_3 \cos\theta_3 - r_4 = 0, \tag{b}$$

$$r_2 \sin\theta_2 + r_3 \sin\theta_3 + r_1 = 0. \tag{c}$$

With r_1, r_2, and r_3 given, and for a known input angle, θ_2, these two equations can be solved for the two variables θ_3 and r_4. In this example, Eq. (c) can be solved for the variable θ_3; that is,

Fig. 4.22 Offset slider-crank linkage.

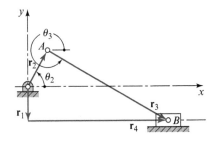

$$\theta_3 = \sin^{-1}\left(\frac{-r_1 - r_2 \sin\theta_2}{r_3}\right). \tag{4.23}$$

Then Eq. (b) can be solved for the variable r_4; that is,

$$r_4 = r_2 \cos\theta_2 + r_3 \cos\theta_3. \tag{4.24}$$

Differentiating Eq. (a) with respect to time gives the velocity equation, which is equivalent to the equation of the velocity polygon in exponential form. After replacing $\dot\theta$ with ω, the result is

$$jr_2\omega_2 e^{j\theta_2} + jr_3\omega_3 e^{j\theta_3} - \dot r_4 e^{j0} = 0. \tag{d}$$

Raven's method, as we have seen, consists of making the trigonometric transformation and separating the result into real and imaginary terms, as was done in obtaining Eqs. (b) and (c) from Eq. (a). When this procedure is carried out for the velocity equation in this example, Eq. (d), we find

$$\omega_3 = -\frac{r_2\cos\theta_2}{r_3\cos\theta_3}\omega_2 \tag{4.25a}$$

and

$$\dot r_4 = -r_2\sin\theta_2\omega_2 - r_3\sin\theta_3\omega_3. \tag{4.25b}$$

We use the same approach to obtain the acceleration equations by first differentiating Eq. (d). Then, separating the result into its real and imaginary components gives two equations that can be solved for the two acceleration unknowns, namely,

$$\alpha_3 = \frac{r_2\sin\theta_2\omega_2^2 - r_2\cos\theta_2\alpha_2 + r_3\sin\theta_3\omega_3^2}{r_3\cos\theta_3}, \tag{4.26}$$

$$\ddot r_4 = -r_2\cos\theta_2\omega_2^2 - r_2\sin\theta_2\alpha_2 - r_3\cos\theta_3\omega_3^2 - r_3\sin\theta_3\alpha_3. \tag{4.27}$$

If the input angular velocity is constant – that is, if $\alpha_2 = 0$ – then the acceleration results are

$$\alpha_3 = \frac{r_2\sin\theta_2\omega_2^2 + r_3\sin\theta_3\omega_3^2}{r_3\cos\theta_3}, \tag{4.28}$$

$$\ddot r_4 = -r_2\cos\theta_2\omega_2^2 - r_3\cos\theta_3\omega_3^2 - r_3\sin\theta_3\alpha_3. \tag{4.29}$$

The same procedure, when carried out for the four-bar linkage, gives results that can be used for computer solutions of both the crank-rocker and drag-link mechanisms. Unfortunately, they cannot be used for other four-bar linkages unless an arrangement is included in the program to cause the solution to stop when an extreme posture is reached.

If the vector loop-closure equation for the four-bar linkage (Fig. 3.18) is written as

$$\mathbf{r}_1 + \mathbf{r}_2 + \mathbf{r}_3 - \mathbf{r}_4 = \mathbf{0}, \tag{e}$$

where the subscripts are the link numbers and where link 2 is the driver having a constant input angular velocity, then the acceleration relations that are obtained by Raven's method are

$$a_3 = \frac{r_2 \cos(\theta_2 - \theta_4)\omega_2^2 + r_3 \cos(\theta_3 - \theta_4)\omega_3^2 - r_4\omega_4^2}{r_3 \sin(\theta_4 - \theta_3)}, \tag{4.30}$$

$$a_4 = \frac{r_2 \cos(\theta_2 - \theta_3)\omega_2^2 - r_4 \cos(\theta_3 - \theta_4)\omega_4^2 + r_3\omega_3^2}{r_4 \sin(\theta_4 - \theta_3)}. \tag{4.31}$$

4.11 Method of Kinematic Coefficients

Again we employ the four-bar linkage as an example of this method of solution. The angular accelerations of links 3 and 4 can be obtained by differentiating Eq. (3) of Example 3.7, with respect to the input angle θ_2; that is,

$$-r_3 \cos \theta_3 {\theta_3'}^2 - r_3 \sin \theta_3 \theta_3'' + r_4 \cos \theta_4 {\theta_4'}^2 + r_4 \sin \theta_4 \theta_4'' = r_2 \cos \theta_2, \tag{4.32}$$

$$-r_3 \sin \theta_3 {\theta_3'}^2 + r_3 \cos \theta_3 \theta_3'' + r_4 \sin \theta_4 {\theta_4'}^2 - r_4 \cos \theta_4 \theta_4'' = r_2 \sin \theta_2, \tag{4.33}$$

where, by definition,

$$\theta_3'' = \frac{d^2\theta_3}{d\theta_2^2} \quad \text{and} \quad \theta_4'' = \frac{d^2\theta_4}{d\theta_2^2},$$

and they are referred to as the *second-order kinematic coefficients* of links 3 and 4. Writing Eqs. (4.32) and (4.33) in matrix form gives

$$\begin{bmatrix} -r_3 \sin \theta_3 & r_4 \sin \theta_4 \\ r_3 \cos \theta_3 & -r_4 \cos \theta_4 \end{bmatrix} \begin{bmatrix} \theta_3'' \\ \theta_4'' \end{bmatrix} = \begin{bmatrix} B_1 \\ B_2 \end{bmatrix}, \tag{4.34}$$

where

$$\begin{aligned} B_1 &= r_2 \cos \theta_2 + r_3 \cos \theta_3 {\theta_3'}^2 - r_4 \cos \theta_4 {\theta_4'}^2, \\ B_2 &= r_2 \sin \theta_2 + r_3 \sin \theta_3 {\theta_3'}^2 - r_4 \sin \theta_4 {\theta_4'}^2, \end{aligned} \tag{4.35}$$

and they are known from position and velocity analyses. Also, note that the (2×2) coefficient matrix on the left-hand side of Eq. (4.34) is the same as the (2×2) coefficient matrix in velocity analysis (Eq. (5) in Example 3.7).[5] This is not a coincidence. The coefficient matrix is the same for all kinematic derivatives – that is, first-order, second-order, third-order kinematic coefficients, and so on. Therefore, this matrix provides a built-in check of all higher-order differentiation. The determinant of the coefficient matrix is as given by Eq. (6) in Example 3.7; that is,

$$\Delta = -r_3 r_4 \sin(\theta_4 - \theta_3). \tag{4.36}$$

---

[5] This coefficient matrix is called the *Jacobian* of the system.

Table 4.2 Summary of second-order kinematic coefficients

	Variable of interest Angle θ_j (use symbol θ_j'' for kinematic coefficient regardless of input)	Variable of interest Magnitude r_j (use symbol r_j'' for kinematic coefficient regardless of input)
Input $\psi = $ angle θ_i	$\alpha_j = \theta_j'' \dot{\psi}^2 + \theta_j' \ddot{\psi}$	$\ddot{r}_j = r_j'' \dot{\psi}^2 + r_j' \ddot{\psi}$
	$\theta_j' = \dfrac{d\theta_j}{d\psi}$ (dimensionless, rad/rad)	$r_j' = \dfrac{dr_j}{d\psi}$ (length, length/rad)
	$\theta_j'' = \dfrac{d^2\theta_j}{d\psi^2}$ (dimensionless, rad/rad^2)	$r_j'' = \dfrac{d^2 r_j}{d\psi^2}$ (length, length/rad^2)
Input $\psi = $ magnitude r_i	$\alpha_j = \theta_j'' \dot{\psi}^2 + \theta_j' \ddot{\psi}$	$\ddot{r}_j = r_j'' \dot{\psi}^2 + r_j' \ddot{\psi}$
	$\theta_j' = \dfrac{d\theta_j}{d\psi}$ (1/length, rad/length)	$r_j' = \dfrac{dr_j}{d\psi}$ (dimensionless, length/length)
	$\theta_j'' = \dfrac{d^2\theta_j}{d\psi^2}$ (1/length^2, rad/length^2)	$r_j'' = \dfrac{d^2 r_j}{d\psi^2}$ (1/length, length/length^2)

The second-order kinematic coefficients for links 3 and 4 can be obtained from Eq. (4.34) by using Cramer's rule, and the results are

$$\theta_3'' = \frac{B_1 \cos\theta_4 + B_2 \sin\theta_4}{r_3 \sin(\theta_4 - \theta_3)} \quad \text{and} \quad \theta_4'' = \frac{B_1 \cos\theta_3 + B_2 \sin\theta_3}{r_4 \sin(\theta_4 - \theta_3)}. \tag{4.37}$$

Note that the second-order kinematic coefficients for the four-bar linkage are nondimensional (rad/rad^2). The angular accelerations of links 3 and 4, obtained from the chain rule, are

$$\alpha_3 = \theta_3'' \omega_2^2 + \theta_3' \alpha_2 \quad \text{and} \quad \alpha_4 = \theta_4'' \omega_2^2 + \theta_4' \alpha_2. \tag{4.38}$$

Table 4.2 summarizes the second-order kinematic coefficients that are related to link j of a planar mechanism (that is, vector $\mathbf{r}_j$) having (a) variable angle θ_j and/or (b) variable magnitude r_j.

To illustrate the method of kinematic coefficients for determining the angular accelerations of links and the accelerations of moving points, consider the following three examples.

Example 4.9

For the four-bar linkage of Example 4.2, determine the angular accelerations of links 3 and 4, and the absolute accelerations of points E and F.

Solution

Recall that the first-order kinematic coefficients from Eq. (8) in Example 3.7 are

$$\theta_3' = 0.271\,8 \text{ rad/rad} \quad \text{and} \quad \theta_4' = 0.526\,5 \text{ rad/rad}. \tag{1}$$

Substituting Eqs. (1) and the known data into Eqs. (4.35) and substituting the results into Eq. (4.37), the second-order kinematic coefficients of links 3 and 4 are

Example 4.9 (cont.)

$$\theta_3'' = 0.013\ 1\ \text{rad/rad}^2 \quad \text{and} \quad \theta_4'' = -0.203\ 0\ \text{rad/rad}^2. \tag{2}$$

Substituting the constant angular velocity, $\omega_2 = 94.25$ rad/s ccw, and Eqs. (1) and (2) into Eqs. (4.38), the angular accelerations of links 3 and 4 are

$$\alpha_3 = 116\ \text{rad/s}^2\ \text{ccw} \quad \text{and} \quad \alpha_4 = 1\ 801\ \text{rad/s}^2\ \text{cw}. \qquad Ans.$$

These results are in good agreement with those obtained from the graphic method of Example 4.2; that is, $\alpha_3 = 113$ rad/s^2 ccw and $\alpha_4 = 1\ 822$ rad/s^2 cw.

To obtain the acceleration of point E, we differentiate Eqs. (13) of Example 3.7 with respect to the input angle θ_2. Therefore, the second-order kinematic coefficients for point E are

$$x_E'' = -r_2 \cos \theta_2 - r_{EB} \cos(\theta_3 - \phi)\theta_3'^2 - r_{EB} \sin(\theta_3 - \phi)\theta_3'', $$
$$y_E'' = -r_2 \sin \theta_2 - r_{EB} \sin(\theta_3 - \phi)\theta_3'^2 + r_{EB} \cos(\theta_3 - \phi)\theta_3''. \tag{3}$$

Substituting Eqs. (1) and (2) and the known data into Eqs. (3) gives

$$x_E'' = 1.206\ 7\ \text{in/rad}^2 \quad \text{and} \quad y_E'' = -3.310\ 8\ \text{in/rad}^2. \tag{4}$$

Differentiating the velocity of point E from Eq. (15) in Example 3.7 with respect to time, the acceleration of point E can be written as

$$\mathbf{A}_E = \left(x_E'' \omega_2^2 + x_E' \alpha_2\right)\hat{\mathbf{i}} + \left(y_E'' \omega_2^2 + y_E' \alpha_2\right)\hat{\mathbf{j}}. \tag{5}$$

Recall that the first-order kinematic coefficients for point E from Eq. (14) in Example 3.7 are

$$x_E' = -3.419\ 1\ \text{in/rad} \quad \text{and} \quad y_E' = 0.926\ 9\ \text{in/rad}. \tag{6}$$

Substituting Eqs. (4) and (6), and the angular velocity and acceleration of the input link, into Eq. (5) gives

$$\mathbf{A}_E = 892.32\hat{\mathbf{i}} - 2\ 448.24\hat{\mathbf{j}}\ \text{ft/s}^2. \qquad Ans.$$

Therefore, the magnitude of the acceleration of point E is $A_E = 2\ 606$ ft/s^2. This result is in good agreement with that in Example 4.2 – that is, $A_E = 2\ 580$ ft/s^2.

The acceleration of point F can be obtained in a similar manner. The first-order and second-order kinematic coefficients for point F are

$$x_F' = -2.626\ 2\ \text{in/rad}, \quad \text{and} \quad y_F' = 3.068\ 6\ \text{in/rad},$$

$$x_F'' = -0.586\ 0\ \text{in/rad}^2, \quad \text{and} \quad y_F'' = -2.551\ 7\ \text{in/rad}^2. \tag{7}$$

Similar to Eq. (5), the acceleration of point F can be written as

$$\mathbf{A}_F = \left(x_F'' \omega_2^2 + x_F' \alpha_2\right)\hat{\mathbf{i}} + \left(y_F'' \omega_2^2 + y_F' \alpha_2\right)\hat{\mathbf{j}}. \tag{8}$$

Example 4.9 (cont.)

Substituting Eqs. (7) and the angular velocity and acceleration of the input into Eq. (8), the acceleration of point F is

$$\mathbf{A}_F = -433\hat{\mathbf{i}} - 1\,886\hat{\mathbf{j}} \text{ ft/s}^2. \hspace{3cm} \textit{Ans.}$$

Therefore, the magnitude of the acceleration of point F is $A_F = 1935$ ft/s^2. This result is in good agreement with the result of Example 4.3; that is, $A_F = 1960$ ft/s^2.

Example 4.10

For the offset slider-crank linkage of Example 3.2 in the posture shown in Fig. 3.8a, link 4 is driven at a constant speed of 10 m/s to the left. Determine the angular accelerations of links 2 and 3, and the acceleration of coupler point D.

Solution

The angular accelerations of links 2 and 3 are obtained by differentiating Eqs. (3) in Example 3.8, with respect to the input displacement, r_4; that is,

$$-r_2 \cos \theta_2 \theta_2'^2 - r_2 \sin \theta_2 \theta_2'' - r_3 \cos \theta_3 \theta_3'^2 - r_3 \sin \theta_3 \theta_3'' = 0,$$
$$-r_2 \sin \theta_2 \theta_2'^2 + r_2 \cos \theta_2 \theta_2'' - r_3 \sin \theta_3 \theta_3'^2 + r_3 \cos \theta_3 \theta_3'' = 0, \hspace{1cm} (1)$$

where the second-order kinematic coefficients of links 2 and 3 are defined as

$$\theta_2'' = \frac{d^2\theta_2}{dr_4^2} \quad \text{and} \quad \theta_3'' = \frac{d^2\theta_3}{dr_4^2}.$$

We recall the first-order kinematic coefficients from Eq. (8) in Example 3.8; they are

$$\theta_2' = -19.856 \text{ rad/m} \quad \text{and} \quad \theta_3' = 5.447 \text{ rad/m}. \hspace{1.5cm} (2)$$

Substituting Eqs. (2) and the known data into Eqs. (1), the second-order kinematic coefficients of links 2 and 3 are

$$\theta_2'' = -246.25 \text{ rad/m}^2 \quad \text{and} \quad \theta_3'' = 162.73 \text{ rad/m}^2. \hspace{1.5cm} (3)$$

The angular accelerations of links 2 and 3 can be written from Table 4.2 as

$$\alpha_2 = \theta_2'' \dot{r}_4^2 + \theta_2' \ddot{r}_4 \quad \text{and} \quad \alpha_3 = \theta_3'' \dot{r}_4^2 + \theta_3' \ddot{r}_4. \hspace{1.5cm} (4)$$

Substituting Eqs. (2) and (3) and the constant input speed $\dot{r}_4 = -10$ m/s into Eq. (4), the angular accelerations of links 2 and 3 are

$$\alpha_2 = -24\,625 \text{ rad/s}^2 \text{ cw} \quad \text{and} \quad \alpha_3 = 16\,273 \text{ rad/s}^2 \text{ ccw}. \hspace{1cm} \textit{Ans.}$$

Example 4.10 (cont.)

To obtain the acceleration of point D, we differentiate Eqs. (12) in Example 3.8 with respect to the input displacement, r_4. The second-order kinematic coefficients for point D are

$$x_D'' = -r_2 \cos\theta_2 \theta_2'^2 - r_2 \sin\theta_2 \theta_2'' - r_{DB} \cos(\theta_3 - \beta)\theta_3'^2 - r_{DB} \sin(\theta_3 - \beta)\theta_3'',$$
$$y_D'' = -r_2 \sin\theta_2 \theta_2'^2 + r_2 \cos\theta_2 \theta_2'' - r_{DB} \sin(\theta_3 - \beta)\theta_3'^2 + r_{DB} \cos(\theta_3 - \beta)\theta_3''. \tag{5}$$

Substituting Eqs. (2) and (3) and the known data into Eqs. (5), we find

$$x_D'' = 6.52 \text{ m/m}^2 \quad \text{and} \quad y_D'' = -12.31 \text{ m/m}^2. \tag{6}$$

Differentiating the velocity of point D, Eq. (14) in Example 3.8, with respect to time, the acceleration of point D can be written as

$$\mathbf{A}_D = \left(x_D'' \dot{r}_4^2 + x_D' \ddot{r}_4\right)\hat{\mathbf{i}} + \left(y_D'' \dot{r}_4^2 + y_D' \ddot{r}_4\right)\hat{\mathbf{j}}. \tag{7}$$

Next, we recall the first-order kinematic coefficients for point D from Eq. (13) in Example 3.8 as

$$x_D' = 1.123 \text{ m/m} \quad \text{and} \quad y_D' = -0.407 \text{ m/m}. \tag{8}$$

Substituting Eqs. (6) and (8) and the constant speed input into Eq. (7), we get

$$\mathbf{A}_D = 652\hat{\mathbf{i}} - 1231\hat{\mathbf{j}} \text{ m/s}^2. \qquad Ans.$$

Therefore, the magnitude of the acceleration of point D is $A_D = 1393 \text{ m/s}^2$.

Example 4.11

Consider the marine steering gear, called *Rapson's slide*, shown in Fig. 4.23. Link 4, O_4B, is the tiller, and link 2, AC, is the actuating rod. In the posture shown, $R_{AD} = R_2 = 4.619$ ft, the velocity of link 2 is a constant 15 ft/s to the left. Determine the angular acceleration of the tiller and the acceleration of point B.

Fig. 4.23 $R_{DO_4} = R_1 = 8$ ft and $R_{BO_4} = 11$ ft.

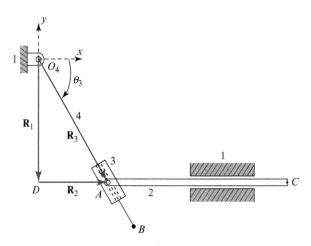

Example 4.11 (cont.)

Solution

The vectors that are chosen for the kinematic analysis of the mechanism are shown in Fig. 4.23. Note that point D is a point of the ground link, 1, such that it lies vertically below point O_4; thus, the input is the magnitude of vector $\mathbf{R}_2$ from point D to point A of link 2. Also, note that the change in the input is negative; that is, the magnitude of the input is decreasing.

With the vectors shown in Fig. 4.23, the vector loop equation for the linkage is

$$\overset{\sqrt{}}{\mathbf{R}_1} + \overset{I\sqrt{}}{\mathbf{R}_2} - \overset{??}{\mathbf{R}_3} = 0, \tag{1}$$

where $\theta_1 = -90^o$, $\theta_2 = 0$, $\theta_3 = \tan^{-1}(R_1/R_2) = -60^o$, and $R_3 = \sqrt{R_1^2 + R_2^2} = 9.238$ ft. The x and y components of Eq. (1) are

$$R_2 - R_3 \cos \theta_3 = 0, \tag{2a}$$

$$-R_1 - R_3 \sin \theta_3 = 0. \tag{2b}$$

Differentiating Eqs. (2) with respect to the input position R_2 gives

$$1 + R_3 \sin \theta_3 \theta_3' - R_3' \cos \theta_3 = 0, \tag{3a}$$

$$-R_3 \cos \theta_3 \, \theta_3' - R_3' \sin \theta_3 = 0. \tag{3b}$$

In matrix form, Eqs. (3) can be written as

$$\begin{bmatrix} R_3 \sin \theta_3 & -\cos \theta_3 \\ -R_3 \cos \theta_3 & -\sin \theta_3 \end{bmatrix} \begin{bmatrix} \theta_3' \\ R_3' \end{bmatrix} = \begin{bmatrix} -1 \\ 0 \end{bmatrix}. \tag{4}$$

The determinant of the Jacobian in Eq. (4) is

$$\Delta = \begin{vmatrix} R_3 \sin \theta_3 & -\cos \theta_3 \\ -R_3 \cos \theta_3 & -\sin \theta_3 \end{vmatrix} = -R_3 \sin^2 \theta_3 - R_3 \cos^2 \theta_3 = -R_3. \tag{5}$$

Therefore, $\Delta = -9.238$ ft.

Differentiating Eqs. (3) with respect to the input position, R_2, gives

$$R_3 \cos \theta_3 \theta_3'^2 + R_3 \sin \theta_3 \theta_3'' + 2R_3' \sin \theta_3 \theta_3' - R_3'' \cos \theta_3 = 0, \tag{6a}$$

$$R_3 \sin \theta_3 \theta_3'^2 - R_3 \cos \theta_3 \theta_3'' - 2R_3' \cos \theta_3 \theta_3' - R_3'' \sin \theta_3 = 0. \tag{6b}$$

In matrix form, Eqs. (6) can be written as

$$\begin{bmatrix} R_3 \sin \theta_3 & -\cos \theta_3 \\ -R_3 \cos \theta_3 & -\sin \theta_3 \end{bmatrix} \begin{bmatrix} \theta_3'' \\ R_3'' \end{bmatrix} = \begin{bmatrix} -2R_3' \sin \theta_3 \theta_3' - R_3 \cos \theta_3 \theta_3'^2 \\ 2R_3' \cos \theta_3 \theta_3' - R_3 \sin \theta_3 \theta_3'^2 \end{bmatrix}. \tag{7}$$

Example 4.11 (cont.)

Note that the coefficient matrix in Eq. (7) is the same as the coefficient matrix in Eq. (4). The coefficient matrix does not change with differentiation of the unknown variables.

Substituting the known data into Eq. (4) and using Cramer's rule, the first-order kinematic coefficients are

$$\theta_3' = \frac{\begin{vmatrix} -1 & -\cos\theta_3 \\ 0 & -\sin\theta_3 \end{vmatrix}}{\Delta} = \frac{\sin\theta_3}{-R_3} = +0.093\,75 \text{ rad/ft}, \tag{8a}$$

and

$$R_3' = \frac{\begin{vmatrix} R_3\sin\theta_3 & -1 \\ -R_3\cos\theta_3 & 0 \end{vmatrix}}{\Delta} = \frac{-R_3\cos\theta_3}{-R_3} = +0.500 \text{ ft/ft}. \tag{8b}$$

Substituting the known data into Eq. (7) and using Cramer's rule, the second-order kinematic coefficients are

$$\theta_3'' = \frac{\begin{vmatrix} -2R_3'\sin\theta_3\theta_3' - R_3\cos\theta_3{\theta_3'}^2 & -\cos\theta_3 \\ 2R_3'\cos\theta_3\theta_3' - R_3\sin\theta_3{\theta_3'}^2 & -\sin\theta_3 \end{vmatrix}}{\Delta} = \frac{2R_3'\theta_3'}{-R_3} = -0.010\,15 \text{ rad/ft}^2, \tag{9a}$$

and

$$R_3'' = \frac{\begin{vmatrix} R_3\sin\theta_3 & -2R_3'\sin\theta_3\theta_3' - R_3\cos\theta_3{\theta_3'}^2 \\ -R_3\cos\theta_3 & 2R_3'\cos\theta_3\theta_3' - R_3\sin\theta_3{\theta_3'}^2 \end{vmatrix}}{\Delta} = \frac{-(R_3\theta_3')^2}{-R_3} = +0.081\,18 \text{ ft/ft}^2. \tag{9b}$$

The angular velocity of link 3 is

$$\omega_3 = \theta_3' \dot{R}_2 = (+0.093\,75 \text{ rad/ft})(-15 \text{ ft/s}) = -1.406 \text{ rad/s}. \tag{10}$$

The negative sign indicates that the direction of the angular velocity of link 3 is clockwise. Note that $\dot{R}_2 = -15$ ft/s, since link 2 is moving to the left (the input magnitude R_2 is decreasing). Also, note that link 4 is constrained to rotate with the same angular velocity as link 3. Therefore, the angular velocity of link 4 is

$$\omega_4 = \omega_3 = -1.406 \text{ rad/s (cw)}. \tag{11}$$

The angular acceleration of link 3 is

$$\begin{aligned} \alpha_3 &= \theta_3'\ddot{R}_2 + \theta_3''\dot{R}_2^2 = (+0.093\,75 \text{ rad/ft})(0) + (-0.010\,15 \text{ rad/ft}^2)(-15 \text{ ft/s})^2 \\ &= -2.283 \text{ rad/s}^2. \end{aligned} \tag{12}$$

Example 4.11 (cont.)

The negative sign indicates that link 3 is accelerating in the clockwise direction. Since link 4 is constrained to rotate with link 3, the angular acceleration of link 4 is

$$\alpha_4 = \alpha_3 = -2.283 \text{ rad/s}^2 \text{ (cw)}. \qquad \text{Ans. (13)}$$

The sliding velocity of link 3 with respect to link 4 is

$$V_{A_3/4} = R_3' \dot{R}_2 = (+0.5 \text{ ft/ft})(-15 \text{ ft/s}) = -7.5 \text{ ft/s}. \qquad (14)$$

The negative sign indicates that the direction of the apparent velocity is along link 4, pointing toward the pin O_4; that is, the direction is $\mathbf{V}_{A_3/4} = 7.5 \text{ ft/s} \angle 120°$.
The sliding acceleration of link 3 with respect to link 4 can be written as

$$A_{A_3/4} = R_3' \ddot{R}_2 + R_3'' \dot{R}_2^2$$
$$= (+0.5 \text{ ft/ft})(0) + (+0.081\,18 \text{ ft/ft}^2)(-15 \text{ ft/s})^2 = +18.266 \text{ ft/s}^2. \qquad (15)$$

The positive sign indicates that the direction of the apparent acceleration is along link 4, pointing away from the pin O_4; that is, the direction is $\mathbf{A}_{A_3/4} = 18.266 \text{ ft/s}^2 \angle -60°$.
To determine the acceleration of point B, we draw vector $\mathbf{R}_B$, as shown in Fig. 4.24.

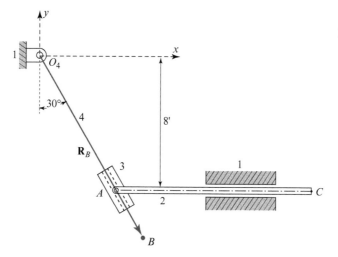

Fig. 4.24 Vector for point B.

The x and y components of the position of point B are

$$x_B = R_B \cos\theta_3 = 5.500 \text{ ft}, \qquad (16a)$$

$$y_B = R_B \sin\theta_3 = -9.526 \text{ ft}. \qquad (16b)$$

Differentiating Eqs. (16) with respect to the input variable, R_2, the first-order kinematic coefficients of point B are

$$x_B' = -R_B \sin\theta_3 \theta_3' = +0.893\,09 \text{ ft/ft}, \qquad (17a)$$

$$y_B' = R_B \cos\theta_3 \theta_3' = +0.515\,63 \text{ ft/ft}. \qquad (17b)$$

Example 4.11 (cont.)

Then, differentiating Eqs. (17) with respect to the input, R_2, gives the second-order kinematic coefficients of point B:

$$x_B'' = -R_B \cos\theta_3 {\theta_3'}^2 - R_B \sin\theta_3 \theta_3'' = -0.14501 \text{ ft/ft}^2, \qquad (18a)$$

$$y_B'' = -R_B \sin\theta_3 {\theta_3'}^2 + R_B \cos\theta_3 \theta_3'' = +0.02791 \text{ ft/ft}^2. \qquad (18b)$$

The velocity of point B can be written as

$$\begin{aligned}
\mathbf{V}_B &= \left(x_B'\hat{\mathbf{i}} + y_B'\hat{\mathbf{j}}\right)\dot{R}_2 \\
&= \left(0.893\,09\,\hat{\mathbf{i}} + 0.515\,63\,\hat{\mathbf{j}}\right)(-15 \text{ ft/s}) = -13.396\,\hat{\mathbf{i}} - 7.734\,\hat{\mathbf{j}} \text{ ft/s} \qquad (19) \\
&= 15.468 \text{ ft/s}\angle{-150°}.
\end{aligned}$$

As a check, the velocity of point B can also be written as

$$V_B = \omega_4 R_{BO_4} = (1.406 \text{ rad/s})(11 \text{ ft}) = 15.466 \text{ ft/s},$$

where the difference is a result of truncation in the value of ω_4 in Eqs. (10) and (11) ($\omega_4 = 1.406\,25$ rad/s).

The acceleration of point B can be written as

$$\begin{aligned}
\mathbf{A}_B &= \left(x_B'\hat{\mathbf{i}} + y_B'\hat{\mathbf{j}}\right)\ddot{R}_2 + \left(x_B''\hat{\mathbf{i}} + y_B''\hat{\mathbf{j}}\right)\dot{R}_2^2 \\
&= \left(0.893\,09\hat{\mathbf{i}} + 0.515\,63\hat{\mathbf{j}} \text{ ft/ft}\right)(0) + \left(-0.145\,01\hat{\mathbf{i}} + 0.027\,91\hat{\mathbf{j}} \text{ ft/ft}^2\right)(-15 \text{ ft/s})^2 \\
&= -32.627\hat{\mathbf{i}} + 6.280\hat{\mathbf{j}} \text{ ft/s}^2 = 33.225 \text{ ft/s}^2\angle169.11°. \qquad Ans.\ (20)
\end{aligned}$$

Note that the inverse tangent function on some calculators initially indicates that the angle is $\angle\mathbf{A}_B = -10.89°$. However, this angle must be transposed into the second quadrant; that is, $\angle\mathbf{A}_B = -10.89° + 180° = 169.11°$.

4.12 Euler–Savary Equation[6]

In Section 4.5 we developed the apparent-acceleration equation, Eq. (4.17). Then, in the examples that followed, we reported that it is very important to carefully choose a point whose apparent path with respect to another body is known so that the radius of curvature of the path, required for the normal component in Eq. (4.14), can be found. This need to know the radius of curvature of the path often dictates the method of approach to such a problem and sometimes

---

[6] Among the most important and most useful references on this subject are [2, Chapter 4], [3, Chapter 5], [4, Chapter 7], [6], and [7, Chapter 4].

even requires the visualization of an equivalent mechanism (as was done in Example 4.7, Fig. 4.18c).[7] It would be convenient if an arbitrary point could be chosen, and the radius of curvature of its path could be calculated. In planar mechanisms, this can be accomplished by the methods that are presented here.

When two rigid bodies move relative to each other with planar motion, an arbitrarily chosen point A of one describes a path, or locus, relative to a coordinate system fixed to the other. At a given instant, there is a point A', attached to the other body, that is the center of curvature of the locus of A. If we take the kinematic inversion of this motion, A' also describes a locus relative to the body containing A, and it so happens that A is the center of curvature of this locus. Each point, therefore, acts as the center of curvature of the path traced by the other, and the two points are called *conjugates* of each other. The distance between these two conjugate points is the magnitude of the radius of curvature of either locus.

Figure 4.25 shows two circles, with centers at C and C', commonly referred to as *osculating circles* [2]. Let us think of the circle with center C' as the fixed centrode and think of the circle with center C as the moving centrode of two bodies experiencing some particular relative planar motion. In actuality, the fixed centrode need not be fixed but is attached to the body that contains the path whose curvature is sought. Also, it is not necessary that the two centrodes be circles; we are interested only in instantaneous values, and, for convenience, we can think of the centrodes as circles matching the curvatures (osculating circles) of the two actual centrodes in the region near their point of contact, I (called the *velocity pole*). The velocity pole is an additional point, coincident with the instantaneous center of velocity.

Recall from Section 3.12 that the instantaneous centers of velocity are a pair of coincident points, labeled I_{jk} and I_{kj}, each attached to one of the bodies, j and k. The velocity pole, I, however, is a third point, not attached to either body, that remains coincident with the changing instantaneous centers of velocity as the motion progresses. Thus, the velocity pole can have a velocity along the two tangent centrodes that is different from the equal velocities of the two instantaneous centers of velocity.

When the bodies containing the two centrodes move relative to each other, the centrodes appear to roll against each other without slip (Section 3.20). Because of these properties, we can think of the two circular centrodes as actually representing the shapes of the two moving bodies, if this helps in visualizing the motion.

If the moving centrode has an angular velocity $\boldsymbol{\omega}$, with respect to the fixed centrode, the velocity[8] of point C is

$$V_C = \omega R_{CI}. \tag{a}$$

Similarly, arbitrary point A, whose conjugate point A' we wish to find, has a velocity of

$$V_A = \omega R_{AI}. \tag{b}$$

---

[7] The concept of equivalent mechanisms is an important topic. For a detailed study, consult [2].

[8] All velocities used in this section are actually apparent velocities, with respect to the coordinate system of the fixed centrode; they are written as absolute velocities to simplify the notation.

Fig. 4.25 Hartmann
construction.

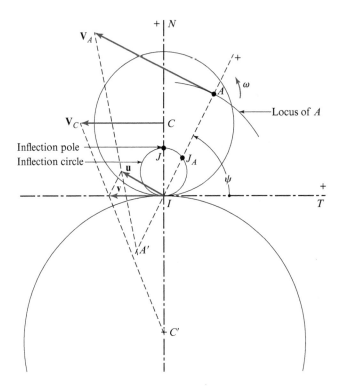

As the motion progresses, the point of contact of the two centrodes, and therefore the location of the velocity pole, I, moves along both centrodes with some velocity (or rate) $\mathbf{v}$. As shown in Fig. 4.25, the pole velocity, $\mathbf{v}$, can be found by connecting a straight line from the terminus of $\mathbf{V}_C$ to point C'. Alternatively, the magnitude of the pole velocity can be obtained from similar triangles, that is

$$v = \frac{R_{IC'}}{R_{CC'}} V_C. \qquad (c)$$

A graphic construction for A', the center of curvature of the locus of the arbitrary point A, is shown in Fig. 4.25 and is called the *Hartmann construction*. First, we find the component, $\mathbf{u}$, of the pole velocity, $\mathbf{v}$, as that component parallel to $\mathbf{V}_A$ or perpendicular to $\mathbf{R}_{AI}$. Then, the intersection of the line AI and a line connecting the termini of the velocities $\mathbf{V}_A$ and $\mathbf{u}$ gives the location of the conjugate point, A'. The radius of curvature, ρ, of the locus of point A is $\rho = R_{AA'}$.

An analytic expression for locating point A' would also be desirable and can be derived from the Hartmann construction. The magnitude of the $\mathbf{u}$ component of the pole velocity is given by

$$u = v \sin \psi, \qquad (d)$$

where ψ is the positive counterclockwise angle measured from the centrode tangent to the line of $\mathbf{R}_{AI}$. Then, noting the similar triangles in Fig. 4.25, we can write the magnitude of the $\mathbf{u}$ component of the pole velocity as

$$u = \frac{R_{IA'}}{R_{AA'}} V_A. \tag{e}$$

Now, equating Eqs. (d) and (e), and substituting Eqs. (a), (b), and (c) into the resulting equation gives

$$u = \frac{R_{IC'} R_{CI}}{R_{CC'}} \omega \sin \psi = \frac{R_{IA'} R_{AI}}{R_{AA'}} \omega. \tag{f}$$

Dividing this equation by $\omega \sin \psi$ and inverting, we obtain

$$\frac{R_{AA'}}{R_{AI} R_{IA'}} \sin \psi = \frac{R_{CC'}}{R_{CI} R_{IC'}} = \frac{\omega}{v}. \tag{g}$$

Next, upon noting that $R_{AA'} = R_{AI} - R_{A'I}$ and $R_{CC'} = R_{CI} - R_{C'I}$, we can reduce this equation to the form

$$\left(\frac{1}{R_{AI}} - \frac{1}{R_{A'I}} \right) \sin \psi = \left(\frac{1}{R_{CI}} - \frac{1}{R_{C'I}} \right). \tag{4.39}$$

This important equation is one form of the *Euler–Savary equation*. If we assume that the radii of curvature of the two centrodes, R_{CI} and $R_{C'I}$, are known, then this equation can be used to determine the position of one of the two conjugate points (A or A') from the position of the other, relative to the velocity pole, I.

Note that in using the Euler–Savary equation, we may arbitrarily choose a positive sense for the centrode tangent ($+T$); the positive centrode normal ($+N$) is then 90° counterclockwise from it. This establishes a positive direction for the line CC' in Fig. 4.25, which may be used in assigning appropriate signs to R_{CI} and $R_{C'I}$. Similarly, an arbitrary positive direction can be chosen for the line AA'. The angle ψ is then taken as positive counterclockwise from the positive centrode tangent to the positive sense of the line AA'. The sense of the line AA' also gives the appropriate signs for R_{AI} and $R_{A'I}$ for Eq. (4.39).

There is a major disadvantage to the above form of the Euler–Savary equation in that the radii of curvature of both centrodes, R_{CI} and $R_{C'I}$, must be known. Usually these are not known, any more than the curvature of the locus itself is known. However, this difficulty can be overcome by seeking another form of the equation.

Let us consider the particular point labeled J in Fig. 4.25. This point is located on the centrode normal at a location defined by

$$\frac{1}{R_{JI}} = \frac{1}{R_{CI}} - \frac{1}{R_{C'I}}. \tag{h}$$

If this particular point is chosen for A in Eq. (4.39), we find that its conjugate point, J', must be located at infinity. The radius of curvature of the path of point J is infinite, and the locus of J therefore has an inflection point at J. The point J is called the *inflection pole*.

Let us now consider whether there are any other points, J_A, of the moving body that also have infinite radii of curvature at the instant considered. If so, then for each of these points, $R_{AA'} = \infty$. Substituting this condition into Eq. (4.39), and with the aid of Eq. (h), we obtain

$$R_{J_A I} = R_{JI} \sin \psi. \tag{4.40}$$

This is the equation of a circle in polar coordinates whose diameter is R_{JI}, as shown in Fig. 4.25. This circle is called the *inflection circle*. Every point on this circle is an inflection point; it has its conjugate point at infinity, and each point therefore has an infinite radius of curvature at the instant indicated. These points are instantaneously moving along straight lines.

Now, with the help of Eq. (4.40), the Euler–Savary equation can be written in the form

$$\frac{1}{R_{AI}} - \frac{1}{R_{A'I}} = \frac{1}{R_{J_A I}}. \tag{4.41}$$

Also, after some further manipulation, the radius of curvature of the path of point A can be written as

$$\rho = R_{AA'} = \frac{R_{AI}^2}{R_{AJ_A}}. \tag{4.42}$$

Either of these two forms of the Euler–Savary equation, Eqs. (4.41) and (4.42), is more useful in practice than Eq. (4.39), since they do not require knowledge of the curvatures of the two centrodes. They do require finding the inflection circle, but the following example shows a graphic procedure for drawing the inflection circle.

Example 4.12

Draw the inflection circle for the coupler link 3 of the slider-crank linkage in the posture shown in Fig. 4.26. Then, use the inflection circle and the Euler–Savary equation to determine the instantaneous radius of curvature of the path of coupler point C.

Fig. 4.26 $R_{AO_2} = 2$ in, $R_{BA} = 2.5$ in, $R_{DA} = 1.25$ in, $R_{CD} = 1$ in, and $\theta_2 = 30°$.

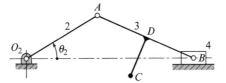

Solution

We begin by locating the velocity pole, I (which is coincident with the instantaneous center of velocity), at the intersection of line O_2A and a line through B perpendicular to its direction of travel (Fig. 4.27). By definition, points B and I must both lie on the inflection circle; hence, we need only one additional point to construct the circle.

Example 4.12 (cont.)

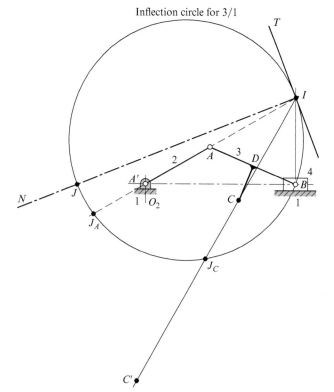

Fig. 4.27 Inflection circle.

The center of curvature of the path of point A is, of course, at O_2, which we now call A'. Taking the positive sense of line AI as being downward and to the left, we measure $R_{AI} = 2.64$ in and $R_{AA'} = -2.00$ in. Then, substituting these values into Eq. (4.42), we obtain

$$R_{AJ_A} = \frac{R_{AI}^2}{R_{AA'}} = \frac{(2.64 \text{ in})^2}{-2.00 \text{ in}} = -3.48 \text{ in}.$$

With this, we lay off 3.48 in from A to locate J_A, which is a third point on the inflection circle. The inflection circle for the motion 3/1 can now be constructed through the three points B, I, and J_A, and the diameter of the circle is measured as

$$R_{JI} = 6.28 \text{ in}. \qquad\qquad Ans.$$

The centrode normal (N) and the centrode tangent (T) can also be drawn, if desired, as shown in Fig. 4.27.

Drawing the ray R_{CI} and continuing to take the positive sense as downward and to the left, we measure $R_{CI} = 3.10$ in and $R_{CJ_C} = -1.75$ in. Substituting these values into Eq. (4.42), the instantaneous radius of curvature of the path of point C is

Example 4.12 (cont.)

$$\rho = R_{CC'} = \frac{R_{CI}^2}{R_{CJ_C}} = \frac{(3.10 \text{ in})^2}{-1.75 \text{ in}} = -5.49 \text{ in},$$ *Ans.*

where the negative sign indicates that C is above C' on the line ICJ_CC'.

4.13 Bobillier Constructions

The Hartmann construction provides one graphic method of finding the conjugate point and the radius of curvature of the path of a moving point, but it requires knowledge of the curvature of the fixed and moving centrodes. It would be desirable to have *graphic* methods for obtaining the inflection circle and the conjugate of a given point without requiring the curvature of the centrodes. Such graphic methods are presented in this section and are called the *Bobillier constructions*.

To understand these constructions, consider the inflection circle, the centrode normal (N), and the centrode tangent (T), as shown in Fig. 4.28. Let us select any two points, A and B, of the moving body that do not lie on a straight line through velocity pole I. Using the Euler–Savary equation, we can find the two corresponding conjugate points, A' and B'. The intersection of the lines AB and $A'B'$ is labeled Q. Then, the straight line drawn through I and Q is called the *collineation axis*. This axis applies only to the two lines AA' and BB', and so is said to belong to these two rays; also, point Q will be located differently on the collineation axis if another pair of points, A and B, is chosen on the same rays. Nevertheless, there is a unique

Fig. 4.28 Bobillier theorem.

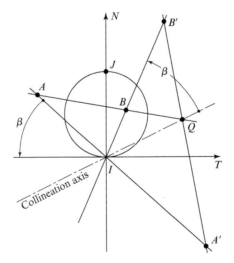

relationship between the collineation axis and the two rays used to define it. This relationship is expressed in *Bobillier's theorem*, which states that *the angle from the collineation axis to the first ray is equal to the angle from the second ray to the centrode tangent.*

In applying the Euler–Savary equation to a planar mechanism, we can usually find two pairs of conjugate points by inspection, and from these we wish to graphically determine the inflection circle. For example, a four-bar linkage with a crank, O_2A, and a follower, O_4B, has A and O_2 as one pair of conjugate points and B and O_4 as another, when we are interested in the motion of the coupler with respect to the frame (3 with respect to 1). Given these two pairs of conjugate points, how do we use the Bobillier theorem to find the inflection circle?

In Fig. 4.29a, let A and A' and B and B' represent the known pairs of conjugate points. Rays constructed through each pair intersect at the velocity pole, I, giving one point on the inflection

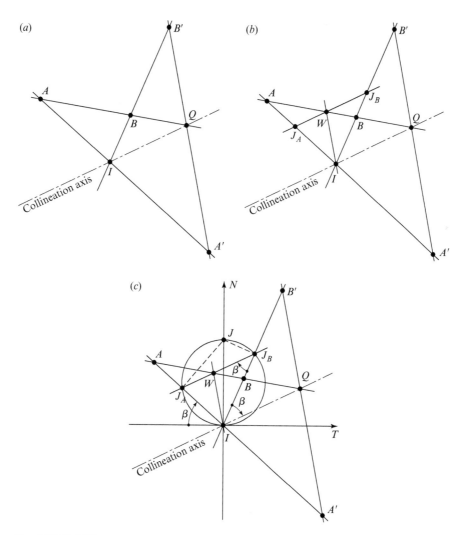

Fig. 4.29 Bobillier construction for locating the inflection circle.

Fig. 4.30 Bobillier construction for locating the conjugate point C'.

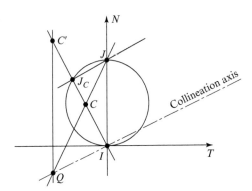

circle. Point Q is located next by the intersection of a ray through A and B with a ray through A' and B'. Then the collineation axis can be drawn as the line IQ.

The next step is shown in Fig. 4.29b. Drawing a straight line through I parallel to $A'B'$, we identify the point W as the intersection of this line with line AB. Now, through W, we draw a second line parallel to the collineation axis. This line intersects AA' at J_A and BB' at J_B, two additional points on the inflection circle for which we are searching.

We could now construct a circle through the three points J_A, J_B, and I, but there is an easy way to do this. Remembering that a triangle inscribed in a semicircle is a right triangle having the diameter as its hypotenuse, we erect a perpendicular to AI at J_A and another perpendicular to BI at J_B. The intersection of these two perpendiculars gives the *inflection pole*, point J, as shown in Fig. 4.29c. Since IJ is the diameter, then the inflection circle, the centrode normal N, and the centrode tangent T can all be easily constructed.

To demonstrate that this construction satisfies the Bobillier theorem, note that the arc from I to J_A is inscribed by the angle that J_AI makes with the centrode tangent. But this same arc is also inscribed by the angle IJ_BJ_A. Therefore, these two angles are equal. But line J_BJ_A was originally constructed parallel to the collineation axis. Therefore, the line IJ_B also makes the same angle, β, with the collineation axis.

Our final problem is to learn how to use the Bobillier theorem to find the conjugate of another arbitrary point, say C, when the inflection circle is given. In Fig. 4.30 we draw a line through point C and the velocity pole, I, and locate the point of intersection of this line with the inflection circle (J_C). This ray serves as one of the two necessary rays to locate the collineation axis. For the other ray, we can use the centrode normal, since J and its conjugate point J', at infinity, are both known. For these two rays, the collineation axis is a line through the velocity pole, I, parallel to the line J_CJ, as we learned in Fig. 4.29c. The balance of the construction is similar to that of Fig. 4.28. Point Q is located by the intersection of a line through J and C with the collineation axis. Then a line through Q and J' at infinity intersects the ray IC at C', the conjugate point for C.

Example 4.13

For the four-bar linkage $A'ABB'$ in the posture shown in Fig. 4.31, use the Bobillier theorem to find the center of curvature of the coupler curve of point C.

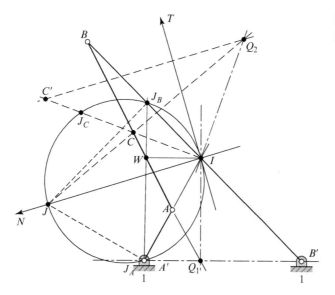

Fig. 4.31 Four-bar linkage.

Solution

Locate the velocity pole I at the intersection of AA' and BB'; also locate Q_1 at the intersection of AB and $A'B'$. Line IQ_1 is the first collineation axis. Through I draw a line parallel to $A'B'$ to locate W on AB. Through W, draw a line parallel to IQ_1 to locate J_A on AA' and J_B on BB'. Then, through J_A, draw a perpendicular to AA'; through J_B, draw a perpendicular to BB'. These perpendiculars intersect at inflection pole J and define the inflection circle, the centrode normal N, and the centrode tangent T.

To obtain the conjugate point of C, we draw ray IC and locate J_C on the inflection circle. The second collineation axis, IQ_2, belonging to the pair of rays IC and IJ, is a line through I parallel to a line (not shown) from J to J_C. Point Q_2 is obtained as the intersection of this collineation axis and a line JC. Now, through Q_2 we draw a line parallel to the centrode normal; its intersection with the ray IC yields C', the center of curvature of the path of C.

4.14 Instantaneous Center of Acceleration

This section defines the *instantaneous center of acceleration* for a planar mechanism. It is important to note that, in general, the instantaneous center of acceleration is not coincident with the instantaneous center of velocity. In other words, the instantaneous center of velocity has acceleration. The instantaneous center of acceleration is defined as *the instantaneous*

Fig. 4.32 (a) Instantaneous center of acceleration; (b) acceleration of point A.

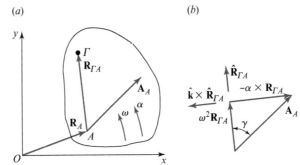

location of a pair of coincident points of two different rigid bodies where the absolute acceler-ations of the two bodies are identical. If we consider a fixed and a moving body, the instantan-eous center of acceleration is the point of the moving body which has zero absolute acceleration at the instant considered.

For the moving plane shown in Fig. 4.32a, assume that ω and α are known and that point A in the plane has a known acceleration, $\mathbf{A}_A$. Let Γ denote the instantaneous center of acceler-ation, a point of zero absolute acceleration whose location is unknown. The acceleration-difference equation can then be written as

$$\mathbf{A}_\Gamma = \mathbf{A}_A - \omega^2 \mathbf{R}_{\Gamma A} + \alpha \times \mathbf{R}_{\Gamma A} = \mathbf{0}. \tag{a}$$

Solving for $\mathbf{A}_A$ gives

$$\mathbf{A}_A = \omega^2 R_{\Gamma A} \hat{\mathbf{R}}_{\Gamma A} - \alpha R_{\Gamma A}(\hat{\mathbf{k}} \times \hat{\mathbf{R}}_{\Gamma A}). \tag{b}$$

Now, recognizing that $\hat{\mathbf{R}}_{\Gamma A}$ is perpendicular to $\hat{\mathbf{k}} \times \hat{\mathbf{R}}_{\Gamma A}$, the two terms on the right-hand side of Eq. (b) are rectangular components of $\mathbf{A}_A$, as shown in Fig. 4.32b. From this figure we can solve for the direction and magnitude of $\mathbf{R}_{\Gamma A}$, that is,

$$\gamma = \tan^{-1} \frac{\alpha}{\omega^2}, \tag{4.43}$$

$$R_{\Gamma A} = \frac{A_A}{\sqrt{\omega^4 + \alpha^2}} = \frac{A_A \cos \gamma}{\omega^2}. \tag{4.44}$$

Equation (4.44) states that the distance $R_{\Gamma A}$ from point A to the instantaneous center of acceleration Γ can be determined from the magnitude of the acceleration, A_A, of any point of the moving plane. Our convention is that the angle γ is defined *from* the line $I\Gamma$ *to* the positive centrode normal.

There are many graphic methods for locating the instantaneous center of acceleration [2,7]. Here, without proof, we present one method (referred to as the four-circle method). In Fig. 4.33, we are given points A and B and their absolute accelerations $\mathbf{A}_A$ and $\mathbf{A}_B$. We extend $\mathbf{A}_A$ and $\mathbf{A}_B$ until they intersect at Q; then we construct a circle through points A, B, and Q. Next, we draw another circle through the termini of $\mathbf{A}_A$ and $\mathbf{A}_B$ and point Q. The second intersection of the two circles locates point Γ, the instantaneous center of acceleration.

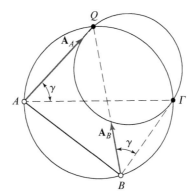

Fig. 4.33 Four-circle method of locating the instantaneous center of acceleration.

4.15 Bresse Circle (or de La Hire Circle)

Another graphic method of locating the instantaneous center of acceleration is by drawing a circle called the Bresse circle. In Section 4.12, we defined the inflection circle as the locus of points that have their conjugate points at infinity, and each therefore has an infinite radius of curvature at the instant under consideration. Therefore, the inflection circle can also be defined as the locus of points with zero normal acceleration. The locus of points with zero tangential acceleration also defines a circle, shown in Fig. 4.34, referred to as the *Bresse circle* or the *de La Hire circle*.

The inflection circle and the Bresse circle intersect at two points, as shown in Fig. 4.34; one point is the velocity pole, I, and the other point is the instantaneous center of acceleration, Γ. Since the velocity pole, I, has acceleration, in general, then it must be discounted as a possible solution. The direction of the acceleration of the velocity pole, I, is along the positive centrode normal, N, and the magnitude can be written as

$$A_I = \omega v, \tag{4.45}$$

where v is the magnitude of the pole velocity [Eq. (*c*) in Section 4.12] and can be written as

$$v = \omega R_{JI}. \tag{4.46}$$

Substituting Eq. (4.46) into Eq. (4.45), the acceleration of I can also be written as

$$A_I = \omega^2 R_{JI}.$$

The angle from the positive centrode normal to the line $I\Gamma$ (that is, the line connecting the instantaneous center of acceleration and the velocity pole) is the angle γ. This angle was defined in Eq. (4.43) as

$$\gamma = \tan^{-1} \frac{\alpha}{\omega^2}.$$

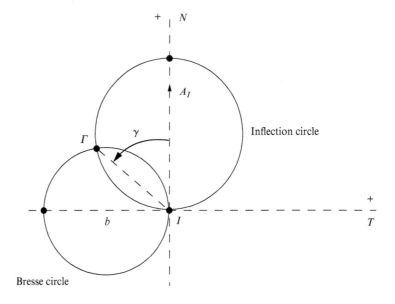

Fig. 4.34 Bresse circle and the instantaneous center of acceleration.

The diameter of the Bresse circle can be written as

$$b = R_{JI}\frac{\omega^2}{\alpha}. \tag{4.47}$$

If the angular acceleration of the moving plane is positive (that is, counterclockwise), then the Bresse circle lies on the negative side of the centrode tangent, T, as shown in Fig. 4.34. Alternatively, if the angular acceleration of the moving plane is negative, then the Bresse circle lies on the positive side of the centrode tangent, T.

In the special case that the angular velocity of the moving plane is constant (that is, $\alpha = 0$), then the diameter of the Bresse circle is infinite (the Bresse circle tends to the centrode normal, N) and the instantaneous center of acceleration is coincident with the inflection pole, J.

Example 4.14

Consider the four-bar linkage of Example 4.1, shown in Fig. 4.8a, and redrawn here as Fig. 4.35. For the given posture (that is, with the crank angle $\theta_2 = 135°$ccw from the ground link, O_2O_4), the angular velocity and angular acceleration of the coupler link, AB, are $\omega_3 = 5$ rad/s ccw and $\alpha_3 = 20$ rad/s^2 cw, respectively. For the instantaneous motion of the coupler link, show: (a) the velocity pole, I, centrode tangent, T, and centrode normal, N; (b) the inflection circle and Bresse circle; and (c) the instantaneous center of acceleration of the coupler link, AB. Then, determine: (d) the radius of curvature of the trajectory of coupler point C where $R_{CB} = 6$ in; (e) the magnitude and direction of the velocity of coupler point C; (f) the magnitude and direction of the angular velocity of the crank; (g) the magnitude and direction of the pole velocity; (h) the magnitude and direction of the acceleration of the velocity pole I; and (i) the magnitude and direction of the acceleration of C.

Example 4.14 (cont.)

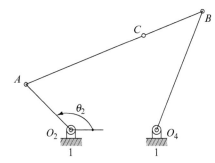

Fig. 4.35 $R_{O_4O_2} = 8$ in, $R_{AO_2} = 6$ in, $R_{BA} = 18$ in, and $R_{BO_4} = 12$ in.

Solution

a. The velocity pole, I, is coincident with the instant center I_{13} as shown in Fig. 4.36. The instant center I_{24} (labeled as point Q) and the collineation axis, IQ, are also shown in Fig. 4.36. From Bobillier's theorem, the angle from the collineation axis to the first ray (say, link 2) is obtained from a scaled drawing as

$$\alpha = 29.08° \text{cw}. \qquad\qquad Ans. \text{ (1)}$$

This is equal to the angle from the second ray (say, link 4) to the centrode tangent, T. Therefore, we draw the centrode tangent, T, and the centrode normal, N, which is 90° counterclockwise from T, as shown in Fig. 4.36.

b. Now, by measuring ray $R_{AI} = 14.14$ in, we can use Eq. (4.42) to locate the inflection point J_A,

$$R_{AJ_A} = \frac{R_{AI}^2}{R_{AA'}} = \frac{(14.14 \text{ in})^2}{6.00 \text{ in}} = 33.31 \text{ in}.$$

Similarly, measuring the ray $R_{BI} = 18.20$ in, we locate the inflection point J_B,

$$R_{BJ_B} = \frac{R_{BI}^2}{R_{BB'}} = \frac{(18.20 \text{ in})^2}{12.00 \text{ in}} = 27.60 \text{ in}.$$

Erecting perpendiculars from these two rays, we locate the inflection pole, J, and draw the inflection circle, as shown in Fig. 4.36. The diameter of the inflection circle is

$$R_{JI} = 19.26 \text{ in}. \qquad\qquad Ans. \text{ (2)}$$

Example 4.14 (cont.)

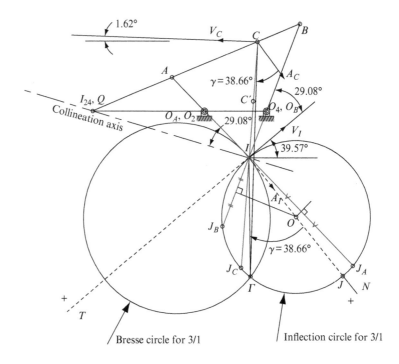

Fig. 4.36 Inflection circle, Bresse circle, and instantaneous center of acceleration.

The diameter of the Bresse circle, from Eq. (4.47), is

$$b = R_{JI}\frac{\omega_3^2}{\alpha_3} = (19.26 \text{ in})\frac{(5 \text{ rad/s})^2}{-20 \text{ rad/s}^2} = -24.08 \text{ in} \qquad Ans. (3)$$

where the negative sign indicates that the Bresse circle lies on the positive side of the centrode tangent.

c. The instantaneous center of acceleration, Γ, of the coupler link is as shown in Fig. 4.36.

d. The radius of curvature of the path of coupler point C (where $R_{CB} = 6$ in) is

$$\rho_C = R_{CC'} = \frac{R_{CI}^2}{R_{CJ_C}} = \frac{(14.67 \text{ in})^2}{28.62 \text{ in}} = 7.52 \text{ in}. \qquad Ans. (4)$$

The center of curvature of the path traced by point C is shown in Fig. 4.36.

e. The velocity of coupler point C is

$$V_C = \omega_3 R_{CI} = (5 \text{ rad/s})(14.65 \text{ in}) = 73.25 \text{ in/s}. \qquad Ans. (5)$$

The orientation of the velocity vector is as shown in Fig. 4.36.

f. The angular velocity of the crank can be written as

$$\omega_2 = \frac{R_{I_{23}I_{13}}}{R_{I_{23}I_{12}}}\omega_3 = \frac{14.14 \text{ in}}{6.00 \text{ in}}(5 \text{ rad/s}) = 11.78 \text{ rad/s ccw}. \qquad Ans. (6)$$

Example 4.14 (cont.)

g. The pole velocity, from Eq. (4.46), is

$$v = \omega_3 R_{JI} = (5 \text{ rad/s})(19.26 \text{ in}) = 96.3 \text{ in/s.} \qquad \textit{Ans. (7)}$$

The direction of the pole velocity is as shown in Fig. 4.36.

h. The acceleration of the velocity pole, I, from Eq. (4.45), is

$$A_I = \omega_3 v = (5 \text{ rad/s})(96.3 \text{ in/s}) = 481.5 \text{ in/s}^2. \qquad \textit{Ans. (8)}$$

The acceleration, A_I, is directed from I toward J, as shown in Fig. 4.36.

i. The acceleration of coupler point C is found by rearranging Eq. (4.44), that is,

$$\begin{aligned} A_C &= R_{\Gamma C}\sqrt{\omega_3{}^4 + \alpha_3{}^2} \\ &= (29.64 \text{ in})\sqrt{(5 \text{ rad/s})^4 + (20 \text{ rad/s}^2)^2} = 948.94 \text{ in/s}^2. \end{aligned} \qquad \textit{Ans. (9)}$$

The direction of the acceleration of point C, from Eq. (4.43), is

$$\gamma = \tan^{-1}\frac{\alpha_3}{\omega_3^2} = \tan^{-1}\frac{-20 \text{ rad/s}^2}{(5 \text{ rad/s})^2} = -38.66°, \qquad \textit{Ans. (10)}$$

which is a clockwise angle from $\mathbf{A}_C$ to the line $C\Gamma$, as shown in Fig. 4.36.

4.16 Radius of Curvature of a Point Trajectory Using Kinematic Coefficients

The radius of curvature of a point trajectory (say, the trajectory of point P) can be written from Eqs. (4.2) and (4.3) or from Eq. (4.14) as

$$\rho = \frac{V_P^2}{A_P^n}. \qquad (4.48)$$

From Eq. (3.28), the speed of point P can be written as

$$V_P = r_P' \dot{\psi}. \qquad (a)$$

The normal component of the acceleration of point P can be written as

$$A_P^n = \mathbf{A}_P \cdot \hat{\mathbf{u}}^n, \qquad (4.49)$$

where the unit normal vector to the point trajectory, see Eq. (3.32), is

$$\hat{\mathbf{u}}^n = \left(\frac{-y'_P}{r'_P}\right)\hat{\mathbf{i}} + \left(\frac{x'_P}{r'_P}\right)\hat{\mathbf{j}}. \tag{b}$$

Taking the time derivative of Eq. (3.27), the acceleration of point P can be written as

$$\mathbf{A}_P = \left(x''_P\dot{\psi}^2 + x'_P\ddot{\psi}\right)\hat{\mathbf{i}} + \left(y''_P\dot{\psi}^2 + y'_P\ddot{\psi}\right)\hat{\mathbf{j}}. \tag{4.50}$$

Substituting Eqs. (b) and (4.50) into Eq. (4.49), the normal component of the acceleration of point P can be written as

$$A_P^n = \left(\frac{x'_P y''_P - y'_P x''_P}{r'_P}\right)\dot{\psi}^2. \tag{4.51}$$

Then, substituting Eqs. (a) and (4.51) into Eq. (4.48), the radius of curvature of the point trajectory at the posture considered can be written as

$$\rho = \frac{r'^3_P}{x'_P y''_P - y'_P x''_P}. \tag{4.52}$$

Sign convention: If the unit normal vector to the point trajectory $\hat{\mathbf{u}}^n$ points toward the center of curvature of the path, then the radius of curvature has a positive value. If the unit normal vector to the point trajectory points away from the center of curvature of the path, then the radius of curvature has a negative value.

The coordinates of the center of curvature of the point trajectory, at the posture under investigation, can be written as

$$x_C = x_P - \rho\left(\frac{y'_P}{r'_P}\right) \quad \text{and} \quad y_C = y_P + \rho\left(\frac{x'_P}{r'_P}\right). \tag{4.53}$$

Example 4.15

Determine the radius of curvature and the center of curvature of the path of point B for Rapson's slide of Example 4.11.

Solution

Here we continue from Example 4.11 in all respects, including the sequence of equation numbers. Therefore, the unit tangent vector to the path of point B is the unit vector pointing in the direction of the velocity vector of point B. The unit tangent vector can be written as

$$\hat{\mathbf{u}}^t = \frac{x'_B\hat{\mathbf{i}} + y'_B\hat{\mathbf{j}}}{r'_B}, \tag{21}$$

Example 4.15 (cont.)

where, using the data of Eqs. (17),

$$r'_B = \pm\sqrt{x'_B{}^2 + y'_B{}^2} = -\sqrt{(0.89309 \text{ ft/ft})^2 + (0.51563 \text{ ft/ft})^2} = -1.03125 \text{ ft/ft}. \qquad (22)$$

Note that the negative sign is chosen, since the input distance R_2 is becoming shorter, and, therefore, the input is negative. Substituting Eqs. (17) and (22) into Eq. (21) gives

$$\hat{\mathbf{u}}^t = \frac{x'_B\hat{\mathbf{i}} + y'_B\hat{\mathbf{j}}}{r'_B} = \frac{(0.89309 \text{ ft/ft})\hat{\mathbf{i}} + (0.51563 \text{ ft/ft})\hat{\mathbf{j}}}{-1.03125 \text{ ft/ft}} = -0.86603\hat{\mathbf{i}} - 0.50000\hat{\mathbf{j}}. \qquad (23)$$

As a check, using Eq. (19), the unit tangent vector can also be written as

$$\hat{\mathbf{u}}^t = \frac{\mathbf{V}_B}{V_B} = \frac{-13.40\,\hat{\mathbf{i}} - 7.73\,\hat{\mathbf{j}} \text{ ft/s}}{15.47 \text{ ft/s}} = -0.86619\hat{\mathbf{i}} - 0.49968\hat{\mathbf{j}}.$$

The unit normal vector is 90° counterclockwise from the unit tangent vector; that is, using Eqs. (17) and (22),

$$\hat{\mathbf{u}}^n = \hat{\mathbf{k}} \times \hat{\mathbf{u}}^t = \hat{\mathbf{k}} \times \frac{X'_B\hat{\mathbf{i}} + Y'_B\hat{\mathbf{j}}}{r'_B} = \frac{-Y'_B\hat{\mathbf{i}} + X'_B\hat{\mathbf{j}}}{r'_B} = 0.50000\hat{\mathbf{i}} - 0.86603\hat{\mathbf{j}}. \qquad (24)$$

The direction of the unit tangent vector and the unit normal vector are shown in Fig. 4.37.

Using Eqs. (17), (18), and (22), the radius of curvature of the path of point B can be found from Eq. (4.52) as

$$\rho_B = \frac{r'_B{}^3}{x'_B y''_B - y'_B x''_B}$$

$$= \frac{(-1.03125 \text{ ft/ft})^3}{(0.89309 \text{ ft/ft})(0.028 \text{ ft/ft}^2) - (0.51563 \text{ ft/ft})(-0.145 \text{ ft/ft}^2)} \qquad Ans. \; (25)$$

$$= -11.0 \text{ ft}.$$

The negative sign indicates that the unit normal vector points away from the center of curvature of the path of point B (Fig. 4.37).

The coordinates of the center of curvature of the path of point B can be found from Eqs. (4.53). Using known values gives

$$x_C = x_B - \rho_B\left[\frac{y'_B}{r'_B}\right] = 5.50 \text{ ft} - (-11.0 \text{ ft})\left[\frac{(0.51563 \text{ ft/ft})}{(-1.03125 \text{ ft/ft})}\right] = 0.00 \text{ ft}, \qquad (26a)$$

$$y_C = y_B + \rho_B\left[\frac{x'_B}{r'_B}\right] = -9.53 \text{ ft} + (-11.0 \text{ ft})\left[\frac{0.89309 \text{ ft/ft}}{-1.03125 \text{ ft/ft}}\right] = 0.00 \text{ ft}. \qquad (26b)$$

Note that Eqs. (25), (26a), and (26b) make intuitive sense, since point B is located on link 4, and link 4 is pinned at the origin, O_4, and, therefore, must rotate about the origin.

Example 4.15 (cont.)

Figure 4.37 shows the directions of the unit tangent vector, $\hat{\mathbf{u}}^t$, and the unit normal vector, $\hat{\mathbf{u}}^n$. Note that the unit normal vector points away from the center of curvature of the path of point B. The center of curvature of the path of point B is coincident with the ground pin O_4. From observation, these answers are corroborated.

Fig. 4.37 Unit tangent and unit normal vectors, and center of curvature of the path of point B.

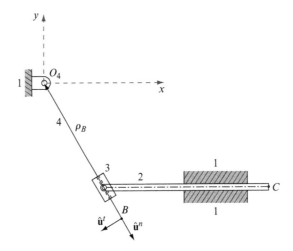

4.17 Cubic of Stationary Curvature

Now let us consider a point fixed in the coupler link of a planar four-bar linkage that generates a path relative to the frame whose radius of curvature, at the instant considered, is ρ. Since the coupler curve is, in general, a tricircular sextic function [1], this radius of curvature changes continuously as the point moves. In certain situations, however, the path has stationary curvature, which means that

$$\frac{d\rho}{ds} = 0, \qquad\qquad (a)$$

where s is the arc distance traveled along the path. The locus of all points in the coupler, or moving plane, which have stationary curvature at the instant considered, is called the *cubic of stationary curvature* or sometimes the *circling-point curve*. We should note that stationary curvature does not necessarily mean constant curvature, but rather that the continually varying radius of curvature is passing through a maximum or a minimum.

Here, we present a quick and easy graphic method for obtaining the cubic of stationary curvature for a coupler link, as described by Hain [2]. Consider the four-bar linkage $A'ABB'$ shown in Fig. 4.38, where A' and B' are the frame pivots. Note that the loci of points A and B have stationary curvature (in fact, constant curvature about centers at A' and B', respectively); hence, A and B must lie on the cubic.

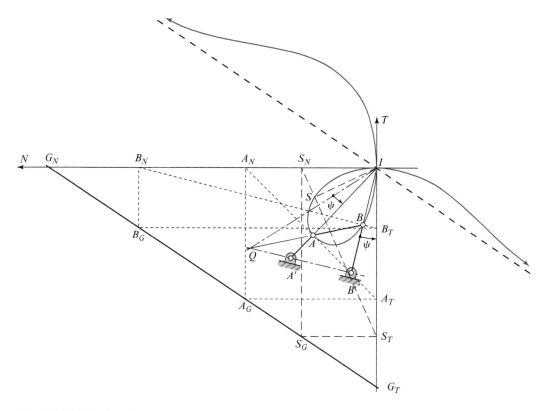

Fig. 4.38 Cubic of stationary curvature.

The first step of the construction is to obtain the centrode normal and the centrode tangent. Since the inflection circle is not needed, we locate the collineation axis, IQ, as indicated and draw the centrode tangent, T, at the angle ψ from line IB', equal to the angle from the collineation axis to line IA'. This construction follows directly from Bobillier's theorem. We also construct the centrode normal, N. At this time it may be convenient to reorient the drawing on the working surface so that the centrode normal lies along the horizontal axis.

Next, we construct a line through A perpendicular to IA and another line through B perpendicular to IB. These lines intersect the centrode normal and centrode tangent at A_N, A_T and B_N, B_T, respectively, as shown in Fig. 4.38. Now, we draw two rectangles, $IA_NA_GA_T$ and $IB_NB_GB_T$; points A_G and B_G define an auxiliary line G that we will use to obtain other points on the cubic of stationary curvature.

Next, we choose any point, S_G, on line G. A ray parallel to N locates S_T, and another ray parallel to T locates S_N. Connecting S_T with S_N and drawing a perpendicular to this line through I locates point S, another point on the cubic of stationary curvature. We repeat this process as often as desired by choosing different points on the line G, and we draw the cubic as a smooth curve through all the points S obtained.

Note that the cubic of stationary curvature has two tangents at the velocity pole I; one is the *centrode-tangent tangent* and the other is the *centrode-normal tangent*. The radius of curvature of the cubic at these tangents is obtained as follows: Extend line G to intersect T at G_T and N at

G_N, as indicated. Then, half the distance IG_T is the radius of curvature of the cubic tangent to the centrode normal, and half the distance IG_N is the radius of curvature of the cubic tangent to the centrode tangent.

A point with interesting properties is the point of intersection of the cubic of stationary curvature with the inflection circle (other than velocity pole I); this point is called *Ball's point*. The point fixed in the coupler that is coincident with Ball's point describes a path that is practically rectilinear for a considerable distance [5]; that is, it describes an excellent straight line near the design position, since it is located at an inflection point of its path and has stationary curvature.

The equation of the cubic of stationary curvature can be written in polar coordinates as

$$\frac{1}{r} = \frac{1}{M \sin \psi} - \frac{1}{N \cos \psi}, \tag{4.54}$$

where r is the distance from the velocity pole, I, to the point on the cubic, measured at an angle ψ from the centrode tangent.[9] The constant parameters M and N are determined using any two points known to lie on the cubic, such as points A and B of Fig. 4.38. Equations for M and N can be written as

$$\frac{1}{M} = \frac{1}{3}\left(\frac{1}{R_{JI}} - \frac{1}{R_{IO_M}}\right) \quad \text{and} \quad \frac{1}{N} = \frac{1}{3R_{JI}}\left(\frac{dR_{JI}}{ds}\right). \tag{4.55}$$

It so happens [8], that M and N are, respectively, the diameters IG_T and IG_N of the circles centered on the centrode tangent and centrode normal whose radii represent the two curvatures of the cubic at the velocity pole I.

The equation of the cubic of stationary curvature can also be expressed in Cartesian coordinates as

$$(x^2 + y^2)\left(\frac{x}{M} - \frac{y}{N}\right) = xy. \tag{4.56}$$

Degenerate forms. From Eq. (4.54), or Eq. (4.56), we observe that the cubic of stationary curvature degenerates to a circle and a straight line when either: (a) N tends to infinity (that is, $1/N$ approaches zero), or (b) M tends to infinity (that is, $1/M$ approaches zero). Consider the following example.

Example 4.16

For the four-bar linkage in the posture shown in Fig. 4.39, link 2 is perpendicular to the frame and the coupler link is parallel to the frame. For the absolute motion of the coupler link, determine the diameter of the inflection circle, the cubic of stationary curvature, and the location of Ball's point. Also, determine the radius of curvature and center of curvature of: (a) the moving centrode, (b) the fixed centrode, and (c) the path of coupler point C, which is midway between pins A and B.

---

[9] For a derivation of this equation see [3] or [4].

Example 4.16 (cont.)

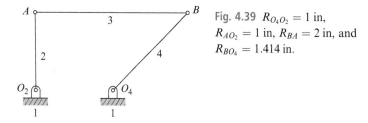

Fig. 4.39 $R_{O_4O_2} = 1$ in,
$R_{AO_2} = 1$ in, $R_{BA} = 2$ in, and
$R_{BO_4} = 1.414$ in.

Solution

The velocity pole, I, is coincident with the instant center I_{13}, as shown in Fig. 4.40. The instant center I_{24} (with label Q) lies at infinity (say, to the right), and the collineation axis IQ is parallel to the coupler link. From Bobillier's theorem, the angle from the collineation axis to the first ray (link 2) is $\alpha = 90°$ ccw. This is equal to the angle from the second ray (link 4) to the centrode tangent, T. Therefore, the centrode tangent, T, and the centrode normal, N (which is $90°$ ccw from the centrode tangent, T), are as shown on Fig. 4.40. Note that link 4 lies along the centrode normal, N.

The location of the inflection point J_A for point A on the coupler link is obtained from the Euler–Savary equation, Eq. (4.42); that is,

$$R_{AJ_A} = \frac{R_{AI}^2}{R_{AA'}} = \frac{(2 \text{ in})^2}{1 \text{ in}} = 4 \text{ in.} \tag{1}$$

The inflection point, J_B, for point B on the coupler link is obtained in a similar manner; that is,

$$R_{BJ_B} = \frac{R_{BI}^2}{R_{BB'}} = \frac{(2\sqrt{2} \text{ in})^2}{\sqrt{2} \text{ in}} = 5.657 \text{ in.} \tag{2}$$

The locations of the inflection points, J_A and J_B, are shown in Fig. 4.40. Note that J_B lies on the centrode normal, N; therefore, this inflection point is coincident with the inflection pole, J. Knowing the centrode normal, N, and the two inflection points, the inflection circle for the motion of link 3 with respect to 1 can now be drawn. The diameter of the inflection circle for the motion 3/1 is

$$R_{JI} = \frac{1}{2} R_{JB} = 2\sqrt{2} \text{ in} = 2.828 \text{ in.} \qquad Ans. \text{ (3)}$$

The inflection circle, the inflection pole J, and the center of the inflection circle (denoted as point O) are shown in Fig. 4.40. Note that the centrode normal, N, is directed from the velocity pole, I, toward the inflection pole, J, and the centrode tangent, T, is $90°$ clockwise from the centrode normal, N.

Example 4.16 (cont.)

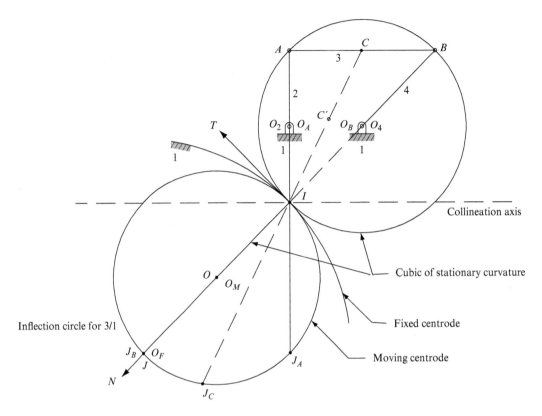

Fig. 4.40 Cubic of stationary curvature and the two centrodes.

The cubic of stationary curvature equation for the coupler link, including pins A and B, can be written from Eq. (4.54) as

$$\frac{1}{R_{AI}} = \frac{1}{M \sin \psi_A} - \frac{1}{N \cos \psi_A} \tag{4a}$$

and also as

$$\frac{1}{R_{BI}} = \frac{1}{M \sin \psi_B} - \frac{1}{N \cos \psi_B}, \tag{4b}$$

where the parameters M and N are as given by Eqs. (4.55). The angle ψ_A is the counterclockwise angle from the centrode tangent, T, to the ray containing point A, and the angle ψ_B is the counterclockwise angle from the centrode tangent, T, to the ray containing point B. From the scale drawing, these angles are measured as

$$\psi_A = -45° \quad \text{and} \quad \psi_B = -90°. \tag{5a}$$

Example 4.16 (cont.)

Note that these measurements can easily be verified from trigonometry. The distances to pin A from the velocity pole, I, and to pin B from the velocity pole, I, are measured consistently as

$$R_{AI} = 2 \text{ in} \quad \text{and} \quad R_{BI} = 2\sqrt{2} \text{ in.} \tag{5b}$$

Substituting Eqs. (5a) into Eq. (4b), gives

$$\frac{1}{R_{BI}} = \frac{1}{-M}. \tag{6}$$

Therefore, using Eq. (5b), we find the parameter

$$M = -R_{BI} = R_{JI} = -2\sqrt{2} \text{ in.} \qquad \textit{Ans.} \tag{7}$$

Note that substituting $\psi_A = -45°$ into Eq. (4a) indicates that $1/N = 0$, and the parameter N tends to infinity.
Ans.

This is a degenerate form of the cubic of stationary curvature. The cubic of stationary curvature degenerates to a straight line (that is, the centrode normal) and a circle of diameter M (the center must lie on the centrode normal). The circle must pass through pins A and B on link 3. Therefore, the center of the circle is coincident with the ground pin O_4.

Substituting the condition $N = \infty$ into Eq. (4.55) gives

$$\frac{1}{3R_{JI}}\left(\frac{dR_{JI}}{ds}\right) = 0. \tag{8a}$$

Since the diameter of the inflection circle is infinite, Eq. (3), the rate of change of the inflection circle diameter must be zero; that is,

$$\frac{dR_{JI}}{ds} = 0. \tag{8b}$$

In other words, the diameter of the inflection circle is at a maximum or a minimum as the four-bar linkage passes through this posture.

Recall that Ball's point is located at the intersection of the inflection circle and the cubic of stationary curvature. Figure 4.40 shows two apparent points of intersection – that is, the velocity pole, I, and the inflection pole, J. Since the velocity pole, I, is not a point fixed in the coupler, it is not a solution. Therefore, Ball's point is coincident with the inflection pole, J.

a. Rearranging Eq. (4.55), the radius of curvature of the moving centrode can be written as

$$\frac{1}{R_{IO_M}} = \frac{1}{R_{JI}} - \frac{3}{M}. \tag{9}$$

Example 4.16 (cont.)

Then, substituting Eq. (7) into this equation gives

$$\frac{1}{R_{IO_M}} = \frac{1}{R_{JI}} - \frac{3}{R_{JI}} = -\frac{2}{R_{JI}}. \tag{10a}$$

Therefore, the radius of curvature of the moving centrode is

$$R_{IO_M} = -\frac{R_{JI}}{2} = \sqrt{2}\text{ in.} \qquad\qquad Ans.\ (10b)$$

The negative signs in these last two equations indicate that the direction from O_M to I is opposite to the direction from I to J. The moving centrode is coincident with (or coalesces with) the inflection circle (Fig. 4.40).

b. The Euler–Savary equation can be written as

$$\frac{1}{R_{JI}} = \frac{1}{R_{IO_F}} - \frac{1}{R_{IO_M}}. \tag{11}$$

Substituting Eq. (10b) into this equation, and rearranging, the radius of curvature of the fixed centrode can be written as

$$R_{IO_F} = -R_{JI} = 2\sqrt{2}\text{ in.} \qquad\qquad Ans.\ (12)$$

The negative sign indicates that the direction from O_F to I is opposite to the direction from I to J. The radius of the osculating circle for the fixed centrode is the same as the diameter of the inflection circle; that is, O_F is coincident with the inflection pole J (Fig. 4.40).

c. The position of the coupler point C is shown in Fig. 4.40. Using complex polar notation, this can be written as

$$\mathbf{R}_{CI} = x_{CI}\hat{\mathbf{i}} + y_{CI}\hat{\mathbf{j}} = jR_{O_2I} + R_{AO_2}e^{j\theta_2} + R_{CA}e^{j\theta_3}, \tag{13}$$

which has real and imaginary components of

$$x_{CI} = R_{AO_2}\cos\theta_2 + R_{CA}\cos\theta_3 \quad \text{and} \quad y_{CI} = R_{O_2I} + R_{AO_2}\sin\theta_2 + R_{CA}\sin\theta_3. \tag{14}$$

Substituting the known data $R_{O_2I} = 1$ in, $R_{AO_2} = 1$ in, and $R_{CA} = 1$ in into Eq. (14) gives

$$x_{CI} = (1\text{ in})\cos\theta_2 + (1\text{ in})\cos\theta_3 \quad \text{and} \quad y_{CI} = (1\text{ in}) + (1\text{ in})\sin\theta_2 + (1\text{ in})\sin\theta_3. \tag{15}$$

For this posture, $\theta_2 = 90°$ and $\theta_3 = 0$; therefore, $x_{CI} = 1$ in and $y_{CI} = 2$ in.
Taking the derivative of Eqs. (14) with respect to the input angle θ_2 gives

$$x'_{CI} = -R_{AO_2}\sin\theta_2 - \theta'_3 R_{CA}\sin\theta_3 \quad \text{and} \quad y'_{CI} = R_{AO_2}\cos\theta_2 + \theta'_3 R_{CA}\cos\theta_3. \tag{16}$$

Next, we use the angular velocity ratio theorem, Eq. (3.30), to find the first-order kinematic coefficient:

$$\theta'_3 = \frac{R_{I_{23}I_{12}}}{R_{I_{23}I_{13}}} = \frac{1\text{ in}}{2\text{ in}} = 0.5\text{ rad/rad.} \tag{17}$$

Example 4.16 (cont.)

Then, using the dimensions given previously and evaluating at the same posture, we find

$$x'_{CI} = -1 \text{ in/rad} \quad \text{and} \quad y'_{CI} = 0.5 \text{ in/rad}, \tag{18}$$

and, from Eq. (3.34b),

$$r'_{CI} = \sqrt{(x'_{CI})^2 + (y'_{CI})^2} = \sqrt{(-1 \text{ in/rad})^2 + (0.5 \text{ in/rad})^2} = 1.118 \text{ in/rad}. \tag{19}$$

Measuring from the centrode tangent and centrode normal axes, the polar coordinates of coupler point C are

$$R_{CI} = \sqrt{(2 \text{ in})^2 + (1 \text{ in})^2} = \sqrt{5} \text{ in} \quad \text{and} \quad \psi_C = -71.57°. \tag{20}$$

The inflection point for point C, from Eq. (4.41), is

$$R_{J_CI} = R_{JI} \sin \psi_C = -2.683 \text{ in}. \tag{21}$$

Then, from Eq. (4.42), the radius of curvature of the path of coupler point C is

$$\rho_C = R_{CC'} = \frac{R_{CI}^2}{R_{CI} - R_{J_CI}} = \frac{(\sqrt{5} \text{ in})^2}{(\sqrt{5} \text{ in}) - (-2.683 \text{ in})} = 1.016 \text{ in}. \quad \textit{Ans.} \tag{22}$$

Finally, measuring from point I, Eqs. (4.53) give the center of curvature of the coupler curve as

$$x_{C'I} = x_{CI} - \rho_C \left(\frac{y'_C}{r'_C}\right) = (1 \text{ in}) - (1.016 \text{ in}) \left(\frac{0.5 \text{ in/rad}}{1.118 \text{ in/rad}}\right) = 0.546 \text{ in}, \quad \textit{Ans.} \tag{23}$$

$$y_{C'I} = y_{CI} - \rho_C \left(\frac{x'_C}{r'_C}\right) = (2 \text{ in}) - (1.016 \text{ in}) \left(\frac{1 \text{ in/rad}}{1.118 \text{ in/rad}}\right) = 1.091 \text{ in}. \quad \textit{Ans.} \tag{24}$$

These results are in good agreement with measurements from Fig. 4.40.

PROBLEMS[10]

4.1 The position vector of a point is defined by the equation

$$\mathbf{R} = \left(100t - \frac{25t^3}{3}\right)\hat{\mathbf{i}} + 250\hat{\mathbf{j}},$$

where R is in millimeters and t is in seconds. Find the acceleration of the point at $t = 2$ s.

---

[10] When assigning problems, the instructor may wish to specify the method of solution to be used, because a variety of approaches are presented in the text.

4.2 A point moves according to the equation

$$\mathbf{R} = \left(42t^2 - 7t^3\right)\hat{\mathbf{i}} + 14t^3\hat{\mathbf{j}},$$

where R is in inches and t is in seconds. Find the acceleration at $t = 3\,\text{s}$.

4.3 The path of a point is described by the equation

$$\mathbf{R} = \left(0.04t^2 + 0.16\right)e^{-j\pi t/10},$$

where R is in inches and t is in seconds. Find the unit tangent vector to the path, the normal and tangential components of the absolute acceleration, and the radius of curvature of the path, at $t = 20\,\text{s}$.

4.4 The motion of a point is described by the equations

$$x = 1.20t\cos\pi t^3 \quad \text{and} \quad y = 0.05t^3\sin 2\pi t,$$

where x and y are in meters and t is in seconds. Find the acceleration of the point at $t = 1.40\,\text{s}$.

4.5 Link 2, in the posture shown, has an angular velocity $\omega_2 = 120\,\text{rad/s}$ ccw and an angular

Fig. P4.5 $R_{AO_2} = 20$ in.

acceleration $\alpha_2 = 4\,800\,\text{rad/s}^2$ ccw. Determine the absolute acceleration of point A.

4.6 The accelerations of points A and B of link 2, which is rotating clockwise, are as given. Determine the angular velocity, the angular acceleration, and the acceleration of the midpoint C of link 2.

Fig. P4.6 $R_{BA} = 500$ mm,
$A_A = 180\,\text{m/s}^2$ and
$A_B = 45\,\text{m/s}^2$.

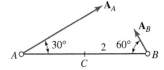

4.7 For the given kinematic data for link 2, find the velocity and acceleration of points B and C.

Fig. P4.7 $R_{BA} = 400$ mm,
$R_{CA} = 250$ mm, $R_{CB} = 200$ mm,
$\omega_2 = 24\,\text{rad/s}$ cw, $V_A = 6\,\text{m/s}$,
$\alpha_2 = 160\,\text{rad/s}^2$ ccw,
$A_A = 120\,\text{m/s}^2$, and $\beta = 15°$.

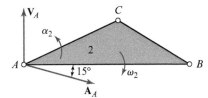

4.8 The angular velocity and angular acceleration of link 2 of the Scott–Russell linkage in the posture shown are $\omega_2 = 20$ rad/s cw and $\alpha_2 = 140$ rad/s^2 cw, respectively. Determine the velocity and acceleration of point B and the angular acceleration of link 3.

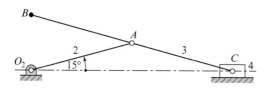

Fig. P4.8 $R_{AO_2} = R_{CA} = R_{BA} = 4$ in.

4.9 In the posture shown in Fig. P4.8, slider 4 is moving to the left with a constant velocity $V_C = 80$ in/s. Find the angular velocity and angular acceleration of link 2.

4.10 If the velocity of point B is constant in the posture shown in Prob. 3.8, determine the acceleration of point A and the angular acceleration of link 3.

4.11 If the angular velocity of crank 2 is constant in the posture shown in Prob. 3.9, determine the angular accelerations of links 3 and 4.

4.12 If the angular velocity of crank 2 is constant in the posture shown in Prob. 3.10, determine the acceleration of point C and the angular accelerations of links 3 and 4.

4.13 If the angular velocity of crank 2 is constant in the posture shown in Prob. 3.11, determine the acceleration of point C and the angular accelerations of links 3 and 4.

4.14 If the angular velocity of crank 2 is constant in the posture shown in Prob. 3.13, determine the accelerations of points C and D and the angular acceleration of link 4.

4.15 If the angular velocity of crank 2 is constant in the posture shown in Prob. 3.14, determine the acceleration of point C and the angular acceleration of link 4.

4.16 If the angular velocity of crank 2 is constant in the posture shown in Prob. 3.16, determine the acceleration of point C and the angular acceleration of link 4.

4.17 If the angular velocity of crank 2 is constant in the posture shown in Prob. 3.17, determine the acceleration of point B and the angular accelerations of links 3 and 6.

4.18 For the four-bar linkage of Prob. 3.18 in the posture shown, determine the angular acceleration of crank 2 to ensure that the angular acceleration of link 4 is zero.

4.19 For the four-bar linkage of Prob. 3.19 in the posture shown, determine the angular acceleration of crank 2 to ensure that the angular acceleration of link 4 is 100 rad/s^2 cw.

4.20 If the angular velocity of crank 2 is constant in the posture shown in Prob. 3.20, determine the acceleration of point C and the angular acceleration of link 3.

4.21 If the angular velocity of crank 2 is constant in the posture shown in Prob. 3.21, determine the acceleration of point C and the angular acceleration of link 3.

4.22 If the angular velocity of crank 2 is constant in the posture shown in Prob. 3.22, determine the accelerations of points B and D.

4.23 If the angular velocity of crank 2 is constant in the posture shown in Prob. 3.23, determine the accelerations of points B and D.

4.24–4.30 The nomenclature for the four-bar linkage is shown in Fig. P4.24; the dimensions and data are given in Table P4.24. Angular velocity ω_2 is constant for each problem; (a negative sign indicates that the direction is clockwise). The dimensions of even-numbered problems are inches and those of odd-numbered problems are millimeters. For each problem, determine $\theta_3, \theta_4, \omega_3, \omega_4, \alpha_3,$ and α_4.

Table P4.24

Problems	r_1	r_2	r_3	r_4	θ_2, deg	ω_2, rad/s
P4.24	4	6	9	10	240	1
P4.25	100	150	250	250	−45	56
P4.26	14	4	14	10	0	10
P4.27	250	100	500	400	70	−6
P4.28	8	2	10	6	40	12
P4.29	400	125	300	300	210	−18
P4.30	16	5	12	12	315	−18

Fig. P4.24

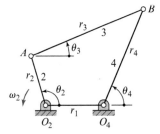

4.31 For the inverted slider-crank linkage in the posture shown, crank 2 has a constant angular velocity of 60 rev/min ccw. Find the velocity and acceleration of point B, and the angular velocity and acceleration of link 4.

Fig. P4.31 $R_{O_4 O_2} = 300$ mm, $R_{AO_2} = 175$ mm, and $R_{BO_4} = 700$ mm.

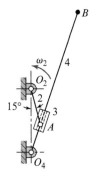

4.32 For the modified Scotch-yoke linkage in the posture shown in Prob. 3.26, determine the acceleration of link 4.

4.33 For the linkage in the posture shown in Prob. 3.27, determine the acceleration of point E.

4.34 For the inverted slider-crank linkage in the posture shown in Prob. 3.24, determine the acceleration of point B and the angular acceleration of link 4.

4.35 For the linkage in the posture shown in Prob. 3.25, determine the acceleration of point B and the angular acceleration of link 3.

4.36 For the linkage in the posture shown in Prob. 3.31, the input angular velocity is constant. Determine the accelerations of points A and B.

4.37 For the mechanism in the posture shown in Prob. 3.32, crank 2 has an angular acceleration of 2 rad/s² ccw. Determine the acceleration of point C_4 and the angular acceleration of link 3.

4.38 For the mechanism in the posture shown in Prob. 3.29, the input angular velocity is constant. Determine the angular accelerations of links 3 and 4.

4.39 For the mechanism in the posture shown in Prob. 3.30, the input angular velocity is constant. Determine the acceleration of point G and the angular accelerations of links 5 and 6.

4.40 Continue Prob. 3.40 and find the second-order kinematic coefficients of links 3 and 4. Assuming an input acceleration of $A_{A_2} = 200$ in/s², find the angular accelerations of links 3 and 4.

4.41 Continue Prob. 3.49 and find the second-order kinematic coefficients of links 3, 4, and 5. Assuming constant angular velocity for link 2, find the angular accelerations of links 3, 4, and 5.

4.42 Continue Prob. 3.50 and find the second-order kinematic coefficients of links 3, 4, and 5. Assuming constant angular velocity for link 2, find the angular accelerations of links 3, 4, and 5.

4.43 Draw the inflection circle for the absolute motion of the coupler link of the double-slider linkage. Select several points on the centrode normal and find their conjugate points. Plot portions of the paths of these points to demonstrate for yourself that the conjugates are indeed the centers of curvature.

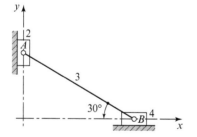

Fig. P4.43 $R_{BA} = 5$ in.

4.44 Draw the inflection circle for the absolute motion of the coupler of the four-bar linkage. Find the center of curvature of the coupler curve of point C and generate a portion of the path of C to verify your findings.

Fig. P4.44 $R_{CA} = 50$ mm, $R_{AO_2} = 18$ mm,
$R_{BO_1} = 70$ mm, $R_{PO_4} = 23.4$ mm, $R_{PC} = 20$ mm,
and $\beta = 51.21°$.

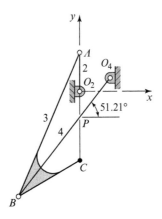

4.45 For the motion of the coupler relative to the frame, find the inflection circle, the centrode normal, the centrode tangent, and the centers of curvature of points C and D of the linkage of Prob. 3.13. Choose points on the coupler coincident with the instantaneous center of velocity and inflection pole, and plot nearby portions of their paths.

4.46 For the four-bar linkage in the posture shown, link 2 is 30° counterclockwise from the ground link, and the angular velocity and angular acceleration of the coupler link are $\omega_3 = 5$ rad/s ccw and $\alpha_3 = 20$ rad/s^2 cw, respectively. For the instantaneous motion of the coupler link, show: (a) the velocity pole, I, the pole tangent, T, and the pole normal, N; (b) the inflection circle and the Bresse circle; and (c) the instantaneous center of acceleration. Then determine: (d) the radius of curvature of the path of coupler point C; (e) the velocity of C; (f) the angular velocity of link 2; (g) the velocity of the pole I; (h) the acceleration of C; and (i) the acceleration of the velocity pole.

Fig. P4.46 $R_{O_4O_2} = 50$ mm,
$R_{AO_2} = 20$ mm, $R_{BA} = 63$ mm,
$R_{BO_4} = 30$ mm, and $R_{CB} = 20$ mm.

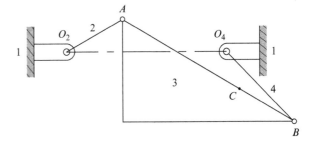

4.47 Consider the double-slider linkage in the posture given in Prob. 3.8. Point B moves with a constant velocity $V_B = 1\,600$ in/s to the left, as shown in the figure. The angular velocity and angular acceleration of coupler link AB are $\omega_3 = 36.6$ rad/s ccw and $\alpha_3 = 1\,340$ rad/s^2 cw, respectively. For the absolute motion of coupler link AB in the specified posture, draw the inflection circle and the Bresse circle. Then determine: (a) the radius of curvature of the path of point C, which is a point of link 3 midway between points A and B; and (b) the velocity of the velocity pole, I. Using the instantaneous center of acceleration, determine: (c) the acceleration of the pole, I; and (d) the accelerations of points A and C.

4.48 For the linkage of Prob. 3.17, link 2 is rotating with an angular velocity $\omega_2 = 15$ rad/s ccw and an angular acceleration $\alpha_2 = 320.93$ rad/s² cw. For the instantaneous motion of the connecting rod 3, find: (a) the inflection circle and the Bresse circle; (b) the location of the instantaneous center of acceleration; (c) the center of curvature of the path traced by the coupler point C; (d) the accelerations of points A, B, and C; and (e) the acceleration of inflection pole J.

4.49 Problem 3.32 shows an epicyclic gear train driven by the arm, link 2, with an angular velocity $\omega_2 = 3.33$ rad/s cw and an angular acceleration $\alpha_2 = 15$ rad/s² ccw. Define point E as a point on the circumference of planet gear 4 horizontal to the right of point B such that the angle $\angle DBE = 90°$. For the absolute motion of gear 4, draw the inflection circle and the Bresse circle on a scale drawing of the epicyclic gear train. Then, determine: (a) the location of the instantaneous center of acceleration of the planet gear; (b) the radii of curvature of the paths of points B and E; (c) the locations of the centers of curvature of the paths of points B and E; and (d) the accelerations of points B and E and pole I.

4.50 On 360×480-mm paper, draw the four-bar linkage full size, placing A' 120 mm up from the lower edge and 140 mm left of the right edge. (Better utilization of the paper is obtained by tilting the frame through about 15° as indicated.) For the coupler link, draw the inflection circle, and the cubic of stationary curvature. Choose a coupler point C coincident with the cubic, and plot a portion of its coupler curve in the vicinity of the cubic. Find the conjugate point C'. Draw a circle through C with center at C', and compare this circle with the actual path of C. Find Ball's point. Locate a point D on the coupler at Ball's point, and plot a portion of its path. Compare the result with a straight line.

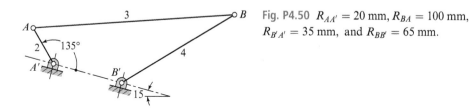

Fig. P4.50 $R_{AA'} = 20$ mm, $R_{BA} = 100$ mm, $R_{B'A'} = 35$ mm, and $R_{BB'} = 65$ mm.

4.51 For the mechanism in the posture shown in Prob. 3.51, the first- and second-order kinematic coefficients are $\theta'_3 = -8.333$ rad/rad, $r'_2 = -5$ in/rad, $\theta''_3 = -8.642$ rad/rad², and $r''_2 = -11.852$ in/rad² (where θ_2 is the input and r_2 is the vector from ground pin O_2 to pin A). Wheel 3 is rolling without slipping on the ground link at point C and sliding in the slot in link 2. The radius of the ground link is $\rho_1 = 3$ in, and the radius of the wheel is $\rho_3 = 0.75$ in. Determine: (a) the unit normal vector to the path of point D; (b) the radius of curvature of the path of this point; and (c) the x and y coordinates of the center of curvature of this path. If the angular velocity of link 2 is a constant $\omega_2 = 30$ rad/s ccw, then determine the acceleration of point D.

4.52 For the mechanism in the posture shown, the first- and second-order kinematic coefficients are $\theta_3' = 1.667$ rad/m, $R_4' = -1.00$ m/m, $\theta_3'' = 2.778$ rad/m², and $R_4'' = -3.333$ m/m². Roller 4 is pinned to link 3 at B and is rolling without slipping on the vertical ground link at C. Determine: (a) the first- and second-order kinematic coefficients of point P; (b) the unit tangent vector and the unit normal vector to the path traced by P; (c) the radius of curvature of this path; and (d) the x and y coordinates of the center of curvature of this path. If the constant velocity of the input is $\mathbf{V}_2 = -3\ \hat{\mathbf{i}}$ m/s, determine the acceleration of P.

Fig. P4.52 $\mathbf{R}_2 = \mathbf{R}_{AO} = 0.6\,\hat{\mathbf{i}}$ m,
$\mathbf{R}_4 = \mathbf{R}_{BO} = 0.6\,\hat{\mathbf{j}}$ m, $R_{PA} = 1.2$ m,
and $\rho_4 = 0.1125$ m.

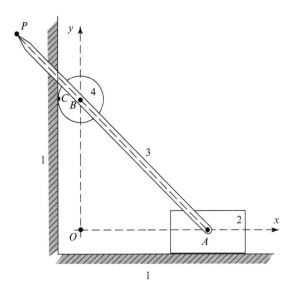

4.53 For the rack-and-pinion mechanism in the posture shown, the first and second-order kinematic coefficients are $R_2' = -16$ in/rad, $R_4' = -13.856$ in/rad, $R_2'' = 55.424$ in/rad², and $R_4'' = 56$ in/rad² (where $\mathbf{R}_2$ is the vector from the ground pin O_2 to the point of contact, C, of link 3). Determine: (a) the first- and second-order kinematic coefficients of point B; (b) the unit tangent vector and the unit normal vector to the path traced by point B; (c) the radius of curvature of this path; and (d) the x and y coordinates of the center of curvature of this path. If the constant angular velocity of the input link is $\omega_2 = -12\,\hat{\mathbf{k}}$ rad/s, determine the acceleration of point B.

Fig. P4.53 $\rho_3 = 4$ in.

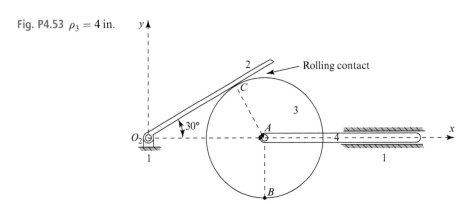

4.54 For the gear train in Prob. 3.54, the angular velocity and acceleration of input gear 2 are $\omega_2 = 77$ rad/s ccw and $\alpha_2 = 5$ rad/s² cw, respectively. Determine: (a) the second-order kinematic coefficients of gear 3 and rack 5; (b) the angular accelerations of gear 3 and link 4; and (c) the acceleration of the rack.

4.55 For the gear train in Prob. 3.55, the angular velocity and acceleration of the input arm 2 are $\omega_2 = 50$ rad/s cw and $\alpha_2 = 15$ rad/s² cw, respectively. Using the method of kinematic coefficients, determine the angular accelerations of gears 3, 4, and 5.

4.56 For the mechanism in the posture shown, link 4 is rolling without slipping on the ground at point C, and the first-order and second-order kinematic coefficients are $\theta_3' = -1.341$ rad/rad, $R_4' = -61.94$ mm/rad, $\theta_3'' = -2.475$ rad/rad², and $R_4'' = -147.44$ mm/rad² (where R_4 is the vector from the origin to pin B, which connects links 3 and 4). Determine the radius of curvature of the path of point D, and the x and y coordinates of the center of curvature of this path. If the constant input angular velocity of link 2 is $\omega_2 = -9\ \hat{\mathbf{k}}$ rad/s, determine the acceleration of point D.

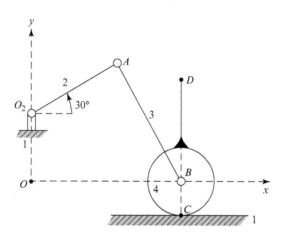

Fig. P4.56 $R_{O_2} = 20$ mm, $R_{AO_2} = 30$ mm, $R_{BA} = 40$ mm, $R_B = 45.34$ mm, $R_{CB} = 10$ mm, and $R_{DB} = 30$ mm.

4.57 For the linkage in the posture shown, the first-order and second-order kinematic coefficients are $\theta_3' = -3.0$ rad/rad, $R_4' = 3.464$ in/rad, $\theta_3'' = -13.856$ rad/rad², and $R_4'' = 6$ in/rad² (where R_4 is the vector from ground pivot O_2 to pin B). Determine the radius of curvature of the path of point C, and the x and y coordinates of the center of curvature of this path. If the input angular velocity of link 2 is a constant $\omega_2 = 22$ rad/s ccw, determine the acceleration of point C.

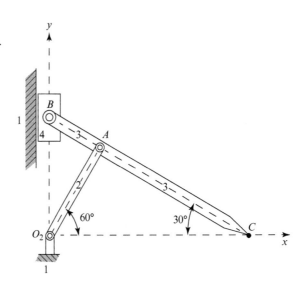

Fig. P4.57 $R_{AO_2} = 1.732$ in,
$R_{BA} = 1$ in, and $R_{CA} = 3$ in.

4.58 For the mechanism in the posture shown, the first-order and second-order kinematic coefficients are $\theta_3' = -3.464$ rad/rad, $\theta_4' = 1$ rad/rad, $\theta_3'' = 5.464$ rad/rad^2, and $\theta_4'' = 7.732$ rad/rad^2, respectively. Line AC in link 3 is parallel to the x-axis. The circular wheel, link 4, is rolling on the ground link at point E and rolling on link 3 at point B. Determine: (a) the first- and second-order kinematic coefficients of point C; (b) the unit tangent vector and the unit normal vector to the path traced by point C; (c) the radius of curvature of this path; and (d) the x and y coordinates of the center of curvature of this path. If the input angular velocity of link 2 is a constant $\omega_2 = 15\,\hat{k}$ rad/s, determine the acceleration of point C.

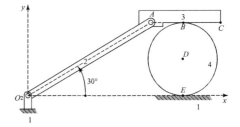

Fig. P4.58 $R_{AO_2} = 100$ mm, $R_{BA} = 25$ mm,
$R_{CA} = 50$ mm, and $\rho_4 = 25$ mm.

4.59 For the linkage in the posture shown, the first-order and second-order kinematic coefficients are $\theta_3' = \theta_4' = 0.5$ rad/rad, $\theta_3'' = \theta_4'' = 0$, $R_{34}' = 0$, and $R_{34}'' = -2$ in/rad^2 (where $\mathbf{R}_{34}$ is the vector from point B fixed in link 3 to point C fixed in link 4). Determine: (a) the radius of curvature of the path of point B; and (b) the center of curvature of the path of this point. If the input angular velocity of link 2 is a constant $\omega_2 = 10$ rad/s cw, determine the acceleration of point B.

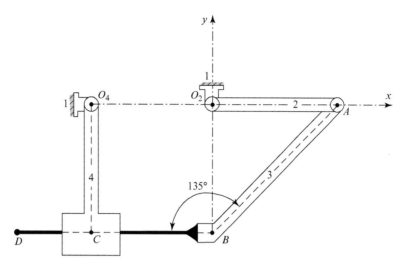

Fig. P4.59 $R_{AO_2} = 4$ in, $R_{BA} = 5.657$ in, $R_{CB} = R_{O_2O_4} = 4$ in, and $R_{CO_4} = 4$ in.

4.60 For the mechanism in the posture shown, the first-order and second-order kinematic coefficients are $\theta_3' = -4.333$ rad/rad, $\theta_4' = 0$, $R_4' = 650$ mm/rad, $\theta_3'' = 0$, $\theta_4'' = 0.813$ rad/rad², and $R_4'' = 121.875$ mm/rad² (where the rotation of link 2 is the input and $\mathbf{R}_4$ is the vector from point O_4 to point C on link 3). The angle between the line AB in link 3 and the line O_2A is a right angle. Determine: (a) the first- and second-order kinematic coefficients of point B; (b) the unit tangent vector and the unit normal vector to the path traced by this point; (c) the radius of curvature of the path traced by this point; and (d) the x and y coordinates of the center of curvature of the path traced by this point. If the input angular velocity of link 2 is a constant $\omega_2 = 15\,\hat{\mathbf{k}}$ rad/s, determine the acceleration of point B.

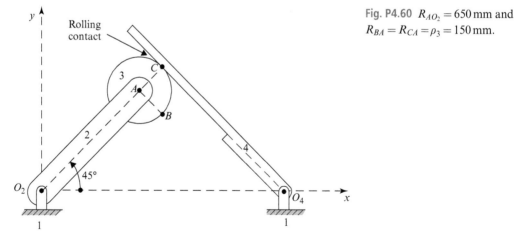

Fig. P4.60 $R_{AO_2} = 650$ mm and $R_{BA} = R_{CA} = \rho_3 = 150$ mm.

4.61 For the linkage of Prob. 3.61 in the posture shown, the first-order and second-order kinematic coefficients are $\theta_3' = \theta_4' = 1$ rad/rad, $R_{34}' = -300$ mm/rad,

$\theta_3'' = \theta_4'' = +2.309 \text{ rad/rad}^2$, and $R_{34}'' = -519.625 \text{ mm/rad}^2$ (where $\mathbf{R}_{34}$ is the vector from O_4 to point B fixed in link 3). Determine the first- and second-order kinematic coefficients of the coupler point C. Then, determine: (a) the unit tangent vector and the unit normal vector to the path traced by point C; (b) the radius of curvature of this path; and (c) the x and y coordinates of the center of curvature of this path. If the angular velocity of input link 2 is a constant $\boldsymbol{\omega}_2 = 15\,\hat{\mathbf{k}}$ rad/s, determine the acceleration of point C.

4.62 For the mechanism in the posture shown, the first-order and second-order kinematic coefficients are $\theta_3' = 2.165 \text{ rad/rad}$, $\theta_4' = -7.143 \text{ rad/rad}$, $\theta_3'' = -9.369 \text{ rad/rad}^2$, and $\theta_4'' = 26.784 \text{ rad/rad}^2$. Determine: (a) the first- and second-order kinematic coefficients for point C; (b) the radius of curvature of the path of this point; and (c) the x and y coordinates of the center of curvature of this path. If the input angular velocity of link 2 is a constant $\omega_2 = 50 \text{ rad/s ccw}$, then determine the acceleration of point C.

Fig. P4.62 $R_{AO_2} = 20$ in, $R_{BA} = 16$ in, $R_{CA} = 13.856$ in, and $\rho_4 = 5.60$ in.

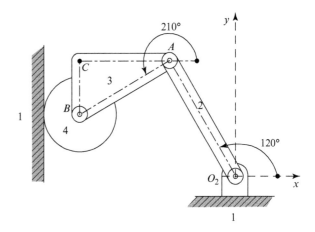

REFERENCES

[1] Beyer, R., 1931. *Technische Kinematik*, Leipzig: Johann Ambrosius Barth (Lithoprint publication, 1948, Ann Arbor, MI: J. W. Edwards).

[2] Hain, K., 1967. *Applied Kinematics*, 2nd ed., trans. T. P. Goodman et al., New York: McGraw-Hill.

[3] Hall, A. S., Jr., 1961. *Kinematics and Linkage Design*, Englewood Cliffs, NJ: Prentice-Hall. (This book is a classic on the theory of mechanisms and contains many useful examples).

[4] Hartenberg, R. S., and J. Denavit, 1964. *Kinematic Synthesis of Linkages*, New York: McGraw-Hill.

[5] Hirschhorn, J., 1962. *Kinematics and Dynamics of Plane Mechanisms*, New York: McGraw-Hill.

[6] de Jonge, A. E. R., 1943. A brief account of modern kinematics, *ASME Trans.*, 65: 663–683.

[7] Rosenauer, N., and A. H. Willis, 1953. *Kinematics of Mechanisms*, Sydney: Associated General Publications; republished New York: Dover, 1967.

[8] Tao, D. C., 1964. *Applied Linkage Synthesis*, Reading, MA: Addison-Wesley.

5 Multi-Degree-of-Freedom Mechanisms

5.1 Introduction

In Chapters 2, 3, and 4, we concentrated entirely on problems that exhibit a single degree of freedom, and can be analyzed by specifying the motion of a single input variable. This was justifiable, since, by far, the vast majority of practical mechanisms are designed to have only one degree of freedom so that they can be driven by a single power source. However, there are mechanisms that have multiple degrees of freedom and can only be analyzed if more than one input motion is given. In this chapter, we will look at how our methods can be used to find the positions, velocities, and accelerations of these mechanisms.

Consider, for example, the planar five-bar linkage shown in Fig. 5.1. The Kutzbach criterion, Eq. (1.1), indicates that this linkage has a mobility of two and, therefore, requires two input motions to provide a unique output motion. This linkage is operated by rotating cranks 2 and 3 independently and can, therefore, produce a wide variety of motions for the two coupler links 4 and 5.

A practical application of the five-bar linkage is to position the end-effector of a robotic manipulator (for example, the General Electric Model P80 robotic manipulator shown in Fig. 5.2).

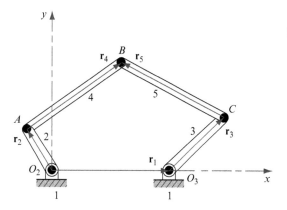

Fig. 5.1 Vectors for the five-bar linkage.

Fig. 5.2 General Electric
Model P80 robotic
manipulator.

Fig. 5.3 A pantograph.

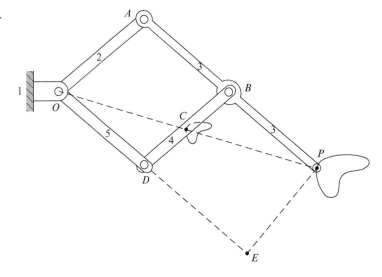

Another common practical application of a five-bar linkage is the pantograph linkage shown in Fig. 1.26. A variation of the pantograph is the linkage shown in Fig. 5.3, where the path of point P is a magnified copy of the path of point C.

If the two input rotations of Fig. 5.1 are interconnected by gears that have rolling contact with each other, as shown in Fig. 5.4, then the resulting mechanism has only a single degree-of-freedom and is commonly referred to as a *geared five-bar* mechanism. Such a mechanism is

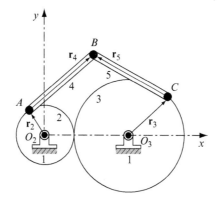

Fig. 5.4 Geared five-bar
mechanism.

often found in machinery, since it can provide more complex motions than the well-known
planar four-bar linkage that was investigated in previous chapters.

Example 5.1

For the five-bar linkage in the posture shown in Fig. 5.5, the postures of the two input links are
$\theta_2 = 120°$ and $\theta_3 = 45°$. Determine the postures of the two coupler links, θ_4 and θ_5.

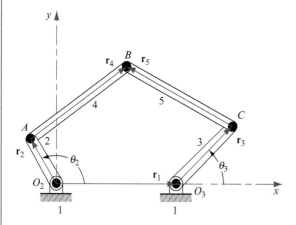

Fig. 5.5 $r_1 = R_{O_3O_2} = 6$ in, $r_2 = R_{AO_2} = 2.5$ in,
$r_3 = R_{CO_3} = 4$ in, $r_4 = R_{BA} = 6$ in, and
$r_5 = R_{BC} = 6$ in.

Solution

The vectors for the kinematic analysis of this linkage are shown in Fig. 5.5. The postures of links
4 and 5 are found from a scale drawing:

$$\theta_4 = 36.5° \quad \text{and} \quad \theta_5 = 151.1°. \qquad \qquad Ans.$$

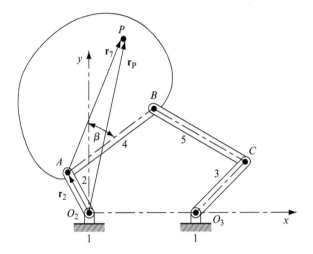

Fig. 5.6 Vectors for coupler point P of link 4.

For a point fixed in coupler link 4 or coupler link 5 to follow a unique path, control of two independent input motions is required. Still, there is an extremely wide variety of curves that can be generated by a coupler point of the planar five-bar linkage. For the purposes of illustration, consider the arbitrary point P of coupler link 4, as shown in Fig. 5.6.

Example 5.2

Consider a continuation of Example 5.1 to include the coupler point P shown in Fig. 5.6. The location of point P is given by $r_7 = R_{PA} = 8$ in and $\beta = 30°$. For the given posture of the linkage, the problem is to determine the absolute position coordinates of point P.

Solution

After Fig. 5.5 of Example 5.1 is completed to scale, vector $\mathbf{r}_7$ can be drawn at the orientation $\theta_7 = \theta_4 + \beta = 66.5°$. The absolute coordinates of point P can then be measured, and the results are

$$x_P = 1.95 \text{ in} \quad \text{and} \quad y_P = 9.50 \text{ in.} \qquad \textit{Ans.}$$

5.2   Algebraic Posture Analysis

The analytic approaches for the solution of the posture analysis of multi-degree-of-freedom mechanisms are parallel to the methods presented in Chapter 2. When sufficient input data are given to represent the positions of all degrees of freedom, the loop-closure equations can be formulated. For planar mechanisms, these equations can be separated into real and imaginary parts (horizontal and vertical components), and these allow solution for two unknowns per loop. We will demonstrate this by continuing with the planar five-bar linkage of Example 5.1 and Example 5.2.

Example 5.3

For the planar five-bar linkage of Fig. 5.5, the loop-closure equation is

$$\overset{\sqrt{I}}{\mathbf{r}_2} + \overset{\sqrt{?}}{\mathbf{r}_4} - \overset{\sqrt{?}}{\mathbf{r}_5} - \overset{\sqrt{I}}{\mathbf{r}_3} - \overset{\sqrt{\sqrt{}}}{\mathbf{r}_1} = \mathbf{0}, \tag{1}$$

where θ_2 and θ_3 are the input angles, marked as given above the appropriate terms in Eq. (1). For the given posture of the linkage, the problem is to determine: (a) the unknown coupler angles θ_4 and θ_5; and (b) the absolute position coordinates of coupler point P.

Solution

In complex polar notation, Eq. (1) becomes

$$r_2 e^{j\theta_2} + r_4 e^{j\theta_4} - r_5 e^{j\theta_5} - r_3 e^{j\theta_3} - r_1 = 0. \tag{2}$$

Separating the real and imaginary parts of this equation gives

$$r_2 \cos\theta_2 + r_4 \cos\theta_4 - r_5 \cos\theta_5 - r_3 \cos\theta_3 - r_1 = 0, \tag{3}$$

$$r_2 \sin\theta_2 + r_4 \sin\theta_4 - r_5 \sin\theta_5 - r_3 \sin\theta_3 = 0. \tag{4}$$

The solution to these equations can be determined either analytically or numerically. For example, using the Newton–Raphson iterative procedure (Section 2.9), the postures of the two coupler links are evaluated as

$$\theta_4 = 36.447° \quad \text{and} \quad \theta_5 = 151.084°. \qquad \qquad Ans.$$

These verify the graphic results determined in Example 5.1, and have higher accuracy.

The vectors defining the location of coupler point P are shown in Fig. 5.6. The vector equation for the location of this point can be written as

$$\overset{??}{\mathbf{r}_P} = \overset{\sqrt{I}}{\mathbf{r}_2} + \overset{\sqrt{C}}{\mathbf{r}_7}, \tag{5}$$

where the constraint equation for the orientation of vector $\mathbf{r}_7$ is

$$\theta_7 = \theta_4 + \beta. \tag{6}$$

From Eq. (5), the absolute position coordinates of point P are

$$x_P = r_2 \cos\theta_2 + r_7 \cos\theta_7, \tag{7}$$

$$y_P = r_2 \sin\theta_2 + r_7 \sin\theta_7. \tag{8}$$

Substituting the given data into these equations, the position of point P (for the given posture of the linkage) is

$$x_P = (2.5 \text{ in}) \cos 120° + (8 \text{ in}) \cos 66.447° = 1.947 \text{ in}, \qquad Ans.$$

$$y_P = (2.5 \text{ in}) \sin 120° + (8 \text{ in}) \sin 66.447° = 9.499 \text{ in}. \qquad Ans.$$

These verify the graphic results determined in Example 5.2 and have higher accuracy.

5.3   Velocity Analysis; Velocity Polygons

Velocity analysis is performed quite easily by the velocity polygon methods of Sections 3.3–3.8 once the velocities of each of the independent inputs are specified. No new methods are required for the utilization of velocity polygons when a mechanism has more than one degree of freedom. However, input velocities must be given for each degree of freedom. An example is presented here to demonstrate how the graphic methods can be extended.

Example 5.4

For the five-bar linkage of Examples 5.1–5.3, in the posture shown in Figs. 5.5–5.7a, suppose that the constant angular velocities of links 2 and 3 are $\omega_2 = 10$ rad/s ccw and $\omega_3 = 5$ rad/s cw, respectively. Determine: (a) the angular velocities of coupler links 4 and 5; and (b) the velocity of coupler point P.

Solution

First, the velocities of points A and C, respectively, are

$$V_A = V_{AO_2} = \omega_2 R_{AO_2} = (10 \text{ rad/s})(2.5 \text{ in}) = 25.0 \text{ in/s},$$

$$V_C = V_{CO_3} = \omega_3 R_{CO_3} = (5 \text{ rad/s})(4.0 \text{ in}) = 20.0 \text{ in/s}.$$

From these, we can write two vector equations for the velocity of point B:

$$\mathbf{V}_B = \overset{\sqrt{}}{\mathbf{V}_A} + \overset{?\sqrt{}}{\mathbf{V}_{BA}} = \overset{\sqrt{}}{\mathbf{V}_C} + \overset{?\sqrt{}}{\mathbf{V}_{BC}}.$$

Solving these equations with the velocity polygon shown in Fig. 5.7b locates the velocity image point B.

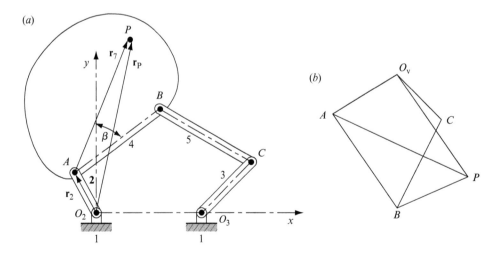

Fig. 5.7 (a) Linkage posture; (b) velocity polygon.

Example 5.4 (cont.)

Now we can measure the velocities V_{BA} and V_{BC} and obtain the angular velocities of coupler links 4 and 5; that is,

$$\omega_4 = \frac{V_{BA}}{R_{BA}} = \frac{35.34 \text{ in/s}}{6.0 \text{ in}} = 5.89 \text{ rad/s cw,} \qquad Ans.$$

$$\omega_5 = \frac{V_{BC}}{R_{BC}} = \frac{30.60 \text{ in/s}}{6.0 \text{ in}} = 5.10 \text{ rad/s ccw.} \qquad Ans.$$

Once the velocities of points A and B are known, we can construct the velocity image of coupler link 4 as explained in Section 3.4 and obtain the velocity of coupler point P; that is,

$$\mathbf{V}_P = 38 \text{ in/s} \angle -55.5°. \qquad Ans.$$

5.4   Instantaneous Centers of Velocity

If we attempt to use the method of instantaneous centers of velocity for planar mechanisms with multiple degrees of freedom, we encounter a complication. We find that the locations of the secondary instant centers cannot be determined from the geometry of the mechanism by the methods of Sections 3.13–3.15 alone. In fact, in addition, their locations depend on the ratios of the independent input velocities [1]. One method for finding the secondary instant centers is demonstrated in the following example.

Example 5.5

For the five-bar linkage in the posture of Examples 5.1–5.4 (Figs. 5.5–5.7a) suppose that links 2 and 3 are rotating with constant angular velocities $\omega_2 = 10 \text{ rad/s ccw}$ and $\omega_3 = 5 \text{ rad/s cw}$, respectively. Using the method of instant centers, determine the angular velocities of coupler links 4 and 5, and the velocity of coupler point P.

Solution

The five primary instant centers for the five-bar linkage, denoted $I_{12}, I_{24}, I_{45}, I_{35},$ and I_{13}, are the centers of the five pin joints, and are shown in Fig. 5.8.

From the Aronhold–Kennedy theorem (Section 3.14), we know that the secondary instant center I_{23}, which relates the two known velocities, ω_2 and ω_3, must lie on the line containing the instant centers I_{12} and I_{13}.

Furthermore, from the angular velocity ratio theorem, Eq. (3.28), we know that the ratio of the angular velocity of link 2 to the angular velocity of link 3 can be written as

Example 5.5 (cont.)

Fig. 5.8 Primary instant
centers.

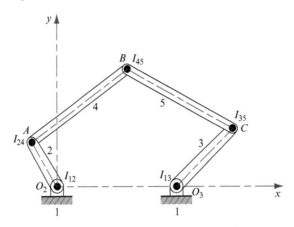

$$\frac{\omega_2}{\omega_3} = \frac{R_{I_{23}I_{13}}}{R_{I_{23}I_{12}}} = \frac{10.0 \text{ rad/s}}{-5.0 \text{ rad/s}} = -2.0. \tag{1}$$

From Section 3.17 we know that, if the instant center I_{23} is located between the two absolute instant centers, I_{12} and I_{13}, then the angular velocity ratio is negative. Similarly, if the instant center I_{23} is outside the two absolute instant centers I_{12} and I_{13}, then the angular velocity ratio is positive. Note, however, that a sign convention is not necessary in Eq. (1), since a directed line is used for measuring the locations of the instant centers.

In this example, Eq. (1) gives

$$R_{I_{23}I_{13}} = -2.0 R_{I_{23}I_{12}}. \tag{2}$$

But we also know from Fig. 5.8, using positive distances to the right, that

$$R_{I_{23}I_{13}} = R_{I_{23}I_{12}} + R_{I_{12}I_{13}} = R_{I_{23}I_{12}} - 6.0 \text{ in,} \tag{3}$$

and solving Eqs. (2) and (3) simultaneously gives

$$-2.0 R_{I_{23}I_{12}} = R_{I_{23}I_{12}} - 6.0 \text{ in,}$$

$$R_{I_{23}I_{12}} = 2.0 \text{ in.} \tag{4}$$

Therefore, instant center I_{23} is located 2.0 in to the right of instant center I_{12}, as shown on a scaled drawing of the linkage in Fig. 5.9.

The remaining secondary instant centers can be obtained directly from the Aronhold–Kennedy theorem. The Kennedy circle, used to help locate the secondary instant centers, is also shown in Fig. 5.9. Secondary instant center I_{14}, for example, must lie on the line containing the instant centers I_{12} and I_{24}. Similarly, secondary instant center I_{15} must lie on the line containing the instant centers I_{13} and I_{35}. The locations of all secondary instant centers that lie within the limits of the page are shown in Fig. 5.9. The location of the instant center I_{34} lies outside of the page, but it is indicated by two dashed lines. Note that it is not necessary to find the locations of all secondary instant centers in order to determine the unknown angular velocities ω_4 and ω_5.

Example 5.5 (cont.)

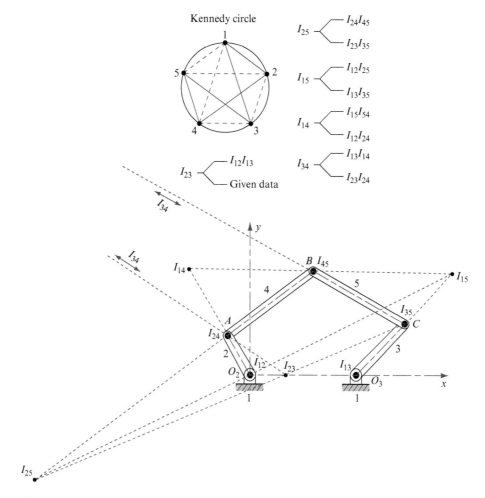

Fig. 5.9 Secondary instant centers.

Choosing the upward direction along each line of centers as positive, the distances between instant centers I_{12} and I_{24} and between instant centers I_{13} and I_{35} are known from the lengths of links 2 and 3, respectively; these are

$$R_{I_{24}I_{12}} = 2.50 \text{ in} \quad \text{and} \quad R_{I_{35}I_{13}} = 4.00 \text{ in}. \tag{5}$$

Distances between other instant centers are also measured from the scaled drawing as

$$R_{I_{24}I_{14}} = -4.24 \text{ in}, \quad R_{I_{25}I_{12}} = -13.41 \text{ in}, \quad R_{I_{25}I_{15}} = -26.29 \text{ in}, \quad \text{and} \quad R_{I_{35}I_{15}} = -3.92 \text{ in}.$$

Example 5.5 (cont.)

Substituting these distances into Eq. (3.28), angular velocity ratios can be written as

$$\frac{\omega_4}{\omega_2} = \frac{R_{I_{24}I_{12}}}{R_{I_{24}I_{14}}} = \frac{2.50 \text{ in}}{-4.24 \text{ in}} = -0.590, \tag{6}$$

$$\frac{\omega_5}{\omega_2} = \frac{R_{I_{25}I_{12}}}{R_{I_{25}I_{15}}} = \frac{-13.41 \text{ in}}{-26.29 \text{ in}} = 0.510, \tag{7}$$

$$\frac{\omega_5}{\omega_3} = \frac{R_{I_{35}I_{13}}}{R_{I_{35}I_{15}}} = \frac{4.00 \text{ in}}{-3.92 \text{ in}} = -1.020. \tag{8}$$

Then, substituting the angular velocity of link 2 into Eqs. (6) and (7), the angular velocities of links 4 and 5 are

$$\omega_4 = -0.590\omega_2 = -0.590(10 \text{ rad/s}) = -5.90 \text{ rad/s cw}, \qquad Ans.$$

$$\omega_5 = 0.510\omega_2 = 0.510(10 \text{ rad/s}) = 5.10 \text{ rad/s ccw}. \qquad Ans.$$

As a check, we can substitute the angular velocity of link 3 into Eq. (8) to verify that the angular velocity of link 5 is

$$\omega_5 = -1.020\omega_3 = -1.020(-5 \text{ rad/s}) = 5.10 \text{ rad/s ccw}. \qquad Ans.$$

Using the method of instant centers, the magnitude of the velocity of point P can be written as

$$V_P = \omega_4 R_{PI_{14}}. \tag{9}$$

From the scale drawing (Fig. 5.10), the distance from the absolute instant center of link 4 (I_{14}) to point P is measured as $R_{PI_{14}} = 6.46$ in. Substituting this value into Eq. (9), the magnitude of the velocity of point P is

$$V_P = (|-5.90 \text{ rad/s}|)(6.46 \text{ in}) = 38.11 \text{ in/s}. \qquad Ans.$$

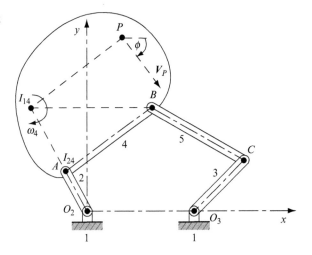

Fig. 5.10 Absolute instant center for coupler link 4.

Example 5.5 (cont.)

The direction of the velocity of point P is measured from Fig. 5.10 as

$$\phi = -55.5°. \hspace{4cm} Ans.$$

Note that these results are in good agreement with those given in Example 5.4.

A special case of the five-bar linkage is when instant center I_{23} is stationary on the line of centers $I_{12}I_{13}$; then links 2 and 3 can be replaced by two circular gears, as shown in Fig. 5.4. This mechanism, referred to as a *geared five-bar mechanism*, has a mobility of 1 (the input is the rotation of either link 2 or link 3). The two gears – that is, links 2 and 3 – are in rolling contact at instant center I_{23}. The ratio of the angular velocity of gear 3 to the angular velocity of gear 2 (also referred to as the *gear ratio*) is the reciprocal of the ratio of Eq. (1).

5.5 First-Order Kinematic Coefficients

If we require higher accuracy in solutions for velocities, we look for an analytic or numeric technique. The method of kinematic coefficients, which was presented in Section 3.11, provides such a technique.

When dealing with mechanisms having multiple degrees of freedom, however, the method of kinematic coefficients raises a new complication in notation. It requires an additional subscript to keep track of the multiple input motions. For example, the time rate of change of a dependent variable, such as θ_i, the angular velocity of link i, can be written as $\omega_i = d\theta_i/dt$. However, since there are multiple degrees of freedom, the dependent variable θ_i is a function of two or more independent variables, say θ_j and θ_k; that is, $\theta_i = \theta_i(\theta_j, \theta_k)$. In this case, the time rate of change of the dependent variable θ_i can be written as

$$\omega_i = \frac{d\theta_i}{dt} = \frac{\partial\theta_i}{\partial\theta_j}\frac{d\theta_j}{dt} + \frac{\partial\theta_i}{\partial\theta_k}\frac{d\theta_k}{dt}. \tag{5.1}$$

Similarly, a point, P, on one of the moving links has a position that is a function of both independent variables, $\mathbf{r}_P = \mathbf{r}_P(\theta_j, \theta_k)$. Therefore, the velocity of point P can be written as

$$\mathbf{V}_P = \frac{d\mathbf{r}_P}{dt} = \frac{\partial\mathbf{r}_P}{\partial\theta_j}\frac{d\theta_j}{dt} + \frac{\partial\mathbf{r}_P}{\partial\theta_k}\frac{d\theta_k}{dt}. \tag{5.2}$$

Also, since $\mathbf{r}_P(\theta_j, \theta_k)$ has coordinates such as $\mathbf{r}_P = x_P\hat{\mathbf{i}} + y_P\hat{\mathbf{j}} + z_P\hat{\mathbf{k}}$, the velocity of point P can be written as

$$\begin{aligned}
\mathbf{V}_P &= \left(\frac{\partial x_P}{\partial\theta_j}\hat{\mathbf{i}} + \frac{\partial y_P}{\partial\theta_j}\hat{\mathbf{j}} + \frac{\partial z_P}{\partial\theta_j}\hat{\mathbf{k}}\right)\frac{d\theta_j}{dt} + \left(\frac{\partial x_P}{\partial\theta_k}\hat{\mathbf{i}} + \frac{\partial y_P}{\partial\theta_k}\hat{\mathbf{j}} + \frac{\partial z_P}{\partial\theta_k}\hat{\mathbf{k}}\right)\frac{d\theta_k}{dt} \\
&= \left(\frac{\partial x_P}{\partial\theta_j}\frac{d\theta_j}{dt} + \frac{\partial x_P}{\partial\theta_k}\frac{d\theta_k}{dt}\right)\hat{\mathbf{i}} + \left(\frac{\partial y_P}{\partial\theta_j}\frac{d\theta_j}{dt} + \frac{\partial y_P}{\partial\theta_k}\frac{d\theta_k}{dt}\right)\hat{\mathbf{j}} + \left(\frac{\partial z_P}{\partial\theta_j}\frac{d\theta_j}{dt} + \frac{\partial z_P}{\partial\theta_k}\frac{d\theta_k}{dt}\right)\hat{\mathbf{k}}.
\end{aligned} \tag{5.3}$$

If we wish to continue to use the prime notation, where a first-order kinematic coefficient carries the symbol $\theta'_{ij} = \partial\theta_i/\partial\theta_j$ or $\mathbf{r}'_{Pj} = \partial\mathbf{r}_P/\partial\theta_j$, we see that an additional subscript is required to indicate with respect to which independent variable the derivative is taken. With this additional subscript, Eq. (5.1) becomes

$$\omega_i = d\theta_i/dt = \theta'_{ij}\omega_j + \theta'_{ik}\omega_k. \tag{5.4}$$

Similarly, Eq. (5.2) becomes

$$\mathbf{V}_P = d\mathbf{r}_P/dt = \mathbf{r}'_{Pj}\omega_j + \mathbf{r}'_{Pk}\omega_k, \tag{5.5}$$

and Eq. (5.3) becomes

$$\mathbf{V}_P = \left(x'_{Pj}\hat{\mathbf{i}} + y'_{Pj}\hat{\mathbf{j}} + z'_{Pj}\hat{\mathbf{k}}\right)\omega_j + \left(x'_{Pk}\hat{\mathbf{i}} + y'_{Pk}\hat{\mathbf{j}} + z'_{Pk}\hat{\mathbf{k}}\right)\omega_k$$

or

$$\mathbf{V}_P = \left(x'_{Pj}\omega_j + x'_{Pk}\omega_k\right)\hat{\mathbf{i}} + \left(y'_{Pj}\omega_j + y'_{Pk}\omega_k\right)\hat{\mathbf{j}} + \left(z'_{Pj}\omega_j + z'_{Pk}\omega_k\right)\hat{\mathbf{k}}. \tag{5.6}$$

Although we find need for this additional subscript, first-order kinematic coefficients do provide insight into the velocity analysis of multi-degree-of-freedom mechanisms, just as they did in Section 3.11 for single-degree-of-freedom mechanisms. This is shown in the following example.

Example 5.6

Consider the five-bar linkage in the posture of Example 5.1 (Fig. 5.5), and with links 2 and 3 rotating with the same constant angular velocities as in Example 5.5. Using the method of kinematic coefficients, determine the angular velocities of the coupler links 4 and 5, and the velocity of the coupler point P.

Solution

We begin the analysis by writing the loop-closure equation as we did in Example 5.3, and separating the real and imaginary parts as in Eqs. (3) and (4) of that example. As we demonstrated there, those equations can be solved for the position variables, θ_4 and θ_5.

Next, we take the partial derivatives of Eqs. (3) and (4) of Example 5.3 with respect to independent variable θ_2; that is,

$$-r_2\sin\theta_2 - r_4\sin\theta_4\theta'_{42} + r_5\sin\theta_5\theta'_{52} = 0, \tag{1a}$$

$$r_2\cos\theta_2 + r_4\cos\theta_4\theta'_{42} - r_5\cos\theta_5\theta'_{52} = 0. \tag{1b}$$

Rearranging these two equations and expressing them in matrix form gives

$$\begin{bmatrix} -r_4\sin\theta_4 & r_5\sin\theta_5 \\ r_4\cos\theta_4 & -r_5\cos\theta_5 \end{bmatrix} \begin{bmatrix} \theta'_{42} \\ \theta'_{52} \end{bmatrix} = \begin{bmatrix} r_2\sin\theta_2 \\ -r_2\cos\theta_2 \end{bmatrix}. \tag{2}$$

Example 5.6 (cont.)

Table 5.1 **Input angles, coupler angles, and first-order
kinematic coefficients**

θ_2	θ_3	θ_4	θ_5	θ'_{42}	θ'_{43}	θ'_{52}	θ'_{53}
deg	deg	deg	deg	rad/rad	rad/rad	rad/rad	rad/rad
40	85	93.419	142.993	−0.533	0.743	−0.440	0.128
50	80	84.547	138.843	−0.513	0.703	−0.291	0.065
60	75	76.200	136.526	−0.466	0.674	−0.134	0.016
70	70	68.479	135.944	−0.412	0.659	0.012	−0.019
80	65	61.343	136.872	−0.360	0.654	0.138	−0.044
90	60	54.690	139.054	−0.316	0.658	0.242	−0.062
100	55	48.401	142.263	−0.281	0.667	0.327	−0.077
110	50	42.360	146.315	−0.254	0.683	0.397	−0.091
120	45	36.447	151.084	−0.237	0.705	0.456	−0.109
130	40	30.526	156.510	−0.230	0.737	0.508	−0.136
140	35	24.383	162.646	−0.241	0.793	0.564	−0.185
150	30	17.496	169.884	−0.306	0.927	0.663	−0.311

Now we can solve this matrix equation for the first-order kinematic coefficients θ'_{42} and θ'_{52}. Using the data obtained so far, the results are $\theta'_{42} = -0.237$ and $\theta'_{52} = 0.456$, which are both dimensionless (rad/rad).

Similarly, taking the partial derivative of Eqs. (3) and (4) of Example 5.3 with respect to independent variable θ_3, and rearranging the resulting equations into matrix form, gives

$$\begin{bmatrix} -r_4 \sin\theta_4 & r_5 \sin\theta_5 \\ r_4 \cos\theta_4 & -r_5 \cos\theta_5 \end{bmatrix} \begin{bmatrix} \theta'_{43} \\ \theta'_{53} \end{bmatrix} = \begin{bmatrix} -r_3 \sin\theta_3 \\ r_3 \cos\theta_3 \end{bmatrix}. \tag{3}$$

We can solve this matrix equation for the first-order kinematic coefficients θ'_{43} and θ'_{53}. The results are $\theta'_{43} = 0.705$ rad/rad and $\theta'_{53} = -0.109$ rad/rad.

Table 5.1 presents the angular variables, θ_4 and θ_5, and the first-order kinematic coefficients of the coupler links, for input angles in the range $40° \le \theta_2 \le 150°$ and $85° \ge \theta_3 \ge 30°$.

As indicated in Section 4.12, it is not a coincidence that the (2×2) coefficient matrices in Eqs. (2) and (3) are identical. This is always true for all sets of derivative equations determined by differentiating loop-closure equations. This coefficient matrix is called the *Jacobian* of the system. In this example, the determinant of the Jacobian is

$$\Delta = r_4 r_5 \sin(\theta_4 - \theta_5). \tag{4}$$

At the current posture this determinant has a value of $\Delta = -32.72$ in^2 and does not cause numeric difficulty. However, we note that this determinant becomes zero when $\theta_5 = \theta_4$ or when $\theta_5 = \theta_4 \pm 180°$. These occur when the coupler links are either fully extended or folded over each other – that is, when the five-bar linkage is in the posture of a quadrilateral with the two coupler links aligned. The kinematic coefficients become indeterminate in such a posture.

Example 5.6 (cont.)

The angular velocities of links 4 and 5 can be determined from Eq. (5.4). At the posture shown in Fig. 5.1 the values are

$$\omega_4 = \theta'_{42}\omega_2 + \theta'_{43}\omega_3 = -0.237(10.0 \text{ rad/s}) + 0.705(-5.0 \text{ rad/s}) = -5.90 \text{ rad/s (cw)}, \quad Ans.$$

$$\omega_5 = \theta'_{52}\omega_2 + \theta'_{53}\omega_3 = 0.456(10.0 \text{ rad/s}) - 0.109(-5.0 \text{ rad/s}) = 5.10 \text{ rad/s ccw}. \quad Ans.$$

These confirm the results found in Examples 5.4 and 5.5.

Differentiating constraint Eq. (6) of Example 5.3 with respect to the two independent inputs, the first-order kinematic coefficients of the orientation of vector $\mathbf{r}_7$ are

$$\theta'_{72} = \theta'_{42} \quad \text{and} \quad \theta'_{73} = \theta'_{43}. \tag{5}$$

Partially differentiating Eqs. (7) and (8) of Example 5.3 with respect to the input variable θ_2 and using Eq. (5), the first-order kinematic coefficients of point P are

$$x'_{P2} = -r_2 \sin\theta_2 - r_7 \sin\theta_7\theta'_{42}, \tag{6a}$$

$$y'_{P2} = r_2 \cos\theta_2 + r_7 \cos\theta_7\theta'_{42}. \tag{6b}$$

Similarly, partially differentiating Eqs. (7) and (8) of Example 5.3 with respect to the input variable θ_3 and using Eq. (5), the first-order kinematic coefficients of point P are

$$x'_{P3} = -r_7 \sin\theta_7\theta'_{43}, \tag{7a}$$

$$y'_{P3} = r_7 \cos\theta_7\theta'_{43}. \tag{7b}$$

Substituting the specified data into Eqs. (6) and (7), the first-order kinematic coefficients of point P are $x'_{P2} = -0.429\,4$ in/rad, $y'_{P2} = -2.006\,6$ in/rad, $x'_{P3} = -5.168\,1$ in/rad, and $y'_{P3} = 2.252\,8$ in/rad.

Finally, substituting these values and the given angular velocities into Eq. (5.6), the velocity of point P is

$$\mathbf{V}_P = [(-0.4294 \text{ in})(10 \text{ rad/s}) + (-5.168\,14 \text{ in})(-5 \text{ rad/s})]\hat{\mathbf{i}}$$
$$+ [(-2.0066 \text{ in})(10 \text{ rad/s}) + (2.252\,84 \text{ in})(-5 \text{ rad/s})]\hat{\mathbf{j}}.$$

That is,

$$\mathbf{V}_P = 21.546 \text{ in/s}\,\hat{\mathbf{i}} - 31.330 \text{ in/s}\,\hat{\mathbf{j}} = 38.024 \text{ in/s}\,\angle{-55.483°}. \quad Ans.$$

The velocity of point P is shown in Fig. 5.11. Note that these results are in good agreement with, and have higher accuracy than, those obtained in Examples 5.4 and 5.5.

Table 5.2 indicates the first-order kinematic coefficients of point P and the magnitude and direction of the velocity of point P for the range of input angles used in Table 5.1.

Example 5.6 (cont.)

Table 5.2 **First-order kinematic coefficients and velocities of coupler point P**

θ_2	θ_3	x'_{P2}	y'_{P2}	x'_{P3}	y'_{P3}	V_P	ϕ
deg	deg	in	in	in	in	in/s	deg
40	85	1.9543	4.2650	−4.9588	−3.2721	73.811	53.081
50	80	1.8180	3.3120	−5.1124	−2.3350	62.609	45.681
60	75	1.4177	2.2909	−5.1816	−1.5054	50.330	37.209
70	70	0.9102	1.3410	−5.2151	−0.7774	39.200	26.183
80	65	0.4201	0.5017	−5.2332	−0.1227	30.885	10.503
90	60	0.0191	−0.2342	−5.2391	0.4870	26.816	−10.260
100	55	−0.2610	−0.8859	−5.2303	1.0735	27.506	−31.146
110	50	−0.4107	−1.4715	−5.2054	1.6553	31.766	−46.367
120	45	−0.4294	−2.0066	−5.1681	2.2528	38.024	−55.483
130	40	−0.3144	−2.5116	−5.1347	2.9020	45.583	−60.380
140	35	−0.0397	−3.0379	−5.1568	3.6942	55.053	−62.539
150	30	0.5534	−3.8178	−5.4662	5.0096	71.258	−62.535

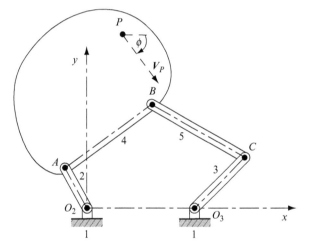

Fig. 5.11 Velocity of coupler point P.

5.6 Method of Superposition

Equations (5.4)–(5.6) suggest that, for mechanisms with more than one degree of freedom, another method of solution for velocities is to solve the problem multiple times, with all except one of the inputs considered inactive (frozen or locked) during each solution, and then to sum the results to find the total solution with all inputs active. This method is called the method of superposition, and it is valid for velocity analysis since the velocity equations are linear, that is, since all dependent velocities are linear combinations of the input velocities. The procedure may be best understood through an example.

Example 5.7

Let us again perform the velocity analysis of Example 5.3, this time using the method of superposition. There are two cases that must be considered: case (a), where link 3 is temporarily considered frozen whereas link 2 has an angular velocity of $\omega_2 = 10$ rad/s ccw; and case (b), where link 2 is temporarily considered frozen whereas link 3 has an angular velocity of $\omega_3 = 5$ rad/s cw.

Solution

The angular velocity of link 4 can be written as a linear combination of case (a) and case (b). Using the angular velocity ratio theorem, this gives

$$\omega_4 = \left(\frac{R_{I_{24}^2 I_{12}^2}}{R_{I_{24}^2 I_{14}^2}}\right)\omega_2 + \left(\frac{R_{I_{34}^3 I_{13}^3}}{R_{I_{34}^3 I_{14}^3}}\right)\omega_3, \tag{1}$$

where the superscripts in the instant center labels denote the input variable that is moving whereas all others are temporarily considered frozen. Similarly, the angular velocity of link 5 can be written as

$$\omega_5 = \left(\frac{R_{I_{25}^2 I_{12}^2}}{R_{I_{25}^2 I_{15}^2}}\right)\omega_2 + \left(\frac{R_{I_{35}^3 I_{13}^3}}{R_{I_{35}^3 I_{15}^3}}\right)\omega_3. \tag{2}$$

Comparing Eqs. (1) and (2) with Eq. (5.4), the first-order kinematic coefficients of links 4 and 5 can be written as

$$\theta'_{42} = \frac{R_{I_{24}^2 I_{12}^2}}{R_{I_{24}^2 I_{14}^2}}, \quad \theta'_{43} = \frac{R_{I_{34}^3 I_{13}^3}}{R_{I_{34}^3 I_{14}^3}}, \quad \theta'_{52} = \frac{R_{I_{25}^2 I_{12}^2}}{R_{I_{25}^2 I_{15}^2}}, \quad \text{and} \quad \theta'_{53} = \frac{R_{I_{35}^3 I_{13}^3}}{R_{I_{35}^3 I_{15}^3}}. \tag{3}$$

Note that a sign convention is not necessary if directed lines are used when measuring the relative locations of instant centers. However, if a sign convention is preferred, then each first-order kinematic coefficient is negative if the relative instant center I_{ij}^k is between the absolute instant centers I_{1i}^k and I_{1j}^k and positive if the relative instant center I_{ij}^k is outside the absolute instant centers I_{1i}^k and I_{1j}^k.

Case (a). Link 3 is temporarily considered frozen. The linkage is temporarily regarded as a four-bar linkage as shown in Fig. 5.12, with input from the rotation of link 2 alone.

From the Aronhold–Kennedy theorem, the secondary instant center I_{14}^2 is the point of inter-section of the line containing instant centers I_{15}^2 and I_{45}^2 and the line containing instant centers I_{12}^2 and I_{24}^2. Similarly, the secondary instant center I_{25}^2 is the point of intersection of the line containing instant centers I_{24}^2 and I_{45}^2 and the line containing instant centers I_{12}^2 and I_{15}^2.

Using the convention that lines proceed positive to the right, the distances between the instant centers, measured from the scaled drawing of Fig. 5.12, are

$$R_{I_{24}^2 I_{12}^2} = -2.50 \text{ in}, \quad R_{I_{24}^2 I_{14}^2} = 10.56 \text{ in}, \quad R_{I_{25}^2 I_{12}^2} = -7.76 \text{ in}, \quad \text{and} \quad R_{I_{25}^2 I_{15}^2} = -17.03 \text{ in}. \tag{4}$$

Example 5.7 (cont.)

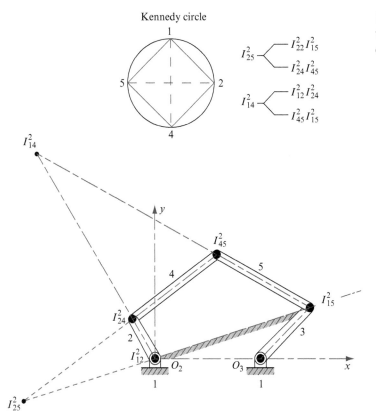

Fig. 5.12 Case (a): four-bar linkage with link 3 temporarily considered frozen.

Case (b). Link 2 is temporarily considered frozen. The linkage is temporarily regarded as a four-bar linkage, as shown in Fig. 5.13, with input from the rotation of link 3 alone.

From the Aronhold–Kennedy theorem, the secondary instant center I^3_{34} is the point of intersection of the line containing instant centers I^3_{13} and I^3_{14} and the line containing instant centers I^3_{35} and I^3_{45}. Similarly, the secondary instant center I^3_{15} is the point of intersection of the line containing instant centers I^3_{14} and I^3_{45} and the line containing instant centers I^3_{13} and I^3_{35}.

Continuing with the convention that lines proceed positive to the right, the distances between the instant centers, measured from the scaled drawing of Fig. 5.13, are

$$R_{I^3_{34}I^3_{13}} = 18.06 \text{ in}, \quad R_{I^3_{34}I^3_{14}} = 25.62 \text{ in}, \quad R_{I^3_{35}I^3_{13}} = 4.00 \text{ in}, \quad \text{and} \quad R_{I^3_{35}I^3_{15}} = -36.67 \text{ in.} \quad (5)$$

Substituting Eqs. (4) and (5) into Eqs. (3), the first-order kinematic coefficients of links 4 and 5 are

$$\theta'_{42} = \frac{R_{I^2_{24}I^2_{12}}}{R_{I^2_{24}I^2_{14}}} = \frac{-2.50 \text{ in}}{10.56 \text{ in}} = -0.237, \quad \theta'_{43} = \frac{R_{I^3_{34}I^3_{13}}}{R_{I^3_{34}I^3_{14}}} = \frac{18.06 \text{ in}}{25.62 \text{ in}} = 0.705, \quad (6a)$$

Example 5.7 (cont.)

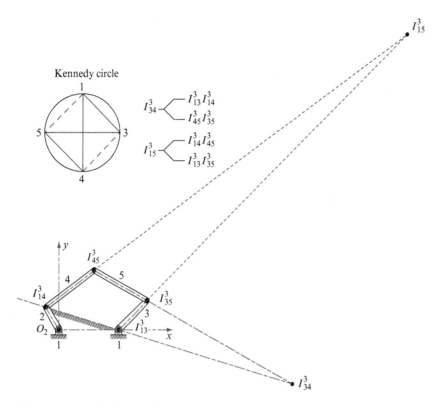

Fig. 5.13 Case (*b*): four-bar linkage with link 2 temporarily considered frozen.

$$\theta'_{52} = \frac{R_{I_{25}^2 I_{12}^2}}{R_{I_{25}^2 I_{15}^2}} = \frac{-7.76 \text{ in}}{-17.03 \text{ in}} = 0.455, \quad \theta'_{53} = \frac{R_{I_{35}^3 I_{13}^3}}{R_{I_{35}^3 I_{15}^3}} = \frac{4.00 \text{ in}}{-36.67 \text{ in}} = -0.109. \quad (6b)$$

Note that the values of the first-order kinematic coefficients of links 4 and 5, given by Eqs. (6), are all nondimensional (rad/rad) and are all in good agreement with the values found analytically in Example 5.6.

Substituting Eqs. (6) into Eq. (5.4), the angular velocities of links 4 and link 5, respectively, are

$$\omega_4 = -0.237(10 \text{ rad/s}) + 0.705(-5 \text{ rad/s}) = -5.90 \text{ rad/s cw}, \qquad Ans.$$

$$\omega_5 = 0.455(10 \text{ rad/s}) - 0.109(-5 \text{ rad/s}) = 5.10 \text{ rad/s ccw}. \qquad Ans.$$

Note that the answers for the angular velocities of coupler links 4 and 5 from the graphic methods (Examples 5.4 and 5.5) are in good agreement with the answers obtained from the analytic methods in Examples 5.6 and 5.7.

5.7   Acceleration Analysis; Acceleration Polygons

If the accelerations of each of the independent inputs are known, then an acceleration analysis can be performed by the acceleration polygon method of Sections 4.3–4.8. An example is presented here to demonstrate how the graphic method for single-degree-of-freedom mechanisms can be extended to multi-degree-of-freedom mechanisms.

Example 5.8

For the five-bar linkage of Example 5.1 in the posture shown in Fig. 5.5, links 2 and 3 are rotating with constant angular velocities of $\omega_2 = 10\ \text{rad/s ccw}$ and $\omega_3 = 5\ \text{rad/s cw}$, respectively. Determine the angular accelerations of coupler links 4 and 5.

Solution

The velocities are determined, using this input data, in the velocity polygon of Example 5.4, which is repeated here as Fig. 5.14b. Since the angular velocities of links 2 and 3 are constant, we have $\alpha_2 = \alpha_3 = 0$, and there remain only the normal components of acceleration for points A and C. Hence,

$$A_A = A^n_{AO_2} = \omega_2^2 R_{AO_2} = (10\ \text{rad/s})^2(2.5\ \text{in}) = 250\ \text{in/s}^2,$$

$$A_C = A^n_{CO_3} = \omega_3^2 R_{CO_3} = (5\ \text{rad/s})^2(4\ \text{in}) = 100\ \text{in/s}^2.$$

Next, two acceleration-difference equations are written to relate the acceleration of point B to those of points A and C; that is,

$$A_B = \overset{\sqrt{\sqrt{}}}{A_A} + \overset{\sqrt{\sqrt{}}}{A^n_{BA}} + \overset{?\sqrt{}}{A^t_{BA}} = \overset{\sqrt{\sqrt{}}}{A_C} + \overset{\sqrt{\sqrt{}}}{A^n_{BC}} + \overset{?\sqrt{}}{A^t_{BC}}. \tag{1}$$

Using results from the velocity polygon (Fig. 5.7b), the magnitudes of the two normal components in Eq. (1) are

$$A^n_{BA} = \frac{V^2_{BA}}{R_{BA}} = \frac{(35.34\ \text{in/s})^2}{6\ \text{in}} = 208.15\ \text{in/s}^2$$

and

$$A^n_{BC} = \frac{V^2_{BC}}{R_{BC}} = \frac{(30.60\ \text{in/s})^2}{6\ \text{in}} = 156.06\ \text{in/s}^2.$$

With these, the terms of Eq. (1) can be constructed graphically as shown in Fig. 5.14c.

The angular accelerations of links 4 and 5 can now be determined from the two tangential components scaled from the acceleration polygon; namely,

$$\alpha_4 = \frac{A^t_{BA}}{R_{BA}} = \frac{1.2\ \text{in/s}^2}{6\ \text{in}} = 0.2\ \text{rad/s}^2\ \text{cw}, \qquad\qquad Ans.$$

Example 5.8 (cont.)

Fig. 5.14 (*a*) Linkage posture; (*b*) velocity polygon; (*c*) acceleration polygon.

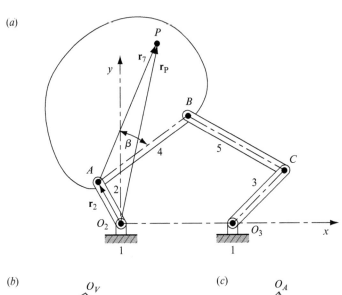

(*a*)

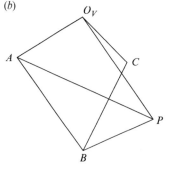

(*b*)

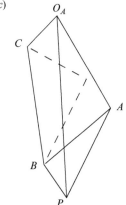

(*c*)

$$\alpha_5 = \frac{A^t_{BC}}{R_{BC}} = \frac{222.7 \text{ in/s}^2}{6 \text{ in}} = 37.1 \text{ rad/s}^2 \text{ ccw.} \qquad Ans.$$

Once the accelerations of points *A* and *B* are known, we can construct the acceleration image of coupler link 4 as explained in Section 4.4, and obtain the acceleration of coupler point *P*; that is,

$$\mathbf{A}_P = 472 \text{ in/s}^2 \angle -88.2°. \qquad Ans.$$

5.8   Second-Order Kinematic Coefficients

For higher accuracy in the solution of acceleration problems, we look for an analytic or numeric technique. The method of kinematic coefficients, which was presented in Section 4.11, provides such a technique.

As with first-order kinematic coefficients, we use subscripts to signify each independent variable when we have multiple degrees of freedom. For example, when we have the second-order kinematic coefficient of the dependent variable θ_i with respect to the two independent input variables θ_j and θ_k, we use the notation

$$\theta''_{ijk} = \frac{\partial^2 \theta_i}{\partial \theta_j \partial \theta_k}. \tag{5.7}$$

Using this notation, the derivative of Eq. (5.4) with respect to time (that is, the angular acceleration of link i) can be written as

$$\alpha_i = \frac{d^2\theta_i}{dt^2} = \frac{d\omega_i}{dt} = \theta''_{ijj}\omega_j^2 + 2\theta''_{ijk}\omega_j\omega_k + \theta''_{ikk}\omega_k^2 + \theta'_{ij}\alpha_j + \theta'_{ik}\alpha_k. \tag{5.8}$$

Differentiating Eq. (5.6) with respect to time, the acceleration of point P can be written as

$$\mathbf{A}_P = \left(x''_{Pjj}\omega_j^2 + 2x''_{Pjk}\omega_j\omega_k + x''_{Pkk}\omega_k^2 + x'_{Pj}\alpha_j + x'_{Pk}\alpha_k\right)\hat{\mathbf{i}}$$
$$+ \left(y''_{Pjj}\omega_j^2 + 2y''_{Pjk}\omega_j\omega_k + y''_{Pkk}\omega_k^2 + y'_{Pj}\alpha_j + y'_{Pk}\alpha_k\right)\hat{\mathbf{j}}. \tag{5.9}$$

The angle for the direction of the acceleration of point P can be written as

$$\psi = \tan^{-1}\left(\frac{A_P^y}{A_P^x}\right). \tag{5.10}$$

Substituting the components of Eq. (5.9) into this equation, the direction of $\mathbf{A}_P$ can be written as

$$\psi = \tan^{-1}\left(\frac{y''_{Pjj}\omega_j^2 + 2y''_{Pjk}\omega_j\omega_k + y''_{Pkk}\omega_k^2 + y'_{Pj}\alpha_j + y'_{Pk}\alpha_k}{x''_{Pjj}\omega_j^2 + 2x''_{Pjk}\omega_j\omega_k + x''_{Pkk}\omega_k^2 + x'_{Pj}\alpha_j + x'_{Pk}\alpha_k}\right). \tag{5.11}$$

For the special case where the input velocities ω_j and ω_k are both constant – that is, where the input accelerations are $\alpha_j = 0$ and $\alpha_k = 0$ – Eq. (5.11) can be written as

$$\psi = \tan^{-1}\left(\frac{y''_{Pjj}\omega_j^2 + 2y''_{Pjk}\omega_j\omega_k + y''_{Pkk}\omega_k^2}{x''_{Pjj}\omega_j^2 + 2x''_{Pjk}\omega_j\omega_k + x''_{Pkk}\omega_k^2}\right). \tag{5.12}$$

Even with these additional subscripts, second-order kinematic coefficients can be used for acceleration analysis of multi-degree-of-freedom mechanisms, almost exactly as they were in Section 4.11 with a single degree of freedom. Let us demonstrate this by again solving Example 5.8 using second-order kinematic coefficients.

Example 5.9

For the five-bar linkage of Example 5.1, in the posture shown in Fig. 5.5, links 2 and 3 are rotating with constant angular velocities of $\omega_2 = 10$ rad/s ccw and $\omega_3 = 5$ rad/s cw, respectively. Determine the angular accelerations of coupler links 4 and 5 and the acceleration of coupler point P.

Example 5.9 (cont.)

Solution

Taking the partial derivative of Eqs. (1) of Example 5.6 with respect to the input angle θ_2, and writing this in matrix form, gives

$$
\begin{bmatrix} -r_4 \sin\theta_4 & r_5 \sin\theta_5 \\ r_4 \cos\theta_4 & -r_5 \cos\theta_5 \end{bmatrix} \begin{bmatrix} \theta_{422}'' \\ \theta_{522}'' \end{bmatrix} = \begin{bmatrix} r_2 \cos\theta_2 + r_4 \cos\theta_4 \theta_{42}'^2 - r_5 \cos\theta_5 \theta_{52}'^2 \\ r_2 \sin\theta_2 + r_4 \sin\theta_4 \theta_{42}'^2 - r_5 \sin\theta_5 \theta_{52}'^2 \end{bmatrix}.
\tag{1}
$$

We can solve this matrix equation for the second-order kinematic coefficients θ_{422}'' and θ_{522}''. Using the data obtained so far, the results are $\theta_{422}'' = 0.139$ rad/rad^2 and $\theta_{522}'' = 0.208$ rad/rad^2.

We also take the partial derivatives of Eqs. (1) of Example 5.6 with respect to the input angle θ_3; writing these in matrix form gives

$$
\begin{bmatrix} -r_4 \sin\theta_4 & r_5 \sin\theta_5 \\ r_4 \cos\theta_4 & -r_5 \cos\theta_5 \end{bmatrix} \begin{bmatrix} \theta_{423}'' \\ \theta_{523}'' \end{bmatrix} = \begin{bmatrix} r_4 \cos\theta_4 \theta_{42}' \theta_{43}' - r_5 \cos\theta_5 \theta_{52}' \theta_{53}' \\ r_4 \sin\theta_4 \theta_{42}' \theta_{43}' - r_5 \sin\theta_5 \theta_{52}' \theta_{53}' \end{bmatrix}.
\tag{2}
$$

Then, solving this matrix equation gives the second-order kinematic coefficients θ_{423}'' and θ_{523}''. Note that taking the partial derivatives of Eqs. (3) of Example 5.6 with respect to the input angle θ_2 gives the same result. For the current geometry and posture of the linkage, the results are $\theta_{423}'' = 0.131$ rad/rad^2 and $\theta_{523}'' = -0.206$ rad/rad^2.

Next, we partially differentiate Eq. (3) of Example 5.6 with respect to the input angle θ_3, which gives

$$
\begin{bmatrix} -r_4 \sin\theta_4 & r_5 \sin\theta_5 \\ r_4 \cos\theta_4 & -r_5 \cos\theta_5 \end{bmatrix} \begin{bmatrix} \theta_{433}'' \\ \theta_{533}'' \end{bmatrix} = \begin{bmatrix} -r_3 \cos\theta_3 + r_4 \cos\theta_4 \theta_{43}'^2 - r_5 \cos\theta_5 \theta_{53}'^2 \\ -r_3 \sin\theta_3 + r_4 \sin\theta_4 \theta_{43}'^2 - r_5 \sin\theta_5 \theta_{53}'^2 \end{bmatrix}.
\tag{3}
$$

Then, we can solve this matrix equation for the second-order kinematic coefficients θ_{433}'' and θ_{533}''. For the current geometry and posture of the linkage, the results are $\theta_{433}'' = -0.038$ rad/rad^2 and $\theta_{533}'' = -0.173$ rad/rad^2.

Table 5.3 presents the second-order kinematic coefficients of the two coupler links for the same range of input angles given in Table 5.1.

Substituting the numeric values of the first-order and second-order kinematic coefficients into Eqs. (5.8), the angular accelerations of the coupler links (for the posture shown in Fig. 5.5) are

$$
\alpha_4 = -0.150 \text{ rad/s}^2 \text{ (cw)} \quad \text{and} \quad \alpha_5 = 37.075 \text{ rad/s}^2 \text{ ccw}. \qquad \textit{Ans.}
$$

Differentiating Eq. (5) of Example 5.6 with respect to the two independent inputs, the second-order kinematic coefficients of vector $\mathbf{r}_7$ are

$$
\theta_{722}'' = \theta_{422}'', \quad \theta_{723}'' = \theta_{423}'', \quad \text{and} \quad \theta_{733}'' = \theta_{433}''.
\tag{4}
$$

Differentiating Eqs. (6) and (7) of Example 5.6 with respect to the input angle θ_2, and using Eqs. (5) of Example 5.6 and Eq. (4) of this example, the second-order kinematic coefficients of point P are

Example 5.9 (cont.)

Table 5.3 **Second-order kinematic coefficients of coupler links**

θ_2	θ_3	θ''_{422}	θ''_{423}	θ''_{433}	θ''_{522}	θ''_{523}	θ''_{533}
deg	deg	rad/rad^2	rad/rad^2	rad/rad^2	rad/rad^2	rad/rad^2	rad/rad^2
40	85	−0.135	−0.263	−0.016	0.535	−0.472	−0.156
50	80	0.095	−0.236	−0.075	0.686	−0.430	−0.214
60	75	0.215	−0.177	−0.107	0.701	−0.361	−0.244
70	70	0.254	−0.112	−0.114	0.635	−0.294	−0.251
80	65	0.249	−0.055	−0.106	0.537	−0.242	−0.245
90	60	0.225	−0.005	−0.088	0.436	−0.208	−0.233
100	55	0.196	0.038	−0.068	0.346	−0.190	−0.217
110	50	0.167	0.081	−0.049	0.269	−0.188	−0.198
120	45	0.139	0.131	−0.038	0.208	−0.206	−0.173
130	40	0.104	0.208	−0.050	0.168	−0.259	−0.128
140	35	0.034	0.371	−0.144	0.174	−0.404	−0.002
150	30	−0.281	0.987	−0.752	0.434	−1.006	0.635

$$x''_{P22} = -r_2\cos\theta_2 - r_7\cos\theta_7\theta'^2_{42} - r_7\sin\theta_7\theta''_{422}, \quad (5)$$

$$y''_{P22} = -r_2\sin\theta_2 - r_7\sin\theta_7\theta'^2_{42} + r_7\cos\theta_7\theta''_{422}. \quad (6)$$

Similarly, differentiating Eqs. (6) and (7) of Example 5.6 with respect to input angle θ_3, and using Eqs. (5) of Example 5.6 and Eq. (4) of this example, the second-order kinematic coefficients of point P are

$$x''_{P23} = -r_7\cos\theta_7\theta'_{42}\theta'_{43} - r_7\sin\theta_7\theta''_{423}, \quad (7)$$

$$y''_{P23} = -r_7\sin\theta_7\theta'_{42}\theta'_{43} + r_7\cos\theta_7\theta''_{423}, \quad (8)$$

$$x''_{P33} = -r_7\cos\theta_7\theta'^2_{43} - r_7\sin\theta_7\theta''_{433}, \quad (9)$$

$$y''_{P33} = -r_7\sin\theta_7\theta'^2_{43} + r_7\cos\theta_7\theta''_{433}. \quad (10)$$

Substituting the results of Eqs. (1) and (3) of Example 5.6 and Eqs. (1)–(3) into Eqs. (5)–(10), the second-order kinematic coefficients of point P are

$$x''_{P22} = -2.5\cos 120° - 8\cos 66.447°(-0.237)^2 - 8\sin 66.447°(+0.139) = +0.054 \text{ in/rad}^2,$$

$$y''_{P22} = -2.5\sin 120° - 8\sin 66.447°(-0.237)^2 + 8\cos 66.447°(+0.139) = -2.133 \text{ in/rad}^2,$$

$$x''_{P23} = -8\cos 66.447°(-0.237)(+0.705) - 8\sin 66.447°(0.131) = -0.429 \text{ in/rad}^2,$$

Example 5.9 (cont.)

$$y''_{P23} = -8\sin 66.447°(-0.237)(+0.705) + 8\cos 66.447°(0.131) = +1.642 \text{ in/rad}^2,$$

$$x''_{P33} = -8\cos 66.447°(+0.705)^2 - 8\sin 66.447°(-0.038) = -1.311 \text{ in/rad}^2,$$

$$y''_{P33} = -8\sin 66.447°(+0.705)^2 + 8\cos 66.447°(-0.038) = -3.762 \text{ in/rad}^2.$$

Then, substituting these results and the specified input angular velocities and angular accelerations into Eq. (5.9), the acceleration of point P is

$$\mathbf{A}_P = 15.51\,\hat{\mathbf{i}} - 471.57\,\hat{\mathbf{j}} \text{ in/s}^2.$$

Therefore, the magnitude and the direction of the acceleration of point P, respectively, are

$$A_P = \sqrt{(15.51 \text{ in/s}^2)^2 + (-471.57 \text{ in/s}^2)^2} = 471.82 \text{ in/s}^2$$

and

$$\psi = \tan^{-1}\left(\frac{-471.57 \text{ in/s}^2}{15.51 \text{ in/s}^2}\right) = -88.12°.$$

The direction of the acceleration of point P is shown in Fig. 5.15.

Table 5.4 presents the second-order kinematic coefficients of point P and the magnitude and direction of the acceleration of point P for the same range of input angles given in Table 5.1.

Note that the answers here are in good agreement with, and have higher accuracy than, the answers obtained from the acceleration polygon (Example 5.8).

Fig. 5.15 Acceleration of coupler point P.

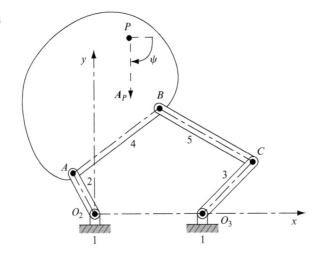

Example 5.9 (cont.)

Table 5.4 **Second-order kinematic coefficients and acceleration of coupler point** *P*

θ_2	θ_3	x''_{P22}	y''_{P22}	x''_{P23}	y''_{P23}	x''_{P33}	y''_{P33}	A_P	ψ
deg	deg	in	in	in	in	in	in	in/s^2	deg
40	85	0.236	−2.914	0.013	3.805	2.537	−3.612	766.98	−83.583
50	80	−1.425	−4.147	0.517	3.406	2.188	−3.342	850.35	−99.446
60	75	−2.417	−4.316	0.656	2.811	1.837	−3.256	836.00	−108.217
70	70	−2.666	−3.992	0.569	2.281	1.417	−3.302	766.03	−112.089
80	65	−2.402	−3.547	0.393	1.896	0.926	−3.404	679.65	−112.157
90	60	−1.870	−3.130	0.197	1.653	0.384	−3.511	599.38	−109.199
100	55	−1.232	−2.764	0.005	1.530	−0.184	−3.600	535.02	−103.873
110	50	−0.578	−2.436	−0.193	1.519	−0.758	−3.673	490.69	−96.725
120	45	0.054	−2.133	−0.429	1.642	−1.311	−3.762	471.82	−88.117
130	40	0.677	−1.875	−0.783	1.999	−1.794	−3.981	497.37	−78.265
140	35	1.423	−1.826	−1.520	2.969	−1.991	−4.761	646.57	−67.780
150	30	3.318	−3.321	−4.290	7.007	−0.210	−9.128	1470.05	−59.072

The following example shows how the method of kinematic coefficients can be applied to the kinematic analysis of a two-degrees-of-freedom mechanism with rolling contact.

Example 5.10

For the mechanism in the posture shown in Fig. 5.16, the planet gear, *j*, is in rolling contact with the ring gear, *h*, and the sun gear, *k*. Link *i* (the arm) has constant angular velocity $\omega_i = 10$ rad/s ccw, and gear *h* has angular velocity $\omega_h = 5$ rad/s cw and angular acceleration

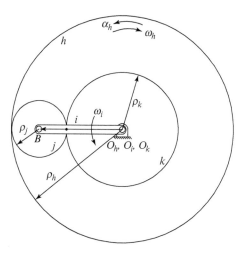

Fig. 5.16 $\rho_h = 8$ in, $\rho_j = 2$ in, and $\rho_k = 4$ in.

Example 5.10 (cont.)

$\alpha_h = 15$ rad/s^2 ccw. Determine: (a) the first- and second-order kinematic coefficients of gears j and k; and (b) the angular velocity and acceleration of gears j and k.

Solution

For the rolling contact between planet gear j and ring gear h, the constraint equation (see Section 2.19) can be written in the form

$$\rho_j(\varDelta\theta_j - \varDelta\theta_i) = +\rho_h(\varDelta\theta_h - \varDelta\theta_i). \tag{1}$$

Similarly, for the rolling contact between planet gear j and sun gear k, the constraint equation can be written as

$$\rho_j(\varDelta\theta_j - \varDelta\theta_i) = -\rho_k(\varDelta\theta_k - \varDelta\theta_i). \tag{2}$$

These two equations can be written in matrix form, with the independent variables on the right, as follows

$$\begin{bmatrix} \rho_j & 0 \\ \rho_j & \rho_k \end{bmatrix} \begin{bmatrix} \varDelta\theta_j \\ \varDelta\theta_k \end{bmatrix} = \begin{bmatrix} \rho_h & \rho_h - \rho_j \\ 0 & \rho_j + \rho_k \end{bmatrix} \begin{bmatrix} \varDelta\theta_h \\ \varDelta\theta_i \end{bmatrix}. \tag{3}$$

a. Taking partial derivatives of Eq. (3) with respect to the first independent variable, we find equations for the following first-order kinematic coefficients

$$\begin{bmatrix} \rho_j & 0 \\ \rho_j & \rho_k \end{bmatrix} \begin{bmatrix} \theta'_{jh} \\ \theta'_{kh} \end{bmatrix} = \begin{bmatrix} \rho_h & \rho_h - \rho_j \\ 0 & \rho_j + \rho_k \end{bmatrix} \begin{bmatrix} 1 \\ 0 \end{bmatrix} = \begin{bmatrix} \rho_h \\ 0 \end{bmatrix}. \tag{4}$$

The determinant of the Jacobian matrix is

$$\varDelta = \rho_j\rho_k = 8 \text{ in}^2,$$

and Cramer's rule gives the solution to Eq. (4) as

$$\theta'_{jh} = \rho_h/\rho_j = 4.0 \text{ rad/rad} \quad \text{and} \quad \theta'_{kh} = -\rho_h/\rho_k = -2.0 \text{ rad/rad}. \qquad \textit{Ans. (5)}$$

Taking partial derivatives of Eq. (3) with respect to the other independent variable, we find equations for the remaining first-order kinematic coefficients:

$$\begin{bmatrix} \rho_j & 0 \\ \rho_j & \rho_k \end{bmatrix} \begin{bmatrix} \theta'_{ji} \\ \theta'_{ki} \end{bmatrix} = \begin{bmatrix} \rho_h & \rho_h - \rho_j \\ 0 & \rho_j + \rho_k \end{bmatrix} \begin{bmatrix} 0 \\ 1 \end{bmatrix} = \begin{bmatrix} \rho_h - \rho_j \\ \rho_j + \rho_k \end{bmatrix}. \tag{6}$$

Cramer's rule gives the solutions to Eqs. (6) as

$$\theta'_{ji} = (\rho_h - \rho_j)/\rho_j = 3.0 \text{ rad/rad} \quad \text{and} \quad \theta'_{ki} = (-\rho_h + 2\rho_j + \rho_k)/\rho_k = 0.0. \qquad \textit{Ans. (7)}$$

Example 5.10 (cont.)

The second-order kinematic coefficients can be found by taking partial derivatives of Eqs. (5) and (7) with respect to each of the independent variables. Since the right-hand sides of these equations are constants, we find the second-order kinematic coefficients to be

$$\theta''_{jhh} = \theta''_{khh} = \theta''_{jhi} = \theta''_{khi} = \theta''_{jii} = \theta''_{kii} = 0.0. \qquad \textit{Ans.}$$

b. The angular velocities of links j and k are

$$\omega_j = \theta'_{jh}\omega_h + \theta'_{ji}\omega_i$$
$$= (4.0 \text{ rad/rad})(-5.0 \text{ rad/s}) + (3.0 \text{ rad/rad})(10.0 \text{ rad/s}) = 10.0 \text{ rad/s ccw}, \qquad \textit{Ans.}$$

$$\omega_k = \theta'_{kh}\omega_h + \theta'_{ki}\omega_i$$
$$= (-2.0 \text{ rad/rad})(-5.0 \text{ rad/s}) + (0.0 \text{ rad/rad})(10.0 \text{ rad/s}) = 10.0 \text{ rad/s ccw}. \ \textit{Ans.}$$

The angular accelerations of links j and k are

$$\alpha_j = \theta''_{jhh}\omega_h^2 + 2\theta''_{jhi}\omega_h\omega_i + \theta''_{jii}\omega_i^2 + \theta'_{jh}\alpha_h + \theta'_{ji}\alpha_i$$
$$= 0 + 0 + 0 + (4.0 \text{ rad/rad})(15 \text{ rad/s}^2) + (3.0 \text{ rad/rad})(0) = 60.0 \text{ rad/s}^2 \text{ ccw}, \qquad \textit{Ans.}$$

$$\alpha_k = \theta''_{khh}\omega_h^2 + 2\theta''_{khi}\omega_h\omega_i + \theta''_{kii}\omega_i^2 + \theta'_{kh}\alpha_h + \theta'_{ki}\alpha_i$$
$$= 0 + 0 + 0 + (-2.0 \text{ rad/rad})(15 \text{ rad/s}^2) + 0 = -30.0 \text{ rad/s}^2 \text{ (cw)}. \qquad \textit{Ans.}$$

5.9 Path Curvature of a Coupler Point Trajectory

After the first and second-order kinematic coefficients of a multi-degree-of-freedom mechanism are known, a study of the curvature of the path traced by a coupler point, which was discussed in Section 4.17, can be undertaken. The radius of curvature of the path traced by coupler point P was given by Eq. (4.52). However, a modified version of this equation is required for multi-degree-of-freedom linkages.

The position of point P can be written as

$$\mathbf{r}_P = x_P\hat{\mathbf{i}} + y_P\hat{\mathbf{j}}, \qquad (5.13)$$

and the velocity $\mathbf{V}_P$ of point P can be determined by differentiating this equation with respect to time. Recall that the velocity of point P can be written as in Eq. (5.6).

The unit vector tangent to the path of point P is given by Eq. (3.30); that is,

$$\hat{\mathbf{u}}_P^t = \frac{\mathbf{V}_P}{V_P}. \qquad (5.14)$$

For the case where x_P and y_P are functions of two independent input variables, θ_j and θ_k, by substituting Eq. (5.6) into this equation, the unit tangent vector can be written as

$$\hat{\mathbf{u}}_P^t = \frac{\left(x'_{Pj}\omega_j + x'_{Pk}\omega_k\right)\hat{\mathbf{i}} + \left(y'_{Pj}\omega_j + y'_{Pk}\omega_k\right)\hat{\mathbf{j}}}{\sqrt{\left(x'_{Pj}\omega_j + x'_{Pk}\omega_k\right)^2 + \left(y'_{Pj}\omega_j + y'_{Pk}\omega_k\right)^2}}. \tag{5.15}$$

Then, the unit normal vector is $90°$ counterclockwise from the unit tangent vector; that is,

$$\hat{\mathbf{u}}_P^n = \hat{\mathbf{k}} \times \hat{\mathbf{u}}_P^t = \frac{-\left(y'_{Pj}\omega_j + y'_{Pk}\omega_k\right)\hat{\mathbf{i}} + \left(x'_{Pj}\omega_j + x'_{Pk}\omega_k\right)\hat{\mathbf{j}}}{\sqrt{\left(x'_{Pj}\omega_j + x'_{Pk}\omega_k\right)^2 + \left(y'_{Pj}\omega_j + y'_{Pk}\omega_k\right)^2}}. \tag{5.16}$$

The normal acceleration of point P is

$$A_P^n = \mathbf{A}_P \cdot \hat{\mathbf{u}}_P^n = A_P^x \hat{\mathbf{u}}_P^{nx} + A_P^y \hat{\mathbf{u}}_P^{ny}. \tag{5.17}$$

Substituting Eqs. (5.9) and (5.16) into Eq. (5.17), and performing the dot product operation, the normal acceleration of point P can be written as

$$A_P^n = \frac{\left(y''_{Pjj}\omega_j^2 + 2y''_{Pjk}\omega_j\omega_k + y''_{Pkk}\omega_k^2 + y'_{Pj}a_j + y'_{Pk}\alpha_k\right)\left(x'_{Pj}\omega_j + x'_{Pk}\omega_k\right) - \left(x''_{Pjj}\omega_j^2 + 2x''_{Pjk}\omega_j\omega_k + x''_{Pkk}\omega_k^2 + x'_{Pj}a_j + x'_{Pk}\alpha_k\right)\left(y'_{Pj}\omega_j + y'_{Pk}\omega_k\right)}{\sqrt{\left(x'_{Pj}\omega_j + x'_{Pk}\omega_k\right)^2 + \left(y'_{Pj}\omega_j + y'_{Pk}\omega_k\right)^2}}. \tag{5.18}$$

For the special case where links j and k have constant angular velocities, then Eq. (5.18) can be written as

$$A_P^n = \frac{\left(y''_{Pjj}\omega_j^2 + 2y''_{Pjk}\omega_j\omega_k + y''_{Pkk}\omega_k^2\right)\left(x'_{Pj}\omega_j + x'_{Pk}\omega_k\right) - \left(x''_{Pjj}\omega_j^2 + 2x''_{Pjk}\omega_j\omega_k + x''_{Pkk}\omega_k^2\right)\left(y'_{Pj}\omega_j + y'_{Pk}\omega_k\right)}{\sqrt{\left(x'_{Pj}\omega_j + x'_{Pk}\omega_k\right)^2 + \left(y'_{Pj}\omega_j + y'_{Pk}\omega_k\right)^2}}. \tag{5.19}$$

Substituting Eqs. (5.6) and (5.19) into Eq. (4.48) and simplifying, the radius of curvature of the path traced by coupler point P can be written as

$$\rho_P = \frac{\left[\left(x'_{Pj}\omega_j + x'_{Pk}\omega_k\right)^2 + \left(y'_{Pj}\omega_j + y'_{Pk}\omega_k\right)^2\right]^{3/2}}{\left(y''_{Pjj}\omega_j^2 + 2y''_{Pjk}\omega_j\omega_k + y''_{Pkk}\omega_k^2\right)\left(x'_{Pj}\omega_j + x'_{Pk}\omega_k\right) - \left(x''_{Pjj}\omega_j^2 + 2x''_{Pjk}\omega_j\omega_k + x''_{Pkk}\omega_k^2\right)\left(y'_{Pj}\omega_j + y'_{Pk}\omega_k\right)}. \tag{5.20}$$

If we define the input angular velocity ratio $\eta = \omega_k/\omega_j$, then Eq. (5.20) can be written as

$$\rho_P = \frac{\left[\left(x'_{Pj} + x'_{Pk}\eta\right)^2 + \left(y'_{Pj} + y'_{Pk}\eta\right)^2\right]^{3/2}}{\begin{array}{l}\left(y''_{Pjj} + 2y''_{Pjk}\eta + y''_{Pkk}\eta^2\right)\left(x'_{Pj} + x'_{Pk}\eta\right)\\ -\left(x''_{Pjj} + 2x''_{Pjk}\eta + x''_{Pkk}\eta^2\right)\left(y'_{Pj} + y'_{Pk}\eta\right)\end{array}}. \tag{5.21}$$

The sign here has the same meaning that it had in earlier chapters. A negative value for the radius of curvature magnitude indicates that the unit normal vector is pointing away from the center of curvature of the path traced by coupler point P.

The coordinates of the center of curvature of the path traced by point P, as given by Eq. (4.53), are written here as

$$x_C = x_P + \rho_P u_P^{nx} \quad \text{and} \quad y_C = y_P + \rho_P u_P^{ny}, \tag{5.22}$$

where x_P and y_P are found as indicated in Example 5.3, and u_P^{nx} and u_P^{ny} are given by the $\hat{\mathbf{i}}$ and $\hat{\mathbf{j}}$ components of Eq. (5.16).

Example 5.11

Continuing with the five-bar linkage of Examples 5.1–5.9 in the posture shown in Fig. 5.5, links 2 and 3 are rotating with constant angular velocities of $\omega_2 = 10$ rad/s ccw and $\omega_3 = 5$ rad/s cw, respectively. Find the unit tangent vector and unit normal vector to the path of point P, the radius of curvature, and the Cartesian coordinates of the center of curvature of the path of point P.

Solution

The velocity of coupler point P, obtained from the first-order kinematic coefficients with the same input data as in Example 5.6, is

$$\mathbf{V}_P = 21.546\hat{\mathbf{i}} - 31.330\hat{\mathbf{j}} \text{ in/s} = 38.024 \text{ in/s} \angle{-55.483°}. \tag{1}$$

Substituting the velocity components of Eq. (1) into Eq. (5.14), the unit tangent vector is

$$\hat{\mathbf{u}}^t = \frac{21.546 \text{ in/s}\,\hat{\mathbf{i}} - 31.330 \text{ in/s}\,\hat{\mathbf{j}}}{38.02 \text{ in/s}} = 0.567\hat{\mathbf{i}} - 0.824\hat{\mathbf{i}}. \qquad\qquad Ans.$$

Then, from Eq. (5.16), the unit normal vector is

$$\hat{\mathbf{u}}^n = 0.824\hat{\mathbf{i}} + 0.567\hat{\mathbf{j}}. \qquad\qquad Ans.$$

The directions of the unit tangent vector and unit normal vector are shown in Fig. 5.17.

Substituting data from Examples 5.6 and 5.9 for the first- and second-order kinematic coefficients and the specified input angular velocities into Eq. (5.20), the radius of curvature of the path of point P is

$$\rho_P = \frac{1445.81 \text{ in}^3/\text{s}^3}{-254.44 \text{ in}^2/\text{s}^3} = -5.682 \text{ in}. \qquad\qquad Ans.$$

Example 5.11 (cont.)

Fig. 5.17 Unit tangent
and unit normal
vectors.

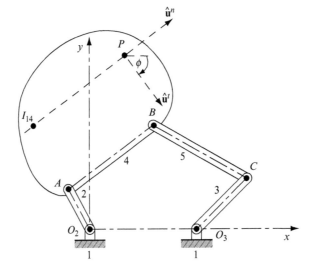

Fig. 5.18 Center of
curvature of the path
traced by point P.

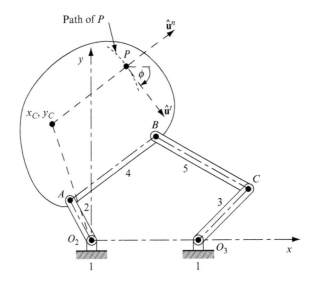

The negative sign indicates that the center of curvature is located in the direction of the negative unit normal vector $\hat{\mathbf{u}}^n$ from point P (Fig. 5.18).

Finally, substituting the answers from Example 5.3 and the above results into Eqs. (5.22), the coordinates of the center of curvature of the path of point P are

$$x_C = 1.947 \text{ in} + (-5.682 \text{ in})(0.824) = -2.735 \text{ in},$$

$$y_C = 9.499 \text{ in} + (-5.682 \text{ in})(0.567) = 6.277 \text{ in}.$$

Example 5.11 (cont.)

Table 5.5 **Radius of curvature and coordinates of the center of curvature**

θ_2	θ_3	ρ_P	x_c	y_c
deg	deg	in	in	in
40	85	−10.350	5.784	2.067
50	80	−8.062	4.052	3.559
60	75	−5.340	2.247	5.595
70	70	−3.014	1.005	7.557
80	65	−1.667	0.551	8.821
90	60	−1.214	0.524	9.271
100	55	−1.481	0.408	9.031
110	50	−2.671	−0.364	8.130
120	45	−5.682	−2.735	6.279
130	40	−13.603	−9.496	2.157
140	35	−51.321	−42.794	−15.556
150	30	57.190	53.984	33.525

The center of curvature of the path of point P is shown in Fig. 5.18. This figure also shows the unit tangent vector $\hat{\mathbf{u}}^t$ and the unit normal vector $\hat{\mathbf{u}}^n$ to the path traced by point P.

Table 5.5 presents the radius of curvature of the path traced by point P and the Cartesian coordinates of the center of curvature of this path for the range of input angles given in Table 5.1.

5.10 Finite Difference Method

The finite difference method can be used with the examples of this chapter to check: (a) the first-order kinematic coefficients of links 4 and 5; and (b) the second-order kinematic coefficients of links 4 and 5.

In general, the first-order kinematic coefficient of link i with respect to link j can be written as

$$\theta'_{ij} = \frac{\partial \theta_i}{\partial \theta_j} \approx \frac{\Delta \theta_i}{\Delta \theta_j}.$$

Keep in mind that the definition of a *partial* derivative implies that the symbol $\Delta \theta_j$ in the denominator corresponds to the change of an independent input θ_j while all other independent inputs are held frozen or locked during the change of $\Delta \theta_j$. However, data such as those in Tables 5.1 and 5.2, for example, do not correspond to this restriction since they were derived with both $\Delta \theta_2$ and $\Delta \theta_3$ changing simultaneously. Therefore, the first-order kinematic coefficient of link i with respect to input j *cannot* be directly approximated by the following formula:

$$\theta'_{ij} \neq \frac{\Delta\theta_i}{\Delta\theta_j} = \frac{(\theta_i)_A - (\theta_i)_B}{(\theta_j)_A - (\theta_j)_B},$$

where subscripts A denote the value *after* and B denote the value *before* the value that is to be checked by the finite difference method. In situations such as Tables 5.1 and 5.2, however, where the total change $\Delta\theta_i$ is the result of separate changes of more than one independent variable, say $\Delta\theta_j$ and $\Delta\theta_k$, the proper procedure is to write the finite change in the angular displacement of link i as

$$\Delta\theta_i \approx \theta'_{ij}\Delta\theta_j + \theta'_{ik}\Delta\theta_k$$

or as

$$\left[(\theta_i)_A - (\theta_i)_B\right] \approx \theta'_{ij}\left[(\theta_j)_A - (\theta_j)_B\right] + \theta'_{ik}\left[(\theta_k)_A - (\theta_k)_B\right]. \tag{5.23}$$

The following example will illustrate the use of these equations to check the first-order kinematic coefficients of Example 5.6.

Example 5.12

Use finite differences to check the first-order kinematic coefficients of links 4 and 5 of the five-bar linkage of Example 5.6.

Solution

By Eq. (5.23) the finite change in the angular displacement of link 4 can be written as

$$\Delta\theta_4 = \left[(\theta_4)_A - (\theta_4)_B\right] \approx \theta'_{42}\left[(\theta_2)_A - (\theta_2)_B\right] + \theta'_{43}\left[(\theta_3)_A - (\theta_3)_B\right], \tag{1}$$

and the finite change in the angular displacement of link 5 can be written as

$$\Delta\theta_5 = \left[(\theta_5)_A - (\theta_5)_B\right] \approx \theta'_{52}\left[(\theta_2)_A - (\theta_2)_B\right] + \theta'_{53}\left[(\theta_3)_A - (\theta_3)_B\right]. \tag{2}$$

For the input angles $\theta_2 = 120°$ and $\theta_3 = 45°$ (row nine of Table 5.1), the first-order kinematic coefficients are

$$\theta'_{42} = -0.237, \quad \theta'_{43} = 0.705, \quad \theta'_{52} = 0.455, \quad \text{and} \quad \theta'_{53} = -0.109.$$

Substituting these values into Eqs. (1) and (2), the finite changes in the angular displacements of links 4 and 5, respectively, are

$$\Delta\theta_4 = [30.526° - 42.360°] \approx -0.237[130° - 110°] + 0.705[40° - 50°], \tag{3}$$

$$\Delta\theta_5 = [156.510° - 146.315°] \approx 0.455[130° - 110°] - 0.109[40° - 50°]. \tag{4}$$

Note that the answers given by the two separate calculations in Eqs. (3) and (4) are in reasonable agreement with each other. If the increments of the two input displacements are decreased, then even closer agreement between the two sets of calculations is obtained. The general conclusion, therefore, is that the first-order kinematic coefficients of the five-bar linkage as indicated by Table 5.1 are correct.

From Eqs. (5.7), the second-order kinematic coefficients of link i with respect to the input angles θ_j and θ_k can be written as

$$\theta''_{ijk} = \frac{\partial^2 \theta_i}{\partial \theta_j \partial \theta_k} = \frac{\partial \theta'_{ij}}{\partial \theta_k}.$$

Remembering the directional nature of finite differences, the finite change in the first-order kinematic coefficients of link i with respect to link j can be written as

$$\Delta\theta'_{ij} = \left[\left(\theta'_{ij} \right)_A - \left(\theta'_{ij} \right)_B \right] \approx \theta''_{ijj} \left[\left(\theta_j \right)_A - \left(\theta_j \right)_B \right] + \theta''_{ijk} \left[\left(\theta_k \right)_A - \left(\theta_k \right)_B \right], \tag{5.24}$$

where subscript A denotes the value after and B denotes the value before the value that is to be checked by the finite difference method.

The following example demonstrates the value of these equations in checking the second-order kinematic coefficients.

Example 5.13

Use finite differences to check the second-order kinematic coefficients of links 4 and 5 of the five-bar linkage of Example 5.9.

Solution

The second-order kinematic coefficients can be checked by writing the finite change in the first-order kinematic coefficients of links 4 and 5 with respect to the input rotations of links 2 and 3 [Eq. (5.24)] as

$$\Delta\theta'_{42} = \left[\left(\theta'_{42} \right)_A - \left(\theta'_{42} \right)_B \right] \approx \theta''_{422} \left[\left(\theta_2 \right)_A - \left(\theta_2 \right)_B \right] + \theta''_{423} \left[\left(\theta_3 \right)_A - \left(\theta_3 \right)_B \right],$$

$$\Delta\theta'_{43} = \left[\left(\theta'_{43} \right)_A - \left(\theta'_{43} \right)_B \right] \approx \theta''_{432} \left[\left(\theta_2 \right)_A - \left(\theta_2 \right)_B \right] + \theta''_{433} \left[\left(\theta_3 \right)_A - \left(\theta_3 \right)_B \right],$$

$$\Delta\theta'_{52} = \left[\left(\theta'_{52} \right)_A - \left(\theta'_{52} \right)_B \right] \approx \theta''_{522} \left[\left(\theta_2 \right)_A - \left(\theta_2 \right)_B \right] + \theta''_{523} \left[\left(\theta_3 \right)_A - \left(\theta_3 \right)_B \right],$$

and

$$\Delta\theta'_{53} = \left[\left(\theta'_{53} \right)_A - \left(\theta'_{53} \right)_B \right] \approx \theta''_{532} \left[\left(\theta_2 \right)_A - \left(\theta_2 \right)_B \right] + \theta''_{533} \left[\left(\theta_3 \right)_A - \left(\theta_3 \right)_B \right],$$

where $\theta''_{432} = \theta''_{423}$ and $\theta''_{532} = \theta''_{523}$. From row nine of Tables 5.1 and 5.3, the values in these equations are

$$\Delta\theta'_{42} = [(-0.230) - (-0.254)] \approx 0.139 \left[\frac{130° - 110°}{57.296°/\text{rad}} \right] + 0.131 \left[\frac{40° - 50°}{57.296°/\text{rad}} \right],$$

$$\Delta\theta'_{43} = [0.737 - 0.683] \approx 0.131 \left[\frac{130° - 110°}{57.296°/\text{rad}} \right] + (-0.038) \left[\frac{40° - 50°}{57.296°/\text{rad}} \right],$$

$$\Delta\theta'_{52} = [0.508 - 0.397] \approx 0.208 \left[\frac{130° - 110°}{57.296°/\text{rad}} \right] + (-0.206) \left[\frac{40° - 50°}{57.296°/\text{rad}} \right],$$

Example 5.13 (cont.)

$$\Delta\theta_{53}' = [(-0.136) - (-0.091)] \approx (-0.206)\left[\frac{130° - 110°}{57.296°/\text{rad}}\right] + (-0.173)\left[\frac{40° - 50°}{57.296°/\text{rad}}\right].$$

Note that the answers given by the two separate calculations in each of these equations are in reasonable agreement with each other. The general conclusion, therefore, is that the second-order kinematic coefficients of links 4 and 5 of the five-bar linkage as given by Table 5.3 are correct.

PROBLEMS[1]

5.1 Slotted links 2 and 3 are driven independently at constant speeds of $\omega_2 = 30$ rad/s cw and $\omega_3 = 20$ rad/s cw, respectively. Find the absolute velocity and acceleration of the center of the pin P carried in the two slots.

Fig. P5.1 $R_B^x = 4$ in, $R_B^y = -1$ in, $\theta_2 = 60°$, and $\theta_3 = 135°$.

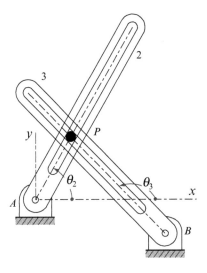

5.2 For the five-bar linkage in the posture shown, the angular velocity of link 2 is 15 rad/s cw and the angular velocity of link 5 is 15 rad/s cw. Determine the angular velocity of link 3 and the apparent velocity $V_{B_4/5}$.

---

[1] When assigning problems, the instructor may wish to specify the method of solution to be used, since a variety of approaches are presented in the text.

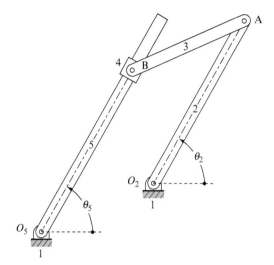

Fig. P5.2 $\mathbf{R}_{O_2O_5} = 8\ \text{in}\angle23.1°$, $\mathbf{R}_{AO_2} = 12\ \text{in}\angle\theta_2$, $\theta_2 = 60°$, $R_{BA} = 8\ \text{in}$, and $\theta_5 = 60°$.

5.3 For Fig. P5.2 in the posture shown in, the angular velocity of link 2 is $\omega_2 = 25$ rad/s ccw and the apparent velocity $V_{B_4/5}$ is 200 in/s upward along link 5. Determine the angular velocities of links 3 and 5.

5.4 For Prob. 5.2, assuming that the two given input velocities are constant, determine the angular acceleration of link 3 at the instant indicated.

5.5 Link 2 rotates at a constant angular velocity of 10 rad/s ccw, while the sliding block 3 slides toward point A on link 2 at a constant rate of 0.125 m/s. Find the absolute velocity and absolute acceleration of point P of block 3.

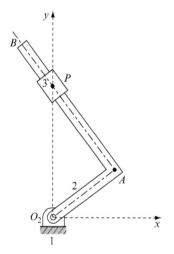

Fig. P5.5 $\mathbf{R}_{AO_2} = 75\ \text{mm}\ \angle36.87°$, $R_{BA} = 150\ \text{mm}$, and $R_{PA} = 100\ \text{mm}$.

5.6 For Prob. 5.5, determine the value of the sliding velocity $V_{P_3/2}$ that minimizes the absolute velocity of point P of block 3. In addition, find the value of $V_{P_3/2}$ that minimizes the absolute acceleration of point P of block 3.

5.7 The left two-link planar robot is attempting to transfer a small object labeled P to the similar right robot. At the posture indicated, $\theta_2 = 45°$ and $\theta_{3/2} = -15°$. (Note that $\theta_{3/2} = \theta_3 - \theta_2$ is given since that is the angle controlled by the motor in joint A.) Determine θ_4 and $\theta_{5/4}$ to allow the right robot to take possession of object P.

Fig. P5.7 $R_{O_4 O_2} = 40$ in, $R_{A O_2} =$ $R_{B O_4} = 12$ in, and $R_{PA} = R_{PB} = 16$ in.

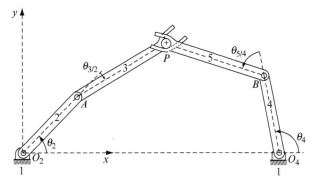

5.8 For the transfer of the object described in Prob. 5.7 it is necessary that the velocities of point P of the two robots match. If the two input velocities of the first robot are $\omega_2 = 10$ rad/s cw and $\omega_{3/2} = 15$ rad/s ccw, what angular velocities must be used for ω_4 and $\omega_{5/4}$?

5.9 For the transfer of the object described in Prob. 5.7 it is necessary that the velocities of point P of the two robots match. If the two input velocities of the first robot are $\omega_2 = 10$ rad/s cw and $\omega_{3/2} = 10$ rad/s ccw, what angular velocities must be used for ω_4 and $\omega_{5/4}$?

5.10 For the transfer of the object described in Prob. 5.7 it is necessary that the velocities of point P of the two robots match. If the two input velocities of the first robot are $\omega_2 = 10$ rad/s cw and $\omega_{3/2} = 0$, what angular velocities must be used for ω_4 and $\omega_{5/4}$?

5.11 To successfully transfer an object between two robots, as described in Probs. 5.7 and 5.8, it is helpful if the accelerations are also matched at point P. Assuming that the two input accelerations are $\alpha_2 = \alpha_3 = 0$ at this instant for the robot on the left, what angular accelerations must be given to the two input joints of the robot on the right to achieve this?

5.12 The circular cam (link 2) is driven at a constant angular velocity $\omega_2 = 15$ rad/s ccw. Link 3 is rotating at a constant angular velocity $\omega_3 = 5$ rad/s cw, causing slipping at point C. For the posture shown, where $\theta_2 = 135°$, determine: (a) the first- and second-order kinematic coefficients of the mechanism; (b) the angular velocity and acceleration of link 4; and (c) the velocity of slipping at point C.

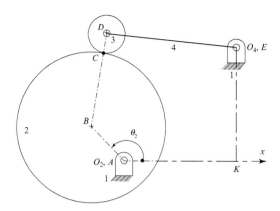

Fig. P5.12 $R_{KO_2} = R_{O_4K} = 60$ mm, $R_{BO_2} = 25$ mm, $R_{CB} = \rho_2 = 40$ mm, $R_{CD} = \rho_3 = 10$ mm, and $R_{DO_4} = 70$ mm.

5.13 Tracing point C of the pantograph linkage is required to follow a prescribed curve; that is, its two independent input variables are x_C and y_C. Then, point P, which carries a pen, traces a similar curve; that is, the outputs of the linkage are the x_P and y_P components of the motion of the pen. Determine the first-order kinematic coefficients of point P.

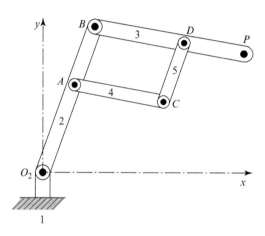

Fig. P5.13 $R_{AO_2} = R_{CA} = R_{DB} = 12$ in and $R_{BA} = R_{DC} = R_{PD} = 8$ in.

5.14 The mechanism has rolling contact at point A, but there can be slipping at point B. Link 2 has a constant angular velocity of $\omega_2 = 20$ rad/s ccw and link 3 has an angular velocity $\omega_3 = 5$ rad/s cw and an angular acceleration $\alpha_3 = 2$ rad/s² ccw. Determine: (a) the first- and second-order kinematic coefficients of link 5; and (b) the angular velocity and angular acceleration of link 5.

Fig. P5.14 $\mathbf{R}_{O_5O_2} = 50\hat{\mathbf{i}} + 75\hat{\mathbf{j}}$ mm,
$R_{CO_2} = 55$ mm, $R_{KO_5} = 20$ mm, $\rho_3 = 40$ mm,
and $\rho_4 = 15$ mm.

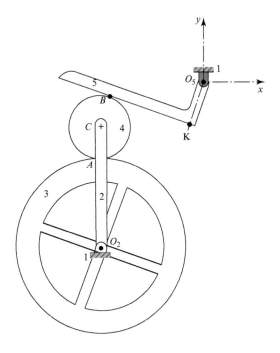

5.15 The mechanism has rolling contact between links 4 and 5 at point C. For the current posture $\theta_2 = 60°$, link 2 has constant angular velocity $\omega_2 = 50$ rad/s ccw, and link 5 has angular velocity $\omega_5 = 25$ rad/s cw and angular acceleration $\alpha_5 = 20$ rad/s^2 ccw. Determine: (a) the first- and second-order kinematic coefficients for links 3 and 4; and (b) the angular velocities and angular accelerations of links 3 and 4.

Fig. P5.15 $\mathbf{R}_{O_5O_2} = 15\hat{\mathbf{i}} - 9.5\hat{\mathbf{j}}$ in, $R_{AO_2} = 15$ in,
$R_{BA} = 15$ in, $\rho_4 = 3$ in, and $\rho_5 = 6.5$ in.

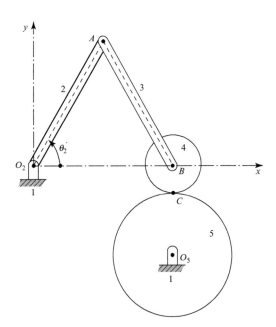

5.16 The mechanism has rolling contact between rack 3 and gear 4 at point F, between gears 4 and 5 at point C, and between gears 5 and 6 at point E. Link 2 has constant angular velocity $\omega_2 = 50$ rad/s ccw, and link 6 has angular velocity $\omega_6 = 25$ rad/s cw and angular acceleration $\alpha_6 = 20$ rad/s² ccw. Determine: (a) the first- and second-order kinematic coefficients for gear 5 and the rack; and (b) the velocity and acceleration of the rack and the angular velocities and accelerations of links 4 and 5.

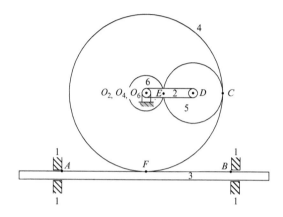

Fig. P5.16 $R_{FA} = R_{BF} = 20$ in,
$\rho_4 = R_{CO_4} = 18$ in, $\rho_5 = R_{CD} = 7$ in,
$\rho_6 = R_{EO_2} = 4$ in, and $R_{DO_2} = 11$ in.

REFERENCES

[1] Pennock, G. R., 2008. Curvature theory for a two-degree-of-freedom planar linkage. *Mech. Mach. Theory*, 43(5):525–548.

Design of Mechanisms

6   Cam Design

6.1   Introduction

In the previous chapters we learned how to analyze the kinematic characteristics of a given mechanism. We were given the design of a mechanism, and we studied ways to determine its mobility, its posture, its velocity, and its acceleration, and we even discussed its suitability for given types of tasks. However, we have said little about how the mechanism is designed – that is, how the sizes and shapes of the links are chosen by the designer.

The next several chapters introduce this *design* point of view as it relates to mechanisms. We will find ourselves looking more at individual types of machine components and learning when and why such components are chosen and how they are sized. In this chapter, which is devoted to the design of cams, we assume that we know the task to be accomplished. However, we do not know but we seek techniques to help discover the size and shape of the cam to perform this task.

Of course, there is the creative step of deciding whether a cam should be used in the first place, as opposed to a gear train, a linkage, or some other mechanical device. This question often cannot be answered on the basis of scientific principles alone; it requires experience and imagination, and involves such factors as economics, marketability, reliability, maintenance, esthetics, ergonomics, ability to manufacture, and suitability for the task. These aspects are not well studied by a general scientific approach; they require human judgment of factors that are often not easily reduced to numbers or formulae. There is usually not a single "right" answer, and generally these questions cannot be answered by this or any other text or reference book.

On the other hand, this is not to say that there is no place for a general science-based approach in design situations. Most mechanical design is based on repetitive analysis. Therefore, in this chapter and in several of the upcoming chapters, we will use the principles of analysis that were presented in Part I. Also, we will use the governing analysis equations to help in our choices of part sizes and shapes and to help us assess the quality of our design as we proceed. It is important to point out that the forthcoming chapters are still based on the laws of mechanics. The primary shift for Part II is that the component dimensions are often the unknowns of the problem, whereas the input and output speeds, for example, may be given information. In this chapter we will discover how to determine a cam contour, or profile, which delivers a specified motion characteristic.

6.2   Classification of Cams and Followers

A *cam* is a machine element used to drive another element, called a *follower*, through a specified motion by direct contact. Cam-and-follower mechanisms are simple and inexpensive, have few moving parts, and occupy very little space. Furthermore, follower motions having almost any desired characteristics are not difficult to design. For these reasons, cam mechanisms are used extensively in modern machinery.

The versatility and flexibility in the design of cam systems are among their more attractive features, yet this also leads to a wide variety of shapes and forms, and the need for terminology to distinguish them.

Sometimes, cams are classified according to their basic shapes. Figure 6.1 shows four different types of cams:

a. a *plate cam*, also called a *disk cam* or a *radial cam*
b. a *wedge* cam
c. a *cylindric cam* or *barrel cam*
d. an *end cam* or *face cam*.

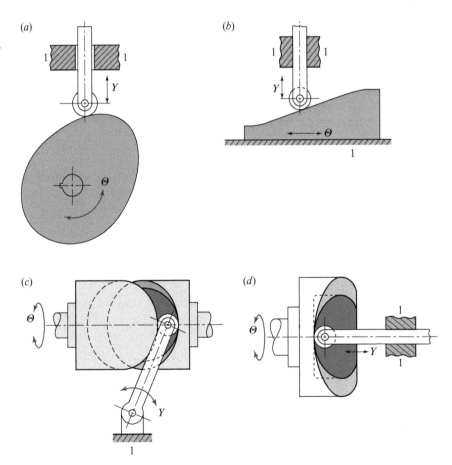

Fig. 6.1 (*a*) Plate cam, (*b*) wedge cam, (*c*) barrel cam, (*d*) face cam.

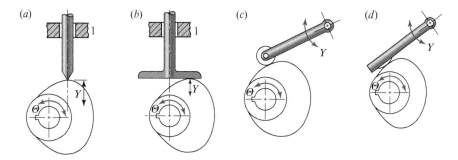

Fig. 6.2 Plate cams with (*a*) an offset reciprocating knife-edge follower; (*b*) a reciprocating flat-face follower; (*c*) an oscillating roller follower; (*d*) an oscillating curved-shoe follower.

The least common of these in practical applications is the wedge cam, because of its need for a reciprocating motion rather than a continuous input motion. By far the most common is the plate cam. For this reason, most of the remainder of this chapter specifically addresses plate cams, although the concepts presented pertain universally.

Cam systems can also be classified according to the basic shape of the follower. Figure 6.2 shows plate cams actuating four different types of followers:

a. a *knife-edge* follower
b. a *flat-face* follower
c. a *roller* follower
d. a *spheric-face* or *curved-shoe* follower.

Note that the follower face is usually chosen to have a simple geometric shape, and the motion is achieved by careful design of the shape of the cam to mate with it. This is not always the case, and examples of *inverse cams*, where the follower is machined to a complex shape, can be found.

Another method of classifying cams is according to the characteristic output motion produced between the follower and the frame. Thus, some cams have *reciprocating* (translating) followers, as in Figs. 6.1*a*, 6.1*b*, and 6.1*d* and Figs. 6.2*a* and 6.2*b*, whereas others have *oscillating* (rotating) followers, as in Figs. 6.1*c*, 6.2*c*, and 6.2*d*. Further classification of reciprocating followers distinguishes whether the centerline of the follower stem is *offset* relative to the center of the cam, as in Fig. 6.2*a*, or *radial*, as in Fig. 6.2*b*.

In all cam systems, the designer must ensure that the follower maintains contact with the cam at all times. This can be accomplished by depending on gravity, by the inclusion of a suitable spring, or by a mechanical constraint. In Fig. 6.1*c*, the follower is constrained by a groove. Figure 6.3*a* shows an example of a *constant-breadth* cam, where two contact points between the cam and the follower provide the constraint. Mechanical constraint can also be introduced by employing *dual* or *conjugate* cams in an arrangement such as that shown in Fig. 6.3*b*. Here, each cam has its own roller, but the rollers are mounted on a common follower.

Throughout this chapter we use the symbol Θ to represent the total motion of the cam, and the symbol Y to represent the total displacement of the follower. To investigate the design of

Fig. 6.3 (*a*) Constant-breadth cam with a reciprocating flat-face follower; (*b*) conjugate cams with an oscillating roller follower.

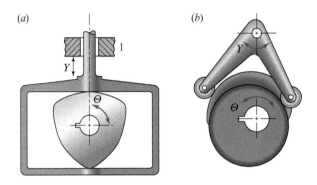

Fig. 6.4 Displacement diagram for a cam.

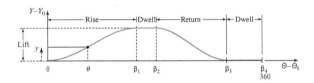

cams in general, we denote the known input variable by $\Theta(t)$ and the output variable by Y. A review of Figs. 6.1, 6.2, and 6.3 reveals the definitions of Θ and Y for various types of cams. These figures show that the input, Θ, is an angle for most cams, but it can be a distance, as in Fig. 6.1*b*. Also, the output, Y, is a translation distance for a reciprocating follower but it is an angle for an oscillating follower.

6.3   Displacement Diagrams

Despite the wide variety of cam types used and their differences in form, they also have certain features in common that allow a systematic approach to their design. Usually a cam system is a single-degree-of-freedom device. It is driven by a known input motion, usually a shaft that rotates at constant speed, and it is intended to produce a certain desired periodic output motion for the follower.

During the rotation of a cam through one cycle, the follower executes a series of events as demonstrated in graphic form in the *displacement diagram* of Fig. 6.4. In such a diagram, the abscissa represents one cycle of the input, Θ (usually one revolution of the cam), and is drawn to any convenient scale. The ordinate represents the follower travel, Y, and, for a reciprocating follower, is usually drawn at full scale to help in the layout of the cam. On a displacement diagram it is possible to identify a portion of the graph called the *rise*, where the displacement of the follower is away from the cam center. The total rise is called the *lift*. The *return* is the portion in which the displacement of the follower is toward the cam center. Portions of the cycle during which the follower is at rest are referred to as *dwells*.

For a particular motion segment (say, segment number k) of the total cam, the rotation lies in the range starting with Θ_k and extending by θ within the segment to

$$\Theta_{k+1} = \Theta_k + \beta_k, \qquad\qquad (a)$$

where β_k is the total cam displacement for segment k.

Similarly, the displacement of the follower is in the range starting from $Y_k + y(0)$ and extending by $y(\theta)$ within the segment to

$$Y_{k+1} = Y_k + y(\beta_k). \qquad\qquad (b)$$

Thus, within a segment we can write

$$y = y(\theta). \qquad\qquad (6.1)$$

Many of the essential features of a displacement diagram, such as the total lift and the placement and duration of dwells, are usually dictated by the requirements of the application. There are, however, many possible choices of follower motions that might be used for the rise and return segments, and some are preferable to others, depending on the situation. One of the key steps in the design of a cam is the choice of suitable forms for these motions. Once the motions have been chosen – that is, once the exact relationship is set between the input Θ and the output Y – the displacement diagram can be constructed precisely and is a graphic representation of the functional relationship

$$Y = Y(\Theta).$$

This equation has stored within it the exact nature of the shape of the final cam: the necessary information for its layout and manufacture, and also the important characteristics that determine the quality of its dynamic performance. Before looking further at these topics, however, we will describe graphic methods of constructing the displacement diagrams for the following rise motions and the similar return motions: uniform motion, parabolic motion, simple harmonic motion, and cycloidal motion.

The displacement diagram for *uniform motion* is a straight line with constant slope. Thus, for constant input speed, the velocity of the follower is also constant. This motion is not useful for the full lift because of the sharp corners produced at the junctions with neighboring segments of the displacement diagram. It is often used, however, between other curve segments that eliminate the corners (called *modified uniform motion*).

Parabolic motion is one possible example of modified uniform motion, and the displacement diagram for this motion is shown in Fig. 6.5a. The central portion of the diagram, bounded by the cam angle β_2 and with lift L_2, is uniform motion. The two end segments, with cam angles β_1 and β_3, and with corresponding lifts L_1 and L_3, are shaped to deliver *parabolic motion* to the follower. Soon we shall learn that these produce constant acceleration of the follower. Figure 6.5a shows a graphic method for matching the slopes of parabolic motions with that of uniform motion. With the cam angles $\beta_1, \beta_2, \beta_3$, and the total lift, L, known, the individual lifts L_1, L_2, and L_3 are determined by locating the midpoints of the β_1 and β_3 segments and constructing a straight line as indicated. Figure 6.5b shows a graphic construction for a parabola to fit within a given rectangular boundary defined, first by L_1 and β_1, and then by L_3 and β_3. The abscissa and ordinate are divided into a convenient but equal number of increments and numbered as indicated. The construction of each point of the parabola then follows that indicated by dashed lines for increment 3.

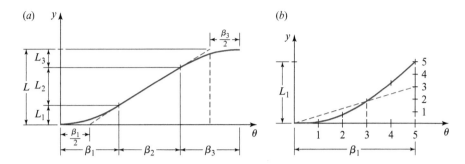

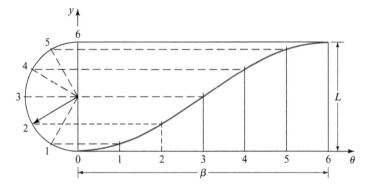

In the layout of an actual cam, if this is done graphically, a great many increments are usually used to obtain good accuracy. At the same time, the drawing is made to a large scale, perhaps ten times full size, and then reduced to actual size by a pantographic method. However, for clarity in reading, the figures in this chapter are shown with fewer increments to define the curves and demonstrate the graphic techniques.

The displacement diagram for *simple harmonic motion* is shown in Fig. 6.6. The graphic construction makes use of a semicircle having a diameter equal to the lift L of the segment. The semicircle and abscissa are divided into an equal number of increments and the construction then follows that indicated by dashed lines for increment number 2.

Cycloidal motion obtains its name from the geometric curve called a cycloid. As shown on the left-hand side of Fig. 6.7, a circle of radius $L/2\pi$, where L is the lift of the segment, makes exactly one revolution by rolling along the ordinate from the origin, $y = 0$, to full lift, $y = L$. Point P of the circle, originally located at the origin, traces a cycloid, as shown. As the circle rolls without slip, the graph of the vertical displacement, y, of the point versus the rotation angle, θ, gives the displacement diagram shown at the right of Fig. 6.7. We find it much more convenient for graphic purposes to draw the circle only once, using either the origin, O, or point B as its center. After dividing the circle and the abscissa into an equal number of increments and numbering them as indicated, we project each point of the circle horizontally until it intersects the ordinate; next, from the ordinate, we project parallel to the diagonal OB to obtain the corresponding point on the displacement diagram.

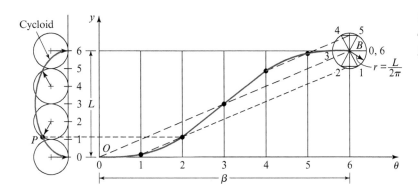

Fig. 6.7 Cycloidal motion displacement diagram; graphic construction.

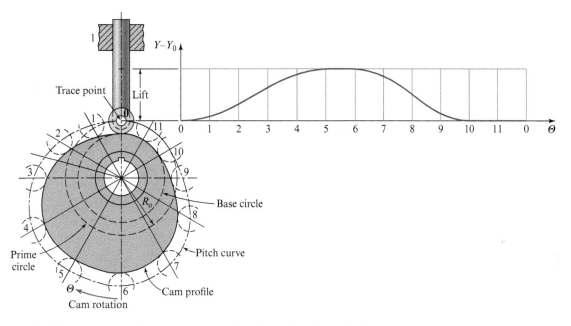

Fig. 6.8 Cam nomenclature. The cam surface is developed by holding the cam stationary and rotating the follower from station 0 through stations 1, 2, 3, and so on.

6.4   Graphic Layout of Cam Profiles

Let us now study the problem of determining the exact shape of the cam profile required to deliver a specified follower motion. We assume here that the required motion has been completely defined – graphically, analytically, or numerically – as discussed in later sections. Thus, a complete displacement diagram can be drawn to scale for the entire cam rotation. The requirement now is to lay out the proper cam profile to achieve the follower motion represented by this displacement diagram.

We explain the procedure using the case of a plate cam with a radial roller follower, as shown in Fig. 6.8. Let us first note some additional nomenclature shown in this figure.

The *trace point* is a theoretic point of the follower, useful primarily in graphic constructions; it corresponds to the tip of a fictitious knife-edge follower. It is located at the center of a roller follower or along the surface of a flat-face follower.

The *pitch curve* is the locus generated by the trace point as the follower moves with respect to the cam. For a knife-edge follower, the pitch curve and cam profile are identical. For a roller follower they are separated by the radius of the roller.

The *prime circle* is the smallest circle that can be drawn tangent to the pitch curve with center on the cam rotation axis. The radius of this circle is denoted R_0.

The *base circle* is the smallest circle tangent to the cam profile centered on the cam rotation axis. For a roller follower, it is smaller than the prime circle by the radius of the roller; for a knife-edge or flat-face follower, it is identical to the prime circle.

In constructing the cam profile, we employ the principle of kinematic inversion. We imagine the sheet of paper on which we are working to be fixed to the cam, and we note that, as the cam rotates, the follower appears to rotate (with respect to the paper) *opposite to the direction of cam rotation*. As shown in Fig. 6.8, we divide the prime circle into a number of increments and assign station numbers to these increments. Dividing the displacement-diagram abscissa into corresponding increments, we transfer $Y-Y_0$ distances from the displacement diagram directly onto the cam layout, measured along the radial lines from the prime circle to the trace point. The smooth curve through these points is the *pitch curve*. For the case of a roller follower, as in this example, we simply draw the roller in its proper location at each increment and then construct the cam profile as a smooth curve tangent to all of these roller locations.

Figure 6.9 shows how the method of construction is modified for an offset roller follower. We begin by constructing an *offset circle*, using a radius equal to the offset distance. After identifying station numbers around the prime circle, the centerline of the follower is constructed for each station, making it tangent to the offset circle. The roller centers for each station are established by transferring $Y-Y_0$ distances from the displacement diagram directly to these follower centerlines, always measuring positive outward from the prime circle. An alternative procedure is to identify points $0', 1', 2'$, and so on, on a single follower centerline and then to rotate them about the cam center to the corresponding follower centerline locations. In either case, the roller circle locations are drawn next and a smooth curve tangent to all roller locations is the required cam profile.

Figure 6.10 shows the construction for a plate cam with a reciprocating flat-face follower. The pitch curve is constructed using a method similar to that used for the roller follower in Fig. 6.8. Instead of roller locations, however, a line representing the flat face of the follower is constructed at each station. The cam profile is a smooth curve drawn tangent to all the follower location lines. It may be helpful to extend each straight line representing a location of the follower face to form a series of triangles. If these triangles are lightly shaded, as suggested in the illustration, it may be easier to draw the cam profile inside all the shaded triangles and tangent to the inner sides of the triangles. Note that the cam profile need not pass through the points 0, 1, 2, 3, etc. constructed from the displacement diagram.

Figure 6.11 shows the layout of the profile of a plate cam with an oscillating roller follower. In this case, to develop the cam profile, we must rotate the fixed pivot center of the follower opposite to the direction of cam rotation. To perform this inversion, first, a circle is drawn

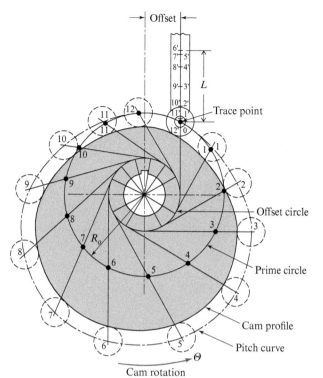

Fig. 6.9 Graphic layout of a plate-cam profile with an offset reciprocating roller follower.

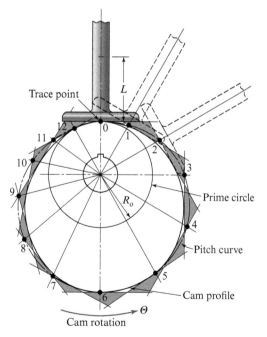

Fig. 6.10 Graphic layout of a plate-cam profile with a reciprocating flat-face follower.

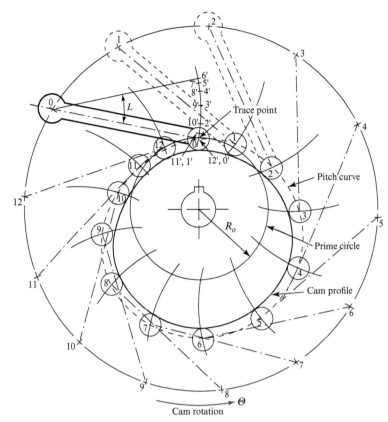

Fig. 6.11 Graphic layout of a plate-cam profile with an oscillating roller follower.

about the cam-shaft center through the fixed pivot of the follower. This circle is then divided and given station numbers to correspond to locations on the displacement diagram. Next, arcs are drawn about each of these centers, all with equal radii corresponding to the length of the follower.

In the case of an oscillating follower, the ordinate values $Y–Y_0$ of the displacement diagram represent angular movements of the follower. If the ordinate scale of the displacement diagram is properly chosen initially, however, and if the total lift of the follower is a reasonably small angle, then ordinate distances from the displacement diagram at each station can be transferred directly to the corresponding arc traveled by the roller center using dividers and measuring positive outward along the arc from the prime circle to locate the trace point for each station. Finally, a circle representing the roller location is drawn with center at the trace point for each station, and the cam profile is constructed as a smooth curve tangent to each of these roller locations.

From the examples presented in this section, it should be clear that each different type of cam-and-follower system requires its own graphic method of construction to determine the cam profile from the displacement diagram. The examples presented here are not intended to be exhaustive of those possible, but they show the general approach. They also demonstrate and reinforce the discussion of the previous section; it is now clear that much of the detailed shape of the cam itself results directly from the displacement diagram. Although different types of cams and followers have different shapes for the same displacement diagram, once a few parameters are chosen, such as the prime-circle radius, which determines the size of a cam, the remainder of its shape results directly from the motion requirements specified in the displacement diagram.

6.5   Kinematic Coefficients of Follower

We have seen that, regardless of the type of cam or the type of follower, the displacement diagram is plotted with the cam input angle, Θ, as the abscissa and the follower output displacement, $Y-Y_0$, as the ordinate. This diagram is made up of a number of segments. In each segment, the abscissa is designated as θ and the ordinate as y. The displacement diagram is, therefore, a graph representing some mathematical function relating the input, Θ, and output, Y, motions of the cam system. In general terms, each segment of this relationship is given by Eq. (6.1):

$$y = y(\theta).$$

Additional graphs can be plotted representing derivatives of y with respect to the input θ – that is, the kinematic coefficients of the follower. The first-order kinematic coefficient of a segment is denoted

$$y'(\theta) = \frac{dy}{d\theta} \tag{6.2}$$

and represents the slope of the displacement diagram at each input position, θ, in the segment. The combined graph of first-order kinematic coefficients of all segments, although it may seem to be of little practical value, is a measure of the "steepness" of the displacement diagram throughout the cycle. We will find later that this is closely related to the mechanical advantage of the cam system and manifests itself in such things as the pressure angle (Section 6.10). If we consider a wedge cam (Fig. 6.1b) with a knife-edge follower (Fig. 6.2a), the displacement diagram itself is of the same shape as the corresponding cam. In such a case we can visualize that difficulties will occur if any segment of the cam is too steep – that is, if the first-order kinematic coefficient y' of any segment has too high a value.

The second-order kinematic coefficient (that is, the second derivative of y with respect to input variable θ) of each segment is also significant. The second-order kinematic coefficient of a segment is denoted

$$y''(\theta) = \frac{d^2y}{d\theta^2}. \tag{6.3}$$

Although it is not as easy to visualize the reason, the second-order kinematic coefficient is very closely related to the curvature of the cam at locations along its profile. Recall that curvature is the reciprocal of the radius of curvature, [Section 4.1, Eq. (g)]. Therefore, if y''

becomes large, the radius of curvature becomes small. In particular, if the second-order kinematic coefficient becomes infinite, then the radius of curvature becomes zero; that is, the cam profile at such a position becomes pointed. This is a highly unsatisfactory condition from the point of view of contact stress between the cam and the follower, and very quickly causes surface damage.

The third-order kinematic coefficient of a segment, denoted

$$y'''(\theta) = \frac{d^3y}{d\theta^3},\qquad(6.4)$$

can also be plotted if desired. Although it is not easy to describe geometrically, this demonstrates the rate of change of y'' with respect to input variable θ. We will see that the third-order kinematic coefficient can also be controlled when choosing the detailed shape of the displacement diagram.

Example 6.1

Derive equations to describe segments of a displacement diagram of a plate cam that rises with parabolic motion from a dwell to another dwell such that the total lift is L and the total cam angle is β. Plot these segments of the displacement diagram and the first-, second-, and third-order kinematic coefficients with respect to input variable θ. The abscissa of each segment graph should be normalized so that the ratio θ/β ranges from $\theta/\beta = 0$ at the left boundary to $\theta/\beta = 1$ at the right boundary of the segment.

Solution

As shown in Fig. 6.5a, two parabolic segments are required, meeting at a common tangent taken here at midrange. For the first segment of the motion, we choose the general equation of a parabola; that is,

$$y = A\theta^2 + B\theta + C.\qquad(1)$$

The first three derivatives of Eq. (1), with respect to input variable θ, are

$$y' = 2A\theta + B,\qquad(2)$$

$$y'' = 2A,\qquad(3)$$

$$y''' = 0.\qquad(4)$$

To properly match the position and slope with those of the preceding dwell, at $\theta = 0$ we must have $y(0) = y'(0) = 0$. Thus, Eqs. (1) and (2) require that $B = C = 0$. Looking next at the inflection point, at $\theta = \beta/2$, we require $y = L/2$. Substituting these conditions into Eq. (1) and rearranging gives

$$A = \frac{2L}{\beta^2}.$$

Example 6.1 (cont.)

Therefore, the displacement equation for the first segment of the *parabolic motion* becomes

$$y = 2L\left(\frac{\theta}{\beta}\right)^2. \tag{6.5a}$$

Differentiating this equation with respect to input variable θ, the first-, second-, and third-order kinematic coefficients, respectively, are

$$y' = \frac{4L}{\beta}\left(\frac{\theta}{\beta}\right), \tag{6.5b}$$

$$y'' = \frac{4L}{\beta^2}, \tag{6.5c}$$

$$y''' = 0. \tag{6.5d}$$

The maximum value for the first-order kinematic coefficient (that is, the maximum slope of y) occurs at the midpoint, where $\theta = \beta/2$. Substituting this value into Eq. (6.5b), the maximum value for the first-order kinematic coefficient is

$$y'_{\text{max}} = \frac{2L}{\beta}. \tag{5}$$

For the second segment of the parabolic motion, we return to the general Eqs. (1)–(4) for a parabola. Substituting the conditions that $y = L$ and $y' = 0$ at $\theta = \beta$ into Eqs. (1) and (2) gives

$$L = A\beta^2 + B\beta + C, \tag{6}$$

$$0 = 2A\beta + B. \tag{7}$$

Since the slope must match that of the first parabola at $\theta = \beta/2$, then from Eqs. (2) and (5) we have

$$\frac{2L}{\beta} = 2A\frac{\beta}{2} + B. \tag{8}$$

Solving Eqs. (6), (7), and (8) simultaneously gives

$$A = -\frac{2L}{\beta^2}, \quad B = \frac{4L}{\beta}, \quad C = -L. \tag{9}$$

Substituting these constants into Eq. (1), the displacement equation for the second segment of the parabolic motion can be written as

$$y = L\left[1 - 2\left(1 - \frac{\theta}{\beta}\right)^2\right]. \tag{6.6a}$$

Example 6.1 (cont.)

Also, substituting Eqs. (9) into Eqs. (2), (3), and (4), the first-, second-, and third-order kinematic coefficients for the second segment of the parabolic motion, respectively, are

$$y' = \frac{4L}{\beta}\left(1 - \frac{\theta}{\beta}\right), \tag{6.6b}$$

$$y'' = -\frac{4L}{\beta^2}, \tag{6.6c}$$

$$y''' = 0. \tag{6.6d}$$

The displacement diagram and the first-, second-, and third-order kinematic coefficients for this example of *full-rise parabolic motion* are shown in Fig. 6.12.

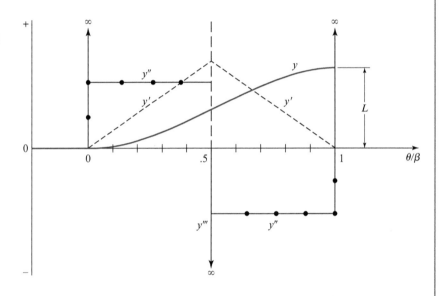

Fig. 6.12 Displacement diagram and derivatives for full-rise parabolic motion, Eqs. (6.5) and (6.6).

The previous discussion relates to the kinematic coefficients of the follower motion. These coefficients are derivatives with respect to the input variable, θ, and relate to the geometry of the cam system. Let us now consider the derivatives of the follower motion with respect to time. First, we assume that the time variation of the input motion, $\Theta(t)$, and, therefore, $\theta(t)$, is known. The angular velocity, $\omega = d\theta/dt$, the angular acceleration, $\alpha = d^2\theta/dt^2$, and the next derivative (often called angular jerk or second angular acceleration) $\dot{\alpha} = d^3\theta/dt^3$, are all assumed to be known. Usually, a plate cam is driven by a constant-speed input shaft. In this case, ω is a known constant, $\theta = \omega t$, and $\alpha = \dot{\alpha} = 0$. During start-up of the cam system, however, this is not the case, and we will consider the more general situation first.

From the general equation of the displacement diagram, for chosen segment number k, we can write from Eqs. (*a*) and (*b*) in Section 6.3

$$y = Y - Y_k = y(\theta) \quad \text{and} \quad \theta = \Theta - \Theta_k = \theta(t). \tag{6.7}$$

Therefore, we can differentiate to find the time derivatives of the follower motion. The velocity of the follower, for example, is given by

$$\dot{y} = \frac{dy}{dt} = \left(\frac{dy}{d\theta}\right)\left(\frac{d\theta}{dt}\right),$$

which, using the first-order kinematic coefficient, can be written as

$$\dot{y} = y'\omega. \tag{6.8}$$

Similarly, the acceleration and the jerk (the third time derivative) of the follower can be written, respectively, as

$$\ddot{y} = \frac{d^2y}{dt^2} = y''\omega^2 + y'\alpha \tag{6.9}$$

and

$$\dddot{y} = \frac{d^3y}{dt^3} = y'''\omega^3 + 3y''\omega\alpha + y'\dot{\alpha}. \tag{6.10}$$

When the cam-shaft speed is constant, then $\alpha = 0$ and Eqs. (6.8) through (6.10) reduce to

$$\dot{y} = y'\omega, \quad \ddot{y} = y''\omega^2, \quad \dddot{y} = y'''\omega^3. \tag{6.11}$$

For this reason, it has become somewhat common to refer to the graphs of the kinematic coefficients y', y'', y''', such as those shown in Fig. 6.12, as the "velocity," "acceleration," and "jerk" curves for a given segment of the motion. These are only appropriate names for a constant-speed cam and then only when scaled by ω, ω^2, and ω^3, respectively.[1] However, it is helpful to use these names for the kinematic coefficients when considering the physical implications of a certain choice of displacement diagram. For example, considering the parabolic motion of Fig. 6.12, it is intuitively meaningful to say that the "velocity" of the follower rises linearly to a maximum at the midpoint $\theta = \beta/2$ and then decreases linearly to zero. The "acceleration" of the follower is zero during the initial dwell and changes abruptly (that is, a step change) to a constant positive value upon beginning the rise. There are two more step changes in the "acceleration" of the follower – namely, at the midpoint and at the end of the rise. At each of these three step changes in the "acceleration" of the follower, the "jerk" of the follower becomes infinite.

[1] Accepting the word "velocity" literally, for example, leads to consternation when it is discovered that, for a plate cam with a reciprocating follower, the units of "velocity" y' are distance per radian. Multiplying these units by radians per second, the units of ω, gives units of distance per second for $\dot{y}$, however.

6.6   High-Speed Cams

Continuing with our discussion of parabolic motion, let us consider briefly the implications of the "acceleration" curve segments of Fig. 6.12 on the dynamic performance of the cam system. Any real follower, of course, has at least some mass and, when this is multiplied by acceleration, exerts an inertia force (Chapter 12). Therefore, the "acceleration" curve of Fig. 6.12 can also be thought of as indicating the inertia force of the follower, which, in turn, is felt at the follower bearings and at the contact point on the cam surface. An "acceleration" curve with abrupt changes (that is, where the "jerk" becomes infinite), such as those demonstrated for parabolic motion, exerts abruptly changing contact stresses at the bearings and on the cam surface and will lead to noise, surface wear, and early failure. Thus, it is very important in choosing and joining segments of a displacement diagram to ensure that the first- and second-order kinematic coefficients (that is, the "velocity" and "acceleration" curves) are continuous – that is, that they contain no step changes.

Sometimes in low-speed cam applications compromises are made with the "velocity" and "acceleration" relationships. It is sometimes simpler to employ a reverse procedure and design the cam shape first, obtaining the displacement diagram as a second step. Such cams are sometimes composed of a combination of curves, such as straight lines and circular arcs, which are readily produced by machine tools. Two examples are the *circle-arc cam* and the *tangent cam* shown in Fig. 6.13. The design approach is by iteration. A trial cam is designed and its kinematic characteristics are found. The process is then repeated until a cam with acceptable characteristics is obtained. Points *A*, *B*, *C*, and *D* of the circle-arc cam and the tangent cam are points of tangency or blending points. It is worth noting, as with the previous parabolic-motion example, that the acceleration changes abruptly at each of the blending points because of the instantaneous change in the radius of curvature of the cam profile.

Although cams with discontinuous acceleration characteristics have sometimes been accepted to reduce costs in low-speed applications, such cams have invariably exhibited major difficulties at some later time when the speed of the machine was raised to increase the productivity of the application. For any high-speed cam application, it is extremely important that not only the displacement and "velocity" curves, but also the "acceleration" curve, be

Fig. 6.13 (*a*) Circle-arc cam; (*b*) tangent cam.

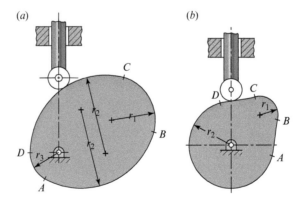

made continuous for the entire motion cycle. No discontinuities should be allowed within or at the junctions of different segments of a cam.

As confirmed by Eq. (6.11), the importance of continuous derivatives becomes more serious as the cam-shaft speed is increased. The higher the speed, the greater the need for smooth curves. At very high speeds it might also be desirable to require that jerk, which is related to rate of change of force, and perhaps even higher derivatives, be made continuous as well. In many applications, however, this is not necessary.

There is no simple answer as to how high a speed one must have before considering the application to require high-speed design techniques. The answer depends not only on the mass of the follower, but also on the stiffness of the return spring, the materials used, the flexibility of the follower, and many other factors [9]. Further analysis techniques on cam dynamics are presented in Sections 6.11–6.16. Still, with the methods presented here, it is not difficult to achieve continuous displacement diagrams with continuous derivatives. Therefore, it is recommended that this be undertaken as standard practice. Cycloidal motion cams, for example, are no more difficult to manufacture than parabolic-motion cams, and there is no good reason for use of the latter. The circle-arc cam and the tangent cam may be easy to produce, but, with modern machining methods, cutting more complex cam shapes is not expensive and is recommended.

6.7 Standard Cam Motions

Example 6.1 in Section 6.5 gave a detailed derivation of the equations for parabolic motion and its first three derivatives [Eqs. (6.5) and (6.6)]. Then, in Section 6.6, reasons were provided for avoiding the use of parabolic motion in high-speed cam systems. The purpose of this section is to present equations for a number of standard types of displacement curve segments that can be used to address most high-speed cam motion requirements. The derivations parallel those of Example 6.1 and are not presented.

The displacement equation and the first-, second-, and third-order kinematic coefficients for a *full-rise simple harmonic motion* segment are

$$y = \frac{L}{2}\left(1 - \cos\frac{\pi\theta}{\beta}\right), \tag{6.12a}$$

$$y' = \frac{\pi L}{2\beta}\sin\frac{\pi\theta}{\beta}, \tag{6.12b}$$

$$y'' = \frac{\pi^2 L}{2\beta^2}\cos\frac{\pi\theta}{\beta}, \tag{6.12c}$$

$$y''' = -\frac{\pi^3 L}{2\beta^3}\sin\frac{\pi\theta}{\beta}. \tag{6.12d}$$

The displacement diagram and the first-, second-, and third-order kinematic coefficients for a full-rise simple harmonic motion segment are shown in Fig. 6.14. Unlike parabolic motion,

Fig. 6.14 Displacement diagram and derivatives for a full-rise simple harmonic motion segment, Eqs. (6.12).

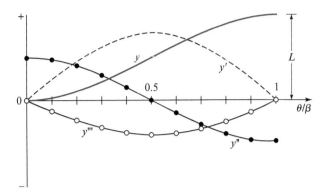

Fig. 6.15 Displacement diagram and derivatives for a full-rise cycloidal motion segment, Eqs. (6.13).

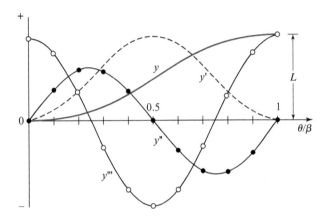

simple harmonic motion exhibits no discontinuity at the inflection point, but it does contain nonzero "accelerations" at its two boundaries.

The displacement equation and the first-, second-, and third-order kinematic coefficients for a *full-rise cycloidal motion* segment are

$$y = L\left(\frac{\theta}{\beta} - \frac{1}{2\pi}\sin\frac{2\pi\theta}{\beta}\right), \tag{6.13a}$$

$$y' = \frac{L}{\beta}\left(1 - \cos\frac{2\pi\theta}{\beta}\right), \tag{6.13b}$$

$$y'' = \frac{2\pi L}{\beta^2}\sin\frac{2\pi\theta}{\beta}, \tag{6.13c}$$

$$y''' = \frac{4\pi^2 L}{\beta^3}\cos\frac{2\pi\theta}{\beta}. \tag{6.13d}$$

The displacement diagram and the first-, second-, and third-order kinematic coefficients for a full-rise cycloidal motion segment are shown in Fig. 6.15. Note that all derivatives at the boundaries of this segment have zero values. Note also that this is the only standard motion segment with zeroes for all derivatives at the boundaries. However, the peak "velocity," "acceleration," and "jerk" values are higher than those for a simple harmonic motion segment.

The displacement equation and the first-, second-, and third-order kinematic coefficients for a *full-rise eighth-order polynomial motion* segment are

$$y = L\left[6.09755\left(\frac{\theta}{\beta}\right)^3 - 20.78040\left(\frac{\theta}{\beta}\right)^5 + 26.73155\left(\frac{\theta}{\beta}\right)^6 - 13.60965\left(\frac{\theta}{\beta}\right)^7 + 2.56095\left(\frac{\theta}{\beta}\right)^8\right],$$

(6.14a)

$$y' = \frac{L}{\beta}\left[18.29265\left(\frac{\theta}{\beta}\right)^2 - 103.90200\left(\frac{\theta}{\beta}\right)^4 + 160.38930\left(\frac{\theta}{\beta}\right)^5 - 95.26755\left(\frac{\theta}{\beta}\right)^6 \right.$$
$$\left. + 20.48760\left(\frac{\theta}{\beta}\right)^7\right],$$

(6.14b)

$$y'' = \frac{L}{\beta^2}\left[36.58530\left(\frac{\theta}{\beta}\right) - 415.60800\left(\frac{\theta}{\beta}\right)^3 + 801.94650\left(\frac{\theta}{\beta}\right)^4 - 571.60530\left(\frac{\theta}{\beta}\right)^5 \right.$$
$$\left. + 143.41320\left(\frac{\theta}{\beta}\right)^6\right],$$

(6.14c)

$$y''' = \frac{L}{\beta^3}\left[36.58530 - 1246.82400\left(\frac{\theta}{\beta}\right)^2 + 3207.78600\left(\frac{\theta}{\beta}\right)^3 - 2858.02650\left(\frac{\theta}{\beta}\right)^4 \right.$$
$$\left. + 860.47920\left(\frac{\theta}{\beta}\right)^5\right].$$

(6.14d)

The displacement diagram and the first-, second-, and third-order kinematic coefficients for the full-rise motion segment formed from an eighth-order polynomial are shown in Fig. 6.16. Equations (6.14) have seemingly awkward coefficients because they have been specially derived

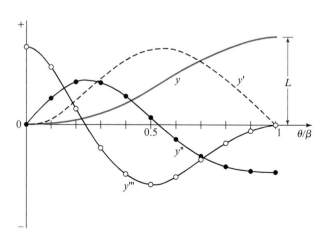

Fig. 6.16 Displacement diagram and derivatives for a full-rise eighth-order polynomial motion segment, Eqs. (6.14).

to have many "nice" properties [5]. Among these, Fig. 6.16 shows not only that several of the kinematic coefficients are zero at both ends of the segment, but also that the "acceleration" characteristics are nonsymmetric. Also, the peak values of "acceleration" are kept as small as possible (that is, the magnitudes of the positive and negative peak "accelerations" are equal).

The displacement diagrams of simple harmonic, cycloidal, and eighth-order polynomial motion segments look quite similar at first glance. Each rises through a lift of L in a cam rotation angle of β, and each begins and ends with zero slope. For these reasons, they are all referred to as *full-rise* motion segments. However, their "acceleration" curves are quite different. A simple harmonic motion segment has nonzero "acceleration" at both boundaries; a cycloidal motion segment has zero "acceleration" at both boundaries; and an eighth-order polynomial motion segment has one zero and one nonzero "acceleration" at its two boundaries. This variety provides the selections necessary when matching these curves with neighboring curves of different types.

Full-return motion segments of the same three types are shown in Figs. 6.17–6.19.

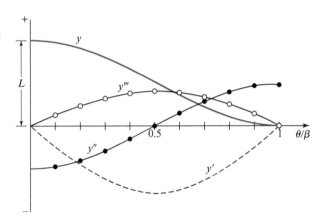

Fig. 6.17 Displacement diagram and derivatives for a full-return simple harmonic motion segment, Eqs. (6.15).

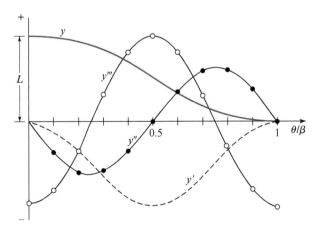

Fig. 6.18 Displacement diagram and derivatives for a full-return cycloidal motion segment, Eqs. (6.16).

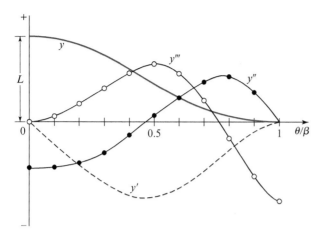

Fig. 6.19 Displacement diagram and derivatives for a full-return eighth-order polynomial motion segment, Eqs. (6.17).

The displacement equation and the first-, second-, and third-order kinematic coefficients for a *full-return simple harmonic motion* segment are

$$y = \frac{L}{2}\left(1 + \cos\frac{\pi\theta}{\beta}\right), \tag{6.15a}$$

$$y' = -\frac{\pi L}{2\beta}\sin\frac{\pi\theta}{\beta}, \tag{6.15b}$$

$$y'' = -\frac{\pi^2 L}{2\beta^2}\cos\frac{\pi\theta}{\beta}, \tag{6.15c}$$

$$y''' = \frac{\pi^3 L}{2\beta^3}\sin\frac{\pi\theta}{\beta}. \tag{6.15d}$$

For a *full-return cycloidal motion* segment, the displacement equation and the first-, second-, and third-order kinematic coefficients are

$$y = L\left(1 - \frac{\theta}{\beta} + \frac{1}{2\pi}\sin\frac{2\pi\theta}{\beta}\right), \tag{6.16a}$$

$$y' = -\frac{L}{\beta}\left(1 - \cos\frac{2\pi\theta}{\beta}\right), \tag{6.16b}$$

$$y'' = -\frac{2\pi L}{\beta^2}\sin\frac{2\pi\theta}{\beta}, \tag{6.16c}$$

$$y''' = -\frac{4\pi^2 L}{\beta^3}\cos\frac{2\pi\theta}{\beta}. \tag{6.16d}$$

For a *full-return eighth-order polynomial motion* segment, the displacement equation and the first-, second-, and third-order kinematic coefficients are

$$y = L\left[1.00000 - 2.63415\left(\frac{\theta}{\beta}\right)^2 + 2.78055\left(\frac{\theta}{\beta}\right)^5 + 3.17060\left(\frac{\theta}{\beta}\right)^6 - 6.87795\left(\frac{\theta}{\beta}\right)^7 \right. \quad (6.17a)$$

$$\left. + 2.56095\left(\frac{\theta}{\beta}\right)^8\right],$$

$$y' = -\frac{L}{\beta}\left[5.26830\frac{\theta}{\beta} - 13.90275\left(\frac{\theta}{\beta}\right)^4 - 19.02360\left(\frac{\theta}{\beta}\right)^5 + 48.14565\left(\frac{\theta}{\beta}\right)^6 - 20.48760\left(\frac{\theta}{\beta}\right)^7\right],$$

$$(6.17b)$$

$$y'' = -\frac{L}{\beta^2}\left[5.26830 - 55.61100\left(\frac{\theta}{\beta}\right)^3 - 95.11800\left(\frac{\theta}{\beta}\right)^4 + 288.87390\left(\frac{\theta}{\beta}\right)^5 - 143.41320\left(\frac{\theta}{\beta}\right)^6\right],$$

$$(6.17c)$$

$$y''' = \frac{L}{\beta^3}\left[166.83300\left(\frac{\theta}{\beta}\right)^2 + 380.47200\left(\frac{\theta}{\beta}\right)^3 - 1444.36950\left(\frac{\theta}{\beta}\right)^4 + 860.47920\left(\frac{\theta}{\beta}\right)^5\right]. \quad (6.17d)$$

Polynomial displacement equations of much higher order, and meeting many more conditions than those presented here, are also in common use. Automated procedures for determining the coefficients have been developed by Stoddart [8], who also indicates how the choice of coefficients can be made to compensate for elastic deformation of the follower system under dynamic conditions. Such cams are referred to as *polydyne cams*.

In addition to the full-rise and full-return motions presented above, it is also useful to have a selection of standard *half-rise* and *half-return* motion segments available. These are curves for which one segment boundary has a nonzero slope and can be used to blend with uniform motion. The displacement diagrams and the first-, second-, and third-order kinematic coefficients for *half-rise simple harmonic motion* segments, sometimes called *half-harmonic rise motion* segments, are shown in Fig. 6.20. The equations corresponding to Fig. 6.20a are

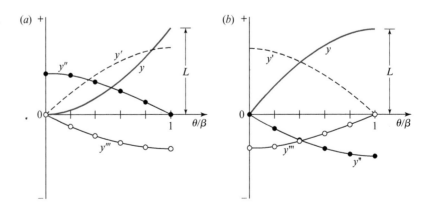

Fig. 6.20 Displacement diagram and derivatives for half-rise simple harmonic motion segments: (*a*) Eqs. (6.18); (*b*) Eqs. (6.19).

$$y = L\left(1 - \cos\frac{\pi\theta}{2\beta}\right), \tag{6.18a}$$

$$y' = \frac{\pi L}{2\beta}\sin\frac{\pi\theta}{2\beta}, \tag{6.18b}$$

$$y'' = \frac{\pi^2 L}{4\beta^2}\cos\frac{\pi\theta}{2\beta}, \tag{6.18c}$$

$$y''' = -\frac{\pi^3 L}{8\beta^3}\sin\frac{\pi\theta}{2\beta}. \tag{6.18d}$$

The displacement equation and the first-, second-, and third-order kinematic coefficients corresponding to the half-rise simple harmonic motion segments of Fig. 6.20b are

$$y = L\sin\frac{\pi\theta}{2\beta}, \tag{6.19a}$$

$$y' = \frac{\pi L}{2\beta}\cos\frac{\pi\theta}{2\beta}, \tag{6.19b}$$

$$y'' = -\frac{\pi^2 L}{4\beta^2}\sin\frac{\pi\theta}{2\beta}, \tag{6.19c}$$

$$y''' = -\frac{\pi^3 L}{8\beta^3}\cos\frac{\pi\theta}{2\beta}. \tag{6.19d}$$

The curves for *half-return simple harmonic motion* segments are shown in Fig. 6.21. The equations corresponding to Fig. 6.21a are

$$y = L\cos\frac{\pi\theta}{2\beta}, \tag{6.20a}$$

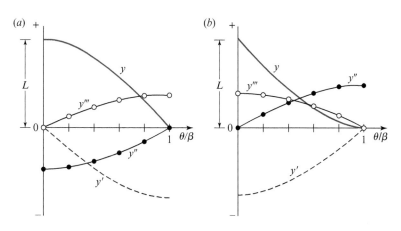

Fig. 6.21 Displacement diagram and derivatives for half-return simple harmonic motion segments: (a) Eqs. (6.20); (b) Eqs. (6.21).

$$y' = -\frac{\pi L}{2\beta} \sin \frac{\pi\theta}{2\beta}, \tag{6.20b}$$

$$y'' = -\frac{\pi^2 L}{4\beta^2} \cos \frac{\pi\theta}{2\beta}, \tag{6.20c}$$

$$y''' = \frac{\pi^3 L}{8\beta^3} \sin \frac{\pi\theta}{2\beta}. \tag{6.20d}$$

The displacement equation and the first-, second-, and third-order kinematic coefficients corresponding to the half-return simple harmonic motion segments of Fig. 6.21b are

$$y = L\left(1 - \sin \frac{\pi\theta}{2\beta}\right), \tag{6.21a}$$

$$y' = -\frac{\pi L}{2\beta} \cos \frac{\pi\theta}{2\beta}, \tag{6.21b}$$

$$y'' = \frac{\pi^2 L}{4\beta^2} \sin \frac{\pi\theta}{2\beta}, \tag{6.21c}$$

$$y''' = \frac{\pi^3 L}{8\beta^3} \cos \frac{\pi\theta}{2\beta}. \tag{6.21d}$$

In addition to the half-harmonics, half-cycloidal motion segments are also useful, since their "accelerations" are zero at both segment boundaries. The displacement diagrams and first-, second-, and third-order kinematic coefficients for *half-rise cycloidal motion* segments are shown in Fig. 6.22. The equations corresponding to Fig. 6.22a are

Fig. 6.22 Displacement diagram and derivatives for half-rise cycloidal motion segments: (*a*) Eqs. (6.22); (*b*) Eqs. (6.23).

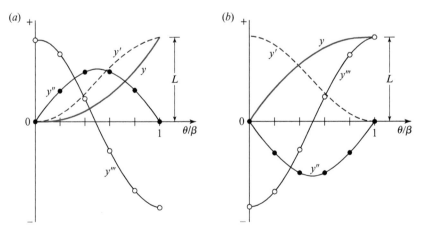

$$y = L\left(\frac{\theta}{\beta} - \frac{1}{\pi} \sin \frac{\pi\theta}{\beta}\right), \tag{6.22a}$$

$$y' = \frac{L}{\beta}\left(1 - \cos \frac{\pi\theta}{\beta}\right), \tag{6.22b}$$

$$y'' = \frac{\pi L}{\beta^2} \sin \frac{\pi\theta}{\beta}, \tag{6.22c}$$

$$y''' = \frac{\pi^2 L}{\beta^3} \cos \frac{\pi\theta}{\beta}. \tag{6.22d}$$

The displacement equation and the first-, second-, and third-order kinematic coefficients corresponding to the *half-rise cycloidal* motion segments of Fig. 6.22b are

$$y = L\left(\frac{\theta}{\beta} + \frac{1}{\pi} \sin \frac{\pi\theta}{\beta}\right), \tag{6.23a}$$

$$y' = \frac{L}{\beta}\left(1 + \cos \frac{\pi\theta}{\beta}\right), \tag{6.23b}$$

$$y'' = -\frac{\pi L}{\beta^2} \sin \frac{\pi\theta}{\beta}, \tag{6.23c}$$

$$y''' = -\frac{\pi^2 L}{\beta^3} \cos \frac{\pi\theta}{\beta}. \tag{6.23d}$$

The curves for *half-return cycloidal motion* segments are shown in Fig. 6.23. The equations corresponding to Fig. 6.23a are

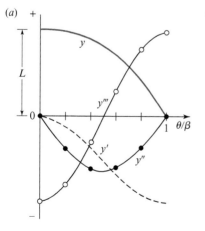 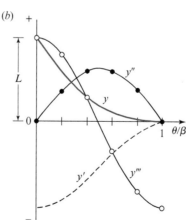

Fig. 6.23 Displacement diagram and derivatives for half-return cycloidal motion segments: (a) Eqs. (6.24); (b) Eqs. (6.25).

$$y = L\left(1 - \frac{\theta}{\beta} + \frac{1}{\pi} \sin \frac{\pi\theta}{\beta}\right), \tag{6.24a}$$

$$y' = -\frac{L}{\beta}\left(1 - \cos \frac{\pi\theta}{\beta}\right), \tag{6.24b}$$

$$y'' = -\frac{\pi L}{\beta^2} \sin \frac{\pi\theta}{\beta}, \tag{6.24c}$$

$$y''' = -\frac{\pi^2 L}{\beta^3} \cos \frac{\pi\theta}{\beta}. \tag{6.24d}$$

The displacement equation and the first-, second-, and third-order kinematic coefficients corresponding to the half-return cycloidal motion segments of Fig. 6.23b are

$$y = L\left(1 - \frac{\theta}{\beta} - \frac{1}{\pi} \sin \frac{\pi\theta}{\beta}\right), \tag{6.25a}$$

$$y' = -\frac{L}{\beta}\left(1 + \cos \frac{\pi\theta}{\beta}\right), \tag{6.25b}$$

$$y'' = \frac{\pi L}{\beta^2} \sin \frac{\pi\theta}{\beta}, \tag{6.25c}$$

$$y''' = \frac{\pi^2 L}{\beta^3} \cos \frac{\pi\theta}{\beta}. \tag{6.25d}$$

We will see shortly how the "standard" segment graphs and equations presented in this section can greatly reduce the analytic effort involved in designing the full displacement diagram for a high-speed cam. First, however, let us summarize several important features of the segment graphs of Figs. 6.14–6.23:

1. Each graph shows one segment of a full displacement diagram.
2. The total lift for that segment is labeled L for each, and the total cam travel is labeled β.
3. The abscissa of each segment graph is normalized so that the ratio θ/β ranges from $\theta/\beta = 0$ at the left boundary to $\theta/\beta = 1$ at the right boundary of the segment.
4. The scales used in plotting the graphs are not depicted but are consistent for all full-rise and full-return curves, and for all half-rise and half-return curves. Thus, in judging the suitability of one curve compared with another, the magnitudes of the "accelerations," for example, can be compared. For this reason, when other factors are equivalent, simple harmonic motion should be used where possible and, in order to keep "accelerations" small, cycloidal motion should be avoided except where necessary.

Finally, it should be noted that the standard cam motion segments presented in this section do not form an exhaustive set. The set presented here is sufficient for most practical applications. However, cams with good dynamic characteristics can also be formed from a wide variety of other possible motion segment curves. A much more extensive set can be found, for example, in the text by Chen [2].

6.8 Matching Derivatives of Displacement Diagrams

In the previous section, a great many equations were presented that might be used to represent different segments of the displacement diagram of a cam. In this section, we will study how these can be joined together to form the motion specification for a complete cam. The procedure is one of solving for proper values of L and β for each segment so that:

1. The motion requirements of the particular application are met.
2. The displacement diagram, as well as the diagrams of the first- and second-order kinematic coefficients, are continuous across the boundaries of the merged segments. The diagram of the third-order kinematic coefficient may be allowed discontinuities if necessary, but must not become infinite; that is, the "acceleration" curve may contain discontinuities in slope (corners) but not discontinuities in value (jumps).
3. The maximum magnitudes of the "velocity" and "acceleration" peaks are kept as low as possible, consistent with the first two conditions.

The procedure may best be understood through an example.

Example 6.2

A plate cam with a reciprocating follower is to be driven by a constant-speed motor at 150 rev/min. The follower is to start from a dwell, accelerate to a uniform velocity of 25 in/s, maintain this velocity for 1.25 in of rise, decelerate to the top of the lift, return, and then dwell for 0.10 s. The total lift is to be 3.00 in. Determine the complete specifications of the displacement diagram.

Solution

The speed of the input shaft is

$$\omega = 150 \text{ rev/min} = 15.70796 \text{ rad/s.} \tag{1}$$

Using Eq. (6.8), the first-order kinematic coefficient (that is, the slope of the uniform "velocity" segment) is

$$y' = \frac{\dot{y}}{\omega} = \frac{25 \text{ in/s}}{15.70796 \text{ rad/s}} = 1.59155 \text{ in/rad.} \tag{2}$$

Since this "velocity" is held constant for 1.25 in of rise, the total cam rotation in this segment is

$$\beta_2 = \frac{L_2}{y'} = \frac{1.25 \text{ in}}{1.59155 \text{ in/rad}} = 0.78540 \text{ rad} = 45.000°. \tag{3}$$

Example 6.2 (cont.)

Similarly, from Eq. (1), the total cam rotation during the final dwell is

$$\beta_5 = 0.10\text{s}\ (15.707\,96\,\text{rad/s}) = 1.570\,796\,\text{rad} = 90.000°. \tag{4}$$

Note that several digits of accuracy higher than usual are utilized here and are recommended as standard practice when matching cam motion derivatives. Any inaccuracies in the L and β values result in discontinuities in the smoothness of derivatives at the junctures of the segments and discontinuities in force, as was explained in Section 6.6.

From these results and the given information, we can sketch the beginnings of the displacement diagram, not necessarily working to scale, but to visualize the motion requirements. This gives the general shapes shown by the heavy segments of the curves of Fig. 6.24a. The lighter segments of the displacement curve are not yet accurately known, but can be sketched by lightly outlining a smooth curve for visualization. Working from this approximate curve, we can also sketch the general nature of the derivative curves. From the changing slope of the displacement diagram we

Fig. 6.24 (a) Displacement diagram, in; (b) "velocity" diagram, in/rad; (c) "acceleration" diagram, in/rad².

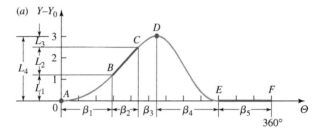

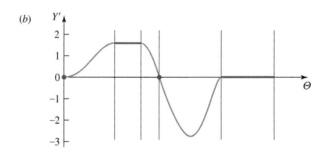

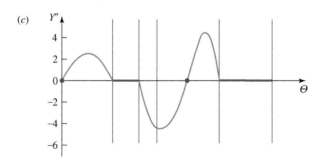

Example 6.2 (cont.)

sketch the "velocity" curve (Fig. 6.24b), and from the changing slope of this curve we sketch the "acceleration" curve (Fig. 6.24c). At this stage, no attempt is made to produce accurate curves drawn to scale, but only to provide an idea of the desired curve shapes.

Next, using the sketches of Fig. 6.24, we compare these desired motion curves with the various standard curve segments of Figs. 6.14–6.23 to choose an appropriate set of equations for each segment of the cam. In segment AB, for example, we find that Fig. 6.22a is the only standard motion curve segment available with half-rise characteristics, an appropriate slope curve, and the necessary zero "acceleration" at both boundaries of the segment. Thus, we choose the half-rise cycloidal motion of Eq. (6.22) for this segment of the cam. There are two sets of choices possible for the segments CD and DE. One set might be the choice of Fig. 6.22b matched with Fig. 6.18. However, to keep the peak "accelerations" low and to keep the "jerk" curves as smooth as possible, we choose Fig. 6.20b matched with Fig. 6.19. Thus, for segment CD, we will use the half-rise harmonic motion of Eq. (6.19), and for segment DE, we choose the eighth-order polynomial return motion of Eq. (6.17).

Choosing the curve types, however, is not sufficient to fully specify the segment characteristics. We must also find values for the unknown parameters of the segment equations; these are $L_1, L_3, \beta_1, \beta_3,$ and β_4. We do this by equating the kinematic coefficients at each nonzero segment boundary. For example, to match the "velocities" at point B we must equate the first-order kinematic coefficient from Eq. (6.22b) at the AB segment right boundary (that is, at $\theta_1/\beta_1 = 1$) with the first-order kinematic coefficient of the BC segment; that is,

$$y'_B = \frac{2L_1}{\beta_1} = \frac{L_2}{\beta_2} = \frac{1.25 \text{ in}}{0.785\,40 \text{ rad}} = 1.591\,55 \text{ in/rad}$$

or

$$L_1 = (0.795\,77 \text{ in/rad})\beta_1. \tag{5}$$

Similarly, to match the "velocities" at point C, we equate the first-order kinematic coefficient of segment BC with the first-order kinematic coefficient of Eq. (6.19b) at the CD segment left boundary (that is, at $\theta_3/\beta_3 = 0$). This gives

$$y'_C = \frac{L_2}{\beta_2} = \frac{\pi L_3}{2\beta_3} = 1.591\,55 \text{ in/rad}$$

or

$$L_3 = (1.013\,21 \text{ in/rad})\beta_3. \tag{6}$$

To match the "accelerations" (that is, the curvatures) at point D, we equate the second-order kinematic coefficient of Eq. (6.19c) at the CD segment right boundary (that is, at $\theta_3/\beta_3 = 1$) with the second-order kinematic coefficient of Eq. (6.17c) at the DE segment left boundary (that is, at $\theta_4/\beta_4 = 0$). This gives

Example 6.2 (cont.)

$$y''_D = -\frac{\pi^2 L_3}{4\beta_3^2} = -5.268\,30\frac{L_4}{\beta_4^2},$$

where the total lift is $L_4 = 3$ in. Substituting Eq. (6) and the total lift into this result and rearranging gives

$$\beta_3 = 0.158\,18\beta_4^2. \tag{7}$$

Finally, for geometric compatibility, we have the constraints

$$L_1 + L_3 = L_4 - L_2 = 1.750 \text{ in}, \tag{8}$$

and, considering Eqs. (3) and (4),

$$\beta_1 + \beta_3 + \beta_4 = 2\pi - \beta_2 - \beta_5 = 3.926\,99 \text{ rad}. \tag{9}$$

Solving the five equations – that is, Eqs. (5)–(9) – simultaneously for the five unknowns, $L_1, L_3, \beta_1, \beta_3,$ and β_4, provides the proper values for the remaining parameters. In summary, the segment parameters are:

$$
\begin{aligned}
L_1 &= 1.183\,1 \text{ in}, & \beta_1 &= 1.486\,74 \text{ rad} = 85.184°, \\
L_2 &= 1.250\,0 \text{ in}, & \beta_2 &= 0.785\,40 \text{ rad} = 45.000°, \\
L_3 &= 0.566\,9 \text{ in}, & \beta_3 &= 0.559\,51 \text{ rad} = 32.058°, \\
L_4 &= 3.000\,0 \text{ in}, & \beta_4 &= 1.880\,74 \text{ rad} = 107.758°, \\
L_5 &= 0.000\,0 \text{ in}, & \beta_5 &= 1.570\,80 \text{ rad} = 90.000°. \quad \textit{Ans.}
\end{aligned}
$$

At this time, if desired, an accurate layout of the displacement diagram and the kinematic coefficients can be made to replace the sketches. The curves of Fig. 6.24 are drawn to scale using these values.

6.9   Plate Cam with Reciprocating Flat-Face Follower

Once the displacement diagram of a cam system has been completely determined, as described in Section 6.8, the layout of the actual cam shape can be attempted, as explained in Section 6.4. In laying out the cam, however, we may find the need for a few more parameters, depending on the type of cam and follower – for example, the prime-circle radius, any offset distance, the roller radius, and so on. Also, as we will see, each different type of cam-and-follower system may be subject to certain further problems unless these remaining parameters are properly chosen.

In this section, we study the problems that may be encountered in the design of a plate cam with a reciprocating flat-face follower. The geometric parameters of such a system that must yet

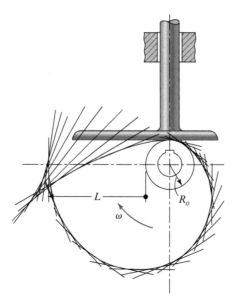

be chosen are the prime-circle radius, R_0, the offset (eccentricity), ε, of the follower stem, and the width of the follower face.

Figure 6.25 shows the layout of a plate cam with a radial reciprocating flat-face follower. In this case, the displacement chosen was a full-rise cycloidal motion segment with $L_1 = 4.000$ in during $\beta_1 = 90°$ of cam rotation, followed by a full-return cycloidal motion segment during the remaining $\beta_2 = 270°$ of cam rotation. The layout procedure of Fig. 6.10 was followed to develop the cam shape, and the radius chosen for the prime circle was $R_0 = 1.000$ in. Obviously, there is a problem, since the resulting cam profile intersects itself. During machining, part of the cam shape is lost, and, when in operation, the intended cycloidal motion is not fully achieved. Such a cam is said to be *undercut*.

Why did undercutting occur in this example and how can it be avoided? It resulted from attempting to achieve too great a lift in too little cam rotation with too small a cam. One possible cure for this problem is to decrease the desired lift, L_1, or to increase the cam rotation angle, β_1. However, this is not possible while still achieving the original design specifications. Another cure is to continue with the same displacement characteristics but to increase the prime-circle radius, R_0, to avoid undercutting. This does produce a larger cam, but with sufficient increase it does overcome the undercutting difficulty.

The minimum value of R_0 that avoids undercutting can be found by developing an equation for the radius of curvature of the cam profile. We start by writing the loop-closure equation using the vectors shown in Fig. 6.26. Using complex polar notation, the loop-closure equation for a general plate cam with a reciprocating flat-face follower is

$$\mathbf{R} = re^{j(\Theta+\varphi)} + j\rho = j(R_0 + Y) + s. \qquad (a)$$

Recall that the symbol Θ represents the total rotation of a general plate cam.

Fig. 6.26 Vectors for a plate-cam profile
with reciprocating flat-face follower.

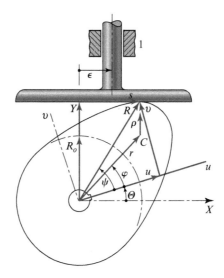

We have carefully chosen the vectors so that point C is located at the instantaneous center of curvature, and ρ is the radius of curvature corresponding to the current contact point. The line along the u axis which separates the angles Θ and φ is fixed on the cam and is horizontal for the cam posture $\Theta = 0$. The value of Y_0 is zero. The angle Θ specifies the rotation of the cam, and the u and v axes rotate with the cam.

Separating Eq. (a) into real and imaginary parts, respectively, gives

$$r\cos(\Theta + \varphi) = s, \tag{b}$$

$$r\sin(\Theta + \varphi) + \rho = R_0 + Y. \tag{c}$$

Since point C is the center of curvature, the magnitudes of r, φ, and ρ do not change for a small increment in cam rotation;[2] that is,

$$\frac{dr}{d\Theta} = \frac{d\varphi}{d\Theta} = \frac{d\rho}{d\Theta} = 0.$$

Therefore, differentiating Eq. (a) with respect to the cam rotation angle, Θ, gives

$$jre^{j(\Theta+\varphi)} = jY' + s', \tag{d}$$

where $Y' = dY/d\Theta = dy/d\theta = y'$ and $s' = ds/d\Theta = ds/d\theta$. Separating Eq. ($d$) into real and imaginary parts, the first-order kinematic coefficients are

$$-r\sin(\Theta + \varphi) = s', \tag{e}$$

$$r\cos(\Theta + \varphi) = y'. \tag{f}$$

[2] The values of r, φ, and ρ are not truly constant but are currently at stationary values; their higher derivatives are nonzero.

Equating Eqs. (*b*) and (*f*), we find the location of the trace point along the surface of the follower as

$$s = y'. \tag{6.26}$$

Differentiating this equation with respect to the cam rotation angle, Θ, we find

$$s' = y''. \tag{g}$$

Substituting Eq. (*g*) into Eq. (*e*) and then substituting the result into Eq. (*c*), the radius of curvature of the cam profile can be written as

$$\rho = R_0 + Y + y''. \tag{6.27}$$

We should carefully note the importance of Eq. (6.27); it states that the radius of curvature of the cam profile can be obtained for each cam rotation angle, Θ, directly from the displacement equations, *before laying out the cam profile*. All that is needed is the choice of the prime-circle radius, R_0, and values for the displacement, Y, and the second-order kinematic coefficient, y''.

We can use Eq. (6.27) to select a value for R_0 that will avoid undercutting. When undercutting occurs, the radius of curvature of the cam profile switches sign from positive to negative. On the verge of undercutting, the cam will come to a point, and the radius of curvature will be zero for some value of the cam rotation angle, Θ. However, we can choose R_0 large enough that this is never the case. In fact, to avoid high contact stresses, we may wish to ensure that ρ is everywhere larger than some specified value, $\rho_{\min}$. To do this, from Eq. (6.27), we require that

$$\rho = R_0 + Y + y'' > \rho_{\min}.$$

Since R_0 and Y are always positive, the critical situation occurs at or near the posture where the second-order kinematic coefficient, y'', has its largest negative value. Denoting this minimum value of y'' as $y''_{\min}$ and remembering that Y corresponds to the same posture, defined by cam angle Θ, we have the condition

$$R_0 > \rho_{\min} - Y - y''_{\min}, \tag{6.28}$$

which must be satisfied. This can easily be checked once the displacement equations have been established, and an appropriate value of R_0 can be chosen before the cam layout is attempted.

Returning now to Fig. 6.26, we see that Eq. (6.26) can also be of value. This equation states that the length of travel of the point of contact on either side of the cam rotation center corresponds precisely with the plot of the first-order kinematic coefficient. Thus, the minimum face width for a flat-face follower must extend at least $Y'_{\max}$ to the right and $-Y'_{\min}$ to the left of the cam-shaft center to maintain contact; that is,

$$\text{Face width} > Y'_{\max} - Y'_{\min}. \tag{6.29}$$

Example 6.3

Assuming that the displacement characteristics in Example 6.2 are to be achieved by a plate cam with a reciprocating flat-face follower, determine the minimum face width and the minimum prime-circle radius to ensure that the radius of curvature of the cam is everywhere greater than $\rho_{min} = 0.25$ in.

Solution

From Fig. 6.24b, the maximum value of the first-order kinematic coefficient occurs in the segment BC and is

$$y'_{max} = \frac{L_2}{\beta_2} = \frac{1.250\,0\ \text{in}}{0.785\,40\ \text{rad}} = 1.592\ \text{in/rad}. \tag{1}$$

The minimum occurs in segment DE at approximately $\theta/\beta_4 = 0.5$. From Eq. (6.17b), the minimum value of the first-order kinematic coefficient is approximately

$$y'_{min} \approx y'(0.5) = -2.812\ \text{in/rad}. \tag{2}$$

Substituting Eqs. (1) and (2) into Eq. (6.29), the minimum face width is

$$\text{Face width} > (1.592\ \text{in}) - (-2.812\ \text{in}) = 4.404\ \text{in}. \qquad \textit{Ans.}$$

Therefore, the follower would be positioned with 1.592 in to the right and 2.812 in to the left of the cam rotation axis, and some appropriate additional allowance may be added on each side.

The largest negative minimum value of the second-order kinematic coefficient occurs at D and can be obtained from Eq. (6.19c) at $\theta/\beta_3 = 1$. That is,

$$y''_{min} = -\frac{\pi^2 L_3}{4\beta_3^2} = -\frac{\pi^2(0.5669\ \text{in})}{4(0.559\,51\ \text{rad})^2} = -4.468\,18\ \text{in/rad}^2.$$

Substituting this result and the known parameters into Eq. (6.28), the minimum prime-circle radius is

$$R_0 > 0.250\ \text{in} - (-4.468\ \text{in}) - 3.000\ \text{in} = 1.718\ \text{in}. \qquad \textit{Ans.}$$

From this calculation we would choose the actual prime-circle radius as, say, $R_0 = 1.75$ in.

We see that the eccentricity of the flat-face follower stem does not affect the geometry of the cam. This eccentricity is usually chosen to avoid high bending stress in the follower. Also, there may be a higher load in the follower during the working stroke, say the lift stroke, than during the return motion. In such a case, the eccentricity may be chosen to locate the follower stem more centrally over the contact point during the lift portion of the motion cycle.

Looking again at Fig. 6.26, we can write another loop-closure equation; that is,

$$ue^{j\Theta} + ve^{j(\Theta+\pi/2)} = j(R_0 + Y) + s,$$

where we recall that u and v denote the coordinates of the contact point in a coordinate system attached to the cam. Dividing this equation by $e^{j\Theta}$ gives

$$u + jv = j(R_0 + Y)e^{-j\Theta} + se^{-j\Theta}.$$

Using Eq. (6.26), the real and imaginary parts of this equation can be written as

$$u = (R_0 + Y)\sin\Theta + y'\cos\Theta, \tag{6.30a}$$

$$v = (R_0 + Y)\cos\Theta - y'\sin\Theta. \tag{6.30b}$$

These two equations give the coordinates of the cam profile and provide an alternative to the graphic layout procedure of Fig. 6.10. They can be used to generate a table of numeric rectangular coordinate data from which the cam can be machined. Polar coordinate equations for this same curve are

$$R = \sqrt{(R_0 + Y)^2 + (y')^2} \tag{6.31a}$$

and

$$\psi = \frac{\pi}{2} - \Theta - \tan^{-1}\frac{y'}{R_0 + Y}. \tag{6.31b}$$

6.10   Plate Cam with Reciprocating Roller Follower

Figure 6.27 shows a plate cam with a reciprocating roller follower. We see that three geometric parameters remain to be chosen after the displacement diagram is completed and before the

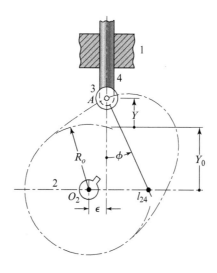

Fig. 6.27 Vectors for a plate cam with reciprocating roller follower.

cam layout is finalized. These three parameters are the radius of the prime circle, R_0, the eccentricity, ε, and the radius of the roller, R_r. There are also two potential problems to be considered when choosing these parameters. One problem is undercutting and the other is an excessive pressure angle.

Pressure angle is the angle between the axis of motion of the follower stem and the line of action of the force exerted by the cam onto the roller follower – that is, the normal to the pitch curve through the trace point. The pressure angle is labeled ϕ in Fig. 6.27. Only the component of force along the line of motion of the follower is useful in overcoming the output load; the perpendicular component and, therefore, the angle ϕ should be kept low to reduce sliding friction between the follower and its guideway and to ease bending of the follower stem. Too high a pressure angle increases the deleterious effect of friction and may cause the translating follower to chatter or perhaps even to jam. Cam pressure angles of up to about 30° or 35° are about the largest that can be used without difficulty.

In Fig. 6.27 we see that the normal to the pitch curve intersects the horizontal axis at point I_{24} – that is, at the instantaneous center of velocity between the cam 2 and the follower stem 4. Since the follower stem is translating, all points of the follower stem have velocities equal to that of the instant center I_{24}. This velocity must also be equal to the velocity of the coincident point of link 2; that is,

$$V_{I_{24}} = \dot{y} = \omega R_{I_{24}O_2}.$$

Dividing this equation by the angular velocity of the cam, ω [Eq. (6.11)], the first-order kinematic coefficient is

$$y' = \frac{\dot{y}}{\omega} = R_{I_{24}O_2}.$$

This first-order kinematic coefficient can also be expressed in terms of the eccentricity of the follower stem and the pressure angle of the cam as

$$y' = \varepsilon + (Y_0 + Y)\tan\phi, \tag{a}$$

where, as shown in Fig. 6.27, the vertical distance from the cam axis to the prime circle is

$$Y_0 = \sqrt{R_0^2 - \varepsilon^2}. \tag{b}$$

Substituting Eq. (b) into Eq. (a) and rearranging, the pressure angle of the cam can be written as

$$\phi = \tan^{-1}\left(\frac{y' - \varepsilon}{Y + \sqrt{R_0^2 - \varepsilon^2}}\right). \tag{6.32}$$

From this equation, we observe that, once the displacement equations and the first-order kinematic coefficient have been determined, the two parameters, R_0 and ε, can be adjusted to seek a suitable pressure angle. We also note that the pressure angle is continuously changing as the cam rotates, and therefore we are particularly interested in studying its extreme values.

Let us first consider the effect of eccentricity. From Eq. (6.32), we observe that increasing ε either increases or decreases the magnitude of the numerator, depending on the sign of the first-order kinematic coefficient y'. Thus, a small eccentricity, ε, can be used to reduce the pressure angle, ϕ, during the rise motion when y' is positive, but only at the expense of an increased pressure angle during the return motion when y' is negative. Still, since the magnitudes of the forces are usually greater during rise, it is common practice to offset the follower to take advantage of this reduction in pressure angle.

A much more significant effect can be made in reducing the pressure angle by increasing the prime-circle radius, R_0. To study this effect, let us take the conservative approach and assume a radial follower – that is, where there is no eccentricity. Substituting $\varepsilon = 0$ into Eq. (6.32), the equation for the pressure angle reduces to

$$\phi = \tan^{-1}\left(\frac{y'}{Y + R_0}\right). \tag{6.33}$$

To find the extremum values of the pressure angle, it is possible to differentiate this equation with respect to the cam rotation angle and equate it to zero, thus finding the values of the rotation angle, Θ, that yield the maximum and the minimum pressure angles. This is a tedious process, however, and can be avoided by using the nomogram of Fig. 6.28. This nomogram was produced by searching out on a computer the maximum value of ϕ from Eq. (6.33) for each of the standard full-rise and full-return motion curves of Section 6.7. With the nomogram, it is possible to use the known values of L and β for each segment of the displacement diagram and to read directly the maximum pressure angle occurring in that motion segment for a particular choice of R_0. Alternatively, a desired maximum pressure angle can be chosen and a corresponding minimum value of R_0 can be determined. The process is best shown by an example.

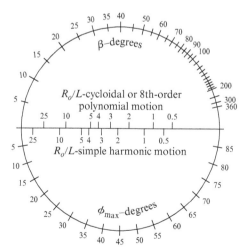

Fig. 6.28 Nomogram relating maximum pressure angle, ϕ_{max}, to prime-circle radius, R_0, lift, L, and segment angle, β, for radial reciprocating roller-follower cams with full-rise or full-return simple harmonic, cycloidal, or eighth-order polynomial motion.

Example 6.4

Assuming that the displacements determined in Example 6.2 are to be achieved by a plate cam with a reciprocating radial roller follower, determine the minimum prime-circle radius that ensures that the pressure angle is everywhere no larger than $30°$.

Solution

Each segment of the displacement diagram can be checked in succession using the nomogram of Fig. 6.28.

For segment AB of Fig. 6.24, we have half-rise cycloidal motion with $L_1 = 1.183$ in and $\beta_1 = 85.184°$. Since this is a half-rise curve, whereas the nomogram of Fig. 6.28 was developed only for full-rise curves, it is necessary to double both L_1 and β_1, thus imagining that the curve is full-rise. This gives $L_1^* = 2.366$ in and $\beta_1^* = 170°$. Next, connecting a straight line from $\beta^* = 170°$ to $\phi_{max} = 30°$, we read from the upper scale on the central axis of the nomogram a value of $R_0/L_1^* \approx 0.75$, from which

$$R_0 \geq 0.75(2.366 \text{ in}) = 1.775 \text{ in}. \tag{1}$$

The segment BC need not be checked because the maximum pressure angle for this segment occurs at the boundary B and cannot be greater than that for segment AB.

Segment CD has half-rise harmonic motion with $L_3 = 0.567$ in and $\beta_3 = 32.058°$. Again, since this is a half-rise curve, these values are doubled and $L_3^* = 1.134$ in and $\beta_3^* = 64°$ are used instead. Then, from the nomogram, we find $R_0^*/L_3^* \approx 2.15$, from which

$$R_0^* \geq 2.15(1.134 \text{ in}) = 2.438 \text{ in}.$$

However, here we must be careful. This value is the radius of a fictitious prime circle for which the horizontal axis of our fictional doubled "full-rise" harmonic curve would have $y = 0$. This is not the R_0 we seek, since our imagined full-harmonic curve has a nonzero Y^* value at its base. Referring to Fig. 6.24, we find

$$Y_3^* = Y_D - 2L_3 = 3.000 - 1.134 = 1.866 \text{ in}.$$

The appropriate value of R_0 for this segment is

$$R_0 \geq 2.438 - 1.866 = 0.572 \text{ in}. \tag{2}$$

Next we check the segment DE, which has eighth-order polynomial motion with $L_4 = 3.000$ in and $\beta_4 = 107.758°$. Because this is a full-return motion curve with $y = 0$ at its base, no adjustments are necessary for use of the nomogram. We find $R_0/L_4 \approx 1.3$ and

$$R_0 \geq 1.3(3.000 \text{ in}) = 3.900 \text{ in}. \tag{3}$$

To ensure that the pressure angle does not exceed $30°$ throughout all segments of the cam motion, we must choose the prime-circle radius to be at least as large as the maximum of these discovered values, Eqs. (1), (2), and (3). Recognizing our inability to read the nomogram with great precision, we might choose a yet larger value, such as

$$R_0 = 4.000 \text{ in}. \qquad\qquad Ans.$$

Example 6.4 (cont.)

Once a final value has been chosen, we can use Fig. 6.28 again to find the actual maximum pressure angle in each segment of the motion:

$$AB: \quad \frac{R_0}{L_1^*} = \frac{4.000}{2.366} = 1.691 \qquad \beta_1^* = 170° \qquad \phi_{max} = 18°,$$

$$CD: \quad \frac{R_0^*}{L_3^*} = \frac{5.866}{1.134} = 5.173 \qquad \beta_3^* = 64° \qquad \phi_{max} = 14°,$$

$$DE: \quad \frac{R_0}{L_4} = \frac{4.000}{3.000} = 1.333 \qquad \beta_4 = 108° \qquad \phi_{max} = 29°.$$

Although the prime circle has been sized to give a satisfactory pressure angle, the follower may still not complete the desired motion. It is still possible that the curvature of the pitch curve is too sharp and that the cam profile may be undercut. Figure 6.29a shows a portion of a cam pitch curve and two cam profiles generated by two different-size rollers. It is clear from this figure that a small roller moving on the given pitch curve generates a satisfactory cam profile. The cam profile generated by the larger roller, however, intersects itself and is said to be *undercut*. The result, after machining, is a pointed cam that does not produce the desired motion. Still, if the prime circle and thus the cam size is increased enough, even the larger roller generates a cam profile that operates satisfactorily.

In Fig. 6.29b we see that the cam profile becomes pointed when the roller radius R_r is equal to the magnitude[3] of the radius of curvature of the pitch curve. Therefore, to achieve some chosen

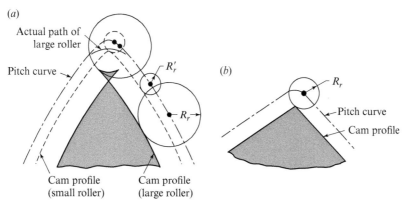

(a)
Actual path of large roller
Pitch curve
R_r'
R_r
Cam profile (small roller)
Cam profile (large roller)

(b)
R_r
Pitch curve
Cam profile

Fig. 6.29 (a) Undercut; (b) pointed cam profile.

---

[3] Remember that radius of curvature can have either a positive or a negative value (Section 4.16). However, here we are only concerned with its magnitude – that is, its absolute value.

minimum size, ρ^*, for the radius of curvature of the cam profile, the radius of curvature of the pitch curve must be of greater magnitude than this value by the radius of the roller; that is,

$$|\rho| \geq \rho^* + R_r. \tag{c}$$

Recall Section 4.16 where the radius of curvature of a point trajectory was given by Eq. (4.52).

In concept, it is possible to search out the minimum value $|\rho|_{min}$ for a particular choice of the displacement equation, Y, and a particular prime-circle radius, R_0. However, since it would be

Fig. 6.30 Minimum radius of curvature of radial reciprocating roller-follower cams with full-rise or full-return simple harmonic motion, Eqs. (6.12) or (6.15). (From [3].)

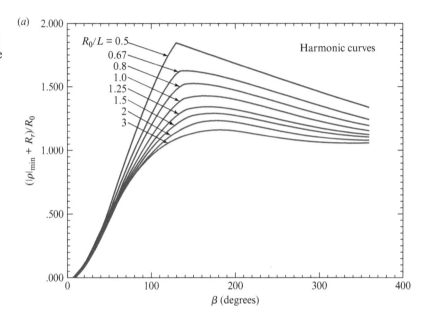

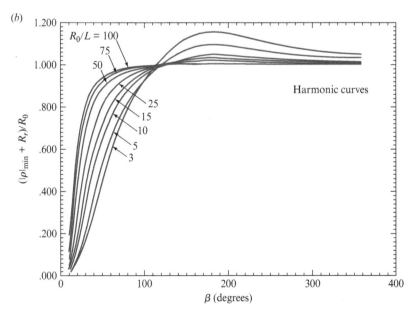

burdensome to perform such a search for each new cam design, the minimum-size radius of curvature (normalized with respect to R_0) has been sought out by computer for each of the standard cam motions of Section 6.7. The results are presented graphically in Figs. 6.30–6.34. Each of these figures shows a graph of $(|\rho|_{min} + R_r)/R_0$ versus β for one type of standard motion curve with various ratios of R_0/L. Since we have already chosen the displacement equations and have found a suitable value of R_0, each segment of the cam can now be checked to find its minimum radius of curvature.

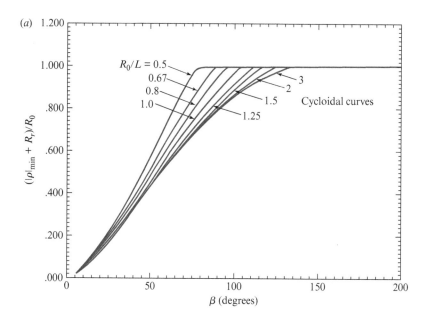

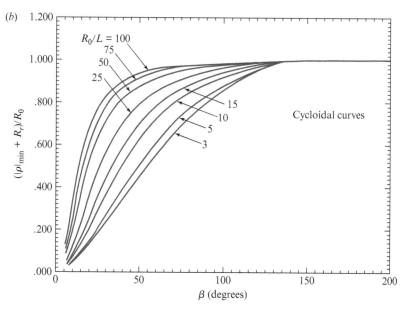

Fig. 6.31 Minimum radius of curvature of radial reciprocating roller-follower cams with full-rise or full-return cycloidal motion, Eqs. (6.13) or (6.16). (From [3].)

Fig. 6.32 Minimum radius of curvature of radial reciprocating roller-follower cams with full-rise or full-return eighth-order polynomial motion, Eqs. (6.14) or (6.17). (From [3].)

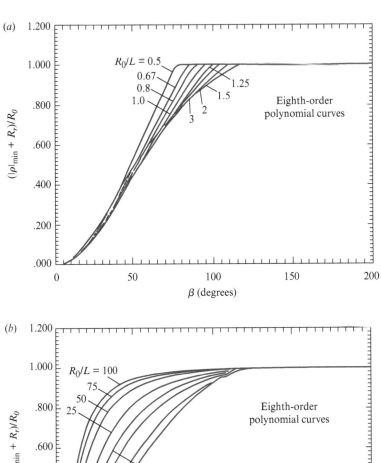

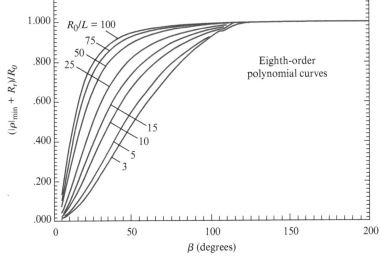

Saving even more effort, it is not necessary to check those segments of the cam where the second-order kinematic coefficient y'' remains positive throughout the segment, such as the half-rise motions of Eqs. (6.18) and (6.22) or the half-return motions of Eqs. (6.21) and (6.25).

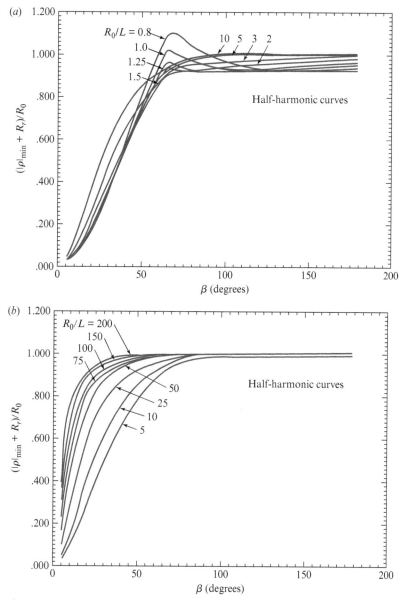

Fig. 6.33 Minimum radius of curvature of radial reciprocating roller-follower cams with half-harmonic motion, Eqs. (6.19) or (6.20). (From [3].)

Assuming that the "acceleration" curve is continuous, the minimum radius of curvature of the cam cannot occur in these segments. For each of these segments,

$$|\rho|_{\min} = R_0 - R_r. \tag{6.34}$$

Fig. 6.34 Minimum radius of curvature of radial reciprocating roller-follower cams with half-cycloidal motion, Eqs. (6.23) or (6.24). (From [3].)

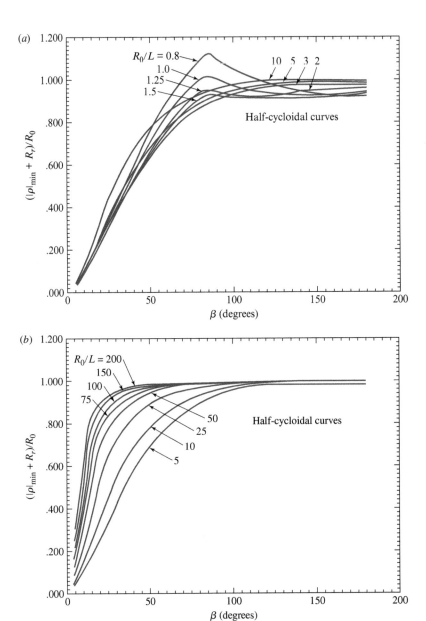

(a)

(b)

Example 6.5

Assuming that the displacement characteristics of Example 6.2 are to be achieved by a plate cam with a reciprocating radial roller follower, determine the minimum radius of curvature of the cam profile if a prime-circle radius of $R_0 = 4.000$ in (from Example 6.4) and a roller radius of $R_r = 0.500$ in are used.

Example 6.5 (cont.)

Solution

For the segment AB of Fig. 6.24, we find from Eq. (6.34) that

$$\rho|_{\min} = R_0 - R_r = 4.000 \text{ in} - 0.500 \text{ in} = 3.500 \text{ in.} \tag{1}$$

For the segment CD, we have half-harmonic rise motion with $L_3 = 0.5669$ in and $\beta_3 = 32.058°$, from which we find

$$\frac{R_0^*}{L_3} = \frac{4.000 \text{ in} + (L_1 + L_2)}{L_3} = \frac{6.433 \text{ in}}{0.567 \text{ in}} = 11.346,$$

where R_0^* was adjusted by $Y_3 = (L_1 + L_2)$, since the curves of Fig. 6.33 were plotted for the condition where $Y = 0$ at the base of each motion segment. Now, using Fig. 6.33b, we find $(\rho|_{\min} + R_r)/R_0^* \approx 0.66$. Therefore,

$$\rho|_{\min} \approx 0.66 R_0^* - R_r = 0.66(6.433 \text{ in}) - 0.500 \text{ in} = 3.746 \text{ in}, \tag{2}$$

where, again, the adjusted value of R_0^* is used.

For the segment DE, we have eighth-order polynomial motion with $L_4 = 3.0000$ in and $\beta_4 = 107.758°$, from which $R_0/L_4 = 1.33$. Using Fig. 6.32a, we find $(\rho|_{\min} + R_r)/R_0 \approx 0.80$, and

$$\rho|_{\min} \approx 0.80 R_0 - R_r = 0.80(4.000 \text{ in}) - 0.500 \text{ in} = 2.700 \text{ in.} \tag{3}$$

Choosing the smallest value among Eqs. (1), (2), and (3), the minimum size of the radius of curvature of the entire cam profile is

$$\rho|_{\min} = 2.700 \text{ in.} \qquad\qquad Ans.$$

To machine the cam to the proper shape, we need the coordinates of the cam surface in the rotating uv coordinate system attached to the cam. The rectangular coordinates of the follower center (the pitch curve) of a plate cam with a reciprocating roller follower in the rotating uv coordinate system are

$$u = \left(\sqrt{R_0^2 - \varepsilon^2} + Y \right) \sin \Theta + \varepsilon \cos \Theta, \tag{d}$$

$$v = \left(\sqrt{R_0^2 - \varepsilon^2} + Y \right) \cos \Theta - \varepsilon \sin \Theta. \tag{e}$$

Differentiating these with respect to the rotation angle of the cam, the first-order kinematic coefficients of the follower center are

$$u' = y' \sin \Theta + \left(\sqrt{R_0^2 - \varepsilon^2} + Y \right) \cos \Theta - \varepsilon \sin \Theta, \qquad (f)$$

$$v' = y' \cos \Theta - \left(\sqrt{R_0^2 - \varepsilon^2} + Y \right) \sin \Theta - \varepsilon \cos \Theta, \qquad (g)$$

and, differentiating again, the second-order kinematic coefficients of the follower center are

$$u'' = y'' \sin \Theta + 2y' \cos \Theta - \left(\sqrt{R_0^2 - \varepsilon^2} + Y \right) \sin \Theta - \varepsilon \cos \Theta, \qquad (h)$$

$$v'' = y'' \cos \Theta - 2y' \sin \Theta - \left(\sqrt{R_0^2 - \varepsilon^2} + Y \right) \cos \Theta + \varepsilon \sin \Theta. \qquad (i)$$

We note that positive (counterclockwise) rotation of the cam causes the center point of the follower to increment in a clockwise direction around the pitch curve of the cam in the rotating uv coordinate system. This defines the positive sense of the unit tangent vector $\hat{\mathbf{u}}^t$ for the pitch curve, and the unit normal vector is then given by $\hat{\mathbf{u}}^n = \hat{\mathbf{k}} \times \hat{\mathbf{u}}^t$ in the rotating coordinate system. Since this unit normal points away from the center of curvature of the pitch curve, the radius of curvature of the pitch curve has a negative value.

To normalize Eqs. (f) and (g), we define

$$w' = +\sqrt{u'^2 + v'^2} \qquad (6.35)$$

or

$$w' = +\left[(y' - \varepsilon)^2 + \left(\sqrt{R_0^2 - \varepsilon^2} + Y \right)^2 \right]^{1/2}, \qquad (j)$$

and with this we find the unit tangent,

$$\hat{\mathbf{u}}^t = \left(\frac{u'}{w'} \right) \hat{\mathbf{i}} + \left(\frac{v'}{w'} \right) \hat{\mathbf{j}}, \qquad (k)$$

and the unit normal,

$$\hat{\mathbf{u}}^n = \left(\frac{-v'}{w'} \right) \hat{\mathbf{i}} + \left(\frac{u'}{w'} \right) \hat{\mathbf{j}}, \qquad (l)$$

in the rotating uv coordinate system attached to the cam.

The coordinates of the point of contact between the cam and the roller follower in the rotating coordinate system of the cam can now be written as

$$u_{\text{cam}} = u + R_r \left(\frac{v'}{w'} \right) \quad \text{and} \quad v_{\text{cam}} = v - R_r \left(\frac{u'}{w'} \right). \qquad (6.36)$$

The radius of curvature of the pitch curve can be written from Eq. (4.52) as

$$\rho = \frac{w'^3}{u'v'' - v'u''}. \tag{6.37}$$

Remembering that the radius of curvature of the pitch curve is expected to have a negative value, as explained earlier, the radius of curvature of the cam profile is

$$\rho_{\text{cam}} = \rho + R_r. \tag{6.38}$$

Note that Eq. (6.38) still yields a negative value, although smaller than Eq. (6.37), since the unit normal vector still points away from the center of curvature of the cam.

The unit vector in the direction of motion of the follower (the Y direction), when expressed in the rotating uv coordinate system, is

$$\hat{\mathbf{u}}^Y = \sin\Theta\hat{\mathbf{i}} + \cos\Theta\hat{\mathbf{j}}. \tag{m}$$

The pressure angle can be written from Fig. 6.27 with the aid of Eqs. (l) and (m) as

$$\cos\phi = \hat{\mathbf{u}}^n \cdot \hat{\mathbf{u}}^Y = -\left(\frac{v'}{w'}\right)\sin\Theta + \left(\frac{u'}{w'}\right)\cos\Theta. \tag{6.39}$$

Example 6.6

A plate cam with a reciprocating radial roller follower is to be designed such that the displacement of the follower is

$$Y = 15(1 - \cos 2\Theta) \text{ mm}.$$

The prime-circle radius is to be $R_0 = 40$ mm, and the roller is to have a radius of $R_r = 12$ mm. The cam is to rotate counterclockwise. For the cam rotation angle $\Theta = 30°$, determine:

a. the coordinates of the point of contact between the cam and the follower in the rotating coordinate system;
b. the radius of curvature of the cam profile; and
c. the pressure angle of the cam.

Solution

At the cam rotation angle $\Theta = 30°$, the lift of the follower, found from the specified displacement equation, is

$$Y = 15(1 - \cos 2\Theta) \text{ mm} = 7.500 \text{ mm}. \tag{1}$$

By differentiating the displacement equation with respect to cam rotation angle, we find

$$y' = 30\sin 2\Theta \text{ mm/rad} = 25.981 \text{ mm/rad}, \tag{2a}$$

Example 6.6 (cont.)

$$y'' = 60 \cos 2\Theta \text{ mm/rad}^2 = 30.000 \text{ mm/rad}^2. \tag{2b}$$

From Eqs. (d) and (e), with offset $\varepsilon = 0$, the coordinates of the roller center in the rotating coordinate system attached to the cam are

$$u = (R_0 + Y) \sin \Theta = (55 - 15 \cos 2\Theta) \sin \Theta \text{ mm} = 23.750 \text{ mm}, \tag{3a}$$

$$v = (R_0 + Y) \cos \Theta = (55 - 15 \cos 2\Theta) \cos \Theta \text{ mm} = 41.136 \text{ mm}. \tag{3b}$$

Then, from Eqs. (f) and (g), the first-order kinematic coefficients of the trace point are

$$\begin{aligned} u' &= y' \sin \Theta + (R_0 + Y) \cos \Theta \\ &= (25.981 \text{ mm/rad}) \sin \Theta + (47.500 \text{ mm}) \cos \Theta = 54.127 \text{ mm/rad}, \end{aligned} \tag{4a}$$

$$\begin{aligned} v' &= y' \cos \Theta - (R_0 + Y) \sin \Theta \\ &= (25.981 \text{ mm/rad}) \cos \Theta - (47.500 \text{ mm}) \sin \Theta = -1.250 \text{ mm/rad}, \end{aligned} \tag{4b}$$

and, from Eqs. (h) and (i), the second-order kinematic coefficients of the trace point are

$$\begin{aligned} u'' &= y'' \sin \Theta + 2y' \cos \Theta - (R_0 + Y) \sin \Theta \\ &= (30.000 \text{ mm/rad}^2) \sin \Theta + (51.962 \text{ mm/rad}) \cos \Theta - (47.500 \text{ mm}) \sin \Theta \\ &= 36.250 \text{ mm/rad}^2, \end{aligned} \tag{5a}$$

$$\begin{aligned} v'' &= y'' \cos \Theta - 2y' \sin \Theta - (R_0 + Y) \cos \Theta \\ &= (30.000 \text{ mm/rad}^2) \cos \Theta - (51.962 \text{ mm/rad}) \sin \Theta - (47.500 \text{ mm}) \cos \Theta \\ &= -41.136 \text{ mm/rad}^2. \end{aligned} \tag{5b}$$

From Eq. (6.35) we have

$$\begin{aligned} w' &= +\sqrt{u'^2 + v'^2} \\ &= +\sqrt{(54.127 \text{ mm/rad})^2 + (-1.250 \text{ mm/rad})^2} = +54.141 \text{ mm/rad}. \end{aligned} \tag{6}$$

a. Substituting Eqs. (3), (4), and (6), and the given dimensions into Eqs. (6.36), the coordinates of the point of contact between the cam and the roller follower, in the rotating coordinate system, are

$$u_{\text{cam}} = 23.750 \text{ mm} + (12 \text{ mm}) \left(\frac{-1.250 \text{ mm/rad}}{54.141 \text{ mm/rad}} \right) = 23.413 \text{ mm}, \qquad \textit{Ans.}$$

$$v_{\text{cam}} = 41.136 \text{ mm} + (12 \text{ mm}) \left(\frac{-54.127 \text{ mm/rad}}{54.141 \text{ mm/rad}} \right) = 29.139 \text{ mm}. \qquad \textit{Ans.}$$

Example 6.6 (cont.)

b. From Eq. (6.37), using Eqs. (4)–(6), the radius of curvature of the pitch curve is

$$\rho = \frac{(54.141 \text{ mm/rad})^3}{(54.127 \text{ mm/rad})(-41.136 \text{ mm/rad}^2) - (-1.250 \text{ mm/rad})(36.250 \text{ mm/rad}^2)}$$
$$= -72.775 \text{ mm},$$

and we note that this value is negative, as explained earlier. With this, Eq. (6.38) gives the radius of curvature of the cam profile as

$$\rho_{\text{cam}} = \rho + R_r = -72.775 \text{ mm} + 12.0 \text{ mm} = -60.775 \text{ mm}, \qquad \textit{Ans.}$$

which we note is still negative. This confirms that the center of curvature is still further inward and that there is no undercutting at this location on the cam profile.

c. From Eq. (6.39), we find the pressure angle for this posture,

$$\cos \phi = -\left(\frac{-1.250 \text{ mm/rad}}{54.141 \text{ mm/rad}}\right) \sin 30° + \left(\frac{54.127 \text{ mm/rad}}{54.141 \text{ mm/rad}}\right) \cos 30°; \quad \phi = 28.68°. \qquad \textit{Ans.}$$

In this and the previous section, we have considered problems that result from poor choice of the prime-circle radius for a plate cam with a reciprocating follower. Although the equations are different for oscillating followers or other types of cams, a similar approach can be used to guard against undercutting [6] and severe pressure angle [4]. Similar equations can also be developed for cam profile data [7]. An extensive survey of the cam design literature has been compiled by Chen [1]. We present one more example here to show a general approach for the design of cams that require a vector analysis in addition to the previous equations.

Example 6.7

A plate cam with an oscillating roller follower, shown in Fig. 6.35*a*, is to be designed such that the cam rotates counterclockwise, and the displacement of the follower follows the equation

$$y = 0.5(1 - \cos 2\Theta) \text{ rad.}$$

The prime-circle radius is $R_0 = 1.500$ in, the roller radius is $R_r = 0.500$ in, the distance between the center of the cam-shaft and the follower pivot is $R_{O_4 O_2} = r_1 = 3.000$ in, and the length of the

Example 6.7 (cont.)

Fig. 6.35 Plate cam with   (a)
oscillating roller follower.

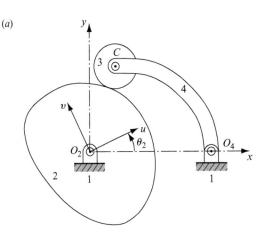

(b)

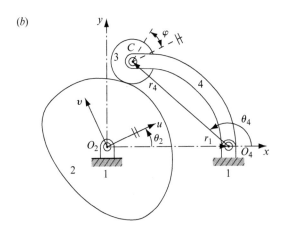

follower arm is $R_{CO_4} = r_4 = 2.598$ in. For the cam rotation angle of $\Theta = \theta_2 = 45°$, determine:

a. the coordinates of the point of contact between the cam and the follower;
b. the radius of curvature of the cam profile; and
c. the pressure angle of the cam.

Solution

Let us designate the rotation of the oscillating follower by the angle θ_4 as shown in Fig. 6.35b. With the specified dimensions, the angle of the follower with zero displacement must be $\theta_4 = 150° = 5\pi/6$ rad. Combining this with the specified displacement gives the complete equation for the rotation of the follower:

Example 6.7 (cont.)

$$\theta_4 = 5\pi/6 - y = 5\pi/6 - 0.5(1 - \cos 2\Theta) \text{ rad.} \tag{1}$$

Note that a negative sign is used since increasing follower displacement, as shown by Fig. 6.35b, is in the clockwise direction.

From Fig. 6.35b, the coordinates of the trace point, C, in the fixed coordinate system are

$$x = r_1 + r_4 \cos \theta_4, \tag{2a}$$

$$y = r_4 \sin \theta_4. \tag{2b}$$

These coordinates can be transformed to the moving coordinate system of the cam by the relations

$$u = x \cos \Theta + y \sin \Theta, \tag{3a}$$

$$v = -x \sin \Theta + y \cos \Theta. \tag{3b}$$

Therefore, substituting Eqs. (2) into Eqs. (3), the coordinates of the trace point, C, in the moving coordinate system are

$$u = (r_4 \sin \theta_4) \sin \Theta + (r_1 + r_4 \cos \theta_4) \cos \Theta,$$

$$v = (r_4 \sin \theta_4) \cos \Theta - (r_1 + r_4 \cos \theta_4) \sin \Theta,$$

which can be written as

$$u = r_1 \cos \Theta + r_4 \cos(\theta_4 - \Theta), \tag{4a}$$

$$v = -r_1 \sin \Theta + r_4 \sin(\theta_4 - \Theta). \tag{4b}$$

These are the equations of the pitch curve.

Differentiating Eqs. (4) with respect to the rotation angle, Θ, of the cam, the first- and second-order kinematic coefficients of the trace point, C, are

$$u' = -r_1 \sin \Theta - r_4 \sin(\theta_4 - \Theta)[\theta_4' - 1], \tag{5a}$$

$$v' = -r_1 \cos \Theta + r_4 \cos(\theta_4 - \Theta)[\theta_4' - 1], \tag{5b}$$

$$u'' = -r_1 \cos \Theta - r_4 \cos(\theta_4 - \Theta)[\theta_4' - 1]^2 - r_4 \theta_4'' \sin(\theta_4 - \Theta), \tag{6a}$$

$$v'' = +r_1 \sin \Theta - r_4 \sin(\theta_4 - \Theta)[\theta_4' - 1]^2 + r_4 \theta_4'' \cos(\theta_4 - \Theta), \tag{6b}$$

where the first- and second-order kinematic coefficients of the follower arm, from Eq. (1), are

$$\theta_4' = -\sin 2\Theta \quad \text{and} \quad \theta_4'' = -2 \cos 2\Theta. \tag{7}$$

Example 6.7 (cont.)

The coordinates of the point of contact between the cam and the follower in the rotating coordinate system are determined in the same way as for Eqs. (6.35) and (6.36).

At the cam rotation angle $\Theta = 45°$, from Eqs. (1) and (7) we have

$$\theta_4 = 2.118 \text{ rad} = 121.35°, \quad \theta_4' = -1.000 \text{ rad/rad}, \quad \text{and} \quad \theta_4'' = 0.000. \tag{8}$$

Substituting Eqs. (8) into Eqs. (4), the coordinates of the trace point, C, are

$$u = (3.000 \text{ in}) \cos 45° + (2.598 \text{ in}) \cos(121.35° - 45°) = 2.734 \text{ in}, \tag{9a}$$

$$v = -(3.000 \text{ in}) \sin 45° + (2.598 \text{ in}) \sin(121.35° - 45°) = 0.403 \text{ in}. \tag{9b}$$

Substituting Eqs. (8) and the given geometry into Eqs. (5), the first-order kinematic coefficients of the trace point are

$$u' = -(3.000 \text{ in}) \sin 45° - (2.598 \text{ in}) \sin(121.35° - 45°)[-2.000] = 2.928 \text{ in/rad}, \tag{10a}$$

$$v' = -(3.000 \text{ in}) \cos 45° + (2.598 \text{ in}) \cos(121.35° - 45°)[-2.000] = -3.347 \text{ in/rad}, \tag{10b}$$

and from Eq. (6.35) we have

$$w' = +\sqrt{(2.928 \text{ in/rad})^2 + (-3.347 \text{ in/rad})^2} = 4.447 \text{ in/rad}. \tag{11}$$

a. Substituting Eqs. (9)–(11) into Eqs. (6.36), the coordinates of the contact point between the cam and the follower, in the rotating coordinate system, are

$$u_{\text{cam}} = (2.734 \text{ in}) + (0.5 \text{ in}) \left(\frac{-3.347 \text{ in/rad}}{4.447 \text{ in/rad}} \right) = 2.358 \text{ in}, \qquad Ans.$$

$$v_{\text{cam}} = (0.403 \text{ in}) - (0.5 \text{ in}) \left(\frac{2.928 \text{ in/rad}}{4.447 \text{ in/rad}} \right) = 0.074 \text{ in}. \qquad Ans.$$

b. In the rotating coordinate system of the cam, the direction of motion of the roller center, C, is defined by the angle φ (Fig. 6.35*b*), which can be written as

$$\varphi = \theta_4 - 90° - \theta_2 = 121.35° - 90° - 45° = -13.65°. \tag{12}$$

Therefore, from Eq. (6.39), the pressure angle of the cam can be written as

$$\cos \phi = \left(-\frac{v'}{w'} \right) \cos \varphi + \left(\frac{u'}{w'} \right) \sin \varphi.$$

Substituting Eqs. (10), (11), and (12) into this equation, the pressure angle (at the cam rotation angle $\theta_2 = 45°$) is

$$\phi = 54.83°. \qquad Ans.$$

Example 6.7 (cont.)

c. Substituting Eqs. (8) and the given geometry into Eqs. (6), the second-order kinematic coefficients of the follower center are

$$u'' = -(3.000 \text{ in}) \cos 45° - (2.598 \text{ in}) \cos(121.35° - 45°)[-2.000]^2$$
$$= -4.573 \text{ in/rad}^2, \tag{13a}$$

$$v'' = (3.000 \text{ in}) \sin 45° - (2.598 \text{ in}) \sin(121.35° - 45°)[-2.000]^2$$
$$= -7.977 \text{ in/rad}^2. \tag{13b}$$

Then, substituting Eqs. (10), (11), and (13) into Eq. (6.37), the radius of curvature of the pitch curve at this location is

$$\rho = \frac{(4.447 \text{ in/rad})^3}{(2.928 \text{ in/rad})(-7.977 \text{ in/rad}^2) - (-3.347 \text{ in/rad})(-4.573 \text{ in/rad}^2)}$$
$$= -2.275 \text{ in.}$$

The negative sign indicates that the unit normal vector to the pitch curve, through the follower center, points away from the center of curvature, as explained earlier. Therefore, from Eq. (6.38), the radius of curvature of the cam profile is

$$\rho_{\text{cam}} = \rho + R_r = -2.275 \text{ in} + 0.500 \text{ in} = -1.775 \text{ in.} \qquad \textit{Ans.}$$

6.11 Rigid and Elastic Cam Systems

Figure 6.36*a* is a cross-sectional view showing the overhead valve arrangement in an automobile engine. In analyzing the dynamics of this or any other cam system, we expect to determine the contact force at the cam surface, the spring force, and the cam-shaft torque, all for a complete rotation of the cam. In one method of analysis, all parts of the cam-follower train, consisting of the push rod, the rocker arm, and the valve stem, together with the cam-shaft, are assumed to be rigid. If this is a valid assumption, and if the speed of the cam-follower train is moderate, then such an analysis usually produces quite satisfactory results. In any event, such a rigid-body analysis should always be attempted as the first step.

Sometimes the speeds are so high or the members are so elastic (perhaps because of extreme lengths or slenderness) that an elastic-body analysis must be used. This fact is usually discovered when troubles are encountered with the system. Such troubles are usually evidenced by noise, chatter, unusual wear, poor product performance, or perhaps fatigue failure of some of the parts. In other cases, laboratory investigation of the operation of a prototype system may reveal substantial differences between the theoretical and the observed performance.

Fig. 6.36 An overhead valve arrangement for an automotive engine.

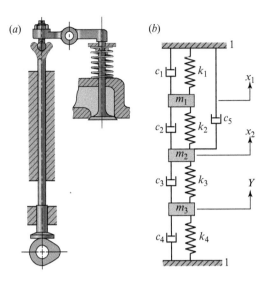

Figure 6.36b is a mathematical model of an elastic-body cam system. Here, m_3 represents the mass of the cam and a portion of the cam-shaft. The motion machined into the cam profile is the coordinate $Y(\Theta)$, a function of the cam-shaft angle Θ. Masses m_1 and m_2 and stiffnesses k_2 and k_3 are lumped characteristics of the follower train. The stiffness of the follower retaining spring is k_1 and the bending stiffness of the cam-shaft is k_4. The dashpots c_i ($i = 1, 2, 3, 4$, and 5) are inserted to represent the effects of friction, which, in the analysis, may indicate either viscous damping or sliding friction or any combination of the two. The system of Fig. 6.36b is a rather sophisticated one requiring the solution of three simultaneous differential equations. This is not presented here; instead, we focus our attention on simpler systems.

6.12 Dynamics of an Eccentric Cam

An eccentric cam is the name given to a circular plate cam with the cam-shaft mounted off-center. The distance e between the center of the disk and the center of the shaft is the eccentricity. Figure 6.37a shows a simple reciprocating-follower eccentric cam system. It consists of an eccentric plate cam, a flat-face follower mass, and a retaining spring of stiffness k. The coordinate Y designates the motion of the follower, as long as the cam remains in contact with the follower. We arbitrarily select $Y = 0$ at the bottom of the stroke. Then, for constant shaft speed, the kinematic quantities of interest are

$$Y = e - e \cos \omega t, \quad \dot{Y} = \omega e \sin \omega t, \quad \ddot{Y} = \omega^2 e \cos \omega t, \tag{6.40}$$

where ωt is the cam angle Θ.

To make a rigid-body analysis, we assume no friction and construct a free-body diagram of the follower. In Fig. 6.37b, F_{23} is the cam contact force and F_S is the spring force. In general,

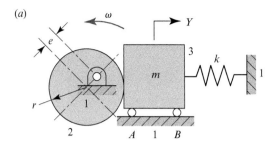

(a)

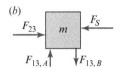

(b)

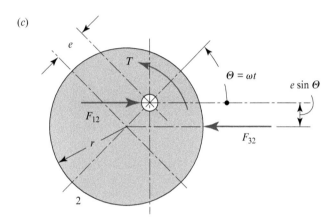

(c)

Fig. 6.37 (a) Eccentric plate cam with flat-face follower; (b) free-body diagram of the follower; (c) free-body diagram of the cam.

F_{23} and F_S do not have the same line of action, so a pair of frame forces, $F_{13,A}$ and $F_{13,B}$, is exerted at bearings A and B.

Before writing the equation of motion, let us investigate the spring force in more detail. Recall that the *spring stiffness*, k, also called the *spring rate*, is defined as the amount of force necessary to deform the spring a unit distance. Thus, the units of k are usually expressed in newtons per meter or pounds per inch. The purpose of the spring is to keep or retain the follower in contact with the cam. Thus, the spring should exert some force even at the bottom of the stroke, where it is extended the most. This force, called the *preload P*, is the force exerted by the spring when $Y = 0$. Thus, $P = k\delta$, where δ is the deformation through which the spring must be compressed to assemble it.

Summing forces on the follower mass in the Y direction gives

$$\sum F^Y = F_{23} - k(Y + \delta) = m\ddot{Y}. \qquad (a)$$

Note that the contact force F_{23} can have only a positive value. Rearranging this equation gives

$$m\ddot{Y} + kY = F_{23} - k\delta \quad \text{or} \quad m\ddot{Y} + kY = F_{23} - P. \qquad (b)$$

Fig. 6.38 (*a*) Plot of displacement, velocity, acceleration, and contact force for an eccentric cam system; (*b*) graph of torque components and total cam-shaft torque.

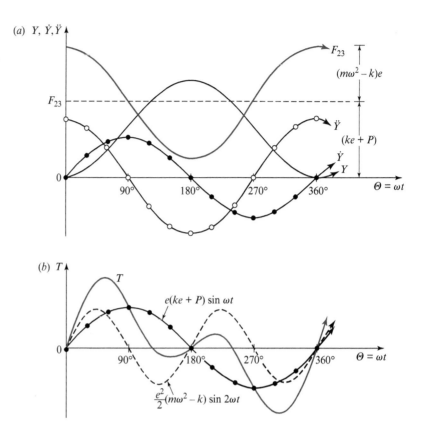

Then, substituting the first and third equations of Eqs. (6.40) into this equation, and rearranging, the contact force can be written as

$$F_{23} = \left(m\omega^2 - k\right)e\cos\omega t + (ke + P). \qquad (6.41)$$

This equation and Fig. 6.38*a* demonstrate that the contact force, F_{23}, consists of a constant term $ke + P$ with a cosine wave superimposed on it. The maximum occurs at $\Theta = 0°$ and the minimum occurs at $\Theta = 180°$. The cosine, or variable, component has an amplitude that depends upon the square of the cam-shaft speed. Thus, as the speed increases, this term increases at a greater rate. At some speed, the contact force could become zero at or near $\Theta = 180°$. When this happens, there is usually some impact between the cam and the follower, resulting in clicking, rattling, or very noisy operation. In effect the sluggishness, or inertia, of the follower prevents it from remaining in contact with the cam. The result is often called *jump* or *float*. The noise occurs when contact is reestablished. Of course, the purpose of the retaining spring is to prevent this. Since the contact force consists of a cosine wave superimposed on a constant term, all we must do to prevent jump is to move, or elevate, the cosine wave away from the zero position. To do this we can increase the constant term $ke + P$, by increasing the preload, P, or the spring rate, k, or both.

Having learned that jump begins at $\cos \omega t = -1$ with zero contact force (that is, $F_{23} = 0$), we can solve Eq. (6.41) for the jump speed. The result is

$$\omega = \sqrt{\frac{2ke + P}{me}}. \tag{6.42}$$

Using the same procedure, we find that jump will not occur for preload in the range

$$P > e(m\omega^2 - 2k). \tag{6.43}$$

Figure 6.37c is a free-body diagram of the cam. The torque, T, applied by the shaft onto the cam, is

$$T = F_{23}e \sin \omega t.$$

Substituting Eq. (6.41) into this equation gives

$$T = \left[(m\omega^2 - k)e \cos \omega t + (ke + P) \right] e \sin \omega t,$$

which, using a trigonometric identity, can be written as

$$T = e(ke + P) \sin \omega t + \tfrac{1}{2} e^2 (m\omega^2 - k) \sin 2\omega t. \tag{6.44}$$

A plot of this equation is presented in Fig. 6.38b. Note that the torque consists of a double-frequency component, whose amplitude is a function of the cam velocity squared, superimposed on a single-frequency component, whose amplitude is independent of velocity. In this example, the area enclosed by the torque-displacement diagram in the positive T direction is the same as in the negative T direction. This means that the energy required to drive the follower in the forward direction is recovered when the follower returns. A flywheel, or inertia, on the cam-shaft can be used to store and release this fluctuating energy requirement. Of course, if an external load is connected in some manner to the follower system, the energy required to drive this load will raise the torque curve in the positive direction and increase the area in the positive loop of the T curve.

Example 6.8

A cam-and-follower mechanism similar to Fig. 6.37a has the cam machined so that it moves the follower to the right through a distance of 40 mm with parabolic motion in 120° of cam rotation, dwells for 30°, then returns with parabolic motion to the starting posture in the remaining cam angle. The spring rate is 5 kN/m, and the mechanism is assembled with a 35 N preload. The follower mass is 18 kg. Assume no friction. (a) Without computing numeric values, sketch approximate graphs of the displacement, the acceleration, and the cam contact force, all versus cam angle for the full cycle of events from $\Theta = 0°$ to $\Theta = 360°$ of cam rotation. On this graph, indicate where jump or liftoff is most likely to begin. (b) At what speed would jump begin?

Solution

a. The cam contact force can be written from Eq. (*a*) as

$$F = kY + P + m\ddot{Y}, \tag{1}$$

Example 6.8 (cont.)

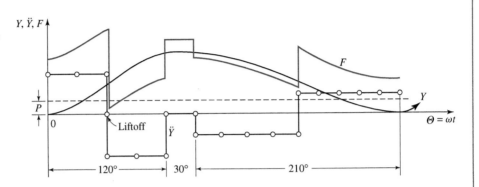

Fig. 6.39 Plots of displacement, acceleration, and contact force.

which is composed of the term $m\ddot{Y}$, which varies with the acceleration, the term kY, which varies with the displacement, and the constant term, P. Figure 6.39 shows the displacement diagram of the cam motion described, along with the acceleration, $\ddot{Y}$, and the contact force, F. Note that jump will first occur at $\Theta = \omega t = 60°$, since this is the first posture where F approaches zero.

b. Liftoff will occur at the midpoint of the rise where (for $\beta_1 = 120°$) the cam angle is $\Theta = \beta_1/2 = 60°$ when the acceleration becomes negative. The acceleration at this posture [Eq. (6.6c)] is

$$\ddot{Y} = -\frac{4L_1\omega^2}{\beta_1^2} = -\frac{4(0.040 \text{ m})\omega^2}{[120°(\pi \text{ rad}/180°)]^2} = (-0.0365 \text{ m/rad}^2)\omega^2.$$

Substituting this value, $P = 35$ N, and $kY = (5000 \text{ N/m})(0.020 \text{ m}) = 100$ N into Eq. (1) with $F = 0$ gives

$$0 = 100 \text{ N} + 35 \text{ N} + (18 \text{ kg})(-0.0365 \text{ m/rad}^2)\omega^2.$$

Then, rearranging this equation gives

$$\omega = \sqrt{\frac{100 \text{ N} + 35 \text{ N}}{(18 \text{ kg})(0.0365 \text{ m/rad}^2)}} = 14.3 \text{ rad/s} = 137 \text{ rev/min}. \qquad Ans.$$

6.13 Effect of Sliding Friction

Let F_μ be the force of sliding (Coulomb) friction as defined by Eq. (11.10). Since the friction force is opposite in sense to the velocity, let us define a sign function as follows:

(a)

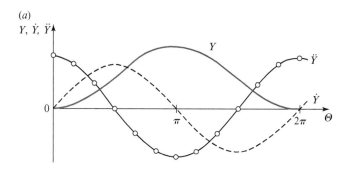

(b)

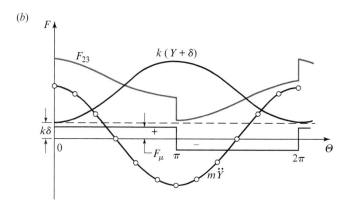

Fig. 6.40 Effect of sliding friction on a cam system with harmonic motion: (a) graph of displacement, velocity, and acceleration for one motion cycle; (b) graph of force components $F_\mu, k\delta, k(Y + \delta)$, and the resultant contact force F_{23}.

$$\operatorname{sgn} Z = \begin{cases} +1 & \text{for } Z \geq 0, \\ -1 & \text{for } Z < 0. \end{cases} \qquad (6.45)$$

With this notation, Eq. (a) of Section 6.12 can be written as

$$\sum F^Y = F_{23} - F_\mu \operatorname{sgn} \dot{Y} - k(Y + \delta) - m\ddot{Y} = 0$$

or

$$F_{23} = F_\mu \operatorname{sgn} \dot{Y} + k(Y + \delta) + m\ddot{Y}. \qquad (6.46)$$

This equation is plotted for simple harmonic motion with no dwells in Fig. 6.40. By studying both parts of this diagram, we note that F_μ is positive when $\dot{Y}$ is positive, and we see how F_{23} is obtained by graphically summing the four component curves.

6.14 Dynamics of Disk Cam with Reciprocating Roller Follower

In Chapter 11 we present a static force analysis of a cam system incorporating a reciprocating roller follower. In this section we present an analytic approach to a similar problem in which sliding friction is included. The geometry of such a system is shown in Fig. 6.41a. In the analysis to follow here, the effect of follower weight on bearings B and C is neglected.

Fig. 6.41 (a) Plate cam driving a reciprocating roller-follower system; (b) free-body diagram of the follower system.

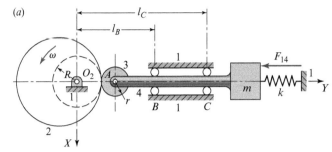

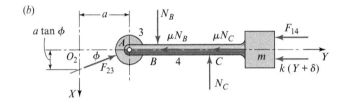

Figure 6.41b is a free-body diagram of the follower and roller. If $Y(\Theta)$ is any motion machined into the cam, and $\Theta = \omega t$ is the cam angle, at $Y = 0$ the follower is at the bottom of its stroke and so $Y_0 = R_{AO_2} = R + r$. Therefore,

$$a = Y_0 + Y = R + r + Y. \qquad (6.47)$$

In Fig. 6.41b the roller contact force forms an angle, ϕ, the pressure angle, with the Y axis. Since the direction of the force $\mathbf{F}_{23}$ is the same as the normal to the contacting surfaces, the intersection of this line with the X axis is the common instant center of the cam and follower. This means that the velocity of this point is the same whether it is considered as a point on the follower or a point on the cam. Therefore,

$$\dot{Y} = a\omega \tan \phi,$$

and so

$$\tan \phi = \frac{\dot{Y}}{a\omega} = \frac{Y'}{a}, \qquad (6.48)$$

where Y' is the first-order kinematic coefficient of the cam motion. Note that this agrees identically with Eq. (6.33).

In the following analysis, the two bearing reactions are N_B and N_C, the coefficient of sliding friction is μ, and δ is the precompression of the retaining spring. Summing forces in the X and Y directions gives

$$\sum F_{13,4}^X = -F_{23}^X + N_B - N_C = 0 \qquad (a)$$

and

$$\sum F_{13,4}^Y = F_{23}^Y - \mu \operatorname{sgn} \dot{Y}(N_B + N_C) - F_{14} - k(Y + \delta) - m\ddot{Y} = 0. \qquad (b)$$

A third equation is obtained by taking moments about pin A:

$$\sum M_A = -N_B(l_B - a) + N_C(l_C - a) = 0. \tag{c}$$

With the help of Eq. (6.48), these three equations can be solved for the unknowns F_{23}, N_B, and N_C.

First, we solve Eq. (*c*) for N_C. This gives

$$N_C = N_B \frac{l_B - a}{l_C - a}. \tag{d}$$

Next we substitute Eq. (*d*) into Eq. (*a*) and solve for the bearing reaction N_B. The result is

$$N_B = \frac{l_C - a}{l_C - l_B} F_{23}^X. \tag{e}$$

Substituting $F_{23}^X = F_{23}^Y \tan \phi$ into this equation, and using Eq. (6.48), gives

$$N_B = \frac{F_{23}^Y(l_C - a) \tan \phi}{l_C - l_B} = \frac{(l_C - a)Y'}{(l_C - l_B)a} F_{23}^Y. \tag{6.49}$$

Next, we substitute Eqs. (*d*) and (6.49) into the friction term of Eq. (*b*):

$$\mu \operatorname{sgn} \dot{Y}(N_B + N_C) = Y' \operatorname{sgn} \dot{Y} \left[\frac{l_C + l_B - 2a}{(l_C - l_B)a} \right] F_{23}^Y. \tag{f}$$

Substituting this result back into Eq. (*b*) and solving for F_{23}^Y gives

$$F_{23}^Y = \frac{F_{14} + k(Y + \delta) + m\ddot{Y}}{1 - Y' \operatorname{sgn} \dot{Y} \left[\dfrac{l_C + l_B - 2a}{(l_C - l_B)a} \right]}. \tag{6.50}$$

For computer or calculator solution, a simple computation for the sgn function is

$$\operatorname{sgn} Z = \frac{Z}{|Z|}. \tag{6.51}$$

Finally, the cam-shaft torque is

$$T = -a \tan \phi F_{23}^Y = -Y' F_{23}^Y. \tag{6.52}$$

The equations in this section require the kinematic expressions for the appropriate rise and return motions that were developed in Section 6.7.

6.15 Dynamics of Elastic Cam Systems

Figure 6.42 shows the effect of follower elasticity upon the actual measured displacement and velocity of a follower system driven by a cycloidal cam. To understand what has happened, we

Fig. 6.42 Oscilloscope traces of the measured displacement and velocity of a dwell–rise–dwell–return cam-and-follower system machined for cycloidal motion. The zero axis of the displacement diagram has been translated downward to obtain a larger diagram in the space available.

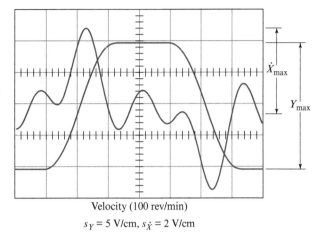

Velocity (100 rev/min)

$s_Y = 5$ V/cm, $s_{\dot{X}} = 2$ V/cm

Fig. 6.43 (a) Undamped model of a cam-and-follower system; (b) free-body diagram of the follower mass; (c) displacement diagram.

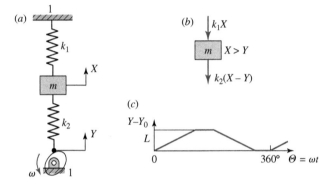

must compare these diagrams with the theoretic ones earlier in the chapter. Although the effect of elasticity is more pronounced for velocity, it is usually the variation of the displacement, especially at the top of rise, that causes the most trouble in practical situations. These difficulties are usually evidenced by poor or unreliable product quality when elastic systems are used in manufacturing or assembly lines, and they result in noise, unusual wear, and fatigue failure.

A complete analysis of elastic cam systems requires a good background in vibration analysis. To avoid the necessity for this background while still developing a basic understanding, we use an extremely simplified cam system with a linear-motion cam. It must be observed, however, that such a cam system would never be used for an actual high-speed application.

In Fig. 6.43a, k_1 is the stiffness of the retaining spring, m is the lumped mass of the follower, and k_2 represents the stiffness of the follower. Since the follower is usually a rod or a lever, k_2 is many times greater than k_1.

Spring k_1 is assembled with a preload. The coordinate X of the follower motion is chosen at the equilibrium position of the mass after spring k_1 is assembled. Thus, k_1 and k_2 exert equal and opposite preload forces on the mass of the follower. Assuming no friction, the free-body diagram of the mass is as shown in Fig. 6.43b. To determine the directions of the forces the coordinate X, representing the actual motion of the follower, has been assumed to be larger than the coordinate Y, representing the theoretic follower motion machined into the cam. However, the same result is obtained if Y is assumed to be larger than X.

Using Fig. 6.43b, we find the equation of motion to be

$$\sum F = -k_1 X - k_2 (X - Y) - m\ddot{X} = 0 \qquad (a)$$

or

$$\ddot{X} + \frac{k_1 + k_2}{m} X = \frac{k_2}{m} Y. \qquad (6.53)$$

This is the differential equation for the motion of the follower. It can be solved as shown in Chapter 13 when the function Y is specified. This equation can be solved piecewise for each cam segment; that is, the ending conditions for one segment of motion must be used as the beginning or starting conditions for the next segment.

Let us analyze the first segment of motion using uniform motion, as shown in Fig. 6.43c. First we use the notation

$$\omega_n = \sqrt{\frac{k_1 + k_2}{m}}. \qquad (6.54)$$

We should not confuse ω_n with the angular velocity of the cam ω. The quantity ω_n is called the *undamped natural frequency*. The units of ω_n and ω are both radians per second.

Equation (6.53) can now be written as

$$\ddot{X} + \omega_n^2 X = \frac{k_2 Y}{m}. \qquad (6.55)$$

The solution to this equation is

$$X = A \cos \omega_n t + B \sin \omega_n t + \frac{k_2 Y}{m\omega_n^2}, \qquad (b)$$

where, for the linear rise segment of duration β_1, the cam motion is

$$Y = \frac{L}{\beta_1} \Theta = \frac{L\omega t}{\beta_1}. \qquad (6.56)$$

Of course, Eq. (6.56) is valid only during the rise segment. We can verify Eq. (b) as the solution by substituting it and its second derivative into Eq. (6.55).

The first derivative of Eq. (b) is

$$\dot{X} = -A\omega_n \sin \omega_n t + B\omega_n \cos \omega_n t + \frac{k_2 \dot{Y}}{m\omega_n^2}. \qquad (c)$$

For $t = 0$ at the beginning of rise with $X = \dot{X} = 0$, we find from Eqs. (b) and (c) that

$$A = 0 \quad \text{and} \quad B = -\frac{k_2 \dot{Y}}{m\omega_n^3}.$$

Fig. **6.44** Displacement diagram of a uniform motion cam mechanism showing the follower response.

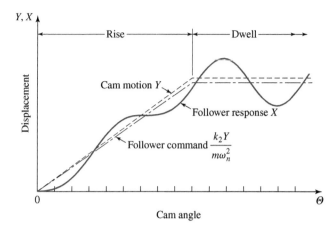

Thus, Eq. (*b*) becomes

$$X = \frac{k_2}{m\omega_n^2} \left(Y - \frac{\dot{Y}}{\omega_n} \sin \omega_n t \right). \tag{6.57}$$

This equation is plotted in Fig. 6.44. Note that the motion consists of a negative sine term superimposed upon a ramp representing the uniform rise. Because of the additional compression of spring k_2 during the rise, the ramp term $k_2 Y / m\omega_n^2$, called the follower command in Fig. 6.44, becomes less than the intended cam rise motion Y.

After the end of rise, Eqs. (6.55)–(6.57) are no longer valid, and a new segment of motion, a dwell, begins. The follower response for this era is shown in Fig. 6.44, but we will not go through the solution here.

Equation (6.57) shows that the vibration amplitude $\dot{Y}/\omega_n$ can be reduced by making ω_n large, and Eq. (6.54) demonstrates that this can be done by increasing k_2, which means that a very rigid follower should be used.

6.16 Unbalance, Spring Surge, and Windup

Unbalance.   As shown in Fig. 6.45*a*, a disk cam produces unbalance because its mass is not symmetric about the axis of rotation. This means that two sets of vibratory forces exist, one caused by the eccentric cam mass and the other caused by the reaction of the follower against the cam. By keeping these effects in mind during design, the engineer can do much to guard against difficulties during operation.

Figures 6.45*b* and 6.45*c* show that face and cylindric cams have good balance characteristics. For this reason, these are good choices when high-speed operation is involved.

Spring Surge.   Texts on spring design demonstrate that helical springs may themselves vibrate because of the mass of the spring coils. When serious vibrations exist, a clear wave motion can be seen traveling along the axis of the spring. This vibration within the retaining

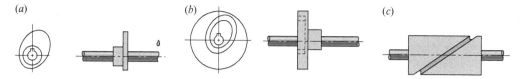

Fig. 6.45 (*a*) A disk cam is inherently unbalanced; (*b*) a face cam is usually well balanced; (*c*) a cylindric cam has good balance.

spring, called *spring surge*, has been photographed with high-speed cameras and the results have been exhibited in slow motion. For example, poorly designed automotive valve springs operating near their critical frequency permit the valve to open for short intervals during the period the valve is intended to be closed. Such conditions result in very poor operation of the engine and rapid fatigue failure of the springs themselves.

Windup.   Figure 6.38*b* is a plot of cam-shaft torque, showing that the shaft exerts torque on the cam during a portion of the cycle and that the cam exerts torque on the shaft during another portion of the cycle. This varying torque requirement may cause the shaft to twist, or wind up, as the torque increases during follower rise. Also during this period, the angular cam velocity and the follower velocity are slowed. Near the end of rise, the energy stored in the shaft by the windup is released, causing both the follower velocity and acceleration to rise above the intended values. The resulting kick may produce follower jump or impact. This effect is most pronounced when heavy loads are being moved by the follower, when the follower moves at high speed, and when the shaft is flexible.

In most cases a flywheel must be employed in a cam system to provide for varying torque requirements. Cam-shaft windup can be prevented to a large extent by mounting the flywheel as close as possible to the cam. Mounting it a long distance from the cam may actually make matters worse.

PROBLEMS

6.1 The reciprocating radial roller follower of a plate cam is to rise 50 mm with simple harmonic motion in 180° of cam rotation and return with simple harmonic motion in the remaining 180°. If the roller radius is 9.5 mm and the prime-circle radius is 50 mm, construct the displacement diagram, the pitch curve, and the cam profile for clockwise cam rotation.

6.2 A plate cam with a reciprocating flat-face follower has the same motion as in Prob. 6.1. The prime-circle radius is 50 mm, and the cam rotates counterclockwise. Construct the displacement diagram and the cam profile, offsetting the follower stem by 20 mm in the direction that reduces the bending of the follower during rise.

6.3 Construct the displacement diagram and the cam profile for a plate cam with an oscillating radial flat-face follower that rises through 30° with cycloidal motion in 150° of counterclockwise cam rotation, then dwells for 30°, returns with cycloidal motion in 120°, and dwells for 60°. Determine the necessary length for the follower face, allowing 0.20 in clearance at the free end. The prime-circle radius is 1.20 in, and the follower pivot is 4.80 in to the right.

6.4 A plate cam with an oscillating roller follower is to produce the same motion as in Prob. 6.3. The prime-circle radius is 2.40 in, the roller radius is 0.40 in, the length of the follower is 4 in, and it is pivoted 5 in to the right of the cam rotation axis. The cam rotation is clockwise. Determine the maximum pressure angle.

6.5 For full-rise simple harmonic motion, write the equations for the velocity and the jerk at the midpoint of the motion. Also, determine the acceleration at the beginning and the end of the motion.

6.6 For full-rise cycloidal motion, determine the values of θ for which the acceleration is maximum and minimum. What are the formulae for the accelerations at these positions? Find the equations for the velocity and the jerk at the midpoint of the motion.

6.7 A plate cam with a reciprocating follower is to rotate clockwise at 400 rev/min. The follower is to dwell for 60° of cam rotation, after which it is to rise to a lift of 62.5 mm. During 25 mm of the return motion, it must have a constant velocity of –1.0 m/s. Standard cam motions from Section 6.7 should be used for high-speed operation. Determine the corresponding lifts and cam rotation angles for each segment of the cam.

6.8 Repeat Prob. 6.7 except with a dwell for 20° of cam rotation.

6.9 If the cam of Prob. 6.7 is driven at constant speed, determine the time of the dwell and the maximum and minimum velocity and acceleration of the follower for the cam cycle.

6.10 A plate cam with an oscillating follower is to rise through 20° in 60° of cam rotation, dwell for 45°, then rise through an additional 20°, return, and dwell for 60° of cam rotation. Assuming high-speed operation, standard cam motions from Section 6.7 are to be used. Determine the lifts and cam rotation angles for each segment of the cam.

6.11 Determine the maximum velocity and acceleration of the follower for Prob. 6.10, assuming that the cam is driven at a constant speed of 600 rev/min.

6.12 The boundary conditions for a polynomial cam motion are as follows: for $\theta = 0$, $y = 0$, and $y' = 0$, whereas for $\theta = \beta$, $y = L$, and $y' = 0$. Determine the appropriate displacement equation and the first three derivatives of this equation with respect to cam rotation. Sketch the corresponding diagrams.

6.13 Determine the minimum face width using 2.5 mm allowances at each end and determine the minimum radius of curvature for the cam of Prob. 6.2.

6.14 Determine the maximum pressure angle and the minimum radius of curvature for the cam of Prob. 6.1.

6.15 A radial reciprocating flat-face follower is to have the motion described in Prob. 6.7. Determine the minimum prime-circle radius if the radius of curvature of the cam is not to be less than 12.5 mm. Using this prime-circle radius, what is the minimum length of the follower face using allowances of 4 mm on each side?

6.16 Graphically construct the cam profile of Prob. 6.15 for clockwise cam rotation.

6.17 A radial reciprocating roller follower is to have the motion described in Prob. 6.7. Using a prime-circle radius of 500 mm, determine the maximum pressure angle and the maximum roller radius that can be used without undercutting.

6.18 Graphically construct the cam profile of Prob. 6.17 using a roller radius of 20 mm. Cam rotation is to be clockwise.

6.19 A plate cam rotates at 300 rev/min and drives a reciprocating radial roller follower through a full rise of 3 in in 180° of cam rotation. Find the minimum radius of the prime circle if simple harmonic motion is used and the pressure angle is not to exceed 25°. Find the maximum acceleration of the follower.

6.20 Repeat Prob. 6.19 except that the motion is cycloidal.

6.21 Repeat Prob. 6.19 except that the motion is eighth-order polynomial.

6.22 Using a roller diameter of 0.8 in, determine whether the cam of Prob. 6.19 is undercut.

6.23 Equations (6.30) and (6.31) describe the profile of a plate cam with a reciprocating flat-face follower. If such a cam is to be cut on a milling machine with cutter radius R_c, determine similar equations for the center of the cutter.

6.24 Write computer programs for each of the displacement equations of Section 6.7.

6.25 Write a computer program to plot the cam profile for Prob. 6.2.

6.26 A plate cam with an offset reciprocating roller follower has a dwell of 60° and then rises in 90° to another dwell of 120°, after which it returns in 90° of cam rotation. The radius of the base circle is 1.6 in, the radius of the roller follower is 0.6 in, and the follower offset is 0.8 in. For the rise motion $60° \le \Theta \le 150°$, the equation of the displacement (the lift) is

$$y = 1.60 \left[\frac{\theta}{\pi} + \sin \theta \right],$$

where y is in inches and θ is the cam rotation angle in radians. (a) Find equations for the first- and second-order kinematic coefficients of the lift y for this rise motion. (b) Sketch the displacement diagram and the first- and second-order kinematic coefficients for the follower motion described. Comment on the suitability of this rise motion in the context of the other displacements specified. At the cam rotation angle $\Theta = 120°$, determine the following: (c) the location of the point of contact between the cam and follower, expressed in the moving Cartesian coordinate system attached to the cam; (d) the radius of curvature of the pitch curve and the radius of curvature of the cam surface; and (e) the pressure angle of the cam. Is this pressure angle acceptable?

6.27 A plate cam with an offset reciprocating roller follower is to be designed using the input, the rise and fall, and the output motion shown in Table P6.27. The radius of the base circle is 1.20 in, the radius of the roller follower is 0.50 in, and the follower offset (eccentricity) is 0.60 in.

Comment on the suitability of the motions specified. At the cam rotation angle $\Theta = 50°$, determine the following: (a) the first-, second-, and third-order kinematic coefficients of the lift curve; (b) the coordinates of the point of contact between the roller follower and the cam surface, expressed in the Cartesian coordinate system rotating with the cam; (c) the radius of curvature of the pitch curve; (d) the unit tangent and the unit normal vectors to the pitch curve; and (e) the pressure angle of the cam.

Table P6.27 Displacement information for a plate cam with a reciprocating roller follower

Cam angle Θ (deg)	Lift L (in)	Output Y
0–20	0	Dwell
20–110	1.00	Full-rise simple harmonic motion
110–120	0	Dwell
120–200	0.20	Full-rise cycloidal motion
200–270	0	Dwell
270–360	1.20	Full-return cycloidal motion

6.28 A plate cam with a radial reciprocating roller follower is to be designed using the input, the rise and fall, and the output motion shown in Table P6.28. The base circle diameter is 75 mm and the diameter of the roller is 25 mm. Displacements are as specified in Table P6.28

Table P6.28 Displacement information for a plate cam with a reciprocating roller follower

Cam angle Θ (deg)	Lift L (mm)	Output Y
0–90	75	Cycloidal rise
90–105	0	Dwell
105–195	75	Cycloidal fall
195–210	0	Dwell
210–270	50	Simple harmonic rise
270–285	0	Dwell
285–345	50	Simple harmonic fall
345–360	0	Dwell

Plot the lift curve (displacement diagram) and the profile of the cam. (a) Comment on the lift curves at appropriate positions of the cam (for example, when the cam rotation angle is $\Theta = 0°$, $\Theta = 45°$, $\Theta = 180°$, $\Theta = 210°$, $\Theta = 225°$, and $\Theta = 300°$). (b) Identify on your cam profile the location(s) and the value(s) of the largest pressure angle. Would this pressure angle cause difficulties for a practical cam-follower system? (c) Identify on your cam profile the location(s) of any discontinuities in position, velocity, acceleration, and/or jerk. Are these discontinuities acceptable (why or why not)? (d) Identify on your cam profile any regions of positive radius of curvature of the cam profile. Are these regions acceptable (why or why not)? (e) For the values given in Table P6.28, what design changes would you suggest to improve this cam design?

6.29 Continue using the same displacement information and the same design parameters as in Prob. 6.28. Use a spreadsheet to determine and plot the following for a complete rotation of the cam: (a) the first-order kinematic coefficients of the follower center; (b) the second-order kinematic coefficients of the follower center; (c) the third-order kinematic coefficients of the follower center; (d) the lift curve (displacement diagram); (e) the radius of curvature of the cam surface; and (f) the pressure angle of the cam-follower system. Is the pressure angle suitable for a practical cam-follower system?

6.30 The cam rotation angle, the rise and fall, and the output motion of a disk cam with a reciprocating roller follower are given in Table P6.30. The diameter of the base circle of the cam is 3.60 in, the diameter of the roller follower is 1.20 in, and the follower eccentricity is 0.80 in.

Table P6.30 **Displacement information for a plate cam with a reciprocating roller follower**

Cam angle Θ (deg)	Lift L (in)	Output Y
0–45	0	Dwell
45–120	1.40	Full-rise simple harmonic motion
120–130	0	Dwell
130–180	0.60	Full-rise cycloidal motion
180–210	0	Dwell
210–290	0.80	Full-return simple harmonic motion
290–310	0	Dwell
310–360	1.20	Full-return cycloidal motion

Sketch the displacement diagram and its first two derivatives. At the cam rotation angle $\Theta = 230°$, determine: (a) the first- and second-order kinematic coefficients of the displacement diagram; (b) the coordinates of the point of contact between the cam and the roller follower, expressed in the moving Cartesian coordinate system attached to the cam; (c) the radius of curvature of the cam profile; and (d) the pressure angle of the cam. Is this pressure angle acceptable for this cam-follower system?

6.31 The cam angle, the rise and fall, and the output motion of a disk cam with a reciprocating roller follower are as given in Table P6.31. The diameter of the base circle of the cam is

Table P6.31 **Displacement information for a plate cam with a reciprocating roller follower**

Cam angle Θ (deg)	Lift L (mm)	Output Y
0–5	0	Dwell
5–115	50	Full-rise cycloidal motion
115–120	0	Dwell
120–180	85	Full-rise simple harmonic motion
180–210	0	Dwell
210–310	120	Full-return eighth-order polynomial motion
310–325	0	Dwell
325–360	15	Full-return simple harmonic motion

240 mm, the diameter of the roller follower is 60 mm, and the eccentricity (offset) of the roller follower is 70 mm.

Sketch the displacement diagram and its first two derivatives. At the cam rotation angle $\Theta = 235°$, determine: (a) the first- and second-order kinematic coefficients of the displacement diagram; (b) the coordinates of the point of contact between the cam and the roller follower, expressed in the rotating Cartesian coordinate system attached to the cam; (c) the radius of the curvature of the cam profile; and (d) the pressure angle of the cam.

6.32 The cam angle, the rise and fall, and the output motion of a disk cam with a reciprocating roller follower are as given in Table P6.32. The diameter of the base circle of the cam is 7.20 in, the diameter of the roller follower is 3.20 in, and the eccentricity of the roller follower is 1.60 in.

Table P6.32 Displacement information for a disk cam and a reciprocating roller follower

Cam angle Θ (deg)	Lift L (in)	Output Y
0–40	0	Dwell
40–100	2.40	Half-rise simple harmonic motion
100–180	0.80π	Half-rise cycloidal motion
180–260	0	Dwell
260–360	$2.40 + 0.80\pi$	Full-return cycloidal motion

Part *I*: Sketch the displacement diagram and its first two derivatives. At the cam rotation angle $\Theta = 85°$, determine: (a) the first-, second-, and third-order kinematic coefficients of the displacement diagram; (b) the radius of curvature of the cam surface; (c) the unit tangent and normal vectors to the cam at the point of contact with the follower; (d) the coordinates of the point of contact between the cam and the follower expressed in the moving Cartesian coordinate system attached to the cam; and (e) the pressure angle of the cam.

Part *II*: Repeat the problem for the cam angle $\Theta = 120°$.

6.33 The cam angle, the rise and fall, and the output motion of a disk cam with a reciprocating roller follower are as given in Table P6.33. The diameter of the base circle of the cam is 70 mm, the diameter of the roller follower is 30 mm, and the follower eccentricity is 10 mm.

Sketch the displacement diagram and its first two derivatives. At the cam angle $\Theta = 150°$, determine: (a) the first-, second-, and third-order kinematic coefficients of the displacement diagram; (b) the radius of the curvature of the cam surface; (c) the unit tangent and normal vectors to the cam at the point of contact with the follower; (d) the coordinates of the point of contact between the cam and the follower expressed in the moving Cartesian coordinate system attached to the cam; and (e) the pressure angle of the cam.

Table P6.33 **Displacement information for a disk cam and a reciprocating roller follower**

Cam angle Θ (deg)	Lift L (mm)	Output Y
0–30	0	Dwell
30–90	30	Full-rise simple harmonic motion
90–120	0	Dwell
120–180	20	Full-rise cycloidal motion
180–210	0	Dwell
210–270	20	Full-return cycloidal motion
270–300	0	Dwell
300–360	30	Full-return simple harmonic motion

6.34 The cam angle, the rise and fall, and the output motion of a disk cam with a reciprocating roller follower are given in Table P6.34. The diameter of the base circle of the cam is 3 in, the diameter of the roller follower is 1 in, and the follower eccentricity is 0.8 in.

Table P6.34 **Displacement information for a disk cam and a reciprocating roller follower**

Cam angle Θ (deg)	Lift L (in)	Output Y
0–60	0	Dwell
60–180	3.6	Full-rise cycloidal motion
180–240	0	Dwell
240–360	3.6	Full-return cycloidal motion

Sketch the displacement diagram and its first two derivatives. At the cam angle $\Theta = 300°$, determine: (a) the first- and second-order kinematic coefficients of the displacement diagram; (b) the coordinates of the point of contact between the cam and the roller follower expressed in the moving Cartesian coordinate system attached to the cam; (c) the radius of curvature of the cam surface; and (d) the pressure angle of the cam.

6.35 The cam rotation angle, the rise and fall, and the output motion of a disk cam with a reciprocating roller follower are as given in Table P6.35. The diameter of the base circle of the cam is 70 mm, the diameter of the follower roller is 30 mm, and the follower eccentricity is 10 mm.

Sketch the displacement diagram and its first two derivatives. At the cam rotation angle $\Theta = 230°$, determine: (a) the first-, second-, and third-order kinematic coefficients of the displacement diagram; (b) the radius of the curvature of the cam surface; (c) the unit tangent and normal vectors to the cam at the point of contact with the follower; (d) the coordinates of the point of contact between the cam and the follower. Express your answers in the moving Cartesian coordinate system attached to the cam. Also, (e) determine the pressure angle of the cam.

Table P6.35 **Displacement information for a disk cam and a reciprocating roller follower**

Cam angle Θ (deg)	Lift L (mm)	Output Y
0–30	0	Dwell
30–90	30	Full-rise simple harmonic motion
90–120	0	Dwell
120–180	20	Full-rise cycloidal motion
180–210	0	Dwell
210–270	20	Full-return cycloidal motion
270–300	0	Dwell
300–360	30	Full-return simple harmonic motion

6.36 In Fig. P6.36a, the mass m is constrained to move only in the vertical direction. The circular cam has an eccentricity of 50 mm, a speed of 20 rad/s, and a weight of 35.5 N. Neglecting friction, find the angle $\Theta = \omega t$ at the instant the cam-follower jumps.

Fig. P6.36   (a)   (b)

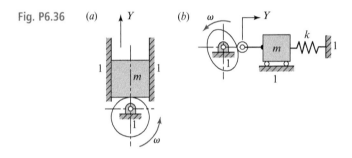

6.37 In Fig. P6.36a, the mass m is driven up and down by the eccentric cam and has a weight of 45 N. The cam eccentricity is 25 mm. Assume no friction. (a) Derive the equation for the contact force. (b) Find the cam velocity, ω, corresponding to the beginning of the cam-follower jump.

6.38 In Fig. P6.36a, the slider has a weight of 5.5 lb. The cam is a simple eccentric, and causes the slider to rise 1 in with no friction. At what cam speed in revolutions per minute will the slider first lose contact with the cam? Sketch a graph of the contact force at this speed for 360° of cam rotation.

6.39 The cam-and-follower system in Fig. P6.36b has $k = 5.5$ lb/in, $w = mg = 2$ lb, $Y = 0.6$ in$(1 - \cos \omega t)$, and $\omega = 60$ rad/s. The retaining spring is assembled with a preload of 4.25 lb. (a) Compute the maximum and minimum values of the contact force. (b) If the follower is found to jump off the cam, compute the angle $\Theta = \omega t$ corresponding to the beginning of jump.

6.40 Figure P6.36b shows the model of a cam-and-follower system. The motion machined into the cam is to move the mass to the right through a distance of 50 mm with parabolic motion in 150° of cam rotation, dwell for 30°, return to the starting position with simple harmonic motion, and dwell for the remaining 30° of cam rotation. There is no friction or damping. The spring rate is 7.10 kN/m, and the preload is 26.6 N, corresponding to the $Y = 0$ position. The weight of the mass is 160 N. (a) Sketch a displacement diagram showing the follower motion for the entire 360° of cam rotation. Without computing numeric values, superimpose graphs of the acceleration and cam contact force onto the same axes. Show where jump is most likely to begin. (b) At what speed in revolutions per minute would jump begin?

6.41 A cam-and-follower mechanism is shown in abstract form in Fig. P6.36b. The cam is cut so that it causes the mass to move to the right a distance of 1 in with harmonic motion in 150° of cam rotation, dwell for 30°, then return to the starting position in the remaining 180° of cam rotation, also with harmonic motion. The spring is assembled with a 5 lb preload and it has a rate of 25 lb/in. The follower weight is 38.5 lb. Compute the cam speed in revolutions per minute at which jump would begin.

6.42 Lever OAB is driven by a cam cut to give the roller a rise of 25 mm with parabolic motion and parabolic return with no dwells. The lever and roller are to be assumed weightless, and there is no friction. Calculate the jump speed if $l = 125$ mm and the mass B weighs 22.2 N.

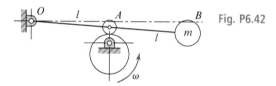

Fig. P6.42

6.43 A cam-and-follower system similar to the one of Fig. 6.41 uses a plate cam driven at a speed of 600 rev/min and employs simple harmonic rise and parabolic return motions. The events are: rise in 150°, dwell for 30°, and return in 180°. The retaining spring has a rate $k = 80$ lb/in with a precompression of 0.5 in. The follower has a weight of 3.5 lb. The external load is related to the follower motion Y by the equation $F_{14} = 13 - 430Y$, where Y is in inches and F_{14} is in pounds. Dimensions corresponding to Fig. 6.41 are $R = 0.80$ in, $r = 0.20$ in, $l_B = 2.40$ in, and $l_C = 3.60$ in. Using a rise of $L = 0.80$ in and assuming no friction, plot the displacement, cam-shaft torque, and radial component of the cam force for one complete revolution of the cam.

6.44 Repeat Prob. 6.43 with a speed of 900 rev/min, and $F_{14} = 25 + 60.5Y$ lb, where Y is in inches and the coefficient of sliding friction is $\mu = 0.025$.

6.45 A plate cam drives a reciprocating roller follower through the distance $L = 30$ mm with parabolic motion in 120° of cam rotation, dwells for 30°, and returns with cycloidal motion in 120°, followed by a dwell for the remaining cam angle. The external load on the follower is $F_{14} = 160$ N during the rise, and is zero during the dwells and the return. In the notation of Fig. 6.41, $R = 75$ mm, $r = 25$ mm, $l_B = 150$ mm, $l_C = 200$ mm, and $k = 26.6$ kN/m. The spring is assembled with a preload of 166 N when the follower is at the bottom of its stroke. The weight of the follower is 8 N, and the cam velocity is

140 rad/s. Assuming no friction, plot the displacement, the torque exerted on the cam by the shaft, and the radial component of the contact force exerted by the roller against the cam surface for one complete cycle of motion.

6.46 Repeat Prob. 6.45 if friction exists with $\mu = 0.04$ and the cycloidal return takes place in 180°.

REFERENCES

[1] Chen, F. Y. 1977. A survey of the state of the art of cam system dynamics, *Mech. Mach. Theory,* 12 (3): 201–224.

[2] Chen, F. Y. 1982. *The Mechanics and Design of Cam Mechanisms*, Oxford: Pergamon Press.

[3] Ganter, M. A., and J. J. Uicker, Jr. 1979. Design charts for disk cams with reciprocating roller followers, *J. Mech. Des. ASME Trans. B*, 101(3): 465–470.

[4] Kloomak, M., and R. V. Mufley. 1955. Plate cam design: Pressure angle analysis, *Prod. Eng.*, 26: 155–171.

[5] Kloomak, M., and R. V. Mufley. 1955. Plate cam design: With emphasis on dynamic effects, *Prod. Eng.*, 26: 178–182.

[6] Kloomak, M., and R. V. Mufley. 1955. Plate cam design: Radius of curvature, *Prod. Eng.*, 26: 186–201.

[7] Molian, S. 1968. *The Design of Cam Mechanisms and Linkages*, London: Constable.

[8] Stoddart, D. A. 1953. Polydyne cam design. *Mach. Design*, 25(1): 121–135; (2): 146–154; (3): 149–164.

[9] Tesar, D., and G. K. Matthew. 1976. *The Dynamic Synthesis, Analysis, and Design of Modeled Cam Systems*, Lexington, MA: Heath.

7   Spur Gears

Gears are machine elements used to transmit rotary motion between two shafts, usually with a constant speed ratio. In this chapter, we will discuss the case where the axes of the two shafts are parallel, and the teeth are straight and parallel to the axes of rotation of the shafts; such gears are called *spur gears*.

7.1   Terminology and Definitions

A pair of spur gears in mesh is shown in Fig. 7.1. The *pinion* is a name given to the smaller of the two mating gears; the larger is often called the *gear* or the *wheel*. The pair of gears, chosen to work together, is often called a *gearset*.

The terminology of gear teeth is shown in Fig. 7.2, where most of the following definitions are shown.

The *pitch circle* is a theoretical circle on which all calculations are based. The pitch circles of a pair of mating gears are tangent to each other, and it is these pitch circles that were pictured in earlier chapters as rolling against each other without slip.

The *diametral pitch P* is the ratio of the number of teeth on the gear to its pitch diameter; that is,

$$P = \frac{N}{2R},\tag{7.1}$$

where N is the number of teeth, and R is the pitch circle radius. Note that the diametral pitch cannot be directly measured on the gear itself. Also, note that, as the value of the diametral pitch becomes larger, the teeth become smaller; this is shown clearly in Fig. 7.3. The diametral pitch is used to indicate the tooth size in US customary units and usually has units of teeth per inch. A pair of mating gears have the same diametral pitch.

The *module m* is the ratio of the pitch diameter of the gear to its number of teeth; that is,

$$m = \frac{2R}{N}.\tag{7.2}$$

Fig. 7.1 Pair of spur gears in mesh. (Courtesy of Bill Oxford/iStock via Getty Images.)

Fig. 7.2 Gear tooth terminology.

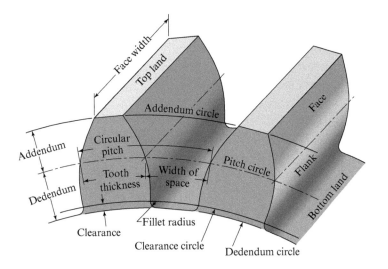

The module is the usual unit for indicating tooth size in International System (SI) units, and it customarily has units of millimeters per tooth. Note that the module is the reciprocal of the diametral pitch, and the relationship can be written as

$$m = \frac{25.4 \ (\text{mm/in})}{P \ (\text{teeth/in})} = \frac{25.4}{P} \ \text{mm/tooth}.$$

Also note that metric gears should not be interchanged with US gears because their standards for tooth sizes are not the same.

The *circular pitch*, p, is the distance from one tooth to the adjacent tooth, measured along the pitch circle. Therefore, it can be determined from

$$p = \frac{2\pi R}{N}. \tag{7.3}$$

Circular pitch is related to the previous definitions, depending on the units, by

$$p = \frac{\pi}{P} = \pi m. \tag{7.4}$$

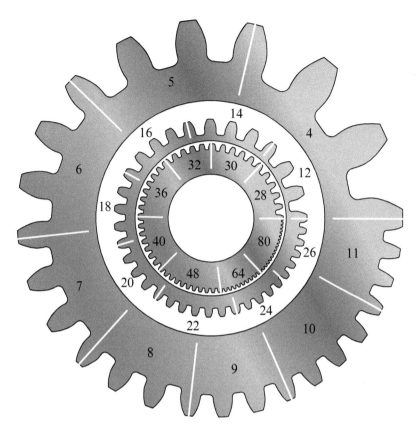

Fig. 7.3 Tooth sizes in teeth per inch for various diametral pitches. (Courtesy of Gleason Cutting Tools Corp., Loves Park, IL.)

The *addendum*, a, is the radial distance from the pitch circle to the top land of each tooth. The *dedendum*, d, is the radial distance from the pitch circle to the bottom land of each tooth. The *whole depth* is the sum of the addendum and dedendum.

The *clearance*, c, is the amount by which the dedendum of a gear exceeds the addendum of the mating gear.

The *backlash* is the amount by which the width of a tooth space exceeds the thickness of the engaging tooth measured along the pitch circles.

7.2 Fundamental Law of Toothed Gearing

Gear teeth mating with each other to produce rotary motion are similar to a cam and follower. When the tooth profiles (or cam and follower profiles) are shaped to produce a constant angular velocity ratio between the two shafts, then the two mating surfaces are said to be *conjugate*. It is possible to specify an arbitrary profile for one tooth and then to find a profile for the mating tooth so that the two surfaces are conjugate. One possible choice for such conjugate profiles is the *involute* profile, which, with few exceptions, is in universal use for gear teeth.

Fig. 7.4

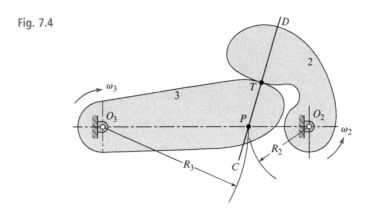

A single pair of mating gear teeth must be shaped such that, as they pass through their entire period of contact, the ratio of the angular velocity of the driven gear to that of the driving gear – that is, the first-order kinematic coefficient – *must remain constant*. This is the fundamental criterion that governs the choice of the tooth profiles. If this were not true in gearing, very serious vibration and impact problems would result, even at low speeds.

In Section 3.17 we learned the angular velocity ratio theorem, which states that the first-order kinematic coefficient of any mechanism is inversely proportional to the segments into which the common instant center cuts the line of centers. In Fig. 7.4, two profiles are in contact at point T; let profile 2 represent the driver and profile 3 represent the driven element. The normal to the surfaces, CD, is called the *line of action*. The normal to the profiles at the point of contact, T, intersects the line of centers, O_2O_3, at the instant center of velocity. In gearing, this instant center is referred to as the *pitch point* and usually carries the label P.

Designating the pitch circle radii of the two gear profiles as R_2 and R_3, from the angular velocity ratio theorem, Eq. (3.28), we see that

$$\frac{\omega_2}{\omega_3} = \frac{R_3}{R_2} \tag{7.5}$$

This equation is frequently used to define the fundamental law of gearing, which states that, as gears go through their mesh, *the pitch point must remain stationary on the line of centers* so that the speed ratio remains constant. This means that the line of action for every new instantaneous point of contact must always pass through the stationary pitch point, P. Thus, the problem of finding a conjugate profile for a given shape is to find a mating shape that satisfies the fundamental law of gearing.

It should not be assumed that any shape or profile is satisfactory just because a conjugate profile can be found. Although, theoretically, conjugate curves might be found, the practical problems of reproducing these curves from steel gear blanks or other materials while using existing machinery still exist. In addition, the sensitivity of the law of gearing to small dimensional changes of the shaft center distance caused either by misalignment or by large forces must also be considered. Finally, the tooth profile selected must be one that can be reproduced quickly and economically in very large quantities. A major portion of this chapter is devoted to explaining how the involute curve profile fulfills these requirements.

7.3   Involute Properties

An *involute* curve is the path generated by a tracing point of a taut cord as the cord is unwrapped from a cylinder called the *base cylinder*. This is shown in Fig. 7.5, where T is the tracing point. Note that the cord, AT, is normal to the involute at T, and the distance AT is the instantaneous value of the radius of curvature. As the involute is generated from its origin, T_0, to T_1, the radius of curvature varies continuously; it is zero at T_0 and increases continuously to T_1. Thus, the cord is the generating line, and it is always normal to the involute.

If the two mating tooth profiles both have the shapes of involute curves, the condition that the pitch point, P, remain stationary is satisfied. This is shown in Fig. 7.6, where two gear blanks with fixed centers O_2 and O_3 are shown having base cylinders with respective radii of

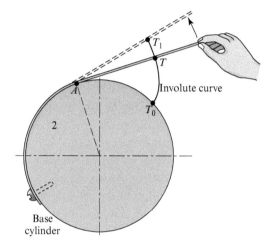

Fig. 7.5 Involute curve.

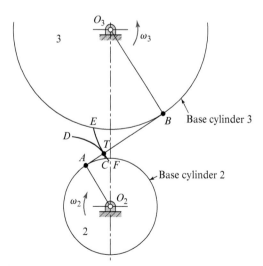

Fig. 7.6 Conjugate involute curves.

Fig. 7.7 Involute action.

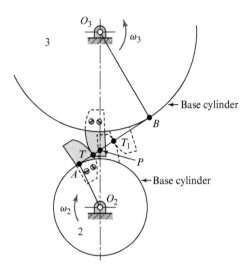

O_2A and O_3B. We now imagine that a cord is wound clockwise abound the base cylinder of gear 2, pulled taut between points A and B, and wound counterclockwise around the base cylinder of gear 3. If the two base cylinders are now rotated in opposite directions to keep the cord tight, a tracing point, T, traces out the involutes CD on gear 2 and EF on gear 3. The two involutes thus generated simultaneously by the single tracing point, T, are conjugate profiles.

Next imagine that the involutes of Fig. 7.6 are scribed on plates and that the plates are cut along the scribed curves and then bolted to the respective cylinders in the same postures. The result is shown in Fig. 7.7. The cord can now be removed, and if gear 2 is moved clockwise, gear 3 is caused to move counterclockwise by the cam-like action of the two curved plates. The path of contact is the line AB, formerly occupied by the cord. Since line AB is the generating line for each involute, it is normal to both profiles at all points of contact. Also, it always occupies the same position since it is always tangent to both base cylinders. Therefore, point P is the pitch point. Point P does not move; therefore, the involute curves are conjugate curves and satisfy the fundamental law of gearing.

7.4   Interchangeable Gears; Gear Standards

A *tooth system* is the name given to a *standard*[1] that specifies the relationships among addendum, dedendum, clearance, tooth thickness, and fillet radius to attain interchangeability of gears of

[1] Standards are defined by the American Gear Manufacturers Association (AGMA) and the American National Standards Institute (ANSI). The AGMA standards may be quoted or extracted in their entirety, provided that an appropriate credit line is included – for example, "Extracted from AGMA Information Sheet – Strength of Spur, Helical, Herringbone, and Bevel Gear Teeth (AGMA 225.01) with permission of the publisher, the American Gear Manufacturers Association, 1500 King Street, Suite 201, Alexandria, VA 22314." The AGMA standards are given in US customary units. Standards in SI units are published by the International Standards Organization (ISO); ISO Central

Table 7.1 Standard gear tooth sizes

Standard diametral pitches, P US customary units, teeth/in	
Coarse pitch	1, 1¼, 1½, 1¾, 2, 2½, 3, 4, 5, 6, 8, 10, 12, 14, 16, 18
Fine pitch	20, 24, 32, 40, 48, 64, 72, 80, 96, 120, 150, 200
Standard modules, m SI units, mm/tooth	
Preferred	1, 1.25, 1.5, 2, 2.5, 3, 4, 5, 6, 8, 10, 12, 16, 20, 25, 32, 40, 50
Next choice	1.125, 1.375, 1.75, 2.25, 2.75, 3.5, 4.5, 5.5, 7, 9, 11, 14, 18, 22, 28, 36, 45

Table 7.2 Standard tooth systems for spur gears

System	Pressure angle, ϕ (deg)	Addendum, a	Dedendum, d
Full depth	20	$1/P$ or $1m$	$1.25/P$ or $1.25m$
Full depth	22½	$1/P$ or $1m$	$1.25/P$ or $1.25m$
Full depth	25	$1/P$ or $1m$	$1.25/P$ or $1.25m$
Stub teeth	20	$0.8/P$ or $0.8m$	$1/P$ or $1m$

different tooth numbers but of the same pressure angle and the same diametral pitch or module. We should be aware of the advantages and disadvantages of such a tooth system so that we can choose the best gears for a given design and have a basis for comparison if we depart from a standard tooth profile.

For a pair of spur gears to properly mesh, they must share the same pressure angle and the same tooth size, as specified by the choice of the diametral pitch or module. The numbers of teeth and the pitch diameters of the two gears in mesh need not match, but are chosen to give the desired speed ratio as demonstrated in Eq. (7.5).

The sizes of the teeth used are chosen by selecting the diametral pitch, P, or module, m. Standard cutters are generally available for the sizes listed in Table 7.1. Once the diametral pitch or module is chosen, the remaining dimensions of the tooth are set by the standards in Table 7.2. Tables 7.1 and 7.2 contain the standards for the spur gears most in use today, and they include values for both SI and US customary units.

Let us clarify the design choices by an example.

Example 7.1

Two parallel shafts, separated by a distance (commonly referred to as the *center distance*) of 3.5 in, are to be connected by a gear pair so that the output shaft rotates at 40% of the speed of the input shaft. Design a gearset to fit this situation.

Secretariat Chemin de Blandonnet 8 CP 401 – 1214 Vernier, Geneva, Switzerland. Both the AGMA and ISO standards have been used extensively in Chapters 7 and 8.

Example 7.1 (cont.)

Solution

The center distance can be written as $R_2 + R_3 = 3.5$ in, and substituting the given information into Eq. (7.5) we have $\omega_3/\omega_2 = R_2/R_3 = 0.40$. Then, substituting the second equation into the first equation and rearranging, we find that $R_2 = 1.0$ in and $R_3 = 2.5$ in. Next, we must choose the size of the teeth by picking a value for the diametral pitch or module. From Eq. (7.1), we find the numbers of teeth on the two gears to be $N_2 = 2PR_2 = 2P$ and $N_3 = 2PR_3 = 5P$. The choice of P or m for tooth size is often iterative. First, we might choose a value of $P = 6$ teeth/in; this gives the numbers of teeth as $N_2 = 12$ teeth and $N_3 = 30$ teeth; if we choose $P = 10$ teeth/in, then we get $N_2 = 20$ teeth and $N_3 = 50$ teeth. At this time, either choice appears acceptable, and we choose $P = 10$ teeth/in. However, this choice of P (or m) must later be checked for possible undercutting, as we will study in Section 7.7; for contact ratio, which we will study in Section 7.8; and for strength and wear of the teeth [4].

7.5   Fundamentals of Gear Tooth Action

To demonstrate fundamentals, we now proceed, step by step, through the actual graphic layout of a pair of spur gears. The dimensions used are those of Example 7.1, assuming standard 20° full-depth involute tooth form as specified in Table 7.2. The various steps, in the correct order, are shown in Figs. 7.8 and 7.9 and are as follows.

Fig. 7.8 Start of layout of a pair of spur gears.

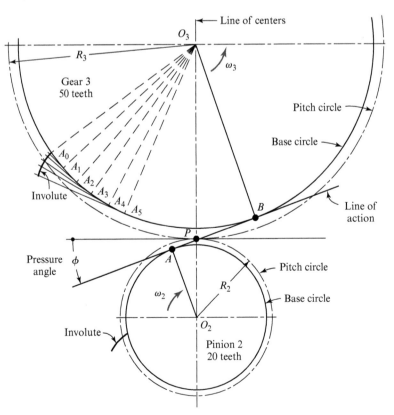

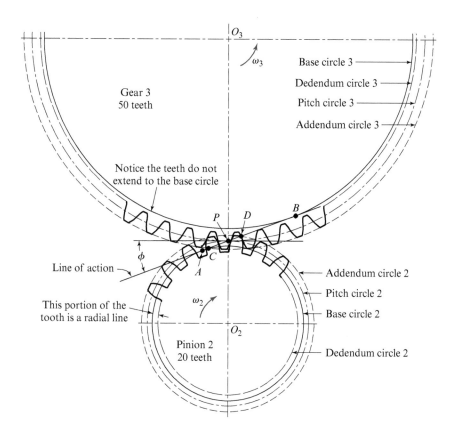

Fig. 7.9 Continued layout of the pair of spur gears.

Step 1. Calculate the two pitch circle radii, $R_2 = 1.0$ in, and $R_3 = 2.5$ in, as in Example 7.1; identify O_2 and O_3 as the two shaft centers (Fig. 7.8); and draw the two pitch circles tangent to each other.

Step 2. Draw the common tangent to the pitch circles perpendicular to the line of centers and through the pitch point P (Fig. 7.8). Draw the *line of action* at an angle equal to the *pressure angle* $\phi = 20°$ from the common tangent. This line of action corresponds to the generating line discussed in Section 7.3; it is always normal to the involute curves and always passes through the pitch point. It is called the line of action, since the point of contact between the gear teeth always lies on this line, or the pressure line, since, assuming no friction, the resultant tooth force acts along this line.

Step 3. Through the centers of the two gears, draw the two perpendiculars, O_2A and O_3B, to the line of action (Fig. 7.8). Draw the two *base circles* with radii of $r_2 = O_2A$ and $r_3 = O_3B$; these correspond to the base cylinders of Section 7.3.

Step 4. From Table 7.2, with $P = 10$ teeth/in, the addendum for both of the gears is found to be

$$a = \frac{1}{P} = \frac{1}{10 \text{ teeth/in}} = 0.10 \text{ in.}$$

Adding this to each of the pitch circle radii, draw the two addendum circles that define the top lands of the teeth on each gear. Carefully identify and label point C where the addendum circle of gear 3 intersects the line of action (Fig. 7.9). Similarly, identify and label point D where the addendum circle of gear 2 intersects the line of action.

Visualizing the rotation of the two gears in the directions shown, we see that contact is not possible before point C, since the teeth of gear 3 are not of sufficient height; thus, C is the first point of contact between the teeth. Similarly, the teeth of gear 2 are too short to allow further contact after reaching point D; thus, contact between one or more pairs of mating teeth begins at C, continues to D, and then ceases. When contact of a pair of teeth ceases at D, one or more pairs of trailing teeth must still be in contact in the span CD and take on the sharing of the transmitted load.

Steps 1–4 are critical for verifying the choice of any gear pair. We will continue with the diagram shown in Fig. 7.9 when we check for interference, undercutting, and contact ratio in later sections. However, to complete our visualization of gear tooth action, let us first proceed to the construction of the complete involute tooth shapes as shown in Fig. 7.8.

Step 5. From Table 7.2, the dedendum for each gear is found to be

$$d = \frac{1.25}{P} = \frac{1.25}{10 \text{ teeth/in}} = 0.125 \text{ in.}$$

Subtracting this from each of the pitch circle radii, draw the two dedendum circles that define the bottom lands of the teeth on each gear (Fig. 7.9). Note that the dedendum circles often lie quite close to the base circles; however, they have distinctly different meanings. In this example, the dedendum circle of gear 3 is larger than its base circle, and the dedendum circle of pinion 2 is smaller than its base circle. However, this is not always the case.

Step 6. Generate an involute curve on each base circle as shown for gear 3 in Fig. 7.8. This is done by first dividing a portion of the base circle into a series of equal small increments, A_0, A_1, A_2, and so on. Next, the radial lines $O_3 A_0, O_3 A_1, O_3 A_2$, and so on, are constructed, and tangents to the base circle are drawn perpendicular to each of these. The involute begins at A_0. The second point is obtained by striking an arc, with center A_1 and radius $A_0 A_1$, up to the tangent line through A_1. The next point is found by striking a similar arc with center at A_2, and so on. This construction is continued until the involute curve is generated far enough to meet the addendum circle of gear 3. If the dedendum circle lies inside of the base circle, as is true for pinion 2 of this example, then, except for the fillet, the involute curve is extended inward to the dedendum circle by a radial line; this portion of the curve is not involute.

Step 7. Using cardboard or preferably a sheet of clear plastic, cut a template for the involute curve and mark on it the center point of the corresponding gear. Note that two templates are needed, since the involute curves are different for gears 2 and 3.

Step 8. Calculate the circular pitch using Eq. (7.4):

$$p = \frac{\pi}{P} = \frac{\pi}{10 \text{ teeth/in}} = 0.314\,16 \text{ in/tooth.}$$

This distance from one tooth to the next is now marked along the pitch circle, and the template is used to draw the involute portion of each tooth (Fig. 7.9). The width of a tooth and that of a tooth space are each equal to half of the circular pitch or $(0.314\,16 \text{ in/tooth})/2 = 0.157\,08 \text{ in/tooth.}$

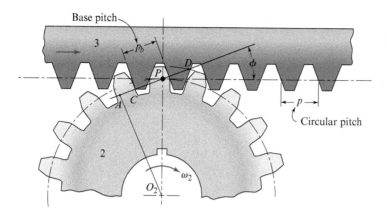

Fig. 7.10 Involute pinion and rack.

These distances are marked along the pitch circle, and the same template is turned over and used to draw the opposite sides of the teeth. The portion of the tooth space between the clearance and the dedendum may be used for a fillet radius. The top and bottom lands are now drawn as circular arcs along the addendum and dedendum circles to complete the tooth profiles. The same process is performed on the other gear using the other template.

Remember that steps 5–8 are not necessary for the proper design of a gearset. They are only included here to help us to visualize the relation between real tooth shapes and the theoretical properties of the involute curve.

Involute Rack. We may imagine a *rack* as a spur gear having an infinitely large pitch diameter. Therefore, in theory, a rack is infinitely long and has an infinite number of teeth, and its base circle is located with its center an infinite distance from the pitch point. For involute teeth, the curves on the sides of the teeth of a rack become straight lines making an angle with the line of centers equal to the pressure angle. The addendum and dedendum distances are the same as those given in Table 7.2. Figure 7.10 shows an involute rack in mesh with the pinion of the previous example.

Base Pitch. Corresponding sides of involute teeth are parallel curves. The *base pitch* is the constant and fundamental distance between these curves – that is, the distance from one tooth to the next – measured along the common normal to the tooth profiles, which is the line of action (Fig. 7.10). The base pitch, p_b, and the circular pitch, p, are related as follows:

$$p_b = p \cos \phi. \tag{7.6}$$

The base pitch is a much more fundamental measurement, as we will see later.

Internal Gear. Figure 7.11 depicts the pinion of the preceding example in mesh with an *internal*, or *annular*, gear. With internal contact, both centers are on the same side of the pitch point. Thus, the locations of the addendum and dedendum circles of an internal gear are reversed with respect to the pitch circle; the addendum circle of the internal gear lies *inside* the pitch circle, whereas the dedendum circle lies *outside* the pitch circle. The base circle lies inside the pitch circle as with an external gear, but is now near the addendum circle. Otherwise, Fig. 7.11 is constructed in the same manner as was Fig. 7.9.

Fig. 7.11 Involute pinion
and internal gear.

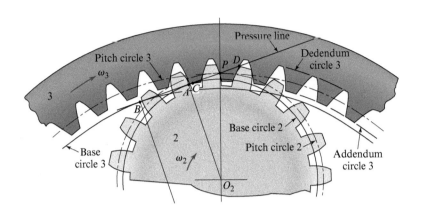

7.6   Manufacture of Gear Teeth

There are many ways of manufacturing the teeth of gears; for example, they can be made by
sand casting, shell molding, investment casting, permanent-mold casting, die casting, or
centrifugal casting. They can be formed by the powder-metallurgy process, or a single bar of
aluminum can be formed by extrusion and then sliced into gears. Gears that carry large loads in
comparison with their sizes are usually made of steel and are cut with either *form cutters* or
generating cutters. In form cutting, the cutter is of the exact shape of the tooth space. With
generating cutters, a tool having a shape different from the tooth space is moved through
several cuts relative to the gear blank to obtain the proper shape for the teeth.

Probably the oldest method of cutting gear teeth is *milling*. A form-milling cutter corresponding to the shape of the tooth space, such as that shown in Fig. 7.12a, is used to machine
one tooth space at a time, as shown in Fig. 7.12b, after which the gear is indexed through one
circular pitch to the next posture. Theoretically, with this method, a different cutter is required
for each gear to be cut, since, for example, the shape of the tooth space in a 25-tooth gear is
different from the shape of the tooth space in, say, a 24-tooth gear. Actually, the change in
tooth space shape is not very large, and eight form cutters can be used to cut any gear in the
range from 12 teeth to a rack with reasonable accuracy. Of course, a separate set of form
cutters is required for each pitch.

Shaping is a highly favored method of *generating* gear teeth. The cutting tool may be either a
rack cutter or a pinion cutter. The operation is explained by reference to Fig. 7.13. For shaping,
the reciprocating cutter is first fed into the gear blank until the pitch circles are tangent. Then,
after each cutting stroke, the gear blank and the cutter roll slightly on their pitch circles. When
the blank and cutter have rolled by a distance equal to the circular pitch, one tooth has been
generated and the cutting continues with the next tooth until all teeth have been cut. Shaping of
an internal gear with a pinion cutter is shown in Fig. 7.14.

Hobbing is another method of generating gear teeth, which is quite similar to shaping them
with a rack cutter. However, hobbing is done with a special tool called a *hob*, a cylindric cutter
with one or more helical threads quite like a screw-thread tap; the threads have straight sides
like a rack. A number of different gear hobs are displayed in Fig. 7.15. A view of the hobbing of

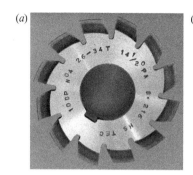

Fig. 7.12 Manufacture of gear teeth by a form cutter: (*a*) a single-tooth involute hob; (*b*) machining of a single tooth space. (Courtesy of Glenn McKechnie / CC-BY-SA-2.0.)

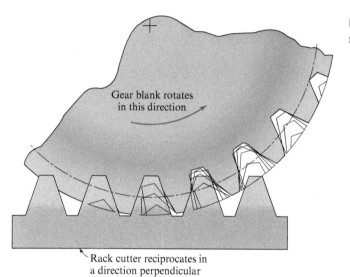

Gear blank rotates in this direction

Rack cutter reciprocates in a direction perpendicular to this page

Fig. 7.13 Shaping of involute teeth with a rack cutter.

Fig. 7.14 Shaping of an internal gear with a pinion cutter. (Courtesy of Gleason Works, Rochester, NY.)

Fig. 7.15 A variety of involute gear hobs. (Courtesy of Gleason Works, Rochester, NY.)

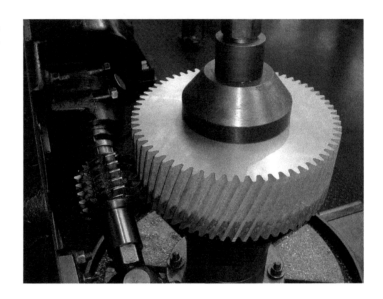

Fig. 7.16 The hobbing of a gear. (Courtesy of Kolossos / CC-BY-SA-3.0.)

a gear is shown in Fig. 7.16. The hob and the gear blank are both rotated continuously at the proper angular velocity ratio, and the hob is fed slowly across the face of the blank to cut the full thickness of the teeth.

Following the cutting process, grinding, lapping, shaving, and burnishing are often used as final finishing processes when tooth profiles of very high accuracy and surface finish are desired.

7.7   Interference and Undercutting

Figure 7.17 shows the pitch circles of the same gears used for discussion in Section 7.5. Let us assume that pinion 2 is the driver and that it is rotating clockwise.

We saw in Section 7.5 that, for involute teeth, contact always takes place along the line of action, AB. Contact begins at point C, where the addendum circle of the driven gear crosses the line of action. Thus, initial contact is on the tip of the driven gear tooth and on the flank of the pinion tooth.

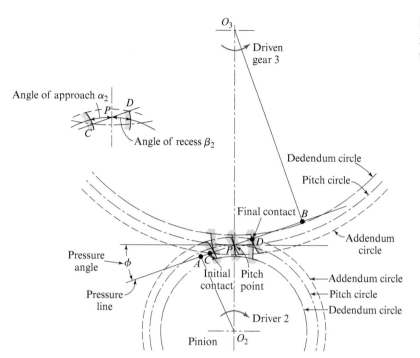

Fig. 7.17 Approach and recess phases of gear tooth action.

As the pinion tooth drives the gear tooth, contact approaches the pitch point, *P*. Near the pitch point, contact slides *up* the flank of the pinion tooth and *down* the face of the gear tooth. At the pitch point, contact is at the pitch circles; note that *P* is the instant center of velocity, and therefore the motion must be rolling with no slip at that point. Note also that this is the only location where the motion is true rolling.

As the teeth recede from the pitch point, the point of contact continues to travel in the same direction as before along the line of action. Contact continues to slide *up* the face of the pinion tooth and *down* the flank of the gear tooth. The last point of contact occurs at the tip of the pinion and the flank of the gear tooth, at the intersection, *D*, of the line of action and the addendum circle of the pinion.

The *approach* phase of the motion is the interval between the initial contact at point *C* and the pitch point, *P*. The *angles of approach* are the angles through which the two gears rotate as the point of contact progresses from *C* to *P*. However, reflecting on the unwrapping cord analogy of Fig. 7.6, we see that the distance *CP* is equal to a length of cord unwrapped from the base cylinder of the pinion during the approach phase of the motion. Similarly, an equal amount of cord has wrapped onto the driven gear during that same phase. Thus, the angles of approach for the pinion and the gear, in radians, are

$$\alpha_2 = \frac{CP}{r_2} \quad \text{and} \quad \alpha_3 = \frac{CP}{r_3}. \tag{7.7}$$

The *recess* phase of the motion is the interval during which contact progresses from the pitch point, *P*, to final contact at point *D*. The *angles of recess* are the angles through which the two

Fig. 7.18 Interference in gear tooth action.

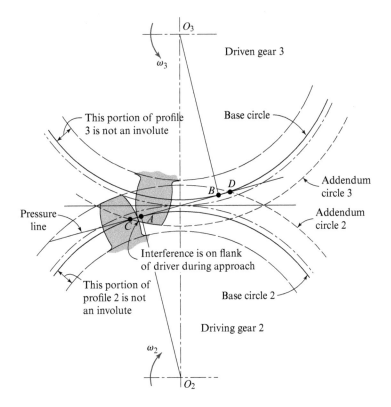

gears rotate as the point of contact progresses from P to D. Again, from the unwrapping cord analogy, we find these angles, in radians, to be

$$\beta_2 = \frac{PD}{r_2} \quad \text{and} \quad \beta_3 = \frac{PD}{r_3}. \tag{7.8}$$

If the teeth come into contact such that they are not conjugate, this is called *interference*. Consider Fig. 7.18; shown here are two 16-tooth $14\tfrac{1}{2}°$ pressure angle gears[2] with full-depth involute teeth. The driver, gear 2, turns clockwise. As with previous figures, the points labeled A and B indicate the points of tangency of the line of action with the two base circles, whereas the points labeled C and D indicate the initial and final points of contact. Note that the points C and D are now *outside* of points A and B. This indicates interference.

The interference is explained as follows. Contact begins when the tip of the driven gear 3 tooth contacts the flank of the driving tooth. In this case, the flank of the driving tooth first tries to make contact with the driven tooth at point C, and this occurs *before* the involute portion of the driving tooth comes within range. In other words, contact occurs before the two teeth become tangent. The actual effect is that the nontangent tip of the driven gear tooth interferes with and digs into the flank of the driving tooth.

---

[2] Such gears were part of an older standard and are now obsolete. They are chosen here only to show an example of interference.

In this example, a similar effect also occurs as the teeth leave contact. Contact should end at or before point B. Since, for this example, contact does not end until point D, the effect is for the nontangent tip of the driving tooth to interfere with and dig into the flank of the driven tooth.

When gear teeth are produced by a generating process, interference is automatically eliminated, since the cutting tool removes the interfering portion of the tooth flank. This effect is called *undercutting*. If undercutting is at all pronounced, the undercut tooth is considerably weakened. Thus, the effect of eliminating interference by a generation process is merely to substitute another problem for the original.

The importance of the problem of teeth that have been weakened by undercutting cannot be emphasized too strongly. Of course, interference can be eliminated by using more teeth on the gears. However, if the gears are to transmit a given amount of power, more teeth can be used only by increasing the pitch diameter. This makes the gears larger, which is seldom desirable. It also increases the pitch-line velocity which makes the gears noisier and somewhat reduces the power transmitted, although not in direct proportion. In general, however, the use of more teeth to eliminate interference or undercutting is seldom an acceptable solution.

Another method of reducing interference and the resulting undercutting is to employ a larger pressure angle. The larger pressure angle creates smaller base circles, so that a greater portion of the tooth profile has an involute shape. In effect, this means that fewer teeth can be used; as a result, gears with larger pressure angle are often smaller.

Of course, the use of standard gears is far less expensive than manufacturing specially made nonstandard gears. However, as indicated in Table 7.2, gears with larger pressure angles can be found without deviating from the standards.

One more way to eliminate interference is to use gears with shorter teeth. If the addendum distance is reduced, then points C and D move inward. One way to do this is to purchase standard gears, and then grind the top lands of the teeth to a new addendum distance. This, of course, makes the gears nonstandard and causes concern about repair or replacement, but it can be effective in eliminating interference. Again, careful study of Table 7.2 indicates that this is also possible by use of standard $20°$ *stub tooth* gears.

7.8   Contact Ratio

The zone of action of meshing gear teeth is shown in Fig. 7.19, where tooth contact begins and ends at the intersections of the two addendum circles with the line of action. As always, initial contact occurs at C, and final contact occurs at D. Tooth profiles drawn through these points intersect the base circle at points c and d. Thinking back to our analogy of the unwrapping cord of Fig. 7.6, the distance CD, measured along the line of action, is equal to the arc length cd, measured along the base circle.

Consider a situation in which the arc length cd, or distance CD, is exactly equal to the base pitch, p_b, of Eq. (7.6). This means that one tooth and its tooth space spans the entire arc cd. In other words, when a tooth is just ending contact at D, the following tooth is just beginning its contact at C. Therefore, during the tooth action from C to D, there is exactly one pair of teeth in contact at all times.

Fig. 7.19

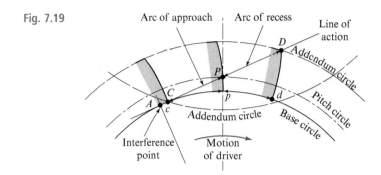

Next, consider a situation for which the arc length cd, or distance CD, is greater than the base pitch – but not much greater, say $cd = 1.1\,p_b$. This means that when one pair of teeth is just leaving contact at D, the following pair, already in contact, has already passed C. Thus, for a short time, there are two pairs of teeth in contact, one in the vicinity of C, and the other nearing D. As meshing proceeds, the earlier pair reaches D and ceases contact, leaving only a single pair of teeth in contact, until the situation repeats itself with the next pair of teeth.

Because of the nature of this tooth action, with one, two, or even more pairs of teeth in contact simultaneously, it is convenient to define the term contact ratio as

$$m_c = \frac{CD}{p_b}. \tag{7.9}$$

This is a value for which the next lower integer indicates the average number of pairs of teeth in contact. Thus, a contact ratio of $m_c = 1.35$, for example, implies that there is always at least one tooth in contact and there are two teeth in contact 35% of the time.

The minimum acceptable value of the contact ratio for smooth operation of meshing gears is $1.2 \le m_c \le 1.4$ and the recommended range of the contact ratio for most spur gearsets is $m_c > 1.4$.

The distance CD is quite convenient to measure if we are working graphically by making a drawing like Fig. 7.20 or Fig. 7.9. However, distances CP and PD can also be found analytically. From triangles O_3BC and O_3BP, we can write

$$CP = \sqrt{(R_3 + a)^2 - (R_3 \cos \phi)^2} - R_3 \sin \phi. \tag{7.10}$$

Similarly, from triangles O_2AD and O_2AP, we have

$$PD = \sqrt{(R_2 + a)^2 - (R_2 \cos \phi)^2} - R_2 \sin \phi. \tag{7.11}$$

The contact ratio is then obtained by substituting the sum of Eqs. (7.10) and (7.11) into Eq. (7.9).

We should note, however, that Eqs, (7.10) and (7.11) are only valid for the conditions where

$$CP \le R_2 \sin \phi \quad \text{and} \quad PD \le R_3 \sin \phi, \tag{7.12}$$

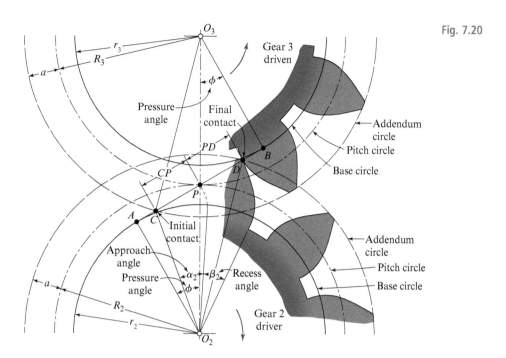

Fig. 7.20

since proper contact cannot begin before point A or end after point B. If either of these inequalities is not satisfied, then the gear teeth have interference and undercutting results.

7.9 Varying Center Distance

Figure 7.21a shows a pair of meshing gears having 20° full-depth involute teeth. Since both sides of the teeth are in contact, the center distance $R_{O_3O_2}$ cannot be reduced without jamming or deforming the teeth. However, Fig. 7.21b shows the same pair of gears, but mounted with a slightly increased distance $R_{O_3O_2}$ between the shaft centers, as might happen through the accumulation of tolerances of surrounding parts. Clearance, or *backlash*, now exists between the teeth, as shown.

When the center distance is increased, the base circles of the two gears do not change; they are fundamental to the shapes of the gears, once manufactured. However, review of Fig. 7.6 indicates that the same involute tooth shapes still touch as conjugate curves, and the fundamental law of gearing is still satisfied. However, the larger center distance results in an increase of the pressure angle and larger pitch circles passing through a new adjusted pitch point.

In Fig. 7.21b we see that the triangles $O_2A'P'$ and $O_3B'P'$ are still similar to each other, although they are both modified by the change in pressure angle. Also, distances O_2A' and O_3B' are the base circle radii and have not changed. Therefore, the ratio of the new pitch radii, O_2P' and O_3P', and the new velocity ratio remain the same as in the original design.

Another effect of increasing the center distance, observable in Fig. 7.21, is the shortening of the path of contact. The original path of contact CD in Fig. 7.21a is shortened to $C'D'$ in

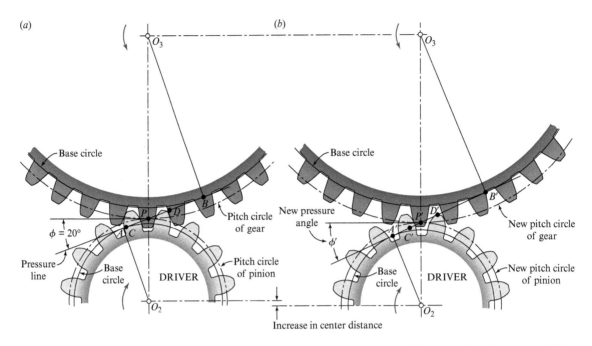

Fig. 7.21 Effect of increased center distance on the action of involute gearing; mounting the gears at (*a*) normal center distance, and (*b*) increased center distance.

Fig. 7.21*b*. The contact ratio, Eq. (7.9), is also reduced when the path of contact $C'D'$ is shortened. Since a contact ratio of less than unity would imply periods during which no teeth would be in contact at all, the center distance must never be increased larger than that corresponding to a contact ratio of unity $(C'D' = p_b)$.

7.10 Involutometry

The study of the geometry of the involute curve is called *involutometry*. In Fig. 7.22, a base circle with center at O is used to generate the involute BC. Line AT is the generating line, ρ is the instantaneous radius of curvature of the involute, and r is the radius to point T on the curve. If we designate the radius of the base circle r_b, line AT has the same length as the arc distance AB and so

$$\rho = r_b(\alpha + \varphi), \qquad (a)$$

where α is the angle between the radius OT and the radius to the origin of the involute OB, and φ is the angle between the radius of the base circle OA and the radius OT. Since OAT is a right triangle,

$$\rho = r_b \tan \varphi. \qquad (7.13)$$

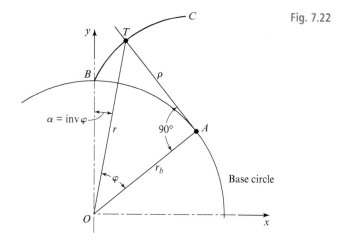

Fig. 7.22

Solving Eqs. (*a*) and (7.13) simultaneously to eliminate ρ and r_b gives

$$\alpha = \tan \varphi - \varphi,$$

which can be written as

$$\text{inv}\varphi = \tan \varphi - \varphi \tag{7.14}$$

and defines the *involute function*. The angle φ in this equation is the variable involute angle, measured in radians. Once φ is known, inv φ can readily be determined from Eq. (7.14). The inverse problem, where inv φ is given and φ is to be found, is more difficult. One approach is to expand Eq. (7.14) in an infinite series and to employ the first several terms to obtain a numeric approximation. Another approach is to use a root finding technique [2, Chapter 4]. Here, we refer to Table A.6 of Appendix A, in which the value of the involute function is tabulated, and the angle φ can be determined directly, in degrees.

Referring again to Fig. 7.22, we see that

$$r = \frac{r_b}{\cos \varphi}. \tag{7.15}$$

To demonstrate the use of the relations obtained earlier, let us determine the tooth dimensions of Fig. 7.23. Here, only the portion of the tooth extending above the base circle has been drawn, and the thickness of the tooth, t_p, at the pitch circle (point A), equal to half of the circular pitch, is given. The problem is to determine the tooth thickness at some other point, say point T. The various quantities shown in Fig. 7.23 are identified as follows:

r_b = radius of the base circle;
r_p = radius of the pitch circle;
r = radius at which the tooth thickness is to be determined;

Fig. 7.23

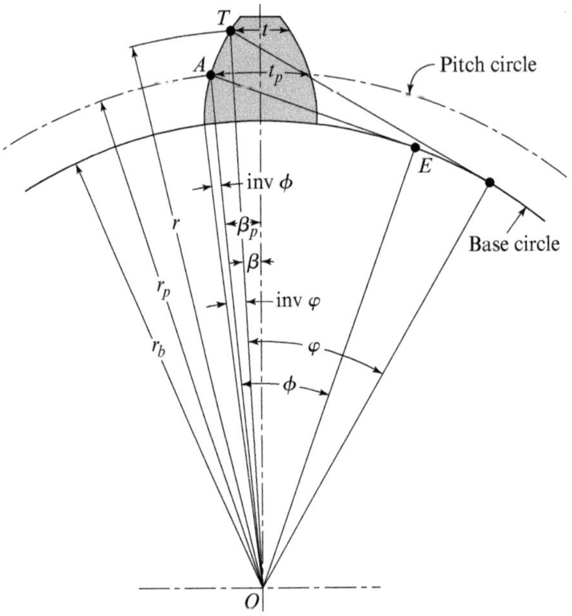

t_p = tooth thickness at the pitch circle;

t = tooth thickness to be determined;

ϕ = pressure angle corresponding to the pitch circle radius r_p;

φ = involute angle corresponding to point T;

β_p = angular half-tooth thickness at the pitch circle; and

β = angular half-tooth thickness at point T.

The half-tooth thicknesses at points A and T are

$$\frac{t_p}{2} = \beta_p r_p \quad \text{and} \quad \frac{t}{2} = \beta r, \tag{b}$$

so that

$$\beta_p = \frac{t_p}{2r_p} \quad \text{and} \quad \beta = \frac{t}{2r}. \tag{c}$$

From these we can write

$$\text{inv}\varphi - \text{inv}\phi = \beta_p - \beta = \frac{t_p}{2r_p} - \frac{t}{2r}. \tag{d}$$

The tooth thickness at point T is obtained by solving Eq. (d) for t:

$$t = 2r\left(\frac{t_p}{2r_p} + \text{inv}\phi - \text{inv}\varphi\right). \tag{7.16}$$

Example 7.2

A gear has 22 teeth cut full depth with pressure angle $\phi = 20°$ and diametral pitch $P = 2$ teeth/in. Find the thickness of the teeth at the base circle and at the addendum circle.

Solution

By the equations of Section 7.1 and Table 7.2, we find the radius of the pitch circle is $r_p = R = N/2P = 5.500$ in, the circular pitch is $p = 1.571$ in/tooth, the addendum is $a = 0.500$ in, and the dedendum is $d = 0.625$ in.

From Eq. (7.15), the radius of the base circle can be written as

$$r_b = r_p \cos\phi = (5.500 \text{ in}) \cos 20° = 5.168 \text{ in.}$$

The thickness of the tooth at the pitch circle is

$$t_p = \frac{p}{2} = \frac{1.571 \text{ in/tooth}}{2} = 0.785\,5 \text{ in.}$$

Converting the tooth pressure angle into radians gives $\phi = 20° = 0.349\,066$ rad. Then the involute function from Eq. (7.14) is

$$\text{inv}\phi = \tan 0.349\,066 - 0.349\,066 = 0.014\,904 \text{ rad.}$$

The involute angle at the base circle, from Eq. (7.15), is $\varphi_b = 0$. Therefore, the involute function is

$$\text{inv } \varphi_b = 0.$$

Substituting these results into Eq. (7.16), the tooth thickness at the base circle is

$$t_b = 2r_b \left[\frac{t_p}{2r_p} + \text{inv}\phi - \text{inv}\varphi_b \right]$$

$$= 2(5.168 \text{ in}) \left[\frac{0.785\,5 \text{ in}}{2(5.500 \text{ in})} + 0.014\,904 \text{ rad} - 0 \right] = 0.892 \text{ in.} \qquad \textit{Ans.}$$

The radius of the addendum circle is $r_a = r_p + a = 5.500 + 0.500 = 6.000$ in. Therefore, the involute pressure angle corresponding to this radius, from Eq. (7.15), is

$$\varphi = \cos^{-1}\left(\frac{r_b}{r}\right) = \cos^{-1}\left(\frac{5.168 \text{ in}}{6.000 \text{ in}}\right) = 30.53° = 0.532\,908 \text{ rad.}$$

Thus, the involute function is

$$\text{inv}\varphi = \tan 0.532\,908 - 0.532\,908 = 0.056\,922 \text{ rad.}$$

Substituting these results into Eq. (7.16), the tooth thickness at the addendum circle is

$$t_a = 2r_a \left[\frac{t_p}{2r_p} + \text{inv}\phi - \text{inv}\varphi \right]$$

$$= 2(6.000 \text{ in}) \left[\frac{0.785\,5 \text{ in}}{2(5.500 \text{ in})} + 0.014\,904 - 0.056\,922 \right] = 0.353 \text{ in.} \qquad \textit{Ans.}$$

Note that the tooth thickness at the base circle is more than double the tooth thickness at the addendum circle.

7.11 Nonstandard Gear Teeth

In this section we will examine the effects obtained by deviating from the specified standards and modifying such things as pressure angle, tooth depth, addendum, or center distance. Some of these modifications do not eliminate interchangeability; all of them are discussed with the intent of obtaining improved performance. Still, making such modifications probably means increased cost, since modified gears will not be available and must be specially machined for the particular application. Of course, this is also necessary at the time of any future repair or design modification.

The designer is often under great pressure to produce a design using gears that are small and yet able to transmit a large amount of power. Consider, for example, a gearset that must have a 4:1 velocity ratio. If the smallest pinion that will carry the load has a pitch diameter of 2 in, the mating gear has a pitch diameter of 8 in, making the overall space required for the two gears more than 10 in. On the other hand, if the pitch diameter of the pinion can be reduced by only ¼ in, the pitch diameter of the gear is reduced by a full 1 in and the overall size of the gearset is reduced by 1¼ in. This reduction assumes considerable importance when it is realized that the associated machine elements, such as shafts, bearings, and enclosure, are also reduced in size.

If a tooth of a certain pitch is required to carry the load, the only method of decreasing the pinion diameter is to use fewer teeth. However, we have already seen that problems involving interference, undercutting, and contact ratio are encountered when the tooth numbers are made too small. Thus, three principal reasons for employing nonstandard gears are: (a) to eliminate undercutting, (b) to prevent interference, and (c) to maintain a reasonable contact ratio. Note also that if a pair of gears is manufactured of the same material, the pinion is the weaker and is subject to greater wear, since each of its teeth are in contact a greater portion of the time. Therefore, any undercutting weakens the tooth that is already weaker. Thus, another objective of nonstandard gears is to gain a better balance of strength between the pinion and the gear.

As an involute curve is generated from its base circle, its radius of curvature becomes larger and larger. Near the base circle, the radius of curvature is quite small, being theoretically zero at the base circle. Contact near this region of sharp curvature should be avoided if possible because of the difficulty in obtaining good cutting accuracy in areas of small curvature, and since contact stresses are likely to be very high. Nonstandard gears present the opportunity to design to avoid these sensitive areas.

Clearance Modification.   A larger fillet radius at the root of the tooth increases the fatigue strength of the tooth and provides extra depth for shaving the tooth profile. Since interchangeability is not lost, the dedendum is sometimes increased to $1.300/P$ or $1.400/P$ ($1.300m$ or $1.400m$) to obtain space for a larger fillet radius.

Center-Distance Modification.   When gears of low tooth numbers are paired with each other or with larger gears, reduction in interference and improvement in the contact ratio can be obtained by increasing the center distance to greater than standard. Although such a system changes the tooth proportions and the pressure angle of the gears, the resulting tooth shapes can be generated with rack cutters (or hobs) of standard pressure angles or with standard pinion shapers. Before introducing this system, however, it is of value to develop certain additional relations about the geometry of gears.

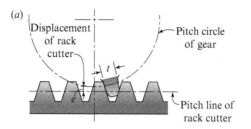 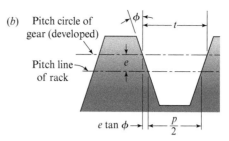

Fig. 7.24

The first new relation is for finding the thickness of a tooth that is cut by a rack cutter (or hob) when the pitch line of the rack cutter is displaced or offset by a distance, e, from the pitch circle of the gear being cut. What we are doing here is moving the rack cutter away from the center of the gear being cut. Stated another way, suppose the rack cutter does not cut as deeply into the gear blank and the teeth are not cut to full depth. This produces teeth that are thicker than the standard, and this thickness will now be determined. Figure 7.24a shows the problem, and Fig. 7.24b shows the solution. The increase of tooth thickness at the pitch circle is $2e\tan\phi$, so that

$$t = 2e\tan\phi + \frac{p}{2}, \tag{7.17}$$

where ϕ is the pressure angle of the rack cutter and t is the thickness of the modified gear tooth measured on its pitch circle.

Now suppose that two gears of different tooth numbers have both been cut with the cutter offset from their pitch circles as in the previous paragraph. Since the teeth of both have been cut with offset cutters, they will mate at a modified pressure angle and with modified pitch circles and consequently modified center distances. The word modified is used here in the sense of being nonstandard. Our problem is to determine the radii of these modified pitch circles and the value of the modified pressure angle.

In the following notation the word *standard* refers to values that would have been obtained had the usual, or standard, systems been employed to obtain the dimensions:

ϕ = pressure angle of generating rack cutter;
ϕ' = modified pressure angle at which gears will mate;
R_2 = standard pitch radius of pinion;
R'_2 = modified pitch radius of pinion when meshing with given gear;
R_3 = standard pitch radius of gear;
R'_3 = modified pitch radius of gear when meshing with given pinion;
t_2 = thickness of pinion tooth at standard pitch radius R_2;
t'_2 = thickness of pinion tooth at modified pitch radius R'_2;
t_3 = thickness of gear tooth at standard pitch radius R_3; and
t'_3 = thickness of gear tooth at modified pitch radius R'_3.

From Eq. (7.16), the thickness of a gear tooth at the standard pitch radius and at the modified pitch radius can be written, respectively, as

$$t_2' = 2R_2'\left(\frac{t_2}{2R_2} + \text{inv}\phi - \text{inv}\phi'\right) \tag{a}$$

and

$$t_3' = 2R_3'\left(\frac{t_3}{2R_3} + \text{inv}\phi - \text{inv}\phi'\right). \tag{b}$$

Note that the sum of these two thicknesses must be the new circular pitch. Therefore, using Eq. (7.3), we can write

$$t_2' + t_3' = p' = \frac{2\pi R_2'}{N_2}. \tag{c}$$

Since the pitch diameters of a pair of mating gears are proportional to their tooth numbers, then

$$R_3 = \frac{N_3}{N_2}R_2 \quad \text{and} \quad R_3' = \frac{N_3}{N_2}R_2'. \tag{d}$$

Substituting Eqs. (a), (b), and (d) into Eq. (c) and rearranging gives

$$\text{inv}\phi' = \frac{N_2(t_2' + t_3') - 2\pi R_2}{2R_2(N_2 + N_3)} + \text{inv}\phi. \tag{7.18}$$

This equation gives the modified pressure angle ϕ' at which a pair of gears will operate when the tooth thicknesses on their standard pitch circles are modified to t_2' and t_3'.

Although the base circle of a gear is fundamental to its shape and fixed once the gear is generated, gears have no pitch circles until a pair of them are brought into contact. Bringing a pair of gears into contact creates a pair of pitch circles that are tangent to each other at the modified pitch point. Throughout this discussion, the idea of a pair of so-called standard pitch circles has been used to define a certain point on the involute curves. These standard pitch circles, as we have seen, are the ones that come into existence when gears are paired *if the gears are not modified from the standard dimensions*. On the other hand, the base circles are fixed circles that are not changed by tooth modifications. The base circle remains the same whether the tooth dimensions are changed or not, so we can determine the base circle radius using either the standard pitch circle or the new pitch circle. Thus, from Eq. (7.15), we can write

$$R_2 \cos \phi = R_2' \cos \phi'.$$

Therefore, the modified pitch radius of the pinion can be written as

$$R_2' = \frac{R_2 \cos \phi}{\cos \phi'}. \tag{7.19}$$

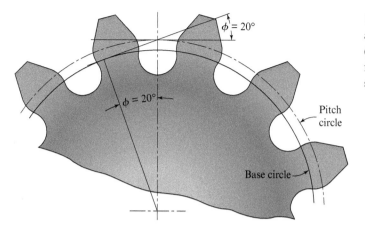

Fig. 7.25 Standard 20° pressure angle, 1 tooth/in diametral pitch (25.4 mm/tooth module), 12-tooth full-depth involute gear showing undercut.

Similarly, the modified pitch radius of the gear can be written as

$$R'_3 = \frac{R_3 \cos \phi}{\cos \phi'}. \tag{7.20}$$

Equations (7.19) and (7.20) give the values of the actual pitch radii when the two gears with modified teeth are brought into mesh without backlash. The new center distance is, of course, the sum of these radii.

All the necessary relations have now been developed to create nonstandard gears with changes in the center distance. The use of these relations is now demonstrated by an example.

Figure 7.25 is a drawing of a 20° pressure angle, 1 tooth/in diametral pitch (25.4 mm/tooth module), 12-tooth pinion generated with a rack cutter to full depth with a standard clearance of $0.250/P$ $(0.250m)$. In the 20° full-depth system, interference is severe when the number of teeth is less than 14. The resulting undercutting is evident in the drawing.

In an attempt to eliminate undercutting, improve the tooth action, and increase the contact ratio, suppose that this pinion were not cut to full depth; suppose instead that the rack cutter were only allowed to cut to a depth for which its addendum passes through the interference point A of the pinion being cut – that is, the point of tangency of the 20° line of action and the base circle – as shown in Fig. 7.26. From Eq. (7.15) we know that

$$r_2 = R_2 \cos \phi. \tag{e}$$

Then, from Fig. 7.26, the depth of the cut would be offset from the standard by

$$e = a + r_2 \cos \phi - R_2. \tag{f}$$

Substituting Eq. (e) into Eq. (f), the offset can be written as

$$e = a + R_2 \cos^2 \phi - R_2 = a - R_2 \sin^2 \phi. \tag{7.21}$$

If the offset is any less than this, then the rack will cut below the interference point, A, and will result in undercutting.

Fig. 7.26 Offset of a rack cutter to cause its addendum line to pass through the interference point A.

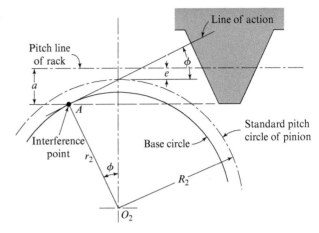

Example 7.3

A 12-tooth pinion with pressure angle $\phi = 20°$ and diametral pitch $P = 1$ tooth/in is to be mated with a standard 40-tooth gear. If the pinion were cut to full depth, then Eq. (7.9) shows that the contact ratio would be 1.41, but there would be undercutting, as indicated in Fig. 7.25. Instead, let the 12-tooth pinion be cut from a larger blank using center-distance modifications. Determine the cutter offset, the modified pressure angle, the modified pitch radii of the pinion and the gear, the modified center distance, the modified outside radii of the pinion and gear, and the contact ratio. Has the contact ratio increased significantly?

Solution

Designating the pinion as subscript 2 and the gear as 3, then with $P = 1$ tooth/in and $\phi = 20°$, the following values are determined:

$$p = 3.142 \text{ in/tooth}, R_2 = 6 \text{ in}, R_3 = 20 \text{ in}, N_2 = 12 \text{ teeth},$$
$$N_3 = 40 \text{ teeth, and } t_3 = 1.571 \text{ in}.$$

For a standard rack cutter, from Table 7.2, the addendum is $a = 1/P = 1.000$ in. From Eq. (7.21), the rack cutter will be offset by

$$e = 1.0 - 6.0 \sin^2 20° = 0.298 \text{ in}. \qquad\qquad Ans.$$

Then, the thickness of the pinion tooth at the 6 in pitch circle, using Eq. (7.17), is

$$t_2' = 2e \tan \phi + \frac{p}{2} = 2(0.298 \text{ in}) \tan 20° + \frac{3.142 \text{ in}}{2} = 1.788 \text{ in}.$$

The pressure angle at which this (and only this) gearset will operate is determined from Eq. (7.18); that is,

$$\text{inv}\phi' = \frac{N_2(t_2' + t_3') - 2\pi R_2}{2R_2(N_2 + N_3)} + \text{inv}\phi$$
$$= \frac{12(1.788 \text{ in} + 1.571 \text{ in}) - 2\pi(6.000 \text{ in})}{2(6.000 \text{ in})(12 + 40)} + \text{inv } 20° = 0.019\,08 \text{ rad}.$$

Example 7.3 (cont.)

From Appendix A, Table A.6, we find that the new pressure angle is

$$\phi' = 21.65°. \hspace{3cm} Ans.$$

Using Eqs. (7.19) and (7.20), the modified pitch radii are found to be

$$R_2' = \frac{R_2 \cos \phi}{\cos \phi'} = \frac{(6.000 \text{ in}) \cos 20°}{\cos 21.65°} = 6.066 \text{ in}, \hspace{1.5cm} Ans.$$

$$R_3' = \frac{R_3 \cos \phi}{\cos \phi'} = \frac{(20.000 \text{ in}) \cos 20°}{\cos 21.65°} = 20.220 \text{ in}. \hspace{1cm} Ans.$$

So the modified center distance is

$$R_2' + R_3' = 6.066 + 20.220 = 26.286 \text{ in}. \hspace{1.5cm} Ans.$$

Note that the center distance has not increased as much as the offset of the rack cutter. Standard clearance of $0.25/P$ results from the standard dedendums equal to $1.25/P$, as indicated in Table 7.2. So, the root radii of the two gears are:

$$\text{Root radius of pinion} = 6.298 - 1.250 = 5.048 \text{ in};$$
$$\text{Root radius of gear} = 20.000 - 1.250 = 18.750 \text{ in};$$
$$\text{Sum of root radii} = 23.798 \text{ in}.$$

The difference between this sum and the center distance is the working depth plus twice the clearance. Since the clearance is 0.250 in for each gear, the working depth is

$$\text{Working depth} = 26.286 - 23.798 - 2(0.250) = 1.988 \text{ in}.$$

The outside radius of each gear is the sum of the root radius, the clearance, and the working depth; that is,

$$\text{Outside radius of pinion} = 5.048 + 0.250 + 1.988 = 7.286 \text{ in}; \hspace{1cm} Ans.$$
$$\text{Outside radius of gear} = 18.750 + 0.250 + 1.988 = 20.988 \text{ in}. \hspace{1cm} Ans.$$

The result is shown in Fig. 7.27, and the pinion is seen to have a stronger-looking form than the one of Fig. 7.25. Undercutting has been completely eliminated.

The contact ratio can be obtained from Eqs. (7.9)–(7.11). The following quantities are needed:

$$\text{Outside radius of pinion} = R_2' + a = 7.286 \text{ in},$$
$$\text{Outside radius of gear} = R_3' + a = 20.988 \text{ in},$$
$$r_2 = R_2 \cos \phi = (6.000 \text{ in}) \cos 20° = 5.638 \text{ in},$$
$$r_3 = R_3 \cos \phi = (20.000 \text{ in}) \cos 20° = 18.794 \text{ in},$$
$$p_b = p \cos \phi = (3.141\ 6 \text{ in/tooth}) \cos 20° = 2.952 \text{ in/tooth}.$$

Example 7.3 (cont.)

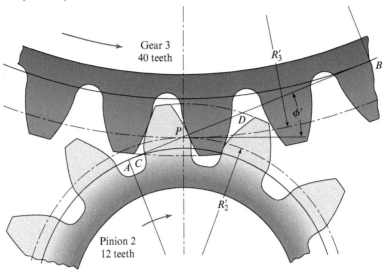

Fig. 7.27

Therefore, for Eqs. (7.10) and (7.11), we have

$$CP = \sqrt{\left(R_3' + a\right)^2 - r_3^2} - R_3' \sin\phi$$
$$= \sqrt{(20.988 \text{ in})^2 - (18.794 \text{ in})^2} - (20.220 \text{ in}) \sin 21.65° = 1.883 \text{ in},$$

$$PD = \sqrt{\left(R_2' + a\right)^2 - r_2^2} - R_2' \sin\phi$$
$$= \sqrt{(7.286 \text{ in})^2 - (5.638 \text{ in})^2} - (6.066 \text{ in}) \sin 21.65° = 2.377 \text{ in}.$$

Finally, from Eq. (7.9), the contact ratio is

$$m_c = \frac{CP + PD}{p_b} = \frac{1.883 \text{ in} + 2.377 \text{ in}}{2.952 \text{ in/tooth}} = 1.443 \text{ teeth avg.} \qquad \textit{Ans.}$$

Therefore, the contact ratio has increased only slightly (approximately 2% increase). The modification, however, is justified because of the elimination of undercutting which results in a substantial improvement in the strength of the teeth.

Long-and-Short-Addendum System.   It often happens in the design of machinery that the center distance between a pair of gears is fixed by some other design consideration or feature of the machine. In such a case, modifications to obtain improved performance cannot be made by varying the center distance.

In the previous section, we saw that improved action and tooth shape can be obtained by backing the rack cutter away from the gear blank during forming of the teeth. The effect of this withdrawal is to create the active tooth profile farther away from the base circle. Examination of Fig. 7.27 indicates that more dedendum could be used on the gear (not the pinion) before the

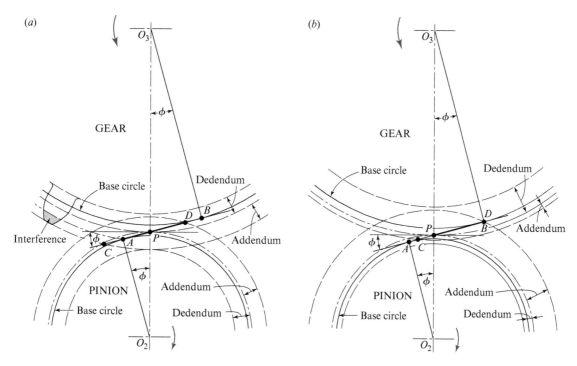

Fig. 7.28 Comparison of standard gears and gears cut by the long-and-short-addendum system: (*a*) gear and pinion with standard addendum and dedendum; (*b*) gear and pinion with long-and-short-addendum.

interference point is reached. If the rack cutter is advanced into the gear blank by a distance equal to the withdrawal from the pinion blank, more of the gear dedendum is used, and at the same time, the center distance is not changed. This is called the *long-and-short-addendum system*.

In the long-and-short-addendum system, there are no changes in the pitch circles and consequently none in the pressure angle. The effect is to move the contact region away from the pinion center toward the gear center, thus shortening the approach action and lengthening the recess action.

The characteristics of the long-and-short-addendum system can be explained by reference to Fig. 7.28. Figure 7.28*a* shows a conventional (standard) set of gears having a dedendum equal to the addendum plus the clearance. Interference exists, and the tip of the gear tooth will have to be relieved as shown or the pinion will be undercut. This is indicated, since the addendum circle crosses the line of action at *C*, outside of the tangency or interference point *A*; hence, the distance *AC* is a measure of the degree of interference.

To eliminate the undercutting or interference, the pinion addendum may be enlarged, as in Fig. 7.28*b*, until the addendum circle of the pinion passes through the interference point (point *B*) of the gear. In this manner we shall be using all of the gear tooth profile. The same whole depth may be retained; hence, the dedendum of the pinion may be reduced by the same amount that the addendum is increased. This means that we must also increase the gear dedendum and decrease the dedendum of the mating pinion. With these changes, the path of contact is the line

CD of Fig. 7.28*b*. It is longer than the path *AD* of Fig. 7.28*a*, and so the contact ratio is higher. Note, too, that the base circles, the pitch circles, the pressure angle, and the center distance have not changed. Both gears can be cut with standard cutters by advancing the cutter into the gear blank for this modification by a distance equal to the amount of withdrawal from the pinion blank. Finally, note that the blanks from which the pinion and gear are cut must now be of different diameters than the standard blanks.

The tooth dimensions for the long-and-short-addendum system can be determined using the equations developed in the previous sections.

A less obvious advantage of the long-and-short-addendum system is that more recess action than approach action is obtained. The approach action of gear teeth is analogous to pushing a piece of chalk across a blackboard; the chalk screeches. But, when the chalk is pulled across a blackboard, analogous to the recess action, it glides smoothly. Thus, recess action is always preferable because of the smoothness and the lower frictional forces.

7.12 Parallel-Axis Gear Trains

Mechanisms arranged in combinations so that the driven member of one mechanism is the driver for another mechanism are called *mechanism trains*. With certain exceptions, to be explored here, the analysis of such trains can proceed in serial fashion by using the methods developed in the previous chapters.

In Chapter 3, we learned that the first-order *kinematic coefficient* is the term used to describe the ratio of the velocity of the driven member to that of the driving member. Thus, for example, in a four-bar linkage with link 2 as the driving or input member and link 4 as the driven or output member, we have

$$\theta'_{42} = \frac{d\theta_4}{d\theta_2} = \frac{d\theta_4/dt}{d\theta_2/dt} = \frac{\omega_4}{\omega_2}, \tag{a}$$

where it is noted that, as in Section 5.5 and after, we adopt the second subscript to explicitly indicate the number of the driving or input member. This second subscript is important in this and following sections, since many mechanism trains have more than one degree of freedom.

In this section, where we deal with serially connected gear trains, we prefer to write Eq. (*a*) as

$$\theta'_{LF} = \frac{d\theta_L}{d\theta_F} = \frac{d\theta_L/dt}{d\theta_F/dt} = \frac{\omega_L}{\omega_F}, \tag{7.22}$$

where ω_L is the angular velocity of the *last* gear and ω_F is the angular velocity of the *first* gear in the train, since, usually, the last gear is the output and is the driven gear, and the first is the input and driving gear.

The term θ'_{LF} in Eq. (7.22) is the first-order kinematic coefficient, called the *speed ratio* by some or the *train value* by others. Equation (7.22) is often written in the more convenient form:

$$\omega_L = \theta'_{LF}\omega_F. \tag{7.23}$$

Next, we consider a pinion 2 driving a gear 3. The speed of the driven gear is

$$\omega_3 = \pm \frac{R_2}{R_3} \omega_2 = \pm \frac{N_2}{N_3} \omega_2, \tag{b}$$

where, for each gear, R is the radius of the pitch circle, N is the number of teeth, and ω is either the angular velocity or the angular displacement completed during a chosen time interval.

For parallel-shaft gearing, the directions can be tracked by following the vector sense – that is, by specifying that angular velocity is positive when counterclockwise as seen from a chosen side. For parallel-shaft gearing, we shall use the following sign convention: If the last gear of a parallel-shaft gear train rotates with the same sense as the first gear, then θ'_{LF} is positive; if the last gear rotates in the opposite sense to the first gear, then θ'_{LF} is negative. This sign convention approach is not as easy, however, when the gear shafts are not parallel, as in bevel, crossed-helical, or worm gearing (Chapter 8). In such cases, it is often simpler to track the directions by visually inspecting a sketch of the train.

The gear train shown in Fig. 7.29 is made up of five gears in series. Applying Eq. (b) three times, we find the speed of gear 6 to be

$$\omega_6 = -\frac{R_5}{R_6} \frac{R_4}{R_5} \frac{R_2}{R_3} \omega_2 = -\frac{N_5}{N_6} \frac{N_4}{N_5} \frac{N_2}{N_3} \omega_2. \tag{c}$$

Here, we note that gear 5 is an *idler*; that is, its tooth numbers cancel in Eq. (c), and hence the only purpose served by gear 5 is to change the direction of rotation of gear 6. We further note that gears 5, 4, and 2 are drivers, whereas gears 6, 5, and 3 are driven members. Thus, Eq. (7.22) can also be written

$$\theta'_{LF} = \pm \frac{\text{product of driving tooth numbers}}{\text{product of driven tooth numbers}}. \tag{7.24}$$

Note also that, since they are proportional, pitch radii can be used in Eq. (7.24) just as well as tooth numbers.

In speaking of gear trains, it is convenient to describe a train having only one gear on each axis as a *simple* gear train. A *compound* gear train, then, is one that has two or more gears on one or more axes, such as the train shown in Fig. 7.29. Another example of a compound gear train is shown in Fig. 7.30, which shows a transmission for a small- or medium-size truck that has four speeds forward and one in reverse.

The compound gear train shown in Fig. 7.31 is composed of bevel, helical, and spur gears (Chapter 8). The helical gears are crossed, so their direction of rotation depends upon their hand.

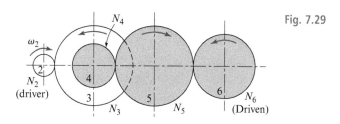

Fig. 7.29

Fig. 7.30 A truck transmission with gears having diametral pitch of 7 teeth/in and pressure angle of 22½°.

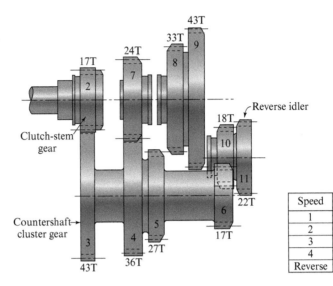

Speed	Drive
1	2-3-6-9
2	2-3-5-8
3	2-3-4-7
4	Straight through
Reverse	2-3-6-10-11-9

Fig. 7.31 A gear train composed of bevel, crossed-helical, and spur gears.

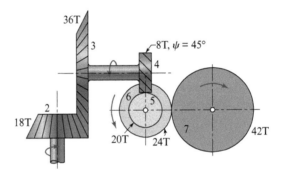

Fig. 7.32 A reverted gear train.

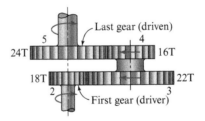

A *reverted* gear train is one in which the first and last gears have collinear axes of rotation, such as the one shown in Fig. 7.32. This produces a compact arrangement and is used in such applications as speed reducers, clocks (to connect the hour hand to the minute hand), and machine tools. As an exercise, it is suggested that you seek out a suitable set of diametral pitches for each pair of gears shown in Fig. 7.32 so that the first and last gears have the same axis of rotation with all gears properly engaged.

7.13 Determining Tooth Numbers

When notable power is transmitted through a speed-reduction unit, the speed ratio of the last pair of meshing gears is usually chosen larger than that of the first gear pair, since the torque is greater at the low-speed end. In a given amount of space, more teeth can be used on gears of lesser pitch; hence, a greater speed reduction can be obtained at the high-speed end.

Without examining the problem of tooth strength, suppose we wish to use two pairs of gears in a train to obtain an overall kinematic coefficient of $\theta'_{LF} = 1/12$. Let us also impose the restriction that the tooth numbers must not be less than 15 and that the reduction in the first pair of gears should be about twice that of the second pair. This means that the overall kinematic coefficient is

$$\theta'_{52} = \frac{N_4}{N_5}\frac{N_2}{N_3} = \frac{1}{12}, \tag{a}$$

where N_2/N_3 is the coefficient of the first gear pair, and N_4/N_5 is that of the second pair. Since the kinematic coefficient of the first pair should be half that of the second, Eq. (a) can be written as

$$\left(\frac{N_4}{N_5}\right)\left(\frac{N_4}{2N_5}\right) \approx \frac{1}{12}, \tag{b}$$

or

$$\frac{N_4}{N_5} \approx \sqrt{\frac{1}{6}} = 0.408\,248, \tag{c}$$

to six decimal places. The following tooth numbers are seen to be close:

$$\frac{15}{37}, \quad \frac{16}{39}, \quad \frac{18}{44}, \quad \frac{20}{49}, \quad \frac{22}{54}, \quad \frac{24}{59}.$$

Of these, $N_4/N_5 = 20/49$ is the closest approximation, but note that

$$\theta'_{52} = \left(\frac{N_4}{N_5}\right)\left(\frac{N_2}{N_3}\right) = \left(\frac{20}{49}\right)\left(\frac{20}{98}\right) = \frac{400}{4\,802} = \frac{1}{12.005},$$

which is very close to 1/12. On the other hand, the choice of $N_4/N_5 = 18/44$ gives exactly

$$\theta'_{52} = \left(\frac{N_4}{N_5}\right)\left(\frac{N_2}{N_3}\right) = \left(\frac{18}{44}\right)\left(\frac{22}{108}\right) = \frac{396}{4\,752} = \frac{1}{12}.$$

In this case, the reduction in the first gear pair is not exactly twice the reduction in the second gear pair. However, this consideration is usually of only minor importance.

The problem of specifying tooth numbers and the number of pairs of gears to give a kinematic coefficient with a specified degree of accuracy has interested many people, but has no exact solution. Consider, for instance, the problem of specifying a set of gears to have a kinematic coefficient of $\theta'_{LF} = \pi/10$ accurate to eight decimal places, while we know that π is an irrational number and cannot be expressed as a ratio of integers.

7.14 Epicyclic Gear Trains

Figure 7.33 shows an elementary epicyclic gear train together with its schematic diagram, as suggested by Lévai.[3] The train consists of a *central gear*, 2, and an *epicyclic gear*, 4, which produces epicyclic motion for its points by rolling around the periphery of the central gear. An *arm* 3 contains the bearing for the epicyclic gear to maintain the gears in mesh.

Epicyclic trains are also called *planetary* or *sun-and-planet* gear trains. In this nomenclature, gear 2 of Fig. 7.33 is called the *sun gear*, gear 4 is called the *planet gear*, and crank 3 is called the *planet carrier*. Figure 7.34 shows the train of Fig. 7.33 with two redundant planet gears added. This produces better force balance; also, adding more planet gears allows lower forces by more load sharing. However, these additional planet gears do not change the kinematic characteristics at all. For this reason, we generally indicate only a single planet in the illustrations and problems in this chapter, although an actual machine will probably be designed with planets in trios.

The simple epicyclic gear train together with its schematic representation in Fig. 7.35 shows how the motion of the planet gear can be transmitted to another central gear. The second central gear in this case is gear 5, an internal gear. In Fig. 7.35a, internal gear 5 is stationary, but this is not a requirement, as shown in Fig. 7.35b. Figure 7.36 shows a similar arrangement,

Fig. 7.33 (*a*) Elementary epicyclic gear train; (*b*) its schematic diagram.

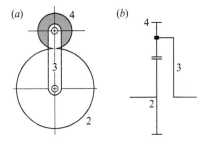

Fig. 7.34 A planetary gearset.

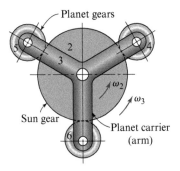

---

[3] Literature devoted to epicyclic gear trains is rather scarce; however, see [3]. For a comprehensive study in the English language, see [1], which lists 104 references.

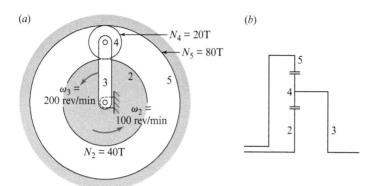

Fig. 7.35 (a) A simple epicyclic gear train; (b) its schematic diagram.

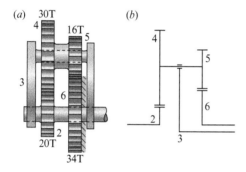

Fig. 7.36 A simple epicyclic gear train with double planet gears.

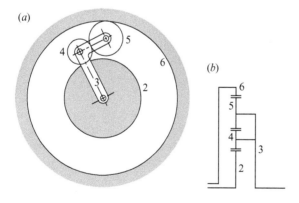

Fig. 7.37 An epicyclic gear train with two planet gears.

with the difference that both central gears are external gears. Note, in Fig. 7.36, that the double planet gears are mounted on a single planet shaft and that each planet gear is in mesh with a separate sun gear rotating at a different speed.

In any case, no matter how many planets are used, only one planet carrier or arm may be used. This principle is shown in Fig. 7.37, in which redundant planets are used, and in Fig. 7.37, where two planets are used to alter the kinematic performance.

Fig. 7.38 All 12 possible epicyclic gear train types according to Lévai [1].

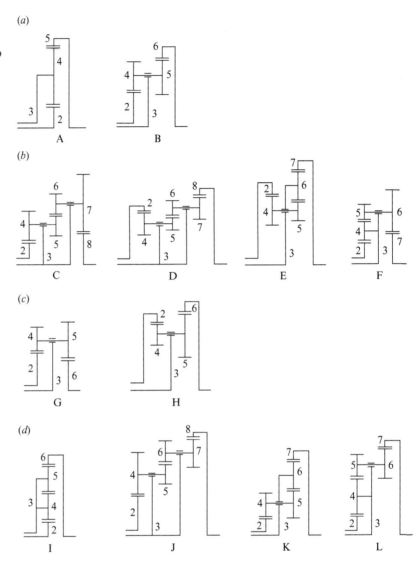

According to Lévai [1], only 12 variations of epicyclic gear trains are possible; they are all shown in schematic form in Fig. 7.38 as Lévai arranged them. In all variations, the arm (the planet carrier) is shown as link number 3. The trains in Fig. 7.38a and Fig. 7.38c are simple trains in which the planet gears mesh with both sun gears. The trains shown in Fig. 7.38b and Fig. 7.38d have planet gear pairs that are partly in mesh with each other and partly in mesh with the sun gears.

7.15 Analysis of Epicyclic Gear Trains by Formula

Figure 7.39 shows a planetary gear train composed of a sun gear 2, an arm or planet carrier 3, and planet gears 4 and 5. Using the apparent angular velocity equation, Eq. (3.10), we can write that the angular velocity of gear 2 as it would appear from a coordinate system fixed to arm 3 is

$$\omega_{2/3} = \omega_2 - \omega_3. \tag{a}$$

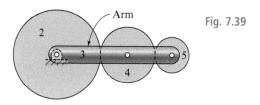

Fig. 7.39

Also, the angular velocity of gear 5 as it would appear from the coordinate system fixed to arm 3 is

$$\omega_{5/3} = \omega_5 - \omega_3. \tag{b}$$

Dividing Eq. (b) by Eq. (a) gives

$$\frac{\omega_{5/3}}{\omega_{2/3}} = \frac{\omega_5 - \omega_3}{\omega_2 - \omega_3}. \tag{c}$$

Equation (c) expresses the ratio of the apparent angular velocity of gear 5 to that of gear 2, with both taken as they would appear from arm 3. This ratio, which is proportional to the tooth numbers, appears the same whether the arm is rotating or not; it is the first-order kinematic coefficient of the gear train. Therefore, the first-order kinematic coefficient can be expressed as

$$\theta'_{52/3} = \frac{\omega_5 - \omega_3}{\omega_2 - \omega_3}. \tag{d}$$

An equation similar to Eq. (d) is all that we need to find the angular velocities in a planetary gear train. It is convenient to express it in the form as it would appear from the arm:

$$\theta'_{LF/A} = \frac{\omega_L - \omega_A}{\omega_F - \omega_A}, \tag{7.25}$$

where:

ω_F = angular velocity of the first gear in the train;
ω_L = angular velocity of the last gear in the train; and
ω_A = angular velocity of the arm.

Note that Eq. (d) and Eq. (7.25) can be expressed entirely in terms of kinematic coefficients; that is,

$$\theta'_{52/3} = \frac{\theta'_5 - \theta'_3}{\theta'_2 - \theta'_3} \tag{e}$$

and

$$\theta'_{LF/A} = \frac{\theta'_L - \theta'_A}{\theta'_F - \theta'_A}, \tag{7.26}$$

where:

θ'_F = kinematic coefficient of the first gear in the train;
θ'_L = kinematic coefficient of the last gear in the train; and
θ'_A = kinematic coefficient of the arm.

The following examples demonstrate the use of the kinematic coefficients presented in Eqs. (7.25) and (7.26).

Example 7.4

Figure 7.36 shows a reverted planetary gear train. Gear 2 is fastened to its shaft and is driven at 250 rev/min in a clockwise direction. Gears 4 and 5 are planet gears that are joined but are free to turn on the shaft carried by the arm. Gear 6 is stationary. Find the speed and direction of rotation of the arm.

Solution

We must first decide which gears to designate the first and last members of the train. Since the speeds of gears 2 and 6 are both given, then either gear may be chosen as the first. The choice makes no difference in the results, but once the decision is made it may not be changed. Here we choose gear 2 as the first; therefore, gear 6 is the last. Thus, choosing counterclockwise as positive gives

$$\omega_F = \omega_2 = -250 \text{ rev}/\min \quad \text{and} \quad \omega_L = \omega_6 = 0 \text{ rev}/\min,$$

and, according to Eq. (7.24), the first-order kinematic coefficient is

$$\theta'_{LF} = \theta'_{62} = \left(-\frac{16}{34}\right)\left(-\frac{20}{30}\right) = \frac{16}{51},$$

where the positive sign results, since there are two external contacts.

Substituting values into Eq. (7.25) gives the first-order kinematic coefficient:

$$\theta'_{LF/A} = \frac{16}{51} = \frac{0 - \omega_3}{-250 \text{ rev}/\min - \omega_3}.$$

Rearranging this equation gives the angular velocity of arm A:

$$\omega_A = \omega_3 = 114.3 \text{ rev}/\min \text{ ccw}. \qquad\qquad Ans.$$

Example 7.5

Consider the simple planetary gear train shown in Fig. 7.40 in which the ring gear h is fixed (that is, the angular velocity of the ring gear is $\omega_h = 0$) and the input is the angular velocity of the

Example 7.5 (cont.)

planet carrier (the arm). The radii of the ring gear h, planet gear j, sun gear k, and arm A are $R_h = 200$ mm, $R_j = 50$ mm, $R_k = 100$ mm, and $R_A = 150$ mm, respectively. If the input angular velocity is $\omega_A = \omega_i = 10$ rad/s ccw, determine the angular velocities of the planet gear j and the sun gear k.

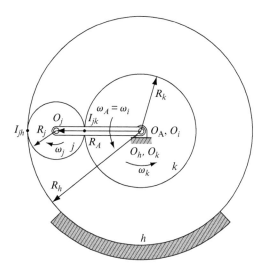

Fig. 7.40 A simple planetary gear train with the ring gear fixed.

Solution

Using Eq. (7.25), the first-order kinematic coefficient of planet gear j, in rolling contact with ring gear h, can be written as

$$\theta'_{jh/A} = \pm \frac{R_h}{R_j} = \frac{\omega_j - \omega_A}{\omega_h - \omega_A}. \qquad (1a)$$

Similarly, the first-order kinematic coefficient of planet gear j, in rolling contact with sun gear k, can be written as

$$\theta'_{jk/A} = \pm \frac{R_k}{R_j} = \frac{\omega_j - \omega_A}{\omega_k - \omega_A}. \qquad (1b)$$

Note that if the ring gear is removed, and the arm is fixed (that is, $\omega_A = 0$), then the planetary gear train reduces to an ordinary gear train, that is, Eq. (1b) reduces to

$$\theta'_{jk/A} = \pm \frac{R_k}{R_j} = \frac{\omega_j}{\omega_k}. \qquad (2)$$

Since there is internal contact between the planet gear and the fixed ring gear, then the positive sign is used in Eq. (1a). Then, rearranging the equation and substituting $\omega_h = 0$ gives

Example 7.5 (cont.)

$$R_j\omega_j = -(R_h - R_j)\omega_A = -R_A\omega_A. \tag{3}$$

Therefore, the angular velocity of the planet gear is

$$\omega_j = -\frac{R_A\omega_A}{R_j} = -\frac{(150 \text{ mm})(10 \text{ rad/s})}{50 \text{ mm}} = -30 \text{ rad/s}. \qquad Ans.$$

The negative sign shows that the direction of the angular velocity of the planet gear is opposite to the direction of the arm and, therefore, is clockwise.

Similarly, since there is external contact between the planet gear and the sun gear, the negative sign is used in Eq. (1b), and the equation can be written as

$$R_k\omega_k = -R_j\omega_j + (R_k + R_j)\omega_A. \tag{4a}$$

The angular velocity of the sun gear can be expressed in terms of the input angular velocity of the arm by substituting Eq. (3) into Eq. (4a) and simplifying; that is,

$$R_k\omega_k = (R_k + R_h)\omega_A = 2R_A\omega_A. \tag{4b}$$

Therefore, the angular velocity of the sun gear is

$$\omega_k = \frac{2R_A\omega_A}{R_k} = \frac{2(150 \text{ mm})(10 \text{ rad/s})}{100 \text{ mm}} = 30 \text{ rad/s}. \qquad Ans.$$

The positive result indicates that the direction of the angular velocity of the sun gear is the same as the arm and, therefore, is counterclockwise.

Example 7.6

Consider the compound planetary gear train that is commonly used in an electric drill or screwdriver, shown in the schematic diagram in Fig. 7.41. The gear and the arm dimensions are $R_1 = 150$ mm, $R_{2A} = R_{4A} = 100$ mm, and $R_3 = R_4 = R_5 = R_6 = 50$ mm. The angular velocity of the input shaft (and arm 2) is $\omega_{2A} = 10$ rad/s ccw (looking from the left) and ring gear 1 is fixed – that is, the angular velocity of the ring gear is $\omega_1 = 0$. Determine the angular velocities of gears 3, 4, 5, and 6.

Solution

This gear train is composed of two simple planetary gear trains in series; that is, the first planetary gear train is composed of ring gear 1, input arm 2, planet gear 3, and sun gear 4, and the second planetary gear train is composed of ring gear 1, arm 4, planet gear 5, and output sun gear 6. Note that the angular velocities of gear 4 and arm 4 are equal in magnitude and direction, since they are rigidly attached. Also, note that gear 4 acts as both the output of the first planetary gear train and

Example 7.6 (cont.)

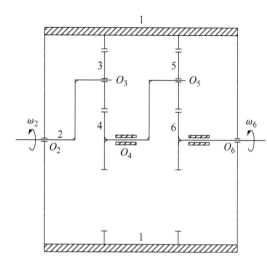

Fig. 7.41 Schematic diagram of a compound planetary gear train.

the input to the second planetary gear train. Using Eq. (7.25), the first-order kinematic coefficient of gear 4 with respect to ring gear 1 can be written as

$$\theta'_{41/2A} = \left(-\frac{R_3}{R_4}\right)\left(\frac{R_1}{R_3}\right) = -\frac{R_1}{R_4} = \frac{\omega_4 - \omega_{2A}}{\omega_1 - \omega_{2A}}. \tag{1a}$$

The correct sign is negative, since gears 4 and 3 are in external contact, and gears 3 and 1 are in internal contact. Substituting the given data into Eq. (1a) gives

$$-\frac{150 \text{ mm}}{50 \text{ mm}} = \frac{\omega_4 - 10 \text{ rad/s}}{0 - 10 \text{ rad/s}}. \tag{1b}$$

Therefore, the angular velocity of gear 4 is

$$\omega_4 = 40 \text{ rad/s}, \qquad\qquad Ans.$$

where the positive result implies that gear 4 is rotating in the same direction as input gear 2 – that is, counterclockwise.

Using Eq. (7.25), the first-order kinematic coefficient of gear 3 with respect to gear 4 can be written as

$$\theta'_{34/2A} = -\frac{R_4}{R_3} = \frac{\omega_3 - \omega_{2A}}{\omega_4 - \omega_{2A}}, \tag{2a}$$

where the negative sign denotes external contact between gears 3 and 4. Therefore, the first-order kinematic coefficient is

$$\theta'_{34/2A} = -\frac{50 \text{ mm}}{50 \text{ mm}} = \frac{\omega_3 - 10 \text{ rad/s}}{40 \text{ rad/s} - 10 \text{ rad/s}}. \tag{2b}$$

Example 7.6 (cont.)

Therefore, the angular velocity of gear 3 is

$$\omega_3 = -20 \text{ rad/s}, \qquad\qquad Ans.$$

where the negative sign indicates that gear 3 is rotating in the direction opposite to input gear 2 – that is, clockwise.

Using Eq. (7.25), the first-order kinematic coefficient of gear 6 with respect to ring gear 1 can be written as

$$\theta'_{61/4A} = \left(-\frac{R_5}{R_6}\right)\left(\frac{R_1}{R_5}\right) = -\frac{R_1}{R_6} = \frac{\omega_6 - \omega_{4A}}{\omega_1 - \omega_{4A}}. \qquad (3a)$$

The correct sign is negative, since gears 6 and 5 are in external contact, and gears 5 and 1 are in internal contact. Substituting the given data into Eq. (3a) gives

$$-\frac{150 \text{ mm}}{50 \text{ mm}} = \frac{\omega_6 - 40 \text{ rad/s}}{0 - 40 \text{ rad/s}}. \qquad (3b)$$

Therefore, the angular velocity of gear 6 is

$$\omega_6 = 160 \text{ rad/s}, \qquad\qquad Ans.$$

where the positive result implies that gear 4 is rotating in the same direction as input gear 2 – that is, counterclockwise.

Using Eq. (7.25), the first-order kinematic coefficient of gear 5 with respect to gear 6 can be written as

$$\theta'_{56/4A} = -\frac{R_6}{R_5} = \frac{\omega_5 - \omega_{4A}}{\omega_6 - \omega_{4A}}, \qquad (4a)$$

where the negative sign denotes external contact between gears 5 and 6. Substituting the known data, the first-order kinematic coefficient is

$$\theta'_{56/4A} = -\frac{50 \text{ mm}}{50 \text{ mm}} = \frac{\omega_5 - 40 \text{ rad/s}}{160 \text{ rad/s} - 40 \text{ rad/s}}. \qquad (4b)$$

Therefore, the angular velocity of gear 5 is

$$\omega_5 = -80 \text{ rad/s}, \qquad\qquad Ans.$$

where the negative sign implies that gear 5 is rotating in the direction opposite to that of input gear 2 – that is, clockwise.

Alternate Solution

The rolling contact constraint between planet gear 3 and ring gear 1 can be written in terms of kinematic coefficients as

Example 7.6 (cont.)

$$\frac{R_3}{R_1} = \frac{50 \text{ mm}}{150 \text{ mm}} = \frac{1}{3} = \frac{\theta'_{12} - \theta'_{22}}{\theta'_{32} - \theta'_{22}}, \tag{1}$$

where the positive sign is chosen, since planet gear 3 has internal contact with ring gear 1.

Since ring gear 1 is fixed, the kinematic coefficient of the ring gear is $\theta'_{12} = 0$ and, by definition, the kinematic coefficient of the input arm 2 is $\theta'_{22} = 1$. Substituting these into Eq. (1), the kinematic coefficient of planet gear 3 with respect to input gear 2 is

$$\theta'_{32} = -2 \text{ rad/rad}, \tag{2}$$

where the negative sign indicates that gear 3 is rotating in the direction opposite to input gear 2 – that is, clockwise. Therefore, the angular velocity of gear 3 is

$$\omega_3 = \theta'_{32}\omega_2 = -20 \text{ rad/s.} \qquad\qquad Ans.$$

The rolling contact constraint between sun gear 4 and planet gear 3 can be written from Eq. (1) as

$$-\frac{R_4}{R_3} = -1 = \frac{\theta'_{32} - \theta'_{22}}{\theta'_{42} - \theta'_{22}}, \tag{3}$$

where the negative sign is chosen since planet gear 3 has external contact with sun gear 4.

Substituting Eq. (2) and $\theta'_{22} = 1$ and the known geometry into Eq. (3), the kinematic coefficient of sun gear 4 with respect to gear 2 is

$$\theta'_{42} = 4 \text{ rad/rad}, \tag{4}$$

where the positive result implies that gear 4 is rotating in the same direction as input gear 2 – that is, counterclockwise. Therefore, the angular velocity of gear 4 is

$$\omega_4 = \theta'_{42}\omega_2 = 40 \text{ rad/s ccw.} \qquad\qquad Ans.$$

The rolling contact constraint between planet gear 5 and ring gear 1 can be written from Eq. (1) as

$$\frac{R_5}{R_1} = \frac{50 \text{ mm}}{150 \text{ mm}} = \frac{1}{3} = \frac{\theta'_{12} - \theta'_{42}}{\theta'_{52} - \theta'_{42}}, \tag{5}$$

where the positive sign is used since planet gear 5 has internal contact with ring gear 1.

Substituting Eq. (4) and $\theta'_{12} = 0$, the kinematic coefficient of planet gear 5 with respect to gear 2 is

$$\theta'_{52} = -8 \text{ rad/rad}, \tag{6}$$

where the negative sign indicates that gear 5 is rotating in the direction opposite to input gear 2 – that is, clockwise. Therefore, the angular velocity of gear 5 is

$$\omega_5 = \theta'_{52}\omega_2 = -80 \text{ rad/s.} \qquad\qquad Ans.$$

Example 7.6 (cont.)

The rolling contact constraint between sun gear 6 and planet gear 5 can be written from Eq. (5) as

$$-\frac{R_6}{R_5} = -1 = \frac{\theta'_{52} - \theta'_{42}}{\theta'_{62} - \theta'_{42}}, \tag{7}$$

where the negative sign is used because of the external contact between planet gear 5 and sun gear 6.

Substituting Eqs. (4) and (6) and the known geometry into Eq. (7), the kinematic coefficient of sun gear 6 with respect to gear 2 is

$$\theta'_{62} = 16 \text{ rad/rad}, \tag{8}$$

where the positive result implies that gear 6 is rotating in the same direction as input gear 2. Therefore, the angular velocity of gear 6 is

$$\omega_6 = \theta'_{62}\omega_2 = 160 \text{ rad/s ccw}. \qquad\qquad Ans.$$

Example 7.7

Consider the planetary gear train shown in the schematic diagram in Fig. 7.42. The radius of the fixed sun gear is $R_1 = 100$ mm; the radius of input gear 2 is $R_2 = 100$ mm; the radius of ring gear 4 is $R_4 = 200$ mm; and the radii of gears 3 and 5 are $R_3 = R_5 = 50$ mm. The angular velocity of the input shaft 2 is $\omega_2 = 10$ rad/s ccw (looking from the left). Determine the angular velocities of gears 3, 4, and 5.

Fig. 7.42 Schematic diagram of a planetary gear train.

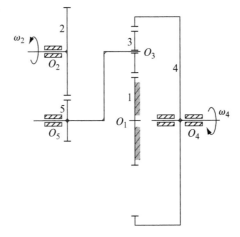

Example 7.7 (cont.)

Solution

The rolling contact constraint between input gear 2 and gear 5 can be written as

$$-\frac{R_2}{R_5} = -\frac{100 \text{ mm}}{50 \text{ mm}} = -2 = \frac{\theta'_{52}}{\theta'_{22}}. \tag{1}$$

Since the kinematic coefficient of the input shaft 2 is $\theta'_{22} = 1$ rad/rad, the kinematic coefficient of gear 5 is

$$\theta'_{52} = -2 \text{ rad/rad}, \tag{2}$$

where the negative sign implies that gear 5 is rotating in the direction opposite to input gear 2 – that is, clockwise. Therefore, the angular velocity of gear 5 is

$$\omega_5 = \theta'_{52}\omega_2 = -20 \text{ rad/s}. \hspace{2cm} Ans.$$

The rolling contact constraint between planet gear 3 and fixed sun gear 1 can be written as

$$-\frac{R_3}{R_1} = \frac{\theta'_{12} - \theta'_{5A2}}{\theta'_{32} - \theta'_{5A2}}, \tag{3}$$

which, for the given conditions, becomes

$$-\frac{50 \text{ mm}}{100 \text{ mm}} = \frac{0 + 2 \text{ rad/rad}}{\theta'_{32} + 2 \text{ rad/rad}}.$$

Therefore, the kinematic coefficient of gear 3 with respect to fixed sun gear 1 is

$$\theta'_{32} = -6 \text{ rad/rad} \tag{4}$$

and the angular velocity of gear 3 is

$$\omega_3 = \theta'_{32}\omega_2 = -60 \text{ rad/s}, \hspace{2cm} Ans.$$

where the negative sign indicates that the direction of ω_3 is clockwise.

The rolling contact constraint between ring gear 4 and planet gear 3 can be written as

$$\frac{R_3}{R_4} = \frac{\theta'_{42} - \theta'_{5A2}}{\theta'_{32} - \theta'_{5A2}}, \tag{5}$$

which, using Eqs. (2) and (4) and the given dimensions, can be written as

$$\frac{50 \text{ mm}}{200 \text{ mm}} = \frac{\theta'_{42} + 2 \text{ rad/rad}}{-6 \text{ rad/rad} + 2 \text{ rad/rad}}.$$

Example 7.7 (cont.)

Therefore, the kinematic coefficient of gear 4 with respect to fixed sun gear 1 is

$$\theta'_{42} = -3 \text{ rad/rad} \tag{6}$$

and the angular velocity of gear 4 is

$$\omega_4 = \theta'_{42}\omega_2 = -30 \text{ rad/s}, \qquad \text{Ans.}$$

where the negative sign indicates that the direction of ω_4 is clockwise.

The kinematic coefficients in Eqs. (2), (4), and (6) define the geometry of the gear train and provide the capability for obtaining the angular velocities (magnitudes and directions) in terms of the input angular velocity ω_2.

7.16 Tabular Analysis of Epicyclic Gear Trains

Another method of determining the rotational speeds of epicyclic gear trains uses the principle of superposition. The total analysis is carried out by finding the apparent rotations of the components with respect to the arm or planet carrier and then summing with the rotation of the components as if all the gears are fixed to the arm. The process is easily carried out in a tabular procedure.

Figure 7.35 shows a planetary gear train composed of a sun gear 2, planet carrier (arm) 3, planet gear 4, and internal gear 5 that is in mesh with the planet gear. Since this gear train has two degrees of freedom, we might reasonably specify the angular velocities of both the sun gear and of the arm, and wish to determine the angular velocity of the internal gear.

The analysis can be carried out in the following three steps:

1. Consider all gears (including the fixed gear, if any) to be locked to the arm, and allow the arm to rotate with angular velocity ω_A. Tabulate the angular velocities of all components under this condition as also equal to ω_A.
2. Free all constraints of step 1, fix the arm, and allow some other gear (such as the sun gear, for example) B to rotate with angular velocity $\omega_{B/A}$. Tabulate the apparent angular velocities of all other gears with respect to the arm as multiples of $\omega_{B/A}$.
3. Add the angular velocities of each gear from steps 1 and 2, and apply the given input velocities in order to find numeric values for ω_A and $\omega_{B/A}$.

A few examples will clarify how this can be done in a convenient tabular procedure.

Example 7.8

Let us assume the tooth numbers shown in Fig. 7.35, and let the angular velocities of the sun gear and the arm be $\omega_2 = 100 \text{ rev/min}$ and $\omega_3 = 200 \text{ rev/min}$, respectively, both in the counterclockwise direction, chosen positive. What is the angular velocity of internal ring gear 5?

Example 7.8 (cont.)

Solution

The solution process described by the three steps enumerated above is demonstrated in Table 7.3.

Table 7.3 Tabular analysis for **Example 7.8** and **Example 7.9**

Step	Gear 2	Arm 3	Gear 4	Gear 5
1. Gears fixed to arm	ω_2	ω_3	ω_3	ω_3
2. Arm fixed	$\omega_{2/3}$	0	$(-40/20)\omega_{2/3}$	$(20/80)(-40/20)\omega_{2/3}$
3. Total	$\omega_3 + \omega_{2/3}$	ω_3	$\omega_3 + (-40/20)\omega_{2/3}$	$\omega_3 + (20/80)(-40/20)\omega_{2/3}$

Next, comparing the given input velocities with the bottom row of columns 2 and 3, we see that, with $\omega_2 = \omega_3 + \omega_{2/3} = 100 \text{ rev/min}$ and $\omega_3 = 200 \text{ rev/min}$, then $\omega_{2/3} = -100 \text{ rev/min}$. Therefore, from the bottom row of column 5, the angular velocity of internal ring gear 5 can be written as

$$\omega_5 = \omega_3 + (20/80)(-40/20)\,\omega_{2/3};$$

that is,

$$\omega_5 = (200 \text{ rev/min}) + (-40/80)(-100 \text{ rev/min}) = 250 \text{ rev/min},$$

where the positive result implies that gear 5 is rotating in the same direction as the positive inputs – that is, counterclockwise.

Example 7.9

What is the angular velocity of internal gear 5 of Fig. 7.35 if gear 2 rotates at 100 rev/min cw while arm 3 rotates at 200 rev/min ccw?

Solution

The analysis procedure is identical to that performed in Table 7.3 for Example 7.8. However, the input velocities have now changed. Still taking counterclockwise as positive, the bottom row of columns 2 and 3 indicate that, since

$$\omega_2 = \omega_3 + \omega_{2/3} = -100 \text{ rev/min} \quad \text{and} \quad \omega_3 = 200 \text{ rev/min},$$

then $\omega_{2/3} = -300 \text{ rev/min}$. Therefore, the bottom row of column 5 indicates

$$\omega_5 = \omega_3 + (20/80)(-40/20)\omega_{2/3} = (200 \text{ rev/min}) + (-40/80)(-300 \text{ rev/min})$$
$$= 350 \text{ rev/min ccw.} \qquad \qquad Ans.$$

Example 7.10

The planetary gear train shown in Fig. 7.43 is called *Ferguson's paradox.*[4] Gear 2 is fixed to the frame. Arm 3 and gears 4, 5, and 6 are free to turn upon their shafts. Gears 2, 4, and 5 have tooth numbers of 100, 101, and 99, respectively, all cut with the same circular pitch (but with slightly different pitch circle radii) so that the thick planet gear 6 meshes with all of them. Find the angular rotations of gears 4 and 5 when arm 3 is given one counterclockwise turn.

Fig. 7.43 Ferguson's paradox.

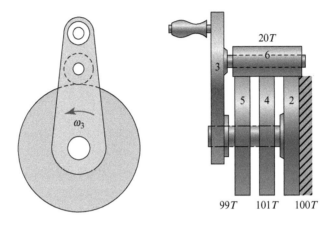

Solution

The solution is shown in Table 7.4.

Table 7.4 **Tabular analysis for Example 7.10**

Step	Gear 2	Arm 3	Gear 4	Gear 5	Gear 6
1. Gears fixed to arm	$\Delta\theta_3$	$\Delta\theta_3$	$\Delta\theta_3$	$\Delta\theta_3$	$\Delta\theta_3$
2. Arm fixed	$\Delta\theta_{2/3}$	0	$(-20/101)(-100/20)$ $\Delta\theta_{2/3}$	$(-20/99)(-100/20)$ $\Delta\theta_{2/3}$	$(-100/20)\Delta\theta_{2/3}$
3. Total	$\Delta\theta_3 + \Delta\theta_{2/3}$	$\Delta\theta_3$	$\Delta\theta_3 + (-20/101)$ $(-100/20)\Delta\theta_{2/3}$	$\Delta\theta_3 + (-20/99)$ $(-100/20)\Delta\theta_{2/3}$	$\Delta\theta_3 + (-100/20)\Delta\theta_{2/3}$

It should be noted that angular displacements are indicated in Table 7.4 rather than angular velocities. This is caused by the nature of the question asked and, recognizing that, during a chosen time interval, $\Delta\theta = \omega\Delta t$ for each of the elements. According to the problem statement, we choose the time interval Δt such that for the arm $\Delta\theta_3 = 1$ rev ccw. Then, for gear 2 to remain stationary, column 2 indicates that $\Delta\theta_3 + \Delta\theta_{2/3} = (1\text{ rev}) + \Delta\theta_{2/3} = 0$, and, therefore, $\Delta\theta_{2/3} = -1$ rev. Finally, from the bottom row of columns 4 and 5, we find that

---

[4] James Ferguson (1710–1776), Scottish physicist and astronomer, first published this device under the title *The Description and Use of a New Machine Called the Mechanical Paradox,* London, 1764.

Example 7.10 (cont.)

$$\Delta\theta_4 = \Delta\theta_3 + (-20/101)(-100/20)\Delta\theta_{2/3} = (1\,\text{rev}) + (100/101)(-1\,\text{rev}) = 1/101\,\text{rev}$$

and

$$\Delta\theta_5 = \Delta\theta_3 + (-20/99)(-100/20)\Delta\theta_{2/3} = (1\,\text{rev}) + (100/99)(-1\,\text{rev}) = -1/99\,\text{rev}.$$

Thus, when arm 3 is turned 1 rev ccw, gear 4 rotates 1/101 rev ccw, whereas gear 5 turns 1/99 rev cw.   *Ans.*

Example 7.11

The overdrive unit shown in Fig. 7.44 is sometimes used following a standard automotive transmission to further reduce engine speed. The engine speed (after the transmission) corresponds to the speed of planet carrier 3, and the drive-shaft speed corresponds to that of gear 5; sun gear 2 is held stationary. Determine the percentage additional reduction in engine speed obtained when the overdrive is active.

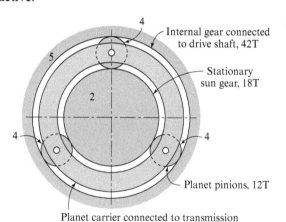

Fig. 7.44 Overdrive unit.

Internal gear connected to drive shaft, 42T

Stationary sun gear, 18T

Planet pinions, 12T

Planet carrier connected to transmission

Solution

The analysis for this problem is indicated in Table 7.5. For gear 2 to remain stationary, column 2 shows that $\omega_3 + \omega_{2/3} = 0$; therefore $\omega_{2/3} = -\omega_3$. Putting this into column 5 yields

$$\omega_5 = \omega_3 + (12/42)(-18/12)\omega_{2/3} = \omega_3 + (-18/42)(-\omega_3) = 1.429\omega_3.$$

Therefore, the percentage reduction in engine speed is

$$(1.429\omega_3 - 1.0\omega_3)/(1.429\omega_3) = 0.300 = 30\%.$$ *Ans.*

Example 7.11 (cont.)

Table 7.5 **Tabular analysis for Example 7.11**

Step	Gear 2	Arm 3	Gear 4	Gear 5
1. Gears fixed to arm	ω_3	ω_3	ω_3	ω_3
2. Arm fixed	$\omega_{2/3}$	0	$(-18/12)\omega_{2/3}$	$(12/42)(-18/12)\omega_{2/3}$
3. Total	$\omega_3 + \omega_{2/3}$	ω_3	$\omega_3 + (-18/12)\omega_{2/3}$	$\omega_3 + (12/42)(-18/12)\omega_{2/3}$

PROBLEMS

7.1 Find the module of a pair of gears having 32 and 84 teeth, respectively, whose center distance is 145 mm.

7.2 Find the number of teeth and the circular pitch of a 150 mm pitch diameter gear whose module is 3 mm/tooth.

7.3 Determine the diametral pitch of a pair of gears having 18 and 40 teeth, respectively, whose center distance is 2.9 in.

7.4 Find the number of teeth and the circular pitch of a gear whose pitch diameter is 8 in if the diametral pitch is 3 teeth/in.

7.5 Find the module and the pitch diameter of a 40-tooth gear whose circular pitch is 8.64 mm/tooth.

7.6 The pitch diameters of a pair of mating gears are 88 mm and 206 mm, respectively. If the module is 2 mm/tooth, how many teeth are there on each gear?

7.7 Find the diametral pitch and the pitch diameter of a gear whose circular pitch is 1.60 in/tooth if the gear has 36 teeth.

7.8 The pitch diameters of a pair of gears are 2.4 in and 4 in, respectively. If their diametral pitch is 10 teeth/in, how many teeth are there on each gear?

7.9 What is the diameter of a 33-tooth gear if its circular pitch is 22 mm/tooth?

7.10 A shaft carries a 30-tooth, 8 mm/tooth module gear that drives another gear at a speed of 480 rev/min. How fast does the 30-tooth gear rotate if the shaft center distance is 240 mm?

7.11 Two gears having an angular velocity ratio of 3:1 are mounted on shafts whose centers are 6 in apart. If the diametral pitch of the gears is 6 teeth/in, how many teeth are there on each gear?

7.12 A gear having a diametral pitch of 8 teeth/in and 21 teeth drives another gear at a speed of 240 rev/min. How fast is the 21-tooth gear rotating if the shaft center distance is 6.25 in?

7.13 A 6 mm/tooth module, 24-tooth pinion is to drive a 36-tooth gear. The gears are cut on the 20° full-depth involute system. Find and tabulate the addendum, dedendum, clearance, circular pitch, base pitch, tooth thickness, pitch circle radii, base circle radii, lengths of paths of approach and recess, and contact ratio.

7.14 A 5 mm/tooth module, 15-tooth pinion is to mate with a 30-tooth internal gear. The gears are 20° full-depth involute. Make a drawing of the gears showing several teeth on each gear. Can these gears be assembled in a radial direction? If not, what remedy should be used?

7.15 A 10 mm/tooth module, 17-tooth pinion, and a 50-tooth gear are paired. The gears are cut on the 20° full-depth involute system. Find the angles of approach and recess of each gear, and the contact ratio.

7.16 A gearset with a diametral pitch of 5 teeth/in has involute teeth with 22½° pressure angle,[5] and 19 and 31 teeth, respectively. They have $1.0/P$ for the addendum and $1.25/P$ for the dedendum. Tabulate the addendum, dedendum, clearance, circular pitch, base pitch, tooth thickness, base circle radii, and contact ratio.

7.17 A gear with a diametral pitch of 3 teeth/in and 22 teeth is in mesh with a rack; the pressure angle is 25°. The addendum and dedendum are $1.0/P$ and $1.25/P$, respectively. Find the lengths of the paths of approach and recess, and determine the contact ratio.

7.18 Repeat Prob. P7.15 using the 25° full-depth system.

7.19 Draw a 12 mm/tooth module, 26-tooth, 20° full-depth involute gear in mesh with a rack. (a) Find the lengths of the paths of approach and recess, and the contact ratio. (b) Draw a second rack in mesh with the same gear but offset 3 mm further away from the gear center. Determine the new contact ratio. Has the pressure angle changed?

7.20–7.24 Shaper gear cutters have the advantage that they can be used for either external or internal gears and also that only a small amount of runout is necessary at the end of the stroke. The generating action of a pinion shaper cutter can easily be simulated by employing a sheet of clear plastic. Figure P7.20 shows one tooth of a 16-tooth pinion cutter with 20° pressure angle as it can be cut from a plastic sheet. To construct the cutter, lay out the tooth on a sheet of drawing paper. Be sure to include the clearance at the top of the tooth. Draw radial lines through the pitch circle spaced at distances equal to one-fourth of the tooth thickness as shown in the figure. Next, fasten the sheet of plastic to the drawing and scribe the cutout, the pitch circle, and the radial lines onto the sheet. Then, remove the sheet and trim the tooth outline with a razor blade. Next, use a small piece of fine sandpaper to remove any burrs.

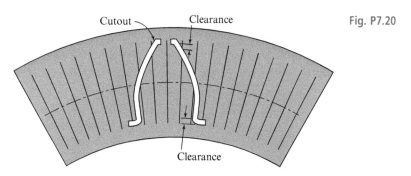

Cutout Clearance Fig. P7.20

Clearance

To generate a gear with the cutter, only the pitch circle and the addendum circle need to be drawn. Divide the pitch circle into spaces equal to those used on the template and construct radial lines through them. The tooth outlines are then obtained by rolling the template pitch circle upon that of the gear and drawing the cutter tooth lightly for each position. The resulting generated tooth upon the gear will be evident. The following problems all employ a standard 1 tooth/in diametral pitch 20° full-depth template constructed as described earlier. In each case, you should generate a few teeth and estimate the amount of undercutting.

Table P7.20–P7.24

Problem	P7.20	P7.21	P7.22	P7.23	P7.24
No. teeth	10	12	14	20	36

7.25 A 2.5 teeth/in diametral pitch gear has 17 teeth, a 20° pressure angle, an addendum of $1.0/P$, and a dedendum of $1.25/P$. Find the thickness of the teeth at the base circle and at the addendum circle. What is the pressure angle corresponding to the addendum circle?

7.26 A 15-tooth pinion has 16 mm/tooth module, 20° full-depth involute teeth. Calculate the thickness of the teeth at the base circle. What are the tooth thickness and the pressure angle at the addendum circle?

7.27 A tooth is 20 mm thick at a pitch circle radius of 200 mm and has a pressure angle of 25°. What is the thickness at the base circle?

7.28 A tooth is 39.3 mm thick at the pitch radius of 400 mm and has a pressure angle of 20°. At what radius does the tooth become pointed?

7.29 A 25° full-depth involute, 2 mm/tooth module pinion has 18 teeth. Calculate the tooth thickness at the base circle. What are the tooth thickness and pressure angle at the addendum circle?

7.30 A nonstandard 10-tooth 3 mm/tooth module involute pinion is to be cut with a 22½° pressure angle.[6] What maximum addendum can be used before the teeth become pointed?

7.31 The accuracy of cutting gear teeth can be measured by fitting hardened and ground pins in diametrically opposite tooth spaces and measuring the distance over the pins. For a 2.5 mm/tooth module 20° full-depth involute system 96-tooth gear: (a) calculate the pin diameter that will contact the teeth at the pitch lines if there is to be no backlash; (b) what should be the distance measured over the pins if the gears are cut accurately?

7.32 A set of interchangeable gears with 4 mm/tooth module is cut on the 20° full-depth involute system. The gears have tooth numbers of 24, 32, 48, and 96. For each gear, calculate the radius of curvature of the tooth profile at the pitch circle and at the addendum circle.

7.33 Calculate the contact ratio of a 17-tooth pinion that drives a 73-tooth gear. The gears are 96-tooth/in diametral pitch and cut on the 20° full-depth involute system.

[6] Such gears came from an older standard and are now obsolete.

7.34 A 25° pressure angle 11-tooth pinion is to drive a 23-tooth gear. The gears have a module of 3 mm/tooth and have involute stub teeth. What is the contact ratio?

7.35 A 22-tooth pinion mates with a 42-tooth gear. The gears have full-depth involute teeth, have a diametral pitch of 16 teeth/in, and are cut with a 17½° pressure angle.[7] Find the contact ratio.

7.36 The center distance of two 24-tooth, 20° pressure angle, full-depth involute spur gears with module of 12 mm/tooth is increased by 3 mm over the standard distance. At what pressure angle do the gears operate?

7.37 The center distance of two 18-tooth, 25° pressure angle, full-depth involute spur gears with module of 8 mm/tooth is increased by 2 mm over the standard distance. At what pressure angle do the gears operate?

7.38 A pair of mating gears have 1 mm/tooth module and are generated on the 20° full-depth involute system. If the tooth numbers are 15 and 50, what maximum addendums may they have if interference is not to occur?

7.39 A set of gears is cut with a 112 mm/tooth circular pitch and a 17½° pressure angle.[8] The pinion has 20 full-depth teeth. If the gear has 240 teeth, what maximum addendum may it have to avoid interference?

7.40 Using the method described for Probs. 7.20–7.24, cut a 25 mm/tooth module 20° pressure angle full-depth involute rack tooth from a sheet of clear plastic. Use a nonstandard clearance of $0.35m$ in order to obtain a stronger fillet. This template can be used to simulate the generating action of a hob. Now, using the variable-center-distance system, generate an 11-tooth pinion to mesh with a 25-tooth gear without interference. Record the values found for center distance, pitch radii, pressure angle, gear blank diameters, cutter offset, and contact ratio. Note that more than one satisfactory solution exists.

7.41 Using the template cut in Prob. 7.40, generate an 11-tooth pinion to mesh with a 44-tooth gear with the long-and-short-addendum system. Determine and record suitable values for gear and pinion addendum and dedendum and for the cutter offset and contact ratio. Compare the contact ratio with that of standard gears.

7.42 A pair of involute spur gears with 9 and 36 teeth are to be cut with a 20° full-depth cutter with module of 8 mm/tooth. (a) Determine the amount that the addendum of the gear must be decreased to avoid interference. (b) If the addendum of the pinion is increased by the same amount, determine the contact ratio.

7.43 A standard 20° pressure angle full-depth involute 25 mm/tooth module 20-tooth pinion drives a 48-tooth gear. The speed of the pinion is 500 rev/min. Using the position of the point of contact along the line of action as the abscissa, plot a curve indicating the sliding velocity at all points of contact. Note that the sliding velocity changes sign when the point of contact passes through the pitch point.

7.44 Find the first-order kinematic coefficient of the gear train. What are the speed and direction of rotation of gear 8?

---

[7] Such gears came from an older standard and are now obsolete.
[8] Such gears came from an older standard and are now obsolete.

Fig. P7.44

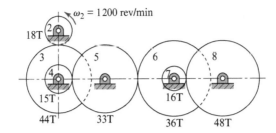

$\omega_2 = 1\,200$ rev/min

18T
15T
44T   33T   36T   48T
16T

7.45 For the given pitch diameters of a set of spur gears forming a train, compute the first-order kinematic coefficient of the train. Determine the speeds and directions of rotation of gears 5 and 7.

Fig. P7.45 $R_2 = 3.5$ in,
$R_3 = 7.5$ in, $R_4 = 4.5$ in,
$R_5 = 15$ in, $R_6 = 4.5$ in,
$R_7 = 8$ in, $\omega_2 = 120$ rev/min.

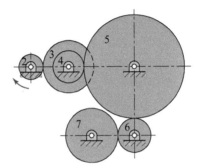

7.46 Use the truck transmission of Fig. 7.28 and an input speed of 3 000 rev/min to find the drive-shaft speed for each forward gear and for the reverse gear.

7.47 Consider the gears in a speed-change gearbox used in machine tool applications. By sliding the cluster gears on shafts B and C, nine speed changes can be obtained. The problem of the machine tool designer is to select tooth numbers for the various gears to produce a reasonable distribution of speeds for the output shaft. The smallest and largest gears are gears 2 and 9, respectively. Using 20 and 45 teeth for these gears, determine a set of suitable tooth numbers for the remaining gears. What are the corresponding speeds of the output shaft? Note that the problem has many solutions.

Fig. P7.47

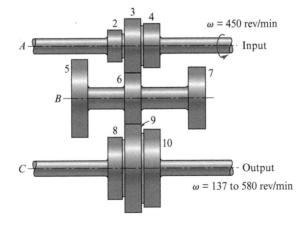

$\omega = 450$ rev/min
Input
Output
$\omega = 137$ to 580 rev/min

7.48 If internal gear 7 rotates at 60 rev/min ccw, determine the speed and direction of rotation of arm 3.

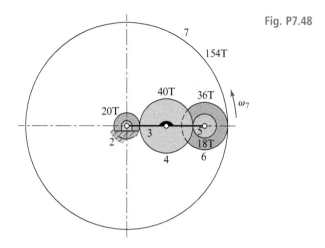

Fig. P7.48

7.49 If the arm in Fig. P7.48 rotates at 300 rev/min ccw, find the speed and direction of rotation of internal gear 7.

7.50 If shaft C is stationary and gear 2 rotates at 800 rev/min ccw, determine the speed and direction of rotation of shaft B.

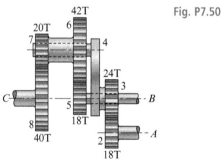

Fig. P7.50

7.51 In Fig. P7.50, shaft B is stationary and shaft C is driven at 380 rev/min ccw. Determine the speed and direction of rotation of shaft A.

7.52 In Prob. Fig. P7.50, determine the speed and direction of rotation of shaft C if: (a) shafts A and B both rotate at 360 rev/min ccw; and (b) shaft A rotates at 360 rev/min cw and shaft B rotates at 360 rev/min ccw.

7.53 Gear 2 is connected to the input shaft, and arm 3 is connected to the output shaft. Determine the speed reduction. What is the sense of rotation of the output shaft? What changes could be made in the train to produce the opposite sense of rotation for the output shaft?

458   **Spur Gears**

Fig. P7.53

108T

6

28T

20T   16T

2   3   5

4

7.54 The Lévai type L train shown in Fig. 7.36 has $N_2 = 16T$, $N_4 = 19T$, $N_5 = 17T$, $N_6 = 24T$, and $N_7 = 95T$. Internal gear 7 is fixed. Find the speed and direction of rotation of the arm if gear 2 is driven at 100 rev/min cw.

7.55 The Lévai type-A train of Fig. 7.36 has $N_2 = 20T$ and $N_4 = 32T$. (a) If the diametral pitch is $P = 4$ teeth/in, find the number of teeth on gear 5 and the crank arm radius. (b) If gear 2 is fixed and internal gear 5 rotates at 10 rev/min ccw, find the speed and direction of rotation of the arm.

7.56 The figure shows a possible arrangement of gears in a lathe headstock. Shaft A is driven by a motor at a speed of 720 rev/min. The three pinions can slide along shaft A to yield the meshes 2 with 5, 3 with 6, or 4 with 8. The gears on shaft C can also slide to mesh either 7 with 9 or 8 with 10. Shaft C is the mandrel shaft. (a) Make a table demonstrating all possible gear arrangements, beginning with the slowest speed for shaft C and ending with the highest, and enter in this table the speeds of shafts B and C. (b) If the gears all have a diametral pitch of $P = 5$ teeth/in, what must be the shaft center distances?

Fig. P7.56 $N_2 = 16T$, $N_3 = 36T$, $N_4 = 25T$, $N_5 = 64T$, $N_6 = 66T$, $N_7 = 17T$, $N_8 = 55T$, $N_9 = 79T$, and $N_{10} = 41T$.

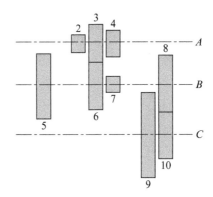

7.57 If shaft A is the output connected to the arm, and shaft B is the input driving gear 2, determine the speed ratio. Can you identify the Lévai type for this train?

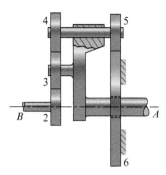

Fig. P7.57 $N_2 = 16T$, $N_3 = 18T$, $N_4 = 16T$, $N_5 = 18T$, and $N_6 = 50T$.

7.58 In Prob. 7.57, shaft B rotates at 100 rev/min cw. Find the speed of shaft A and of gears 3 and 4 about their own axes.

7.59 In the clock mechanism, a pendulum on shaft A drives an anchor (Fig. 1.12c). The pendulum period is such that one tooth of the 30T escapement wheel on shaft B is released every 2 s, causing shaft B to rotate once every minute. Note that the second (to the right) 64T gear is pivoted loosely on shaft D and is connected by a tubular shaft to the hour hand. (a) Show that the train values are such that the minute hand rotates once every hour and that the hour hand rotates once every 12 hours. (b) How many turns does the drum on shaft F make every day?

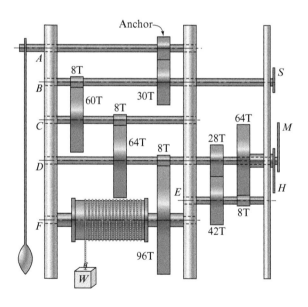

Fig. P7.59 Clockwork mechanism.

REFERENCES

[1] Lévai, Z. L., 1966. *Theory of Epicyclic Gears and Epicyclic Change-Speed Gears*, Budapest: Technical University of Building, Civil, and Transport Engineering.

[2] Mischke, C. R., 1980. *Mathematical Model Building*, Ames, IA: Iowa State University Press.

[3] Pennock, G. R., and J. J. Alwerdt, 2007. Duality between the kinematics of gear trains and the statics of beam systems, *Mech. Mach. Theory* 42(11): 512–526.

[4] Shigley, J. E., and C. R. Mischke, 2001. *Mechanical Engineering Design*, 6th ed., Boston, MA: McGraw-Hill.

8   Helical Gears, Bevel Gears, Worms, and Worm Gears

When rotational motion is to be transmitted between parallel shafts, engineers often prefer to use spur gears, since they are easy to design and very economical to manufacture. However, sometimes the design requirements are such that helical gears are a better choice. This is especially true when the loads are heavy, the speeds are high, or the noise level must be kept low.

When motion is to be transmitted between shafts which are not parallel, spur gears cannot be used; the designer must then choose between crossed-helical, bevel, hypoid, or worm gears. Bevel gears have straight teeth, line contact, and high efficiency. Crossed-helical and worm gears have a much lower efficiency because of their increased sliding action; however, if good engineering is used, crossed-helical and worm gears may be designed with quite acceptable values of efficiency. Bevel and hypoid gears are used for similar applications, and, although hypoid gears have inherently stronger teeth, their efficiency is often much less. Worm gears are used when a very small velocity ratio (first-order kinematic coefficient) is required.

8.1   Parallel-Axis Helical Gears

The shape of the tooth of a helical gear is shown in Fig. 8.1. If a piece of paper is cut into the shape of a parallelogram and wrapped around a cylinder, the angular edge of the paper wraps into a helix. The cylinder plays the same role as the base cylinder of a spur gear in Chapter 7. If the paper is unwound, each point on the angular edge generates an involute curve, as indicated in Section 7.3 for spur gears. The surface obtained when every point on the angular edge of the paper generates an involute is called an *involute helicoid*. If we imagine the strip of paper as unwrapping from a base cylinder on one gear and wrapping up onto the base cylinder of another, then a slanted line on this strip of paper generates two involute helicoids meshing as two tangent tooth shapes.

The initial contact of spur gear teeth, as we saw in Chapter 7, is a line extending across the face of the tooth. The initial contact of helical gear teeth starts as a point and changes into a line as the teeth come into more engagement; in helical gears, however, the line is diagonal across

461

Fig. 8.1 An involute helicoid.

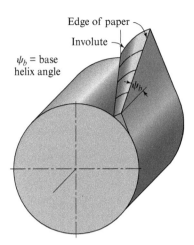

Edge of paper

Involute

ψ_b = base
helix angle

ψ_b

the face of the tooth. It is this gradual engagement of the teeth and the smooth transfer of load from one tooth to another that give helical gears their ability to quietly transmit heavy loads at high speeds.

8.2   Helical Gear Tooth Relations

As shown in Fig. 8.2, two parallel shaft helical gears must have equal pitches and equal helix angles to mesh properly, but must be of opposite hand. Helical gears with the same hand, a right-hand driver and a right-hand driven gear, for example, can be meshed with their axes skewed and are referred to as *crossed-axis gears* (Section 8.7).

Figure 8.3 represents a portion of the top view of a helical rack. Lines *AB* and *CD* are the centerlines of two adjacent helical teeth taken on the pitch plane. The angle ψ is the helix angle and is measured at the pitch diameter unless otherwise specified. The distance *AC* in the plane of rotation of the gear is the *transverse circular pitch*, p_t. The distance *AE* is the *normal circular pitch*, p_n, and is related to the transverse circular pitch as follows:

$$p_n = p_t \cos \psi. \tag{8.1}$$

Distance *AD* is called the *axial pitch*, p_x, and can be written as

$$p_x = \frac{p_t}{\tan \psi}. \tag{8.2}$$

The *normal diametral pitch* P_n can be written as

$$P_n = \frac{\pi}{p_n} = \frac{\pi}{p_t \cos \psi} = \frac{P_t}{\cos \psi}, \tag{8.3}$$

where P_t is the *transverse diametral pitch*.

Fig. 8.2 Pair of helical gears in mesh. Note the opposite hand of the two gears. (Courtesy of Kwhisky/iStock via Getty Images.)

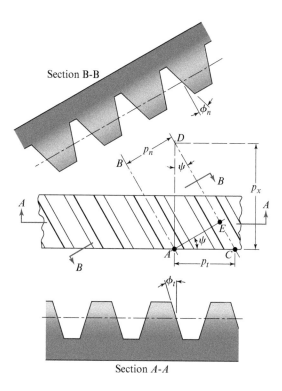

Fig. 8.3 Helical gear tooth relations.

Because of the angularity of the teeth, we must define two different pressure angles, the *normal pressure angle*, ϕ_n, and the *transverse pressure angle*, ϕ_t, both shown in Fig. 8.3. The two pressure angles are related by

$$\tan \phi_n = \tan \phi_t \cos \psi. \tag{8.4}$$

In applying these equations, it is convenient to remember that all equations and relations that are valid for a spur gear apply equally for the transverse plane of a helical gear.

A better picture of the tooth relations can be obtained by an examination of Fig. 8.4. To obtain the geometric relations, a helical gear has been cut by the oblique plane AA at an angle ψ to a

Fig. 8.4

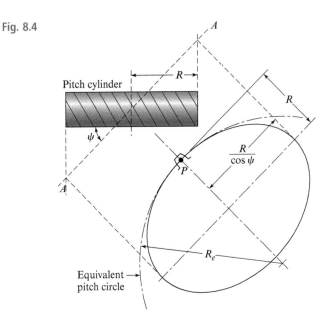

right section. For convenience, only the pitch cylinder of radius R is given. Figure 8.4 shows that the intersection of the AA plane and the pitch cylinder is an ellipse whose radius at the pitch point P is R_e. This is called the *equivalent pitch radius*, and it is the radius of curvature of the pitch surface in the normal cross-section. For the condition that $\psi = 0$, this radius of curvature is $R_e = R$. If we imagine the angle ψ to be gradually increased from 0 to 90°, we see that R_e begins at a value of $R_e = R$ and increases until, when $\psi = 90°$, the value of $R_e = \infty$.

It is demonstrated in a note at the end of this chapter (Section 8.17) that

$$R_e = \frac{R}{\cos^2 \psi},\qquad(8.5)$$

where R is the pitch radius of the helical gear and R_e is the pitch radius of an equivalent spur gear. This equivalence is taken on the normal section of the helical gear.

Let us define the number of teeth on the helical gear as N and that on the equivalent spur gear as N_e. Then,

$$N_e = 2R_e P_n.$$

Using Eqs. (8.3) and (8.5) we can write this as

$$N_e = 2\frac{R}{\cos^2 \psi}\frac{P_t}{\cos \psi} = \frac{N}{\cos^3 \psi}.\qquad(8.6)$$

8.3   Helical Gear Tooth Proportions

Except for fine pitch (normal diametral pitch of 20 teeth/in and finer), there is no generally accepted standard for the proportions of helical gear teeth.

In determining the tooth proportions for helical gears, it is necessary to consider the manner in which the teeth are formed. If the helical gear is hobbed, then tooth proportions are calculated in a plane normal to the tooth. As a general guide, tooth proportions are then often based on a normal pressure angle of $\phi_n = 20°$. Most of the proportions used for spur gears, given in Table 7.2, can then be used. The tooth proportions are calculated by using the normal diametral pitch, P_n. These proportions are suitable for helix angles from 0° to 30°, and all helix angles can be cut with the same hob. Of course, the normal diametral pitch of the hob and the gear are the same.

If the gear is to be cut by a shaper, an alternative set of tooth proportions is used, based on a transverse pressure angle of $\phi_t = 20°$ and the transverse diametral pitch, P_t. For these gears, the helix angles are generally restricted to 15°, 23°, 30°, or 45°; helix angles greater than 45° are not recommended. The normal diametral pitch, P_n, must be used to compute the tooth dimensions; the proportions given in Table 7.2 are usually satisfactory. If the shaper method is used, however, the same cutter cannot be used to cut both spur and helical gears.

8.4   Contact of Helical Gear Teeth

For spur gears, contact between meshing teeth occurs along a line that is parallel to their axes of rotation. As shown in Fig. 8.5, contact between meshing helical gear teeth occurs along a diagonal line. When contact of another tooth is just beginning at A, contact at the other end of the tooth may have already progressed from B to C.

Several kinds of contact ratios are used in evaluating the performance of helical gearsets. The *transverse contact ratio* is designated by m_t and is the contact ratio in the transverse plane. It is obtained exactly as was m_c for spur gears.

The *normal contact ratio*, m_n, is the contact ratio in the normal section. It is also determined exactly as was the contact ratio m_c for spur gears, but the dimensions of equivalent spur gears must be used in the determination. The base helix angle ψ_b and the pitch helix angle ψ, for helical gears, are related by

$$\tan \psi_b = \tan \psi \cos \phi_t. \tag{8.7}$$

Then the transverse and normal contact ratios are related by

$$m_n = \frac{m_t}{\cos^2 \psi_b}. \tag{8.8}$$

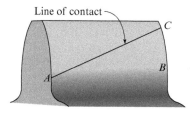

Line of contact

Fig. 8.5 Line of contact on a helical gear tooth.

The *axial contact ratio*, m_x, also called the *face contact ratio*, is the ratio of the face width of the gear to the axial pitch, determined from Eq. (8.2). It is given by

$$m_x = \frac{F}{p_x} = \frac{F \tan \psi}{p_t},\tag{8.9}$$

where F is the face width of the helical gear. Figure 8.5 shows that this contact ratio is greater than unity when another tooth is beginning contact, solely because of the helix angle of the teeth before the previous tooth contact has finished. Note that this face contact ratio, also called *overlap*, has no parallel for spur gears, and note that, because of the helix angle, this face contact ratio can be made greater than unity for helical gears by the choice of face width despite the choice of tooth size. If the face width is made greater than the axial pitch, continuous contact of at least one tooth is assured. This means that fewer teeth may be used on helical pinions than on spur pinions. The overlapping action also results in smoother operation of the gears. Note also that the face contact ratio depends solely on the geometry of a single gear, whereas the transverse and normal contact ratios depend upon the geometry of a pair of mating gears.

The *total contact ratio* is the sum of the face contact ratio m_x and the transverse contact ratio m_t. In a sense this sum gives the average total number of teeth in contact.

8.5   Replacing Spur Gears with Helical Gears

Because of their ability to carry heavy loads at high speed with little noise, it is sometimes desirable to replace a pair of spur gears by parallel shaft helical gears, although the cost may be slightly higher. An example shows the calculations.

Example 8.1

A pair of $20°$ full-depth involute spur gears with 32 and 80 teeth, diametral pitch of 16 teeth/in, and face width of 0.75 in are to be replaced by helical gears. The same hob used for the spur gears is to be used for the helical gears. The shaft center distance and the angular velocity ratio must remain the same. The helix angle is to be as small as possible, and the overlap is to be 1.5 or greater. Determine the helix angle, the numbers of teeth, and the face width of the new helical gears.

Solution

From the spur gear data and Eq. (7.1), the center distance is

$$R_2 + R_3 = \frac{N_2 + N_3}{2P} = \frac{32 \text{ teeth} + 80 \text{ teeth}}{2(16 \text{ teeth/in})} = 3.5 \text{ in.}\tag{1}$$

From Eq. (7.5), the first-order kinematic coefficient, the angular velocity ratio, is

$$\left|\theta'_{3/2}\right| = \left|\frac{\omega_3}{\omega_2}\right| = \frac{R_2}{R_3} = \frac{N_2}{N_3} = \frac{32 \text{ teeth}}{80 \text{ teeth}} = 0.4.\tag{2}$$

Example 8.1 (cont.)

Since the same hob is to be used, the normal diametral pitch, P_n, for the helical gears must also be 16 teeth/in. Since the shaft center distance must remain unchanged, that is,

$$R_2 + R_3 = \frac{N_2 + N_3}{2P_n \cos \psi} = \frac{N_2 + N_3}{2(16 \text{ teeth/in}) \cos \psi} = 3.5 \text{ in}$$

or

$$\cos \psi = \frac{N_2 + N_3}{112 \text{ teeth}} = \frac{(N_2/N_3)N_3 + N_3}{112 \text{ teeth}} = \frac{1.4N_3}{112 \text{ teeth}} = \frac{N_3}{80 \text{ teeth}}. \tag{3}$$

This implies that N_3 must be less than 80 teeth, while Eq. (2) requires that the ratio N_2/N_3 must remain 0.4, or $N_2 = 0.4N_3$.

Since $N_3 = 79$ teeth does not give an integer solution for N_2, the next smallest integer solution (which gives the smallest nonzero helix angle, ψ) is $N_3 = 75$ teeth and $N_2 = 30$ teeth, giving a helix angle of $\psi = 20.364°$. The transverse circular pitch is

$$p_t = \frac{\pi}{P_n \cos \psi} = \frac{\pi}{(16 \text{ teeth/in}) \cos 20.364°} = 0.209 \text{ in/tooth},$$

for which Eq. (8.9) indicates a face width of $F \geq 0.845$ in. Unfortunately, the space available does not allow this increase in face width. Therefore, this solution is not acceptable.

The next integer solution for tooth numbers is $N_3 = 70$ teeth and $N_2 = 28$ teeth, giving a helix angle of $\psi = 28.955°$. The transverse circular pitch is $p_t = p/(P_n \cos \psi) = 0.224$ in/tooth, and the face width is $F \geq 0.607$ in. This is an acceptable solution with face width of less than the original spur gears. The same face width of $F = 0.750$ in can still be used if desired.

8.6   Herringbone Gears

Double-helical or *herringbone gears* comprise teeth having both a right- and a left-hand helix cut on the same gear blank, as shown in Fig. 8.6. One of the primary disadvantages of the single-helical gear is the axial thrust loads that must be accounted for in the design of the bearings. In addition, the desire to obtain a good overlap without an excessively large face width may lead to the use of a comparatively larger helix angle, thus producing even higher

Fig. 8.6 A herringbone gearset. (Courtesy of Matzner Photography, Madison, WI.)

axial thrust loads. These thrust loads are eliminated by the herringbone configuration, since the axial force of the right-hand half is balanced by that of the left-hand half. Thus, with the absence of thrust reactions, helix angles arc usually larger for herringbone gears than for single-helical gears. However, one of the members of a herringbone gearset should always be mounted with some axial play or float to accommodate slight tooth errors and mounting tolerances.

For the efficient transmission of substantial amounts of power at high speeds, herringbone gears are almost universally employed.

8.7   Crossed-Axis Helical Gears

Crossed-axis helical or *spiral gears* are sometimes used when the shaft centerlines are neither parallel nor intersecting. These are essentially nonenveloping worm gears (Section 8.14), since the gear blanks have a cylindrical form with the two cylinder axes skew to each other.

The tooth action of crossed-axis helical gears is quite different from that of parallel-axis helical gears. The teeth of crossed-axis helical gears have only *point contact*. In addition, there is much greater sliding action along the tooth surfaces than for parallel-axis helical gears. For these reasons, they are chosen only to carry small loads. Because of the point contact, however, they need not be mounted accurately; either the center distance or the shaft angle may vary slightly without affecting the amount of contact.

There is no difference between crossed-axis helical gears and other helical gears until they are mounted in mesh. They are manufactured in the same way. Two meshing crossed-axis helical gears usually have the same hand; that is, a right-hand driver goes with a right-hand driven gear. The relation between thrust, hand, and rotation for crossed-axis helical gears is shown in Fig. 8.7.

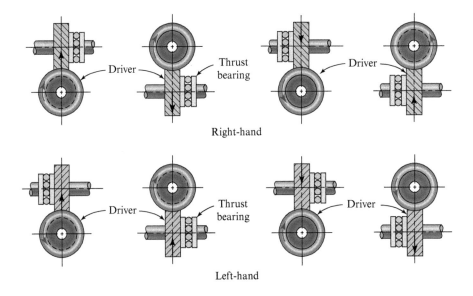

Fig. 8.7 Thrust, rotation, and hand-relations for crossed-axis helical gearing.

Table 8.1 **Tooth proportions for crossed-axis helical gears**

	Driver	Driven	Both
Helix angle	Minimum number of teeth	Helix angle	Normal pressure angle
ψ_2, deg	N_2	ψ_3, deg	ϕ_n, deg
45	20	45	14½
60	9	30	17½
75	4	15	19½
86	1	4	20

Normal diametral pitch $P_n = 1$ teeth/in; working depth $= 2.400$ in; whole depth $= 2.650$ in; addendum $a = 1.200$ in.

For crossed-axis helical gears to mesh properly, they must share the same normal pitch. When tooth sizes are chosen, the normal pitch should always be used. The reason for this is that when different helix angles are used for the driver and the driven gear, the transverse pitches are not the same. The relation between the shaft and helix angles is

$$\Sigma = \psi_2 \pm \psi_3. \tag{8.10}$$

The positive sign is used when both helix angles are of the same hand, and the negative sign is used when they are of opposite hand. Opposite-hand crossed-axis helical gears are used when the shaft angle Σ is small. The first-order kinematic coefficient, the angular velocity ratio between the shafts, is

$$\left| \theta'_{3/2} \right| = \left| \frac{\omega_3}{\omega_2} \right| = \frac{N_2}{N_3} = \frac{R_2 \cos \psi_2}{R_3 \cos \psi_3}. \tag{8.11}$$

Crossed-axis helical gears have the least sliding at the point of contact when the two helix angles are equal. If the two helix angles are not equal, the larger helix angle should be used with the driver if both gears have the same hand.

There is no widely accepted standard for crossed-axis helical gear tooth proportions. Many different combinations of proportions give good tooth action. Since the teeth are in point contact, an effort should be made to obtain a contact ratio of 2 or more. For this reason, crossed-axis helical gears are usually cut with a low pressure angle and deep teeth. The tooth proportions given in Table 8.1 are representative of good design. The driver tooth numbers indicated are the minimum required to avoid undercut. The driven gear should have 20 teeth or more if a contact ratio of 2 is to be obtained.

To illustrate the calculations for a pair of crossed-axis helical gears, consider the following example.

Example 8.2

Two shafts at an angle of 60° are to have a velocity ratio of 1/1.5. The center distance between the shafts is 8.63 in. Design a pair of crossed-axis helical gears for this application.

Example 8.2 (cont.)

Solution

Choosing $\psi_2 = 35°$ for the pinion, Eq. (8.10) then gives $\psi_3 = 25°$ for the gear. Substituting these angles into Eq. (8.11), the first-order kinematic coefficient can be written as

$$\left|\theta'_{3/2}\right| = \left|\frac{\omega_3}{\omega_2}\right| = \frac{R_2 \cos 35°}{R_3 \cos 25°} = \frac{1}{1.5}.$$

Therefore, the pitch radius of the pinion is

$$R_2 = 0.737\,6R_3.$$

This, along with the given shaft center distance, $R_2 + R_3 = 8.63$ in, gives the pitch radius of the pinion, $R_2 = 3.663$ in, and the pitch radius of the gear, $R_3 = 4.967$ in. Choosing a normal diametral pitch of $P_n = 6$ teeth/in, the numbers of teeth on the pinion and the gear, respectively, are

$$N_2 = 2P_n R_2 \cos \psi_2 = 2(6 \text{ teeth/in})(3.663 \text{ in}) \cos 35° = 36 \text{ teeth} \qquad Ans.$$

and

$$N_3 = 2P_n R_3 \cos \psi_3 = 2(6 \text{ teeth/in})(4.967 \text{ in}) \cos 25° = 54 \text{ teeth}. \qquad Ans.$$

8.8   Straight-Tooth Bevel Gears

When rotational motion is transmitted between shafts whose axes intersect, some form of bevel gears is usually used. Bevel gears have pitch surfaces that are cones, with their cone axes matching the two shaft rotation axes, as shown in Fig. 8.8. The gears are mounted so that the apexes of the two pitch cones are coincident with the point of intersection of the shaft axes. These pitch cones roll together without slipping.

Fig. 8.8 The pitch surfaces of bevel gears are cones which have only rolling contact. (Courtesy of Gleason Works, Rochester, NY.)

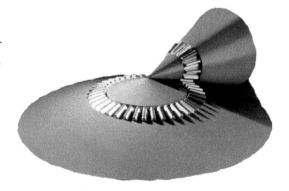

Fig. 8.9 A pair of miter gears in mesh. (Courtesy of Gleason Works, Rochester, NY.)

Although bevel gears can be designed for almost any angle between the shafts, they are often made for an angle of 90°. When the shaft intersection angle is other than 90°, the gears are called *angular bevel gears*. For the special case where the shaft intersection angle is 90° and both gears are of equal size, such bevel gears are called *miter gears*. A pair of miter gears is shown in Fig. 8.9.

For straight-tooth bevel gears, the true shape of a tooth is obtained by taking a spheric section through the tooth, where the center of the sphere is at the common apex, as shown in Fig. 8.8. As the radius of the sphere increases, the same number of teeth is projected onto a larger surface; therefore, the size of the teeth increases as larger spheric sections are taken. We have seen that the action and contact conditions for spur gear teeth may be viewed in a plane taken at right angles to the axes of the spur gears. For bevel gear teeth, the action and contact conditions should properly be viewed on a spheric surface (instead of a plane). We can even think of spur gears as a special case of bevel gears in which the spheric radius is infinite, thus producing a plane surface on which the tooth action is viewed. Figure 8.10 is typical of many straight-tooth bevel gearsets.

It is standard practice to specify the pitch diameter of a bevel gear at the large end of the teeth. In Fig. 8.11, the pitch cones of a pair of bevel gears are drawn, and the pitch radii are given as R_2 and R_3, respectively, for the pinion and the gear. The cone angles γ_2 and γ_3 are defined as the pitch angles, and their sum is equal to the shaft intersection angle, Σ; that is,

$$\Sigma = \gamma_2 + \gamma_3.$$

The first-order kinematic coefficient, the angular velocity ratio between the shafts, is obtained in the same manner as for spur gears and is

$$\left| \theta'_{3/2} \right| = \left| \frac{\omega_3}{\omega_2} \right| = \frac{R_2}{R_3} = \frac{N_2}{N_3}. \tag{8.12}$$

In the kinematic design of bevel gears, the tooth numbers of the two gears and the shaft angle are usually given, and the corresponding pitch angles are to be determined. Although they can

Fig. 8.10 A pair of straight-tooth bevel gears. (Courtesy of Waldemarus/iStock via Getty Images.)

Fig. 8.11 Pitch cones of bevel gears.

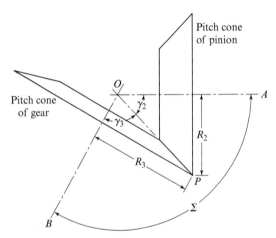

be determined graphically, the analytic approach gives more precise values. From Fig. 8.11, distance OP may be written as

$$OP = \frac{R_2}{\sin \gamma_2} = \frac{R_3}{\sin \gamma_3} \tag{a}$$

so that

$$\sin \gamma_2 = \frac{R_2}{R_3} \sin \gamma_3 = \frac{R_2}{R_3} \sin(\Sigma - \gamma_2) \tag{b}$$

or

$$\sin \gamma_2 = \frac{R_2}{R_3} (\sin \Sigma \cos \gamma_2 - \cos \Sigma \sin \gamma_2). \tag{c}$$

Dividing both sides of this equation by $\cos \gamma_2$ and rearranging gives

$$\tan \gamma_2 = \frac{R_2}{R_3}(\sin \Sigma - \cos \Sigma \tan \gamma_2).$$

Then, rearranging this equation gives

$$\tan \gamma_2 = \frac{\sin \Sigma}{(R_3/R_2) + \cos \Sigma} = \frac{\sin \Sigma}{(N_3/N_2) + \cos \Sigma}. \tag{8.13}$$

Similarly,

$$\tan \gamma_3 = \frac{\sin \Sigma}{(N_2/N_3) + \cos \Sigma}. \tag{8.14}$$

For a shaft angle of $\Sigma = 90°$, the above expressions reduce to

$$\tan \gamma_2 = \frac{N_2}{N_3} \tag{8.15}$$

and

$$\tan \gamma_3 = \frac{N_3}{N_2}. \tag{8.16}$$

The projection of bevel gear teeth onto the surface of a sphere would indeed be a difficult and time-consuming task. Fortunately, an approximation that reduces the problem to that of ordinary spur gears is common. This approximation is called *Tredgold's approximation*, and, as long as the gear has eight or more teeth, it is accurate enough for practical purposes. It is in almost universal use, and the terminology of bevel gear teeth has evolved around it.

In Tredgold's method, a *back cone* is formed of elements that are perpendicular to the elements of the pitch cone at the large end of the teeth, as shown in Fig. 8.12. The length of a back-cone

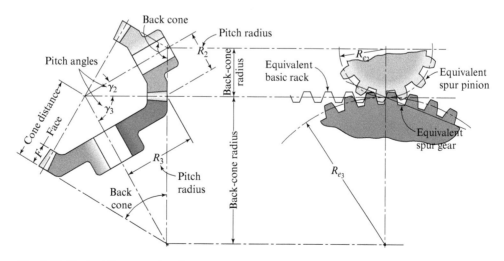

Fig. 8.12 Tredgold's approximation.

element is called the *back-cone radius*. Now an equivalent spur gear is constructed whose pitch radius, R_e, is equal to the back-cone radius. Thus, from a pair of bevel gears, using Tredgold's approximation, we can obtain a pair of equivalent spur gears that are then used to define the tooth profiles. They can also be used to determine the tooth action and the contact conditions, just as for ordinary spur gears, and the results correspond closely to those for the bevel gears.

From the geometry of Fig. 8.12, the equivalent pitch radii are

$$R_{e2} = \frac{R_2}{\cos \gamma_2}, \quad R_{e3} = \frac{R_3}{\cos \gamma_3}. \tag{8.17}$$

The number of teeth on each of the equivalent spur gears is

$$N_e = \frac{2\pi R_e}{p}, \tag{8.18}$$

where p is the circular pitch of the bevel gear measured at the large end of the teeth. Usually, the equivalent spur gears do *not* have integral numbers of teeth.

8.9 Tooth Proportions for Bevel Gears

Practically all straight-tooth bevel gears manufactured today use a 20° pressure angle. It is not necessary to use an interchangeable tooth form, since bevel gears cannot be interchanged. For this reason, the long-and-short-addendum system, described in Section 7.11, is used. The proportions are tabulated in Table 8.2.

Bevel gears are usually mounted on the outboard side of the bearings, since the shaft axes intersect, which means that the effect of shaft deflection is to pull the small end of the teeth away from mesh, causing the larger end to take more of the load. Thus, the load across the tooth is variable; for this reason, it is desirable to design a fairly short tooth. As indicated in

Table 8.2 **Tooth proportions for 20° straight-tooth bevel gears**

Item	Formula			
Working depth	$h_k = 2.0/P$			
Clearance	$c = 0.188/P + 0.002\,\text{in}$			
Addendum of gear	$a_G = \dfrac{0.540}{P} + \dfrac{0.460}{P(m_{90})^2}$			
Gear ratio	$m_G = N_G/N_P$			
Equivalent 90° ratio	$m_{90} = \begin{cases} m_G & \text{when } \Sigma = 90° \\ \sqrt{m_G \dfrac{\cos \gamma_P}{\cos \gamma_G}} & \text{when } \Sigma \neq 90° \end{cases}$			
Face width	$F = \dfrac{1}{3}$ or $F = \dfrac{10}{P}$ whichever is smaller			
Minimum number of teeth	Pinion	16	15	14	13
	Gear	16	17	20	30

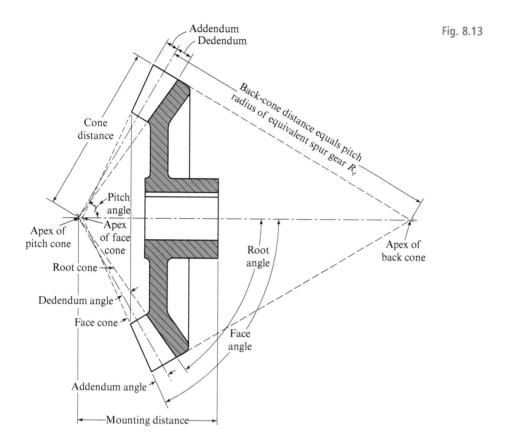

Fig. 8.13

Table 8.2, the face width is usually limited to about one-third of the cone distance. We note also that a short face width simplifies the tooling problems in cutting bevel gear teeth.

Figure 8.13 defines additional terminology characteristic of bevel gears. Note that a constant clearance is maintained by making the elements of the face cone parallel to the elements of the root cone of the mating gear. This explains why the face cone apex is not coincident with the pitch cone apex in Fig. 8.13. This permits a larger fillet than would otherwise be possible.

8.10 Bevel Gear Epicyclic Trains

The bevel gear train shown in Fig. 8.14 is called *Humpage's reduction gear*. Bevel gear epicyclic trains are used quite frequently, and they are similar to spur gear epicyclic trains except that their axes of rotation are not all on parallel shafts. The train of Fig. 8.14 is, in fact, a double epicyclic train, and the spur gear counterpart of each can be found in Fig. 7.36. The analysis of such trains can be done by formulae the same as for spur gear trains, as shown by the following example:

Fig. 8.14 Humpage's reduction gear.

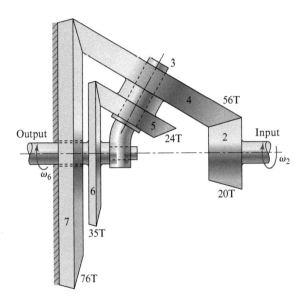

Example 8.3

In the bevel gear train in Fig. 8.14, the input is gear 2, and the output is gear 6, which is connected to the output shaft. Arm 3 turns freely on the output shaft and carries planets 4 and 5. Gear 7 is fixed to the frame. What is the output shaft speed if gear 2 rotates at 2 000 rev/min?

Solution

The problem is solved in two steps. In the first step we consider the train to be made up of gears 2, 4, and 7, and calculate the rotational speed of the arm. Thus,

$$\omega_F = \omega_2 = 2\,000 \text{ rev}/\min \quad \text{and} \quad \omega_L = \omega_7 = 0 \text{ rev}/\min,$$

and, according to Eq. (7.22),

$$\theta'_{LF} = \theta'_{72} = -\left(-\frac{56}{76}\right)\left(-\frac{20}{56}\right) = -\frac{5}{19}.$$

The negative sign is chosen because, if gear 7 were not fixed, it would appear to rotate in the direction opposite to that of gear 2, when viewed from a coordinate system fixed to arm 3.

Substituting into Eq. (7.25) and solving for the angular velocity of arm 3 gives

$$\theta'_{LF/A} = -\frac{5}{19} = \frac{0 - \omega_3}{2\,000 \text{ rev}/\min - \omega_3},$$

$$\omega_A = \omega_3 = 416.67 \text{ rev}/\min.$$

Next, we consider the train as composed of gears 2, 4, 5, and 6. Then $\omega_F = \omega_2 = 2\,000 \text{ rev}/\min$, as before, and $\omega_L = \omega_6$, which is to be determined. The first-order kinematic coefficient of the train is

Example 8.3 (cont.)

$$\theta'_{LF} = \theta'_{62} = -\left(-\frac{24}{35}\right)\left(-\frac{20}{56}\right) = -\frac{12}{49}.$$

Again, the negative sign is chosen because gear 6 appears to rotate in the direction opposite to that of gear 2, when viewed from a coordinate system fixed to arm 3.

Substituting into Eq. (7.25) again and solving for ω_6, with ω_3 known from above, now gives

$$\theta'_{LF/A} = -\frac{12}{49} = \frac{\omega_L - 416.67 \text{ rev}/\min}{2\,000 \text{ rev}/\min - 416.67 \text{ rev}/\min}.$$

Rearranging this equation, the speed of gear 6 (and the output shaft) is

$$\omega_L = \omega_6 = 28.91 \text{ rev}/\min. \qquad\qquad Ans.$$

Because the result is positive, we conclude that the output shaft rotates in the same direction as input shaft 2 with a speed reduction of 2 000:28.91, or 69.18:1.

8.11 Crown and Face Gears

If the pitch cone half-angle of one of a pair of bevel gears is made equal to 90°, the pitch cone becomes a flat surface and the resulting gear is called a *crown gear*. Figure 8.15 shows a crown gear in mesh with a bevel pinion. Note that a crown gear is the counterpart of a rack in spur gearing. The back cone of a crown gear is a cylinder, and the resulting involute teeth have straight sides, as indicated in Fig. 8.15.

A pseudo-bevel gearset can be obtained using a cylindric spur gear for a pinion in mesh with a gear having a planar pitch surface (similar to a crown gear) called a *face gear*. When the axes of the pinion and gear intersect, the face gear is called *on-center*; when the axes do not intersect, the face gear is called *off-center*.

To understand how a spur pinion, with a cylindric rather than conic pitch surface, can properly mesh with a face gear, we must consider how the face gear is formed; it is generated by

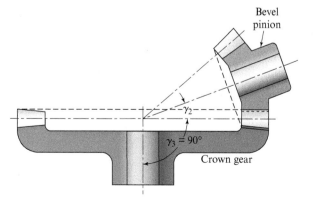

Fig. 8.15 A crown gear and bevel pinion.

a reciprocating cutter that is a replica of the spur pinion. Since the cutter and the gear blank are rotated as if in mesh, the resulting face gear is conjugate to the cutter and, therefore, to the spur pinion. The face width of the teeth on the face gear must be held quite short, however; otherwise the top land will become pointed.

Face gears are not capable of carrying heavy loads, but since the axial mounting posture of the pinion is not critical, they are sometimes more suitable for angular drives than bevel gears.

8.12 Spiral Bevel Gears

Straight-tooth bevel gears are easy to design, simple to manufacture, and give very good results in service if they are mounted accurately and positively. As in the case of spur gears, however, they become noisy at high pitch-line velocities. In such cases, it is often good design practice to use *spiral bevel gears*, which are the bevel counterparts of helical gears. A mating pair of spiral bevel gears is shown in Fig. 8.16. The pitch surfaces and the nature of contact are the same as for straight-tooth bevel gears, except for the differences brought about by the spiral-shaped teeth.

Spiral bevel gear teeth are conjugate to a basic crown rack, which can be generated as shown in Fig. 8.17 using a circular cutter. The spiral angle ψ is measured at the mean radius of the gear. As with helical gears, spiral bevel gears give much smoother tooth action than straight-tooth bevel gears and hence are useful where high speeds are required. To obtain true spiral tooth action, the face contact ratio should be at least 1.25.

Pressure angles used with spiral bevel gears are generally $14\frac{1}{2}$–$20°$, whereas the spiral angle is about 30–$35°$. As far as tooth action is concerned, the spiral may be either right- or left-handed; it makes no difference. However, looseness in the bearings might result in jamming or separating of the teeth, depending on the direction of rotation and the hand of the spiral. Since jamming of the teeth would do the most damage, the hand of the spiral should be such that the teeth tend to separate.

Zerol Bevel Gears.   The Zerol bevel gear is a patented gear that has curved teeth but a $0°$ spiral angle. An example is shown in Fig. 8.18. It has no advantage in tooth action over the

Fig. 8.16 Spiral bevel gears.

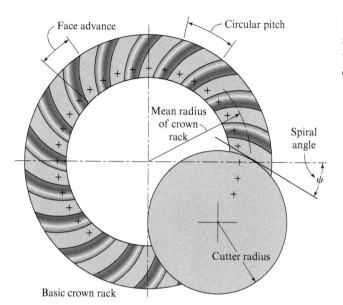

Fig. 8.17 Cutting spiral bevel gear teeth on a basic crown rack.

Fig. 8.18 Zerol bevel gears. (Courtesy of Gleason Works, Rochester, NY.)

straight-tooth bevel gear and is designed simply to take advantage of the cutting machinery used for cutting spiral bevel gears.

8.13 Hypoid Gears

It is frequently desirable, as in the case of rear-wheel drive automotive differentials, to have a gearset similar to bevel gears but where the shafts do not intersect. Such gears are called *hypoid gears*, since, as shown in Fig. 8.19, their pitch surfaces are hyperboloids of revolution.

Fig. 8.19 The pitch surfaces for hypoid gears are hyperboloids of revolution.

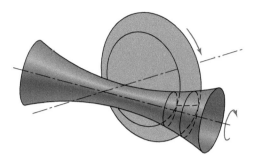

Fig. 8.20 Hypoid gears. (Courtesy of Sergey Ryzhov/Shutterstock.com.)

Figure 8.20 shows a pair of hypoid gears in mesh. The tooth action between these gears is a combination of rolling and sliding along a straight line and has much in common with that of worm gears (Section 8.14).

8.14 Worms and Worm Gears

A *worm* is a machine element having a screw-like thread, and worm teeth are frequently spoken of as *threads*. A worm meshes with a conjugate gear-like member called a *worm gear* or a *worm wheel*. Figure 8.21 shows a worm and worm gear in an application. These gears are used with nonintersecting shafts that are usually at a shaft angle of 90°, but there is no reason why shaft angles other than 90° cannot be used if a design demands it.

Worms in common use have 1–4 teeth and are said to be *single-threaded*, *double-threaded*, and so on. As we will see, there is no definite relation between the number of teeth and the pitch diameter of a worm. The number of teeth on a worm gear is usually much higher, and, therefore, the angular velocity of the worm gear is usually much lower than that of the worm.

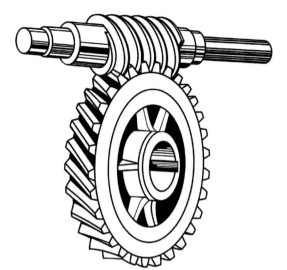

Fig. 8.21 A single-enveloping worm and worm gearset. (Courtesy of migfoto/iStock via Getty Images.)

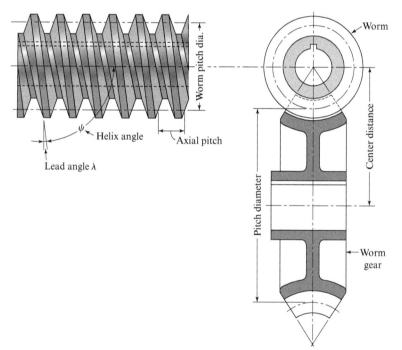

Fig. 8.22 Nomenclature of a single-enveloping worm and worm gearset.

In fact, often, one primary application for a worm and worm gear is to obtain a very large angular velocity reduction – that is, a very low first-order kinematic coefficient. In keeping with this low velocity ratio, the worm gear is usually the driven member of the pair, and the worm is usually the driving member.

A worm gear, unlike a spur or helical gear, has a face that is made concave so that it partially wraps around, or envelops, the worm, as shown in Fig. 8.22. Worms are sometimes designed

Fig. 8.23 Face width of a worm gear.

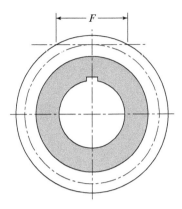

with a cylindric pitch surface, or they may have an hourglass shape, such that the worm also wraps around or partially encloses the worm gear. If an enveloping worm gear is mated with a cylindric worm, the set is said to be *single-enveloping*. When the worm is hourglass-shaped, the worm and worm gearset is said to be *double-enveloping*, since each member partially wraps around the other; such a worm is sometimes called a *Hindley worm*. The nomenclature of a single-enveloping worm and worm gearset is shown in Fig. 8.22.

A worm and worm gear combination is similar to a pair of mating crossed-helical gears except that the worm gear partially envelops the worm. For this reason, they have line contact instead of the point contact found in crossed-helical gears and are thus able to transmit more power. When a double-enveloping worm and worm gearset is used, even more power can be transmitted, at least in theory, since contact is distributed over an area on both tooth surfaces.

In a single-enveloping worm and worm gearset it makes no difference whether the worm rotates on its own axis and drives the gear by a screwing action or whether the worm is translating along its axis and drives the worm gear through rack action. The resulting motion and contact are the same. For this reason, a single-enveloping worm need not be accurately mounted along its axis. However, the worm gear should be accurately mounted along its rotation axis; otherwise, its pitch surface is not properly aligned with the worm axis. In a double-enveloping worm and worm gearset, both members are throated, and therefore both must be accurately mounted in all directions to obtain correct contact.

A mating worm and worm gear with a 90° shaft angle have the same hand of helix, but the helix angles are usually very different. The helix angle on the worm is usually quite large (at least for one or two teeth) and quite small on the worm gear. On the worm, the *lead angle* is the complement of the helix angle, as shown in Fig. 8.22. Because of this, it is customary to specify the lead angle for the worm and specify the helix angle for the worm gear. This is convenient, since these two angles are equal for a 90° shaft angle.

In specifying the pitch of a worm and worm gearset, it is usual to specify the axial pitch of the worm and the circular pitch of the worm gear. These are equal if the shaft angle is 90°. It is common to employ simple fractions, such as ¼, ⅜, ½, ¾, 1, and 1¼ in/tooth, and so on, for the circular pitch of the worm gear; there is no reason, however, why the AGMA standard diametral pitches used for spur gears (Table 7.1) should not also be used for worm gears.

The pitch radius of a worm gear is determined in the same manner as that of a spur gear; that is,

$$R_3 = \frac{N_3 p}{2\pi},$$ (8.19)

where all values are defined in the same manner as for spur gears, but refer to the parameters of the worm gear.

The pitch radius of the worm may have any value, but it should be the same as that of the hob used to cut the worm gear teeth. The relation between the pitch radius of the worm and the center distance, as recommended by AGMA, is

$$R_2 = \frac{(R_2 + R_3)^{0.875}}{4.4},$$ (8.20)

where the quantity $(R_2 + R_3)$ is the center distance in inches. This equation gives proportions that result in good power capacity. The AGMA standard also states that the denominator of Eq. (8.20) may vary from 3.4 to 6.0 without appreciably affecting the power capacity. Equation (8.20) is not required, however; other proportions will also serve well, and, in fact, power capacity may not always be the primary consideration. However, there are many variables in worm gear design, and the equation is helpful in obtaining trial dimensions.

The *lead* of a worm has the same meaning as for a screw thread and is the axial distance through which a point on the helix will move when the worm is turned through one revolution. Thus, in equation form, the lead of the worm is given by

$$l = p_x N_2,$$ (8.21)

where p_x is the axial pitch, and N_2 is the number of teeth (threads) on the worm. The lead and the *lead angle* are related as follows:

$$\lambda = \tan^{-1}\left(\frac{l}{2\pi R_2}\right),$$ (8.22)

where λ is the lead angle, as shown in Fig. 8.22.

The teeth on a worm are usually cut in a milling machine or on a lathe. Worm gear teeth are most often produced by hobbing. Except for clearance at the top of the hob teeth, the worm should be an exact duplicate of the hob in order to obtain conjugate action. This also means that, where possible, the worm should be designed using the dimensions of existing hobs.

The pressure angles used on worms and worm gearsets vary widely and should depend approximately on the value of the lead angle. Good tooth action is obtained if the pressure angle is made large enough to eliminate undercutting of the worm gear tooth on the side at which the contact ends. Recommended values are given in Table 8.3.

A satisfactory tooth depth that has about the right relation to the lead angle is obtained by making the depth a proportion of the normal circular pitch. Using an addendum of $1/P = p_n/\pi$, as for full-depth spur gears, we obtain the following proportions for worms and worm gears:

Table 8.3 **Recommended pressure angles for worm and worm gearsets**

Lead angle, λ (deg)	Pressure angle, ϕ (deg)
0–16	14½
16–25	20
25–35	25
35–45	30

$$\text{addendum} = 1.000/P = 0.318\ 3p_n,$$

$$\text{whole depth} = 2.000/P = 0.636\ 6p_n,$$

$$\text{clearance} = 0.157/P = 0.050\ 7p_n.$$

The face width of the worm gear should be obtained as shown in Fig. 8.23. This makes the face of the worm gear equal to the length of a tangent to the worm pitch circle between its points of intersection with the addendum circle.

8.15 Summers and Differentials

Figure 8.24 shows a variety of mechanisms used as computing devices. Since each of these is a two-degree-of-freedom mechanism, two input motion variables must be defined so that the motions of the remaining elements of the system are determined. The equation below each of these mechanisms indicates that the output parameter is a direct measure of the sum of the two input parameters. For this reason, such a mechanism is referred to as an *adder* or a *summer*.

The spur gear differential of Fig. 8.25 helps in visualizing its action. If planet carrier 2 is held stationary and gear 3 is turned by some amount $\Delta\theta_3$, then gear 8 turns in the opposite direction by an amount $\Delta\theta_8 = -\Delta\theta_3$. If the planet carrier is also allowed to turn, then $\Delta\theta_8 = \Delta\theta_2 - \Delta\theta_3$. It is for this reason that this two-degree-of-freedom mechanism is called a *differential*. Of course, a better force balance is obtained by employing several sets of planets equally spaced about the sun gears; sets of three are usual. Also, you will note in Fig. 8.25 that planets 4, 5, 6, and 7 are identical; by making planets 4 and 7 longer (thicker), they meet with each other and eliminate the need for planets 5 and 6.

If a spur gear differential were used on the driving axles of an automobile, then shafts A and B of Fig. 8.25 would drive the right and left wheels, respectively, whereas arm 2 receives power from a main drive shaft connected to the transmission.

It is interesting that a differential was used in China, long before the invention of the magnetic compass, to indicate geographic direction. In Fig. 8.26 each wheel of a carriage drives a vertical shaft through pin wheels. The right-hand shaft drives the upper pin wheel and the left-hand shaft drives the lower pin wheel of the differential shown in Fig. 8.27. When the cart is driven in a straight line, the upper and lower pin wheels rotate at the same speed but in opposite directions. Thus, the planet gear turns about its own center, but the axle to which it is

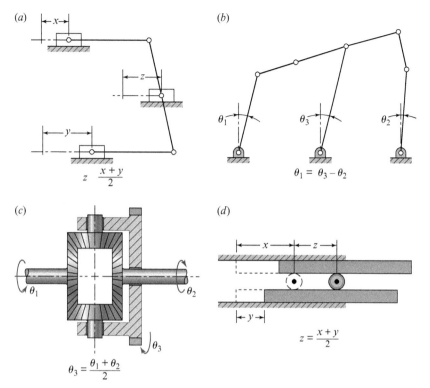

(a)

(b)

$$\theta_1 = \theta_3 - \theta_2$$

$$z \quad \frac{x+y}{2}$$

(c)

$$\theta_3 = \frac{\theta_1 + \theta_2}{2}$$

(d)

$$z = \frac{x+y}{2}$$

Fig. 8.24 Differential mechanisms used for: (a, c, d) adding; (b) subtracting; and (a, c, d) averaging two quantities.

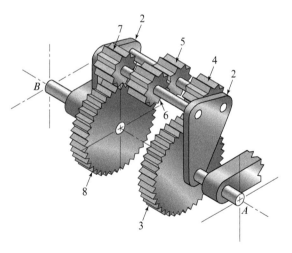

Fig. 8.25 A spur gear differential.

mounted remains stationary and so the figure continues to point in a constant direction. However, when the cart makes a turn, one of the pin wheels rotates faster than the other, causing the planet axle to turn just enough to cause the figure to continue to point in the same geographic direction as before.

Figure 8.28 is a drawing of the ordinary bevel gear automotive differential. The drive shaft pinion and the ring gear are normally hypoid gears (Section 8.13). The ring gear acts as the

Fig. 8.26 The image mounted on this chariot continues to point in a constant geographic direction. (Courtesy of Science & Society Picture Library/Contributor via Getty Images.)

Fig. 8.27 The Chinese differential.

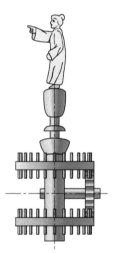

planet carrier, and its speed can be calculated as for a simple gear train when the speed of the drive shaft is given. Gears 5 and 6 are connected, respectively, to each of the rear wheels, and, when the car is traveling in a straight line, these two gears rotate in the same direction with the same speed. Thus, for straight-line motion of the car, there is no relative motion between the planet gears and gears 5 and 6. The planet gears, in effect, serve only as keys to transmit motion from the planet carrier to both wheels.

When the vehicle is making a turn, the wheel on the inside of the curve makes fewer revolutions than the wheel with the larger turning radius. Unless this difference in speed is accommodated in some manner, one or both of the tires must slip to make the turn. The

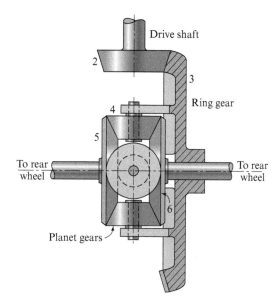

Fig. 8.28 Schematic drawing of a bevel gear automotive rear-axle differential.

Drive shaft

2

3

Ring gear

4

5

To rear wheel

To rear wheel

6

Planet gears

differential permits the two wheels to rotate at different angular velocities while, at the same time, delivering power to both. During a turn, the planet gears rotate about their own axes, thus permitting gears 5 and 6 to revolve at different angular velocities.

The purpose of the differential is to allow different speeds for the two driving wheels. In the usual differential of a rear-wheel-drive passenger car, the torque is divided approximately equally whether the car is traveling in a straight line or on a curve. Sometimes, however, the road conditions are such that the tractive effort developed by the two wheels is unequal. In such a case the total tractive effort is only twice that at the wheel having the least traction, since the differential divides the torque equally. If one wheel happens to be traveling on snow or ice, the total tractive effort possible at that wheel is very small since only a small torque is required to cause the wheel to slip. Thus, the car remains stationary with one wheel spinning and the other having only trivial tractive effort. If the car is in motion and encounters a slippery surface, then all traction as well as control is lost!

Limited-Slip Differential.   It is possible to overcome this disadvantage of the simple bevel gear differential by adding a coupling unit that is sensitive to wheel speeds. The objective of such a unit is to cause more of the torque to be directed to the slower-moving wheel. Such a combination is then called a *limited-slip differential.*

Mitsubishi, for example, utilizes a viscous coupling unit, called a VCU, which is torque-sensitive to wheel speeds. A difference in wheel speeds causes more torque to be delivered to the slower-moving wheel. A large difference, perhaps caused by the spinning of one wheel on ice, causes a much larger share of the torque to be delivered to the nonspinning wheel. The arrangement, as used on the rear axle of an automobile, is shown in Fig. 8.29.

Another approach is to employ Coulomb friction, or clutching action, in the coupling. Such a unit, as with the VCU, is engaged whenever a significant difference in wheel speeds occurs.

Fig. 8.29 Viscous coupling used on the rear axle of the Mitsubishi Galant and Eclipse GSX automobiles.

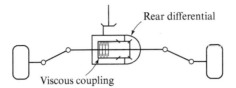

Of course, it is also possible to design a bevel gear differential that is capable of being locked by the driver whenever dangerous road conditions are encountered. This is equivalent to a solid axle and forces both wheels to move at the same speed. It seems obvious that such a differential should not be locked when the tires are on dry pavement, because of excessive tire wear caused by slipping during vehicle turns.

Fig. 8.30 The TORSEN differential used on the drive shaft of Audi automobiles. (Courtesy of Audi of America, Inc., Troy, MI.)

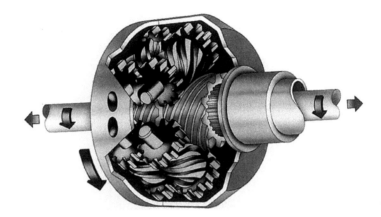

Worm-Gear Differential.   If gears 3 and 8 in Fig. 8.25 are replaced with worm gears, and planet gears 4 and 7 are replaced with mating worm wheels, then the result is a *worm-gear differential*. Of course, planet carrier 2 has to rotate about a new axis perpendicular to axle *AB*, since worm and worm wheel axes act at right angles to each other. Such an arrangement can provide the traction of a locked differential or solid axle without the penalty of restricting small differential movement between the wheels.

The worm-gear differential was invented by Mr. Vernon Gleasman of Pittsford, NY, in 1958 and was developed by Gleason Works as the TORSEN differential, a registered trademark now owned by JTEKT Torsen North America, Inc. The word TORSEN is an acronym for the words "torque-sensing" since the differential can be designed to provide any desired locking value by varying the lead angle of the worm. Figure 8.30 shows a TORSEN differential as used on Audi automobiles.[1]

---

[1] A very nice animation of a TORSEN worm-gear differential can be viewed at www.youtube.com/watch?v=Z9iPqIQ_8iM.

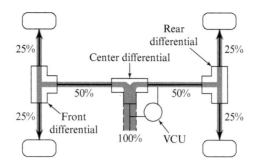

Fig. 8.31 All-wheel drive (AWD) system used on the Mitsubishi Galant, demonstrating the power distribution for straight ahead operation.

8.16 All-Wheel Drive Train

As shown in Fig. 8.31, an all-wheel drive train consists of a center differential, geared to the transmission, driving the ring gears on both the front- and rear-axle differentials. Dividing the thrusting force between all four wheels instead of only two is itself an advantage, since it allows easier handling on curves and in crosswinds.

Dr. Herbert H. Dobbs, Colonel (Ret.), an automotive engineer, states:

One of the major improvements in automotive design is antilocking braking. This provides stability and directional control when stopping by ensuring a balanced transfer of momentum from the car to the road through all wheels. As too many have found out, loss of traction at one of the wheels when braking produces unbalanced forces on the car, which can throw it out of control.

It is equally important to provide such control during starting and acceleration. As with braking, this is not a problem when a car is driven prudently and driving conditions are good. When driving conditions are not good, the problems quickly become manifold. The antilock braking system provides the answer for the deceleration portion of the driving cycle, but the devices generally available to help during the remainder of the cycle are much less satisfactory.[2]

An early solution to this problem, used by Audi, is to electrically lock the center or the rear differential, or both, when driving conditions deteriorate. Locking only the center differential causes half of the power to be delivered to the rear wheels and half to the front wheels. If one of the rear wheels, say, rests on a slippery surface such as ice, the other rear wheel has no traction. But the front wheels still provide traction. So the car has two-wheel drive. If the rear differential is then also locked, the car has three-wheel drive since the rear-wheel drive is then 50–50 distributed.

Another solution is to use a limited-slip differential as the center differential on an AWD vehicle. This then has the effect of distributing most of the driving torque to the front or rear axle depending on which is moving slower. An even better solution is to use limited-slip differentials on both the center and the rear differentials.

Unfortunately, both locking differentials and limited-slip differentials interfere with antilock braking systems. However, they are quite effective during low-speed winter operation.

[2] Dr. Herbert H. Dobbs, Rochester Hills, MI, personal communication.

The most effective solution seems to be the use of TORSEN differentials in an AWD vehicle. Here is what Dr. Dobbs has to say about their use:

If they are cut to preclude any slip, the TORSEN distributes torque proportional to available traction at the driven wheels under all conditions just like a solid axle does, but it never locks up under any circumstances. Both of the driven wheels are always free to follow the separate paths dictated for them by the vehicle's motion, but are constrained by the balancing gears to stay synchronized with each other. All this adds up to a true "TORque SENsing and proportioning" differential, which of course is where the name came from.

The result, particularly with a high performance front-wheel drive vehicle, is remarkable, to say the least. The Army has TORSENS in the High-Mobility Multipurpose Wheeled Vehicle (HMMWV) or "Hummer," which has replaced the Jeep. The only machines in production more mobile off-road than this one have tracks, and it is very capable on highway as well. It is fun to drive. The troops love it. Beyond that, Teledyne has an experimental "FAst Attack Vehicle" with TORSENS front, center, and rear. I've driven that machine over 50 mi/h on loose sand washes at Hank Hodges' Nevada Automotive Center, and it handled like it was running on dry pavement. The constant redistribution of torque to where traction was available kept all wheels driving and none digging!

8.17 Note

The equation of an ellipse with its center at the origin of an xy coordinate system with a and b as its semi-major and semi-minor axes, respectively, is

$$\frac{x^2}{a^2} + \frac{y^2}{b^2} = 1. \tag{a}$$

Also, the formula for radius of curvature is

$$\rho = \frac{\left[1 + (dy/dx)^2\right]^{3/2}}{d^2y/dx^2}. \tag{b}$$

Using these two equations, it is not difficult to find the radius of curvature corresponding to $x = 0$, $y = b$. The result is

$$\rho = a^2/b. \tag{c}$$

Then, referring to Fig. 8.3, we substitute $\rho = R_e$, $a = R/\cos\psi$, and $b = R$ into Eq. (c) and obtain Eq. (8.5).

PROBLEMS

8.1 A pair of parallel-axis helical gears has $14\frac{1}{2}°$ normal pressure angle, module of 4 mm/ tooth, and $45°$ helix angle. The pinion has 15 teeth, and the gear has 24 teeth. Calculate the transverse and normal circular pitches, the normal module, the pitch radii, and the equivalent tooth numbers.

8.2 A pair of parallel-axis helical gears are cut with a 20° normal pressure angle and a 30° helix angle. They have module of 1.5 mm/tooth and have 16 and 40 teeth, respectively. Find the transverse pressure angle, the normal circular pitch, the axial pitch, and the pitch radii of the equivalent spur gears.

8.3 A parallel-axis helical gearset is made with a 20° transverse pressure angle and a 35° helix angle. The gears have module of 2.5 mm/tooth and have 15 and 25 teeth, respectively. If the face width is 19 mm, calculate the base helix angle and the axial contact ratio.

8.4 A pair of helical gears is to be cut for parallel shafts whose center distance is to be about 88 mm to give a velocity ratio of approximately 1.8. The gears are to be cut with a standard 20° pressure angle hob whose module is 3 mm/tooth. Using a helix angle of 30°, determine the transverse values of the diametral and circular pitches and the tooth numbers, pitch radii, and center distance.

8.5 A 16-tooth helical pinion is to run at 1 800 rev/min and drive a helical gear on a parallel shaft at 400 rev/min. The centers of the shafts are to be spaced 275 mm apart. Using a helix angle of 23° and a pressure angle of 20°, determine the values for the tooth numbers, pitch radii, normal circular pitch and module, and the face width.

8.6 The catalog description of a pair of helical gears is as follows: 14½° normal pressure angle, 45° helix angle, module of 3 mm/tooth, 25 mm face width, and normal module of 203 mm/tooth. The pinion has 12 teeth and a 38 mm pitch diameter, and the gear has 32 teeth and a 100 mm pitch diameter. Both gears have full-depth teeth, and they may be purchased either right- or left-handed. If a right-hand pinion and left-hand gear are placed in mesh, find the transverse contact ratio, the normal contact ratio, the axial contact ratio, and the total contact ratio.

8.7 In a medium-size truck transmission a 22-tooth clutch-stem gear meshes continuously with a 41-tooth countershaft gear. The data indicate normal module of 289 mm/tooth, 18½° normal pressure angle, 23½° helix angle, and a 28 mm face width. The clutch-stem gear is cut with a left-hand helix, and the countershaft gear is cut with a right-hand helix. Determine the normal and total contact ratios if the teeth are cut full-depth with respect to the normal module.

8.8 A helical pinion is right-handed, has 12 teeth, has a 60° helix angle, and is to drive another gear at a velocity ratio of 3.0. The shafts are at a 90° angle, and the normal module of the gears is 3.13 mm/tooth. Find the helix angle and the number of teeth on the mating gear. What is the shaft center distance?

8.9 A right-hand helical pinion is to drive a gear at a shaft angle of 90°. The pinion has 6 teeth and a 75° helix angle and is to drive the gear at a velocity ratio of 6.5. The normal module of the gear is 2.08 mm/tooth. Calculate the helix angle and the number of teeth on the mating gear. Also determine the pitch radius of each gear.

8.10 Gear 2 rotates clockwise and drives gear 3 counterclockwise at a velocity ratio of 2:1. Use a normal module of 5 mm/tooth, a shaft angle of 50°, a shaft center distance of about 250 mm, and the same helix angle for both gears. Find the tooth numbers, the helix angles, and the exact shaft center distance.

Fig. P8.10

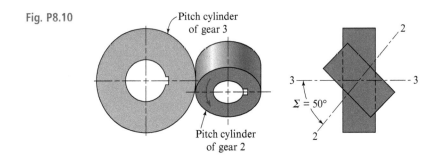

8.11 A pair of straight-tooth bevel gears is to be manufactured for a shaft angle of 90°. If the driver has 18 teeth and the velocity ratio is 3:1, what are the pitch angles?

8.12 A pair of straight-tooth bevel gears has a velocity ratio of 1.5 and a shaft angle of 75°. What are the pitch angles?

8.13 A pair of straight-tooth bevel gears is to be mounted at a shaft angle of 120°. The pinion and gear are to have 15 and 33 teeth, respectively. What are the pitch angles?

8.14 A pair of straight-tooth bevel gears with module of 12.5 mm/tooth have 19 teeth and 28 teeth, respectively. The shaft angle is 90°. Determine the pitch diameters, pitch angles, addendum, dedendum, and face width of the equivalent spur gears.

8.15 A pair of straight-tooth bevel gears with module of 3 mm/tooth have 17 teeth and 28 teeth, respectively, and a shaft angle of 105°. For each gear, calculate the pitch radius, pitch angle, addendum, dedendum, face width, and equivalent number of teeth. Make a sketch of the two gears in mesh. Use standard tooth proportions as for a 90° shaft angle.

8.16 A worm having 4 teeth and a lead of 25 mm drives a worm gear at a velocity ratio of 7.5. Determine the pitch diameters of the worm and worm gear for a center distance of 44 mm.

8.17 Specify a suitable worm and worm-gear combination for a velocity ratio of 60 and a center distance of 162.50 mm. Use an axial module of 12.50 mm/tooth.

8.18 A triple-threaded worm drives a worm gear having 40 teeth. The axial module is 31 mm/tooth, and the pitch diameter of the worm is 44 mm. Calculate the lead and lead angle of the worm. Find the helix angle and pitch diameter of the worm gear.

8.19 A triple-threaded worm with a lead angle of 20° and an axial module of 10 mm/tooth drives a worm gear with a velocity reduction of 15 to 1. Determine the following for the worm gear: (a) the number of teeth, (b) the pitch radius, (c) the helix angle, (d) the pitch radius of the worm, and (e) the center distance.

8.20 The gear train shown in Fig. P8.20 consists of bevel gears, spur gears, and a worm and worm gear. The bevel pinion is mounted on a shaft that is driven by a V-belt on pulleys. If pulley 2 rotates at 1 200 rev/min in the direction indicated, find the speed and direction of rotation of gear 9.

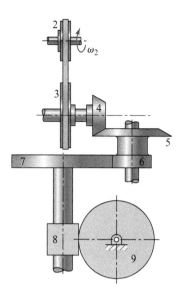

Fig. P8.20 $R_2 = 3$ in, $R_3 = 5$ in, $N_4 = 18T$, $N_5 = 38T$, $N_6 = 20T$, $N_7 = 48T$, Worm 8 3T-R.H., and $N_9 = 36T$.

8.21 The marine reduction differential shown in Fig. P8.21 has bevel gear 2 driven by the engine shaft A. Bevel planets 3 mesh with fixed crown gear 4 and are pivoted on the spider (arm), which is connected to propeller shaft B. Find the percentage speed reduction.

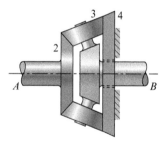

Fig. P8.21 $N_2 = 36T$, $N_3 = 21T$, and $N_4 = 52T$.

8.22 The tooth numbers for the automotive rear-axle differential shown in Fig. 8.28 are $N_2 = 17T$, $N_3 = 54T$, $N_4 = 11T$, and $N_5 = N_6 = 16T$. The drive shaft turns at 1 200 rev/min. What is the speed of the right wheel if it is jacked up and the left wheel is resting on the road surface?

8.23 A vehicle using the rear-axle differential shown in Fig. 8.28 turns to the right at a speed of 48 km/h on a curve of 24 m radius. Use the same tooth numbers as in Prob. 8.22. The tire diameter is 375 mm. Use 1.50 m as the distance between treads. Calculate the speed of each rear wheel and the speed of the ring gear.

9   Synthesis of Linkages

In previous chapters we have concentrated on the analysis of mechanisms where the dimensions of the links are known. By kinematic *synthesis* we mean the design or creation of a new mechanism to yield a desired set of motion characteristics. Because of the very large number of techniques available, this chapter presents only a few of the more useful approaches to show applications of the planar theory.[1]

9.1   Type, Number, and Dimensional Synthesis

There are three general stages in the design of a new mechanism: type synthesis, number synthesis, and dimensional synthesis. Although these three steps may not be consciously followed, they are always present in the creation of any new device.

Type synthesis refers to the choice of the kind of mechanism to be used; it might be a linkage, a geared system, a cam system, or even belts and pulleys. This beginning stage of the total design process usually involves the consideration of design and construction factors, such as manufacturing processes, materials, space, safety, and economics. The study of kinematics is usually only slightly involved in type synthesis.

Number synthesis deals with finding a satisfactory number of links and number and types of joints to obtain the desired mobility (Section 1.6). Number synthesis is the second stage in the design process and follows type synthesis.

The third stage in design, determining the detailed dimensions of the individual links, is called *dimensional synthesis*. This is the subject of the balance of this chapter.

9.2   Function Generation, Path Generation, and Body Guidance

A frequent requirement in mechanism design is that of causing an output body to rotate, oscillate, or reciprocate according to a specified function of time or function of the input

---

[1] Some of the most useful works on planar kinematic synthesis in the English language are included in the References at the end of this chapter. Extensive bibliographies may be found in [6] and [8].

motion. This is the first type of synthesis and is called *function generation*. A simple example is that of synthesizing a four-bar linkage to generate the function $y = f(x)$. In this case, x represents the motion (crank angle) of the input crank, and the linkage is to be designed so that the motion (angle) of the output rocker approximates the desired function y. A few other examples of function generation are as follows:

1. In a conveyor line the output body of a mechanism must move at the constant velocity of the conveyor while some operation is performed – for example, capping a bottle – and then it must return, pick up the next cap, and repeat.
2. The output body of a mechanism must pause or stop during its motion cycle to provide time for another event. The following event might be a sealing, stapling, or fastening operation of some kind.
3. The output body of a mechanism must rotate according to a specified velocity function, since it must be synchronized with another mechanism that requires a particular rotating motion.

Recall that in Chapter 2 we defined the location of a point to be described by the term *position*. However, the location and orientation of a rigid body and the arrangement of a mechanism or system of rigid bodies are each described by the term *posture*. This terminology continues consistently throughout this and subsequent chapters.

A second type of synthesis problem is called *path generation*. This refers to a problem in which the position of a coupler point must follow a path having a prescribed shape. Common requirements are that a portion of the path be circular, elliptic, or a straight line. Sometimes it is required that the path cross over itself, as in a figure-eight.

The third general class of synthesis problems is called *body guidance*. Here we are interested in moving a rigid body from one posture to another. The problem may call for a simple translation or a combination of translation and rotation. In the construction industry, for example, a heavy part such as a scoop or a bulldozer blade must be moved through a series of prescribed postures.

The general problem of dimensional synthesis is to design a mechanism that will guide a moving rigid body through N finitely separated postures in a single plane (where $N = 2, 3, 4, \ldots$). In general, the N postures are specified; that is, the design problem dictates the N postures to which the rigid body must move. The desire is to synthesize a planar, single-degree-of-freedom mechanism that will achieve the postures (locations and orientations) of the rigid body for each of the N postures. For example, the synthesis of a planar four-bar linkage might be the first attempt. If the rigid body that is to be guided through the N postures is the coupler link, then the synthesis problem is one of body guidance. If the rigid body is the input or output crank, then the synthesis problem is one of function generation.

In this chapter, we present a general geometric approach that can be used to synthesize a four-bar linkage for $N = 2, 3, 4$, or 5 finitely separated postures of the coupler link. We will see that the number of four-bar linkages that can be found to guide a rigid body through two finitely separated postures is ∞^6; for three finitely separated postures this number is ∞^4; for four finitely separated postures there are ∞^1 solutions; and for five finitely separated postures there are only a small finite number of solutions. In general, it is not possible to synthesize a four-bar linkage for N greater than five finitely separated postures of a rigid body unless the postures share special geometric relationships.

9.3   Two Finitely Separated Postures of a Rigid Body ($N = 2$)

Let us start with the study of simple two-posture synthesis methods.

Two-Posture Synthesis of a Slider-Crank Linkage.
The central slider-crank linkage of Fig. 9.1a has a stroke, $B_1 B_2$, equal to twice the crank radius, r_2. As shown, the extreme positions of points B_1 and B_2, also called limit postures of the slider, are determined by constructing circular arcs around O_2 of length $r_3 - r_2$ and $r_3 + r_2$, respectively. The two dimensions r_2 and r_3 can be found from these two measured lengths.

In general, the central slider-crank linkage must have r_3 larger than r_2. However, the special case of $r_3 = r_2$ results in the *isosceles slider-crank linkage*, in which the slider reciprocates along an axis through O_2, and the stroke is four times the crank radius. All points on the coupler of the isosceles slider-crank linkage generate elliptic paths. The paths generated by points on the coupler of the central slider-crank linkage of Fig. 9.1a, which is not isosceles, are not elliptic; however, they are always symmetric about the sliding axis $O_2 B$.

The linkage of Fig. 9.1b is called the *general* or *offset slider-crank linkage*. Note that the limit postures of the slider can be found in the same manner as explained above for the central slider-crank linkage. Certain special effects can be obtained by changing the offset distance, e. For example, stroke $B_1 B_2$ is always greater than twice the crank radius. Also, the crank angle required to execute the forward stroke is different from that for the return stroke. This feature can be used to synthesize quick-return mechanisms where a slower working stroke may be desired to reduce the power requirement (Section 1.7), or a quicker return stroke may be desired to increase productivity.

Two-Posture Synthesis of a Crank-Rocker Linkage.
The limit postures of the rocker in a crank-rocker linkage are shown by points B_1 and B_2 in Fig. 9.2. Note that the crank and the

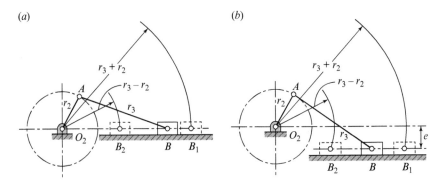

Fig. 9.1 (a) Central slider-crank linkage; (b) general or offset slider-crank linkage.

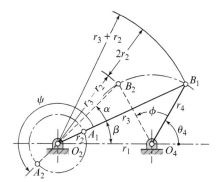

Fig. 9.2 Two limit postures of the crank-rocker linkage.

coupler form a single straight line at each of the limit postures. Also, note that the two dimensions r_2 and r_3 can be determined from measurements made in these two postures in the same way as for the slider-crank linkage.

In this case, the crank executes angle ψ while the rocker moves from B_1 to B_2 through angle ϕ. Note on the return stroke that the rocker swings from B_2 back to B_1 through the angle, $-\phi$, but the crank moves through the angle $360° - \psi$.

There are many cases in which a crank-rocker linkage is superior to a cam-and-follower system. Among the advantages over cam systems are the smaller forces involved, the elimination of the retaining spring, and the tighter clearances obtainable by the use of revolute joints.

If the direction of rotation of the input crank is chosen so that $\psi > 180°$ in Fig. 9.2, then we can define $\alpha = \psi - 180°$, and an equation for the advance-to-return ratio (Section 1.7) can be written as

$$Q = \frac{180° + \alpha}{180° - \alpha}. \tag{9.1}$$

A problem which frequently arises in the synthesis of crank-rocker linkages is to obtain the dimensions or geometry that will cause the linkage to generate a specified output displacement, ϕ, when the advance-to-return ratio Q is specified.[2]

To synthesize a crank-rocker linkage for specified values of ϕ and α, we locate an arbitrary point, O_4, in Fig. 9.3a and choose any desired rocker length, r_4; we note that this choice does not change the solution but only sets the scale of the figure. Then, we draw the two postures of link 4, that is, $O_4 B_1$ and $O_4 B_2$, separated by the given angle, ϕ. Through B_1, we construct any line X. Then, through B_2, we construct line Y at the given angle, α, to line X. The intersection of these two lines defines the location of the crank pivot, O_2. Since the orientation of line X was chosen arbitrarily, there are an infinite number of solutions to this problem.

Next, as shown in Fig. 9.3a, distance $B_2 C$ is $2r_2$, or twice the crank length. So, we can bisect this distance to find r_2. Then, the coupler length is found from $r_3 = O_2 B_1 - r_2$. The completed linkage is shown in Fig. 9.3b.

---

[2] The method described here appears in [9] and in [14]. Both [8] and [15] describe another method that yields different results.

Fig. 9.3 Synthesis of a
four-bar linkage to
generate a specified
rocker displacement, ϕ,
with a specified advance-
to-return ratio, Q.

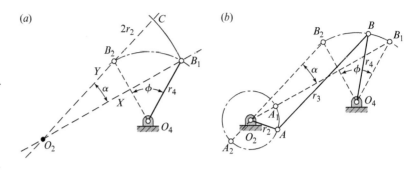

Fig. 9.4 Two finitely separated
postures of a rigid body.

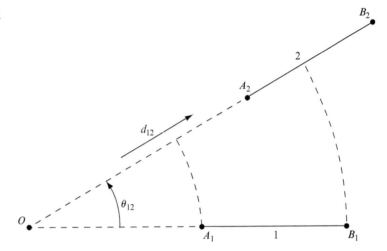

Pole of a Finite Displacement.

For two finitely separated postures of a rigid body in planar motion, the moving body can be represented by the line AB, as shown in Fig. 9.4. In the first posture the body is denoted as A_1B_1, and in the second posture the body is denoted as A_2B_2. The displacement of the body from posture 1 to posture 2 can be described by a rotation, θ_{12}, and a translation, d_{12}, as shown in Fig. 9.4, although this may not be how the body actually moves.

Note that a unique point can be located as the center of rotation of the rigid body displacement. The center of rotation is the point of intersection of the perpendicular bisector of the line connecting A_1 to A_2 and the perpendicular bisector of the line connecting B_1 to B_2. The center of rotation is henceforth referred to as the *pole* for the finite displacement and denoted as P_{12}, as shown in Fig. 9.5.

The angle of rotation of the body AB is denoted

$$\angle A_1 P_{12} A_2 = \angle B_1 P_{12} B_2 = \theta_{12} = 2\phi_{12}. \tag{9.2}$$

Therefore, the angles

$$\angle A_1 P_{12} A^* = \angle A^* P_{12} A_2 = \phi_{12},$$
$$\angle B_1 P_{12} B^* = \angle B^* P_{12} B_2 = \phi_{12}, \tag{9.3}$$

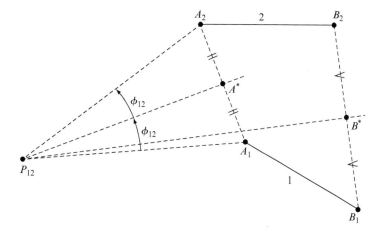

Fig. 9.5 A single pivot to guide the rigid body through a finite displacement.

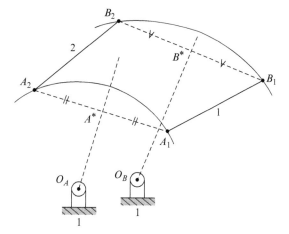

Fig. 9.6 A possible four-bar linkage.

where A^* and B^* are the midpoints of the lines A_1A_2 and B_1B_2, respectively. Note that the pole P_{ij} for the finite displacement of the body from posture i to posture j is analogous to the instantaneous center of velocity for an infinitesimal displacement of a rigid body (Section 3.12). The following two statements are important: (a) the pole for a finite displacement is located at the point of the moving body having zero displacement; and (b) for the purpose of calculating displacements, the finite displacement of the body can be considered a rotation around the pole.

A better solution than a single pivot to guide the moving body between two postures is to use a four-bar linkage, since a four-bar linkage is more stable and can support a larger load than a single pivot. A possible four-bar linkage solution is shown in Fig. 9.6. The body AB would be attached to the coupler link; the previous or new points A and B could be used as the two coupler pivots; ground pivot O_A can be chosen as any point on the perpendicular bisector A^*P_{12}, and ground pivot O_B can be chosen as any point on the perpendicular bisector B^*P_{12}.

The total number of possible four-bar linkages that can guide the body through two finitely separated postures can be obtained from the fact that there are ∞^2 choices of locations for pin A, ∞^2 choices of locations for pin B, ∞^1 choices for ground pivot O_A, and ∞^1 choices for ground pivot O_B along the perpendicular bisector lines. Therefore, there are a total of ∞^6 possible four-bar linkages that can guide the body through the two specified finitely separated postures.

It is important to note the relationships between the signs of the rotation angles and the order of their subscripts. The rotation angle through which the body turns in moving from posture 1 to posture 2 is denoted $2\phi_{12}$ [Eq. (9.2)]. Similarly, the rotation angle through which the body turns in moving from posture 2 back to posture 1 is denoted $2\phi_{21}$. Therefore, we can write that

$$2\phi_{12} + 2\phi_{21} = 0°, 360° \quad \text{or} \quad \phi_{12} + \phi_{21} = 0°, 180°. \tag{9.4}$$

Note that Eqs. (9.4) are valid when the two angles are measured in the same direction. If the angles are measured in opposite directions, then

$$\phi_{21} = -\phi_{12}. \tag{9.5}$$

An expression for the average velocity of an arbitrary point fixed in the body, say point A, can be written as

$$(V_A)_{\text{avg}} = \frac{\Delta s_A}{\Delta t}, \tag{9.6}$$

where the arc distance $\Delta s_A = A_1 A_2$. The relationship between the distance, Δs_A, and the angle, ϕ_{12}, can be obtained as follows. Consider the triangle $A_1 P_{12} A^*$ of Fig. 9.5; then,

$$\sin \phi_{12} = \frac{A_1 A^*}{P_{12} A_1},$$

which can be written as

$$A_1 A^* = (P_{12} A_1) \sin \phi_{12}. \tag{9.7}$$

Note that

$$A_1 A^* = \frac{A_1 A_2}{2} = \frac{1}{2} \Delta s_A. \tag{9.8}$$

Setting Eq. (9.8) equal to Eq. (9.7) and rearranging, the displacement can be written as

$$\Delta s_A = 2 A_1 A^* = 2(P_{12} A_1) \sin \phi_{12}. \tag{9.9}$$

Then, substituting Eq. (9.9) into Eq. (9.6) and rearranging, the average velocity of point A can be written as

$$(V_A)_{\text{avg}} = \frac{2 \sin \phi_{12}}{\Delta t} (P_{12} A_1). \tag{9.10}$$

From this equation, the magnitude of the instantaneous velocity of point A in the body can be written as

$$V_A = \lim_{\Delta t \to 0} \left(\frac{2 \sin \phi_{12}}{\Delta t} \right) P_{12} A_1. \tag{9.11}$$

Recall that the magnitude of the instantaneous velocity of point A (Chapter 3) can also be written as

$$V_A = \omega R, \tag{9.12}$$

where ω is the angular velocity of the body, and $R = P_{12}A_1$ is the distance of point A from the pole (center of rotation). Comparing Eqs. (9.11) and (9.12), we note that the angular velocity of the body can be written as

$$\omega = \lim_{\Delta t \to 0} \left(\frac{2 \sin \phi_{12}}{\Delta t} \right). \tag{9.13}$$

9.4   Three Finitely Separated Postures of a Rigid Body ($N = 3$)

Three finitely separated postures of a body, denoted 1, 2, and 3, can be specified by: (a) the poles P_{12}, P_{23}, and P_{31}; and (b) the rotation angle from posture 1 to posture 2 denoted $2\phi_{12}$, the rotation angle from posture 2 to posture 3 denoted $2\phi_{23}$, and the rotation angle from posture 3 to posture 1 denoted $2\phi_{31}$.

We can obtain the locations of the three poles in the same manner as before. We choose an arbitrary line in the body, say the line DE shown in Fig. 9.7. For the body in its first posture, the line is denoted D_1E_1; in the second posture the line is denoted D_2E_2; and in the third posture the line is denoted D_3E_3. The intersection of the perpendicular bisectors of the lines D_1D_2 and E_1E_2 is the pole P_{12}. The intersection of the perpendicular bisectors of the lines D_2D_3 and E_2E_3 is the pole P_{23}. Finally, the intersection of the perpendicular bisectors of the lines D_3D_1 and E_3E_1 is the pole P_{31}. The three poles for the three finitely separated postures of the body are shown in Fig. 9.7.

We connect poles P_{12} and P_{23} by a straight line; we connect poles P_{23} and P_{31} by a straight line; and we connect poles P_{31} and P_{12} by a straight line. These three lines form a triangle called the *pole triangle*. An understanding of the geometry of the pole triangle (for example, the

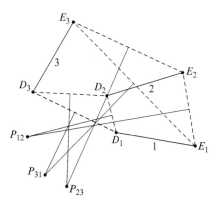

Fig. 9.7 Three poles for three postures of rigid body DE.

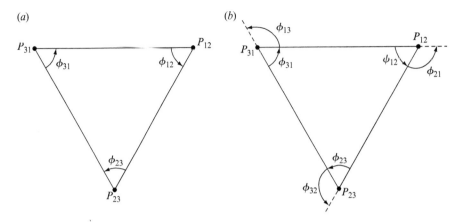

Fig. 9.8 (a) Interior angles of a pole triangle; (b) exterior angles of a pole triangle.

interior angles and the exterior angles) is essential for synthesizing a mechanism that can guide the body through the three finitely separated postures.

For convenience, the finite displacements of the rigid body in Fig. 9.8a are shown counterclockwise. For the pole triangle shown, the sides are labeled so that side 1 is the side with the common subscript 1 for the two poles ($P_{31}P_{12}$), side 2 is the side with the common subscript 2 for the two poles ($P_{12}P_{23}$), and side 3 is the side with the common subscript 3 for the two poles ($P_{23}P_{31}$).

Interior Angles of a Pole Triangle.

The order of the subscripts for the interior angles of a pole triangle are defined as follows: (a) The interior angle about the pole P_{12} from side 1 to side 2 of the pole triangle is the angle ϕ_{12} (positive ccw); (b) the interior angle about the pole P_{23} from side 2 to side 3 of the pole triangle is the angle ϕ_{23} (positive ccw); and (c) the interior angle about the pole P_{31} from side 3 to side 1 of the pole triangle is the angle ϕ_{31} (positive ccw).

Recall that these interior angles are half the rotation angles for the body; that is,

$$\phi_{12} = \frac{\theta_{12}}{2}, \quad \phi_{23} = \frac{\theta_{23}}{2}, \quad \text{and} \quad \phi_{31} = \frac{\theta_{31}}{2}. \tag{9.14}$$

Also note that, consistent with Eq. (9.14), the sum of the three interior angles of the pole triangle must be

$$\phi_{12} + \phi_{23} + \phi_{31} = 180°. \tag{9.15}$$

Exterior Angles of a Pole Triangle.

Note the following relationships, again consistent with Eq. (9.14):

$$\phi_{12} + \phi_{21} = 180°, \quad \phi_{23} + \phi_{32} = 180°, \quad \text{and} \quad \phi_{31} + \phi_{13} = 180°, \tag{9.16}$$

which are valid when the angles are measured in the same direction, as shown in Fig. 9.8b. Also, note the relationships

$$\phi_{12} = -\phi_{21}, \quad \phi_{23} = -\phi_{32}, \quad \text{and} \quad \phi_{31} = -\phi_{13}, \tag{9.17}$$

which are valid when the angles are measured in opposite directions, consistent with Eq. (9.5).

Example 9.1

Given the pole triangle shown in Fig. 9.9, where the length of side 1 is $P_{31}P_{12} = 2$ in and the interior angles are $\phi_{12} = \theta_{12}/2 = 53.13°$ ccw and $\phi_{31} = \theta_{31}/2 = 36.87°$ ccw, and also given that the location of point D fixed in the moving body in posture 1, that is, point D_1, is midway between the poles P_{31} and P_{12} and 1.5 in vertically below this line, then determine the location of point D when the rigid body is in posture 2 and in posture 3; that is, find points D_2 and D_3.

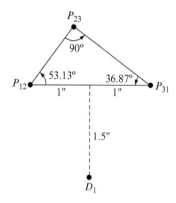

Fig. 9.9 Pole triangle and point D_1.

Solution

Note that the interior angle $\phi_{23} = \theta_{23}/2 = 90°$ ccw; therefore, the pole triangle is a 3:4:5 right-angled triangle, and the lengths of the other two sides are $P_{12}P_{23} = 1.2$ in for side 2 and $P_{23}P_{31} = 1.6$ in for side 3.

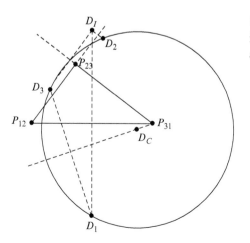

Fig. 9.10 Location of point D in the three positions (D_1, D_2, and D_3).

Example 9.1 (cont.)

A straightforward approach for finding the locations of points D_2 and D_3 is to use the so-called *image point* (also referred to as the *image pole*). The image point, denoted here D_I, is defined as that point that will give D_i $(i = 1, 2, 3)$ when reflected about the side (or the extended side) of the pole triangle with the common subscript i. In other words, D_1 is the reflection of D_I about side 1 $(P_{31}P_{12})$, D_2 is the reflection of D_I about side 2 $(P_{12}P_{23})$, and D_3 is the reflection of D_I about side 3 $(P_{23}P_{31})$. Points D_1, D_2, and D_3 are shown in Fig. 9.10.

The three points D_1, D_2, and D_3 must also lie on the circumference of a circle, that is, we know that it is always possible to find a circle that passes through the three locations defined by any arbitrary point fixed in the body as it travels through three finitely separated postures. This makes it possible to design a four-bar linkage to guide a rigid body through three finitely separated postures since pins A and B on the coupler link must be located on circular arcs.

The center of the circle that passes through the three points D_1, D_2, and D_3 is called a *center point* and is denoted D_C. All points such as D_C make up the center point system (or center system), which will be explained in more detail later. The center point, D_C, is the point of intersection of the perpendicular bisectors of the lines D_2D_3 and D_3D_1. Since the perpendicular bisector of the line D_1D_2 must also pass through the center point, D_C, this can be used to check the accuracy of the previous constructions. Also note that, consistent with the definition of a pole, the perpendicular bisector of the line D_iD_j must pass through the pole P_{ij}.

The geometric relationships can be written as follows:

$$\angle D_1P_{12}D_C = \angle D_CP_{12}D_2 = \phi_{12},$$
$$\angle D_2P_{23}D_C = \angle D_CP_{23}D_3 = \phi_{23}, \qquad (9.18)$$
$$\angle D_3P_{31}D_C = \angle D_CP_{31}D_1 = \phi_{31}.$$

Example 9.2

Assume we are given the pole triangle and the center point E_C, as shown in Fig. 9.11. The length of the side $P_{31}P_{12}$ is 2 in, the interior angles $\phi_{12} = \theta_{12}/2 = 25°$ ccw and $\phi_{31} = \theta_{31}/2 = 40°$ ccw, the distance from P_{12} to center point E_C is 1 in, and the angle $\angle P_{31}P_{12}E_C = 45°$cw, as shown in Fig. 9.11.

Fig. 9.11 Pole triangle and center point E_C.

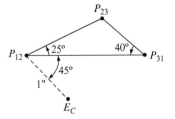

Example 9.2 (cont.)

Find the locations of point E when the body is in postures 1, 2, and 3 (that is, find E_1, E_2, and E_3), and determine the radius of the circle that passes through these three points.

Solution

The interior angle from posture 1 to posture 2 can be written as

$$\phi_{12} = \angle P_{31}P_{12}P_{23} = 25° \text{ ccw,} \tag{1a}$$

and the interior angle from posture 3 to posture 1 can be written as

$$\phi_{31} = \angle P_{23}P_{31}P_{12} = 40° \text{ ccw.} \tag{1b}$$

Therefore, the interior angle from posture 2 to posture 3 is

$$\phi_{23} = \angle P_{12}P_{23}P_{31} = 115° \text{ ccw.} \tag{1c}$$

The geometric relationships are

$$
\begin{aligned}
\angle E_1 P_{12} E_C &= \angle E_C P_{12} E_2 = \phi_{12}, \\
\angle E_2 P_{23} E_C &= \angle E_C P_{23} E_3 = \phi_{23}, \\
\angle E_3 P_{31} E_C &= \angle E_C P_{31} E_1 = \phi_{31}.
\end{aligned}
\tag{2}
$$

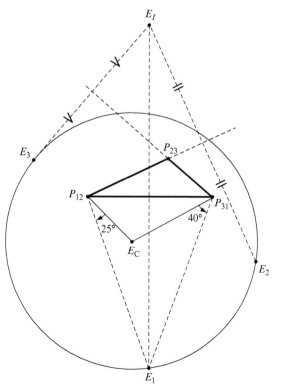

Fig. 9.12 Pole triangle and points E_1, E_2, and E_3.

Example 9.2 (cont.)

Two of these three equations can be used to locate E_1; that is, the first and the third equations can be written as

$$\angle E_C P_{12} E_1 = -\phi_{12} = 25°\text{cw},$$
$$\angle E_C P_{31} E_1 = \quad \phi_{31} = 40°\text{ ccw}. \tag{3}$$

Therefore, the point of intersection of these two lines is E_1, as shown in Fig. 9.12. The location of point E in positions 2 and 3 can be obtained from a similar procedure; that is, to locate E_2 use the first two parts of Eqs. (2), and to locate E_3 use the second and third parts. An alternative procedure is to find the image point E_I and then reflect this point about the common side 2 and the common side 3 of the pole triangle to find E_2 and E_3, respectively.

A circle can be drawn with center E_C and passing through points E_1, E_2, and E_3. The radius of this circle is measured as

$$E_C E_1 = E_C E_2 = E_C E_3 = 1.9 \text{ in.}$$

A special case, which is important in synthesis, is when the radius of the circle is infinite, that is when the circle degenerates to a straight line. This implies that point E is moving on a straight line through the three finitely separated positions. In such a case, point E would be suitable for a prismatic joint. An example of this special case is presented next.

Example 9.3

We are given the pole triangle and the point D_1, as shown in Fig. 9.13. The length of the side $P_{31} P_{12}$ is 2 in, and the interior angles are $\phi_{12} = 30°$ ccw, $\phi_{23} = 90°$ ccw, and $\phi_{31} = 60°$ ccw. The location of point D_1 is midway between the poles P_{31} and P_{12} and 1 in below this line, as shown in Fig. 9.13. Find the location of point D when the rigid body is in postures 2 and 3; that is, find the locations of D_2 and D_3.

Fig. 9.13 Pole triangle and point D_1.

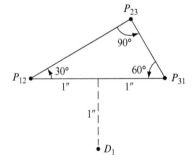

Example 9.3 (cont.)

Solution

The pole triangle is a $1:2:\sqrt{3}$ right-angled triangle; therefore, the lengths of the remaining two sides are $P_{12}P_{23} = \sqrt{3}$ in and $P_{23}P_{31} = 1$ in. Reflecting point D_1 across the side $P_{31}P_{12}$ gives the image point D_I. Then, reflecting D_I across the side $P_{12}P_{23}$ gives point D_2 and, finally, reflecting D_I across the side $P_{23}P_{31}$ gives point D_3. The points D_1, D_2, and D_3 are shown in Fig. 9.14.

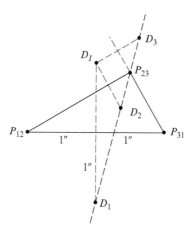

Fig. 9.14 Locations of points D_1, D_2, and D_3.

Note that the three points D_1, D_2, and D_3 lie on a straight line. Also note that this line passes through the pole P_{23}; that is, the rotation angle

$$\angle D_2 P_{23} D_3 = 2\phi_{23} = 2\angle P_{12}P_{23}P_{31} = 180° \text{ccw}.$$

The following will clarify why the path of point D, in this example, is a straight line.

Circumscribing Circle.
An important geometric property of a triangle is its circumscribing circle. A circle can always be drawn such that the three vertices of the triangle lie on the circumference of this circle, referred to as the *circumscribing circle*. The center of the circumscribing circle is the intersection of the perpendicular bisectors of the three sides of the triangle. The center of the circumscribing circle is denoted point O, as shown in Fig. 9.15.

The reflection of point O across side 1 ($P_{31}P_{12}$) is denoted O_1, that across side 2 ($P_{12}P_{23}$) is denoted O_2, and that across side 3 ($P_{23}P_{31}$) is denoted O_3. Poles P_{31} and P_{12} lie on the circumference of a circle with the same radius as the circumscribing circle and with center O_1; poles P_{12} and P_{23} lie on the circumference of a circle with the same radius as the circumscribing circle and with center O_2; and poles P_{23} and P_{31} lie on the circumference of a circle with the same radius as the circumscribing circle and with center O_3. These three circles also intersect at a unique point which is referred to as the *orthocenter* and is denoted as point H, as shown in Fig. 9.16.

Fig. 9.15 Circumscribing circle of
a pole triangle.

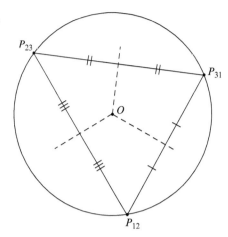

Fig. 9.15 Circumscribing circle of
a pole triangle.

Fig. 9.16 Orthocenter.

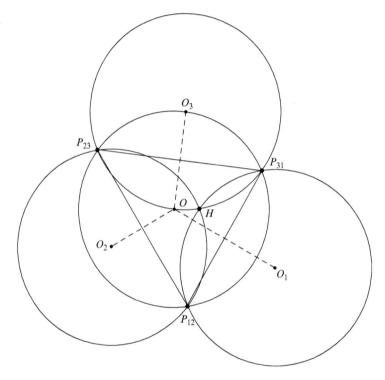

The orthocenter of a triangle is defined as the point of intersection of the lines drawn through the vertices of the triangle and perpendicular to the opposite sides. Therefore, if the pole triangle is an acute-angled triangle then the orthocenter lies inside the triangle. If the pole triangle is a right-angled triangle, then the orthocenter is coincident with a pole and is located at the apex of the right angle. If the pole triangle has an obtuse angle, then the orthocenter lies outside the triangle.

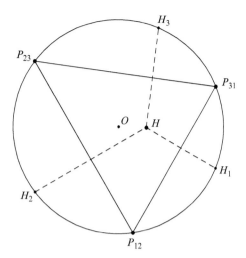

Fig. 9.17 Reflections of the orthocenter.

The reflection of the orthocenter H about side 1 ($P_{31}P_{12}$) is denoted H_1, the reflection of point H about side 2 ($P_{12}P_{23}$) is denoted H_2, and the reflection of point H about side 3 ($P_{23}P_{31}$) is denoted H_3. Note that the reflections of the orthocenter lie on the circumscribing circle of the pole triangle, as shown in Fig. 9.17.

Example 9.4

For the pole triangle given in Example 9.3, find the center of the circumscribing circle, point O, and draw the circumscribing circle. Then locate the points O_1, O_2, and O_3, and draw the circles with the same radii as the circumscribing circle and with centers at O_1, O_2, and O_3. Finally, determine the locus of all points in the body having three positions on straight lines.

Solution

Since the pole triangle from Example 9.3 is a right-angled triangle, the center of the circumscribing circle, point O, must lie at the midpoint of the hypotenuse. Note that the hypotenuse is side 1 of the pole triangle, with common subscript 1; therefore, point O_1 is coincident with O. Reflecting point O across side 2 of the pole triangle with the common 2 subscript gives O_2, and reflecting point O across side 3 of the pole triangle with the common 3 subscript gives O_3, as shown in Fig. 9.18.

Since point O_1 is coincident with O, then the circle with center at O_1 and the same radius as the circumscribing circle is coincident with the circumscribing circle. Also note that the given point D_1 in Example 9.3 lies on this circle; therefore, points D_2 and D_3 must lie on the circles with centers at O_2 and O_3. The conclusion is that the locus of points having three positions on a straight line are the circles with the same radii as the circumscribing circle and centers at O_1, O_2, and O_3. These are the only points in the body that can travel on straight lines through the three finitely separated postures.

This explains why point D in Example 9.3 travels on a straight line.

Example 9.4 (cont.)

Fig. 9.18 Locus of points having three positions on a straight line.

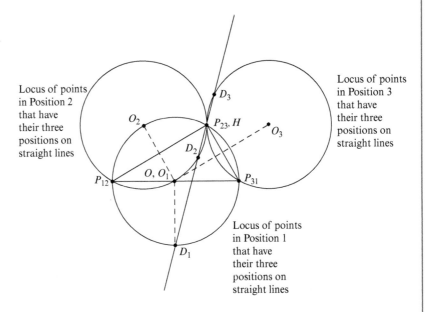

It is interesting to note that, for three infinitesimally separated postures of the body, the three circles coalesce and become the inflection circle (Section 4.12). Recall that the inflection circle is the locus of all inflection points in the body – that is, points that instantaneously travel on straight lines.

In designing a four-bar linkage to guide the coupler link through three finitely separated postures there are ∞^4 possible solutions, that is, ∞^2 choices for crankpin A in the coupler link and ∞^2 choices for crankpin B in the coupler link. Based on the choices of A and B, the locations of the ground pivots O_A and O_B are uniquely determined.

9.5   Four Finitely Separated Postures of a Rigid Body ($N = 4$)

For four finitely separated postures of a rigid body, there are six poles: (a) pole P_{12} for the finite displacement from posture 1 to posture 2; (b) pole P_{13} for the finite displacement from posture 1 to posture 3; (c) pole P_{14} for the finite displacement from posture 1 to posture 4; (d) pole P_{23} for the finite displacement from posture 2 to posture 3; (e) pole P_{24} for the finite displacement from posture 2 to posture 4; and (f) pole P_{34} for the finite displacement from posture 3 to posture 4.

The four finitely separated positions of an arbitrary point fixed in the moving body do not, in general, lie on the circumference of a circle. However, there are some points of the moving body whose four positions do lie on the circumference of a circle; these points are important in kinematic synthesis of a mechanism to guide the body through four given postures and are referred to as *circle points*. Circle points are suitable for the crankpins of a four-bar linkage with the ground pivots at the centers of the circles; these are called *center points*. For four finitely separated postures of a plane, all possible circle points lie on a curve that is referred to as the *circle point curve* and all center points lie on a curve that is referred to as the *center point curve*. These two curves are each third-degree [1] or cubic curves; that is, the curves can be described by third-order polynomials.

To design a four-bar linkage to guide a rigid body through four finitely separated postures, we only need to focus on four of the six poles. In other words, the circle point curve and the center point curve can be obtained by considering only four of the six poles. If the solution is not satisfactory then we can choose a different combination of four poles and repeat the synthesis procedure. The geometry involved is that of a pole quadrilateral referred to as an *opposite pole quadrilateral*.

Two poles that do not contain a common numeric subscript are referred to as *opposite poles*. There are three pairs of *opposite poles*, namely, (P_{12}, P_{34}), (P_{13}, P_{24}), and (P_{14}, P_{23}). Two poles that do contain a common numeric subscript are referred to as *adjacent poles*. There are 12 pairs of adjacent poles, namely, (P_{12}, P_{13}), (P_{12}, P_{14}), (P_{12}, P_{23}), (P_{12}, P_{24}), (P_{13}, P_{14}), (P_{13}, P_{23}), (P_{13}, P_{34}), (P_{14}, P_{24}), (P_{14}, P_{34}), (P_{23}, P_{24}), (P_{23}, P_{34}), and (P_{24}, P_{34}). There are a total of three opposite pole quadrilaterals; they have diagonals that connect opposite poles. Therefore, the sides of an opposite pole quadrilateral are lines connecting adjacent poles. The three possible opposite pole quadrilaterals are: (a) (P_{13}, P_{12}), (P_{12}, P_{24}), (P_{24}, P_{34}), and (P_{34}, P_{13}); (b) (P_{14}, P_{12}), (P_{12}, P_{23}), (P_{23}, P_{34}), and (P_{34}, P_{14}); and (c) (P_{14}, P_{13}), (P_{13}, P_{23}), (P_{23}, P_{24}), and (P_{24}, P_{14}). These are shown in Fig. 9.19.

Theorem. *If four positions of a point fixed in a rigid body lie on the circumference of a circle then the center of that circle (that is, the center point) views the opposite sides of an opposite pole quadrilateral under angles which are equal or differ by* 180°.

The converse is also true, namely: *If a point (say, E_C) views the opposite sides of an opposite pole quadrilateral under angles that are equal or differ by* 180°, *then that point is the center of a circle which passes through the four positions of point E of the body.*

The procedure to find circle points is shown in Fig. 9.20; it is as follows. Consider one of the three opposite pole quadrilaterals, say the opposite pole quadrilateral (P_{13}, P_{12}), (P_{12}, P_{24}), (P_{24}, P_{34}), and (P_{34}, P_{13}). Then, choose one pair of opposite sides of this quadrilateral, say the opposite sides (P_{12}, P_{24}) and (P_{13}, P_{34}). Draw a circle with the side (P_{12}, P_{24}) as a chord; that is, perpendicularly bisect the side (P_{12}, P_{24}), and choose the center of the circle as a point on this

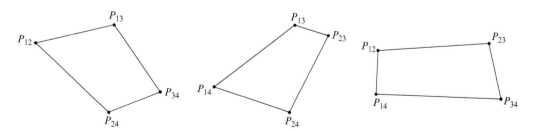

Fig. 9.19 Three possible opposite pole quadrilaterals.

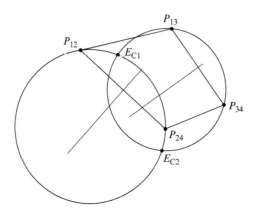

Fig. 9.20 Circle with side (P_{12}, P_{24}) as a chord.

bisector. Choose a convenient radius R, and draw the circle as shown in Fig. 9.20. The center of this circle can be chosen on the left side of the line (P_{12}, P_{24}) or on the right side of the line. Left and right are defined as follows: If we stand on pole P_{12}, for example, and look toward pole P_{24}, that is, we are looking from 1 to 4, as we ignore the common 2 subscript. So, as shown in Fig. 9.20, the center of the circle has been chosen on the right side of the line.

Now we draw a circle with the opposite side (P_{13}, P_{34}) as a chord. The radius, r, of this circle must be chosen in the ratio

$$r = \frac{P_{13}P_{34}}{P_{12}P_{24}} R. \tag{9.19}$$

The center of this circle must be chosen on the same side of the line (P_{13}, P_{34}) as the center of the first circle was chosen relative to the line (P_{12}, P_{24}). So, we stand on pole P_{13} and look toward pole P_{34}, that is, looking from 1 to 4, as we ignore the common 3 subscript, consistent with the previous procedure. The center must again be chosen on the right of the line, as shown in Fig. 9.20.

In general, these two circles will intersect in two points, denoted here as E_{C_1} and E_{C_2}, and both are possible center points. These points both satisfy the theorem; that is, they both view the opposite sides of the opposite pole quadrilateral under angles that are equal or differ by $180°$. The relationships can be written as

$$\angle P_{12}E_{C_1}P_{24} = \angle P_{13}E_{C_1}P_{34},$$
$$\angle P_{12}E_{C_2}P_{24} = \angle P_{13}E_{C_2}P_{34}, \tag{9.20}$$

or

$$\angle P_{12}E_{C_1}P_{24} = \angle P_{13}E_{C_1}P_{34} \pm 180°,$$
$$\angle P_{12}E_{C_2}P_{24} = \angle P_{13}E_{C_2}P_{34} \pm 180°. \tag{9.21}$$

The second of Eqs. (9.20) and the first of Eqs. (9.21) can be verified to be true for the example shown in Fig. 9.20.

By choosing different values for R (and r) and repeating the above procedure, another set of center points can be obtained. A curve can then be drawn through these points and is called the *center point curve.* Points on this curve are all suitable candidates for fixed pivots of the four-bar linkage. Note that all six poles, by definition, must lie on the center point curve.

Example 9.5

The locations of five of the six poles for four finitely separated postures of the coupler link of a planar four-bar linkage are as shown in Fig. 9.21. Draw an opposite pole quadrilateral. Draw the center point curve, which can be used in the synthesis of a four-bar linkage. For the specified fixed pivots of the four-bar linkage, O_A and O_B, shown in Fig. 9.21, locate the corresponding circle points A and B in the first three postures of the linkage (that is, A_1B_1, A_2B_2, and A_3B_3). Then, specify the lengths of the two links O_AA and O_BB and the length of the coupler link AB of the synthesized four-bar linkage $O_A, A, B,$ and O_B. Is the synthesized four-bar linkage a Grashof four-bar linkage or a non-Grashof four-bar linkage?

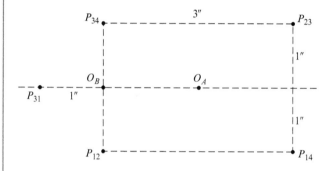

Fig. 9.21 Four finitely separated postures of a coupler link.

Solution

First, an opposite pole quadrilateral is drawn. The four sides of the opposite pole quadrilateral are $(P_{12}, P_{14}), (P_{14}, P_{34}), (P_{34}, P_{23})$, and (P_{23}, P_{12}), as shown in Fig. 9.22. Note that this is only one of three possible opposite pole quadrilaterals. However, it is the only one that can be drawn from the five poles given. Pole P_{24} is not known; therefore, pole P_{31} cannot be used to draw an opposite pole quadrilateral.

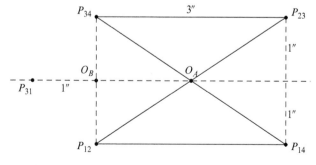

Fig. 9.22 An opposite pole quadrilateral.

The procedure to draw the center point curve is as follows:

1. Choose a side (P_{12}, P_{14}) and the opposite side (P_{34}, P_{23}) of the opposite pole quadrilateral. From the property of similar triangles,

Example 9.5 (cont.)

$$\frac{R}{r} = \frac{P_{12}P_{14}}{P_{23}P_{34}} = \frac{3 \text{ in}}{3 \text{ in}} = 1, \text{ that is, } R = r, \tag{1}$$

where R is the radius of the circle with center on the perpendicular bisector of $P_{12}P_{14}$, and r is the radius of the circle with center on the perpendicular bisector of $P_{34}P_{23}$.

2. Consider the side (P_{12}, P_{14}) of the opposite pole quadrilateral. Standing at pole P_{12}, we look at pole P_{14}; we are looking from 2 to 4, ignoring the common subscript 1. We draw a circle with radius R and with center, say, to our right on the perpendicular bisector of (P_{12}, P_{14}).

3. Now we consider the opposite side (P_{23}, P_{34}) of the opposite pole quadrilateral. Standing at pole P_{23}, we look at pole P_{34}; looking from 2 to 4, ignoring the common subscript 3. We draw a circle with radius $r = R$ and with center, again to our right on the perpendicular bisector of (P_{23}, P_{34}).

4. Identify the two points of intersection of the two circles in steps 2 and 3 as two center points.

5. Choose a different value for R (and r) and follow the above procedure (steps 2, 3, and 4) to obtain another set of center points.

6. Repeating steps 2–5 several times, we draw a curve through these center points.

This results in the center point curve shown in Fig. 9.23.

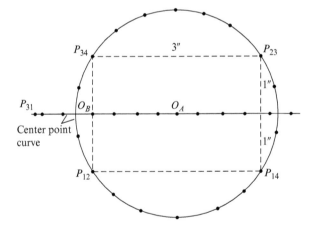

Fig. 9.23 Center point curve.

Note that any two points on this curve are suitable as fixed pivots of the four-bar linkage. Note, also, that the center point curve for this example is the straight horizontal line passing through O_A and O_B and the circle circumscribing the opposite pole quadrilateral. This is an overly simplified (and degenerate) case; that is, the cubic curve here consists of a straight line and a circle. Still, we note that the center point curve passes through the four poles of the opposite pole quadrilateral (and the given fifth pole). As mentioned previously, the poles (by definition) must always lie on the center point curve; this helps to confirm the graphic construction.

Example 9.5 (cont.)

To locate circle points A and B corresponding to the given center points O_A and O_B, consider the pole triangle formed by the poles P_{12}, P_{23}, and P_{31}; the interior angles are

$$\phi_{12} = 101.3° \text{ cw}, \quad \phi_{23} = 19.6° \text{ cw}, \quad \text{and} \quad \phi_{31} = 59.1° \text{ cw}. \tag{2}$$

Consider the center point, that is, the fixed pivot O_A; then we find the angles

$$\angle O_A P_{12} A_1 = -\phi_{12} = 101.3° \text{ ccw} \quad \text{and} \quad \angle O_A P_{31} A_1 = +\phi_{31} = 59.1° \text{ cw}. \tag{3}$$

We draw a line through pole P_{12} such that it makes an angle of 101.3° ccw from line $O_A P_{12}$. Then, we draw a line through pole P_{31} such that it makes an angle of 59.1° cw from line $O_A P_{31}$. The intersection of these two lines gives point A_1. Note that point A_1 is coincident with pole P_{31}. Next, we reflect A_1 about the side 1 $(P_{12}P_{31})$ to find image point A_I. Then, we reflect image point A_I across side 2 $(P_{12}P_{23})$ and side 3 $(P_{23}P_{31})$ to give points A_2 and A_3, respectively. The locations of points A_1, A_2, and A_3 are as shown in Fig. 9.24. Note that the image point A_I is coincident with point A_1.

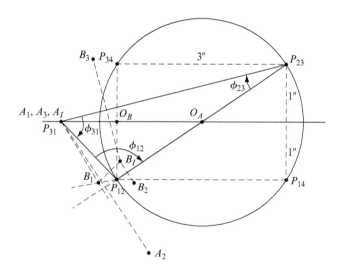

Fig. 9.24 Four-bar linkage.

Similarly, we consider the center point, that is, the fixed pivot O_B, where the angles are

$$\angle O_B P_{12} B_1 = -\phi_{12} = 101.3° \text{ ccw} \quad \text{and} \quad \angle O_B P_{31} B_1 = +\phi_{31} = 59.1° \text{ cw}. \tag{4}$$

Then, we follow the same procedure as above to obtain points B_1, B_2, and B_3. The locations of points B_1, B_2, and B_3, are as shown in Fig. 9.24.

The lengths of the links of the synthesized four-bar linkage $O_A A B O_B$ are measured and found to be

$$AB = 1.25 \text{ in} = p, \quad O_A O_B = 1.5 \text{ in} = q, \quad O_A A = 2.5 \text{ in} = l, \quad \text{and} \quad O_B B = 1.2 \text{ in} = s.$$

Example 9.5 (cont.)

From the Grashof criterion, Eq. (1.6), a planar four-bar linkage has a crank if and only if

$$s + l \leq p + q,$$

where l is the length of the longest link, s is the length of the shortest link, and p and q are the lengths of the remaining two links. From the measurements

$$s + l = 1.2 \text{ in} + 2.5 \text{ in} = 3.7 \text{ in} \quad \text{and} \quad p + q = 1.25 \text{ in} + 1.5 \text{ in} = 2.75 \text{ in}.$$

Therefore, the synthesized planar four-bar linkage is a non-Grashof four-bar linkage and does not have a continuously rotating crank.

Example 9.6

For a rigid body in plane motion, posture 2 coincides with posture 1, and posture 4 coincides with posture 3. The locations of the six poles for four finitely separated postures of the moving body are as shown in Fig. 9.25. Draw the center point curve that can be used in the synthesis of a four-bar linkage to guide the rigid body through the four finitely separated postures.

Fig. 9.25 Four finitely separated postures of a rigid body.

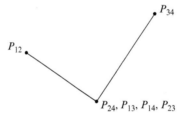

Solution

Following the procedure that was outlined in the previous example, the center point curve is as shown in Fig. 9.26.

Fig. 9.26 Center point curve.

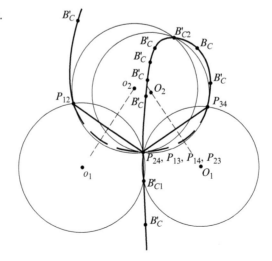

9.6   Five Finitely Separated Postures of a Rigid Body ($N = 5$)

It may be possible to guide a rigid body through five specified finitely separated postures using a four-bar linkage. However, to do this, we must find two points belonging to the rigid body and having their five positions on the circumferences of two circles. If such circle points exist, they must lie at the intersection of two circle point curves. For example, one could plot the circle point curve that is associated with postures 1, 2, 3, and 4 and then plot the circle point curve that is associated with postures 1, 2, 3, and 5. The points of intersection of these two circle point curves are called *Burmester points* [8], and they are possible locations for the coupler pivots of the four-bar linkage. If the two center points corresponding to the two chosen coupler pivots lie in locations where there are no practical obstructions to establishing these center points as ground pivots for the linkage, then a workable design may be possible.

We probably agree that it is not likely that these particular points will also happen to fall on the circle point curve of yet another group of four postures of the body. Therefore, it is generally *not* possible to synthesize a planar four-bar linkage that passes through more than five arbitrarily prescribed postures.

It is worth noting that a four-bar linkage designed using the methods presented in this section may not be workable even though the theory has been applied correctly. Since we are dealing with finitely separated postures of the coupler link, and we are not exerting any control over intermediate postures, it is possible that the designed four-bar linkage may not pass through the desired motion because of intermediate limiting postures.

9.7   Precision Postures; Structural Error; Chebyshev Spacing

The synthesis examples in the preceding sections are of the body guidance type. However, it was pointed out that linkages of the function generation type can also be synthesized by kinematic inversion if we consider the motion of the input with respect to the output. That is, if x is the orientation of the input, and y is the orientation of the output, then, for function generation, we are trying to find the dimensions of a linkage for which the input–output relationship fits a given functional relationship,

$$y = f(x). \tag{a}$$

In general, however, a mechanism has only a limited number of design parameters, a few link lengths, starting angles for the input and output, and a few more. Therefore, except for very special cases, a linkage synthesis problem usually has no exact solution over its entire range of travel.

In the preceding sections, we have chosen to work with two, three, four, or even five postures of the linkage, called *precision postures*, and to seek a linkage that exactly satisfies the desired requirements at these few chosen postures. Our implicit assumption has been that if the design fits the specifications at these few postures, then it will probably deviate only slightly from the desired motion between the precision postures, and that the deviation will probably be acceptably small. *Structural error* is defined as the theoretical difference between the function produced by the synthesized linkage and the function originally prescribed. For many function generation problems, the structural error in a four-bar linkage solution can be held to less than 4% [4]. We should note, however, that structural error usually exists even with no *graphic error* resulting from a graphic solution process and even with no *mechanical error*, which stems from imperfect manufacturing tolerances.

Of course, the amount of structural error in the solution can be affected by the choice of the precision postures. One objective of linkage design is to select a set of precision postures for use in the synthesis procedure that keeps this structural error acceptably small.

Although it is not perfect, a very good trial for the distribution of these precision postures is called *Chebyshev* spacing [6]. For N precision postures in the range $x_0 \le x \le x_{N+1}$, Chebyshev spacing is given by

$$x_j = \frac{1}{2}(x_{N+1} + x_0) - \frac{1}{2}(x_{N+1} - x_0)\cos\frac{(2j-1)\pi}{2N}, \qquad j = 1, 2, \ldots, N. \qquad (9.22)$$

As an example, suppose we wish to design a linkage to generate the function

$$y = x^{0.8} \qquad (b)$$

over the range $1 \le x \le 3$ using three precision postures.

From Eq. (9.22), the three values of x_j are

$$x_1 = \frac{1}{2}(3+1) - \frac{1}{2}(3-1)\cos\frac{(2-1)\pi}{2(3)} = 2 - \cos\frac{\pi}{6} = 1.134,$$

$$x_2 = 2 - \cos\frac{3\pi}{6} = 2.000,$$

$$x_3 = 2 - \cos\frac{5\pi}{6} = 2.866.$$

From Eq. (*b*), we find the corresponding values of y to be

$$y_1 = 1.106, \quad y_2 = 1.741, \quad \text{and} \quad y_3 = 2.322.$$

Chebyshev spacing of the precision postures is also easily determined using the graphic approach shown in Fig. 9.27. As shown in Fig. 9.27a, a circle is first constructed whose diameter is equal to the range Δx, given by

Fig. 9.27 Graphic construction for Eq. (9.22): (*a*) $N = 5$; (*b*) $N = 3$.

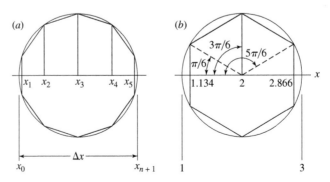

$$\Delta x = x_{N+1} - x_0. \tag{c}$$

Next, we inscribe a regular polygon having $2N$ sides in this circle, with its first side placed perpendicular to the x axis. Perpendiculars dropped from each jth vertex now intersect the diameter Δx at the precision posture value of x_j. Figure 9.27b shows the construction for this numeric example of Eq. (9.22).

It should be noted that Chebyshev spacing is a good approximation of precision postures that usually reduce structural error in a design. Depending on the accuracy requirements of the problem, it may be satisfactory. If additional accuracy is required, then by plotting a curve of structural error versus x, we can usually determine visually the adjustments to be made in the choice of precision postures for another trial.

Before closing this section, however, we should note two more problems that can arise to confound the designer in choosing precision postures for synthesis. These are called *branch defect* and *order defect* [19]. Branch defect refers to a possible completed design that meets all of the prescribed requirements at each of the precision postures but that cannot be moved continuously between these postures without being taken apart and reassembled. Order defect refers to a completed linkage design that can reach all of the precision postures, but not in the desired order.

9.8   Overlay Method

Synthesis of a function-generator mechanism, say, using the overlay method, is one of the easiest and quickest of all methods of synthesis in use. It is not always possible to obtain a solution, and sometimes the accuracy may be rather poor. Theoretically, however, one can approximate as many precision postures as are desired in the process.

Let us design a function-generator linkage to generate the function

$$y = x^{0.8}, \quad 1 \le x \le 3. \tag{a}$$

Suppose we choose six precision postures of the linkage for this example and use uniform spacing of the output. Table 9.1 indicates the rounded values of x and y and the corresponding angles selected for the input, ψ, and output, ϕ.

Table 9.1 Rounded values of input and output angles

Posture	x	Input angle ϕ, deg	y	Output angle ψ, deg
1	1	0	1	0
2	1.363	21.8	1.281	14.4
3	1.748	44.9	1.563	28.8
4	2.150	69.0	1.845	43.2
5	2.569	94.1	2.127	57.6
6	3.000	120.0	2.408	72.0

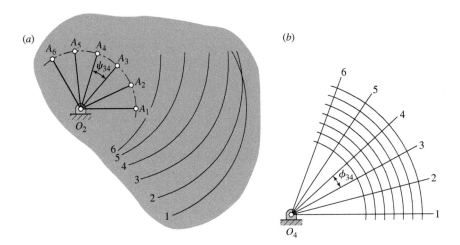

Fig. 9.28.

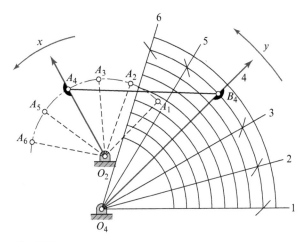

Fig. 9.29.

The first step in the synthesis is shown in Fig. 9.28a. We use a sheet of tracing paper and construct the input, O_2A, in all six postures. This requires an arbitrary choice for the length of O_2A. Also, on this sheet, we choose another arbitrary length for the coupler AB and draw arcs numbered 1–6 using points A_1 to A_6, respectively, as centers.

Now, on another sheet of paper, we construct the output, whose length is unknown, in all six postures, as shown in Fig. 9.28b. Through point O_4 we draw a number of arbitrarily spaced arcs intersecting the lines O_41, O_42, and so on; these represent possible lengths of the output rocker.

As the final step, we lay the tracing over the drawing and manipulate it in an effort to find a fit. In this case, a good fit is found, and the result is shown in Fig. 9.29.

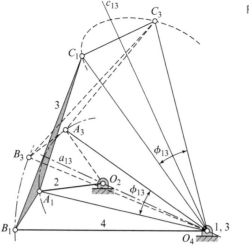

Fig. 9.30.

9.9 Coupler-Curve Synthesis[3]

In this section, we synthesize a four-bar linkage so that a tracing point on the coupler traces a specified path when the linkage is moved. Then, in Sections 9.10–9.14, we discover that paths having certain characteristics are particularly useful, for example, in synthesizing linkages having dwells of the output for certain periods of rotation of the input.

In synthesizing a linkage to generate a specified path, we can choose up to six precision points along the path. If the synthesis is successful, the tracing point will pass through each precision position. Because of branch or order defects, the final result may or may not approximate the desired path.

Two postures of a four-bar linkage are shown in Fig. 9.30. Link 2 is the input; it is connected at A to coupler 3, containing the tracing point C, and connected to output link 4 at B. Two postures of the linkage are shown by subscripts 1 and 3. Points C_1 and C_3 are two positions of the tracing point on the path to be generated. In this example, C_1 and C_3 have been especially selected so that the perpendicular bisector c_{13} passes through O_4. Note, for this selection of points, that the angle $\angle C_1 O_4 C_3$ is equal to the angle $\angle A_1 O_4 A_3$, as indicated in Fig. 9.30.

The advantage of making these two angles equal is that when the linkage is finally synthesized, the triangles $C_3 A_3 O_4$ and $C_1 A_1 O_4$ are congruent. Thus, if the tracing point is made to pass through C_1 on the path, it will also pass through C_3.

To synthesize a linkage so that the coupler point will pass through four precision positions, we locate any four points, $C_1, C_2, C_3,$ and C_4, on the desired path (Fig. 9.31). Choosing C_1 and C_3, say, we first locate O_4 anywhere on the perpendicular bisector, c_{13}. Then, with O_4 as a center and using any radius R, we construct a circular arc. Next, with centers at C_1 and C_3, and any other radius, r, we strike arcs to intersect the arc of radius R. These two intersections define points A_1 and A_3 on the input link. We construct the perpendicular

---

[3] The methods presented here were devised and presented by Hain in [8].

Fig. 9.31.

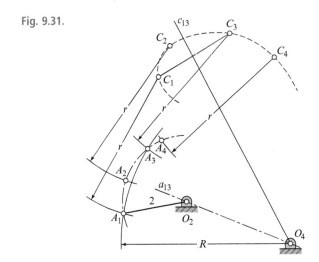

bisector, a_{13}, to A_1A_3 and note that it passes through O_4. We locate O_2 anywhere on a_{13}. This provides an opportunity to choose a convenient length for the input link. Now we use O_2 as a center and draw the crank circle through A_1 and A_3. Points A_2 and A_4 on this circle are obtained by striking arcs of radius r again about C_2 and C_4. This completes the first phase of the synthesis; we have located O_2 and O_4 relative to the desired path and hence defined distance O_2O_4. We have also defined the length of the input link and located its positions relative to the four precision points on the path.

Our next task is to locate point B, the point of attachment of the coupler and output link. Any one of the four locations of B can be used; in this example we use the B_1 position.

Before beginning the final step, we note that the linkage is now defined. Four arbitrary decisions were made: the location of O_4, the radii R and r, and the location of O_2. Thus ∞^4 solutions are possible.

Referring to Fig. 9.32, we locate point 2 by making triangles $C_2A_2O_4$ and C_1A_12 congruent. We locate point 4 by making $C_4A_1O_4$ and C_1A_14 congruent. Points 4, 2, and O_4 define a circle whose center is B_1. So we find B_1 at the intersection of the perpendicular bisectors of O_42 and O_44. Note that the procedure used causes points 1 and 3 to coincide with O_4. With B_1 located, the links can be drawn in place and the mechanism tested to see how well it traces the desired path.

To synthesize a linkage to generate a path through five precision points, we can make two point reductions. We begin by choosing five points, C_1 to C_5, on the path to be traced. We choose two pairs of these for reduction purposes. In Fig. 9.33, we choose the pairs C_1C_5 and C_2C_3. Other pairs that could have been chosen are

$$C_1C_5, C_2C_4; \quad C_1C_5, C_3C_4; \quad C_1C_4, C_2C_3; \quad C_2C_5, C_3C_4.$$

We construct the perpendicular bisectors c_{23} and c_{15} of the lines connecting each pair. These intersect at point O_4. Note that O_4 can, therefore, be located conveniently by a judicious choice of the pairs to be used as well as by the choice of the positions of the points C_i on the path.

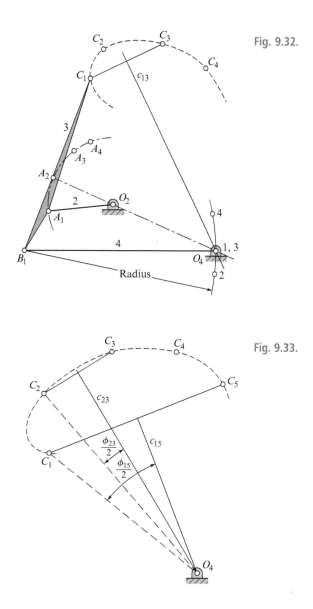

Fig. 9.32.

Fig. 9.33.

The next step is best performed using a sheet of tracing paper as an overlay. We secure the tracing paper to the drawing and mark upon it the center O_4, the perpendicular bisector c_{23}, and another line from O_4 to C_2. Such an overlay is shown in Fig. 9.34a with the line O_4C_2 designated O_4C_2'. This defines the angle $\phi_{23}/2$. Now we rotate the overlay about point O_4 until the perpendicular bisector coincides with c_{15} and repeat for point C_1. This defines the angle $\phi_{15}/2$ and the corresponding line, O_4C_1'.

Now, we pin the overlay at point O_4, using a thumbtack, and rotate it until a good position is found. It is helpful to set the compass to some convenient radius, r, and draw circles about each point C_i. The intersection of these circles with lines O_4C_1' and O_4C_2' on the overlay, and with each other, will reveal which areas will be worthwhile to investigate (Fig. 9.34b).

(a)   (b)

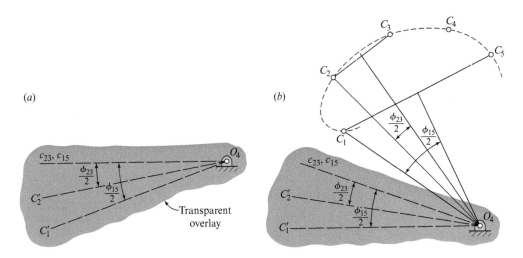

Fig. 9.34.

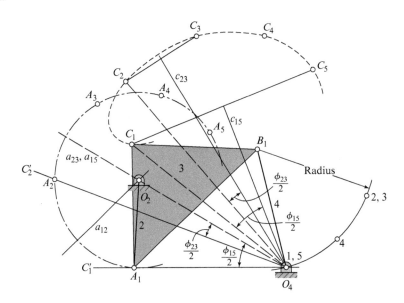

Fig. 9.35.

The final steps in the solution are shown in Fig. 9.35. Having located a good position for the overlay, we transfer the three lines to the drawing and remove the overlay. Now, we draw a circle of radius r to intersect $O_4 C'_1$ and locate point A_1. Another arc of the same radius r from point C_2 intersects $O_4 C'_2$ at point A_2. With A_1 and A_2 located, we draw the perpendicular bisector a_{12}; it intersects the perpendicular bisector a_{23} at O_2, giving us the length of the input rocker. A circle through A_1 about O_2 contains all the design positions of point A; we use the same radius r and locate A_3, A_4, and A_5 on arcs about C_3, C_4, and C_5.

We have now located everything except point B_1, and this is determined as before. Note that points 2 and 3 are coincident (called a *double point*) because of the choice of point O_4 on the perpendicular bisector c_{23}. To locate this double point, we strike an arc from C_1 of radius C_2O_4. Then we strike another arc from A_1 of radius A_2O_4. These intersect at the double point 2,3. To locate point 4, we strike an arc from C_1 of radius C_4O_4, and another from A_1 of radius A_4O_4. Note that point O_4 and the double point 1,5 are coincident, since the synthesis is based on inversion on the O_4B_1 position. Points O_4, 4, and double point 2,3 lie on a circle whose center is B_1, as shown in Fig. 9.35. The linkage is completed by drawing the coupler link and the follower link in the first design posture.

9.10 Cognate Linkages; Roberts–Chebyshev Theorem

One of the remarkable properties of the planar four-bar linkage is that there is not just one but three four-bar linkages that generate the same coupler curve. This was discovered by Roberts[4] in 1875 and by Chebyshev in 1878 and hence is known as the Roberts–Chebyshev theorem. Although mentioned in an English publication in 1954 [7], it did not appear in the American literature until 1958, when it was presented, independently and almost simultaneously, by Hartenberg and Denavit of Northwestern University [10] and by Hinkle of Michigan State University [12].

In Fig. 9.36, let O_1ABO_2 be the original four-bar linkage with coupler point P attached to AB. The remaining two linkages defined by the Roberts–Chebyshev theorem were termed *cognate linkages* by Hartenberg and Denavit. Each of the cognate linkages is shown in Fig. 9.36; one is $O_1A_1C_1O_3$ and uses short dashes for showing the links, and the other is $O_2B_2C_2O_3$ and uses long dashes. The construction is evident by observing that there are four similar triangles, each containing the angles $\alpha, \beta,$ and γ, and three different parallelograms.

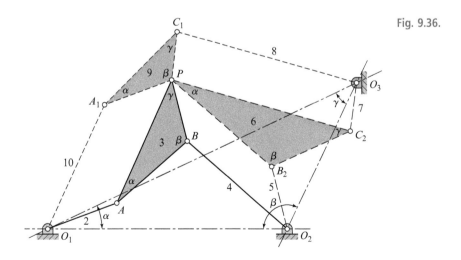

Fig. 9.36.

---

[4] Samuel Roberts (1827–1913) was a British mathematician; not to be confused with Richard Roberts (1789–1864), the British inventor of the approximate straight-line generator (Fig. 1.24b).

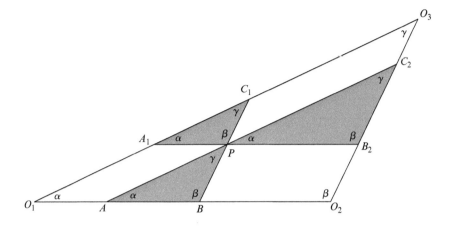

Fig. 9.37 The
Cayley diagram.

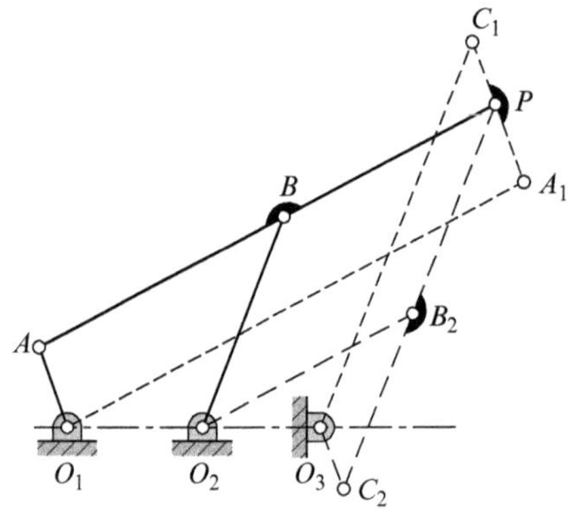

Fig. 9.38.

A good way to obtain the dimensions of the two cognate linkages is to imagine that the frame connections O_1, O_2, and O_3 can be unfastened. Then, imagine that O_1, O_2, and O_3 are "pulled" away from each other until a straight line is formed by the crank, coupler, and follower of each linkage. If we were to do this for Fig. 9.36, then we would obtain Fig. 9.37. Note that the frame distances are now incorrect, but all the movable links are of the correct lengths. Given any four-bar linkage and its coupler point, one can create a drawing similar to Fig. 9.37 and obtain the dimensions of the other two cognate linkages. This approach was discovered by A. Cayley and is called the *Cayley diagram*.[5]

If the tracing point P is on the straight line AB or its extensions, a figure like Fig. 9.37 is of little help since all three linkages are compressed into a single straight line. An example is shown in Fig. 9.38, where O_1ABO_2 is the original linkage having a coupler point, P, on an extension of AB. To find the cognate linkages, locate O_3 on an extension of O_1O_2 in the same

---

[5] Arthur Cayley (1821–1895). See [2]. In Cayley's time, a four-bar linkage was described as a three-bar linkage because the idea of a kinematic chain had not yet been conceived.

ratio as AB is to BP. Then, construct, in order, the parallelograms O_1A_1PA, O_2B_2PB, and $O_3C_1PC_2$.

Hartenberg and Denavit demonstrated that the angular-velocity relations between the links in Fig. 9.36 are

$$\omega_9 = \omega_2 = \omega_7, \qquad \omega_{10} = \omega_3 = \omega_5, \qquad \omega_8 = \omega_4 = \omega_6. \tag{9.23}$$

They also observed that if crank 2 is driven at a constant angular velocity and if the velocity relationships are to be preserved during generation of the coupler curve, the cognate linkages must be driven at variable angular velocities.

9.11 Freudenstein's Equation

In Fig. 9.39, we replace the links of a four-bar linkage by position vectors and write the loop-closure equation

$$\mathbf{r}_1 + \mathbf{r}_2 + \mathbf{r}_3 + \mathbf{r}_4 = \mathbf{0}. \tag{a}$$

In complex polar notation, Eq. (a) is written as

$$r_1 e^{j\theta_1} + r_2 e^{j\theta_2} + r_3 e^{j\theta_3} + r_4 e^{j\theta_4} = 0. \tag{b}$$

From Fig. 9.39, we see that $\theta_1 = 180° = \pi$ radians, from which $e^{j\theta_1} = -1$. Therefore, if Eq. (b) is transformed into complex rectangular form, and if the real and the imaginary components are separated, we obtain the two algebraic equations

$$-r_1 + r_2 \cos\theta_2 + r_3 \cos\theta_3 + r_4 \cos\theta_4 = 0, \tag{c}$$

$$r_2 \sin\theta_2 + r_3 \sin\theta_3 + r_4 \sin\theta_4 = 0. \tag{d}$$

The coupler angle θ_3 can be eliminated, and the output angle θ_4 can be expressed in terms of the input angle θ_2 by the following procedure. Moving all terms except those involving θ_3 to the right-hand side and squaring both sides gives

$$r_3^2 \cos^2\theta_3 = (r_1 - r_2 \cos\theta_2 - r_4 \cos\theta_4)^2, \tag{e}$$

$$r_3^2 \sin^2\theta_3 = (-r_2 \sin\theta_2 - r_4 \sin\theta_4)^2. \tag{f}$$

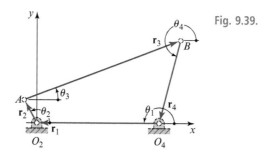

Fig. 9.39.

Now, expanding the right-hand sides and adding the two equations gives

$$r_3^2 = r_1^2 + r_2^2 + r_4^2 - 2r_1r_2 \cos\theta_2 - 2r_1r_4 \cos\theta_4 + 2r_2r_4(\cos\theta_2 \cos\theta_4 + \sin\theta_2 \sin\theta_4). \quad (g)$$

Making the substitution $(\cos\theta_2 \cos\theta_4 + \sin\theta_2 \sin\theta_4) = \cos(\theta_2 - \theta_4)$, and then dividing by the factor $2r_2r_4$, and rearranging again gives

$$\frac{r_3^2 - r_1^2 - r_2^2 - r_4^2}{2r_2r_4} + \frac{r_1}{r_4} \cos\theta_2 + \frac{r_1}{r_2} \cos\theta_4 = \cos(\theta_2 - \theta_4). \quad (h)$$

Freudenstein [4] writes this equation in the form

$$K_1 \cos\theta_2 + K_2 \cos\theta_4 + K_3 = \cos(\theta_2 - \theta_4), \quad (9.24)$$

where

$$K_1 = \frac{r_1}{r_4}, \quad (9.25)$$

$$K_2 = \frac{r_1}{r_2}, \quad (9.26)$$

$$K_3 = \frac{r_3^2 - r_1^2 - r_2^2 - r_4^2}{2r_2r_4}. \quad (9.27)$$

We have already learned graphic methods for synthesizing a four-bar linkage so that the motion of the output is coordinated with that of the input. Freudenstein's equation, Eq. (9.24), enables us to perform this same task by analytic means. Thus, suppose we wish the output of a four-bar linkage to occupy the postures $\psi_1, \psi_2,$ and ψ_3 corresponding to the postures $\phi_1, \phi_2,$ and ϕ_3 of the input. In Eq. (9.24) we simply replace θ_2 with ϕ_i, and θ_4 with ψ_i, and write the equation three times, once for each posture. This gives

$$\begin{aligned} K_1 \cos\phi_1 + K_2 \cos\psi_1 + K_3 &= \cos(\phi_1 - \psi_1), \\ K_1 \cos\phi_2 + K_2 \cos\psi_2 + K_3 &= \cos(\phi_2 - \psi_2), \\ K_1 \cos\phi_3 + K_2 \cos\psi_3 + K_3 &= \cos(\phi_3 - \psi_3). \end{aligned} \quad (i)$$

Equations (i) are then solved simultaneously for the three unknowns, $K_1, K_2,$ and K_3. Then, a length, say r_1, is selected for one of the links, and Eqs. (9.25)–(9.27) are solved for the dimensions of the other three links. The method is best shown by an example.

Example 9.7

Synthesize a function generator to follow the equation

$$y = \frac{1}{x} \quad \text{over the range} \quad 1 \le x \le 2$$

using three precision postures.

Example 9.7 (cont.)

Solution

Choosing Chebyshev spacing, we find, from Eq. (9.22), the values of x and the corresponding values of y to be

$$x_1 = 1.067, \qquad y_1 = 0.937,$$
$$x_2 = 1.500, \qquad y_2 = 0.667,$$
$$x_3 = 1.933, \qquad y_3 = 0.517.$$

We must now choose starting angles for the input and output and also total displacement angles for each. These are arbitrary decisions and may or may not result in a good linkage design in the sense that the structural errors between the precision postures may be large or the transmission angles may be poor. Sometimes, in such a synthesis, it is found that one of the pivots must be disconnected to move from one precision posture to another. Generally, some trial-and-error work may be necessary to discover the best choices of starting angles and total displacement angles.

Here, for the input, we choose a starting angle of $\phi_{min} = 30°$ and a total displacement angle of $\Delta\phi = \phi_{max} - \phi_{min} = 90°$. For the output, we choose a starting angle of $\psi_{min} = 240°$ and again choose a displacement of $\Delta\psi = \psi_{max} - \psi_{min} = 90°$ total travel. With these choices made, the first and last rows of Table 9.2 can be completed.

Next, to obtain the values of ϕ and ψ corresponding to the precision postures, we write

$$\phi = ax + b, \qquad \psi = cy + d, \tag{1}$$

and use the data in the first and last rows of Table 9.2 to evaluate the constants a, b, c, and d. When this is done, we find Eqs. (1) are

$$\phi = 90°x - 60°, \qquad \psi = -180°y + 420°. \tag{2}$$

These equations can now be used to compute the data for the remaining rows in Table 9.2 and to determine the scales of the input and output links of the synthesized linkage.

Table 9.2 **Precision postures**

Posture	x	Input angle ϕ, deg	y	Output angle ψ, deg
–	1.000	30.00	1.000	240.00
1	1.067	36.03	0.937	251.34
2	1.500	75.00	0.667	300.00
3	1.933	113.97	0.517	326.94
–	2.000	120.00	0.500	330.00

Now we take the values of ϕ and ψ from the second line of Table 9.2 and substitute them for θ_2 and θ_4 in Eq. (9.24). Then, if we repeat this for the third and fourth lines, we have the three equations

Example 9.7 (cont.)

$$K_1 \cos 36.03° + K_2 \cos 251.34° + K_3 = \cos(36.03° - 251.34°),$$
$$K_1 \cos 75.00° + K_2 \cos 300.00° + K_3 = \cos(75.00° - 300.00°), \tag{3}$$
$$K_1 \cos 113.97° + K_2 \cos 326.94° + K_3 = \cos(113.97° - 326.94°).$$

When the trigonometric operations are carried out, we have

$$0.8087K_1 - 0.3200K_2 + K_3 = -0.8160,$$
$$0.2588K_1 + 0.5000K_2 + K_3 = -0.7071, \tag{4}$$
$$-0.4062K_1 + 0.8381K_2 + K_3 = -0.8389.$$

Upon solving Eqs. (4), we obtain

$$K_1 = 0.4032, \quad K_2 = 0.4032, \quad K_3 = -1.0130.$$

Using $r_1 = 1.000$ units, we obtain, from Eq. (9.25),

$$r_4 = \frac{r_1}{K_1} = \frac{1.000}{0.4032} = 2.480 \text{ units.} \qquad\qquad Ans.$$

Similarly, from Eqs. (9.26) and (9.27), we learn that

$$r_2 = 2.480 \text{ units} \quad \text{and} \quad r_3 = 0.917 \text{ units.} \qquad\qquad Ans.$$

The result is the crossed four-bar linkage shown in Fig. 9.40.

Fig. 9.40. The crossed four-bar linkage.

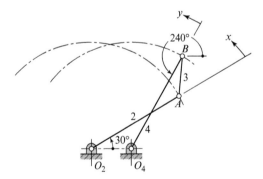

Freudenstein offers the following suggestions, which are helpful in synthesizing such function generators.

1. The total displacement angles of the input and output links should be less than 120°.
2. Avoid the generation of symmetric functions, such as $y = x^2$, over a symmetric range, such as $-1 \le x \le 1$.
3. Avoid the generation of functions having abrupt changes in slope.

9.12 Analytic Synthesis Using Complex Algebra

Another very powerful approach to the synthesis of planar linkages takes advantage of the concept of precision postures and the operations available through the use of complex algebra. Basically, as with Freudenstein's equation in the previous section, the idea is to write complex algebraic equations describing the final linkage in each of its precision postures.

Since links do not change lengths during the motion, the magnitudes of these complex vectors do not change from one posture to the next, but their angles vary. By writing equations at several precision postures, we obtain a set of simultaneous equations that may be solved for the unknown magnitudes and angles.

The method is very flexible and much more general than is shown here. More complete coverage is given in texts such as that by Erdman, Sandor, and Kota [3]. However, the fundamental ideas and some of the operations are clarified here by an example.

Example 9.8

In this example, we wish to design a mechanical strip-chart recorder. The concept of the final design is shown in Fig. 9.41. We assume that the signal to be recorded is available as a shaft rotation having a range of $0 \le \phi \le 90°$ clockwise. This rotation is to be converted to a straight-line motion of the recorder pen over a range of $0 \le s \le 4$ to the right, with a linear relationship between changes of ϕ and s.[6]

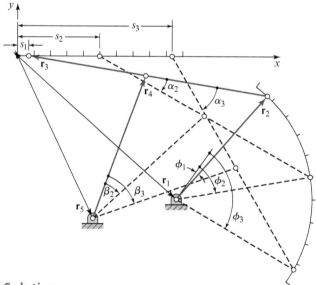

Fig. 9.41 Three-posture synthesis of chart-recorder linkage using complex algebra.

Solution

We choose three accuracy postures for our design approach. Using Chebyshev spacing over the range to reduce structural error and taking counterclockwise angles as positive, the three accuracy postures are given by Eq. (9.22):

---

[6] A similar problem is solved graphically in [11].

Example 9.8 (cont.)

$$\phi_1 = -6° = -0.104\,72 \text{ rad}, \qquad s_1 = 0.267\,95 \text{ in},$$
$$\phi_2 = -45° = -0.785\,40 \text{ rad}, \qquad s_2 = 2.000\,00 \text{ in}, \qquad (1)$$
$$\phi_3 = -84° = -1.466\,08 \text{ rad}, \qquad s_3 = 3.732\,05 \text{ in}.$$

First we tackle the design of the dyad consisting of the input crank, $\mathbf{r}_2$, and the coupler link, $\mathbf{r}_3$. Taking these to be complex vectors for the mechanism in its first accuracy posture, and designating the unknown location of the fixed pivot by the complex vector $\mathbf{r}_1$, we write a loop-closure equation at each of the three precision postures:

$$\mathbf{r}_1 + \mathbf{r}_2 + \mathbf{r}_3 = s_1,$$
$$\mathbf{r}_1 + \mathbf{r}_2 e^{j(\phi_2 - \phi_1)} + \mathbf{r}_3 e^{j\alpha_2} = s_2, \qquad (2)$$
$$\mathbf{r}_1 + \mathbf{r}_2 e^{j(\phi_3 - \phi_1)} + \mathbf{r}_3 e^{j\alpha_3} = s_3,$$

where the angles, α_j, represent the angular displacements of the coupler link from its first posture. Next, by subtracting the first of these equations from each of the others and rearranging, we obtain

$$\left[e^{j(\phi_2 - \phi_1)} - 1 \right] \mathbf{r}_2 + \left[e^{j\alpha_2} - 1 \right] \mathbf{r}_3 = s_2 - s_1,$$
$$\left[e^{j(\phi_3 - \phi_1)} - 1 \right] \mathbf{r}_2 + \left[e^{j\alpha_3} - 1 \right] \mathbf{r}_3 = s_3 - s_1. \qquad (3)$$

Here, we note that we have two complex equations in two complex unknowns, $\mathbf{r}_2$ and $\mathbf{r}_3$, except we note that the coupler displacement angles, α_j, which appear in the coefficients, are also unknowns. Thus, we have more unknowns than equations and we are free to specify additional conditions or additional data for the problem. Making estimates based on crude sketches of our contemplated design, therefore, we make the following arbitrary decisions:

$$\alpha_2 = -20° = -0.349\,07 \text{ rad},$$
$$\alpha_3 = -50° = -0.872\,66 \text{ rad}. \qquad (4)$$

Collecting the data from Eqs. (1) and (4), substituting into Eqs. (3), and evaluating, we find

$$-(0.222\,85 + j0.629\,32)\mathbf{r}_2 - (0.060\,31 + j0.342\,02)\mathbf{r}_3 = 1.732\,05,$$
$$-(0.792\,09 + j0.978\,15)\mathbf{r}_2 - (0.357\,21 + j0.766\,04)\mathbf{r}_3 = 3.464\,10,$$

which we can now solve for the two complex unknowns:

$$\mathbf{r}_2 = 2.153\,26 + j2.448\,60 = 3.261 \text{ in}\angle 48.67°, \qquad \textit{Ans.}$$

$$\mathbf{r}_3 = -5.725\,48 + j0.952\,04 = 5.804 \text{ in}\angle 170.56°. \qquad \textit{Ans.}$$

Then, using the first of Eqs. (2), we solve for the location of the fixed pivot:

$$\mathbf{r}_1 = s_1 - \mathbf{r}_2 - \mathbf{r}_3$$
$$= 3.840\,17 - j3.400\,64 = 5.129 \text{ in}\angle -41.53°. \qquad \textit{Ans.}$$

Example 9.8 (cont.)

Thus far, we have completed the design of the dyad, which includes the input crank. Before proceeding, we should note that an identical procedure could have been used for the design of a slider-crank linkage, for one dyad of a four-bar linkage used for any path generation or motion generation problem, or for a variety of other applications. Our total design is not yet completed, but we should note the general applicability of the procedures covered to other linkage synthesis problems.

Continuing with our design of the recorder linkage, however, we now must find the postures and dimensions of the dyad, $\mathbf{r}_4$ and $\mathbf{r}_5$ of Fig. 9.41. As shown, we choose to connect the moving pivot of the output crank at the midpoint of the coupler link to minimize its mass and to keep dynamic forces low. Thus, we can write another loop-closure equation including the rocker at each of the three precision postures:

$$
\begin{aligned}
\mathbf{r}_5 + \mathbf{r}_4 \quad\; + 0.5\mathbf{r}_3 \quad\;\; &= s_1, \\
\mathbf{r}_5 + \mathbf{r}_4 e^{j\beta_2} + 0.5\mathbf{r}_3 e^{j\alpha_2} &= s_2, \\
\mathbf{r}_5 + \mathbf{r}_4 e^{j\beta_3} + 0.5\mathbf{r}_3 e^{j\alpha_3} &= s_3.
\end{aligned}
\tag{5}
$$

Substituting the known data into these equations and rearranging, we obtain

$$
\begin{aligned}
\mathbf{r}_5 + \quad\quad \mathbf{r}_4 + (-3.130\,69 + j0.476\,02) &= \mathbf{0}, \\
\mathbf{r}_5 + \left(e^{j\beta_2}\right)\mathbf{r}_4 + (-4.527\,25 + j1.426\,52) &= \mathbf{0}, \\
\mathbf{r}_5 + \left(e^{j\beta_3}\right)\mathbf{r}_4 + (-5.207\,45 + j2.499\,03) &= \mathbf{0}.
\end{aligned}
\tag{6}
$$

These are three simultaneous complex equations in only two vector unknowns, $\mathbf{r}_4$ and $\mathbf{r}_5$, and thus we are not free to choose the rotation angles, β_2 and β_3, arbitrarily. For Eqs. (6) to have consistent nontrivial solutions, it is necessary that the determinant of the matrix of coefficients be zero. Thus, β_2 and β_3 must be chosen such that

$$
\begin{vmatrix}
1 & 1 & (-3.130\,69 + j0.476\,02) \\
1 & e^{j\beta_2} & (-4.527\,25 + j1.426\,52) \\
1 & e^{j\beta_3} & (-5.207\,45 + j2.499\,03)
\end{vmatrix} = 0,
\tag{7}
$$

which expands to

$$
(-2.076\,76 + j2.023\,01)e^{j\beta_2} + (1.396\,56 - j0.950\,50)e^{j\beta_3} + (0.680\,20 - j1.072\,51) = 0.
$$

Solving this for $e^{j\beta_2}$ gives

$$
e^{j\beta_2} = (0.573\,82 + j0.101\,28)e^{j\beta_3} + (0.426\,19 - j0.101\,28),
$$

and equating the real and imaginary parts, respectively, gives

$$
\begin{aligned}
\cos\beta_2 &= 0.573\,82\cos\beta_3 - 0.101\,28\sin\beta_3 + 0.426\,19, \\
\sin\beta_2 &= 0.101\,28\cos\beta_3 + 0.573\,82\sin\beta_3 - 0.101\,28.
\end{aligned}
\tag{8}
$$

Example 9.8 (cont.)

These equations can now be squared and added to eliminate the unknown β_2. The result, after rearrangement, is a single equation in β_3:

$$0.468\,59\cos\beta_3 - 0.202\,56\sin\beta_3 - 0.468\,57 = 0. \tag{9}$$

This equation can be solved by substituting the tangent of the half-angle identities

$$x = \tan\frac{\beta_3}{2}, \quad \cos\beta_3 = \frac{1-x^2}{1+x^2}, \quad \sin\beta_3 = \frac{2x}{1+x^2}, \tag{10}$$

$$0.468\,59(1-x^2) - 0.202\,56(2x) - 0.468\,57(1+x^2) = 0.$$

This reduces to the quadratic equation

$$-0.937\,16x^2 - 0.405\,12x + 0.000\,02 = 0,$$

for which the roots are

$$x = -0.432\,66 \quad \text{or} \quad x = 0.000\,27,$$

and, from Eq. (10),

$$\beta_3 = -46.79° \quad \text{or} \quad \beta_3 = 0.03°.$$

Guided by our sketch of the desired design, we choose the first of these roots, $\beta_3 = -46.79°$. Then, returning to Eqs. (8), we find the value $\beta_2 = -26.76°$, and finally, from Eqs. (6), we find the final solution:

$$\mathbf{r}_4 = 1.299\,71 + j3.410\,92 = 3.650 \text{ in}\angle 69.14°, \qquad Ans.$$

$$\mathbf{r}_5 = 1.830\,98 - j3.886\,94 = 4.297 \text{ in}\angle -64.78°. \qquad Ans.$$

Note that this second part of the solution, solving for the rocker $\mathbf{r}_4$, is also a very general approach that could be used to design a crank to go through three given precision postures in a variety of other problems. Although only a specific case is presented here, the approach arises repeatedly in linkage design.

Of course, before we finish, we should evaluate the quality of our solution by analysis of the linkage we have designed. This was performed here using the equations of Chapter 2 to find the locations of the coupler point for 20 equally spaced crank increments spanning the given range of motion. As expected, there is structural error; the coupler curve of the recording pen tip is not exactly straight, and the displacement increments are not perfectly linear over the range of travel of the pen. However, the solution is quite good; the deviation from a straight line is less than 0.020 in, or 0.5% of the travel, and the linearity between the input crank rotation and coupler point travel is better than 1% of the travel. As expected, the structural error follows a regular pattern and vanishes at the three precision postures. The transmission angle remains larger than

Example 9.8 (cont.)

$70°$ throughout the range; thus, no problems with force transmission are expected. Although the design might be improved slightly using additional precision postures, the present solution seems excellent and the additional effort does not appear worthwhile.

9.13 Synthesis of Dwell Linkages

One of the most interesting uses of coupler curves having straight-line or circular-arc segments is in the synthesis of linkages having a substantial dwell during a portion of their operating cycle. Using segments of coupler curves, it is not difficult to synthesize linkages having a dwell at either or both of the extremes of their motion or at an intermediate posture.

In Fig. 9.42a, a coupler curve having an approximately elliptic shape is selected from the Hrones and Nelson atlas [13] so that a substantial portion of the curve approximates a circular arc. The four-bar linkage that generates this coupler curve is not shown in the figure, but is specified in the atlas. Connecting link 5 is given a length equal to the radius of this arc. Thus, in Fig. 9.42a, points $D_1, D_2,$ and D_3 are stationary as coupler point C moves through positions $C_1, C_2,$ and C_3. Also, the same purpose can be accomplished with a slider, as shown in Fig. 9.42b; link 6 dwells as point C travels along the straight portion of the coupler curve. The length of output link 6 and the location of the frame pivot, O_6, depend upon the desired angle of oscillation of this link. The frame pivot should also be positioned for a desirable transmission angle.

When segments of circular arcs are desired for a coupler curve, an organized method of searching the Hrones and Nelson atlas [13] can be employed. The overlay, shown in Fig. 9.43, is made on a sheet of tracing paper and can be fitted over the paths in the atlas very quickly. It reveals immediately the radius of curvature of the segment, the location of pivot point D, and the displacement angle of the connecting link.

(a)

(b)

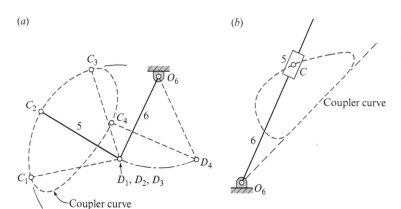

Fig. 9.42 Link 6 dwells as point C travels: (a) the circular-arc path $C_1C_2C_3$; (b) on the straight portion of the coupler curve.

Fig. 9.43 Overlay for use with the Hrones and Nelson atlas [13].

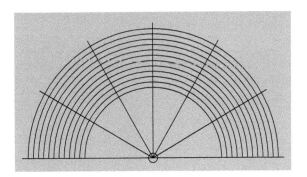

Fig. 9.44 Link 6 dwells: (*a*) at each end of its swing; (*b*) in the central portion of its swing.

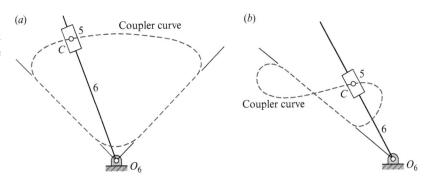

Figure 9.44 shows two ideas for dwell linkages employing a slider. A coupler curve having a straight-line segment is used in each, and the pivot point, O_6, is placed on an extension of this line.

The arrangement shown in Fig. 9.44*a* has a dwell at both extremes of the motion of link 6. A practical design of this linkage may be difficult to achieve, however, since link 6 has a high velocity when the slider is near the pivot, O_6.

The slider linkage of Fig. 9.44*b* uses a figure-eight coupler curve having a straight-line segment to produce an intermediate dwell linkage. Pivot O_6 must be located on an extension of the straight-line segment, as shown.

9.14 Intermittent Rotary Motion

The *Geneva wheel*, or *Maltese cross*, is a cam-like mechanism that provides intermittent rotary motion and is widely used in both low-speed and high-speed machinery. Although originally developed as a stop to prevent overwinding of watches, it is now used extensively in automatic machinery – for example, where a spindle, turret, or worktable must be indexed. It is also used in motion-picture projectors to provide the intermittent advance of the film.

A drawing of a six-slot Geneva mechanism is shown in Fig. 9.45. Note that the centerlines of the slot and crank are mutually perpendicular at engagement and at disengagement. The crank, which usually rotates at a uniform angular velocity, carries a roller to engage with the slots.

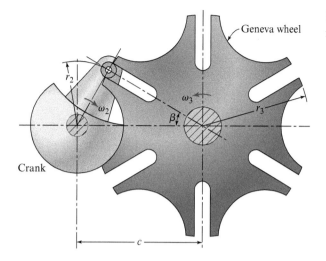

Fig. 9.45 Geneva mechanism.

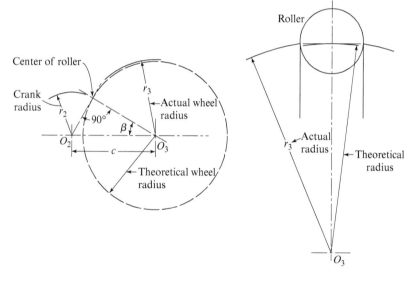

Fig. 9.46 Design of a Geneva wheel.

During one revolution of the crank, the Geneva wheel rotates a fractional part of a revolution, the amount of which is dependent upon the number of slots. The circular segment attached to the crank effectively locks the wheel against rotation when the roller is not in engagement and also locates the wheel for correct engagement of the roller with the next slot.

The design of a Geneva mechanism is initiated by specifying the crank radius, the roller diameter, and the number of slots. At least three slots are necessary, but most problems can be solved with wheels having 4–12 slots. The design procedure is shown in Fig. 9.46. The angle β is half the angle subtended by adjacent slots; that is,

$$\beta = \frac{360°}{2n},$$

(a)

Fig. 9.47.

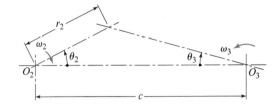

where n is the number of slots in the wheel. Then, defining r_2 as the crank radius, we have

$$c = \frac{r_2}{\sin \beta}, \tag{b}$$

where c is the center distance. Note, too, from Fig. 9.46, that the actual Geneva wheel radius is more than that which would be obtained by a zero-diameter roller. This is because of the difference between the sine and the tangent of the angle subtended by the roller, measured from the wheel center.

After the roller has entered the slot and is driving the wheel, the geometry is that of Fig. 9.47. Here, θ_2 is the crank angle, and θ_3 is the wheel angle. They are related trigonometrically by

$$\tan \theta_3 = \frac{\sin \theta_2}{(c/r_2) - \cos \theta_2}. \tag{c}$$

We can determine the angular velocity of the wheel for any value of θ_2 by differentiating Eq. (c) with respect to time. This produces

$$\omega_3 = \frac{(c/r_2)\cos \theta_2 - 1}{1 + (c^2/r_2^2) - 2(c/r_2)\cos \theta_2} \omega_2. \tag{9.28}$$

The maximum wheel velocity occurs when the crank angle is zero. Substituting $\theta_2 = 0$, therefore, gives

$$(\omega_3)_{\text{max}} = \frac{r_2}{c - r_2} \omega_2. \tag{9.29}$$

The angular acceleration of the wheel with constant input crank speed, obtained by differentiating Eq. (9.28) with respect to time, is

$$\alpha_3 = \frac{(c/r_2)\sin \theta_2 \left(1 - c^2/r_2^2\right)}{\left[1 + (c/r_2)^2 - 2(c/r_2)\cos \theta_2\right]^2} \omega_2^2. \tag{9.30}$$

The angular acceleration reaches a maximum when the crank angle is

$$\theta_2 = \cos^{-1}\left\{\pm \sqrt{\left[\frac{1 + (c/r_2)^2}{4(c/r_2)}\right]^2 + 2} - \frac{1 + (c/r_2)^2}{4(c/r_2)}\right\}. \tag{9.31}$$

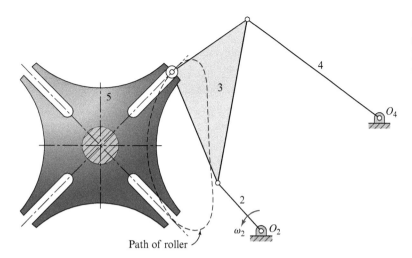

Fig. 9.48 Geneva wheel activated by a four-bar linkage with link 2 as the crank.

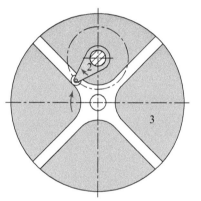

Fig. 9.49 Inverse Geneva mechanism.

This occurs when the roller has advanced about 30% into the slot.

Several methods have been employed to reduce the wheel angular acceleration that reduces inertia forces and the consequent wear on the sides of the slot. Among these is the idea of using a curved slot. This can reduce the acceleration but also increases the deceleration and consequently the wear on the other side of the slot.

Another method uses the Hrones and Nelson atlas [13] for synthesis. The idea is to place the roller on the coupler of a four-bar linkage. During the period in which it drives the wheel, the path of the roller should be curved, and the roller should have a low value of angular acceleration. Figure 9.48 shows one solution and includes the path taken by the roller. This is the type of path that is sought while leafing through the atlas.

The inverse Geneva mechanism of Fig. 9.49 enables the wheel to rotate in the same direction as the crank and requires less radial space. The locking device is not shown, but this can be a circular segment attached to the crank, as before, which locks by coming into contact with a built-up rim on the periphery of the wheel.

PROBLEMS

9.1 A function varies from 0 to 10. Find the Chebyshev spacing for two, three, four, five, and six precision postures.

9.2 Determine the link lengths of a slider-crank linkage to have a stroke of 24 in and an advance-to-return ratio of 1.20.

9.3 Determine a set of link lengths for a slider-crank linkage such that the stroke is 400 mm and the advance-to-return ratio is 1.25.

9.4 The rocker of a crank-rocker linkage is to have a length of 20 in and swing through a total angle of 45° with an advance-to-return ratio of 1.25. Determine a suitable set of dimensions for r_1, r_2, and r_3.

9.5 A crank-rocker linkage is to have a rocker of 1.80 m length and a rocking angle of 75°. If the advance-to-return ratio is to be 1.32, what is a suitable set of link lengths for the remaining three links?

9.6 Design a crank and coupler to drive rocker 4 such that slider 6 reciprocates through a distance of 400 mm with an advance-to-return ratio of 1.20. Use $a = r_4 = 400$ mm and $r_5 = 600$ mm with $\mathbf{r}_4$ vertical at midstroke. Record the location of O_2 and dimensions r_2 and r_3.

Fig. P9.6.

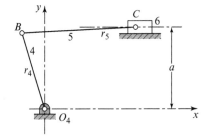

9.7 Design a crank and rocker for a six-bar linkage such that the slider in Fig. P9.6 reciprocates a distance of 32 in with an advance-to-return ratio of 1.12; use $a = r_4 = 48$ in and $r_5 = 72$ in. Locate O_4 such that rocker 4 is vertical when the slider is at midstroke. Find suitable coordinates for O_2 and lengths for r_2 and r_3.

9.8 Two postures of a folding seat used in the aisles of buses to accommodate extra passengers are shown. Design a four-bar linkage to support the seat so that it will lock in the open posture and fold to a stable closing posture along the side of the aisle.

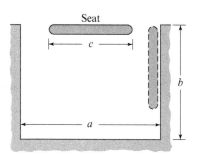

Fig. P9.8 $a = 500$ mm,
$b = 400$ mm, $c = 300$ mm.

9.9 Design a spring-operated four-bar linkage to support a heavy lid, such as the trunk lid of an automobile. The lid is to swing through an angle of 80° from the closed to the open posture. The springs are to be mounted so that the lid will be held closed against a stop, and they should also hold the lid in a stable open posture without the use of a stop.

9.10 Synthesize a linkage to move AB from posture 1 to posture 2 and return.

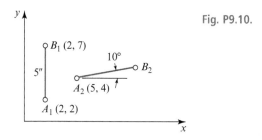

Fig. P9.10.

9.11 Synthesize a linkage to move AB successively through postures 1, 2, and 3.

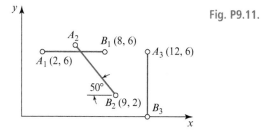

Fig. P9.11.

9.12–9.21[7] Figure P9.12 shows a function-generator linkage in which the motion of rocker 2 corresponds to x and the motion of rocker 4 to the function $y = f(x)$. Use four precision points with Chebyshev spacing, and synthesize a linkage to generate each function in the table. Plot a curve of the desired function and a curve of the actual function that the linkage generates. Compute the maximum error between them in percent.

Fig. P9.12.

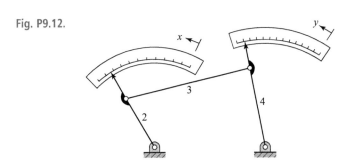

Table P9.12–P9.31

Problem number	Function, $y =$	Range
P9.12, P9.22	$\log_{10} x$	$1 \le x \le 2$
P9.13, P9.23	$\sin x$	$0 \le x \le \pi/2$
P9.14, P9.24	$\tan x$	$0 \le x \le \pi/4$
P9.15, P9.25	e^x	$0 \le x \le 1$
P9.16, P9.26	$1/x$	$1 \le x \le 2$
P9.17, P9.27	$x^{1.5}$	$0 \le x \le 1$
P9.18, P9.28	x^2	$0 \le x \le 1$
P9.19, P9.29	$x^{2.5}$	$0 \le x \le 1$
P9.20, P9.30	x^3	$0 \le x \le 1$
P9.21, P9.31	x^2	$-1 \le x \le 1$

9.22–9.31 Repeat Probs. 9.12–9.21 using the overlay method.

9.32   The figure shows a coupler curve generated by a four-bar linkage (not shown). Link 5 is to be attached to the coupler point, and link 6 is to be a rotating member with O_6 as the frame connection. In this problem, we wish to find a coupler curve from the Hrones and Nelson atlas or by precision postures, such that, for an appreciable distance, point C moves through an arc of a circle. Link 5 is then proportioned so that D lies at the center of curvature of this arc. The result is then called a *hesitation motion*, since link 6 hesitates in its rotation for the period during which point C traverses the approximate circular arc. Make a drawing of the complete linkage, and

---

[7] Solutions for these problems were among the earliest computer work in kinematic synthesis and results are reported in [5].

plot the first-order kinematic coefficient ϕ_6', of link 6 for 360° of displacement of the input link.

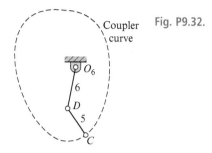

Coupler curve   Fig. P9.32.

9.33 Synthesize a four-bar linkage to obtain a coupler curve having an approximate straight-line segment. Then, using the suggestion included in Figs. 9.42*b* or 9.44*b*, synthesize a dwell motion. Using an input crank angular velocity of unity, plot the first-order kinematic coefficient ϕ_6' of rocker 6 versus the input crank displacement.

9.34 Synthesize a dwell mechanism using the idea suggested in Fig. 9.42*a* and the Hrones and Nelson atlas. Rocker 6 is to have a total angular displacement of 60°. Using this displacement as the abscissa, plot the first-order kinematic coefficient ϕ_6', of the motion of the rocker to show the dwell motion.

REFERENCES

[1] Beyer, R., 1964. *Kinematic Synthesis of Linkages,* trans. H. Kuenzel, New York: McGraw-Hill.

[2] Cayley, A., 1876. On three-bar motion, *Proc. Lond. Math. Soc.,* 7:136–166.

[3] Erdman, A. G., G. N. Sandor, and S. Kota 2001. *Mechanism Design: Analysis and Synthesis,* vol. I, 4th ed., Englewood Cliffs, NJ: Prentice-Hall.

[4] Freudenstein, F., 1955. Approximate synthesis of four-bar linkages. *ASME Trans.* 77:853–861.

[5] Freudenstein, F., 1958. Four-bar function generators, *Mach. Design,* 30(24): 119–123.

[6] Freudenstein, F., and G. N. Sandor, 1985. Kinematics of mechanism, in *Mechanical Design and Systems Handbook,* 2nd ed., edited by H. A. Rothbart, New York: McGraw-Hill, 4-56–4-68.

[7] Grodzinski, P., and E. M'Ewan, 1954. Link mechanisms in modern kinematics, *Proc. Inst. Mech. Eng.,* 168(37):877–896.

[8] Hain, K., 1967. *Applied Kinematics,* 2nd ed., edited by H. Kuenzel, T. P. Goodman, trans. D. P. Adams et al., New York: McGraw-Hill, 639–727.

[9] Hall, A. S., Jr., 1961. *Kinematics and Linkage Design,* Englewood Cliffs, NJ: Prentice-Hall.

[10] Hartenberg, R. S., and J. Denavit, 1958. The fecund four-bar, *Trans. 5th Conf. Mech.,* West Lafayette, IN: Purdue University, 194.

[11] Hartenberg, R. S., and J. Denavit, 1964. *Kinematic Synthesis of Linkages,* New York: McGraw-Hill.

[12] Hinkle, R. T., 1958. Alternate four-bar linkages, *Prod. Eng.,* 29:4.

[13] Hrones, J. A., and G. L. Nelson, 1951. *Analysis of the Four-Bar Linkage,* Cambridge, MA: Technology Press.

[14] Soni, A. H., 1974. *Mechanism Synthesis and Analysis,* New York: McGraw-Hill.

[15] Tao, D. C., 1967. *Fundamentals of Applied Kinematics,* Reading, MA: Addison-Wesley.

10   Spatial Mechanisms and Robotics

10.1 Introduction

The large majority of mechanisms in use today have *planar* motion, that is, the motions of all points produce paths that lie in a single plane or in parallel planes. This means that all motions can be seen in true size and shape from a single viewing direction and that graphic methods of analysis require only a single view. If the coordinate system is chosen with the x and y axes parallel to the plane(s) of motion, then all z values remain constant, and the problem can be solved, either graphically or analytically, with only two-dimensional methods. Although this is usually the case, it is not a necessity. Mechanisms having three-dimensional point paths do exist and are called *spatial mechanisms*. Another special category, called *spherical mechanisms*, have point paths that lie on concentric spherical surfaces.

Recall that these definitions were raised in Chapter 1; however, almost all of the examples presented in previous chapters have dealt only with planar mechanisms. This is justified because of their very extensive use in practical applications. Although a few nonplanar mechanisms, such as universal shaft couplings and bevel gears, have been known for centuries, it is only relatively recently that kinematicians have made substantial progress in developing design procedures for spatial mechanisms. It is probably not a coincidence, given the greater difficulty of the mathematic manipulations, that the emergence of such tools has awaited the development and availability of computers.

Although we have concentrated so far on mechanisms with planar motion, a brief review demonstrates that most of the previous theory has been derived in sufficient generality for both planar and spatial motion. The focus has been planar, since planar motion can be more easily visualized and requires less tedious computations than three-dimensional applications. However, most of the theory extends directly to spatial mechanisms. This chapter reviews some of the previous techniques, including examples with spatial motion. This chapter also introduces new mathematical tools and solution techniques that were not required for planar motion.

In Section 1.6 we learned that the mobility of a mechanism can be obtained from the Kutzbach criterion. The three-dimensional form of this criterion was given in Eq. (1.3); that is,

$$m = 6(n - 1) - 5j_1 - 4j_2 - 3j_3 - 2j_4 - 1j_5, \tag{10.1}$$

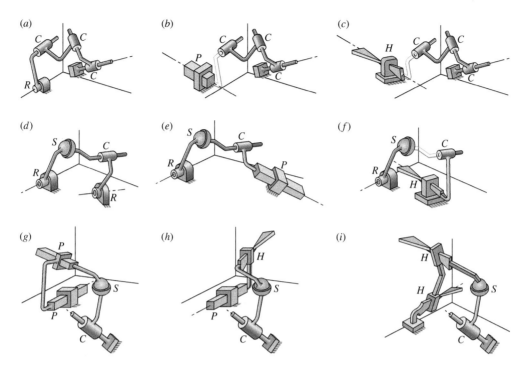

Fig. 10.1 (a) *RCCC*, (b) *PCCC*, (c) *HCCC*, (d) *RSCR*, (e) *RSCP*, (f) *RSCH*, (g) *PPSC*, (h) *PHSC*, (i) *HHSC*.

where n is the number of links, and each j_k is the number of joints having k degrees of freedom. There are many possible combinations of n and j_k that produce a mobility of $m = 1$. One possibility has $n = 7$, $j_1 = 7$, and $j_2 = j_3 = j_4 = j_5 = 0$. Harrisberger [10] called such a combination a mechanism type; in particular, he called this the $7j_1$ type. Other combinations of j_k, of course, produce different types of mechanisms. For example, the $3j_1 + 2j_2$ type has $n = 5$ links, and the $1j_1 + 2j_3$ type has $n = 3$ links.

Each mechanism type contains a finite number of *kinds* of mechanisms; there are as many kinds of mechanisms of each type as there are ways of arranging the different types of joints between the links. In Table 1.1 we saw that three of the six lower pairs, the revolute, R, the prismatic, P, and the helical, H, have one degree of freedom. Thus, using any seven of these lower pairs, we obtain 36 kinds of type $7j_1$ mechanisms. Altogether, Harrisberger lists 435 kinds of mechanisms that satisfy the Kutzbach criterion with a mobility of $m = 1$. Not all of these types, or kinds, however, are likely to have much practical value. Consider, for example, the $7j_1$ type with all revolute joints, each connected in series in a single loop.[1]

For mechanisms having mobility $m = 1$, Harrisberger selected nine kinds from two types that appeared to him to be useful; these are shown in Fig. 10.1. They are all spatial four-bar linkages having four joints with either rotating or sliding input and output. The designations in

---

[1] The only application known to the authors for this type is its use in the front landing gear mechanism of the Boeing 727 aircraft.

the legend, such as the *RSCH* four-bar linkage in Fig. 10.1*f*, identify the kinematic pair types (Table 1.1) beginning with the input and proceeding through the coupler and the output back to the frame. Thus, for this linkage, the input is connected to the frame by a revolute pair, *R*, and to the coupler by a spheric pair, *S*, the coupler is connected to the output by a cylindric pair, *C*, and the output is connected to the frame by a helical pair, *H*. The freedoms of these pairs, from Table 1.1 are $R = 1, S = 3, C = 2$, and $H = 1$.

The linkages of Fig. 10.1*a* through Fig. 10.1*c* are described by Harrisberger as type 1, or of the $1j_1 + 3j_2$ type. The remaining linkages of Fig. 10.1 are described as type 2, or of the $2j_1 + 1j_2 + 1j_3$ type. All have $n = 4$ links and have a mobility of $m = 1$.

10.2 Exceptions to the Mobility Criterion

Curiously, the most common and most useful spatial mechanisms that have been discovered date back many years and are, in fact, exceptions to the Kutzbach criterion. As shown by the example in Fig. 1.6, geometric conditions sometimes occur that are not accounted for in the Kutzbach criterion and lead to apparent exceptions. As a case in point, consider that every planar mechanism, once constructed, truly exists in three dimensions. Yet, a planar four-bar linkage has $n = 4$ and is of type $4j_1$; thus Eq. (10.1) predicts a mobility of $m = 6(4 - 1) - 5(4) = -2$ (implying redundant constraints). However, we know that the mobility is, in fact, $m = 1$. The special geometric conditions in this case lie in the fact that all revolute axes remain parallel and are all perpendicular to the plane of motion. The Kutzbach criterion does not consider such geometric conditions and can produce false predictions.

Three more *RRRR* four-bar linkages are exceptions to the Kutzbach criterion, namely: the spheric four-bar linkage; the wobble-plate mechanism; and the Bennett linkage. As with the planar four-bar linkage, the Kutzbach criterion for these mechanisms predicts $m = -2$, and yet they are truly of mobility $m = 1$. The spheric four-bar linkage is shown in Fig. 10.2; the axes of all four revolute joints intersect at the center of a sphere. The links may be regarded as great-circle arcs existing on the surface of the sphere; what had been link lengths are now spheric angles. By properly proportioning these angles, it is possible to design all of the spheric counterparts of the planar four-bar linkage such as the spheric crank-rocker linkage and the

Fig. 10.2 Spheric four-bar linkage.

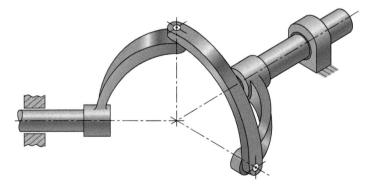

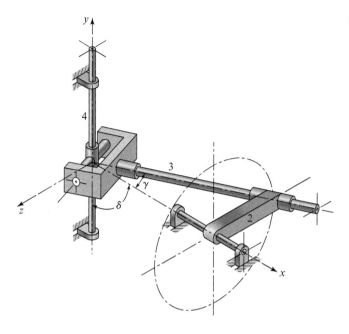

Fig. 10.3 Wobble-plate mechanism.

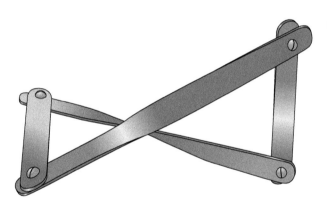

Fig. 10.4 Bennett linkage.

spheric drag-link linkage. The spheric four-bar linkage is easy to design and manufacture and hence is one of the most useful of all spatial linkages. The Hooke or Cardan joint, which is the basis of a universal shaft coupling, is a special case of a spheric four-bar linkage with input and output cranks that subtend equal spheric angles.

The wobble-plate mechanism is shown in Fig. 10.3. Note that all of the revolute axes intersect at the origin; thus, it is also a spheric mechanism. Note also that input crank 2 rotates and output shaft 4 oscillates; also, when $\delta = 90°$, the mechanism is called a *spheric-slide oscillator*, and if $\gamma > \delta$, the output shaft rotates.

The Bennett linkage [2], shown in Fig. 10.4, is probably one of the less practical of the known spatial four-bar linkages. However, it has stimulated development of kinematic theory for spatial mechanisms [1]. In this linkage, opposite links are of equal lengths and are twisted by

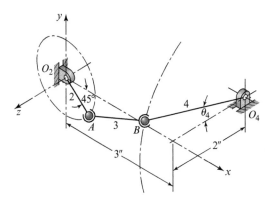

Fig. 10.5 *RSSR* four-bar linkage.

equal amounts. The sines of the twist angles, α_1 and α_2, are proportional to the link lengths, a_1 and a_2, according to the equation

$$\frac{\sin \alpha_1}{a_1} = \pm \frac{\sin \alpha_2}{a_2}.$$

Two more exceptions to the Kutzbach mobility criterion are the Goldberg (Michael, not Rube!) five-bar *RRRRR* linkage and the Bricard six-bar *RRRRRR* linkage [8].[2] Again, it is doubtful whether these linkages have much practical value.

Harrisberger and Soni [11] identified all spatial linkages having one general constraint, that is, that have mobility of $m = 1$, but for which the Kutzbach criterion predicts $m = 0$. They identified eight types and 212 kinds and found seven new mechanisms that may have useful applications.

The spatial four-bar *RSSR* linkage, shown in Fig. 10.5, is an important and useful linkage. Since $n = 4$, $j_1 = 2$, and $j_3 = 2$, the Kutzbach criterion of Eq. (10.1) predicts a mobility of $m = 2$. Although this might appear at first glance to be another exception, upon closer examination we find that the second degree of freedom actually exists; it is the freedom of the coupler to spin about the axis between the two spheric joints. Since this degree of freedom does not affect the input–output kinematic relationship, it is called an *idle* degree of freedom. This extra freedom does no harm if the mass of the coupler is truly distributed along its axis; in fact, it may be an advantage, since the spin of the coupler about its axis may equalize wear on the two ball-and-socket joints. If the mass center of the coupler lies off-axis, however, then this second freedom is not idle dynamically and can cause quite erratic performance of the linkage at high speed.

Note that all mechanisms that defy the Kutzbach criterion always predict a mobility less than the actual mobility. This is always the case; the Kutzbach criterion always predicts a lower limit on the mobility; the reason for this is mentioned in Section 1.6. The argument for the development of the Kutzbach equation, Eq. (10.1), came from counting the freedoms for

---

[2] For pictures of these see [12].

motion of all links before any connections are made, less the numbers of these presumably eliminated by connecting various types of joints. Yet, when there are special geometric conditions, such as intersections or parallelism between joint axes, the criterion counts each joint as eliminating its own share of freedoms, although two (or more) of them may eliminate the same freedom(s). Thus, the exceptions arise from the false assumption of independence among the constraints of the joints.

Let us carry this thought one step further. When two (or more) of the constraint conditions eliminate the same motion freedom, the problem is said to have *redundant constraints*. Under these conditions, the same redundant constraints also determine how the forces are shared where the motion freedom is eliminated. Thus, when we come to analyzing forces, we find that there are too many constraints (equations) relating the number of unknown forces. The force analysis problem is then said to be *overconstrained*, and we find that there are statically indeterminate forces in the same number as the error in the predicted mobility.

There is an important lesson buried in this argument:[3] Whenever there are redundant constraints on the motion, there are an equal number of statically indeterminate force components in the mechanism. Despite the higher simplicity in the design equations of planar linkages, for example, we should consider the force effects of these redundant constraints. All of the out-of-plane force and moment components become statically indeterminate. Slight machining tolerance errors or misalignments of axes can cause indeterminate stresses with cyclic loading as the mechanism is operated. The engineer needs to address the question: What effects will these stresses have on the fatigue life of the members of the mechanism?

On the other hand, as pointed out by Phillips [18], when motion is only occasional and loads are not high, this might be an ideal design decision.[4] If errors are small, the additional indeterminate forces may be small and such designs are tolerated although they may seem to exhibit friction effects and wear in the joints. As errors become larger, however, we may find that binding becomes a major issue.

The very existence of the large number of planar linkages in the world testifies that such effects can be tolerated with appropriate tolerances on manufacturing errors, good lubrication, and proper fits between mating joint elements. Still, too few mechanical designers truly understand that the root of the problem can be eliminated by removing the redundant constraints in the first place. Thus, even in planar motion mechanisms, for example, we can make use of ball-and-socket or cylindric joints that, when properly located, can help to relieve statically indeterminate force and moment components.

10.3 Spatial Posture Analysis Problem

Like planar mechanisms, a spatial mechanism is usually connected to form one or more closed loops. Thus, following methods similar to those of Section 2.7, loop-closure equations can be written that define the kinematic relationships of the mechanism. A variety of different

---

[3] A detailed discussion of this entire topic forms one main theme of an excellent two-volume set by Phillips [18].
[4] See [18, section 20.16].

Table 10.1 Classification of solutions of the vector tetrahedron equation

Case	Unknowns	Known quantities		Degree of polynomial
		Vectors	Scalars	
1	r, θ_r, ϕ_r	$\mathbf{C}$		1
2a	r, θ_r, s	$\mathbf{C}, \hat{\omega}_r, \hat{s}$	ϕ_r	2
2b	r, θ_r, θ_s	$\mathbf{C}, \hat{\omega}_r, \hat{\omega}_s$	ϕ_r, s, ϕ_s	4
2c	θ_r, ϕ_r, s	$\mathbf{C}, \hat{s}$	r	2
2d	$\theta_r, \phi_r, \theta_s$	$\mathbf{C}, \hat{\omega}_s$	r, s, ϕ_s	2
3a	r, s, t	$\mathbf{C}, \hat{r}, \hat{s}, \hat{t}$		1
3b	r, s, θ_t	$\mathbf{C}, \hat{r}, \hat{s}, \hat{\omega}_t$	t, ϕ_t	2
3c	r, θ_s, θ_t	$\mathbf{C}, \hat{r}, \hat{\omega}_s, \hat{\omega}_t$	s, ϕ_s, t, ϕ_t	4
3d	$\theta_r, \theta_s, \theta_t$	$\mathbf{C}, \hat{\omega}_r, \hat{\omega}_s, \hat{\omega}_t$	$r, \phi_r, s, \phi_s, t, \phi_t$	8

mathematical tools can be used to solve the spatial posture problem, including vectors [3], dual numbers [17], quaternions [22], and transformation matrices [21]. In vector notation, the closure of a spatial mechanism, such as the four-bar *RSSR* linkage shown in Fig. 10.5, can be defined by a loop-closure equation of the form

$$r + s + t + C = 0. \tag{10.2}$$

This equation is called the *vector tetrahedron equation*, since the individual vectors can be thought of as defining four of the six edges of a tetrahedron.

The vector tetrahedron equation is three-dimensional and hence can be solved for three scalar unknowns. These can be either distances or angles and can exist in any combination in vectors $\mathbf{r}$, $\mathbf{s}$, and $\mathbf{t}$. Vector $\mathbf{C}$ is the sum of all known vectors in the loop. Using spheric coordinates, each of the vectors $\mathbf{r}$, $\mathbf{s}$, and $\mathbf{t}$ can be expressed as a magnitude and two angles. Vector $\mathbf{r}$, for example, is defined once its magnitude, r, and two angles, θ_r and ϕ_r, are known. Thus, in Eq. (10.2), any three of the nine quantities $r, \theta_r, \phi_r, s, \theta_s, \phi_s, t, \theta_t$, and ϕ_t can be unknowns that must be found from the vector equation. Chace has solved these nine cases by first reducing each to a polynomial [3]. He classifies the solutions depending upon whether the three unknowns occur in one, two, or three separate vectors, and he tabulates the forms of the solutions as shown in Table 10.1.

In Table 10.1, the unit vectors $\hat{\omega}_r$, $\hat{\omega}_s$, and $\hat{\omega}_t$ are axes about which the angles ϕ_r, ϕ_s, and ϕ_t are measured. In case 1, vectors $\mathbf{s}$ and $\mathbf{t}$ are not needed; therefore, they are dropped from the equation. The three unknowns are all in vector $\mathbf{r}$. In cases 2a, 2b, 2c, and 2d, vector $\mathbf{t}$ is not needed and is dropped; the three unknowns are shared by vectors $\mathbf{r}$ and $\mathbf{s}$. Cases 3a, 3b, 3c, and 3d have single unknowns in each of the vectors $\mathbf{r}$, $\mathbf{s}$, and $\mathbf{t}$.

One advantage of the Chace vector tetrahedron solutions is that, since they provide known forms for the solutions of the nine cases, we can write a set of nine program modules for numeric evaluation. Eight of the nine cases have been reduced to explicit closed-form solutions for the unknowns and therefore can be quickly evaluated. Only case 3d, involving the solution of an eighth-order polynomial, must be solved by numeric iteration.

Although the vector tetrahedron equation and its nine case solutions can be used to solve most practical spatial problems, we recall from Section 10.1 that the Kutzbach criterion predicts the existence of up to seven j_1 joints in a single-loop, one-degree-of-freedom mechanism. For example, the seven-link $7R$ linkage has one input and six unknown joint variables [6]. Using the vector form of the loop-closure equation, it is not possible to determine the values for the six unknowns since the vector equation is equivalent to only three scalar equations, and does not fully describe three-dimensional rotation. Therefore, other mathematical tools, such as dual numbers, quaternions, or transformation matrices, must be used. In general, such problems require numeric techniques for their evaluation [6]. Anyone attempting the solution of such problems by hand-algebraic techniques quickly appreciates that *posture* analysis, not velocity or acceleration analysis, is the most difficult problem in kinematics.

Solving the polynomials of the Chace vector tetrahedron equation turns out to be equivalent to finding the intersections of straight lines or circles with various surfaces of revolution. In general, such problems can often be solved easily and quickly by the methods of descriptive geometry. The graphic approach has the added advantage that the geometric nature of the problem is not concealed in a multiplicity of mathematic operations. Graphic methods performed by hand are not nearly as accurate as an analytic or a numeric approach. However, graphic methods performed by computer software can give the same accuracy as an analytic or numeric approach.

Example 10.1

The dimensions of the links of the spatial *RSSR* four-bar linkage shown in Fig. 10.5 are $R_{AO_2} = 1$ in, $R_{BA} = 3.5$ in, and $R_{BO_4} = 4$ in. Note that the planes of rotation of the input and output are known. For the input angle $\theta_2 = -45°$, determine the postures of coupler 3 and output link 4.

Graphic Solution

Once we replace link 4 by vector $\boldsymbol{R}_{BO_4}$, then the only unknown is the angle θ_4, since the magnitude of the vector and the plane of rotation are given. Similarly, once link 3 is replaced by vector $\boldsymbol{R}_{BA}$, then the magnitude is known, and there are only two unknown spheric coordinate angular direction variables for this vector. The loop-closure equation is

$$(\boldsymbol{R}_{BA}) + (-\boldsymbol{R}_{BO_4}) + (\boldsymbol{R}_{AO_2} - \boldsymbol{R}_{O_4O_2}) = \boldsymbol{0}. \tag{1}$$

We identify this as case 2*d* in Table 10.1, requiring the solution of a second-degree (quadratic) polynomial and hence it yields two solutions.

This problem can be solved graphically using two orthographic views, the frontal and profile views. If we imagine, in Fig. 10.5, that the coupler is disconnected from the output crank at B and allowed to occupy any possible posture with respect to A, then B of link 3 must lie on the surface of a sphere of known radius with its center at A. With joint B still disconnected, the locus of B of link 4 is a circle of known radius about O_4 in a plane parallel to the yz plane. Therefore, to solve this problem, we need to find the two points of intersection of a circle and a sphere.

Example 10.1 (cont.)

The solution is shown in Fig. 10.6, where the subscripts F and P denote projections on the frontal and profile planes, respectively. First, we locate points O_2, A, and O_4 in both views. In the profile view, we draw a circle of radius $R_{BO_4} = 4$ in about center O_{4P}; this is the locus of point B_4. This circle appears in the frontal view as the vertical line $M_F O_{4F} N_F$.

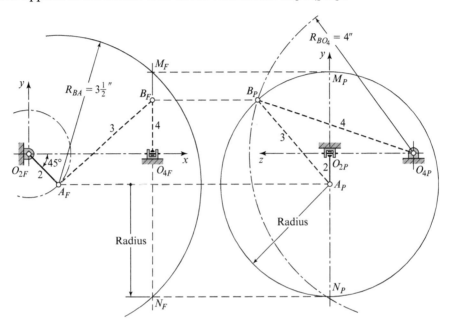

Fig. 10.6 Graphic posture analysis.

Next, in the frontal view, we construct the outline of a sphere with A_F as center and coupler length $R_{BA} = 3.5$ in as its radius. If $M_F O_{4F} N_F$ is regarded as the trace of a plane normal to the frontal view plane, the intersection of this plane with the sphere appears as the full circle in the profile view, having diameter $M_P N_P = M_F N_F$. The circular arc of radius R_{BO_4} intersects the circle at two points, yielding two solutions. One of the points is chosen as B_P and is projected back to the frontal view to locate B_F. Links 3 and 4 are drawn next as the dashed lines shown in the profile and frontal views of Fig. 10.6.

Measuring the $x, y,$ and z projections from the graphic solution, we can write the vectors for each link:

$$\boldsymbol{R}_{O_4 O_2} = 3.00\hat{\boldsymbol{i}} - 2.00\hat{\boldsymbol{k}} \text{ in,} \qquad\qquad Ans.$$

$$\boldsymbol{R}_{AO_2} = 0.71\hat{\boldsymbol{i}} - 0.71\hat{\boldsymbol{j}} \text{ in,} \qquad\qquad Ans.$$

$$\boldsymbol{R}_{BA} = 2.30\hat{\boldsymbol{i}} + 1.95\hat{\boldsymbol{j}} + 1.77\hat{\boldsymbol{k}} \text{ in,} \qquad\qquad Ans.$$

$$\boldsymbol{R}_{BO_4} = 1.22\hat{\boldsymbol{j}} + 3.81\hat{\boldsymbol{k}} \text{ in.} \qquad\qquad Ans.$$

Example 10.1 (cont.)

These components were obtained from a true-size graphic solution; better accuracy is possible, of course, by making the drawings two or four times actual size.

Analytic Solution

The given information establishes that

$$\boldsymbol{R}_{O_4O_2} = 3.000\hat{\boldsymbol{i}} - 2.000\hat{\boldsymbol{k}} \text{ in,}$$

$$\boldsymbol{R}_{AO_2} = \cos(-45°) + \sin(-45°) = 0.707\hat{\boldsymbol{i}} - 0.707\hat{\boldsymbol{j}} \text{ in,}$$

$$\boldsymbol{R}_{BA} = R^x_{BA}\hat{\boldsymbol{i}} + R^y_{BA}\hat{\boldsymbol{j}} + R^z_{BA}\hat{\boldsymbol{k}},$$

$$\boldsymbol{R}_{BO_4} = -4.000\sin\theta_4\,\hat{\boldsymbol{j}} + 4.000\cos\theta_4\hat{\boldsymbol{k}} \text{ in.}$$

The same loop-closure equation as in Eq. (1) must be solved. Substituting the given information gives

$$\left(R^x_{BA}\hat{\boldsymbol{i}} + R^y_{BA}\hat{\boldsymbol{j}} + R^z_{BA}\hat{\boldsymbol{k}}\right) + \left(4.000\sin\theta_4\hat{\boldsymbol{j}} - 4.000\cos\theta_4\hat{\boldsymbol{k}}\right)$$
$$+ \left(0.707\hat{\boldsymbol{i}} - 0.707\hat{\boldsymbol{j}} - 3.000\hat{\boldsymbol{i}} + 2.000\hat{\boldsymbol{k}}\right) = \boldsymbol{0}.$$

Then, separating this equation into its $\hat{\boldsymbol{i}}$, $\hat{\boldsymbol{j}}$, and $\hat{\boldsymbol{k}}$ components and rearranging gives the following three equations:

$$R^x_{BA} = 2.293 \text{ in,}$$

$$R^y_{BA} = -4.000\sin\theta_4 + 0.707 \text{ in,}$$

$$R^z_{BA} = 4.000\cos\theta_4 - 2.000 \text{ in.}$$

Next, we square and add these equations, and, remembering that $R_{BA} = 3.500$ in, we obtain

$$(2.293 \text{ in})^2 + (-4.000\sin\theta_4 + 0.707 \text{ in})^2 + (4.000\cos\theta_4 - 2.000 \text{ in})^2 = (3.500 \text{ in})^2.$$

Expanding this equation and rearranging gives

$$-5.656\sin\theta_4 - 16.000\cos\theta_4 + 13.508 \text{ in}^2 = 0.$$

We now have a single equation with only one unknown, namely, θ_4. However, it contains both sine and cosine terms (a transcendental equation). A standard technique for dealing with this situation is to convert the equation into a second-order (quadratic) polynomial by using the tangent of half-angle identities:

$$\sin\theta_4 = \frac{2\tan(\theta_4/2)}{1 + \tan^2(\theta_4/2)} \quad \text{and} \quad \cos\theta_4 = \frac{1 - \tan^2(\theta_4/2)}{1 + \tan^2(\theta_4/2)}.$$

Example 10.1 (cont.)

Substituting these identities into the above equation and multiplying by the common denominator gives a single quadratic equation in one unknown, $\tan(\theta_4/2)$, that is,

$$-11.312 \, \tan(\theta_4/2) - 16.000\left[1 - \tan^2(\theta_4/2)\right] + 13.508\left[1 + \tan^2(\theta_4/2)\right] = 0,$$

or

$$29.508 \, \tan^2(\theta_4/2) - 11.312 \tan(\theta_4/2) - 2.492 = 0.$$

This quadratic equation can now be solved for two roots, which are

$$\tan \frac{\theta_4}{2} = \begin{cases} 0.5398; & \theta_4 = -123.28° \\ -0.1565; & \theta_4 = 162.21° \end{cases}.$$

Of these two roots, we choose the second as the solution of interest. Back-substituting into the previous conditions, we find numeric values for the four vectors, namely:

$$\boldsymbol{R}_{O_4O_2} = 3.000\hat{\boldsymbol{i}} - 2.000\hat{\boldsymbol{k}} \text{ in,} \qquad\qquad Ans.$$

$$\boldsymbol{R}_{AO_2} = 0.707\hat{\boldsymbol{i}} - 0.707\hat{\boldsymbol{j}} \text{ in,} \qquad\qquad Ans.$$

$$\boldsymbol{R}_{BA} = 2.293\hat{\boldsymbol{i}} + 1.930\hat{\boldsymbol{j}} + 1.809\hat{\boldsymbol{k}} \text{ in,} \qquad\qquad Ans.$$

$$\boldsymbol{R}_{BO_4} = 1.223\hat{\boldsymbol{j}} + 3.809\hat{\boldsymbol{k}} \text{ in.} \qquad\qquad Ans.$$

These results match those of the graphic solution within graphic error.

We should note from this example that we have solved for the output angle θ_4 and the current values of the four vectors for the given position of the input crank angle θ_2. However, have we really solved for the postures of all links? The answer is no! As indicated above, we have not determined how the coupler link may have rotated about its axis $\boldsymbol{R}_{BA}$ between the two spheric joints. This is still unknown and explains how we were able to solve this two-degree-of-freedom system without specifying another input angle for the idle freedom. Depending upon our motivation, the above solution may be sufficient; however, it is important to note that the above vectors are not sufficient to solve for this additional unknown variable.

The spheric four-bar *RRRR* linkage shown in Fig. 10.2 is also case *2d* of the vector tetrahedron equation and, for a specified posture of the input link, can be solved in the same manner as the example shown.

10.4 Spatial Velocity and Acceleration Analyses

Once the postures of all links of a spatial mechanism are known, the velocities and accelerations can then be determined using the methods of Chapters 3 and 4. In planar mechanisms, the angular velocity and acceleration vectors are always perpendicular to the plane of motion.

This considerably reduces the effort required in the solution process for both graphic and analytic approaches. In spatial problems, however, these vectors may be skew in space. Otherwise, the equations and the methods of analysis are the same. The following example shows the differences.

Example 10.2

For the spatial *RSSR* four-bar linkage in the posture shown in Fig. 10.7, the constant angular velocity of link 2 is $\omega_2 = 40\hat{k}$ rad/s. Find the angular velocities and angular accelerations of links 3 and 4 and the velocity and acceleration of point B for the specified posture.

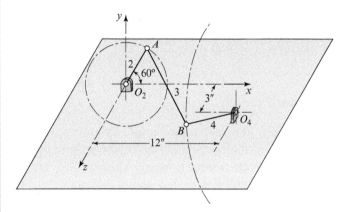

Fig. 10.7 $R_{AO_2} = 4$ in, $R_{BA} = 15$ in, and $R_{BO_4} = 10$ in.

Graphic Posture Solution

The graphic posture analysis follows exactly the procedure given in Section 10.3 and Example 10.1. The solution is shown in Fig. 10.8.

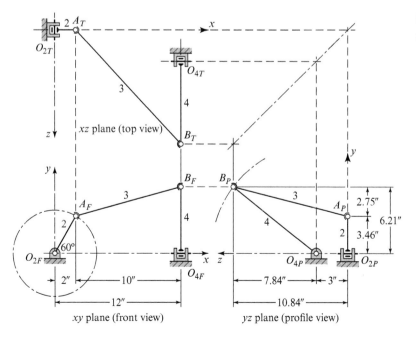

Fig. 10.8 Graphic posture analysis.

Example 10.2 (cont.)

Analytic Posture Solution

The analytic posture solution also follows the procedure given in Section 10.3 and Example 10.1. For the given input angle, the results are:

$$\boldsymbol{R}_{O_4O_2} = 12.000\hat{i} + 3.000\hat{k} \text{ in,}$$

$$\boldsymbol{R}_{AO_2} = 2.000\hat{i} + 3.464\hat{j} \text{ in,}$$

$$\boldsymbol{R}_{BA} = 10.000\hat{i} + 2.746\hat{j} + 10.838\hat{k} \text{ in,}$$

$$\boldsymbol{R}_{BO_4} = 6.210\hat{j} + 7.838\hat{k} \text{ in.}$$

Graphic Velocity and Acceleration Solutions

The determination of the velocities and accelerations of a spatial mechanism by graphic means can be conducted in the same manner as for a planar mechanism. However, the posture information as well as the velocity and acceleration vectors often do not appear in their true lengths in the front, top, and profile views, but are foreshortened. This means that spatial motion problems usually require the use of auxiliary views where the vectors appear in true lengths.

The velocity solution for this example is shown in Fig. 10.9 with notation corresponding to that used in many texts on descriptive geometry. The subscripts F, T, and P designate the front, top, and profile views, and the subscripts 1 and 2 indicate the first and second auxiliary views, respectively.

The steps to obtain the solution to the velocity problem are as follows:

1. We first construct to scale the front, top, and profile views of the linkage and designate each point.
2. Next, we calculate $\mathbf{V}_A$ as above and place this vector in position with its origin at A in the three views. The velocity $\mathbf{V}_A$ shows in true length in the frontal view. We designate its terminus as point a_F and project this point into the top and profile views to find a_T and a_P.
3. The magnitude of the velocity $\mathbf{V}_B$ is unknown, but its direction is perpendicular to $\boldsymbol{R}_{BO_4}$, and, once the problem is solved, it will show in true size in the profile view. Therefore, we construct a line in the profile view that originates at point A_P (which becomes the origin of our velocity polygon) and corresponds in direction to that of $\mathbf{V}_B$ (perpendicular to $R_{B_PO_{4P}}$). We then choose any point d_P on this line and project it into the front view (d_F) and top view (d_T) to establish the line of action of $\mathbf{V}_B$ in those views.
4. The equation to be solved is

$$V_B = V_A + V_{BA}, \tag{1}$$

Example 10.2 (cont.)

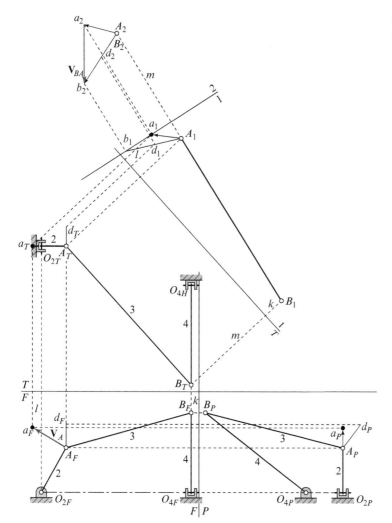

Fig. 10.9 Graphic velocity analysis.

where $\mathbf{V}_A$ and the directions of $\mathbf{V}_B$ and $\mathbf{V}_{BA}$ are known. We note that $\mathbf{V}_{BA}$ must lie in a plane perpendicular (in space) to $\mathbf{R}_{BA}$, but its magnitude and that of $\mathbf{V}_B$ are unknown. Vector $\mathbf{V}_{BA}$ must originate at the terminus of $\mathbf{V}_A$ and must lie in a plane perpendicular to $\mathbf{R}_{BA}$; it terminates by intersecting the line Ad or its extension. To find the plane perpendicular to $\mathbf{R}_{BA}$, we start by constructing the first auxiliary view, which shows vector $\mathbf{R}_{BA}$ in true length, so we construct the edge view of plane 1 parallel to $A_T B_T$ and project $\mathbf{R}_{BA}$ to this plane. In this projection, we note that the distances k and l are the same in this first auxiliary view as in the frontal view. The first auxiliary view of AB is $A_1 B_1$, which is true length. We also project points a and d into this view, but the remaining links need not be projected.

5. In this step, we construct a second auxiliary view, plane 2, such that the projection of AB upon it is a point. Then, all lines drawn parallel to this plane are perpendicular to $\mathbf{R}_{BA}$. The edge

Example 10.2 (cont.)

view of such a plane must be perpendicular to A_1B_1 extended. In this example, it is convenient to choose this plane so that it contains point a; therefore, we construct the edge view of plane 2 through point a_1 perpendicular to A_1B_1 extended. Now we project points $A, B, a,$ and d into this plane. Note that the distances – m, for example – of points from plane 1 must be the same in the top view and in the second auxiliary view.

6. We now extend the line A_1d_1 until it intersects the edge view of plane 2 at b_1, and we find projection b_2 of this point into plane 2 where the projector meets line A_2d_2 extended. Now both points a and b lie in plane 2, and any line drawn in plane 2 is perpendicular to $\mathbf{R}_{BA}$. Therefore, line ab is $\mathbf{V}_{BA}$, and the second auxiliary view of that line shows its true length. Line A_2b_2 is the projection of $\mathbf{V}_B$ on the second auxiliary plane, but not in its true length, since point A is not in plane 2.

7. To simplify the reading of the drawing, step 7 is not shown; those who follow the first six steps carefully should have no difficulty with the seventh. We can project the three vectors into the top, front, and profile views. Velocity V_B can be measured from the profile view where it appears in true length. The result is

$$V_B = 200\hat{j} - 158\hat{k} \text{ in/s.} \qquad\qquad Ans.$$

When all vectors have been projected into these three views, we can measure their $x, y,$ and z projections directly.

8. If we assume that the idle freedom of spin is not active and, therefore, that the angular velocity vector ω_3 is perpendicular to $\mathbf{R}_{BA}$ and appears in true length in the second auxiliary view,[5] then the magnitudes of the angular velocities can be determined from the equations

$$\omega_3 = \frac{V_{BA}}{R_{BA}} = \frac{242 \text{ in/s}}{15 \text{ in}} = 16.13 \text{ rad/s,}$$

$$\omega_4 = \frac{V_{BO_4}}{R_{BO_4}} = \frac{255 \text{ in/s}}{10 \text{ in}} = 25.50 \text{ rad/s.}$$

Therefore, we can draw the angular velocity vectors in the views where they appear in their true lengths and project them into the views where their vector components can be measured. The results are

$$\omega_3 = -7.69\hat{i} + 13.71\hat{j} + 3.63\hat{k} \text{ rad/s,} \qquad\qquad Ans.$$

$$\omega_4 = 25.50\hat{i} \text{ rad/s.} \qquad\qquad Ans.$$

---

[5] This is the same assumption we use in the analytic solution when we write the equation $\omega_3 \cdot \mathbf{R}_{BA} = \mathbf{0}$.

Example 10.2 (cont.)

The steps to obtain the solution to the acceleration problem, also not shown, are followed in an identical manner, using the same two auxiliary planes. The equation to be solved is

$$A^n_{BO_4} + A^t_{BO_4} = A^n_{AO_2} + A^t_{AO_2} + A^n_{BA} + A^t_{BA}, \tag{2}$$

where the vectors $A^n_{BO_4}$, $A^n_{AO_2}$, $A^t_{AO_2}$, and A^n_{BA} can be determined now that the velocity analysis has been completed. Also, we see that $A^t_{BO_4}$ and A^t_{BA} have known directions. Therefore, the solution can proceed exactly as for the velocity polygon; the only difference in approach is that there are more known vectors. The final results are

$$\alpha_3 = -527\hat{i} + 619\hat{j} + 329\hat{k} \text{ rad/s}^2, \qquad\qquad Ans.$$

$$\alpha_4 = -865\hat{i} \text{ rad/s}^2, \qquad\qquad Ans.$$

$$A_B = 2\,750\hat{j} - 10\,450\hat{k} \text{ in/s}^2. \qquad\qquad Ans.$$

Analytic Velocity and Acceleration Solutions

From the given information and the constraints, the angular velocities and accelerations can be written as

$$\omega_2 = 40\hat{k} \; rad/s, \qquad\qquad \alpha_2 = 0,$$
$$\omega_3 = \omega_3^x\hat{i} + \omega_3^y\hat{j} + \omega_3^z\hat{k}, \qquad\qquad \alpha_3 = \alpha_3^x\hat{i} + \alpha_3^y\hat{j} + \alpha_3^z\hat{k},$$
$$\omega_4 = \omega_4\hat{i}, \qquad\qquad \alpha_4 = \alpha_4\hat{i}.$$

First, we find the velocity of point A as the velocity difference from point O_2. Thus,

$$V_A = \omega_2 \times R_{AO_2} = (40\hat{k} \text{ rad/s}) \times (2.000\hat{i} + 3.464\hat{j} \text{ in}) = -138.560\hat{i} + 80.000\hat{j} \text{ in/s}. \tag{3}$$

Similarly, for link 3, the velocity difference is

$$V_{BA} = (\omega_3^x\hat{i} + \omega_3^y\hat{j} + \omega_3^z\hat{k}) \times (10.000\hat{i} + 2.746\hat{j} + 10.838\hat{k} \text{ in})$$
$$= (10.838\omega_3^y - 2.746\omega_3^z)\hat{i} + (10.000\omega_3^z - 10.838\omega_3^x)\hat{j} \tag{4}$$
$$+ (2.746\omega_3^x - 10.000\omega_3^y)\hat{k} \text{ in}$$

and for link 4 the velocity of point B is

$$V_B = \omega_4 \times R_{BO_4} = (\omega_4\hat{i}) \times (6.210\hat{j} + 7.838\hat{k} \text{ in}) = -(7.838 \text{ in})\omega_4\hat{j} + (6.210 \text{ in})\omega_4\hat{k}. \tag{5}$$

Example 10.2 (cont.)

The next step is to substitute these into the velocity difference equation,

$$V_B = V_A + V_{BA}. \tag{6}$$

After this, we separate the $\hat{i}, \hat{j}$, and $\hat{k}$ components to obtain the three algebraic equations:

$$10.838 \text{ in } \omega_3^y - 2.746 \text{ in } \omega_3^z = 138.560 \text{ in/s}, \tag{7}$$

$$-10.838 \text{ in } \omega_3^x + 10.000 \text{ in } \omega_3^z + 7.838 \text{ in } \omega_4 = -80.000 \text{ in/s}, \tag{8}$$

$$2.746 \text{ in } \omega_3^x - 10.000 \text{ in } \omega_3^y - 6.210 \text{ in } \omega_4 = 0.000. \tag{9}$$

We now have three equations; however, we note that there are four unknowns, ω_3^x, ω_3^y, ω_3^z, and ω_4. This would not occur in most problems, but does here because of the idle freedom of the coupler to spin about its own axis. Since this spin does not affect the input–output relationship, we will find the same result for ω_4 regardless of the rate of spin. One way to proceed, therefore, would be to choose one component of $\boldsymbol{\omega}_3$ and give it a value, thus setting the rate of spin, and then solve for the other unknowns. Another approach is to set the rate of spin about the coupler link axis to zero by requiring that

$$\boldsymbol{\omega}_3 \cdot \mathbf{R}_{BA} = 0,$$
$$R_{BA}^x \omega_3^x + R_{BA}^y \omega_3^y + R_{BA}^z \omega_3^z = 0, \tag{10}$$
$$(10.000 \text{ in})\omega_3^x + (2.746 \text{ in})\omega_3^y + (10.838 \text{ in})\omega_3^z = 0.$$

Equations (7)–(10) can now be solved simultaneously for the four unknowns. The result is

$$\boldsymbol{\omega}_3 = -7.692\hat{i} + 13.704\hat{j} + 3.625\hat{k} \text{ rad/s}, \qquad Ans.$$

$$\boldsymbol{\omega}_4 = -25.468\hat{i} \text{ rad/s}. \qquad Ans.$$

Substituting these angular velocities into Eq. (5), the velocity of point B is

$$V_B = 199.618\hat{j} - 158.156\hat{k} \text{ in/s}. \qquad Ans.$$

Turning next to acceleration analysis, we compute the following components:

$$A_{AO_2}^n = \boldsymbol{\omega}_2 \times (\boldsymbol{\omega}_2 \times \mathbf{R}_{AO_2}) = -3\,200.00\hat{i} - 5\,542.56\hat{j} \text{ in/s}^2, \tag{11}$$

$$A_{AO_2}^t = \boldsymbol{\alpha}_2 \times \mathbf{R}_{AO_2} = \mathbf{0}, \tag{12}$$

$$A_{BA}^n = \boldsymbol{\omega}_3 \times (\boldsymbol{\omega}_3 \times \mathbf{R}_{BA}) = -2\,601.01\hat{i} - 714.26\hat{j} - 2\,818.99\hat{k} \text{ in/s}^2, \tag{13}$$

$$\begin{aligned} A_{BA}^t = \boldsymbol{\alpha}_3 \times \mathbf{R}_{BA} &= \left(\alpha_3^x\hat{i} + \alpha_3^y\hat{j} + \alpha_3^z\hat{k}\right) \times (10.000\hat{i} + 2.746\hat{j} + 10.838\hat{k} \text{ in}) \\ &= \left(10.838\alpha_3^y - 2.746\alpha_3^z\right)\hat{i} + \left(10.000\alpha_3^z - 10.838\alpha_3^x\right)\hat{j} + \left(2.746\alpha_3^x - 10.000\alpha_3^y\right)\hat{k} \text{ in}, \end{aligned} \tag{14}$$

Example 10.2 (cont.)

$$A^n_{BO_4} = \omega_4 \times (\omega_4 \times R_{BO_4}) = -4\,028.05\hat{j} - 5\,084.03\hat{k} \text{ in/s}^2, \tag{15}$$

$$A^t_{BO_4} = \alpha_4 \times R_{BO_4} = -7.838\alpha_4\hat{j} + 6.210\alpha_4\hat{k} \text{ in.} \tag{16}$$

These quantities are now substituted into the acceleration difference equation,

$$A^n_{BO_4} + A^t_{BO_4} = A^n_{AO_2} + A^t_{AO_2} + A^n_{BA} + A^t_{BA}, \tag{17}$$

and separated into $\hat{i}$, $\hat{j}$, and $\hat{k}$ components. Along with the condition $\alpha_3 \cdot R_{BA} = 0$ for the spin of the idle freedom, this results in four equations in four unknowns as follows:

$$10.838 \text{ in } \alpha_3^y - 2.746 \text{ in } \alpha_3^z = 5\,801.01 \text{ in/s}^2,$$
$$-10.838 \text{ in } \alpha_3^x + 10.000 \text{ in } \alpha_3^z + 7.838 \text{ in } \alpha_4 = 2\,228.61 \text{ in/s}^2,$$
$$2.746 \text{ in } \alpha_3^x - 10.000 \text{ in } \alpha_3^y - 6.210 \text{ in } \alpha_4 = -2\,265.04 \text{ in/s}^2,$$
$$10.000 \text{ in } \alpha_3^x + 2.746 \text{ in } \alpha_3^y + 10.838 \text{ in } \alpha_3^z = 0.00.$$

The coefficients (on the left-hand side) of these acceleration equations are identical to those of the velocity equations [Eqs. (7)–(10)]. This is always the case. These equations are now solved to give the desired accelerations.

$$\alpha_3 = -526.94\hat{i} + 618.71\hat{j} + 329.43\hat{k} \text{ rad/s}^2, \qquad Ans.$$

$$\alpha_4 = -864.59\hat{i} \text{ rad/s}^2, \qquad Ans.$$

$$A_B = A^n_{BO_4} + A^t_{BO_4} = 2\,748.61\hat{j} - 10\,453.10\hat{k} \text{ in/s}^2. \qquad Ans.$$

10.5 Euler Angles

We saw in Section 3.2 that angular velocity is a vector quantity; hence, like all vector quantities, it has components along any set of rectilinear axes,

$$\omega = \omega^x\hat{i} + \omega^y\hat{j} + \omega^z\hat{k},$$

and it obeys the laws of vector algebra. Unfortunately, we also saw in Fig. 3.2 that finite three-dimensional angular displacements do not behave as vectors. Their order of summation is not arbitrary; that is, they are not commutative in addition. Therefore, they do not follow the rules of vector algebra. The inescapable conclusion is that we cannot find a set of three angles, or angular displacements, that specify the three-dimensional orientation of a rigid body and that also have ω^x, ω^y, and ω^z as their time derivatives.

To clarify the problem further, we visualize a rigid body rotating in space about a fixed point, O, that we take as the origin of a grounded or absolute reference frame, xyz. We also visualize a moving reference system, $x'y'z'$, sharing the same origin but attached to and moving with the rotating body. The $x'y'z'$ system is defined as a set of *body-fixed axes*. Our problem here is to determine some way to specify the orientation of the body-fixed axes with respect to the absolute reference axes, a method that is convenient for use with finite three-dimensional rotations.

Let us assume that the stationary axes are aligned along unit vector directions $\hat{\mathbf{i}}, \hat{\mathbf{j}}$, and $\hat{\mathbf{k}}$ and that the rotated body-fixed axes have directions denoted by $\hat{\mathbf{i}}', \hat{\mathbf{j}}'$, and $\hat{\mathbf{k}}'$. Then, any point in the absolute coordinate system is located by the position vector, $\mathbf{R}$, and in the rotated system by $\mathbf{R}'$. Since both vectors are descriptions of the position of the same point, then we conclude that $\mathbf{R} = \mathbf{R}'$ and

$$R^x\hat{\mathbf{i}} + R^y\hat{\mathbf{j}} + R^z\hat{\mathbf{k}} = R^{x'}\hat{\mathbf{i}}' + R^{y'}\hat{\mathbf{j}}' + R^{z'}\hat{\mathbf{k}}'.$$

Now, taking the vector dot products of this equation with $\hat{\mathbf{i}}, \hat{\mathbf{j}}$, and $\hat{\mathbf{k}}$, respectively, gives the transformation equations between the two sets of axes, namely

$$R^x = \left(\hat{\mathbf{i}} \cdot \hat{\mathbf{i}}'\right) R^{x'} + \left(\hat{\mathbf{i}} \cdot \hat{\mathbf{j}}'\right) R^{y'} + \left(\hat{\mathbf{i}} \cdot \hat{\mathbf{k}}'\right) R^{z'},$$
$$R^y = \left(\hat{\mathbf{j}} \cdot \hat{\mathbf{i}}'\right) R^{x'} + \left(\hat{\mathbf{j}} \cdot \hat{\mathbf{j}}'\right) R^{y'} + \left(\hat{\mathbf{j}} \cdot \hat{\mathbf{k}}'\right) R^{z'}, \qquad (a)$$
$$R^z = \left(\hat{\mathbf{k}} \cdot \hat{\mathbf{i}}'\right) R^{x'} + \left(\hat{\mathbf{k}} \cdot \hat{\mathbf{j}}'\right) R^{y'} + \left(\hat{\mathbf{k}} \cdot \hat{\mathbf{k}}'\right) R^{z'}.$$

But we remember that the dot product of two unit vectors is equal to the cosine of the angle between them. For example, if the angle from $\hat{\mathbf{i}}$ to $\hat{\mathbf{j}}'$ is denoted by $\theta_{ij'}$, then

$$\hat{\mathbf{i}} \cdot \hat{\mathbf{j}}' = \cos\theta_{ij'}. \qquad (b)$$

Thus, the above set of equations becomes

$$R^x = \cos\theta_{ii'} R^{x'} + \cos\theta_{ij'} R^{y'} + \cos\theta_{ik'} R^{z'},$$
$$R^y = \cos\theta_{ji'} R^{x'} + \cos\theta_{jj'} R^{y'} + \cos\theta_{jk'} R^{z'}, \qquad (10.3)$$
$$R^z = \cos\theta_{ki'} R^{x'} + \cos\theta_{kj'} R^{y'} + \cos\theta_{kk'} R^{z'}.$$

These equations can be written, in matrix notation, as

$$\begin{bmatrix} R^x \\ R^y \\ R^z \end{bmatrix} = \begin{bmatrix} \cos\theta_{ii'} & \cos\theta_{ij'} & \cos\theta_{ik'} \\ \cos\theta_{ji'} & \cos\theta_{jj'} & \cos\theta_{jk'} \\ \cos\theta_{ki'} & \cos\theta_{kj'} & \cos\theta_{kk'} \end{bmatrix} \begin{bmatrix} R^{x'} \\ R^{y'} \\ R^{z'} \end{bmatrix}. \qquad (10.4)$$

This means of defining the orientation of the $x'y'z'$ axes with respect to the xyz axes uses what are called *direction cosines* as coefficients in a set of *transformation equations* between the two coordinate systems. Although we will have use for direction cosines, we note the disadvantage that there are nine of them, whereas only three variables can be independent for spatial rotation. The nine direction cosines are not all independent but are related by six *orthogonality conditions*. We would prefer a technique that had only three independent variables, preferably all angles.

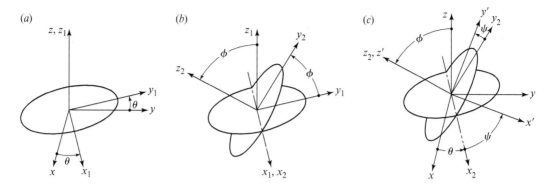

Fig. 10.10 Euler angles.

Three angles, called *Euler angles*, can be used to specify the orientation of the body-fixed axes with respect to the reference axes, as shown in Fig. 10.10. To explain the Euler angles, we begin with the body-fixed axes coincident with the reference axes. We then specify three successive rotations, θ, ϕ, and ψ, which must occur in the specified order and about the specified axes, to move the body-fixed axes into their final orientation.

The first Euler angle describes a rotation through an angle θ and is taken positive counter-clockwise about the positive z axis, as shown in Fig. 10.10a. This rotation turns the xy axes by the angle θ to the new $x_1 y_1$ locations shown, whereas z and z_1 remain fixed and collinear. The transformation equations for this first rotation are

$$\begin{bmatrix} R^x \\ R^y \\ R^z \end{bmatrix} = \begin{bmatrix} \cos\theta & -\sin\theta & 0 \\ \sin\theta & \cos\theta & 0 \\ 0 & 0 & 1 \end{bmatrix} \begin{bmatrix} R^{x_1} \\ R^{y_1} \\ R^{z_1} \end{bmatrix}. \tag{10.5}$$

The second Euler angle describes a rotation through an angle ϕ and is taken positive counterclockwise about the positive x_1 axis, as shown in Fig. 10.10b. This rotation turns the displaced $y_1 z_1$ axes by the angle ϕ to the new $y_2 z_2$ locations shown, whereas x_1 and x_2 remain fixed. The transformation equations for this second rotation are

$$\begin{bmatrix} R^{x_1} \\ R^{y_1} \\ R^{z_1} \end{bmatrix} = \begin{bmatrix} 1 & 0 & 0 \\ 0 & \cos\phi & -\sin\phi \\ 0 & \sin\phi & \cos\phi \end{bmatrix} \begin{bmatrix} R^{x_2} \\ R^{y_2} \\ R^{z_2} \end{bmatrix}. \tag{10.6}$$

The third Euler angle describes a rotation through an angle ψ and is taken positive counter-clockwise about the positive z_2 axis, as shown in Fig. 10.10c. This rotation turns the $x_2 y_2$ axes by the angle ψ to the final $x' y'$ locations shown, whereas z_2 and z' remain fixed. The transformation equations for this third rotation are

$$\begin{bmatrix} R^{x_2} \\ R^{y_2} \\ R^{z_2} \end{bmatrix} = \begin{bmatrix} \cos\psi & -\sin\psi & 0 \\ \sin\psi & \cos\psi & 0 \\ 0 & 0 & 1 \end{bmatrix} \begin{bmatrix} R^{x'} \\ R^{y'} \\ R^{z'} \end{bmatrix}. \tag{10.7}$$

Now, substituting Eq. (10.6) into Eq. (10.5) gives the transformation from the xyz axes to the $x_2 y_2 z_2$ axes:

$$
\begin{bmatrix} R^x \\ R^y \\ R^z \end{bmatrix} = \begin{bmatrix} \cos\theta & -\sin\theta\cos\phi & \sin\theta\sin\phi \\ \sin\theta & \cos\theta\cos\phi & -\cos\theta\sin\phi \\ 0 & \sin\phi & \cos\phi \end{bmatrix} \begin{bmatrix} R^{x_2} \\ R^{y_2} \\ R^{z_2} \end{bmatrix}. \tag{10.8}
$$

Finally, substituting Eq. (10.7) into Eq. (10.8) gives the total transformation from the xyz axes to the $x'y'z'$ axes:

$$
\begin{bmatrix} R^x \\ R^y \\ R^z \end{bmatrix} = \begin{bmatrix} \cos\theta\cos\psi - \sin\theta\cos\phi\sin\psi & -\cos\theta\sin\psi - \sin\theta\cos\phi\cos\psi & \sin\theta\sin\phi \\ \sin\theta\cos\psi + \cos\theta\cos\phi\sin\psi & -\sin\theta\sin\psi + \cos\theta\cos\phi\cos\psi & -\cos\theta\sin\phi \\ \sin\phi\sin\psi & \sin\phi\cos\psi & \cos\phi \end{bmatrix} \begin{bmatrix} R^{x'} \\ R^{y'} \\ R^{z'} \end{bmatrix}. \tag{10.9}
$$

Since the transformation is expressed as a function of only three Euler angles, θ, ϕ, and ψ, it represents the same information as Eqs. (10.3) and (10.4), but without the difficulty of having nine variables (direction cosines) related by six (orthogonality) conditions. Thus, Euler angles have become a useful tool in treating problems involving three-dimensional rotation.

Note, however, that the form of the transformation equations given here depends on using precisely the conventions for the particular set of Euler angles defined in Fig. 10.10. Unfortunately, there appears to be little agreement among different authors regarding how these angles should be defined. A wide variety of other definitions, differing in the axes about which the rotations are to be measured, or in their order, or in the sign conventions for positive values of the angles, are to be found in other references. The differences are not great, but are sufficient to frustrate easy comparison of the formulae derived.

We must also remember that the time derivatives of the Euler angles, that is, $\dot{\theta}, \dot{\phi}$, and $\dot{\psi}$, are *not* the components of the angular velocity, $\boldsymbol{\omega}$, of the body-fixed axes. We see that these angular velocities act about different axes and are expressed in different coordinate systems. From Fig. 10.10, we can see that the angular velocity of the body-fixed coordinate system is

$$
\boldsymbol{\omega} = \dot{\theta}\hat{\boldsymbol{k}} + \dot{\phi}\hat{\boldsymbol{i}}_1 + \dot{\psi}\hat{\boldsymbol{k}}'.
$$

When these different unit vectors are all transformed into the fixed reference frames and then added, the three components of the angular velocity vector can be written as

$$
\begin{aligned}
\omega^x &= \dot{\phi}\cos\theta + \dot{\psi}\sin\theta\sin\phi, \\
\omega^y &= \dot{\phi}\sin\theta - \dot{\psi}\cos\theta\sin\phi, \\
\omega^z &= \dot{\theta} + \dot{\psi}\cos\phi.
\end{aligned} \tag{10.10}
$$

On the other hand, we can also transform the unit vectors into the body-fixed axes; then

$$
\begin{aligned}
\omega^{x'} &= \dot{\theta}\sin\phi\sin\psi + \dot{\phi}\cos\psi, \\
\omega^{y'} &= \dot{\theta}\sin\phi\cos\psi - \dot{\phi}\sin\psi, \\
\omega^{z'} &= \dot{\theta}\cos\phi + \dot{\psi}.
\end{aligned} \tag{10.11}
$$

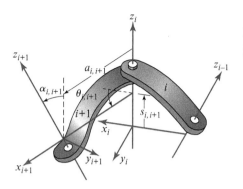

Fig. 10.11 Definitions of Denavit–Hartenberg parameters.

10.6 Denavit–Hartenberg Parameters

The transformation equations of the previous section dealt only with three-dimensional rotations about a fixed point. Yet the same approach can be generalized to include both translations and rotations and thus to treat general spatial motion. This has been done, and much literature dealing with these techniques exists, usually under titles such as "matrix methods." Most modern publications on this approach stem from the work of Denavit and Hartenberg [5]. These authors developed a notation for labeling all single-loop lower pair linkages and also devised a transformation matrix technique for their analysis.

In the Denavit–Hartenberg approach, we start by numbering the joints of the linkage, usually starting with the input joint and numbering consecutively around a kinematic loop to the output joint. If we assume that the linkage has only a single loop with only j_1 joints and a mobility of $m = 1$, then the Kutzbach criterion indicates that there will be $n = 7$ binary links and $j_1 = 7$ joints numbered from $i = 1$ through $i = 7$.[6] Figure 10.11 shows a typical joint of this loop, revolute joint number i in this case, and the two links that it connects.

Next, we identify the motion axis of each of the joints, choose a positive orientation on each, and label each as a z_i axis. Although the z_i axes may be skew in space, it is possible to find a common perpendicular between each consecutive pair and label these as x_i axes such that each x_i axis is the common perpendicular between z_{i-1} and z_i, choosing an arbitrary positive orientation.[7]

Having done this, we can now identify and label y_i axes such that there is a right-handed Cartesian coordinate system $x_i y_i z_i$ associated with each joint of the loop. Careful study of Fig. 10.11 and visualization of the motion allowed by the joint indicates that coordinate system $x_i y_i z_i$ remains fixed to the link carrying joint $i - 1$ and joint i, whereas $x_{i+1} y_{i+1} z_{i+1}$ moves with and remains fixed to the link carrying joint i and joint $i + 1$. Thus, we have a basis for assigning numbers to the links that correspond to the numbers of the coordinate system attached to each link.

---

[6] The treatment presented here deals only with j_1 joints; multi-degree-of-freedom lower pairs may be represented as serial combinations of revolute joints and prismatic joints, however, with fictitious links between them.

[7] When $i = n$, then $i + 1$ must be taken as 1. Similarly, when $i = 1$, then $i - 1$ must be taken as n, because, in the loop, the last joint is also the joint before the first.

The key to the Denavit–Hartenberg approach comes in the standard method they defined for determining the dimensions of the links. The relative location and orientation (posture) of any two consecutive coordinate systems placed as described above can be defined by four parameters, labeled a, α, θ, and s, shown in Fig. 10.11, and defined as follows:

$a_{i,i+1}$ = the distance along x_{i+1} from z_i to z_{i+1}, with the sign taken from the sense of x_{i+1};

$\alpha_{i,i+1}$ = the angle from positive z_i to positive z_{i+1}, taken positive counterclockwise as seen from positive x_{i+1};

$\theta_{i,i+1}$ = the angle from positive x_i to positive x_{i+1}, taken positive counterclockwise as seen from positive z_i; and

$s_{i,i+1}$ = the distance along z_i from x_i to x_{i+1}, with sign taken from the sense of z_i.

When joint i is a revolute, as depicted in Fig. 10.11, then the $a_{i,i+1}, \alpha_{i,i+1}$, and $s_{i,i+1}$ parameters are constants defining the shape of link $i+1$, but $\theta_{i,i+1}$ is a variable; in fact, it serves to measure the joint variable of joint i. When joint i is a prismatic joint, then the $a_{i,i+1}, \alpha_{i,i+1}$, and $\theta_{i,i+1}$ parameters are constants, and $s_{i,i+1}$ serves as the joint variable.

Example 10.3

Find the Denavit–Hartenberg parameters for the Hooke, or Cardan, universal joint shown in Fig. 10.12.

Fig. 10.12 Hooke, or Cardan, universal joint.

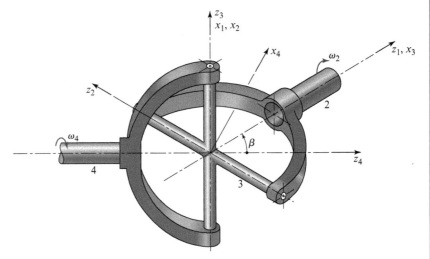

Solution

The linkage has four links and four revolute joints (it is identical to the spheric four-bar linkage shown in Fig. 10.2). First, we label the axes of the four revolute joints as z_1, z_2, z_3, and z_4. Next, we identify and label the common perpendiculars between the joint axes, choosing positive orientations for these, and label them x_1, x_2, x_3, and x_4, as shown. Note that, in the posture shown, x_1 and x_3 appear to lie along z_3 and z_1, respectively; this is only a temporary coincidence and changes as the linkage moves.

Example 10.3 (cont.)

Now, following the definitions given above, we find the values for the parameters:

$$a_{12} = 0, \quad a_{23} = 0, \quad a_{34} = 0, \quad a_{41} = 0, \qquad Ans.$$

$$\alpha_{12} = 90°, \quad \alpha_{23} = 90°, \quad \alpha_{34} = 90°, \quad \alpha_{41} = \beta, \qquad Ans.$$

$$\theta_{12} = \phi_1, \quad \theta_{23} = \phi_2, \quad \theta_{34} = \phi_3, \quad \theta_{41} = \phi_4, \qquad Ans.$$

$$s_{12} = 0, \quad s_{23} = 0, \quad s_{34} = 0, \quad s_{41} = 0, \qquad Ans.$$

where β is the angle between the two shafts.

We note that all a and s distance parameters are zero; this signifies that the linkage is, indeed, spheric, with all motion axes intersecting at a common center. Also, all θ parameters are joint variables since this is an *RRRR* linkage. They have been given the symbols ϕ_i rather than numeric values since they are variables; solutions for their values will be found in the next section.

10.7 Transformation Matrix Posture Analysis

The Denavit–Hartenberg parameters provide a standard method for measuring the important geometric characteristics of a linkage, but the value of this approach does not stop there. Having standardized the placement of the coordinate systems on each link, Denavit and Hartenberg also demonstrated that the transformation equations for the postures of successive coordinate systems can be written in a standard matrix format that uses these parameters.

If we know the position coordinates of a point measured in one of the coordinate systems, say R_{i+1}, then we can find the position coordinates of the same point in the previous coordinate system, R_i, as follows:

$$R_i = T_{i,i+1} R_{i+1}, \tag{10.12}$$

where the transformation $T_{i,i+1}$ has the standard form[8]

$$T_{i,i+1} = \begin{bmatrix} \cos\theta_{i,i+1} & -\cos\alpha_{i,i+1}\,\sin\theta_{i,i+1} & \sin\alpha_{i,i+1}\,\sin\theta_{i,i+1} & a_{i,i+1}\cos\theta_{i,i+1} \\ \sin\theta_{i,i+1} & \cos\alpha_{i,i+1}\,\cos\theta_{i,i+1} & -\sin\alpha_{i,i+1}\,\cos\theta_{i,i+1} & a_{i,i+1}\sin\theta_{i,i+1} \\ 0 & \sin\alpha_{i,i+1} & \cos\alpha_{i,i+1} & s_{i,i+1} \\ 0 & 0 & 0 & 1 \end{bmatrix}, \tag{10.13}$$

---

[8] Note that the first three rows and columns of the T matrix are the direction cosines needed for the rotation between the coordinate systems [for purposes of comparison see Eq. (10.8)]. The fourth column adds the translation terms for the separation of the origins. The fourth row represents a dummy equation, $1 = 1$, but keeps the T matrix square and nonsingular.

and each point position vector is given by

$$R = \begin{bmatrix} x \\ y \\ z \\ 1 \end{bmatrix}. \tag{10.14}$$

Now, by applying Eq. (10.12) recursively from one link to the next, for an n-link single-loop mechanism,

$$R_1 = T_{12}R_2,$$

$$R_1 = T_{12}T_{23}R_3,$$

$$R_1 = T_{12}T_{23}T_{34}R_4,$$

and, in general,

$$R_1 = T_{12}T_{23}\ldots T_{i-1,i}R_i. \tag{10.15}$$

If we agree on a notation in which a product of these T matrices is still denoted as a T matrix,

$$T_{i,i+1}T_{i+1,i+2}\ldots T_{j-1,j} = T_{i,j}, \tag{10.16}$$

then Eq. (10.15) becomes

$$R_1 = T_{1,i}R_i. \tag{10.17}$$

Finally, since link 1 follows link n at the end of the loop,

$$R_1 = T_{12}T_{23}\ldots T_{n-1,n}T_{n,1}R_1,$$

and, since this equation must be true no matter what point we choose for R_1,

$$T_{1,2}T_{2,3}\ldots T_{n-1,n}T_{n,1} = I, \tag{10.18}$$

where I is the 4×4 identity transformation matrix.

This important equation is the transformation matrix form of the *loop-closure equation*. Just as the vector tetrahedron equation, Eq. (10.2), states that the *sum of vectors* around a kinematic loop must equal zero for the loop to close, Eq. (10.18) states that the *product of transformation matrices* around a kinematic loop must equal the identity transformation. Whereas the vector sum ensures that the loop returns to its starting location, the transformation matrix product also ensures that the loop returns to its starting orientation. This becomes critical, for example, in spheric motion problems, and cannot be shown by the vector tetrahedron equation. The following example illustrates this point.

Example 10.4

Develop equations for the positions of all the joint variables of the Hooke, or Cardan, universal joint shown in Fig. 10.12 when the input shaft angle ϕ_1 is specified.

Solution

The Denavit–Hartenberg parameters for this linkage were determined in Example 10.3. Substituting these parameters into Eq. (10.13) gives the individual transformation matrices for each link, namely

$$T_{1,2} = \begin{bmatrix} \cos\phi_1 & 0 & \sin\phi_1 & 0 \\ \sin\phi_1 & 0 & -\cos\phi_1 & 0 \\ 0 & 1 & 0 & 0 \\ 0 & 0 & 0 & 1 \end{bmatrix}, \tag{1}$$

$$T_{2,3} = \begin{bmatrix} \cos\phi_2 & 0 & \sin\phi_2 & 0 \\ \sin\phi_2 & 0 & -\cos\phi_2 & 0 \\ 0 & 1 & 0 & 0 \\ 0 & 0 & 0 & 1 \end{bmatrix}, \tag{2}$$

$$T_{3,4} = \begin{bmatrix} \cos\phi_3 & 0 & \sin\phi_3 & 0 \\ \sin\phi_3 & 0 & -\cos\phi_3 & 0 \\ 0 & 1 & 0 & 0 \\ 0 & 0 & 0 & 1 \end{bmatrix}, \tag{3}$$

$$T_{4,1} = \begin{bmatrix} \cos\phi_4 & -\cos\beta\sin\phi_4 & \sin\beta\sin\phi_4 & 0 \\ \sin\phi_4 & \cos\beta\cos\phi_4 & -\sin\beta\cos\phi_4 & 0 \\ 0 & \sin\beta & \cos\beta & 0 \\ 0 & 0 & 0 & 1 \end{bmatrix}. \tag{4}$$

Although we could now use Eq. (10.18) directly, that is,

$$T_{1,2}T_{2,3}T_{3,4}T_{4,1} = I,$$

the number of computations can be reduced if we first rearrange the equation to the form

$$T_{2,3}T_{3,4} = T_{1,2}^{-1}T_{4,1}^{-1} = (T_{4,1}T_{12})^{-1}. \tag{5}$$

The inverse matrix on the right-hand side can easily be found here by simply transposing the matrix – that is, by switching rows and columns.[9] Therefore, substituting and carrying out the matrix computations, Eq. (5) becomes

---

[9] Although this is true in this special case, where all translation terms are zero, it is not true for all (4×4) transformation matrices.

Example 10.4 (cont.)

$$\begin{bmatrix} \cos\phi_2\cos\phi_3 & \sin\phi_2 & \cos\phi_2\sin\phi_3 & 0 \\ \sin\phi_2\cos\phi_3 & -\cos\phi_2 & \sin\phi_2\sin\phi_3 & 0 \\ \sin\phi_3 & 0 & -\cos\phi_3 & 0 \\ 0 & 0 & 0 & 1 \end{bmatrix}$$

$$= \begin{bmatrix} \cos\phi_1\cos\phi_4 & \cos\phi_1\sin\phi_4 & & \\ -\cos\beta\sin\phi_1\sin\phi_4 & +\cos\beta\sin\phi_1\cos\phi_4 & \sin\beta\sin\phi_1 & 0 \\ \sin\beta\sin\phi_4 & -\sin\beta\cos\phi_4 & \cos\beta & 0 \\ \sin\phi_1\cos\phi_4 & \sin\phi_1\sin\phi_4 & & \\ +\cos\beta\cos\phi_1\sin\phi_4 & -\cos\beta\cos\phi_1\cos\phi_4 & -\sin\beta\cos\phi_1 & 0 \\ 0 & 0 & 0 & 1 \end{bmatrix}. \tag{6}$$

Since corresponding row and column elements on both sides of this equation must be equal, then equating the third-row elements of the third column, we have

$$\cos\phi_3 = \sin\beta\cos\phi_1 \quad\text{and}\quad \sin\phi_3 = \sqrt{1-\sin^2\beta\cos^2\phi_1} = \sigma, \qquad \textit{Ans.}$$

where we define the new symbol σ. Next, equating the first- and second-row elements of column three gives

$$\cos\phi_2 = \frac{\sin\beta}{\sigma} \quad\text{and}\quad \sin\phi_2 = \frac{\cos\beta}{\sigma}. \qquad \textit{Ans.}$$

Once we have solved for the sine and cosine of ϕ_2 and ϕ_3, we can equate the elements of the second column, second row and the first column, second row. Using the known solutions, this gives

$$\cos\phi_4 = \frac{\sin\phi_1}{\sigma} \quad\text{and}\quad \sin\phi_4 = \frac{\cos\beta\cos\phi_1}{\sigma}. \qquad \textit{Ans.}$$

Since both the sine and cosine of each variable are now known, the arc-tangent function gives both the magnitude and the proper quadrant of each angle.

10.8   Matrix Velocity and Acceleration Analyses

The power of the matrix method of posture analysis was demonstrated in the previous section. However, the method is not limited to posture analysis. The same standardized approach has been extended to velocity and acceleration analyses. To see this, let us start by noting that, of the four Denavit–Hartenberg parameters, three are constants describing the link shape and one is the joint variable. Therefore, in the basic transformation of Eq. (10.13), that is

$$T = \begin{bmatrix} \cos\theta & -\cos\alpha\sin\theta & \sin\alpha\sin\theta & a\cos\theta \\ \sin\theta & \cos\alpha\cos\theta & -\sin\alpha\cos\theta & a\sin\theta \\ 0 & \sin\alpha & \cos\alpha & s \\ 0 & 0 & 0 & 1 \end{bmatrix}, \tag{10.19}$$

there is only one variable, and it is either the θ or the s parameter, depending on the type of joint.

If we consider the case where the joint variable is the angle θ, then the derivative of T with respect to this variable is

$$\frac{dT}{d\theta} = \begin{bmatrix} -\sin\theta & -\cos\alpha\cos\theta & \sin\alpha\cos\theta & -a\sin\theta \\ \cos\theta & -\cos\alpha\sin\theta & \sin\alpha\sin\theta & a\cos\theta \\ 0 & 0 & 0 & 0 \\ 0 & 0 & 0 & 0 \end{bmatrix}. \tag{10.20}$$

On the other hand, if the joint variable is the distance s, as in a prismatic joint, then the derivative of T with respect to this variable is

$$\frac{dT}{ds} = \begin{bmatrix} 0 & 0 & 0 & 0 \\ 0 & 0 & 0 & 0 \\ 0 & 0 & 0 & 1 \\ 0 & 0 & 0 & 0 \end{bmatrix}. \tag{10.21}$$

It is interesting that both of these derivatives can be defined by the same formula,

$$\frac{dT_{i,i+1}}{d\phi_i} = Q_i T_{i,i+1}, \tag{10.22}$$

where we understand that when joint i is a revolute, then $\phi_i = \theta_{i,i+1}$, and we use

$$Q_i = \begin{bmatrix} 0 & -1 & 0 & 0 \\ 1 & 0 & 0 & 0 \\ 0 & 0 & 0 & 0 \\ 0 & 0 & 0 & 0 \end{bmatrix}, \tag{10.23}$$

and when joint i is prismatic, then $\phi_i = s_{i,i+1}$, and we use

$$Q_i = \begin{bmatrix} 0 & 0 & 0 & 0 \\ 0 & 0 & 0 & 0 \\ 0 & 0 & 0 & 1 \\ 0 & 0 & 0 & 0 \end{bmatrix}. \tag{10.24}$$

For velocity analysis, we need derivatives with respect to time rather than with respect to the joint variables. Therefore, using Eq. (10.22),

$$\frac{dT_{i,i+1}}{dt} = Q_i T_{i,i+1} \dot{\phi}_i, \tag{10.25}$$

where $\dot{\phi}_i = d\phi_i/dt$. In general, these are unknown variables. However, we will present a general procedure by which these variables can be determined. We start by differentiating the loop-closure conditions, Eq. (10.18), with respect to time. Using the chain rule with Eq. (10.25) to differentiate each factor, we get

$$\sum_{i=1}^{n} T_{12} T_{23} \cdots T_{i-1,i} Q_i T_{i,i+1} \cdots T_{n-1,n} T_{n1} \dot{\phi}_i = 0,$$

and using the more condensed notation of Eq. (10.16), this becomes

$$\sum_{i=1}^{n} T_{1i} Q_i T_{i1} \dot{\phi}_i = 0. \tag{10.26}$$

If we now define the symbol

$$D_i = T_{1i} Q_i T_{i1}, \tag{10.27a}$$

and take note that the loop-closure condition indicates that this is the same as

$$D_i = T_{1i} Q_i T_{1i}^{-1}, \tag{10.27b}$$

then Eq. (10.26) can be written as

$$\sum_{i=1}^{n} D_i \dot{\phi}_i = 0. \tag{10.28}$$

This equation contains all the conditions that the joint variable velocities must obey to fit the mechanism in question and be compatible with each other. Note that once the posture analysis is completed, the D_i matrices can be evaluated from known information. Some of the joint variable velocities, namely, the m input variables, will be given as input information; all other $\dot{\phi}_i$ values may then be determined from Eq. (10.28).

To make the above procedure clear, we continue Example 10.4.

Example 10.5

Given the angular velocity of the input shaft, $\dot{\phi}_1$, determine the angular velocity of the output shaft $\dot{\phi}_4$, for the Hooke, or Cardan, universal joint of Example 10.4.

Solution

From the results of the previous example, we find the following relationships between the input shaft angle, ϕ_1, and the other pair variables,

$$\sin \phi_2 = \frac{\cos \beta}{\sigma}, \qquad \cos \phi_2 = \frac{\sin \beta \sin \phi_1}{\sigma},$$

$$\sin \phi_3 = \sigma, \qquad \cos \phi_3 = \sin \beta \cos \phi_1,$$

$$\sin \phi_4 = \frac{\cos \beta \cos \phi_1}{\sigma}, \qquad \cos \phi_4 = \frac{\sin \phi_1}{\sigma},$$

Example 10.5 (cont.)

where

$$\sigma = \sqrt{1 - \sin^2 \beta \cos^2 \phi_1}.$$

Substituting these relationships into Eq. (10.13) gives the transformation matrices as functions of ϕ_1 alone. These, in turn, can be substituted into Eq. (10.27) to give the derivative operator matrices D_i, which become

$$D_1 = \begin{bmatrix} 0 & -1 & 0 & 0 \\ 1 & 0 & 0 & 0 \\ 0 & 0 & 0 & 0 \\ 0 & 0 & 0 & 0 \end{bmatrix},$$

$$D_2 = \begin{bmatrix} 0 & 0 & -\cos\phi_1 & 0 \\ 0 & 0 & -\sin\phi_1 & 0 \\ \cos\phi_1 & \sin\phi_1 & 0 & 0 \\ 0 & 0 & 0 & 0 \end{bmatrix},$$

$$D_3 = \begin{bmatrix} 0 & \dfrac{\sin\beta\sin\phi_1}{\sigma} & \dfrac{\cos\beta\sin\phi_1}{\sigma} & 0 \\ -\dfrac{\sin\beta\sin\phi_1}{\sigma} & 0 & -\dfrac{\cos\beta\cos\phi_1}{\sigma} & 0 \\ -\dfrac{\cos\beta\sin\phi_1}{\sigma} & \dfrac{\cos\beta\cos\phi_1}{\sigma} & 0 & 0 \\ 0 & 0 & 0 & 0 \end{bmatrix},$$

$$D_4 = \begin{bmatrix} 0 & -\cos\beta & \sin\beta & 0 \\ \cos\beta & 0 & 0 & 0 \\ -\sin\beta & 0 & 0 & 0 \\ 0 & 0 & 0 & 0 \end{bmatrix}.$$

These can now be used in Eq. (10.28). Taking the elements from row 3, column 2, then from row 1, column 3, and finally from row 2, column 1, gives the three equations:

$$(\sin\phi_1)\dot{\phi}_2 + \left(\frac{\cos\beta\cos\phi_1}{\sigma}\right)\dot{\phi}_3 = 0,$$

$$(-\cos\phi_1)\dot{\phi}_2 + \left(\frac{\cos\beta\sin\phi_1}{\sigma}\right)\dot{\phi}_3 + (\sin\beta)\dot{\phi}_4 = 0,$$

$$\left(-\frac{\sin\beta\sin\phi_1}{\sigma}\right)\dot{\phi}_3 + (\cos\beta)\dot{\phi}_4 = -\dot{\phi}_1.$$

Example 10.5 (cont.)

Solving these equations simultaneously gives

$$\dot{\phi}_2 = \frac{-\sin\beta\cos\beta\cos\phi_1}{1 - \sin^2\cos^2\phi_1}\dot{\phi}_1,$$

$$\dot{\phi}_3 = \frac{\sin\beta\sin\phi_1}{\sqrt{1 - \sin^2\beta\cos^2\phi_1}}\dot{\phi}_1,$$

$$\dot{\phi}_4 = \frac{-\cos\beta}{1 - \sin^2\beta\cos^2\phi_1}\dot{\phi}_1. \qquad \text{Ans.}$$

Note that when the input shaft has constant angular velocity, the angular velocity of the output shaft is not constant unless the shafts are in line (that is, unless the shaft angle $\beta = 0$). If β is not equal to zero, the output angular velocity fluctuates as the shaft rotates. Since the shaft angle β is constant, but usually not zero, the maximum value of the output/input velocity ratio, $\dot{\phi}_4/\dot{\phi}_1$, occurs when $\cos\phi_1 = 1$, that is, when $\phi_1 = 0°, 180°, 360°, 540°$, and so on; the minimum value of this velocity ratio occurs when $\cos\phi_1 = 0$. If the speed fluctuation, that is, the difference between the maximum and minimum velocity ratio, is expressed as a percentage and plotted against the shaft angle, β, the result is the graph shown in Fig. 10.13. This curve, showing the percentage speed fluctuation, is useful in evaluating Hooke, or Cardan, universal shaft coupling applications.

Fig. 10.13 Relationship between shaft angle and speed fluctuation.

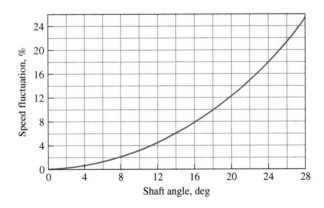

If we want to find the velocity of a point attached to link i, we can differentiate Eq. (10.15), for the position of the point, with respect to time. Using the D_i operator matrices to do this, we are led to defining a set of velocity operator matrices,

$$\omega_1 = 0, \qquad (10.29a)$$

$$\omega_{i+1} = \omega_i + D_i\dot{\phi}_i, \quad i = 1, 2, \ldots, n, \qquad (10.29b)$$

and then the velocity of a point on link i is given by

$$\dot{R}_i = \omega_i T_{1,i} R_i. \tag{10.30}$$

Acceleration analysis follows the same approach as for the velocity analysis. Without showing all of the details, the important equations are presented here; for further details, see [21]. First, we must find a formula for taking the derivative of a D_i matrix. This is

$$\frac{dD_i}{dt} = (\omega_i D_i - D_i \omega_i), \quad i = 1, 2, \ldots, n. \tag{10.31}$$

Differentiating Eq. (10.28) again with respect to time, by means of Eq. (10.31), and rearranging, we get a set of equations relating the pair-variable accelerations:

$$\sum_{i=1}^{n} D_i \ddot{\phi}_i = -\sum_{i=1}^{n} (\omega_i D_i - D_i \omega_i) \dot{\phi}_i. \tag{10.32}$$

These equations can be solved for the dependent $\ddot{\phi}_i$ joint variable acceleration values once the input position, velocity, and acceleration values are known. The solution process is identical to that for Eq. (10.28); in fact, the matrix of coefficients of the unknowns is identical for both sets of equations.

This matrix of coefficients, called the *Jacobian*, is essential for the solution of any set of derivatives of the joint variables. As indicated in Section 3.11, if this matrix becomes singular, there is no unique solution for the velocities (or accelerations) of the joint variables. If this occurs, such a posture of the mechanism is called a *singular* posture; one example would be a dead-center posture.

The derivatives of the velocity operator matrices of Eq. (10.29) give rise to the definition of a set of acceleration operator matrices, that is

$$\alpha_1 = 0, \tag{10.33a}$$

$$\alpha_{i+1} = \alpha_i + D_i \ddot{\phi}_i + (\omega_i D_i - D_i \omega_i) \dot{\phi}_i. \tag{10.33b}$$

From these, the acceleration of a point on link i is given by

$$\ddot{R}_i = (\alpha_i + \omega_i \omega_i) T_{1,i} R_i. \tag{10.34}$$

Much greater detail and more power have been developed using this transformation matrix approach for the kinematic and dynamic analyses of rigid-body systems. These methods [21] go far beyond the scope of this book. Further examples dealing with robotics, however, are presented later in this chapter.

10.9 Generalized Mechanism Analysis Computer Programs

It will probably be observed that the methods taken for the solution of each new problem are quite similar from one problem to the next. However, particularly in three-dimensional analysis, the number and complexity of the calculations make solution by hand a very tedious task. These characteristics indicate that a general computer program might have a broad range of applications and that the development costs for such a program might be justified through repeated use, increased accuracy, relief of drudgery, and elimination of error. General computer programs for the simulation of rigid-body kinematic and dynamic systems have been under development for several decades, and many are available and currently used in industrial settings, particularly in the automotive and aerospace industries.

The first widely available computer program for mechanism analysis was named Kinematic Analysis Method (KAM) and was written and distributed by IBM. It included capabilities for position, velocity, acceleration, and force analysis of both planar and spatial mechanisms, and was developed using the Chace vector tetrahedron equation solutions discussed in Section 10.3. Released in 1964, this program was the first to recognize the need for a general program for mechanical systems exhibiting large geometric movements. Being first, however, it had limitations and has been superseded by more powerful programs, including those described here.

Powerful generalized programs have also been developed using finite element and finite difference methods; NASTRAN and ANSYS are two examples. In the realm of mechanical systems, these programs have been developed primarily for stress analysis, and have excellent capabilities for static- and dynamic-force analysis. These allow the links of the simulated system to deflect under load and are capable of solving statically indeterminate force problems. They are very powerful programs with wide applications in industry. Although they are sometimes used for mechanism analysis, they are limited in their ability to simulate the geometric changes typical of kinematic systems.

The most widely used generalized programs for kinematic and dynamic simulation of three-dimensional rigid-body mechanical systems are ADAMS, DADS, and IMP. The MSC ADAMS® program, standing for Automatic Dynamic Analysis of Mechanical Systems, grew from the research efforts of Chace, Orlandea, and others at the University of Michigan [16] and is now available from MSC Software.[10] DADS, an acronym for Dynamic Analysis and Design System, was originally developed by Haug and others at the University of Iowa [13] and is now available through LMS International.[11] The Integrated Mechanisms Program (IMP) was developed by Uicker, Sheth, and others at the University of Wisconsin, Madison [19]. These and other similar programs are all applicable to single- or multiple-degree-of-freedom systems in both open- and closed-loop configurations. All operate on workstations, and some operate on microprocessors; all can display results with graphic animation. All are capable of solving position, velocity, acceleration, and static- and dynamic-force analyses. All can formulate the dynamic equations of motion and can predict the system response to a given set of initial conditions with prescribed input motions and/or forces that may be functions of time. Some of these programs include multibody collision detection, the ability to simulate impact, elasticity, or control system effects. Other commercial software of this type include Mechanism Dynamics[12] and MSC Working Model®[13] systems.

---

[10] MSC Software Corp., 2 MacArthur Place, Santa Ana, CA 92707, USA.

[11] LMS International, Researchpark Z1, Interleuvenlaan 68, 3001 Leuven, Belgium; LMS CAE Division, 2651 Crosspark Road, Coralville, IA 52241, USA.

[12] Parametric Technologies, Corp., 140 Kendrick Street, Needham, MA 02494, USA.   [13] MSC.Software Corp.

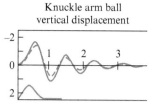

Knuckle arm ball
vertical displacement

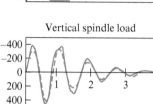

Vertical spindle load

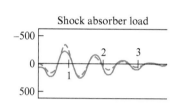

Shock absorber load

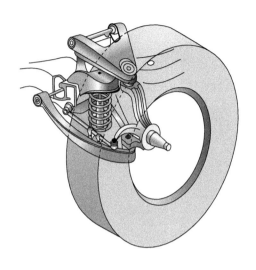

Fig. 10.14 Half-front automotive suspension. (Mechanical Dynamics, Inc., Ann Arbor, MI, and JML Research, Inc., Madison, WI.)

A typical application for any of these programs is the simulation of the half-front automotive suspension shown in Fig. 10.14. The graphs show the comparison of experimental test data and numeric results as a suspension encounters a hole as simulated by both the ADAMS and IMP programs.[14] Simulations of this type have been performed with several of these programs and they compare well with experimental data.

Another type of generalized program available today is intended for kinematic synthesis. The earliest such program was KINSYN (KINematic SYNthesis) [15],[15] which was followed by LINCAGES (Linkage INteractive Computer Assisted Geometrically Enhanced Synthesis) [7],[16] RECSYN (RECtified SYNthesis) [4], and others. These systems are directed toward the kinematic synthesis of planar linkages using methods analogous to those presented in Chapter 9. Users may input their motion requirements through a graphic user interface (GUI); the computer accepts a sketch and provides the required design information on the display screen. Another system of this type is the WATT Mechanism Design Tool from Heron Technologies,[17] a spin-off company from Twente University in the Netherlands. An example, shown in Fig. 10.15, shows a pipe-clamp mechanism designed in about 15 minutes using KINSYN III, developed at MIT under the direction of Dr. R. E. Kaufman, Professor Emeritus, The George Washington University.

---

[14] Simulations of the system shown in Fig. 10.14 were performed using the ADAMS and IMP programs in 1974 for the Strain History Prediction Committee of the Society of Automotive Engineers. Vehicle data and experimental test results were provided by Chevrolet Division, General Motors Corp.

[15] This paper was accompanied by the presentation of a 16-mm motion picture of 28 minute duration. The final version of the program, KINSYN7, was developed and marketed by KINTECH, Inc. until it closed in 1989.

[16] LINCAGES is available through Dr. A. G. Erdman by sending an email to agerdman@me.umn.edu.

[17] Heron Technologies b.v., Reekalf 10, 7908 XG Hoogeveen, the Netherlands.

Fig. 10.15 This pipe-
clamp mechanism. (Courtesy of
Dr. Kaufman.)

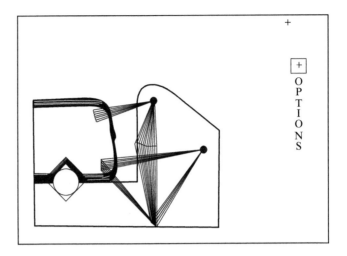

10.10 Introduction to Robotics

In preceding chapters, we have studied methods for analyzing the kinematics of machines. First, we studied planar kinematics at great length, justifying this emphasis by observing that over 90% of all mechanisms in use today have planar motion. Then, in the earlier sections of this chapter, we demonstrated how these methods extend to problems with spatial motion. Although the algebra became lengthier with spatial problems, we determined that there was no significant new block of theory required. We also concluded that, as the mathematics becomes more tedious, there is a role for the computer to relieve the drudgery. Still, we reported only a few problems with both practical application and spatial motion.

However, within the past few decades, advancing technology has given considerable attention to the development of robotic devices. Most are spatial mechanisms and require the attendant, more extensive, calculations. Fortunately, however, robots also carry on-board computing capability and can deal with this added complexity. The one remaining requirement is that the engineers and designers of the robot itself have a clear understanding of their characteristics and have appropriate tools for their analysis. That is the purpose of the remaining sections of this chapter.

The term *robot* is of Slavic origin; it comes from the Czech word *robota*, meaning work or labor, and was introduced into the English vocabulary by Czechoslovakian dramatist Karl Capek in the early part of the twentieth century. This word has been applied to a wide variety of computer-controlled electromechanical systems, from autonomous landrovers to underwater vehicles to teleoperated arms in industrial manipulators. The Robot Institute of America (RIA) defines a robot as *a reprogrammable multifunctional manipulator designed to move material, parts, tools, or specialized devices through variable programmed motions for the performance of a variety of tasks.* Many industrial manipulators bear a strong resemblance in their conceptual design to a human arm; however, this is not always true and is not essential. The key to the above definition is that a robot has flexibility through its programming, and its motion can be adapted to fit a variety of tasks.

10.11 Topological Arrangements of Robotic Arms

Up to this point, we have studied machines with only a few degrees of freedom. With traditional types of machines, it is usually desirable to drive the entire machine from a single source of power; thus they are designed to have mobility of $m = 1$. In keeping with the idea of flexibility of application, however, robots must have more. We know that, if a robot is to reach an arbitrary location in three-dimensional space, it must have mobility of at least $m = 3$ to adjust to proper values of x, y, and z. In addition, if the robot is to be able to manipulate a tool or an object it carries into an arbitrary orientation, upon reaching the desired location, an additional three degrees of freedom are required, giving a desired mobility of at least $m = 6$.

Along with the desire for six or more degrees of freedom, we also strive for simplicity, not only for reasons of good design and reliability, but also to minimize the computing burden in the control of the robot. Therefore, we often find that only three of the freedoms are designed into the robot arm and that another two or three may be included in the wrist. Since the tool, or *end-effector*, often varies with the task to be performed, this portion of the total robot may be made interchangeable; the same basic robot arm might carry any of several different end effectors specially designed for particular tasks.

Based on the first three freedoms of the arm, one common arrangement of joints for a manipulator is an *RRR* linkage, also called an *articulated* configuration. Two examples are the Model T26 Cincinnati Milacron T^3 robot shown in Fig. 10.16 and the Intelledex articulated robot shown in Fig. 10.17.

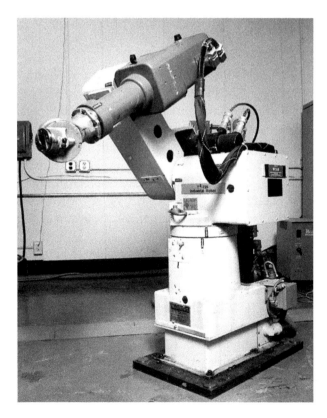

Fig. 10.16 Model T26 Cincinnati Milacron T^3 articulated six-axis robot.

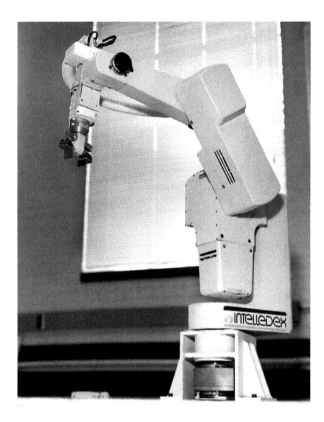

The Selectively Compliant Articulated Robot for Assembly, also called a SCARA robot, is a more recent but popular configuration based on the *RRP* linkage. An example is shown in Fig. 10.18. Although there are other *RRP* robot configurations, note that the SCARA robot has all three joint variable axes parallel (usually parallel to the direction of gravity), restricting its freedom of movement but particularly suiting it to assembly operations.

A manipulator based on the *PPP* chain is shown in Fig. 10.19. It has three mutually perpendicular prismatic joints and, therefore, is called a *Cartesian configuration* or a *gantry* robot. The kinematic analysis of, and the programming for, this style of robot is particularly simple because of the perpendicularity of the three joint axes. It has applications in table-top assembly and in transfer of cargo or material.

The arms depicted up to this point are simple, series-connected open kinematic chains. This is desirable since it reduces their kinematic complexity and eases their design and programming. However, this is not always true; there are robotic arms that include closed kinematic loops in their basic topology. One example is the robot shown in Fig. 10.20.

In fact, much attention has been given to the design of parallel robot topologies. One example is the Gough/Stewart platform [9, 20] shown in Fig. 10.21. Although parallel chains are more complex for analysis and control, they do provide better stiffness, larger payloads, and increased positional accuracy.

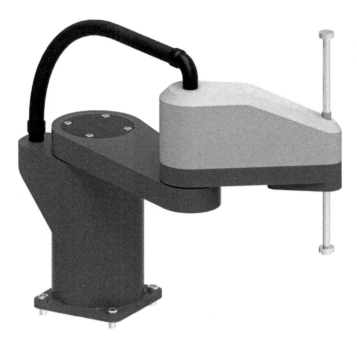

Fig. 10.18 A Selectively Compliant
Articulated Robot for
Assembly (SCARA). (Courtesy of
Boris15/Shutterstock.com.)

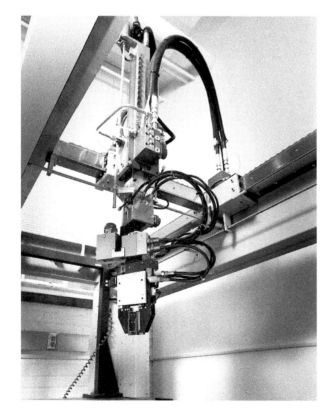

Fig. 10.19 IBM model 7650 gantry robot.

Fig. 10.20 General
Electric model P80
robotic manipulator.

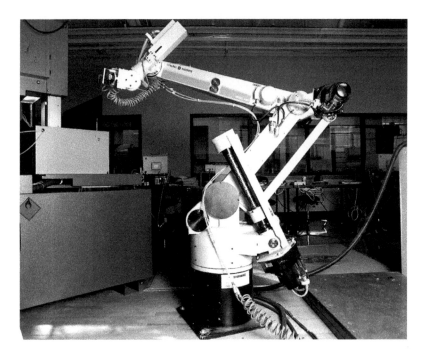

Fig. 10.21 Gough/Stewart
platform manipulator.

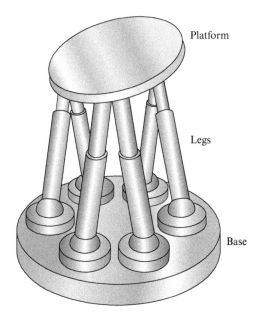

10.12 Forward Kinematics Problem

The kinematic analysis problem addressed here is finding the posture of the end-effector (tool)
of a robot once we are given the geometry of each component and the positions of the several

actuators controlling the degrees of freedom. This problem is commonly referred to as the *forward* or *direct* kinematics problem. For serial robots, the solution can be obtained from the methods on spatial mechanisms presented earlier in this chapter (Section 10.7). First, we should recognize three important characteristics of most robotics problems that are different from the spatial mechanisms of earlier sections in this chapter.

1. Knowledge of the position of a single point at the tip of the tool (often called the *tool point*) is often not sufficient. To have a complete knowledge of the end-effector, we must know the posture of a coordinate system attached to the end-effector. This implies that vector methods are not sufficient and that the matrix methods of Section 10.7 are better suited.
2. Perhaps because of the previous observation, many robot manufacturers supply the Denavit–Hartenberg parameters of Section 10.6 for their particular robot designs. Then, use of the transformation matrix approach is straightforward.
3. All joint variables of a serially connected chain are independent and are degrees of freedom. Thus, in the forward kinematics problem being discussed now, all joint variables are actuator variables and have given values; there are no loop-closure conditions and no dependent joint variable values to be determined.

The conclusion implied by these three observations combined is that finding the posture of the end-effector for given positions of the actuators is a straightforward application of Eq. (10.17). The absolute position of any chosen point R_{n+1} in the end-effector coordinate system attached to link $n+1$ is given by

$$R_1 = T_{1,n+1} R_{n+1}, \tag{10.35}$$

where

$$T_{1,n+1} = T_{1,2} T_{2,3} \ldots T_{n,n+1}, T_{1,n+1}, \tag{10.36}$$

and each T matrix is given by Eq. (10.13) once the Denavit–Hartenberg parameters (including the actuator positions) are known.

Of course, if the robot has $n + 1 = 7$ links ($n = 6$ joints), then symbolic multiplication of these matrices may become lengthy, unless some of the (constant) shape parameters are conveniently set to "nice" values. However, remembering that the robot has computing capability on-board, numeric evaluation of Eq. (10.36) is no great challenge for a particular set of actuator values. Still, because of speed requirements for real-time computer control, it is desirable to simplify these expressions as much as possible before programming. Toward this goal, most robot manufacturers have chosen simplified designs (having "nice" shape parameters) to simplify these expressions. Many have also worked out the final expressions for their particular robots and make these available in the technical documentation.

Example 10.6

For the Microbot model TCM five-axis robot shown in Fig. 10.22, find the transformation matrix T_{16} relating the posture of the end-effector coordinate system to the ground coordinate system when the actuators are set to position values $\phi_1 = 30°$, $\phi_2 = 60°$, $\phi_3 = -30°$, $\phi_4 = \phi_5 = 0°$. Also, find the absolute position of the tool point that has local coordinates of $x_6 = y_6 = 0$, $z_6 = 2.5$ in.

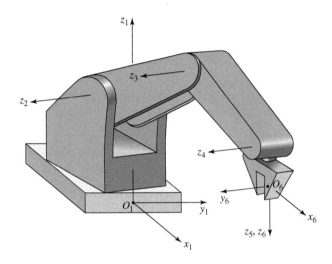

Fig. 10.22 Microbot model TCM five-axis robot.

Solution

The coordinate system axes are shown Fig. 10.22. Note that the $x_6 y_6 z_6$ coordinate system is not based on a joint axis, but was chosen arbitrarily to form a convenient tool coordinate system. Based on these axes and the data in the manufacturer's documentation, we find the Denavit–Hartenberg parameters to be as follows:

$$
\begin{array}{llll}
a_{12} = 0, & \alpha_{12} = 90°, & \theta_{12} = \phi_1 = 30°, & s_{12} = 7.68 \text{ in}, \\
a_{23} = 7.00 \text{ in}, & \alpha_{23} = 0°, & \theta_{23} = \phi_2 = 60°, & s_{23} = 0, \\
a_{34} = 7.00 \text{ in}, & \alpha_{34} = 0°, & \theta_{34} = \phi_3 = -30°, & s_{34} = 0, \\
a_{45} = 0, & \alpha_{45} = 90°, & \theta_{45} = \phi_4 = 0°, & s_{45} = 0, \\
a_{56} = 0, & \alpha_{56} = 0°, & \theta_{56} = \phi_5 = 0°, & s_{56} = 3.80 \text{ in}.
\end{array}
$$

Example 10.6 (cont.)

Using Eqs. (10.13) and (10.16) we can now form the following matrices and matrix products:

$$T_{12} = \begin{bmatrix} 0.866 & 0 & 0.500 & 0 \\ 0.500 & 0 & -0.866 & 0 \\ 0 & 1 & 0 & 7.68 \text{ in} \\ 0 & 0 & 0 & 1 \end{bmatrix},$$

$$T_{23} = \begin{bmatrix} 0.500 & -0.866 & 0 & 3.50 \text{ in} \\ 0.866 & 0.500 & 0 & 6.06 \text{ in} \\ 0 & 0 & 1 & 0 \\ 0 & 0 & 0 & 1 \end{bmatrix}, \quad T_{13} = \begin{bmatrix} 0.433 & -0.750 & 0.500 & 3.03 \text{ in} \\ 0.250 & -0.433 & -0.866 & 1.75 \text{ in} \\ 0.866 & 0.500 & 0 & 13.74 \text{ in} \\ 0 & 0 & 0 & 1 \end{bmatrix},$$

$$T_{34} = \begin{bmatrix} 0.866 & 0.500 & 0 & 6.06 \text{ in} \\ -0.500 & 0.866 & 0 & -3.50 \text{ in} \\ 0 & 0 & 1 & 0 \\ 0 & 0 & 0 & 1 \end{bmatrix}, \quad T_{14} = \begin{bmatrix} 0.750 & -0.433 & 0.500 & 8.28 \text{ in} \\ 0.433 & -0.250 & -0.866 & 4.78 \text{ in} \\ 0.500 & 0.866 & 0 & 17.24 \text{ in} \\ 0 & 0 & 0 & 1 \end{bmatrix},$$

$$T_{45} = \begin{bmatrix} 1 & 0 & 0 & 0 \\ 0 & 0 & -1 & 0 \\ 0 & 1 & 0 & 0 \\ 0 & 0 & 0 & 1 \end{bmatrix}, \quad T_{15} = \begin{bmatrix} 0.750 & 0.500 & 0.433 & 8.28 \text{ in} \\ 0.433 & -0.866 & 0.250 & 4.78 \text{ in} \\ 0.500 & 0 & -0.866 & 17.24 \text{ in} \\ 0 & 0 & 0 & 1 \end{bmatrix},$$

$$T_{56} = \begin{bmatrix} 1 & 0 & 0 & 0 \\ 0 & 1 & 0 & 0 \\ 0 & 0 & 1 & 3.80 \text{ in} \\ 0 & 0 & 0 & 1 \end{bmatrix}, \quad T_{16} = \begin{bmatrix} 0.750 & 0.500 & 0.433 & 9.93 \text{ in} \\ 0.433 & -0.866 & 0.250 & 5.73 \text{ in} \\ 0.500 & 0 & -0.866 & 13.95 \text{ in} \\ 0 & 0 & 0 & 1 \end{bmatrix}.$$

Ans.

The absolute position of the specified tool point is now found from Eq. (10.35):

$$R_1 = \begin{bmatrix} 0.750 & 0.500 & 0.433 & 9.93 \text{ in} \\ 0.433 & -0.866 & 0.250 & 5.73 \text{ in} \\ 0.500 & 0 & -0.866 & 13.95 \text{ in} \\ 0 & 0 & 0 & 1 \end{bmatrix} \begin{bmatrix} 0 \\ 0 \\ 2.5 \text{ in} \\ 1 \end{bmatrix} = \begin{bmatrix} 11.00 \text{ in} \\ 6.36 \text{ in} \\ 11.78 \text{ in} \\ 1 \end{bmatrix}.$$

Ans.

Note that the direct computation of the velocity and acceleration of an arbitrary point on a moving link of the robot can also be accomplished in a straightforward manner by the matrix methods of Section 10.8. In particular, Eqs. (10.29), (10.30), (10.33), and (10.34) are of direct use once the actuator velocities and accelerations are known. A continuation of the previous example makes this clear.

Example 10.7

For the Microbot model TCM robot of Fig. 10.22 in the posture described in Example 10.6, find the instantaneous velocity and acceleration of the same tool point $x_6 = y_6 = 0$, and $z_6 = 2.5$ in if the actuators are given constant velocities of $\dot{\phi}_1 = 0.20$ rad/s, $\dot{\phi}_4 = -0.35$ rad/s, and $\dot{\phi}_2 = \dot{\phi}_3 = \dot{\phi}_5 = 0$.

Solution

Using Eq. (10.27) and the results from Example 10.6, we find

$$D_1 = \begin{bmatrix} 0 & -1 & 0 & 0 \\ 1 & 0 & 0 & 0 \\ 0 & 0 & 0 & 0 \\ 0 & 0 & 0 & 0 \end{bmatrix}, \quad D_4 = \begin{bmatrix} 0 & 0 & -0.866 & 14.93 \text{ in} \\ 0 & 0 & -0.500 & 8.62 \text{ in} \\ 0.866 & 0.500 & 0 & -9.56 \text{ in} \\ 0 & 0 & 0 & 0 \end{bmatrix},$$

and then, from Eq. (10.29), we obtain

$$\omega_1 = 0,$$

$$\omega_2 = \omega_3 = \omega_4 = \begin{bmatrix} 0 & -0.200 \text{ rad/s} & 0 & 0 \\ 0.200 \text{ rad/s} & 0 & 0 & 0 \\ 0 & 0 & 0 & 0 \\ 0 & 0 & 0 & 0 \end{bmatrix},$$

$$\omega_5 = \omega_6 = \begin{bmatrix} 0 & -0.200 \text{ rad/s} & 0.303 \text{ rad/s} & -5.225 \text{ in/s} \\ 0.200 \text{ rad/s} & 0 & 0.175 \text{ rad/s} & -3.017 \text{ in/s} \\ -0.303 \text{ rad/s} & -0.175 \text{ rad/s} & 0 & 3.346 \text{ in/s} \\ 0 & 0 & 0 & 0 \end{bmatrix}.$$

Now, using Eq. (10.30), we find that the velocity of the tool point is

$$\dot{R}_6 = \omega_6 T_{16} R_6 = \begin{bmatrix} -2.93 \text{ in/s} \\ 1.24 \text{ in/s} \\ -1.10 \text{ in/s} \\ 0 \end{bmatrix}, \qquad \textit{Ans.}$$

which in vector notation is

$$V = -2.93\hat{i}_1 + 1.24\hat{j}_1 - 1.10\hat{k}_1 \text{ in/s}. \qquad \textit{Ans.}$$

The accelerations can be obtained in similar fashion. From Eq. (10.33), we obtain

$$\alpha_1 = \alpha_2 = \alpha_3 = \alpha_4 = 0,$$

Example 10.7 (cont.)

$$
\alpha_5 = \alpha_6 = \begin{bmatrix} 0 & 0 & -0.035 \text{ rad/s}^2 & 0.603 \text{ in/s}^2 \\ 0 & 0 & 0.152 \text{ rad/s}^2 & -1.045 \text{ in/s}^2 \\ 0.035 \text{ rad/s}^2 & -0.152 \text{ rad/s}^2 & 0 & 0 \\ 0 & 0 & 0 & 0 \end{bmatrix}.
$$

Finally, using Eq. (10.34), we find that the acceleration of the tool point is

$$
\ddot{R}_6 = (\alpha_6 + \omega_6\omega_6) T_{16} R_6 = \begin{bmatrix} -0.390 \text{ in/s}^2 \\ -0.033 \text{ in/s}^2 \\ 0.088 \text{ in/s}^2 \\ 0 \end{bmatrix}, \qquad Ans.
$$

which in vector notation is

$$
A = -0.390\hat{i}_1 - 0.033\hat{j}_1 + 0.088\hat{k}_1 \text{ in/s}^2. \qquad Ans.
$$

10.13 Inverse Kinematics Problem

The previous section demonstrated the basic procedure for finding the positions, velocities, and accelerations of arbitrary points on the moving links of a serial robot once the positions, velocities, and accelerations of the joint actuators are known. The procedures are straightforward and simply applied if, as is usual, the robot is a simply connected chain and has no closed loops. In this simple case, every joint variable is an independent degree of freedom, and no loop-closure equations need be solved. This avoids a major complication.

However, even though the robot itself might have no closed loops, the manner in which a problem or question is presented can sometimes lead to the same complications. Consider, for example, an open-loop robot that we wish to control so that the tool point follows a specified path; we are given the desired path, but we do not know the actuator values (or, more precisely, the time functions) required to achieve this. This problem cannot be solved by the methods of the previous section. When the joint variables (or their derivatives) are the unknowns of the problem rather than given information, the problem is called the *inverse kinematics* problem. In general, this is a more complex problem to solve, and is commonly encountered in robot applications.

When the inverse kinematics problem arises, we must find a set of equations that describe the given situation and that can be solved. In robotics, this usually means that there are one or more constraint equations, not necessarily defined by closed loops within the robot topology, but perhaps defined by the manner in which the problem is posed. For example, if we are told that the tool point is to travel along a certain path with certain timing, then we are being told the values required for the transformation $T_{1n}T_{1n}T_{1n}T_{1n}$ as known functions of time. Then Eq. (10.36) becomes a set of required constraints that must be satisfied and that must be solved to find the joint actuator values as functions of time. An example will make the procedure clear.

Example 10.8

The Microbot robot of Fig. 10.22, described in Example 10.6, is to be guided such that: (1) the tool point O_6 follows the straight-line path given by

$$R_{O_6}(t) = (4.0 + 1.6t)\hat{i}_1 + (3.0 + 1.2t)\hat{j}_1 + 2.0\hat{k}_1 \text{ in}$$

with t varying from 0.0 to 5.0 s; and (2) the orientation of the end-effector is to remain constant with $\hat{i}_6 = \hat{k}_1$ (vertical) and $\hat{k}_6$ radially outward from the base of the robot. Find expressions for how each of the joint actuators must be driven, as functions of time, to achieve this motion.

Solution

From the problem requirements stated, we can express the required time history of the tool coordinate system by the transformation matrix

$$T_{16}(t) = \begin{bmatrix} 0 & 0.600 & 0.800 & 4.0 + 1.6t \text{ in} \\ 0 & -0.800 & 0.600 & 3.0 + 1.2t \text{ in} \\ 1 & 0 & 0 & 2.0 \text{ in} \\ 0 & 0 & 0 & 1 \end{bmatrix}. \tag{1}$$

We can also multiply out the matrix description for $T_{16}T_{16}T_{16}T_{16}$ from Eq. (10.36); this gives

$$T_{16} = \begin{bmatrix} \cos\phi_1\cos\beta\cos\phi_5 & -\cos\phi_1\cos\beta\sin\phi_5 & \cos\phi_1\sin\beta & 7\cos\phi_1\cos\phi_2 \\ -\sin\phi_1\sin\phi_5 & +\sin\phi_1\cos\phi_5 & & +7\cos\phi_1\cos(\phi_2+\phi_3) \\ & & & +3.8\cos\phi_1\sin\beta \\ \sin\phi_1\cos\beta\cos\phi_5 & -\sin\phi_1\cos\beta\sin\phi_5 & \sin\phi_1\sin\beta & 7\sin\phi_1\cos\phi_2 \\ -\cos\phi_1\sin\phi_5 & -\cos\phi_1\cos\phi_5 & & +7\sin\phi_1\sin(\phi_2+\phi_3) \\ & & & +3.8\sin\phi_1\sin\beta \\ \sin\beta\cos\phi_5 & -\sin\beta\sin\phi_5 & -\cos\beta & 7\sin\phi_2 \\ & & & +7\sin(\phi_2+\phi_3) \\ & & & -3.8\cos\beta \\ 0 & 0 & 0 & 1 \end{bmatrix}, \tag{2}$$

where, to save space, we have adopted the definition

$$\beta = \phi_2 + \phi_3 + \phi_4. \tag{3}$$

Now, to achieve the problem requirements, we must find expressions for the actuator variables ϕ_i that ensure that the elements of Eq. (2) are equal to those of Eq. (1) for all values of time in the range $0 \le t \le 5$ s; this is our required constraint condition. Equating the ratios of the elements of

Example 10.8 (cont.)

the second row, third column with the elements of the first row, third column of Eqs. (2) and (1) gives

$$\tan \phi_1 = \frac{0.600}{0.800} = 0.750.$$

Therefore, the first joint variable must be

$$\phi_1 = \tan^{-1} 0.750 = 36.87°. \qquad \qquad Ans.$$

Similarly, from the ratios of the elements of the third row, second column with the elements of the third row, first column we find

$$\tan \phi_5 = 0,$$

$$\phi_5 = 0° \quad \text{or} \quad \pm 180°.$$

Since the problem gives no preference for the orientation of the end-effector in the horizontal plane, we are free to choose

$$\phi_5 = 0°. \qquad \qquad Ans.$$

Next, from the third row, third column elements, we obtain

$$\cos \beta = 0,$$

$$\beta = \phi_2 + \phi_3 + \phi_4 = \pm 90°.$$

Since the orientation of the end-effector in the horizontal plane is not specified, we arbitrarily choose

$$\beta = \phi_2 + \phi_3 + \phi_4 = 90°. \qquad \qquad (4)$$

Turning now to the elements of the fourth column, and simplifying according to already known results, we obtain two more independent equations:

$$7 \cos \phi_2 + 7 \cos(\phi_2 + \phi_3) = 1.2 + 2t, \qquad \qquad (5)$$

$$7 \sin \phi_2 + 7 \sin(\phi_2 + \phi_3) = -5.68. \qquad \qquad (6)$$

Squaring and adding these two equations gives

$$98 + 98 \cos \phi_3 = 33.7024 + 4.8t + 4t^2. \qquad \qquad (7)$$

Therefore, a solution for ϕ_3 is

$$\phi_3 = \cos^{-1}(-0.656\,10 + 0.048\,98t + 0.040\,816t^2). \qquad \qquad Ans. \ (8)$$

Example 10.8 (cont.)

Since this form admits to multiple values, we take the smallest solution with $\phi_3 \leq 0$ so that the elbow is kept above the path. At time $t = 0$, for example, the value of ϕ_3 is

$$\phi_3 = \cos^{-1}(-0.656\ 10) = -131.00°.$$

Knowing the solution for $\phi_3, \phi_3, \phi_3, \phi_3$, we now rewrite Eqs. (5) and (6) as follows:

$$\sin \phi_2 = \frac{-5.68}{14} - \frac{(1.2 + 2t) \sin \phi_3}{14(1 + \cos \phi_3)}, \tag{9}$$

$$\cos \phi_2 = \frac{(1.2 + 2t)}{14} - \frac{5.68 \sin \phi_3}{14(1 + \cos \phi_3)}. \tag{10}$$

Since the third actuator variable ϕ_3 now has a known solution, either of these equations can be used to find ϕ_2. Actually, both should be evaluated to ensure that the proper quadrant is found. These two equations can be numerically evaluated at any value of time, and they are considered solutions even though they are not explicit, closed-form functions of time.

Similarly, from Eq. (4), we now find the fourth joint variable:

$$\phi_4 = 90° - \phi_2 - \phi_3. \qquad\qquad \textit{Ans. (11)}$$

Graphs of the five actuator variables versus time are shown in Figure 10.23. Note that for the path being followed, the actuator variables follow very smooth but nonlinear curves. Thus, it appears that the actuator velocities and accelerations will not be zero, but should vary within very reasonable limits.

Fig. 10.23 Graphs of the actuator variables versus time.

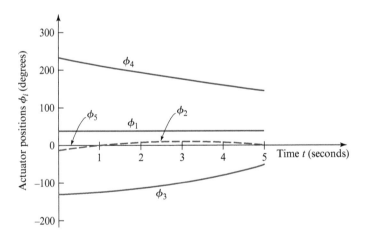

10.14 Inverse Velocity and Acceleration Analyses

The previous section presented the general approach to the inverse posture analysis problem. However, at least for control purposes, it may also be important to find the required actuator variable velocities and accelerations. One way to do this is to directly differentiate the actuator variable solution equations with respect to time. However, another, more general approach is to differentiate our constraint conditions.

The constraint equations are given by Eq. (10.36), and, for convenience, can be written as

$$T_{1,2}T_{2,3}\dots T_{n,n+1} = T_{1,n+1}(t),$$

where, for the inverse kinematics problem, the right-hand side of this equation consists of known functions of time given by the problem specifications, namely, the trajectory (or path) to be followed. Now, using Eq. (10.25) and the D_i differentiation operator matrices of Eq. (10.27b), we can differentiate the left-hand side of the equation, and by direct differentiation with respect to time, we can also differentiate the right-hand side. Thus, we obtain

$$\sum_{i=1}^{n} D_i \dot{\phi}_i T_{1,n+1} = \frac{dT_{1,n+1}(t)}{dt}. \tag{10.37}$$

If we post-multiply both sides of Eq. (10.37) by $T_{1,n+1}^{-1}$, we obtain

$$\sum_{i=1}^{n} D_i \dot{\phi}_i = \frac{dT_{1,n+1}(t)}{dt} T_{1,n+1}^{-1}. \tag{10.38}$$

For convenience, we define the right-hand side by

$$\Omega = \frac{dT_{1,n+1}(t)}{dt} T_{1,n+1}^{-1}, \tag{10.39}$$

where Ω is a new (velocity) operator matrix. Then, substituting Eq. (10.39) into Eq. (10.38) gives

$$\sum_{i=1}^{n} D_i \dot{\phi}_i = \Omega, \tag{10.40}$$

which, as we will see, can be solved for the actuator velocities $\dot{\phi}_i$.

Let us now look more carefully at the nature of Eq. (10.40). Each of the matrices has four rows and four columns. Yet, since they represent velocities in a single chain in three dimensions, there should only be six independent equations, three for translations and three for rotations. Looking at the definitions of the Q_i differentiation operator matrices in Eq. (10.23) and reviewing the form of the D_i matrices in Examples 10.5 and 10.7, we discover which of the elements of these matrices contain the six independent equations. Although the proof is left as an exercise, it can be demonstrated that the D_i matrices always have a bottom row of zeroes and the upper-left three rows and three columns are always skew-symmetric; that is, they are always of the form [21]

$$D_i = \begin{bmatrix} 0 & -c & +b & d \\ +c & 0 & -a & e \\ -b & +a & 0 & f \\ 0 & 0 & 0 & 0 \end{bmatrix}. \tag{10.41}$$

As expected, there are only six independent values ($a, b, c, d, e,$ and f) in each of these matrices. It is now convenient to rearrange these into a single column and to give this column a new symbol:

$$\vec{D}_i = \begin{bmatrix} a \\ b \\ c \\ d \\ e \\ f \end{bmatrix}, \tag{10.42}$$

where the over-arrow indicates that, without loss of generality, the elements of the (4×4) matrix have been rearranged into a (6×1) column matrix. There is actually much more mathematical and physical significance to this (6×1) column vector than is explained here; they form the *Plücker coordinates* or *line coordinates* of the motion axis of the joint variable's freedom for motion [14]. The first (top) three elements form the three-dimensional unit vector for the orientation of the motion axis, whereas the later (bottom) three uniquely locate that axis in the absolute coordinate system.

If we now reorganize all matrices in Eq. (10.40) into this column vector form, it becomes

$$\sum_{i=1}^{n} \vec{D}_i \dot{\phi}_i = \vec{\Omega}, \tag{10.43}$$

which is a set of six simultaneous algebraic equations relating the joint variable velocities. If we write the equations out explicitly, they appear as

$$a_1 \dot{\phi}_1 + a_2 \dot{\phi}_2 + \cdots + a_n \dot{\phi}_n = a,$$
$$b_1 \dot{\phi}_1 + b_2 \dot{\phi}_2 + \cdots + b_n \dot{\phi}_n = b,$$
$$c_1 \dot{\phi}_1 + c_2 \dot{\phi}_2 + \cdots + c_n \dot{\phi}_n = c,$$
$$d_1 \dot{\phi}_1 + d_2 \dot{\phi}_2 + \cdots + d_n \dot{\phi}_n = d$$
$$e_1 \dot{\phi}_1 + e_2 \dot{\phi}_2 + \cdots + e_n \dot{\phi}_n = e,$$
$$f_1 \dot{\phi}_1 + f_2 \dot{\phi}_2 + \cdots + f_n \dot{\phi}_n = f,$$

where a, b, and so on are the elements of $\vec{\Omega}$ taken in the order consistent with Eqs. (10.41) and (10.42). Regrouping the coefficients of this set of equations into matrix form, we get

$$J = \begin{bmatrix} a_1 & a_2 & \cdots & a_n \\ b_1 & b_2 & \cdots & b_n \\ c_1 & c_2 & \cdots & c_n \\ d_1 & d_2 & \cdots & d_n \\ e_1 & e_2 & \cdots & e_n \\ f_1 & f_2 & \cdots & f_n \end{bmatrix}, \tag{10.44}$$

which is the *Jacobian* (Section 10.8). Using this matrix, Eq. (10.43) now becomes

$$J\{\dot{\phi}\} = \vec{\Omega}, \tag{10.45}$$

where $\{\dot{\phi}\}$ is a column of the unknown actuator velocities, $\dot{\phi}_i$, taken in order. This set of equations specifies the constraint conditions that must exist between the actuator velocities to trace the specified path according to the given functions of time.

Assuming that the posture analysis has already been completed, the Ω and D_i matrices are all known functions of time, and this set of equations can now be solved for the actuator velocities as functions of time. If the robot has six independent joint variables, then the solution can be determined by inverting the Jacobian. In this case, the solution is

$$\{\dot{\phi}\} = J^{-1}\vec{\Omega}, \tag{10.46}$$

where J^{-1} is the inverse of the Jacobian of Eq. (10.44).

If the Jacobian is not a square matrix, then two cases are possible: (1) When there are fewer than six joint variables, other solution methods, such as Gaussian elimination, must be used. (2) When there are more than six joint variables, the problem has no unique solution; further equations, perhaps even arbitrary conditions, must be supplied. If the Jacobian is square but singular or if the equations are inconsistent, this shows that there is no finite unique solution for velocities; the problem specifications are not realistic for this robot to perform the task specified. If none of these problems arises, the actuator velocities become known functions of time.

Joint actuator accelerations can be determined by completely analogous methods. Taking the next time derivative of Eq. (10.40), we see that

$$\sum_{i=1}^{n} D_i \ddot{\phi}_i = \frac{d\Omega}{dt} - \sum_{i=1}^{n} (\omega_i D_i - D_i \omega_i)\dot{\phi}_i = A, \tag{10.47}$$

where this equation defines a new matrix A and the ω_i matrices are defined as in Eqs. (10.29). Extracting the same six independent equations, in the same order as before, we get

$$\sum_{i=1}^{n} \vec{D}_i \ddot{\phi}_i = \vec{A}. \tag{10.48}$$

We note that the coefficients of the unknown $\ddot{\phi}_i$ joint variable accelerations are identical to those of Eq. (10.43); thus, we can use the same Jacobian defined in Eq. (10.44). The solution to these equations becomes

$$\{\ddot{\phi}\} = J^{-1}\vec{A}, \tag{10.49}$$

where $\{\ddot{\phi}\}$ is a column of the unknown joint actuator accelerations, $\ddot{\phi}_i$, taken in order. This set of equations specifies the relations that must exist between the joint actuator accelerations required to trace the specified path according to the given functions of time.

As with the direct kinematics problem, once the actuator velocities and accelerations are known, the velocity and acceleration of a point on one of the moving links of the robot can be determined from Eqs. (10.29) through (10.34).

Example 10.9

Continue Example 10.8 and find the velocities required at the actuators as functions of time to achieve the motion described.

Solution

Using Eq. (10.27b) and the solutions for the posture analysis from Example 10.8, we find the D_i matrices. These are

$$
D_1 = \begin{bmatrix} 0 & -1 & 0 & 0 \\ 1 & 0 & 0 & 0 \\ 0 & 0 & 0 & 0 \\ 0 & 0 & 0 & 0 \end{bmatrix}, \quad D_2 = \begin{bmatrix} 0 & 0 & -0.8 & 6.144 \text{ in} \\ 0 & 0 & -0.6 & 4.608 \text{ in} \\ 0.8 & 0.6 & 0 & 0 \\ 0 & 0 & 0 & 0 \end{bmatrix},
$$

$$
D_3 = \begin{bmatrix} 0 & 0 & -0.8 & (5.6 \sin \phi_2 + 6.144) \text{ in} \\ 0 & 0 & -0.6 & (4.2 \sin \phi_2 + 4.608) \text{ in} \\ 0.8 & 0.6 & 0 & -7 \cos \phi_2 \text{ in} \\ 0 & 0 & 0 & 0 \end{bmatrix},
$$

$$
D_4 = \begin{bmatrix} 0 & 0 & -0.8 & 1.6 \text{ in} \\ 0 & 0 & -0.6 & 1.2 \text{ in} \\ 0.8 & 0.6 & 0 & -(2t + 1.2) \text{ in} \\ 0 & 0 & 0 & 0 \end{bmatrix},
$$

$$
D_5 = \begin{bmatrix} 0 & 0 & 0.6 & -1.2 \text{ in} \\ 0 & 0 & -0.8 & 1.6 \text{ in} \\ -0.6 & 0.8 & 0 & 0 \\ 0 & 0 & 0 & 0 \end{bmatrix}. \tag{12}
$$

Using Eq. (10.39) and the trajectory specified in Example 10.8, we also find the Ω matrix.

$$
\Omega = \begin{bmatrix} 0 & 0 & 0 & 1.6 \text{ in/s} \\ 0 & 0 & 0 & 1.2 \text{ in/s} \\ 0 & 0 & 0 & 0 \\ 0 & 0 & 0 & 0 \end{bmatrix}. \tag{13}
$$

From these and Eq. (10.43), we construct the set of equations relating the actuator velocities:

$$
\begin{bmatrix} 0 & 0.6 & 0.6 & 0.6 & 0.8 \\ 0 & -0.8 & -0.8 & -0.8 & 0.6 \\ 1 & 0 & 0 & 0 & 0 \\ 0 & 6.144 \text{ in} & (5.6 \sin \phi_2 + 6.144) \text{ in} & 1.6 \text{ in} & -1.2 \text{ in} \\ 0 & 4.608 \text{ in} & (4.2 \sin \phi_2 + 4.608) \text{ in} & 1.2 \text{ in} & 1.6 \text{ in} \\ 0 & 0 & -7 \cos \phi_2 \text{ in} & -(2t + 1.2) \text{ in} & 0 \end{bmatrix} \begin{bmatrix} \dot{\phi}_1 \\ \dot{\phi}_2 \\ \dot{\phi}_3 \\ \dot{\phi}_4 \\ \dot{\phi}_5 \end{bmatrix} = \begin{bmatrix} 0 \\ 0 \\ 0 \\ 1.6 \text{ in/s} \\ 1.2 \text{ in/s} \\ 0 \end{bmatrix}. \tag{14}
$$

Example 10.9 (cont.)

Here we see the Jacobian for the end-effector, such that the tool point follows the specified trajectory. However, in this example, the Jacobian is not square and cannot be inverted. Note that there are six equations and only five unknowns, namely, the five actuator velocities. This has occurred because this robot has only five degrees of freedom, and, therefore, the end-effector is not capable of arbitrary spatial motion.

Fortunately, however, the specified trajectory is within the capability of this robot. There is a set of solutions for the five unknown actuator velocities that fits all six equations; these solutions are:

$$\dot{\phi}_1 = 0, \qquad\qquad Ans.$$

$$\dot{\phi}_2 = \frac{(2t+1.2) - 7\cos\phi_2}{\varDelta}\ \text{rad/s}, \qquad\qquad Ans.$$

$$\dot{\phi}_3 = \frac{-(2t+1.2)}{\varDelta}\ \text{rad/s}, \qquad\qquad Ans.$$

$$\dot{\phi}_4 = \frac{7\cos\phi_2}{\varDelta}\ \text{rad/s}, \qquad\qquad Ans.$$

$$\dot{\phi}_5 = 0, \qquad\qquad Ans.$$

where

$$\varDelta = -3.5(2t+1.2)\sin\phi_2 - 19.9\cos\phi_2. \qquad\qquad (15)$$

Since the parameter ϕ_2 is known from the posture analysis, the above equations are solutions for the velocity analysis.

10.15 Robot Actuator Force Analysis

Of course, the purpose of moving the tool point along the planned trajectory is to perform some useful function, and this almost certainly requires the expenditure of work or power. This work or power must come from the joint actuators. Therefore, it is very helpful to find the sizes of forces and/or torques that must be exerted by the actuators to perform the given task. Conversely, it is useful to know how much force is produced at the end-effector for a given set of forces or torques applied by the actuators.

If the force or torque capacity of the actuators is less than those demanded by the task attempted, the robot is not capable of performing the task. It will probably not fail catastrophically, but it will deviate from the desired trajectory and perform a different motion than that specified. The purpose of this section is to find a means of evaluating the forces required of the actuators to perform a given trajectory with a given task loading, so that overloading of the actuators can be anticipated and avoided.

The entire subject of force analysis in mechanical systems is covered in much more depth in Chapters 11 and 12. Therefore, the treatment here is limited by simplifying assumptions. It is suited specifically to the study of robots performing specified tasks with specified loads at low speeds with no friction or other losses. If these assumptions do not apply, the more extensive treatments of the later chapters should be employed.

The interaction of the robot with the task being performed produces a set of forces and torques at the end effector. Let us assume that these required forces and torques are known functions of time and are grouped into a given six-element column, $\{F\}$, in the following order:

$$\{F\} = \begin{bmatrix} M^{x_1} \\ M^{y_1} \\ M^{z_1} \\ F^{x_1} \\ F^{y_1} \\ F^{z_1} \end{bmatrix}, \tag{10.50}$$

where the first three elements are the components of the required end-effector torque about the x_1, y_1, z_1 axes, and the final three are the components of the required force along the same axes.

The source of energy for these needs is a set of unknown forces or torques, τ_i, each one applied by the actuator at the ith joint. We arrange these into another column matrix τ in the order of numbering of the joint variables:

$$\tau = \begin{bmatrix} \tau_1 \\ \tau_2 \\ \vdots \\ \tau_n \end{bmatrix}. \tag{10.51}$$

Next we define a small (absolute) displacement of the end-effector against the loads by another six-element column matrix, $\{\delta R\}$, arranged in the same order as the load vector, $\{F\}$:

$$\{\delta R\} = \begin{bmatrix} \delta\theta^{x_1} \\ \delta\theta^{y_2} \\ \delta\theta^{z_1} \\ \delta R^{x_1} \\ \delta R^{y_1} \\ \delta R^{z_1} \end{bmatrix}, \tag{10.52}$$

and yet another column matrix, $\delta\phi$, of displacements of the joint variables, $\delta\phi_i$, during this small displacement.

Now, if the end-effector is to undergo the small displacement $\{\delta R\}$ against the loads $\{F\}$, the energy must come from the torques, τ, of the actuators acting through their small displacements, $\delta\phi$. Therefore, since work input must equal work output, in the absence of other losses, we have

$$\{F\}^T\{\delta R\} = \tau^T \delta\phi, \qquad (a)$$

where the superscript ^{T} signifies the transpose of a matrix. But we also know by integrating Eq. (10.45) over a short time interval δt that the task displacement $\{\delta R\}$ is related to the joint actuator displacements $\{\delta\phi\}$ by

$$\{\delta R\} = J\{\delta\phi\}. \qquad (10.53)$$

Substituting this relationship into Eq. (a) and rearranging, we get

$$\left[\tau^T - \{F\}^T J\right]\delta\phi = 0. \qquad (b)$$

Since we assume that the small displacement $\delta\phi$ is not zero, we can solve for the actuator torques τ by setting the leading factor to zero. This shows that

$$\tau = J^T\{F\}, \qquad (10.54)$$

which is the relationship we are seeking. Under the simplifying conditions assumed above, we have determined that: (a) the Jacobian relates the motions of the actuators to the absolute motion of the end-effector [Eqs. (10.46), (10.49), and (10.53)]; and (b) the transpose of the Jacobian relates the actuator torques to the loads on the end-effector [Eq. (10.54)].

Example 10.10

Continue with Example 10.9 and assume that the end-effector is working against a time-varying load of $10\hat{i}_1 + 5t\hat{k}_1$ lb in addition to a constant torque of $25\hat{k}_1$ in · lb. Find the torques required at the five actuators as functions of time to achieve the motion described.

Solution

From the data given, the tool force vector of Eq. (10.50) is

$$\{F\} = \begin{bmatrix} 0 \\ 0 \\ 25 \text{ in} \cdot \text{lb} \\ 10 \text{ lb} \\ 0 \\ 5t \text{ lb} \end{bmatrix}. \qquad (16)$$

Since the Jacobian was already found in Example 10.9, Eq. (14), we can now use Eq. (10.54) to find the actuator torques directly. We find them to be

$$\tau_1 = 25.0 \text{ in} \cdot \text{lb}, \qquad Ans.$$

$$\tau_2 = 61.4 \text{ in} \cdot \text{lb}, \qquad Ans.$$

Example 10.10 (cont.)

$$\tau_3 = (-35t\cos\phi_2 + 56.0\sin\phi_2 + 61.4) \text{ in} \cdot \text{lb}, \qquad \textit{Ans.}$$

$$\tau_4 = \left(-10t^2 - 6t + 16.0\right) \text{ in} \cdot \text{lb}, \qquad \textit{Ans.}$$

$$\tau_5 = -12.0 \text{ in} \cdot \text{lb}. \qquad \textit{Ans.}$$

Recall that, since ϕ_2 is known from the posture analysis, then the above five equations are solutions for the actuator torques as functions of time.

PROBLEMS

10.1 Use the Kutzbach criterion to determine the mobility of the *SSC* linkage. Identify any idle freedoms and state how they can be removed. What is the nature of the path described by point *B*?

Fig. P10.1 $R_{BA} = R_{O_3O} = 3$ in, $R_{BO_3} = 6$ in, and $\theta_2 = 30°$.

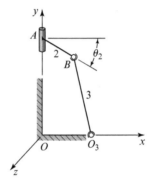

10.2 For the *SSC* linkage shown in Fig. P10.1, express the posture of each link in vector form.
10.3 For the linkage of Fig. P10.1 with $V_A = -2\hat{\mathbf{j}}$ in/s, use vector analysis to find the angular velocities of links 2 and 3 and the velocity of point *B* at the posture specified.
10.4 Solve Prob. 10.3 using graphic techniques.
10.5 For the spheric *RRRR* linkage in the posture shown, use vector algebra to make complete velocity and acceleration analyses at the posture indicated.

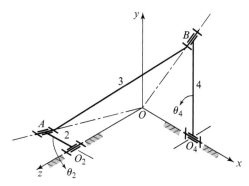

Fig. P10.5 $R_{O_2O} = 175\mathbf{k}$ mm, $R_{O_4O} = 50\hat{\mathbf{i}}$ mm, $R_{AO_2} = -75\hat{\mathbf{i}}$ mm, $R_{BO_4} = 225\hat{\mathbf{j}}$ mm, $R_{BA} = 125\hat{\mathbf{i}} + 225\hat{\mathbf{j}} - 175\mathbf{k}$ mm, and $\omega_2 = -60\mathbf{k}$ rad/s.

10.6 Solve Prob. 10.5 using graphic techniques.

10.7 Solve Prob. 10.5 using transformation matrix techniques.

10.8 Solve Prob. 10.5 except with $\theta_2 = 90°$.

10.9 Determine the advance-to-return ratio for Prob. 10.5. What is the total angle of oscillation of link 4?

10.10 For the spheric *RRRR* linkage shown, determine whether the crank is free to turn through a complete revolution. If so, find the angle of oscillation of link 4 and the advance-to-return ratio.

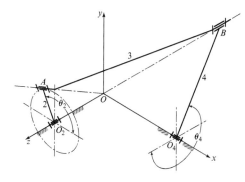

Fig. P10.10 $R_{O_2O} = 6$ in, $R_{O_4O} = 9$ in, $R_{AO_2} = 1.5$ in, $R_{BO_4} = 10.5$ in, $R_{BA} = 16.5$ in, $\theta_2 = 120°$, and $\omega_2 = 30\mathbf{k}$ rad/s.

10.11 Use vector algebra to make complete velocity and acceleration analyses of the linkage of Fig. P10.10 at the posture specified.

10.12 Solve Prob. 10.11 using graphic techniques.

10.13 Solve Prob. 10.11 using transformation matrix techniques.

10.14 Figure P10.14 shows the top, front, and auxiliary views of a spatial slider-crank *RSSP* linkage. In the construction of many such linkages, a provision is made to vary the angle β; thus, the stroke of slider 4 becomes adjustable from zero, when $\beta = 0$, to twice the crank length when $\beta = 90°$. With $\beta = 30°$, use vector algebra to make a complete velocity analysis of the linkage at the given posture.

Fig. P10.14 $R_{AO} = 50$ mm, $R_{BA} = 150$ mm, $\theta_2 = 240°$, and $\omega_2 = 24\hat{\imath}$ rad/s.

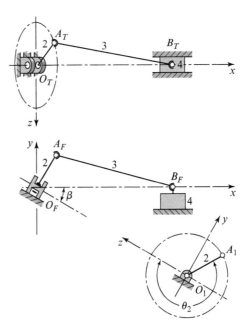

10.15 Solve Prob. 10.14 using graphic techniques.

10.16 Solve Prob. 10.14 using transformation matrix techniques.

10.17 Solve Prob. 10.14 with $\beta = 60°$ using vector algebra.

10.18 Solve Prob. 10.14 with $\beta = 60°$ using graphic techniques.

10.19 Solve Prob. 10.14 with $\beta = 60°$ using transformation matrix techniques.

10.20 Figure P10.20 shows the top, front, and profile views of an *RSRC* crank and oscillating-slider linkage. Link 4, the oscillating slider, is rigidly attached to a round rod that rotates and slides in the two bearings. (a) Use the Kutzbach criterion to find the mobility of this linkage. (b) With crank 2 as the driver, find the total angular and linear travel of link 4. (c) Write the loop-closure equation for this linkage and use vector algebra to solve it for all unknown position data.

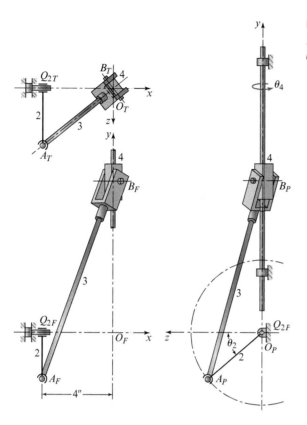

Fig. P10.20 $R_{Q_2O} = 100$ mm,
$R_{AO} = 100$ mm, $R_{BA} = 300$ mm,
$\theta_2 = 40°$, and $\boldsymbol{\omega}_2 = -48\hat{\mathbf{i}}$ rad/s.

10.21 Use vector algebra to find V_B, $\boldsymbol{\omega}_3$, and $\boldsymbol{\omega}_4$ for Prob. 10.20.

10.22 Solve Prob. 10.21 using graphic techniques.

10.23 Solve Prob. 10.21 using transformation matrix techniques.

10.24 For the SCARA robot shown in Fig. P10.24, find the transformation matrix T_{15} relating the posture of the tool coordinate system to the ground coordinate system when the joint actuators are set to the values $\phi_1 = 30°$, $\phi_2 = -60°$, $\phi_3 = 50$ mm, and $\phi_4 = 0$. Also find the absolute position of the tool point which has local coordinates $x_5 = y_5 = 0$, $z_5 = 37.5$ mm.

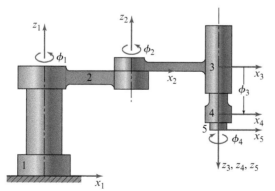

Fig. P10.24 $a_{12} = a_{23} = 250$ mm,
$a_{34} = a_{45} = 0$, $\alpha_{12} = \alpha_{34} = \alpha_{45} = 0$,
$\alpha_{23} = 180°$, $\theta_{12} = \phi_1$, $\theta_{23} = \phi_2$, $\theta_{34} = 0$,
$\theta_{45} = \phi_4$, $s_{12} = 300$ mm, $s_{23} = 0$, $s_{34} = \phi_3$,
and $s_{45} = 50$ mm.

10.25 Solve Prob. 10.24 using symbolic values for the joint variables.

10.26 For the gantry robot shown, find the transformation matrix T_{15} relating the posture of the tool coordinate system to the ground coordinate system when the joint actuators are set to the values $\phi_1 = 18$ in, $\phi_2 = 7.20$ in, $\phi_3 = 2$ in, and $\phi_4 = 0$. Also find the absolute position of the tool point which has coordinates $x_5 = y_5 = 0, z_5 = 1.80$ in.

Fig. P10.26 $a_{12} = a_{23} = a_{34} = a_{45} = 0$,
$\alpha_{12} = 90°, \alpha_{23} = -90°, \alpha_{34} = \alpha_{45} = 0$,
$\theta_{12} = \theta_{23} = 90°, \theta_{34} = 0, \theta_{45} = \phi_4, s_{12} = \phi_1$,
$s_{23} = \phi_2, s_{34} = \phi_3$, and $s_{45} = 2$ in.

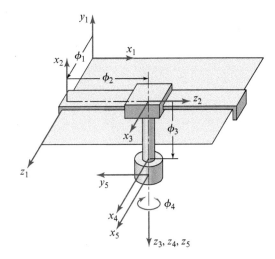

10.27 Repeat Prob. 10.26 using symbolic values for the joint variables.

10.28 For the SCARA robot of Prob. 10.24 in the posture described, find the instantaneous velocity and acceleration of the same tool point, $x_5 = y_5 = 0, z_5 = 37.5$ mm, if the actuators have constant velocities of $\dot{\phi}_1 = 0.20$ rad/s, $\dot{\phi}_2 = -0.35$ rad/s, and $\dot{\phi}_3 = \dot{\phi}_4 = 0$.

10.29 For the gantry robot of Prob. 10.26 in the posture described, find the instantaneous velocity and acceleration of the same tool point, $x_5 = y_5 = 0, z_5 = 1.80$ in, if the actuators have (constant) velocities of $\dot{\phi}_1 = \dot{\phi}_2 = 0, \dot{\phi}_3 = 1.60$ in/s, and $\dot{\phi}_4 = 20$ rad/s.

10.30 The SCARA robot of Prob. 10.24 is to be guided along a path for which the origin of the end-effector O_5 follows the straight line given by

$$R_{O_5}(t) = (40t + 100)\hat{i}_1 + (30t + 75)\hat{j}_1 + 50\hat{k}_1 \text{ mm},$$

with t varying from 0.0 to 5.0 s; the orientation of the end-effector is to remain constant with $\hat{k}_5 = -\hat{k}_1$ (vertically downward) and $\hat{i}_5$ radially outward from the base of the robot. Find expressions for how each of the actuators must be driven, as functions of time, to achieve this motion.

10.31 The gantry robot of Prob. 10.26 is to travel a path for which the origin of the end-effector O_5 follows the straight line given by

$$R_{O_5}(t) = (4.8t + 12)\hat{i}_1 - 6\hat{j}_1 + (3.6t + 9)\hat{k}_1 \text{ in}$$

with t varying from 0.0 to 4.0 s; the orientation of the end-effector is to remain constant with $\hat{k}_5 = -\hat{j}_1$ (vertically downward) and $\hat{i}_5 = \hat{i}_1$. Find expressions for the positions of each of the actuators, as functions of time, for this motion.

10.32 The end-effector of the SCARA robot of Prob. 10.24 is working against a force loading of $44\hat{i}_1 + 22t\hat{k}_1$ N and a constant torque loading of $2.8\hat{k}_1$ N $\cdot$ m as it follows the trajectory described in Prob. 10.30. Find the torques required at the actuators, as functions of time, to achieve the motion described.

10.33 The end effector of the gantry robot of Prob. 10.26 is working against a force loading of $4.5\hat{i}_1 + 2.25t\hat{j}_1$ lb and a constant torque loading of $45\hat{j}_1$ in $\cdot$ lb as it follows the trajectory described in Prob. 10.31. Find the torques required at the actuators, as functions of time, to achieve the motion described.

REFERENCES

[1] Baker, J. E., 1988. The Bennett linkage and its associated quadric surfaces, *Mech. Mach. Theory*, **23** (2):147–156.

[2] Bennett, G. T., 1903. A new mechanism, *Engineering*, **76**:777–778.

[3] Chace, M. A., 1963. Vector analysis of linkages, *J. Eng. Ind. ASME Trans. B* **85**: 289–297.

[4] Chuang, J. C., R. T. Strong, and K. J. Waldron, 1981. Implementation of solution rectification techniques in an interactive linkage synthesis program, *J. Mech. Des. ASME Trans.*, **103**(3):657–664.

[5] Denavit, J., and R. S. Hartenberg, 1955. A kinematic notation for lower-pair mechanisms based on matrices, *J. Appl. Mech. ASME Trans. E* **22**(2):215–221.

[6] Duffy, J., and C. Crane, 1980. A displacement analysis of the general spatial 7-link, 7R mechanisms, *Mech. Mach. Theory* **15**(3):153–169.

[7] Erdman, A. G., and J. E. Gustafson, 1977. LINCAGES: Linkage interactive computer analysis and graphically enhanced synthesis package, ASME Paper No. 77-DTC-5.

[8] Goldberg, M., 1943. New five-bar and six-bar linkages in three dimensions, *ASME Trans.* **65**:649–661.

[9] Gough, V. E., 1956. Contribution to discussion of papers on research in automobile stability, control and tyre performance, *Proc. Auto Div. Inst. Mech. Eng.* **171**:392–394.

[10] Harrisberger, L., 1965. A number synthesis survey of three-dimensional mechanisms, *J. Eng. Ind. ASME Trans. B* **87**(2):213–218.

[11] Harrisberger, L., and A. H. Soni, 1966. A survey of three-dimensional mechanisms with one general constraint, ASME Paper No. 66-MECH-44. (This paper contains 45 references on spatial mechanisms.)

[12] Hartenberg, R. S., and J. Denavit, 1964. *Kinematic Synthesis of Linkages*, New York: McGraw-Hill.

[13] Haug, E. J., 1989. *Computer-Aided Kinematics and Dynamics of Mechanical Systems*, Boston, MA: Allyn Bacon.

[14] Hunt, K. H., 1978. *Kinematic Geometry of Mechanisms*, New York: Oxford University Press.

[15] Kaufman, R. E., 1971. KINSYN: An interactive kinematic design system, *Trans. 3rd World Cong. on Theory of Mach. and Mech.*, Dubrovnik, Yugoslavia.

[16] Orlandea, N., M. A. Chace, and D. A. Calahan, 1977. A sparsity-oriented approach to the dynamic analysis and design of mechanical systems, parts I and II, *J. Eng. Ind. ASME Trans.*, **99**:773–784.

[17] Pennock, G. R., and A.T. Yang, 1985. Application of dual-number matrices to the inverse kinematics problem of robot manipulators, *J. Mech. Trans. Automation Design ASME Trans.* **107** (2):201–208.

[18] Phillips, J., 1984/1990. *Freedom of Machinery*, vols 1–2, New York: Cambridge University Press.

[19] Sheth, P. N., and J. J. Uicker, Jr., 1972. IMP (Integrated Mechanisms Program): A computer-aided design analysis system for mechanisms and linkages, *J. Eng. Ind. ASME Trans.*, **94**:454–464.

[20] Stewart, D, 1965. A platform with six degrees of freedom, *Proc. Inst. Mech. Eng. (UK)* **180**: 371–386.

[21] Uicker, J. J., B. Ravani, and P. N. Sheth, 2013. *Matrix Methods in the Design Analysis of Mechanisms and MultiBody Systems*, New York: Cambridge University Press.

[22] Yang, A. T., and F. Freudenstein, 1964. Application of dual-number and quaternion algebra to the analysis of spatial mechanisms, *J. Appl. Mech. ASME Trans. E* **86**: 300–308.

Part III

Dynamics of Machines

11   Static Force Analysis

11.1   Introduction

In our studies of kinematic analysis in the previous chapters, we limited ourselves to consideration of the geometry of the motions and of the relationships between displacement and time. The forces required to produce those motions or the motion that would result from the application of a given set of forces were not considered. We are now ready for a study of the dynamics of machines and systems. Such a study is usually simplified by starting with the statics of such systems.

In the design of a machine, consideration of only those effects that are described by units of *length* and *time* is a tremendous simplification. It frees the mind from the complicating influence of many other factors that ultimately enter into the problem, and it permits our attention to be focused on the primary goal: designing a mechanism to obtain a desired motion. That was the problem of *kinematics*, which was the focus of the previous chapters of this book.

The fundamental units in kinematic analysis are length and time; in dynamic analysis, the fundamental units are *length*, *time*, and *force*.

Forces are transmitted between machine members through mating surfaces – that is, from a gear to a shaft or from one gear through meshing teeth to another gear, from a connecting rod through a bearing to a lever, from a V-belt to a pulley, from a cam to a follower, or from a brake drum to a brake shoe. It is necessary to know the magnitudes of these forces for a variety of reasons. The distribution of these forces at the boundaries of mating surfaces must be reasonable, and their intensities must remain within the working limits of the materials composing the surfaces. For example, if the force operating on a sleeve bearing becomes too high, it will squeeze out the oil film and cause metal-to-metal contact, overheating, and rapid failure of the bearing. If the forces between gear teeth are too large, the oil film may be squeezed out from between them; this could result in rough motion, noise, vibration, flaking and spalling of the metal, and eventual failure. In our study of dynamics, we are interested principally in determining the magnitudes, directions, and locations of the forces, but not in sizing the members on which they act. The determination of the sizes of machine members is the subject of books usually titled "machine design" or "mechanical design," such as [5].

Some of the key terms used in this phase of our study are defined as follows.

Force Our earliest perception of force probably arose from our desire to push, lift, or pull various objects. Thus, force is the action of one body acting on another. Our intuitive concept of force includes such properties as *magnitude*, *direction*, and *point of application*, and these are commonly referred to as the *characteristics* of the force.

Matter Matter is a material substance. If it is closed and retains its shape, it is called a body.

Mass Newton defined the mass of a body as the *quantity of matter*. He recognized the fact that all bodies possess some inherent property that signifies the amount of matter, and that is different from weight. Thus, a lunar rock has a certain fixed amount of matter, although its Moon weight is different from its Earth weight. The amount of matter, or quantity of substance, is called the *mass* of the rock.

Inertia Inertia is the property of mass that causes it to resist any effort to change its motion.

Weight Weight is the force that results from gravity acting upon a mass.

Particle A particle is a body whose dimensions may be neglected (Section 2.1). The dimensions may or may not be small; if the body is in translation, then the motion of all points of the particle are equal, and it may be considered to be located at a single point. A particle is not a point, however, in the sense that a particle can consist of matter and can have mass, whereas a point cannot.

Rigid body All real bodies are either elastic or plastic and deform, although perhaps only slightly, when acted upon by forces. When the deformation of a body is small enough to be neglected, such a body is frequently assumed to be *rigid* (that is, incapable of deformation) to simplify the analysis. This assumption of rigidity is the key step that allows the treatment of kinematics without consideration of forces (Section 1.3). Without this simplifying assumption of rigidity, forces and motions are interdependent, and kinematic and dynamic analyses require simultaneous solution.

Deformable body The rigid-body assumption cannot be maintained when internal stresses and strains caused by applied forces and/or moments are to be analyzed. If stress is to be found, we must admit to the existence of strain; thus, we must consider the body to be capable of deformation, even though small. If deformations are small in comparison to the gross dimensions and motion of the body, we can still treat the body as rigid while performing the kinematic (motion) analysis, but must consider it deformable while stresses are determined. (The only place in this text where bodies are considered deformable is in the study of buckling in Sections 11.14–11.18). Using the additional assumption that forces and stresses remain within the elastic range, such analysis is frequently called *elastic-body analysis*.

11.2 Newton's Laws

Newton's three laws [3] can be stated as follows:

Law 1 If all the forces acting on a body are balanced, that is, they sum to zero, then the body either remains at rest or continues to move in a straight line at a uniform velocity.

Law 2 If the forces acting on a body are not balanced, the body experiences an acceleration proportional to, and in the direction of, the resultant force.

Law 3 When two bodies interact, a pair of reaction forces come into existence between them; these forces have equal magnitudes and opposite senses.

Newton's first two laws can be summarized by the equation

$$\sum \mathbf{F}_{ij} = m_j \mathbf{A}_{G_j}, \tag{11.1}$$

which is called the *equation of motion* for body j. In this equation, $\mathbf{A}_{G_j}$ is the absolute acceleration of the center of mass G_j of body j that has mass m_j, and this acceleration is produced by the sum of all forces $\mathbf{F}_{ij}$ acting on the body, that is, all values of the index i. Both $\mathbf{F}_{ij}$ and $\mathbf{A}_{G_j}$ are vector quantities.

In general, this chapter focuses on the first and third of Newton's laws; Chapter 12 concentrates on the second law. For all three laws, the issue of units is critical to successful solution of physical problems; therefore, this topic is addressed in the next section.

11.3 Systems of Units

An important use of Eq. (11.1) occurs in the standardization of systems of units. Let us employ the following symbols to designate the units associated with physical quantities:

length L
time T
mass M
force F.

These symbols are each to stand for any unit we may choose to use for that respective quantity. Thus, possible choices for L are inches, feet, miles, millimeters, meters, or kilometers. In our use here, the symbols L, T, M, and F do not have numeric values; however, they can be substituted into an equation as if they did. The equality sign then indicates that the units on one side of the equation must be equivalent to those on the other. Making the indicated entries into Eq. (11.1), for example, gives

$$F = MLT^{-2}, \tag{11.2}$$

since acceleration $\mathbf{A}$ has units of length divided by time squared. Equation (11.2) expresses an equivalence relationship among the four units: force, mass, length, and time. We are free to choose units to be used for three of these, and then the units used for the fourth depend on those used for the chosen three. For this reason, the three units chosen are called *basic units*, whereas the fourth is called the *derived unit*.

When force, length, and time are chosen as basic units, then mass is the derived unit, and the system is called a *gravitational system of units*. When mass, length, and time are chosen as basic units, then force is the derived unit, and the system is called an *absolute system of units*.

In some English-speaking countries, the US customary foot–pound–second (fps) system and the inch–pound–second (ips) system are two gravitational systems of units that are still in use by some practicing engineers.[1] In the fps system, the (derived) unit of mass is

$$M = FT^2/L = \text{lb} \cdot \text{s}^2/\text{ft}. \tag{11.3}$$

Thus, force, length, and time are the three basic units in the US customary (fps) system. The unit of force is the pound, abbreviated lb.[2] When using the US customary system, we note in Eq. (11.3) that the derived unit of mass in the (fps) gravitational system is the $\text{lb} \cdot \text{s}^2/\text{ft}$, sometimes referred to as a slug; there is no abbreviation for slug. Since there is no known use of the term slug outside of academia, this text does not use this term. The proper unit of mass in the (fps) system is the $\text{lb} \cdot \text{s}^2/\text{ft}$, with no abbreviation.

The derived unit of mass in the (ips) gravitational system is

$$M = FT^2/L = \text{lb} \cdot \text{s}^2/\text{in}. \tag{11.4}$$

Note that this unit of mass has not been given an official name.

Note that lb is a unit of force, not a unit of mass. It is improper to use the term lb when defining a quantity of mass, although this mistake is frequently heard. Some incorrectly refer to mass and weight as though these terms are interchangeable; they mistakenly refer to a quantity of mass which weighs 1 lb as 1 lb of mass. This is improper and is not used in this text. When using the fps or the ips gravitational systems of units, the proper units of mass are the $\text{lb} \cdot \text{s}^2/\text{ft}$ or the $\text{lb} \cdot \text{s}^2/\text{in}$.

The International System (SI)[3] of units is an absolute system. The three basic units are the meter (m), the kilogram mass (kg), and the second (s). The unit of force is derived and is called a *newton* (N) to distinguish it from the kilogram, which, as indicated, is a unit of mass. The dimensions of the newton are

$$F = ML/T^2 = \text{kg} \cdot \text{m/s}^2 = \text{N}. \tag{11.5}$$

The weight of a body is the force exerted upon it by gravity. Designating the gravitational force called weight as **W** and the acceleration due to gravity as **g**, Eq. (11.1) becomes

$$\mathbf{W} = m\mathbf{g}.$$

In the (fps) system, the magnitude of standard gravity is $g = 32.174 \text{ ft/s}^2$, which for most cases is approximated as 32.2 ft/s^2. Therefore, the weight of a mass of one $\text{lb} \cdot \text{s}^2/\text{ft}$, under standard gravity, is

$$W = mg = \left(1 \text{ lb} \cdot \text{s}^2/\text{ft}\right)\left(32.2 \text{ ft/s}^2\right) = 32.2 \text{ lb}.$$

---

[1] Some prefer to use gravitational systems; this helps to explain some of the resistance to the use of the International System (SI), which is an absolute system. However, most US engineering companies with international markets have already required this change.

[2] The abbreviation lb for pound comes from the Latin word *Libra*, the balance, the seventh sign of the zodiac, which is represented as a pair of scales.

[3] SI stands for *Le Système Internationale d'Unités* (French).

In the ips system, standard gravity is 386.088 in/s^2, or approximately 386 in/s^2. Thus, in standard gravity, a unit mass weighs

$$W = mg = \left(1 \text{ lb} \cdot \text{s}^2/\text{in}\right)\left(386 \text{ in/s}^2\right) = 386 \text{ lb}.$$

With SI units, the magnitude of standard gravity is 9.80665 m/s^2, or approximately 9.81 m/s^2. Therefore, the weight of a mass of one kilogram, under standard gravity, is

$$W = mg = (1 \text{ kg})\left(9.81 \text{ m/s}^2\right) = 9.81 \text{ kg} \cdot \text{m/s}^2 = 9.81 \text{ N}.$$

It is convenient to remember that the weight of a large apple is approximately 1 N.

11.4 Applied and Constraint Forces

A great many, if not most, of the forces with which we are concerned in mechanical equipment occur through direct physical or mechanical contact. However, examples of forces that may be applied without physical contact are electrical, magnetic, and gravitational forces. When two or more bodies are connected to form a group or system, the forces acting on this system from outside are called *applied* (or *external*) forces. The action and reaction forces between any two of the connected bodies are called *constraint* (or *internal*) forces. These constraint forces constrain the connected bodies to behave in a specific manner defined by the nature of the connection.

As shown in the next section, the constraint forces of action and reaction at a mechanical contact occur in equal but opposite pairs and thus have no *net* force effect on the system of bodies being considered. Such pairs of constraint forces, although they clearly exist and may be large, are usually not considered further when both the action and the reaction forces act on bodies of the system being considered. However, when we consider a body, or a system of bodies, to be isolated from its surroundings, only one of each pair of constraint forces acts on the system being considered at a point of contact on the separation boundary. The other constraint force, the reaction, is left acting on the surroundings. When we isolate the system being considered, these constraint forces at points of separation must be clearly identified, since they are essential to further study of the isolated system.

As indicated earlier, the *characteristics* of a force are its *magnitude*, its *direction*, and its *point of application*. The direction of a force includes the concept of a line along which the force acts, and a *sense*. Therefore, a force may be directed either positively or negatively along its line of action.

The notation used for force vectors is shown in Fig. 11.1. Boldface letters are used for force vectors and lightface letters for their magnitudes. Thus, the components of a force vector are

$$\mathbf{F} = F^x\hat{\mathbf{i}} + F^y\hat{\mathbf{j}} + F^z\hat{\mathbf{k}}. \tag{a}$$

Consistent with the previous chapter, the directions of the components of a force vector are indicated by superscripts, not subscripts.

Fig. 11.1 Rectangular
components of a force vector.

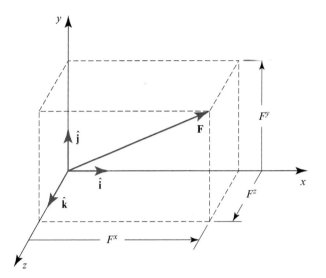

Fig. 11.2 (*a*) $\mathbf{R}_{BA}$ is a position-difference
vector, whereas $\mathbf{F}$ and $\mathbf{F}'$ are force
vectors; (*b*) force couple formed by
$\mathbf{F}$ and $\mathbf{F}'$.

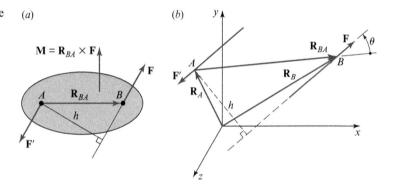

Two equal and opposite forces along two parallel but *noncollinear* straight lines in a body cannot be combined to produce a single (null) resultant force on the body. Any two such forces acting on the body constitute a *couple*. For example, consider the two forces, $\mathbf{F}$ and $\mathbf{F}'$, shown in Fig. 11.2*a*. The *arm of the couple* is the perpendicular distance between their lines of action, denoted here as h, and the *plane of the couple* is the plane containing the two lines of action. The free vector $\mathbf{M}$ is the *moment* of the couple formed by $\mathbf{F}$ and $\mathbf{F}'$. This vector is directed normal to the plane of the couple and the sense is in accordance with the right-hand rule. The magnitude of the moment is the product of the arm h and the magnitude of one of the forces, say F. Thus,

$$M = hF. \tag{11.6}$$

In vector form, the moment, $\mathbf{M}$, is the cross-product of the position-difference vector $\mathbf{R}_{BA}$ and the force vector $\mathbf{F}$, and so it is defined by the equation

$$\mathbf{M} = \mathbf{R}_{BA} \times \mathbf{F}. \tag{11.7}$$

Some interesting properties of couples can be determined by examination of Fig. 11.2b. Here, **F** and **F**$'$ are two equal-sized, opposite, and parallel forces. We can choose any point on each line of action and define the positions of these points by vectors **R**$_A$ and **R**$_B$. Then the "relative" position, or position-difference vector (Section 2.3), is

$$\mathbf{R}_{BA} = \mathbf{R}_B - \mathbf{R}_A. \tag{b}$$

The moment of the couple is the sum of the moments of the two forces and is

$$\mathbf{M} = \mathbf{R}_A \times \mathbf{F}' + \mathbf{R}_B \times \mathbf{F}. \tag{c}$$

But **F**$' = -$**F**, and, therefore, using Eq. (b), Eq. (c) can be written as

$$\mathbf{M} = (\mathbf{R}_B - \mathbf{R}_A) \times \mathbf{F} = \mathbf{R}_{BA} \times \mathbf{F}. \tag{d}$$

Equation (d) indicates the following:

1. The value of the moment of the couple is independent of the choice of the reference point about which the moments are taken, since the vector **R**$_{BA}$ is the same for all positions of the origin.
2. Since **R**$_A$ and **R**$_B$ define any choices of points on the lines of action of the two forces, vector **R**$_{BA}$ is not restricted to perpendicularity with **F** and **F**$'$. This is a very important characteristic of the vector product, since it demonstrates that the value of the moment is independent of how **R**$_{BA}$ is chosen. The magnitude of the moment can be obtained as follows: Resolve **R**$_{BA}$ into two components **R**$_{BA}^n$ and **R**$_{BA}^t$, perpendicular and parallel to **F**, respectively. Then,

$$\mathbf{M} = \left(\mathbf{R}_{BA}^n + \mathbf{R}_{BA}^t\right) \times \mathbf{F}.$$

But **R**$_{BA}^n$ is the perpendicular between the two lines of action, and **R**$_{BA}^t$ is parallel to **F**. Therefore, **R**$_{BA}^t \times$ **F** $= \mathbf{0}$, and

$$\mathbf{M} = \mathbf{R}_{BA}^n \times \mathbf{F} \tag{e}$$

is the moment of the couple. Since $R_{BA}^n = R_{BA} \sin \theta$, where θ is the angle between **R**$_{BA}$ and **F**, the magnitude of the moment is

$$M = (R_{BA}\sin \theta)F = hF. \tag{f}$$

3. The moment vector **M** is independent of any particular origin or line of application and is thus a *free vector*.
4. The forces of a couple can be rotated together within their plane, keeping their magnitudes and the distance between their lines of action constant, or they can be translated to any parallel plane, without changing the magnitude, direction, or sense of the moment vector.
5. Two couples are equal if they have the same moment vectors, regardless of the forces or moment arms. This means that it is the vector product of the two that is significant and not the individual values.

11.5 Free-Body Diagrams

The term "body" as used here may consist of an entire machine, several connected parts of a machine, a group of parts, a single part, or a portion of a machine part. A *free-body diagram* is a sketch or drawing of a body or group of bodies, isolated from the rest of the machine and its surroundings, upon which the forces and moments are shown in action. It is sometimes desirable to include on the diagram the known magnitudes and directions as well as other pertinent information.

The diagram so obtained is called "free," since the machine part, or parts, or portion of the system has been freed (or isolated) from the remaining machine elements, and their effects have been replaced by forces and moments. If the free-body diagram is of a machine part or system of parts, the forces shown on it are the applied (external) forces and moments exerted by adjacent or connecting parts that are not part of this free-body diagram. If the diagram is of a portion of a part, the forces and moments shown acting on the cut (separation) surface are constraint (internal) forces and moments.

The construction and presentation of clear, detailed, and neatly drawn free-body diagrams represent the heart of engineering communication. This is true, since they report a part of the thinking process, whether they are actually placed on paper or not, and since the construction of these diagrams is the *only* way the results of thinking can be communicated to others. It is important to acquire the habit of drawing free-body diagrams no matter how simple the problem may appear. They are a means of storing ideas while concentrating on the next step in the solution of a problem. Construction of, or, at the very minimum, sketching of free-body diagrams speeds up the problem-solving process and dramatically decreases the chance of mistakes.

The advantages of using free-body diagrams can be summarized as follows:

1. They make it easier to translate words, thoughts, and ideas into physical models.
2. They assist in seeing and understanding all facets of a problem.
3. They show the constraint forces and moments acting on the system.
4. They help in planning an approach to a problem.
5. They make mathematical relations easier to see or formulate.
6. Their use makes it easy to keep track of progress and helps in making simplifying assumptions.
7. They are useful for storing methods of solution for future reference.
8. They assist our memory and make it easier to present and explain our work to others.

In analyzing the forces in a machine, we almost always need to separate the machine into its individual components or subsystems and construct free-body diagrams indicating the forces that act on each. Many of these components are connected to each other by kinematic pairs (joints). Accordingly, Fig. 11.3 shows the constraint forces acting between the elements of the six lower pairs (Section 1.4) when all friction forces are assumed to be zero. Note that these are not complete free-body diagrams in that they show only the forces on the mating pair surfaces and do not show the forces where the partial links have been separated from the remaining portions of the links.

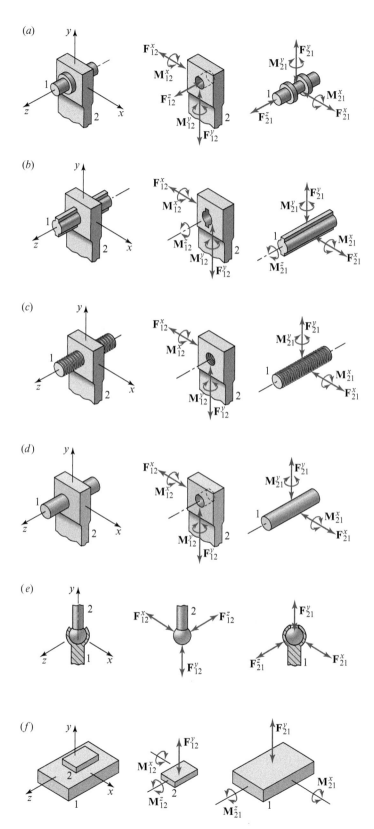

Fig. 11.3 (*a*) Revolute or turning pair; (*b*) prism or prismatic pair; (*c*) screw or helical pair; (*d*) cylinder or cylindric pair; (*e*) sphere or globular pair; (*f*) flat or planar pair.

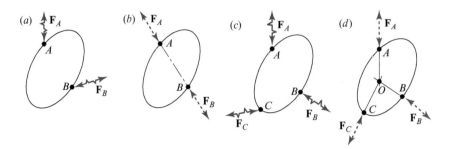

Fig. 11.4 (*a*) Two-force member is not in equilibrium; (*b*) two-force member is in equilibrium; (*c*) three-force member is not in equilibrium; (*d*) three-force member is in equilibrium.

Upon careful examination of Fig. 11.3 we see that there is no component of a constraint force or moment transmitted along an axis where motion is possible – that is, along with the direction of a joint variable. This is consistent with the assumption of no friction. If one of these pair elements were disposed to transmit a force, or a moment, to its mating element in the direction of the joint variable, in the absence of friction, the tendency would result in motion of the joint variable rather than in the transmission of a force or moment. Similarly, in the case of higher pairs, in the absence of friction the constraint forces are always normal to the contacting surfaces.

The notation shown in Fig. 11.3 is used consistently throughout this chapter. The force that link i exerts onto link j is denoted $\mathbf{F}_{ij}$, whereas the reaction to this force is denoted $\mathbf{F}_{ji}$ and is the force from link j acting back onto link i.

Note that, when the sense of a force or moment is unknown, the diagram showing this force or moment is drawn with arrowheads at both ends (as in Fig. 11.3). When the orientation is also unknown, the diagram either shows a broken (zig-zag) line for the shaft of the arrow (as in Fig. 11.4*a*), or it shows multiple components for the force or moment (as in Fig. 11.1). These symbolic conventions are used throughout the remainder of this text.

11.6 Conditions for Equilibrium

A body or group of bodies is said to be in equilibrium if all the forces and moments exerted on the system are in balance. In such a situation, Newton's laws, as expressed in Eq. (11.1), indicate that no acceleration results. This may imply that no motion takes place, meaning that all velocities are zero. If so, the system is said to be in *static equilibrium* [4]. On the other hand, no acceleration may imply that velocities do exist but remain constant; the system is then said to be in *dynamic equilibrium*. In either case, Eq. (11.1) demonstrates that a system of bodies is in equilibrium if and only if:

1. *The vector sum of all forces acting upon it is zero.*
2. *The vector sum of the moments of all forces acting about an arbitrary axis is zero.*

Mathematically, these two statements are expressed as

$$\sum \mathbf{F} = \mathbf{0} \quad \text{and} \quad \sum \mathbf{M} = \mathbf{0}. \tag{11.8}$$

Note how these statements are a result of Newton's first and third laws, it being understood that a body is composed of a collection of particles.

Many problems have forces acting in a single plane. When this is true, it is convenient to choose this as the xy plane. Then, Eqs. (11.8) can be simplified by taking the components in that plane as follows:

$$\sum F^x = 0, \quad \sum F^y = 0, \quad \text{and} \quad \sum M = 0,$$

where the z direction for the moment M component is implied by the fact that all forces act only in the xy plane.

11.7 Two- and Three-Force Members

A *member* is the name given to a rigid body subjected only to forces – that is, with no applied moment(s). The free-body diagram of Fig. 11.4a shows a two-force member, that is an arbitrarily shaped body acted upon by two forces, $\mathbf{F}_A$ and $\mathbf{F}_B$, acting at points A and B, respectively. The forces are shown with two arrowheads and with zig-zag lines signifying that the orientations of these forces are not yet known, and with arrowheads at both ends to remind us that the magnitude and sense of each is also not yet known.

Assuming that the free-body diagram is complete, that is, no other forces are active on the body, then the first of Eqs. (11.8) gives

$$\sum \mathbf{F} = \mathbf{F}_A + \mathbf{F}_B = \mathbf{0}.$$

This requires that $\mathbf{F}_A$ and $\mathbf{F}_B$ have *equal magnitudes, parallel lines of action*, and *opposite senses*. Using the second of Eqs. (11.8) about either point A or point B demonstrates that $\mathbf{F}_A$ and $\mathbf{F}_B$ must also share the *same line of action*; otherwise their moments cannot sum to zero. Thus, for any two-force member, with no applied moments, we have learned that the forces must be oppositely directed along the unique line of action defined by the points A and B, as shown in Fig. 11.4b. When the magnitude of one force becomes known, the magnitude of the second force must be equal to that of the first force but of the opposite sense (simply stated, the two forces must be equal, opposite, and collinear).

Note that in order for a rigid body with two active forces and an applied moment to be in static equilibrium, the two forces cannot be coaxial; that is, they must be equal, opposite, and parallel. Also, because of the applied moment, the body is not referred to as a member.

Next consider a three-force member – that is, a body with three active forces and no applied moments, as shown in Fig. 11.4c. Here, we also assume that the three forces, $\mathbf{F}_A$, $\mathbf{F}_B$, and $\mathbf{F}_C$, are coplanar; that is, they are all known to act in the plane defined by the three points, A, B, and C. Suppose further that we also know two of the lines of action, say, those of $\mathbf{F}_A$ and $\mathbf{F}_B$, and that they intersect at some point, O. Since neither $\mathbf{F}_A$ nor $\mathbf{F}_B$ exerts any moment about point O, applying $\sum \mathbf{M} = \mathbf{0}$ demonstrates that the moments of each of the three forces (including $\mathbf{F}_C$) about point O must be zero. Thus, the lines of action of all three forces must intersect at the common point O; that is, the three forces acting on the body must be *concurrent*, as shown in Fig. 11.4d. This explains why, in two dimensions, a three-force member can be solved for only two force magnitudes, even though there are three scalar equations implied in Eqs. (11.8); the moment equation has already been satisfied once the lines of action become known. If one of the three force magnitudes is known, then the other two magnitudes can be found by using $\sum \mathbf{F} = \mathbf{0}$. This is demonstrated in the following example.

Example 11.1

The four-bar linkage in the posture shown in Fig. 11.5a has an external load $\mathbf{P} = 120\text{ lb}\angle 220°$ acting on link 4 at point Q. For the given posture, find the internal reaction forces and the crank torque $\mathbf{T}_{12}$ to hold the linkage in static equilibrium.

Assumptions

1. Gravitational forces are small in comparison with applied force $\mathbf{P}$ and can be ignored.
2. Frictional forces can be neglected. (Frictional forces are addressed in Section 11.9.)

Graphic Solution

1. Draw the linkage and the given force to scale, as shown in Fig. 11.5a.
2. Draw the free-body diagrams of links 2, 3, and 4, as shown in Figs. 11.5b, 11.5c, and 11.5d. At this stage, unknown information, such as applied forces and torques on each link, can be sketched lightly on the free-body diagrams, but will be redrawn to scale once they become known.
3. From the free-body diagram, observe that link 3 is a two-force member. Thus, according to the above discussion, the lines of action of $\mathbf{F}_{23}$ and $\mathbf{F}_{43}$ are shown along the axis defined by points A and B in Fig. 11.5c. Since action and reaction forces must be equal and opposite, these also define the lines of action of $\mathbf{F}_{32}$ on link 2 in Fig. 11.5b and $\mathbf{F}_{34}$ on link 4 in Figs. 11.5d and 11.5e.
4a. Measure the moment arms of forces $\mathbf{P}$ and $\mathbf{F}_{34}$ about O_4, which are found to be 2.38 in and 8.63 in, respectively. Then, assuming that F_{34} is positive to the right, and taking counterclockwise moments as positive, Eqs. (11.8) gives

$$\sum M_{O_4} = (2.38\text{ in})(120\text{ lb}) - (8.63\text{ in})F_{34} = 0.$$

The solution is $F_{34} = 33.09\text{ lb}$, where the fact that the result is positive confirms the sense assumed for $\mathbf{F}_{34}$. If F_{34} were assumed positive to the left, then the solution would give $F_{34} = -33.09\text{ lb}$.

Example 11.1 (cont.)

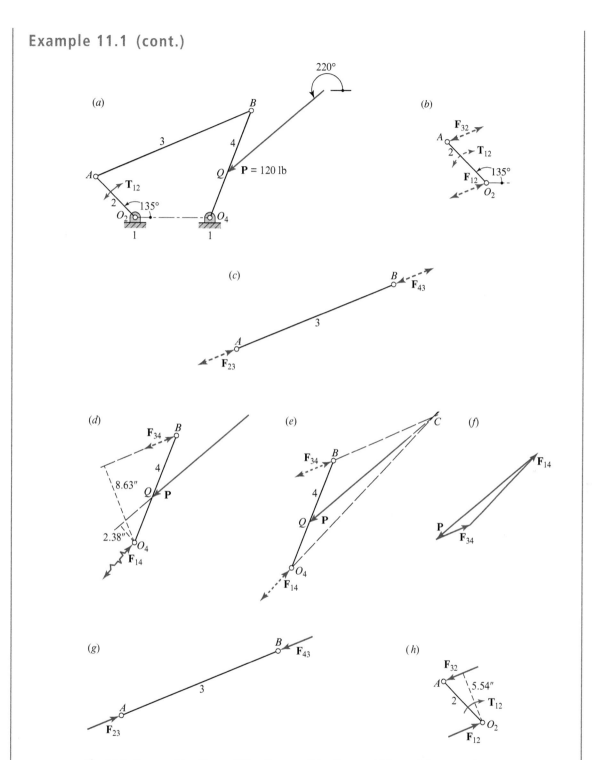

Fig. 11.5 $R_{AO_2} = 6$ in, $R_{BA} = 18$ in, $R_{O_4O_2} = 8$ in, $R_{BO_4} = 12$ in, and $R_{QO_4} = 5$ in.

Example 11.1 (cont.)

4*b*. As an alternative approach to step 4*a*, note that link 4 is a three-force member; therefore, when the lines of action of **P** and **F**$_{34}$ are extended, they intersect at the concurrency point *C* and define the line of action of **F**$_{14}$, as shown in Fig. 11.5*e*.

5. No matter whether step 4*a* or step 4*b* is used, the force polygon for link 4 shown in Fig. 11.5*f* can be used to graphically solve the equation

$$\sum \mathbf{F} = \mathbf{P} + \mathbf{F}_{34} + \mathbf{F}_{14} = \mathbf{0}.$$

Note that step 4*b* avoids the need for assuming a positive sense for F_{34} or for use of the moment equation of step 4*a*; these are satisfied by use of the concurrency point *C*. The force polygon gives $F_{34} = 33.09$ lb and $F_{14} = 89.0$ lb with orientations and senses as indicated.

6. From the free-body diagram of link 3, Fig. 11.5*g*, and the nature of action and reaction forces, note that $\mathbf{F}_{23} = -\mathbf{F}_{43} = \mathbf{F}_{34}$. For a two-force member, it is important to note whether the member is in tension or in compression. In the case of a deformable body, a member in tension, if overloaded, would fail due to yielding; however, if the member is in compression, it could fail either by yielding or by buckling. Hence, buckling becomes an important topic for deformable bodies and is addressed in Sections 11.14–11.18. In this example, link 3 is in compression.

7. Show the known force $\mathbf{F}_{32} = -\mathbf{F}_{23}$ on the free-body diagram of link 2 in Fig. 11.5*h*. Since link 2 is subjected to two forces and a torque, $\mathbf{F}_{12}$ must be equal, opposite, and parallel to (but not collinear with) $\mathbf{F}_{32}$. Alternatively, from the summation of forces on the free-body diagram of link 2 in Fig. 11.5*h*, $\mathbf{F}_{12} = -\mathbf{F}_{32}$. Measure the moment arm of $\mathbf{F}_{32}$ about point O_2; it is found to be 5.54 in. Taking counterclockwise moments as positive gives

$$\sum M_{O_2} = (5.54 \text{ in})F_{32} - T_{12} = 0.$$

Therefore, the crank torque to maintain static equilibrium is

$$T_{12} = (5.54 \text{ in})(33.09 \text{ lb}) = 183.3 \text{ in} \cdot \text{lb}, \qquad\qquad Ans.$$

where the positive sign confirms the clockwise sense shown in Fig. 11.5*h*.

Analytic Solution

1. First, use Eqs. (2.25)–(2.30) to perform a posture analysis of the linkage. The posture of each link is shown in Fig. 11.6*a*. In addition, note that link 3 is a two-force member, assume that F_{34} is positive to the right, and find

$$\mathbf{R}_{AO_2} = 6.0 \text{ in}\angle 135° = -4.242 \, 64\hat{\mathbf{i}} + 4.242 \, 64\hat{\mathbf{j}} \text{ in},$$

$$\mathbf{R}_{BO_4} = 12.0 \text{ in}\angle 68.65° = 4.368 \, 18\hat{\mathbf{i}} + 11.176 \, 72\hat{\mathbf{j}} \text{ in},$$

$$\mathbf{R}_{QO_4} = 5.0 \text{ in}\angle 68.65° = 1.820 \, 08\hat{\mathbf{i}} + 4.656 \, 96\hat{\mathbf{j}} \text{ in},$$

Example 11.1 (cont.)

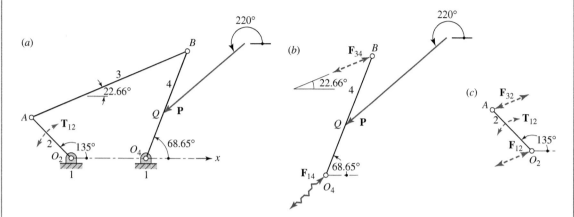

Fig. 11.6 Analytic solution.

$$\mathbf{F}_{34} = F_{34}\angle22.66° = 0.922\,81F_{34}\hat{\mathbf{i}} + 0.385\,26F_{34}\hat{\mathbf{j}},$$

$$\mathbf{P} = 120\,\text{lb}\angle220° = -91.925\,33\hat{\mathbf{i}} - 77.134\,51\hat{\mathbf{j}}\,\text{lb}.$$

2. Referring to Fig. 11.6b, sum moments about point O_4. Thus,

$$\sum\mathbf{M}_{O_4} = \mathbf{R}_{QO_4} \times \mathbf{P} + \mathbf{R}_{BO_4} \times \mathbf{F}_{34} = \mathbf{0}.$$

Perform the cross-product operations, showing the first term is $\mathbf{R}_{QO_4} \times \mathbf{P} = 287.701\,61\hat{\mathbf{k}}$ in · lb, whereas the second term is $\mathbf{R}_{BO_4} \times \mathbf{F}_{34} = -8.631\,10F_{34}\hat{\mathbf{k}}$ in · lb. Substitute these values into the above equation and solve, giving

$$\mathbf{F}_{34} = 33.333\,13\,\text{lb}\angle22.66° = 30.760\,05\hat{\mathbf{i}} + 12.841\,98\hat{\mathbf{j}}\,\text{lb}. \qquad \textit{Ans.}$$

3. Next, determine the force $\mathbf{F}_{14}$ from the equation

$$\sum\mathbf{F} = \mathbf{P} + \mathbf{F}_{34} + \mathbf{F}_{14} = \mathbf{0}.$$

Solve, giving

$$\mathbf{F}_{14} = 61.165\,28\hat{\mathbf{i}} + 64.292\,53\hat{\mathbf{j}}\,\text{lb} = 88.739\,62\,\text{lb}\angle46.43°. \qquad \textit{Ans.}$$

4. Next, from the free-body diagram of link 2 (Fig. 11.6c), write

$$\sum\mathbf{M}_{O_2} = \mathbf{T}_{12} + \mathbf{R}_{AO_2} \times \mathbf{F}_{32} = \mathbf{0}.$$

Using $\mathbf{F}_{32} = -\mathbf{F}_{34} = -30.760\,05\hat{\mathbf{i}} - 12.841\,98\hat{\mathbf{j}}\,\text{lb}$, find

$$\mathbf{T}_{12} = -184.987\,72\hat{\mathbf{k}}\,\text{in · lb}. \qquad \textit{Ans.}$$

Fig. 11.7 (*a*) Balanced revolute joint connection; (*b*) unbalanced revolute joint connection.

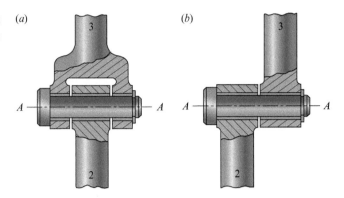

Note that the graphic solution is in good agreement with the analytic result. Also note that in both the graphic and analytic solutions, the free-body diagram of the frame, link 1, is not drawn, since it is not helpful in the solution. If drawn, force $\mathbf{F}_{21} = -\mathbf{F}_{12}$ would be shown at O_2, a force $\mathbf{F}_{41} = -\mathbf{F}_{14}$ at O_4, and reaction moment $\mathbf{M}_{21} = -\mathbf{T}_{12}$ would be shown. In addition, a minimum of two more force components and, perhaps, a moment would be shown where link 1 is anchored to keep it stationary. Since this would introduce at least three additional unknowns and since Eqs. (11.8) would not allow the solution for more than three, drawing the free-body diagram of the frame is of no benefit in the solution process. The only case for which this is necessary is when these anchoring forces, called "shaking forces," and moments are sought as part of the solution. This is discussed in more detail in Section 12.7.

In the preceding example, it was assumed that all forces act in the same plane. For the connecting link 3, for example, it was assumed that the line of action of the forces and the centerline of the link are coincident. A careful designer sometimes takes extreme measures to approach these conditions as closely as possible. Note that if the pin connections are arranged as shown in Fig. 11.7*a*, such conditions are obtained theoretically. If, on the other hand, the connections are made as shown in Fig. 11.7*b*, the pin itself as well as the link have moments acting upon them. If the forces are not in the same plane, then moments exist proportional to the distance between the force planes.

Example 11.2

For the slider-crank linkage in the posture shown in Fig. 11.8*a*, an external load $P = 100$ lb acts horizontally on link 4 at point Q, which is 1 in above the centerline. Block 4 is 8 in wide and 3 in high, with pin B centrally located. Determine the torque $\mathbf{T}_{12}$ that must be applied to link 2 to hold the linkage in static equilibrium.

Assumptions

1. Gravitational forces are small in comparison with the applied force $\mathbf{P}$ and can be ignored.
2. Frictional forces can be neglected.

Graphic Solution

1. Draw the linkage and the given force to scale, as shown in Fig. 11.8*a*. (The size scale shown here is about 13 in/in and the force scale shown is about 140 lb/in. The numeric values shown

Example 11.2 (cont.)

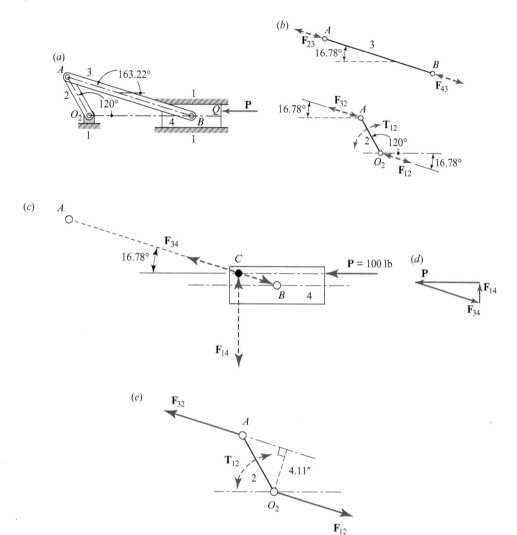

Fig. 11.8 $R_{AO_2} = 6$ in and $R_{BA} = 18$ in.

below, however, were obtained from larger figures with a size scale of 2 in/in and a force scale of 10 lb/in.)

2. Draw the free-body diagram of each moving link, as shown in Figs. 11.8b and 11.8c.
3. Observe that link 3 is a two-force member. Show the lines of action of F_{23} and F_{43} along the axis defined by points A and B in Fig. 11.8b. Also show the lines of action of F_{34} on link 4 in Fig. 11.8c and F_{32} on link 2 in Fig. 11.8e.
4. For link 4, neither the magnitude nor the location of the frame reaction force F_{14} is known as yet. However, the orientation of this force must be perpendicular to the sliding surface, since friction is neglected. Also, link 4 is a three-force member. Therefore, when the lines of action of

Example 11.2 (cont.)

P and **F**$_{34}$ are extended, they intersect at concurrency point C and define the line of action of **F**$_{14}$, as shown in Fig. 11.8c.

5. Draw the force polygon for link 4, as shown in Fig. 11.8d, to graphically solve the equation

$$\sum \mathbf{F} = \mathbf{P} + \mathbf{F}_{34} + \mathbf{F}_{14} = \mathbf{0}.$$

From this polygon, measure

$$F_{34} = 105 \text{ lb} \quad \text{and} \quad F_{14} = 31 \text{ lb},$$

and also recognize the sense of each.

6. From the free-body diagram of link 3, Fig. 11.8b, and the nature of action and reaction forces, note that **F**$_{23}$ = −**F**$_{43}$ = **F**$_{34}$. Also note that link 3 is in compression; therefore, if link 3 is a deformable body, the possibility of buckling should be investigated (Sections 11.14–11.18).

7. From the free-body diagram of link 2, Fig. 11.8e, and the nature of action and reaction forces, note that **F**$_{32}$ = −**F**$_{23}$. Also, from the summation of forces, **F**$_{12}$ = −**F**$_{32}$. When the moment arm of **F**$_{32}$ about point O_2 is measured, it is found to be 4.11 in. Taking counterclockwise moments as positive gives

$$\sum M_{O_2} = (4.11 \text{ in})F_{32} + T_{12} = 0.$$

Therefore, the torque acting on link 2 to maintain static equilibrium is

$$T_{12} = -(4.11 \text{ in})(105 \text{ lb}) = -432 \text{ in} \cdot \text{lb (cw)}, \qquad \textit{Ans.}$$

where the negative sign indicates a clockwise sense for the torque T_{12}.

Analytic Solution

1. From a posture analysis of the linkage with

$$\mathbf{R}_{AO_2} = 6.0 \text{ in}\angle 120° = -3.000\,00\hat{\mathbf{i}} + 5.196\,15\hat{\mathbf{j}} \text{ in}$$

the posture of link 3 is obtained, see Fig. 11.9a.

2. Write **F**$_{14}$ = $F_{14}\hat{\mathbf{j}}$ and

$$\mathbf{F}_{34} = F_{34}\angle -16.78° = 0.957\,42F_{34}\hat{\mathbf{i}} - 0.288\,70F_{34}\hat{\mathbf{j}}.$$

Then, considering Fig. 11.9b, the forces **F**$_{34}$ and **F**$_{14}$ are determined from the equation

$$\sum \mathbf{F} = \mathbf{P} + \mathbf{F}_{34} + \mathbf{F}_{14} = \mathbf{0}.$$

This equation can be written as

$$-100.00\hat{\mathbf{i}} \text{ lb} + 0.957\,42F_{34}\hat{\mathbf{i}} - 0.288\,70F_{34}\hat{\mathbf{j}} + F_{14}\hat{\mathbf{j}} = \mathbf{0}.$$

Example 11.2 (cont.)

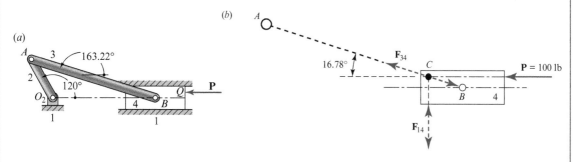

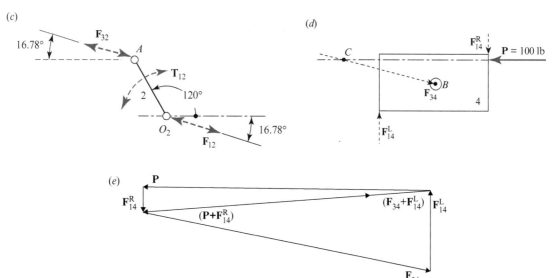

Fig. 11.9 Analytic solution.

Then, equating the $\hat{\mathbf{i}}$ *and* $\hat{\mathbf{j}}$ components, respectively, gives

$$F_{34} = \frac{100.00 \text{ lb}}{0.957\,42} = 104.45 \text{ lb},$$

$$F_{14} = 0.288\,70 F_{34} = 30.15 \text{ lb}.$$

These forces can be written as

$$\mathbf{F}_{34} = 100.00\,\hat{\mathbf{i}} - 30.15\,\hat{\mathbf{j}} \text{ lb} = 104.45 \text{ lb} \angle -16.78°,$$

$$\mathbf{F}_{14} = 30.15\hat{\mathbf{j}} \text{ lb}.$$

3. Next, from the free-body diagram of link 2 (Fig. 11.9*c*), write

$$\sum \mathbf{M}_{O_2} = \mathbf{T}_{12} + \mathbf{R}_{AO_2} \times \mathbf{F}_{32} = \mathbf{0}.$$

Example 11.2 (cont.)

Using $\mathbf{F}_{32} = -\mathbf{F}_{34} = -100.00\hat{\mathbf{i}} + 30.15\hat{\mathbf{j}}$ lb and solving, find

$$\mathbf{T}_{12} = -429.17\hat{\mathbf{k}} \text{ in} \cdot \text{lb (cw).} \qquad\qquad Ans.$$

Note that the graphic solution is in good agreement with the analytic result.

The graphic and the analytic solutions both encounter difficulties in any of the following three situations: (a) if block 4 were of lesser width; (b) if force P were applied at a location even higher on link 4; or (c) if the connecting rod were at a shallower angle, as it would be for input crank angles between $135°$ and $225°$. If any of these situations occur, then concurrency point C lies to the left of link 4, and the location of force F_{14} cannot be as shown in Fig. 11.9b.

In such a case, block 4 would tip counterclockwise unless prevented by its top-right corner being constrained by the upper sliding surface; the single force F_{14}, would then be replaced by a couple $(F_{14}^{\mathrm{L}},$ and $F_{14}^{\mathrm{R}})$ at the lower-left and top-right corners of block 4, as shown in Fig. 11.9d.

Block 4 would then become a four-force member, that is, $P, F_{34}, F_{14}^{\mathrm{L}},$ and F_{14}^{R}. The method of analysis will be studied in the following section, and this example will be concluded, as shown in the force polygon of Fig. 11.9e.

11.8 Four- and More Force Members

When all free-body diagrams of a system have been constructed, at least in planar problems, we usually find that one or more represent either two- or three-force members, and the techniques of the previous section can be used for their solution. We start by seeking out such members and using them to establish known lines of action for other unknown forces. In doing this, we are implicitly enforcing the summation-of-moments equation for such members. Then, by noting that action and reaction forces share the same line of action, we proceed to establish more lines of action. Finally, upon reaching a body where one or more forces and all lines of action are known, we apply the summation-of-forces equations to find the magnitudes and senses of the unknown forces. Proceeding from body to body in this way, we find solutions for the unknown forces.

It is always good practice to replace multiple known forces on the same free-body diagram by a single force representing the sum of known forces. This sometimes simplifies the diagrams to reveal two- and three-force members that are otherwise not noticed.

Sometimes it is helpful to combine two links, or even three links, and to draw a free-body diagram of their combination. This approach can sometimes be used to eliminate unknown forces on the individual links, which combine with their reactions as pairs of constraint forces within the combined assembly.

In some problems, we find that the solution cannot proceed in this fashion – for example, when two- or three-force members cannot be found, and lines of action remain unknown. It is then wise to count the number of unknown quantities (magnitudes and directions) to be determined for each free-body diagram. In planar problems, it is clear that Eqs. (11.8), the equilibrium conditions, cannot be solved for more than three unknowns on any single free-body diagram.

In planar problems, the most general case of a system of forces that is solvable is one in which the three unknowns are the magnitudes of three forces. If all other forces are combined into a single force, we then have a four-force member. In order for such a member to be in static equilibrium, the resultant of any two of these forces must be equal, opposite, and collinear with the resultant of the other two forces.

The following example demonstrates the solution of a four-force member.

Example 11.3

A cam with a reciprocating roller follower is shown in Fig. 11.10a. The follower is held in contact with the cam by a spring pushing downward at C with a spring force of $F_C = 12$ N at this particular posture. Also, an external load $F_E = 35$ N acts on the follower at E in the direction indicated. For the posture shown, determine the follower pin force at A and the bearing reactions at B and D.

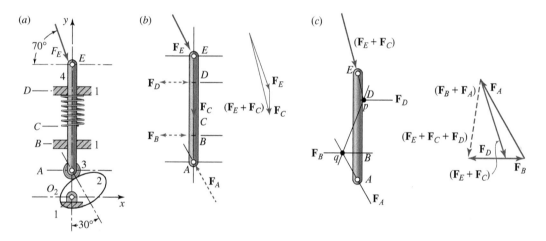

Fig. 11.10 (a) $R_{AO_2} = 15$ mm, $R_{BO_2} = 30$ mm, $R_{CA} = 25$ mm, $R_{DB} = 30$ mm, and $R_{EA} = 60$ mm; (b) free-body diagram of link 4, and force polygon; (c) free-body diagram of reduced link 4, and force polygon.

Assumptions

1. Gravitational forces are small in comparison with applied force **P** and can be ignored.
2. Frictional forces can be neglected.

Solution

The free-body diagram of the follower (link 4) is shown in Fig. 11.10b. Note that the follower is a five-force member ($\mathbf{F}_A$, $\mathbf{F}_B$, $\mathbf{F}_C$, $\mathbf{F}_D$, and $\mathbf{F}_E$). However, forces $\mathbf{F}_C$ and $\mathbf{F}_E$ are known, and their sum is obtained from the force polygon in Fig. 11.10b. The free-body diagram of link 4 is redrawn to indicate this in Fig. 11.10c. The resultant force ($\mathbf{F}_E + \mathbf{F}_C$) is shown with its point of application at

Example 11.3 (cont.)

E, where the original lines of action of $\mathbf{F}_E$ and $\mathbf{F}_C$ intersect, and with its line of action dictated by the direction of their vector sum. This placement of the point of application at E is not arbitrary, but is required. Since the original forces, $\mathbf{F}_E$ and $\mathbf{F}_C$, each have no moment about point E, their resultant must have no moment about this point. The result of combining these two forces is that the free-body diagram of link 4 in Fig. 11.10c is now reduced to a four-force member with one known force, the resultant $(\mathbf{F}_E + \mathbf{F}_C)$, and three unknown force magnitudes $(F_A, F_B,$ and $F_D)$.

In a similar manner, if the magnitude of $\mathbf{F}_D$ were known, it could be added to $(\mathbf{F}_E + \mathbf{F}_C)$ to produce a new resultant, $(\mathbf{F}_E + \mathbf{F}_C + \mathbf{F}_D)$, which would act through the intersection point p.

Now, consider the moment equation about point q, the intersection of $\mathbf{F}_A$ and $\mathbf{F}_B$. If we now write $\sum \mathbf{M}_q = \mathbf{0}$, we see that this equation can only be satisfied if the resultant, $(\mathbf{F}_E + \mathbf{F}_C + \mathbf{F}_D)$, has no moment about point q and thus has pq as its line of action. This is the basis for the solution shown in Fig. 11.10c. The direction of the line of action pq of the resultant, $(\mathbf{F}_E + \mathbf{F}_C + \mathbf{F}_D)$, is used in the force polygon to determine the magnitude of force $\mathbf{F}_D$; the solution gives $F_D = 29.2$ N. The force polygon is then completed by finding $\mathbf{F}_A$ and $\mathbf{F}_B$, whose lines of action are known. The solutions are $F_A = 51.8$ N and $F_B = 43.1$ N when rounded to three significant figures.

Note that this approach defines a general concept, useful in either the graphic or the analytic approach. When there are three unknown force magnitudes on a member, we choose a point such as q where the lines of action of two of the unknown forces intersect and write the moment equation about that point, that is, $\sum \mathbf{M}_q = \mathbf{0}$. This equation will have only one unknown remaining and can be solved directly. Only then should $\sum \mathbf{F} = \mathbf{0}$ be written, since the member has then been reduced to two unknowns.

11.9 Friction Force Models

Over the years, there has been much interest in the subjects of friction and wear, and many papers and books have been devoted to these subjects. It is not our purpose to explore the mechanics of friction here, but to present classical simplifications that have been used to analyze the performance of mechanical devices with friction. The results of any such analysis may not be theoretically ideal, but they do correspond closely to experimental measurements, so that reliable decisions can be made regarding a design and its operating performance.

Consider two bodies constrained to remain in contact with each other, such as the surfaces of block 3 and link 2 shown in Fig. 11.11a. A force, $\mathbf{F}_{43}$, may be exerted on block 3 by link 4, tending to cause block 3 to slide relative to link 2. Without the presence of friction between the interface surfaces of links 2 and 3, block 3 cannot transmit the component of force $\mathbf{F}_{32}^t$ tangential to the surfaces. Instead of transmitting this component of force onto link 2, block 3 slides in the direction of this unbalanced force component; equilibrium is not possible without friction unless the line of action of $\mathbf{F}_{32}$ and its reaction $\mathbf{F}_{23}$ are normal to the surface.

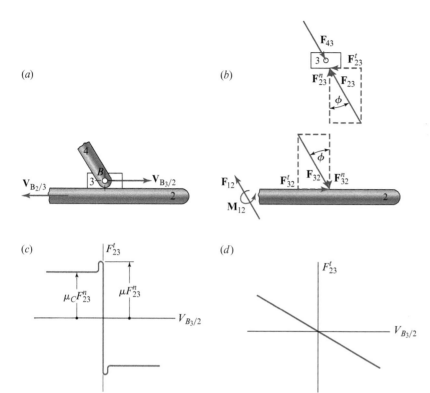

Fig. 11.11 (a) Physical system; (b) free-body diagrams; (c) static and Coulomb friction models; (d) viscous friction model.

With friction, however, a resisting force $\mathbf{F}_{23}^t$ is developed at the contact surface, as shown in the free-body diagrams of Fig. 11.11b. This friction force component $\mathbf{F}_{23}^t$ acts in addition to the usual constraint force component $\mathbf{F}_{23}^n$ across the surface of the sliding joint; together, these two force components form the total constraint force $\mathbf{F}_{23}$ that balances the constraint force $\mathbf{F}_{43}$ when block 3 is in static equilibrium. Of course, the reaction force components, $\mathbf{F}_{32}^n$ and $\mathbf{F}_{32}^t$ (or their sum), also act simultaneously onto link 2, as shown in the other free-body diagram of Fig. 11.11b. The force component $\mathbf{F}_{23}^t$ and its reaction $\mathbf{F}_{32}^t$ are called *friction forces*.

Depending on the materials of links 2 and 3, there is a limit to the size of the force component F_{23}^t that can be transmitted by friction while maintaining static equilibrium. This limit is expressed by the relationship

$$F_{23}^t \leq \mu F_{23}^n, \tag{11.9}$$

where μ, referred to as the *coefficient of static friction*, is a characteristic property of the contacting materials. Values of the coefficient μ have been determined experimentally for many material combinations and are found in many engineering handbooks; for example [2].

If constraint force $\mathbf{F}_{43}$ is tipped too much, so that its tangential component and therefore the friction force component F_{23}^t becomes too large to satisfy the inequality of Eq. (11.9), static equilibrium is not possible, and block 3 then slides relative to link 2 with an apparent velocity $\mathbf{V}_{B_3/2}$. When sliding takes place, the friction force takes on the value

$$F_{23}^t = \mu_c F_{23}^n, \tag{11.10}$$

where μ_c is the *coefficient of sliding friction.* This model for approximating sliding friction is called *Coulomb friction,* and we shall use this term to refer to the relationship of Eq. (11.10). The coefficient μ_c can also be found experimentally and for most material combinations is slightly less than μ, the coefficient of static friction.

To summarize what we have discussed so far, Fig. 11.11c shows a graph of the magnitude of the friction force, F_{23}^t, versus the apparent sliding velocity, $V_{B_3/2}$. Here it can be seen that when the apparent sliding velocity is zero, the friction force, F_{23}^t, can have any magnitude between μF_{23}^n and $-\mu F_{23}^n$. When the velocity is not zero, the magnitude of the friction force, F_{23}^t, drops slightly to the value $\mu_c F_{23}^n$ and has a sense opposing that of the apparent sliding motion, $V_{B_3/2}$.

Looking again at the total force $\mathbf{F}_{23}$ in Fig. 11.11b, we see that it is inclined at an angle to the surface normal and is equal and opposite to $\mathbf{F}_{43}$ whenever the system is in equilibrium. Thus, the angle ϕ is given by

$$\tan \phi = \frac{F_{23}^t}{F_{23}^n},$$

$$\tan \phi \leq \frac{\mu F_{23}^n}{F_{23}^n} = \mu,$$

$$\phi \leq \tan^{-1}\mu.$$

When $\mathbf{F}_{43}$ is tipped so that block 3 is on the verge of sliding, then

$$\phi = \tan^{-1}\mu. \tag{11.11}$$

This limiting value of the angle ϕ, called the *friction angle,* defines the maximum angle through which the force $\mathbf{F}_{23}$ can tip from the surface normal before equilibrium is not possible and sliding begins (commonly referred to as *impending motion*). Note that the limiting value of the friction angle does not depend on the magnitude of the force $\mathbf{F}_{23}$, but only on the coefficient of friction of the materials involved.

Note also that, in the above discussion, we have treated only the effects of friction between surfaces that permit sliding motion. We might ask how this can be extended to treat rotational motion, such as in a revolute joint. Since good low-friction rotational bearings are not expensive, and since they can be easily lubricated to reduce friction, this is usually not a problem. Extensions of the above ideas are known for rotational bearings [1]. However, since the results are usually not affected by more than a very small percentage, the additional difficulty does not often warrant their use.

Although the static friction or Coulomb friction models are often used and often do represent good approximations for friction forces in mechanical equipment, they are not the only models. Sometimes – for example, when representing a machine or its dynamic effects by its differential equation of motion – it is more convenient to analyze the machine's performance using a different approximation for friction forces, called *viscous friction* or *viscous damping.* As shown in the graph of Fig. 11.11d, this model assumes a linear relationship between the magnitude of the friction force and the sliding velocity. This viscous friction model is especially useful when the dynamic analysis of a machine leads to the use of one or more differential equations – for example, in vibration analysis (Sections 13.8–13.10). The nonlinear relationship of static and/or Coulomb friction, shown in Fig. 11.11c, leads to a nonlinear differential equation that is more difficult to solve.

Whether the friction effect is represented by the static, Coulomb, or viscous friction models, it is important to recognize the sense of the friction force. As a mnemonic device, the rule is often stated that "friction opposes motion," as shown by the free-body diagram of link 3 in Fig. 11.11b, where the sense of $\mathbf{F}_{23}^t$ is opposite to that of $\mathbf{V}_{B_3/2}$. This rule of thumb is not wrong if it is applied carefully, but it can be dangerous and even misleading. It will be noted in Fig. 11.11a that there are two motions that might be thought of, namely, $\mathbf{V}_{B_3/2}$ and $\mathbf{V}_{B_2/3}$; there are also two friction forces, $\mathbf{F}_{23}^t$ and $\mathbf{F}_{32}^t$. Careful examination of Fig. 11.11b indicates that $\mathbf{F}_{23}^t$ opposes the sense of $\mathbf{V}_{B_3/2}$, whereas $\mathbf{F}_{32}^t$ opposes the sense of $V_{B_2/3}$. In machine systems, particularly where both sides of a sliding joint are in motion, it is very important to understand *which* friction force "opposes" *which* motion.

Of course, in static force analysis, the assumption throughout is that forces are found under conditions of static equilibrium; therefore, there should be no motion. When we say that "friction opposes motion," we are speaking of *impending motion*; that is, we are assuming that the system is on the verge of moving and we are speaking of the movement that would begin if a small change in force were to perturb the equilibrium.

11.10 Force Analysis with Friction

When we begin a force analysis with friction, it is recommended to solve the problem first without friction. The purpose is to find the sense of each of the normal force components. We will demonstrate this procedure in the following two examples.

Example 11.4

Repeat Example 11.2 (Fig. 11.8a) with a coefficient of static friction between the slider and the ground of $\mu = 0.25$.

Assumptions

1. The impending motion of sliding block 4 is to the left.
2. Friction can be neglected in the revolute joints.

Solution

The normal force component $\mathbf{F}_{14}^n$ without friction was obtained in Example 11.2 and was determined (Fig. 11.8d) to act vertically upward. (If this previous example had not already been solved to provide the direction of $\mathbf{F}_{14}^n$, it may be necessary to solve this problem without friction to provide it.)

Since the impending motion of the slider is to the left, that is, $\mathbf{V}_{B_4/1}$ is about to slide to the left, then the friction force, $\mathbf{F}_{14}^t$, must act to the right. We can redraw the free-body diagram of link 4 (Fig. 11.8c) and include the friction force as shown in Fig. 11.12a. Because of static friction, the

Example 11.4 (cont.)

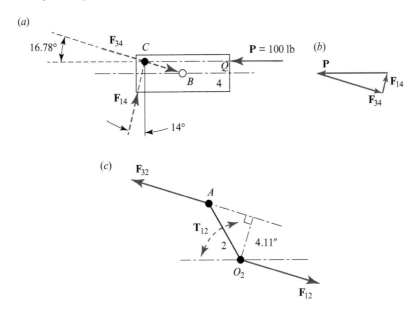

Fig. 11.12 Graphic solution.

line of action of $\mathbf{F}_{14}$ is shown tipped through the friction angle ϕ, which is calculated from Eq. (11.11):

$$\phi = \tan^{-1}0.25 = 14°.$$

In deciding the direction of tip of the angle ϕ, it is necessary to know both the sense of the normal force (upward) and the sense of the friction force (horizontal to the right) at the point of contact. This explains why the recommended procedure is to first solve the problem without friction.

Once the new line of action of the force $\mathbf{F}_{14}$ is known, the solution proceeds exactly as in Example 11.2. The graphic solution, with friction effects, is shown in Fig. 11.12b, where it is found that

$$F_{34} = 97 \text{ lb} \quad \text{and} \quad F_{14} = 29 \text{ lb}.$$

Note that the line of action of $\mathbf{F}_{34}$ has not changed, but that the magnitude is less than in the frictionless case. Therefore, less torque is required on link 2 to maintain the linkage in static equilibrium. We find that

$$T_{12} = (4.11 \text{ in})(97 \text{ lb}) = 399 \text{ in} \cdot \text{lb cw}. \qquad \textit{Ans.}$$

Note that, if the impending motion of sliding block 4 were to the right, then a greater torque would be required on link 2 to maintain the linkage in static equilibrium.

Example 11.4 (cont.)

Note also that we have $F_{14}^n = (29\text{ lb})\cos 14° = 28.14\text{ lb}$, whereas in Example 11.2 the force without friction was $F_{14} = 30.15\text{ lb}$. This indicates why it would be totally wrong to multiply the value of F_{14} without friction by μ to obtain F_{14}^t and then to include it by superposition; friction changes both the normal and the tangential components of all forces.

Finally, note that the point of application of force F_{14} is located 3.76 in to the left of point B. Fortunately, this is still within the width of block 4; therefore the solution is valid. However, if the width of block 4 were not 8 in, but only 6 in, then the point of application of force $\mathbf{F}_{14}$ would fall outside of the width of block 4. In this case, block 4 would not be in equilibrium, but would tip counterclockwise until prevented by contact with the ground at both the top-right and bottom-left corners. Block 4 would then become a four-force member with two unknown forces, $\mathbf{F}_{14}^R$ at the top-right corner and $\mathbf{F}_{14}^L$ at the bottom-left corner, each tipped by the friction angle, ϕ, but tipped in opposite directions since one normal component is downward and one is upward, and a new force polygon would be required for the solution for block 4 under these conditions.

Example 11.5

Repeat Example 11.3 (Fig. 11.10a) with a coefficient of static friction between links 1 and 4 at both sliding bearings B and D of $\mu = 0.15$. Determine the minimum force necessary at follower pin A to hold the system in static equilibrium.

Assumptions

Friction in all other joints is negligible.

Solution

First, we solve the problem without friction to find the sense of the normal force components, $\mathbf{F}_B^n$ and $\mathbf{F}_D^n$. This was performed in Example 11.3, where $\mathbf{F}_B$ and $\mathbf{F}_D$ were both found in Fig. 11.10c.

The next step is to determine the direction of the impending motion of the follower, link 4. The problem statement asks for the *minimum* force at A necessary for static equilibrium; this implies that if $\mathbf{F}_A$ were any smaller, the system would move downward. Thus, the impending motion of link 4 is downward for both velocities, $\mathbf{V}_{D_4/1}$ and $\mathbf{V}_{B_4/1}$. Therefore, the two friction forces from link 1 onto link 4 at B and D must both act upward.

Next, we redraw the free-body diagram of link 4 (Fig. 11.10c) to include the friction forces as shown in Fig. 11.13a. Here, the points of application of forces $\mathbf{F}_B$ and $\mathbf{F}_D$ are on the left and right side of link 4, respectively, at the points of contact with the bearings, not at the centerline of link 4. However, because of static friction, the lines of action of $\mathbf{F}_B$ and $\mathbf{F}_D$ are each shown tipped through the angle ϕ, which is calculated from Eq. (11.11):

$$\phi = \tan^{-1} 0.15 = 8.5°.$$

Example 11.5 (cont.)

Fig. 11.13 (a) Free-body diagram of link 4 with static friction; (b) force polygon.

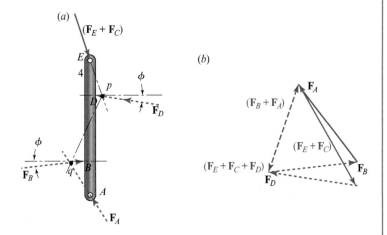

In deciding the direction of tip of the angles ϕ, it is necessary to know the sense of each friction force (upward) and the sense of each normal force (toward the right at B and toward the left at D). This explains why the solution without friction must be done first.

Once the new lines of action of the forces $\mathbf{F}_B$ and $\mathbf{F}_D$ are known, the solution proceeds exactly as in Example 11.3. The graphic solution, with friction effects, is shown in Fig. 11.13b, where it is found that

$$F_B = 39.1 \text{ N}, \quad F_D = 34.7 \text{ N}, \quad \text{and} \quad F_A = 39.1 \text{ N}. \qquad Ans.$$

The normal components of the two forces at B and D in this example are $F_B^n = 37.7$ N and $F_D^n = 33.6$ N. Note that these values are quite different from $F_B = 43.1$ N and $F_D = 29.2$ N, determined without friction in Example 11.3. This should be a warning, whether working graphically or analytically, that it is incorrect to simply multiply the frictionless normal forces by the coefficient of friction to find the friction forces and then to add these to the frictionless solution. All forces may (and usually do) change magnitude when friction is included, and the problem must be completely reworked from the beginning with friction included. The effects of static or Coulomb friction *cannot* be added by superposition.

We note, also, that if the problem statement had asked for the maximum force at A, the impending motion of link 4 would have been upward and the friction forces downward on this link. This would have reversed the tilt of the two lines of action for both $\mathbf{F}_B$ and $\mathbf{F}_D$ and would have totally changed the final results. In practice, if the actual value of the force $\mathbf{F}_A$ is between these minimum and maximum values, equilibrium is maintained, and the values of other constraint forces is between the two extreme values found by this type of analysis. Also, if the value of $\mathbf{F}_A$ is slowly increased from the minimum to the maximum value, the follower remains in equilibrium until the maximum value is reached, and then begins to move; this discontinuous action is sometimes referred to as "stiction."

11.11 Spur- and Helical-Gear Force Analysis

Gear systems are typically designed to operate at constant speed, and accelerations do not usually occur, except at startup. Therefore, it is appropriate to analyze their forces in this chapter under these conditions. Recall that, when there are no accelerations, this implies that velocities may exist, but that they remain constant; this is then referred to as dynamic equilibrium (Section 11.6). Figure 11.14a shows a pinion with center O_2 rotating clockwise at a constant speed of ω_2 and driving a gear with center O_3 at a speed of ω_3. As discussed in Chapter 7, the reactions between the teeth occur along the pressure line AB, tipped by the pressure angle ϕ from the common tangent to the pitch circles. Free-body diagrams of the pinion and the gear are shown in Fig. 11.14b. The action of the pinion on the gear is indicated by the force $\mathbf{F}_{23}$ acting at the pitch point along the pressure line.[4] Since the gear is supported by its shaft, then from $\sum\mathbf{F} = \mathbf{0}$, an equal and opposite force $\mathbf{F}_{13}$ must act at the centerline of the shaft. A similar analysis of the pinion indicates that the same observations are true. In each

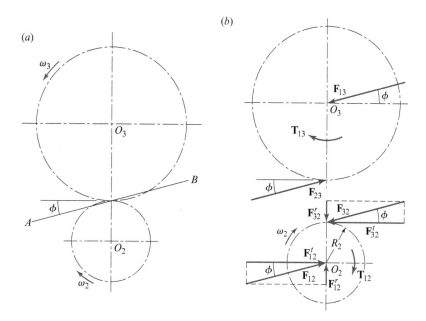

Fig. 11.14 Forces on spur gears.

[4] It is true that treating the force $\mathbf{F}_{23}$ in this manner ignores the possible effects of friction forces between the meshing gear teeth. Justification for this comes in four forms: (1) The actual point of contact is continually varying during the meshing cycle but always remains near the pitch point, which is the instantaneous center of velocity; thus, the relative motion between the teeth is close to true rolling motion, with only a very small amount of slip. (2) The fact that the friction forces are continually changing in both magnitude and direction throughout the meshing cycle, and that the total force is often shared by more than one tooth in contact, makes a more exact analysis impractical. (3) The machined surfaces of the teeth and the fact that they are usually well-lubricated produces a very small coefficient of friction. (4) Experimental data indicate that gear efficiencies are usually very high, often approaching 99%, demonstrating that any errors produced by ignoring friction are quite small.

case, the forces are equal in magnitude, opposite in direction, parallel, and in the same plane. On either gear, therefore, they form a couple.

Note that the free-body diagram of the pinion indicates the forces resolved into components. Here we employ the superscripts r and t to indicate the radial and tangential directions with respect to the pitch circle. It is expedient to use the same superscripts for the components of the force, $\mathbf{F}_{12}$, that the shaft exerts on the pinion. The moment of the couple formed by $\mathbf{F}_{32}^t$ and $\mathbf{F}_{12}^t$ is in equilibrium with the torque, $\mathbf{T}_{12}$, that must be applied by the shaft to drive the gearset. When the pitch radius of the pinion is designated R_2, then $\sum \mathbf{M}_{O_2} = \mathbf{0}$ indicates that the magnitude of the torque is

$$T_{12} = R_2 F_{32}^t. \tag{11.12}$$

Note that the radial force component $\mathbf{F}_{32}^r$ serves no purpose as far as the transmission of power is concerned. For this reason, the tangential component $\mathbf{F}_{32}^t$ is frequently called the *transmitted* force.

In applications involving gears, the power transmitted and the shaft speeds are often specified. Remembering that power is the product of force times velocity, or torque times angular velocity, we can find the relation between power and the transmitted force. Using the symbol P to denote power, we obtain

$$P = T_{12}\omega_2 = R_2 F_{32}^t \omega_2. \tag{11.13a}$$

Therefore, the transmitted force can be written as

$$F_{32}^t = \frac{P}{R_2\omega_2}. \tag{11.13b}$$

In application of these formulae, it is often necessary to remember that the US customary unit used for power is horsepower, abbreviated hp, where $1\ \text{hp} = 33\,000\ \text{ft}\cdot\text{lb}/\text{min}$, and in SI units power is measured in watts (or in kilowatts), abbreviated W (or kW), where $1\ \text{W} = 1\ \text{N}\cdot\text{m/s}$.

Once the transmitted force is known, the following relations for spur gears are evident from Fig. 11.14b:

$$F_{32}^r = F_{32}^t \tan\phi \quad \text{and} \quad F_{32} = \frac{F_{32}^t}{\cos\phi}.$$

In the treatment of forces on helical gears, it is convenient to determine the axial force, work with it independently, and treat the remaining force components the same as for spur gears. Figure 11.15 is a drawing of a helical gear with a portion of the face removed to show the tooth contact force and its components acting at the pitch point. The gear is imagined to be driven clockwise under load. The driving gear has been removed from the figure and its effect replaced by the forces shown acting on the teeth.

The resultant force, $\mathbf{F}$, is shown divided into three components, $\mathbf{F}^a$, $\mathbf{F}^r$, and $\mathbf{F}^t$, which are the axial, radial, and tangential components, respectively. The tangential force component is the transmitted force and is the force that is effective in transmitting torque. When the transverse pressure angle is designated ϕ^t and the helix angle ψ, the following relations are evident from Fig. 11.15:

$$\mathbf{F} = \mathbf{F}^a + \mathbf{F}^r + \mathbf{F}^t, \tag{11.14}$$

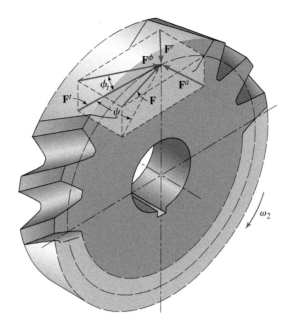

Fig. 11.15 Force components on a helical gear at the tooth contact point.

$$F^a = F^t \tan \psi, \tag{11.15}$$

$$F^r = F^t \tan \phi^t. \tag{11.16}$$

It is also expedient to make use of the resultant of $\mathbf{F}^r$ and $\mathbf{F}^t$. We shall designate this force $\mathbf{F}^\phi$, defined by the equation

$$\mathbf{F}^\phi = \mathbf{F}^r + \mathbf{F}^t. \tag{11.17}$$

Example 11.6

A gear train is composed of three helical gears with shaft centers in line, as shown in Fig. 11.16a. The driver is a right-hand helical gear having a pitch radius of 2 in, a transverse pressure angle of 20°, and a helix angle of 30°. An idler gear in the train has the teeth cut left-handed and has a pitch radius of 3.25 in. The idler transmits no power to its shaft. The driven gear in the train has the teeth cut right-handed and has a pitch radius of 2.50 in. If the transmitted force is 600 lb, find the shaft forces acting on each gear.

Assumptions

Gravitational and friction forces can be neglected.

Solution

First, we consider only the axial forces, as previously suggested. For each mesh, the axial component of the reaction, from Eq. (11.15), is

$$F^a = F^t \tan \psi = (600 \text{ lb}) \tan 30° = 346 \text{ lb}.$$

Example 11.6 (cont.)

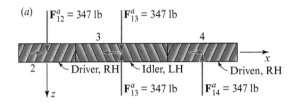

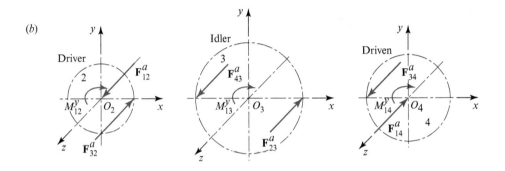

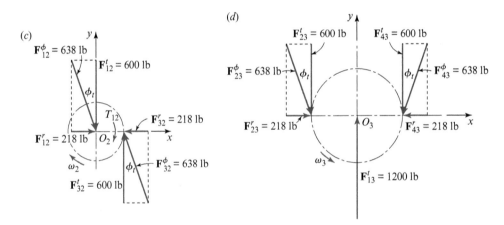

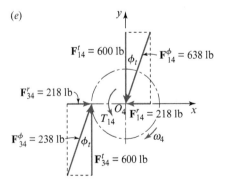

Fig. 11.16 (*a,b*) Axial forces; (*c*) free-body diagram of driver gear 2; (*d*) free-body diagram of idler gear 3; (*e*) free-body diagram of driven gear 4.

Example 11.6 (cont.)

Figure 11.16*a* is a top view of the three gears, looking down on the plane formed by the three axes of rotation. For each gear, rotation is considered to be about the z axis for this problem. In Fig. 11.16*b*, free-body diagrams of each of the three gears are drawn in projection, and the three coordinate axes are shown. As indicated, the idler exerts an axial force $\mathbf{F}_{32}^a$ on the driver. This is resisted by the axial shaft force $\mathbf{F}_{12}^a$. Forces $\mathbf{F}_{12}^a$ and $\mathbf{F}_{32}^a$ form a couple that is resisted by the moment $\mathbf{M}_{12}^y$. Note that this moment is negative (clockwise) about the y axis, and the magnitude of this moment is

$$M_{12}^y = -R_2F_{12}^a = -(2.00 \text{ in})(346 \text{ lb}) = -692 \text{ in} \cdot \text{lb}.$$

Turning our attention next to the idler, we see from Figs. 11.16*a* and 11.16*b* that the net axial force on the shaft of the idler is zero. The axial component of the force from the driver onto the idler is $\mathbf{F}_{23}^a$ and that from the driven gear onto the idler is $\mathbf{F}_{43}^a$. These two force components are equal and opposite and form a couple that is resisted by the moment on the shaft of the idler of magnitude

$$M_{13}^y = -2R_3F_{23}^a = -2(3.25 \text{ in})(346 \text{ lb}) = -2\,249 \text{ in} \cdot \text{lb}.$$

The driven gear has the axial force component $\mathbf{F}_{34}^a$ acting along its pitch line caused by the helix angle of the idler. This is resisted by the axial shaft reaction, $\mathbf{F}_{14}^a$. As indicated, these forces form a couple that is resisted by the moment $\mathbf{M}_{14}^y$. The magnitude of this moment is

$$M_{14}^y = -R_4F_{34}^a = -(2.50 \text{ in})(346 \text{ lb}) = -865 \text{ in} \cdot \text{lb}.$$

It is emphasized that the three resisting moments, $\mathbf{M}_{12}^y, \mathbf{M}_{13}^y$, and $\mathbf{M}_{14}^y$, are caused solely by the axial components of the reaction forces between the gear teeth. These produce static bearing reactions and have no effect on the amount of power transmitted.

Now that all of the reactions due to the axial components have been determined, we turn our attention to the remaining force components and examine their effects as if they were operating independently of the axial forces. Free-body diagrams showing the force components in the plane of rotation for the driver, idler, and driven gears are shown in Figs. 11.16*c*, 11.16*d*, and 11.16*e*, respectively. These force components can be obtained graphically as shown, or by applying Eqs. (11.12)–(11.16). Equation (11.12) gives the input torque,

$$T_{12} = R_2F_{32}^t = (2.00 \text{ in})(600 \text{ lb}) = 1\,200 \text{ in} \cdot \text{lb} = 100 \text{ ft} \cdot \text{lb},$$

and the output torque,

$$T_{14} = R_4F_{34}^t = (2.50 \text{ in})(600 \text{ lb}) = 1\,500 \text{ in} \cdot \text{lb} = 125 \text{ ft} \cdot \text{lb}.$$

It is not necessary to combine the components to determine the resultant forces, since the components are exactly those that are desired to proceed with machine design.

Example 11.6 (cont.)

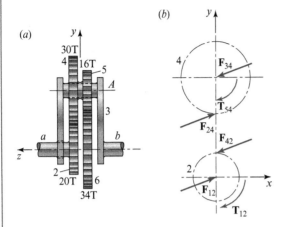

(a)

(b)

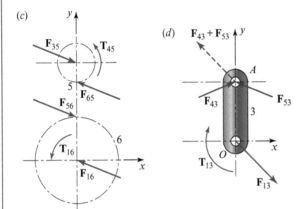

(c)

(d)

Fig. 11.17 (a) Planetary gear train with tooth numbers; (b) free-body diagrams of gears 2 and 4; (c) free-body diagrams of gears 5 and 6; (d) free-body diagram of planet carrier arm 3.

Example 11.7

For the reverted planetary gear train shown in Fig. 11.17a, input shaft 2 is driven by a torque $T_{12} = -100\hat{k}$ in·lb. Shafts 2 and 3 are not connected but rotate about the same axis. Gear 6 is fixed to frame 1 (not shown). All gears have a diametral pitch of 10 teeth per inch and a pressure angle of $20°$. Perform a force analysis on the parts of the train and compute the magnitude and direction of the output torque delivered by shaft 3.

Assumptions

The forces act in a single plane, and gravitational forces, friction forces, and centrifugal forces on the planet gears can be neglected.

Example 11.7 (cont.)

Solution

The pitch radii of the gears are $R_2 = (20 \text{ teeth})/(2 \cdot 10 \text{ teeth/in}) = 1.00 \text{ in}$, $R_4 = 1.50 \text{ in}$, $R_5 = 0.80 \text{ in}$, and $R_6 = 1.70 \text{ in}$. The distance between centers of meshing gear pairs is $(R_2 + R_4) = (1.00 \text{ in} + 1.50 \text{ in}) = 2.50 \text{ in} = (R_5 + R_6)$. The transmitted force is $F_{42}^t = T_{12}/R_2 = (100 \text{ in} \cdot \text{lb})/(1 \text{ in}) = 100 \text{ lb}$. Therefore, $F_{42} = F_{42}^t / \cos \phi = (100 \text{ lb})/ \cos 20° = 106 \text{ lb}$. The free-body diagram of gear 2 is shown in Fig. 11.17b. In vector form, the results are

$$\mathbf{F}_{12} = -\mathbf{F}_{42} = 106 \text{ lb} \angle 20°.$$

Figure 11.17b also shows the free-body diagram of gear 4. The forces are

$$\mathbf{F}_{24} = -\mathbf{F}_{34} = 106 \text{ lb} \angle 20°,$$

where $\mathbf{F}_{34}$ is the force of planet arm 3 onto gear 4. Gears 4 and 5 are connected to each other but turn freely on the planet arm shaft. Thus, $\mathbf{T}_{54}$ is the torque exerted by gear 5 onto gear 4. This torque is $T_{54} = (R_4)F_{24}^t = (1.50 \text{ in})(100 \text{ lb}) = 150 \text{ in} \cdot \text{lb cw}$.

Turning next to the free-body diagram of gear 5 in Fig. 11.17c, we first find $F_{65}^t = T_{45}/R_5 = (150 \text{ in} \cdot \text{lb})/(0.800 \text{ in}) = 188 \text{ lb}$. Therefore, $F_{65} = (188 \text{ lb})/ \cos 20° = 200 \text{ lb}$. In vector form, the results for gear 5 are summarized as

$$\mathbf{F}_{65} = -\mathbf{F}_{35} = 200 \text{ lb} \angle 160°, \qquad \mathbf{T}_{45} = 150 \hat{\mathbf{k}} \text{ in} \cdot \text{lb}.$$

For gear 6, shown in Fig. 11.17c, we have

$$\mathbf{F}_{16} = -\mathbf{F}_{56} = 200 \text{ lb} \angle 160°,$$

$$\mathbf{T}_{16} = R_6 F_{56}^t \hat{\mathbf{k}} = (1.70 \text{ in})(200 \cos 20° \text{lb})\hat{\mathbf{k}} = 319 \hat{\mathbf{k}} \text{ in} \cdot \text{lb}.$$

Note that $\mathbf{F}_{16}$ and $\mathbf{T}_{16}$ are the force and torque, respectively, exerted by the frame on gear 6.

The free-body diagram of arm 3 is shown in Fig. 11.17d. As noted earlier, the forces are assumed to act in a single plane. The two forces $\mathbf{F}_{43}$ and $\mathbf{F}_{53}$ are

$$\mathbf{F}_{43} = -\mathbf{F}_{34} = 106 \text{ lb} \angle 20°, \qquad \mathbf{F}_{53} = -\mathbf{F}_{35} = 200 \text{ lb} \angle 160°,$$

and can be summed to

$$\mathbf{F}_{43} + \mathbf{F}_{53} = 137 \text{ lb} \angle 130.2°.$$

We now find the output shaft reaction to be

$$\mathbf{F}_{13} = -(\mathbf{F}_{43} + \mathbf{F}_{53}) = 137 \text{ lb} \angle -49.8°.$$

Using $\mathbf{R}_{AO} = 2.50 \hat{\mathbf{j}} \text{ in}$ and the equation

$$\sum \mathbf{M}_O = \mathbf{T}_{13} + \mathbf{R}_{AO} \times (\mathbf{F}_{43} + \mathbf{F}_{53}) = \mathbf{0},$$

we find $\mathbf{T}_{13} = -221 \hat{\mathbf{k}} \text{ in} \cdot \text{lb}$. Therefore, the output shaft torque is

$$\mathbf{T}_{13} = 221 \hat{\mathbf{k}} \text{ in} \cdot \text{lb}. \qquad\qquad Ans.$$

Fig. 11.18 Force components
on a straight-tooth bevel gear
at the tooth contact point.

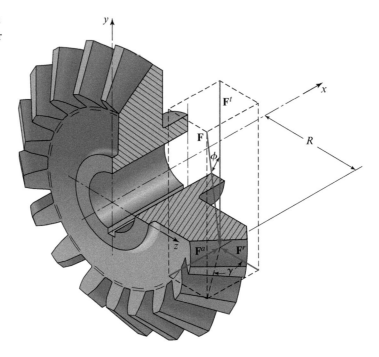

Fig. 11.18 Force components on a straight-tooth bevel gear at the tooth contact point.

11.12 Straight-Tooth Bevel-Gear Force Analysis

In determining the forces on bevel gear teeth, it is customary to use the forces that would occur at the mid-thickness of the tooth on the pitch cone. The resultant tangential force probably occurs somewhere between the midpoint and the large end of the tooth, but there will be only a small error in making this approximation. The tangential or transmitted force is then given by

$$F^t = \frac{T}{R}, \tag{11.18}$$

where R is the mid-radius of the pitch cone, as shown in Fig. 11.18, and T is the shaft torque.

Figure 11.18 also shows all components of the resultant force acting at the midpoint of the tooth. The following relationships can be derived by inspection of Fig. 11.18:

$$\mathbf{F} = \mathbf{F}^a + \mathbf{F}^r + \mathbf{F}^t, \tag{11.19}$$

$$F^r = F^t \tan\phi \cos\gamma, \tag{11.20}$$

$$F^a = F^t \tan\phi \sin\gamma. \tag{11.21}$$

Note that the axial force $\mathbf{F}^a$ results in a couple on the shaft.

Example 11.8

The bevel pinion shown in Fig. 11.19 rotates at 600 rev/min in the direction indicated and transmits 5 hp to the gear. The mounting distances are shown, together with the locations of the bearings on each shaft. Bearings A and C are capable of taking both radial and axial loads, whereas bearings B and D are designed to receive only radial loads. The teeth of the gears have a $20°$ pressure angle. Find the components of the forces that the bearings exert on the shafts in the $x, y,$ and z directions.

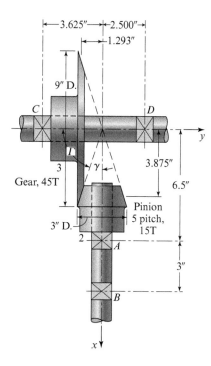

Fig. 11.19 Bevel gearset and bearing locations.

Solution

The pitch angles for the pinion and gear are

$$\gamma = \tan^{-1}\left(\frac{1.5\ \text{in}}{4.5\ \text{in}}\right) = 18.4°,$$

$$\Gamma = \tan^{-1}\left(\frac{4.5\ \text{in}}{1.5\ \text{in}}\right) = 71.6°.$$

The radii to the midpoints of the teeth are shown on the figure and are $R_2 = 1.293$ in and $R_3 = 3.875$ in for the pinion and gear, respectively.

Let us now determine the forces acting on the pinion. Using Eq. (11.13b), we find the transmitted force to be

Example 11.8 (cont.)

$$F_{32}^t = \frac{P}{R_2 \omega_2} = \frac{(5 \text{ hp})(33\,000 \text{ ft} \cdot \text{lb}/\min/\text{hp})(12 \text{ in}/\text{ft})}{(1.293 \text{ in})(600 \text{ rev}/\min)(2\pi \text{ rad}/\text{rev})} = 406 \text{ lb}.$$

This force acts in the negative z direction – that is, into the plane of Fig. 11.19. The radial and axial components of $\mathbf{F}_{32}$ are obtained from Eqs. (11.20) and (11.21),

$$F_{32}^r = 406 \text{ lb} \tan 20° \cos 18.4° = 140 \text{ lb},$$
$$F_{32}^a = 406 \text{ lb} \tan 20° \sin 18.4° = 46.6 \text{ lb},$$

where $\mathbf{F}_{32}^r$ acts in the positive y direction and $\mathbf{F}_{32}^a$ in the positive x direction. Thus,

$$\mathbf{F}_{32} = 46.6\hat{\mathbf{i}} + 140\hat{\mathbf{j}} - 406\hat{\mathbf{k}} \text{ lb}.$$

The torque applied to the pinion shaft must be

$$\mathbf{T}_{12} = -\mathbf{R}_2 \times \mathbf{F}_{32}^t = -(-1.293\hat{\mathbf{j}} \text{ in}) \times (-406\hat{\mathbf{k}} \text{ lb}) = -525\hat{\mathbf{i}} \text{ in} \cdot \text{lb}.$$

A free-body diagram of the pinion and shaft is shown schematically in Fig. 11.20a. The dimensions, the torque $\mathbf{T}_{12}$, and the force $\mathbf{F}_{32}$ are known, and the problem is to determine the bearing reactions $\mathbf{F}_A$ and $\mathbf{F}_B$. To determine $\mathbf{F}_B$, we sum moments about A:

$$\sum \mathbf{M}_A = \mathbf{T}_{12} + \mathbf{R}_{BA} \times \mathbf{F}_B + \mathbf{R}_{PA} \times \mathbf{F}_{32} = 0, \tag{1}$$

where the two position-difference vectors are

$$\mathbf{R}_{BA} = 3.0\hat{\mathbf{i}} \text{ in} \quad \text{and} \quad \mathbf{R}_{PA} = -2.625\hat{\mathbf{i}} - 1.293\hat{\mathbf{j}} \text{ in}.$$

Therefore, the second and third terms of Eq. (1), respectively, are

$$\mathbf{R}_{BA} \times \mathbf{F}_B = (3.0 \text{ in } \hat{\mathbf{i}}) \times (F_B^y \hat{\mathbf{j}} + F_B^z \hat{\mathbf{k}})$$
$$= -(3.0 \text{ in}) F_B^z \hat{\mathbf{j}} + (3.0 \text{ in}) F_B^y \hat{\mathbf{k}}, \tag{2}$$

$$\mathbf{R}_{PA} \times \mathbf{F}_{32} = (-2.625\hat{\mathbf{i}} - 1.293\hat{\mathbf{j}} \text{ in}) \times (46.6\hat{\mathbf{i}} + 140\hat{\mathbf{j}} - 406\hat{\mathbf{k}} \text{ lb})$$
$$= 525\hat{\mathbf{i}} - 1\,066\hat{\mathbf{j}} - 307\hat{\mathbf{k}} \text{ in} \cdot \text{lb}. \tag{3}$$

Substituting the value of $\mathbf{T}_{12}$ and Eqs. (2) and (3) into Eq. (1) and solving, the bearing reaction at B is

$$\mathbf{F}_B = 102\hat{\mathbf{j}} - 355\hat{\mathbf{k}} \text{ lb}. \qquad\qquad Ans.$$

The magnitude of the bearing reaction at B is 370 lb.

Example 11.8 (cont.)

Next, to determine the bearing reaction at A, we write

$$\sum \mathbf{F} = \mathbf{F}_{32} + \mathbf{F}_A + \mathbf{F}_B = \mathbf{0}.$$

Then, solving this equation, the bearing reaction at A is

$$\mathbf{F}_A = -46.6\hat{\mathbf{i}} - 242\hat{\mathbf{j}} + 761\hat{\mathbf{k}} \text{ lb.} \qquad\qquad \textit{Ans.}$$

The magnitude of the bearing reaction at A is 798 lb. The results are shown in Fig. 11.20b. A similar procedure is used for the gear shaft. The results are displayed in Fig. 11.20c.

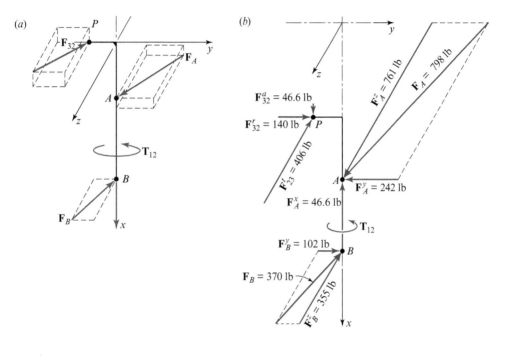

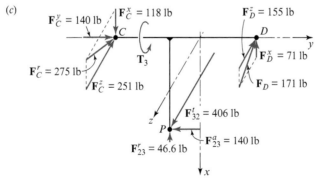

Fig. 11.20 (*a*) Free-body diagrams of pinion and input shaft; (*b*) force results and component values on pinion and input shaft; (*c*) free-body diagrams of gear and output shaft with force results and component values.

11.13 Method of Virtual Work

So far in this chapter, we have learned to analyze problems involving the static equilibrium of mechanical systems by the application of Newton's laws. Another fundamentally different approach to force-analysis problems is based on the principle of virtual work, first proposed by the Swiss mathematician J. Bernoulli in the eighteenth century.

The method is based on an energy balance of the system that requires that the net change in internal energy during a small displacement must be equal to the difference between the work input to the system and the work output, including the work done against friction, if any. Thus, for a system of rigid bodies in equilibrium under a system of applied forces, if given an arbitrary small displacement from equilibrium, the net change in the internal energy, denoted here by dU, is equal to the work dW done on the system:

$$dU = dW. \qquad (11.22)$$

Work and change in internal energy are positive when work is done on the system, giving it increased internal energy, and negative when energy is lost from the system, such as when it is dissipated through friction. If the system has no friction or other dissipation losses, energy is conserved, and the net change in internal energy during a small displacement from equilibrium is equal to the work done by any external forces and/or moments.

Of course, such a method requires that we know how to calculate the work done by each force during the small *virtual displacement* chosen. If some force $\mathbf{F}$ acts at a point of application Q that undergoes a small displacement $d\mathbf{R}_Q$, then the work done by this force on the system is given by

$$dW = \mathbf{F} \cdot d\mathbf{R}_Q, \qquad (11.23)$$

where dW is a scalar value, having units of work or energy, and is positive for work done onto the system and negative for work output from the system.

The displacement considered is called a *virtual* displacement, since it need not be one that truly happens for the physical machine. It need only be a small displacement that is hypothetically possible and consistent with the constraints imposed on the system. The small displacement relationships of the system can be determined through the principles of kinematics covered in Part I of this text, and the input-to-output force relationships of the machine can therefore be determined. This will become clear in the following example.

Example 11.9

Using the method of virtual work, determine the input crank torque, $\mathbf{T}_{12}$, required for equilibrium of the four-bar linkage of Example 11.1 subjected to the external load $\mathbf{P}$ as shown in Fig. 11.5a.

Assumptions

Friction effects and the weights of the links can be neglected.

Solution

From Example 11.1, we recall that the position-difference vector from O_4 to Q and the applied force at Q, respectively, are

Example 11.9 (cont.)

$$\mathbf{R}_{QO_4} = 1.84\hat{\mathbf{i}} + 4.65\hat{\mathbf{j}} \text{ in,} \tag{1}$$

$$\mathbf{P} = 120 \text{ lb}\angle 220° = -91.93\hat{\mathbf{i}} - 77.13\hat{\mathbf{j}} \text{ lb.} \tag{2}$$

Figure 11.21 presents a scale drawing of the linkage at the posture in question. If a small virtual displacement $d\theta_2$ is considered for the input crank 2, this results in a small displacement $d\theta_4$ of the output crank 4, along with a small displacement $d\mathbf{R}_{QO_4}$ of point Q, as shown.

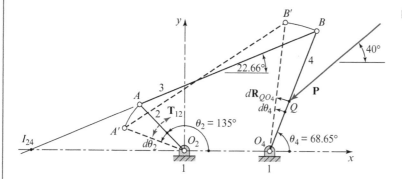

Fig. 11.21 Virtual work solution.

Assuming that these virtual displacements are small and happen in a short time increment, dt, the velocity difference equation, Eq. (3.3), can be multiplied by dt to yield the relationship between $d\mathbf{R}_{QO_4}$ and $d\theta_4$, as follows:

$$\begin{aligned} d\mathbf{R}_{QO_4} &= d\theta_4\hat{\mathbf{k}} \times \mathbf{R}_{QO_4} = (d\theta_4\hat{\mathbf{k}} \text{ rad}) \times (1.84\hat{\mathbf{i}} + 4.65\hat{\mathbf{j}} \text{ in}) \\ &= (-4.65\hat{\mathbf{i}} + 1.84\hat{\mathbf{j}} \text{ in})d\theta_4. \end{aligned} \tag{3}$$

Since the input torque, $\mathbf{T}_{12}$, and the applied force, $\mathbf{P}$, are the only forces doing work during the chosen displacement, Eq. (11.22) can be written for this problem as

$$dU = T_{12}\hat{\mathbf{k}} \cdot d\theta_2\hat{\mathbf{k}} + \mathbf{P} \cdot d\mathbf{R}_{QO_4} = 0.$$

Substituting Eqs. (2) and (3) into this equation gives

$$T_{12}d\theta_2 + (-91.93\hat{\mathbf{i}} - 77.13\hat{\mathbf{j}} \text{ lb}) \cdot (-4.65\hat{\mathbf{i}} + 1.84\hat{\mathbf{j}} \text{ in})d\theta_4 = 0.$$

Then, dividing by $d\theta_2$, rearranging, and recognizing the first-order kinematic coefficient, we find

$$T_{12} = (-285.6 \text{ in} \cdot \text{lb})\frac{d\theta_4}{d\theta_2} = -285.6\theta_4' \text{ in} \cdot \text{lb.} \tag{4}$$

By dividing the numerator and denominator of this angular displacement ratio by dt, we recognize that this is equal to the angular-velocity ratio ω_4/ω_2, which can be determined from

Example 11.9 (cont.)

the angular-velocity ratio theorem of Eq. (3.28).[5] Using the location of the instant center I_{24} and measurements from Fig. 11.21, we find

$$\theta_4' = \frac{d\theta_4}{d\theta_2} = \frac{\omega_4}{\omega_2} = \frac{R_{I_{24}I_{12}}}{R_{I_{24}I_{14}}} = \frac{14.65 \text{ in}}{22.65 \text{ in}} = 0.646\,8 \text{ rad/rad}.$$

Finally, substituting this result into Eq. (4) gives the input torque,

$$\mathbf{T}_{12} = \left(-285.6\hat{\mathbf{k}} \text{ in} \cdot \text{lb}\right)\left(0.646\,8 \text{ rad/rad}\right) = -184.7\hat{\mathbf{k}} \text{ in} \cdot \text{lb}. \qquad \textit{Ans.}$$

Note that the solution is in good agreement with the analytic solution in Example 11.1.

One primary advantage of the method of virtual work for force analysis over the other methods demonstrated above comes in problems where input-to-output force relationships are sought. Note that the constraint forces are not required in this solution technique, since both the action and reaction forces are internal to the system; both move through identical displacements, and thus their virtual work contributions cancel each other. This is not true for friction forces where the displacements are different for the action and reaction force components, with the work difference representing the energy dissipated. Otherwise, constraint forces need not be considered, since they cause no net virtual work.

Summary of Force-Analysis Procedures

The following is a summary of the procedures for the methods of static force analysis that have been presented up to this point in the chapter.

1. We start by drawing complete free-body diagrams of each member in the mechanism. It is good practice to combine all known forces on a free-body diagram into a single vector. We classify each member; that is, we identify two-force members, three-force members, four-force members, and so on. We then commence with the member with the lowest number of forces. That is, we draw the two-force members first, then we draw the three-force members, and, finally, we draw any four-force members. If a member is acted upon by more than four forces, then either (a) it can be reduced to one of the above cases, or (b) it has more than three unknowns and is not yet solvable.
2. Using the definitions of two-force, three-force, and four-force members, we apply the following rules:
   a. For a two-force member, the two forces must be equal, opposite, and collinear. Note that for a link with two forces and a moment, the forces are equal, opposite, and parallel.

---

[5] This technique of treating small displacement ratios as velocity ratios or first-order kinematic coefficients can often be extremely helpful because, as we remember from Section 3.8, velocity relationships lead to linear equations rather than the nonlinear relations implied by position or displacement. Velocity polygons and instant-center methods are also helpful.

b. For a three-force member, the three forces must intersect at a single point.

c. For a four-force member, the resultant of any two forces must be equal, opposite, and collinear with the resultant of the other two forces.

3. We draw force polygons for three- and four-force members, clearly stating the scale of each force polygon. Note that, in general, it is less confusing if force polygons are not superimposed on the mechanism diagram. To draw a force polygon for a three-force member, we must have only two unknown pieces of information, both of which are force magnitudes. For a four-force member, we may have three unknown pieces of information, all of which are force magnitudes.

4. If desirable, multiple links can be grouped together and treated as a single free-body diagram.

11.14 Introduction to Buckling

Throughout previous chapters and sections of this text, we have assumed that all bodies are completely rigid and cannot deform. In fact, we noted in Section 1.4 that this assumption of rigidity is mandatory to allow the separation of the study of kinematics from the treatment of dynamics. But we also noted that this assumption of rigidity prevents the study of strain or deformation and therefore of stress and strength.

Sections 11.14–11.18 deviate from the assumption of rigidity. In these sections we consider a nonrigid, large deformation effect called *buckling*, which is the sudden failure of a long, slender member subjected to compressive loading. When a member buckles, it experiences severe bending and loses its strength. This loss in strength can be very dramatic, and for this reason a buckling failure in a structure can occur without warning and is often catastrophic. Figure 11.22 shows the buckling collapse of steel beams in the Cortland St. subway station in New York. Here we can see that buckling is a very important problem. It is also a fascinating topic in mechanical design and is the subject of the following four sections.

11.15 Euler Column Formula

Consider a long, slender member of length L with fixed–free end conditions, as shown in Fig. 11.23. Assume that the member is subject to the central compressive load P shown in the figure. The first problem is to derive an equation for the critical value of the load, P, that is, the load that will cause the member to become unstable. This load is referred to as the *critical load*.

For small deflections and for homogeneous and isotropic materials, the curvature of the member caused by the compressive load P can be written as

$$\kappa = \frac{d^2y}{dx^2} = \frac{M}{EI}, \tag{11.24}$$

where d^2y/dx^2 is the second derivative of the transverse deflection, M is the bending moment at an arbitrary location x along the member, E is Young's modulus, and I is the second moment

Fig. 11.22 A catastrophic failure due to buckling in the Cortland St. subway station in New York, NY (circa September 2001).

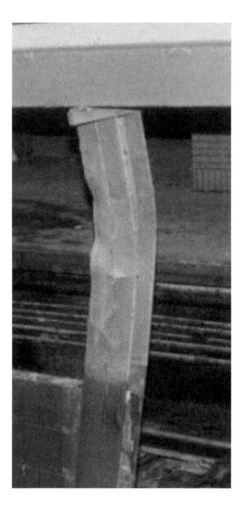

Fig. 11.23 A long, slender member subjected to a central compressive load, P.

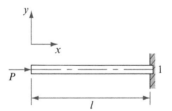

of area (sometimes called the *area moment of inertia*) of the member cross-section at location x. The bending moment can be written as

$$M = -Py, \tag{11.25}$$

where y is the transverse deflection of the member at location x.

Young's modulus E is the modulus of elasticity of the material of the member; it has units of stress (that is, lb/in^2 or Pa) and is a measure of the stiffness of the material. Consider a typical

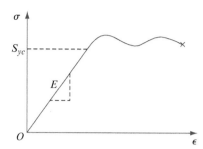

Fig. 11.24 Stress versus strain for a mild steel bar in tension.

plot of stress, σ, versus strain, ε, for a mild steel bar in tension (Fig. 11.24). Hooke's law states that, in the elastic region, the stress in a mild steel bar is proportional to the strain, with a proportionality constant of the modulus of elasticity. This can be written as

$$E = \sigma/\varepsilon.$$

As a rule of thumb, the modulus of elasticity for a steel alloy is

$$E = 30 \text{ Mpsi} = 30 \times 10^6 \text{ lb/in}^2 \quad \text{or} \quad E = 207 \text{ GPa} = 207 \times 10^9 \text{ N/m}^2,$$

and for stainless steel is

$$E = 27.5 \text{ Mpsi} \quad \text{or} \quad E = 190 \text{ GPa}.$$

Substituting Eq. (11.25) into Eq. (11.24), the second derivative of the transverse deflection of the member can be written as

$$\frac{d^2y}{dx^2} = -\frac{Py}{EI}.$$

Rearranging this equation gives a second-order differential equation for the member:

$$\frac{d^2y}{dx^2} + \left(\frac{P}{EI}\right)y = 0. \tag{11.26}$$

The solution to this differential equation, that is, the transverse deflection of the member, is

$$y = A\sin\left(\sqrt{\frac{P}{EI}}\,x\right) + B\cos\left(\sqrt{\frac{P}{EI}}\,x\right), \tag{11.27}$$

where A and B are the constants of integration and are obtained from the support conditions at the ends of the member.

Pinned–pinned end conditions. Consider the case where both ends of the member are pinned, as shown in Fig. 11.25 (also referred to in civil engineering as *rounded–rounded ends*).

Fig. 11.25 A member with pinned–pinned ends.

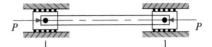

The end conditions are: (a) at $x = 0$, the deflection is $y = 0$, and (b) at $x = L$, the deflection is $y = 0$. Substituting $y = 0$ at $x = 0$ into Eq. (11.27) gives

$$0 = A \sin 0 + B \cos 0.$$

Therefore, one constant of integration is $B = 0$. Substituting this into Eq. (11.27) gives

$$y = A \sin \left(\sqrt{\frac{P}{EI}} x \right). \tag{11.28}$$

Then, substituting $y = 0$ at $x = L$ into this equation gives

$$0 = A \sin \left(\sqrt{\frac{P}{EI}} L \right).$$

There are two solutions to this equation. Either (a) the constant of integration $A = 0$, which is a trivial solution (it implies that the deflection is $y = 0$ for all values of x) and can be ignored, or (b)

$$\sin \left(\sqrt{\frac{P}{EI}} L \right) = 0.$$

The solution to this equation is

$$\sqrt{\frac{P_{cr}}{EI}} L = n\pi (n = 1, 2, 3, \ \ldots), \tag{11.29}$$

where P_{cr} is a compressive load that places the member in a condition of unstable equilibrium and is referred to as the *critical load*. The first critical load (that is, the lowest or minimum value of P_{cr}) is the most important from physical considerations. For $n = 1$, the critical load is

$$P_{cr} = \left(\frac{\pi^2}{L^2} \right) EI. \tag{11.30}$$

This equation is commonly referred to as *Euler's column formula* for a long, slender member with pinned–pinned ends. Note an interesting result, that the strength of the material is not a factor; that is, the critical load depends only on: (a) the length, L; (b) the second moment of area, I; and (c) Young's modulus, E. Therefore, if the member is steel, then using a stronger steel (with a higher compressive yield strength, S_{yc}) does not help, since all steel alloys have essentially the same modulus of elasticity E (= 30 Mpsi or 207 GPa).

Substituting Eq. (11.30) into Eq. (11.28), the transverse deflection of the member at the critical load can be written as

Fig. 11.26 Deflection curve is a half-sine wave.

Fig. 11.27 Deflection curve is a full-sine wave.

$$y = A \sin\left(\frac{\pi x}{L}\right), \tag{11.31}$$

which indicates that the deflection curve of the member (for pinned–pinned ends) is a half-sine wave, as shown in Fig. 11.26.

Values of $n > 1$ in Eq. (11.29) result in deflection curves that cross the x axis at points of inflection and are multiples of half-sine waves. For $n = 2$, the deflection curve is a full-sine wave as shown in Fig. 11.27.

11.16 Critical Unit Load

Next, we hope to obtain an expression for the critical load per unit cross-sectional area of the member, referred to as the *critical unit load*, which has units of stress or strength. The second moment of area (also called the *area moment of inertia*[6]) of the member cross-section in a plane perpendicular to the axis of the member can be written as

$$I = Ak^2,$$

where k is called the *radius of gyration*.[7] Rearranging this equation, the radius of gyration of the member cross-section can be written as

$$k = \sqrt{\frac{I}{A}}.$$

From this we define a nondimensional parameter called the *slenderness ratio*,

$$S_r = \frac{L}{k}. \tag{11.32}$$

The slenderness ratio, rather than the actual length, L, is commonly used to classify members in compression into length categories.

---

[6] Note that the *area* moment of inertia, shown here, is quite different from the *mass* moment of inertia of Section 12.3. Note the dimensions of the fourth power of length compared to mass length squared.

[7] Note that this is *not* the same as the term with similar name defined in Eq. (12.10).

Substituting this definition into Eq. (11.30), the first *critical load* can be written as

$$P_{cr} = \frac{\pi^2 EA}{S_r^2}.$$

The *critical unit load* is defined by dividing this critical load by the cross-sectional area,

$$\frac{P_{cr}}{A} = \frac{\pi^2 E}{S_r^2}. \tag{11.33}$$

The critical unit load, rather than the strength of the material from which it is made – that is, the yield strength or the ultimate strength – represents the strength of a particular member under compression.

Including the pinned–pinned end conditions presented above, there are four common pairs of end conditions for a member. These are: (a) fixed–fixed ends, (b) fixed–pinned ends, (c) pinned–pinned ends, and (d) fixed–free ends. The critical unit load for each of these four cases can be written in the form of Eq. (11.33) using an end-condition constant, C; that is,

$$\frac{P_{cr}}{A} = C\frac{\pi^2 E}{S_r^2}. \tag{11.34}$$

This equation is commonly referred to as the Euler column formula for different end conditions.

To determine the end-condition constant, C, the right-hand side of Eq. (11.34) can be written as

$$C\frac{\pi^2 E}{S_r^2} = \frac{\pi^2 E}{(L_{eff}/k)^2},$$

where L_{eff} denotes the effective length of the member. The effective length is the length of the half-sine wave for the equivalent member. Therefore, by rearranging this equation, the end-condition constant can be expressed in terms of the effective length of the member as $C = (L/L_{eff})^2$. Also, using the American Institute of Steel Construction (AISC) recommended column end-condition effective length factor, $\alpha = L_{eff}/L$, the end-condition constant can be written as

$$C = 1/\alpha^2.$$

Case (*a*): Fixed–fixed ends are shown in Fig. 11.28. For the case of a member with fixed–fixed ends, the effective length of the member is $L_{eff} = 0.5L$. Substituting this value into the above equations, the end-condition constant is $C = 4$.

Case (*b*): Fixed–pinned ends are shown in Fig. 11.29. For the case of a member with fixed–pinned ends, the effective length of the member is $L_{eff} = 0.707L$. Using this value, the above equations give the end-condition constant $C = 2$.

Case (*c*): Pinned–pinned ends are shown in Fig. 11.25. The case of a member with pinned–pinned ends is the case used for the above equations. The effective length of the member is $L_{eff} = L$, and the end-condition constant is $C = 1$.

Fig. 11.28 Fixed–fixed ends.

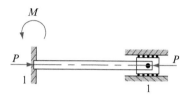

Fig. 11.29 Fixed–pinned ends.

Fig. 11.30 Fixed–free ends.

Case (*d*): Fixed–free ends are shown in Fig. 11.30. For the case of a member with fixed–free ends, the effective length of the member is $L_{\text{eff}} = 2L$. With this value in the above equations, the end-condition constant is $C = 1/4$.

When analyzing buckling in links of a planar mechanism, we often find that the end conditions are pinned–pinned for bending in the *xy* plane. Recall that the end-condition constant for a pinned–pinned link is $C = 1$. Note, however, that if bending of such a link is assumed to be out of the *xy* plane then the end conditions of the link are not known. It is common for out-of-plane bending to assume that the ends of the link are fixed–fixed. In such a case, the end-condition constant for a fixed–fixed link is $C = 4$.

11.17 Critical Unit Load and Slenderness Ratio

A typical plot of critical unit load versus slenderness ratio of a long, slender member subjected to an axial compressive load P is shown in Fig. 11.31.

Note that there are two distinct curves shown in Fig. 11.31: (a) Euler's column formula, and (b) *Johnson's parabolic equation.*[8] The criterion for using Euler's column formula, Eq. (11.34), is

---

[8] Proposed by Prof. J. B. Johnson in 1898 and, independently, by Prof. A. Ostenfeld of Copenhagen.

Fig. 11.31 Critical unit load versus slenderness ratio.

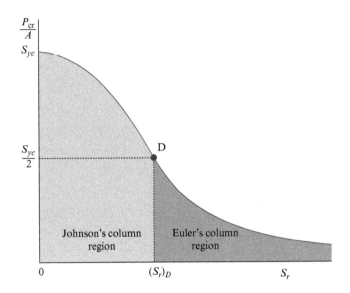

$$S_r > (S_r)_D, \tag{11.35}$$

where $(S_r)_D$ is the slenderness ratio at the point of tangency, D, between Euler's column formula and Johnson's parabolic equation. The critical unit load at the point of tangency is

$$\frac{P_{cr}}{A} = 0.5 S_{yc}.$$

Substituting this equation into Eq. (11.34) and rearranging, the slenderness ratio at the point of tangency – that is, where the slenderness ratio is $S_r = (S_r)_D$ – can be written as

$$(S_r)_D = \pi \sqrt{\frac{2CE}{S_{yc}}}. \tag{11.36}$$

If the slenderness ratio of the member is less than this value, $S_r < (S_r)_D$, then the criterion that is most commonly used for buckling is Johnson's parabolic equation.

11.18 Johnson's Parabolic Equation

The equation of the parabola in the left side of Fig. 11.31 can be written as

$$\frac{P_{cr}}{A} = a + b S_r^2. \tag{11.37}$$

The focus here is to determine the constants a and b. There are two known data points in Fig. 11.31: (a) for the slenderness ratio $S_r = 0$, the critical unit load is $P_{cr}/A = S_{yc}$; and (b) for the slenderness ratio $S_r = (S_r)_D$, the critical load is $P_{cr}/A = 0.5 S_{yc}$.

Substituting the first of these data points into Eq. (11.37), we find the coefficient

$$a = S_{yc}.$$

Substituting this result into Eq. (11.37), the critical unit load can be written as

$$\frac{P_{cr}}{A} = S_{yc} + bS_r^2. \tag{11.38}$$

Then, substituting the second data point and Eq. (11.36) into Eq. (11.38) gives

$$0.5S_{yc} = S_{yc} + b\pi^2 \frac{2CE}{S_{yc}}.$$

Finally, rearranging this equation, the coefficient b can be written as

$$b = -\frac{S_{yc}^2}{4CE\pi^2}.$$

Substituting this result into Eq. (11.38), Johnson's parabolic equation for the critical unit load can be written as

$$\frac{P_{cr}}{A} = S_{yc} - \frac{1}{CE}\left(\frac{S_{yc}S_r}{2\pi}\right)^2. \tag{11.39}$$

Therefore, the critical load for a member using Johnson's parabolic equation is

$$P_{cr} = A\left[S_{yc} - \frac{1}{CE}\left(\frac{S_{yc}S_r}{2\pi}\right)^2\right]. \tag{11.40}$$

Example 11.10

For the four-bar linkage in the posture shown in Fig. 11.32, a vertical load P is acting at point E on the horizontal coupler link 3 – that is, the link parallel to the x axis. The coupler link is pinned to the vertical links 2 and 4 at points A and B. The links are made of a steel alloy with a compressive yield strength $S_{yc} = 60$ kpsi and a modulus of elasticity $E = 30$ Mpsi. Links 2 and 4 have hollow circular cross-sections with 2 in outside diameters and 0.25 in wall thicknesses. Assume that the value for the end-condition constant for both vertical links is $C = 1$. The known lengths are $R_{AO} = 8$ ft, $R_{BA} = 6$ ft, $R_{BC} = 5$ ft, and $R_{EA} = 4$ ft. Determine: (a) the radii of gyration for links 2 and 4, (b) the values of the slenderness ratios of links 2 and 4 at the point of tangency between Euler's column formula and Johnson's parabolic equation, (c) the critical loads for links 2 and 4, and (d) the maximum value of load P if the factor of safety guarding against buckling for links 2 and 4 are both specified as $N = 2$.

Solution

a. The cross-sectional area of both links 2 and 4 is

$$A = \frac{\pi}{4}\left(D^2 - d^2\right) = \frac{\pi}{4}\left[(2\text{ in})^2 - (1.5\text{ in})^2\right] = 1.374\text{ in}^2. \tag{1}$$

Example 11.10 (cont.)

Fig. 11.32 Four-bar linkage
subjected to a vertical load P.

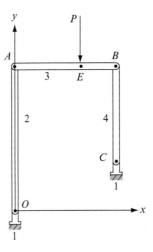

The area moment of inertia for the hollow circular cross-section of links 2 and 4 is (Appendix A, Table A.4)

$$I = \frac{\pi}{64}\left(D^4 - d^4\right) = \frac{\pi}{64}\left[(2 \text{ in})^4 - (1.5 \text{ in})^4\right] = 0.537 \text{ in}^4. \tag{2}$$

The radius of gyration for links 2 and 4 is

$$k = \sqrt{\frac{I}{A}} = \sqrt{\frac{0.537 \text{ in}^4}{1.374 \text{ in}^2}} = 0.625 \text{ in}. \qquad Ans.\ (3)$$

b. From the given end conditions for links 2 and 4, that is, pinned–pinned ends, the end-condition constant is $C = 1.0 = (1/\alpha)^2$ and $\alpha = 1.0 = L/L_{\text{eff}}$. This indicates that the effective length, L_{eff}, is equal to the actual length, L, for both link 2 and link 4. Therefore, the slenderness ratio of link 2 is

$$(S_r)_2 = \frac{(L_{\text{eff}})_2}{k} = \frac{(8 \text{ ft}) \ 12 \text{ in/ft}}{0.625 \text{ in}} = 153.60, \qquad Ans.\ (4a)$$

and the slenderness ratio of link 4 is

$$(S_r)_4 = \frac{(L_{\text{eff}})_4}{k} = \frac{(5 \text{ ft}) 12 \text{ in/ft}}{0.625 \text{ in}} = 96.00. \qquad Ans.\ (4b)$$

The slenderness ratio of either link at the point of tangency is

$$(S_r)_D = \pi\sqrt{\frac{2CE}{S_{yc}}} = \pi\sqrt{\frac{2 \cdot 1.0\left(30 \times 10^6 \text{ psi}\right)}{60 \times 10^3 \text{ psi}}} = 99.35. \qquad Ans.\ (5)$$

Example 11.10 (cont.)

c. The critical load is determined by first determining whether each link is considered an Euler column or a Johnson column. Note that the slenderness ratios given by Eqs. (4a), (4b), and (5) are

$$(S_r)_2 = 153.60, (S_r)_4 = 96.00, \text{ and } (S_r)_D = 99.35.$$

The criterion for using Euler's column formula is given in Eq. (11.35) as $S_r > (S_r)_D$. Therefore, because $153.60 > 99.35$, the conclusion is that link 2 is regarded as an Euler column. The critical unit load for link 2, from Euler's column formula, Eq. (11.34), can be written as

$$\frac{(P_{cr})_2}{A} = \frac{C\pi^2 E}{(S_r)_2^2}. \tag{6}$$

Substituting the known values into Eq. (6) gives

$$\frac{(P_{cr})_2}{1.374 \text{ in}^2} = \frac{(1.0)\pi^2 (30 \times 10^6 \text{ psi})}{(53.610)^2}.$$

Therefore, the critical load for link 2 is

$$(P_{cr})_2 = 17\ 244 \text{ lb.} \qquad\qquad Ans.\ (7)$$

The criterion for using Johnson's parabolic equation is $S_r < (S_r)_D$. Therefore, since $96.00 < 99.35$, Johnson's parabolic equation must be used for link 4. The critical load for link 4 from Johnson's parabolic equation is given by Eq. (11.40); that is,

$$(P_{cr})_4 = A\left[S_{yc} - \frac{1}{CE}\left(\frac{S_{yc}S_r}{2\pi}\right)^2 \right].$$

Substituting the known values into this equation gives

$$(P_{cr})_4 = (1.374 \text{ in}^2)\left[(60 \times 10^3 \text{ psi}) - \frac{1}{1.0(30 \times 10^6 \text{ psi})}\left(\frac{(60 \times 10^3 \text{ psi})96.00}{2\pi}\right)^2 \right].$$

Therefore, the critical load for link 4 is

$$(P_{cr})_4 = 43\ 950 \text{ lb.} \qquad\qquad Ans.\ (8)$$

d. For the given factor of safety guarding against buckling, an equation can be written for the applied load on either of the two side links as follows:

$$P_{app} = \frac{P_{cr}}{N} = \frac{P_{cr}}{2}.$$

The following two cases must be compared.

Example 11.10 (cont.)

Case 1: When the load acting at point A from link 3 onto link 2 is a maximum, the maximum applied load at point A can be written from Eq. (7) as

$$(P_{\text{app}})_2 = \frac{(P_{\text{cr}})_2}{2} = \frac{17\,244\,\text{lb}}{2} = 8622\,\text{lb}.$$

To determine the load on link 2 from the applied load P, we take moments about point B. Then, the applied load P necessary to cause buckling of link 2 is

$$(P)_2 = \frac{R_{AB}(P_{\text{app}})_2}{R_{EB}} = \frac{6\,\text{ft}\,(8622\,\text{lb})}{2\,\text{ft}} = 25\,866\,\text{lb}. \tag{9a}$$

Case 2: When the load acting at point B from link 3 onto link 4 is a maximum, the maximum applied load at point B can be written from Eq. (8) as

$$(P_{\text{app}})_4 = \frac{(P_{\text{cr}})_4}{2} = \frac{43\,950\,\text{lb}}{2} = 21\,975\,\text{lb}.$$

To determine the load on link 4 from the applied load P, we take moments about point A. Then, the applied load P necessary to cause buckling of link 4 is

$$(P)_4 = \frac{R_{BA}(P_{\text{app}})_4}{R_{EA}} = \frac{6\,\text{ft}\,(21\,975\,\text{lb})}{4\,\text{ft}} = 32\,963\,\text{lb}. \tag{9b}$$

To ensure a factor of safety of at least 2 on both members, we must take the minimum of the applied loads from cases 1 and 2. That is, in comparing Eqs. (9a) and (9b), since $(P)_2 < (P)_4$ (that is, since $25\,866\,\text{lb} < 32\,963\,\text{lb}$), the maximum load P to guard against buckling of either link with a factor of safety of 2 is

$$P < 25\,866\,\text{lb}. \qquad\qquad Ans.$$

Example 11.11

A force P is applied vertically upward on the horizontal link 4 that is pinned to the ground at point B and pinned to links 2 and 3 at point A, as shown in Fig. 11.33. Link 2 is fixed to the ground at point D. Assume that links 2 and 4 are in static equilibrium. The distances are $R_{BA} = 1\,\text{m}$ and $R_{CA} = 3\,\text{m}$, and links 2 and 4 are made from a steel alloy with a compressive yield strength $S_{yc} = 415\,\text{MPa}$ and a modulus of elasticity $E = 207\,\text{GPa}$. The length of the vertical link 2 is $R_{DA} = 2.75\,\text{m}$, and this link has a hollow circular cross-section with an outside diameter of 50 mm and a wall thickness of 6 mm. Using the theoretic value for the end-condition constant for link 2, determine: (a) the value of the slenderness ratio at the point of tangency between Euler's column formula and Johnson's parabolic formula; (b) the critical load acting on link 2 and the critical unit load on link 2; and (c) the force P for the factor of safety of link 2 to guard against

Example 11.11 (cont.)

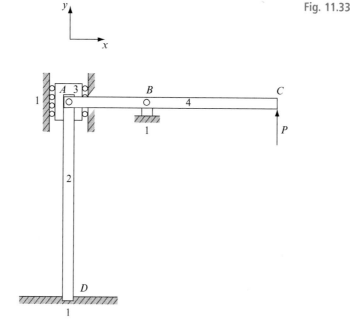

Fig. 11.33

buckling to be $N = 2$. (d) If link 2 is replaced with a solid circular cross-section of the same material with a diameter of 40 mm and force $P = 70$ kN, then determine the maximum length of link 2 for the factor of safety to guard against buckling of the link to be $N = 2$.

Solution

a. From the fixed–pinned end conditions of link 2, the end-condition constant for this link is $C = 2$. Therefore, the slenderness ratio of link 2 at the point of tangency is

$$(S_r)_D = \pi\sqrt{\frac{2CE}{S_{yc}}} = \pi\sqrt{\frac{2 \cdot 2(207 \times 10^9 \text{ Pa})}{415 \times 10^6 \text{ Pa}}} = 140.3. \qquad \textit{Ans. (1)}$$

b. The cross-sectional area of link 2 can be written as

$$A = \frac{\pi}{4}\left(D^2 - d^2\right) = \frac{\pi}{4}\left[(0.050 \text{ m})^2 - (0.038 \text{ m})^2\right] = 0.000\,829 \text{ m}^2.$$

The second moment of area of link 2 can be written as (Appendix A, Table A.4)

$$I = \frac{\pi}{64}\left(D^4 - d^4\right) = \frac{\pi}{64}\left[(0.050 \text{ m})^4 - (0.038 \text{ m})^4\right] = 0.2044 \times 10^{-6} \text{ m}^4.$$

Example 11.11 (cont.)

The radius of gyration of link 2 can be written as

$$k = \sqrt{\frac{I}{A}} = \sqrt{\frac{0.2044 \times 10^{-6} \text{ m}^4}{0.000\,829 \text{ m}^2}} = 0.0157 \text{ m}.$$

The slenderness ratio of link 2 is

$$S_r = \frac{L}{k} = \frac{2.75 \text{ m}}{0.0157 \text{ m}} = 175.2. \tag{2}$$

To determine the critical load on link 2, we must first determine whether this link is an Euler column or a Johnson column. The criterion for using Johnson's parabolic equation is $S_r < (S_r)_D$. From Eqs. (1) and (2), we have $S_r > (S_r)_D$, that is, $175.2 > 140.3$. Therefore, link 2 is an Euler column.

The critical load on link 2, from the Euler column equation, is

$$P_{cr} = A\left[\frac{C\pi^2 E}{S_r^2}\right] = (0.000\,829 \text{ m}^2)\left[\frac{2\pi^2(207 \times 10^9 \text{ Pa})}{175.2^2}\right] = 110.35 \text{ kN}. \qquad Ans.$$

The critical unit load on link 2 is defined as

$$\frac{P_{cr}}{A} = \frac{110.35 \text{ kN}}{0.000\,829 \text{ m}^2} = 133.1 \text{ MPa}. \qquad Ans.$$

c. Links 2 and 4 are in static equilibrium. The free-body diagram of link 4 is shown in Fig. 11.34.

Fig. 11.34 Free-body diagram of link 4.

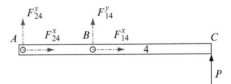

Taking moments about pin B gives

$$\mathbf{R}_{CB} \times \mathbf{P} = \mathbf{R}_{AB} \times \mathbf{F}_{24}^y.$$

Substituting the given data into this equation gives

$$P = \frac{(1 \text{ m})F_{24}^y}{2 \text{ m}} = \frac{F_{24}^y}{2}. \tag{3}$$

Link 2 is in compression as shown in Fig. 11.35.

The reaction force $P_A^y = F_{42}^y$ onto link 2 is opposite to the reaction force F_{24}^y; therefore, the magnitude of P_A^y is equal to the magnitude of F_{24}^y. Equation (3) can be written as

Example 11.11 (cont.)

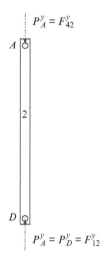

$P_A^y = F_{42}^y$

Fig. 11.35 Link 2 is in compression.

A

2

D

$P_A^y = P_D^y = F_{12}^y$

$$P = \frac{P_A^y}{2}. \tag{4}$$

The factor of safety guarding against buckling for link 2 is defined as

$$N = \frac{P_{cr}}{P_A^y}, \tag{5}$$

where $P_A^y = F_{42}^y$ is the compressive load at point D on link 2, as shown in Fig. 11.35. Therefore, rearranging Eq. (5), the force P_A^y is

$$P_A^y = \frac{P_{cr}}{N} = \frac{110.35 \text{ kN}}{2} = 55.175 \text{ kN}. \tag{6}$$

Finally, substituting Eq. (6) into Eq. (4) gives

$$P = \frac{P_A^y}{2} = \frac{55.175 \text{ kN}}{2} = 27.588 \text{ kN}. \qquad\qquad Ans.$$

d. Link 2 is a solid circular column with a factor of safety $N = 2$ and $P = 70$ kN. Rearranging Eq. (4), the compressive force is

$$P_A^y = 2P = 2(70 \text{ kN}) = 140 \text{ kN}.$$

Rearranging Eq. (5), the critical load applied on link 2 is

$$P_{cr} = P_A^y \cdot N = 140 \text{ kN} \cdot 2 = 280 \text{ kN}.$$

First, assume the link is an Euler column. Then from Euler's column formula, Eq. (11.34), the critical load applied on link 2 is

Example 11.11 (cont.)

$$P_{cr} = A\left[\frac{C\pi^2 E}{S_r^2}\right].$$ (7)

The cross-sectional area of link 2 is

$$A = \frac{\pi}{4}(D^2) = \frac{\pi}{4}(0.040 \text{ m})^2 = 0.001\,26 \text{ m}^2.$$

The second moment of area of the cross-section of link 2 is

$$I = \frac{\pi}{64}(D^4) = \frac{\pi}{64}(0.040 \text{ m})^4 = 0.126 \times 10^{-6} \text{ m}^4.$$

The radius of gyration of the cross-section of link 2 is

$$k = \sqrt{\frac{I}{A}} = \sqrt{\frac{0.126 \times 10^{-6} \text{ m}^4}{0.001\,26 \text{ m}^2}} = 0.010 \text{ m}.$$

Rearranging Eq. (7), the slenderness ratio of link 2 is

$$(S_r)_{\text{Euler}} = \sqrt{\frac{C\pi^2 EA}{P_{cr}}} = \sqrt{\frac{2\pi^2(207 \times 10^9 \text{ Pa})(0.001\,26 \text{ m}^2)}{280 \times 10^3 \text{ N}}} = 135.6.$$

Next, assume the link is a Johnson column. The critical unit load can be written from Johnson's parabolic equation, Eq. (11.39), as

$$\frac{P_{cr}}{A} = \left[S_{yc} - \frac{1}{CE}\left(\frac{S_{yc}S_r}{2\pi}\right)^2\right].$$ (8)

Rearranging Eq. (8), the slenderness ratio can be written as

$$(S_r)_{\text{Johnson}} = \frac{2\pi}{S_{yc}}\sqrt{\left(S_{yc} - \frac{P_{cr}}{A}\right)CE}.$$

Substituting the known data into this equation gives

$$(S_r)_{\text{Johnson}} = \frac{2\pi}{(415 \times 10^6 \text{ Pa})}\sqrt{\left[(415 \times 10^6 \text{ Pa}) - \frac{280\,000 \text{ N}}{0.001\,26 \text{ m}^2}\right]2(207 \times 10^9 \text{ Pa})}$$
$$= 135.3.$$

The slenderness ratio at the point of tangency can be written as

$$(S_r)_D = \pi\sqrt{\frac{2CE}{S_{yc}}}.$$

Example 11.11 (cont.)

Since the material properties are the same, this gives the same result as Eq. (1), that is $(S_r)_D = 140.3$. Therefore, since $(S_r)_{\text{Johnson}} < (S_r)_{\text{Euler}} < (S_r)_D$, the link is a Johnson column. Using the slenderness ratio for the case of a Johnson column; that is, $S_r = (S_r)_{\text{Johnson}} = 135.3$, and rearranging Eq. (2), the maximum possible length of link 2 is

$$L = kS_r = (0.010 \text{ m})135.3 = 1.353 \text{ m}. \qquad\qquad Ans.$$

Example 11.12

A load P_C is acting on the horizontal link 2 at point C of a simple truss, as shown in Fig. 11.36. Link 2 is pinned to the ground at point O_2 and to the inclined link 3 at point B, whereas link 3 is pinned to the ground at point O_3. The lengths are $R_{CO_2} = 5$ ft, $R_{BO_2} = 4$ ft, $R_{O_3O_2} = 3$ ft, and $R_{BO_3} = 5$ ft. The two links each have a solid circular cross-section with a diameter of $D = 2$ in. Also, the two links are made from a steel alloy with a compressive yield strength $S_{yc} = 60$ kpsi, a tensile yield strength $S_{yt} = 50$ kpsi, and a modulus of elasticity $E = 30$ Mpsi.

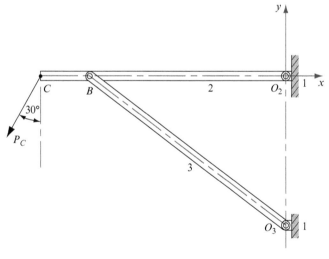

Fig. 11.36 A simple pinned truss.

Part 1: Using the theoretic value for the end-condition constants for the two links, determine: (a) the value of the slenderness ratio at the point of tangency between Euler's column formula and Johnson's parabolic formula for link 3; (b) the critical load and the critical unit load acting on link 3; and (c) the maximum load P_C for the factor of safety to guard against buckling of link 3 to be $N = 1$.

Part 2: If the load acting at point C is specified as $P_C = 24\,000$ lb, determine: (a) the factor of safety for link 2; (b) the factor of safety for link 3; and (c) which link is most likely to fail first. Briefly explain your answer.

Example 11.12 (cont.)

Solution
Part 1.

a. Links 2 and 3 are in static equilibrium. The free-body diagram of link 2 is shown in Fig. 11.37. Note that the link is in tension due to the external load P_C.

Fig. 11.37 Free-body diagram of link 2.

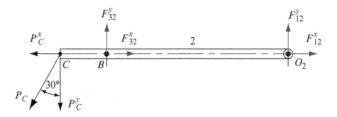

Taking moments about the pin O_2 gives the y component of the load at pin B as

$$F_{32}^y = \frac{5 \text{ ft}}{4 \text{ ft}} P_C^y \quad \text{or} \quad F_{23}^y = -\frac{5}{4} P_C^y.$$

The free-body diagram of link 3 is as shown in Fig. 11.38.

Fig. 11.38 Free-body diagram of link 3.

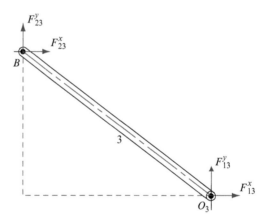

Taking moments about the pin O_3 gives

$$F_{23}^x = -\frac{4 \text{ ft}}{3 \text{ ft}} F_{23}^y = -\frac{4}{3}\left(-\frac{5}{4} P_C^y\right) = \frac{5}{3} P_C^y.$$

The magnitude of the force at pin B can be written as

$$F_{23} = \sqrt{\left(\frac{5}{3} P_C^y\right)^2 + \left(-\frac{5}{4} P_C^y\right)^2} = 5P_C^y\sqrt{\frac{1}{9} + \frac{1}{16}} = 5P_C^y\sqrt{\frac{16+9}{144}} = \frac{25}{12} P_C^y. \tag{1}$$

Example 11.12 (cont.)

The direction of this force at pin B – that is, the line of action – is

$$\angle F_{23} = \tan^{-1}\left(\frac{F_{23}^y}{F_{23}^x}\right) = \tan^{-1}\left(\frac{-5P_C^y/4}{5P_C^y/3}\right) = \tan^{-1}\left(\frac{-3}{4}\right) = -36.87°.$$

Since link 3 is a two-force member, forces F_{23} and F_{13} must be equal, opposite, and collinear; that is, $F_{23} = F_{13}$, and link 3 is in compression because of the internal load P_B (Fig. 11.39). Therefore, it is possible that link 3 could fail because of buckling.

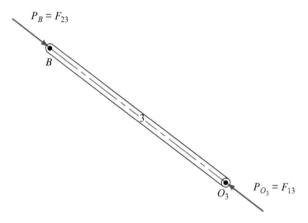

Fig. 11.39 Link 3 is in compression.

We must check link 3 for the factor of safety to guard against buckling to be $N = 1$. The cross-sectional area of link 3 is

$$A = \frac{\pi}{4}D^2 = \frac{\pi}{4}(2\text{ in})^2 = 3.1416\text{ in}^2.$$

The second moment of area of link 3 can be written as

$$I = \frac{\pi}{64}D^4 = \frac{\pi}{64}(2\text{ in})^4 = 0.7854\text{ in}^4.$$

The radius of gyration of link 3 is

$$k = \sqrt{\frac{I}{A}} = \sqrt{\frac{0.785\,4\text{ in}^4}{3.141\,6\text{ in}^2}} = 0.500\text{ in}.$$

The slenderness ratio of link 3 is

$$S_r = \frac{L}{k} = \frac{5\text{ ft}\cdot 12\text{ in/ft}}{0.500\text{ in}} = 120. \tag{2}$$

From the pinned–pinned end conditions, the end-condition constant for link 3 is $C = 1.0$. Therefore, the slenderness ratio of link 3 at the point of tangency is

Example 11.12 (cont.)

$$(S_r)_D = \pi \sqrt{\frac{2CE}{S_{yc}}} = \pi \sqrt{\frac{2 \cdot 1.0 (30 \times 10^6 \text{ psi})}{60 \times 10^3 \text{ psi}}} = 99.35. \qquad \textit{Ans. (3)}$$

b. To determine the critical load on link 3, we must first determine whether this link is an Euler column or a Johnson column. The criterion for using Johnson's parabolic equation is $S_r < (S_r)_D$. From Eqs. (2) and (3), we have $120 > 99.35$; that is, $S_r > (S_r)_D$. Therefore, link 3 is an Euler column. The critical load on link 3 is

$$P_{cr} = A \left[\frac{C\pi^2 E}{S_r^2} \right] = (3.141\ 6 \text{ in}^2) \left[\frac{1.0\pi^2 (30 \times 10^6 \text{ psi})}{120^2} \right] = 64\,596 \text{ lb.} \qquad \textit{Ans. (4)}$$

The critical unit load is

$$\frac{P_{cr}}{A} = \frac{64\,596 \text{ lb}}{3.141\ 6 \text{ in}^2} = 20\,562 \text{ lb/in}^2. \qquad \textit{Ans.}$$

c. If we identify $P_B = F_{23}$ as the compressive load at point B on link 3 (Fig. 11.40), then, from the given factor of safety of $N = 1$ guarding against buckling for link 3, the applied load at pin B on link 3 can be written as

Fig. **11.40** The compressive load acting on link 3.

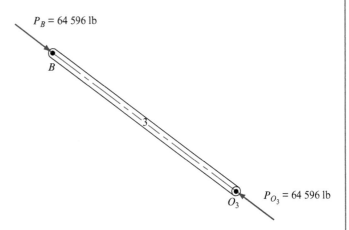

$P_B = 64\ 596$ lb

$P_{O_3} = 64\ 596$ lb

$$P_B = \frac{P_{cr}}{N} = \frac{P_{cr}}{1} = 64\,596 \text{ lb.} \qquad (5)$$

Therefore, load $P_B (= F_{23})$ is equal to the critical load P_{cr}, as shown in Fig. 11.40. Note that a compressive load $P_B = 64\,596$ lb is the limiting load on link 3 to avoid buckling, with a factor of safety of $N = 1$.

Example 11.12 (cont.)

Rearranging Eq. (1), the vertical component of the load at point C is

$$P_C^y = \frac{12}{25} F_{23} = \frac{12}{25} P_B. \tag{6}$$

Substituting Eq. (5) into Eq. (6), the vertical component of the load at point C is

$$P_C^y = \frac{12}{25} (64\,596 \text{ lb}) = 31\,006 \text{ lb}.$$

Finally, the load acting at point C is

$$P_C = \frac{P_C^y}{\cos 30°} = \frac{31\,006 \text{ lb}}{0.866} = 35\,803 \text{ lb}. \qquad Ans.$$

Note that the load $P_C = 35\,803$ lb is the load that ensures that the factor of safety to guard against buckling of link 3 is $N = 1$. A load $P_C > 35\,803$ lb would cause link 3 to buckle; that is, the factor of safety would be $N < 1$.

Part 2.

a. The x and y components of the load $P_C = 24\,000$ lb are

$$P_C^x = P_C \sin 30° = (24\,000 \text{ lb})0.500 = 12\,000 \text{ lb},$$

$$P_C^y = P_C \cos 30° = (24\,000 \text{ lb})0.866 = 20\,785 \text{ lb}. \tag{7}$$

Link 2 is subject to the tensile load P_C^x, which creates a tensile stress in the link. The factor of safety guarding against yielding of link 2 is defined as

$$N = \frac{S_{yt}}{\sigma},$$

where the normal stress due to the tensile load can be written as

$$\sigma = \frac{P_C^x}{A} = \frac{12\,000 \text{ lb}}{3.1416 \text{ in}^2} = 3\,819.7 \text{ lb/in}^2.$$

Therefore, the factor of safety guarding against yielding of link 2 is

$$N = \frac{50\,000 \text{ psi}}{3\,819.7 \text{ psi}} = 13.1. \qquad Ans. \text{ (8)}$$

Note that this relatively high factor of safety indicates safety against tensile failure.

b. Substituting Eq. (7) into Eq. (6), the compressive force F_{23} exerted on link 3 is

$$F_{23} = \frac{25}{12} P_C^y = \frac{25}{12} (20\,785 \text{ lb}) = 43\,302 \text{ lb}, \tag{9}$$

Example 11.12 (cont.)

where force F_{23} is equal to force P_B. For link 3, the factor of safety to guard against buckling can be calculated by substituting Eqs. (4) and (9) into Eq. (5); that is,

$$N = \frac{P_{cr}}{P_B} = \frac{P_{cr}}{F_{23}} = \frac{64\,596\ lb}{43\,302\ lb} = 1.49. \hspace{2cm} Ans.\ (10)$$

c. Note that the factor of safety guarding against yielding of link 2 [the tensile member; Eq. (8)] is much higher than the factor of safety guarding against buckling of link 3 [the compressive member; Eq. (10)]. Therefore, the prediction is that link 3 will fail before link 2. *Ans.*

Example 11.13

A vertically downward force $F = 15\,000$ lb is applied at point B on member 2, which is pinned to the ground at point O_2 and to the vertical member 3 at point A, as shown in Fig. 11.41. The length of member 2, that is R_{BO_2}, is 60 in, and the ground pin O_3 is located at a distance $b = 30$ in along the x axis from the ground pin O_2. Members 2 and 3 are made from a steel alloy with a compressive yield strength $S_{yc} = 60$ kpsi and a modulus of elasticity $E = 30$ Mpsi. For buckling in the plane, member 3 behaves as a pinned–pinned member. Therefore, the theoretic end-condition constant is $C = 1$. For buckling out of the plane, member 3 behaves as a fixed–fixed member with a theoretic end-condition constant $C = 4$.

Fig. 11.41 The structure for buckling analysis.

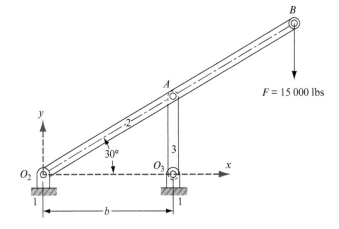

Part A: Assuming that the slenderness ratio of member 3 is such that Euler's column formula is valid, determine whether the following statements are true or false; give a brief explanation for each.

1. For members with a slenderness ratio greater than 10, it is safe to assume that the members designed for failure against yielding are automatically safe against buckling failure.

Example 11.13 (cont.)

2. With all other design parameters unchanged, a member with fixed–free end conditions has a larger factor of safety guarding against buckling than the same member with pinned–pinned end conditions.
3. Two members with the same length, cross-section, and modulus of elasticity, but with different compressive yield strengths, have the same critical buckling load.

Part B: Assume that member 3 has a hollow rectangular cross-section as shown in Fig. 11.42a. Initially, it is not known whether member 3 will be attached to the ground at pin O_3 as in either Case I or Case II shown in Fig. 11.42b. Using symbolic arguments (no numeric calculations are required), determine whether Case I or Case II leads to a higher factor of safety guarding against buckling. Assume the possibility of buckling failure in the plane of Fig. 11.41.

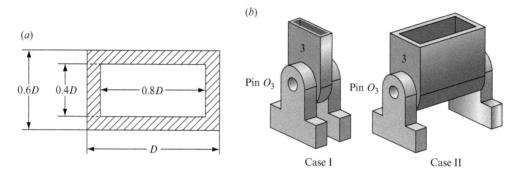

Fig. 11.42 (a) Cross-section of member 3 (dimensions in inches); (b) Case I and Case II.

Part C: Assuming the worst-case scenario, that is, the case with the lower factor of safety guarding against buckling, design the cross-section of member 3, that is, determine the dimension D, such that member 3 has a factor of safety guarding against buckling failure of $N = 2$. Note that designing the cross-section for the case with the lower factor of safety automatically ensures a safe design for the case with the higher factor of safety.

Part D: Determine the factor of safety of member 3 guarding against buckling failure. Use the dimensions of the rectangular cross-section obtained in Part C. Assume that member 3 is attached as in Case I and that there is the possibility of buckling out of the plane.

Part E: Assume that member 3 is attached as shown in Case II and that there is the possibility of buckling in the plane. The problem is to determine a new value for b; that is, to change the location of pin O_3. The distance b cannot be less than 5 in and cannot be greater than 45 in. Using the dimensions of the rectangular cross-section obtained in Part C, determine an expression for the factor of safety guarding against buckling failure as a function of the distance b. Then, plot a graph of N versus b. Using this graph, determine the minimum value of b above which the member is safe against buckling failure.

Example 11.13 (cont.)

Solution
Part A:

1. For members with a slenderness ratio greater than 10, it is safe to assume that members designed for failure against yielding are automatically safe against buckling failure.
   FALSE.   *Ans.*
2. With all other design parameters unchanged, a member with fixed–free end conditions has a larger factor of safety than the same member with pinned–pinned end conditions.
   FALSE.   *Ans.*
3. Two members with the same length, cross-section, and modulus of elasticity but with different compressive yield strengths have the same value of critical buckling load.
   TRUE.   *Ans.*

Part B:

For Case I, the second moment of area for member 3 can be written as

$$I_{\text{Case I}} = \frac{0.6D(D)^3}{12} - \frac{0.4D(0.8D)^3}{12} = \frac{0.3952D^4}{12} = 0.03293D^4.$$

For Case II, the second moment of area for member 3 can be written as

$$I_{\text{Case II}} = \frac{D(0.6D)^3}{12} - \frac{0.8D(0.4D)^3}{12} = \frac{0.1648D^4}{12} = 0.01373D^4. \tag{1}$$

Therefore, the second moment of area of Case I is greater than the second moment of area of Case II; that is,

$$I_{\text{Case I}} > I_{\text{Case II}}. \tag{2}$$

Since the cross-sectional area is the same in both cases, Eq. (2) implies that the radius of gyration of Case I is greater than the radius of gyration of Case II; that is,

$$k_{\text{Case I}} > k_{\text{Case II}}. \tag{3}$$

Since the length of the member is the same in both cases, Eq. (3) implies that the slenderness ratio in Case I is less than the slenderness ratio in Case II; that is,

$$(S_{\text{r}})_{\text{Case I}} < (S_{\text{r}})_{\text{Case II}}.$$

Therefore, for either Euler's column formula or Johnson's parabolic equation, the critical load of Case I is greater than the critical load of Case II; that is,

$$(P_{\text{cr}})_{\text{Case I}} > (P_{\text{cr}})_{\text{Case II}}.$$

Since the applied load is the same in both cases, the factor of safety guarding against buckling for Case I is greater than the factor of safety guarding against buckling for Case II; that is,

$$(N)_{\text{Case I}} > (N)_{\text{Case II}}. \qquad\qquad \textit{Ans.}$$

Example 11.13 (cont.)

This implies that, if the cross-sectional area of the member is designed for Case II, then the member will automatically be safe against buckling for Case I.

Part C:
The cross-sectional area of member 3, that is, a hollow rectangular cross-section, can be written as

$$A = D(0.6D) - 0.8D(0.4D) = 0.28D^2.$$

From Eq. (1), the second moment of area for member 3 can be written as $I = 0.01373D^4$. The radius of gyration of member 3 can be written as

$$k = \sqrt{\frac{I}{A}} = \sqrt{\frac{0.01373D^4}{0.28D^2}} = 0.2215D. \tag{4}$$

Since $b = 30$ in, the length of member 3 is

$$L = b\tan 30^\circ = 17.321 \text{ in.}$$

Therefore, the slenderness ratio of the member is

$$S_r = \frac{L}{k} = \frac{17.321 \text{ in}}{0.2215D} = \frac{78.199 \text{ in}}{D}.$$

The slenderness ratio at the point of tangency can be written as

$$(S_r)_D = \pi\sqrt{\frac{2CE}{S_{yc}}}.$$

From the specified fixed–fixed end conditions, the end-condition constant is $C = 1$. Therefore, the slenderness ratio at the point of tangency is

$$(S_r)_D = \pi\sqrt{\frac{2 \cdot 1\left(30 \times 10^6 \text{ psi}\right)}{60 \times 10^3 \text{ psi}}} = 99.3.$$

To determine the applied force at point A, we take moments about point O_2, which can be written as

$$(60 \text{ in})\cos 30^\circ F = bP_{\text{app}}.$$

Therefore, the applied force at point A is

$$P_{\text{app}} = \frac{(60 \text{ in})\cos 30^\circ(15000 \text{ lb})}{30 \text{ in}} = 25981 \text{ lb.}$$

The factor of safety guarding against buckling is defined as

$$N = \frac{P_{\text{cr}}}{P_{\text{app}}}.$$

Example 11.13 (cont.)

Since $N = 2$, the critical load is

$$P_{cr} = P_{app}N = (25\,981\text{ lb})2 = 51\,962\text{ lb.}$$

Note: At this point, it is not known whether Euler's column formula or Johnson's parabolic equation is the proper equation to determine the dimensions of the cross-section of the member. If we assume that Johnson's parabolic equation is the proper equation, then the critical load is

$$P_{cr} = A\left[S_{yc} - \frac{1}{CE}\left(\frac{S_{yc}S_r}{2\pi}\right)^2\right].$$

Substituting the known values into this equation gives

$$51\,962\text{ lb} = 0.28D^2\left[(60 \times 10^3\text{ psi}) - \frac{1}{1(30 \times 10^6\text{ psi})}\left(\frac{(60 \times 10^3\text{ psi})78.199}{2\pi D}\right)^2\right].$$

Rearranging this equation gives

$$(16\,800\text{ psi})D^2 = 51\,962\text{ lb} + 5\,204\text{ lb} = 57\,166\text{ lb.}$$

Solving this equation, the dimension D is

$$D = 1.845\text{ in.}\qquad\qquad\qquad Ans.$$

As a check, using Eq. (4) the slenderness ratio is

$$S_r = \frac{L}{k} = \frac{17.321\text{ in}}{0.221\,5(1.845\text{ in})} = 42.384 < (S_r)_D = 99.3.$$

Therefore, Johnson's parabolic equation is proper for this case.

Part D:
Since the dimensions of the rectangular cross-section are now known, the cross-sectional area of member 3 can be written as

$$A = 0.28D^2 = 0.9531\text{ in}^2.$$

From Eq. (1), the moment of inertia of the hollow rectangular member can be written as

$$I = 0.013\,73D^4 = 0.159\,1\text{ in}^4.$$

From Eq. (4), the radius of gyration for the member can be written as

$$k = 0.221\,5D = 0.408\,7\text{ in.}$$

Since $b = 30$ in, the length of the member is

$$L = b\tan 30^o = 17.310\text{ in.}$$

Example 11.13 (cont.)

The slenderness ratio of the member is

$$S_r = \frac{L}{k} = \frac{17.310 \text{ in}}{0.4087 \text{ in}} = 42.354.$$

Since the member is attached as shown in Case I, and there is the possibility of buckling out of the plane of Fig. 11.41, the end-condition constant is $C = 4$. Since $S_r < (S_r)_D$, Johnson's parabolic equation is appropriate, and the critical load is

$$P_{cr} = A\left[S_{yc} - \frac{1}{CE}\left(\frac{S_{yc}S_r}{2\pi}\right)^2\right],$$

which gives

$$P_{cr} = 0.9531 \text{ in}^2\left[60\,000 \text{ psi} - \frac{1}{4(30 \times 10^6 \text{ psi})}\left(\frac{60\,000 \text{ psi} \cdot 42.353}{2\pi}\right)^2\right] = 55\,886 \text{ lb}.$$

Since the applied load $P_{app} = 25\,981$ lb, the factor of safety is

$$N = \frac{55\,886 \text{ lb}}{25\,981 \text{ lb}} = 2.15. \qquad\qquad Ans.$$

Part E:

Let the distance b between pins O_2 and O_3 be varied between 5 in and 45 in. The goal is to determine the minimum value of b such that the member does not fail in buckling. Assume that member 3 is located at an unknown distance b from the pin O_2. Then the force P_{app} acting at point A can be obtained by taking moments about O_2, that is,

$$P_{app} = \frac{(60 \text{ in} \cos 30^o)15\,000 \text{ lb}}{b} = \frac{779\,420 \text{ in} \cdot \text{lb}}{b}. \qquad (5)$$

As in Part D, the cross-sectional area of member 3 is $A = 0.9531$ in^2, the moment of inertia is $I = 0.1591$ in^4, and the radius of gyration of the member is $k = 0.4087$ in.

The length of the member is

$$L = b \tan 30^o = 0.5774b.$$

The slenderness ratio of the member is

$$S_r = \frac{L}{k} = \frac{0.5774b}{0.4087 \text{ in}} = (1.413 \text{ in}^{-1})b.$$

Note: Since b can vary between 5 in and 45 in, the maximum value of the slenderness ratio of the member is $1.07 \le S_r \le 63.58$, which is still less than $(S_r)_D = 99.3$. This implies that Johnson's parabolic equation is valid for all possible values of b. Therefore, the critical load can be written as

Example 11.13 (cont.)

$$P_{cr} = A \left[S_{yc} - \frac{1}{CE} \left(\frac{S_{yc}S_r}{2\pi} \right)^2 \right].$$

Substituting the known data into this equation, the critical load can be written as

$$P_{cr} = 0.953\,1 \text{ in}^2 \left[\left(60 \times 10^3 \text{ psi} \right) - \frac{1}{1\left(30 \times 10^6 \text{ psi} \right)} \left(\frac{\left(60 \times 10^3 \text{ psi} \right)\left(1.413 \text{ in}^{-1} \right)b}{2\pi} \right)^2 \right],$$

or as

$$P_{cr} = 57\,186 \text{ lb} - (5.784 \text{ psi})b^2. \tag{6}$$

The factor of safety guarding against buckling is defined as

$$N = \frac{P_{cr}}{P_{app}}. \tag{7}$$

Substituting Eqs. (5) and (6) into Eq. (7), the factor of safety guarding against buckling can be written as

$$N = \frac{57\,186 \text{ lb} - (5.784 \text{ psi})b^2}{779\,420 \text{ in} \cdot \text{lb}/b} = \frac{(57\,186 \text{ lb})b - (5.784 \text{ psi})b^3}{779\,420 \text{ in} \cdot \text{lb}}$$

or as

$$N = \left(0.073\,37 \text{ in}^{-1} \right)b - \left(7.421 \times 10^{-6} \text{ in}^{-3} \right)b^3.$$

Fig. 11.43 Plot of the factor of safety N versus dimension b.

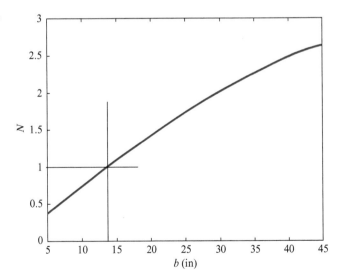

Example 11.13 (cont.)

The member will fail in buckling for all values of b for which the value of N is less than 1. A plot of N versus b is shown in Fig. 11.43. We observe from this plot that the factor of safety guarding against buckling is less than 1 for values of b less than 13.9 in. This implies failure in the buckling mode.   *Ans.*

Example 11.14

A solid aluminum member is subjected to the concentric axial load $P = 3\,\text{kN}$, as shown in Fig. 11.44. The length of the member is $L = 180\,\text{mm}$, and the cross-section is rectangular with dimensions $b = 20\,\text{mm}$ and $h = 4\,\text{mm}$. The aluminum has a modulus of elasticity $E = 75\,\text{GPa}$ and a compressive yield strength $S_{yc} = 150\,\text{MPa}$. Use the AISC recommended end-condition constant for fixed–pinned end conditions.

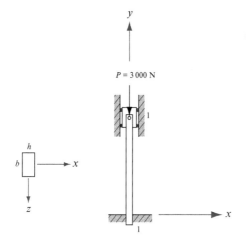

Fig. 11.44 A pinned–fixed member; $h = 4\,\text{mm}, b = 20\,\text{mm}$.

$P = 3\,000\,\text{N}$

a. Is the member an Euler column or a Johnson column?
b. Determine the maximum load P_{cr} that can be applied before buckling occurs.
c. Determine the factor of safety guarding against buckling of the member.

Solution

The cross-sectional area of the member is

$$A = bh = (0.020\ \text{m})(0.004\ \text{m}) = 0.000\,080\ \text{m}^2.$$

The second moment of area (Table A.4, Appendix A) is

$$I = \frac{bh^3}{12} = \frac{(0.020\ \text{m})(0.004\ \text{m})^3}{12} = 1.067 \times 10^{-10}\text{m}^4.$$

Example 11.14 (cont.)

Note here that the second moment of area has the smallest value allowable, since the member will buckle in the weakest plane. For this example, the member will buckle in the plane of Fig. 11.44.

The radius of gyration of the member is

$$k = \sqrt{\frac{I}{A}} = \sqrt{\frac{1.067 \times 10^{-10} \text{ m}^4}{0.000\,080 \text{ m}^2}} = 1.155 \text{ mm}.$$

Using the AISC recommended end-condition effective length factor for fixed–pinned end conditions (that is, $\alpha = L_{\text{eff}}/L = 0.707$, Fig. 11.29), the end-condition constant is

$$C = \frac{1}{\alpha^2} = \frac{1}{0.707^2} = 2.$$

a. The slenderness ratio of the member is

$$S_r = \frac{L}{k} = \frac{0.180 \text{ m}}{0.001\,155 \text{ m}} = 155.84.$$

The slenderness ratio at the point of tangency is

$$(S_r)_D = \pi\sqrt{\frac{2EC}{S_{yc}}} = \pi\sqrt{\frac{2(75 \times 10^9 \text{ Pa})2}{150 \times 10^6 \text{ Pa}}} = 140.5.$$

Since the slenderness ratio $S_r \geq (S_r)_D$, the member is an Euler column.   *Ans.*

b. Using Euler's column formula, the critical load is

$$P_{\text{cr}} = A\left[\frac{C\pi^2 E}{S_r^2}\right] = (0.000\,080 \text{ m}^2)\left[\frac{2\pi^2(75 \times 10^9 \text{ Pa})}{155.84^2}\right] = 4877 \text{ N}.$$ *Ans.*

c. The factor of safety guarding against buckling is

$$N = \frac{P_{\text{cr}}}{P} = \frac{4877 \text{ N}}{3000 \text{ N}}1.63$$ *Ans.*

Therefore, the member should not fail in buckling.

Example 11.15

For the simple truss shown in Fig. 11.45, a vertical load $P = 10$ kN is acting at point A in the negative y direction on the horizontal member 2, which is pinned to the vertical wall at point D.

Example 11.15 (cont.)

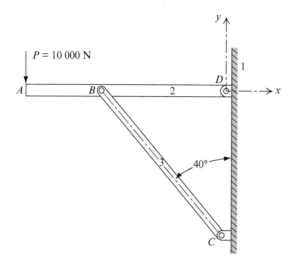

P = 10 000 N

A   B   2   D   x

40°

3

C

Fig. 11.45 A simple truss subjected to a vertical load P. $R_{DC} = 5.0$ m.

Assume that member 3 is pinned to member 2 at point B and also to the vertical wall at point C. Use the theoretic value for the end-condition constant of member 3, and assume that this member is steel with a yield strength $S_{yc} = 370$ MPa and a modulus of elasticity $E = 207$ GPa.

a. Assume that member 3 has a solid circular cross-section with a constant diameter d. Determine the diameter d to ensure that the member has a factor of safety guarding against buckling of $N = 2$. Does Euler's column formula or Johnson's parabolic equation give the correct diameter?

b. Assume that member 3 has a square cross-section with a constant width t. Determine the width, t, to ensure that the member has a factor of safety guarding against buckling of $N = 2$. Does Euler's column formula or Johnson's parabolic equation give the correct width?

Solution

First, determine the magnitude of the compressive load that is acting on member 3. The unknown lengths can be determined from the given geometry. The length of member 3 can be obtained from the trigonometric relation

$$L_3 = (5.0 \text{ m})/\cos 40° = 6.527 \text{ m}.$$

The distance from point B to point D is

$$R_{BD} = (5.0 \text{ m}) \tan 40° = 4.196 \text{ m}.$$

The force acting on member 3 can be obtained by summing moments about pin D; that is,

$$(7.0 \text{ m})(10\,000 \text{ N}) - (4.196 \text{ m})F_B^y = 0.$$

Example 11.15 (cont.)

Therefore, the vertical force acting on member 3 is

$$F_B^y = 16\,683 \text{ N}.$$

The total compressive force P acting on member 3 can then be obtained from the geometry; that is,

$$P = \frac{16\,683 \text{ N}}{\cos 40°} = 21\,778 \text{ N}.$$

a. Member 3 has a circular cross-section with a constant diameter d. Therefore, from Table A.4 of Appendix A, the cross-sectional area and second moment of area of the member are

$$A = \frac{\pi d^2}{4} \quad \text{and} \quad I = \frac{\pi d^4}{64}.$$

The radius of gyration of the member is

$$k = \sqrt{\frac{I}{A}} = \sqrt{\frac{\pi d^4 / 64}{\pi d^2 / 4}} = \frac{d}{4}.$$

The end conditions of member 3 are pinned–pinned. Using the theoretic value for the end-condition constant (that is, $C = 1$ or $\alpha = 1$) gives $L_{\text{eff}} = L$. Therefore, the slenderness ratio of member 3 is

$$S_{\text{r}} = \frac{L_{\text{eff}}}{k} = \frac{6.527 \text{ m}}{d/4} = \frac{26.108 \text{ m}}{d}.$$

The factor of safety guarding against buckling is defined as

$$N = \frac{P_{\text{cr}}}{P}.$$

Rearranging this equation gives the critical load as

$$P_{\text{cr}} = NP = 2(21\,778 \text{ N}) = 43\,556 \text{ N}.$$

The slenderness ratio at the point of tangency is

$$(S_{\text{r}})_D = \pi\sqrt{\frac{2E}{S_{yc}}} = \pi\sqrt{\frac{2\left(207 \times 10^9 \text{ Pa}\right)}{370 \times 10^6 \text{ Pa}}} = 105.1.$$

The diameter d can be obtained from Euler's column formula; that is,

$$\frac{P_{\text{cr}}}{A} = \frac{\pi^2 E}{S_{\text{r}}^2}.$$

Example 11.15 (cont.)

Substituting known values gives

$$\frac{43\,556\text{ N}}{\pi d^2/4} = \frac{\pi^2\left(207\times10^9\text{ Pa}\right)}{(26.108\text{ m}/d)^2}.$$

Solving for the diameter d gives

$$d = 0.066\text{ m}. \tag{1}$$

Check: The diameter d can also be obtained from Johnson's parabolic equation; that is,

$$\frac{P_{\text{cr}}}{A} = S_{yc} - \frac{1}{E}\left(\frac{S_{yc}S_r}{2\pi}\right)^2. \tag{2}$$

Substituting the known values into Eq. (2) gives

$$\frac{43\,556\text{ N}}{(\pi d^2/4)} = \left(370\times10^6\text{ Pa}\right) - \frac{1}{207\times10^9\text{ Pa}}\left(\frac{\left(370\times10^6\text{ Pa}\right)(26.108\text{ m}/d)}{2\pi}\right)^2.$$

Solving this for diameter d gives

$$d = 0.176\text{ m}. \tag{3}$$

Now we must check which answer is valid; that is, Eq. (1) or (3). We recall that the slenderness ratio is

$$S_r = \frac{26.108\text{ m}}{d}.$$

First, we check to determine whether Euler's column result, Eq. (1), is valid. The slenderness ratio is

$$(S_r)_{\text{Euler}} = \frac{26.108\text{ m}}{0.066\text{ m}} = 395.6.$$

Since the slenderness ratio $(S_r)_{\text{Euler}}$ is greater than $(S_r)_D$, that is, since $395.6 > 105.1$, Euler's column formula gives a valid result. The diameter of the member is

$$d = 0.066\text{ m} = 66\text{ mm}. \qquad Ans.$$

Next we check to make sure that Johnson's parabolic equation result is not valid. The slenderness ratio from this result is

$$(S_r)_{\text{Johnson}} = \frac{26.108\text{ m}}{0.176\text{ m}} = 148.3.$$

Since the slenderness ratio $(S_r)_{\text{Johnson}}$ is greater than $(S_r)_D$; that is, since $148.3 > 105.1$, Johnson's parabolic equation result is *not* valid.

Example 11.15 (cont.)

b. Member 3 has a square cross-section with constant width t. Therefore, from Table A.4 of Appendix A, the cross-sectional area and second moment of area of the member are

$$A = t^2 \quad \text{and} \quad I = t^4/12.$$

The radius of gyration of the member is

$$k = \sqrt{\frac{I}{A}} = \sqrt{\frac{t^4/12}{t^2}} = 0.289t.$$

The slenderness ratio of member 3 is

$$S_r = \frac{L_{eff}}{k} = \frac{6.527 \text{ m}}{0.289t} = \frac{22.6 \text{ m}}{t}.$$

First, t can be found using Euler's column formula; that is,

$$\frac{P_{cr}}{A} = \frac{\pi^2 E}{S_r^2}.$$

Substituting the known values into this equation gives

$$\frac{43\,556 \text{ N}}{t^2} = \frac{\pi^2 \left(207 \times 10^9 \text{ Pa}\right)}{\left(22.6 \text{ m}/t\right)^2}.$$

Solving for the width t gives

$$t = 0.057 \text{ m}.$$

The width t can also be obtained from the Johnson parabolic equation; that is,

$$\frac{P_{cr}}{A} = S_{yc} - \frac{1}{E}\left(\frac{S_{yc} S_r}{2\pi}\right)^2.$$

Substituting the known values into this equation gives

$$\frac{43\,556 \text{ N}}{t^2} = \left(370 \times 10^6 \text{ Pa}\right) - \frac{1}{\left(207 \times 10^9 \text{ Pa}\right)}\left(\frac{\left(370 \times 10^6 \text{ Pa}\right)22.6 \text{ m}/t}{2\pi}\right)^2.$$

Therefore, the width t is

$$t = 0.152 \text{ m}.$$

To check which answer is valid, we recall that the slenderness ratio is

$$S_r = \frac{22.6 \text{ m}}{t}.$$

Example 11.15 (cont.)

First, check to see whether the use of Euler's column formula is valid:

$$(S_r)_{Euler} = \frac{22.6 \text{ m}}{0.057 \text{ m}} = 396.5.$$

Since the slenderness ratio $(S_r)_{Euler}$ is greater than $(S_r)_D$, that is, since $396.5 > 105.1$, Euler's column formula gives a valid result. The correct width of the member is

$$t = 0.057 \text{ m} = 57 \text{ mm.} \qquad\qquad Ans.$$

Next we check to make sure that Johnson's parabolic equation result is not valid. The slenderness ratio from this result is

$$(S_r)_{Johnson} = \frac{22.6 \text{ m}}{0.152 \text{ m}} = 148.7.$$

Since the slenderness ratio $(S_r)_{Johnson}$ is greater than $(S_r)_D$, that is, since $148.7 > 105.1$, use of Johnson's parabolic equation is *not* valid.

PROBLEMS

11.1 Figure P11.1 shows four linkages and the external forces and torques exerted on or by the linkages. Sketch the free-body diagram of each part of each linkage. Do not attempt to show the magnitudes of the forces, except roughly, but do sketch them in their proper locations and orientations.

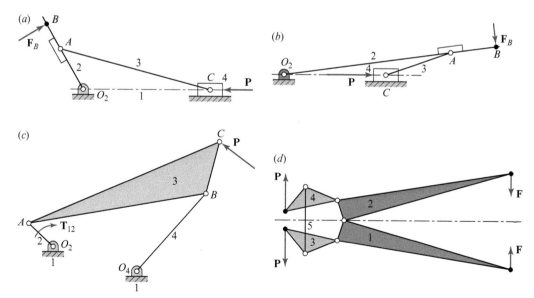

Fig. P11.1

11.2 If force **P** = 200 lb, determine the torque, T_{12}, that must be applied to crank 2 to maintain the linkage in static equilibrium at the posture shown.

Fig. P11.2 R_{AO_2} = 3 in and
R_{BA} = 14 in.

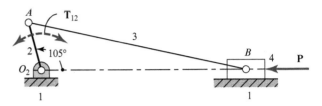

11.3 If T_{12} = 900 in · lb cw for the linkage shown in Fig. P11.2, determine the force **P** to maintain static equilibrium at the posture shown

11.4 If force **P** = 400 N, determine the forces acting on the ground and the torque, T_{12}, to maintain static equilibrium for the four-bar linkage at the posture shown in Fig. P11.4a.

Fig. P11.4 R_{AO_2} = 88 mm, R_{BA} =
R_{BO_4} = 150 mm,
R_{CO_4} = 100 mm,
R_{DO_4} = 175 mm, and
$R_{O_2O_4}$ = 50 mm.

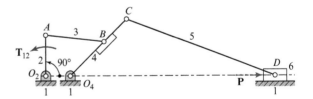

11.5 If force **P** = 400 N, determine the torque that must be applied to link 2 of the linkage shown in Fig. P11.4b to maintain static equilibrium at the posture shown.

11.6 Sketch a complete free-body diagram of each link and determine the force, **P**, to maintain static equilibrium at the posture shown.

Fig. P11.6 R_{AO_2} = 4 in, R_{BA} = 6 in,
R_{BO_4} = 5 in, R_{CO_4} = 8 in, R_{CD} = 16 in,
$R_{O_2O_4}$ = 2.4 in, and T_{12} = 800 in·lb.

11.7 Determine the torque, T_{12}, required to drive slider 6 of Fig. P11.7 against a load of **P** = 444 N at a crank angle of θ = 30°.

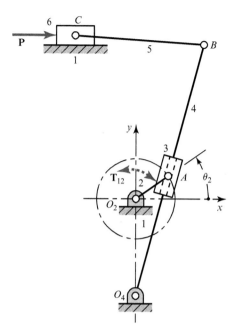

Fig. P11.7 $R_{AO_2} = 62.5$ mm, $R_{O_2O_4} = 150$ mm, $R_{CO_2}^y = 250$ mm, $R_{BO_4} = 400$ mm, and $R_{BC} = 200$ mm.

11.8 Sketch complete free-body diagrams of each link and determine the torque, $\mathbf{T}_{12}$, that must be applied to link 2 to maintain static equilibrium at the posture shown.

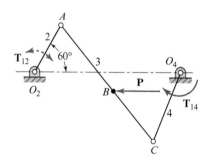

Fig. P11.8 $R_{AO_2} = 8$ in, $R_{BA} = 16$ in, $R_{CA} = R_{O_4O_2} = 28$ in, $R_{CO_4} = 14$ in; $P = 75$ lb and $T_{14} = 400$ in·lb cw.

11.9 Sketch free-body diagrams of each link, and show all the forces acting. Find the magnitude and direction of the torque that must be applied to link 2 at the posture shown to drive the linkage against the forces shown.

Fig. P11.9 $R_{AO_2} = 100$ mm,
$R_{CA} = 350$ mm, $R_{O_4O_2} = 350$ mm, $R_{CO_4} = 250$ mm, $R_{DO_4} = 175$ mm, $R_{BA} = 350$ mm, $R_{BC} = 200$ mm; $P_B = 450$ N and $P_D = 900$ N.

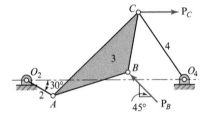

11.10 Figure P11.10 shows a four-bar linkage with external forces applied at points B and C. Draw a free-body diagram of each link, and show all the forces acting on each. Find the torque that must be applied to link 2 to maintain static equilibrium at the posture shown.

Fig. P11.10 $R_{AO_2} = 3$ in, $R_{CA} = 12$ in, $R_{O_4O_2} = 16$ in, $R_{CO_4} = R_{BA} = 8$ in, $R_{BC} = 6$ in; $P_B = 100$ lb and $P_C = 360$ lb.

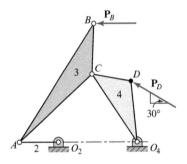

11.11 Draw a free-body diagram of each member of the linkage, and find the magnitudes and the directions of all forces and moments. Compute the magnitude and direction of the torque that must be applied to link 2 to maintain static equilibrium at the posture indicated.

Fig. P11.11 $R_{AO_2} = 100$ mm,
$R_{CA} = 250$ mm, $R_{O_4O_2} = R_{CO_4} = 200$ mm, $R_{DO_4} = 150$ mm, $R_{DC} = 100$ mm, $R_{BA} = 350$ mm, $R_{BC} = 125$ mm; $P_B = 500$ N and $P_D = 750$ N.

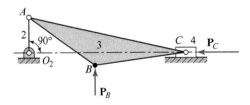

11.12 Determine the magnitude and direction of the torque that must be applied to link 2 to maintain static equilibrium at the posture shown.

Fig. P11.12 $R_{AO_2} = 75$ mm,
$R_{CA} = 350$ mm, $R_{BA} = 175$ mm, $R_{BC} = 200$ mm; $P_B = 225$ N and $P_C = 450$ N.

11.13 Figure P11.13a shows a Figee floating crane with lemniscate boom configuration, and Fig. P11.13b shows a schematic diagram of the crane with dimensions given in the legend. The lifting capacity is 16 T (where 1 T = 1 metric ton = 1 000 kg) including the grab, which is about 10 T. The maximum outreach is 30 m, which corresponds to the position $\theta_2 = 49°$. The minimum outreach is 10.5 m at $\theta_2 = 132°$. For the maximum outreach posture and a grab load of 10 T (under standard gravity), find the bearing reactions at A, B, O_2, and O_4, as well as the torque, $\mathbf{T}_{12}$, at O_2. Notice that the photograph shows a counterweight on link 2; neglect this weight and also the weights of the members.

(a)

(b)

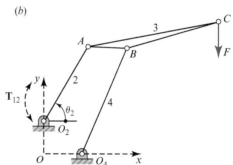

Fig. P11.13 (a) Photograph (Courtesy of Zoonar RF/Zoonar via Getty Images), and (b) $R_{AO_2} = 14.7$ m; $R_{BA} = 6.5$ m; $R_{BO_4} = 19.3$ m; $R_{CA} = 22.3$ m, $R_{CB} = 16$ m; and $\mathbf{R}_{O_2O_4} = -6.4\hat{\mathbf{i}} + 5.3\hat{\mathbf{j}}$ m.

11.14 Repeat Prob. 11.13 for the minimum outreach posture.

11.15 Repeat Prob. 11.7, assuming coefficients of Coulomb friction $\mu_c = 0.20$ between links 1 and 6 and $\mu_c = 0.10$ between links 3 and 4. Determine the torque, $\mathbf{T}_{12}$, necessary to drive the system, including friction, against the load **P**.

11.16 Repeat Prob. 11.12, assuming a coefficient of static friction $\mu = 0.15$ between links 1 and 4. Determine the torque, $\mathbf{T}_{12}$, necessary to overcome friction.

11.17 In each case shown, pinion 2 is the driver, gear 3 is an idler, the gears have module of 4 and 20° pressure angle. For each case, sketch the free-body diagram of gear 3 and show all forces acting. For (a), pinion 2 rotates at 600 rev/min and transmits 13.4 kW to the gearset. For (b) and (c), pinion 2 rotates at 900 rev/min and transmits 18.6 kW to the gearset.

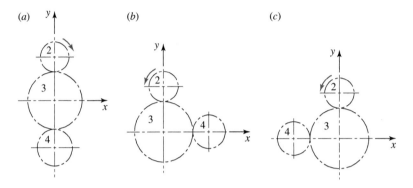

Fig. P11.17 $N_2 = 18$T, $N_3 = 36$T, $N_4 = 20$T.

11.18 A 15-tooth spur pinion has a module of 5 and 20° pressure angle, rotates at 600 rev/min, and drives a 60-tooth gear. The drive transmits 18 kW. Construct a free-body diagram of each gear, showing upon it the tangential and radial components of the forces and their proper directions.

11.19 A 16-tooth pinion on shaft 2 rotates at 1 720 rev/min and transmits 37.5 kW to the double-reduction gear train. All gears have 20° pressure angle. Find the magnitude and direction of the radial force that each bearing exerts against the shaft.

Fig. P11.19 $R_{BA} = 200$ mm, $R_{CA} = R_{DB} = 50$ mm, $m_2 = 3$ mm/tooth, $N_2 = 16$ T, $N_3 = 24$ T, $m_4 = 4$ mm/tooth, $N_4 = 36$ T, $N_5 = 64$ T.

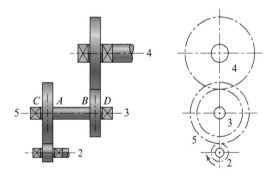

11.20 Solve Prob. 11.17 if each pinion has right-hand helical teeth with a 30° helix angle and a 20° pressure angle. All gears in the train are helical, and the normal module is 4 mm/ tooth for each case.

11.21 Analyze the gear shaft of Example 11.8, and find the bearing reactions $\mathbf{F}_C$ and $\mathbf{F}_D$.

11.22 In each of the bevel gear drives shown, bearing A takes both thrust load and radial load, whereas bearing B takes only radial load. The teeth are cut with a 20° pressure angle. Compute the bearing loads for each case.

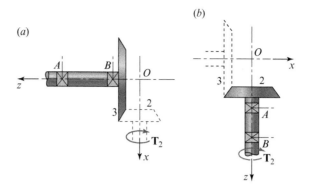

(b)

(a)

Fig. P11.22 (a) $R_{BO} = 26.25$ mm, $R_{AB} = 50$ mm, $R_2 = 17.25$ mm, 16T, $R_3 = 34.5$ mm, 32T, $m = 2.5$ mm/tooth, $\mathbf{T}_2 = -20\hat{\imath}$ N·m; and (b) $R_{AO} = 62.5$ mm, $R_{BA} = 50$ mm, $R_2 = 32$ mm, 18T, $R_3 = 42.5$ mm, 24T, $m = 4$ mm/tooth, $\mathbf{T}_2 = -27\hat{k}$ N·m.

11.23 The figure shows a gear train composed of a pair of helical gears and a pair of straight-tooth bevel gears. Shaft 4 is the output of the train and delivers 4.5 kW to the load at a speed of 370 rev/min. All gears have pressure angles of 20°. If bearing E is to take both thrust load and radial load, whereas bearing F is to take only radial load, determine the force that each bearing exerts against shaft 4.

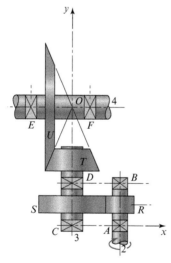

Fig. P11.23 $R_{AR} = R_{BR} =$ 21 mm, $N_R = 15$, $m = 2$ mm/tooth, $\psi = 30°$, $N_S = 35$, $R_{TD} = 12$ mm, $N_T = 20$ T, $m = 3$ mm/tooth, $F = 24$ mm, $N_U = 40$ T, $R_{EO} = 42$ mm, $R_{FO} = 18$ mm.

11.24 Using the data of Prob. 11.23, find the forces exerted by bearings C and D onto shaft 3. Which of these bearings should take the thrust load if the shaft is to be loaded in compression?

11.25 Use the method of virtual work to solve the slider-crank linkage of Prob. 11.2.

11.26 Use the method of virtual work to solve the four-bar linkage of Prob. 11.5.

11.27 Use the method of virtual work to analyze the crank-shaper linkage of Prob. 11.7. Given that the load remains constant at $\mathbf{P} = 4\hat{\imath}$ kN, find and plot a graph of the crank torque, T_{12}, for all postures in the cycle using increments of 30° for the input crank.

11.28 Use the method of virtual work to solve the four-bar linkage of Prob. 11.10.

11.29 A car (link 2) that weighs 9.0 kN is slowly backing a 4.5 kN trailer (link 3) up a 30° ramp. The car wheels are of 325 mm radius, and the trailer wheels have 250 mm radius; the center of the hitch ball is also 325 mm above the roadway. The centers of mass of the car and trailer are located at G_2 and G_3, respectively, and gravity acts vertically downward. The weights of the wheels and friction in the bearings are considered negligible. Assume that there are no brakes applied on the car or on the trailer and that the car has front-wheel drive. Determine the loads on each of the wheels and the minimum coefficient of static friction between the driving wheels and the road to avoid slipping.

Fig. P11.29 $R_{FR} = 2\,000$ mm,
$R_{G_2F}^x = 800$ mm,
$R_{G_2F}^y = 475$ mm,
$R_{BR}^x = 900$ mm,
$R_{BT}^u = -1\,250$ mm,
$R_{G_3T}^v = 350$ mm.

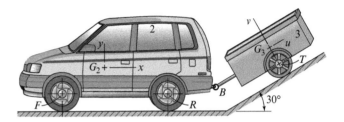

11.30 Repeat Prob. 11.29, assuming that the car has rear-wheel drive rather than front-wheel drive.

11.31 The low-speed disk cam with oscillating flat-faced follower is driven at a constant shaft speed. The displacement curve for the cam has a full-rise cycloidal motion, defined by Eq. (6.13) with parameters $L = 30°$, $\beta = 150°$, and a prime-circle radius $R_0 = 1.20$ in; the instant pictured is at $\theta_2 = 112.5°$. A force of $F_C = 1.8$ lb is applied at point C and continues at 45° from the face of the follower as shown. Use the virtual work approach to determine the torque $\mathbf{T}_{12}$ required on the crankshaft at the instant shown to produce this motion.

Fig. P11.31 $R_{O_2O_3} = 2$ in,
$R_B = 1.68$ in, and $R_C = 6$ in.

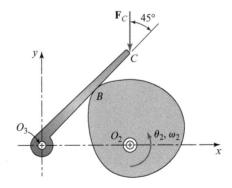

11.32 Repeat Prob. 11.31 for the entire lift portion of the cycle, finding $\mathbf{T}_{12}$ as a function of θ_2.

11.33 Disk 3 of radius R is being slowly rolled under a pivoted bar 2 driven by an applied torque, T. Assume a coefficient of static friction of μ between the disk and ground and that all other joints are frictionless. A force, $\mathbf{F}$, is acting vertically downward on the bar at a distance d from the pivot O_2. Assume that the weights of the links are negligible in comparison to $\mathbf{F}$. Find an equation for the torque, T, required as a function of the distance $x = R_{CO_2}$, and an equation for the final distance, x, that is reached when friction no longer allows further movement.

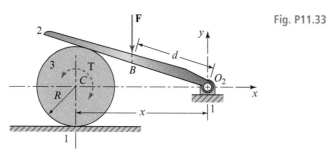

Fig. P11.33

11.34 For the linkage in the posture shown, an external torque $\mathbf{T}_{14} = -5.5\hat{\mathbf{k}}$ N·m is acting on link 4 about O_4, and a horizontal force $\mathbf{F}_C$ is acting at point C on link 3 to hold the linkage in static equilibrium. Assume that gravity is acting into the plane, and the effects of friction can be neglected. (a) Draw free-body diagrams of links 2, 3, and 4. (b) Determine the magnitudes, directions, and locations of all internal reaction forces. (c) Determine the magnitude and direction of the external force acting at point C.

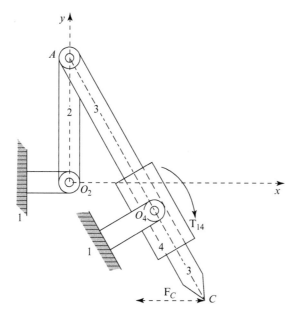

Fig. P11.34 $R_{AO_4} = 100$ mm, $R_{AO_2} = 65$ mm, and $R_{CA} = 150$ mm.

11.35 A horizontal force $\mathbf{F}_C = 5.6$ lb is acting at point C on link 4 and an external torque $\mathbf{T}_{12}$ is acting on link 2. Link 3 is inclined at $30°$ to the horizontal. The coefficient of friction between link 4 and the ground link is $\mu = 0.3$, and the coefficient of friction between link 2 and the ground link is not specified, but is large enough that there is no slip at E. The impending motion of link 4 is to the right. Assume that gravity is acting into the plane of the figure. Block 4 is 2.40 in wide by 1.60 in high, with pin B centrally located. Point C is 0.20 in above the centerline. (a) Determine the magnitude and direction of the external torque, $\mathbf{T}_{12}$, necessary to overcome the force, $\mathbf{F}_C$. (b) Determine the magnitudes, the directions, and the locations of the internal reaction forces. (c) Determine whether link 4 is slipping or tipping on the ground link 1.

Fig. P11.35 $\mathbf{R}_{BA} = 7.20$ in
$\angle 30°$ and $R_{EA} = 1.20$ in.

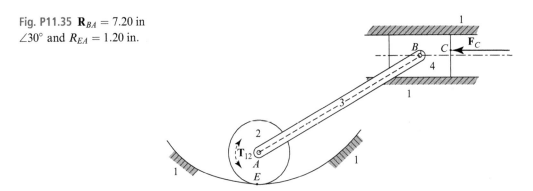

11.36 For the linkage in the posture shown, a constant external torque $\mathbf{T}_{12} = 20\hat{\mathbf{k}}$ N·m is acting on link 2 about the shaft O_2, and a horizontal external force $\mathbf{P}$ is acting on link 5 to hold the linkage in static equilibrium. Assume that gravity is acting into the plane and effects of friction in the linkage can be neglected. (a) Draw free-body diagrams for links 2, 3, 4, 5, and 6. (b) Determine the magnitudes, directions, and locations of the internal reaction forces. (c) Determine the magnitude and direction of the external force $\mathbf{P}$.

Fig. P11.36 $\mathbf{R}_{O_2O} = 175\hat{\mathbf{i}} + 125\hat{\mathbf{j}}$ mm,
$R_{AO_2} = 125$ mm, $R_{BA} = 68.8$ mm,
$R_{CD} = 150$ mm, and $R_{BD} = 70.5$ mm.
Blocks 5 and 6 are each 50 mm by
25 mm, with the pins centrally located.

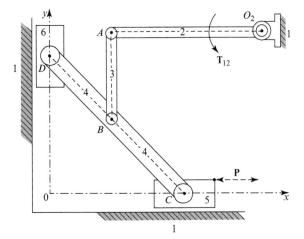

11.37 For the linkage in the posture shown, a torque $T_{12} = 2.16\hat{k}$ in·lb is acting on link 2 at the crankshaft O_2. A force P is applied to point D on link 4 at an angle of $45°$ to hold the linkage in static equilibrium. Gravity is acting into the plane and friction in the linkage can be neglected. Determine the magnitude and the direction of the force P. Determine the magnitudes, directions, and locations of the internal reaction forces in the linkage. Is link 4 slipping or tipping on the ground link?

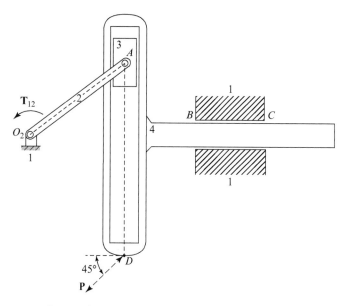

Fig. P11.37 $R_{AO_2} = 1.60\hat{i} + 1.20\hat{j}$ in, $R_{BO_2} = 2.80\hat{i}$ in, $R_{DO_2} = 1.60\hat{i} - 2.00\hat{j}$ in, and $R_{CB} = 1.20\hat{i}$ in.

11.38 For the linkage in the posture shown, a force $P = 45$ N is applied on link 4. A torque T_{12} is acting on link 2 at the crankshaft O_2 to hold the linkage in static equilibrium. Gravity is acting into the plane and friction in the linkage can be neglected. Is the impending motion of link 3 slipping or tipping in link 4? Is the impending motion of link 4 slipping or tipping on the ground link? Determine the magnitudes, directions, and locations of the internal reaction forces in the linkage. Determine the magnitude and the direction of the torque T_{12}.

Fig. P11.38 $\mathbf{R}_{AO_2} = 125$ mm$\angle -60°$, $R_{CO_2}^y = -165$ mm, and $R_{EO_2}^y = -230$ mm.

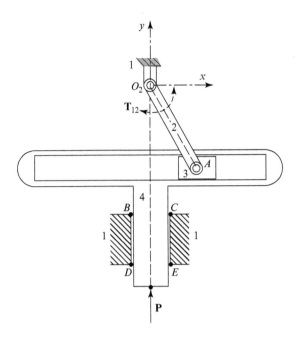

11.39 Links 2 and 3 are pinned together at B, and a constant vertical load $P = 180\,000$ lb is applied at B. Link 2 is fixed in the ground at A, and link 3 is pinned to the ground at C. The length of link 2 is 32 in and it has a 0.6 in solid square cross-section. The length of link 3 is 20 in and it has a solid circular cross-section with diameter D. Both links are made from a steel with a modulus of elasticity $E = 30 \times 10^6$ psi and a compressive yield strength $S_{yc} = 29\,000$ psi. Using the theoretic values for the end-condition constants of each link, determine: (a) the slenderness ratio, the critical load, and the factor of safety guarding against buckling of link 2; and (b) the minimum diameter D_{min} of link 3 if the static factor of safety guarding against buckling is to be $N = 2$.

Fig. P11.39 Two links AB and BC pinned at B.

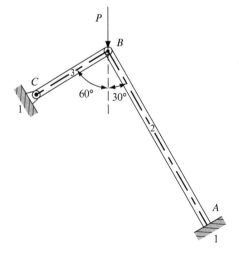

11.40 The horizontal link 2 is subjected to the load $F = 33\,720$ lb at C as shown. The link is supported by the solid circular aluminum link 3. The lengths of the links are $L_2 = R_{CA} = 200$ in, $R_{BA} = 120$ in, and $L_3 = R_{BD} = 120$ in. End D of link 3 is fixed in the ground, and the opposite end, B, is pinned to link 2 (that is, the effective length of the link is $L_{\text{eff}} = 0.5L_2$). For aluminum, the yield strength is $S_{yc} = 53\,660$ psi, and the modulus of elasticity is $E = 30 \times 10^6$ psi. Determine the diameter, d, of the solid circular cross-section of link 3 to ensure that the static factor of safety is $N = 2.5$.

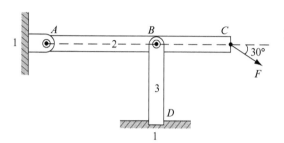

Fig. P11.40 Link BD supporting the link AC.

11.41 The horizontal link 2 is subjected to the inclined load $F = 1\,800$ lb at C as shown. The link is supported by a solid circular cross-section link 3 whose length $L_3 = BD = 200$ in. End D of vertical link 3 is fixed in the ground link, and end B supports link 2 (that is, the effective length of the link is $L_{\text{eff}} = 0.5L_3$). Link 3 is steel with a compressive yield strength $S_{yc} = 53\,660$ psi and a modulus of elasticity $E = 30 \times 10^6$ psi. Determine the diameter, d, of link 3 to ensure that the factor of safety guarding against buckling is $N = 2.5$. Also, classify the following statements as true or false and briefly give your reasons. (a) The slenderness ratio at the point of tangency between Euler's column formula and Johnson's parabolic equation does not depend on the geometry of the column. (b) Under the same loading conditions, a link with pinned–pinned ends gives a higher factor of safety against buckling than an identical link with fixed–fixed ends. (c) If the slenderness ratio $S_r = (S_r)_D$ at the point of tangency, then the critical unit load does not depend on the yield strength of the column material.

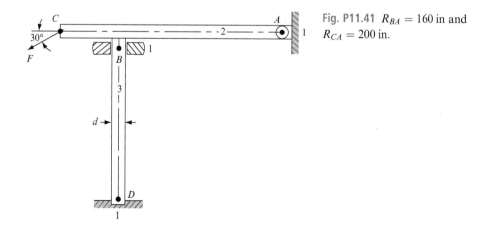

Fig. P11.41 $R_{BA} = 160$ in and $R_{CA} = 200$ in.

11.42 Load $\mathbf{P}_A$ is acting at A and load $\mathbf{P}_B$ is acting at B of the horizontal link 3. Link 3 is pinned to the vertical link 2 at O and link 2 is fixed in the ground link 1 at D. The lengths are $AO = 1.20$ m, $OB = 0.60$ m, and $DO = 1.80$ m. Both links have solid circular cross-sections with diameter $D = 50$ mm and are made from a steel alloy with compressive yield strength $S_{yc} = 604$ MPa, tensile yield strength $S_{yt} = 533$ MPa, and modulus of elasticity $E = 213$ GPa. Assuming that links 2 and 3 are in static equilibrium and using the theoretic value for the end-condition constant for link 2, determine: (a) the magnitude of the force $\mathbf{P}_B$ that is acting as shown at B if $P_A = 133\,200$ N; (b) the critical load, the critical unit load, and the factor of safety to guard against buckling for link 2; and (c) the diameter of a solid circular cross-section for link 2 that will ensure the factor of safety guarding against buckling of the link is $N = 4$.

Fig. P11.42 Link AB supporting two loads.

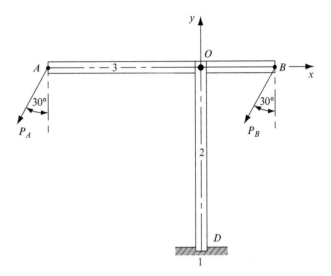

11.43 The horizontal link 2 is subjected to the load $P = 1\,125$ lb and is supported by the vertical link 3, which has a constant circular cross-section. The lengths are $AC = 200$ in, $AB = 160$ in, and $DB = L_3 = 200$ in. For vertical link 3, end D is fixed in the ground link, and end B supports link 2 (that is, the effective length of link 3 is $L_{\text{eff}} = 0.5\ L_3$). The yield strength and the modulus of elasticity for the aluminum link 3 are $S_y = 52\,000$ psi and $E = 29.1 \times 10^6$ psi, respectively. Determine the diameter, d, of link 3 to ensure that the static factor of safety guarding against buckling is $N = 2.5$.

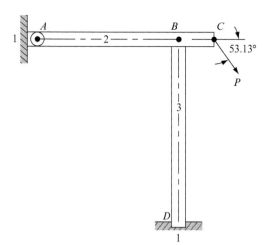

Fig. P11.43 Link 3 supporting link AC.

11.44 Horizontal link 2 is pinned to the vertical wall at A and pinned to link 3 at B. The opposite end of link 3 is pinned to the wall at C. A vertical force $P = 5600$ lb is acting on link 2 at B. Link 2 has a 0.80 in by 1.20 in solid rectangular cross-section, and link 3 has a 1.60 in by 1.60 in solid square cross-section. The length of link 3 is $BC = 48$ in, and $\angle ABC = 30°$. The two links are made from a steel alloy with tensile yield strength $S_{yt} = 26700$ psi, compressive yield strength $S_{yc} = 28800$ psi, and modulus of elasticity $E = 29.1 \times 10^6$ psi. Using the theoretic value for the end-condition constant for link 3, determine: (a) the value of the slenderness ratio at the point of tangency between Euler's column formula and Johnson's parabolic formula; (b) the critical load and the factor of safety guarding against buckling of link 3; and (c) the minimum width of the square cross-section of link 3 for the factor of safety to guard against buckling to be $N = 1$.

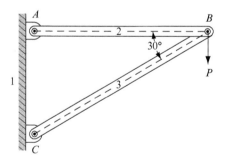

Fig. P11.44 Simple truss.

11.45 Link $BC = 48$ in and 1.0 in square cross-section is fixed in the vertical wall at C and pinned at B to a circular steel cable, AB, with diameter $d = 0.8$ in. Distance $AC = 28$ in. The mass, m, of a container, suspended from pin B, produces a gravitational load at B, which results in the moment at point C in the wall $M_C = 72000$ in · lb ccw. The yield strength and modulus of elasticity of the steel cable AB and the steel link BC are

$S_y = 52000$ lb/in^2 and $E = 29.1 \times 10^6$ lb/in^2, respectively. Given that $mg = 4400$ lb, determine: (a) the tension in the cable AB and the factor of safety guarding against tensile failure; (b) the compressive load acting in link BC; and (c) the factor of safety guarding against buckling of link BC. (Use the theoretic value of the end-condition constant assuming that the link has fixed–pinned ends.) If $M_C = 90000$ in·lb ccw, then determine the maximum weight of a container that can be suspended from pin B before buckling of link BC begins (that is, the factor of safety guarding against buckling failure is $N = 1$).

Fig. P11.45 Structure supporting a load.

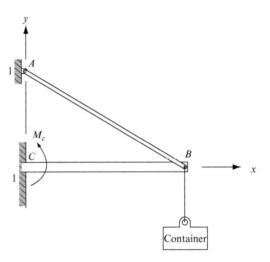

11.46 A vertically upward force, F, is applied at C of the horizontal link 4, which is pinned to the ground at B and pinned to the vertical links 2 and 3 at A. Lengths $AB = 0.6$ m and $AC = 1.5$ m and the three links are made from a steel alloy with a compressive yield strength $S_{yc} = 426$ MPa and a modulus of elasticity $E = 213$ GPa. Link

Fig. P11.46 Link AC supporting a load F.

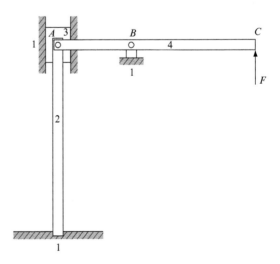

2 has a hollow circular cross-section with an outside diameter $D = 50$ mm, wall thickness $t = 6.25$ mm, and length $L = 1.5$ m. Using the theoretic value for the end-condition constant for link 2, determine: (a) the value of the slenderness ratio at the point of tangency between Euler's column formula and Johnson's parabolic formula; (b) the critical load acting on link 2; and (c) the force, F, for the factor of safety of link 2 to guard against buckling to be $N = 1$. (d) If link 2 has a solid circular cross-section with diameter $D = 75$ mm and $F = 88.8$ kN, then determine the maximum length of link 2 for the factor of safety to guard against buckling to be $N = 2$.

11.47 A force F is acting at C perpendicular to link 2, end A is pinned to the ground, and supporting link 3 is pinned to link 2 at B and pinned to the ground at D. The lengths are $AC = 8$ in and $AD = 6$ in. Link 3 has a circular cross-section with diameter $D = 0.20$ in. Both links are a steel alloy with a compressive yield strength $S_{yc} = 59\,100$ psi and modulus of elasticity $E = 29 \times 10^6$ psi. Using the theoretic value for the end-condition constant for link 3, determine: (a) the slenderness ratio at the point of tangency between Euler's column formula and Johnson's parabolic formula; (b) the critical load and the critical unit load acting on the link; and (c) the force, F, for the factor of safety to guard against buckling to be $N = 1$. If force $F = 676$ lb, then for link 3 determine: (d) the critical load for the factor of safety to guard against buckling $N = 1$; and (e) the slenderness ratio.

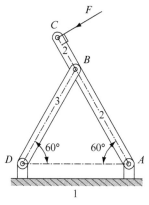

Fig. P11.47 Simple truss.

11.48 For the four-bar linkage in the posture shown, the torque acting on link 2 at the crankshaft O_2 is $\mathbf{T}_{12} = -8\,925\hat{\mathbf{k}}$ N·m. There is also a torque, $\mathbf{T}_{14}$, acting on link 4 at crankshaft O_4 to hold the linkage in static equilibrium. The length of coupler link 3 is $AB = 150$ mm, and the cross-section is rectangular with width $3t$ and thickness $4t$ (into the plane). The coupler link is a steel alloy with compressive yield strength $S_{yc} = 426$ MPa and a modulus of elasticity $E = 213$ GPa. If the critical load for the coupler link is $P_{cr} = 666\,000$ N, and the effects of gravity are ignored, then determine: (a) the factor of safety guarding against buckling; (b) the slenderness ratio at the point of tangency between Euler's column formula and Johnson's parabolic equation; and (c) the numeric value of the parameter t. (d) If the rectangular cross-section is replaced by a circular cross-section of the same material and diameter $d = 5$ mm, then determine the factor of safety guarding against buckling of the coupler link.

Fig. P11.48 Four-bar linkage.

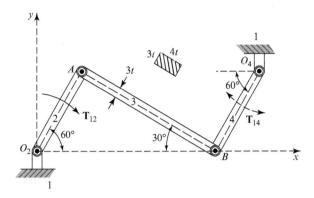

11.49 The single-cylinder engine is in static equilibrium due to the external force $\mathbf{F} = -3375\hat{\mathbf{i}}$ lb acting at point D. The cross-section of the connecting rod (link 3) of length $L_3 = 8$ in is rectangular with width $4t$ and thickness $2t$, and the material is a steel alloy with compressive yield strength $S_{yc} = 28\,800$ psi and modulus of elasticity $E = 29.1 \times 10^6$ psi. Assume for practical purposes that the thickness ($2t$) of link 3 must be greater than 0.20 in and that buckling will occur in the xy plane. The effects of gravity and friction in the engine can be ignored. Determine: (a) the slenderness ratio of link 3 in terms of the unknown thickness parameter, t; (b) the slenderness ratio of link 3 at the point of tangency between Euler's column formula and Johnson's parabolic equation; and (c) the thickness parameter, t, to ensure that the factor of safety guarding against buckling for link 3 is $N = 5$.

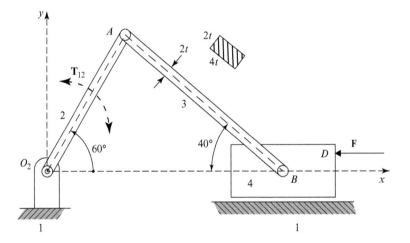

Fig. P11.49 Single-cylinder engine.

11.50 Vertical link 2 is rigidly fixed to the ground (at A) and pinned to horizontal link 3 at B. Link 4 is pinned to link 3 at H, and the weight of link 4 is 440 lb. Vertical link 5 is pinned to link 3 at D and to the ground at O_5. Link 2 has a solid circular cross-section with diameter $d_2 = 1.0$ in, compressive yield strength $S_{yc} = 52\,100$ psi, and modulus of elasticity $E = 28.9 \times 10^6$ psi. The factor of safety guarding against buckling, for link 2, is $N = 2.5$, and the end-condition constant $C = 2$. Assume that gravity is acting vertically downward, and links 2, 3, and 5 are massless compared to link 4. (a) For link 2 determine: (1) the slenderness ratio; and (2) the slenderness ratio at the point of tangency between Euler's column formula and Johnson's parabolic formula. (b) Determine the critical unit load acting on link 2. (c) If the diameter of link 2 is increased to $d_2 = 3.60$ in, then is link 2 an Euler column or a Johnson column?

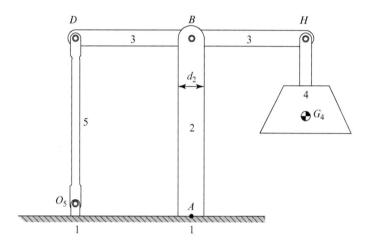

Fig. P11.50 $AB = 100$ in and $DB = BH = 60$ in.

REFERENCES

[1] Beer, F. P., E. R. Johnston, and E. R. Eisenberg, 2007. *Vector Mechanics for Engineers, Statics*, 8th ed., New York: McGraw-Hill.

[2] Neale, M. J. (ed.), 1975. *Tribology Handbook*, London: Butterworths.

[3] Newton, I., 1687. *Principia*, trans. A. Motte, 1729. London: B. Motte.

[4] Plesha, M. E., G. L. Gray, and F. Constanzo, 2013. *Engineering Mechanics: Statics*, 2nd ed., New York: McGraw-Hill.

[5] Shigley, J. E., and C. R. Mischke, 2001. *Mechanical Engineering Design*, 6th ed., New York: McGraw-Hill.

12 Dynamic Force Analysis

12.1 Introduction

In Chapter 11, we studied the forces in machine systems in which all forces on the bodies were in balance, and therefore the systems were in either static or dynamic equilibrium. However, in real machines this is seldom, if ever, the case except when the machine is stopped. We learned in Chapter 4 that although the input crank of a machine may be driven at constant speed, this does not mean that all points of the input crank have constant velocity vectors or that other links of the machine operate at constant speeds. In general, there will be accelerations, and therefore machines with moving parts having mass are not balanced.

Of course, techniques for static force analysis are important, not only because stationary structures must be designed to withstand their imposed loads, but also because they introduce concepts and approaches that can be built upon and extended to nonequilibrium situations. That introduces the purpose of this chapter: to learn how much acceleration will result from a system of unbalanced forces and also to learn how these *dynamic* forces can be assessed for systems that are not in balance.

12.2 Centroid and Center of Mass

We recall from Section 11.2 that Newton's laws set forth the relationships between the net unbalanced force on a *particle*, its mass, and its acceleration. For Chapter 11, since we were only studying systems in equilibrium, we made use of that relationship for entire rigid bodies, arguing that they are made up of collections of particles and that the action and the reaction forces between the particles balance each other. In Chapter 12, we must be more careful. We must remember that each of these particles may have acceleration, and that the accelerations of the particles may all be different from each other. Important questions include the following: Which of these many particle accelerations are we to use? And why?

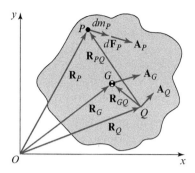

Fig. 12.1 Particle P of mass dm_P.

Referring to the rigid body shown in Fig. 12.1, we consider an infinitesimal particle, with mass dm_P, say at point P. Equation (11.1) tells us that the net unbalanced force, $d\mathbf{F}_P$, on this particle is proportional to its mass and its absolute acceleration, $\mathbf{A}_P$, that is,

$$d\mathbf{F}_P = \mathbf{A}_P dm_P. \qquad (a)$$

Our task now is to sum these effects, that is, to integrate over all particles of the body and to put the result in a usable form for rigid bodies other than single particles. As we did in Chapter 11, we can conclude that the action and the reaction forces between particles of the body balance each other and therefore cancel in the process of the summation. The only net forces remaining are the constraint forces, those whose reactions are on some other body than this one. Thus, integrating Eq. (a) over all particles of mass in our rigid body, we obtain

$$\sum \mathbf{F} = \int \mathbf{A}_P dm_P. \qquad (b)$$

We find it difficult to perform this integration, since each particle has a different acceleration. However, if we know (or can find) the acceleration of one particular point of the body, say point Q, and also the angular velocity $\boldsymbol{\omega}$ and angular acceleration $\boldsymbol{\alpha}$ of the body, then we can express the various accelerations of all other particles of the body in terms of their acceleration differences from that of point Q. Thus, from Eqs. (4.5), we write

$$\mathbf{A}_P = \mathbf{A}_Q + \boldsymbol{\omega} \times \left(\boldsymbol{\omega} \times \mathbf{R}_{PQ}\right) + \boldsymbol{\alpha} \times \mathbf{R}_{PQ}. \qquad (c)$$

Substituting this equation into Eq. (b), the sum of the forces can be written as

$$\sum \mathbf{F} = \mathbf{A}_Q \int dm_P + \boldsymbol{\omega} \times \left(\boldsymbol{\omega} \times \int \mathbf{R}_{PQ} dm_P\right) + \boldsymbol{\alpha} \times \int \mathbf{R}_{PQ} dm_P, \qquad (d)$$

where we have factored out of the integrals all quantities that are equal for all particles of the body.

We recognize the first integral of Eq. (d); the summation of all particle masses gives the total mass of the body:

$$\int dm_P = m. \qquad (12.1)$$

The second and third terms of Eq. (*d*) both contain another integral, which is not as easy to recognize. However, because of its frequent appearance in the study of mechanics, a particular point *G* having a location given by the position vector $\mathbf{R}_G$ is defined by the integral equation:

$$\int \mathbf{R}_P dm_P = m\mathbf{R}_G. \tag{12.2}$$

This point *G* is called the *centroid*, or the *center of mass* or, simply, the *mass center* of the body. More will be said about this special point later.

Substituting Eqs. (12.1) and (12.2) into Eq. (*d*) gives

$$\sum \mathbf{F} = m\mathbf{A}_Q + \boldsymbol{\omega} \times \left[\boldsymbol{\omega} \times (m\mathbf{R}_{GQ}) \right] + \boldsymbol{\alpha} \times (m\mathbf{R}_{GQ}),$$

which can be written as

$$\sum \mathbf{F} = m \left[\mathbf{A}_Q + \boldsymbol{\omega} \times (\boldsymbol{\omega} \times \mathbf{R}_{GQ}) + \boldsymbol{\alpha} \times \mathbf{R}_{GQ} \right].$$

Finally, recognizing the sum of the three acceleration terms as the acceleration of the mass center, we can write

$$\sum \mathbf{F} = m\mathbf{A}_G. \tag{12.3}$$

This important equation is the integrated form of Newton's law for a particle, now extended to a rigid body. Note that it is the same equation as Eq. (11.1). However, careful derivation of the acceleration term was not presented there, since we were treating static problems where accelerations were zero.

We have now answered the question raised earlier in this section. Recognizing that each particle of a rigid body may have a different acceleration, which one should be used? Equation (12.3) indicates clearly that the absolute acceleration of the center of mass of the body is the proper acceleration to be used in Newton's law. That particular point, and no other, is the proper point for which Newton's law for a rigid body has the same form as for a single particle.

In engineering problems, we frequently find that forces are distributed in some manner along a line, over an area, or over a volume. The resultant of these distributed forces is usually not too difficult to find. To have the equivalent effect, this resultant must act at the centroid of the system. Thus, *the centroid of a system is a point at which a system of distributed forces may be considered concentrated with exactly the same effect* [1].

Instead of a system of forces, we may have a distributed mass, as in the above derivation. Then, by *center of mass* we mean *the point at which the mass may be considered concentrated* and that the effect is the same.

In Fig. 12.2*a*, a series of particles with masses are shown located at various positions along a line. The location of the center of mass *G* is given by

$$\bar{x} = \frac{m_1 x_1 + m_2 x_2 + m_3 x_3}{m_1 + m_2 + m_3} = \frac{\sum m_i x_i}{\sum m_i}. \tag{12.4}$$

(a)   (b)

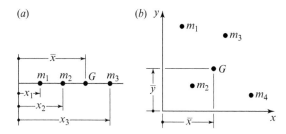

Fig. 12.2 Particles of mass distributed: (a) along a line; (b) over a plane.

(a)   (b)   (c)

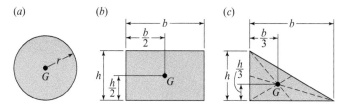

Fig. 12.3 Center of mass location for: (a) a right circular solid; (b) a rectangular solid; (c) a triangular prism.

In Fig. 12.2b, the mass particles m_i are located at various positions $\mathbf{R}_i$ over an area. The location of the center of mass G is now given by

$$\mathbf{R}_G = \frac{m_1\mathbf{R}_1 + m_2\mathbf{R}_2 + m_3\mathbf{R}_3 + m_4\mathbf{R}_4}{m_1 + m_2 + m_3 + m_4} = \frac{\sum m_i\mathbf{R}_i}{\sum m_i}. \tag{12.5}$$

This procedure can also be extended to particles distributed over a volume by simply using Eq. (12.5) and treating position vectors $\mathbf{R}_i$ as three-dimensional rather than two-dimensional.

When the mass is distributed continuously along a line, over a plane, or over a volume, the concept of summation in Eq. (12.5) is replaced by integration over infinitesimal particles of mass. The result is

$$\mathbf{R}_G = \frac{1}{m}\int \mathbf{R}\,dm, \tag{12.6}$$

where m is the total mass of the body considered. It is from this definition that Eq. (12.2) was obtained.

When mass is uniformly distributed along a line, over a plane, or volume, the center of mass can often be found by symmetry. Figure 12.3 shows the locations of the centers of mass for a circular solid, a rectangular solid, and a triangular solid. Each is assumed to have constant thickness and uniform density distribution.

When a body is of a more irregular shape, the center of mass can often still be determined by considering it a combination of simpler subshapes, as demonstrated in the following example.

Example 12.1

The composite shape with dimensions in millimeters shown in Fig. 12.4 consists of a rectangular solid, less a circular through-hole, plus a triangular plate of a different thickness. Determine the center of mass of this shape, which is made of a homogeneous material with density ρ kg/(mm)^3.

Fig. 12.4 Composite shape; dimensions are in millimeters.

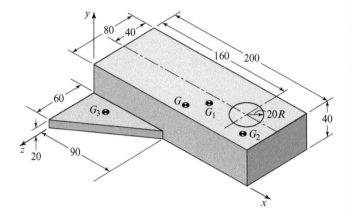

Solution

We can find the masses and the centers of mass of each of the three subshapes with the help of the formulae in Appendix A. For the rectangular solid subshape, we obtain[1]

$$m_1 = (80 \text{ mm})(200 \text{ mm})(40 \text{ mm})\left[\rho \text{ kg/(mm)}^3\right] = 640\,000\,\rho \text{ kg},$$

$$\mathbf{R}_{G_1} = 100\hat{\mathbf{i}} + 20\hat{\mathbf{j}} - 40\hat{\mathbf{k}} \text{ mm}.$$

For the hole we treat the mass as negative, giving

$$m_2 = \pi(20 \text{ mm})^2(40 \text{ mm})\left[-\rho \text{ kg/(mm)}^3\right] = -50\,265\,\rho \text{ kg},$$

$$\mathbf{R}_{G_2} = 160\hat{\mathbf{i}} + 20\hat{\mathbf{j}} - 40\hat{\mathbf{k}} \text{ mm}.$$

Finally, for the triangular subshape, we find

$$m_3 = 0.5(60 \text{ mm})(90 \text{ mm})(20 \text{ mm})\left[\rho \text{ kg/(mm)}^3\right] = 54\,000\,\rho \text{ kg},$$

$$\mathbf{R}_{G_3} = 30\hat{\mathbf{i}} + 10\hat{\mathbf{j}} + 20\hat{\mathbf{k}} \text{ mm}.$$

---

[1] Note that in this example we use SI units. US customary units, pounds, slugs, or other units might have been used. However, this is not emphasized here, because the density cancels in the use of Eq. (12.5).

Example 12.1 (cont.)

Since we are completing a process of integration, we can combine the values of the subshapes; that is, from Eq. (12.5) we can write

$$\mathbf{R}_G = \frac{m_1\mathbf{R}_1 + m_2\mathbf{R}_2 + m_3\mathbf{R}_3}{m_1 + m_2 + m_3}.$$

This gives the following result:

$$\mathbf{R}_G = 89.4\hat{\mathbf{i}} + 19.2\hat{\mathbf{j}} - 35.0\hat{\mathbf{k}} \text{ mm.} \qquad\qquad Ans.$$

12.3 Mass Moments and Products of Inertia

Another problem that often arises when forces are distributed over an area or volume is that of calculating their moment about a specified point or axis of rotation. Sometimes the force intensity varies according to its distance from the point or axis of rotation. Although we will save a more thorough derivation of these equations until Section 12.11 and what follows, we point out here that such problems always give rise to integrals of the form $\int (\text{distance})^2 dm$.

In three-dimensional problems, three such integrals are defined as follows:[2]

$$I^{xx} = \int (\hat{\mathbf{i}} \times \mathbf{R}) \cdot (\hat{\mathbf{i}} \times \mathbf{R}) dm = \int \left[(R^y)^2 + (R^z)^2 \right] dm,$$

$$I^{yy} = \int (\hat{\mathbf{j}} \times \mathbf{R}) \cdot (\hat{\mathbf{j}} \times \mathbf{R}) dm = \int \left[(R^z)^2 + (R^x)^2 \right] dm,$$

$$I^{zz} = \int (\hat{\mathbf{k}} \times \mathbf{R}) \cdot (\hat{\mathbf{k}} \times \mathbf{R}) dm = \int \left[(R^x)^2 + (R^y)^2 \right] dm. \qquad (12.7)$$

These three integrals are called the *mass moments of inertia* (or *second moments of mass*) of the body. Another three similar integrals are

$$I^{xy} = I^{yx} = \int (\hat{\mathbf{i}} \times \mathbf{R}) \cdot (\hat{\mathbf{j}} \times \mathbf{R}) dm = -\int (R^x R^y) dm,$$

$$I^{yz} = I^{zy} = \int (\hat{\mathbf{j}} \times \mathbf{R}) \cdot (\hat{\mathbf{k}} \times \mathbf{R}) dm = -\int (R^y R^z) dm,$$

---

[2] It should be carefully noted here that these integrals are not the same as those called *area moments of inertia* (or *second moments of area*), which are integrals over dA, a differential area, rather than integrals over dm, a differential mass. These other integrals also arise in problems involving forces distributed over an area or volume, but are different. In two-dimensional problems of constant thickness, however, they are easily related, since dA times the thickness times the mass density yields dm for the integral.

$$I^{zx} = I^{xz} = \int (\hat{\mathbf{k}} \times \mathbf{R}) \cdot (\hat{\mathbf{i}} \times \mathbf{R}) \, dm = -\int (R^z R^x) \, dm, \tag{12.8}$$

and these three integrals are called the *mass products of inertia* of the body. Sometimes it is convenient to arrange these mass moments of inertia and mass products of inertia into a symmetric square array or matrix called the *inertia tensor* of the body [3]:

$$\mathbf{I} = \begin{bmatrix} I^{xx} & I^{xy} & I^{xz} \\ I^{yx} & I^{yy} & I^{yz} \\ I^{zx} & I^{zy} & I^{zz} \end{bmatrix}. \tag{12.9}$$

A careful look at the above integrals shows that they represent the mass distribution of the body with respect to the coordinate system about which they are determined, but that they change if evaluated in a different coordinate system. To keep their meaning direct and simple, we assume that the coordinate system chosen for each body is attached to the body in a convenient location and orientation. Therefore, for rigid bodies, the mass moments and products of inertia are constant properties of the body and its mass distribution, and they do not change when the body moves; they do, however, depend on the coordinate system chosen.

An interesting property of these integrals is that it is always possible to choose the coordinate system so that its origin is located at the center of mass of the body and oriented such that all of the mass products of inertia become zero. Such a choice of the coordinate axes of the body is called its *principal axes*, and the corresponding values of Eqs. (12.7) are called the *principal mass moments of inertia*. A variety of simple geometric solids, the orientations of their principal axes, and formulae for their principal mass moments of inertia are included in Appendix A, Table A.5.

If we note that mass moments of inertia have units of mass times distance squared, it seems natural to define a radius value for the body as

$$I_G = mk^2 \quad \text{or} \quad k = \sqrt{\frac{I_G}{m}}. \tag{12.10}$$

This distance k is called the *radius of gyration* of the body, and it is always calculated or measured from the center of mass of the body about one of the principal axes. For three-dimensional motion of a body there are three radii of gyration, k^x, k^y, and k^z, associated with the three mass moments of inertia, I^{xx}, I^{yy}, and I^{zz}.

It is often necessary to determine the mass moments and products of inertia of bodies that are composed of several simpler subshapes for which formulae are known, such as those given in Table A.5. The easiest method of finding these is to compute the mass moments about the principal axes of each subshape, then to shift the origin of each to the mass center of the composite body, and then to sum the results. This requires that we develop methods of redefining mass moments and products of inertia when the axes are translated to a new posture. The form of the *transfer*, or *parallel-axis theorem* for a mass moment of inertia, is written

$$I = I_G + md^2, \tag{12.11}$$

where I_G is one of the principal mass moments of inertia about a known principal axis, and I is the mass moment of inertia about a parallel axis at distance d from that principal axis. Equation (12.11) must only be used for *translation* of inertia axes starting from a principal axis. Also, the rotation of these axes results in the introduction of product-of-inertia terms. More will be said on general transformations of inertia axes in Section 12.11.

Example 12.2

Figure 12.5 shows a connecting rod made of ductile iron with density of $0.260 \, \text{lb/in}^3$. Find the mass moment of inertia about the z axis.

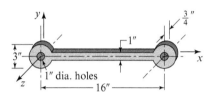

Fig. 12.5 Connecting rod.

Solution

First we find the mass moment of inertia of each of the annular cylinders at the ends of the rod and of the central prismatic bar, each taken about its own mass center. Then we use the parallel-axis theorem to transfer each to the z axis.

The mass of each hollow cylinder is

$$m_{\text{cyl}} = \rho \pi (r_o^2 - r_i^2) l$$
$$= \frac{0.260 \, \text{lb/in}^3}{386 \, \text{in/s}^2} \pi \left[(1.5 \, \text{in})^2 - (0.75 \, \text{in})^2 \right]$$
$$= 0.003 \, 17 \, \text{lb} \cdot \text{s}^2/\text{in}.$$

The mass of the central prismatic bar is

$$m_{\text{bar}} = \rho w h l = \frac{0.260 \, \text{lb/in}^3}{386 \, \text{in/s}^2} (13 \, \text{in})(1 \, \text{in})(0.75 \, \text{in})$$
$$= 0.006 \, 57 \, \text{lb} \cdot \text{s}^2/\text{in}.$$

From Table A.5 we find the mass moment of inertia of each cylinder and the bar as

$$I_{\text{cyl}} = \frac{m_{\text{cyl}} (r_o^2 + r_i^2)}{2}$$
$$= \frac{0.003 \, 17 \, \text{lb} \cdot \text{s}^2/\text{in} \left[(1.5 \, \text{in})^2 + (0.5 \, \text{in})^2 \right]}{2}$$
$$= 0.003 \, 96 \, \text{in} \cdot \text{lb} \cdot \text{s}^2$$

Example 12.2 (cont.)

and

$$I_{\text{bar}} = \frac{m_{\text{bar}}\left(w^2 + h^2\right)}{12}$$

$$= \frac{0.006\,57 \text{ lb} \cdot \text{s}^2/\text{in}\left[(1 \text{ in})^2 + (13 \text{ in})^2\right]}{12}$$

$$= 0.093\,08 \text{ in} \cdot \text{lb} \cdot \text{s}^2.$$

Then, using Eq. (12.11), to transfer the axes, the mass moment of inertia about the z axis is

$$I^{zz} = I_{\text{cyl}} + \left(I_{\text{bar}} + m_{\text{bar}}d_{\text{bar}}^2\right) + \left(I_{\text{cyl}} + m_{\text{cyl}}d_{\text{cyl}}^2\right)$$

$$= (0.003\,96 \text{ in} \cdot \text{lb} \cdot \text{s}^2) + (0.093\,08 \text{ in} \cdot \text{lb} \cdot \text{s}^2) + (0.006\,57 \text{ lb} \cdot \text{s}^2/\text{in})(8 \text{ in})^2$$

$$+ (0.003\,96 \text{ in} \cdot \text{lb} \cdot \text{s}^2) + (0.003\,17 \text{ lb} \cdot \text{s}^2/\text{in})(16 \text{ in})^2$$

$$= 1.333 \text{ in} \cdot \text{lb} \cdot \text{s}^2. \qquad\qquad\qquad Ans.$$

It is noted that only one mass moment of inertia, I^{zz}, was requested or found in this example. This does not mean that I^{xx} and I^{yy} are zero, but rather that they are likely not needed for further analysis. In problems with only planar motion, only I^{zz} is needed since I^{xx} and I^{yy} are used only with rotations out of the xy plane. These other mass moments and products of inertia are used in Sections 12.10 and later, where we treat problems with spatial motion, and they are found in identical fashion.

12.4 Inertia Forces and d'Alembert's Principle

Next let us consider a moving rigid body of mass m acted upon by a system of forces, say, $\mathbf{F}_1$, $\mathbf{F}_2$, and $\mathbf{F}_3$, as shown in Fig. 12.6a. We designate the center of mass of the body as point G, and we find the resultant of the system of forces from the equation

$$\sum \mathbf{F} = \mathbf{F}_1 + \mathbf{F}_2 + \mathbf{F}_3.$$

In the general case, the line of action of this resultant is *not* through the mass center but may be displaced by some distance, shown in Fig. 12.6a as distance h. We demonstrated in Eq. (12.3) that the effect of this unbalanced force system is to produce an acceleration of the center of mass of the body, that is

$$\sum \mathbf{F} = m\mathbf{A}_G. \qquad\qquad\qquad (12.12)$$

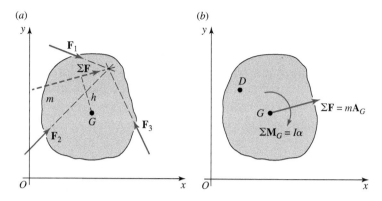

Figure 12.6 (a) Unbalanced system of forces; (b) accelerations which result from the forces.

In a similar way, the unbalanced moment effect of this resultant force about the center of mass causes angular acceleration of the body that obeys the equation

$$\sum \mathbf{M}_G = I_G \boldsymbol{\alpha}. \tag{12.13}$$

However, this equation is restricted to use in taking moments about the center of mass, G; that is, it cannot be used for taking moments about an arbitrary point in the body.

The quantity $\sum \mathbf{F}$ is the resultant of all external forces acting upon the body, and $\sum \mathbf{M}_G$ is the sum of all applied external moments and the moments of all externally applied forces about point G. The mass moment of inertia is designated I_G, signifying that it must be taken with respect to the mass center, G.

Equation (12.12) indicates that when an unbalanced system of forces acts upon a rigid body, the mass center of the body experiences a rectilinear acceleration, $\mathbf{A}_G$, in the same direction as the resultant force, $\sum \mathbf{F}$. Also, Eq. (12.13) indicates that the body experiences an angular acceleration, $\boldsymbol{\alpha}$, in the same direction as the resultant moment, $\sum \mathbf{M}_G$, caused by the moments of the forces about the mass center. This situation is shown in Fig. 12.6b. If the forces and moments are known, Eqs. (12.12) and (12.13) may be used to determine the resulting acceleration pattern; that is, the resulting motion of the body.

As pointed out, Eq. (12.13) is restricted to the sum of moments taken about the center of mass. Suppose we wish to take moments about the arbitrary point D shown in Fig. 12.6b; then the sum of the moments about point D can be written as

$$\sum \mathbf{M}_D = I_G \boldsymbol{\alpha} + \mathbf{R}_{GD} \times m \mathbf{A}_G. \tag{a}$$

Then, from Section 4.3, the acceleration of the center of mass can be expressed in terms of the acceleration of point D as

$$\sum \mathbf{M}_D = I_G \boldsymbol{\alpha} + \mathbf{R}_{GD} \times m \left(\mathbf{A}_D - \omega^2 \mathbf{R}_{GD} + \boldsymbol{\alpha} \times \mathbf{R}_{GD} \right). \tag{b}$$

Simplifying and then rearranging this equation gives

$$\sum \mathbf{M}_D = \left(I_G + m R_{GD}^2 \right) \boldsymbol{\alpha} + \mathbf{R}_{GD} \times m \mathbf{A}_D. \tag{c}$$

Then, according to Eq. (12.11), the parallel-axis theorem, Eq. (c) can be written as

$$\sum \mathbf{M}_D = I_D \boldsymbol{\alpha} + \mathbf{R}_{GD} \times m\mathbf{A}_D. \tag{12.14}$$

Comparing the right-hand side of this equation with the right-hand side of Eq. (a), we note a pattern, namely, the point used for computing the acceleration is the same point as used for the axis of the second moment of mass. This equation accomplishes our goal that the sum of moments can now be taken about an arbitrary point D; it is no longer restricted to the sum of moments about the center of mass.

During engineering design, however, the motions of the machine members are often specified in advance by other machine requirements. The problem, then, is this: given the motions of the machine parts, what forces are required to produce these motions? The problem requires: (1) a kinematic analysis to determine the translational and rotational accelerations of the various links; and (2) definitions of the actual shapes, dimensions, and material specifications to determine the centroids and mass moments of inertia of the links. In the examples presented here, only the results of kinematic analysis are included, since methods of finding these have been presented in Chapter 4. The selection of the materials, shapes, and many of the dimensions of machine parts form the subject of machine design and are also not further discussed here.

In the dynamic analysis of machines, the acceleration vectors are usually known or found by the methods of Chapter 4; therefore, an alternative form of Eqs. (12.12) and (12.13) is often convenient in determining the forces required to produce these known accelerations. Thus, we can write

$$\sum \mathbf{F} + (-m\mathbf{A}_G) = \mathbf{0} \tag{12.15}$$

and

$$\sum \mathbf{M}_G + (-I_G \boldsymbol{\alpha}) = \mathbf{0}, \tag{12.16a}$$

or

$$\sum \mathbf{M}_D + (-I_D \boldsymbol{\alpha}) = \mathbf{0}. \tag{12.16b}$$

These are vector equations applying to the planar motion of a rigid body. Equation (12.15) states that the vector sum of all external forces acting upon the body plus the fictitious force $-m\mathbf{A}_G$ sum to zero. This new fictitious force, $-m\mathbf{A}_G$, is called an *inertia force*. It has the same line of action as the absolute acceleration, $\mathbf{A}_G$, but is opposite in sense. Equation (12.16) states that the sum of all external moments and the moments of all external forces acting upon the body about an axis through G (or D) and perpendicular to the plane of motion plus the fictitious moment $-I_G \boldsymbol{\alpha}$ (or $-I_D \boldsymbol{\alpha}$) sum to zero. This new fictitious moment, $-I_G \boldsymbol{\alpha}$ (or $-I_D \boldsymbol{\alpha}$), is called an *inertia moment*. The inertia moment is opposite in sense to the angular acceleration vector, $\boldsymbol{\alpha}$, of the body. We recall that Newton's first law states that a body perseveres in its state of uniform motion except when compelled to change by impressed forces; in other words, bodies resist any change in their motions. In a sense we can picture the fictitious inertia force

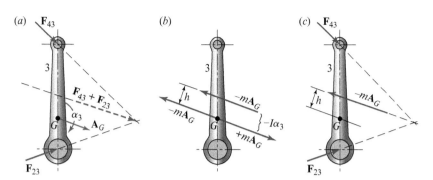

Fig. 12.7 (a) Unbalanced forces and resulting accelerations; (b) inertia force and inertia couple; (c) inertia force offset from center of mass.

and inertia moment vectors as resistances of a body to the change of motion required by the net unbalanced forces.

Equations (12.15) and (12.16) are known as *d'Alembert's principle*, since he[3] was the first to call attention to the fact that addition of the inertia force and inertia moment to the real system of forces and moments enables a solution from the equations of static equilibrium. We note that the equations can also be written

$$\sum \mathbf{F} = \mathbf{0} \quad \text{and} \quad \sum \mathbf{M} = \mathbf{0}, \tag{12.17}$$

where it is understood that both the external and the inertia forces and moments are to be included in the summations. Equations (12.17) are useful, since they permit us to take the summation of moments about any axis perpendicular to the plane of motion.

D'Alembert's principle is summarized as follows: The vector sum of all external forces and inertia forces acting upon a system of rigid bodies is zero. The vector sum of all external moments and inertia moments acting upon a system of rigid bodies is also separately zero.

When a graphic solution by a force polygon is desired, the sum of the forces and the sum of the moments given by Eqs. (12.17) can be combined. For example, consider Fig. 12.7*a*, where link 3 is acted upon by the external forces $\mathbf{F}_{23}$ and $\mathbf{F}_{43}$. The resultant $\mathbf{F}_{23} + \mathbf{F}_{43}$ produces an acceleration of the center of mass, $\mathbf{A}_G$, and an angular acceleration of the link, $\boldsymbol{\alpha}_3$, since the line of action of the resultant does not pass through the center of mass. Representing the inertia moment $-I_G\boldsymbol{\alpha}_3$ as a couple, as shown in Fig. 12.7*b*, we intentionally choose the two forces of this couple to be $\pm m\mathbf{A}_G$. For the moment of the couple to be $-I_G\alpha_3$, the distance between the forces of the couple must be

$$h = \frac{I_G\alpha_3}{mA_G}. \tag{12.18}$$

Because of this particular choice for the couple, one force of the couple exactly balances the inertia force itself and leaves only a single force, as shown in Fig. 12.7*c*. This force includes the

---

[3] Jean leRond d'Alembert (1717–1783) was a French mathematician, physicist, and philosopher.

combined effects of the inertia force and the inertia moment, yet appears as only a single inertia force offset by the distance, h, to include the effect of the inertia moment.

To illustrate the graphic approach to dynamic force analysis, we present the following example.

Example 12.3

For the elliptic trammel linkage shown in Fig. 12.8a, determine the force $\mathbf{F}_A$ that ensures that point A has the constant velocity $\mathbf{V}_A = 12.6\hat{\mathbf{j}}$ ft/s. The linkage is in the horizontal plane with gravity normal to the plane of motion. The weight and mass moment of inertia of coupler link 3 are $w_3 = 2.20$ lb and $I_{G_3} = 0.047\,9$ in·lb·s², respectively.

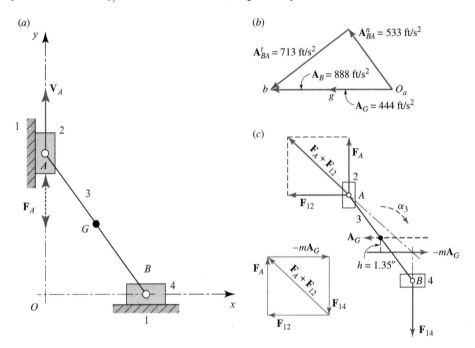

Fig. 12.8 (a) $R_{BA} = 10$ in, $R_{CA} = 5$ in, $R_{AO} = 8$ in, and $R_{BO} = 6$ in; (b) acceleration polygon; (c) free-body diagram and force polygon.

Assumptions

The masses of links 2 and 4 and friction in the linkage are negligible.

Solution

A kinematic analysis provides the acceleration information shown in the polygon of Fig. 12.8b. The acceleration of the mass center of the coupler link is $A_G = 444$ ft/s², and the angular acceleration of this link is

Example 12.3 (cont.)

$$\alpha_3 = \frac{A_{BA}^t}{R_{BA}} = \frac{(713 \text{ ft/s}^2)(12 \text{ in/ft})}{10 \text{ in}} = 856 \text{ rad/s}^2.$$

The mass of the coupler link is

$$m_3 = \frac{w_3}{g} = \frac{2.20 \text{ lb}}{32.17 \text{ ft/s}^2(12 \text{ in/ft})} = 0.005\,70 \text{ lb} \cdot \text{s}^2/\text{in}.$$

Substituting the known information into Eq. (12.18), the offset distance is

$$h = \frac{(0.0479 \text{ in} \cdot \text{lb} \cdot \text{s}^2)(856 \text{ rad/s}^2)}{(0.005\,70 \text{ lb} \cdot \text{s}^2/\text{in})(444 \text{ ft/s}^2)(12 \text{ in/ft})} = 1.35 \text{ in}.$$

The free-body diagram of the assembly of links 2, 3, and 4 and the resulting force polygon are shown in Fig. 12.8c. Note that the inertia force, $-m\mathbf{A}_G$, is offset from G by the distance h on the proper side of (below) G to include a counterclockwise moment $-I_G\alpha_3$ (opposite to α_3) about G, and with the inertia force in the opposite sense to $\mathbf{A}_G$. The constraint reaction at B is $\mathbf{F}_{14}$ and, with no friction, is vertical. The forces at A are the horizontal constraint reaction, $\mathbf{F}_{12}$, and the vertical actuating force, $\mathbf{F}_A$. Recognizing this as a four-force member, as demonstrated in Section 11.8, we find the concurrency point at the intersection of the offset inertia force and $\mathbf{F}_{14}$, with the directions of both being known. The line of action of the total force, $\mathbf{F}_{12} + \mathbf{F}_A$, at A must pass through this point of concurrency. This fact permits construction of the force polygon, where the unknown forces, $\mathbf{F}_{12}$ and $\mathbf{F}_A$, having known directions, are found as components of $\mathbf{F}_{12} + \mathbf{F}_A$. The actuating force $\mathbf{F}_A$ is found by measurement to be

$$\mathbf{F}_A = 27\hat{\mathbf{j}} \text{ lb}. \qquad\qquad Ans.$$

To illustrate the analytic approach to dynamic force analysis of a mechanism, we present another example.

Example 12.4

Determine the reaction forces and the driving torque for the four-bar linkage in the posture shown in Fig. 12.9. The input crank is operating at a constant angular velocity $\omega_2 = 48$ rad/s ccw, and the applied force acting at point C is $\mathbf{F}_C = -0.8\hat{\mathbf{j}}$ kN. The masses of links 3 and 4 are $m_3 = 1.5$ kg and $m_4 = 5$ kg, respectively, and the mass moments of inertia of links 2, 3, and 4 are $I_{G_2} = 0.025$ kg $\cdot$ m^2, $I_{G_3} = 0.012$ kg $\cdot$ m^2, and $I_{G_4} = 0.054$ kg $\cdot$ m^2, respectively.

Assumptions

Gravity and friction effects are negligible.

Example 12.4 (cont.)

Fig. 12.9 $R_{AO_2} = 60$ mm, $R_{O_4O_2} = 100$ mm, $R_{BA} = 220$ mm, $R_{BO_4} = 150$ mm, $R_{CO_4} = R_{CB} = 120$ mm, $R_{G_3A} = 90$ mm, and $R_{G_4O_4} = 90$ mm.

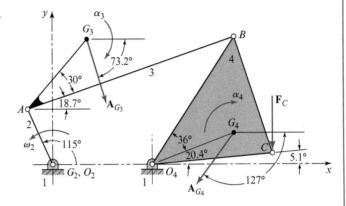

Solution

The information given in the figure caption in vector form is:

$$\mathbf{R}_{AO_2} = 60 \text{ mm} \angle 115° = -25.4\hat{\mathbf{i}} + 54.4\hat{\mathbf{j}} \text{ mm},$$

$$\mathbf{R}_{G_3A} = 90 \text{ mm} \angle 48.7° = 59.4\hat{\mathbf{i}} + 67.6\hat{\mathbf{j}} \text{ mm},$$

$$\mathbf{R}_{BA} = 220 \text{ mm} \angle 18.7° = 208.4\hat{\mathbf{i}} + 70.5\hat{\mathbf{j}} \text{ mm},$$

$$\mathbf{R}_{BO_4} = 150 \text{ mm} \angle 61.5° = 71.6\hat{\mathbf{i}} + 131.8\hat{\mathbf{j}} \text{ mm},$$

$$\mathbf{R}_{G_4O_4} = 90 \text{ mm} \angle 25.5° = 81.2\hat{\mathbf{i}} + 38.7\hat{\mathbf{j}} \text{ mm},$$

$$\mathbf{R}_{CO_4} = 120 \text{ mm} \angle 5.1° = 119.5\hat{\mathbf{i}} + 10.7\hat{\mathbf{j}} \text{ mm},$$

$$\mathbf{R}_{O_4O_2} = 100 \text{ mm} \angle 0° = 100.0\hat{\mathbf{i}} \text{ mm}.$$

A kinematic analysis of the linkage gives: $\alpha_3 = -119\hat{\mathbf{k}}$ rad/s^2, $\alpha_4 = -625\hat{\mathbf{k}}$ rad/s^2, $\mathbf{A}_{G_2} = \mathbf{0}$, $\mathbf{A}_{G_3} = 162$ m/s$^2 \angle -73.2° = 46.8\hat{\mathbf{i}} - 155\hat{\mathbf{j}}$ m/s^2, and $\mathbf{A}_{G_4} = 104$ m/s$^2 \angle -127° = -62.6\hat{\mathbf{i}} - 83.1\hat{\mathbf{j}}$ m/s^2. Next, we calculate the inertia forces and inertia moments using Eqs (12.12) and (12.13):

$$-m_2\mathbf{A}_{G_2} = \mathbf{0},$$

$$-m_3\mathbf{A}_{G_3} = -(1.5 \text{ kg})(46.8\hat{\mathbf{i}} - 155\hat{\mathbf{j}} \text{ m/s}^2) = -70.2\hat{\mathbf{i}} + 233\hat{\mathbf{j}} \text{ N},$$

$$-m_4\mathbf{A}_{G_4} = -(5.0 \text{ kg})(-62.6\hat{\mathbf{i}} - 83.1\hat{\mathbf{j}} \text{ m/s}^2) = 313\hat{\mathbf{i}} + 415\hat{\mathbf{j}} \text{ N},$$

$$-I_{G_2}\alpha_2 = \mathbf{0},$$

$$-I_{G_3}\alpha_3 = -(0.012 \text{ kg} \cdot \text{m}^2)(-119\hat{\mathbf{k}} \text{ rad/s}^2) = 1.43\hat{\mathbf{k}} \text{ N} \cdot \text{m},$$

and

$$-I_{G_4}\alpha_4 = -(0.054 \text{ kg} \cdot \text{m}^2)(-625\hat{\mathbf{k}} \text{ rad/s}^2) = 33.8\hat{\mathbf{k}} \text{ N} \cdot \text{m}.$$

Example 12.4 (cont.)

Since the solution here is analytic, we do not need to calculate offset distances using Eq. (12.18).

For our next step, we make sketches of the free-body diagrams of link 3 and link 4. From the two free-body diagrams, we note that there are a total of six unknown force components. Since there are a total of six scalar equations available, two components of the forces and one for the sum of the moments about the mass center of each link, then the problem is solvable. However, these six equations must be solved simultaneously to find the six unknowns. In an effort to decouple the equations, an alternative strategy is to take the sum of the moments about different points in the links.

Considering the free-body diagram of link 3, we formulate the summation of moments about point A:

$$\sum \mathbf{M}_A = \mathbf{R}_{G_3 A} \times (-m_3 \mathbf{A}_{G_3}) + (-I_{G_3} \boldsymbol{\alpha}_3) + \mathbf{R}_{BA} \times \mathbf{F}_{43} = \mathbf{0}. \tag{1}$$

Also, considering the free-body diagram of link 4, we formulate the summation of moments about point O_4 :

$$\sum \mathbf{M}_{O_4} = \mathbf{R}_{G_4 O_4} \times (-m_4 \mathbf{A}_{G_4}) + (-I_{G_4} \boldsymbol{\alpha}_4) + \mathbf{R}_{C O_4} \times \mathbf{F}_C + \mathbf{R}_{B O_4} \times \mathbf{F}_{34} = \mathbf{0}. \tag{2}$$

Remembering that $\mathbf{F}_{43} = -\mathbf{F}_{34}$, the in-plane components of $\mathbf{F}_{34}$ are the only two unknowns, and they are shared by these two equations.

The cross-product terms of these equations are

$$\mathbf{R}_{G_3 A} \times (-m_3 \mathbf{A}_{G_3}) = (59.4\hat{\mathbf{i}} + 67.6\hat{\mathbf{j}} \text{ mm}) \times (70.2\hat{\mathbf{i}} + 233\hat{\mathbf{j}} \text{ N}) = 18.6\hat{\mathbf{k}} \text{ N} \cdot \text{m},$$

$$\mathbf{R}_{BA} \times \mathbf{F}_{43} = (208.4\hat{\mathbf{i}} + 70.5\hat{\mathbf{j}} \text{ mm}) \times (-F_{34}^x \hat{\mathbf{i}} - F_{34}^y \hat{\mathbf{j}} \text{ N})$$
$$= (70.5 F_{34}^x - 208.4 F_{34}^y)\hat{\mathbf{k}} \text{ N} \cdot \text{m},$$

and

$$\mathbf{R}_{G_4 O_4} \times (-m_4 \mathbf{A}_{G_4}) = (81.2\hat{\mathbf{i}} + 38.7\hat{\mathbf{j}} \text{ mm}) \times (313\hat{\mathbf{i}} + 415\hat{\mathbf{j}} \text{ N}) = 21.6\hat{\mathbf{k}} \text{ N} \cdot \text{m},$$

$$\mathbf{R}_{C O_4} \times \mathbf{F}_C = (119.5\hat{\mathbf{i}} + 10.7\hat{\mathbf{j}} \text{ mm}) \times (-800\hat{\mathbf{j}} \text{ N}) = -95.6\hat{\mathbf{k}} \text{ N} \cdot \text{m},$$

$$\mathbf{R}_{B O_4} \times \mathbf{F}_{34} = (71.6\hat{\mathbf{i}} + 131.8\hat{\mathbf{j}} \text{ mm}) \times (F_{34}^x \hat{\mathbf{i}} + F_{34}^y \hat{\mathbf{j}} \text{ N}) = (-131.8 F_{34}^x + 71.6 F_{34}^y)\hat{\mathbf{k}} \text{ N} \cdot \text{m}.$$

Then, substituting these values into Eqs. (1) and (2) gives

$$\sum \mathbf{M}_A = 18.6\hat{\mathbf{k}} + 1.43\hat{\mathbf{k}} + (70.5 F_{34}^x - 208.4 F_{34}^y)\hat{\mathbf{k}} = \mathbf{0}$$

and

$$\sum \mathbf{M}_{O_4} = 21.6\hat{\mathbf{k}} + 33.8\hat{\mathbf{k}} - 95.6\hat{\mathbf{k}} + (-131.8 F_{34}^x + 71.6 F_{34}^y)\hat{\mathbf{k}} = \mathbf{0}.$$

Example 12.4 (cont.)

Rearranging terms gives two equations in two unknowns, that is,

$$70.5F_{34}^x - 208.4F_{34}^y = -20.0 \text{ N} \cdot \text{m}$$

and

$$-131.8F_{34}^x + 71.6F_{34}^y = 40.2 \text{ N} \cdot \text{m}.$$

Solving simultaneously, the two components of the force vector $\mathbf{F}_{34}$ are:

$$F_{34}^x = -0.310\,16 \text{ kN} = -310.16 \text{ N} \text{ and } F_{34}^y = -0.008\,84 \text{ kN} = -8.84 \text{ N}.$$

Therefore, the force vector is

$$\mathbf{F}_{34} = -310.16\hat{\mathbf{i}} - 8.84\hat{\mathbf{j}} \text{ N} = 310.29 \text{ N}\angle -178.4°. \qquad Ans.$$

Next, summing forces on link 4 yields the equation

$$\sum \mathbf{F}_{i4} = \mathbf{F}_{14} + \mathbf{F}_{34} + \mathbf{F}_C + (-m_4\mathbf{A}_{G_4}) = \mathbf{0},$$

and, upon solving, we find the force vector

$$\begin{aligned}
\mathbf{F}_{14} &= -\mathbf{F}_{34} - \mathbf{F}_C - (-m_4\mathbf{A}_{G_4}) \\
&= -(-310.16\hat{\mathbf{i}} - 8.84\hat{\mathbf{j}} \text{ N}) - (-800\hat{\mathbf{j}} \text{ N}) - (313\hat{\mathbf{i}} + 415\hat{\mathbf{j}} \text{ N}) \\
&= -2.84\hat{\mathbf{i}} + 393.84\hat{\mathbf{j}} \text{ N} = 393.85 \text{ N}\angle 90.41°.
\end{aligned}$$

Similarly, summing forces on link 3 yields

$$\sum \mathbf{F}_{i3} = \mathbf{F}_{23} + \mathbf{F}_{43} + (-m_3\mathbf{A}_{G_3}) = \mathbf{0}$$

and solving this, the force vector is

$$\begin{aligned}
\mathbf{F}_{23} &= -(-\mathbf{F}_{34}) - (-m_3\mathbf{A}_{G_3}) \\
&= -(310.16\hat{\mathbf{i}} + 8.84\hat{\mathbf{j}} \text{ N}) - (-70.2\hat{\mathbf{i}} + 233\hat{\mathbf{j}} \text{ N}) \\
&= -239.96\hat{\mathbf{i}} - 241.84\hat{\mathbf{j}} \text{ N} = 340.69 \text{ N}\angle -134.78°.
\end{aligned}$$

Now, for link 2, we have

$$\sum \mathbf{F}_{i2} = \mathbf{F}_{12} + \mathbf{F}_{32} + (-m_2\mathbf{A}_{G_2}) = \mathbf{0}.$$

Rearranging this gives

$$\mathbf{F}_{12} = -\mathbf{F}_{32} = \mathbf{F}_{23} = -239.96\hat{\mathbf{i}} - 241.84\hat{\mathbf{j}} \text{ N} = 340.69 \text{ N}\angle -134.78°. \qquad Ans.$$

Summing moments on link 2 about O_2 gives

$$\sum \mathbf{M}_{O_2} = \mathbf{R}_{AO_2} \times \mathbf{F}_{32} + \mathbf{T}_{12} + (-I_{G_2}\boldsymbol{\alpha}_2) = \mathbf{0},$$

Example 12.4 (cont.)

and solving for the driving torque gives

$$\mathbf{T}_{12} = -\mathbf{R}_{AO_2} \times \mathbf{F}_{32} = -\left(-25.4\hat{\mathbf{i}} + 54.4\hat{\mathbf{j}} \text{ mm}\right) \times \left(239.96\hat{\mathbf{i}} + 241.84\hat{\mathbf{j}} \text{ N}\right) = 19.2\hat{\mathbf{k}} \text{ N} \cdot \text{m}. \quad Ans.$$

12.5 Principle of Superposition

Linear systems are those in which effect is proportional to cause. This means that the response, or output, of a linear system is directly proportional to the drive, or input, to the system. An example of a linear system is a spring, where the deflection (output) is directly proportional to the force (input) exerted on the spring.

The *principle of superposition* may be used to solve problems involving linear systems by considering each of the inputs to the system separately. If the system is linear, the responses to each of these inputs can be summed or superimposed on each other to determine the total response of the system. Thus the principle of superposition states that, *for a linear system, the individual responses to several disturbances, or driving functions, can be superimposed on each other to obtain the total response of the system.*

The principle of superposition does not apply to nonlinear systems. Some examples of nonlinear systems, where superposition may not be used, are systems with static or Coulomb friction, systems with clearances or backlash, or systems with springs that change stiffness as they are deflected.

We have now reviewed all the principles necessary for making a complete dynamic force analysis of a planar mechanism. The steps in using the principle of superposition for making such an analysis are summarized as follows:

1. Perform a kinematic analysis of the mechanism. Locate the center of mass of each link, and find the acceleration of each mass center; also, find the angular acceleration of each link.
2. If the inertia forces are attached to all links simultaneously, along with other applied forces and moments, then often there are no two- or three-force members, and it may be difficult to find the lines of action of unknown constraint forces. Instead of doing this, it is sometimes convenient to ignore the masses and applied forces and moments on all but one or two links and to leave other links as two- or three-force members. By choosing in this manner, a solution may become possible for the constraint forces caused by the masses or applied forces and moments being considered, but without those caused by the masses and applied forces and moments being ignored.
3. Those masses and applied forces and moments considered in step 2 can be ignored while a solution is obtained for additional constraint force components caused by some of the previously ignored masses or applied forces and moments. This process can be repeated until constraint force components caused by all masses and all applied forces and moments are found.

4. The results of steps 2 and 3 can be vectorially added to obtain the resultant forces and moments on each link caused by the combined effects of all masses and all applied forces and moments.

This process of superposition is demonstrated in the following example.

Example 12.5

Perform a complete dynamic force analysis of the four-bar linkage in the posture shown in Fig. 12.10a, using the principle of superposition. The input crank is operating at the constant angular velocity $\omega_2 = 60$ rad/s ccw, and the applied force is $\mathbf{F}_C = 40\hat{\mathbf{i}}$ lb. The weights of links 3 and 4 are $w_3 = 7.13$ lb and $w_4 = 3.42$ lb, respectively, and the mass moments of inertia of links 2, 3, and 4 are $I_{G_2} = 0.25$ in·lb·s^2, $I_{G_3} = 0.625$ in·lb·s^2, and $I_{G_4} = 0.037$ in·lb·s^2, respectively.

Fig. 12.10 (a) $R_{AO_2} = 3$ in,
$\theta_2 = 135°$, $R_{O_4O_2} = 14$ in,
$R_{BA} = 20$ in, $\theta_3 = 22.3°$,
$R_{BO_4} = 10$ in, $\theta_4 = 75.85°$,
$R_{CO_4} = 8$ in, $\beta = 36.9°$, $R_{CB} = 6$ in,
$R_{G_3A} = 10$ in, $R_{G_4O_4} = 5.69$ in,
$\gamma = 18.4°$, $F_C = 40$ lb; (b)
$A_A = 900$ ft/s^2, $A_{G_3} = 758$ ft/s^2,
$A_B = 615$ ft/s^2, $A_{G_4} = 349$ ft/s^2,
$A_C = 492$ ft/s^2.

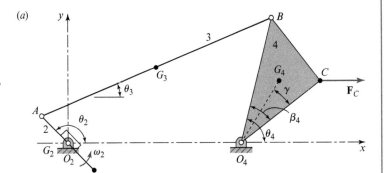

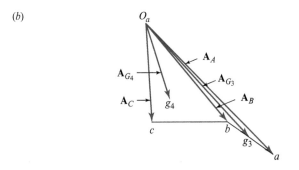

Assumptions

Gravity and friction effects are negligible.

Solution

Step 1. All of the details of a complete kinematic analysis of the linkage are not included here, but the resulting acceleration polygon is shown in Fig. 12.10b (the numeric results are shown on the acceleration polygon). From the methods of Chapter 4, the angular accelerations of links 3 and 4 are

$$\alpha_3 = 148 \text{ rad/s}^2 \text{ ccw} \quad \text{and} \quad \alpha_4 = 604 \text{ rad/s}^2 \text{ cw}.$$

Example 12.5 (cont.)

Step 2. Since the center of mass of link 2 is located at bearing O_2, the force analysis is concerned primarily with links 3 and 4. Free-body diagrams of links 4 and 3 are shown in Figs. 12.11 and 12.12, respectively. Note that these diagrams are arranged in pseudo-equation form to emphasize the concept of superposition. Thus, in each figure, the forces in (a) plus those in (b) and (c) produce the results shown in (d). The action and reaction forces are also correlated; for example, $\mathbf{F}'_{34}$ in Fig. 12.11a is equal to $-\mathbf{F}'_{43}$ in Fig. 12.12a and so on. The following analysis is not difficult, but it is complex; therefore, it is important to read it slowly and to examine the figures carefully, detail by detail.

We start by considering the effects of the mass of link 4 alone, while ignoring the masses of links 2 and 3. Using Figs. 12.11a and 12.12a, we make the following calculations:

$$I_{G_4}\alpha_4 = \left(0.037 \text{ in} \cdot \text{lb} \cdot \text{s}^2\right)\left(604 \text{ rad/s}^2\right) = 22.3 \text{ in} \cdot \text{lb cw},$$

$$m_4 A_{G_4} = \frac{3.42 \text{ lb}}{32.17 \text{ ft/s}^2}\left(349 \text{ ft/s}^2\right) = 37.1 \text{ lb},$$

$$h_4 = \frac{I_{G_4}\alpha_4}{m_4 A_{G_4}} = \frac{22.3 \text{ in} \cdot \text{lb}}{37.1 \text{ lb}} = 0.602 \text{ in}.$$

Now, we position the inertia force $-m_4 A_{G_4} = 37.1$ lb on the free-body diagram of link 4 opposite in direction to A_{G_4} and offset from G_4 by the distance $h_4 = 0.602$ in. The direction of the offset is to the right of G_4 so that the inertia force $-m_4 A_{G_4}$ produces a counterclockwise inertia moment $-I_{G_4}\alpha_4$ about G_4, opposite to the clockwise sense of α_4. The line of action of $\mathbf{F}'_{34}$ is along link 3 for part (a) of our superposition approach, since link 3 is a two-force member in this step. The intersection of the inertia force $-m_4 A_{G_4}$ and the line of action of $\mathbf{F}'_{34}$ gives the point of concurrency and establishes the line of action of $\mathbf{F}'_{14}$. The force polygon for part (a) for link 4 can be constructed and the magnitudes of $\mathbf{F}'_{34}$ and $\mathbf{F}'_{14}$ can be determined. These values are shown in the legend to Fig. 12.11.

Note that the forces $\mathbf{F}'_{43}$ and $\mathbf{F}'_{23}$ are now known as shown in Fig. 12.12a.

Step 3. Now, we consider the effects of the mass of link 3 alone. Using Fig. 12.12b, we make the following calculations:

$$I_{G_3}\alpha_3 = \left(0.625 \text{ in} \cdot \text{lb} \cdot \text{s}^2\right)\left(148 \text{ rad/s}^2\right) = 92.5 \text{ in} \cdot \text{lb ccw},$$

$$m_3 A_{G_3} = \frac{7.13 \text{ lb}}{32.17 \text{ ft/s}^2}\left(758 \text{ ft/s}^2\right) = 168 \text{ lb},$$

$$h_3 = \frac{I_{G_3}\alpha_3}{m_3 A_{G_3}} = \frac{92.5 \text{ in} \cdot \text{lb}}{168 \text{ lb}} = 0.550 \text{ in}.$$

Now, we position the inertia force $-m_3 A_{G_3} = 168$ lb on the free-body diagram of link 3 opposite in direction to A_{G_3} and offset from G_3 by the distance $h_3 = 0.550$ in. The direction of the offset

Example 12.5 (cont.)

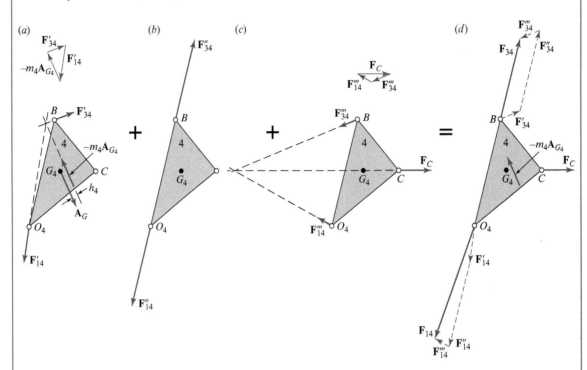

Fig. 12.11 Free-body diagrams of link 4 showing superposition of forces: (a) $h_4 = 0.602$ in, $F'_{34} = 24.3$ lb, and $F'_{14} = 44.3$ lb; (b) $F''_{34} = F''_{14} = 94.8$ lb; (c) $F'''_{34} = 25$ lb and $F'''_{14} = 19.3$ lb; (d) $F_{34} = 94.3$ lb and $F_{14} = 132$ lb.

is to the left of G_3 so that the inertia force $-m_3 \mathbf{A}_{G_3}$ produces a clockwise inertia moment $-I_{G_3} \alpha_3$ about G_3, opposite to the counterclockwise sense of α_3. The line of action of $\mathbf{F}''_{43}$ is along link 4, for part (b) of our superposition approach, since link 4 is a two-force member in this step. The intersection of the line of action of $\mathbf{F}''_{43}$ and the inertia force $-m_3 \mathbf{A}_{G_3}$ gives the point of concurrency. Thus, the line of action of $\mathbf{F}''_{23}$ is known, and the force polygon for link 3 can be constructed. The resulting magnitudes of $\mathbf{F}''_{43}$ and $\mathbf{F}''_{23}$ are included in the legend of Fig. 12.12.

Note that the forces $\mathbf{F}''_{34}$ and $\mathbf{F}''_{14}$ are now known, as shown in Fig. 12.11b. The results of the static force analysis with the applied force $F_C = 40$ lb as the specified loading, but with no inertia forces, are as shown in Fig. 12.11c.

Step 4. Recognizing that link 3 is again a two-force member, the force polygon in Fig. 12.11c determines the values of the forces acting on link 4. From these forces, the magnitudes and directions of the forces $\mathbf{F}''_{43}$ and $\mathbf{F}''_{23}$ acting on link 3 can be determined, as shown in Fig. 12.12c.

Step 5. Now we perform the superposition of the results obtained in steps 2–4; this is accomplished by the vector additions shown in part (d) of each figure. The analysis is completed by taking the resultant force $\mathbf{F}_{23}$ from Fig. 12.12d and applying the negative value, $\mathbf{F}_{32}$, to link 2. This is shown in the free-body diagram of link 2, Fig. 12.13. There is no inertia force or inertia moment on link 2 because the center of mass is stationary and the angular velocity is constant.

Example 12.5 (cont.)

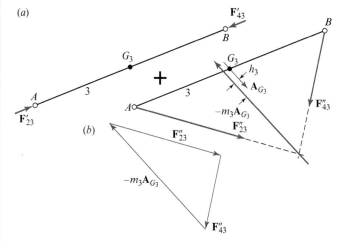

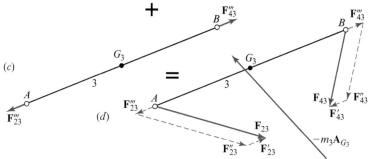

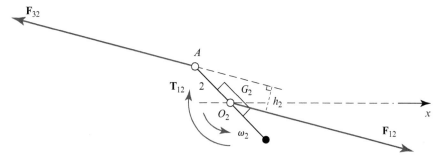

Fig. 12.12 Free-body diagrams of link 3 showing superposition of forces: (a) $F'_{23} = F'_{43} = 24.3$ lb; (b) $h_3 = 0.550$ in, $F''_{23} = 145$ lb, and $F''_{43} = 94.8$ lb; (c) $F'''_{23} = F'''_{43} = 25$ lb; (d) $F_{23} = 145$ lb and $F_{43} = 94.3$ lb.

Fig. 12.13 Free-body diagram of link 2 showing $F_{32} = F_{12} = 145$ lb and $T_{12} = 226$ in·lb cw.

The offset distance h_2 is measured as $h_2 = 1.56$ in. Therefore, the external torque applied to link 2 is

$$T_{12} = h_2 F_{32} = (1.56 \text{ in})(145 \text{ lb}) = 226 \text{ in·lb cw.}$$

Note that this torque is opposite in sense to the direction of rotation of link 2; this is not true for the entire cycle of operation, but it can occur at particular crank angles. The torque must sometimes be in the reverse direction to maintain the constant angular velocity of link 2.

12.6   Planar Rotation about a Fixed Center

The previous sections have dealt with the case of dynamic forces for a rigid body having general planar motion. It is important to emphasize that the equations and methods of analysis investigated in these sections are general and apply to *all* problems with planar motion. It will be interesting now to study the application of these methods to a special case, that of a rigid body rotating about a fixed point [4].

Let us consider a rigid body constrained as shown in Fig. 12.14*a* to rotate with an angular velocity, ω, about some fixed point, O, not coincident with its center of mass, G. A system of forces (not shown) is applied to the body, causing it to undergo an angular acceleration, α. This motion of the body implies that the mass center, G, has normal and tangential components of acceleration, $\mathbf{A}_{GO}^n$ and $\mathbf{A}_{GO}^t$, whose magnitudes are $R_{GO}\omega^2$ and $R_{GO}\alpha$, respectively, as shown in Fig. 12.14*b*. Thus, if we resolve the resultant external force into normal and tangential components, these components must have magnitudes

$$F^n = mR_{GO}\omega^2 \quad \text{and} \quad F^t = mR_{GO}\alpha, \tag{a}$$

in accordance with Eq. (12.12). In addition, Eq. (12.13) states that an external moment must exist to create the angular acceleration and that the magnitude of this moment must be $M_G = I_G\alpha$.

If we sum the moments of these forces about O, we have

$$\sum M_O = I_G\alpha + R_{GO}(mR_{GO}\alpha) = (I_G + mR_{GO})\alpha. \tag{b}$$

Observe that the quantity in parentheses in Eq. (*b*) is identical in form with Eq. (12.11) and transfers the moment of inertia to an axis through O that is parallel to an axis through the center of mass. Therefore, Eq. (*b*) can be written in the form

$$\sum M_O = I_O\alpha.$$

Equations (12.15) and (12.16*b*) then become

$$\sum \mathbf{F} - m\mathbf{A}_G = \mathbf{0} \tag{12.19}$$

and

$$\sum \mathbf{M}_O - I_O\boldsymbol{\alpha} = \mathbf{0}, \tag{12.20}$$

by including the inertia force $-m\mathbf{A}_G$ and inertia moment $-I_O\alpha$, as shown in Fig. 12.14*c*. We observe particularly that the system of forces does *not* reduce to a single couple because of the existence of the inertia force component, $-m\mathbf{R}_{GO}\omega^2$, which has no moment arm about point, O. Thus, both Eqs. (12.19) and (12.20) are necessary.

A special case arises when the angular velocity of the body is constant, that is, the angular acceleration of the body is $\boldsymbol{\alpha} = \mathbf{0}$. Then, the external moment, $\sum \mathbf{M}_O$, is zero, and the only force is, from Fig. 12.14*c*, the inertia force, $-m\mathbf{R}_{GO}\omega^2$. A second special case exists under starting conditions, when $\omega = 0$ but α is not zero. Under these conditions, the only inertia force is $-mR_{GO}\boldsymbol{\alpha}$, and the system reduces to a single couple.

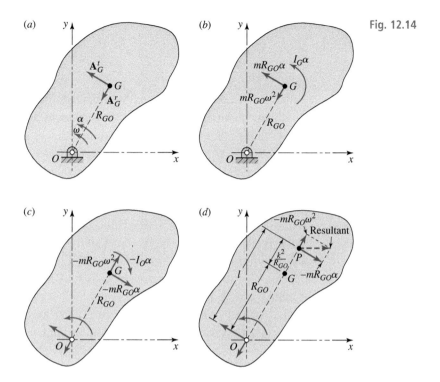

Fig. 12.14

When a rigid body has only translation, the resultant inertia force and the resultant external force have the same line of action, which passes through the center of mass of the body. When a rigid body has both angular velocity and angular acceleration, again, the resultant inertia force and the resultant external force have the same line of action; however, in this case, this line does *not* pass through the center of mass. Let us now locate a point on this offset line of action.

The resultant inertia force passes through some point, P, on the line OG or its extension (Fig. 12.14d). This force can be resolved into two components: the component, $-m\mathbf{R}_{GO}\omega^2$, acting along OG; and the component, $-m\mathbf{R}_{GO}\boldsymbol{\alpha}$, acting perpendicular to OG but not through point G. The distance from O to P, designated as l, can be determined by equating the moment of the component $-mR_{GO}\alpha$ through P (Fig. 12.14d) to the sum of the inertia moment and the moment of the inertia forces which act through G (Fig. 12.14c). Thus, taking moments about O in each figure, we have

$$(-mR_{GO}\alpha)l = -I_G\alpha + (-mR_{GO}\alpha)R_{GO}$$

or

$$l = \frac{I_G}{mR_{GO}} + R_{GO}. \tag{c}$$

Substituting Eq. (12.10) into Eq. (*c*) and eliminating mass gives

$$l = \frac{k^2}{R_{GO}} + R_{GO}. \tag{12.21}$$

Point *P*, located by Eq. (12.21), is called the *center of percussion*. As indicated, the resultant inertia force passes through *P*, and consequently, the inertia force has zero moment about the center of percussion. If an external force is applied at *P*, perpendicular to *OG*, an angular acceleration, α, results, but the bearing reaction at *O* is zero except for the component caused by the inertia force $-mR_{GO}\omega^2$. It is the usual practice in shock-testing machines to apply the force at the center of percussion in order to eliminate the tangential bearing reaction that is otherwise caused by the externally applied shock load.

As another example, suppose we consider the impact of a baseball against a bat. If we crudely approximate the bat as a cylindric rod of length *L*, then its center of mass is located at approximately $R_G = L/2$, and its radius of gyration is approximately $k = L/\sqrt{12}$. Substituting these values into Eq. (12.21) gives the location of the center of percussion as $l = 2L/3$. Hitting the ball at that point produces no dynamic reaction force on the hands and is commonly referred to as the "sweet spot" in both baseball and tennis. Equation (12.21) shows that the location of the center of percussion is independent of the values of ω and α. If the axis of rotation is coincident with the center of mass, $R_{GO} = 0$ and Eq. (12.21) shows that $l = \infty$. Under these conditions there is no resultant inertia force but rather a resultant inertia couple, $-I_G\alpha$.

12.7   Shaking Forces and Moments

Of special interest to the designer are the forces transmitted to the frame or foundation of a machine owing to the inertia of the moving links. When these forces vary in magnitude or direction, they tend to shake or vibrate the machine (and the frame); consequently, such effects are called *shaking forces* and *shaking moments*.

If we consider some machine, say a four-bar linkage, with links 2, 3, and 4 as the moving members and link 1 as the frame, then taking the entire group of moving parts as a system, not including the frame, and drawing a free-body diagram, we can immediately write

$$\sum \mathbf{F} = \mathbf{F}_{12} + \mathbf{F}_{14} + (-m_2\mathbf{A}_{G_2}) + (-m_3\mathbf{A}_{G_3}) + (-m_4\mathbf{A}_{G_4}) = \mathbf{0}.$$

Using $\mathbf{F}_S$ as a symbol for the resulting shaking force, we define this as equal to the resultant of all the reaction forces on the ground link 1,

$$\mathbf{F}_S = \mathbf{F}_{21} + \mathbf{F}_{41}.$$

Therefore, from the previous equation, we have

$$\mathbf{F}_S = (-m_2\mathbf{A}_{G_2}) + (-m_3\mathbf{A}_{G_3}) + (-m_4\mathbf{A}_{G_4}). \tag{12.22}$$

Generalizing from this example, we can write the shaking force for any machine as

$$\mathbf{F}_S = \sum \left(-m_j \mathbf{A}_{G_j} \right). \tag{12.23}$$

This makes sense since, if we consider a free-body diagram of the entire machine *including the frame*, all other applied and constraint forces have equal and opposite reaction forces, and these cancel within the free-body system. Only the inertia forces, having no reactions, are ultimately external to the system and remain unbalanced.[4] These are not balanced by reaction forces and produce unbalanced shaking effects between the frame and the surface on which it is mounted. These are the forces that require that the machine be fastened down to prevent it from moving.

A similar derivation can be made for unbalanced moments. Using the symbol $\mathbf{M}_S$ for the shaking moment, we take the summation of moments about the coordinate system origin and find

$$\mathbf{M}_S = \sum \left[\mathbf{R}_{G_j} \times \left(-m_j \mathbf{A}_{G_j} \right) \right] + \sum \left(-I_{G_j} \boldsymbol{\alpha}_j \right). \tag{12.24}$$

12.8 Complex-Algebraic Approach

In this section, we investigate the dynamic force analysis of a mechanism using the complex-algebraic approach. The general procedure that is adopted is as follows:

1. Perform a complete kinematic analysis.
2. Draw a free-body diagram of each link of the mechanism.
3. Write the dynamic force equations for each moving link.
4. Corresponding to each input angle, compute the following items:
    a. the axial and the transverse forces exerted on each moving link;
    b. the force components exerted on the ground bearings (here referred to as the bearing loads);
    c. the inertia moment; and
    d. the shaking moment on the foundation.

To demonstrate the complex-algebraic approach, we consider the four-bar linkage shown in Fig. 12.15.

For a specified angular velocity and acceleration of input link 2, the angular velocities and accelerations of coupler link 3 and output link 4 are determined following the methods of Chapters 3 and 4. The accelerations of the mass centers of links 2, 3, and 4 are now expressed in coordinates attached to and aligned along the respective links.

---

[4] The same arguments can be made about gravitational, magnetic, electric, or other noncontacting-force fields. However, because these are usually static forces, it is not customary to include them in the definition of shaking force.

Fig. 12.15 Four-bar linkage.

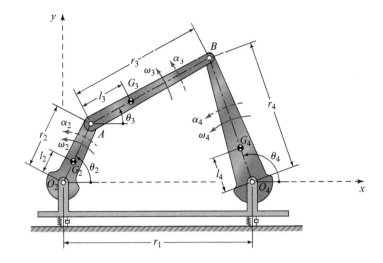

The acceleration of the mass center of the input link 2 can be written as

$$\mathbf{A}_{G_2} = I_2\left(-\omega_2^2 + j\alpha_2\right)e^{j\theta_2} = \left(A_{G_2}^{x_2} + jA_{G_2}^{y_2}\right). \tag{12.25a}$$

Therefore, the x_2 and y_2 components of the acceleration, respectively, are

$$A_{G_2}^{x_2} = -I_2\omega_2^2 \quad \text{and} \quad A_{G_2}^{y_2} = -I_2\alpha_2. \tag{12.25b}$$

The acceleration of the mass center of the coupler link 3 can be written as

$$\mathbf{A}_{G_3} = r_2\left(-\omega_2^2 + j\alpha_2\right)e^{j\theta_2} + I_3\left(-\omega_3^2 + j\alpha_3\right)e^{j\theta_3}.$$

or as

$$\mathbf{A}_{G_3} = \left[r_2\left(-\omega_2^2 + j\alpha_2\right)e^{j\eta_{23}} + I_3\left(-\omega_3^2 + j\alpha_3\right)\right]e^{j\theta_3}, \tag{12.26}$$

where $\eta_{23} = \theta_2 - \theta_3$. The acceleration of the mass center of the coupler link can also be written as

$$\mathbf{A}_{G_3} = \left(A_{G_3}^{x_3} + jA_{G_3}^{y_3}\right)e^{j\theta_3}. \tag{12.27}$$

Equating Eqs. (12.26) and (12.27), the x_3 and y_3 components of the acceleration are

$$A_{G_3}^{x_3} = r_2\left(-\omega_2^2\cos\eta_{23} - \alpha_2\sin\eta_{23}\right) - I_3\omega_3^2, \tag{12.28a}$$

$$A_{G_3}^{y_3} = r_2\left(-\omega_2^2\sin\eta_{23} + \alpha_2\cos\eta_{23}\right) - I_3\alpha_3. \tag{12.28b}$$

The acceleration of the mass center of the output link 4 can be written as

$$\mathbf{A}_{G_4} = l_4\left(-\omega_4^2 + j\alpha_4\right)e^{j\theta_4} = \left(A_{G_4}^{x_4} + jA_{G_4}^{y_4}\right)e^{j\theta_4}. \tag{12.29a}$$

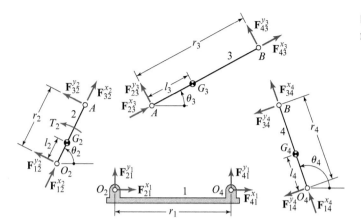

Fig. 12.16 Free-body diagrams for the four-bar linkage.

Therefore, the x_4 and y_4 components of the acceleration, respectively, are

$$A_{G_4}^{x_4} = -l_4\omega_4^2 \quad \text{and} \quad A_{G_4}^{y_4} = l_4\alpha_4. \tag{12.29b}$$

The free-body diagrams of the four links are presented in Fig. 12.16.
For the input link, the sum of the forces gives

$$\left[(F_{12}^{x_2} + F_{32}^{x_2}) + j(F_{12}^{y_2} + F_{32}^{y_2})\right]e^{j\theta_2} = m_2\left(A_{G_2}^{x_2} + jA_{G_2}^{y_2}\right)e^{j\theta_2}, \tag{12.30a}$$

and the sum of the moments about the bearing O_2 gives

$$r_2 F_{32}^{y_2} + T_2 = \left(I_{G_2} + m_2 l_2^2\right)\alpha_2. \tag{12.30b}$$

For the coupler link, the sum of the forces gives

$$\left[(F_{23}^{x_3} + F_{43}^{x_3}) + j(F_{23}^{y_3} + F_{43}^{y_3})\right]e^{j\theta_3} = m_3\left(A_{G_3}^{x_3} + jA_{G_3}^{y_3}\right)e^{j\theta_3}, \tag{12.31a}$$

and the sum of the moments about the mass center G_3 gives

$$F_{43}^{y_3}(r_3 - l_3) - F_{23}^{y_3}l_3 = I_{G_3}\alpha_3. \tag{12.31b}$$

For the output link, the sum of the forces gives

$$\left[(F_{14}^{x_4} + F_{34}^{x_4}) + j(F_{14}^{y_4} + F_{34}^{y_4})\right]e^{j\theta_4} = m_4\left(A_{G_4}^{x_4} + jA_{G_4}^{y_4}\right)e^{j\theta_4}, \tag{12.32a}$$

and the sum of the moments about the bearing O_4 gives

$$r_4 F_{34}^{y_4} = \left(I_{G_4} + m_4 l_4^2\right)\alpha_4. \tag{12.32b}$$

For the revolute joint A, the sum of the forces gives

$$\left(F_{32}^{x_2} + jF_{32}^{y_2}\right)e^{j\theta_2} + \left(F_{23}^{x_3} + jF_{23}^{y_3}\right)e^{j\theta_3} = 0, \tag{12.33a}$$

and for the revolute joint B, the sum of the forces gives

$$\left(F_{43}^{x_3} + jF_{43}^{y_3}\right)e^{j\theta_3} + \left(F_{34}^{x_4} + jF_{34}^{y_4}\right)e^{j\theta_4} = 0. \tag{12.33b}$$

For the bearing O_2, the sum of the forces gives

$$\left(F_{21}^{x_1} + jF_{21}^{y_1}\right) + \left(F_{12}^{x_2} + jF_{12}^{y_2}\right)e^{j\theta_2} = 0, \tag{12.33c}$$

and for the bearing O_4, the sum of the forces gives

$$\left(F_{41}^{x_1} + jF_{41}^{y_1}\right) + \left(F_{14}^{x_4} + jF_{14}^{y_4}\right)e^{j\theta_4} = 0. \tag{12.33d}$$

To obtain solutions for the dynamic force analysis, we proceed as follows. From Eq. (12.32b), the transverse force acting at point B on the output link 4 is

$$F_{34}^{y_4} = \left(I_{G_4} + m_4 l_4^2\right)\frac{\alpha_4}{r_4}. \tag{12.34}$$

Equating the real and imaginary parts of Eq. (12.30a), the x_2 and y_2 components of the forces, respectively, are

$$F_{12}^{x_2} + F_{32}^{x_2} = m_2 A_{G_2}^{x_2}, \tag{12.35a}$$

$$F_{12}^{y_2} + F_{32}^{y_2} = m_2 A_{G_2}^{y_2}. \tag{12.35b}$$

Equating the real and imaginary parts of Eq. (12.31a), the x_3 and y_3 components of the forces, respectively, are

$$F_{23}^{x_3} + F_{43}^{x_3} = m_3 A_{G_3}^{x_3}, \tag{12.36a}$$

$$F_{23}^{y_3} + F_{43}^{y_3} = m_3 A_{G_3}^{y_3}. \tag{12.36b}$$

Equating the real and imaginary parts of Eq. (12.32a), the x_4 and y_4 components of the forces, respectively, are

$$F_{14}^{x_4} + F_{34}^{x_4} = m_4 A_{G_4}^{x_4}, \tag{12.37a}$$

$$F_{14}^{y_4} + F_{34}^{y_4} = m_4 A_{G_4}^{y_4}. \tag{12.37b}$$

Substituting Eq. (12.36b) into Eq. (12.31b), the transverse force acting at revolute joint B on the coupler link is

$$F_{43}^{y_3} = \frac{1}{r_3}\left(I_{G_3}\alpha_3 + m_3 l_3 A_{G_3}^{y_3}\right). \tag{12.38a}$$

Then from Eq. (12.36b), the transverse force acting at revolute joint A on the coupler link is

$$F_{23}^{y_3} = m_3 A_{G_3}^{y_3} - F_{43}^{y_3}. \tag{12.38b}$$

Equation (12.33*b*) can be written as

$$\left(F_{43}^{x_3} + jF_{43}^{y_3}\right)e^{j\eta_{34}} + \left(F_{34}^{x_4} + jF_{34}^{y_4}\right) = 0, \tag{12.39a}$$

or as

$$\left(F_{43}^{x_3} + jF_{43}^{y_3}\right) + \left(F_{34}^{x_4} + jF_{34}^{y_4}\right)e^{-j\eta_{34}} = 0, \tag{12.39b}$$

where $\eta_{34} = \theta_3 - \theta_4$. Equating the imaginary parts of Eq. (12.39*a*), and equating the imaginary parts of Eq. (12.39*b*), respectively, gives

$$F_{43}^{x_3}\sin\eta_{34} + F_{43}^{y_3}\cos\eta_{34} + F_{34}^{y_4} = 0, \tag{12.40a}$$

$$F_{43}^{y_3} - \left(F_{34}^{x_4}\sin\eta_{34} - F_{34}^{y_4}\cos\eta_{34}\right) = 0. \tag{12.40b}$$

Then, from these two equations we have

$$F_{43}^{x_3} = -\left(F_{43}^{y_3}\cot\eta_{34} + F_{34}^{y_4}\csc\eta_{34}\right), \tag{12.41a}$$

$$F_{34}^{x_4} = F_{43}^{y_3}\csc\eta_{34} + F_{34}^{y_4}\cot\eta_{34}. \tag{12.41b}$$

Then, from Eqs. (12.36*a*) and (12.37*a*), the axial forces acting on links 3 and 4, respectively, are

$$F_{23}^{x_3} = m_3 A_{G_3}^{x_3} - F_{43}^{x_3}, \tag{12.42a}$$

$$F_{14}^{x_4} = m_4 A_{G_4}^{x_4} - F_{34}^{x_4}. \tag{12.42b}$$

Rewriting Eq. (12.33*a*) as

$$F_{32}^{x_2} + jF_{32}^{y_2} = -\left(F_{23}^{x_3} + jF_{23}^{y_3}\right)e^{-j\eta_{23}}, \tag{12.43}$$

and equating the real and imaginary parts, respectively, gives

$$F_{32}^{x_2} = -F_{23}^{x_3}\cos\eta_{23} - F_{23}^{y_3}\sin\eta_{23}, \tag{12.44a}$$

$$F_{32}^{y_2} = F_{23}^{x_3}\sin\eta_{23} - F_{23}^{y_3}\cos\eta_{23}. \tag{12.44b}$$

The input torque (sometimes referred to as the inertia torque) can now be obtained by writing Eq. (12.30*b*) as

$$T_2 = -r_2 F_{32}^{y_2} + \left(I_{G_2} + m_2 l_2^2\right)\alpha_2. \tag{12.45}$$

If the four-bar linkage is in *steady-state operation*, that is, if the input angular velocity is constant, then substituting $\alpha_2 = 0$ into Eq. (12.45) gives

$$T_2 = -r_2 F_{32}^{y_2}. \tag{12.46}$$

From Eqs. (12.35) and (12.37) we have

$$F_{12}^{x_2} = m_2 A_{G_2}^{x_2} - F_{32}^{x_2} \quad \text{and} \quad F_{14}^{x_4} = m_4 A_{G_4}^{x_4} - F_{34}^{x_4}, \tag{12.47a}$$

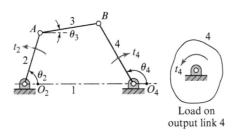

Fig. 12.17 Output torque of the four-bar linkage.

$$F_{12}^{y_2} = m_2 A_{G_2}^{y_2} - F_{32}^{y_2} \quad \text{and} \quad F_{14}^{y_4} = m_4 A_{G_4}^{y_4} - F_{34}^{y_4}. \tag{12.47b}$$

If we write Eqs. (12.33c) and (12.33d), respectively, as

$$\left(F_{21}^{x_1} + jF_{21}^{y_1}\right) = -\left(F_{12}^{x_2} + jF_{12}^{y_2}\right)e^{j\theta_2},$$

$$\left(F_{41}^{x_1} + jF_{41}^{y_1}\right) = -\left(F_{14}^{x_4} + jF_{14}^{y_4}\right)e^{j\theta_4},$$

then the x and y components of the loads at bearing O_2, respectively, are

$$F_{21}^{x_1} = -F_{12}^{x_2}\cos\theta_2 + F_{12}^{y_2}\sin\theta_2 \quad \text{and} \quad F_{21}^{y_1} = -F_{12}^{x_2}\sin\theta_2 - F_{12}^{y_2}\cos\theta_2, \tag{12.48a}$$

and the x and y components of the loads at bearing O_4, respectively, are

$$F_{41}^{x_1} = -F_{14}^{x_4}\cos\theta_4 + F_{14}^{y_4}\sin\theta_4 \quad \text{and} \quad F_{41}^{y_1} = -F_{14}^{x_4}\sin\theta_4 - F_{14}^{y_4}\cos\theta_4. \tag{12.48b}$$

Finally, the dynamic shaking moment can be written as

$$M_S = r_1 F_{41}^{y_1}.$$

Substituting Eq. (12.48b) into this equation gives

$$M_S = -r_1\left(F_{14}^{x_4}\sin\theta_4 + F_{14}^{y_4}\cos\theta_4\right). \tag{12.49}$$

Let us consider the special case where the four-bar linkage is in static equilibrium because of the output torque, t_4, shown in Fig. 12.17, that is, the torque exerted on the load by the output crankshaft on the bearing O_4. The free-body diagrams are as shown in Fig. 12.18.

From the dynamic force analysis, we can now determine the following as functions of the input angle:

1. the axial and the transverse force components exerted on the three moving links;
2. the bearing loads;
3. the inertia torque; and
4. the shaking moment.

From Eqs. (12.30), the sum of the forces on link 2 and the sum of the moments about the bearing O_2 are

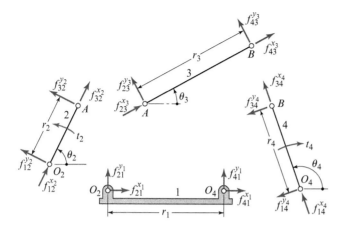

Fig. 12.18 Free-body diagrams for the four-bar linkage.

$$\left[\left(f_{12}^{x_2} + f_{32}^{x_2}\right) + j\left(f_{12}^{y_2} + f_{32}^{y_2}\right)\right]e^{j\theta_2} = 0, \tag{12.50a}$$

$$r_2 f_{32}^{y_2} + t_2 = 0. \tag{12.50b}$$

From Eqs. (12.31), the sum of the forces on link 3 and the sum of the moments about the mass center G_3 are

$$\left[\left(f_{23}^{x_3} + f_{43}^{x_3}\right) + j\left(f_{23}^{y_3} + f_{43}^{y_3}\right)\right]e^{j\theta_3} = 0, \tag{12.51a}$$

$$r_3 f_{43}^{y_3} = 0. \tag{12.51b}$$

From Eq. (12.32), the sum of the forces on link 4 and the sum of the moments about the bearing O_4 are

$$\left[\left(f_{14}^{x_4} + f_{34}^{x_4}\right) + j\left(f_{14}^{y_4} + f_{34}^{y_4}\right)\right]e^{j\theta_4} = 0, \tag{12.52a}$$

$$r_4 f_{34}^{y_4} + t_4 = 0. \tag{12.52b}$$

For the revolute joint A, a sum of the forces gives

$$\left(f_{32}^{x_2} + jf_{32}^{y_2}\right)e^{j\theta_2} + \left(f_{23}^{x_3} + jf_{23}^{y_3}\right)e^{j\theta_3} = 0, \tag{12.53a}$$

and for the revolute joint B, a sum of the forces gives

$$\left(f_{43}^{x_3} + jf_{43}^{y_3}\right)e^{j\theta_3} + \left(f_{34}^{x_4} + jf_{34}^{y_4}\right)e^{j\theta_4} = 0. \tag{12.53b}$$

For bearing O_2, a sum of the forces gives

$$\left(f_{21}^{x_1} + jf_{21}^{y_1}\right) + \left(f_{12}^{x_2} + jf_{12}^{y_2}\right)e^{j\theta_2} = 0, \tag{12.54a}$$

and for bearing O_4, a sum of the forces gives

$$\left(f_{41}^{x_1} + jf_{41}^{y_1}\right) + \left(f_{14}^{x_4} + jf_{14}^{y_4}\right)e^{j\theta_4} = 0. \tag{12.54b}$$

From Eqs. (12.51a), the axial and transverse forces at bearing A on the coupler link are

$$f_{23}^{x_3} = -f_{43}^{x_3} \quad \text{and} \quad f_{23}^{y_3} = -f_{43}^{y_3}. \tag{12.55}$$

From Eq. (12.52b), the transverse force acting at bearing B on output link 4 is

$$f_{34}^{x_4} = -\frac{x_4}{r_4}. \tag{12.56}$$

Substituting Eq. (12.55) into Eq. (12.53a), and rearranging, gives

$$f_{32}^{x_2} + jf_{32}^{y_2} = -f_{23}^{x_3} e^{-j\eta_{23}}.$$

Equating the real and imaginary parts of this equation, respectively, gives

$$f_{32}^{x_2} = f_{43}^{x_3} \cos \eta_{23} \quad \text{and} \quad f_{32}^{y_2} = -f_{43}^{y_3} \sin \eta_{23}. \tag{12.57}$$

Equation (12.53b) can be written as

$$f_{43}^{x_3} e^{j\eta_{34}} + \left(f_{34}^{x_4} + jf_{34}^{y_4} \right) = 0.$$

Equating the real and imaginary parts of this equation, respectively, gives

$$f_{43}^{x_3} = -f_{23}^{x_3} = -\frac{t_4}{r_4} \frac{1}{\sin \eta_{34}} \quad \text{and} \quad f_{43}^{x_3} \cos \eta_{34} + f_{34}^{x_4} = 0. \tag{12.58}$$

Rearranging these two equations, the axial force on the output link can be written as

$$f_{34}^{x_4} = \frac{t_4}{r_4} \cot \eta_{34}. \tag{12.59}$$

From Eqs. (12.50a) and (12.52a) we have

$$f_{12}^{x_2} = -f_{32}^{x_2} \quad \text{and} \quad f_{12}^{y_2} = -f_{32}^{y_2}, \tag{12.60a}$$

$$f_{14}^{x_4} = -f_{34}^{x_4} \quad \text{and} \quad f_{14}^{y_4} = -f_{34}^{y_4}. \tag{12.60b}$$

Adding Eqs. (12.53a) and (12.54a) and with the aid of Eqs. (12.60a), we obtain

$$f_{21}^{x_1} + jf_{21}^{y_1} = f_{43}^{x_3} e^{j\theta_3}.$$

Equating the real and imaginary parts of this equation, respectively, gives

$$f_{21}^{x_1} = f_{43}^{x_3} \cos \theta_3 \quad \text{and} \quad f_{21}^{y_1} = f_{43}^{x_3} \sin \theta_3. \tag{12.61}$$

Adding Eqs. (12.53b) and (12.54b) and with the aid of Eqs. (12.60b), we obtain

$$f_{41}^{x_1} + jf_{41}^{y_1} = f_{43}^{x_3} e^{j\theta_3}.$$

Equating the real and imaginary parts of this equation, respectively, gives

$$f_{41}^{x_1} = f_{43}^{x_3} \cos \theta_3 \quad \text{and} \quad f_{41}^{y_1} = f_{43}^{x_3} \sin \theta_3. \tag{12.62}$$

Substituting Eq. (12.58) into Eq. (12.57), the transverse force acting at point A on the input link can be written as

$$f_{32}^{x_2} = \frac{t_4}{r_4}\frac{\sin\eta_{23}}{\sin\eta_{34}}.$$

Then, substituting this equation into Eq. (12.50b), and rearranging, the input torque can be written as

$$t_2 = -\frac{r_2}{r_4}\frac{\sin\eta_{23}}{\sin\eta_{34}}t_4. \tag{12.63}$$

Recall that the mechanical advantage of a four-bar linkage is defined as the size of the ratio of the output torque to the input torque (Sections 1.10 and 3.20). Therefore, the mechanical advantage can be written from Eq. (12.63), as

$$MA = \frac{t_4}{t_2} = -\frac{r_4\sin\eta_{34}}{r_2\sin\eta_{23}}.$$

Note that this result is consistent with Eq. (3.35).

Finally, the shaking moment can be written with the aid of Eq. (12.62) as

$$m_S = r_1 f_{41}^{y_1} = r_1 f_{43}^{x_3}\sin\theta_3.$$

Substituting Eq. (12.58) into this equation, the shaking moment can be written as

$$m_S = \frac{r_1}{r_4}\frac{\sin\theta_3}{\sin\eta_{34}}t_4. \tag{12.64}$$

If friction in the mechanism is ignored, then the total load, which is denoted here by an asterisk superscript, can be determined as the sum of the static load and the dynamic load; this is the principle of superposition. Therefore, the load torque is

$$T_4^* = t_4 + T_4. \tag{12.65}$$

The axial and transverse loads at bearing O_2, respectively, are

$$(F_{12}^{x_2})^+ = -f_{32}^{x_2} + F_{12}^{x_2}, \; (F_{12}^{y_2})^* = -f_{32}^{y_2} + F_{12}^{y_2}. \tag{12.66}$$

For the input link, the axial and transverse loads at revolute joint A, respectively, are

$$(F_{32}^{x_2})^* = f_{32}^{x_2} + F_{32}^{x_2}, \; (F_{32}^{y_2}) = f_{32}^{y_2} + F_{32}^{y_2}. \tag{12.67a}$$

For the coupler link, the axial and transverse loads at revolute joint A, respectively, are

$$(F_{23}^{x_3})^* = -f_{43}^{x_3} + F_{23}^{x_3}, \; (F_{23}^{y_3})^* = F_{23}^{y_3}. \tag{12.67b}$$

For the coupler link, the axial and transverse loads at revolute joint B, respectively, are

$$(F_{43}^{x_3})^* = -f_{43}^{x_3} + F_{43}^{x_3}, \; (F_{43}^{y_3})^* = F_{43}^{y_3}. \tag{12.68a}$$

For the output link, the axial and transverse loads at revolute joint B, respectively, are

$$\left(F_{34}^{x_4}\right)^* = -f_{34}^{x_4} + F_{34}^{x_4}, \ \left(F_{34}^{y_4}\right)^* = f_{34}^{y_4} + F_{34}^{y_4}. \tag{12.68b}$$

The axial and transverse loads at bearing O_4, respectively, are

$$\left(F_{14}^{x_4}\right)^* = -f_{34}^{x_4} + F_{14}^{x_4}, \ \left(F_{34}^{y_4}\right)^* = -f_{34}^{y_4} + F_{34}^{y_4}. \tag{12.69}$$

The x and y components of the load at bearing O_2, respectively, are

$$\left(F_{21}^{x_1}\right)^* = f_{21}^{x_1} + F_{21}^{x_1} \quad \text{and} \quad \left(F_{21}^{y_1}\right)^* = f_{21}^{y_1} + F_{21}^{y_1}. \tag{12.70a}$$

Therefore, the magnitude and the orientation, respectively, are

$$(F_{21})^* = \sqrt{\left(F_{21}^{x_1}\right)^{*2} + \left(F_{21}^{y_1}\right)^{*2}} \quad \text{and} \quad \psi_2 = \tan_2^{-1}\left(\frac{\left(F_{21}^{y_1}\right)^*}{\left(F_{21}^{x_1}\right)^*}\right). \tag{12.70b}$$

The x and y components of the load at bearing O_4, respectively, are

$$\left(F_{41}^{x_1}\right)^* = f_{41}^{x_1} + F_{41}^{x_1} \quad \text{and} \quad \left(F_{41}^{y_1}\right)^* = f_{41}^{y_1} + F_{41}^{y_1}. \tag{12.71a}$$

Therefore, the magnitude and orientation, respectively, are

$$(F_{41})^* = \sqrt{\left(F_{41}^{x_1}\right)^{*2} + \left(F_{41}^{y_1}\right)^{*2}} \quad \text{and} \quad \psi_4 = \tan_2^{-1}\left(\frac{\left(F_{41}^{y_1}\right)^*}{\left(F_{41}^{x_1}\right)^*}\right). \tag{12.71b}$$

The total driving torque to the input link 2 is

$$T_2^* = T_2 + t_2,$$

where the inertia input torque, T_2, and the static input torque, t_2, are given by Eqs. (12.45) and (12.63), respectively.

The total shaking moment is

$$M_2^* = M_2 + m_2,$$

where the dynamic shaking moment, M_2, and the static shaking moment, m_2, are given by Eqs. (12.49) and (12.64), respectively.

For stable operation we could vary the link lengths, l_1, l_2, and l_3, such that the power $P = T_2^* \omega_2$, discussed in the following section, is maximized and the shaking moment is minimized. For a better suspension system we should try to minimize the vertical components of the loads at the two ground bearings, that is, $\left(F_{21}^{y_1}\right)^*$ and $\left(F_{41}^{y_1}\right)^*$.

12.9 Equation of Motion from Power Equation

This section presents a method of obtaining the equation of motion of a mechanism based on energy considerations. We first write the power equation for the mechanism and then express this equation in terms of kinematic coefficients, as discussed in Chapters 3 and 4.

The work-energy equation for a mechanism can be written as

$$W = \Delta E + \Delta U + W_f, \tag{12.72}$$

where:

W = the net work of the mechanism – that is, the work input less the work output;
ΔE = the change in the kinetic energy of the moving links;
ΔU = the change in the potential energy stored in the mechanism; and
W_f = the energy dissipated through damping and friction.

Differentiating Eq. (12.72) with respect to time gives

$$P = \frac{dW}{dt} = \frac{dE}{dt} + \frac{dU}{dt} + \frac{dW_f}{dt}, \tag{12.73}$$

where P is the *net power* of the mechanism. Equation (12.73) is commonly referred to as the *power equation*. The net power is a scalar quantity and can be written as

$$P = Q\dot{\psi}, \tag{12.74}$$

where Q is the generalized input force (that is, the force or the torque acting on the input link), and $\dot{\psi}$ is the generalized input velocity (that is, either the rectilinear or the angular velocity of the input). If the product of these two signed quantities is positive (that is, if the signed quantities are acting in the same direction), then power is going into the system. If the product is negative (that is, if the two signed quantities are acting in opposite directions), then power is being removed from the system.

The kinetic energy of a link in the mechanism, say link j, can be written as

$$E_j = \frac{1}{2} m_j V_{G_j}^2 + \frac{1}{2} I_{G_j} \omega_j^2. \tag{12.75}$$

Recall from Section 3.12 that the velocity of the mass center of link, j, can be written in terms of first-order kinematic coefficients as

$$\mathbf{V}_{G_j} = \left(x'_{G_j} \hat{\mathbf{i}} + y'_{G_j} \hat{\mathbf{j}} \right) \dot{\psi}, \tag{12.76a}$$

and the angular velocity of link j can be written as

$$\omega_j = \theta'_j \dot{\psi}. \tag{12.76b}$$

Substituting Eqs. (12.76) into Eq. (12.75), the kinetic energy of link j can be written as

$$E_j = \frac{1}{2} m_j \left(x'^2_{G_j} + y'^2_{G_j} \right) \dot{\psi}^2 + \frac{1}{2} I_{G_j} \theta'^2_j \dot{\psi}^2. \tag{12.77}$$

Differentiating this equation with respect to time, the time rate of change in the kinetic energy of link j can be written as

$$\frac{dE_j}{dt} = A_j \dot{\psi}\ddot{\psi} + B_j \dot{\psi}^3, \tag{12.78}$$

Fig. 12.19 Potential energy due to elevation.

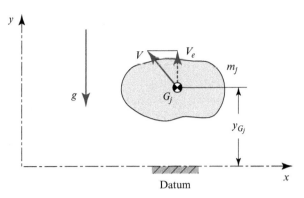

where

$$A_j = m_j\left(x'^2_{G_j} + y'^2_{G_j}\right) + I_{G_j}\theta'^2_j, \qquad (12.79a)$$

$$B_j = m_j\left(x'_{G_j}x''_{G_j} + y'_{G_j}y''_{G_j}\right) + I_{G_j}\theta'_j\theta''_j. \qquad (12.79b)$$

Comparing Eq. (12.79a) and Eq. (12.77), we see that the kinetic energy can be written as

$$E_j = \frac{1}{2}A_j\dot{\psi}^2. \qquad (12.80)$$

From Eqs. (12.79a) and (12.79b), we also note the relationship between the coefficients, that is,

$$B_j = \frac{1}{2}\frac{dA_j}{d\psi}. \qquad (12.81)$$

For a mechanism with n links and with link 1 fixed, the time rate of change in kinetic energy can be written, from Eq. (12.78), as

$$\frac{dE}{dt} = \sum_{j=2}^{n} A_j\dot{\psi}\ddot{\psi} + \sum_{j=2}^{n} B_j\dot{\psi}^3. \qquad (12.82)$$

Next we consider the potential energy stored in link j of the mechanism:

1. *Due to gravity.* If we assume that the gravitational force acts in the negative y direction, then the potential energy due to gravity is

$$U_{gj} = m_j g y_{G_j}, \qquad (12.83)$$

where m_j is the mass of the link, g is the gravitational constant, and y_{G_j} is the height of the mass center above a datum arbitrarily chosen at the origin, as shown in Fig. 12.19.

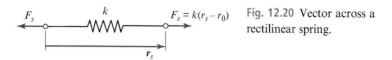

Fig. 12.20 Vector across a rectilinear spring.

Differentiating Eq. (12.83) with respect to time, the rate of change in the gravitational potential energy of link j can be written as

$$\frac{dU_{gj}}{dt} = m_j g y'_{G_j} \dot{\psi},$$ (12.84a)

where the first-order kinematic coefficient is

$$y'_{G_j} = \frac{dy_{G_j}}{d\psi}.$$ (12.84b)

Therefore, the time rate of change in the gravitational potential energy stored in the mechanism is

$$\frac{dU_g}{dt} = \sum_{j=2}^{n} m_j g y'_{G_j} \dot{\psi}.$$ (12.85)

2. *Due to a rectilinear spring.* If the mechanism contains a spring element, then the potential energy stored in the spring is

$$U_s = \frac{1}{2} k (r_s - r_0)^2,$$ (12.86)

where k is the spring rate, r_s is the actual length of the spring, and r_0 is the free length of the spring, as shown in Fig. 12.20. Differentiating Eq. (12.86) with respect to time, the time rate of change in the potential energy stored in the spring can be written as

$$\frac{dU_s}{dt} = k (r_s - r_0) r'_s \dot{\psi},$$ (12.87a)

where the first-order kinematic coefficient of the spring is

$$r'_s = \frac{dr_s}{d\psi}.$$ (12.87b)

Next we consider the dissipative effects due to the following:

1. A *viscous damper.* The work done in overcoming the damping effect is

$$W_c = C\dot{r}_c \Delta r_c,$$ (12.88)

where C is the damping coefficient and r_c is the length of the vector across the damper, as shown in Fig. 12.21. Differentiating Eq. (12.88) with respect to time, the time rate of change of the dissipative effect can be written

Fig. 12.21 Vector across a
damper.

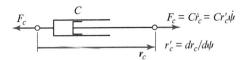

Fig. 12.22 Vector to the point
of contact.

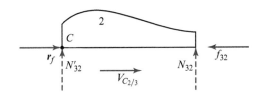

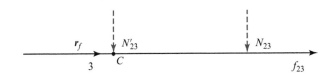

$$\frac{dW_c}{dt} = Cr'^2_c \dot{\psi}^2, \qquad\qquad (12.89a)$$

where the first-order kinematic coefficient of the damper is

$$r'_c = \frac{dr_c}{d\psi}. \qquad\qquad (12.89b)$$

2. *Coulomb friction.* The work done in overcoming Coulomb friction is

$$W_f = \mu N \Delta r_f, \qquad\qquad (12.90)$$

where μ is the coefficient of Coulomb friction, N is the normal force between the surfaces on the contacting links, and $\mathbf{r}_f$ is the vector along the surface to the point of contact shown as point C in Fig. 12.22. Differentiating Eq. (12.90) with respect to time, the time rate of change of the dissipative effect can be written as

$$\frac{dW_f}{dt} = \mu N \left| r'_f \dot{\psi} \right|, \qquad\qquad (12.91a)$$

where the first-order kinematic coefficient is

$$r'_f = \frac{dr_f}{d\psi}. \qquad\qquad (12.91b)$$

Substituting Eqs. (12.74), (12.82), (12.85), (12.87a), (12.89a), and (12.91a) into Eq. (12.73), the net power can be written as

$$Q\dot{\psi} = \sum_{j=2}^{n} A_j \dot{\psi}\ddot{\psi} + \sum_{j=2}^{n} B_j \dot{\psi}^3 + \sum_{j=2}^{n} m_j g y'_{G_j} \dot{\psi} + k(r_s - r_0)r'_s \dot{\psi} + Cr'^2_c \dot{\psi}^2 + \mu N \left| r'_f \dot{\psi} \right|. \tag{12.92}$$

We note that each term in this equation contains the generalized input velocity, $\dot{\psi}$. If we divide through by this common factor, then we obtain the *equation of motion* for the mechanism, that is,

$$Q = \sum_{j=2}^{n} A_j \ddot{\psi} + \sum_{j=2}^{n} B_j \dot{\psi}^2 + \sum_{j=2}^{n} m_j g y'_{G_j} + k(r_s - r_0)r'_s + Cr'^2_c \dot{\psi} + \mu N \left| r'_f \right|. \tag{12.93}$$

This result is also valid when the generalized input velocity is zero – that is, when the input link is instantaneously stopped or when the mechanism is starting from rest. In this case, Eq. (12.93) reduces to

$$Q = \sum_{j=2}^{n} A_j \ddot{\psi} + \sum_{j=2}^{n} m_j g y'_{G_j} + k(r_s - r_0)r'_s + \mu N \left| r'_f \right|, \tag{12.94}$$

and for this case there is no necessity to calculate $\sum_{j=2}^{n} B_j$ or the first-order kinematic coefficient of the viscous damper.

We will consider two cases for the input motion: (1) input rotation, that is, $\dot{\psi} = \dot{\theta}_i$; and (2) input translation, that is, $\dot{\psi} = \dot{r}_i$.

1. If the input motion is rotation, then Eq. (12.93) can be written as

$$T_i = \sum_{j=2}^{n} A_j \ddot{\theta}_i + \sum_{j=2}^{n} B_j \dot{\theta}_i^2 + \sum_{j=2}^{n} m_j g y'_{G_j} + k(r_s - r_0)r'_s + Cr'^2_c \dot{\theta}_i + \mu N \left| r'_f \right|, \tag{12.95}$$

where T_i is the torque acting at the input. Note that the coefficient $\sum_{j=2}^{n} A_j$ must have units of moment of inertia and is referred to as the *equivalent mass moment of inertia* of the system and is denoted I_{EQ}. This moment of inertia, if thought of as concentrated on the input link alone, would have the same kinetic energy as the entire mechanism. From Eq. (12.80), the kinetic energy could then be written as

$$E = \frac{1}{2} I_{EQ} \dot{\theta}^2. \tag{12.96}$$

The equation of motion, that is Eq. (12.95), could then be written as

$$T_i = I_{EQ} \ddot{\theta}_i + \frac{1}{2} \frac{dI_{EQ}}{d\theta_i} \dot{\theta}_i^2 + \sum_{j=2}^{n} m_j g y'_{G_j} + k(r_s - r_0)r'_s + Cr'^2_c \dot{\theta}_i + \mu N \left| r'_f \right|. \tag{12.97}$$

2. If the input motion is translation, then Eq. (12.93) can be written as

$$F_i = \sum_{j=2}^{n} A_j \ddot{r}_i + \sum_{j=2}^{n} B_j r_i^2 + \sum_{j=2}^{n} m_j g y'_{G_j} + k(r_s - r_0) r'_s + C r'^2_c \dot{r}_i + \mu N \left| r'_f \right|, \tag{12.98}$$

where F_i is the force acting at the input. For this case, the coefficient $\sum_{j=2}^{n} A_j$ has units of mass and is referred to as the *equivalent mass* of the system and denoted as m_{EQ}. This mass, if thought of as concentrated on the input link alone, would have the same kinetic energy as the entire mechanism. From Eq. (12.80) the kinetic energy could then be written as

$$E = \frac{1}{2} m_{EQ} \dot{r}^2. \tag{12.99}$$

The equation of motion, that is Eq. (12.98), could then be written as

$$F_i = m_{EQ} \ddot{r}_i + \frac{1}{2} \frac{dm_{EQ}}{dr_i} \dot{r}_i^2 + \sum_{j=2}^{n} m_j g y'_{G_j} + k(r_s - r_0) r'_s + C r'^2_c \dot{r}_i + \mu N \left| r'_f \right|. \tag{12.100}$$

Example 12.6

Determine the torque, T_2, that must be applied to input link 2 of the parallelogram four-bar linkage in the posture shown in Fig. 12.23. The angular velocity and angular acceleration of link 2 are $\omega_2 = 10 \text{ rad/s ccw}$ and $\alpha_2 = 10 \text{ rad/s}^2 \text{ cw}$, respectively. The masses of the links are $m_2 = 2.5 \text{ kg}$, $m_3 = 5 \text{ kg}$, and $m_4 = 2.5 \text{ kg}$, and the mass moments of inertia of the links are $I_{G_2} = I_{G_4} = 0.25 \text{ kg} \cdot \text{m}^2$ and $I_{G_3} = 4.5 \text{ kg} \cdot \text{m}^2$, and the center of mass of each link is coincident with its geometric center. The free length of the spring is $r_0 = 0.6 \text{ m}$, the spring rate is $k = 500 \text{ N/m}$, and the damping constant is $C = 10 \text{ N} \cdot \text{s/m}$.

Fig. 12.23 $r_1 = 0.8 \text{ m}$, $r_2 = 0.4 \text{ m}$,
$r_3 = 0.8 \text{ m}$, $r_4 = 0.4 \text{ m}$,
$r_s = R_{EB} = 0.8 \text{ m}$, and
$r_c = R_{DA} = 0.8 \text{ m}$.

Assumptions

Gravity acts vertically downward and Coulomb friction can be ignored.

Solution

From a kinematic analysis of the four-bar linkage as presented in Sections 3.12 and 4.12, the first- and second-order kinematic coefficients of links 3 and 4 are

Example 12.6 (cont.)

$$\theta_3' = 0, \quad \theta_4' = 1 \text{ rad/rad}, \quad \theta_3'' = 0, \quad \text{and} \quad \theta_4'' = 0. \tag{1}$$

Recall that the angular velocities and angular accelerations of links 3 and 4 can be written as

$$\omega_3 = \theta_3'\omega_2 \quad \text{and} \quad \omega_4 = \theta_4'\omega_2$$

$$\alpha_3 = \theta_3''\omega_2^2 + \theta_3'\alpha_2 \quad \text{and} \quad \alpha_4 = \theta_4''\omega_2^2 + \theta_4'\alpha_2. \tag{2}$$

Substituting Eqs. (1) and the input angular velocity and angular acceleration into Eqs. (2) gives

$$\omega_3 = 0, \quad \omega_4 = 10 \text{ rad/s ccw}, \quad \alpha_3 = 0, \quad \text{and} \quad \alpha_4 = 10 \text{ rad/s}^2 \text{ cw}. \tag{3}$$

The x and y components of the position of the mass center G_2 with respect to the ground pivot O_2 are

$$x_{G_2} = \frac{1}{2}r_2\cos\theta_2 \quad \text{and} \quad y_{G_2} = \frac{1}{2}r_2\sin\theta_2. \tag{4}$$

Differentiating these equations twice with respect to θ_2 and substituting $\theta_2 = 270°$, the first- and second-order kinematic coefficients of the mass center of link 2 are

$$x_{G_2}' = 0.2 \text{ m/rad}, \ y_{G_2}' = 0, \ x_{G_2}'' = 0, \ y_{G_2}'' = 0.2 \text{ m/rad}^2. \tag{5}$$

The x and y components of the position of the mass center G_3 with respect to the ground pivot O_2 can be written as

$$x_{G_3} = r_2\cos\theta_2 + \frac{1}{2}r_3\cos\theta_3 \quad \text{and} \quad y_{G_3} = r_2\sin\theta_2 + \frac{1}{2}r_3\sin\theta_3. \tag{6}$$

Differentiating these equations twice with respect to θ_2 and substituting $\theta_2 = 270°$ and $\theta_3 = 0°$, the first- and second-order kinematic coefficients of the mass center of link 3 are

$$x_{G_3}' = 0.4 \text{ m/rad}, \ y_{G_3}' = 0, \ x_{G_3}'' = 0, \ y_{G_3}'' = 0.4 \text{ m/rad}^2. \tag{7}$$

The first- and second-order kinematic coefficients of the mass center of link 4 are the same as for the mass center of link 2, that is,

$$x_{G_4}' = x_{G_2}' = 0.2 \text{ m/rad}, \ y_{G_4}' = y_{G_2}' = 0, \ x_{G_4}'' = x_{G_2}'' = 0, \ y_{G_4}'' = y_{G_2}'' = 0.2 \text{ m/rad}^2. \tag{8}$$

To obtain the first-order kinematic coefficient for the spring, the vector loop-closure equation, shown in Fig. 12.24, can be written as

$$\mathbf{r}_4 + \mathbf{r}_s - \mathbf{r}_6 = \mathbf{0}. \tag{9}$$

Example 12.6 (cont.)

Fig. 12.24 Vector loop for the spring.

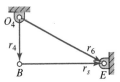

The two scalar equations are

$$r_4 \cos \theta_4 + r_s \cos \theta_s - r_6 \cos \theta_6 = 0,$$

$$r_4 \sin \theta_4 + r_s \sin \theta_s - r_6 \sin \theta_6 = 0. \qquad (10)$$

Differentiating Eqs. (10) with respect to input θ_2 gives

$$-r_4 \sin \theta_4 \theta_4' - r_s \sin \theta_s \theta_s' + r_s' \cos \theta_s = 0, \qquad (11a)$$

$$r_4 \cos \theta_4 \theta_4' + r_s \cos \theta_s \theta_s' + r_s' \sin \theta_s = 0. \qquad (11b)$$

Substituting $\theta_s = 0°$, $\theta_4 = 270°$, and $\theta_4' = 1$ rad/rad into Eq. (11a), the first-order kinematic coefficient for the spring is

$$r_s' = -r_4 \theta_4' = -0.4 \text{ m/rad}. \qquad (12)$$

To obtain the first-order kinematic coefficient for the damper, the vector loop-closure equation, shown in Fig. 12.25, can be written as

$$\mathbf{r}_2 + \mathbf{r}_c - \mathbf{r}_8 = \mathbf{0}. \qquad (13)$$

Fig. 12.25 Vector loop for the viscous damper.

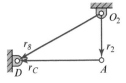

The two scalar equations are

$$r_2 \cos \theta_2 + r_c \cos \theta_c - r_8 \cos \theta_8 = 0, \qquad (14a)$$

$$r_2 \sin \theta_2 + r_c \sin \theta_c - r_8 \sin \theta_8 = 0. \qquad (14b)$$

Differentiating Eqs. (14) with respect to input θ_2 gives

$$-r_2 \sin \theta_2 - r_c \sin \theta_c \theta_c' + r_c' \cos \theta_c = 0, \qquad (15a)$$

$$r_2 \cos \theta_2 + r_c \cos \theta_c \theta_c' + r_c' \sin \theta_c = 0. \qquad (15b)$$

Example 12.6 (cont.)

Substituting $\theta_2 = 270°$ and $\theta_c = 180°$ into Eqs. (15a), the first-order kinematic coefficient for the damper is

$$r'_c = r_2 = 0.4 \text{ m/rad.} \tag{16}$$

The equivalent mass moment of inertia for the four-bar linkage is

$$I_{EQ} = \sum_{j=2}^{4} A_j.$$

From Eq. (12.79a), the equivalent mass moment of inertia can be written as

$$I_{EQ} = m_2\left(x'^2_{G_2} + y'^2_{G_2}\right) + I_{G_2} + m_3\left(x'^2_{G_3} + y'^2_{G_3}\right) + I_{G_3}\theta'^2_3 + m_4\left(x'^2_{G_4} + y'^2_{G_4}\right) + I_{G_4}\theta'^2_4.$$

Substituting the known data and the kinematic coefficients into this equation gives

$$I_{EQ} = 0.1 + 0.25 + 0.8 + 0.1 + 0.25 = 1.5 \text{ kg}\cdot\text{m}^2. \tag{17}$$

From Eq. (12.79b), we can write

$$\sum_{j=2}^{4} B_j = m_2\left(x'_{G_2}x''_{G_2} + y'_{G_2}y''_{G_2}\right) + m_3\left(x'_{G_3}x''_{G_3} + y'_{G_3}y''_{G_3}\right) + I_{G_3}\theta'_3\theta''_3$$

$$+ m_4\left(x'_{G_4}x''_{G_4} + y'_{G_4}y''_{G_4}\right) + I_{G_4}\theta'_4\theta''_4. \tag{18}$$

For the given posture, the kinematic coefficients are given by Eqs. (1), (5), (7), and (8). Substituting these kinematic coefficients into Eq. (18) gives

$$\sum_{j=2}^{4} B_j = 0. \tag{19}$$

Then, substituting the known data, the kinematic coefficients, and Eqs. (17) and (19) into Eq. (12.97), the equation of motion can be written

$$T_2 = \left(1.5 \text{ kg}\cdot\text{m}^2\right)\ddot{\theta}_2 + k(r_s - r_0)r'_s + Cr'^2_c\dot{\theta}_2.$$

Finally, substituting the known data and Eqs. (12) and (16) into this equation, the input torque is

$$T_2 = -15 \text{ N}\cdot\text{m} + (500 \text{ N/m})(0.8 \text{ m} - 0.6 \text{ m})(-0.4 \text{ m/rad})$$
$$+ (10 \text{ N}\cdot\text{s/m})(0.4 \text{ m/rad})^2(10 \text{ rad/s})$$

$$T_2 = -39 \text{ N}\cdot\text{m} = 39 \text{ N}\cdot\text{m cw.} \qquad\qquad Ans.$$

Example 12.6 (cont.)

Since the answer is negative, the input torque acts in the direction opposite to the input angular velocity, that is, clockwise. Note that this is necessary to achieve the clockwise input angular acceleration specified. Also note that a clockwise input torque of $T_2 = 24\,\text{N}\cdot\text{m}$ would yield a zero input acceleration at the instant under study.

12.10 Measuring Mass Moments of Inertia

Sometimes the shapes of machine parts are very complex, and it is extremely tedious and time-consuming to calculate their mass moment(s) of inertia. Consider, for example, the problem of finding the mass moment of inertia of an automobile body about a vertical axis through its center of mass. For such problems, it is usually possible to determine the mass moment of inertia of a body by observing its dynamic behavior caused by a known rotational disturbance.

Many bodies – connecting rods, for example – are shaped so that their masses can be assumed to lie in a single plane. Once such bodies have been weighed and their mass centers located, they can be suspended like a pendulum and caused to oscillate. The mass moment of inertia of such a body can then be computed from an observation of its period or frequency of oscillation. For best experimental results, the part should be suspended with the pivot located close to, but not coincident with, the center of mass. It is not usually necessary to drill a hole to suspend the body; for example, a spoked wheel or a gear can be suspended on a knife edge at its rim.

When the body shown in Fig. 12.26a is displaced through angle θ, a gravity force mg acts at G. Summing moments about the pivot, O, gives

$$\sum M_O = -mg(r_G \sin\theta) - I_O \ddot{\theta} = 0. \qquad (a)$$

We intend that the pendulum be displaced only through a small angle, so that $\sin\theta$ can be replaced by θ. Equation (a) can then be written as

$$\ddot{\theta} + \frac{mgr_G}{I_O}\theta = 0. \qquad (b)$$

Fig. 12.26 (*a*) Simple pendulum; (*b*) torsional pendulum.

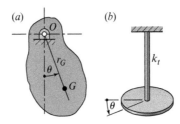

This second-order, linear differential equation has the well-known solution

$$\theta = c_1 \sin \sqrt{\frac{mgr_G}{I_O}} + c_2 \cos \sqrt{\frac{mgr_G}{I_O}}, \tag{c}$$

where c_1 and c_2 are constants of integration.

We start the pendulum motion by displacing it through a small angle θ_0 and releasing it with no initial velocity from this posture. Thus, at time $t = 0$, we obtain $\theta = \theta_0$ and $\dot{\theta} = 0$. Substituting these conditions into Eq. (c) and its first time derivative enables us to evaluate the two constants. They are $c_1 = 0$ and $c_2 = \theta_0$. Therefore,

$$\theta = \theta_0 \cos \sqrt{\frac{mgr_G}{I_O}} t. \tag{12.101}$$

Since a cosine function repeats itself every $360°$ or 2π radians, the period of the motion is

$$\tau = 2\pi \sqrt{\frac{I_O}{mgr_G}}. \tag{d}$$

Therefore, the mass moment of inertia of the body about the pivot, O, is

$$I_O = mgr_G \left(\frac{\tau}{2\pi}\right)^2. \tag{12.102}$$

This equation indicates that the body must be weighed to get mg; the distance, r_G, must be measured, and then the pendulum must be suspended and oscillated so that the period, τ, can be observed; Eq. (12.102) can then be evaluated to give I_O about O. If the moment of inertia about the mass center is desired, it can be obtained by using the parallel-axis theorem, Eq. (12.11).

Figure 12.26b shows how the mass moment of inertia can be determined without actually weighing the body. The body is connected to a slender rod or wire at the mass center. The torsional stiffness, k_t, of the rod or wire is defined as the torque necessary to twist the rod through a unit angle. If the body of Fig. 12.26b is turned through a small angle, θ, and released, the equation of motion becomes

$$\ddot{\theta} + \frac{k_t}{I_G} \theta = 0. \tag{e}$$

This is similar to Eq. (b) and, with the same starting conditions, has the solution

$$\theta = \theta_0 \cos \sqrt{\frac{k_t}{I_G}} t. \tag{12.103}$$

The period of oscillation is then

$$\tau = 2\pi \sqrt{\frac{I_G}{k_t}}, \tag{f}$$

Fig. 12.27 Trifilar pendulum.

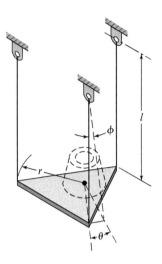

and

$$I_G = k_t \left(\frac{\tau}{2\pi}\right)^2. \tag{12.104}$$

The torsional stiffness is often known or can be computed from a knowledge of the length and diameter of the rod or wire and its material. Then, the oscillation of the body can be observed and Eq. (12.104) used to compute the mass moment of inertia, I_G. Alternatively, when the torsional stiffness, k_t, is unknown, a body with known mass moment of inertia, I_G, can be mounted and Eq. (12.104) can be used to determine k_t.

A *trifilar pendulum*, also called a *three-string torsional pendulum*, shown in Fig. 12.27, provides a very accurate method of measuring mass moment of inertia. Three strings of equal length support a lightweight platform and are equally spaced about its center. A round platform serves just as well as the triangular one shown. The part whose mass moment of inertia is to be determined is carefully placed on the platform so that the center of mass of the object coincides with the platform center. The platform is then made to oscillate, and the number of oscillations is counted over a specified period of time ([1], p. 129).

The notation for the three-string torsional pendulum analysis is as follows:

m = mass of the part;
m_p = mass of the platform;
I_G = mass moment of inertia of the part;
I_p = mass moment of inertia of the platform;
r = platform radius;
θ = platform angular displacement;
l = string length;
ϕ = string angular displacement; and
z = vertical axis through the center of the platform.

We begin by writing the summation of moments about the z axis. This gives

$$\sum M^z = -r(m + m_p)g \sin\phi - (I_G + I_p)\ddot{\theta} = 0. \tag{g}$$

Since we assume small displacements, the sine of an angle can be approximated by the angle itself. Therefore,

$$\phi = \frac{r}{l}\theta, \tag{h}$$

and Eq. (g) becomes

$$\ddot{\theta} + \frac{(m + m_p)gr^2}{(I_G + I_p)l}\theta = 0. \tag{i}$$

This equation can be solved in the same manner as Eq. (b). The result is

$$I_G + I_p = \frac{(m + m_p)gr^2}{l}\left(\frac{\tau}{2\pi}\right)^2. \tag{12.105}$$

This equation can be used with an empty platform to find I_p. With m_p and I_p known, the equation can then be used to find I_G of the part being measured.

12.11 Transformation of Inertia Axes

A brief review will confirm that all examples presented so far have been limited to planar motion. It is now time to extend our study to include spatial motion problems. The basic principles are not new, but the problems presented may seem more complex because of our difficulty in visualizing in three dimensions. In addition, our previous treatment of angular motion was not presented in enough detail to deal with three-dimensional rotations.

Up to this point, we have continually used what are called the *principal mass moments of inertia*. In a coordinate system with origin at the center of mass and with axes aligned in the principal axis directions, all *mass products of inertia* [Eqs. (12.8)] are zero. This choice of axes has been used to simplify the form of the equations from that required for other choices.

In Eq. (12.11), the parallel-axis formula, we demonstrated how translation of axes can be performed. However, up to now, we have said little of how inertia properties can be rotated to a coordinate system that is not parallel to the principal axes. This is the purpose of this section.

Let us assume that the principal axes are aligned along unit vector directions $\hat{\mathbf{i}}$, $\hat{\mathbf{j}}$, and $\hat{\mathbf{k}}$, and that the rotated axes desired have directions denoted by $\hat{\mathbf{i}}'$, $\hat{\mathbf{j}}'$, and $\hat{\mathbf{k}}'$. Then, any point in the principal axis coordinate system is located by the position vector $\mathbf{R}$ and by $\mathbf{R}'$ in the rotated system. Both are descriptions of the same point position, $\mathbf{R}' = \mathbf{R}$, and

$$R^{x'}\hat{\mathbf{i}}' + R^{y'}\hat{\mathbf{j}}' + R^{z'}\hat{\mathbf{k}}' = R^x\hat{\mathbf{i}} + R^y\hat{\mathbf{j}} + R^z\hat{\mathbf{k}}.$$

Now, by taking dot products of this equation with $\hat{\mathbf{i}}'$, $\hat{\mathbf{j}}'$, and $\hat{\mathbf{k}}'$, respectively, we find the transformation equations between the two sets of axes. They are

$$
\begin{aligned}
R^{x'} &= \left(\hat{\mathbf{i}}' \cdot \hat{\mathbf{i}}\right)R^x + \left(\hat{\mathbf{i}}' \cdot \hat{\mathbf{j}}\right)R^y + \left(\hat{\mathbf{i}}' \cdot \hat{\mathbf{k}}\right)R^z, \\
R^{y'} &= \left(\hat{\mathbf{j}}' \cdot \hat{\mathbf{i}}\right)R^x + \left(\hat{\mathbf{j}}' \cdot \hat{\mathbf{j}}\right)R^y + \left(\hat{\mathbf{j}}' \cdot \hat{\mathbf{k}}\right)R^z, \\
R^{z'} &= \left(\hat{\mathbf{k}}' \cdot \hat{\mathbf{i}}\right)R^x + \left(\hat{\mathbf{k}}' \cdot \hat{\mathbf{j}}\right)R^y + \left(\hat{\mathbf{k}}' \cdot \hat{\mathbf{k}}\right)R^z.
\end{aligned}
\tag{a}
$$

But we may remember that the dot product of two unit vectors is equal to the cosine of the angle between them. For example, if the angle between $\hat{\mathbf{i}}'$ and $\hat{\mathbf{j}}$ is denoted by $\theta_{i'j}$, then

$$\hat{\mathbf{i}}' \cdot \hat{\mathbf{j}} = \cos\theta_{i'j}.$$

Thus, Eqs. (a) become

$$
\begin{aligned}
R^{x'} &= \cos\theta_{i'i}R^x + \cos\theta_{i'j}R^y + \cos\theta_{i'k}R^z, \\
R^{y'} &= \cos\theta_{j'i}R^x + \cos\theta_{j'j}R^y + \cos\theta_{j'k}R^z, \\
R^{z'} &= \cos\theta_{k'i}R^x + \cos\theta_{k'j}R^y + \cos\theta_{k'k}R^z.
\end{aligned}
\tag{b}
$$

By substituting these into the definitions of mass moments of inertia in Eqs. (12.7), expanding, and then performing the integration, we get

$$
\begin{aligned}
I^{x'x'} &= \cos^2\theta_{i'i}I^{xx} + \cos^2\theta_{i'j}I^{yy} + \cos^2\theta_{i'k}I^{zz} \\
&\quad + 2\cos\theta_{i'i}\cos\theta_{i'j}I^{xy} + 2\cos\theta_{i'i}\cos\theta_{i'k}I^{xz} + 2\cos\theta_{i'j}\cos\theta_{i'k}I^{yz},
\end{aligned}
$$

$$
\begin{aligned}
I^{y'y'} &= \cos^2\theta_{j'i}I^{xx} + \cos^2\theta_{j'j}I^{yy} + \cos^2\theta_{j'k}I^{zz} \\
&\quad + 2\cos\theta_{j'i}\cos\theta_{j'j}I^{xy} + 2\cos\theta_{j'i}\cos\theta_{j'k}I^{xz} + 2\cos\theta_{j'j}\cos\theta_{j'k}I^{yz},
\end{aligned}
\tag{12.106}
$$

$$
\begin{aligned}
I^{z'z'} &= \cos^2\theta_{k'i}I^{xx} + \cos^2\theta_{k'j}I^{yy} + \cos^2\theta_{k'k}I^{zz} \\
&\quad + 2\cos\theta_{k'i}\cos\theta_{k'j}I^{xy} + 2\cos\theta_{k'i}\cos\theta_{k'k}I^{xz} + 2\cos\theta_{k'j}\cos\theta_{k'k}I^{yz}.
\end{aligned}
$$

Similarly, for the mass products of inertia, starting from Eqs. (12.8), we get

$$
\begin{aligned}
I^{x'y'} &= -\cos\theta_{i'i}\cos\theta_{j'i}I^{xx} - \cos\theta_{i'j}\cos\theta_{j'j}I^{yy} - \cos_{i'k}\cos_{j'k}I^{zz} \\
&\quad + \left(\cos_{i'i}\cos\theta_{j'j} + \cos\theta_{i'j}\cos\theta_{j'i}\right)I^{xy} \\
&\quad + \left(\cos_{i'j}\cos\theta_{j'k} + \cos\theta_{i'k}\cos\theta_{j'j}\right)I^{yz} \\
&\quad + \left(\cos_{i'k}\cos\theta_{j'i} + \cos\theta_{i'i}\cos\theta_{j'k}\right)I^{zx}
\end{aligned}
\tag{12.107a}
$$

$$
\begin{aligned}
I^{y'z'} &= -\cos\theta_{j'i}\cos\theta_{k'i}I^{xx} - \cos\theta_{j'j}\cos\theta_{k'j}I^{yy} - \cos\theta_{j'k}\cos\theta_{k'k}I^{zz} \\
&\quad + \left(\cos\theta_{j'i}\cos\theta_{k'j} + \cos\theta_{j'j}\cos\theta_{k'i}\right)I^{xy} \\
&\quad + \left(\cos\theta_{j'j}\cos\theta_{k'k} + \cos\theta_{j'k}\cos\theta_{k'j}\right)I^{yz} \\
&\quad + \left(\cos\theta_{j'k}\cos\theta_{k'i} + \cos\theta_{j'i}\cos\theta_{k'k}\right)I^{zx}
\end{aligned}
\tag{12.107b}
$$

$$
\begin{aligned}
I^{z'x'} &= -\cos\theta_{k'i}\cos\theta_{i'i}I^{xx} - \cos\theta_{k'j}\cos\theta_{i'j}I^{yy} - \cos\theta_{k'k}\cos\theta_{i'k}I^{zz} \\
&\quad + \left(\cos\theta_{k'i}\cos\theta_{ij} + \cos\theta_{k'j}\cos\theta_{i'i}\right)I^{xy} \\
&\quad + \left(\cos\theta_{k'j}\cos\theta_{i'k} + \cos\theta_{k'k}\cos\theta_{i'j}\right)I^{yz} \\
&\quad + \left(\cos\theta_{k'k}\cos\theta_{i'i} + \cos\theta_{k'i}\cos\theta_{i'k}\right)I_{zx}.
\end{aligned}
\tag{12.107c}
$$

Once we have rotated away from the principal axes, or if we have started with other than the principal axes, then the situation is different. The parallel-axis formulae of Eq. (12.11) are no longer sufficient for the translation of axes. Let us assume that the mass moments and products of inertia are known in one coordinate system and are desired in another coordinate system, which is translated from the first by the equations

$$
\begin{aligned}
R^{x''} &= R^{x'} + d^{x'}, \\
R^{y''} &= R^{y'} + d^{y'}, \\
R^{z''} &= R^{z'} + d^{z'}.
\end{aligned}
\tag{12.108}
$$

Then, substituting into the definition of I^{xx} of Eq. (12.7), for example, and integrating:

$$
\begin{aligned}
I^{x''x''} &= \int \left[\left(R^{y''} \right)^2 + \left(R^{z''} \right)^2 \right] dm \\
&= \int \left[\left(R^{y'} + d^{y'} \right)^2 + \left(R^{z'} + d^{z'} \right)^2 \right] dm \\
&= \int \left[\left(R^{y'} \right)^2 + \left(R^{z'} \right)^2 \right] dm + 2 \left(\int R^{y'} dm \right) d^{y'} \\
&\quad + 2 \left(\int R^{z'} dm \right) d^{z'} + \left(d^{y'} \right)^2 \int dm + \left(d^{z'} \right)^2 \int dm \\
&= I^{x'x'} + 2mR_G^{y'} d^{y'} + 2mR_G^{z'} d^{z'} + m \left(d^{y'} \right)^2 + m \left(d^{z'} \right)^2.
\end{aligned}
$$

Proceeding in this manner through each of Eqs. (12.7) and (12.8), we find

$$
\begin{aligned}
I^{x''x''} &= I^{x'x'} + 2mR_G^{y'} d^{y'} + 2mR_G^{z'} d^{z'} + m \left(d^{y'} \right)^2 + m \left(d^{z'} \right)^2, \\
I^{y''y''} &= I^{y'y'} + 2mR_G^{z'} d^{z'} + 2mR_G^{x'} d^{x'} + m \left(d^{z'} \right)^2 + m \left(d^{x'} \right)^2, \\
I^{z''z''} &= I^{z'z'} + 2mR_G^{x'} d^{x'} + 2mR_G^{y'} d^{y'} + m \left(d^{x'} \right)^2 + m \left(d^{y'} \right)^2, \\
I^{x''y''} &= I^{x'y'} + mR_G^{x'} d^{y'} + mR_G^{y'} d^{x'} + md^{x'} d^{y'}, \\
I^{y''z''} &= I^{y'z'} + mR_G^{y'} d^{z'} + mR_G^{z'} d^{y'} + md^{y'} d^{z'}, \\
I^{z''x''} &= I^{z'x'} + mR_G^{z'} d^{x'} + mR_G^{x'} d^{z'} + md^{z'} d^{x'}.
\end{aligned}
\tag{12.109}
$$

Example 12.7

A circular steel disk, called a swashplate, is fastened at its center at an angle of $30°$ to a steel shaft, as shown in Fig. 12.28. Using $0.282 \ \text{lb/in}^3$ as the density of steel, find the mass moments and products of inertia of the combined disk and shaft about the combined center of mass and aligned along the axes of the shaft.

Example 12.7 (cont.)

Fig. 12.28 Shaft with swashplate.

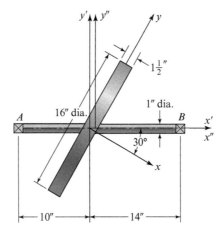

Solution

First, we calculate the mass of the disk and the principal mass moments of inertia of the disk using the equations of Table A.5, in Appendix A. These are

$$m = \frac{0.282 \text{ lb/in}^3}{0.386 \text{ in/s}^2}\, \pi (8 \text{ in})^2 (1.5 \text{ in}) = 0.220 \text{ lb} \cdot \text{s}^2/\text{in},$$

$$I^{xx} = \frac{(0.220 \text{ lb} \cdot \text{s}^2/\text{in})(8 \text{ in})^2}{2} = 7.04 \text{ in} \cdot \text{lb} \cdot \text{s}^2,$$

$$I^{yy} = I^{zz} = \frac{(0.220 \text{ lb} \cdot \text{s}^2/\text{in})(8 \text{ in})^2}{4} = 3.52 \text{ in} \cdot \text{lb} \cdot \text{s}^2.$$

From Fig. 12.28 we now find the direction cosines between the coordinate systems. These are

$$\begin{aligned}
\cos\theta_{i'i} &= 0.866, & \cos\theta_{i'j} &= 0.500, & \cos\theta_{i'k} &= 0.000, \\
\cos\theta_{j'i} &= -0.500, & \cos\theta_{j'j} &= 0.866, & \cos\theta_{j'k} &= 0.000, \\
\cos\theta_{k'i} &= 0.000, & \cos\theta_{k'j} &= 0.000, & \cos\theta_{k'k} &= 1.000.
\end{aligned}$$

Substituting these values into Eqs. (12.106) and (12.107) gives the mass moments and products of inertia of the disk aligned with the axes of the shaft. These are

$$I^{x'x'} = (0.866)^2(7.04 \text{ in} \cdot \text{lb} \cdot \text{s}^2) + (0.500)^2(3.52 \text{ in} \cdot \text{lb} \cdot \text{s}^2) = 6.16 \text{ in} \cdot \text{lb} \cdot \text{s}^2,$$

$$I^{y'y'} = (-0.500)^2(7.04 \text{ in} \cdot \text{lb} \cdot \text{s}^2) + (0.866)^2(3.52 \text{ in} \cdot \text{lb} \cdot \text{s}^2) = 4.40 \text{ in} \cdot \text{lb} \cdot \text{s}^2,$$

$$I^{z'z'} = I^{zz} = 3.52 \text{ in} \cdot \text{lb} \cdot \text{s}^2,$$

Example 12.7 (cont.)

$$I^{x'y'} = -(0.866)(-0.500)(7.04 \text{ in} \cdot \text{lb} \cdot \text{s}^2) - (0.500)(0.866)(3.52 \text{ in} \cdot \text{lb} \cdot \text{s}^2)$$
$$= 1.52 \text{ in} \cdot \text{lb} \cdot \text{s}^2,$$

$$I^{y'z'} = I^{z'x'} = 0.$$

Next, we find the mass and the principal mass moments of inertia of the shaft using the equations of Table A.5. These are

$$m = \frac{0.282 \text{ lb/in}^3}{386 \text{ in/s}^2} \pi (0.500 \text{ in})^2 (24 \text{ in}) = 0.013\,8 \text{ lb} \cdot \text{s}^2/\text{in},$$

$$I^{xx} = \frac{(0.013\,8 \text{ lb} \cdot \text{s}^2/\text{in})(0.500 \text{ in})^2}{2} = 0.001\,72 \text{ in} \cdot \text{lb} \cdot \text{s}^2,$$

$$I^{yy} = \frac{(0.013\,8 \text{ lb} \cdot \text{s}^2/\text{in})\left[3(0.500 \text{ in})^2 + (24 \text{ in})^2\right]}{12} = 0.662 \text{ in} \cdot \text{lb} \cdot \text{s}^2.$$

The mass products of inertia of the shaft are zero because the axes used are principal axes. Translating the values for the shaft to the same axes as the disk, Eq. (12.7) gives

$$I^{x'x'} = 0.001\,72 \text{ in} \cdot \text{lb} \cdot \text{s}^2,$$

$$I^{y'y'} = I^{z'z'} = (0.662 \text{ in} \cdot \text{lb} \cdot \text{s}^2) + (0.013\,8 \text{ lb} \cdot \text{s}^2/\text{in})(2 \text{ in})^2 = 0.717 \text{ in} \cdot \text{lb} \cdot \text{s}^2.$$

The values of the disk and the shaft are now both in the same coordinate system and are combined to give

$$I^{x'x'} = (6.16 \text{ in} \cdot \text{lb} \cdot \text{s}^2) + (0.001\,72 \text{ in} \cdot \text{lb} \cdot \text{s}^2) = 6.16 \text{ in} \cdot \text{lb} \cdot \text{s}^2,$$

$$I^{y'y'} = (4.40 \text{ in} \cdot \text{lb} \cdot \text{s}^2) + (0.717 \text{ in} \cdot \text{lb} \cdot \text{s}^2) = 5.12 \text{ in} \cdot \text{lb} \cdot \text{s}^2,$$

$$I^{z'z'} = (3.52 \text{ in} \cdot \text{lb} \cdot \text{s}^2) + (0.717 \text{ in} \cdot \text{lb} \cdot \text{s}^2) = 4.24 \text{ in} \cdot \text{lb} \cdot \text{s}^2,$$

$$I^{x'y'} = (1.52 \text{ in} \cdot \text{lb} \cdot \text{s}^2) + 0 = 1.52 \text{ in} \cdot \text{lb} \cdot \text{s}^2,$$

$$I^{y'z'} = I^{z'x'} = 0.$$

Finally, all values are translated to the combined center of mass using Eqs. (12.109):

$$m = (0.220 \text{ lb} \cdot \text{s}^2/\text{in}) + (0.013\,8 \text{ lb} \cdot \text{s}^2/\text{in}) = 0.234 \text{ lb} \cdot \text{s}^2/\text{in},$$

Example 12.7 (cont.)

$$R_G^{x'} = \frac{(0.220 \text{ lb} \cdot \text{s}^2/\text{in})(0) + (0.0138 \text{ lb} \cdot \text{s}^2/\text{in})(2.0 \text{ in})}{0.235 \text{ lb} \cdot \text{s}^2/\text{in}} = 0.127 \text{ in},$$

$$R_G^{y'} = R_G^{z'} = 0,$$

$$d^{x'} = -0.127 \text{ in},$$

$$d^{y'} = d^{z'} = 0,$$

$$I^{x''x''} = 6.16 \text{ in} \cdot \text{lb} \cdot \text{s}^2, \qquad\qquad Ans.$$

$$I^{y''y''} = (5.12 \text{ in} \cdot \text{lb} \cdot \text{s}^2) + 2(0.235 \text{ lb} \cdot \text{s}^2/\text{in})(0.118 \text{ in})(-0.118 \text{ in}) + (0.234 \text{ in})(-0.118 \text{ in})^2$$
$$= 5.12 \text{ in} \cdot \text{lb} \cdot \text{s}^2, \qquad\qquad Ans.$$

$$I^{z''z''} = (4.24 \text{ in} \cdot \text{lb} \cdot \text{s}^2) + 2(0.234 \text{ lb} \cdot \text{s}^2/\text{in})(0.118 \text{ in})(-0.118 \text{ in}) + 4(0.234 \text{ in} \cdot \text{lb} \cdot \text{s}^2)(-0.118 \text{ in})^2$$
$$= 4.24 \text{ in} \cdot \text{lb} \cdot \text{s}^2, \qquad\qquad Ans.$$

$$I^{x''y''} = 1.52 \text{ in} \cdot \text{lb} \cdot \text{s}^2, \qquad\qquad Ans.$$

$$I^{y''z''} = I^{z''x''} = 0. \qquad\qquad Ans.$$

12.12 Euler's Equations of Motion

In Section 12.2, a detailed derivation found the form of Newton's law for a rigid body by integrating over all particles of the body. This demonstrated that the center of mass of the body is the single unique point of the body where Newton's law has the same form as that for a single particle. Repeating Eq. (12.12), this is

$$\sum \mathbf{F}_{ij} = m_j \mathbf{A}_{G_j}. \tag{12.110}$$

Taking much less care with derivation, we then gave the "equivalent" rotational form, Eq. (12.13), as

$$\sum \mathbf{M}_{Gij} = I_{G_j} \boldsymbol{\alpha}_j.$$

Although this equation is valid for all problems with only planar motion, this equation is oversimplified and is *not* always valid for problems with spatial motion. Our task now is to more carefully derive an appropriate equation that *is* valid for problems that include three-dimensional effects.

We start, as before, with Newton's law written for a single differential particle at location P on a rigid body. This can be written as

$$d\mathbf{F} = \mathbf{A}_P dm,$$

where $d\mathbf{F}$ is the net unbalanced force on the particle, $\mathbf{A}_P$ is the absolute acceleration of the particle, and dm is the mass of the particle. Next, we find the net unbalanced moment that this particle contributes to the body by taking the moment of its net unbalanced force about the center of mass of the body, point G:

$$\mathbf{R}_{PG} \times d\mathbf{F} = \mathbf{R}_{PG} \times \mathbf{A}_P dm$$
$$= \mathbf{R}_{PG} \times [\mathbf{A}_G + \boldsymbol{\omega} \times (\boldsymbol{\omega} \times \mathbf{R}_{PG}) + \boldsymbol{\alpha} \times \mathbf{R}_{PG}] dm.$$

On the left-hand side of this equation is the net unbalanced moment contribution of this single particle. If we rearrange the terms on the right-hand side, the equation becomes

$$d\mathbf{M} = \mathbf{R}_{PG} \times \mathbf{A}_G dm + \mathbf{R}_{PG} \times [\boldsymbol{\omega} \times (\boldsymbol{\omega} \times \mathbf{R}_{PG})] dm + \mathbf{R}_{PG} \times (\boldsymbol{\alpha} \times \mathbf{R}_{PG}) dm. \qquad (a)$$

We now recall the following vector identity for the triple cross-product of three arbitrary vectors, say $\mathbf{A}$, $\mathbf{B}$, and $\mathbf{C}$:

$$\mathbf{A} \times (\mathbf{B} \times \mathbf{C}) = (\mathbf{A} \cdot \mathbf{C})\mathbf{B} - (\mathbf{A} \cdot \mathbf{B})\mathbf{C}.$$

Using this identity on the second term on the right-hand side of Eq. (a), we have

$$d\mathbf{M} = \mathbf{R}_{PG} \times \mathbf{A}_G dm + \mathbf{R}_{PG} \times (\boldsymbol{\omega} \cdot \mathbf{R}_{PG})\boldsymbol{\omega} dm - \mathbf{R}_{PG} \times (\boldsymbol{\omega} \cdot \boldsymbol{\omega})\mathbf{R}_{PG} dm - \mathbf{R}_{PG} \times (\boldsymbol{\alpha} \times \mathbf{R}_{PG}) dm.$$
$$(b)$$

Now, we hope to integrate this equation over all particles of the body. We will look at this term by term. The integral of the moment contributions of the individual particles gives the net externally applied unbalanced moment on the body:

$$\int d\mathbf{M} = \sum \mathbf{M}_{Gij}. \qquad (c)$$

Recognizing that $\mathbf{A}_G$ is common for all particles, we see that the first term on the right-hand side of Eq. (b) integrates to

$$\int \mathbf{R}_{PG} \times \mathbf{A}_G dm = \int \mathbf{R}_{PG} dm \times \mathbf{A}_G = \mathbf{0}, \qquad (d)$$

where we have taken advantage of the definition of the center of mass and noted that $\int \mathbf{R}_{PG} dm = m\mathbf{R}_{GG} = \mathbf{0}$.

The second term on the right-hand side of Eq. (b) is expanded according to its components along the coordinate axes and integrated. The integrals will be recognized as the mass moments and products of inertia of the body. Thus,

$$\int \mathbf{R}_{PG} \times (\boldsymbol{\omega} \cdot \mathbf{R}_{PG})\boldsymbol{\omega} dm = \left[-(I_G^{yy} - I_G^{zz})\omega^y \omega^z - I_G^{yz}(\omega^y \omega^y - \omega^z \omega^z) + (I_G^{xy} \omega^z - I_G^{xz} \omega^y)\omega^x \right] \hat{\mathbf{i}}$$
$$+ \left[-(I_G^{zz} - I_G^{xx})\omega^z \omega^x - I_G^{zx}(\omega^z \omega^z - \omega^x \omega^x) + (I_G^{yz} \omega^x - I_G^{yx} \omega^z)\omega^y \right] \hat{\mathbf{j}}$$
$$+ \left[-(I_G^{xx} - I_G^{yy})\omega^x \omega^y - I_G^{xy}(\omega^x \omega^x - \omega^y \omega^y) + (I_G^{zx} \omega^y - I_G^{zy} \omega^x)\omega^z \right] \hat{\mathbf{k}}.$$
$$(e)$$

Since $\mathbf{R}_{PG} \times \mathbf{R}_{PG} = \mathbf{0}$ for every particle, the third term on the right-hand side of Eq. (b) integrates to

$$(\boldsymbol{\omega} \cdot \boldsymbol{\omega}) \int \mathbf{R}_{PG} \times \mathbf{R}_{PG}\,dm = \mathbf{0}. \qquad (f)$$

The final term on the right-hand side of Eq. (b) is also expanded according to its components along the coordinate axes and integrated. Thus,

$$\int \mathbf{R}_{PG} \times (\boldsymbol{\alpha} \times \mathbf{R}_{PG})\,dm = \left(I_G^{xx}\alpha^x - I_G^{xy}\alpha^y - I_G^{xz}\alpha^z\right)\hat{\mathbf{i}} + \left(I_G^{yx}\alpha^x - I_G^{yy}\alpha^y - I_G^{yz}\alpha^z\right)\hat{\mathbf{j}}$$
$$+ \left(I_G^{zx}\alpha^x - I_G^{zy}\alpha^y - I_G^{zz}\alpha^z\right)\hat{\mathbf{k}}. \qquad (g)$$

When we now substitute Eqs. (c)–(g) into Eq. (b) and equate components in the $\hat{\mathbf{i}}$, $\hat{\mathbf{j}}$, and $\hat{\mathbf{k}}$ directions, we obtain the following three equations:

$$\sum M_{Gij}^x = I_G^{xx}\alpha^x - I_G^{xy}\alpha^y - I_G^{xz}\alpha^z + \left(I_G^{xy}\omega^z - I_G^{zx}\omega^y\right)\omega^x - \left(I_G^{yy} - I_G^{zz}\right)\omega^y\omega^z - I_G^{yz}\left(\omega^y\omega^y - \omega^z\omega^z\right),$$
$$\sum M_{Gij}^y = -I_G^{xy}\alpha^x + I_G^{yy}\alpha^y - I_G^{yz}\alpha^z + \left(I_G^{yz}\omega^x - I_G^{xy}\omega^z\right)\omega^y - \left(I_G^{zz} - I_G^{xx}\right)\omega^z\omega^x - I_G^{zx}\left(\omega^z\omega^z - \omega^x\omega^x\right),$$
$$\sum M_{Gij}^z = -I_G^{xz}\alpha^x - I_G^{yz}\alpha^y + I_G^{zz}\alpha^z + \left(I_G^{zx}\omega^y - I_G^{yz}\omega^x\right)\omega^z - \left(I_G^{xx} - I_G^{yy}\right)\omega^x\omega^y - I_G^{xy}\left(\omega^x\omega^x - \omega^y\omega^y\right).$$
$$(12.111)$$

This set of equations is the most general case of the moment equation. It allows for angular velocities and angular accelerations in three dimensions. Its only restriction is that the summation of moments and the mass moments and products of inertia must be taken about the mass center of the body.

If we further require that the mass moments of inertia be measured around the principal axes of the body, then the mass products of inertia are zero. The above equations then reduce to

$$\sum M_{Gij}^x = I_G^{xx}\alpha^x - \left(I_G^{yy} - I_G^{zz}\right)\omega^y\omega^z,$$
$$\sum M_{Gij}^y = I_G^{yy}\alpha^y - \left(I_G^{zz} - I_G^{xx}\right)\omega^z\omega^x, \qquad (12.112)$$
$$\sum M_{Gij}^z = I_G^{zz}\alpha^z - \left(I_G^{xx} - I_G^{yy}\right)\omega^x\omega^y.$$

This important set of equations is called *Euler's equations of motion*. It should be emphasized that they govern any three-dimensional rotational motion of a rigid body, but that the principal axes of inertia must be used for the $x, y,$ and z component directions.[5]

Example 12.8

Consider the swashplate which was analyzed in Example 12.7. The shaft is supported by radial bearings at A and B as shown in Fig. 12.29. At the instant shown, the shaft is rotating at a speed of 1 500 rev/min cw, which is decreasing at a rate of 100 rev/min/s. Calculate the bearing reactions.

---

[5] It should now be recognized that when we restricted ourselves to motions in the xy plane, it was a special case of Eq. (12.112) that was used in Eq. (12.13) and what followed.

Example 12.8 (cont.)

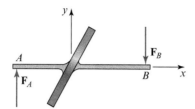

Fig. 12.29 Shaft with swashplate.

Solution

From the choice of axes shown in Fig. 12.29 and the data specified in the problem statement, the angular velocity and acceleration of the shaft, respectively, are

$$\boldsymbol{\omega} = -(1\,500\ \text{rev/min})(2\pi\ \text{rad/rev})/(60\ \text{s/min})\hat{\mathbf{i}} = -157\hat{\mathbf{i}}\ \text{rad/s},$$

$$\boldsymbol{\alpha} = +(100\ \text{rev/min/s})(2\pi\ \text{rad/rev})/(60\ \text{s/min})\hat{\mathbf{i}} = +10.5\hat{\mathbf{i}}\ \text{rad/s}^2.$$

The mass moments and products of the swashplate inertia were found in Example 12.7. Substituting these values into Eqs. (12.111) gives

$$\sum M^x = \left(6.16\ \text{in}\cdot\text{lb}\cdot\text{s}^2\right)\left(10.5\ \text{rad/s}^2\right) = 64.7\ \text{in}\cdot\text{lb},$$

$$\sum M^y = -\left(1.52\ \text{in}\cdot\text{lb}\cdot\text{s}^2\right)\left(10.5\ \text{rad/s}^2\right) = -16\ \text{in}\cdot\text{lb},$$

$$\sum M^z = -\left(1.52\ \text{in}\cdot\text{lb}\cdot\text{s}^2\right)\left(-157\ \text{rad/s}\right)^2 = -37\,500\ \text{in}\cdot\text{lb},$$

or

$$\sum \mathbf{M} = 64.7\hat{\mathbf{i}} - 16.0\hat{\mathbf{j}} - 37\,500\hat{\mathbf{k}}\ \text{in}\cdot\text{lb}.$$

The $\sum M^x$ component is the result of the shaft torque causing the deceleration. The other two components are provided by forces at the bearings, A and B. From these we find, by taking moments about the mass center, that

$$\sum M^y = (10\ \text{in})\left(F_A^z\right) - (14\ \text{in})\left(F_B^z\right) = -16.0\ \text{in}\cdot\text{lb},$$

$$\sum M^z = -(10\ \text{in})\left(F_A^y\right) + (14\ \text{in})\left(F_B^y\right) = -37\,500\ \text{in}\cdot\text{lb}.$$

Remembering that $\left(F_A^y\right) + \left(F_B^y\right) = 0$ and $\left(F_A^z\right) + \left(F_B^z\right) = 0$ for equilibrium, we find that

$$F_A^z = -F_B^z = (-16.0\ \text{in}\cdot\text{lb})/(24\ \text{in}) = -0.667\ \text{lb} \approx 0,$$

$$F_A^y = -F_B^y = (37\,500\ \text{in}\cdot\text{lb})/(24\ \text{in}) = 1\,560\ \text{lb}.$$

Thus the two bearing reaction forces on the shaft are

$$\mathbf{F}_A = 1\,560\hat{\mathbf{j}}\ \text{lb}, \qquad\qquad Ans.$$

$$\mathbf{F}_B = -1\,560\hat{\mathbf{j}}\ \text{lb}. \qquad\qquad Ans.$$

Example 12.8 (cont.)

Note that these bearing reactions are caused entirely by the z component of the inertia moment. This, in turn, is caused by the angular velocity of the shaft and not by its angular acceleration; the bearing reaction loads are still quite large even when the shaft is operating at constant speed.

Note also that this example exhibits only planar motion, but if it had been analyzed by the methods of Section 12.6, we would have found *no* net bearing reaction forces! This should be sufficient warning of the danger of ignoring the third dimension in dynamic force analysis, based solely on the justification that the *motion* is planar.

12.13 Impulse and Momentum

If we consider that both force and acceleration may be functions of time, we can multiply both sides of Eq. (12.110) by dt and integrate between two chosen times, t_1 and t_2. This results in

$$\sum \int_{t_1}^{t_2} \mathbf{F}_{ij}(t)dt = m_j \int_{t_1}^{t_2} \mathbf{A}_{G_j}(t)dt = m_j \mathbf{V}_{G_j}(t_2) - m_j \mathbf{V}_{G_j}(t_1), \qquad (a)$$

where $\mathbf{V}_{G_j}(t_2)$ and $\mathbf{V}_{G_j}(t_1)$ are the velocities of the centers of mass at times t_2 and t_1, respectively.

The product of the mass and the velocity of the center of mass of a moving body is called its *momentum*. Momentum is a vector quantity and is given the symbol $\mathbf{L}$. Thus, for body j, the momentum is

$$\mathbf{L}_j(t) = m_j \mathbf{V}_{G_j}(t). \qquad (12.113)$$

Using this definition, Eq. (*a*) becomes

$$\sum \int_{t_1}^{t_2} \mathbf{F}_{ij}(t)dt = \mathbf{L}_j(t_2) - \mathbf{L}_j(t_1) = \Delta \mathbf{L}_j. \qquad (12.114)$$

The integral of a force over an interval of time is called the *impulse* of the force. Therefore, Eq. (12.114) expresses the *principle of impulse and momentum*, which is that *the total impulse on a rigid body is equal to its change in momentum during the same time interval.* Conversely, *any change in momentum of a rigid body over a time interval is due to the total impulse of the external forces on that body.*

If we now make the time interval between t_1 and t_2 infinitesimal, divide Eq. (12.114) by this interval, and take the limit, it becomes

$$\sum \mathbf{F}_{ij} = \frac{d\mathbf{L}_j}{dt}. \qquad (12.115)$$

Therefore, *the resultant external force acting upon a rigid body is equal to the time rate of change of its momentum.* If there is no net external force acting on a body, then there can be no change in its momentum. Thus, with no net external force,

$$\frac{d\mathbf{L}_j}{dt} = \mathbf{0} \quad \text{or} \quad \mathbf{L}_j = \text{constant}, \tag{12.116}$$

is a statement of the law of *conservation of momentum.*

12.14 Angular Impulse and Angular Momentum

When a body translates, its motion is described in terms of its velocity and acceleration, and we are interested in the forces that produce this motion. When a body rotates, we are interested in the moments, and the motion is described by angular velocity and angular acceleration. Similarly, in dealing with rotational motion, we must consider terms such as *angular impulse* or *moment of impulse*, and *angular momentum*, also called *moment of momentum.*

Angular momentum is a vector quantity and is given the symbol $\mathbf{H}$. For a particle P of mass dm at location $\mathbf{R}_P$ with momentum $d\mathbf{L}$, the moment of momentum, $d\mathbf{H}$, of this single particle is defined, as implied by its name, as follows:

$$d\mathbf{H} = \mathbf{R}_P \times d\mathbf{L} = \mathbf{R}_P \times \mathbf{V}_P dm. \tag{a}$$

Analogous to the development in Section 12.12, we can express this equation as

$$\begin{aligned} d\mathbf{H} &= \mathbf{R}_P \times (\mathbf{V}_G + \boldsymbol{\omega} \times \mathbf{R}_{PG})dm \\ &= \mathbf{R}_P \times \mathbf{V}_G dm + \mathbf{R}_P \times (\boldsymbol{\omega} \times \mathbf{R}_{PG})dm \\ &= \mathbf{R}_P \times \mathbf{V}_G dm + \mathbf{R}_G \times (\boldsymbol{\omega} \times \mathbf{R}_{PG})dm + \mathbf{R}_{PG} \times (\boldsymbol{\omega} \times \mathbf{R}_{PG})dm. \end{aligned}$$

This equation can now be integrated over all particles of a rigid body:

$$\int d\mathbf{H} = \int \mathbf{R}_P dm \times \mathbf{V}_G + \mathbf{R}_G \times \left(\boldsymbol{\omega} \times \int \mathbf{R}_{PG} dm \right) + \int \mathbf{R}_{PG} \times (\boldsymbol{\omega} \times \mathbf{R}_{PG})dm. \tag{b}$$

The left-hand side of this equation integrates to give the total angular momentum of the body about the origin of coordinates:

$$\int d\mathbf{H} = \mathbf{H}_O. \tag{c}$$

Recognizing the definition of the center of mass, we see that the first two terms on the right-hand side of Eq. (*b*) become

$$\int \mathbf{R}_P dm \times \mathbf{V}_G = m\mathbf{R}_G \times \mathbf{V}_G = \mathbf{R}_G \times \mathbf{L}, \tag{d}$$

$$\mathbf{R}_G \times \left(\boldsymbol{\omega} \times \int \mathbf{R}_{PG} dm\right) = \mathbf{R}_G \times (\boldsymbol{\omega} \times \mathbf{R}_{GG}) = \mathbf{0}. \tag{e}$$

As in Section 12.12, the final term of Eq. (b) must be expanded according to its components along the coordinate axes and integrated. Thus,

$$\int \mathbf{R}_{PG} \times (\boldsymbol{\omega} \times \mathbf{R}_{PG}) dm = \left(I_G^{xx}\omega^x - I_G^{xy}\omega^y - I_G^{xz}\omega^z\right)\hat{\mathbf{i}} + \left(-I_G^{yx}\omega^x - I_G^{yy}\omega^y - I_G^{yz}\omega^z\right)\hat{\mathbf{j}}$$
$$+ \left(-I_G^{zx}\omega^x - I_G^{zy}\omega^y - I_G^{zz}\omega^z\right)\hat{\mathbf{k}}. \tag{f}$$

Upon substituting Eqs. (c)–(f) into Eq. (b), we obtain

$$\begin{aligned}
\mathbf{H}_O = \mathbf{R}_G \times \mathbf{L} & \\
&+ \left(I_G^{xx}\omega^x - I_G^{xy}\omega^y - I_G^{xz}\omega^z\right)\hat{\mathbf{i}} \\
&+ \left(-I_G^{yx}\omega^x - I_G^{yy}\omega^y - I_G^{yz}\omega^z\right)\hat{\mathbf{j}} \\
&+ \left(-I_G^{zx}\omega^x - I_G^{zy}\omega^y - I_G^{zz}\omega^z\right)\hat{\mathbf{k}}.
\end{aligned} \tag{g}$$

We can recognize that, if we had chosen a coordinate system with origin at the center of mass of the body, then $\mathbf{R}_G = \mathbf{0}$, and the above equation gives the angular momentum as

$$\mathbf{H}_G = H_G^x \hat{\mathbf{i}} + H_G^y \hat{\mathbf{j}} + H_G^z \hat{\mathbf{k}}, \tag{12.117}$$

where

$$\begin{aligned}
H_G^x &= +I_G^{xx}\omega^x - I_G^{xy}\omega^y - I_G^{xz}\omega^z, \\
H_G^y &= -I_G^{yx}\omega^x + I_G^{yy}\omega^y - I_G^{yz}\omega^z, \\
H_G^z &= -I_G^{zx}\omega^x - I_G^{zy}\omega^y + I_G^{zz}\omega^z.
\end{aligned} \tag{12.118}$$

However, when moments are taken about other than the center of mass, then the moment of momentum about an arbitrary point, O, is

$$\mathbf{H}_O = \mathbf{H}_G + \mathbf{R}_{GO} \times \mathbf{L}. \tag{12.119}$$

Now, as we did in Eq. (a), we can take the cross-product between the position vector of a particle and the terms of Newton's laws. Using an inertial coordinate system,

$$\sum \mathbf{R}_P \times d\mathbf{F} = \mathbf{R}_P \times \mathbf{A}_P dm. \tag{h}$$

If we consider that both force and acceleration may be functions of time, we can multiply both sides of this equation by dt and integrate between t_1 and t_2. The result is

$$\sum \int_{t_1}^{t_2} \mathbf{R}_P \times d\mathbf{F}(t) dt = \mathbf{R}_P \times \mathbf{V}_P(t_2) dm - \mathbf{R}_P \times \mathbf{V}_P(t_1) dm.$$

After integrating over all particles of a rigid body, this yields

$$\sum \int_{t_1}^{t_2} \mathbf{M}_{ij}(t) dt = \mathbf{H}(t_2) - \mathbf{H}(t_1) = \Delta\mathbf{H}. \tag{12.120}$$

The integral on the left-hand side of this equation is the net result of all external moment impulses that occur on the body in the time interval t_1 to t_2, and is called the *angular impulse*. On the right is the change in the moment of angular momentum that occurs over the same time interval as a result of the angular impulse.

If we now make the interval between t_1 and t_2 infinitesimal, divide Eq. (12.120) by this time interval, and take the limit, it becomes

$$\sum \mathbf{M}_{ij} = \frac{d\mathbf{H}}{dt}. \tag{12.121}$$

Therefore, *the resultant external moment acting upon a rigid body is equal to the time rate of change of its angular momentum.* If there is no net external moment acting on a body, then there can be no change in its angular momentum. Thus, with no net external moment,

$$\frac{d\mathbf{H}}{dt} = \mathbf{0} \quad \text{or} \quad \mathbf{H} = \text{constant} \tag{12.122}$$

is a statement of the *law of conservation of angular momentum*.

It is wise now to review this section mentally, and to take careful note of how the coordinate axes may be selected for a particular application. This review will show that, stated or not, all results which were derived from Newton's law depend on the use of an inertial coordinate system. For example, the angular velocity used in Eq. (12.118) must be taken with respect to an absolute coordinate system so that the derivative of Eq. (12.121) contains the required absolute angular acceleration terms. Yet this seems contradictory, since the integration performed in finding the mass moments and products of inertia must often be done in a coordinate system attached to the body itself.

If the coordinate axes are chosen stationary, then the moments and products of inertia used in finding the angular momentum must be transformed to an inertial coordinate system, using the methods of Section 12.10, and therefore become functions of time. This complicates the use of Eqs. (12.120), (12.121), and (12.122). For this reason, it is usually preferable to choose the x, y, and z components to be directed along axes fixed in the moving body.

Equations (12.120), (12.121), and (12.122), being vector equations, can be expressed in any coordinate system, including body-fixed axes, and are still correct. This has the great advantage that the mass moments and products of inertia are constants for a rigid body. However, it must be kept in mind that the $\hat{\mathbf{i}}$, $\hat{\mathbf{j}}$, and $\hat{\mathbf{k}}$ unit vectors are moving. They are functions of time in Eq. (12.117), and they have derivatives when using Eqs. (12.121) or (12.122). Therefore, the components of Eqs. (12.121), for example, cannot be found directly by differentiating Eqs. (12.118), but will be

$$\begin{aligned}
\sum M_G^x &= I_G^{xx}\alpha^x - I_G^{xy}\alpha^y - I_G^{xz}\alpha^z + \omega^y H^z - \omega^z H^y, \\
\sum M_G^y &= -I_G^{yx}\alpha^x + I_G^{yy}\alpha^y - I_G^{yz}\alpha^z + \omega^z H^x - \omega^x H^z, \\
\sum M_G^z &= -I_G^{zx}\alpha^x - I_G^{zy}\alpha^y + I_G^{zz}\alpha^z + \omega^x H^y - \omega^y H^x.
\end{aligned} \tag{12.123}$$

Here, we must be careful to find the *components* of $\boldsymbol{\omega}$ and $\boldsymbol{\alpha}$ along the axes of the body, even though they are absolute angular velocity and acceleration terms. In addition, we must remember when results are interpreted that these moment components are expressed along the body axes.

Example 12.9

Figure 12.30 illustrates an offset flywheel on a turntable, which is a problem typical of situations occurring in the design or analysis of machines in which gyroscopic forces must be considered. A round plate designated body 2 rotates about its central axis with a constant angular velocity $\boldsymbol{\omega}_2 = 5\hat{\mathbf{k}}$ rad/s. Mounted on this revolving round plate are two bearings, A and B, which retain a shaft and a mass, 3, rotating at an angular velocity with respect to the rotating plate, $\boldsymbol{\omega}_{3/2} = 350\hat{\mathbf{i}}$ rad/s. The rotating body 3 has a mass of 4.5 kg, a radius of gyration of 50 mm about its spin axis, and center of mass located at G. Calculate the bearing reactions at A and B.

Fig. 12.30 Offset flywheel on a turntable. $R_{BA} = 150$ mm, $R_{GA} = 50$ mm, $R_{GO_2} = 150$ mm.

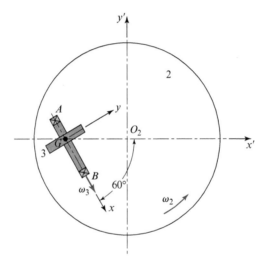

Assumptions

The weight of the shaft is negligible; body 3 rotates at a constant angular velocity; and bearing B can only support a radial load.

Solution

We begin by choosing the local body coordinate system shown by axes x and y in Fig. 12.30. The origin is at point G, the center of mass of body 3, and the axes are principal axes of inertia. Note that this coordinate system is fixed to body 2, not body 3, and it is not stationary. Since body 3 is radially symmetric about this x axis, its mass properties remain constant.

From the given data and a kinematic analysis, we have

$$I_{G_3}^{xx} = m_3\left(k_3^x\right)^2 = (4.5 \text{ kg})(0.050 \text{ m})^2 = 0.01125 \text{ kg} \cdot \text{m}^2,$$

$$\boldsymbol{\omega}_2 = 5\hat{\mathbf{k}}' \text{ rad/s} = 5\hat{\mathbf{k}} \text{ rad/s},$$

Example 12.9 (cont.)

$$\boldsymbol{\omega}_3 = \boldsymbol{\omega}_2 + \boldsymbol{\omega}_{3/2} = 350\hat{\mathbf{i}} + 5\hat{\mathbf{k}} \text{ rad/s},$$

$$\boldsymbol{\alpha}_2 = \boldsymbol{\alpha}_3 = \mathbf{0},$$

$$\mathbf{R}_{GO_2} = -(150\cos 60°)\hat{\mathbf{i}} - (150\sin 60°)\hat{\mathbf{j}} \text{ mm}$$
$$= -0.075\hat{\mathbf{i}} - 0.130\hat{\mathbf{j}} \text{ m},$$

$$\mathbf{V}_G = \boldsymbol{\omega}_2 \times \mathbf{R}_{GO_2} = 0.650\hat{\mathbf{i}} - 0.375\hat{\mathbf{j}} \text{ m/s},$$

$$\mathbf{R}_{AG} = -0.050\hat{\mathbf{i}} \text{ m},$$

$$\mathbf{R}_{BG} = 0.100\hat{\mathbf{i}} \text{ m},$$

and the time rates of change of the unit vectors are

$$\frac{d\hat{\mathbf{i}}}{dt} = \boldsymbol{\omega}_2 \times \hat{\mathbf{i}} = 5\hat{\mathbf{j}} \text{ rad/s},$$

$$\frac{d\hat{\mathbf{j}}}{dt} = \boldsymbol{\omega}_2 \times \hat{\mathbf{j}} = -5\hat{\mathbf{i}} \text{ rad/s},$$

$$\frac{d\hat{\mathbf{k}}}{dt} = \boldsymbol{\omega}_2 \times \hat{\mathbf{k}} = \mathbf{0}.$$

From Eq. (12.113), we find the momentum, which is,

$$\mathbf{L}_3 = m_3\mathbf{V}_G = (4.5 \text{ kg})(0.650\hat{\mathbf{i}} - 0.375\hat{\mathbf{j}} \text{ m/s})$$
$$= 2.93\hat{\mathbf{i}} - 1.69\hat{\mathbf{j}} \text{ kg} \cdot \text{m/s}.$$

From Eq. (12.115), we find the dynamic force effects attributable to the motion of the center of mass. Remembering that $\hat{\mathbf{i}}$ and $\hat{\mathbf{j}}$ are moving and have derivatives, this equation gives

$$\sum \mathbf{F}_{i3} = \frac{d\mathbf{L}_3}{dt},$$

$$\mathbf{F}_{A_{23}} + \mathbf{F}_{B_{23}} = 2.93\frac{d\hat{\mathbf{i}}}{dt} - 1.69\frac{d\hat{\mathbf{j}}}{dt} = 8.45\hat{\mathbf{i}} + 14.65\hat{\mathbf{j}},$$

$$\begin{aligned} F_{A_{23}}^x + F_{B_{23}}^x &= 8.45 \text{ N}, \\ F_{A_{23}}^y + F_{B_{23}}^y &= 14.65 \text{ N}, \\ F_{A_{23}}^z + F_{B_{23}}^z &= 0. \end{aligned} \quad (1)$$

Example 12.9 (cont.)

From Eqs. (12.118) and (12.117) we find the angular momentum, which is,

$$H^x_{G_3} = (0.011\,25\ \text{kg}\cdot\text{m}^2)(350\ \text{rad/s}) = 3.94\ \text{kg}\cdot\text{m}^2/\text{s},$$
$$H^y_{G_3} = 0,$$
$$H^z_{G_3} = I^{zz}_G(5\ \text{rad/s}) = 5I^{zz}_G\ \text{kg}\cdot\text{m}^2/\text{s},$$
$$\mathbf{H}_{G_3} = 3.94\hat{\mathbf{i}} + 5I^{zz}_G\hat{\mathbf{k}}\ \text{kg}\cdot\text{m}^2/\text{s}.$$

We can find the dynamic moment effects from Eq. (12.121); these are

$$\sum\mathbf{M}_{G_3} = \frac{d\mathbf{H}_{G_3}}{dt},$$

$$\mathbf{R}_{AG}\times\mathbf{F}_{A_{23}} + \mathbf{R}_{BG}\times\mathbf{F}_{B_{23}} = 3.94\frac{d\hat{\mathbf{i}}}{dt} + 5I^{zz}_G\frac{d\hat{\mathbf{k}}}{dt}\ \text{N}\cdot\text{m},$$

$$(-0.050\hat{\mathbf{i}}\ \text{m})\times\mathbf{F}_{A_{23}} + (0.100\hat{\mathbf{i}}\ \text{m})\times\mathbf{F}_{B_{23}} = 3.94(5\hat{\mathbf{j}})\ \text{N}\cdot\text{m}.$$

Equating the $\hat{\mathbf{j}}$ and $\hat{\mathbf{k}}$ components, respectively, gives

$$(0.050\ \text{m})F^z_{A_{23}} - (0.100\ \text{m})F^z_{B_{23}} = 19.7\ \text{N}\cdot\text{m},$$
$$-(0.050\ \text{m})F^y_{A_{23}} + (0.100\ \text{m})F^y_{B_{23}} = 0. \tag{2}$$

Since bearing B can only support a radial load, $F^x_{B_{23}} = 0$. Therefore, solving Eqs. (1) and (2) simultaneously, we find

$$F^x_{A_{23}} = 8.45\ \text{N},$$

$$F^y_{A_{23}} = 9.77\ \text{N},$$

$$F^z_{A_{23}} = 131\ \text{N},$$

$$F^y_{B_{23}} = 4.88\ \text{N},$$

$$F^z_{B_{23}} = -131\ \text{N}.$$

Therefore, the total bearing reactions at A and B are

$$\mathbf{F}_{A_{23}} = 8.45\hat{\mathbf{i}} + 9.77\hat{\mathbf{j}} + 131\hat{\mathbf{k}}\ \text{N}, \qquad\qquad Ans.$$

$$\mathbf{F}_{B_{23}} = 4.88\hat{\mathbf{j}} - 131\hat{\mathbf{k}}\ \text{N}. \qquad\qquad Ans.$$

We note that the effect of the gyroscopic couple is to lift the front bearing upward and to push the rear bearing against the plate. We can also note that there is no x component in $\sum M_{23}$; this justifies our assumption that $\boldsymbol{\alpha}_3 = \mathbf{0}$, even though there is no motor controlling that shaft.

The general moment equation is given by Eq. (12.121) and can be written as

$$\sum \mathbf{M}_O = \dot{\mathbf{H}}_O, \tag{12.124}$$

where O denotes either a fixed point or the center of mass of the system. Recall that, in the derivation, the time derivative of the angular momentum is taken with respect to an absolute coordinate system. In many problems it is helpful to express the time derivative of the angular momentum in terms of components measured relative to a moving coordinate system xyz that has an angular velocity, $\mathbf{\Omega}$. The relation between the time derivative of the angular momentum in the moving xyz system and the absolute time derivative can be written as

$$\dot{\mathbf{H}}_O = \left(\frac{dH_O}{dt}\hat{\mathbf{H}}_O\right)_{xyz} + \mathbf{\Omega} \times \mathbf{H}_O. \tag{12.125}$$

The first term on the right-hand side represents that part of $\dot{\mathbf{H}}_O$ caused by the change in the magnitude of $\mathbf{H}_O$, and the cross-product term represents that part caused by the change in direction of $\mathbf{H}_O$. An application of this equation to a dynamic problem is demonstrated in the following example.

Example 12.10

Bevel gear A (with radius $R = 8$ in) is rolling around the fixed horizontal bevel gear B, as shown in Fig. 12.31. Gear A also rotates about the shaft OG with angular velocity ω_1. The shaft OG ($R_{GO} = l = 16$ in), attached to a vertical shaft by the clevis pin O, is initially at rest when it is given a constant angular acceleration, $\dot{\Omega}$. If the weight of gear A is 20 lb and the weight of shaft OG is negligible, determine the tangential force and the normal force between the two gears as a function of the angle θ.

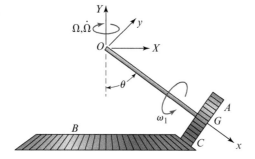

Fig. 12.31 Meshing bevel gears.

Solution

The coordinate system shown by the x and y axes in Fig. 12.31 is fixed to shaft OG and it is not stationary. The origin of this system is chosen at fixed point O, and the axes are principal axes of inertia. Since gear A is assumed to be symmetric about the x axis, its mass properties remain constant.

Example 12.10 (cont.)

The mass moments of inertia of gear A about point O are

$$I_O^{xx} = \frac{1}{2}mR^2 = \frac{1}{2}\left(\frac{20\,\text{lb}}{386\,\text{in/s}^2}\right)(8\,\text{in})^2 = 1.658\,\text{in}\cdot\text{lb}\cdot\text{s}^2, \tag{1}$$

$$I_O^{yy} = I_O^{zz} = I_G^{yy} + ml^2 = 1/4mR^2 + ml^2$$
$$= \left(\frac{20\,\text{lb}}{386\,\text{in/s}^2}\right)\left(\frac{(8\,\text{in})^2}{4} + (16\,\text{in})^2\right) = 14.093\,\text{in}\cdot\text{lb}\cdot\text{s}^2, \tag{2}$$

and, since these are principal axes, the products of inertia are

$$I_O^{xy} = I_O^{yz} = I_G^{xz} = 0. \tag{3}$$

The angular velocity and angular acceleration of gear A, respectively, are

$$\boldsymbol{\omega} = (\omega_1 + \Omega\cos\theta)\hat{\mathbf{i}} - \Omega\sin\theta\hat{\mathbf{j}}\,\text{rad/s}, \tag{4}$$

$$\boldsymbol{\alpha} = (\dot{\omega}_1 + \dot{\Omega}\cos\theta)\hat{\mathbf{i}} - \dot{\Omega}\sin\theta\hat{\mathbf{j}} + \omega_1\Omega\sin\theta\hat{\mathbf{k}}\,\text{rad/s}^2. \tag{5}$$

The velocity of the point of contact, C, between the two gears can be written as

$$\mathbf{V}_C = \mathbf{V}_O + \boldsymbol{\omega}\times\mathbf{R}_{CO},$$

where $\mathbf{R}_{CO} = 16\hat{\mathbf{i}} - 8\hat{\mathbf{j}}$ in. Since the velocities of points O and C are both zero then this equation can be written, with the aid of Eq. (4), as

$$\boldsymbol{\omega}\times\mathbf{R}_{CO} = [-(8\,\text{in})(\omega_1 + \Omega\cos\theta) + (16\,\text{in})\Omega\sin\theta]\hat{\mathbf{k}} = \mathbf{0}.$$

Rearranging this equation, the angular velocity of gear A relative to shaft OG is

$$\omega_1 = \Omega(2\sin\theta - \cos\theta). \tag{6}$$

Differentiating this equation with respect to time gives

$$\dot{\omega}_1 = \dot{\Omega}(2\sin\theta - \cos\theta). \tag{7}$$

Substituting Eqs. (6) and (7) into Eqs. (4) and (5), respectively, gives

$$\boldsymbol{\omega} = \Omega\sin\theta(2\hat{\mathbf{i}} - \hat{\mathbf{j}}), \tag{8}$$

$$\boldsymbol{\alpha} = \dot{\Omega}\sin\theta(2\hat{\mathbf{i}} - \hat{\mathbf{j}}) + \Omega^2\sin\theta(2\sin\theta - \cos\theta)\hat{\mathbf{k}}. \tag{9}$$

Example 12.10 (cont.)

From Eq. (12.125), the time rate of change of the angular momentum of gear A can be written as

$$\dot{\mathbf{H}}_O = \left(\frac{dH_O}{dt}\hat{\mathbf{H}}_O\right)_{xyz} + \boldsymbol{\omega} \times \mathbf{H}_O. \tag{10}$$

From Eqs. (12.117) and (12.118), the angular momentum of gear A about point O can be written as

$$\mathbf{H}_O = I_O^{xx}\omega^x\hat{\mathbf{i}} + I_O^{yy}\omega^y\hat{\mathbf{j}} + I_O^{zz}\omega^z\hat{\mathbf{k}}. \tag{11}$$

Substituting Eqs. (1), (2), and (8) into this equation gives

$$\begin{aligned}\mathbf{H}_O &= \left(1.658 \text{ in} \cdot \text{lb} \cdot \text{s}^2\right)(2\Omega\sin\theta)\hat{\mathbf{i}} + \left(14.093 \text{ in} \cdot \text{lb} \cdot \text{s}^2\right)(-\Omega\sin\theta)\hat{\mathbf{j}} \\ &= \Omega\sin\theta\left(3.316\hat{\mathbf{i}} - 14.093\hat{\mathbf{j}} \text{ in} \cdot \text{lb} \cdot \text{s}^2\right).\end{aligned} \tag{12}$$

Differentiating Eq. (11) with respect to time gives

$$\left(\frac{dH_O}{dt}\hat{\mathbf{H}}_O\right)_{xyz} = I_O^{xx}\alpha^x\hat{\mathbf{i}} + I_O^{yy}\alpha^y\hat{\mathbf{j}} + I_O^{zz}\alpha^z\hat{\mathbf{k}}.$$

Substituting Eqs. (1), (2), and (9) into this equation gives

$$\begin{aligned}\left(\frac{dH_O}{dt}\hat{\mathbf{H}}_O\right)_{xyz} &= \left(1.658 \text{ in} \cdot \text{lb} \cdot \text{s}^2\right)(2\dot{\Omega}\sin\theta)\hat{\mathbf{i}} + \left(14.093 \text{ in} \cdot \text{lb} \cdot \text{s}^2\right)(-\dot{\Omega}\sin\theta)\hat{\mathbf{j}} \\ &\quad + \left(14.093 \text{ in} \cdot \text{lb} \cdot \text{s}^2\right)\left[\Omega^2\sin\theta(2\sin\theta - \cos\theta)\right]\hat{\mathbf{k}},\end{aligned}$$

which can be written as

$$\left(\frac{dH_O}{dt}\hat{\mathbf{H}}_O\right)_{xyz} = \sin\theta\left[3.316\dot{\Omega}\hat{\mathbf{i}} - 14.093\dot{\Omega}\hat{\mathbf{j}} + 14.093\Omega^2(2\sin\theta - \cos\theta)\hat{\mathbf{k}}\right] \text{ in} \cdot \text{lb} \cdot \text{s}^2. \tag{13}$$

Substituting Eqs. (8), (12), and (13) into Eq. (10) gives

$$\dot{\mathbf{H}}_O = \sin\theta\left[3.316\dot{\Omega}\hat{\mathbf{i}} - 14.093\dot{\Omega}\hat{\mathbf{j}} + \Omega^2(3.316\sin\theta - 14.093\cos\theta)\hat{\mathbf{k}}\right] \text{ in} \cdot \text{lb} \cdot \text{s}^2. \tag{14}$$

From the free-body diagram of the shaft and the gear shown in Fig. 12.32, the sum of the moments about point O can be written as

$$\begin{aligned}\sum\mathbf{M}_O &= M_O^y\hat{\mathbf{j}} + \mathbf{R}_{GO} \times (20 \text{ lb})\left(\cos\theta\hat{\mathbf{i}} - \sin\theta\hat{\mathbf{j}}\right) + \mathbf{R}_{GO} \times \left(N\hat{\mathbf{j}} + f\hat{\mathbf{k}}\right) \\ &= -(8 \text{ in})f\hat{\mathbf{i}} + \left[M_O^y - (16 \text{ in})f\right]\hat{\mathbf{j}} + \left[-(320 \text{ in} \cdot \text{lb})\sin\theta + (16 \text{ in})N\right]\hat{\mathbf{k}},\end{aligned} \tag{15}$$

where f is the tangential force between the two gears at the point of contact C.

Example 12.10 (cont.)

Fig. 12.32 Free-body diagram of shaft OG and gear A. $W_A = 20$ lb.

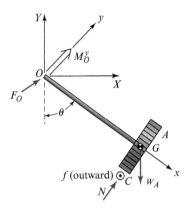

Equating the $\hat{\mathbf{i}}$ components of Eqs. (14) and (15) gives.

$$-(8 \text{ in})f = (3.316 \text{ in} \cdot \text{lb} \cdot \text{s}^2) \sin \theta \, \dot{\Omega}.$$

Therefore, the tangential force between the two gears is

$$f = -(0.415 \text{ lb} \cdot \text{s}^2) \sin \theta \, \dot{\Omega}. \qquad \qquad Ans.$$

Equating the $\hat{\mathbf{j}}$ components of Eqs. (14) and (15) gives

$$M_O^y - (16 \text{ in})f = -(14.093 \text{ in} \cdot \text{lb} \cdot \text{s}^2) \sin \theta \, \dot{\Omega}.$$

Substituting the tangential force into this equation, and rearranging, the moment about point O can be written as

$$M_O^y = -(20.725 \text{ in} \cdot \text{lb} \cdot \text{s}^2) \sin \theta \, \dot{\Omega}.$$

Finally, equating the $\hat{\mathbf{k}}$ components of Eqs. (14) and (15) gives

$$-(320 \text{ in} \cdot \text{lb}) \sin \theta + (16 \text{ in})N = \Omega^2 \sin \theta (3.316 \sin \theta - 14.093 \cos \theta) \text{ in} \cdot \text{lb} \cdot \text{s}^2.$$

Rearranging this equation, the normal reaction force can be written as

$$N = 20 \sin \theta \text{ lb} + \Omega^2 \sin \theta (0.207 \sin \theta - 0.881 \cos \theta) \text{ lb} \cdot \text{s}^2. \qquad Ans.$$

Note that the first term on the right-hand side of this equation represents the static gravitational effect, and the second term represents two opposing inertial effects. If gear A did not mesh with the fixed gear B, then the angular velocity ω_1 would be zero, and the centripetal acceleration of point G would decrease the normal reaction force N. Due to ω_1, however, a gyroscopic moment is created that has the effect of forcing gear A downward. Note that if $\theta = \tan^{-1}(0.881/0.207) = 76.78°$, then the two dynamic effects counterbalance each other and the normal reaction force reduces to its static value of $N = 20 \sin \theta$ lb.

Problems[6]

12.1 The 19 mm wide by 9 mm thick steel bell crank has $R_{AO} = 75$ mm and $R_{BO} = 150$ mm arms, and has a 25 mm diameter by 12 mm thick roller at its right end. It is used as an oscillating cam follower. Using $8\,120$ kg/m^3 for the density of steel, find the mass moment of inertia of the lever about an axis through O.

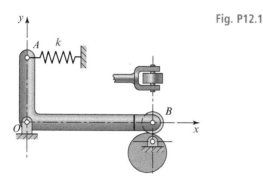

Fig. P12.1

12.2 A 0.20 in by 2 in by 12 in steel bar has two round steel disks, each 2 in diameter and 0.80 in long, welded to one end. A small hole is drilled 1 in from the other end. Using 0.268 lb/in^3 for the density of steel, find the mass moment of inertia of the bar about an axis through the hole.

Fig. P12.2

12.3 Determine the reaction forces at the joints and the external torque applied to input link 2 of the four-bar linkage in the posture shown. For the constant angular velocity $\omega_2 = 180\hat{k}$ rad/s, the known kinematics are: $\alpha_3 = 4950\hat{k}$ rad/s^2, $\alpha_4 = -8\,900\hat{k}$ rad/s^2, $\mathbf{A}_{G_3} = 1\,896\hat{i} + 225\hat{j}$ m/s^2, and $\mathbf{A}_{G_4} = 684\hat{i} + 225\hat{j}$ m/s^2. The masses and mass moments of inertia of the links are: $m_3 = 0.319$ kg, $m_4 = 0.351$ kg, $I_{G_2} = 0.002\,86$ kg·m^2, $I_{G_3} = 0.001\,71$ kg·m^2, and $I_{G_4} = 0.001\,24$ kg·m^2.

---

[6] Unless instructed otherwise, solve all problems without friction and without gravitational loads.

Fig. P12.3 $R_{AO_2} = 75$ mm, $\theta_2 = 180°$, $R_{O_4O_2} = 175$ mm, $R_{BA} = 200$ mm, $\theta_3 = 36.9°$, $R_{BO_4} - 150$ mm, $\theta_4 = 126.9°$, $R_{G_3A} = 100$ mm, and $R_{G_4O_4} = 75$ mm.

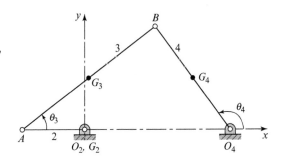

12.4 Determine the reaction forces at the joints and the external torque applied to input link 2 of the four-bar linkage in the posture shown. For the constant angular velocity $\omega_2 = 200\hat{k}$ rad/s, the known kinematics are $\alpha_3 = -6\,500\hat{k}$ rad/s^1, $\alpha_4 = -240\hat{k}$ rad/s^1, $A_{G_3} = -948\hat{i} + 79\hat{j}$ m/s^2, and $A_{G_4} = -240\hat{i} - 633\hat{j}$ m/s^2. The masses and mass moments of inertia of the links are: $m_3 = 1.193$ kg, $m_4 = 3.024$ kg, $I_{G_2} = 0.002\,65$ kg·m^2, $I_{G_3} = 0.006\,73$ kg·m^2, and $I_{G_4} = 0.058\,94$ kg·m^2.

Fig. P12.4 $R_{AO_2} = 50$ mm, $\theta_2 = -45°$, $R_{O_4O_2} = 325$ mm, $R_{BA} = 425$ mm, $\theta_3 = 31.1°$, $R_{BO_4} = 200$ mm, $\theta_4 = 67.2°$, $R_{G_3A} = 212$ mm, and $R_{G_4O_4} = 100$ mm.

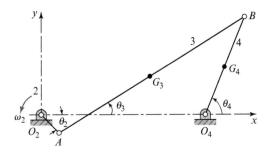

12.5 Determine the reaction forces at the joints and the crank torque of the slider-crank linkage in the posture shown. For the constant angular velocity $\omega_2 = 210\hat{k}$ rad/s, the known kinematics are: $\alpha_3 = -7\,670\hat{k}$ rad/s^2, $A_{G_3} = -2\,346\hat{i} - 1\,463\hat{j}$ m/s^2, and $A_{G_4} = -2\,355\hat{i}$ m/s^2. The masses and mass moments of inertia of the links are: $m_3 = 1.53$ kg, $m_4 = 1.287$ kg, $I_{G_2} = 0.039$ kg·m^2, and $I_{G_3} = 0.012$ kg·m^2.

Fig. P12.5 $R_{AO_2} = 75$ mm, $\theta_2 = 45°$, $R_{BA} = 300$ mm, and $R_{G_3A} = 112$ mm.

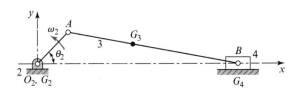

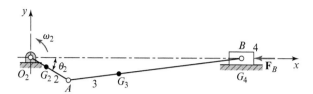

Fig. P12.6 $R_{AO_2} = 75$ mm, $\theta_2 = -30°$, $R_{BA} = 300$ mm, $R_{G_2O_2} = 31$ mm, and $R_{G_3A} = 88$ mm.

12.6 Determine the reaction forces at the joints and the crank torque of the slider-crank linkage in the posture shown if the external force acting through pin B of the piston is $\mathbf{F}_B = -3\,550\hat{\mathbf{i}}$ N. For the constant angular velocity $\boldsymbol{\omega}_2 = 160\hat{\mathbf{k}}$ rad/s, the known kinematics are: $\boldsymbol{\alpha}_3 = -3\,090\hat{\mathbf{k}}$ rad/s^2, $\mathbf{A}_{G_2} = 792$ m/s$^2\angle150°$, $\mathbf{A}_{G_3} = 1\,839$ m/s$^2\angle158.3°$, and $\mathbf{A}_{G_4} = 1\,884$ m/s$^2\angle180°$. The masses and mass moments of inertia of the links are: $m_2 = 0.428$ kg, $m_3 = 1.58$ kg, $m_4 = 1.125$ kg, $I_{G_2} = 0.000\,41$ kg$\cdot$m^2, and $I_{G_3} = 0.012$ kg$\cdot$m^2.

12.7 Determine the reaction forces at the joints and the torque applied to input link 2 of the four-bar linkage in the posture when $\theta_2 = 53°$. For the constant angular velocity $\boldsymbol{\omega}_2 = 12\hat{\mathbf{k}}$ rad/s, the known kinematics are: $\theta_3 = -2.26°$, $\theta_4 = 13.04°$, $\alpha_3 = -85.6\hat{\mathbf{k}}$ rad/s^2, $\alpha_4 = -172\hat{\mathbf{k}}$ rad/s^2, $\mathbf{A}_{G_3} = 321$ ft/s$^2\angle79.4°$, and $\mathbf{A}_{G_4} = 326$ ft/s$^2\angle-90°$. The weights and mass moments of inertia of the links are: $w_2 = 11.4$ lb, $w_3 = 145$ lb, $w_4 = 48$ lb, $I_{G_2} = 20.7$ in$\cdot$lb$\cdot$s^2, $I_{G_3} = 38.3$ in$\cdot$lb$\cdot$s^2, and $I_{G_4} = 4.6$ in$\cdot$lb$\cdot$s^2.

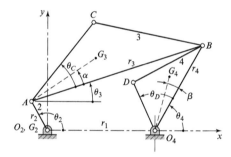

Fig. P12.7 $R_{AO_2} = 12$ in, $R_{O_4O_2} = 36$ in, $R_{BA} = 60$ in, $R_{BO_4} = 32$ in, $\theta_C = 33°$, $R_{CA} = 34$ in, $\theta_D = 53°$, $R_{DO_4} = 16$ in, $\alpha = 16°$, $R_{G_2O_2} = 0$. $R_{G_3A} = 26$ in. $\beta = 17°$, and $R_{G_4O_4} = 18$ in.

12.8 Solve Prob. 12.7 with an external force $\mathbf{F}_D = 2\,700\hat{\mathbf{i}}$ lb acting at point D.

12.9 Make a complete kinematic and dynamic analysis of the four-bar linkage of Prob. 12.7 using the data in the figure caption but in the posture when $\theta_2 = 170°$. The constant angular velocity of input link 2 is $\omega_1 = 12$ rad/s ccw, and the external force at point D is $\mathbf{F}_D = 2\,012$ lb$\angle64.3°$.

12.10 Repeat Prob. 12.9 in the posture when $\theta_2 = 200°$. The constant angular velocity of the input link 2 is $\omega_2 = 12$ rad/s ccw, the external force at point C is $\mathbf{F}_C = 1\,910$ lb$\angle45°$, and there is no external force at point D.

12.11 Make a complete dynamic analysis of the four-bar linkage of Prob. 12.7, but in the posture when $\theta_2 = 270°$. The constant angular velocity of the input link 2 is $\omega_2 = 18$ rad/s ccw, and the external force at point D is $\mathbf{F}_D = 2\,012$ lb$\angle64.3°$. The known kinematics are: $\theta_3 = 46.6°$, $\theta_4 = 80.5°$, $\alpha_3 = 178$ rad/s^{2}cw, $\alpha_4 = 256$ rad/s^2 cw, $\mathbf{A}_{G_3} = 373$ ft/s$^2\angle22.7°$, and $\mathbf{A}_{G_4} = 397$ ft/s$^2\angle-7.5°$.

12.12 Make a complete dynamic analysis of the four-bar linkage in the posture when $\theta_2 = 90°$, $\theta_3 = 23.9°$, and $\theta_4 = 91.7°$. For the constant angular velocity $\omega_2 = 32$ rad/s ccw, the known kinematics are: $\alpha_3 = 221$ rad/s^2 ccw, $\alpha_4 = 122$ rad/s^2 ccw, $\mathbf{A}_{G_3} = 295$ ft/s$^2\angle-105°$, and $\mathbf{A}_{G_3} = 109$ ft/s$^2\angle-116°$. There is an external force $\mathbf{F}_C = 142$ lb$\angle-18°$ acting at point C. The weights and mass moments of inertia of the links are $w_2 = 1.1$ lb, $w_3 = 8.8$ lb, $w_4 = 3.3$ lb, $I_{G_2} = 0.045$ in $\cdot$ lb $\cdot$ s^2, $I_{G_3} = 0.100$ in $\cdot$ lb $\cdot$ s^2, and $I_{G_4} = 0.021$ in $\cdot$ lb $\cdot$ s^2.

Fig. P12.12 $R_{AO_2} = 4.8$ in, $R_{O_4O_2} = 12$ in, $R_{BA} = 12.8$ in, $R_{BO_4} = 10$ in, $\theta_C = 15°$, $R_{CA} = 14.4$ in, $\theta_D = 0°$, $R_{DO_4} = 0$, $R_{G_2O_2} = 0$, $\alpha = 8°$, $R_{G_3A} = 8$ in, $\beta = 0°$, and $R_{G_4O_4} = 5$ in.

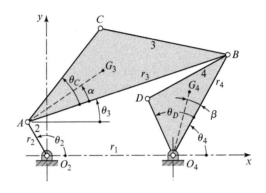

12.13 Make a complete kinematic and dynamic analysis of the four-bar linkage in Prob. 12.12, but in the posture when $\theta_2 = -100°$.

12.14 Make a complete kinematic and dynamic analysis of the four-bar linkage in Prob. 12.12, but in the posture when $\theta_2 = -60°$.

12.15 Make a complete kinematic and dynamic analysis of the offset slider-crank linkage in the posture when $\theta_2 = 120°$ and the constant input angular velocity is $\omega_2 = 18$ rad/s cw. The dimensions, the weights, and the mass moments of inertia of the links are shown in Table P12.15. The external forces acting at points B and C are $\mathbf{F}_B = -450\hat{\mathbf{i}}$ lb and $\mathbf{F}_C = -225\hat{\mathbf{i}}$ lb, respectively.

Fig. P12.15

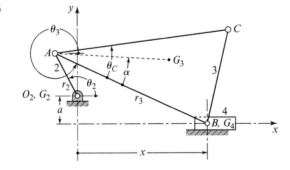

Table P12.15

Problem	R_{AO_2}	a	θ_3	R_{BA}	θ_C	R_{CA}	α	R_{G_3A}
	(in)	(in)	(deg)	(in)	(deg)	(in)	(deg)	(in)
P12.15	4.0	2.40	−22.69	15.2	32	16.0	22	10.4
P12.17	4.0	0.00	−11.09	18.0	0	18.0	0	8.0
P12.19	10.0	0.32	−10.35	50.0	−38	40.0	−18	30.0

Problem	w_2	w_3	w_4	I_{G_2}	I_{G_3}
	(lb)	(lb)	(lb)	(in·lb·s²)	(in·lb·s²)
P12.15	5.5	16.3	5.5	0.045	0.122
P12.17	3.3	7.7	2.6	0.090	0.540
P12.19	22	308	110	18	75.8

12.16 Perform a kinematic and dynamic analysis of the offset slider-crank linkage of Prob. 12.15 for a complete rotation of the crank. The forces are $\mathbf{F}_B = -225\hat{\mathbf{i}}$ lb and $\mathbf{F}_C = \mathbf{0}$ when the velocity of link 4 is to the right, and the forces are $\mathbf{F}_B = \mathbf{F}_C = \mathbf{0}$ when the velocity of link 4 is to the left. Plot the crank torque $\mathbf{T}_{12}$ versus the crank angle, θ_2.

12.17 Make a complete dynamic analysis of the slider-crank linkage shown in Fig. P12.15 in the posture when $\theta_2 = 120°$ and the constant input angular velocity is $\omega_2 = 24$ rad/s cw. The dimensions, the known kinematics, and the weights and mass moments of inertia of the links are shown in Table P12.15. The constant crank torque is $T_{12} = 540$ in · lb.

12.18 Repeat Prob. 12.17 in the posture where $\theta_2 = -120°$. The known kinematics are: $\theta_3 = 11.09°$, $R_B = 15.7$ in, $\alpha_3 = 112$ rad/s² cw, $\mathbf{A}_B = 117\hat{\mathbf{i}}$ ft/s², and $\mathbf{A}_{G_3} = 105\hat{\mathbf{i}} + 92\hat{\mathbf{j}}$ ft/s².

12.19 Make a complete kinematic and dynamic analysis of the offset slider-crank linkage shown in Fig. P12.15 in the posture when $\theta_2 = 120°$ and the constant input angular velocity is $\omega_2 = 6$ rad/s ccw. The dimensions, the weights, and the mass moments of inertia of the links are shown in Table P12.15. The external forces at points B and C are $\mathbf{F}_B = -11\,250\hat{\mathbf{i}}$ lb and $\mathbf{F}_C = 18\,000$ lb∠−60°, respectively.

12.20 Find the driving torque and the reaction forces at the joints for the crossed linkage in the posture shown. For the constant angular velocity $\omega_2 = 10$ rad/s ccw, the known kinematics are: $\omega_3 = 1.43$ rad/s cw, $\omega_4 = 11.43$ rad/s cw, $\alpha_3 = \alpha_4 = 84.8$ rad/s² ccw, and $\mathbf{A}_{G_3} = 7.78\hat{\mathbf{i}} + 7.37\hat{\mathbf{j}}$ m/s². The external force at point C is $\mathbf{F}_C = -100\hat{\mathbf{i}}$ N. The masses and mass moments of inertia of the links are $m_3 = 1.8$ kg, $I_{G_2} = I_{G_4} = 0.007$ kg · m², and $I_{G_3} = 0.055$ kg · m².

Fig. P12.20 $R_{AO_2} = 150$ mm, $\theta_2 = 60°$, $R_{O_4O_2} = 450$ mm, $R_{BA} = 450$ mm, $\theta_3 = -38.2°$, $R_{BO_4} - 150$ mm, $\theta_4 - -98.2°$, $R_{CA} = 600$ mm, and $R_{G_3A} = 300$ mm.

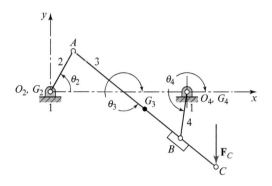

12.21 Find the driving torque and the reaction forces at the joints for Prob. 12.20 under the same dynamic conditions but with crank 4 as the driver of the linkage.

12.22 Using the same force $\mathbf{F}_C$ as in Prob. 12.20, compute the crank torque and the reaction forces at the joints in the posture when $\theta_2 = 210°$. For the constant angular velocity $\omega_2 = 10$ rad/s ccw, the known kinematics are $\theta_3 = 14.7°$, $\theta_4 = 164.7°$, $\omega_3 = 4.73$ rad/s ccw, $\omega_4 = 5.27$ rad/s cw, $\alpha_3 = \alpha_4 = 10.39$ rad/s cw, and $\mathbf{A}_{G_3} = 12.5$ m/s$^2 \angle 20.85°$.

12.23 Make a kinematic and dynamic analysis of the linkage for a complete rotation of the crank with the constant angular velocity $\omega_2 = 10$ rad/s ccw. The external force at point C is $\mathbf{F}_C = -2.22\hat{\mathbf{i}} + 3.85\hat{\mathbf{j}}$ kN for $90° \le \theta_2 \le 300°$, and $\mathbf{F}_C = \mathbf{0}$ otherwise. The masses and mass moments of inertias of the links are $m_3 = 100$ kg, $m_4 = 93.6$ kg, $I_{G_3} = 25.1$ kg·m^2, and $I_{G_4} = 29.3$ kg·m^2.

Fig. P12.23 $R_{AO_2} = 0.4$ m, $R_{O_4O_2} = R_{BA} = 1$ m, $R_{BO_4} = 1.4$ m, $a = 0.4$ m, $b = 0.1$ m, $R_{G_3A} = 0.8$ m, and $R_{G_4O_4} = 0.5$ m.

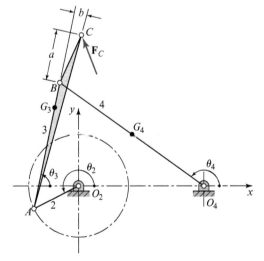

12.24 The motor is geared to a shaft on which a flywheel is mounted. The mass moments of inertia of the parts are: flywheel, $I = 0.303$ kg·m^2; flywheel shaft, $I = 0.00172$ kg·m^2; gear (225 mm D), $I = 0.01909$ kg·m^2; pinion (50 mm D), $I = 0.000387$ kg·m^2; and motor, $I = 0.00959$ kg·m^2. If the motor has a starting torque of 8.33 N·m, determine the angular acceleration of the flywheel shaft at the instant the motor is started.

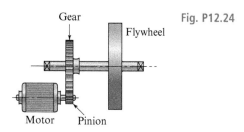

Fig. P12.24

12.25 The disk cam of Prob. 11.31 is driven at the constant input shaft speed $\omega_2 = 20$ rad/s ccw. Both the cam and the follower have been balanced so that the centers of mass of each are located at their respective fixed pivots. The weight and radius of gyration of the cam are 0.165 lb and 1.20 in, respectively, and the weight and radius of gyration of the follower are 0.066 lb and 1.40 in, respectively. At the instant shown, determine the torque $\mathbf{T}_{12}$ required on the camshaft to produce this motion.

12.26 Repeat Prob. 12.25 with the constant shaft speed $\omega_2 = 40$ rad/s ccw.

12.27 A rotating drum is pivoted at O_2 and is decelerated by the double-shoe brake mechanism. The mass and radius of gyration of the drum are 104 kg and 142 mm, respectively. The brake is actuated by the force $\mathbf{P} = -444\hat{\mathbf{j}}$ N. Assume that the contact points between the two shoes and the drum are C and D, where the coefficients of Coulomb friction are $\mu = 0.300$. Determine the angular deceleration of the drum and the reaction force, $\mathbf{F}_{12}$, at the fixed pivot, O_2.

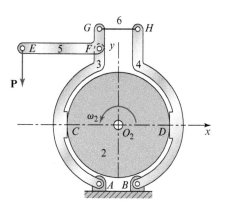

Fig. P12.27 $R_{BA} = 150$ mm, $R^y_{AO_2} = 225$ mm,
$R_{CO_2} = R_{DO_2} = 200$ mm, $R^y_{HO_2} = 400$ mm,
$R_{EF} = 300$ mm, $R_{GF} = 75$ mm, and $R_{HG} = 150$ mm.

12.28 The length of link 4 is 8 in, symmetric about O_4, and the ground bearing is midway between E and G_2. Link 2 is in translation with a velocity $\mathbf{V}_{G_2} = 4.59\hat{\mathbf{j}}$ in/s and acceleration $\mathbf{A}_{G_2} = -14\hat{\mathbf{j}}$ in/s², and the acceleration of the mass center of link 3 is $\mathbf{A}_{G_3} = +42.1\hat{\mathbf{i}} + 17.3\hat{\mathbf{j}}$ in/s². The kinematic coefficients of the linkage are $\theta'_3 = -0.288$ rad/in, $\theta''_3 = -0.238$ rad/in², $R'_{43} = +2$ in/in, and $R''_{43} = +1$ in/in² (where $\mathbf{R}_{43}$ is the vector from G_3 to G_4). A moment M_{12} acts on the input link 2 and a torque $\mathbf{T}_4$ acts on link 4. The weights and second moments of mass of the links are

$w_2 = w_4 = 1.1$ lb, $w_3 = 2.2$ lb, $I_{G_2} = I_{G_4} = 18$ in · lb · s^2, and $I_{G_3} = 45$ in · lb · s^2. Gravity is in the negative z direction. Determine the internal reaction forces, the moment M_{12}, and the torque $\mathbf{T}_4$

Fig. P12.28 $\theta_4 = 150°$, $\mathbf{P} = 9$ lb$\angle 120°$, $R_{G_2O_4} = 6$ in, and $R_{EG_2} = 8$ in.

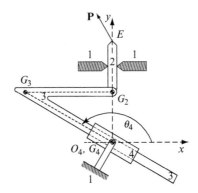

12.29 The first- and second-order kinematic coefficients for the elliptic trammel linkage are $\theta_3' = -0.050$ rad/in, $\theta_3'' = -0.00433$ rad/in^2, $R_4' = -1.732$ in/in, and $R_4'' = -0.200$ in/in^2. A linear spring with a free length $L = 20$ in and spring constant $K = 14$ lb/in is attached between O and A. A viscous damper with damping coefficient $C = 0.253$ lb · s/in is connected between the ground and link 4. The weights and mass moments of inertia of the links are $w_2 = 1.65$ lb, $w_3 = 4.4$ lb, $w_4 = 3.3$ lb, $I_{G_2} = 2.25$ in · lb · s^2, $I_{G_3} = 9$ in · lb · s^2, and $I_{G_4} = 3.15$ in · lb · s^2. The velocity, acceleration, and force acting on the input link 2 are $\mathbf{V}_{A_2} = -200\hat{\mathbf{j}}$ in/s, $\mathbf{A}_{A_2} = -800\hat{\mathbf{j}}$ in/s^2, and $\mathbf{F} = -45\hat{\mathbf{j}}$ lb, respectively. If gravity acts in the negative y direction, determine: (a) the first-order kinematic coefficients of the spring and the viscous damper; (b) the equivalent mass of the linkage; and (c) the horizontal force, P acting on link 4.

Fig. P12.29 $R_{AO} = 34.64$ in, $R_{BA} = 40$ in, and $R_{G_3G_2} = 20$ in.

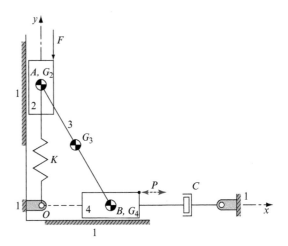

12.30 The input crank of the four-bar linkage is rotating with the constant angular velocity $\omega_2 = 10$ rad/s ccw. The angular acceleration of link 3 and the acceleration of the mass center of link 3 are $\alpha_3 = 84.8$ rad/s^2 ccw and $\mathbf{A}_{G_3} = 7.75\hat{\mathbf{i}} + 7.38\hat{\mathbf{j}}$ m/s^2, respectively. The masses and second moments of mass of the links are $m_2 = 1.5$ kg, $m_3 = 4.5$ kg, $m_4 = 1.5$ kg, $I_{G_2} = 0.007$ kg·m^2, $I_{G_3} = 0.055$ kg·m^2, and $I_{G_4} = 0.007$ kg·m^2. The spring has stiffness $k = 2.13$ kN/m and free length $R_0 = 0.112$ m. The viscous damper has a damping coefficient $C = 44.4$ N·s/m. The external force acting at point C is $\mathbf{F}_C = -555\hat{\mathbf{j}}$ N and gravity is in the negative y direction. Determine the equivalent mass moment of inertia of the linkage and the driving torque $\mathbf{T}_2$.

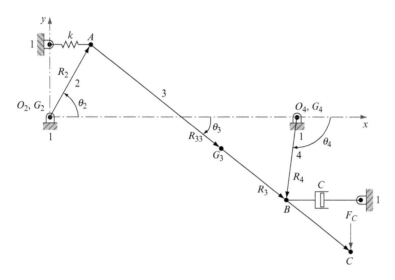

Fig. P12.30 $R_{AO_2} = 150$ mm, $R_{O_4O_2} = 450$ mm, $R_{BA} = 450$ mm, $R_{BO_4} = 150$ mm, $\theta_2 = 60°$, $\theta_3 = -38.20°$, $\theta_4 = -98.20°$, $R_{CA} = 600$ mm, and $R_{G_3A} = 300$ mm.

12.31 For the Scotch-yoke linkage in the posture shown, the angular position, velocity, and acceleration of the input link 2 are $\theta_2 = -150°$, $\omega_2 = 15\hat{\mathbf{k}}$ rad/s, and $\alpha_2 = 2\hat{\mathbf{k}}$ rad/s^2, respectively. The accelerations of the centers of mass of the links are $\mathbf{A}_{G_2} = -216\hat{\mathbf{i}} + 452\hat{\mathbf{j}}$ in/s^2, $\mathbf{A}_{G_3} = -432\hat{\mathbf{i}} + 904\hat{\mathbf{j}}$ in/s^2, and $\mathbf{A}_{G_4} = 904\hat{\mathbf{j}}$ in/s^2. The weights and mass moments of inertia of the links are $w_2 = 11$ lb, $w_3 = 11$ lb, $w_4 = 33$ lb, $I_{G_2} = 0.18$ in·lb·s^2, $I_{G_3} = 1.08$ in·lb·s^2, and $I_{G_4} = 0.72$ in·lb·s^2. Gravity is in the negative y direction, an external force $\mathbf{P} = 28.1\hat{\mathbf{j}}$ lb is acting on link 4, and an unknown torque, $\mathbf{T}_2$ is acting on link 2. Determine the internal reaction forces and the torque, $\mathbf{T}_2$. Indicate the point(s) of contact between link 4 and the ground link.

Fig. P12.31 $a = 70$ in, $b = 20$ in, $c = 15$ in, $R_{G_3O_2} = 40$ in.

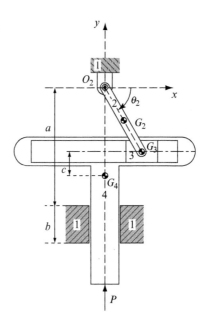

12.32 For the parallelogram four-bar linkage in the posture shown, the angular velocity and acceleration of the input link 2 are $\omega_2 = 2$ rad/s ccw and $\alpha_2 = 1$ rad/s² ccw, respectively. The first- and second-order kinematic coefficients of the center of mass of link 3 are $x'_{G_3} = -5.64$ in/rad, $y'_{G_3} = -5.64$ in/rad, $x''_{G_3} = 5.64$ in/rad², and $y''_{G_3} = -5.64$ in/rad². The weights and second moments of mass of the links are: $w_2 = w_4 = 1.1$ lb, $w_3 = 2.2$ lb, $I_{G_2} = I_{G_4} = 0.11$ in·lb·s², and $I_{G_3} = 0.275$ in·lb·s². Gravity is in the negative z direction, the force $\mathbf{F}_C = 22.5\hat{\mathbf{i}}$ lb acts at point C, and the torque $\mathbf{T}_4 = 90\hat{k}$ in·lb acts on link 4. Determine: (a) the acceleration of the mass center of link 3; (b) the internal reaction forces $\mathbf{F}_{23}$ and $\mathbf{F}_{43}$; and (c) the torque $\mathbf{T}_2$.

Fig. P12.32 $\theta_2 = 135°$.
$R_{BO_2} = R_{AO_4} = 8$ in,
$R_{BA} = R_{O_2O_4} = 12$ in, and
$R_{CB} = 4$ in.

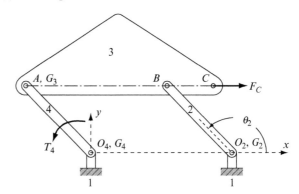

12.33 For the mechanism in the posture shown, the distance $R_{OG_3} = 100$ in and the velocity and acceleration of the input link 2 are $\mathbf{V}_{G_2} = -400\hat{\mathbf{i}}$ in/s and $\mathbf{A}_{G_2} = 400\hat{\mathbf{i}}$ in/s², respectively. The first- and second-order kinematic coefficients of link 3 are $\theta_3' = 0.005$ rad/in, and $\theta_3'' = -0.0000866$ rad/in², respectively. The weights and second moments of mass of links 2 and 3 are $w_2 = 6.6$ lb, $w_3 = 11$ lb, $I_{G_2} = 2.475$ in·lb·s², and $I_{G_3} = 67.5$ in·lb·s². Gravity is in the negative y direction, force $\mathbf{F}_C = -2.25\hat{\mathbf{i}}$ lb acts at point C, and the line of action of the unknown force, P, acting on link 2 is parallel to the x axis. Determine: (a) the internal reaction forces; (b) force P; and (c) the point(s) of contact of link 2 with the ground link.

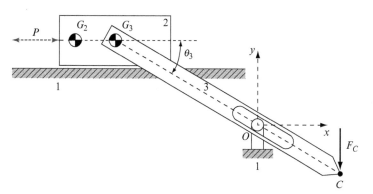

Fig. P12.33 Link 2 is 60 in by 30 in, $R_{G_2G_3} = 20$ in, and $R_{CG_3} = 140$ in.

12.34 For the slider-crank linkage in the posture when $\theta_2 = 45°$, the angular velocity and acceleration of the input link 2 are $\boldsymbol{\omega}_2 = 100\hat{\mathbf{k}}$ rad/s and $\boldsymbol{\alpha}_2 = 10\hat{\mathbf{k}}$ rad/s², respectively. The postures, velocities, and accelerations of links 3 and 4 are provided in Table P12.34. Gravity is in the negative y direction and the masses and mass moments of inertia of the links are: $m_3 = 1.53$ kg, $m_4 = 1.29$ kg, $I_{G_2} = 0.039$ N·m·s², and $I_{G_3} = 0.012$ N·m·s². The stiffness and unstretched length of the spring are $K_S = 3552$ N/m and $r_0 = 75$ mm, respectively. The damping coefficient of the viscous damper is $C = 178$ N·s/m. An external force, F_B, acts horizontally at pin B on link 4, and the motor torque is $\mathbf{T}_2 = 6.67\hat{\mathbf{k}}$ N·m. Determine: (a) the first- and second-order kinematic coefficients of the linkage; (b) the equivalent mass moment of inertia; and (c) the external force $\mathbf{F}_B$.

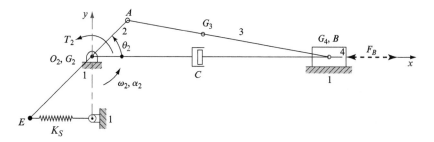

Fig. P12.34 $R_{BA} = 300$ mm, and $R_{EG_2} = 125$ mm.

Table P12.34

θ_2	θ_3	R_{AO_2}	R_{G_3A}	ω_3	V_{B_4}	α_3	A_{B_4}
(deg)	(deg)	(mm)	(mm)	(rad/s)	(m/s)	(rad/s^2)	(m/s^2)
45	−10.18	75	112.5	−17.96	−6.256	1 736.2	−0.534

12.35 For the mechanism in the posture shown, $\theta_3 = 150°$ and massless link 4 is rolling on the ground link. The first- and second-order kinematic coefficients of links 3 and 4 are $\theta_3' = 0$, $\theta_4' = -0.025$ rad/in, $\theta_3'' = -0.000\,625$ rad/in², and $\theta_4'' = 0.001\,875$ rad/in². The velocity and acceleration of the mass center of input link 2 are $\mathbf{V}_{G_2} = 280\hat{\mathbf{i}}$ in/s and $\mathbf{A}_{G_2} = -80\hat{\mathbf{i}}$ in/s², respectively. The free length and spring rate of the spring, the damping constant of the viscous damper, and the weights and mass moments of inertia of links 2 and 3 are provided in Table P12.35. Gravity is in the negative y direction and a horizontal external force P acts on link 2. Determine: (a) the first-order kinematic coefficients of the spring and the damper; (b) the first- and second-order kinematic coefficients of the mass center of link 3; (c) the equivalent mass; (d) the equation of motion; and (e) the force P acting on link 2.

Table P12.35

R_0	k	C	w_2	w_3	I_{G_2}	I_{G_3}
(in)	(lb/in)	(lb·s/in)	(lb)	(lb)	(in·lb·s^2)	(in·lb·s^2)
120	225	135	2.64	1.76	880	352

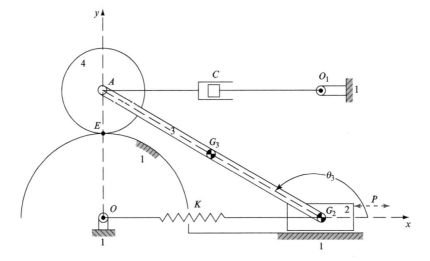

Fig. P12.35 $R_{AE} = 40$ in, $R_{G_2A} = 240$ in, and $R_{EO} = 80$ in.

12.36 For the mechanism in the posture shown, the first- and second-order kinematic coefficients of links 3 and 4 are $\theta'_3 = -0.125$ rad/rad, $R'_4 = 1.299$ m/rad, $\theta''_3 = 0$, and $R''_4 = 0.094$ m/rad^2. The constant angular velocity of input link 2, which rolls without slipping on the inclined plane, is $\omega_2 = 20$ rad/s cw. The free length of the spring is 3 m, the spring rate is $k = 25$ N/m, and the damping constant of the viscous damper is $C = 15$ N·s/m. The masses and mass moments of inertia of links 2 and 4 are: $m_2 = 7$ kg, $m_4 = 4$ kg, $I_{G_2} = 18$ kg·m^2, and $I_{G_4} = 22$ kg·m^2. The mass of link 3 is negligible compared to the masses of links 2 and 4, and gravity is in the negative y direction. Determine: (a) the first- and second-order kinematic coefficients of the mass centers of links 2 and 4; (b) the equivalent mass moment of inertia; and (c) the magnitude and direction of the torque acting on link 2.

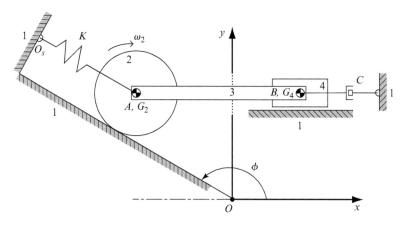

Fig. P12.36 $\rho_2 = 1.5$ m, $R_{BA} = R_{G_4 G_2} = 6$ m, $\phi = 150°$, and $R_{AO_S} = 2.5$ m.

12.37 The two-throw opposed-crank crankshaft is mounted in bearings at A and G. Each crank has an eccentric mass of 0.15 kg, which may be considered located at a radius of 50 mm from the axis of rotation, and at the center of each throw (points C and E). It is proposed to locate masses at B and F to reduce the bearing reactions, caused by the rotating eccentric cranks, to zero. Determine the magnitudes of these masses if they are to be mounted 75 mm from the axis of rotation.

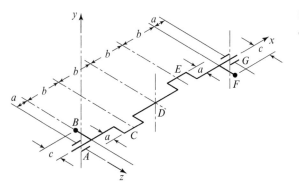

Fig. P12.37 $a = 50$ mm, $b = 150$ mm, $c = 75$ mm.

12.38 The two-throw crankshaft is mounted in bearings at A and F, with the cranks spaced 90° apart. Each crank may be considered to have an eccentric mass of 2.69 kg at the center of the throw and 50 mm from the axis of rotation. It is proposed to eliminate the rotating bearing reactions, caused by the cranks, by mounting additional correction masses on 75-mm arms at points B and E. Calculate the magnitudes and angular locations of these masses.

Fig. P12.38 $a = 50$ mm, $b = 150$ mm, $c = 75$ mm.

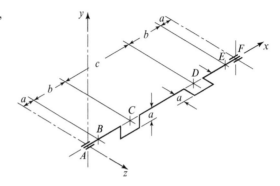

12.39 Solve Prob. 12.38 when the angle between the two throws is reduced from 90° to 0°.

12.40 The connecting rod of mass 3.55 kg is pivoted on a knife edge and caused to oscillate as a pendulum. The rod is observed to complete 64.5 oscillations in 1 min. Determine the mass moment of inertia of the rod about the center of mass.

Fig. P12.40 $a = 44$ mm D, $b = 56$ mm, $c = 350$ mm.

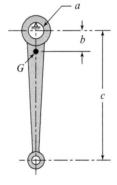

12.41 The gear is suspended on a knife edge at the rim and caused to oscillate as a pendulum. The period of oscillation is observed to be 1.08 s. Assume that the center of mass and the axis of rotation are coincident. If the weight of the gear is 41 lb, find the mass moment of inertia and the radius of gyration of the gear.

Fig. P12.41 $r = 8$ in.

12.42 The wheel is mounted on a shaft in bearings with very low frictional resistance to rotation. At one end of the shaft and on the outboard side of the bearings is connected a rod with a weight, W_b, secured to its end. W_b is displaced from equilibrium, and the assembly is permitted to oscillate. If the weight of the pendulum arm is neglected, show that the mass moment of inertia of the wheel can be written as

$$I = W_b l \left(\frac{\tau^2}{4\pi^2} - \frac{l}{g} \right).$$

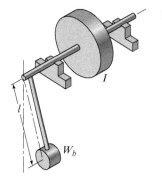

Fig. P12.42

12.43 If the weight of the pendulum arm, W_a, is not neglected in Prob. 12.42, but is assumed to be uniformly distributed over the length, l, show that the mass moment of inertia of the wheel can be written as

$$I = l \left[\frac{\tau^2}{4\pi^2} \left(W_b + \frac{W_a}{2} \right) - \frac{l}{g} \left(W_b + \frac{W_a}{3} \right) \right],$$

where W_a is the weight of the arm.

12.44 Wheel 2 is a round disk that rotates about a vertical axis, z, through its center. The wheel carries a pin, B, at a distance R from the axis of rotation of the wheel, about which link 3 is free to rotate. The center of mass, G, of link 3 is located at a distance r from the vertical axis through B, and link 3 has a weight W_3 and a mass moment of inertia I_G about its own mass center. The wheel rotates at an angular velocity ω_2 with link 3 fully extended. Develop an expression for the angular velocity, ω_3, that link 3 would acquire if the wheel were suddenly stopped.

Fig. P12.44

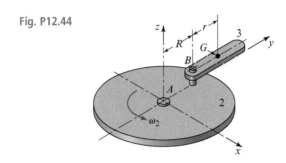

12.45 Repeat Prob. 12.44, but assume that the wheel rotates with link 3 radially inward. Under these conditions, is there a value for the distance, r, for which the resulting angular velocity, ω_3, is zero?

12.46 A planetary gear-reduction unit that utilizes 3.5 module spur gears is cut on the 20° full-depth system. All parts are steel with a density $1.98(10)^{-6}$ kg/m³. The arm is rectangular and is 100 mm wide by 350 mm long with a 100 mm diameter central hub and two 75 mm diameter planetary hubs. The segment separating the planet gears is a 12 mm by 100 mm diameter cylinder. The inertia of the gears can be obtained by treating them as cylinders equal in diameter to their respective pitch circles. The input to the reducer is driven with 18.65 kW at 600 rev/min. The mass moment of inertia of the resisting load is 6.35 kg · m². Calculate the bearing reactions on the input, output, and planetary shafts. As a designer, what forces would you use in designing the mounting bolts? Why?

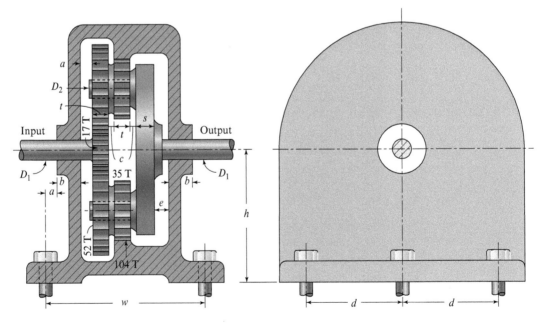

Fig. P12.46 $h = 269$ mm, $w = 338$ mm, $d = 200$ mm, $D_1 = 40.63$ D, $D_2 = 37.5$ mm D, $a = 25$ mm, $b = 50$ mm, $c = 12.5$ mm, $e = 31.25$ mm, $s = 37.5$ mm, $t = 31.88$ mm.

12.47 It frequently happens in motor-driven machinery that the greatest torque is exerted when the motor is first turned on, because of the fact that some motors are capable of delivering more starting torque than running torque. Analyze the bearing reactions of Prob. 12.46 again, but this time use a starting torque equal to 250% of the full-load torque. Assume a normal-load torque and a speed of zero. How does this starting condition affect the forces on the mounting bolts?

12.48 The gear-reduction unit of Prob. 12.46 is running at $600\ \text{rev/min}$ when the motor is suddenly turned off, without changing the resisting-load torque. Solve Prob. 12.46 for this condition.

12.49 The differential gear train has gear 1 fixed and is driven by rotating shaft 5 at $500\ \text{rev/min}$ in the direction shown. Gear 2 has fixed bearings constraining it to rotate about the positive y axis, which remains vertical; this is the output shaft. Gears 3 and 4 have bearings connecting them to the ends of the carrier arm which is integral with shaft 5. The pitch diameters of gears 1 and 2 are both $200\ \text{mm}$, while the pitch diameters of gears 3 and 4 are both $150\ \text{mm}$. All gears have $20°$ pressure angles and are each $18\ \text{mm}$ thick, and all are made of steel with density $81.3\ \text{kg/m}^3$. The mass of shaft 5 and all gravitational loads are negligible. The output shaft torque loading is $\mathbf{T} = 0.925\hat{\mathbf{j}}\ \text{N}\cdot\text{m}$ as shown. Note that the coordinate axes shown rotate with the input shaft 5. Determine the driving torque required, and the forces and moments in each of the bearings. (*Hint*: It is reasonable to assume through symmetry that $F_{13}^t = F_{14}^t$. It is also necessary to recognize that only compressive loads, not tension, can be transmitted between gear teeth.)

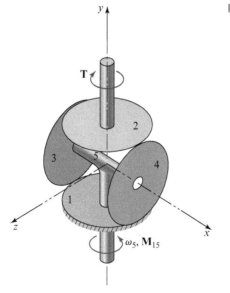

Fig. P12.49 Differential gear train.

12.50 Arms 2 and 3 of the flyball governor are pivoted to block 6, which remains at the height shown but is free to rotate around the y axis. Block 7 also rotates about, and is free to slide along, the y axis. Links 4 and 5 are pivoted at both ends between the two arms and block 7. The two balls at the ends of links 2 and 3 weigh $15.5\ \text{N}$ each, and all other weights are negligible in comparison; gravity acts in the $-\hat{\mathbf{j}}$ direction. The spring between links 6 and 7 has a stiffness of 178 N/m and would be unloaded if block 7 were at a height of $\mathbf{R}_D = 275\hat{\mathbf{j}}$ mm. All moving links rotate about the y axis with angular velocities of $\omega\hat{\mathbf{j}}$. Make a graph of the height R_D versus the rotational speed ω in rev/min, assuming that the changes in speed are slow.

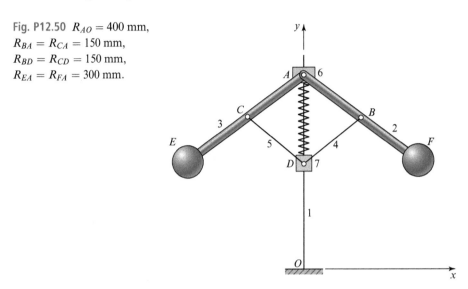

Fig. P12.50 $R_{AO} = 400$ mm,
$R_{BA} = R_{CA} = 150$ mm,
$R_{BD} = R_{CD} = 150$ mm,
$R_{EA} = R_{FA} = 300$ mm.

12.51 For the mechanism in the posture shown, the pin at the center of the wheel is sliding in the slot in link 2, and wheel 3 is rolling without slipping on the ground link. The constant angular velocity of input link 2 is $\omega_2 = 1.50\hat{\mathbf{k}}$ rad/s, the angular acceleration of the wheel is $\alpha_3 = -62.354\,\hat{\mathbf{k}}$ rad/s^2, and the acceleration of the center of mass of the wheel is $\mathbf{A}_{G_3} = 124.7\,\hat{\mathbf{i}}$ in/s^2. The weights and mass moments of inertia of link 2 and the wheel are $w_2 = 33$ lb, $w_3 = 55$ lb, $I_{G_2} = 0.945$ in $\cdot$ lb $\cdot$ s^2, and $I_{G_3} = 0.144$ in $\cdot$ lb $\cdot$ s^2, respectively. Gravity acts in the negative y direction. The force acting at point C is $\mathbf{F}_C = 16.9\hat{\mathbf{i}}$ lb, and there is a torque, $\mathbf{T}_{12}$, acting on link 2 at the crankshaft, O_2. Determine the torque, $\mathbf{T}_{12}$, and the minimum coefficient of friction between the wheel and the ground.

Fig. P12.51 $\theta_2 = -30°$, $\beta = 60°$,
$R_{G_3 O_2} = 8$ in, $R_{C O_2} = 14$ in, and
$\rho_3 = 2$ in.

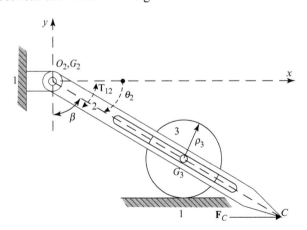

12.52 For the mechanism in the posture shown, $\theta_2 = 30°$, $R_{AO_2} = 1700$ mm, the position of the mass center of block 3 is 100 mm below pin A, and the first- and second-order kinematic coefficients of link 3 are $R_3' = 1440$ mm/rad and $R_3'' = 1667$ mm/rad^2 (where $\mathbf{R}_3$ is the vertical vector from point C fixed in link 3 to the x axis). The angular velocity and acceleration of the input link 2 are $\boldsymbol{\omega}_2 = 12\,\hat{\mathbf{k}}$ rad/s and $\boldsymbol{\alpha}_2 = 15\,\hat{\mathbf{k}}$ rad/s^2, respectively. The masses and mass moments of inertia of links 2 and 3 are $m_2 = 8.93$ kg, $m_3 = 6.94$ kg, $I_{G_2} = 2.09$ kg·m^2, and $I_{G_3} = 1.8$ kg·m^2, respectively. The torque acting on link 2 at the crankshaft O_2 is $\mathbf{T}_{12} = -50\,\hat{\mathbf{k}}$ N·m. Gravity is in the negative y direction. Determine the internal reaction forces and the force $\mathbf{P}$ acting at point B on link 3. Is link 3 sliding or tipping on the ground link in this posture?

12.53 For the mechanism in the posture shown, $\theta_2 = 60°$, the constant angular velocity of

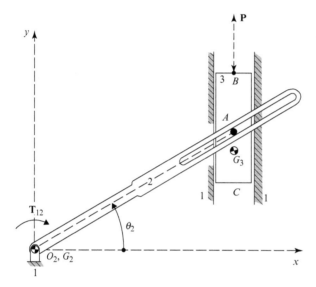

Fig. P12.52 Block 3 is 600 mm by 200 mm and pin A is centrally located.

the input link 2 is $\boldsymbol{\omega}_2 = 26\,\hat{\mathbf{k}}$ rad/s, and the accelerations of the centers of mass of links 3 and 5 are $\mathbf{A}_{G_3} = -338\,\hat{\mathbf{i}} - 586\,\hat{\mathbf{j}}$ in/s^2, and $\mathbf{A}_{G_5} = 280\hat{\mathbf{i}} - 1045\,\hat{\mathbf{j}}$ in/s^2, respectively. The weights and mass moments of inertia of the links are: $w_2 = w_3 = w_4 = w_5 = 15.4$ lb and $I_{G_2} = I_{G_3} = I_{G_4} = I_{G_5} = 0.117$ in·lb·s^2, respectively. The input torque is $\mathbf{T}_{12} = 4.5\,\hat{\mathbf{k}}$ in·lb, and gravity is in the negative z direction. The link lengths, the posture variables, and the first- and second-order kinematic coefficients of links 3, 4, and 5 are shown in Table P12.53. Determine the internal reaction forces between links 3 and 5 and between links 2 and 5 and the torque acting on link 4.

Table P12.53

O_2A	O_2C	CD	AD	AB	O_4B	O_2O_4	θ_3	θ_4	θ_5
(in)	(in)	(in)	(in)	(in)	(in)	(in)	(deg)	(deg)	(deg)
1.0	1.6	1.8	1.2	2.0	2.4	3.0	41.2	114.5	4.0

θ_3'	θ_4'	θ_5'	θ_3''	θ_4''	θ_5''
(rad/rad)	(rad/rad)	(rad/rad)	(rad/rad^2)	(rad/rad^2)	(rad/rad^2)
−0.425	0.140	−0.188	0.333	0.563	1.810

Fig. P12.53 The angle $\beta = 45°$.

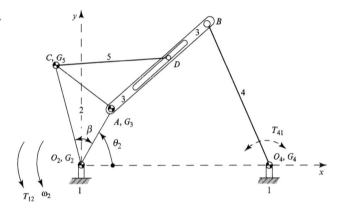

12.54 For the linkage in the posture shown, $\theta_2 = 45°$, the angular velocity and acceleration of the input link 2 are $\omega_2 = -5\hat{\mathbf{k}}$ rad/s and $\alpha_2 = 3\hat{\mathbf{k}}$ rad/s^2, respectively. The first- and second-order kinematic coefficients are $R_{AO_2}' = -0.26$ m/rad, $R_{CA}' = 0.37$ m/rad, $R_{AO_2}'' = -0.45$ m/rad^2, $R_{CA}'' = -0.37$ m/rad^2, $x_C' = -3.5$ m/rad, $y_C' = 0.35$ m/rad, $x_C'' = -0.35$ m/rad^2, and $y_C'' = -0.35$ m/rad^2. Gravity is in the negative y direction. The distances, the free length and spring rate of the rectilinear spring, the damping constant of the viscous damper, and the masses and second moments of mass about the mass centers of links 2, 3, and 4 are shown in Table P12.54. Determine: (a) the first-order kinematic coefficients of the spring and the damper; (b) the kinetic energy of the linkage; and (c) the torque $\mathbf{T}_{12}$.

Table P12.54

R_{CO_2}	R_{AO_2}	R_{CD}	R_{S0}	K	C	m_2	m_3	m_4	I_{G_2}	I_{G_3}	I_{G_4}
(mm)	(mm)	(mm)	(mm)	(N/ m)	(N · s/m)	(kg)	(kg)	(kg)	(kg · m^2)	(kg · m^2)	(kg · m^2)
500	448	500	600	0.031	0/009 32	4.8	49	39	3.04	7.10	6.10

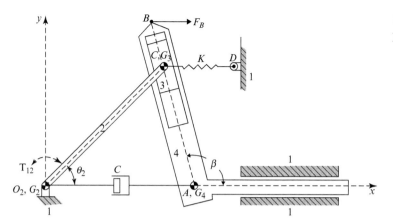

Fig. P12.54 $\beta = 105°$,
$\mathbf{F}_B = 250\hat{\mathbf{i}}$ N.

12.55 For the elliptic trammel linkage in the posture shown, $R_A = 86.60$ in, the velocity and acceleration of the input link 2 are $\mathbf{V}_{A_2} = -120\hat{\mathbf{j}}$ in/s and $\mathbf{A}_{A_2} = 280\hat{\mathbf{j}}$ in/s^2, respectively. The kinematic coefficients, the weights, and the second moments of mass about the mass centers of the links are shown in Table P12.55. The stiffness and free length of the spring are $K = 22.2$ lb/in and $R_0 = 20$ in, respectively, and the coefficient of the viscous damper is $C = 0.084$ lb · s/in. The only friction in the linkage is between links 1 and 4, where the coefficient of friction is $\mu = 0.35$. The normal force at the point of contact between links 1 and 4 is $\mathbf{F}_{41}^n = 112\hat{\mathbf{j}}$ lb, and gravity is vertically downward. Determine: (a) the first-order kinematic coefficients of the spring and the damper; (b) the kinetic energy of the linkage; and (c) the external force $\mathbf{P}$ acting at point H.

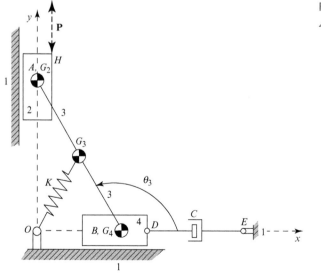

Fig. P12.55 $\theta_3 = 120°$,
$AG_3 = G_3B = 50$ in.

Table P12.55

θ_3'	θ_3''	R_4'	R_4''	w_2	w_3	w_4	I_{G_2}	I_{G_3}	I_{G_4}
(rad/in)	(rad/in²)	(in/in)	(in/in²)	(lb)	(lb)	(lb)	(in·lb·s²)	(in·lb·s²)	(in·lb·s²)
−0.020	−0.000 69	−1.732	−0.080	7.7	13.2	17.6	22.5	65.2	15.8

12.56 For the linkage in the posture shown, the angular velocity and acceleration of the input link 2 are $\boldsymbol{\omega}_2 = -5\ \hat{\mathbf{k}}$ rad/s and $\boldsymbol{\alpha}_2 = 3\ \hat{\mathbf{k}}$ rad/s², respectively, and the angular acceleration and the acceleration of the mass center of link 3 are $\boldsymbol{\alpha}_3 = 15.75\ \hat{\mathbf{k}}$ rad/s² and $\mathbf{A}_{G_3} = 0.24\ \hat{\mathbf{i}} - 0.38\ \hat{\mathbf{j}}$ m/s², respectively. The masses and second moments of mass of the links are $m_2 = m_3 = 5.94$ kg, $m_4 = 0.99$ kg, $I_{G_2} = I_{G_3} = 5.07$ kg·m², and $I_{G_4} = 0.506$ kg·m². The torque acting on link 4 is $\mathbf{T}_{14} = -1.965\ \hat{\mathbf{k}}$ N·m and there is an unknown torque $\mathbf{T}_{12}$ acting on link 2. Gravity is in the negative y direction. Determine: (a) the magnitude, direction, and location of the reaction force between links 3 and 4; and (b) the magnitude and direction of the torque $\mathbf{T}_{12}$.

Fig. P12.56 $O_2 A = 50$ mm, $AO_4 = 100$ mm, and $AG_3 = 60$ mm.

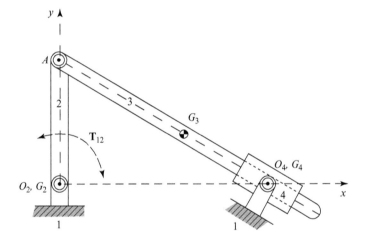

12.57 For the mechanism in the posture shown, a force $F_2 = 3.4$ lb is applied at point A on the input link 2, which causes the link to move up the inclined plane with a velocity $V_{G_2} = 20$ in/s. The first- and second-order kinematic coefficients (where $R_3 = R_{G_2 O_3}$) are shown in Table P12.57. A linear spring with a free length $R_{S0} = 0.40$ in and a spring constant $K = 14$ lb/in is attached to link 2 at point B and has a current length of 0.75 in. A viscous damper with a damping coefficient $C = 0.45$ lb·s/in is attached to link 2 at point D. The weights and mass moments of inertia of links 2 and 3 are $w_2 = 6.6$ lb, $w_3 = 11$ lb, $I_{G_2} = 0.000\,315$ in·lb·s², and $I_{G_3} = 0.000\,63$ in·lb·s². Gravity is in the negative y direction. Determine the first- and second-order kinematic coefficients of the mass center of link 2, and the first-order kinematic coefficients of the spring and damper. Write the equation of motion for the mechanism in symbolic form and then determine the acceleration of link 2.

Table P12.57

θ_3' (rad/in)	R_3', (in/in)	θ_3'' (rad/in²)	R_3'' (in/in²)
–0.31	–0.500	–0.113	0.268

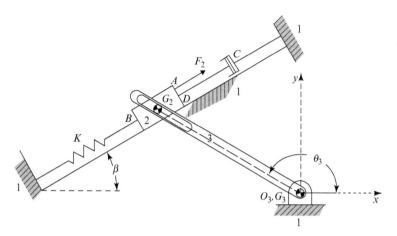

Fig. P12.57 $\theta_3 = 150°$, $\beta = 30°$, $G_3G_2 = 2.80$ in, and $BG_2 = 0.40$ in.

12.58 For the mechanism in the posture shown, the constant angular velocity of the input link 2 is $\boldsymbol{\omega}_2 = 3\hat{\mathbf{k}}$ rad/s. The free length of the linear spring attached between ground pin O_1 and pin A is 500 mm and a viscous damper is attached between pin O and link 3. The torque acting on link 2 is $\mathbf{T}_{12} = -1.48\hat{\mathbf{k}}$ N · m, there is a horizontal force $\mathbf{P}$ acting at pin A, and gravity is in the negative y direction. The known data (where $\mathbf{R}_2$ is the vector from bearing O_2 to the mass center G_3 and $\mathbf{R}_3$ is the vector from ground pin O to the mass center G_3) are shown in Table P12.58. Determine: (a) the first-order kinematic coefficients of the spring and the damper; (b) the potential energy of the spring; (c) the equivalent mass moment of inertia; and (d) the force $\mathbf{P}$.

Table P12.58

R_2'	R_3'	R_2''	R_3''	m_2	m_3	I_{G_2}	I_{G_3}	K	C
(mm/rad)	(mm/rad)	(mm/rad²)	(mm/rad²)	(kg)	(kg)	(kg · m²)	(kg · m²)	(N/m)	(N · s/m)
52	104	150	130	6.95	11.91	0.744	2.233	14.7	73.3

Fig. P12.58 $\theta_2 = -60°$, $R_{AO_2} = 120$ mm,
$R_{G_3O_2} = 90$ mm, and $R_{G_2O_2} = 45$ mm.

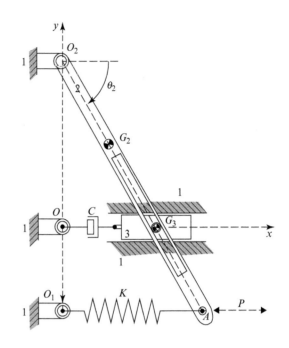

12.59 For the linkage in the posture shown, the kinematic coefficients of link 3 are $\theta_3' = 0.25$ rad/rad and $\theta_3'' = 0.60$ rad/rad^2. The angular velocity and acceleration of the input link 2 are $\boldsymbol{\omega}_2 = -5$ $\hat{\mathbf{k}}$ rad/s and $\boldsymbol{\alpha}_2 = 3$ $\hat{\mathbf{k}}$ rad/s^2, respectively. The free length and stiffness of the horizontal spring are $R_{S0} = 1.2$ in and $K = 11.25$ lb/in, respectively. The damping coefficient of the viscous damper is $C = 0.844$ lb·s/in. The weights and mass moments of inertia of the links are $w_2 = w_3 = 13.2$ lb, $w_4 = 2.2$ lb, $I_{G_2} = I_{G_3} = 45$ in·lb·s^2, and $I_{G_4} = 36$ in·lb·s^2. The input torque is $\mathbf{T}_{12} = -90$ $\hat{\mathbf{k}}$ in·lb and there is a horizontal force, $\mathbf{F}$, acting at the mass center of link 3. Gravity is in the negative y direction. Determine: (a) the first-order kinematic coefficients of the horizontal spring and the viscous damper; (b) the kinetic energy of this linkage; and (c) the force, $\mathbf{F}$.

Fig. P12.59 $O_2A = 2$ in,
$AO_S = 2.08$ in, $AO_4 = 3.2$ in,
$AD = 4$ in, and $AG_3 = 2.4$ in.

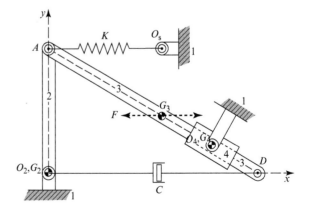

12.60 For the mechanism in the posture shown, the velocity and acceleration of the input link 2 down the slope EA are $V_A = 200\ \text{mm/s}$ and $A_A = 600\ \text{mm/s}^2$, respectively. The wheel, link 4, is rolling without slipping on the ground link. The free length of the linear spring is $R_{SO} = 200\ \text{mm}$ and the damping coefficient of the viscous damper is $C = 0.155\ \text{N}\cdot\text{s/m}$. Gravity acts vertically downward. The dimensions, first- and second-order kinematic coefficients (R_{44} is the vector from point O to pin B), masses, and second moments of mass about the mass centers of links 2, 3, and 4 are shown in Table P12.60. Determine: (a) the first-order kinematic coefficients of the spring and the damper; (b) the kinetic energy of the mechanism; and (c) the stiffness of the spring.

Table P12.60

EA	AB	BD	ρ_4	θ_3'	θ_3''	R_{44}'	R_{44}''	m_2	m_3	m_4	I_{G_2}	I_{G_3}	I_{G_3}
(mm)	(mm)	(mm)	(mm)	(rad/m)	(rad/m^2)	(m/m)	(m/m^2)	(kg)	(kg)	(kg)	(kg·m^2)	(kg·m^2)	(kg·m^2)
250	600	350	150	−2.892	14.4	2.0	−10	2.4	9.7	19.4	0.033	0.059	0.044

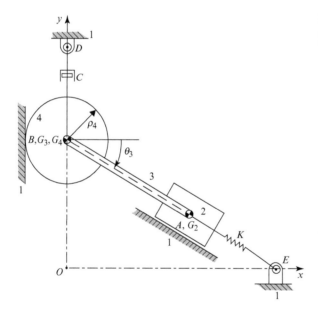

Fig. P12.60 $\theta_3 = -30°$, planar mechanism.

REFERENCES

[1] Beer, F. P., E. R. Johnston, and W. E. Clausen, 2007. *Vector Mechanics for Engineers, Dynamics*, 8th ed., New York: McGraw-Hill.

[2] Fisher, F. E., and H. H. Alford, 1977. *Instrumentation for Mechanical Analysis*, University of Michigan Summer Conferences, Ann Arbor, MI.

[3] Gary, G. L., F. Costanzo, and M. E. Plesha, 2013. *Engineering Mechanics, Dynamics*, 2nd ed., New York: McGraw-Hill.

[4] Meriam, J. L., and L. G. Kraige, 2007. *Engineering Mechanics, Dynamics*, 6th ed., Hoboken, NJ: Wiley.

13 Vibration Analysis

The existence of vibrating elements in a mechanical system produces unwanted noise, high stresses, wear, poor reliability, and, frequently, premature failure of one or more of the parts. The moving parts of all machines are inherently vibration producers, and for this reason engineers must expect vibrations to exist in the devices they design. But there is a great deal they can do during the design of the system to anticipate a vibration problem and to minimize its undesirable effects.

Sometimes it is necessary to build a vibratory system into a machine – a vibratory conveyor, for example. Under these conditions the engineer must understand the mechanics of vibration to obtain an optimal design.

13.1 Differential Equations of Motion

Any motion that exactly repeats itself after a certain interval of time is a periodic motion and is called a *vibration*. Vibrations may be either free or forced. A mechanical element is said to have a *free vibration* if the periodic motion continues after the cause of the original disturbance is removed, but if a vibratory motion persists because of the continuing existence of a disturbing force, then it is called a *forced vibration*. Any free vibration of a mechanical system eventually ceases because of loss of energy. In vibration analysis, we often take account of these energy losses using a single factor called the *damping factor*. Thus, a heavily damped system is one in which the vibration decays rapidly. The *period* of a vibration is the time for a single cycle of the motion after which repetition begins; the *frequency* is the number of cycles or periods occurring in unit time. The *natural frequency* is the frequency of a free vibration. If the forcing frequency becomes equal to the natural frequency of a system, then *resonance* is said to occur.

We shall also use the terms *steady-state vibration* to indicate that a motion is repeating itself exactly in each successive cycle and *transient vibration* to indicate a vibration-like motion that is changing in character. If a periodic force operates on a mechanical system, the resulting motion is transient in character when the force first begins to act, but after an interval of time the transient decays, owing to damping, and the resulting motion is termed a *steady-state vibration*.

The word *response* is frequently used in discussing vibratory systems. The words *response*, *behavior*, and *performance* have roughly the same meaning when used in dynamic analysis.

Thus, we can apply an external force having a sine-wave relationship with time to a vibrating system to determine how the system "responds," or "behaves," when the frequency of the force is varied. A plot using the vibration amplitude along the ordinate and the forcing frequency along the abscissa is then described as a *performance* or *response* curve for the system. Sometimes it is useful to apply arbitrary input disturbances or forces to a system. These may not resemble the force characteristics that a real system receives in use at all, yet the response of the system to these chosen disturbances can provide much useful information about the system.

Vibration analysis is sometimes called *elastic-body analysis* or *deformable-body analysis*, since, as we shall see, it is elasticity in a mechanical system that allows vibration. When a rotating shaft has a torsional vibration, this means that a mark on the circumference at one end of the shaft is successively ahead of and then behind a corresponding mark on the other end of the shaft. In other words, torsional vibration of a shaft is the alternate twisting and untwisting of the rotating material and requires elasticity for its existence. We shall begin our study of vibration by assuming that elastic parts have no mass and that heavy parts are absolutely rigid, that is, they have no elasticity. Of course these assumptions are never true, and so, in the course of our studies, we must also learn to correct for the effects of making these assumptions.

Figure 13.1 shows an idealized vibrating system having a mass m guided to move only in the x direction. The mass is connected to a fixed frame through the spring k, and the dashpot, c. The assumptions made are as follows:

1. The spring and the dashpot are massless.
2. The mass is absolutely rigid.
3. All damping is concentrated in the dashpot.

It turns out that a great many mechanical systems can be analyzed quite accurately using these assumptions.

The elasticity of the system of Fig. 13.1 is completely represented by the spring. The *stiffness* is designated as k and is defined by the equation

$$k = \frac{F}{x},\qquad (13.1)$$

where F is the force required to deflect the spring a distance x.

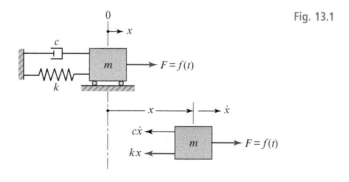

Fig. 13.1

Similarly, the friction, or damping, is assumed to be entirely viscous damping (later we shall examine other kinds of friction) and is designated using a *coefficient of viscous damping, c*. Thus,

$$c = \frac{F}{\dot{x}}, \tag{13.2}$$

where F is the force required to move the mass at a velocity of $\dot{x}$.

Note that we are using the notation for time derivatives that is customary in vibration theory. Thus,

$$\dot{x} = \frac{dx}{dt} \quad \text{and} \quad \ddot{x} = \frac{d^2x}{dt^2}.$$

This is called *Newton's notation*, since Newton was the first to use it. It is used only when derivatives are with respect to time.

The vibrating system of Fig. 13.1 has one degree of freedom, since the position of the mass is completely defined by a single coordinate. An external force, $F = f(t)$, is shown acting upon the mass. Thus, this system is classified as a forced, single-degree-of-freedom system with damping.

The equation of motion of the system is written by displacing the mass in the positive direction, giving it a positive velocity, and then summing all forces, including the inertia force. As shown in Fig. 13.1, there are three forces acting: a spring force, kx, acting in the negative direction, a damping force, $c\dot{x}$, also acting in the negative direction, and an external force, $f(t)$, acting in the positive direction. Summing these forces together with the inertia force gives

$$\sum F = -kx - c\dot{x} + f(t) + (-m\ddot{x}) = 0$$

or

$$m\ddot{x} + c\dot{x} + kx = f(t). \tag{13.3}$$

Equation (13.3) is an important equation in dynamic analysis, and we shall eventually solve it for many specialized conditions. It is a linear differential equation of the second order.

Consider next the idealized torsional vibrating system of Fig. 13.2. Here, a disk having a mass moment of inertia, I, is mounted upon the end of a weightless shaft having a torsional spring constant, k, defined by

$$k = \frac{T}{\theta}, \tag{13.4}$$

Fig. 13.2

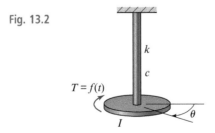

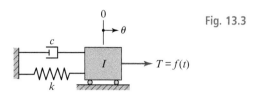

Fig. 13.3

where T is the torque necessary to produce an angular deflection, θ, of the shaft. In a similar manner, the torsional viscous damping coefficient is defined by

$$c = \frac{T}{\dot\theta}. \tag{13.5}$$

Note that we are using the same symbols for denoting these torsional parameters as for rectilinear ones. The nature of the problem or of the differential equation reveals whether the system is rectilinear or torsional, and so there is no confusion in this usage.

Next, designating an external torque forcing function by $T = f(t)$, we find that the differential equation for the torsional system is

$$\sum M = -k\theta - c\dot\theta + f(t) + \left(-I\ddot\theta\right) = 0$$

or

$$I\ddot\theta + c\dot\theta + k\theta = f(t), \tag{13.6}$$

which is of the same form as Eq. (13.3). Thus, with appropriate substitutions, the solution of Eq. (13.6) is the same as that of Eq. (13.3). Note, too, that this means we can *simulate* a torsional system, as in Fig. 13.3, merely by substituting torsional notation for the rectilinear notation.

It has been said that mathematics is a human invention and, therefore, that it can never perfectly describe nature [1]. However, the linear differential equation frequently does such an excellent job of simulating the action of a mechanical system that one wonders whether differential equations are not an exception to this statement. Perhaps people just happened to discover them, as they have other curiosities of nature.

Example 13.1

Figure 13.4*a* shows a vibrating system in which a time-dependent displacement, $y = y(t)$, excites a spring–mass system through a viscous dashpot. Write the differential equation of this system.

(a)   (b)   Fig. 13.4 (a) System; (b) forces acting on the mass.

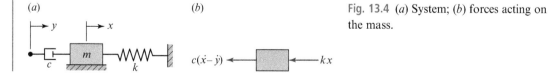

Example 13.1 (cont.)

Solution

We begin by assuming an arbitrary relationship between the coordinates and their first derivatives, say, $x > y$ and $\dot{x} > \dot{y}$. We also assume senses for x and $\dot{x}$, say, $x > 0$ and $\dot{x} > 0$. Regardless of the assumptions made, the resulting differential equation is the same. With these assumptions, the forces acting on the mass are as shown in Fig. 13.4b. Note that these external forces act in the negative direction because of the arbitrary assumptions made. Summing the forces in the x direction and including the inertia force gives

$$\sum F = -kx - c(\dot{x} - \dot{y}) + (-m\ddot{x}) = 0. \tag{1}$$

Such equations are usually rearranged with the response derivative terms on the left-hand side and the forcing term on the right, that is,

$$m\ddot{x} + c\dot{x} + kx = c\dot{y}. \qquad\qquad Ans.$$

13.2 A Vertical Model

The system of Fig. 13.1 moves in the horizontal direction, and, as a result, gravity has no effect on its motion. Let us now turn the system to the vertical direction and, also incidentally, make the damping and the external forces zero. This yields the model of Fig. 13.5. We choose the origin of the coordinate system to correspond with the position of the mass where the spring has no force, and we choose the symbol x to denote the vertical displacement. Thus, when the mass is given a positive displacement, the spring force is made up of two components. One of these is $k\delta_{st}$, where $\delta_{st} = W/k$ is the distance the spring is deflected when the weight is suspended from it. The other component is kx. Summing the forces acting on the mass gives

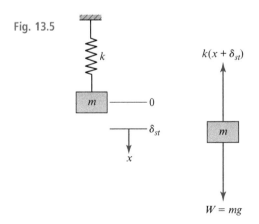

Fig. 13.5

$$\sum F = -k(x + \delta_{\text{st}}) + W + (-m\ddot{x}) = 0.$$

Rearranging this equation gives

$$m\ddot{x} + kx = W - k\delta_{\text{st}},$$

and, because of the definition $\delta_{\text{st}} = W/k$, this equation reduces to the homogeneous form

$$m\ddot{x} + kx = 0. \tag{13.7}$$

Note that Eq. (13.3) is identical to Eq. (13.7) if the damping force, $c\dot{x}$, and the external force, $f(t)$, are made zero.

13.3 Solution of the Differential Equation

If Eq. (13.7) is arranged in the form

$$\ddot{x} = -\frac{k}{m}x, \tag{a}$$

then we can see that the function used for x has to be of the same form as the negative of its second derivative. One such function is

$$x = A \sin bt, \tag{b}$$

for which

$$\dot{x} = Ab \cos bt \tag{c}$$

and

$$\ddot{x} = -Ab^2 \sin bt. \tag{d}$$

Substituting Eqs. (b) and (d) into Eq. (a) produces

$$-Ab^2 \sin bt = -\frac{k}{m}A \sin bt. \tag{e}$$

Then dividing both sides by $(-A \sin bt)$ leaves

$$b^2 = \frac{k}{m}.$$

Thus, Eq. (b) is a solution, provided that $b = \sqrt{k/m}$. By an exactly parallel development, it is clear that

$$x = B \cos bt \tag{f}$$

is also a solution to Eq. (a). A general solution is obtained by adding Eqs. (b) and (f), that is,

$$x = A \sin \sqrt{\frac{k}{m}} t + B \cos \sqrt{\frac{k}{m}} t. \tag{g}$$

Mathematically, the constants A and B are the constants of integration, since, theoretically, we could integrate Eq. (a) twice to get the solution. Unfortunately, most differential equations cannot be solved in this manner, and we must resort to cleverness to obtain the solution. Physically, the constants A and B represent the manner in which the vibration is started, or, to put this another way, the state of the motion at the instant $t = 0$.

We define the quantity

$$\omega_n = \sqrt{\frac{k}{m}}, \tag{13.8}$$

which is called the *natural frequency* of the system; it has units of radians per second.

The solution, Eq. (g), can now be written in the form

$$x = A \sin \omega_n t + B \cos \omega_n t. \tag{13.9}$$

As an example of the meaning of constants A and B, suppose we move the mass by a distance x_0 and release it with zero velocity at the instant $t = 0$. Then the starting conditions are

$$\text{at } t = 0, \quad x = x_0, \quad \dot{x} = 0.$$

Substituting the first of these into Eq. (13.9) gives

$$x_0 = A(0) + B(1)$$

or

$$B = x_0.$$

Next, taking the first time derivative of Eq. (13.9) gives

$$\dot{x} = A\omega_n \cos \omega_n t - B\omega_n \sin \omega_n t, \tag{h}$$

and substituting the other condition into this equation yields

$$0 = A\omega_n(1) - B\omega_n(0)$$

or

$$A = 0.$$

Then, substituting A and B into Eq. (13.9) gives the result

$$x = x_o \cos \omega_n t. \tag{13.10}$$

If we start the motion with the initial conditions that at $t = 0$, $x = 0$, $\dot{x} = v_0$, the result is

$$x = \frac{v_0}{\omega_n} \sin \omega_n t. \tag{13.11}$$

But, if we start the motion with at $t = 0$, $x = x_0$, $\dot{x} = v_0$, the result is

$$x = \frac{v_0}{\omega_n} \sin \omega_n t + x_0 \cos \omega_n t, \tag{13.12}$$

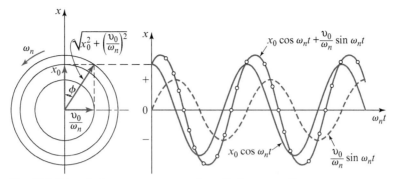

Fig. 13.6 Use of phasors to generate time-displacement diagrams.

which is the most general form of the solution. We can substitute Eq. (13.12) together with its second derivative into the differential equation and so demonstrate its validity.

The three solutions [Eqs. (13.10), (13.11), and (13.12)] are represented graphically in Fig. 13.6 using *phasors* to generate the trigonometric functions. Phasors are not vectors in the classic sense, since phasors can be manipulated in ways that are not defined for vectors. They are complex numbers, however, and they may be added and subtracted just as vectors are.

The ordinate of the graph of Fig. 13.6 is displacement, x, and the abscissa can be considered as the time axis or as the angular displacement, $\omega_n t$, of the phasors for a given time after the motion has commenced. The phasors x_0 and v_0/ω_n are shown in their initial positions, and as time passes, these rotate counterclockwise with an angular velocity of ω_n and generate the displacement curves shown. Figure 13.6 shows that the phasor x_0 starts from a maximum positive displacement and the phasor v_0/ω_n starts from a zero displacement. These, therefore, are very special, and the most general form is that given by Eq. (13.12), in which motion begins at some intermediate point.

In Eq. (13.8), the quantity $\omega_n = \sqrt{k/m}$ was defined as the natural frequency of the system. For most systems, the natural frequency is a constant since the mass and the spring constant do not vary. Since one cycle is equal to 2π radians, the period of a free vibration is

$$\tau = \frac{2\pi}{\omega_n} = 2\pi\sqrt{\frac{m}{k}}, \tag{13.13}$$

where τ is usually expressed in seconds per cycle. Although the natural units for ω_n are radians per second, frequency can also be defined as the reciprocal of the period and this gives

$$f = \frac{\omega_n}{2\pi} = \frac{1}{2\pi}\sqrt{\frac{k}{m}}, \tag{13.14}$$

where f is expressed in hertz $\left(\text{cycles/s or s}^{-1}\right)$ and is abbreviated as Hz.[1]

[1] Named after German scientist Heinrich Rudolf Hertz (1857–1894).

Fig. 13.7 Phase relationship of displacement, velocity, and acceleration phasors.

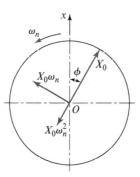

A study of Fig. 13.6 indicates that one should also be able to express the motion by the equation[2]

$$x = X_0 \cos(\omega_n t - \phi),$$
(13.15)

where X_0 and ϕ are the constants of integration whose values depend upon the initial conditions. These constants can be obtained directly from the trigonometry of Fig. 13.6 and are

$$X_0 = \sqrt{x_0^2 + \left(\frac{v_0}{\omega_n}\right)^2} \quad \text{and} \quad \phi = \tan^{-1}\frac{v_0/\omega_n}{x_0}.$$
(13.16)

Equation (13.15) can now be written in the form

$$x = \sqrt{x_0^2 + \left(\frac{v_0}{\omega_n}\right)^2} \cos(\omega_n t - \phi).$$
(13.17)

This is a particularly convenient form of the equation since the coefficient is the *amplitude* of the vibration. The amplitude is the maximum displacement of the mass. The angle ϕ is called the *phase angle*, and it denotes the angular lag of the motion with respect to the cosine function.

The velocity and acceleration are obtained by successively differentiating Eq. (13.15), that is,

$$\dot{x} = -X_0\omega_n \sin(\omega_n t - \phi) \quad \text{and} \quad \ddot{x} = -X_0\omega_n^2 \cos(\omega_n t - \phi),$$

where the velocity amplitude is $X_0\omega_n$, and the amplitude of the acceleration is $X_0\omega_n^2$. The displacement, velocity, and acceleration phasors are shown in Fig. 13.7 to clarify their phase relationships. All three phasors maintain fixed relative phase angles as they rotate together at the constant angular velocity ω_n. The velocity leads the displacement by $90°$ and the acceleration is $180°$ out of phase with the displacement.

---

[2] The motion can also be expressed in the form

$$x = X_0 \sin(\omega_n t + \psi),$$

where X_0 and ψ are the constants of integration. This is probably not a good way of expressing it, however, because in the study of forced vibration it might imply that the output can lead the input, which, of course, is not possible.

13.4 Step Input Forcing

Let us again consider the vibrating system of Fig. 13.1, this time adding to the system a force F applied to the mass that, in general, might be a function of time. As a start, however, let us assume that this force is constant and acting in the positive x direction. As before, we consider the damping to be zero. For this condition Eq. (13.3) is written

$$\ddot{x} + \frac{k}{m}x = \frac{F}{m}. \tag{13.18}$$

The general solution to this equation is

$$x = A\cos\omega_n t + B\sin\omega_n t + \frac{F}{k}, \tag{13.19}$$

where A and B are the constants of integration and where the natural frequency is

$$\omega_n = \sqrt{\frac{k}{m}}, \tag{a}$$

as before. We note that the quantity F/k has units of length. The physical interpretation is that a force F applied to a spring of stiffness k would produce a deformation of the spring of magnitude F/k.

It is interesting to consider starting the motion when the system is at rest. Thus, with the system in equilibrium at $x = 0$, we apply a constant force F at the instant $t = 0$ and observe the behavior of the system. Since the system is motionless at the instant the force is applied, the starting conditions are $x = 0$ and $\dot{x} = 0$ when $t = 0$. Substituting these conditions into Eq. (13.19) produces the following values for the two constants of integration:

$$A = -\frac{F}{k} \quad \text{and} \quad B = 0.$$

The equation for the motion is obtained by substituting these back into Eq. (13.19). This gives

$$x = \frac{F}{k}(1 - \cos\omega_n t). \tag{13.20}$$

This equation is plotted in Fig. 13.8 to show how the system behaves. Figure 13.8 shows that the application of a constant force F produces a vibration of amplitude F/k about a position of

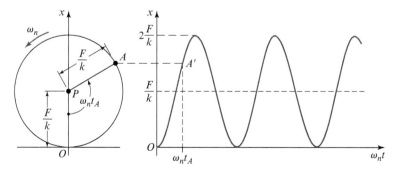

Fig. 13.8 Response of an undamped vibrating system to a constant force.

equilibrium displaced a distance F/k from the origin. This is evident from Eq. (13.20) since it contains a positive constant term,

$$x_1 = \frac{F}{k},$$

and a negative trigonometric term,

$$x_2 = -\frac{F}{k}\cos\omega_n t.$$

For $t = 0$ these two terms become

$$x_1 = \frac{F}{k} \quad \text{and} \quad x_2 = -\frac{F}{k}.$$

On the phase diagram (Fig. 13.8) the motion is represented by the phasor PA of length F/k rotating counterclockwise at ω_n rad/s. This phasor starts from the posture PO when $t = 0$, rotates about P as a center, and generates the circle of radius F/k.

The velocity and acceleration are obtained by successive differentiation of Eq. (13.20) and are

$$\dot{x} = \frac{F}{k}\omega_n \sin\omega_n t, \tag{13.21}$$

$$\ddot{x} = \frac{F}{k}\omega_n^2 \cos\omega_n t. \tag{13.22}$$

These can be represented as phasors too, as shown in Fig. 13.9. Their starting postures are determined by calculating their values for $\omega_n t = 0$. Velocity–time and acceleration–time graphs can also be plotted for these phasors, employing the same methods used to obtain the displacement–time plot of Fig. 13.8.

Figure 13.10 shows the results when the force is just as suddenly removed. If, for example, the force is removed at time $t_B = 5\pi/\omega_n$ when the phasor occupies the posture PB and the displacement diagram has progressed to B', the resulting motion has an amplitude about the zero axis of $2F/k$. The diagram shows an amplitude of F/k about the zero axis if the force is removed at time $t_A = 9\pi/4\omega_n$ at position A' on the displacement diagram. Finally, note that if

Fig. 13.9 Starting postures of the velocity and acceleration phasors.

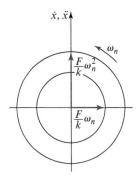

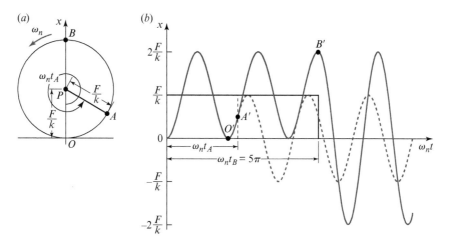

Fig. 13.10 (a) Phase diagram; (b) displacement diagram.

the force is removed at $t = 2\pi/\omega_n$ at point O' on the displacement diagram, the resulting motion is zero.

13.5 Phase-Plane Representation

The phase-plane method is a graphic method of solving transient vibration problems that is quite easy to understand and use. The method eliminates the necessity for solving differential equations, some of which can be very difficult, and even enables solutions to be obtained when the functions involved are not expressed in algebraic form. Engineers must concern themselves as much with transient disturbances and motions of machine parts as with steady-state motions. The phase-plane method presents the physics of the problem with so much clarity that it serves as an excellent vehicle for the study of mechanical transients.

Before introducing the details of the phase-plane method, it is of value to demonstrate how the displacement–time and the velocity–time relations are generated by a single rotating phasor. We have already observed that a free undamped vibrating system has a displacement equation which can be expressed in the form

$$x = X_0 \cos(\omega_n t - \phi), \tag{13.23}$$

and that its velocity is

$$\dot{x} = -X_0 \omega_n \sin(\omega_n t - \phi). \tag{13.24}$$

The displacement given by Eq. (13.23) can be represented by the projection on a vertical axis of a phasor of length X_0 rotating at ω_n rad/s in the counterclockwise direction (Fig. 13.11a). The angle $(\omega_n t - \phi)$ in this example is measured from the vertical axis. Similarly, the velocity can be represented on the same vertical axis by the projection of another phasor of length $X_0 \omega_n$

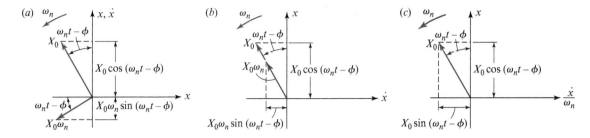

Fig. 13.11

rotating at the same angular velocity, but leading X_0 by a phase angle of 90° as shown in Fig. 13.11a. Therefore, the angular location of the velocity phasor is measured from the horizontal axis. If we take the coordinate system containing the velocity phasor and rotate it backward (clockwise) through an angle of 90°, then the velocity and displacement phasors become collinear and their angular locations can be measured from the same vertical axis. This step has been taken in Fig. 13.11b, where the displacement is still measured along the same vertical axis. But, having rotated the coordinate system in which the velocity is measured, we see that the velocity is obtained by projecting the $X_0\omega_n$ phasor to the horizontal axis. Thus, we can now measure velocities on a separate axis from displacements. Note, too, that the direction of positive velocities is to the right.

Our final step is taken by noting that the velocity phasor differs in length from the displacement phasor by the constant factor, ω_n. Thus, instead of plotting velocities, if we plot the quantity $\dot{x}/\omega_n$, then we have a quantity that is proportional to velocity. This step has been taken in Fig. 13.11c, where the horizontal axis is designated the $\dot{x}/\omega_n$ axis. With this change the projection of the phasor X_0 on the x axis gives the displacement and its projection on the $\dot{x}/\omega_n$ axis gives a quantity that is directly proportional to the velocity.

The displacement–time and the velocity–time graphs of a free undamped vibration have been plotted in Fig. 13.12 to demonstrate how the vector diagram is related to them. A point A on the phase diagram corresponds with A' on the displacement diagram and with A'' on the velocity diagram. Note that quantities obtained from the velocity plot must be multiplied by ω_n to obtain the actual velocity.

It is possible to arrive at these same conclusions in a different manner. If Eq. (13.23) is squared, we obtain

$$x^2 = X_0^2 \cos^2(\omega_n t - \phi). \tag{a}$$

Next, dividing Eq. (13.24) by ω_n and squaring the result gives

$$\left(\frac{\dot{x}}{\omega_n}\right)^2 = X_0^2 \sin^2(\omega_n t - \phi). \tag{b}$$

Then, adding Eqs. (a) and (b) gives

$$x^2 + \left(\frac{\dot{x}}{\omega_n}\right)^2 = X_0^2. \tag{13.25}$$

(a)

(b)

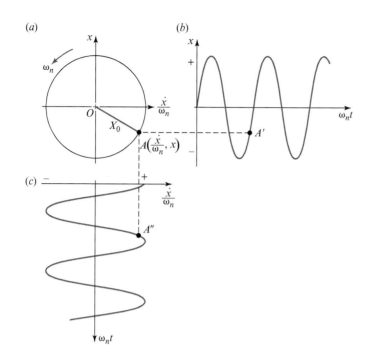

(c)

Fig. 13.12 (a) Phase diagram; (b) displacement diagram; (c) velocity diagram.

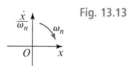

Fig. 13.13

This is the equation of a circle having the amplitude X_0 as its radius and with its center at the origin of a coordinate system having the axes x and $\dot{x}/\omega_n$. Thus, Eq. (13.25) describes the circle of Fig. 13.12a, where OA is the amplitude, X_0, of the motion. The coordinate system $x, \dot{x}/\omega_n$ defines the position of points in a region called the *phase plane*.

The phase plane is widely used in the solution of nonlinear differential equations. Such equations occur frequently in the study of vibrations and feedback-control systems. When the phase plane is used to solve nonlinear differential equations, it is customary to arrange the axes as shown in Fig. 13.13, with ω_n considered positive in the clockwise direction instead of the counterclockwise direction as we use it here. This arrangement of the axes seems more logical to some, since $\dot{x}/\omega_n$ is a function of x. However, for the analysis of transient disturbances to mechanical systems and for analyzing cam mechanisms, the arrangement of the axes as in Fig. 13.12 is more appropriate, and this is the one we shall employ throughout this chapter. The reason for this is that we employ the phase-plane method not as an end in itself, as is done when it is used for the solution of nonlinear problems, but as a tool to obtain transient response.

13.6 Phase-Plane Analysis

As a first example of the use of the phase-plane method, we consider the vibrating system of Fig. 13.14. This is a spring–mass system having a spring attached to a frame that can be repositioned at will. To begin the motion, we consider the system as initially at rest and then, at $t = 0$, we suddenly move the frame a distance x_1 to the right. The effect of this sudden motion is to compress the spring an amount x_1 and to shift the equilibrium position of the mass a distance x_1 to the right. For $t < 0$, the mass is in equilibrium at the position shown in Fig. 13.14. For $t > 0$, the equilibrium position of the mass is a distance x_1 to the right of the position shown. If the motion of the frame occurs instantaneously, the initial conditions are

$$\text{at } t = 0, \quad x = -x_1 \quad \text{and} \quad \dot{x} = 0, \tag{a}$$

where the motion is now measured from the shifted equilibrium position. If these initial conditions are used to evaluate the constants of integration of Eq. (13.15), we obtain

$$X_0 = -x_1 \quad \text{and} \quad \phi = 0$$

so that

$$x = -x_1 \cos \omega_n t. \tag{b}$$

Next, recall from Eq. (13.20) that the application of a constant force to an undamped spring–mass system gives as the equation of motion

$$x = \frac{F}{k} - \frac{F}{k} \cos \omega_n t. \tag{c}$$

By rearranging this, we obtain

$$x - \frac{F}{k} = -\frac{F}{k} \cos \omega_n t. \tag{d}$$

If we let $x_1 = F/k$, then Eq. (d) becomes

$$x - x_1 = -x_1 \cos \omega_n t. \tag{e}$$

But, if the origin of x is shifted a distance x_1, then Eq. (e) is the same as Eq. (b). Thus, shifting of the frame of a vibrating system through a distance x_1 is equivalent to adding a constant force $F = kx_1$ acting upon the mass. This problem is shown in Fig. 13.15, where the earlier origin is O and the new origin is O_1. These are separated, then, by the distance x_1. At the instant $t = 0$, the origin O is shifted to O_1, initiating the motion. A phasor, O_1A, equal in magnitude to x_1, begins its rotation from the posture O_1O. Rotating at ω_n rad/s, the projection of this phasor on

Fig. 13.14

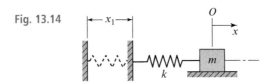

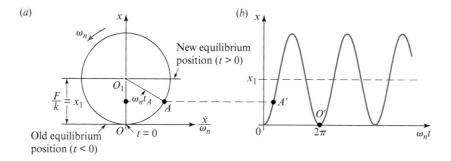

Fig. 13.15 (a) Phase-plane diagram; (b) displacement–time diagram.

the x axis describes the displacement of the mass. This phasor then continues to rotate until something else happens to the system. Arriving at point A, it has traversed an angle $\omega_n t_A$, and the corresponding point on the displacement–time diagram is A'.

The phase-plane method permits us to move the origin in any manner we choose and at any instant in time that we choose. In the example above, the origin was shifted a distance x_1, and we have seen that this is equivalent to the sudden application of a force $F = kx_1$ applied to the mass in the positive direction. Suppose that we permit the phasor to make one complete revolution of 360°. It then generates one complete cycle of displacement and returns to point O. If, at this instant, we shift the frame back to its original position, we might reasonably ask: What are the displacement and velocity at the instant of making this shift? The phase-plane diagram shows that both the displacement and velocity are zero at the end of a cycle. The displacement and velocity are exactly the quantities we require to determine the motion of the system in the next era. Since these are both zero, there is no motion after returning the frame to its original position. It will be recalled that these are exactly the results that we obtained in analyzing square-wave forcing functions when the force endured for an integral number of cycles. Thus, the phasor generates the displacement diagram through the angle $\omega_n t = 2\pi$ in this case; and at this point, we return the frame to its original position. The initial conditions are now $x = 0, \dot{x} = 0$; consequently, the mass stops its motion completely.

Let us now create a vibration by shifting the frame from $x = 0$ to $x = x_1$, waiting a period of time Δt, then shifting the frame back to $x = 0$. This constitutes a square-wave forcing function, or a step disturbance to the system, and the phase-plane and displacement diagrams for such a motion are shown in Fig. 13.16. The motion can be described in three eras. During the first, from t_0 to t_1, everything is at rest. At $t = t_1$, the frame suddenly moves to the right a distance $x = x_1$, compressing the spring. The duration of the second era is from t_1 to $t_2(\Delta t)$, and at time t_2, the frame suddenly returns to its original position. The third era represents time when $t \geq t_2$. The position of the frame during these three eras is shown in the displacement diagram. The displacement diagram also describes the motion of the mass, and the phase-plane diagram explains why it moves as it does. Between t_0 and t_1, the mass is at rest and nothing happens. At time t_1, the frame shifts from O to O_1 on the phase-plane diagram, which is the distance x_1. If we designate the original frame position as the origin of the $x, \dot{x}/\omega_n$ system, then the conditions at $t = t_1$ are $x = +x_1$ and $\dot{x} = 0$. However, the motion of the mass about O is equal to its motion about O_1 plus the distance from O to O_1. Therefore,

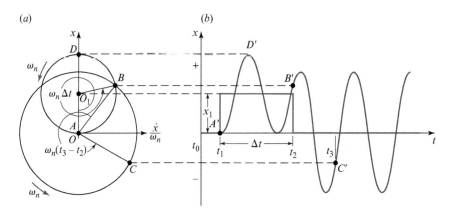

Fig. 13.16 A step disturbance.

$$x = -x_1 \cos \omega_n t' + x_1 = x_1(1 - \cos \omega_n t'). \qquad (f)$$

Also

$$\dot{x} = x_1 \omega_n \sin \omega_n t', \qquad (g)$$

where t' is understood to begin at time t_1. As shown in Fig. 13.16b, the mass vibrates about the position $x = x_1$ during this era. The vibration begins at point A on the phase-plane diagram and continues as a free vibration for the time Δt. During this period, the phasor O_1A rotates counterclockwise with angular velocity ω_n rad/s until at time t_2 it occupies the posture O_1B. At this instant the third era begins when the frame suddenly returns through the distance x_1 to its original position. Thus the starting conditions for the third era are the same as the ending conditions for the second and are

$$x_2 = x_1(1 - \cos \omega_n \Delta t) \quad \text{and} \quad \dot{x}_2 = x_1 \omega_n \sin \omega_n \Delta t. \qquad (h)$$

Substituting these conditions into Eq. (13.12) and rearranging gives the equation of motion for the third era as

$$x = x_1(1 - \cos \omega_n \Delta t) \cos \omega_n t'' + x_1 \sin \omega_n \Delta t \sin \omega_n t'', \qquad (i)$$

where t'' is the time measured from the start of the third era. Equation (i) can be transformed into an equivalent expression containing only a single trigonometric term and a phase angle as in Eq. (13.15), but we shall not do so here. The third era, as we have seen, begins at point B on the phase-plane diagram (B' on the displacement diagram); the motion of point B as it moves about a circle with center at O in the counterclockwise direction describes the motion of the mass. Thus, at instant t_3, the phasor OB has rotated through the angle $\omega_n(t_3 - t_2)$ and is located at C. The corresponding point on the displacement diagram is C''.

The extension of the phase-plane method to any number of steps, taken in either or both the positive or negative x direction, should now be apparent. In each case the starting conditions for the next era are taken equal to the ending conditions for the previous era. The equations of motion should be written for each era with time counted from the start of that era. You can now understand that if the third era of Fig. 13.16 had begun at point D on the phase-plane diagram, instead of at point B, then the resulting motion would have an amplitude twice as large as that in the second era.

13.7 Transient Disturbances

Any action that destroys the equilibrium of a vibratory system may be called a *disturbance* to that system. A *transient disturbance* is any action that endures for only a relatively short period of time. The analyses in the several preceding sections have dealt with transient disturbances having a stepwise relationship to time. Since all machine parts have elasticity and inertia, forces do not come into existence instantaneously in real life.[3] Consequently, we usually expect to encounter forcing functions that vary smoothly with time. Although the step forcing function is not true to nature, it is our purpose in this section to demonstrate how the step function is used with the phase-plane method to obtain a good approximation of the vibration of a system excited by a "natural" disturbance.

The procedure is to plot the disturbance as a function of time, to divide this into steps, and then to use the steps successively to make a phase-plane plot. The resulting displacement and velocity diagrams can then be obtained by graphically projecting points from the phase-plane diagram, as previously explained. It turns out that very accurate results can frequently be obtained using only a small number of steps. Of course, as in any graphic solution, better results are obtained when a larger number of steps are employed and when the work is plotted to a larger scale. It is difficult to set up general rules for selecting the size of the steps to be used. For slowly vibrating systems and for relatively smooth forcing functions, the step width can be quite large, but even a slowly varying system requires shorter steps if the forcing function has numerous sharp peaks and valleys, that is, if it has a great deal of frequency content. For smooth forcing functions and slowly vibrating systems, a step size such that the phasor sweeps out an angle of 180° is probably about the largest that one should use. It is a good idea to check the step size during the construction of the phase-plane diagram. Too great a step size causes a discontinuity in the slope of two curves at the point of adjacency of the two. If this occurs, then the step can immediately be decreased and the procedure resumed.

Figure 13.17 shows how to find the heights of the steps. The first step for the forcing function in Fig. 13.17 has been given a duration of Δt_1 and a height of h_1. This height is obtained by constructing a horizontal line across the force curve such that the areas of the two shaded triangles are equal. For a step size of Δt_1, the phasor sweeps out an angle of $\omega_n \Delta t_1$. Thus, the angular rotation of the phasor is obtained simply by multiplying the step size in seconds by the natural frequency of the system.

Figure 13.18 shows a four-step forcing function that we assume has been deduced from a natural function. The widths and heights of the steps have been plotted to scale and are Δt_1 and h_1 for the first step, Δt_2 and h_2 for the second step, and so on. Note that the fourth step has a

---

[3] If a force could truly be applied instantaneously, then, according to Newton, an acceleration would occur instantaneously. It would still take at least one instant to integrate this to become a change in velocity, and another instant to integrate into a change in position. Finally, according to Hooke's law, our elastic body would display a force.

Fig. 13.17 Finding the heights of the steps.

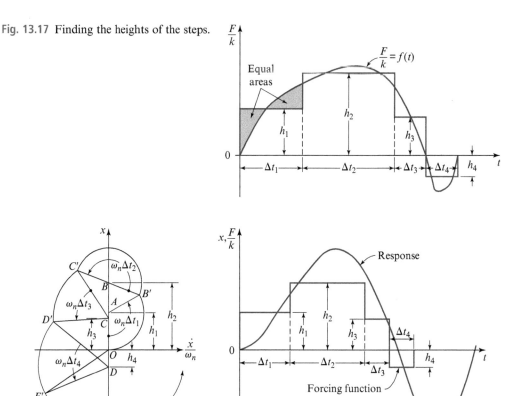

Fig. 13.18 Construction of the phase-plane and displacement diagrams.

negative h_4. The motion begins at $t = 0$ when a phasor of length h_1 starts rotating about A as a center from the initial posture AO. This phasor rotates through an angle $\omega_n \Delta t_1$ and generates the portion of the response curve contained in the interval Δt_1. At the end of this period of time the second step begins with the center of rotation of the phasor shifting from A to B. The length of the phasor for the duration of the second step is the distance BB'. This phasor then rotates through the angle $\omega_n \Delta t_2$ about a center at B until it arrives at the posture BC'. At this instant, the third step begins. The center of rotation shifts to C, and the phasor CC' rotates through the angle $\omega_n \Delta t_3$. At the end of the fourth step the phasor has arrived at E'. The center of rotation now shifts to the origin O, and the motion continues as a free vibration of amplitude OE' until something else (not shown) happens to the system.

Example 13.2

Measurements on a mechanical vibrating system show that the mass has a weight of 16.90 lb and that the springs combine to give an equivalent spring rate of 30 lb/in. A transient force resembling the first half-cycle of a sine wave operates on the system. The maximum value of the force is 10 lb, and it is applied for 0.120 s. Determine the response.

Example 13.2 (cont.)

Assumption

Since the system is observed to vibrate quite freely, damping is neglected.

Solution

The natural frequency of the system is

$$\omega_n = \sqrt{\frac{k}{m}} = \sqrt{\frac{(30 \text{ lb/in})}{(16.90 \text{ lb})/(386 \text{ in/s}^2)}} = 26.2 \text{ rad/s}.$$

The period and frequency are

$$\tau = \frac{2\pi}{\omega_n} = \frac{2\pi \text{ rad/cycle}}{26.2 \text{ rad/s}} = 0.240 \text{ s/cycle} \quad \text{and} \quad f = \frac{1}{\tau} = \frac{1}{0.240 \text{ s/cycle}} = 4.17 \text{ Hz}.$$

We first replace the entire forcing function by a single step disturbance. The height of the step should be the time average of the forcing function for the duration of the step. The average ordinate of a half-cycle of the sine wave is

$$h_1 = \frac{1}{\pi} \int_0^\pi \left[\frac{F_{\max}}{k} \sin bt \right] d(bt) = \frac{2F_{\max}}{\pi k} = 0.637 \frac{10 \text{ lb}}{30 \text{ lb/in}} = 0.212 \text{ in}. \tag{1}$$

This is the only value required to solve the problem. The graphic solution is shown in Fig. 13.19, together with the force and the step, all plotted to the same time scale. As shown, the amplitude of a vibration resulting from a single step is $X_0 = 2h_1 = 0.424$ in.

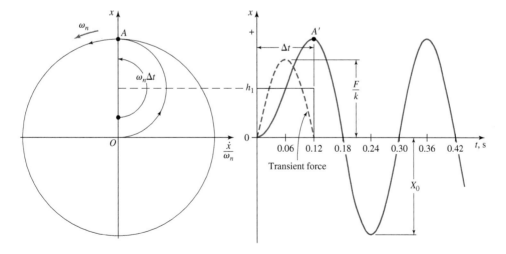

Fig. 13.19 Response determined with only a single step of force.

Example 13.2 (cont.)

To check the accuracy of a single step, we shall solve the example again using three force steps. In Fig. 13.20, the first step has a height of h_1 and is taken from O to B' with a radius AO; the second has a height h_2 and is from B' to C' with a radius BB'; and the third has a height h_3 and is from C' to D' with a radius CC'. The vibration is a free motion after point D'. These steps need not be of equal size, but it is convenient in this example to make them so. For the three steps of equal size, the ordinate of each part of the half-cycle of the sine wave of Eq. (1) can be written as

$$h_1 = h_3 = \frac{1}{\pi/3} \int_0^{\pi/3} \left[\frac{F_{max}}{k} \sin bt\right] d(bt) = \frac{3}{2\pi}\left(\frac{F_{max}}{k}\right) = \frac{3}{2\pi}\left(\frac{10\,\text{lb}}{30\,\text{lb/in}}\right) = 0.159\,\text{in},$$

$$h_2 = \frac{1}{\pi/3} \int_{\pi/3}^{2\pi/3} \left[\frac{F_{max}}{k} \sin bt\right] d(bt) = \frac{3}{\pi}\left(\frac{F_{max}}{k}\right) = \frac{3}{\pi}\left(\frac{10\,\text{lb}}{30\,\text{lb/in}}\right) = 0.318\,\text{in}.$$

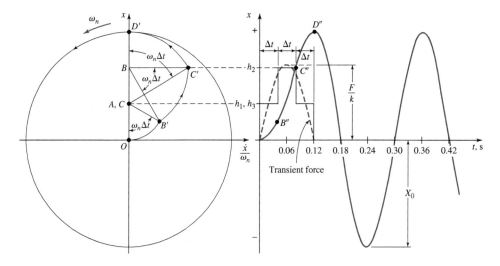

Fig. 13.20 Response determined in three force steps.

The construction and results are shown in Fig. 13.20. Note that the amplitude $(X_0 = h_1 + h_2 + h_3 = 0.636\,\text{in})$ is larger than that for a single step. However, it is doubtful whether the extra labor of a five-step solution is justified.

13.8 Free Vibration with Viscous Damping

Combining all the friction in a system and representing it as purely viscous damping is the method most widely used in the simulation of vibration in mechanical systems. This is primarily because it results in a linear differential equation that is convenient to solve, not necessarily because this truly represents the response of the real system. The differential equation for an unforced, viscously damped system was derived in Eq. (13.3), and, with no externally applied force, this is

$$m\ddot{x} + c\dot{x} + kx = 0.$$

Dividing by m gives

$$\ddot{x} + \frac{c}{m}\dot{x} + \frac{k}{m}x = 0. \tag{13.26}$$

Inspection of this equation indicates that the time variation of x and its derivatives must be alike if the relation is to be satisfied. The exponential function satisfies this requirement, so it is not unreasonable to guess that the solution might be of the form

$$x = Ae^{st}, \tag{a}$$

where A and s are values still to be determined. The first and second time derivatives of Eq. (a) are

$$\dot{x} = Ase^{st} \quad \text{and} \quad \ddot{x} = As^2e^{st}.$$

Substituting these into Eq. (13.26) and dividing out the common factors gives

$$s^2 + \frac{c}{m}s + \frac{k}{m} = 0. \tag{b}$$

Thus, Eq. (a) is a solution of Eq. (13.26), provided that the quantity s is selected to satisfy Eq. (b). The roots of Eq. (b) are

$$s = -\frac{c}{2m} \pm \sqrt{\left(\frac{c}{2m}\right)^2 - \frac{k}{m}}, \tag{c}$$

and so Eq. (a) can be written

$$x = Ae^{s_1 t} + Be^{s_2 t}, \tag{d}$$

where s_1 and s_2 are the two roots of Eq. (c), and A and B are the constants of integration.

The value of damping that makes the radical of Eq. (c) zero has special significance; we shall call it the *critical damping coefficient* and designate it by the symbol c_c. With critical damping the radical is zero, and so

$$\frac{c_c}{2m} = \sqrt{\frac{k}{m}} = \omega_n \quad \text{or} \quad c_c = 2m\omega_n. \tag{13.27}$$

It is also convenient to define a *damping ratio* ζ, which is the ratio of the actual to the critical damping. Thus,

$$\zeta = \frac{c}{c_c} = \frac{c}{2m\omega_n}. \tag{13.28}$$

After some algebraic manipulation, Eq. (*c*) can be written as

$$s = \left(-\zeta \pm \sqrt{\zeta^2 - 1}\right)\omega_n. \tag{e}$$

For the case in which the actual damping is larger than the critical, $\zeta > 1$ and the radical is real. This is analogous to movement of the mass in a very thick viscous fluid, such as molasses, and there is no vibration. Although this case does have application in certain mechanical systems, because of space limitations we shall not pursue it further here. It is not difficult to find that, if the mass is displaced and released in a system with more than critical damping, the mass returns slowly to its position of equilibrium without overshooting.

When the damping is less than critical, $\zeta < 1$ and the radical of Eq. (*e*) is imaginary. It is then written as

$$s = \left(-\zeta \pm j\sqrt{1 - \zeta^2}\right)\omega_n, \tag{f}$$

where we have used the notation $j = \sqrt{-1}$. Substituting these roots into Eq. (*d*) gives

$$x = e^{-\zeta\omega_n t}\left(Ae^{j\sqrt{1-\zeta^2}\omega_n t} + Be^{-j\sqrt{1-\zeta^2}\omega_n t}\right). \tag{g}$$

Since x must be real, A and B are complex conjugates, and this equation can be written as

$$x = e^{-\zeta\omega_n t}[(A + B)\cos\omega_d t + j(A - B)\sin\omega_d t], \tag{h}$$

where ω_d is the natural frequency of the damped vibration, and

$$\omega_d = \sqrt{1 - \zeta^2}\,\omega_n. \tag{13.29}$$

It is expedient to transform Eq. (*h*) into a form having only a single trigonometric term. Making this transformation gives

$$x = X_0 e^{-\zeta\omega_n t}\cos(\omega_d t - \phi), \tag{13.30}$$

where X_0 and ϕ are the new constants of integration. It is clear that Eq. (13.30) reduces to Eq. (13.15) if the damping is zero. The constants of integration can also be found in exactly the same manner.

Equation (13.30) is the product of a trigonometric function and a decreasing exponential function. The resulting motion is therefore oscillatory with an exponentially decreasing amplitude, as shown in Fig. 13.21. The frequency does *not* depend upon amplitude and is less than that of an undamped system, as is indicated by the factor $\sqrt{1 - \zeta^2}$ in Eq. (13.29).

13.9 Damping Obtained by Experiment

The classic method of obtaining the damping coefficient is by an experiment in which the system is disturbed in some manner – say, by striking it with a hammer – and then recording the

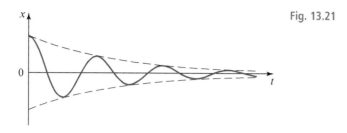

Fig. 13.21

decaying response by means of a strain gauge, rotary potentiometer, solar cell, or other transducer. If the friction is primarily viscous, the result resembles Fig. 13.21. The rate of decay can then be measured and, by means of the analysis to follow, the viscous damping coefficient can be calculated.

The recording also provides a qualitative indication of the predominating friction. In many mechanical systems, recordings reveal that the first portion of the decay is curved as in Fig. 13.21, but the later part decays at a linear rate instead. A linear rate of decay signifies Coulomb (sliding) friction. Still another type of decay sometimes found is curved at the beginning but then flattens out for small amplitudes and requires a much greater time to decay to zero. This kind of damping is proportional to the square of the velocity. Most mechanical systems have several kinds of friction present, and the investigator is usually interested only in the predominant kind. For this reason he or she should analyze that portion of the decay record in which the amplitudes are closest to those actually experienced in normal operation of the system.

Perhaps the best method of obtaining an average damping coefficient is to use a number of cycles of decay, if they can be obtained, rather than a single cycle. If we take any response curve, such as that of Fig. 13.21, and measure the amplitude of the nth and also of the $(n + N)$th cycle, then these measurements are taken when the cosine term of Eq. (13.30) is approximately unity; so

$$x_n = X_0 e^{-\zeta \omega_n t_n} \quad \text{and} \quad x_{n+N} = X_0 e^{-\zeta \omega_n (t_n + N\tau)},$$

where τ is the period of vibration and N is the number of cycles of motion between the amplitude measurements. The *logarithmic decrement*, δ_N, is defined as the natural logarithm of the ratio of these two amplitudes and is

$$\delta_N = \ln \frac{x_n}{x_{n+N}} = \ln \frac{X_0 e^{-\zeta \omega_n t_n}}{X_0 e^{-\zeta \omega_n (t_n + N\tau)}} = \ln e^{\zeta \omega_n N\tau} = \zeta \omega_n N\tau. \tag{13.31}$$

The period of the vibration is

$$\tau = \frac{2\pi}{\omega_d} = \frac{2\pi}{\omega_n \sqrt{1 - \zeta^2}}. \tag{13.32}$$

Therefore, the logarithmic decrement can be written in the form

$$\delta = \frac{\delta_N}{N} = \frac{2\pi\zeta}{\sqrt{1 - \zeta^2}}. \tag{13.33}$$

Measurements of many damping ratios indicate that a value of $\zeta < 0.20$ can be expected for most machine systems, with a value of $\zeta = 0.10$ or less being the most probable. For this range of values, the radical in Eq. (13.33) can be taken as approximately unity, giving

$$\delta \approx 2\pi\zeta \tag{13.34}$$

as an approximate formula.

Example 13.3

Let the vibrating system of Example 13.2 have a dashpot attached that exerts a force of 0.25 lb on the mass when the mass has a velocity of 1 in/s. Find the critical damping constant, the logarithmic decrement, and the ratio of two consecutive maxima.

Solution

The data from Example 13.2 are repeated here for convenience:

$$W = 16.90 \text{ lb}, \quad k = 30 \text{ lb/in}, \quad \omega_n = 26.2 \text{ rad/s}.$$

The critical damping constant is

$$c_c = 2m\omega_n = 2\left(\frac{16.90 \text{ lb}}{386 \text{ in/s}^2}\right)(26.2 \text{ rad/s}) = 2.29 \text{ lb}\cdot\text{s/in}. \qquad Ans.$$

The viscous damping constant is $c = F/\dot{x} = 0.25 \text{ lb}\cdot\text{s/in}$. Therefore, the damping ratio is

$$\zeta = \frac{c}{c_c} = \frac{0.25 \text{ lb}\cdot\text{s/in}}{2.29 \text{ lb}\cdot\text{s/in}} = 0.109.$$

The logarithmic decrement is obtained from Eq. (13.33):

$$\delta = \frac{2\pi\zeta}{\sqrt{1-\zeta^2}} = \frac{2\pi(0.109)}{\sqrt{1-(0.109)^2}} = 0.689. \qquad Ans.$$

The ratio of two consecutive maxima is

$$\frac{x_{n+1}}{x_n} = e^{-\delta} = e^{-0.689} - 0.502. \qquad Ans.$$

Therefore, for this amount of damping, each peak amplitude is approximately 50% of that of the previous cycle.

13.10 Phase-Plane Representation of Damped Vibration

We have seen that when undamped mechanical systems are subjected to transient forces, the resulting motion (mathematically, at least) endures forever without decreasing in amplitude.

We know, however, that energy losses always exist, although sometimes only in minute amounts, and that these losses eventually cause vibration to cease. In the case of vibrating systems known to be acted upon by transient forces, it is often desirable to introduce additional damping as one means of decreasing the number of cycles of vibration. For this reason, the phase-plane solution of damped vibration is particularly important.

We have seen that the phase-plane diagram of an undamped free vibration is generated by a phasor of constant length rotating at a constant angular velocity. In the case of a damped free vibration, the phase-plane diagram is generated by a phasor whose length is decreasing exponentially with time. Thus, the phase-plane trajectory for such a motion is a spiral instead of a circle.

It turns out that a simple spiral will give incorrect values for the velocity if plotted on the same $x, \dot{x}/\omega_n$ axes as used for undamped free vibration. The reason for this is that the phase relationship between the velocity and displacement is not 90° as it is for an undamped system. Instead, for a damped system, the phase angle depends upon the amount of damping present.

An easy way to obtain a phase-plane diagram for damped motion is to tilt or rotate the $\dot{x}/\omega_n$ axis through angle σ, as shown in Fig. 13.22. The angle σ is called the *damping phase angle*, and the following analysis demonstrates how this angle comes about.

Taking the time derivative of Eq. (13.30) gives

$$\dot{x} = -X_0\zeta\omega_n e^{-\zeta\omega_n t}\cos\left(\sqrt{1-\zeta^2}\,\omega_n t - \phi\right) - X_0\sqrt{1-\zeta^2}\,\omega_n e^{-\zeta\omega_n t}\sin\left(\sqrt{1-\zeta^2}\,\omega_n t - \phi\right)$$

or

$$\frac{\dot{x}}{\omega_n} = -X_0 e^{-\zeta\omega_n t}\left[\zeta\cos\left(\sqrt{1-\zeta^2}\,\omega_n t - \phi\right) + \sqrt{1-\zeta^2}\sin\left(\sqrt{1-\zeta^2}\,\omega_n t - \phi\right)\right]. \qquad (a)$$

Designating $\sin\sigma = \zeta$ and $\cos\sigma = \sqrt{1-\zeta^2}$ and noting the trigonometric identity

$$\sin(\alpha + \beta) = \sin\alpha\cos\beta + \cos\alpha\sin\beta,$$

we see that Eq. (*a*) can be written as

$$\frac{\dot{x}}{\omega_n} = -X_0 e^{-\zeta\omega_n t}\sin\left(\sqrt{1-\zeta^2}\,\omega_n t - \phi + \sigma\right). \qquad (13.35)$$

As shown in Fig. 13.22, the $\dot{x}/\omega_n$ axis is rotated counterclockwise through the angle σ when plotting the phase-plane diagram. The velocity diagram is then obtained by projecting

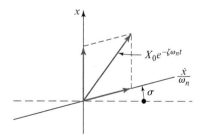

Fig. 13.22 Damping phase angle σ is applied only to the velocity.

Fig. 13.23 Spiral template for
$\zeta = 0.15$.

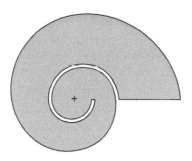

perpendicular to the $\dot{x}/\omega_n$ axis. If the damping is zero, note that $\sqrt{1 - \zeta^2} = 1$, $e^{-\zeta\omega_n t} = 1$, and $\sigma = 0$, and Eqs. (13.30) and (13.35) reduce to the same set of equations that we employed to develop the phase-plane analysis of undamped motion.

In analyzing a damped system that is acted upon by transient forces by the phase-plane method, it is very helpful to construct transparent spiral templates, as shown in Fig. 13.23. These are made from a sheet of clear plastic that is placed over a drawing of the spiral so that the spiral and its center can be scribed on the plastic using the point of a pair of dividers. The scribed curve can then be trimmed with a pair of manicurist's scissors or a sharp razor blade. Since only one template is needed for each damping ratio, these can be stored for use in future problems. Do not forget to label each template with the value of the damping ratio for which it is constructed.

The spirals can be plotted directly from a table calculated using Eq. (13.30), but this is tedious. A more rapid method, which is accurate enough for graphic purposes, is to approximate each quadrant of the spiral by a circular arc. To do this it is necessary only to calculate the change in the length of the phasor in a quarter of a rotation. From Eq. (13.30), the length of the phasor is $x = X_0 c^{-\zeta\omega_n t}$ and its angular velocity is $\omega_n \sqrt{1 - \zeta^2}$. Therefore, for a 90° rotation,

$$\omega_n \sqrt{1 - \zeta^2}\, t = \frac{\pi}{2} \quad \text{or} \quad \omega_n t = \frac{\pi}{2\sqrt{1 - \zeta^2}}. \tag{b}$$

Thus, the length, after rotation through 90°, is

$$x_{90°} = X_0 e^{-\zeta\pi/2\sqrt{1-\zeta^2}}. \tag{c}$$

To demonstrate an example of the construction of the spirals, let us employ a damping ratio $\zeta = 0.15$ and begin with a phasor 2.50 in long. Then, after 90° of rotation, the length, from Eq. (c), is

$$x_{90°} = (2.50 \text{ in})e^{-0.15\pi/2\sqrt{1-(0.15)^2}} = 1.98 \text{ in.}$$

The construction is shown in Fig. 13.24 and is explained as follows: Construct the axes 1 and 2 at right angles to each other and lay off 2.50 in to A and 1.98 in to B on axes 1 and 2, respectively. Draw two lines through the origin at 45° to the 1,2 axes. The perpendicular bisector of AB intersects one of these 45° lines at P_1; using P_1 as a center, strike an arc from A to B. Line $P_1 B$ crosses the other 45° line defining the center P_2 of arc BC. This process is continued in the same fashion with the point P_3 defining the center of the arc CD, and so on.

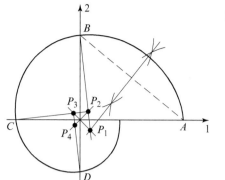

Fig. 13.24

The template is used by placing its center at the same point in the phase-plane diagram from which a circular arc would be constructed for undamped motion. The template is then turned until the spiral coincides with the point from which the spiral diagram is to be started.

Example 13.4

A vibrating system has an 80 lb mass mounted upon springs with an equivalent spring rate of 360 lb/in. A dashpot is included in the system and is estimated to produce 15% of critical damping. A transient force on the system is assumed to act in three steps as follows: 720 lb for 0.0508 s, −270 lb for 0.0635 s, and 180 lb for 0.0381 s. The negative sign on the second force step simply means that its direction is opposite to the direction of the first and third steps. Determine the response of the system assuming no motion when the force function begins to act.

Solution

The undamped natural frequency is

$$\omega_n = \sqrt{\frac{k}{m}} = \sqrt{\frac{(360\ \text{lb/in})}{(80\ \text{lb})/(386\ \text{in/s}^2)}} = 41.7\ \text{rad/s},$$

and so the damped natural frequency is

$$\omega_d = \omega_n\sqrt{1 - \zeta^2} = (41.7\ \text{rad/s})\sqrt{1 - (0.15)^2} = 41.2\ \text{rad/s}.$$

The period of the motion is $\tau = 2\pi/\omega_d = (2\pi\ \text{rad/cycle})/(41.2\ \text{rad/s}) = 0.152\ \text{s/cycle}$. The F/k values are

$$\frac{F_1}{k} = \frac{720\ \text{lb}}{360\ \text{lb/in}} = 2.00\ \text{in}, \quad \frac{F_2}{k} = \frac{-270\ \text{lb}}{360\ \text{lb/in}} = -0.75\ \text{in}, \quad \frac{F_3}{k} = \frac{180\ \text{lb}}{360\ \text{lb/in}} = 0.50\ \text{in}.$$

These three steps are plotted to scale in Fig. 13.25. The angular duration of each step for the phase-plane diagram is obtained by multiplying the damped natural frequency by the time of each step and converting to degrees. This gives

Example 13.4 (cont.)

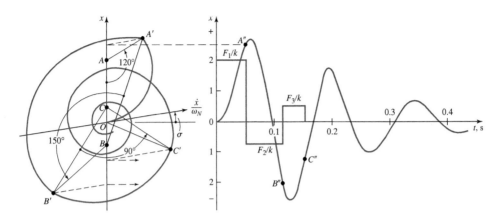

Fig. 13.25 Phase-plane diagram.

Step 1: $\omega_d \Delta t_1 \dfrac{180°}{\pi \text{ rad}} = (41.2 \text{ rad/s})(0.0508 \text{ s}) \dfrac{180°}{\pi \text{ rad}} = 120°$

Step 2: $\omega_d \Delta t_2 \dfrac{180°}{\pi \text{ rad}} = (41.2 \text{ rad/s})(0.0635 \text{ s}) \dfrac{180°}{\pi \text{ rad}} = 150°$

Step 3: $\omega_d \Delta t_3 \dfrac{180°}{\pi \text{ rad}} = (41.2 \text{ rad/s})(0.0381 \text{ s}) \dfrac{180°}{\pi \text{ rad}} = 90°.$

The $\dot{x}/\omega_n$ axis must be rotated $\sigma = \sin^{-1}(0.15) = 8.63°$ counterclockwise from the horizontal direction, as indicated in the phase-plane diagram.

The construction of Fig. 13.25 is explained as follows: We project F_1/k to A on the x axis. A phasor AO, with center at A, begins rotating at the instant $t = 0$. This phasor rotates through an angle of $120°$ to AA' while exponentially decreasing in length. At A', the step function changes the origin of the phasor to B, and it rotates $150°$ about this point to B', again exponentially decreasing in length. At B' the origin again shifts, this time to C, and the phasor rotates $90°$ to C', again exponentially decreasing. Now the origin shifts to O, and the phasor continues its rotation and its decreasing magnitude with O as the origin until the motion becomes too small to follow. Note that points on the spiral diagram are projected parallel to the $\dot{x}/\omega_n$ axis to the x axis, and then horizontally to the displacement diagram. The displacement diagram shows the resulting response. The maximum amplitude occurs on the first positive half-cycle and is 2.72 in. In a similar manner, the velocity diagram, although not shown, would be projected perpendicular to the $\dot{x}/\omega_n$ axis. Points on the spiral would project parallel to the x axis until they intersect the $\dot{x}/\omega_n$ axis.

13.11 Response to Periodic Forcing

Differential equations of the form

$$m\ddot{x} + kx = ky(\omega t), \tag{13.36}$$

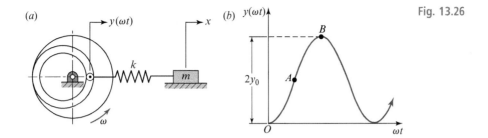

Fig. 13.26

where $y(\omega t)$ is a periodic forcing function, occur frequently in mechanical systems. Let us define $y(\omega t + n\pi) = y(\omega t)$, where n is an integer; we shall then be concerned in this section only with small values of n, say, n less than 8. An example of Eq. (13.36) is shown in Fig. 13.26 in which a roller riding in the groove of a face cam generates the motion from point O (Fig. 13.26b) of

$$y(\omega t) = y_0(1 - \cos \omega t). \tag{13.37}$$

This, then, is the motion applied to the left end of a spring, k, connected to a sliding mass, m, with friction assumed to be zero. Note, in Fig. 13.26b, that we choose Eq. (13.37) to represent the driving function and that we have assumed that the roller is in its extreme left position when time is zero and cam rotation begins.

Alternatively, we could have chosen point A in Fig. 13.26b to correspond to the start of the motion. For this choice the driving function would have been

$$y_A(\omega t) = y_0 \sin \omega t.$$

One of the reasons for introducing this problem is to show an interesting relationship: Fig. 13.26a represents a single-degree-of-freedom vibrating system, but when Eq. (13.36) or a similar equation is solved, the solution often looks more like that of a two-degree-of-freedom system. Novices are sometimes quite surprised when they first view this solution and they are unable to explain what is happening.

Substituting Eq. (13.37) into Eq. (13.36) gives the equation to be solved as

$$m\ddot{x} + kx = ky_0(1 - \cos \omega t). \tag{13.38}$$

The complementary solution is

$$x' = A \cos \omega_n t + B \sin \omega_n t,$$

and the particular solution is

$$x'' = C + D \cos \omega t. \tag{a}$$

The second derivative of this equation is

$$\ddot{x}'' = -D\omega^2 \cos \omega t. \tag{b}$$

By substituting Eqs. (*a*) and (*b*) into Eq. (13.38), we find that $C = y_0$ and

$$D = -\frac{ky_0}{k - m\omega^2} - \frac{y_0}{1 - (\omega/\omega_n)^2}.$$

Therefore, the complete solution is

$$x = A \cos \omega_n t + B \sin \omega_n t + y_0 \left[1 - \frac{\cos \omega t}{1 - (\omega/\omega_n)^2} \right]. \tag{13.39}$$

The velocity is

$$\dot{x} = -A\omega_n \sin \omega_n t + B\omega_n \cos \omega_n t + \frac{y_0 \omega \sin \omega t}{1 - (\omega/\omega_n)^2}. \tag{13.40}$$

Since we are starting from rest, where $x(0) = \dot{x}(0) = 0$, Eqs. (13.39) and (13.40) give

$$A = \frac{(\omega/\omega_n)^2 y_0}{1 - (\omega/\omega_n)^2} \quad \text{and} \quad B = 0.$$

Equation (13.39) then becomes

$$x = \frac{(\omega/\omega_n)^2 y_0 \cos \omega_n t}{1 - (\omega/\omega_n)^2} - \frac{y_0 \cos \omega t}{1 - (\omega/\omega_n)^2} + y_0. \tag{13.41}$$

The first term on the right-hand side of Eq. (13.41) is called the *starting transient*. Note that this is a vibration at the natural frequency, ω_n, not at the forcing frequency, ω. The usual physical system contains a certain amount of friction, which, as we shall see in the sections to follow, causes this term to die out after a period of time. The second and third terms on the right represent the *steady-state* solution and these denote another component of the vibration at the forcing frequency, ω. Thus, the total motion is composed of a constant plus two vibrations of differing amplitudes and frequencies. If Fig. 13.26*a* were a cam-and-follower system, spring k would probably be quite stiff; consequently, ω_n would be quite large compared with ω. This would cause the amplitude of the starting transient to be comparatively small, and a phase-plane plot would resemble Fig. 13.27. Here we can see that the motion is formed by the sum of a fixed phasor of length y_0 and two phasors of different lengths and different rotational velocities.

A computer solution for the special case $\omega/\omega_n = 1/3$ is shown in Fig. 13.28. Figure 13.28*a* is a plot of the forcing function, and Fig. 13.28*b* is the response given by Eq. (13.41). This case is of interest since there are short dwells at the extremes of the motion. One can verify the existence of these dwells by constructing a plot similar to Fig. 13.27 or by substituting $\omega/\omega_n = 1/3$ into Eq. (13.41) and investigating the solution in the region for which ωt is near zero or π.

Examination of Eq. (13.41) for $\omega_n \to \infty$ indicates that $\omega/\omega_n = 0$, and the solution becomes

$$x = y_0(1 - \cos \omega t),$$

which is the rigid-body solution. If we replace the spring with a rigid member, then k and, therefore, ω_n, become very large, and so the mass exactly follows the cam motion.

Fig. 13.27

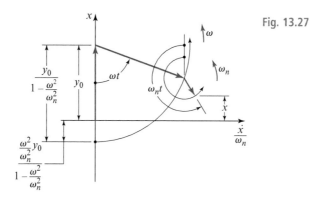

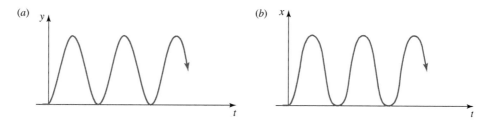

Fig. 13.28 (*a*) Forcing function at frequency ω and (*b*) response. The amplitude scales for both figures are the same.

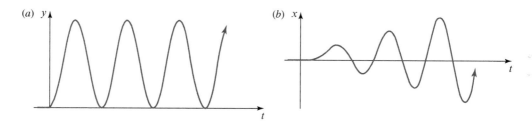

Fig. 13.29 (*a*) Forcing function and (*b*) response at resonance; the amplitude scale was increased so the plot stays within reasonable bounds.

Resonance is defined by the case in which $\omega = \omega_n$. Rearranging Eq. (13.41) to

$$x = y_0 \frac{(\omega/\omega_n)^2 \cos \omega_n t - \cos \omega t + 1 - (\omega/\omega_n)^2}{1 - (\omega/\omega_n)^2},$$

we see that $x = 0/0$ for $\omega/\omega_n = 1$. However, l'Hôpital's rule enables us to find the solution; thus,

$$\lim_{\omega \to \omega_n} (x) = y_0 \left(1 - \cos \omega_n t + \frac{\omega_n t}{2} \sin \omega_n t\right).$$

A plot of this equation, obtained by numerically solving the differential equation, is shown in Fig. 13.29. Note that the sine term grows and very quickly dominates the response.

13.12 Harmonic Forcing

We have seen that the action of any temporary forcing function on a damped system is to create a vibration, but the vibration decays when the applied force is removed. A machine element that is connected to or is a part of any moving machinery is often subject to forces that vary periodically with time. Since all metal machine parts have both mass and elasticity, the opportunity for vibration always exists. Many machines do operate at fairly constant speeds and uniform output, and it is not difficult to see that vibratory forces may exist that have a fairly constant amplitude over a period of time. Of course, these varying forces do change in magnitude when the machine speed or output changes, but there is a rather broad class of vibration problems that can be analyzed and corrected using the assumption of a periodically varying force of constant amplitude. Sometimes these forces exhibit a time characteristic that is very similar to that of a sine wave. At other times, they are quite complex and have to be analyzed as a Fourier series. Such a series is the sum of a number of sine and cosine terms, and the resultant motion is the sum of the responses to the individual terms. In this text, we shall study only the motion resulting from the application of a single sinusoidal force.

In Section 13.11, we examined an undamped system subjected to a periodic force and discovered that the solution contains components at two frequencies. One of these components exhibits the forcing frequency, whereas the other occurs at the natural frequency. Real systems always have damping present, and this causes the component at the (damped) natural frequency to become insignificant after a period of time; the motion that remains contains only the forcing frequency and is termed the *steady-state motion*.

To illustrate steady-state motion, we shall solve the equation

$$m\ddot{x} + c\dot{x} + kx = F_0 \cos \omega t. \tag{13.42}$$

The solution contains two parts, as before. The first, the transient or complementary, part is the solution to the homogeneous equation

$$m\ddot{x} + c\dot{x} + kx = 0. \tag{13.43}$$

This we solved in Section 13.8, and we obtained Eq. (*h*),

$$x' = e^{-\zeta \omega_n t}(C_1 \cos \omega_d t + C_2 \sin \omega_d t), \tag{13.44}$$

where C_1 and C_2 are the constants of integration. Since the forcing is harmonic, the particular solution is obtained by assuming a solution of the form

$$x'' = A \cos \omega t + B \sin \omega t, \tag{13.45}$$

where ω is the frequency of the forcing function, and where A and B are constants yet to be determined, but are not the constants of integration. Taking successive time derivatives of Eq. (13.45) produces

$$\dot{x}'' = -A\omega \sin \omega t + B \cos \omega t, \tag{a}$$

$$\ddot{x}'' = -A\omega^2 \cos \omega t - B\omega^2 \sin \omega t, \tag{b}$$

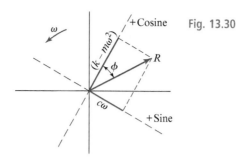

Fig. 13.30

and substituting Eqs. (13.45), (a), and (b) into the differential equation, Eq. (13.42), yields

$$\left(kA + c\omega B - m\omega^2 A\right)\cos \omega t + \left(kB + c\omega A - m\omega^2 B\right)\sin \omega t = F_0 \cos \omega t.$$

Next, separating the terms, we obtain

$$\left(k - m\omega^2\right)A + c\omega B = F_0,$$
$$-c\omega A + \left(k - m\omega^2\right)B = 0.$$

(c)

Solving Eqs. (c) for A and B and substituting the results into Eq. (13.45), we obtain

$$x'' = \frac{F_0}{\left(k - m\omega^2\right)^2 + c^2\omega^2}\left[\left(k - m\omega^2\right)\cos \omega t + c\omega \sin \omega t\right].$$

(d)

Now, when we create the phase diagram (Fig. 13.30) and plot the amplitudes of the two trigonometric terms, we find the resultant is

$$R = \sqrt{\left(k - m\omega^2\right)^2 + c^2\omega^2},$$

(e)

lagging the direction of the positive cosine by a phase angle of

$$\phi = \tan^{-1}\left(\frac{c\omega}{k - m\omega^2}\right).$$

(13.46)

Substituting Eq. (e) into Eq. (d) and dividing out common factors gives

$$x'' = \frac{F_0 \cos(\omega t - \phi)}{\sqrt{\left(k - m\omega^2\right)^2 + c^2\omega^2}},$$

(13.47)

and so the complete solution is

$$x = x' + x'' = e^{-\zeta\omega_n t}\left(C_1\cos \omega_d t + C_2 \sin \omega_d t\right) + \frac{F_0 \cos(\omega t - \phi)}{\sqrt{\left(k - m\omega^2\right)^2 + c^2\omega^2}}.$$

(13.48)

The first two terms of Eq. (13.48) enclosed in parentheses are called *transient* terms because the exponential factor causes those terms to decay in a short period of time. When this happens, only the final term remains, and, since this has no end, it is called the *steady-state solution*.

When we are interested only in the steady-state solution, we write the solution in the form of Eq. (13.47), omitting the prime marks on x.

Now we write Eq. (13.47) in the form

$$x = X \cos(\omega t - \phi), \tag{13.49}$$

and find the successive derivatives to be

$$\dot{x} = -\omega X \sin(\omega t - \phi), \tag{f}$$

$$\ddot{x} = -\omega^2 X \cos(\omega t - \phi). \tag{g}$$

These are then substituted into Eq. (13.42), and the forcing term is brought to the left-hand side of the equation. This gives

$$-m\omega^2 X \cos(\omega t - \phi) - c\omega X \sin(\omega t - \phi) + kX \cos(\omega t - \phi) - F_0 \cos \omega t = 0. \tag{h}$$

The amplitudes of the trigonometric terms are identified as:

$m\omega^2 X$ = inertia force;
$c\omega X$ = damping force;
kX = spring force; and
F_0 = excitation force.

Our next step is to plot these phasors on the phase diagram. We are interested only in the steady-state solution; it has no end, and so, although the phasors have a definite phase relation with each other, they can exist anywhere on the diagram at the instant we examine them.

We begin Fig. 13.31a by selecting one of the directions, say, the direction of $\cos \omega t$. The direction of $\cos(\omega t - \phi)$ lags that of $\cos \omega t$ by the angle ϕ. These automatically define the direction of $\sin(\omega t - \phi)$, lagging $\cos(\omega t - \phi)$ by 90°. Having defined this set of directions, we lay off each phasor in the proper direction, being sure not to forget the signs.

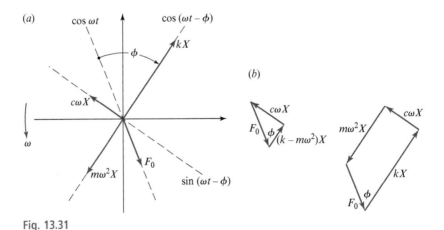

Fig. 13.31

Since Eq. (*h*) states that the sum of these four phasors is zero, we can construct a phasor or vector polygon of them. This has been done in Fig. 13.31*b*; from the figure, we see

$$\left[(k - m\omega^2)^2 + c^2\omega^2\right]X^2 = F_0^2$$

or

$$X = \frac{F_0}{\sqrt{(k - m\omega^2)^2 + c^2\omega^2}} \qquad (i)$$

and

$$\phi = \tan^{-1}\left(\frac{c\omega}{k - m\omega^2}\right). \qquad (13.50)$$

Consequently, Eq. (13.49) becomes

$$x = \frac{F_0 \cos(\omega t - \phi)}{\sqrt{(k - m\omega^2)^2 + c^2\omega^2}}. \qquad (13.51)$$

These equations can be simplified by introducing the symbols

$$\omega_n = \sqrt{\frac{k}{m}}, \quad \zeta = \frac{c}{c_c}, \quad c_c = 2m\omega_n.$$

They can then be written in the form

$$\frac{X}{F_0/k} = \frac{1}{\sqrt{\left(1 - \omega^2/\omega_n^2\right)^2 + \left(2\zeta\omega/\omega_n\right)^2}} \qquad (13.52)$$

and

$$\phi = \tan^{-1}\left(\frac{2\zeta\omega/\omega_n}{1 - \omega^2/\omega_n^2}\right), \qquad (13.53)$$

which is a dimensionless form that is convenient for analysis. It is interesting to note that the quantity F_0/k is the deflection that a spring of rate k would experience if acted upon by the force F_0.

If various values are selected for the frequency and damping ratio, curves can be plotted showing the effects upon the motion of varying these quantities. This has been done in Figs. 13.32 and 13.33. Note that the amplitude is very large for small damping ratios near resonance (that is, where ω/ω_n is near unity). Note, too, that the peak amplitudes occur slightly before resonance is reached. Figure 13.33 also indicates that the phase angle increases very rapidly near resonance.

13.13 Forcing Caused by Unbalance

Quite frequently (Chapter 15) the exciting force in a machine is due to the rapid rotation of an unbalanced mass. In this case, the differential equation is written

Fig. 13.32 Relative displacement of a damped forced system as a function of frequency for various damping ratios.

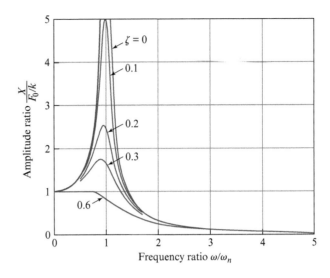

Fig. 13.33 Relationship of the phase angle to frequency for different damping ratios.

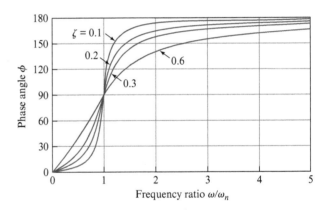

$$m\ddot{x} + c\dot{x} + kx = m_u e\omega^2 \cos\omega t, \tag{13.54}$$

where m = vibrating mass; m_u = unbalanced mass; and e = eccentricity. The solution to this equation is the same as that for Eq. (13.51) with F_0 replaced by $m_u e\omega^2$. Thus, the motion of the vibrating mass is

$$x = \frac{m_u e\omega^2 \cos(\omega t - \phi)}{\sqrt{(k - m\omega^2)^2 + c^2\omega^2}}. \tag{13.55}$$

If we make the same transformation we used to obtain Eq. (13.52), we get

$$\frac{mX}{m_u e} = \frac{\omega^2/\omega_n^2}{\sqrt{\left[1 - (\omega^2/\omega_n^2)\right]^2 + (2\zeta\omega/\omega_n)^2}}. \tag{13.56}$$

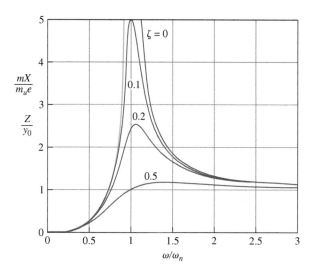

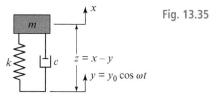

Fig. 13.35

This is plotted in Fig. 13.34. In one case, the main mass is excited by a rotating unbalanced mass m_u. The plot gives the nondimensional absolute response of the mass, m. In another case, as shown in Fig. 13.35, the mass is excited by moving the far end of a spring and dashpot, which is connected to the mass, with harmonic motion. The plot then gives the nondimensional response Z/y_0 relative to the end of the spring–dashpot combination. This plot shows that for high speeds where the frequency ratio is greater than unity, we can reduce the amplitude only by reducing the mass and eccentricity of the rotating unbalance.

13.14 Relative Motion

In many cases – in vibration-measuring instruments, for example – it is useful to know the response z of a system relative to another moving system y. Thus, in Fig. 13.35, we might wish to learn the relative response, $z = x - y$, instead of the absolute response, x.

The differential equation for this system is

$$m\ddot{x} + c\dot{x} + kx = ky + c\dot{y}. \tag{13.57}$$

Note that since $\dot{z} = \dot{x} - \dot{y}$ and $\ddot{z} = \ddot{x} - \ddot{y}$, then, written in terms of z and y, Eq. (13.57) becomes

$$m\ddot{z} + c\dot{z} + kz = -m\ddot{y}$$

or

$$m\ddot{z} + c\dot{z} + kz = m\omega^2 y_0 \cos \omega t. \qquad (13.58)$$

So the solution is the same as that of Eq. (13.54), and we can write

$$\frac{Z}{y_0} = \frac{\omega^2/\omega_n^2}{\sqrt{\left[1 - (\omega^2/\omega_n^2)\right]^2 + (2\zeta\omega/\omega_n)^2}}. \qquad (13.59)$$

The response chart of Fig. 13.34 also applies to the relative response of such a system.

13.15 Isolation

The investigations so far enable the engineer to design mechanical equipment to minimize vibration and other dynamic problems, but since machines are inherently vibration generators, in many cases, it is impractical to eliminate all such motion. In such cases, a more practical and economic vibration approach is that of reducing as much as possible the difficulties that the vibratory motion causes. This may take any of several directions, depending upon the nature of the problem. For example, everything economically feasible may have been undertaken to eliminate vibrations in a machine, but the residuals may still be strong and may cause objectionable noise by transmitting these vibrations to the base structure. In another case, a piece of equipment, such as a computer monitor, which is not of itself a vibration generator, may receive objectionable vibrations from another source. Both of these problems can be solved by isolating the equipment from its support. In analyzing these problems, we may be interested, therefore, in the *isolation of forces* or in the *isolation of motions*.

In Fig. 13.36a, the exciting-force amplitude is constant; however, in Fig. 13.36c it depends upon ω since it is due to a rotating unbalance. In both parts (a) and (c), the support is to be isolated from the vibrating mass, but in part (b) the support is vibrating. In parts (a) and (c) the forces acting against the support must be transmitted through the spring and dashpot, since

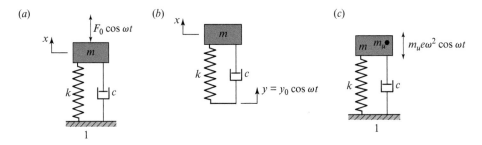

Fig. 13.36 (a) Exciting-force amplitude is constant; (b) support is vibrating; (c) exciting force depends upon ω.

these are the only connections. The spring and damping forces are phasors that are always at right angles to each other, and so the force transmitted through these connections is

$$F_{tr} = \sqrt{(kX^2) + (c\omega X)^2} = kX\sqrt{1 + \left(\frac{2\zeta\omega}{\omega_n}\right)^2}.$$ (13.60)

The *transmissibility*, T, is a nondimensional ratio that defines the percentage of the exciting force transmitted to the frame. For example, for the system of Fig. 13.36a, X is given by Eq. (13.52), and so the transmissibility is

$$T = \frac{F_{tr}}{F_0} = \frac{\sqrt{1 + (2\zeta\omega/\omega_n)^2}}{\sqrt{\left[1 - (\omega/\omega_n)^2\right]^2 + (2\zeta\omega/\omega_n)^2}}.$$ (13.61)

Figure 13.37 is a plot of: (1) the force transmissibility versus frequency ratio for a system in which a steady-state periodic forcing function is applied directly to the mass; and (2) the amplitude-magnification factor for a system in which the far end of a spring and dashpot are driven by a sinusoidal displacement of amplitude y_0 (shown in Fig. 13.36b). The plots give the nondimensional absolute response of the mass and show that the isolator can be designed so that ω/ω_n is large and the damping is small.

Finally, the transmissibility for the system of Fig. 13.36c, using Eqs. (13.56) and (13.60), is

$$T = \frac{F_{tr}/k}{(m_u/m)e} = \frac{(\omega/\omega_n)^2\sqrt{1 + (2\zeta\omega/\omega_n)^2}}{\sqrt{\left[1 - (\omega/\omega_n)^2\right]^2 + (2\zeta\omega/\omega_n)^2}}.$$ (13.62)

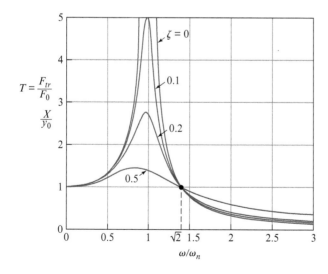

Fig. 13.37 Plot of transmissibility versus frequency ratio for various damping ratios.

Fig. 13.38 Plot of transmissibility versus frequency ratio for different damping ratios.

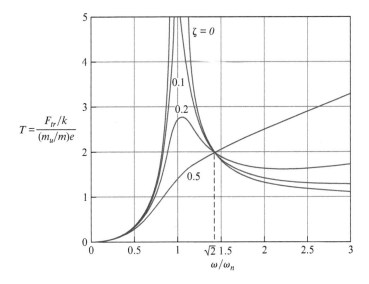

In the plot of Fig. 13.38, for large values of ω/ω_n, the damping actually serves to increase the transmissibility. Therefore, the only practical solution, in general, is to balance the equipment as perfectly as possible to minimize the product $m_u e$.

We choose the complex-operator method for the solution of the system of Fig. 13.36b. Note that this system is the same as that of Section 13.14, except that now we are interested in the absolute motion of the mass. We begin by defining the forcing function as

$$y = y_0 e^{j\omega t}.$$

Then the time derivative is

$$\dot{y} = j\omega y_0 e^{j\omega t}.$$

Also, assuming a solution in the form

$$x = X e^{j\omega t}$$

gives velocity and acceleration of the form

$$\dot{x} = j\omega X e^{j\omega t} \quad \text{and} \quad \ddot{x} = -\omega^2 X e^{j\omega t}.$$

Substituting these into Eq. (13.57) and dividing through by the exponential produces

$$-m\omega^2 X + jc\omega X + kX = ky_0 + jc\omega y_0,$$

and so

$$\frac{X}{y_0} = \frac{k + jc\omega}{(k - m\omega^2) + jc\omega}. \tag{a}$$

The product of two complex values is a new quantity whose magnitude is the product of the magnitudes of the original complex values and whose angle is the sum of the angles of the original complex values. Complex values can also be divided; in this case, we divide their magnitudes and subtract their angles.

In this case we are not interested in the phase, and so Eq. (*a*) becomes

$$\frac{X}{y_0} = \frac{\sqrt{k^2 + c^2\omega^2}}{\sqrt{(k - m\omega^2)^2 + c^2\omega^2}}.$$

Finally, dividing both numerator and denominator by *m*, we obtain

$$\frac{X}{y_0} = \frac{\sqrt{1 + (2\zeta\omega/\omega_n)^2}}{\sqrt{\left[1 - (\omega/\omega_n)^2\right]^2 + (2\zeta\omega/\omega_n)^2}}, \tag{13.63}$$

which matches Eq. (13.61) and is plotted in Fig. 13.37.

13.16 Rayleigh's Method

For any single-degree-of-freedom system, the quantity W/k represents the static deflection y_0 of a spring of stiffness k due to a weight W acting upon it. Rearranging the natural frequency equation by substituting $m = W/g$ gives

$$\omega_n = \sqrt{\frac{k}{m}} = \sqrt{\frac{g}{W/k}} = \sqrt{\frac{g}{y_0}}. \tag{13.64}$$

This equation is, incidentally, very useful for determining natural frequencies of mechanical systems, since static deflection can usually be measured quite easily.

The static deflection of a cantilever beam caused by a concentrated weight W on the free end is [3, table E-9, pp. 1179–1186]

$$y_0 = \frac{Wl^3}{3EI},$$

where *l* is the length of the cantilever, *E* is its modulus of elasticity, and *I* is the second moment of area of its cross-section. Substituting this value of y_0 into Eq. (13.64) produces

$$\omega_n = \sqrt{\frac{g}{y_0}} = \sqrt{\frac{3EIg}{Wl^3}},$$

so that the natural frequency in hertz is

$$f = \frac{1}{2\pi} \sqrt{\frac{3EIg}{Wl^3}}.$$

In our investigation of the mechanics of vibration we have been concerned with systems whose motions can be described using a single coordinate. In the cases of vibrating beams and rotating shafts, there may be many masses involved or the mass may be distributed. A coupled set of differential equations must be written, with one equation for each mass or each element of mass of the system, and these equations must be solved simultaneously if we are to obtain the equations of motion of a multimass or distributed mass system. Although a number of optional approaches are available, here we shall present an energy method, which is due to Lord Rayleigh [4], because of its importance in the study of vibration.

It is probable that Eq. (13.64) first suggested to Rayleigh the idea of employing the static deflection to find the natural frequency of a system. If we consider a freely vibrating system without damping, then, during its motion, no energy is added to the system, nor is any lost. Yet when the mass has velocity, kinetic energy exists, and when the spring is compressed or extended, potential energy exists. Since no energy is created or destroyed, the maximum kinetic energy of a system must be the same as its maximum potential energy. This is the basis of Rayleigh's method; a mathematical statement of it is

$$E_{\max} = V_{\max}, \tag{13.65}$$

where E denotes the kinetic energy and V denotes the potential energy in the system.

In order to see how it works, let us apply the method to a simple cantilever beam. We begin by assuming that the motion, y, of any point along the beam is harmonic and of the form

$$y = y_0 \sin \omega_n t. \tag{a}$$

The potential energy is maximum when a spring is fully extended or compressed and occurs when $\sin \omega_n t = 1$. That is,

$$V_{\max} = \frac{1}{2} W y_0. \tag{b}$$

The velocity of the weight is given by

$$\dot{y} = y_0 \omega_n \cos \omega_n t. \tag{c}$$

The kinetic energy reaches a maximum when the velocity is maximum – that is, when $\cos \omega_n t = 1$. The maximum kinetic energy is

$$E_{\max} = \frac{1}{2} m \dot{y}_{\max}^2 = \frac{1}{2} \frac{W}{g} (y_0 \omega_n)^2. \tag{d}$$

Applying Eq. (13.65),

$$\frac{1}{2} \frac{W}{g} (y_0 \omega_n)^2 = \frac{1}{2} W y_0 \quad \text{or} \quad \omega_n = \sqrt{\frac{g}{y_0}}, \tag{e}$$

which is identical to Eq. (13.64).

It is true in distributed or multimass systems that the dynamic deflection curves are not the same as the static deflection curves. The importance of the method, however, is that any *reasonable* deflection curve can be used in the process. The static deflection curve is a reasonable one, and hence it gives a good approximation. Rayleigh also shows that the correct curve always gives the lowest value for the natural frequency (a lower bound), although we shall not demonstrate this fact here. It is sufficient to know that, if many different deflection curves are assumed, the one giving the lowest natural frequency is the best.

Rayleigh's method is now applied to a multimass system composed of weights W_1, W_2, W_3, and so on, by assuming, as before, a deflection of each mass according to the equation

$$y_i = y_{0i} \sin \omega_n t. \tag{f}$$

The maximum deflections are therefore $y_{0_1}, y_{0_2}, y_{0_3}$, and so on, and the maximum velocities are $y_{0_1}\omega_n, y_{0_2}\omega_n, y_{0_3}\omega_n$, and so on. The maximum potential energy for the system is

$$V_{\max} = \frac{1}{2}W_1 y_{01} + \frac{1}{2}W_2 y_{02} + \frac{1}{2}W_2 y_{03} + \cdots = \frac{1}{2}\sum W_i y_{0i}. \tag{g}$$

The maximum kinetic energy for the system is

$$E_{\max} = \frac{W_1}{2g}y_{01}^2\omega_n^2 + \frac{W_2}{2g}y_{02}^2\omega_n^2 + \frac{W_3}{2g}y_{03}^2\omega_n^2 + \cdots = \frac{\omega_n^2}{2g}\sum W_i y_{0i}^2. \tag{h}$$

Applying Rayleigh's principle and solving for the frequency yields

$$\omega_n = \sqrt{\frac{g\sum W_i y_{0i}}{\sum W_i y_{0i}^2}} \quad \text{or} \quad f = \frac{1}{2\pi}\sqrt{\frac{g\sum W_i y_{0i}}{\sum W_i y_{0i}^2}}. \tag{13.66}$$

Equation (13.66), commonly called the Rayleigh–Ritz equation, can be applied to a beam consisting of several masses or to a rotating shaft having masses in the form of gears, pulleys, flywheels, and the like, mounted upon it. If the speed of the rotating shaft should become equal to the natural frequency given by Eq. (13.66), then violent vibrations occur. For this reason, it is common practice to designate this frequency as the *critical speed*.

13.17 First and Second Critical Speeds of a Shaft

This section presents an approach to obtain exact solutions to the first and second critical speeds of a rotating shaft using the method of *influence coefficients*.

The deflection at location i of a shaft due to a load of unit magnitude at location j is denoted a_{ij} and is referred to as an *influence coefficient*. Note the reciprocal relationship between the definition of an influence coefficient and the definition of spring stiffness, that is,

$$a_{ij} = \frac{1}{k_{ij}}. \tag{13.67}$$

So, for a shaft of negligible mass supporting two disks with mass (for example, gears, pulleys, or flywheels), the influence coefficients are as follows:

a_{11} = the deflection at location 1 due to a load of unit magnitude at location 1;
a_{12} = the deflection at location 1 due to a load of unit magnitude at location 2;
a_{21} = the deflection at location 2 due to a load of unit magnitude at location 1;
a_{22} = the deflection at location 2 due to a load of unit magnitude at location 2.

Maxwell's Reciprocity Theorem.   This theorem states that the deflection at location i caused by a unit load at location j is equal to the deflection at location j caused by a unit load at location i:

$$a_{ij} = a_{ji}. \tag{13.68}$$

The total deflection at location 1 caused by both disks can be written, by superposition, as

$$y_1 = a_{11}W_1 + a_{12}W_2. \tag{13.69}$$

Similarly, the total deflection at location 2 caused by both disks can be written as

$$y_2 = a_{21}W_1 + a_{22}W_2. \tag{13.70}$$

Therefore, for a rotating shaft with two masses, the total deflections at locations 1 and 2 can be written from Eqs. (13.69) and (13.70) as

$$y_1 = a_{11}\left(m_1 y_1 \omega^2\right) + a_{12}\left(m_2 y_2 \omega^2\right)$$

and

$$y_2 = a_{21}\left(m_1 y_1 \omega^2\right) + a_{22}\left(m_2 y_2 \omega^2\right).$$

Rearranging these two equations and writing them in matrix form gives

$$\begin{bmatrix} a_{11}m_1\omega^2 - 1 & a_{12}m_2\omega^2 \\ a_{21}m_1\omega^2 & a_{22}m_2\omega^2 - 1 \end{bmatrix} \begin{bmatrix} y_1 \\ y_2 \end{bmatrix} = \begin{bmatrix} 0 \\ 0 \end{bmatrix}. \tag{13.71}$$

A nontrivial solution requires that the determinant of the (2×2) coefficient matrix must be zero, that is,

$$\left(a_{11}m_1\omega^2 - 1\right)\left(a_{22}m_2\omega^2 - 1\right) - \left(a_{12}m_2\omega^2\right)\left(a_{21}m_1\omega^2\right) = 0,$$

which can be written as

$$m_1 m_2 (a_{11}a_{22})\omega^4 - (a_{11}m_1 + a_{22}m_2)\omega^2 + 1 - m_1 m_2 (a_{12}a_{21})\omega^4 = 0.$$

Rearranging this equation and dividing by ω^4, we have

$$1/\omega^4 - (a_{11}m_1 + a_{22}m_2)1/\omega^2 + m_1 m_2 (a_{11}a_{22} - a_{12}a_{21}) = 0. \tag{13.72}$$

Equation (13.72) is commonly referred to as the *frequency equation*. If we substitute $Z = 1/\omega^2$, then we have a quadratic equation in Z, that is,

$$Z^2 - (a_{11}m_1 + a_{22}m_2)Z + m_1 m_2 (a_{11}a_{22} - a_{12}a_{21}) = 0. \tag{13.73}$$

Then, the two solutions for Z can be written as

$$Z_1, Z_2 = \left(\frac{a_{11}m_1 + a_{22}m_2}{2}\right) \pm \sqrt{\left(\frac{a_{11}m_1 + a_{22}m_2}{2}\right)^2 - m_1 m_2 (a_{11}a_{22} - a_{12}a_{21})}. \qquad (13.74)$$

This equation is commonly referred to as the *exact* equation for the first two critical speeds of the system.

This procedure can easily be extended to a rotating shaft supporting n disks. In this case, Eq. (13.71) contains an $(n \times n)$ matrix, and Eq. (13.73) is an nth order polynomial in $Z = 1/\omega^2$. The n critical speeds of the system can then be determined from this polynomial.

Dunkerley's Method. The Dunkerley equation is an approximation for the first critical speed of a rotating shaft supporting several mass disks. Note that the coefficient of Z in Eq. (13.73) is the sum of the roots of the polynomial; therefore, for two mass disks, this is,

$$\frac{1}{\omega_1^2} + \frac{1}{\omega_2^2} = a_{11}m_1 + a_{22}m_2. \qquad (13.75)$$

This is a single equation in two unknowns, ω_1 and ω_2. To obtain an approximation to the first critical speed, assume that

$$\frac{1}{\omega_1^2} \gg \frac{1}{\omega_2^2}. \qquad (13.76)$$

Therefore, Eq. (13.75) can be written as

$$\frac{1}{\omega_1^2} = a_{11}m_1 + a_{22}m_2. \qquad (13.77)$$

From Eq. (13.77), the Dunkerley approximation for the first critical speed can be written as

$$\omega_1 = \frac{1}{\sqrt{a_{11}m_1 + a_{22}m_2}}. \qquad (13.78)$$

Note that the Dunkerley approximation always underestimates the first critical speed (a lower bound). However, in general, the Rayleigh–Ritz equation [Eq. (13.66)] overestimates the first critical speed (an upper bound).

Example 13.5

A rotating steel shaft that is 4 ft long and has 0.5 in diameter is simply supported in two bearings as shown in the Fig. 13.39. Two 80 lb gears are attached to the shaft. One gear is 12 in to the right of the left bearing, and the other is 12 in to the left of the right bearing. The mass of the shaft is negligible compared to the masses of the gears. The influence coefficients are known to be $a_{11} = a_{22} = 5.504 \times 10^{-5}$ in/lb and $a_{12} = a_{21} = 4.278 \times 10^{-5}$ in/lb. Determine the first critical speed of the system using (*a*) the Rayleigh equation, (*b*) the exact equation, and (*c*) the Dunkerley equation.

Example 13.5 (cont.)

Fig. 13.39 Two gears on a rotating shaft.

Solution

a. Rayleigh's equation [Eq. (13.66)] can be written as

$$\omega_1^2 = \frac{g(W_1 y_1 + W_2 y_2)}{W_1 y_1^2 + W_2 y_2^2}. \tag{1}$$

Since the two gears have the same weight, $W_1 = W_2$, Eq. (1) can be written

$$\omega_1^2 = \frac{g(y_1 + y_2)}{y_1^2 + y_2^2}. \tag{2}$$

Note from symmetry that the deflection is $y_1 = y_2 = y$; therefore, Eq. (2) can be written

$$\omega_1^2 = \frac{g}{y}. \tag{3}$$

The total deflections at locations 1 and 2 from Eqs. (13.69) and (13.70) are

$$y_1 = a_{11} W_1 + a_{12} W_2, \tag{4}$$

$$y_2 = a_{21} W_1 + a_{22} W_2. \tag{5}$$

Substituting the given data into Eq. (4) or Eq. (5) gives

$$y = y_1 = y_2 = 80 \text{ lb}(a_{11} + a_{12}).$$

Therefore, the deflection is

$$y = 80 \text{ lb}(5.504 + 4.278) \times 10^{-5} \text{ in/lb} = 782.56 \times 10^{-5} \text{ in.} \tag{6}$$

Substituting Eq. (6) into Eq. (3) gives

$$\omega_1^2 = \frac{386 \text{ in/s}^2 \times 10^5}{782.56 \text{ in}} = 49\,325 \text{ rad}^2/\text{s}^2.$$

Therefore, the first critical speed by Rayleigh's method is

$$\omega_1 = 222 \text{ rad/s.} \qquad\qquad Ans.$$

b. From Eq. (13.74), the exact equation for the first two critical speeds of the system is

$$Z_1, Z_2 = \left(\frac{a_{11} m_1 + a_{22} m_2}{2}\right) \pm \sqrt{\left(\frac{a_{11} m_1 + a_{22} m_2}{2}\right)^2 - m_1 m_2 (a_{11} a_{22} - a_{12} a_{21})}. \tag{7}$$

Example 13.5 (cont.)

Substituting the given data into this equation gives

$$Z_1, Z_2 = 1.141 \times 10^{-5}\, s^2 \pm \sqrt{(1.141^2 - 0.516) \cdot 10^{-10}}\, s^2,$$

which can be written as

$$\frac{1}{\omega_1^2}, \frac{1}{\omega_2^2} = (1.141 \pm 0.886) \cdot 10^{-5}\, s^2$$

or as

$$\omega_1^2 = 49\,334\ \text{rad}^2/s^2 \quad \text{and} \quad \omega_2^2 = 392\,157\ \text{rad}^2/s^2. \tag{8}$$

Therefore, the first and second critical speeds from the exact equation are

$$\omega_1 = 222\ \text{rad/s} \quad \text{and} \quad \omega_2 = 626\ \text{rad/s}. \qquad Ans.$$

Note that for this example the answer for the first critical speed from the Rayleigh equation is the same as the answer from the exact equation.

c. The Dunkerley equation for the rotating shaft supporting the two mass disks is

$$\frac{1}{\omega_1^2} = a_{11}m_1 + a_{22}m_2. \tag{9}$$

Substituting the given data into this equation gives

$$\frac{1}{\omega_1^2} = 2\left(\frac{80\ \text{lb}}{386\ \text{in/s}^2}\right)5.504 \cdot 10^{-5}\ \text{in/lb} = 2.281 \cdot 10^{-5}\, s^2. \tag{10}$$

Therefore, the first critical speed for the system is

$$\omega_1 = 209\ \text{rad/s} \qquad Ans.$$

Note that this answer is less than the first critical speed obtained from the exact equation.

Example 13.6

The rotating shaft shown in Fig. 13.40 supports two flywheels, each of mass m. The first critical speed of the system has been determined experimentally to be 300 rad/s and the first critical speed of the shaft alone is 600 rad/s. If one of the flywheels is removed from the shaft, then determine the first critical speed of the system.

Solution

The first natural frequency of the shaft carrying the two flywheels can be written as

Example 13.6 (cont.)

Fig. 13.40 Two flywheels on a rotating shaft.

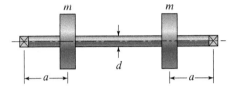

$$\frac{1}{\omega_1^2} = \frac{1}{\omega_{11}^2} + \frac{1}{\omega_{22}^2} + \frac{1}{\omega_{ss}^2} \tag{1a}$$

or as

$$\frac{1}{\omega_1^2} = ma_{11} + ma_{22} + \frac{1}{\omega_{ss}^2}, \tag{1b}$$

where ω_{ss} is the critical speed of the shaft alone.

From symmetry we note that the influence coefficient $a_{11} = a_{22}$; therefore Eq. (1b) can be written as

$$2ma_{11} = \frac{1}{\omega_1^2} - \frac{1}{\omega_{ss}^2}. \tag{2}$$

Substituting the known critical speeds into Eq. (2) gives

$$2ma_{11} = \frac{1}{(300 \text{ rad/s})^2} - \frac{1}{(600 \text{ rad/s})^2} = \frac{1}{120\,000} \text{ s}^2. \tag{3}$$

For the massless shaft with only one flywheel, the first natural frequency can be written from Eq. (1a) as

$$\frac{1}{\omega_1^2} = \frac{1}{\omega_{11}^2} + \frac{1}{\omega_{ss}^2} \tag{4}$$

or as

$$\frac{1}{\omega_1^2} = ma_{11} + \frac{1}{\omega_{ss}^2}. \tag{5}$$

Substituting Eq. (3) into Eq. (5) gives

$$\frac{1}{\omega_1^2} = \frac{1}{2} \left[\frac{1}{120\,000} \text{ s}^2 \right] + \frac{1}{\omega_{ss}^2}. \tag{6}$$

Example 13.6 (cont.)

Finally, substituting the critical speed of the shaft $\omega_{ss} = 600$ rad/s into Eq. (6) gives

$$\frac{1}{\omega_1^2} = \frac{1}{2}\left[\frac{1}{120\,000}\,\text{s}^2\right] + \frac{1}{(600\text{ rad/s})^2} = 6.944 \cdot 10^{-6}\,\text{s}^2. \tag{7}$$

Therefore, the first critical speed of the system is

$$\omega_1 = 379.5 \text{ rad/s}. \qquad\qquad Ans.$$

In the investigation of rectilinear vibrations to follow, we will see that there are as many natural frequencies in a multimass vibrating system as there are degrees of freedom. Therefore, a three-mass torsional system has three degrees of freedom and, consequently, three natural frequencies. Equation (13.66) gives only the first or lowest of these frequencies.

It is probable that the critical speed can be determined by Eq. (13.66), for most cases, within about 5%. This is because of the assumption that the static and dynamic deflection curves are identical. Bearings, couplings, belts, and so on, all have effect upon the spring constant and the damping. Shaft vibration usually occurs over an appreciable range of shaft speed, and for this reason Eq. (13.66) is accurate enough for many engineering purposes.

As a general rule, shafts have many discontinuities – such as shoulders, keyways, holes, and grooves – to locate and secure various gears, pulleys, and other shaft-mounted masses. These diametral changes usually require deflection analysis using numeric integration. For such problems, Simpson's rule is easy to apply using a computer, a calculator, or by hand calculations [3, pp. 142–143].

13.18 Torsional Systems

Figure 13.41 shows a shaft supported on bearings at A and B with two masses connected at the ends. The masses represent rotating machine parts – an engine and its flywheel, for example. We wish to study the possibility of free vibration of the system when it rotates at constant angular velocity. To investigate the motion of each mass, it is necessary to picture a reference system fixed to the shaft and rotating with the shaft at the same angular velocity. Then, we can measure the angular displacement of either mass by finding the instantaneous angular location

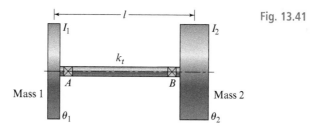

Fig. 13.41

of a mark on the mass relative to one of the rotating axes. Thus, we define θ_1 and θ_2 as the angular displacements of mass 1 and mass 2, respectively, with respect to the rotating axes.

Now, assuming no damping, Eq. (13.6) is written for each mass,

$$I_1\ddot{\theta}_1 + k_t(\theta_1 - \theta_2) = 0,$$
$$I_2\ddot{\theta}_2 + k_t(\theta_2 - \theta_1) = 0,$$

(a)

where the angle $(\theta_1 - \theta_2)$ represents the total twist of the shaft. A solution is assumed in the form

$$\theta_i = \gamma_i \sin \omega_n t.$$

Then, the acceleration is

$$\ddot{\theta}_i = -\gamma_i \omega_n^2 \sin \omega_n t.$$

If these equations are substituted into Eqs. (a), we obtain

$$\left(k_t - I_1\omega_n^2\right)\gamma_1 - k_t\gamma_2 = 0,$$
$$-k_t\gamma_1 + \left(k_t - I_2\omega_n^2\right)\gamma_2 = 0,$$

(b)

which can be written in matrix form as

$$\begin{bmatrix} k_t - I_1\omega_n^2 & -k_t \\ -k_t & k_t - I_2\omega_n^2 \end{bmatrix} \begin{bmatrix} \gamma_1 \\ \gamma_2 \end{bmatrix} = \begin{bmatrix} 0 \\ 0 \end{bmatrix}.$$

For a nontrivial solution, the determinant of the coefficient matrix must be zero; that is

$$\left(k_t - I_1\omega_n^2\right)\left(k_t - I_2\omega_n^2\right) - k_t^2 = 0,$$

so that

$$\omega_n^2\left(\omega_n^2 - k_t\frac{I_1 + I_2}{I_1 I_2}\right) = 0$$

and

$$\omega_{n1} = 0 \quad \text{and} \quad \omega_{n2} = \sqrt{k_t\frac{I_1 + I_2}{I_1 I_2}}.$$

(13.79)

Substituting $\omega_{n1} = 0$ into Eqs. (b), we find that $\gamma_1 = \gamma_2$. Therefore, for this case, the masses rotate together without any relative displacement and there is no vibration.

Substituting the value of ω_{n2} into Eqs. (b), it is determined that

$$\gamma_2 = -\frac{I_1}{I_2}\gamma_1,$$

indicating that the motion of the second mass is opposite to that of the first and proportional to I_1/I_2. Figure 13.42 is a graphic representation of the vibration. Here, l is the distance between

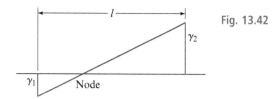

Fig. 13.42

the two masses, and γ_1 and γ_2 are the instantaneous angular displacements plotted to scale. If a line is drawn connecting the ends of the angular displacements, this line crosses the axis at a *node*, which is a location on the shaft having zero angular displacement. It is convenient to designate the configuration of the vibrating system as a *mode* of vibration. This system has two modes of vibration, corresponding to the two frequencies, although we have seen that one of them is degenerate. A system of n masses would have n modes corresponding to n different frequencies.

For multimass torsional systems, the *Holzer tabulation method* [2] can be used to find all the natural frequencies. It is easy to use and avoids solving the simultaneous differential equations.

PROBLEMS

13.1 Derive the differential equation of motion for each system and write the formula for the natural frequency ω_n for each system.

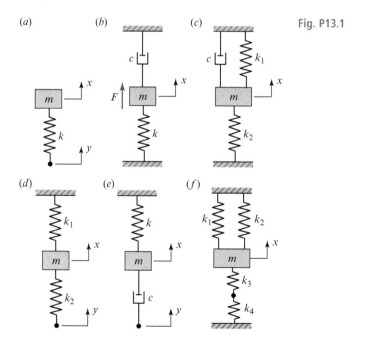

Fig. P13.1

13.2 Evaluate the constants of integration of the solution to the differential equation for an undamped free system, using the following sets of starting conditions: (a) $x = x_0, \dot{x} = 0$; (b) $x = 0, \dot{x} = v_0$; (c) $x - x_0, \ddot{x} = a_0$; and (d) $x = x_0, \dddot{x} = b_0$. For each case, transform the solution to a form containing a single trigonometric term.

13.3 A system like Fig. 13.5 has $w = 2.2$ lb and an equation of motion $x = 0.8 \cos(8\pi t - \pi/4)$ in. Determine: (a) spring constant k; (b) static deflection δ_{st}; (c) period; (d) frequency in hertz; and (e) velocity, acceleration, and spring force at $t = 0.20$ s. Plot a phase diagram to scale showing the displacement, velocity, acceleration, and spring-force phasors at this instant.

13.4 The weight W_1 drops through the distance h and collides with W_2 with plastic impact (a coefficient of restitution of zero). Derive the differential equation of motion of the system and determine the amplitude of the resulting motion of W_2.

Fig. P13.4
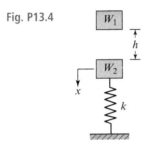

13.5 The vibrating system has $k_1 = k_3 = 5$ lb/in, $k_2 = 10$ lb/in, and $W = 9$ lb. What is the natural frequency in hertz?

Fig. P13.5
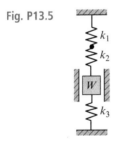

13.6 Weight $W = 66.6$ N is connected to a pivoted rod which is assumed to be weightless but rigid. A spring having a rate of $k = 10.65$ kN/m is connected to the center of the rod and holds the system in static equilibrium at the posture shown. Assuming that the rod can vibrate with a small amplitude, determine the period of the motion.

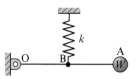

Fig. P13.6 $\ell = R_{AO} = 300$ mm, $R_{BO} = 150$ mm.

13.7 The upside-down pendulum of length l is retained by two springs connected a distance a from the pivot. The springs have been positioned such that the pendulum is in static equilibrium when it is in the vertical posture. (a) For small amplitudes, find the natural frequency of this system. (b) Find the ratio l/a at which the system becomes unstable.

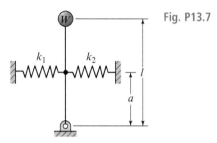

Fig. P13.7

13.8 Write the differential equation for the system, and find the natural frequency. Find the response x if y is a step input of height y_0. Find the relative response $z = x - y$ to this step input.

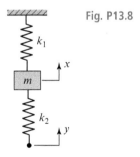

Fig. P13.8

13.9 An undamped vibrating system consists of a spring whose scale is 200 lb/in and a weight of 2.64 lb. A step force $F = 11.25$ lb is exerted on the weight for 0.040 s. (a) Write the equations of motion of the system for the era in which the force acts and for the era that follows. (b) What are the amplitudes in each era? (c) Sketch a time plot of the displacement.

13.10 A round shaft whose torsional spring constant is k_t N·m/rad connects two wheels having mass moments of inertia I_1 and I_2. Show that the system is likely to vibrate torsionally with a frequency of

$$\omega_n = \sqrt{\frac{k_t(I_1 + I_2)}{I_1 I_2}}.$$

Fig. P13.10

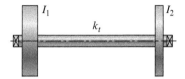

13.11 A motor is connected to a flywheel by a 15.6 mm diameter steel shaft 900 mm long. Using the methods of this chapter, it can be demonstrated that the torsional spring rate of the shaft is 77.7 N·m/rad. The mass moments of inertia of the motor and flywheel are 2.66 and 6.22 kg·m², respectively. The motor is turned on for 2 s, and during this period it exerts a constant torque of 22.2 N·m on the shaft. (a) What speed in revolutions per minute does the shaft attain? (b) What is the natural frequency of vibration of the system? (c) Assuming no damping, what is the amplitude of the vibration of the system in degrees during the first era? During the second era?

Fig. P13.11

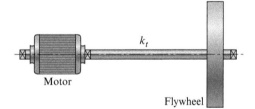

13.12 The mass of a vibrating system is 4.5 kg, and it has a natural frequency of 1 Hz. Using the phase-plane method, plot the response of the system to the given forcing function. What is the final amplitude of the motion?

Fig. P13.12

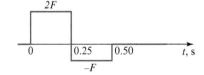

13.13 An undamped vibrating system has a spring scale of 34.75 kN/m and a mass of 22.5 kg. Find the response and the final amplitude of vibration of the system if it is acted upon by the given forcing function. Use the phase-plane method.

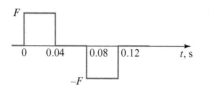

Fig. P13.13 $F = 54.3$ kN.

13.14 A vibrating system has a spring rate of $k = 71$ kN/m and a mass of $m = 36$ kg. Plot the response of this system to the given forcing function using: (a) three steps; and (b) six steps.

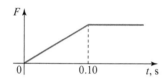

Fig. P13.14 $F_{max} = 888$ N.

13.15 What is the value of the coefficient of critical damping for a spring–mass–damper system in which $k = 315$ lb/in and $w = 88$ lb? If the actual damping is 20% of critical, what is the natural frequency of the system? What is the period of the damped system? What is the value of the logarithmic decrement?

13.16 A vibrating system has a spring scale of $k = 19.7$ lb/in and a weight of 33 lb. When disturbed, it is observed that the amplitude decays to one-fourth of its original value in 4.80 s. Find the damping coefficient and the damping factor.

13.17 A vibrating system has $k = 53.3$ kN/m, $m = 40.5$ kg, and damping equal to 20% of critical. (a) What is the damped natural frequency ω_d of the system? (b) What are the period and the logarithmic decrement?

13.18 Solve Prob. 13.14 using damping equal to 15% of critical.

13.19 A damped vibrating system has an undamped natural frequency of 10 Hz and a mass of 360 kg. The damping ratio is 0.15. Using the phase-plane method, determine the response of the system to the given forcing function.

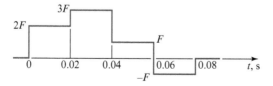

Fig. P13.19 $F = 4.44$ kN.

13.20 A vibrating system has a spring rate of 533 kN/m, a damping factor of 9 770 N·s/m, and a weight of 3550 N. It is excited by a harmonically varying force $F_0 = 444$ N at a frequency of 435 cycles per minute. (a) Calculate the amplitude of the forced vibration and the phase angle between the vibration and the force. (b) Plot several cycles of the displacement–time and force–time diagrams.

13.21 A spring-mounted mass has $k = 2960 \, \text{lb/in}$, $c = 54.3 \, \text{lb} \cdot \text{s/in}$, and $w = 795 \, \text{lb}$. This system is excited by a force having an amplitude of 100 lb at a frequency of 4.80 Hz. Find the amplitude and phase angle of the resulting vibration and plot several cycles of the force–time and displacement–time diagrams.

13.22 When a 2700 kg press is mounted upon structural-steel floor beams, it causes them to deflect 18.8 mm. If the press has a reciprocating unbalance of 1865 N and it operates at a speed of 80 rev/min, how much of the force will be transmitted from the floor beams to other parts of the building? Assume no damping. Can this mounting be improved?

13.23 Four vibration mounts are used to support a 900 lb machine that has a rotating unbalance of 28 in · lb and runs at 300 rev/min. The vibration mounts have damping equal to 30% of critical. What must the spring constant of the mounting be if 20% of the exciting force is transmitted to the foundation? What is the resulting amplitude of motion of the machine?

13.24 A 24 in long steel shaft is simply supported by two bearings at A and C. Flywheels 1 and 2 are attached to the shaft at locations B and D, respectively. Flywheel 1 at location B weighs 11 lb, flywheel 2 at location D weighs 4.5 lb, and the weight of the shaft can be neglected. The known stiffness coefficients are $k_{11} = 140 \, \text{lb/in}$, $k_{12} = 280 \, \text{lb/in}$, and $k_{22} = 224 \, \text{lb/in}$. Determine (a) the first and second critical speeds of the shaft using the exact solution and (b) the first critical speed using the Dunkerley and (c) Rayleigh–Ritz approximations. (d) If flywheel 2 is then placed at location B and flywheel 1 is placed at location D, determine the first critical speed of the new system using the Dunkerley approximation.

Fig. P13.24 Steel shaft simply supported by two rolling element bearings. $R_{BA} = 10 \, \text{in}$, $R_{CB} = 10 \, \text{in}$, and $R_{DC} = 4 \, \text{in}$.

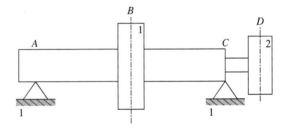

13.25 The first critical speeds of a rotating shaft with two mass disks, obtained from three different mathematical techniques, are 110 rad/s, 112 rad/s, and 100 rad/s, respectively. (a) Which values correspond to the first critical speed of the shaft from the exact solution, the Dunkerley approximation, and the Rayleigh–Ritz approximation? (b) If the influence coefficients are $a_{11} = a_{22} = 0.0178 \, \text{in/lb}$ and the masses of the two disks are the same, that is, $m_1 = m_2 = m$, then use the Dunkerley approximation to calculate the mass m. (c) If the influence coefficients are $a_{11} = a_{22} = 0.0178 \, \text{in/lb}$ and the weights of the two disks are specified as $w_1 = w_2 = w = 1.1 \, \text{lb}$, use the Rayleigh–Ritz approximation to calculate the influence coefficient a_{12}.

13.26 A steel shaft is simply supported by two rolling element bearings at A and B. The length of the shaft is 58 in and two flywheels with weight 67.5 lb are attached to the shaft at the locations shown. One flywheel is 14 in to the right of the left bearing at A and the other flywheel is 14 in to the left of the right bearing at B. The weight of the shaft can be neglected. The influence coefficients are specified as $a_{11} = 1.135 \times 10^{-5}$ in/lb and $a_{21} = 0.833 \times 10^{-5}$ in/lb. (a) Determine the first and second critical speeds of the shaft using the exact solution. Determine the first critical speed of the shaft using: (b) the Dunkerley approximation and (c) the Rayleigh–Ritz equation.

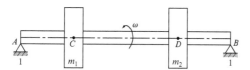

Fig. P13.26 Steel shaft simply supported by two rolling element bearings.

13.27 A steel shaft is simply supported by two rolling element bearings at A and C. The length of the shaft is 24 in and two flywheels are attached to the shaft at the locations B and D as shown. The flywheel at location B weighs 888 lb and the flywheel at location D weighs 400 lb. The weight of the shaft can be neglected. It was determined that with flywheel 1 alone the first critical speed of the shaft is 800 rad/s and with flywheel 2 alone the first critical speed of the shaft is 1 200 rad/s. (a) Determine the first critical speed for the two-mass system. (b) If the two flywheels are interchanged (that is, flywheel 2 is placed at location B and flywheel 1 is placed at location D), determine the first critical speed of the new system using the Dunkerley approximation.

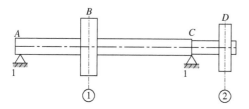

Fig. P13.27 Steel shaft simply supported by two rolling element bearings. $R_{BA} = 8$ in, and $R_{CD} = 4$ in.

13.28 A steel shaft, which is 1.250 m in length, is simply supported by two bearings at B and D. Flywheels 1 and 2 are attached to the shaft at A and C, respectively. The flywheel at location A weighs 67 N, the flywheel at location C weighs 133 N, and the weight of the shaft can be neglected. The stiffness coefficients are specified as $k_{11} = 0.0444$ N/m and $k_{22} = 0.071$ N/m. (a) Determine the first critical speed for the two-mass system using the Dunkerley approximation. (b) If the two flywheels are interchanged (that is, flywheel 2 is placed at location A and flywheel 1 is placed at location C), determine the first critical speed of the new system.

Fig. P13.28 Steel shaft simply supported by
two rolling element bearings. $R_{BA} = 250$ mm,
and $R_{CD} = 500$ mm.

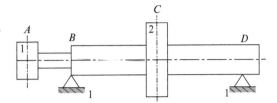

13.29   The weights of two gears rigidly attached to a shaft at two different locations, denoted
        as 1 and 2, are $W_1 = 45$ lb and $W_2 = 79$ lb, respectively. The shaft is rotating
        counterclockwise with a constant operating speed $\omega = 100$ rad/s. From a deflection
        analysis of the shaft, the known influence coefficients are $a_{11} = 6.22 \times 10^{-6}$ in/lb,
        $a_{12} = 8.89 \times 10^{-6}$ in/lb, and $a_{22} = 16 \times 10^{-6}$ in/lb. Neglecting gravity and the mass
        of the shaft, determine the first and second critical speeds of the shaft using the exact
        equation. Is the operating speed of the shaft acceptable? Determine the first critical speed
        of the shaft using: (a) the Rayleigh–Ritz method; and (b) the Dunkerley approximation.

13.30   The weights of the two masses m_1 and m_2 which are rigidly attached to the rotating shaft
        are 140 N and 60 N, respectively. The shaft is rotating counterclockwise with a constant
        angular velocity $\omega = 100$ rad/s. From a deflection analysis, the influence coefficients for
        the shaft are $a_{11} = 20 \times 10^{-6}$ mm/N, $a_{22} = 120 \times 10^{-6}$ mm/N, and $a_{12} = a_{21} =$
        40×10^{-6} mm/N. Neglecting the mass of the shaft, determine the first and second
        critical speeds of the shaft using the exact equation. Is the operating speed of the shaft
        acceptable? Determine the first critical speed of the shaft using the Rayleigh–Ritz
        equation and the Dunkerley approximation. Determine the first critical speed of the
        shaft if the mass m_1 is moved to location 2 and the mass m_2 is moved to location 1.

Fig. P13.30 Rotating shaft
with two mass disks.
$R_{BA} = 100$ mm,
$R_{CB} = 70$ mm, and
$R_{CD} = 150$ mm.

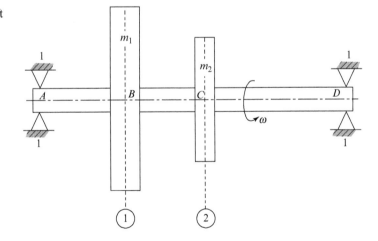

13.31   The first and second critical speeds of a rotating shaft with two flywheels rigidly
        attached are $\omega_1 = 375$ rad/s and $\omega_2 = 615$ rad/s. The weights of the flywheels are
        $W_1 = 14.6$ lb and $W_2 = 18$ lb and the known influence coefficients of the shaft are

$a_{11} = 5.52 \times 10^{-6}$ in/lb and $a_{21} = 2.48 \times 10^{-4}$ in/lb. Determine the influence coefficient a_{22}.

13.32 A shaft rotating with a constant angular velocity is simply supported at A and B. Gears C and D are rigidly attached to the shaft at locations 1 and 2, respectively. The weight of gear C at location 1 is 2.93 lb and the weight of gear D at location 2 is 5.18 lb. The known influence coefficients of the shaft are $a_{11} = 1.21 \times 10^{-3}$ in/lb, $a_{12} = 0.94 \times 10^{-3}$ in/lb, and $a_{22} = 1.40 \times 10^{-3}$ in/lb. Determine the first critical speed of the shaft using: (a) the Dunkerley approximation; and (b) the Rayleigh–Ritz method. If gear C is moved to location 2 and gear D is moved to location 1 then calculate the new first critical speed using: (c) the Dunkerley approximation; and (d) the Rayleigh–Ritz method.

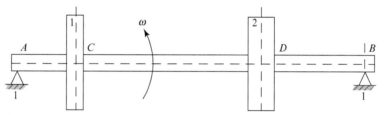

Fig. P13.32 Rotating shaft with two gears. $R_{CA} = 3.20$ in, $R_{DC} = 14$ in, and $R_{DB} = 7.20$ in.

13.33 A shaft is simply supported at A and B and is rotating with constant angular velocity ω. Two identical flywheels C and D of unknown mass are rigidly attached to the shaft (at locations 1 and 2). The known influence coefficients are $a_{11} = 0.23 \times 10^{-6}$ m/N, $a_{12} = 0.166 \times 10^{-6}$ m/N, and $a_{22} = 0.30 \times 10^{-6}$ m/N. The first critical speed of the shaft, obtained from the exact equation, is 135 rad/s. Determine: (a) the masses of the two flywheels, and (b) the second critical speed of the shaft.

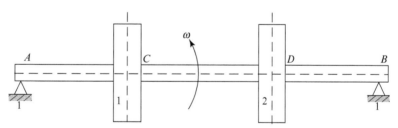

Fig. P13.33 Rotating shaft with two identical flywheels. $R_{CA} = 175$ mm, $R_{DC} = 238$ mm, and $R_{DB} = 175$ mm.

13.34 The 24-in long shaft is simply supported by the bearings at A and D. Flywheel 1 at location B weighs 6.75 lb and flywheel 2 at location C weighs 3.38 lb. The weight of the shaft can be neglected. The stiffness coefficients of the shaft are $k_{11} = 9.0 \times 10^6$ lb/in and $k_{22} = 45.0 \times 10^6$ lb/in. (a) Using the Dunkerley approximation, determine the first critical speed of the shaft. (b) If the flywheels are interchanged (that is, if flywheel 2 is at location B and flywheel 1 is at location C), then use the Dunkerley approximation to determine the first critical speed of the new system.

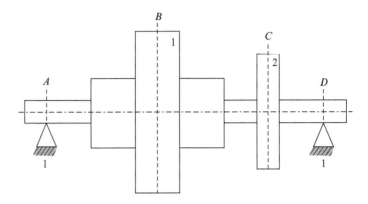

Fig. P13.34 Shaft simply supported at A and D. $R_{BA} = 9.60$ in and $R_{CD} = 4.80$ in.

13.35 Three identical flywheels, each weighing 84.4 N, are rigidly attached to a rotating shaft. The known influence coefficients of the shaft are: $a_{11} = 12.66 \times 10^{-4}$ mm/N, $a_{22} = 60.47 \times 10^{-4}$ mm/N, $a_{33} = 35.16 \times 10^{-4}$ mm/N, $a_{12} = 4.50 \times 10^{-4}$ mm/N, $a_{13} = 0.84 \times 10^{-4}$ mm/N, and $a_{23} = 5.62 \times 10^{-4}$ mm/N. Use the Rayleigh–Ritz method to determine the first critical speed of the shaft.

13.36 The first and second critical speeds of a rotating shaft supporting two identical gears, each with a weight of 13.2 lb, are $\omega_1 = 500$ rad/s and $\omega_2 = 1120$ rad/s, respectively. If the influence coefficient $a_{11} = 2a_{22}$, then determine the numeric values of the influence coefficients of the shaft.

13.37 The shaft is simply supported by the bearings at A and D. Flywheel 1 at location B weighs 24.86 N, flywheel 2 at location C weighs 79.48 N, and the weight of the shaft can be neglected. The influence coefficients of the shaft are $a_{11} = 10.13 \times 10^{-8}$ mm/N, $a_{12} = 3.94 \times 10^{-8}$ mm/N, and $a_{22} = 55.13 \times 10^{-8}$ mm/N. Determine the first critical speed of the shaft using: (a) the Dunkerley approximation; and (b) the Rayleigh–Ritz approximation.

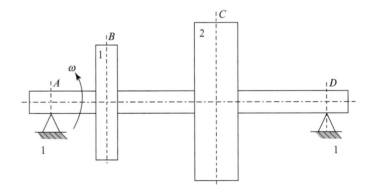

Fig. P13.37 Shaft simply supported at A and D. $R_{BA} = 60$ mm, $R_{CB} = 120$ mm, and $R_{CD} = 120$ mm.

13.38 A shaft with negligible mass is simply supported by two bearings. When a gear with a weight of 15.4 lb is attached to the shaft at location 1, the first critical speed is measured as 1 200 rad/s. After a second identical gear is attached to the shaft at location 2, the first and second critical speeds of the shaft are measured as 500 rad/s and 1 800 rad/s, respectively. Using the exact solution to the first and second critical speeds of a rotating shaft, determine the four influence coefficients of this shaft.

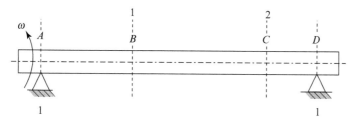

Fig. P13.38 Simply supported shaft. $R_{BA} = 4$ in, $R_{CB} = 7.2$ in, and $R_{DC} = 2.4$ in.

13.39 The 1.80 m long shaft is simply supported by the bearings at B and D. Flywheel 1 at location A weighs 40 N and flywheel 2 at location C weighs 71 N. The weight of the shaft can be neglected. The stiffness coefficients of the shaft are $k_{11} = 622$ N/m, $k_{12} = 3730$ N/m, and $k_{22} = 2486$ N/m. Determine the first and second critical speeds of the shaft using the exact equation. Then determine the first critical speed of the shaft using: (a) the Rayleigh–Ritz method; and (b) the Dunkerley approximation.

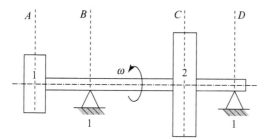

Fig. P13.39 Shaft simply supported at B and D. $R_{BA} = 0.050$ m, $R_{CB} = 0.085$ m, and $R_{CD} = 0.045$ m.

13.40 The steel shaft is simply supported by two bearings at A and D. Two gears are rigidly attached to the shaft at locations B and C. Gear 1 at location B weighs 33.76 lb, gear 2 at location C weighs 20.25 lb, and the weight of the shaft can be neglected. The stiffness coefficients are $k_{11} = 141$ lb/in, $k_{12} = 281$ lb/in, and $k_{22} = 225$ lb/in. Determine the first critical speed of the shaft using: (a) the Dunkerley approximation; and (b) the Rayleigh–Ritz method. Determine the first and second critical speeds of the shaft using the exact solution. If gear 2 is now placed at location B and gear 1 is placed at location C, then determine the first critical speed of the new system using the Dunkerley approximation.

Fig. P13.40 Shaft simply supported at A and D. $R_{BA} = 16$ in, $R_{CB} = 6$ in, and $R_{DC} = 16$ in.

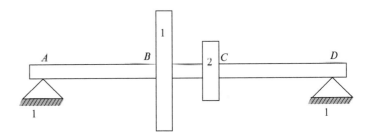

13.41 The first critical speeds of a rotating shaft with two mass disks, obtained from three different mathematical techniques, are 150 rad/s, 152 rad/s, and 153 rad/s, respectively. The known influence coefficients are $a_{11} = a_{22} = 0.0533$ in/lb. (a) Specify which value corresponds to the first critical speed from the exact solution, the Rayleigh–Ritz method, and the Dunkerley approximation. (b) If the masses of the two disks are identical then use the Dunkerley approximation to calculate the mass of each disk. (c) If the weights of the two disks are specified as $w_1 = w_2 = 0.220$ lb, then use the Rayleigh–Ritz method to calculate the influence coefficient a_{12}.

REFERENCES

[1] Bridgman, P. W., 1946. *The Logic of Modern Physics*, New York: Macmillan.
[2] Sankar, T. S., and R. B. Bhat, 1986. Vibration and control of vibration, in J. *Standard Handbook of Machine Design*, edited by E. Shigley and C. R. Mischke, New York: McGraw-Hill, pp. 38.22–38.27.
[3] Shigley, J. E., and C. R. Mischke, 2001. *Mechanical Engineering Design*, 6th ed. New York: McGraw-Hill, pp. 142-3.
[4] Strutt, J. W. (Baron Rayleigh), 1945 (republished). *Theory of Sound*, New York: Dover.

14   Dynamics of Reciprocating Engines

The purpose of this chapter is to apply fundamentals – kinematic and dynamic analysis – in a complete investigation of a particular class of machines. The reciprocating engine has been selected for this purpose, since it has reached a high state of development and is of more general interest than most other machines. For our purposes, however, another type of machine involving interesting dynamic situations would serve just as well. The primary objective of the chapter is to demonstrate methods of applying fundamentals to the dynamic analysis of machines.

14.1   Engine Types

The descriptions and characteristics of all the engine types that have been conceived and constructed over the years would fill many books. Here, our purpose is to describe very briefly a few of the engine types that are currently in general use. This exposition is not intended to be complete. Furthermore, since you, the reader, are anticipated to be mechanically inclined and generally familiar with internal combustion engines, the primary purpose of this section is merely to record things that you already know and to furnish nomenclature for the remainder of the chapter.

In this chapter, we classify engines according to their intended application, the combustion cycle used, and the number and arrangement of cylinders. Thus, we refer to aircraft engines, automotive engines, marine engines, and stationary engines, for example, all so named because of the applications for which they are designed. Similarly, one might have in mind an engine designed on the basis of the *Otto cycle*, in which the fuel and air are mixed before compression and in which combustion takes place with no excess air, or on the basis of the *Diesel cycle*, in which the fuel is injected near the end of compression and combustion takes place with a good deal of excess air. The Otto-cycle engine uses quite volatile fuels and ignition is by spark, but the Diesel-cycle engine operates on fuels of lower volatility and ignition occurs because of compression.

Both Diesel- and Otto-cycle engines may use either a *two-stroke cycle* or a *four-stroke cycle*, depending upon the number of piston strokes comprising the complete combustion cycle. In the four-stroke cycle the exhaust ports open near the end of each power stroke to permit exhaust gases to flow out during the exhaust stroke. Next, the exhaust ports close and inlet ports open and permit entry of a fuel–air mixture during the suction stroke. Then, all ports are closed and the piston compresses the fuel–air mixture. Finally, the fuel–air mixture is ignited, causing an expansion or power stroke, and the cycle begins again. Note that each four-stroke cycle requires two revolutions of the crank. In contrast, many outboard or lawn mower engines use a two-stroke cycle or are simply two-stroke engines. Note that a two-stroke engine has only an expansion stroke and a compression stroke and that both occur during each revolution of the crank.

The four-stroke engine uses four piston strokes in a single combustion cycle corresponding to two revolutions of the crank. The events corresponding to the four strokes are: (1) the suction, or input stroke; (2) the compression stroke; (3) the expansion, or power stroke; and (4) the exhaust stroke.

Multi-cylinder engines are also broadly classified according to how the cylinders are arranged with respect to each other and the crankshaft. Thus, an *in-line* engine is one in which all cylinder axes form a single plane containing the crankshaft, and, in that plane, the pistons are all on the same side of the crankshaft. Figure 14.1 is a schematic drawing of a three-cylinder in-line engine with the three cranks spaced at 120° of crank rotation. A firing-order diagram for four-stroke operation is included for interest.

Figure 14.2 shows a cutaway view of a five-cylinder in-line engine of a recent passenger car. We can see here that the plane formed by the centerlines of the cylinders is not mounted vertically, to keep the height of the engine smaller to fit low-profile styling requirements. Figure 14.2 also includes a view of the relative location of the camshaft and the overhead valves.

A V-type engine uses two banks of one or more in-line cylinders, all connected to a single crankshaft. Figure 14.3 shows some typical crank arrangements. The pistons in the right and left banks of Figs. 14.3*a* and 14.3*b* may be in the same plane, but those in Fig. 14.3*c* are in different planes.

If the V angle is increased to 180°, the result is called an *opposed-piston engine*. An opposed engine may have the two piston axes coincident or offset, and the rods may connect to the same crank throw or to separate crank throws 180° apart.

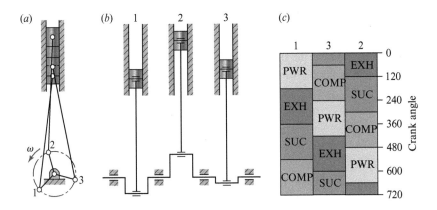

Fig. 14.1 Three-cylinder in-line engine: (*a*) front view; (*b*) side view; (*c*) firing order.

Fig. 14.2 Five-cylinder in-line engine from an Audi Coupe Quattro.
(Courtesy of Audi of America, Inc., Troy, MI.)

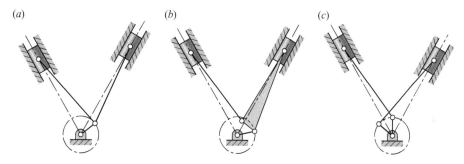

Fig. 14.3 Crank arrangements of different V engines: (*a*) single crank throw per pair of
cylinders – connecting rods interlock with each other and are of *fork-and-blade* design; (*b*)
single crank throw per pair of cylinders – the *master connecting rod* carries a bearing for the
articulated rod; (*c*) separate crank throws connect to staggered rods and pistons.

A *radial* engine is one having multiple cylinder axes arranged radially about the crank center.
Radial engines use a master connecting rod for one cylinder, and the remaining pistons are
connected to the master rod by articulated rods, somewhat the same as for the V engine of
Fig. 14.3*b*.

Figures 14.4–14.6 show, respectively, the piston–connecting-rod assembly, the crankshaft,
and the block of an internal combustion engine. These are included as typical of engine design
to demonstrate the forms of important parts of an engine. The following information also gives
a general idea of the performance and design characteristics of typical engines, together with
the sizes of parts used in them.

Fig. 14.4 Piston–connecting-rod assembly for an internal combustion engine. (Courtesy of JazzIRT/E+ via Getty Images.)

Fig. 14.5 Cast crankshaft for an internal combustion engine. (Courtesy of Saibo / CC-BY-SA-3.0.)

The reader will note that all quantities throughout this chapter are quoted in SI units rather than US customary units. Because of the international nature of the automotive and related industries, engines today are all designed in SI units, even in the United States. Although units of horsepower and gallons are still seen in advertisements, the power of an engine is measured in kilowatts (kW), and engine displacement is quoted in liters (L)[1] by knowledgeable engineers throughout the industry. A cross-sectional view of an automotive engine is shown in Fig. 14.7.

The GMC V6 truck engines were manufactured in four displacements, and they include one model, a V12 (11.5 L), which is described as a "twin six" since many of the V6 parts are interchangeable with it. Typical performance curves for the truck engine are illustrated in Figs. 14.8–14.10.

---

[1] The unit for volume of an engine or cylinder is the liter. The international symbol for liter is lower-case "l," which can easily be confused with the numeral "1." Therefore, the symbol "L" is used throughout this chapter.

Fig. 14.6 Block for an internal combustion engine. (Courtesy of Bill Pugliano/Stringer via Getty Images News.)

Fig. 14.7 Cross-sectional view of an internal combustion engine. (Courtesy of ineb1599/iStock via Getty Images.)

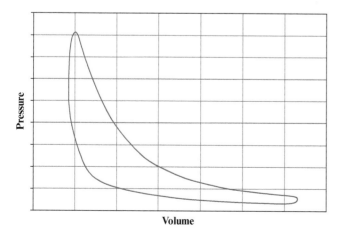

Fig. 14.8 Typical indicator diagram for a 6.57 L V6 truck engine; conditions unknown.

Fig. 14.9 Pressure-crank-angle curve for the 6.57 L V6 truck engine.

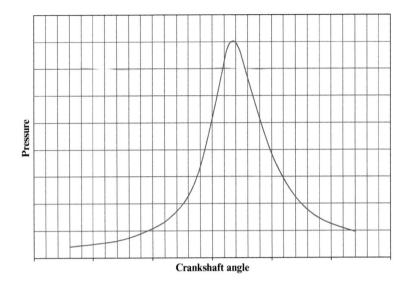

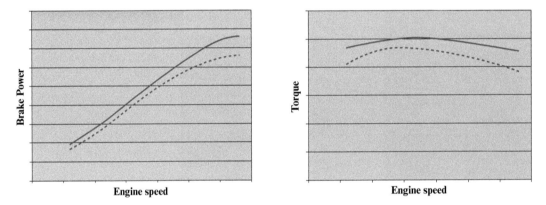

Fig. 14.10 Power and torque characteristics of the 6.57 L V6 truck engine. The solid curve is the net output as installed; the dashed curve is the maximum output without accessories. The maximum torque occurs at low engine speed.

The specifications for the truck engine are as follows: 60° V design; bore = 123.8 mm; stroke = 90.4 mm; connecting rod length = 182.6 mm; compression ratio = 7.50:1; cylinders are numbered 1, 3, and 5 from front to rear on the left bank, and 2, 4, and 6 on the right bank; firing order is 1, 6, 5, 4, 3, 2; the crank arrangement is shown in Fig. 14.11.

14.2 Indicator Diagrams

Experimentally, an instrument called an *engine indicator* is used to measure the variation in pressure within a cylinder. This instrument constructs a graph, during operation of the engine,

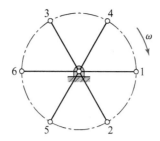

Fig. 14.11 Front view of a V6 engine showing crank arrangement and direction of rotation.

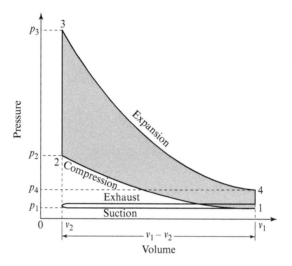

Fig. 14.12 An ideal indicator diagram for a four-stroke engine.

which is known as an *indicator diagram*. Known constraints of the indicator make it possible to study the diagram and determine the relationship between the gas pressure and the crank angle for the particular running conditions in existence at the time the diagram is recorded.

When an engine is in the design stage, it is necessary to estimate a diagram from theoretic considerations. From such an approximation, a pilot model of the proposed engine is designed and built, and the actual indicator diagram is taken and compared with the theoretically estimated one. This provides much useful information for the design of the production model.

An indicator diagram for the ideal air-standard cycle is shown in Fig. 14.12 for a four-stroke engine. During compression, the cylinder volume changes from v_1 to v_2 and the cylinder pressure changes from p_1 to p_2. The relationship, at any point of this stroke, is given by the polytropic gas law as

$$pv^k = p_1 v_1^k = \text{constant}. \tag{14.1}$$

In an actual indicator card, the corners at points 2 and 3 are rounded, and the line joining these points is curved. This is explained by the fact that combustion is not instantaneous, and ignition occurs before the end of the compression stroke. An actual card is also rounded at points 4 and 1, since the valves do not operate simultaneously.

The polytropic exponent, k in Eq. (14.1) is often taken to be about 1.30 for both compression and expansion, although differences probably exist.

The relationship between the power developed and the dimensions of the engine is given by

$$P_b = \frac{p_b lan \left(1\,000 \text{ N/m}^2/\text{kPa}\right)}{(1\,000 \text{ N} \cdot \text{m/s/kW})(1\,000 \text{ mm/m})(60 \text{ s/min})} = \frac{p_b lan}{60\,000}, \tag{14.2}$$

where

P_b = brake power per cylinder (kW);
p_b = brake mean effective pressure (kPa);
l = length of the stroke (mm);
a = piston area $\left(m^2\right)$;
n = number of working strokes per minute.

The amount of power that can be obtained from one liter of piston displacement varies considerably, depending upon the engine type. For typical automotive engines it ranges from about 25 to 45 kW/L, with a current average of perhaps 32. On the other hand, many marine diesel engines have ratios varying from 4.5 to 9 kW/L. About the best that can be done in designing a new engine is to use standard references to discover what others have done with the same types of engines, and then choose a value that seems to be reasonably attainable.

For many engines, the bore-to-stroke ratio varies from about 0.75 to 1.00. The tendency in automotive-engine design seems to be toward shorter-stroke engines to reduce engine height. Decisions on bore–stroke ratio and power per unit displacement volume are helpful in solving Eq. (14.2) to obtain suitable dimensions once the power, speed, and number of cylinders have been decided upon.

The ratio of the brake mean effective pressure, p_b, to the indicated mean effective pressure, p_i, obtained experimentally from an indicator card, is the mechanical efficiency, e_m:

$$e_m = \frac{p_b}{p_i}. \tag{14.3}$$

Differences between a theoretic and an experimentally determined indicator diagram can be accounted for by applying a correction called a *card factor*. The card factor is defined by

$$f_c = \frac{p_i}{p_i'}, \tag{14.4}$$

where p_i' is the theoretic indicated mean effective pressure, and the card factor is usually about 0.90 to 0.95.

Since the *compression ratio* (Fig. 14.12) is defined as

$$R = \frac{v_1}{v_2}, \tag{14.5}$$

the work done during compression is

$$U_c = \int_{v_2}^{v_1} p\,dv = p_1 v_1^k \int_{v_2}^{v_1} \frac{dv}{v^k} = \frac{p_1 v_1}{k-1} \left(R^{k-1} - 1\right). \tag{a}$$

The displacement volume can be written as

$$v_1 - v_2 = v_1 - \frac{v_1}{R} = \frac{v_1(R-1)}{R}. \tag{b}$$

Substituting v_1 from Eq. (b) into Eq. (a), the work done during compression can be written as

$$U_c = \frac{p_1(v_1 - v_2)}{k-1} \frac{R^k - R}{R-1}. \tag{c}$$

The work done during expansion is the area under the curve between points 3 and 4 of Fig. 14.12. This is determined in the same manner; the result is

$$U_e = \frac{p_4(v_1 - v_2)}{k-1} \frac{R^k - R}{R-1}. \tag{d}$$

The net amount of work accomplished in a cycle is the difference between Eqs. (c) and (d), and it must be equal to the product of the theoretical indicated mean effective pressure and the displacement volume. That is,

$$U = U_e - U_c = \frac{p_4(v_1 - v_2)}{k-1} \frac{R^k - R}{R-1} - \frac{p_1(v_1 - v_2)}{k-1} \frac{R^k - R}{R-1} = p_i'(v_1 - v_2). \tag{14.6}$$

If the exponent is the same for expansion as for compression, Eq. (14.6) can be solved to give

$$p_4 = p_1'(k-1)\frac{R-1}{R^k - R} + p_1. \tag{e}$$

Substituting the theoretical indicated mean effective pressure, p_i' [Eq. (14.4)] into Eq. (e) gives

$$p_4 = (k-1)\frac{R-1}{R^k - R}\frac{p_i}{f_c} + p_1. \tag{14.7}$$

Equations (14.1) and (14.7) can be used to create the theoretical indicator diagram. The corners are then rounded off so that the pressure at point 3 is made about 75% of that given by Eq. (14.1). As a check, the area of the diagram can be measured and divided by the displacement volume; the result should equal the indicated mean effective pressure.

14.3 Dynamic Analysis: General

Recall that Chapters 11 and 12 present both graphic and analytic methods for finding the static and dynamic forces in a mechanism. The balance of this chapter is devoted to an analysis of the dynamics of a single cylinder of an operating engine. To simplify this work, it is assumed that the engine is running at a constant crankshaft speed and that gravitation forces and friction forces can be neglected in comparison with dynamic force effects. The gas forces and inertia forces are determined separately in the upcoming sections; then, these forces are combined (Section 14.7) using the principle of superposition (Section 12.5) to obtain the total bearing forces and crankshaft torque.

The subject of engine balancing is treated in Chapter 15.

14.4 Gas Forces

Analytic or graphic force analysis methods can be used to solve this gas-force problem. The advantage of the analytic methods is that they can be programmed for automatic computation throughout a cycle. In contrast, the graphic technique must be repeated for each crank posture until a complete cycle of operation (720° for a four-stroke engine) is completed.

From Chapter 12 we recall that the dynamic force analysis will require the acceleration of the piston. Recall from Chapters 3 and 4 that the equations for the approximate position [Eq. (3.16a)] and the acceleration [Eq. (4.22a)] of the piston can be written as

$$x = l - \frac{r^2}{4l} + r\left(\cos \omega t + \frac{r}{4l} \cos 2\omega t\right), \tag{14.8a}$$

$$\ddot{x} = -r\omega^2 \left(\cos \omega t + \frac{r}{l} \cos 2\omega t\right) - r\alpha \left(\sin \omega t + \frac{r}{2l} \sin 2\omega t\right). \tag{14.8b}$$

For constant crank angular velocity, Eq. (14.8b) reduces to

$$\ddot{x} = -r\omega^2 \left(\cos \omega t + \frac{r}{l} \cos 2\omega t\right). \tag{14.8c}$$

We assume that the moving parts of the engine are massless so that gravity, inertia forces, and torques are zero. We also assume that the effects of friction can be neglected. These assumptions allow us to trace the force effects of the gas pressure from the piston to the crankshaft without the complicating effects of other forces. These other effects, except for friction, will be added later, using superposition.

Referring to Fig. 14.13, we designate a gas-force vector $\mathbf{P}$ as defined or obtained using the methods of Section 14.2. Reactions caused by this force are designated using a single prime. Thus, $\mathbf{F}'_{14}$ is the force from the cylinder wall acting against the piston and $\mathbf{F}'_{34}$ is the force of the connecting rod acting onto the piston at the piston pin. The force polygon in Fig. 14.13 shows the relation among $\mathbf{P}$, $\mathbf{F}'_{14}$, and $\mathbf{F}'_{34}$. Thus, we have

$$\mathbf{F}'_{14} = P \tan \phi \hat{\mathbf{j}}. \tag{14.9}$$

The quantity $\tan \phi$ appears frequently in expressions throughout the chapter; therefore, it is convenient to develop an expression for $\tan \phi$ in terms of the crank angle, $\theta = \omega t$. From Eqs. (a) and (d) in Section 3.9 this quantity can be written as

Fig. 14.13 Graphic analysis of forces resulting from gas pressure.

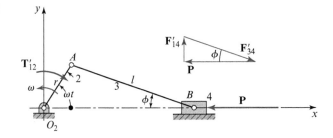

$$\tan \phi = \frac{(r/l) \sin \omega t}{\cos \phi} = \frac{(r/l) \sin \omega t}{\sqrt{1 - [(r/l) \sin \omega t]^2}}. \qquad (a)$$

Recall that the binomial expansion for a variable x can be written as

$$(1 - x)^n = 1 - nx + \frac{n(n-1)x^2}{2!} - \frac{n(n-1)(n-2)x^3}{3!} + \frac{n(n-1)(n-2)(n-3)x^4}{4!} - \cdots \qquad (b)$$

Therefore, the denominator on the right-hand side of Eq. (a) can be written as

$$\frac{1}{\sqrt{1 - [(r/l) \sin \omega t]^2}} = \left\{1 - [(r/l) \sin \omega t]^2\right\}^{-\frac{1}{2}} \approx 1 + \frac{r^2}{2l^2} \sin^2 \omega t, \qquad (c)$$

where only the first two terms of the expansion have been retained, see Section 3.9. Substituting Eq. (c) into Eq. (a) we have

$$\tan \phi = \frac{r}{l} \sin \omega t \left(1 + \frac{r^2}{2l^2} \sin^2 \omega t\right). \qquad (14.10)$$

Then, substituting this equation into Eq (14.9), the force from the cylinder wall acting against the piston is

$$\mathbf{F}'_{14} = P \frac{r}{l} \sin \omega t \left(1 + \frac{r^2}{2l^2} \sin^2 \omega t\right) \hat{\mathbf{j}}. \qquad (d)$$

The trigonometry of Fig. 14.13 indicates that the *wrist pin* (piston pin) bearing force has a magnitude of

$$F'_{34} = \frac{P}{\cos \phi} = \frac{P}{\sqrt{1 - [(r/l) \sin \omega t]^2}} = P\left(1 + \frac{r^2}{2l^2} \sin^2 \omega t\right), \qquad (e)$$

or, in vector notation,

$$\mathbf{F}'_{34} = P\hat{\mathbf{i}} - F'_{14}\hat{\mathbf{j}} = P\hat{\mathbf{i}} - P \tan \phi \hat{\mathbf{j}}. \qquad (14.11)$$

By taking moments about the crank center, we find the torque, $\mathbf{T}'_{21}$, delivered by the crank to the shaft is the product of the force, $\mathbf{F}'_{14}$, and the piston coordinate, x, that is

$$\mathbf{T}'_{21} = F'_{14}x\hat{\mathbf{k}}.$$

Substituting Eqs. (d) and (14.8a) into this equation, the torque is

$$\mathbf{T}'_{21} = P\frac{r}{l} \sin \omega t \left(1 + \frac{r^2}{2l^2} \sin^2 \omega t\right)\left[l - \frac{r^2}{4l} + r\left(\cos \omega t + \frac{r}{4l} \cos 2\omega t\right)\right]\hat{\mathbf{k}}. \qquad (f)$$

Multiplying these terms, and (with only a very small error) neglecting second or higher powers of (r/l) gives

$$\mathbf{T}'_{21} = Pr \sin \omega t \left(1 + \frac{r}{l} \cos \omega t\right) \hat{\mathbf{k}}. \tag{14.12}$$

This is the torque delivered to the crankshaft by the gas force; the counterclockwise direction is positive.

14.5 Equivalent Masses

Problems 12.5 and 12.6, at the end of Chapter 12, focused on the slider-crank linkage, which is the basis for most engine mechanisms. The dynamics of this linkage were analyzed using the methods presented in Chapter 12. In this chapter we are concerned with the same problem. However, here we will show certain simplifications that are customarily used to reduce the complexity of the algebraic solution process. These simplifications are approximations and do introduce certain errors. In this chapter we demonstrate the simplifications and comment on the errors that they introduce.

In analyzing the inertia forces caused by the connecting rod of an engine, it is often convenient to picture a portion of the mass of the rod as concentrated at the crankpin, A, with the remaining portion lumped at the wrist pin, B (Fig. 14.14). The reason for this is that the crankpin moves on a circle and the wrist pin on a straight line, and both of these motions are quite easy to analyze. However, the center of mass, G_3, of the connecting rod is somewhere between the crankpin and the wrist pin, and its motion is more complex and consequently more difficult to determine accurately in algebraic form.

The mass of the connecting rod m_3, is correctly located at the centroid, G_3. However, we can mentally divide this mass into two mass particles; one particle, m_{3B}, is concentrated at the wrist pin B, whereas the other particle, m_{3P}, is concentrated at the center of percussion, P (Section 12.6), for oscillation of the rod about point B. The separation of these masses is dynamically equivalent to the original rod if: (1) the total mass is identical, (2) the position of the centroid G_3 is unchanged, and (3) the mass moment of inertia is the same. Writing these three conditions in equation form gives

$$m_3 = m_{3B} + m_{3P}, \tag{a}$$

$$m_{3B}l_B = m_{3P}l_P, \tag{b}$$

Fig. 14.14

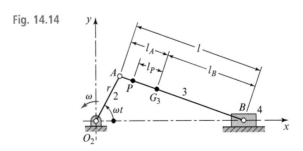

and

$$I_G = m_{3B}l_B^2 + m_{3P}l_P^2. \qquad (c)$$

These three equations can be solved to give the mass particles, m_{3B} and m_{3P}, and the location of the center of percussion, l_P. The procedure is as follows. Solving Eqs. (a) and (b) simultaneously gives the mass particles at the crankpin A and the wrist pin B:

$$m_{3B} = m_3 \frac{l_P}{l_B + l_P} \quad \text{and} \quad m_{3P} = m_3 \frac{l_B}{l_B + l_P}. \qquad (14.13)$$

Then substituting these equations into Eq. (c) gives

$$I_G = m_3 \frac{l_P}{l_B + l_P} l_B^2 + m_3 \frac{l_B}{l_B + l_P} l_P^2 = m_3 l_P l_B \qquad (d)$$

or

$$l_P l_B = \frac{I_G}{m_3}. \qquad (e)$$

This equation demonstrates that the two distances, l_P and l_B, are dependent on each other. Therefore, if l_B is known, then l_P becomes set by Eq. (e). The result can be shown to be purely geometric by substituting $I_G = m_3 k_G^2$ into the equation, and rearranging, to give

$$l_P = \frac{k_G^2}{l_B}. \qquad (14.14)$$

Note that this answer is in complete agreement with Eq. (12.20).

In the usual connecting rod, the center of percussion, P, is close to the crankpin A, and it is usual to assume that they are coincident. Thus, if we let $l_P = l_A$, Eqs. (14.13) reduce to

$$m_{3B} = \frac{m_3 l_A}{l} \quad \text{and} \quad m_{3A} = \frac{m_3 l_B}{l}. \qquad (14.15)$$

We note again that the equivalent masses, obtained by Eqs. (14.15), are not exact because of the assumption made, but they are close enough for ordinary connecting rods. This approximation, however, would not be valid for the master connecting rod of a radial engine, since the crankpin end has bearings for all of the other connecting rods as well as its own bearing.

For estimating and checking purposes, about two-thirds of the mass of the usual connecting rod is concentrated at the crankpin A, and the remaining one-third is concentrated at wrist pin B.

Figure 14.15 shows an engine linkage in which the mass of the crank m_2 is not balanced, as evidenced by the fact that the center of gravity, G_2, is located outward along the crank, a distance r_G from the axis of rotation. In the inertia force analysis, another simplification is made by replacing this by an equivalent mass m_{2A}, at the crankpin. For equivalence, this requires

$$m_2 r_G = m_{2A} r \quad \text{or} \quad m_{2A} = m_2 \frac{r_G}{r}. \qquad (14.16)$$

Fig. 14.15

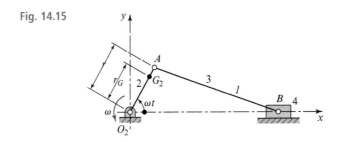

Fig. 14.16

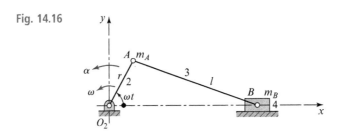

14.6 Inertia Forces

Using the methods of the previous section, we begin by locating equivalent masses at the crankpin and at the wrist pin. Thus,

$$m_A = m_{2A} + m_{3A}, \tag{14.17}$$

$$m_B = m_{3B} + m_4. \tag{14.18}$$

Equation (14.17) states that the rotating mass, m_A, located at the crankpin is composed of the equivalent mass, m_{2A}, of the crank, and the mass, m_{3A}, of a portion of the connecting rod. Of course, if the crank is balanced (Chapter 15), including balancing the m_{3A} portion of the connecting rod, all of its mass is located at the axis of rotation, and m_A is then zero. Equation (14.18) indicates that the reciprocating mass, m_B, located at the wrist pin is composed of the equivalent mass, m_{3B}, of the other portion of the connecting rod and the mass, m_4, of the piston assembly.

Figure 14.16 shows the (unbalanced) slider-crank linkage with masses m_A and m_B located at crankpin A and wrist pin B, respectively.

If we designate the angular velocity and angular acceleration of the crank as ω and α, respectively, the position vector of the crankpin A relative to the origin, O_2, is

$$\mathbf{R}_A = r\cos\omega t\,\hat{\mathbf{i}} + r\sin\omega t\,\hat{\mathbf{j}}. \tag{a}$$

Differentiating this equation twice with respect to time, the acceleration of crankpin A is

$$\mathbf{A}_A = \left(-r\alpha\sin\omega t - r\omega^2\cos\omega t\right)\hat{\mathbf{i}} + \left(r\alpha\cos\omega t - r\omega^2\sin\omega t\right)\hat{\mathbf{j}}. \tag{14.19}$$

The inertia force of the rotating mass is then

$$-m_A\mathbf{A}_A = m_A r\left(\alpha \sin \omega t + \omega^2 \cos \omega t\right)\hat{\mathbf{i}} + m_A r\left(-\alpha \cos \omega t + \omega^2 \sin \omega t\right)\hat{\mathbf{j}}. \qquad (14.20)$$

Since the analysis is usually performed for a constant angular velocity (that is, $\alpha = 0$), then Eq. (14.20) reduces to

$$-m_A\mathbf{A}_A = m_A r\omega^2\left(\cos \omega t\,\hat{\mathbf{i}} + \sin \omega t\,\hat{\mathbf{j}}\right). \qquad (14.21)$$

The acceleration of the piston is known from Eq. (14.18b); therefore, the inertia force of the reciprocating mass is

$$-m_B\mathbf{A}_B = \left[m_B r\omega^2\left(\cos \omega t + \frac{r}{l}\cos 2\omega t\right) + m_B r\alpha\left(\sin \omega t + \frac{r}{2l}\sin 2\omega t\right)\right]\hat{\mathbf{i}}. \qquad (14.22)$$

For constant angular velocity of the crank, Eq. (14.8c), Eq. (14.22) can be written as

$$-m_B\mathbf{A}_B = m_B r\omega^2\left(\cos \omega t + \frac{r}{l}\cos 2\omega t\right)\hat{\mathbf{i}}. \qquad (14.23)$$

Adding Eqs. (14.21) and (14.23) gives the total inertia force for all of the moving parts (for constant angular velocity). The components in the x and y directions are

$$F^x = (m_A + m_B)r\omega^2 \cos \omega t + \left(m_B \frac{r}{l}\right)r\omega^2 \cos 2\omega t \qquad (14.24)$$

and

$$F^y = m_A r\omega^2 \sin \omega t. \qquad (14.25)$$

It is customary to refer to the portion of the force occurring at the frequency ω rad/s as the *primary inertia force* and the portion occurring at 2ω rad/s as the *secondary inertia force*. We note that the x component, which is in the direction of the cylinder axis, has a primary part varying directly with the crankshaft speed and a secondary part varying at twice the crankshaft speed. On the other hand, the y component, perpendicular to the cylinder axis, has only a primary part, and it therefore varies directly with the crankshaft speed.

We proceed now to a determination of the inertia torque. As shown in Fig. 14.17, the inertia force caused by the mass at the crankpin A has no moment about O_2 and, therefore, produces no inertia torque. Consequently, we need consider only the inertia force given by Eq. (14.23) caused by the reciprocating mass, m_B.

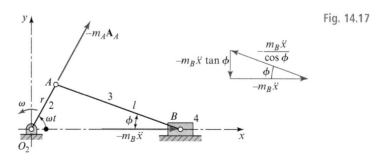

Fig. 14.17

From the force polygon of Fig. 14.17, the inertia torque exerted by the engine on the crankshaft is

$$\mathbf{T}_{21}'' = -(-m_B A_B \tan \phi) x \hat{\mathbf{k}}. \tag{b}$$

Substituting the expressions for x, A_B, and $\tan \phi$, as given by Eqs. (14.18a), (14.18c), and (14.10), into Eq. (b), the inertia torque can be written as

$$\mathbf{T}_{21}'' = -m_B r \omega^2 \left(\cos \omega t + \frac{r}{l} \cos 2\omega t \right) \left[\frac{r}{l} \sin \omega t \left(1 + \frac{r^2}{2l^2} \sin^2 \omega t \right) \right] \left[l - \frac{r^2}{4l} + r \left(\cos \omega t + \frac{r}{4l} \cos 2\omega t \right) \right] \hat{\mathbf{k}}. \tag{c}$$

Terms that are proportional to the second and higher powers of r/l are now neglected in performing the indicated products. Therefore, Eq. (c) can be written as

$$\mathbf{T}_{21}'' = -m_B r^2 \omega^2 \sin \omega t \left(\frac{r}{2l} + \cos \omega t + \frac{3r}{2l} \cos 2\omega t \right) \hat{\mathbf{k}}. \tag{d}$$

Substituting the identities

$$2 \sin \omega t \cos \omega t = \sin 2\omega t \tag{e}$$

and

$$2 \sin \omega t \cos 2\omega t = \sin 3\omega t - \sin \omega t \tag{f}$$

into Eq. (d), and rearranging, gives

$$\mathbf{T}_{21}'' = \frac{m_B}{2} r^2 \omega^2 \left(\frac{r}{2l} \sin \omega t - \sin 2\omega t - \frac{3r}{2l} \sin 3\omega t \right) \hat{\mathbf{k}}. \tag{14.26}$$

Equation (14.26) is the inertia torque exerted by the engine on the crankshaft in the positive direction. A clockwise, or negative, inertia torque of the same magnitude is, of course, exerted by the frame onto the engine.

The assumed distribution of the connecting-rod mass results in a moment of inertia that is greater than the true value. Consequently, the torque given by Eq. (14.26) is not the exact value. In addition, terms proportional to the second- and higher-order powers of r/l were neglected in simplifying Eq. (c). These two approximations are of about the same magnitude and are quite small for ordinary connecting rods having $r/l \leq 1/4$.

14.7 Bearing Loads in a Single-Cylinder Engine

The designer of a reciprocating engine must know the values of the forces acting upon the bearings and how these forces vary during a cycle of operation. These are necessary to proportion and select the bearings properly and are also needed for the design of other engine parts. This section is an investigation of the force exerted by the piston against the cylinder wall and the forces acting against the piston pin and against the crankpin. The main bearing forces

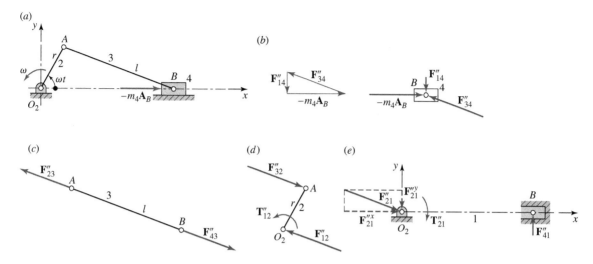

Fig. 14.18 Graphic analysis of forces resulting from mass m_4.

are investigated in a later section, since they depend upon the action within the cylinders of the engine.

The resultant total bearing loads are made up of the following components:

1. gas-force components, designated by a single prime;
2. inertia force due to the mass m_4 of the piston assembly, designated by a double prime;
3. inertia force of that part m_{3B} of the connecting rod assigned to the wrist-pin end, designated by a triple prime; and
4. connecting-rod inertia force of that part m_{3A} at the crankpin end, designated by a quadruple prime.

Equations for the gas-force components have been determined in Section 14.4, and references are made to them in finding the total bearing loads.

Figure 14.18 is a graphic analysis of the forces in the engine linkage with zero gas force and subjected only to inertia force resulting from the mass, m_4, of the piston assembly. Figure 14.18a shows the posture of the linkage selected for analysis, and the inertia force, $-m_4 \mathbf{A}_B$, is shown acting upon the piston. In Fig. 14.18b the free-body diagram of the piston is shown together with the polygon from which the forces are obtained. Figures 14.18c–14.18e, respectively, show the free-body diagrams showing forces acting upon the connecting rod, crank, and frame.

In Fig. 14.18e we note that the torque $\mathbf{T}_{21}''$ balances the couple formed by the forces $\mathbf{F}_{41}''$ and $\mathbf{F}_{21}''{}^y$. However, force $\mathbf{F}_{21}''{}^x$ at the crank center remains unopposed by any other force. This is a very important observation that we shall reserve for discussion in a separate section.

The following forces are of interest to us:

1. the force $\mathbf{F}_{41}''$ of the piston against the cylinder wall;
2. the force $\mathbf{F}_{34}''$ of the connecting rod against the wrist pin;

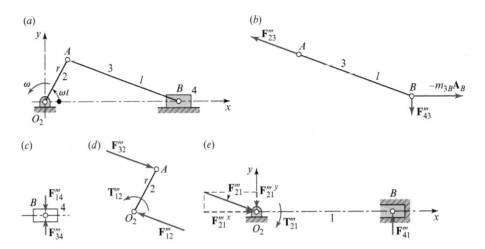

Fig. 14.19 Graphic analysis of the forces resulting from mass m_{3B}.

3. the force $\mathbf{F}''_{32}$ of the connecting rod against the crankpin; and
4. the force $\mathbf{F}''_{12}$ of the engine block against the crank.

By methods similar to those presented earlier in the chapter, the analytic expressions are determined to be:

$$\mathbf{F}''_{41} = -m_4\ddot{x}\tan\phi\hat{\mathbf{j}}, \tag{14.27}$$

$$\mathbf{F}''_{34} = m_4\ddot{x}\hat{\mathbf{i}} - m_4\ddot{x}\tan\phi\hat{\mathbf{j}}, \tag{14.28}$$

$$\mathbf{F}''_{32} = -\mathbf{F}''_{34}, \tag{14.29}$$

$$\mathbf{F}''_{12} = -\mathbf{F}''_{32} = \mathbf{F}''_{34}, \tag{14.30}$$

where m_4 is the mass of the piston assembly, $\ddot{x}$ is the acceleration of the piston [Eq. (14.8b)], and the quantity $\tan\phi$ can be written in terms of the crank angle [Eq. (14.10)].

Figure 14.19 shows only those forces that result because of that part m_{3B} of the mass of the connecting rod that is assumed to be concentrated at the wrist-pin center. Thus, Fig. 14.19b is a free-body diagram of the connecting rod showing the inertia force, $-m_{3B}\mathbf{A}_B$, acting at wrist pin B.

We note that it is incorrect when finding the bearing loads to add m_{3B} and m_4 together and then to compute a resultant inertia force, although such a procedure would seem to be simpler. The reason for this is that m_4 is the mass of the piston assembly and the corresponding inertia force acts on the piston side of the wrist pin. But m_{3B} is part of the connecting-rod mass, and hence its inertia force acts on the connecting-rod side of the wrist pin. Thus, adding the two would yield correct results for the crankpin load and the force of the piston against the cylinder wall but would give *incorrect* results for the wrist-pin load.

The forces on the piston pin, the crank, and the frame resulting from the inertia of mass m_{3B} are shown in Figs. 14.19c, 14.19d, and 14.19e, respectively. The equations for these forces for a crank having uniform angular velocity are determined to be

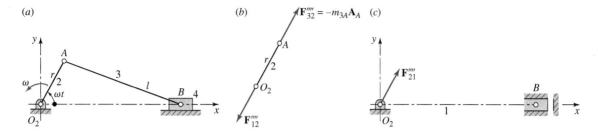

Fig. 14.20 Graphic analysis of the forces resulting from mass m_{3A}.

$$\mathbf{F}_{41}''' = -m_{3B}\ddot{x}\tan\phi\hat{\mathbf{j}},\tag{14.31}$$

$$\mathbf{F}_{34}''' = \mathbf{F}_{41}''',\tag{14.32}$$

$$\mathbf{F}_{32}''' = -m_{3B}\ddot{x}\hat{\mathbf{i}} + m_{3B}\ddot{x}\tan\phi\hat{\mathbf{j}},\tag{14.33}$$

$$\mathbf{F}_{12}''' = -\mathbf{F}_{32}'''.\tag{14.34}$$

Figure 14.20 shows the forces that result from that part m_{3A} of the connecting-rod mass that is assumed concentrated at the crankpin end. Whereas a counterweight attached to the crank balances the reaction at O_2, it cannot make F_{32}''' zero. Thus, the crankpin force exists no matter whether or not the rotating mass of the connecting rod is balanced. This force is

$$\mathbf{F}_{32}'''' = m_{3A}r\omega^2\left(\cos\omega t\hat{\mathbf{i}} + \sin\omega t\hat{\mathbf{j}}\right).\tag{14.35}$$

The last step is to sum the above individual component equations to obtain the resultant bearing loads. For example, the total force of the piston against the cylinder wall is found by summing Eqs.(14.9), (14.27), and (14.31), with due regard for subscripts and signs. When simplified, the result is

$$\mathbf{F}_{41} = \mathbf{F}_{41}' + \mathbf{F}_{41}'' + \mathbf{F}_{41}''' = -[P + (m_{3B} + m_4)\ddot{x}]\tan\phi\hat{\mathbf{j}}.\tag{14.36}$$

The forces on the wrist pin, the crankpin, and the crankshaft are determined in a similar manner and are

$$\mathbf{F}_{34} = (m_4\ddot{x} + P)\hat{\mathbf{i}} - [(m_{3B} + m_4)\ddot{x} + P]\tan\phi\hat{\mathbf{j}},\tag{14.37}$$

$$\mathbf{F}_{32} = [m_{3A}r\omega^2\cos\omega t - (m_{3B} + m_4)\ddot{x} - P]\hat{\mathbf{i}} + \{m_{3A}r\omega^2\sin\omega t +$$
$$[(m_{3B} + m_4)\ddot{x} + P]\tan\phi\}\hat{\mathbf{j}},\tag{14.38}$$

$$\mathbf{F}_{21} = \mathbf{F}_{32}.\tag{14.39}$$

The torque delivered by the crankshaft to the load is called the *crankshaft torque*, and it is the negative of the moment of the couple formed by the forces $\mathbf{F}_{41}$ and $\mathbf{F}_{21}^y$. Therefore, it is obtained from the equation

$$\mathbf{T}_{21} = -F_{41}x\hat{\mathbf{k}} = [(m_{3B} + m_4)\ddot{x} + P]x\tan\phi\hat{\mathbf{k}}.\tag{14.40}$$

14.8   Shaking Forces of Engines

The inertia force caused by the reciprocating masses is shown acting in the positive direction in Fig. 14.21a. In Fig. 14.21b, the forces acting upon the engine block caused by these inertia forces are shown. The resultant forces are $\mathbf{F}_{21}$, the force exerted by the crankshaft on the main bearings, and a positive couple formed by the forces $\mathbf{F}_{41}$ and $\mathbf{F}^y_{21}$. The force $\mathbf{F}^x_{21} = -m_B\mathbf{A}_B$ is termed the *shaking force*, and the couple $T = xF_{41}$ is called the *shaking moment*. As indicated by Eqs. (14.23) and (14.26), the magnitude and the direction of this force and couple change with ωt; consequently, the shaking force induces rectilinear vibration of the block in the x direction, and the shaking moment induces a torsional vibration of the block about the crank center.

A graphic representation of the inertia force is possible if Eq. (14.23) is rearranged as

$$F = m_B r \omega^2 \cos \omega t + m_B r \omega^2 \frac{r}{l} \cos 2\omega t, \qquad (14.41)$$

where we write $F = F^x_{21}$ for simplicity of notation. The first term of Eq. (14.41) is represented by the x projection of a vector of length $m_B r\omega^2$ rotating at ω rad/s. This is the primary part of the inertia force. The second term is similarly represented by the x projection of a vector of length $m_B r\omega^2 (r/l)$ rotating at 2ω rad/s; this is the secondary part. Such a diagram is shown in Fig. 14.22 for $r/l = \frac{1}{4}$. The total inertia or shaking force is the algebraic sum of the horizontal projections of these two vectors.

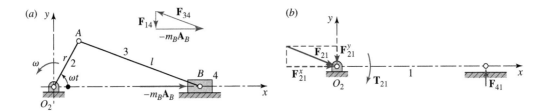

Fig. 14.21 Dynamic forces due to the reciprocating masses; the primes are omitted for clarity.

Fig. 14.22 Circle diagram illustrating inertia forces. The total shaking force is $OA' + OB'$.

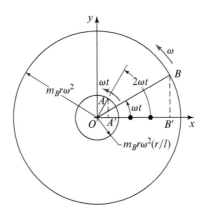

14.9 Computation Hints

This section contains suggestions for using a computer or programmable calculator in solving the dynamics of an engine mechanism. Many of the ideas, however, are also useful for readers using nonprogrammable calculators, as well as for checking purposes.

Indicator Diagrams.   It would be very convenient if a subprogram for computing the gas forces could be devised and the results used directly in a main program to compute all the resultant bearing forces and crankshaft torques. Unfortunately, the theoretical indicator diagram must be manipulated by hand to obtain a reasonable approximation to the experimental data. This manipulation can be done graphically or with a computer having a graphic display. The procedure is illustrated by the following example.

Example 14.1

Determine the pressure-versus-piston-displacement relation for a six-cylinder engine having a displacement of 2.30 L, a compression ratio of 8, and a brake power of 42.5 kW at 2 400 rev/min. Use a mechanical efficiency of 75%, a card factor of 0.85, a suction pressure of 100 kPa, and a polytropic exponent of 1.30.

Solution

Rearranging Eq. (14.2), we find the brake mean effective pressure as follows:

$$p_b = \frac{(42.5\ \text{kW})(1\,000\ \text{N}\cdot\text{m/s/kW})(60\text{s/min})}{(0.002\,30\ \text{m}^3)(2\,400\ \text{rev/min})(0.5\ \text{cycle/rev})(1\,000\text{N/m}^2/\text{kPa})} = 923.9\ \text{kPa}.$$

Then, from Eq. (14.3), the indicated mean effective pressure is

$$p_i = \frac{p_b}{e_m} = \frac{923.9\ \text{kPa}}{0.75} = 1\,232\ \text{kPa}.$$

Next, we determine the pressure, p_4, on the theoretical diagram of Fig. 14.12. Substituting the information into Eq. (14.7) gives

$$p_4 = (1.3 - 1)\left(\frac{8 - 1}{8^{1.3} - 8}\right)\frac{1\,232\ \text{kPa}}{0.85} + 100\ \text{kPa} = 355\ \text{kPa}.$$

The volume difference, $v_1 - v_2$, in Fig. 14.12 is the volume swept out by each piston. Therefore,

$$v_1 - v_2 = la = \frac{2.30\ \text{L}}{6\ \text{cyl}} = 0.383\ \text{L/cyl}.$$

Substituting this displacement volume and the compression ratio $R = 8$ into Eq. (b) of Section 14.2 gives

$$v_1 = \frac{R(v_1 - v_2)}{R - 1} = \frac{8(0.383\ \text{L})}{8 - 1} = 0.438\ \text{L}.$$

Example 14.1 (cont.)

Therefore,

$$v_2 = 0.438 \text{ L} - 0.383 \text{ L} = 0.055 \text{ L}.$$

The percentage clearance C is

$$C = \frac{v_2}{v_1 - v_2}(100\%) = \frac{0.055 \text{ L}}{0.383 \text{ L}}(100\%) = 14.4\%.$$

Expressing volumes as percentages of the displacement volume enables us to write Eq. (14.1) in the form

$$p_x(X + C)^k = p_1(100 + C)^k,$$

where X is the percentage of piston travel measured from the head end of the stroke. Thus, the formula

$$p_x = p_1\left(\frac{100 + C}{X + C}\right)^k = 1\,232 \text{ kPa}\left(\frac{114.4}{X + 14.4}\right)^{1.3} \tag{1}$$

is used to compute the pressure during the compression stroke for any piston position between $X = 0$ and $X = 100\%$. For the expansion stroke Eq. (14.1) becomes

$$p_x = p_4\left(\frac{100 + C}{X + C}\right)^k = 355 \text{ kPa}\left(\frac{114.4}{X + 14.4}\right)^{1.3}. \tag{2}$$

Fig. 14.23 The diagram was rounded by hand. The peak cylinder pressure is about 75% of the maximum computed pressure at the beginning of the expansion stroke.

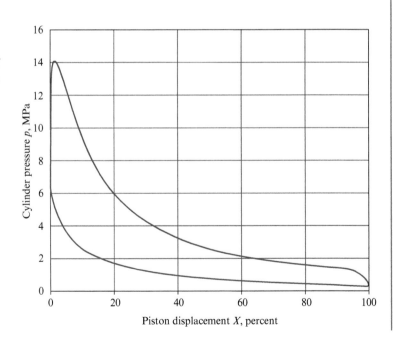

Example 14.1 (cont.)

Equations (1) and (2) are easy to program for machine computation. The results can be displayed and recorded, or printed, for graphic use. Alternatively, the results can be displayed for hand calculation.

Figure 14.23 shows the plotted results of the computation using $\Delta X = 5\%$ increments. Note particularly how the results are rounded to obtain a smooth indicator diagram. This rounding, of course, produces results that are not exactly duplicated in subsequent trials. The greatest differences occur in the vicinity of point B.

Force Analysis.   In a computer analysis the values of pressure are read from a diagram like Fig. 14.23. Most analysts prefer to tabulate these data; a table is usually constructed with the first column containing values of the crank angle, ωt. For a four-stroke engine, values of this angle are entered from $0°$ to $720°$.

Values of piston position, x, corresponding to each crank angle, ωt, are obtained from Eq. (3.15). Then the corresponding piston displacement, X, in percentage is obtained from the equation

$$X = \frac{r + l - x}{2r}(100\%). \qquad (14.42)$$

Some care must be taken in tabulating X and the corresponding pressures. Then, the gas forces corresponding to each value of ωt are computed using the piston area.

The remainder of the analysis is perfectly straightforward; we use Eqs. (14.8c), (14.10), and (14.36)–(14.40), in that order.

PROBLEMS

14.1 A one-cylinder, four-stroke engine has a compression ratio of 7.6 and develops brake power of 2.25 kW at 3 000 rev/min. The crank length is 22 mm with a 60 mm bore. Develop and plot a rounded indicator diagram using a card factor of 0.90, a mechanical efficiency of 72%, a suction pressure of 100 kPa, and a polytropic exponent of 1.30.

14.2 Construct a rounded indicator diagram for a four-cylinder, four-stroke gasoline engine having a 85 mm bore, a 90 mm stroke, and a compression ratio of 6.25. The operating conditions to be used are 22.4 kW at 1 900 rev/min. Use a mechanical efficiency of 72%, a card factor of 0.90, a suction pressure of 100 kPa, and a polytropic exponent of 1.30.

14.3 Construct an indicator diagram for a V6 four-stroke gasoline engine having a 100 mm bore, a 90 mm stroke, and a compression ratio of 8.40. The engine develops 150 kW at 4 400 rev/min. Use a mechanical efficiency of 72%, a card factor of 0.88, a suction pressure of 100 kPa, and a polytropic exponent of 1.30.

14.4 A single-cylinder, two-stroke gasoline engine develops 30 kW at 4 500 rev/min. The engine has an 80 mm bore, a stroke of 70 mm, and a compression ratio of 7.0. Develop a rounded indicator diagram for this engine using a card factor of 0.990, a mechanical efficiency of 65%, a suction pressure of 100 kPa, and a polytropic exponent of 1.30.

14.5 The engine of Prob. 14.1 has a connecting rod 80 mm long and a mass of 0.100 kg, with the mass center 10 mm from the crankpin end. Piston mass is 0.180 kg. Find the bearing reactions and the crankshaft torque during the expansion stroke corresponding to a piston displacement of $X = 30\%$ ($\omega t = 60°$). To find p_e, see the answer to Prob. 14.1 in Appendix B.

14.6 Repeat Prob. 14.5, but do the computations for the compression cycle ($\omega t = 660°$).

14.7 Make a complete force analysis of the engine of Prob. 14.5. Plot a graph of the crankshaft torque versus crank angle for 720° of crank rotation.

14.8 The engine of Prob. 14.3 uses a connecting rod 300 mm long. The masses are $m_{3A} = 0.80$ kg, $m_{3B} = 0.38$ kg, and $m_4 = 1.64$ kg. Find all the bearing reactions and the crankshaft torque for one cylinder of the engine during the expansion stroke at a piston displacement of $X = 30\%$ ($\omega t = 63.2°$). The pressure should be obtained from the indicator diagram, Fig. AP14.3 in Appendix B.

14.9 Repeat Prob. 14.8, but do the computations for the same position in the compression cycle ($\omega t = 656.8°$).

14.10 Additional data for the engine of Prob. 14.4 are $l_3 = 110$ mm, $R_{G_3A} = 15$ mm, $m_4 = 0.24$ kg, and $m_3 = 0.13$ kg. Make a complete force analysis of the engine and plot a graph of the crankshaft torque versus crank angle for 360° of crank rotation.

14.11 The four-stroke engine of Prob. 14.1 has a stroke of 66 mm and a connecting rod length of 183 mm. The mass of the rod is 0.386 kg, and the center of mass is 42 mm from the crankpin. The piston assembly has mass of 0.576 kg. Make a complete force analysis for one cylinder of this engine for 720° of crank rotation. Use 110 kPa for the exhaust pressure and 70 kPa for the suction pressure. Plot a graph to show the variation of the crankshaft torque with the crank angle. Use Fig. 14.23 for the pressures.

15 Balancing

Balancing is defined here as the process of correcting or eliminating unwanted inertia forces and moments in rotating machinery. In previous chapters, we have seen that shaking forces on the frame can vary significantly during a cycle of operation. Such forces can cause vibrations that at times may reach dangerous amplitudes. Even if they are not dangerous, vibrations increase component stresses and subject bearings to repeated loads that may cause parts to fail prematurely by fatigue. Thus, in the design of machinery, it is not sufficient merely to avoid operation near the critical speeds; we must eliminate, or at least reduce, the dynamic forces that produce these vibrations in the first place.

Tolerances used in the manufacture of machinery are typically set as tight as possible without prohibitively increasing the cost of manufacture. In general, however, it is more economic to produce parts that are not perfect and then to subject them to a balancing procedure than it is to produce such perfect parts that no correction is needed. Therefore, each part produced is an individual case, in that no two parts can normally be expected to require the same corrective measures. Thus, determining the unbalance and the required correction is the principal topic in the study of balancing.

15.1 Static Unbalance

Figure 15.1 shows a disk–shaft combination resting on rigid horizontal rails so that the shaft, which is assumed to be perfectly straight, can roll without friction. A reference system xyz is attached to the disk and moves with it (with the z axis as the axis of the shaft). Simple experiments to determine whether the disk is statically unbalanced can be conducted as follows. We can roll the disk gently by hand and permit it to coast until it comes to rest. Then we can mark with chalk the lowest point of the periphery of the disk. After we repeat this four or five times, if the chalk marks are scattered at different locations around the periphery, the disk is in static balance. If all of the chalk marks are clustered, the disk is statically unbalanced, which means that the axis of the shaft and the center of mass of the disk are not coincident. The position of the chalk marks with respect to the xy axes indicates the angular location of the unbalance but *not* the amount.

It is unlikely that any of the marks will be located 180° from the remaining cluster, although it is theoretically possible to obtain static equilibrium with the unbalance above the shaft axis.

Fig. 15.1 Static balancing experiment.

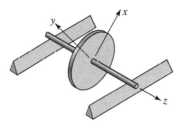

Fig. 15.2 Unbalanced disk and shaft.

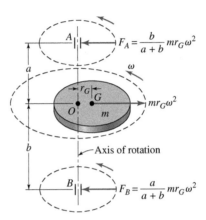

If static unbalance is found to exist, it can be corrected by drilling out material at the chalk mark cluster or by adding mass to the periphery 180° from the cluster. Since the amount of unbalance is unknown, the correction magnitude must be chosen by trial and error.

15.2 Equations of Motion

If an unbalanced disk and shaft of mass m is mounted in bearings and caused to rotate at an angular speed of ω, then an inertia force of magnitude $mr_G\omega^2$ exists (where r_G, that is, the eccentricity, is the distance from the shaft axis, O, to the mass center, G, of the disk), as shown in Fig. 15.2. The figure also shows the force acting on the shaft that produces the rotating bearing reactions F_A and F_B.

To determine the equation of motion of the system, we specify m as the total mass and m_u as the unbalanced mass. We also let k signify the shaft stiffness, a value that describes the amount of force necessary to bend the shaft a unit distance when applied at O. Thus, k has units of pounds per inch or newtons per meter. Let c be the coefficient of viscous damping as defined in Eq. (13.2). Selecting any x coordinate normal to the shaft axis, we can now write

$$\sum F_O = -kx - c\dot{x} - m\ddot{x} + m_u r_G\omega^2 \cos \omega t = 0,$$

or, upon rearranging,

$$m\ddot{x} + c\dot{x} + kx = m_u r_G \omega^2 \cos \omega t. \qquad (a)$$

Recall that the solution to this differential equation was studied in Sections 13.12 and 13.13. From Eq. (13.55), the deflection of the shaft can be written as

$$x = \frac{m_u r_G \omega^2 \cos(\omega t - \phi)}{\sqrt{(k - m\omega^2)^2 + c^2 \omega^2}}, \qquad (b)$$

where ϕ is the phase angle between the force $m_u r_G \omega^2$ and the amplitude X of the shaft vibration. The *phase angle* is given by Eq. (13.50) and is repeated here for convenience:

$$\phi = \tan^{-1}\left(\frac{c\omega}{k - m\omega^2}\right). \qquad (c)$$

Certain simplifications can be made in Eq. (b) to clarify its meaning. First, consider the term $k - m\omega^2$ in the denominator. If this term were zero, the amplitude of x would be very large, since it would be limited only by the damping constant, c, which is usually very small. The value of ω that makes the term $k - m\omega^2$ equal to zero is denoted as ω_n and is called the *natural angular velocity*, *critical speed*, or *natural frequency*. Therefore,

$$\omega_n = \sqrt{\frac{k}{m}}, \qquad (15.1)$$

which agrees with the definition in Section 13.2, Eq. (13.8).

In the study of free or unforced vibrations (Section 13.8), it is found that a certain value of the viscous damping factor, c, results in no vibration at all. This special value is called the *critical coefficient of viscous damping* [Eq. (13.27)], that is,

$$c_c = 2m\omega_n. \qquad (15.2)$$

The *damping ratio* ζ is the ratio of the actual to the critical damping, Eq. (13.28), that is,

$$\zeta = \frac{c}{c_c} = \frac{c}{2m\omega_n}. \qquad (15.3)$$

For most machine systems in which damping is not deliberately introduced, ζ is generally in the range $0.015 \le \zeta \le 0.120$.

Recall that the deflection can be written from Eq. (13.15) or Eq. (13.23) as

$$x = X \cos(\omega t - \phi). \qquad (d)$$

Substituting this equation into Eq. (b), and dividing the numerator and denominator of the result by k, and introducing Eqs. (15.1) and (15.3), we obtain the ratio given by Eq. (13.56), that is,

$$\frac{mX}{m_u r_G} = \frac{(\omega/\omega_n)^2}{\sqrt{(1 - \omega^2/\omega_n^2)^2 + (2\zeta\omega/\omega_n)^2}}, \qquad (15.4)$$

where $r_G = e$. This is the equation for the amplitude ratio of the vibration of a rotating disk–shaft combination. If we neglect damping, that is, set $\zeta = 0$, and let $m = m_u$, then the amplitude of the vibration corresponding to any frequency ratio ω/ω_n can be written as

$$X = r_G \frac{(\omega/\omega_n)^2}{1 - (\omega/\omega_n)^2}. \tag{15.5}$$

Figure 15.3 is a plot of Eq. (15.5) where the amplitude is the ordinate and the frequency ratio is the abscissa. When rotation is just beginning, ω is much less than ω_n and the graph indicates that the amplitude of the vibration is very small. As the shaft speed increases, the amplitude also increases and theoretically becomes infinite at the critical speed. As the shaft goes through the critical speed, the amplitude changes to a negative value and decreases as the shaft speed increases further. In the high-speed range, the disk is rotating about its own center of mass, which has then become coincident with the centerline of the bearings.

In summary, we note the following interesting conclusions:

a. The amplitude approaches infinity when the denominator becomes zero, that is, when the speed of rotation is equal to the natural frequency.
b. Regardless of how small r_G is (even if it is zero), trouble can always be expected when $\omega = \omega_n$.
c. The amplitude switches from positive to negative as the speed passes through the natural frequency.
d. The amplitude never returns to zero, no matter how much the shaft speed is increased, but reaches a limiting value of $-r_G$.

The discussion in Sections 15.1 and 15.2 demonstrates that statically unbalanced rotating systems produce undesirable vibrations and bearing reactions. Using static balancing equipment, the eccentricity can be reduced, but it is impossible in practice to make it exactly zero (see note b, above). When the operating speed must be higher than the natural frequency, a machine should be designed so that, on startup, it passes through the natural frequency quickly to avoid the buildup of dangerous vibrations.

Fig. 15.3 The small figures below the graph indicate the relative locations of points O and G for four typical frequency ratios.

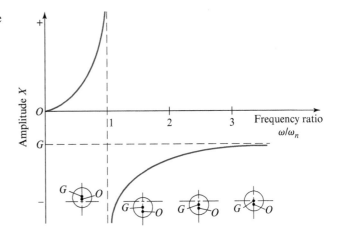

15.3   Static Balancing Machines

The first purpose of a balancing machine is to indicate whether a mechanical part is in balance. If it is out of balance, the balancing machine should indicate the unbalance by showing its *magnitude* and *location*.

Static balancing machines are used only for parts whose axial dimensions are small, such as gears, fans, and impellers, and the machines are often called *single-plane balancers*, since the mass must lie practically in a single plane. In the sections to follow, we discuss balancing in several planes, but it is important to note here that if several disks are to be mounted upon a shaft that is to rotate, the parts should be individually statically balanced before mounting. Although it is possible, instead, to balance the assembly in two planes after the parts are mounted, additional external moments inevitably come into existence when this is done.

Static balancing is essentially a weighing process in which the part is acted upon either by gravity or by inertia force. We have seen that the disk and shaft of the preceding section could be balanced by placing it on parallel rails, rocking it, and permitting it to seek equilibrium. In this case, the location of the unbalance is determined through the force of gravity. Another method of balancing the disk is to rotate it at a predetermined speed. Then, the bearing reactions can be measured and their magnitudes used to indicate the amount of unbalance. Since the part is rotating while the measurements are taken, a stroboscope is used to indicate the location of the required correction.

When machine parts are manufactured in quantity, a balancer is required that measures both the amount and location of the unbalance and gives the correction directly and quickly. Also, time can be saved if it is not necessary to make the part rotate. Such a balancing machine is shown in Fig. 15.4. This machine is essentially a pendulum that can tilt in any direction, as shown by the schematic drawing of Fig. 15.5a.

When an unbalanced specimen is mounted on the balancing machine, the pendulum tilts. The direction of the tilt gives the location of the unbalance, whereas the tilt angle θ (Fig. 15.5b),

Fig. 15.4 Aircraft propeller assembly balancer. (Courtesy of Schenck Ro Tec Corp., Auburn Hills, MI.)

Fig. 15.5 Operation of a
static balancing machine.

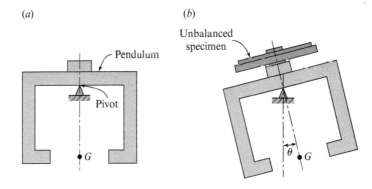

Fig. 15.6 If $m_1 = m_2$ and
$r_1 = r_2$, the rotor is statically
balanced but dynamically
unbalanced.

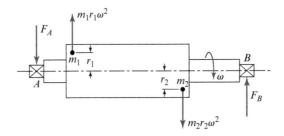

indicates the magnitude of the unbalance. Some damping is employed to reduce oscillations of
the pendulum.

15.4 Dynamic Unbalance

Figure 15.6 shows a long rotor with distributed mass that is mounted in bearings at A and B.
Suppose that two equal masses m_1 and m_2 are placed at opposite ends of the rotor and at equal
distances r_1 and r_2 from the axis of rotation. Since the masses are equal and on opposite sides of
the rotational axis, placing the rotor on rails (as described in Section 15.1) demonstrate that it is
statically balanced in all angular orientations. If the rotor is placed in bearings and caused to
rotate at an angular velocity ω, then m_1 and m_2 create inertia forces of magnitudes $m_1 r_1 \omega^2$ and
$m_2 r_2 \omega^2$, respectively. These inertia forces produce bearing reaction forces $\mathbf{F}_A$ and $\mathbf{F}_B$ that rotate
with the rotor. Therefore, a rotating part may be statically balanced and, at the same time,
dynamically unbalanced.

   For the single-mass system in Fig. 15.7a, both bearing reaction forces are in the same plane
and in the same direction. Therefore, the system is statically unbalanced. For the two-mass
system (with equal masses m) in Fig. 15.7b, the static unbalance creates a couple tending to turn
the shaft end over end. The shaft is in static equilibrium because of the opposite couple formed
by the bearing reactions. Note that the bearing reaction forces $\mathbf{F}_A$ and $\mathbf{F}_B$ are still in the same
plane, however, they now act in opposite directions. Therefore, the system is dynamically
unbalanced. Note also that a rotating mass system may be modeled as either a discrete or a

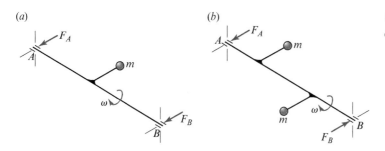

(a)

(b)

Fig. 15.7 (a) Static unbalance; (b) dynamic unbalance.

distributed system. Procedures for static and dynamic balancing of such systems are presented in the following section.

In the general case, accurate distribution of the mass along the axis of a part depends upon the configuration of the part, but inaccuracies occur in machining and also in casting and in forging. Other errors or unbalances may be caused by improper boring, keys, and assembly. It is the designer's responsibility to design a rotating part so that a line joining all mass centers is a straight line coinciding with the axis of rotation. However, perfect parts and perfect assembly are seldom attained; consequently, a line from one end of the part to the other, joining all mass centers, is usually a spatial curve that may only occasionally cross or coincide with the axis of rotation. Before balancing, therefore, a part is usually out of balance, both statically and dynamically. This is the most general kind of unbalance, and when the part is supported by two bearings, we can expect the magnitudes as well as the directions of the rotating bearing reactions to be different.

15.5 Analysis of Unbalance

In this section we will see how to analyze either a discrete mass or a distributed mass unbalanced rotating system and determine proper corrections using graphic methods, vector methods, and numeric methods. This section shows that a system can be balanced either by adding or removing correcting (or balancing) masses.

Graphic Analysis.   The two equations

$$\sum \mathbf{F} = \mathbf{0} \ \text{ and } \ \sum \mathbf{M} = \mathbf{0} \tag{a}$$

are used to determine the amounts and orientations of the corrections. We begin by noting that the magnitude of the inertia force, $m\mathbf{R}\omega^2$, of an eccentric rotating mass is proportional to its $m\mathbf{R}$ product. Thus, for Fig. 15.8a, vector quantities (proportional to the inertia force of each of the three masses) $m_1\mathbf{R}_1$, $m_2\mathbf{R}_2$, and $m_3\mathbf{R}_3$, act in radial directions, as indicated. The first of Eqs. (a) is applied by constructing a force polygon as shown in Fig. 15.8b. Since this polygon requires another vector, $m_c\mathbf{R}_c$, for closure, the magnitude of the correction is $m_c\mathbf{R}_c$, and its orientation is parallel to the vector $m_c\mathbf{R}_c$. The three masses of Fig. 15.8 are assumed to rotate in a single plane, and so this is the correction for a case of static unbalance.

Fig. 15.8 (*a*) Three-mass system rotating in a single plane; (*b*) inertia force polygon.

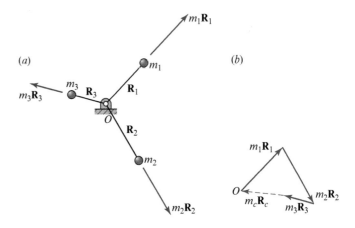

When the rotating masses are in different planes, both of Eqs. (*a*) must be used. Note that taking the sum of moments about a point in one of the correcting planes eliminates that unknown correcting mass. This is illustrated in the following example.

Example 15.1

The end view of a shaft with masses m_1, m_2, and m_3 at radial locations $\mathbf{R}_1$, $\mathbf{R}_2$, and $\mathbf{R}_3$, respectively, is shown in Fig. 15.9*a*. A side view of the shaft showing left and right correcting

Fig. 15.9 Graphic analysis of unbalance.

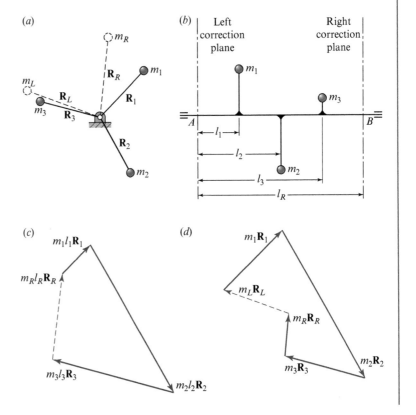

Example 15.1 (cont.)

planes and the axial distances l_1, l_2, and l_3 of the three masses is shown in Fig. 15.9b. Determine the magnitudes and angular orientations of the corrections to be added in each correcting plane.

Solution

Applying the second of Eqs. (a), the sum of the moments of the inertia forces about point A (in order to eliminate the moment of the left correction mass) gives

$$\sum M_A = m_1 l_1 \mathbf{R}_1 + m_2 l_2 \mathbf{R}_2 + m_3 l_3 \mathbf{R}_3 + m_R l_R \mathbf{R}_R = \mathbf{0}. \tag{1}$$

This is a vector equation in which the directions of the vectors can be drawn parallel, respectively, to the vectors $\mathbf{R}_j$ of Fig. 15.9a. Consequently, the moment polygon in Fig. 15.9c can be constructed. The closing vector, $m_R l_R \mathbf{R}_R$, gives the magnitude and orientation of the correction required for the right-hand correcting plane.

Although Fig. 15.9c is called a moment polygon, it is worth noting that the vectors making up this polygon consist of quantities proportional to the moment magnitudes and are drawn in the position vector directions. A true moment polygon would require rotating this polygon 90° ccw from the orientation shown, since each moment vector is properly expressed as $l\hat{\mathbf{k}} \times m\mathbf{R}\omega^2$.

Applying the first of Eqs. (a),

$$\sum F = m_1 \mathbf{R}_1 + m_2 \mathbf{R}_2 + m_3 \mathbf{R}_3 + m_R \mathbf{R}_R + m_L \mathbf{R}_L = \mathbf{0}. \tag{2}$$

By constructing the force polygon of Fig. 15.9d, this equation is solved for the left-hand correction, $m_L \mathbf{R}_L$.

After solution for the $m_R \mathbf{R}_R$ and $m_L \mathbf{R}_L$ products, the separate quantities m_R, R_R, m_L, and R_L can be determined, since, in general, two of these magnitudes are specified in the problem statement or are chosen by the designer.

Vector Analysis. This section will show how to analyze an unbalanced rotating system and determine proper corrections using a vector algebra approach. The following two examples illustrate this method.

Example 15.2

Figure 15.10 represents a rotating system that has been idealized for illustrative purposes. A weightless shaft that is supported in bearings at A and B rotates with an angular velocity $\omega = 100\hat{\mathbf{i}}$ rad/s. Three mass particles, with weights $w_1 = 2$ oz, $w_2 = 1$ oz, and $w_3 = 1.5$ oz, are connected to the shaft and rotate with it, causing unbalance. Determine the magnitudes and orientations of the bearing reactions at A and B.

Example 15.2 (cont.)

Fig. 15.10 $a = 1$ in, $b = 2$ in, $r_1 = 3$ in, $r_2 = 2$ in, and $r_3 = 2.5$ in.

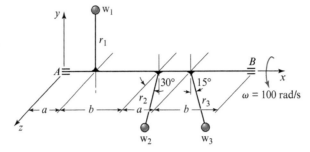

Solution

The inertia force caused by each rotating mass particle is

$$m_1 r_1 \omega^2 = \frac{(2 \text{ oz})(3 \text{ in})(100 \text{ rad/s})^2}{(386 \text{ in/s}^2)(16 \text{ oz/lb})} = 9.72 \text{ lb},$$

$$m_2 r_2 \omega^2 = \frac{(1 \text{ oz})(2 \text{ in})(100 \text{ rad/s})^2}{(386 \text{ in/s}^2)(16 \text{ oz/lb})} = 3.24 \text{ lb},$$

and

$$m_3 r_3 \omega^2 = \frac{(1.5 \text{ oz})(2.5 \text{ in})(100 \text{ rad/s})^2}{(386 \text{ in/s}^2)(16 \text{ oz/lb})} = 6.07 \text{ lb}.$$

These three forces are each parallel to the yz plane, and can be written in vector form as

$$\mathbf{F}_1 = m_1 r_1 \omega^2 \angle \theta_1 = 9.72 \text{ lb} \angle 0° = 9.72 \hat{\mathbf{j}} \text{ lb},$$

$$\mathbf{F}_2 = m_2 r_2 \omega^2 \angle \theta_2 = 3.24 \text{ lb} \angle 120° = -1.62 \hat{\mathbf{j}} + 2.81 \hat{\mathbf{k}} \text{ lb},$$

and

$$\mathbf{F}_3 = m_3 r_3 \omega^2 \angle \theta_3 = 6.07 \text{ lb} \angle 195° = -5.86 \hat{\mathbf{j}} - 1.57 \hat{\mathbf{k}} \text{ lb},$$

where the angles θ_1, θ_2, and θ_3 are measured counterclockwise from the y axis when viewed from the positive x axis. The moments of these forces taken about the bearing at A must be balanced by the moment of the bearing reaction at B. Therefore,

$$\sum \mathbf{M}_A = (1 \hat{\mathbf{i}} \text{ in}) \times (9.72 \hat{\mathbf{j}} \text{ lb}) + (3 \hat{\mathbf{i}} \text{ in}) \times (-1.62 \hat{\mathbf{j}} + 2.81 \hat{\mathbf{k}} \text{ lb})$$
$$+ (4 \hat{\mathbf{i}} \text{ in}) \times (-5.86 \hat{\mathbf{j}} - 1.57 \hat{\mathbf{k}} \text{ lb}) + (6 \hat{\mathbf{i}} \text{ in}) \times (\mathbf{F}_B) = \mathbf{0}.$$

Solving this equation, the bearing reaction at B is

$$\mathbf{F}_B = 3.10 \hat{\mathbf{j}} - 0.36 \hat{\mathbf{k}} \text{ lb} = 3.12 \text{ lb} \angle -6.6°. \qquad Ans.$$

Example 15.2 (cont.)

To find the bearing reaction at A, we repeat the procedure. Taking moments about B gives

$$\sum \mathbf{M}_B = \left(-2\hat{\mathbf{i}} \text{ in}\right) \times \left(-5.86\hat{\mathbf{j}} - 1.57\hat{\mathbf{k}} \text{ lb}\right) + \left(-3\hat{\mathbf{i}} \text{ in}\right) \times \left(-1.62\hat{\mathbf{j}} + 2.81\hat{\mathbf{k}} \text{ lb}\right)$$
$$+\left(-5\hat{\mathbf{i}} \text{ in}\right) \times \left(9.72\hat{\mathbf{j}} \text{ lb}\right) + \left(-6\hat{\mathbf{i}} \text{ in}\right) \times \left(\mathbf{F}_A\right) = \mathbf{0}.$$

Then solving this equation, the bearing reaction at A is

$$\mathbf{F}_A = -5.34\hat{\mathbf{j}} - 0.88\hat{\mathbf{k}} \text{ lb} = 5.41 \text{ lb}\angle 189.4°. \qquad\qquad Ans.$$

Note that these are rotating reactions and that static (stationary) components caused by gravity are not included.

Example 15.3

The angular speed of the system shown in Fig. 15.11 is 750 rev/min. Determine (a) the magnitudes and orientations of the bearing reactions at A and B, and (b) the magnitude and location of a correcting mass to be added at a radius of 0.25 m.

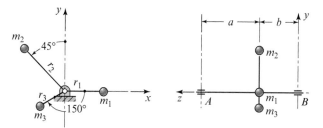

Fig. 15.11 $a = 0.3$ m, $b = 0.2$ m, $r_1 = 0.2$ m, $r_2 = 0.3$ m, $r_3 = 0.15$ m, $m_1 = 12$ kg, $m_2 = 3$ kg, and $m_3 = 10$ kg.

Solution

a. The angular velocity of the system is

$$\omega = (750 \text{ rev/min})(2\pi \text{ rad/rev})/(60 \text{ s/min}) = 78.54 \text{ rad/s}.$$

The inertia forces caused by the rotating masses are

$$F_1 = m_1 r_1 \omega^2 = (12 \text{ kg})(0.2 \text{ m})(78.54 \text{ rad/s})^2 = 14.80 \text{ kN},$$

$$F_2 = m_2 r_2 \omega^2 = (3 \text{ kg})(0.3 \text{ m})(78.54 \text{ rad/s})^2 = 5.55 \text{ kN},$$

and

$$F_3 = m_3 r_3 \omega^2 = (10 \text{ kg})(0.15 \text{ m})(78.54 \text{ rad/s})^2 = 9.25 \text{ kN}.$$

Example 15.3 (cont.)

These forces can be written in vector form as

$$\mathbf{F}_1 = 14.80 \text{ kN} \angle 0° = 14.80 \hat{\mathbf{i}} \text{ kN},$$
$$\mathbf{F}_2 = 5.55 \text{ kN} \angle 135° = -3.93 \hat{\mathbf{i}} + 3.93 \hat{\mathbf{j}} \text{ kN},$$
$$\mathbf{F}_3 = 9.25 \text{ kN} \angle -150° = -8.01 \hat{\mathbf{i}} - 4.63 \hat{\mathbf{j}} \text{ kN}.$$

To find the bearing reaction at B, we take moments about the bearing A. This equation is written as

$$\sum \mathbf{M}_A = \left(-0.3 \hat{\mathbf{k}} \text{ m}\right) \times \left[\left(14.80 \hat{\mathbf{i}} \text{ kN}\right) + \left(-3.93 \hat{\mathbf{i}} + 3.93 \hat{\mathbf{j}} \text{ kN}\right)\right.$$
$$\left. + \left(-8.01 \hat{\mathbf{i}} - 4.63 \hat{\mathbf{j}} \text{ kN}\right)\right] + \left(-0.5 \hat{\mathbf{k}} \text{ m}\right) \times \mathbf{F}_B = \mathbf{0}.$$

Taking the cross-products and rearranging gives

$$\left(0.5 \hat{\mathbf{k}} \text{ m}\right) \times \mathbf{F}_B = -0.21 \hat{\mathbf{i}} - 0.86 \hat{\mathbf{j}} \text{ kN} \cdot \text{m}.$$

Solving this equation for $\mathbf{F}_B$ gives

$$\mathbf{F}_B = -1.72 \hat{\mathbf{i}} + 0.42 \hat{\mathbf{j}} \text{ kN} = 1.77 \text{ kN} \angle 166.28°. \qquad Ans.$$

Summing forces gives

$$\mathbf{F}_A + \mathbf{F}_1 + \mathbf{F}_2 + \mathbf{F}_3 + \mathbf{F}_B = \mathbf{0}.$$

Therefore, the reaction at A is

$$\mathbf{F}_A = -\left(14.80 \hat{\mathbf{i}} \text{ kN}\right) - \left(-3.93 \hat{\mathbf{i}} + 3.93 \hat{\mathbf{j}} \text{ kN}\right)$$
$$- \left(-8.01 \hat{\mathbf{i}} - 4.63 \hat{\mathbf{j}} \text{ kN}\right) - \left(-1.72 \hat{\mathbf{i}} + 0.42 \hat{\mathbf{j}} \text{ kN}\right)$$

or

$$\mathbf{F}_A = -1.14 \hat{\mathbf{i}} + 0.28 \hat{\mathbf{j}} \text{ kN} = 1.17 \angle 166.20° \text{kN}. \qquad Ans.$$

b. Let $\mathbf{F}_C$ be the correcting force. Then for zero bearing reactions

$$\sum \mathbf{F} = \mathbf{F}_1 + \mathbf{F}_2 + \mathbf{F}_3 + \mathbf{F}_C = \mathbf{0}.$$

Thus,

$$\mathbf{F}_C = -\left(14.80 \hat{\mathbf{i}} \text{ kN}\right) - \left(-3.93 \hat{\mathbf{i}} + 3.93 \hat{\mathbf{j}} \text{ kN}\right) - \left(-8.00 \hat{\mathbf{i}} - 4.63 \hat{\mathbf{j}} \text{ kN}\right)$$
$$= -2.86 \hat{\mathbf{i}} + 0.70 \hat{\mathbf{j}} \text{ kN} = 2.94 \text{ kN} \angle 166.25°.$$

Therefore, the correcting mass should be added at $166.25°$. $\qquad Ans.$

At a radius of 0.25 m the correcting mass is

$$m_C = \frac{F_C}{r_C \omega^2} = \frac{2.94(10)^3 \text{ N}}{0.25 \text{ m}(78.54 \text{ rad/s})^2} = 1.91 \text{ kg}. \qquad Ans.$$

Scalar Equations.   For a rotating system with n discrete mass particles and two correcting planes ($j = 1, 2$) we can write the following four scalar equations:

$$\sum_{i=1}^{n} m_i r_i \cos \phi_i + \sum_{j=1}^{2} m_{cj} r_{cj} \cos \phi_{cj} = 0, \tag{15.6a}$$

$$\sum_{i=1}^{n} m_i r_i \sin \phi_i + \sum_{j=1}^{2} m_{cj} r_{cj} \sin \phi_{cj} = 0, \tag{15.6b}$$

$$\sum_{i=1}^{n} z_i m_i r_i \cos \phi_i + \sum_{j=1}^{2} z_{cj} m_{cj} r_{cj} \cos \phi_{cj} = 0, \tag{15.7a}$$

$$\sum_{i=1}^{n} z_i m_i r_i \sin \phi_i + \sum_{j=1}^{2} z_{cj} m_{cj} r_{cj} \sin \phi_{cj} = 0. \tag{15.7b}$$

Note that the equations do not depend on the angular velocity of the rotating system; that is, if the system is balanced for one speed then it is balanced for all speeds.

For a distributed mass system and two correcting planes we can write the following four scalar equations:

$$F_{A_{21}}^x + F_{B_{21}}^x + \omega^2 \sum_{j=1}^{2} m_{cj} r_{cj} \cos \phi_{cj} = 0, \tag{15.8a}$$

$$F_{A_{21}}^y + F_{B_{21}}^y + \omega^2 \sum_{j=1}^{2} m_{cj} r_{cj} \sin \phi_{cj} = 0, \tag{15.8b}$$

$$z_A F_{A_{21}}^x + z_B F_{B_{21}}^x + \omega^2 \sum_{j=1}^{2} z_j m_{cj} r_{cj} \cos \phi_{cj} = 0, \tag{15.9a}$$

$$z_A F_{A_{21}}^y + z_B F_{B_{21}}^y + \omega^2 \sum_{j=1}^{2} z_j m_{cj} r_{cj} \sin \phi_{cj} = 0. \tag{15.9b}$$

A Direct Method of Dynamic Balancing.   To avoid solving four equations in four unknowns, that is Eqs. (15.6) and (15.7) or Eqs. (15.8) and (15.9), we can take moments about one correcting plane and solve for two unknowns. Then we can take moments about the other correcting plane and solve for the remaining two unknowns. Recall that if a rotating system is in dynamic balance, then it is automatically in static balance. However, if the system is in static balance there is no guarantee that it is in dynamic balance (Section 15.4).

The following example illustrates the direct method of dynamic balancing of a distributed mass system.

Example 15.4

The distributed mass system shown in Fig. 15.12 has been tested for unbalance by rotating the rotor at an angular velocity of 100 rad/s. The bearing reactions on the frame at the two bearings A and B are $F_A = 40$ lb and $F_B = 35$ lb, respectively. Determine the magnitudes and orientations of the correcting masses to be removed in the specified planes 1 and 2 to achieve dynamic balance. The correcting masses are to be located at the radii $r_1 = r_2 = r = 4$ in.

Fig. 15.12 $a = 5$ in and $b = 40$ in.

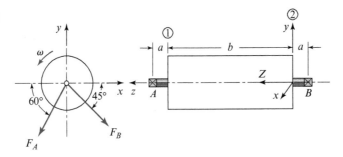

Solution

The x and y components of the forces acting on the frame at the bearings A and B are

$$F^x_{A_{21}} = 40\cos 240°\,\text{lb} = -20\text{ lb}, \tag{1a}$$

$$F^y_{A_{21}} = 40\sin 240°\,\text{lb} = -20\sqrt{3}\text{ lb}, \tag{1b}$$

$$F^x_{B_{21}} = 35\cos(-45°)\text{ lb} = 17.5\sqrt{2}\text{ lb}, \tag{1c}$$

$$F^y_{B_{21}} = 35\sin(-45°)\text{ lb} = -17.5\sqrt{2}\text{ lb}. \tag{1d}$$

Equations (15.8) and (15.9) can be written as

$$m_1 r\omega^2 \cos\phi_1 + m_2 r\omega^2 \cos\phi_2 = -F^x_{A_{21}} - F^x_{B_{21}}, \tag{2a}$$

$$m_1 r\omega^2 \sin\phi_1 + m_2 r\omega^2 \sin\phi_2 = -F^y_{A_{21}} - F^y_{B_{21}}, \tag{2b}$$

$$z_1 m_1 r\omega^2 \cos\phi_1 + z_2 m_2 r\omega^2 \cos\phi_2 = -z_A F^x_{A_{21}} - z_B F^x_{B_{21}}, \tag{3a}$$

$$z_1 m_1 r\omega^2 \sin\phi_1 + z_2 m_2 r\omega^2 \sin\phi_2 = -z_A F^y_{A_{21}} - z_B F^y_{B_{21}}. \tag{3b}$$

Using the direct method, we take moments about one of the two balancing planes. Here we take moments about plane 2 and note that the distances are

$$z_1 = +40\text{ in}, \quad z_2 = 0, \quad z_A = +45\text{ in}, \quad \text{and } z_B = -5\text{ in.} \tag{4}$$

Substituting Eqs. (4) and the known data into Eqs. (2) and (3) gives

$$(4\text{ in})(100\text{ rad/s})^2 (m_1\cos\phi_1 + m_2\cos\phi_2) = -4.749\text{ lb}, \tag{5a}$$

Example 15.4 (cont.)

$$(4 \text{ in})(100 \text{ rad/s})^2(m_1\sin\phi_1 + m_2\sin\phi_2) = 59.390 \text{ lb}, \qquad (5b)$$

$$(40 \text{ in})(4 \text{ in})(100 \text{ rad/s})^2 m_1\cos\phi_1 = 1\,023.744 \text{ in}\cdot\text{lb}, \qquad (6a)$$

$$(40 \text{ in})(4 \text{ in})(100 \text{ rad/s})^2 m_1\sin\phi_1 = 1\,435.102 \text{ in}\cdot\text{lb}. \qquad (6b)$$

Dividing Eq. (6b) by Eq. (6a), the location of the first correcting mass is

$$\phi_1 = \tan^{-1}\left(\frac{1\,435.102 \text{ in}\cdot\text{lb}}{1\,023.744 \text{ in}\cdot\text{lb}}\right) = 54.50°. \qquad \textit{Ans. (7a)}$$

Substituting Eq. (7a) into Eq. (6a) or (6b), the first correcting mass is

$$m_1 = \frac{(1\,435.102 \text{ in}\cdot\text{lb})\left(386 \text{ in/s}^2\right)}{\left(1\,600\,000 \text{ in}^2/\text{s}^2\right)\sin 54.5°} = 0.425 \text{ lb}. \qquad \textit{Ans. (7b)}$$

Substituting Eqs. (7) into Eqs. (5), rearranging, and dividing Eq. (5b) by Eq. (5a), the location of the second correcting mass is

$$\phi_2 = \tan^{-1}\left(\frac{23.535 \text{ lb}}{-30.324 \text{ lb}}\right) = 142.18°. \qquad \textit{Ans. (8a)}$$

Substituting Eq. (8a) into Eq. (5a) or (5b), the second correcting mass is

$$m_2 = \frac{(-30.342 \text{ lb})\left(386 \text{ in/s}^2\right)}{(40\,000 \text{ in/s}^2)\cos 142.2°} = 0.370 \text{ lb}. \qquad \textit{Ans. (8b)}$$

Note that the answers given by Eq. (7a) and (8a) are for adding mass to the system.
  To balance the rotating system by removing mass, we use:

$$\phi_{\text{remove}} = \phi_{\text{add}} \pm 180°.$$

Therefore, using this equation the mass to be removed from plane 1 is

$$m_1 = 0.425 \text{ lb} \text{ at the angle } 54.50° + 180° = 234.50° = -125.50°. \qquad \textit{Ans.}$$

Also, the mass to be removed from plane 2 is

$$m_2 = 0.370 \text{ lb} \text{ at the angle } 142.18° + 180° = 322.18° = -37.82°. \qquad \textit{Ans.}$$

Numeric Analysis.   For this approach, it is convenient to choose the xy plane as the plane of rotation with z as the axis of rotation, as shown in Fig. 15.13. In this manner, the unbalance vectors $m_i\mathbf{R}_i$ and the two correction vectors, $m_L\mathbf{R}_L$ in the left plane and $m_R\mathbf{R}_R$ in the right plane, can be expressed in the two-dimensional polar notation $m\mathbf{R} = mR\angle\theta$. This makes it easy to use the polar–rectangular conversion feature and its inverse, found on many calculators.

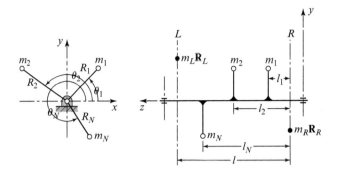

Fig. 15.13 Notation for numeric solution; correction vectors are not shown in the end view.

Note that Fig. 15.13 has $m_1, m_2, \ldots, m_N$ unbalances. Solving Eqs. (b) and (c) we have

$$m_L \mathbf{R}_L = -\sum_{i=1}^{N} \frac{m_i l_i}{l} \mathbf{R}_i, \tag{15.10}$$

$$m_R \mathbf{R}_R = -m_L \mathbf{R}_L - \sum_{i=1}^{N} m_i \mathbf{R}_i. \tag{15.11}$$

These two equations can easily be solved. If a calculator is used, it is suggested that the summation key be employed with each term of the summation entered.

15.6 Dynamic Balancing

The units in which each unbalance is measured have customarily been the ounce · inch (oz · in), the gram · centimeter (g · cm), or the mixed unit of gram · inch (g · in). If correct practice is followed in the use of SI units, however, the most appropriate unit of unbalance is the milligram · meter (mg · m), since prefixes in multiples of 1 000 are preferred in SI; thus, the prefix centi- is not recommended. Furthermore, only one prefix should be used in a compound unit, and, preferably, the first-named quantity should be prefixed. Thus, although both are acceptable in size, neither the gram · centimeter nor the kilogram · millimeter should be used. In this text we use the ounce · inch (oz · in) and the milligram · meter (mg · m) for units of unbalance.

We have seen that static balancing is sufficient for rotating disks, wheels, gears, and the like, where the mass can be assumed to exist in a single rotating plane. In the case of longer-axis machine elements, such as turbine rotors or motor armatures, the unbalanced inertia forces result in a couple whose effect tends to cause the rotor to turn end over end. The purpose of balancing is to determine this unbalanced couple and to add a new couple of equal magnitude in the opposite direction. The new couple is introduced by the addition of masses in two preselected correction planes or by subtracting (drilling out) masses from these two planes. To require balancing, a rotor usually has both static and dynamic unbalance; consequently, the correction masses, their radial orientations, or both are not the same for the two correction masses; also their radial orientations need not be the same for the two correction planes. This

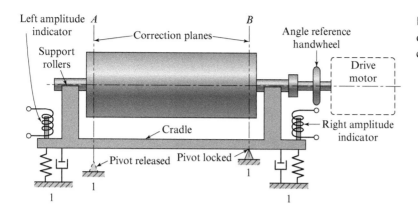

Fig. 15.14 Schematic drawing of a pivoted-cradle balancing machine.

means that the angular separation of the correction masses on the two planes is usually not 180°. Thus, to balance a rotor, we must measure the magnitude and the angular orientation of the correction mass for each of the two correction planes.

Three methods of measuring the corrections for two planes are in general use, the *pivoted-cradle*, the *nodal-point*, and the *mechanical-compensation* methods.

Figure 15.14 schematically illustrates a pivoted-cradle balancing machine with a rotor to be balanced, mounted on half-bearings or rollers attached to a cradle. The right end of the rotor is connected to a drive motor through a universal joint. The cradle can be rocked about either of two points that are adjusted to coincide with the two correction planes on the rotor. The left pivot is shown in the released position and the cradle and rotor are free to rock or oscillate about the right pivot, which is in the locked position. Springs and dashpots are secured at each end of the cradle to provide a single-degree-of-freedom vibrating system. Often, these are made adjustable so that the motor speed can be tuned to match the natural frequency. Amplitude indicators at each end of the cradle are differential transformers, or they may consist of a permanent magnet mounted on the cradle that moves relative to a stationary coil to generate a voltage proportional to the unbalance.

With the pivots located in the two correction planes, the operator can lock either pivot and take readings of the amount and orientation angle of the correction in the unlocked plane. The readings obtained are completely independent of the measurements taken in the other correction plane, since an unbalance in the plane of the locked pivot has no moment about that pivot. With the right pivot locked, an unbalance correctable in the left correction plane causes vibration whose amplitude is measured by the left amplitude indicator. Once this correction is measured, the right pivot is released, the left pivot is locked, and another set of measurements is made for the right correction plane using the right amplitude indicator.

Not shown in the balancing machine of Fig. 15.14 is a sine-wave signal generator that is attached to the drive shaft. If the resulting sine-wave signal generated is compared on a dual-beam oscilloscope with the wave generated by one of the amplitude indicators, a phase difference is found. The angular phase difference is the angular orientation of the unbalance. In a balancing machine, an electronic phasemeter measures the phase angle and gives the result on another meter calibrated in degrees. To locate the correction on the rotor, the angular reference handwheel is turned by hand until the indicated angle is in line with a reference pointer. This places the heavy side of the rotor in a preselected posture and permits the correction to be made.

Fig. 15.15.   (a)

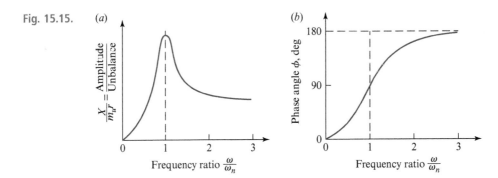

(b)

The relation between the amount of unbalance and the measured amplitude is given by Eq. (15.4). Rearranging that equation, and substituting r for e, the amplitude of the motion can be written as

$$X = \frac{m_u r(\omega/\omega_n)^2}{m\sqrt{\left(1 - \omega^2/\omega_n^2\right)^2 + (2\zeta\omega/\omega_n)^2}},$$ (15.12)

where $m_u r$ is the unbalance, and m is the mass of the cradle and the specimen. This equation demonstrates that the amplitude X is directly proportional to the unbalance, $m_u r$. Figure 15.15a shows a plot of this equation for a particular damping ratio, ζ, and demonstrates that the machine is most sensitive near resonance ($\omega = \omega_n$), since, in this region, the greatest amplitude is recorded for a given unbalance. Damping is deliberately introduced in balancing machines to filter noise and other vibrations that might affect the results. Damping also helps to maintain calibration against effects of temperature and other environmental conditions.

Substituting Eqs. (15.1) and (15.3) into Eq. (c) of Section 15.2 and rearranging, the phase angle can be written in parametric form as

$$\phi = \tan^{-1} \frac{2\zeta\omega/\omega_n}{1 - \omega^2/\omega_n^2}.$$ (15.13)

A plot of this equation for a single damping ratio and for various frequency ratios is shown in Fig. 15.15b. This curve indicates that, at resonance, when the speed, ω, matches the natural frequency, ω_n, the displacement lags the unbalance by the angle $\phi = 90°$. If the top of the rotor is turning away from the operator, the unbalance is horizontal and directly in front of the operator when the displacement is maximum downward. The angular orientation approaches $180°$ as ω is increased above resonance.

15.7 Dynamic Balancing Machines

A dynamic balancing machine for high-speed production is shown in Fig. 15.16.

Nodal-Point Balancing.   Plane separation using a point of zero or minimum vibration is called the *nodal-point method of balancing*. To see how this method is used, examine Fig. 15.17.

Fig. 15.16 Engine crankshaft dynamic balancing machine. (Courtesy of Schenck Ro Tec Corp., Auburn Hills, MI.)

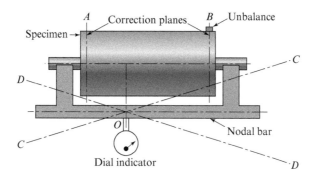

Fig. 15.17 Plane separation by the nodal-point method.

Here, the specimen to be balanced is shown mounted on bearings that are fastened to a nodal bar which experiences the same vibration as the specimen. We assume that the specimen is already balanced in the left-hand correction plane and that an unbalance still exists in the right-hand plane, as shown. Because of this unbalance, a vibration of the entire assembly takes place, causing the nodal bar to oscillate about some point O, occupying first (exaggerated) posture CC and then DD. Point O is easily located by sliding a dial indicator along the nodal bar; a point of zero motion or minimum motion is readily determined. This is the null or nodal point. Its location is the center of oscillation for a center of percussion (Section 12.6) in the right correction plane.

We assumed, at the beginning of this discussion, that no unbalance existed in the left correction plane. However, if unbalance is present, its magnitude is given by the dial indicator located at the nodal point just determined. Thus, by locating the dial indicator at this nodal point, we measure the unbalance in the left plane without any interference from that in the right plane. In a similar manner, another nodal point is found that measures only the unbalance in the right correction plane without any interference from that in the left plane.

In commercial balancing machines employing the nodal-point principle, the plane separation is accomplished in electronic networks. Typical of these is the Micro Dynamic Balancer; a schematic is shown in Fig. 15.18. On this machine, a switching knob selects either correction

Fig. 15.18 Diagram of the electronic circuit in a Micro Dynamic Balancer. (Courtesy of Schenck Ro Tec Corp., Auburn Hills, MI.)

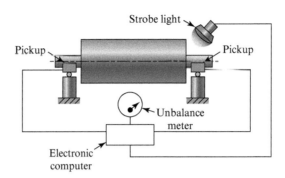

plane and displays the unbalance on a voltmeter, which is calibrated in appropriate unbalance units. The computer contains a filter that eliminates bearing noise and other frequencies not related to the unbalance. A multiplying network is used to give the sensitivity desired and to cause the meter to read in preselected balance units. The strobe light is driven by an oscillator that is synchronized to the rotor speed.

The rotor is driven at a speed that is much greater than the natural frequency of the system, and, since the damping is quite small, the direction of the vibration is horizontal, and the phase angle is approximately 180°. Marked on the right-hand end of the rotor are degrees or numbers that are readable and stationary under the strobe light during rotation of the rotor. Thus, it is only necessary to observe the particular station number of the degree marking under the strobe light to locate the heavy spot. When the switch is shifted to the other correction plane, the meter again reads the amount, and the strobe light illuminates the station. Sometimes as few as five station numbers distributed uniformly around the periphery are adequate for balancing. Rotation such that the top of the rotor moves away from the operator causes the heavy spot to be in a horizontal plane and on the near side of the axis when illuminated by the strobe lamp. A pointer is usually placed here to indicate its location. If, during production balancing, it is found that the phase angle is less than 180°, the pointer can be shifted slightly to indicate the proper position to observe.

Mechanical Compensation. An unbalanced rotating rotor mounted in a balancing machine produces a vibration. Counterforces can be introduced in each correction plane in the balancing machine that exactly balance the forces causing the vibration. The result of introducing these forces is a smooth-running system. Upon stopping, the location and amount of each counterforce can be measured to give the exact correction required. This is called *mechanical compensation.*

When mechanical compensation is used, the speed of the rotor during balancing is not important since the equipment is in calibration for all speeds. The rotor may be driven by a belt, through a universal joint, or it may be self-driven if, for example, it is a gasoline engine. The electronic equipment is simple, no built-in damping is necessary, and the machine is easy to operate, since the unbalances in both correction planes are measured simultaneously and the magnitudes and orientations are read directly.

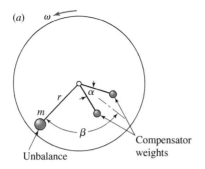

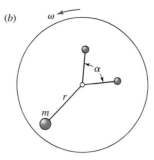

Fig. 15.19 (a) Postures of compensator weights; (b) after compensation.

We can understand how mechanical compensation is applied by examining Fig. 15.19a. Looking at the end of the rotor, we see one of the correction planes with the unbalance to be corrected (represented by *mr*). The correction plane, viewed along the axis of rotation to indicate the unbalance, and the two compensator weights are shown in Fig. 15.19a. All three of these weights are to rotate with the same angular velocity ω, but the postures of the compensator weights relative to one another and their posture relative to the unbalanced weight can be varied by two controls. One of these controls changes the angle α, that is, the angle between the compensator weights. The other control changes the angular posture of the pair of compensator weights relative to the unbalance, that is, the angle β. The knob that changes the angle β is the *orientation control*, and, when the rotor is compensated (balanced) in this plane, a pointer on the knob indicates the exact angular orientation of the unbalance. The knob that changes the angle α is the *amount control*, and it also gives a direct reading when the rotor unbalance is compensated as shown in Fig. 15.19b. The magnitude of the vibration is measured electrically and displayed on a voltmeter. Thus, compensation is found when the controls are manipulated to make the voltmeter read zero.

15.8 Field Balancing with a Programmable Calculator[1]

Field balancing is necessary for very large rotors for which balancing machines are impractical. Despite the fact that high-speed rotors are balanced in the shop during manufacture, it is frequently necessary to rebalance them in the field because of slight deformations brought on by shipping, creep, or high operating temperatures.

It is possible to balance a machine in the field by balancing a single plane at a time. But cross-coupling effects and correction-plane interference often require balancing each end of a rotor two or three times to obtain satisfactory results. Some machines may require as much as an hour to bring them up to full speed, resulting in even more delays in the balancing procedure.

---

[1] The authors are grateful to Prof. W. B. Fagerstrom, University of Delaware, for contributing some of the ideas of this section.

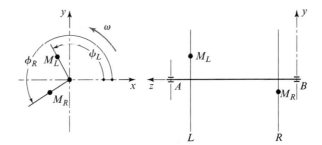

Fig. 15.20 Notation for two-plane field balancing.

Both Rathbone [5] and Thearle [8] have developed methods of two-plane field balancing that can be expressed in complex-number notation and solved using a programmable calculator. The time saved is several hours when compared with graphic methods or analysis with complex numbers using an ordinary scientific calculator.

In the analysis that follows, boldface letters are used to represent complex numbers; for example

$$\mathbf{R} = R\angle\theta = Re^{j\theta} = x + jy.$$

In Fig. 15.20, unknown unbalances $\mathbf{M}_L$ and $\mathbf{M}_R$ are assumed to exist in the left and right correction planes, respectively. The magnitudes of these unbalances are M_L and M_R, and they are located at angles ϕ_L and ϕ_R from the rotating reference, denoted as the xy system. When these unbalances have been found, their negatives are located in the left and right planes to achieve balance.

The rotating unbalances $\mathbf{M}_L$ and $\mathbf{M}_R$ produce disturbances at bearings A and B. Using commercial field-balancing equipment, it is possible to measure the amplitudes and the angular orientations of these disturbances. The notation $\mathbf{X} = X\angle\phi$, with appropriate subscripts, is used to designate these quantities.

In field balancing, three runs or tests are made, as follows:

- *First run:* Measure vector $\mathbf{X}_A = X_A\angle\phi_A$ at bearing A and vector $\mathbf{X}_B = X_B\angle\phi_B$ at bearing B due only to the original unbalances, $\mathbf{M}_L = M_L\angle\phi_L$ and $\mathbf{M}_R = M_R\angle\phi_R$.
- *Second run:* Add trial mass $\mathbf{m}_L = m_L\angle\theta_L$ to the left correction plane, and measure the amplitudes $\mathbf{X}_{AL} = X_{AL}\angle\phi_{AL}$ and $\mathbf{X}_{BL} = X_{BL}\angle\phi_{BL}$ at the left and right bearings (A and B), respectively.
- *Third run:* Remove trial mass $\mathbf{m}_L = m_L\angle\theta_L$. Add trial mass $\mathbf{m}_R = m_R\angle\theta_R$ to the right-hand correction plane and again measure the bearing amplitudes. These results are designated $\mathbf{X}_{AR} = X_{AR}\angle\phi_{AR}$ for bearing A and $\mathbf{X}_{BR} = X_{BR}\angle\phi_{BR}$ for bearing B.

Note in the above runs that the term "trial mass" means the same as a trial unbalance, and a unit distance from the axis of rotation is used.

To develop the equations for the unbalance being sought, we first define *complex stiffness* as the amplitude that would result at either bearing caused by a unit unbalance located at the intersection of the rotating reference mark and one of the correction planes. Thus, we must find the complex stiffnesses $\mathbf{A}_L$ and $\mathbf{B}_L$ caused by a unit unbalance located at the intersection of the rotating reference mark and plane L. In addition, we require the complex stiffnesses $\mathbf{A}_R$ and $\mathbf{B}_R$ caused by a unit unbalance located at the intersection of the rotating reference mark and plane R.

If these stiffnesses were available, we could write the following sets of complex equations:

$$\mathbf{X}_{AL} = \mathbf{X}_A + \mathbf{A}_L \mathbf{m}_L, \quad \mathbf{X}_{BL} = \mathbf{X}_B + \mathbf{B}_L \mathbf{m}_L, \tag{a}$$

$$\mathbf{X}_{AR} = \mathbf{X}_A + \mathbf{A}_R \mathbf{m}_R, \quad \mathbf{X}_{BR} = \mathbf{X}_B + \mathbf{B}_R \mathbf{m}_R. \tag{b}$$

After the three runs are made, the stiffnesses are the only unknowns in these equations. Therefore,

$$\mathbf{A}_L = \frac{\mathbf{X}_{AL} - \mathbf{X}_A}{\mathbf{m}_L}, \quad \mathbf{B}_L = \frac{\mathbf{X}_{BL} - \mathbf{X}_B}{\mathbf{m}_L}, \quad \mathbf{A}_R = \frac{\mathbf{X}_{AR} - \mathbf{X}_A}{\mathbf{m}_R}, \quad \mathbf{B}_R = \frac{\mathbf{X}_{BR} - \mathbf{X}_B}{\mathbf{m}_R}. \tag{15.14}$$

Then, from the definitions of these stiffnesses, we have from the first run

$$\mathbf{X}_A = \mathbf{A}_L \mathbf{M}_L + \mathbf{A}_R \mathbf{M}_R, \quad \mathbf{X}_B = \mathbf{B}_L \mathbf{M}_L + \mathbf{B}_R \mathbf{M}_R. \tag{c}$$

Solving this pair of equations simultaneously gives

$$\mathbf{M}_L = \frac{\mathbf{X}_A \mathbf{B}_R - \mathbf{A}_R \mathbf{X}_B}{\mathbf{A}_L \mathbf{B}_R - \mathbf{A}_R \mathbf{B}_L}, \quad \mathbf{M}_R = \frac{\mathbf{A}_L \mathbf{X}_B - \mathbf{X}_A \mathbf{B}_L}{\mathbf{A}_L \mathbf{B}_R - \mathbf{A}_R \mathbf{B}_L}. \tag{15.15}$$

These equations can be programmed, either in complex polar form or in complex rectangular form. The suggestions that follow were formed assuming a complex rectangular form for the solution.

Since the original data are formulated in polar coordinates, a subprogram should be written to transform the data into rectangular coordinates before storage.

The equations reveal that complex subtraction, division, and multiplication are used often. These operations can be set up as subprograms to be called from the main program. If $\mathbf{A} = a + jb$ and $\mathbf{B} = c + jd$, the formula for complex subtraction is

$$\mathbf{A} - \mathbf{B} = (a - c) + j(b - d). \tag{15.16}$$

For complex multiplication the formula is

$$\mathbf{AB} = (ac - bd) + j(bc + ad), \tag{15.17}$$

and for complex division the formula is

$$\frac{\mathbf{A}}{\mathbf{B}} = \frac{(ac + bd) + j(bc - ad)}{c^2 + d^2}. \tag{15.18}$$

With these subprograms it is an easy matter to program Eqs. (15.14) and (15.15).

As a check on your programming, you may wish to use the following data: $\mathbf{X}_A = 8.6\angle 63°$, $\mathbf{X}_B = 6.5\angle 206°$, $\mathbf{m}_L = 10\angle 270°$, $\mathbf{m}_R = 12\angle 180°$, $\mathbf{X}_{AL} = 5.9\angle 123°$, $\mathbf{X}_{BL} = 4.5\angle 228°$, $\mathbf{X}_{AR} =$

$6.2\angle36°$, $\mathbf{X}_{BR} = 10.4\angle162°$. The correct results are $\mathbf{M}_L = 10.76\angle146.6°$ and $\mathbf{M}_R = 6.20$ $\angle245.4°$.

According to Fagerstrom, the vibration angles used can be expressed in two different systems. The first is the rotating-protractor-stationary-mark system (RPSM). This is the system used in the preceding analysis and is the one a theoretician would prefer. In actual practice, however, it is usually easier to have the protractor stationary and use a rotating mark like a key or keyway. This is called the rotating-mark-stationary-protractor system (RMSP). The only difference between the two systems is in the sign of the vibration angle, but there is no sign change on the trial or correction masses.

15.9 Balancing a Single-Cylinder Engine

The rotating masses in a single-cylinder engine can be balanced using the methods already discussed in this chapter. The reciprocating masses, however, cannot be completely balanced, and so our studies in this section are really concerned with minimizing the unbalance. Although the reciprocating masses cannot be totally balanced using a simple counterweight, it is possible to modify the shaking forces (Section 14.8) by unbalancing the rotating masses. As an example of this, let us add a counterweight opposite the crankpin whose mass exceeds the rotating mass by one-half of the reciprocating mass (from one-half to two-thirds of the reciprocating mass is usually added to the counterweight to alter the balance characteristics in a single-cylinder engine). We designate the mass of the counterweight m_C, substitute this mass into Eq. (14.24), and use a negative sign, since the counterweight is opposite the crankpin; then, the inertia force caused by this counterweight is

$$\mathbf{F}_C = -m_C r\omega^2 \cos\omega t\hat{\mathbf{i}} - m_C r\omega^2 \sin\omega t\hat{\mathbf{j}}. \qquad (a)$$

Note that the balancing mass and the crankpin both have the same radius. Designating by m_A and m_B the masses of the rotating and the reciprocating parts, respectively, as in Chapter 14, then, according to the supposition above, we have

$$m_C = m_A + \frac{1}{2} m_B. \qquad (b)$$

Equation (a) can now be written

$$\mathbf{F}_C = -\left(m_A + \frac{1}{2}m_B\right)r\omega^2 \cos\omega t\hat{\mathbf{i}} - \left(m_A + \frac{1}{2}m_B\right)r\omega^2 \sin\omega t\hat{\mathbf{j}}. \qquad (c)$$

From Eqs. (14.24) and (14.25), the inertia force caused by the rotating and reciprocating masses can be written as

$$\mathbf{F}_{A,B} = F^x\hat{\mathbf{i}} + F^y\hat{\mathbf{j}} = \left[(m_A + m_B)r\omega^2 \cos\omega t + \left(m_B\frac{r}{l}\right)r\omega^2 \cos2\omega t\right]\hat{\mathbf{i}} + m_A r\omega^2 \sin\omega t\hat{\mathbf{j}}. \qquad (d)$$

Adding Eqs. (c) and (d), the resultant inertia force can be written as

$$\mathbf{F} = \left[\frac{1}{2} m_B r\omega^2 \cos\omega t + \left(m_B\frac{r}{l}\right)r\omega^2 \cos2\omega t\right]\hat{\mathbf{i}} - \frac{1}{2} m_B r\omega^2 \sin\omega t\hat{\mathbf{j}}. \qquad (15.19)$$

The first term is called the *primary component*. This component has a magnitude $\frac{1}{2}m_B r\omega^2$ and can be represented as a backward-rotating (clockwise) vector with angular velocity ω. The second term is called the *secondary component*. This component is the x projection of a vector of length $m_B(r/l)r\omega^2$ rotating forward (counterclockwise) with an angular velocity of 2ω.

The maximum inertia force occurs when $\omega t = 0$. Substituting this condition into Eq. (15.19) gives

$$F_{\text{max}} = m_B r\omega^2\left(\frac{r}{l} + \frac{1}{2}\right), \tag{e}$$

since $\cos\omega t = \cos 2\omega t = 1$ and $\sin\omega t = 0$ when $\omega t = 0$. Before the extra counterweight is added, the maximum inertia force is

$$F_{\text{max}} = m_B r\omega^2\left(\frac{r}{l} + 1\right). \tag{f}$$

Therefore, in this instance, the effect of the added counterweight is to reduce the maximum shaking force by 50% of the primary component and to add axial inertia forces where formerly none existed. Equation (15.19) is plotted as a polar diagram in Fig. 15.21 for $r/l = 1/4$. Here, the vector OA rotates counterclockwise at angular velocity of 2ω. The horizontal projection of this vector, OA', is the secondary component. Vector OB, the primary component, rotates clockwise at an angular velocity ω. The total shaking force, F, is shown for the 30° orientation and is the sum of the vectors OB and $BB' = OA'$.

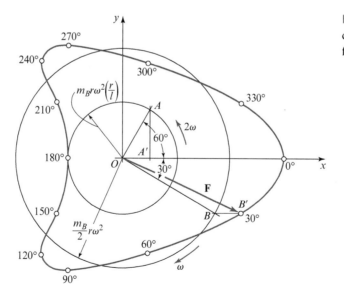

Fig. 15.21 Polar diagram of inertia forces.

Fig. 15.22 View of the crank circle from the negative z axis.

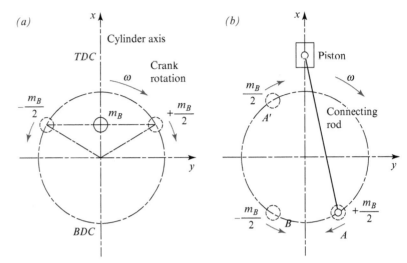

Imaginary-Mass Approach. Stevensen has refined and extended a method of engine balancing that here is called the *imaginary-mass approach*.[2] It is possible that the method is known in some circles as the *virtual-rotor approach* since it uses what might be called a virtual rotor that counter-rotates to accommodate part of the piston effect in a reciprocating engine.

Before going into details, it is necessary to explain a change in the method of viewing the crank circle of an engine. In developing the imaginary-mass approach in this and the following section, we use the right-handed coordinate system of Fig. 15.22a. This might appear to be a left-hand system since the y axis is located clockwise from the x axis, and since positive rotation is shown as clockwise. If you prefer, you can think of this system as a right-hand three-dimensional system viewed from the negative z axis. We adopt this notation since it has been used for so long by the automotive industry.[3]

The imaginary-mass approach uses two fictitious masses, each equal to half the equivalent reciprocating mass at the particular harmonic frequency studied. The purpose of these fictitious masses is to replace the effects of the reciprocating mass. These imaginary masses rotate about the crank center in opposite directions and with equal velocities. They are arranged so that they come together at both the top dead center (TDC) and bottom dead center (BDC), as shown in Fig. 15.22a. The mass $+\frac{1}{2}m_B$ rotates *with* the crank rotation; the other mass rotates opposite to the crank rotation. The mass rotating with the crank is designated in Fig. 15.22a by a plus sign, and the mass rotating in the opposite direction is designated by a minus sign. The center of mass of these two rotating masses always lies on the cylinder axis x.

---

[2] The presentation here is from Prof. E. N. Stevensen, University of Hartford, class notes [7], with his permission. Although some changes have been made to conform to the notation of this text, the material is all Stevensen's. He refers to Maleev [4] and Lichty [2] and states that the method first came to his attention in both of those books.

[3] Readers who are antique automobile buffs will understand this convention because it is the direction in which an antique engine is hand-cranked.

The imaginary-mass approach was conceived remembering that the piston motion and the resulting inertia force can always be represented by a Fourier series. Such a series has an infinite number of terms, each term representing a simple harmonic motion having a known frequency and amplitude. The higher-frequency amplitudes are so small that they can be neglected and hence only a small number of the lower-frequency amplitudes are needed. Also, the odd harmonics (third, fifth, etc.) are not present because of the symmetry of the piston motion.

Each harmonic (the first, second, fourth, and so on) is represented by a pair of imaginary masses. The angular velocities of these masses are $\pm\omega$ for the first harmonic, $\pm 2\omega$ for the second, $\pm 4\omega$ for the fourth, and so on. It is rarely necessary to consider the sixth or higher harmonics.

Stevensen gives the following rule for placing the imaginary masses:

For a given orientation of the cranks the orientations of the imaginary masses are found, first, by determining the angles of travel of each crank from its top dead center and, second, by moving their imaginary masses, one clockwise and the other counterclockwise, by angles equal to the crank angle times the number of the harmonic.

All of these angles must be measured from the same dead-center crank orientation.

Let us apply this approach to the single-cylinder engine, considering only the first harmonic. In Fig. 15.22b, the mass $+\frac{1}{2}m_B$ at A rotates at the angular velocity ω with the crank, whereas mass $-\frac{1}{2}m_B$ at B rotates at the angular velocity $-\omega$ opposite to crank rotation. The imaginary mass at A can be balanced by adding an equal mass at A' to rotate with the crankshaft. However, the mass at B can be balanced neither by the addition nor by the subtraction of mass from any part of the crankshaft, since it is rotating in the opposite direction. When half the mass of the reciprocating parts is balanced in this manner – that is, by adding the mass at A' – the unbalanced part of the first harmonic, caused by the mass at B, causes the engine to vibrate in the plane of rotation equally in all directions like a true unbalanced rotating mass.

It is interesting to find that in a one-cylinder motorcycle engine, a fore-and-aft unbalance is less objectionable than an up-and-down unbalance. For this reason, such engines are overbalanced by using a counterweight whose mass is more than half the reciprocating mass.

It is impossible to balance the second and higher harmonics with masses rotating at crankshaft speeds, since the frequency of the unbalance is higher than that of the crankshaft rotation. Balancing of second harmonics has been accomplished by using shafts geared to run at twice the engine crankshaft speed, as in the case of the 1976 Plymouth Arrow engine, but at the cost of complexity. It is not usually done.

For a convenient reference we summarize the inertia forces in a single-cylinder engine with balancing masses (Table 15.1). Note that the expressions in the final two columns of the table are obtained from Eqs. (14.24) and (14.27). Note that the subscript C is used to designate counterweights (balancing masses) and their radii. Also, note that the effect of the second harmonic is presented as a mass equal to $\frac{1}{4}m_B(r/l)$ reciprocating at a speed of 2ω; see column 2. Since the inertia balancing masses are selected and placed to counterbalance the rotating inertia forces, then the only unbalance that results along the cylinder axis is due to the harmonics of the reciprocating masses. Similarly, the only unbalance across the cylinder axis is due to the first harmonic of the reciprocating masses. The maximum values of the unbalance in these two

Table 15.1 **Single-cylinder engine inertia forces**

Type	Equivalent mass	Radius	Along (x) cylinder axis	Across (y) cylinder
Rotating	m_A	r	$m_A r \omega^2 \cos \omega t$	$m_A r \omega^2 \sin \omega t$
	m_{AC}	r_C	$m_{AC} r_C \omega^2 \cos(\omega t + \pi)$	$m_{AC} r_C \omega^2 \sin(\omega t + \pi)$
Reciprocating	m_B	r	$m_B r \omega^2 \cos \omega t$	0
First harmonic	m_{BC}	r_C	$m_{BC} r_C \omega^2 \cos(\omega t + \pi)$	$m_{BC} r_C \omega^2 \sin(\omega t + \pi)$
Reciprocating Second harmonic	$\dfrac{m_B r}{4l}$	r	$\dfrac{m_B r}{4l}(r)(2\omega)^2 \cos 2\omega t$	0

Fig. 15.23 (*a*) Two-cylinder engine; (*b*) first harmonics.

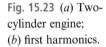
(*a*)

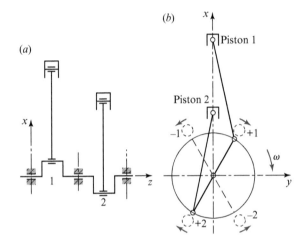
(*b*)

directions can be predetermined in any desired ratio to each other, as indicated previously, and a solution can be obtained for the equivalent mass, m_{BC}, at radius r_C.

If this approach is used to also include the effect of the fourth harmonic, there results an added mass of $(1/16)m_B(r/l)^3$ reciprocating at a speed of 2ω and another mass of $-(1/64)m_B(r/l)^3$ reciprocating at a speed of 4ω, illustrating the decreasing significance of higher harmonics.

15.10 Balancing Multi-Cylinder Engines

To obtain a basic understanding of the balancing problem in multi-cylinder engines, let us consider the two-cylinder in-line engine with three main bearings shown in Fig. 15.23a. The cranks (numbered 1 and 2) are 180° apart, and the rotating parts are already balanced by counterweights. Applying the imaginary-mass approach for the first harmonic results in the diagram of Fig. 15.23b. This diagram shows that masses +1 and +2, rotating clockwise,

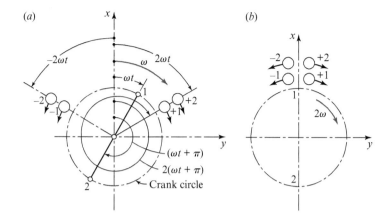

Fig. 15.24 (a) Crank 1 at TDC; (b) extreme, or dead-center, postures.

balance each other, as do masses −1 and −2, rotating counterclockwise. Thus, the first-harmonic forces are inherently balanced for this crank arrangement. These forces, however, are not in the same plane; therefore, unbalanced couples will be set up that tend to rotate the engine about the y axis. The values of these couples can be determined using the force expressions in Table 15.1 together with the coupling distance, since the equations can be applied to each cylinder separately. It is possible to balance the couple resulting from the real rotating masses as well as the imaginary half-masses that rotate with the engine; however, the couple resulting from the half-mass of the first harmonic that is counter-rotating cannot be balanced.

The locations of the imaginary masses for the second harmonic using Stevensen's rule are shown in Fig. 15.24a. This diagram shows that the second-harmonic forces due to the imaginary masses are not balanced. Since the greatest unbalance occurs at the dead-center orientations, the diagrams are usually drawn for this extreme orientation, with crank 1 at TDC, as in Fig. 15.24b. This unbalance causes a vibration in the xz plane having the frequency 2ω. The diagram for the fourth harmonics, not shown, is the same as in Fig. 15.24b, but, of course, the speed is 4ω.

Four-Cylinder Engine.

A four-cylinder in-line engine with cranks spaced 180° apart is shown in Fig. 15.25c. This engine can be treated as two two-cylinder engines placed back to back. Thus, the first-harmonic forces still balance, and, in addition, from Figs. 15.25a and 15.25c, the first-harmonic couples also balance. These couples tend, however, to deflect the center bearing of a three-bearing crankshaft up and down and to bend the center of a two-bearing shaft in the same manner.

Figure 15.25b shows that when cranks 1 and 4 are at TDC, all the masses representing the second harmonic traveling in both directions accumulate at TDC, giving an unbalanced force. The center of mass of all the masses is always on the x axis, and so the unbalanced second harmonics cause a vertical vibration with a frequency of twice the engine speed. This characteristic is typical of all four-cylinder engines with this crank arrangement. Since the masses and forces all act in the same direction, there is no coupling action.

Fig. 15.25 Postures for: (*a*) first harmonics; (*b*) second harmonics. (*c*) Crankshaft with first-harmonic couples added.

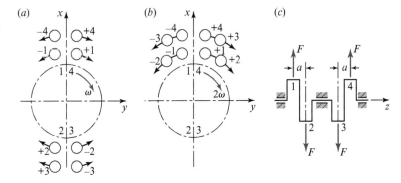

Fig. 15.26 First-harmonic forces.

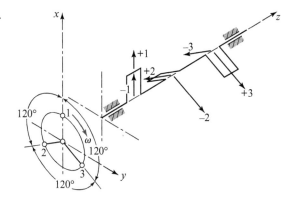

A diagram of the fourth harmonics is identical to that of Fig. 15.25*b*, and the effects are the same, but they have a higher frequency and exert less force.

Three-Cylinder Engine.

A three-cylinder in-line engine with cranks spaced 120° apart is shown in Fig. 15.26. Note that the cylinders are numbered according to the order in which they arrive at TDC. The analysis, using imaginary masses, of the couples of the first-harmonic forces indicates that when crank 1 is at TDC (Fig. 15.26), there is a vertical component of the forces on cranks 2 and 3 equal in magnitude to half of the force on crank 1. The resultant of these two downward components is equivalent to a force downward, equal in magnitude to the force on crank 1 and located halfway between cranks 2 and 3. Thus, a couple is set up with an arm equal to the distance between the center of crank 1 and the centerline between cranks 2 and 3.

At the same time, the horizontal components of the +2 and –2 forces cancel each other, as do the horizontal components of the +3 and –3 forces, as shown in Fig. 15.27. Therefore, no horizontal couple exists. Similar couples are found for both the second and fourth harmonics. Thus, a three-cylinder engine is inherently balanced for forces in the first, second, and fourth harmonics; only the sixth harmonic forces are completely unbalanced. These unbalanced forces tend to create a vibration in the plane of the centerlines of the cylinders, but the magnitude of the forces is very small and can be neglected as far as vibration is concerned.

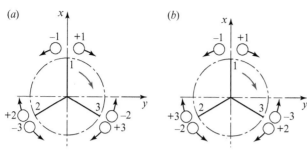

Fig. 15.27 Postures for: (*a*) first harmonics; (*b*) second harmonics; (*c*) fourth harmonics; (*d*) sixth harmonics.

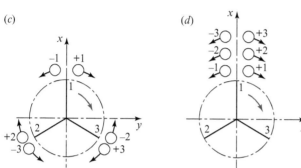

Six-Cylinder Engine.   If a six-cylinder in-line engine is conceived as a combination of two three-cylinder engines placed back to back with parallel cylinders, it has the same inherent balance of the first, second, and fourth harmonics. By virtue of symmetry, the couples of each three-cylinder engine act in opposite directions and balance each other. These couples, although perfectly balanced, tend to bend the crankshaft and crankcase and necessitate the use of rigid construction for high-speed operation. As before, the sixth harmonic forces are completely unbalanced and tend to create a vibration in the vertical plane with a frequency of 6ω. The magnitude of these forces, however, is very small and practically negligible as a source of vibration.

Other Engines.   Taking into consideration the cylinder arrangement and crank spacing permits a great many configurations. For any combination, the balancing situation can be investigated to any harmonic desired by the methods outlined in this section. Particular attention must be paid when analyzing the part of Stevensen's rule that calls for determining the angle of travel from the TDC of the cylinder under consideration and moving the imaginary masses through the appropriate angles from that same TDC. This is especially important when radial and opposed-piston engines are investigated.

As practice problems you may wish to use these methods to confirm the following facts:

1. In a three-cylinder radial engine with one crank and three connecting rods having the same crankpin, the negative masses are inherently balanced for the first-harmonic forces, whereas the positive masses are always located at the crankpin. These two findings are inherently true for all radial engines. Also, since the radial engine has its cylinders in a single plane,

unbalanced couples do not occur. The three-cylinder engine has unbalanced forces in the second and higher harmonics.

2. A two-cylinder opposed-piston engine with a crank spacing of 180° is balanced for forces in the first, second, and fourth harmonics but unbalanced for couples.

3. A four-cylinder in-line engine with cranks at 90° is balanced for forces in the first harmonic but unbalanced for couples. In the second harmonic, it is balanced for both forces and couples.

4. An eight-cylinder in-line engine with the cranks at 90° is inherently balanced for both forces and couples in the first and second harmonics but unbalanced in the fourth harmonic.

5. An eight-cylinder V-engine with cranks at 90° is inherently balanced for forces in the first and second harmonics and for couples in the second harmonic. The unbalanced couples in the first harmonic can be balanced by counterweights that introduce an equal and opposite couple. Such an engine is unbalanced for forces in the fourth harmonic.

15.11 Analytic Technique for Balancing Multi-Cylinder Engines

First-Harmonic Forces.   The x and y components of the resultant of the first-harmonic forces for any multi-cylinder reciprocating engine can be written in the form

$$F_P^x = A \cos \theta + B \sin \theta,$$

$$F_P^y = C \cos \theta + D \sin \theta, \tag{15.20}$$

where

$$A = \sum_{i=1}^{n} P_i \cos (\psi_i - \phi_i) \cos \psi_i, \tag{15.21a}$$

$$B = \sum_{i=1}^{n} P_i \sin (\psi_i - \phi_i) \cos \psi_i, \tag{15.21b}$$

$$C = \sum_{i=1}^{n} P_i \cos (\psi_i - \phi_i) \sin \psi_i, \tag{15.21c}$$

and

$$D = \sum_{i=1}^{n} P_i \sin (\psi_i - \phi_i) \sin \psi_i, \tag{15.21d}$$

where n is the number of cylinders, $P_i = m_i r_i \omega^2$ is the inertia force of piston i, m_i is the effective mass of piston i (Section 14.5), r_i is the length of crank i, ω is the constant angular velocity of the crankshaft, ψ_i is the angle from the x axis to the sliding axis of piston i, and ϕ_i is the angle from the reference line on crank 1 to crank i (by definition, $\phi_1 = 0$). Note that, if the coefficients

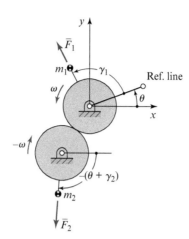

Fig. 15.28 General arrangement of counter-rotating masses for balancing the primary forces.

A, B, C, and D are all zero, then the x and y components of the resultant of the first-harmonic forces are zero; that is, there is no primary shaking force. This is the case, for example, with a six-cylinder in-line engine (see the previous section).

   If the coefficients A, B, C, and D are not all zero, then the first-harmonic forces can always be balanced by a pair of rotating masses or in some special cases by a single mass, as shown in Fig. 15.28. The correcting masses are rotating with the same speed as the crankshaft. The correcting mass m_1 creates a correcting inertia force F_1 at an angle of $(\theta + \gamma_1)$ from the x axis, and the correcting mass m_2 creates a correcting inertia force F_2 at an angle of $-(\theta + \gamma_2)$ from the x axis. For balance, these two correcting forces plus the resultant of the first-harmonic forces (the primary shaking force) must be equal to zero; that is,

$$\begin{aligned} A\cos\theta + B\sin\theta + F_1\cos(\theta+\gamma_1) + F_2\cos[-(\theta+\gamma_2)] &= 0, \\ C\cos\theta + D\sin\theta + F_1\sin(\theta+\gamma_1) + F_2\sin[-(\theta+\gamma_2)] &= 0. \end{aligned} \tag{15.22}$$

Expanding these two equations, in terms of functions of θ, γ_1, and γ_2, and rearranging, gives

$$\begin{aligned} (F_1\cos\gamma_1 + F_2\cos\gamma_2)\cos\theta - (F_1\sin\gamma_1 + F_2\sin\gamma_2)\sin\theta &= -A\cos\theta - B\sin\theta, \\ (F_1\sin\gamma_1 - F_2\sin\gamma_2)\cos\theta + (F_1\cos\gamma_1 - F_2\cos\gamma_2)\sin\theta &= -C\cos\theta - D\sin\theta. \end{aligned} \tag{15.23}$$

To satisfy Eqs. (15.23) for all values of the crank angle θ, the necessary conditions are

$$F_1\cos\gamma_1 + F_2\cos\gamma_2 = -A,$$

$$F_1\sin\gamma_1 + F_2\sin\gamma_2 = B,$$

$$F_1\sin\gamma_1 - F_2\sin\gamma_2 = -C,$$

$$F_1\cos\gamma_1 - F_2\cos\gamma_2 = -D. \tag{15.24}$$

Solving Eqs. (15.24), the forces are

$$F_1 = \frac{1}{2}\sqrt{(D+A)^2 + (C-B)^2},$$

$$F_2 = \frac{1}{2}\sqrt{(D-A)^2 + (C+B)^2}, \tag{15.25}$$

and the orientation angles are given by

$$\tan\gamma_1 = (C-B)/(D+A),$$

$$\tan\gamma_2 = (C+B)/(D-A). \tag{15.26}$$

Second-Harmonic Forces. The x and y components of the resultant of the second-harmonic forces can be written in the form

$$F_S^x = A'\cos 2\theta + B'\sin 2\theta,$$
$$F_S^y = C'\cos 2\theta + D'\sin 2\theta, \tag{15.27}$$

where

$$A' = \sum_{i=1}^{n} Q_i \cos 2(\psi_i - \phi_i)\cos\psi_i,$$

$$B' = \sum_{i=1}^{n} Q_i \sin 2(\psi_i - \phi_i)\cos\psi_i,$$

$$C' = \sum_{i=1}^{n} Q_i \cos 2(\psi_i - \phi_i)\sin\psi_i,$$

and

$$D' = \sum_{i=1}^{n} Q_i \sin 2(\psi_i - \phi_i)\sin\psi_i,$$

where the inertia force $Q_i = m_i(r_i/l_i)r_i\omega^2{}_i = (r_i/l_i)P_i$. We have stated previously that, in general, $r_i/l_i < 1/4$; therefore, $Q_i < P_i$, which indicates that the magnitudes of the second-harmonic forces are much smaller than the magnitudes of the first-harmonic forces.

The second-harmonic forces can always be balanced by a pair of gears rotating at twice the crankshaft speed, as shown in Fig. 15.29. The mass m_1' creates a force F_1' at an orientation angle of $(2\theta + \lambda_1)$ from the x axis, and the mass m_2' creates a force F_2' at an orientation angle of $-(2\theta + \lambda_2)$ from the x axis. For balance, these two forces plus the resultant of the second-harmonic forces must be equal to zero.

Note that the analysis for the second-harmonic forces is exactly the same as for the first-harmonic forces, with A, B, C, and D replaced by A', B', C', and D', θ replaced by 2θ, and the orientation angles γ_1 and γ_2 replaced by the orientation angles λ_1 and λ_2, respectively. The two correcting forces can be written as

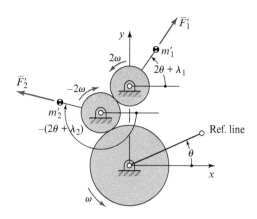

$$F'_1 = \frac{1}{2}\sqrt{(D'+A')^2 + (C'-B')^2}$$

and

$$F'_2 = \frac{1}{2}\sqrt{(D'-A')^2 + (C'+B')^2}.$$

The two orientation angles can be written as

$$\tan\lambda_1 = \frac{(C'-B')}{(D'+A')}$$

and

$$\tan\lambda_2 = \frac{(C'+B')}{(D'-A')}.$$

Example 15.5

Determine the magnitudes and orientations of the forces created by the correcting masses needed to balance the first- and second-harmonic shaking forces of the three-cylinder engine shown in Fig. 15.30.

Solution

The resultant first-harmonic force vector is

$$\mathbf{F}_P = 1.5P\cos\theta\hat{\mathbf{i}} - 1.5P\sin\theta\hat{\mathbf{j}},\tag{1}$$

where

$$P = mr\omega^2.$$

Example 15.5 (cont.)

Fig. 15.30 Three-cylinder arrangement.

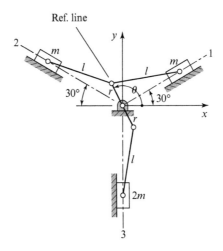

Comparing Eq. (1) with Eqs. (15.20), the coefficients are

$$A = 1.5P, \quad B = 0, \quad C = 0, \quad \text{and} \quad D = -1.5P.$$

Therefore, from Eqs. (15.25), the forces are

$$F_1 = \frac{1}{2}\sqrt{(D+A)^2 + (C-B)^2} = 0$$

and

$$F_2 = \frac{1}{2}\sqrt{(D-A)^2 + (C+B)^2} = 1.5P.$$

Since the first force is zero, there is only one correcting mass, that is, m_2. The orientation angle for this mass, from the second of Eqs. (15.26), is

$$\tan \gamma_2 = (C+B)/(D-A) = 0/-3P = 0.$$

Therefore, since the denominator (the cosine of the angle) is negative, the result is

$$\gamma_2 = 180°. \hspace{4cm} Ans.$$

The conclusion is that the correcting mass m_2 is rotating opposite to the crankshaft (at crankshaft speed) and oriented at an angle $-(\theta + 180°)$. Figure 15.31 shows the orientation of the correcting mass for $\theta = 0°$, that is, the reference line is coincident with the x axis. If the correcting mass m_2 is placed at a radius r_2, then the inertia force is

$$F_2 = m_2 r_2 \omega^2 = 1.5mr\omega^2.$$

Example 15.5 (cont.)

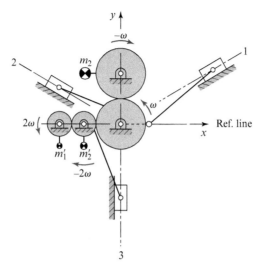

Fig. 15.31 Balancing arrangement for $\theta = 0°$.

Therefore, the product is

$$m_2 r_2 = 1.5mr. \qquad Ans.$$

The second-harmonic force vector is

$$\mathbf{F}_s = 1.5Q\sin 2\theta\,\hat{\mathbf{i}} + 2.5Q\cos 2\theta\,\hat{\mathbf{j}}, \qquad (2)$$

where

$$Q = \frac{mr^2\omega^2}{l}.$$

Comparing Eq. (2) with Eqs. (15.23), the coefficients are

$$A' = 0, \quad B' = 1.5Q, \quad C' = 2.5Q, \quad \text{and} \quad D' = 0.$$

Therefore, the inertia forces, from Eqs. (15.25), are

$$F_1' = \frac{1}{2}\sqrt{(D' + A')^2 + (C' - B')^2} = 0.5Q$$

and

$$F_2' = \frac{1}{2}\sqrt{(D' - A')^2 + (C' + B')^2} = 2.0Q.$$

The forces can be expressed as

$$F_1' = m_1' r_1' (2\omega)^2 = 0.5m\frac{r^2}{l}\omega^2$$

Example 15.5 (cont.)

and

$$F_2' = m_2' r_2' (2\omega)^2 = 2m \frac{r^2}{l} \omega^2.$$

Therefore, the correcting mass m_1' can be placed at a radius r_1' such that

$$m_1' r_1' = 0.125 mr^2 / l, \qquad\qquad Ans.$$

and the correcting mass m_2' can be placed at a radius r_2' such that

$$m_2' r_2' = 0.5 mr^2 / l. \qquad\qquad Ans.$$

From Eq. (15.26a), the first orientation angle is

$$\tan \lambda_1 = (C' - B)/(D' + A') = -1.0Q/0 = \infty.$$

Therefore, $\lambda_1 = 270°$, since the numerator (the sine of the angle) is negative. $\qquad$ *Ans.*
    From Eq. (15.25b), the second orientation angle is

$$\tan \lambda_2 = (C' + B')/(D' - A') = +4.0Q/0 = \infty.$$

Thus, $\lambda_2 = 90°$, since the numerator (the sine of the angle) is positive. $\qquad$ *Ans.*
    The conclusion is that we need an arrangement such as that shown in Fig. 15.31. The mass m_1' is located on a shaft turning in the same direction as the crankshaft at twice the crankshaft speed. The orientation angle is $2\theta + 270°$. The mass m_2' is located on a shaft turning opposite to the crankshaft at twice the crankshaft speed. The orientation angle is $-(2\theta + 90°)$. $\qquad$ *Ans.*

15.12 Balancing of Linkages[4]

The two problems that arise in balancing linkages are balancing the shaking force and balancing the shaking moment. Lowen and Berkof [3] note that very few studies have been reported on the problem of balancing the shaking moment. This problem is discussed further in Section 15.13.
    In force balancing a linkage we must concern ourselves with the position of the total center of mass. If a way can be found to cause this total center of mass to remain stationary, the vector sum of all the frame forces is always zero. Berkof and Lowen [1] have listed five methods of force balancing:

---

[4] Those who wish to investigate this topic in detail should begin with reference [3], in which the entire issue is devoted to the subject of linkage balancing. This issue contains 11 translations on the subject from the German and Russian literature.

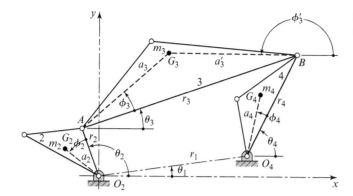

Fig. 15.32 Four-bar linkage showing arbitrary postures of the link masses.

1. The method of static balancing, in which concentrated link masses are replaced by systems of masses that are statically equivalent.
2. The method of principal vectors, in which an analytic expression is obtained for the center of mass and then manipulated to learn how its trajectory can be influenced.
3. The method of linearly independent vectors, in which the center of mass of a mechanism is made stationary, causing the coefficients of the time-dependent terms of the equation describing the trajectory of the total center of mass to vanish.
4. The use of cam-driven masses to keep the total center of mass stationary.
5. The addition of an axially symmetric duplicate mechanism by which the new combined total center of mass is made stationary.

Here we present only the Berkof–Lowen method [1], which employs the method of linearly independent vectors. The method is developed completely for the four-bar linkage, but the final results alone are given for a typical six-bar linkage. The procedure is as follows. First, find the equation that describes the trajectory of the total center of mass of the linkage. This equation will contain certain terms whose coefficients are time-dependent. Then the total center of mass is made stationary by changing the positions of the individual link masses so that the coefficients of all time-dependent terms vanish. To accomplish this, it is necessary to write the equation in such a form that the time-dependent unit vectors contained in the equation are independent.

In Fig. 15.32 a general four-bar linkage is shown having link mass m_2 located at G_2, m_3 located at G_3, and m_4 located at G_4. The total mass M of the four-bar linkage is

$$M = m_2 + m_3 + m_4. \tag{a}$$

We begin by defining the position of the total center of mass of the linkage by the vector $\mathbf{r}_s$,

$$\mathbf{r}_s = \frac{1}{M}(m_2\mathbf{r}_{s_2} + m_3\mathbf{r}_{s_3} + m_4\mathbf{r}_{s_4}), \tag{b}$$

where $\mathbf{r}_{s_2}, \mathbf{r}_{s_3}$, and $\mathbf{r}_{s_4}$ are the vectors that describe the positions of m_2, m_3, and m_4, respectively, in the global xy coordinate system. The coordinates a_i, ϕ_i describe the positions of the mass centers within each link. Thus, from Fig. 15.32,

$$\begin{aligned}
\mathbf{r}_{s_2} &= a_2 e^{j(\theta_2+\phi_2)}, \\
\mathbf{r}_{s_3} &= r_2 e^{j\theta_2} + a_3^{j(\theta_3+\phi_3)}, \\
\mathbf{r}_{s_4} &= r_1 e^{j\theta_1} + a_4^{j(\theta_4+\phi_4)}.
\end{aligned} \tag{c}$$

Substituting Eqs. (c) into (b) gives

$$M\mathbf{r}_s = \left(m_2 a_2 e^{j\phi_2} + m_3 r_2\right)e^{j\theta_2} + \left(m_3 a_3 e^{j\phi_3}\right)e^{j\theta_3} + \left(m_4 a_4 e^{j\phi_4}\right)e^{j\theta_4} + m_4 r_1 e^{j\theta_1}, \tag{d}$$

where we have used the identity $e^{j\alpha}e^{j\beta} = e^{j(\alpha+\beta)}$.

For a four-bar linkage the loop-closure equation can be written as

$$r_2 e^{j\theta_2} + r_3 e^{j\theta_3} - r_4 e^{j\theta_4} - r_1 e^{j\theta_1} = 0. \tag{e}$$

Thus, the time-dependent terms $e^{j\theta_2}$, $e^{j\theta_3}$, and $e^{j\theta_4}$ in Eq. (d) are not independent. To make them so, we solve Eq. (e) for one of the unit vectors, say $e^{j\theta_3}$, and substitute the result into Eq. (d). Thus,

$$e^{j\theta_3} = \frac{1}{r_3}\left(r_1 e^{j\theta_1} - r_2 e^{j\theta_2} + r_4 e^{j\theta_4}\right), \tag{f}$$

and Eq. (d) now becomes

$$M\mathbf{r}_s = \left(m_2 a_2 e^{j\phi_2} + m_3 r_2 - m_3 a_3 \frac{r_2}{r_3} e^{j\phi_3}\right)e^{j\theta_2} + \left(m_4 a_4 e^{j\phi_4} + m_3 a_3 \frac{r_4}{r_3} e^{j\phi_3}\right)e^{j\theta_4} \\
+ \left(m_4 r_1 + m_3 a_3 \frac{r_1}{r_3} e^{j\phi_3}\right)e^{j\theta_1}. \tag{g}$$

Equation (g) shows that the center of mass will be stationary at the position

$$\mathbf{r}_s = \frac{r_1}{r_3 M}\left(m_4 r_3 + m_3 a_3 e^{j\phi_3}\right)e^{j\theta_1} \tag{15.28}$$

if we make the following coefficients of the time-dependent terms vanish:

$$m_2 a_2 e^{j\phi_2} + m_3 r_2 - m_3 a_3 \frac{r_2}{r_3} e^{j\phi_3} = 0, \tag{h}$$

$$m_4 a_4 e^{j\phi_4} + m_3 a_3 \frac{r_4}{r_3} e^{j\phi_3} = 0. \tag{i}$$

But, Eq. (h) can be simplified by locating G_3 from point B instead of point A (Fig. 15.32). Thus,

$$a_3 e^{j\phi_3} = r_3 + a_3' e^{j\phi_3'}.$$

With this substitution, Eq. (h) becomes

$$m_2 a_2 e^{j\phi_2} - m_3 a_3' \frac{r_2}{r_3} e^{j\phi_3'} = 0. \tag{j}$$

Equations (i) and (j) must be satisfied to obtain total force balance. These equations yield the two sets of conditions:

$$m_2 a_2 = m_3 a_3' \frac{r_2}{r_3} \quad \text{and} \quad \phi_2 = \phi_3',$$

$$m_4 a_4 = m_3 a_3 \frac{r_4}{r_3} \quad \text{and} \quad \phi_4 = \phi_3 + \pi. \tag{15.29}$$

A study of these conditions indicates that the mass and its location can be specified in advance for any single link; then, full balance can be obtained by rearranging the mass of the other two links.

The usual problem in balancing a four-bar linkage is that the link lengths r_i are specified in advance because of the functional requirements. For this situation, counterweights can be added to the input and output links to redistribute their masses, while the geometry of the coupler link is undisturbed.

When adding counterweights, the following relations must be satisfied:

$$m_i a_i \angle \phi_i = m_i^\circ a_i^\circ \angle \phi_i^\circ + m_i^* a_i^* \angle \phi_i^*, \tag{15.30}$$

where $m_i^\circ, a_i^\circ, \phi_i^\circ$ are the parameters of the unbalanced linkage, m_i^*, a_i^*, ϕ_i^* are the parameters of the counterweights, and m_i, a_i, ϕ_i are the parameters that result from Eqs. (15.29). A second condition that must generally be satisfied is

$$m_i = m_i^\circ + m_i^*. \tag{15.31}$$

If the solution to a balancing problem can remain as the mass–distance product $m_i^* \, a_i^*$, Eq. (15.31) need not be used and Eq. (15.30) can be solved to yield

$$m_i^* a_i^* = \sqrt{(m_i a_i)^2 + (m_i^\circ a_i^\circ)^2 - 2(m_i a_i)(m_i^\circ a_i^\circ) \cos(\phi_i - \phi_i^\circ)}, \tag{15.32}$$

$$\phi_i^* = \tan^{-1}\left(\frac{m_i a_i \sin \phi_i - m_i^\circ a_i^\circ \sin \phi_i^\circ}{m_i a_i \cos \phi_i - m_i^\circ a_i^\circ \cos \phi_i^\circ} \right). \tag{15.33}$$

Figure 15.33 shows a typical six-bar linkage and the notation. For this, the Berkof–Lowen conditions for total balance are:

$$m_2 \frac{a_2}{r_2} e^{j\phi_2} = m_5 \frac{a_5' \, b_2}{r_5 \, r_2} e^{j(\phi_1 + \alpha_2)} + m_3 \frac{a_3'}{r_3} e^{j\phi_1},$$

$$m_4 \frac{a_4}{r_4} e^{j\phi_4} = m_6 \frac{a_6' \, b_4}{r_6 \, r_4} e^{j\phi_6'} - m_3 \frac{a_3}{r_3} e^{j(\phi_3 + \alpha_4)}, \tag{15.34}$$

$$m_5 \frac{a_5}{r_5} e^{j\phi_5} = -m_6 \frac{a_6}{r_6} e^{j\phi_6}.$$

Similar relations can be devised for other six-bar linkages. For total balance, Eqs. (15.34) demonstrate that a certain mass–geometry relation between links 5 and 6 must be satisfied, after which the masses of any two links and their locations can be specified. Balance is then achieved by a redistribution of the masses of the remaining three movable links.

It is important to note that the addition of counterweights to balance the shaking forces will probably increase the internal bearing forces as well as the shaking moment. Thus, only a partial balance may represent the best compromise between these three effects.

Fig. 15.33 Notation
for a six-bar linkage.

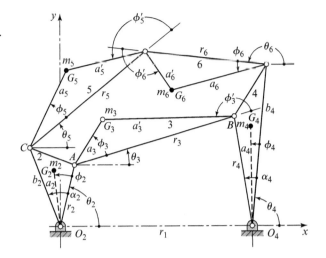

Example 15.6

Table 15.2 is a tabulation of the dimensions, masses, and the locations of the mass centers of a
four-bar linkage having link 2 as the input and link 4 as the output. Complete force balancing is to
be achieved by adding counterweights to the input and the output links. Find the mass–distance
product and the angular orientation of each counterweight.

Table 15.2 **Parameters of an unbalanced four-bar linkage**

Link i	1	2	3	4
r_i, mm	140	50	150	75
a_i^0, mm	–	25	80	40
ϕ_i^0, deg	–	0	15	0
$a_i'^0$, mm	–	–	75.6	–
$\phi_i'^0$, deg	–	–	164.1	–
m_i^0, kg	–	0.046	0.125	0.054

Solution

From Eqs. (15.29) we first find

$$m_2 a_2 = m_3^o a_3'^o \frac{r_2}{r_3} = (0.125 \text{ kg})(75.6 \text{ mm})\frac{50 \text{ mm}}{150 \text{ mm}} = 3.150 \text{ g} \cdot \text{m},$$

$$\phi_2 = \phi_3' = 164.1°,$$

Example 15.6 (cont.)

$$m_4 a_4 = m_4^o a_3^o \frac{r_4}{r_3} = (0.125 \text{ kg})(80 \text{ mm}) \frac{75 \text{ mm}}{150 \text{ mm}} = 5.000 \text{ g} \cdot \text{m},$$

$$\phi_4 = \phi_4^o + 180° = 15° + 180° = 195°.$$

Note that $m_2 a_2$ and $m_4 a_4$ are the mass–distance products *after* the counterweights have been added. Also note that the link 3 parameters are altered. We next compute

$$m_2^o a_2^o = (0.046 \text{ kg})(25 \text{ mm}) = 1.150 \text{ g} \cdot \text{m} \quad \text{and}$$
$$m_4^o a_4^o = (0.054 \text{ kg})(40 \text{ mm}) = 2.160 \text{ g} \cdot \text{m}.$$

Using Eq. (15.32), we compute the mass–distance product for the link 2 counterweight as

$$m_2^* a_2^* = \sqrt{(m_2 a_2)^2 + (m_2^o a_2^o)^2 - 2(m_2 a_2)(m_2^o a_2^o) \cos(\phi_2 - \phi_2^o)}$$
$$= \sqrt{(3.150 \text{ g} \cdot \text{m})^2 + (1.150 \text{ g} \cdot \text{m})^2 - 2(3.150 \text{ g} \cdot \text{m})(1.150 \text{ g} \cdot \text{m}) \cos(164.1° - 0°)}$$
$$= 4.268 \text{ g} \cdot \text{m}. \hspace{4cm} \textit{Ans.}$$

From Eq. (15.33), we find the orientation of this counterweight to be

$$\phi_2^* = \tan^{-1}\left(\frac{m_2 a_2 \sin \phi_2 - m_2^o a_2^o \sin \phi_2^o}{m_2 a_2 \cos \phi_2 - m_2^o a_2^o \cos \phi_2^o}\right)$$
$$= \tan^{-1}\left(\frac{(3.150 \text{ g} \cdot \text{m}) \sin 164.1° - (1.150 \text{ g} \cdot \text{m}) \sin 0°}{(3.150 \text{ g} \cdot \text{m}) \cos 164.1° - (1.150 \text{ g} \cdot \text{m}) \cos 0°}\right) = 168.3°. \hspace{1cm} \textit{Ans.}$$

Using the same procedure for link 4 yields

$$m_4^* a_4^* = 7.11 \text{ g} \cdot \text{m} \quad \text{at} \quad \phi_4^* = 190.5°. \hspace{3cm} \textit{Ans.}$$

Figure 15.34 is a scale drawing of the complete linkage with the two counterweights added to achieve complete force balance.

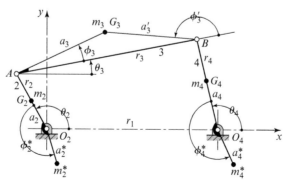

Figure 15.34 Four-bar crank-rocker linkage showing counterweights added to input and output links.

15.13 Balancing of Machines[5]

In Section 15.12 we learned how to force balance a simple linkage using two or more counterweights, depending upon the number of links composing the linkage. Unfortunately, this does not balance the shaking moment and, in fact, may even make it worse because of the addition of the counterweights. If a machine is imagined to be composed of several mechanisms, one might consider balancing the machine by balancing each mechanism separately. But this may not result in the best balance for the machine, since the addition of a large number of counterweights may cause the inertia torque to be completely unacceptable. Furthermore, unbalance of one mechanism may counteract the unbalance in another, eliminating the need for some of the counterweights in the first place.

Stevensen demonstrates that any single harmonic of unbalanced forces, moments of forces, and torques in a machine can be balanced by the addition of six counterweights. They are arranged on three shafts, two per shaft, driven at the constant speed of the harmonic, and have axes parallel, respectively, to each of the three mutually perpendicular axes through the center of mass of the machine. The method is too complex to be included here, but it is worthwhile to look at the overall approach.

Using the methods of this text together with a computer, the rectilinear and angular accelerations of each of the moving mass centers of a machine are computed for postures throughout a cycle of motion. The masses and mass moments of inertia of the machine must also be computed or determined experimentally. Then, the inertia forces, the inertia torques, and the moments of the forces are computed with reference to the three mutually perpendicular coordinate axes through the center of mass of the machine. When these are summed for each posture in the cycle, six functions of time are the result, three for the forces and three for the moments. With the aid of a computer, it is then possible to use numeric harmonic analysis to define the component harmonics of the unbalanced forces parallel to the three axes and of the unbalanced moments about the three axes.

To balance a single harmonic, each component of the unbalance of the machine is represented in the form $A \cos \omega t + B \sin \omega t$ with appropriate subscripts. Six equations of equilibrium are then written, which include the unbalances as well as the effects of the six unknown counterweights. These equations are arranged so that each of the $\sin \omega t$ and $\cos \omega t$ terms is multiplied by parenthetical coefficients. Balance is then achieved by setting the parenthetic terms in each equation equal to zero, much in the same manner as in the preceding section. This results in 12 equations, all linear, in 12 unknowns. With the locations for the balancing counterweights on the three shafts specified, the 12 equations can be solved for the six *mr* products, and the six phase angles needed for the six balancing weights. Stevensen goes on to demonstrate that when fewer than the necessary three shafts are available, it becomes necessary to optimize some effect of the unbalance, such as the motion of a point on the machine.

---

[5] The material for this section is from Stevensen [6]. It is included with the advice and consent of Professor E. N. Stevensen.

PROBLEMS

15.1 Determine the bearing reactions at A and B if the speed of the shaft is 300 rev/min. Also determine the magnitude and orientation of the balancing weight if it is located at a radius of 2 in.

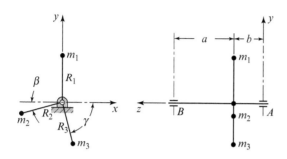

Fig. P15.1 $a = 32$ in, $b = 8$ in, $\beta = 15°$, $\gamma = 75°$, $R_1 = 1$ in, $R_2 = 1.4$ in, $R_3 = 1.6$ in, $w_1 = 4.4$ lb, $w_2 = 3.3$ lb, and $w_3 = 6.6$ lb.

15.2 Three masses are connected to a shaft that rotates in bearings at A and B. Determine the magnitudes and orientations of the bearing reactions if the speed of the shaft is 300 rev/min. Also, determine the magnitude and orientation of a counterweight that is to be located at a radius of 250 mm.

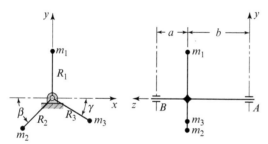

Fig. P15.2 $a = 150$ mm, $b = 300$ mm, $\beta = 45°$, $\gamma = 30°$, $R_1 = 200$ mm, $R_2 = 300$ mm, $R_3 = 150$ mm, $m_1 = 56$ g, $m_2 = 42$ g, and $m_3 = 84$ g.

15.3 Two masses are connected to a rotating shaft and mounted outboard of bearings A and B. If the speed of the shaft is 120 rev/min, what are the magnitudes and orientations of the bearing reactions at A and B? Suppose the system is to be balanced by removing mass at a radius of 125 mm. Determine the magnitude and orientation of the mass to be removed.

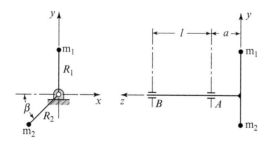

Fig. P15.3 $l = 100$ mm, $a = 50$ mm, $\beta = 45°$, $R_1 = 100$ mm, $R_2 = 150$ mm, $m_1 = 1.8$ kg, and $m_2 = 1.35$ kg.

15.4 If the speed of the shaft is 220 rev/min, calculate the magnitudes and orientations of the bearing reactions at A and B for the two-mass system.

Fig. P15.4 $a = b = 10$ in, $c = 2$ in,
$R_1 = 2.4$ in, $R_2 = 1.6$ in, $w_1 = 4.4$ lb, and
$w_2 = 3.3$ lb.

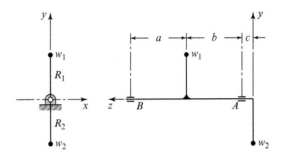

15.5 Determine the bearing reactions at A and B and their orientations if the shaft speed is 100 rev/min.

Fig. P15.5 $a = c = 12$ in,
$b = 24$ in, $R_1 = R_2 = 2.4$ in,
$w_1 = 2.2$ lb, and $w_2 = 6.6$ lb.

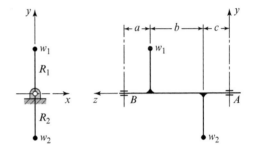

15.6 The rotating shaft shown in Fig. P15.5 supports two masses m_1 and m_2, whose masses are 1.8 kg and 2.25 kg, respectively. The dimensions are $a = 50$ mm, $b = 200$ mm, $c = 75$ mm, $R_1 = 100$ mm, and $R_2 = 75$ mm. Find the magnitudes of the rotating bearing reactions at A and B and their orientations if the shaft speed is 360 rev/min.

15.7 The shaft is to be balanced by placing masses in the correcting planes L and R. Calculate the magnitudes and orientations of the correcting masses.

Fig. P15.7 $a = 25$ mm, $b = e = 200$ mm,
$c = 250$ mm, $d = 225$ mm, $\beta = 30°$,
$\gamma = 60°$, $R_1 = R_3 = 125$ mm,
$R_2 = 100$ mm, $m_1 = m_3 = 112$ g, and
$m_2 = 84$ g.

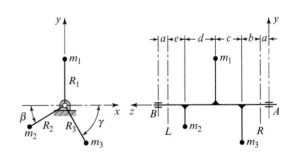

15.8 The shaft of Prob. 15.7 is to be balanced by removing mass from the two correcting planes. Determine the correcting masses and their orientations.

15.9 The shaft shown in Fig. P15.7 is to be balanced by removing weight in the two correcting planes, L and R. The three weights are $w_1 = 0.211$ oz, $w_2 = 0.246$ oz, and $w_3 = 0.160$ oz. The dimensions are $a = 1$ in, $b = 12$ in, $c = 24$ in, $d = 6$ in, $e = 3$ in, $\beta = 30°$, $\gamma = 60°$, $R_1 = 5$ in, $R_2 = 6$ in, and $R_3 = 4$ in. Calculate the magnitudes and orientations of the correcting weights.

15.10 Repeat Prob. 15.9 if weights are to be added in the two correcting planes.

15.11 Solve the two-plane balancing problem as stated in Section 15.8.

15.12 A rotor to be balanced in the field yields an amplitude of 5 units at an angle of 142° at the left-hand bearing and an amplitude of 3 units at an angle of $-22°$ at the right-hand bearing because of unbalance. To correct this, a trial mass of 12 units is added to the left-hand correcting plane at an angle of 210° from the rotating reference. A second run then gives left-hand and right-hand responses of 8 units∠160° and 4 units∠260°, respectively. The first trial mass is then removed and a second mass of 6 units is added to the right-hand correction plane at an angle of $-70°$. The responses to this are 2 units∠74° and 4.5 units∠$-80°$ for the left- and right-hand bearings, respectively. Determine the original unbalances.

15.13 A rotor is rotating with a constant angular velocity $\omega = 50$ rad/s, and is dynamically balanced by two correcting weights. A decision has been made to use planes 1 and 3 instead of planes 1 and 2. Determine the magnitudes and orientations of the new weights $(w_{C1})_{new}$ and $(w_{C3})_{new}$ at the same radii $R_{C1} = 2$ in and $R_{C3} = 6$ in.

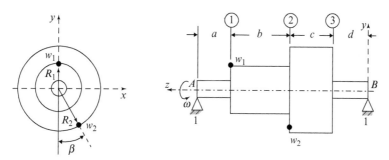

Fig. P15.13 $a = d = 2.4$ in, $b = 4$ in, $c = 3.2$ in, $\beta = 30°$, $R_{C1} = 2$ in, $R_{C2} = 6$ in, $w_{C1} = 6.6$ lb, and $w_{C2} = 4.4$ lb.

15.14 The shaft is rotating with a constant angular velocity $\omega = 80$ rad/s. For dynamic balance, determine the magnitudes and orientations of the correcting masses to be removed from planes 1 and 2 at radii $R_{C1} = R_{C2} = 65$ mm.

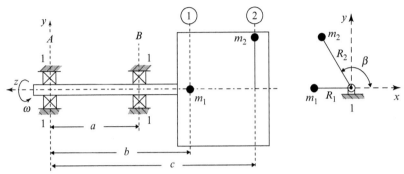

Fig. P15.14 $a = 250$ mm, $b = 400$ mm, $c = 600$ mm, $\beta = 120°$,
$R_1 = 30$ mm, $R_2 = 50$ mm, $m_1 = 4$ kg, and $m_2 = 2$ kg.

15.15 A shaft is simply supported by the bearings at A and B and is rotating with a constant angular velocity $\omega = 50$ rad/s. Determine the magnitudes and orientations of the correcting weights that must be removed in the correcting planes 1 and 2 to dynamically balance the system. The radial distances from the shaft axis are specified as $R_{C1} = R_{C2} = 2.8$ in. Show the orientations of the correcting weights on the right-hand figure.

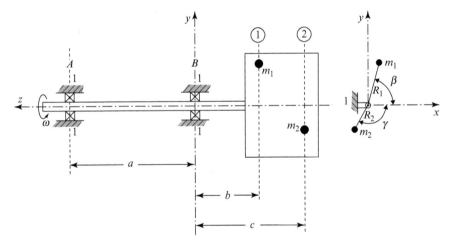

Fig. P15.15 $a = 16$ in, $b = 6$ in, $c = 12$ in, $\beta = 75°$, $\gamma = 120°$,
$R_1 = 2.4$ in, $R_2 = 1.4$ in, $w_1 = 26.4$ lb, and $w_2 = 33$ lb.

15.16 The shaft is simply supported by the bearings at A and B and is rotating with a constant angular velocity $\omega = 75$ rad/s. To dynamically balance the system, determine the magnitudes and orientations of the correcting masses that must be removed from planes 1 and 2 at the radial distances $R_{C1} = R_{C2} = 55$ mm.

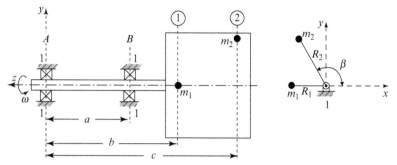

Fig. P15.16 $a = 250$ mm, $b = 400$ mm, $c = 600$ mm, $\beta = 120°$, $R_1 = 25$ mm, $R_2 = 40$ mm, $m_1 = 5$ kg, and $m_2 = 7$ kg.

15.17 Three mass particles are rigidly attached to a simply supported shaft which is rotating with a constant angular velocity $\omega = 15$ rad/s. Determine the magnitudes and orientations of the bearing reaction forces at A and B. Then determine the magnitudes and orientations of the correcting weights that must be added in the two correcting planes 1 and 2 to dynamically balance the system. The correcting weights are to be placed at radial distances $R_{C1} = R_{C1} = 1$ in from the shaft axis.

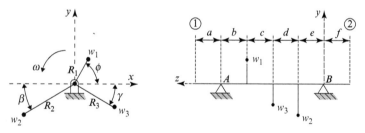

Fig. P15.17 $a = 2$ in, $b = 8.6$ in, $c = 11$ in, $d = 4$ in, $e = 6$ in, $f = 5$ in, $\beta = \gamma = 30°, \phi = 60°$, $R_1 = 0.6$ in, $R_2 = 1.4$ in, $R_3 = 0.8$ in, $w_1 = 6.6$ lb, $w_2 = 2.2$ lb, and $w_3 = 4.4$ lb.

15.18 Three mass particles are rigidly attached to a shaft which is rotating with constant angular velocity $\omega = 350\hat{k}$ rev/min. Determine the magnitudes and orientations of the reaction forces at the bearings A and B. Then determine the correcting masses that must be removed from the correcting planes 1 and 2 to dynamically balance the system if the radial distances of the correcting weights are specified as $R_{C1} = R_{C2} = 75$ mm.

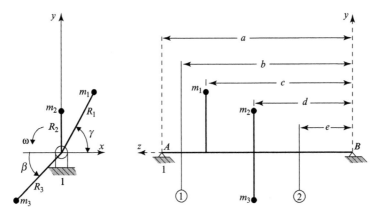

Fig. P15.18 $a = 265$ mm, $b = 235$ mm, $c = 200$ mm, $d = 135$ mm, $e = 75$ mm, $\beta = 45°, \gamma = 60°, R_1 = 95$ mm, $R_2 = 55$ mm, $R_3 = 90$ mm, $m_1 = 4$ kg, $m_2 = 3.15$ kg, and $m_3 = 0.5$ kg.

15.19 The shaft of the three-mass system is simply supported by the bearings at A and B and has a constant speed of 550 rev/min. Using the graphic approach determine the magnitudes and angular orientations of the reaction forces at bearings A and B.

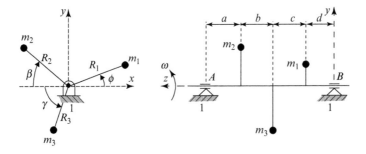

Fig. P15.19 $a = 30$ mm, $b = c = 25$ mm, $d = 20$ mm, $\beta = 40°$, $\gamma = -70°, \phi = 20°, R_1 = 20$ mm, $R_2 = 20$ mm, $R_3 = 15$ mm, $m_1 = 7$ kg, $m_2 = 12$ kg, and $m_3 = 8$ kg.

15.20 The distributed mass system, denoted as body 2, is simply supported in bearings at A and B and has a constant speed of 240 rev/min. The forces from the system acting on the ground at bearings A and B are $(\mathbf{F}_{21})_A = 56.3\hat{i} + 16.9\hat{j}$ lb and $(\mathbf{F}_{21})_B = 18\hat{i} - 28.1\hat{j}$ lb, respectively. Determine the magnitudes and orientations of the correcting weights that must be removed in the correction planes 1 and 2 to ensure moment (dynamic) balance of the system.

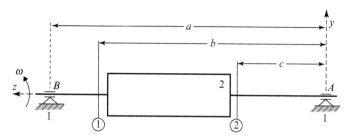

Fig. P15.20 $a = 10$ in, $b = 8$ in, $c = 3$ in, and $R_{C1} = R_{C2} = 1.2$ in.

15.21 The shaft, simply supported by bearings at A and B, has a constant speed of 300 rev/min. Using the graphic approach determine the magnitudes and orientations of the reaction forces at bearings A and B.

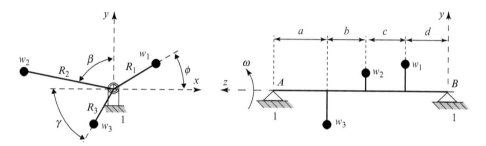

Fig. P15.21 $a = 1.6$ in, $b = c = 0.6$ in, $d = 1$ in, $R_1 = 0.4$ in, $\phi = 30°$, $R_2 = 1.4$ in, $\beta = 80°$, $R_3 = 0.6$ in, $\gamma = 60°$, $w_1 = 37.4$ lb, $w_2 = 8.8$ lb, and $w_3 = 17.6$ lb.

15.22 The angular speed of the continuous mass system, denoted as 2, in the simply supported bearings at A and B is a constant 360 rev/min. The forces from the system acting on the ground at bearings A and B are specified as $(\mathbf{F}_{21})_A = 324\hat{\mathbf{i}} - 351\hat{\mathbf{j}}$ N and $(\mathbf{F}_{21})_B = 280\hat{\mathbf{i}} + 524\hat{\mathbf{j}}$ N, respectively. Determine the magnitudes and orientations of the masses that must be removed in the correcting planes 1 and 2 to ensure dynamic balance of the system.

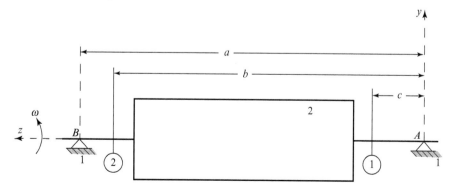

Fig. P15.22 $a = 750$ mm, $b = 675$ mm, $c = 175$ mm, and $R_{C1} = R_{C2} = 50$ mm.

15.23 The constant angular velocity of the two-cylinder engine crankshaft is $\omega = 200$ rad/s counterclockwise. Determine the x and y components of the primary shaking force acting on the crankshaft bearing in terms of the crank angle θ. Then determine the magnitudes and orientations of the correcting weights which must be added at the radial distance $R_C = 1.6$ in from the crankshaft axis. Determine the answers when the reference line (attached to crank 1) is specified at the crank angle $\theta = 30°$, as shown on the figure to the right.

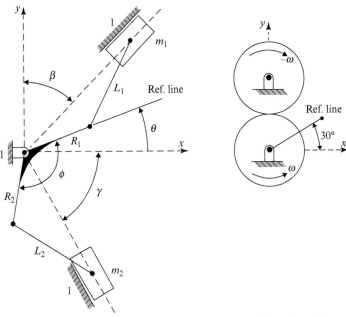

Fig. P15.23 $R_1 = R_2 = R = 3.2$ in, $L_1 = L_2 = L = 6.4$ in, $\beta = 45°$, $\gamma = 60°$, $\phi = 120°$, and $w_1 = w_2 = w = 33$ lb.

15.24 The two-cylinder engine crankshaft is rotating counterclockwise with a constant angular velocity $\omega = 45$ rad/s. Determine the magnitude and direction of the primary shaking force in terms of crank angle θ. If correcting masses are required to balance the primary shaking force then determine: (a) the magnitudes and orientations of the inertial forces created by these correcting masses, and (b) the magnitudes and orientations of the correcting masses if $R_{C1} = R_{C2} = 150$ mm. Determine the answers for parts (a) and (b) when the reference line attached to crank 1 is at the crank angle $\theta = 210°$.

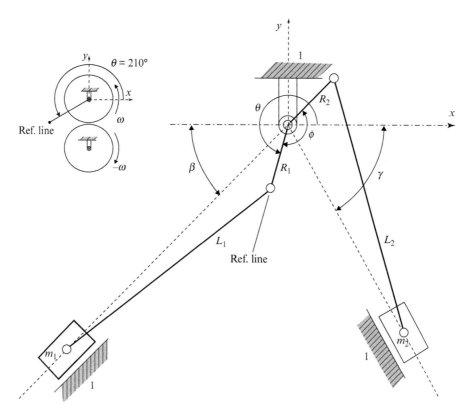

Fig. P15.24 $R_1 = R_2 = R = 75$ mm, $L_1 = L_2 = L = 600$ mm, $\beta = 45°$, $\gamma = 60°$, $\phi = 150°$, and $m_1 = m_2 = m = 6$ kg.

15.25 The crankshaft of the two-cylinder engine is rotating counterclockwise with a constant angular velocity $\omega = \dot\theta = 45$ rad/s. Determine the magnitude and orientation of the primary shaking force in terms of crank angle θ. If correcting weights are required to balance the primary shaking force, then determine (a) the magnitudes and orientations of the inertial forces created by these correcting weights, (b) the magnitudes and orientations of the correcting weights if $R_{C_1} = R_{C_2} = 8$ in, and (c) the answers when the reference line (attached to crank 1) is specified at the crank angle $\theta = 0°$.

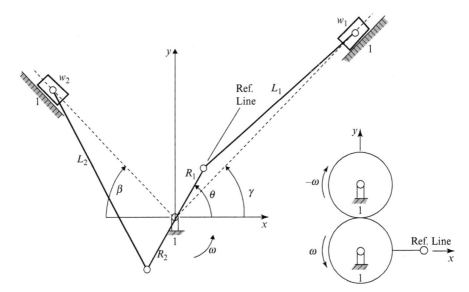

Fig. P15.25 $R_1 = R_2 = R = 4$ in, $L_1 = L_2 = L = 22$ in, $\beta = -45°$, $\gamma = 45°$, and $w_1 = w_2 = w = 11$ lb.

15.26 The crankshaft of the three-cylinder engine is rotating with a constant angular velocity $\omega = 50\hat{k}$ rad/s. Determine the x and y components of the primary shaking force, and the magnitude(s) of the correcting force (or forces) created by the correcting mass (or masses), and the orientation(s) of the correcting force (or forces). Determine the answers when the reference line (attached to crank 1) is specified at the crank angle $\theta = 0°$.

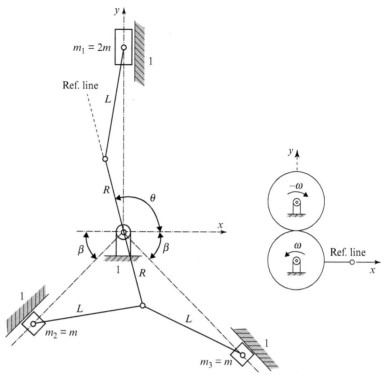

Fig. P15.26 $R_1 = R_2 = R_3 = R = 150$ mm, $L_1 = L_2 = L_3 = L = 750$ mm, $\beta = 45°, m_1 = 2m = 9.9$ kg, and $m_2 = m_3 = m = 4.95$ kg.

15.27 The crankshaft of the two-cylinder engine is rotating counterclockwise with a constant angular speed $\omega = 330$ rev/min. Determine the x and y components of the primary shaking force on the crankshaft bearing, and the magnitudes and orientations of the inertial forces created by the correcting weights that balance the primary shaking force. Determine the magnitude and orientation of the primary shaking force when the reference line (attached to crank 1) is specified at the crank angle $\theta = 0°$.

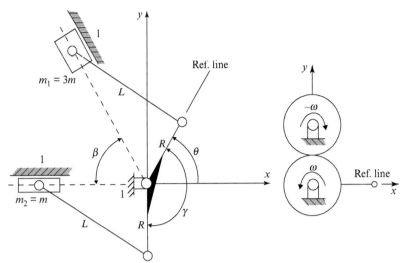

Fig. P15.27 $R_1 = R_2 = R = 4.8$ in, $L_1 = L_2 = L = 18$ in, $\beta = 60°$, $\gamma = 150°$,
$w_1 = 3w = 132$ lb, and $w_2 = w = 44$ lb.

15.28 The two cranks of the two-cylinder engine are oriented at 240° to each other, and the crankshaft is rotating counterclockwise with a constant angular speed $\omega = 690$ rev/min. Both pistons are in the same xy plane. Determine the x and y components of the primary shaking force acting on the ground bearing, and the magnitudes and orientations of the correcting masses that balance the primary shaking force. The radial distances of the correcting masses are $R_{C_1} = R_{C_2} = 400$ mm. Determine the answers when the reference line (attached to crank 1) is specified at the crank angle $\theta = 60°$.

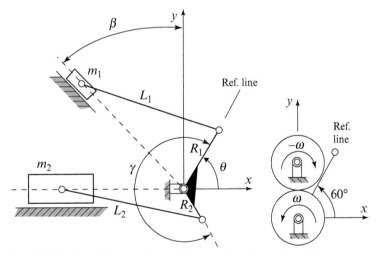

Fig. P15.28 $R_1 = 300$ mm, $R_2 = 150$ mm, $L_1 = L_2 = L = 650$ mm,
$\beta = 45°$, $m_1 = 10$ kg, and $m_2 = 50$ kg.

15.29 The constant angular velocity of the crankshaft of the two-cylinder engine is $\omega = 250\hat{\mathbf{k}}$ rad/s. Determine the x and y components of the primary shaking force in terms of the crank angle θ, and determine the magnitudes and orientations of the correcting weights. The correcting weights are to be added at a radial distance $R_C = 2$ in from the crankshaft axis. Determine the answers when the reference line (attached to crank 1) is specified at the crank angle $\theta = 30°$.

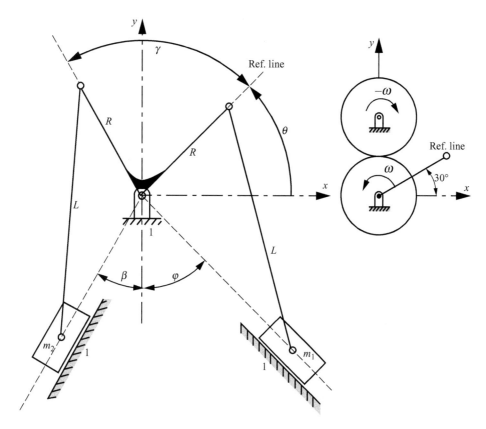

Fig. P15.29 $R_1 = R_2 = R = 6$ in, $L_1 = L_2 = L = 12$ in, $\beta = 30°$, $\gamma = 45°$, $\phi = 75°$, and $w_1 = w_2 = w = 44$ lb.

REFERENCES

[1] Berkof, R. S., and G. G. Lowen, 1969. A new method for completely force balancing simple linkages, *J. Eng. Ind. ASME Trans. B* **91**(1): 21–26.
[2] Lichty, L. C., 1939. *Internal Combustion Engines*, 5th ed., New York: McGraw-Hill.
[3] Lowen, G. G., and R. S. Berkof, 1968. Survey of investigations into the balancing of linkages, *J. Mech.*, **3**(4): 221–232.

[4] Maleev, V. L., 1993. *Internal Combustion Engines*, New York: McGraw-Hill.

[5] Rathbone, T. C., 1929. Turbine vibration and balancing, *ASME Trans.* **51**: 267–284.

[6] Stevensen, E. N., Jr., 1973. Balancing of machines, *J. Eng. Ind. ASME Trans. B* **95**(5): 650–656.

[7] Stevensen, E. N., Jr. Class notes, University of Hartford.

[8] Thearle, E. L., 1934. Dynamic balancing in the field, *ASME Trans.* **56**(10): 745–753.

16   Flywheels, Governors, and Gyroscopes

16.1   Dynamic Theory of Flywheels

A flywheel is an energy storage device. It absorbs mechanical energy by increasing its angular velocity and delivers energy by decreasing its angular velocity. Commonly, a flywheel is used to smooth the flow of energy between a power source and its load. If the load happens to be a punch press, for example, the actual punching operation requires energy for only a small fraction of its motion cycle. As another example, if the power source happens to be a two-cylinder, four-stroke engine, the engine delivers energy during only about half of its motion cycle. Other applications involve using flywheels to absorb braking energy and deliver acceleration energy for automobiles, or to act as energy-smoothing devices for electric utilities as well as solar- and wind-power-generating facilities. Electric railways have long used regenerative braking by absorbing braking energy back into power lines, but newly introduced and stronger materials now make the flywheel feasible for such purposes.

Figure 16.1 is a schematic representation of a flywheel. The flywheel, whose motion is measured by the angular coordinate θ has a mass moment of inertia I. An input torque T_i, corresponding to coordinate θ_i, causes the flywheel speed to increase. A load or output torque T_o, with corresponding coordinate θ_o, absorbs energy from the flywheel and causes it to slow down. If the work entering into the system is considered positive and work output is negative, the equation of motion of the flywheel is

$$\sum M = T_i\left(\theta_i, \dot{\theta}_i\right) - T_o\left(\theta_o, \dot{\theta}_o\right) - I\ddot{\theta} = 0,$$

which can be written as

$$I\alpha = T_i(\theta_i, \omega_i) - T_o(\theta_o, \omega_o). \tag{a}$$

Note that both T_i and T_o may depend for their values on their angular positions, θ_i and θ_o, as well as their angular velocities, ω_i and ω_o. Typically, the torque characteristic depends upon only one variable, either position or velocity. For example, the torque delivered by an induction motor depends on the speed of the motor. In fact, it is common practice for manufacturers of electric motors to publish charts detailing the torque–speed characteristics of their motors.

Fig. 16.1 Representation of
a flywheel.

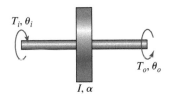

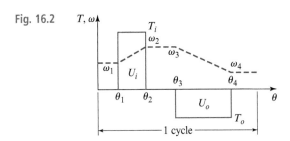

Fig. 16.2

When the input and output torque functions are known, Eq. (*a*) can be solved for the motion of the flywheel using well-known techniques for solving linear and nonlinear differential equations. The resulting equations can easily be solved using the methods of Chapter 13. In this chapter we assume rigid shafting, giving $\theta_i = \theta_o = \theta$. Thus, Eq. (*a*) becomes

$$I\alpha = T_i(\theta, \omega) - T_o(\theta, \omega). \tag{b}$$

When the two torque functions are known, and the starting values of displacement θ and velocity ω are given, Eq. (*b*) can be solved for θ, ω, and α as functions of time. However, typically, we are usually not interested in the instantaneous values of the kinematic parameters. Primarily, we want to know the overall performance of the flywheel. Important questions include: What is the mass moment of inertia? How do we match the power source to the load to get an optimum size of motor or engine? And what are the resulting performance characteristics of the system?

To gain insight into the problem, consider a hypothetical situation where an input power source subjects a flywheel to a constant torque T_i while the output shaft rotates from θ_1 to θ_2. This is positive torque and is plotted upward in Fig. 16.2. Equation (*b*) indicates that a positive acceleration, α, is the result, and so the shaft angular velocity increases from ω_1 to ω_2. As shown in the diagram, the shaft now rotates from θ_2 to θ_3 with zero torque and hence, from Eq. (*b*), with zero angular acceleration; therefore, $\omega_3 = \omega_2$. From θ_3 to θ_4, a load, or output torque, of constant magnitude is applied, causing the shaft to slow from ω_3 to ω_4. Note that the output torque is plotted in the negative direction in accordance with Eq. (*b*).

The energy input to the flywheel is the area of the rectangle between θ_1 and θ_2, or

$$U_i = T_i(\theta_2 - \theta_1). \tag{c}$$

The energy output from the flywheel is the area of the rectangle between θ_3 and θ_4, or

$$U_o = T_o(\theta_4 - \theta_3). \qquad (d)$$

If U_o is greater than U_i, the load uses more energy than has been entered into the flywheel and so ω_4 will be less than ω_1. If U_o is equal to U_i, then ω_4 will equal ω_1, since the gain and loss are equal; we are assuming no friction losses. Finally, if U_i is greater than U_o, then ω_4 will be greater than ω_1.

We can also write these relations in terms of kinetic energy. At $\theta = \theta_1$, the flywheel has an angular velocity of ω_1 rad/s, and so its kinetic energy is

$$U_1 = \tfrac{1}{2}I\omega_1^2. \qquad (e)$$

At $\theta = \theta_2$, the angular velocity is ω_2, and so its kinetic energy is

$$U_2 = \tfrac{1}{2}I\omega_2^2. \qquad (f)$$

Therefore, the change in kinetic energy is

$$U_2 - U_1 = \tfrac{1}{2}I\left(\omega_2^2 - \omega_1^2\right). \qquad (16.1)$$

16.2 Integration Technique

Many of the torque–displacement functions encountered in practical engineering situations are so complex that they must be integrated by approximate methods. Figure 16.3, for example, is a plot of the engine torque for one cycle of motion of a single-cylinder, four-stroke internal combustion engine. Since part of the torque curve is negative, it is clear that the flywheel returns

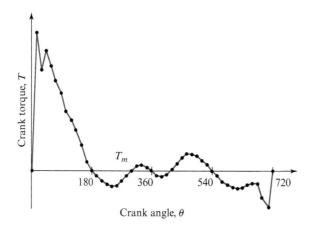

Fig. 16.3 Relation between torque and crank angle for a single-cylinder, four-stroke engine.

part of its energy back to the engine. Approximate integration of this curve for a cycle of 4π rad yields the mean torque, T_m, available to drive the load.

A simple integration routine is Simpson's rule; this approximation can be programmed for any computer and is simple enough for even the crudest programmable calculators. In fact, this routine is often found as a part of the library for many calculators and personal computers. The equation used is

$$\int_{x_0}^{x_n} f(x)dx = \frac{h}{3}(f_0 + 4f_1 + 2f_2 + 4f_3 + 2f_4 + \cdots + 2f_{n-2} + 4f_{n-1} + f_n), \qquad (16.2)$$

where

$$h = (x_n - x_0)/n \text{ with } x_n > x_0,$$

and n is the number of subintervals used, 2, 4, 6, ..., which must be even. If memory is limited, Eq. (16.2) can be solved in two or more steps, say from 0 to $n/2$ and then from $n/2$ to n.

For flywheel applications, it is convenient to define a *coefficient of speed fluctuation* as

$$C_s = \frac{\omega_2 - \omega_1}{\omega}, \qquad (16.3)$$

where ω is the nominal or average angular velocity, given by

$$\omega = \tfrac{1}{2}(\omega_2 + \omega_1). \qquad (16.4)$$

Equation (16.1) can be factored to give the change in kinetic energy as

$$U_2 - U_1 = \tfrac{1}{2}I(\omega_2 - \omega_1)(\omega_2 + \omega_1).$$

Substituting Eqs. (16.3) and (16.4) into this equation gives

$$U_2 - U_1 = C_s I \omega^2, \qquad (16.5)$$

which can be used to obtain an appropriate flywheel inertia corresponding to the change in the kinetic energy.

Example 16.1

Table 16.1 lists numeric values of the torque for the single-cylinder, four-stroke engine (plotted in Fig. 16.3). The nominal speed of the engine is to be 250 rad/s.

a. Integrate the torque–displacement function for one cycle, and find the energy that can be delivered to the load during the cycle.
b. Determine the mean torque T_m (Fig. 16.3).
c. The greatest energy fluctuation occurs approximately between $\theta = 15°$ and $\theta = 150°$ on the T diagram (Fig. 16.3). Using a coefficient of speed fluctuation of $C_S = 0.1$, find a suitable value for the flywheel inertia.
d. Find ω_2 and ω_1.

Example 16.1 (cont.)

Table 16.1 Torque data for the single-cylinder, four-stroke engine

θ_i (deg)	T_i (in · lb)	θ_i (deg)	T_i (in · lb)	θ_i (deg)	T_i (in · lb)	θ_i (deg)	T_i (in · lb)
0	0	180	0	360	0	540	0
15	2 800	195	−107	375	−85	555	−107
30	2 090	210	−206	390	−125	570	−206
45	2 430	225	−280	405	−89	585	−292
60	2 160	240	−323	420	8	600	−355
75	1 840	255	−310	435	126	615	−371
90	1 590	270	−242	450	242	630	−362
105	1 210	285	−126	465	310	645	−312
120	1 066	300	−8	480	323	660	−272
135	803	315	89	495	280	675	−274
150	532	330	125	510	206	690	−548
165	184	345	85	525	107	705	−760

Solution

a. Using $n = 48$ and $h = 4\pi/48$, we enter the data of Table 16.1 into our calculator and compute the integral by Simpson's rule as defined in Eq. (16.2); the amount of energy that can be delivered to the load during the cycle is

$$U = 3\ 490 \text{ in} \cdot \text{lb.} \qquad \qquad \textit{Ans.}$$

b. The mean torque is

$$T_m = 3\ 490 \text{ in} \cdot \text{lb}/4\pi \text{ rad} = 278 \text{ in} \cdot \text{lb.} \qquad \qquad \textit{Ans.}$$

c. The largest positive loop in the torque–displacement diagram occurs between $\theta = 0°$ and $\theta = 180°$. We select this loop as yielding the largest speed change. Subtracting 278 in·lb from the values in Table 16.1 for this loop gives, respectively, −278, 2 522, 1 812, 2 152, 1 882, 1 562, 1 312, 932, 788, 535, 254, −94, and −278 in · lb.

Entering the data into Simpson's rule again, using $n = 12$ and $h = \pi/12$, the change in kinetic energy is

$$U_2 - U_1 = 3\ 663 \text{ in} \cdot \text{lb.}$$

Rearranging Eq. (16.5), the mass moment of inertia of the flywheel can be written as

$$I = \frac{U_2 - U_1}{C_S \omega^2}.$$

Example 16.1 (cont.)

Substituting the known information into this equation, a suitable value for the flywheel inertia is

$$I = \frac{3\,663 \text{ in} \cdot \text{lb}}{0.1(250 \text{ rad/s})^2} = 0.586 \text{ in} \cdot \text{lb} \cdot \text{s}^2. \qquad Ans.$$

d. Equations (16.3) and (16.4) can now be solved simultaneously for ω_2 and ω_1. Substituting appropriate values into these two equations yields

$$\omega_2 = \frac{1}{2}(2 + C_S)\omega = \frac{1}{2}(2 + 0.1)250 \text{ rad/s} = 262.5 \text{ rad/s}, \qquad Ans.$$

$$\omega_1 = 2\omega - \omega_2 = 2(250 \text{ rad/s}) - 262.5 \text{ rad/s} = 237.5 \text{ rad/s}. \qquad Ans.$$

These two speeds occur at $\theta = 180°$ and $\theta = 0$, respectively.

16.3 Multi-Cylinder Engine Torque Summation

Once the torque–displacement relation has been defined for a single cylinder, it is easy to assume these for multi-cylinder engines. If, for example, we wish to find the torque function for a three-cylinder engine, then we examine the firing order shown in Fig. 14.1 and write

$$T_{total} = T_\theta + T_{\theta+240°} + T_{\theta+480°}, \qquad (a)$$

since the torque events are spaced 240° apart. Here, θ is the crank angle of the first cylinder.

Applying this same procedure for a four-cylinder, four-stroke internal combustion engine, along with using the torque values in Table 16.1, gives the results in Table 16.2. Figure 16.4 shows how this torque varies during each 180° of crank rotation.

16.4 Classification of Governors

The availability today of a wide variety of low-cost solid-state electronic devices and transducers makes it possible to regulate the speed of mechanical systems to a finer degree and at less cost than was possible with older all-mechanical governors. The electronic governor also has the advantage that the speed to be regulated can be changed quite easily and at will.

So far in this chapter we have learned that flywheels are used to regulate speed over short intervals of time, such as a single revolution or the duration of an engine cycle. *Governors* are also devices that are used to regulate speed. In contrast to flywheels, however, governors are used to regulate speed over much longer intervals of time; in fact, they are intended to maintain

Table 16.2 Torque data for a four-cylinder, four-stroke engine

θ_i (deg)	T_θ (in·lb)	$T_{\theta+180}$ (in·lb)	$T_{\theta+360}$ (in·lb)	$T_{\theta+540}$ (in·lb)	T_{total} (in·lb)
0	0	0		0	0
15	2 800	−107	−85	−107	2 501
30	2 090	−206	−125	−206	1 553
45	2 430	−280	−89	−292	1 769
60	2 160	−323	8	−355	1 490
75	1 840	−310	126	−371	1 285
90	1 590	−242	242	−362	1 228
105	1 210	−126	310	−312	1 082
120	1 066	−8	323	−272	1 109
135	803	89	280	−274	898
150	532	125	206	−548	315
165	184	85	107	−760	−384

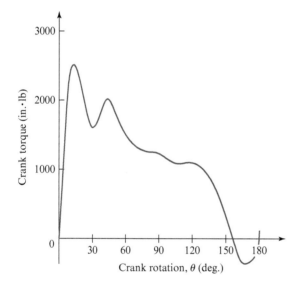

Fig. 16.4 Relation between torque and crank angle for a four-cylinder, four-stroke engine.

a balance between the energy supplied to a moving system and the external load or resistance applied to that system. When the speed of a machine must, over its lifetime, always be kept at the same level, or approximately so, then a shaft-mounted mechanical device may be an appropriate speed regulator. Such governors may be classified as either centrifugal governors or inertia governors.

16.5 Centrifugal Governors

Figure 16.5 shows a simple spring-controlled shaft governor. Masses attached to the bell-crank levers are driven outward by inertia force against springs. The motion of the shaft-mounted sleeve is dependent on the motions of the masses and the ratio of the bell-crank lengths.

Defining the nomenclature for Fig. 16.6, we let

k = spring rate;
P = spring force;
W = sleeve weight acting along the vertical shaft;
r = instantaneous radial position of mass center; and
r_0 = position of mass center at zero spring force.

Then, the spring force at any position r is

$$P = k(r - r_0). \tag{a}$$

Fig. 16.5 Centrifugal shaft governor.

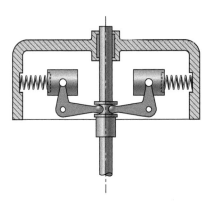

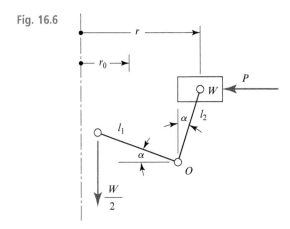

Fig. 16.6

Taking moments about pivot O gives

$$\sum M_O = Pl_2 \cos \alpha + \frac{W}{2}l_1 \cos \alpha + (-m\ddot{r})l_2 \cos \alpha = 0. \qquad (b)$$

Now, with $\ddot{r} = r\omega^2$ and $m = W/g$, this gives

$$\frac{W}{g}r\omega^2 l_2 = Pl_2 + \frac{W}{2}l_1. \qquad (c)$$

The controlling force can be written as

$$F = \frac{W}{g}r\omega^2. \qquad (d)$$

Then, substituting this expression into Eq. (c) and using Eq. (a), the controlling force is

$$F = P + \frac{W}{2}\frac{l_1}{l_2} = k(r - r_0) + \frac{W}{2}\frac{l_1}{l_2}. \qquad (16.6)$$

Equating Eqs. (d) and (16.6) gives

$$\frac{W}{g}r\omega^2 = k(r - r_0) + \frac{W}{2}\frac{l_1}{l_2}.$$

Then, rearranging this equation, the corresponding angular velocity can be written as

$$\omega^2 = \frac{kg}{Wr}\left[r - \left(r_0 - \frac{W}{2k}\frac{l_1}{l_2}\right)\right]. \qquad (16.7)$$

16.6 Inertia Governors

In a centrifugal governor, an increase in speed causes the rotating masses to move radially outward. Thus, it is the radial (normal) component of the acceleration that is primarily responsible for creating the controlling force. In an inertia shaft governor, as shown in Fig. 16.7, the mass pivot is located at point A, very close to the shaft center at O. Thus, the radial component of acceleration is small and much less effective, but a sudden change of speed, producing an angular acceleration, causes the masses to lag and produces tension in the spring. This spring tension produces a transverse (tangential) acceleration force. Consequently, it is the angular acceleration that determines the positions of the masses.

Compared with a centrifugal governor, an inertia governor is more sensitive, since it acts at the very beginning of a speed change.

Fig. 16.7 Inertia shaft governor.

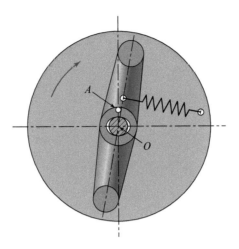

Fig. 16.8 Block diagram of a closed-loop control system.

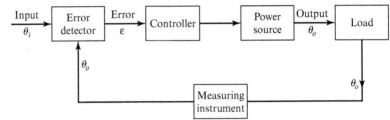

16.7 Mechanical Control Systems

Many mechanical control systems are represented schematically, as shown in Fig. 16.8. Here, θ_i and θ_o represent any set of input and output signals. In the case of a governor, θ_i represents the desired speed and θ_o the actual speed. For control systems in general, the input and output functions can represent force, torque, or displacement, either rectilinear or angular.

The system shown in Fig. 16.8 is called a *closed-loop* or *feedback control system*, since the output value, θ_o, is fed back to the detector at the input to indicate error ε, which is the difference between the input and the output. The objective of the controller is to cause this error to become as close to zero as possible. The mechanical characteristics of the system, that is, the mechanical clearances, friction, inertias, and stiffnesses, sometimes cause the output to differ somewhat from the input, and so it is the designer's responsibility to examine these mechanical effects in an effort to minimize the error for all operating conditions.

The differential equation for a control system is always written as an expression of the dynamic equilibrium of the elements of the system. Thus, for the rotational system of Fig. 16.8, we express mathematically that the inertia torque is equal to the sum of all other torques acting upon the system. If the load has both inertia and viscous damping, then the expression is

$$I\ddot{\theta}_o = -c\dot{\theta}_o + k'f(\varepsilon). \tag{a}$$

The function $f(\varepsilon)$ depends upon the characteristics of the controller and the power source. A very simple system would result if the components had characteristics such that the relation between $f(\varepsilon)$ and ε is linear. For this condition we can make the substitution

$$k'f(\varepsilon) = k\varepsilon. \tag{b}$$

Substituting Eq. (b) into Eq. (a), and rearranging, gives

$$I\ddot{\theta}_o + c\dot{\theta}_o = k\varepsilon. \tag{c}$$

By definition, we have

$$\varepsilon = \theta_i - \theta_o. \tag{d}$$

Therefore, Eq. (c) can be written as

$$I\ddot{\theta}_o + c\dot{\theta}_o + k\theta_o = k\theta_i. \tag{16.8}$$

This second-order differential equation can be simplified by the following substitutions:

$$\omega_n = \sqrt{\frac{k}{I}} \tag{16.9}$$

and

$$\zeta = \frac{c}{2\sqrt{kI}}, \tag{16.10}$$

where I = mass moment of inertia, c = torsional damping coefficient, k = torsional stiffness, ω_n = natural frequency, and ζ = damping ratio, c/c_{cr}.

After a review of this notation for linear systems in Chapter 13, Eq. (16.8) can be written in the form

$$\ddot{\theta}_o + 2\zeta\omega_n\dot{\theta}_o + \omega_n^2\theta_o = \omega_n^2\theta_i. \tag{16.11}$$

16.8 Standard Input Functions

There is a great deal of useful information that can be gained from a mathematical analysis as well as a laboratory analysis of feedback control systems when standard input functions are used to study their performance. Standard input functions result in differential equations that are easier to analyze mathematically than they are if actual operating conditions are used as input. Furthermore, the use of standard inputs makes it possible to compare the performance of different control systems.

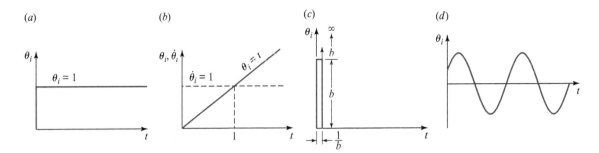

Fig. 16.9 (*a*) Unit-step function, (*b*) unit-step velocity function, (*c*) unit-impulse function, (*d*) harmonic input function.

One of the most useful standard input functions is the *unit-step function* shown in Fig. 16.9*a*. This function is not continuous, and, consequently, we cannot define initial conditions at $t = 0$. It is customary to specify the conditions at $t = 0+$ and $t = 0-$, where the signs indicate the conditions for values of time slightly greater than or slightly less than zero. Thus, for the unit-step function we have

$$\theta_i = 0, \quad \dot{\theta}_i = 0 \quad \text{when } t = 0-,$$

$$\theta_i = 1, \quad \dot{\theta}_i = 0 \quad \text{when } t = 0+.$$

The performance resulting from the application of these input conditions is called the *unit-step response*.

Another standard input function is the *unit-step velocity function*, shown in Fig. 16.9*b*. This function is sometimes called the *unit-slope ramp function* because of the form of its displacement diagram. This function is defined as follows:

$$\theta_i = 0, \quad \dot{\theta}_i = 0 \quad \text{when } t = 0-,$$

$$\theta_i = t, \quad \dot{\theta}_i = 1 \quad \text{when } t = 0+.$$

Also shown in Figs. 16.9*c* and 16.9*d* are the unit-impulse function and the harmonic input function, both of which are useful in studying a system's response.

It sometimes happens in control systems that the input function is relatively constant, but the output is subjected to abruptly varying applied forces or torques. For example, an automatically controlled machine tool may encounter a deeper cut, causing a greater load torque to be exerted on the machine, thus tending to slow it down. The differential equation for this condition is written as

$$I\ddot{\theta}_o + c\dot{\theta}_o + k\theta_o = k\theta_i - T. \tag{a}$$

When this equation is simplified by substituting the values from Eqs. (16.9) and (16.10) and rearranged, it becomes

$$\ddot{\theta}_o + 2\zeta\omega_n\dot{\theta}_o + \omega_n^2\theta_o = \omega_n^2\theta_i - \frac{T}{I}. \tag{16.12}$$

When the solution to Eq. (16.11) is obtained for a given function θ_i, the solution to Eq. (16.12) can be obtained by superposition.

16.9 Solution of Linear Differential Equations

In the analysis of feedback control systems, nth-order linear differential equations are frequently encountered. The general form of these equations is

$$a_n \frac{d^n \theta}{dt^n} + a_{n-1} \frac{d^{n-1} \theta}{dt^{n-1}} + \cdots + a_1 \frac{d\theta}{dt} + a_0 \theta = f(t). \tag{16.13}$$

The function $f(t)$ is the driving or forcing function, and so t and θ are the independent and dependent variables, respectively. The coefficients $a_0, a_1, \ldots, a_n$ are constants and are independent of t and θ.

The solution to equations of the form of Eq. (16.13) is composed of two parts (Chapter 13). The first part is called the *complimentary solution*, and it is the solution to the homogeneous equation

$$a_n \frac{d^n \theta}{dt^n} + a_{n-1} \frac{d^{n-1} \theta}{dt^{n-1}} + \cdots + a_1 \frac{d\theta}{dt} + a_0 \theta = 0. \tag{16.14}$$

The complimentary solution is also called the *transient solution* in the literature of automatic controls since, for a damped stable system, this part of the solution fades away. If the control system should happen to be unstable, then the controlled quantity increases without limit. Note that the transient solution is obtained by solving with a forcing function of zero.

The other part of the solution is called the *particular solution* by mathematicians and the *steady-state solution* by control engineers. It is any specific solution of Eq. (16.13).

The complete solution is the sum of the transient and the steady-state solutions. The transient part contains n arbitrary constants that are evaluated from the initial conditions. The amplitude of the transient solution depends upon both the forcing function and the initial conditions, but all other characteristics of the transient solution are completely independent of the forcing function. We find, for linear systems, that neither the forcing function nor the amplitude have any effect on system stability, this being dependent only on the transient portion of the solution.

The Transient Solution.   The following steps are used to obtain the transient solution:

1. Set the forcing function equal to zero and arrange the equation in the form of Eq. (16.14).
2. Assume a solution of the form

$$\theta = A e^{st}. \tag{16.15}$$

3. Substitute Eq. (16.15) and its derivatives into the differential equation, and simplify to obtain the characteristic equation.
4. Solve the characteristic equation for its roots.
5. Obtain the transient solution by substituting the roots into the assumed solution.

As an example of this procedure, let us solve Eq. (16.11) for the transient solution. Following step 1, we write

$$\ddot{\theta}_o + 2\zeta\omega_n\dot{\theta}_o + \omega_n^2\theta_o = 0, \qquad (a)$$

and for step 2 we have

$$\theta_o = Ae^{st}, \quad \dot{\theta}_o = Ase^{st}, \quad \ddot{\theta}_o = As^2e^{st}. \qquad (b)$$

For step 3, we substitute Eqs. (b) into Eq. (a), which gives

$$As^2e^{st} + 2\zeta\omega_n Ase^{st} + \omega_n^2 Ae^{st} = 0. \qquad (c)$$

The characteristic equation is then obtained by dividing out common factors, that is,

$$s^2 + 2\zeta\omega_n s + \omega_n^2 = 0. \qquad (d)$$

Step 4 is to solve this equation and obtain its roots, namely

$$\begin{aligned} s_1 &= -\zeta\omega_n + \sqrt{\zeta^2 - 1}\omega_n, \\ s_2 &= -\zeta\omega_n - \sqrt{\zeta^2 - 1}\omega_n. \end{aligned} \qquad (e)$$

We shall not consider the situation in which $\zeta > 1$. Consequently, for $\zeta \leq 1$, the radical in Eq. (e) has an imaginary value, and we prefer to write the roots in the form

$$\begin{aligned} s_1 &= -\zeta\omega_n + j\sqrt{1 - \zeta^2}\omega_n, \\ s_2 &= -\zeta\omega_n - j\sqrt{1 - \zeta^2}\omega_n, \end{aligned} \qquad (f)$$

where $j = \sqrt{-1}$.

Finally, for step 5, we substitute the roots back into the assumed solution. Then, after factoring, the transient solution becomes

$$\theta_{o,t} = e^{-\zeta\omega_n t}\left(Ce^{j\sqrt{1-\zeta^2}\omega_n t} + De^{-j\sqrt{1-\zeta^2}\omega_n t}\right), \qquad (g)$$

where C and D are the constants of integration and are two in number, the same as the order of the highest derivative in the differential equation. These cannot be evaluated until the complete solution is found. At that time, they will be determined from the initial conditions. We can, however, transform Eq. (g) into trigonometric form by using Euler's equation [Eq. (2.40)]. The result of this transformation is

$$\theta_{o,t} = e^{-\zeta\omega_n t}\left(A\cos\sqrt{1 - \zeta^2}\omega_n t + B\sin\sqrt{1 - \zeta^2}\omega_n t\right). \qquad (16.16)$$

The common factor $e^{-\zeta\omega_n t}$ of Eq. (16.16) indicates its transient nature, since this term becomes approximately zero for large values of t.

It can sometimes happen that two or more roots of the characteristic equation are identical. For the case of two identical roots, $s_1 = s_2$, the transient solution is

$$\theta_{o,t} = Ee^{s_1 t} + Fte^{s_1 t}. \tag{h}$$

For other special cases, the reader should refer to a text such as [2] that focuses on the analysis of control systems.

The Steady-State Solution.   Here we shall solve Eq. (16.11) for various standard inputs. For the unit-step function, $\theta_i = 1$. Substituting this function into Eq. (16.11) gives

$$\ddot{\theta}_o + 2\zeta\omega_n\dot{\theta}_o + \omega_n^2\theta_o = \omega_n^2. \tag{i}$$

The solution is

$$\theta_{o,s} = 1, \tag{16.17}$$

which can be verified by substituting Eq. (16.17) and its derivatives into Eq. (i).

For the unit-slope ramp function, $\theta_i = t$, substituting this function into Eq. (16.11) gives

$$\ddot{\theta}_o + 2\zeta\omega_n\dot{\theta}_o + \omega_n^2\theta_o = \omega_n^2 t. \tag{j}$$

We assume a solution of the form

$$\theta_{o,s} = At + B. \tag{k}$$

The successive derivatives are

$$\dot{\theta}_{o,s} = A, \quad \ddot{\theta}_{o,s} = 0.$$

Substituting the assumed solution and its derivatives into Eq. (j) gives

$$2\zeta\omega_n A + \omega_n^2(At + B) = \omega_n^2 t.$$

We now arrange the equation into the form

$$\left(2\zeta\omega_n A + \omega_n^2 B\right) + \left(\omega_n^2 A\right)t = \omega_n^2 t,$$

and solve for A and B. The solutions are

$$A = 1, \quad B = -\frac{2\zeta}{\omega_n}.$$

Substituting these into Eq. (k), the steady-state solution for a unit-slope ramp function is

$$\theta_{o,s} = t - \frac{2\zeta}{\omega_n}. \tag{16.18}$$

The Complete Solution. The complete solution is the sum of the transient and the steady-state solutions. Thus,

$$\theta_o = \theta_{o,t} + \theta_{o,s}.$$

Therefore, adding Eqs. (16.16) and (16.17) gives

$$\theta_o = e^{-\zeta \omega_n t}\left(C_1 \cos \sqrt{1 - \zeta^2}\omega_n t + C_2 \sin \sqrt{1 - \zeta^2}\omega_n t \right) + 1. \qquad (16.19)$$

The initial conditions for $t = 0+$ can now be applied in order to evaluate the constants C_1 and C_2. Imposing the condition that $\theta_o = 0$ before the step $(t = 0-)$ gives

$$0 = 1(C_1 \cos 0 + C_2 \sin 0) + 1$$

or

$$C_1 = -1.$$

The second condition to be imposed is that $\dot{\theta}_o = 0$ at $t = 0+$. To apply this condition, it is necessary to take the derivative of Eq. (16.19). This is

$$\dot{\theta}_o = -\zeta \omega_n e^{-\zeta \omega_n t}\left(C_1 \cos \sqrt{1 - \zeta^2}\omega_n t + C_2 \sin \sqrt{1 - \zeta^2}\omega_n t \right)$$

$$+ e^{-\zeta \omega_n t}\left(-C_1\sqrt{1 - \zeta^2}\omega_n \sin \sqrt{1 - \zeta^2}\omega_n t + C_2\sqrt{1 - \zeta^2}\omega_n \cos \sqrt{1 - \zeta^2}\omega_n t \right).$$

Substituting $\dot{\theta}_o = 0$ and $t = 0$ gives

$$0 = -\zeta \omega_n (C_1 \cos 0 + C_2 \sin 0) + 1\left(-C_1\sqrt{1 - \zeta^2}\omega_n \sin 0 + C_2\sqrt{1 - \zeta^2}\omega_n \cos 0 \right).$$

Substituting the value of C_1 and solving yields

$$C_2 = -\frac{\zeta}{\sqrt{1 - \zeta^2}}.$$

Substituting the values of C_1 and C_2 into Eq. (16.19) and rearranging, the complete solution can be written as

$$\theta_o = 1 - e^{-\zeta \omega_n t}\left(\cos \sqrt{1 - \zeta^2}\omega_n t + \sin \sqrt{1 - \zeta^2}\omega_n t \right). \qquad (16.20)$$

If transformed to a single trigonometric term with a phase angle as demonstrated in Chapter 13, the complete solution is

$$\theta_o = 1 - \frac{e^{-\zeta \omega_n t}}{\sqrt{1 - \zeta^2}} \cos\left(\sqrt{1 - \zeta^2}\omega_n t - \phi \right), \qquad (16.21)$$

where

$$\phi = \tan^{-1}\left(\frac{\zeta}{\sqrt{1-\zeta^2}}\right).$$

The complete solution for the unit-slope ramp function input is obtained in a similar manner. The initial conditions to be applied in order to evaluate the constants of integration are

$$\theta_o = 0 \quad \text{and} \quad \dot{\theta}_o = 0 \quad \text{when} \quad t = 0-.$$

The equation to be solved for the integration constants is obtained by adding Eqs. (16.16) and (16.18); it is

$$\theta_o = t - \frac{2\zeta}{\omega_n} + e^{-\zeta\omega_n t}\left(C_1\cos\sqrt{1-\zeta^2}\omega_n t + C_2\sin\sqrt{1-\zeta^2}\omega_n t\right). \tag{16.22}$$

The two constants are found to be

$$C_1 = \frac{2\zeta}{\omega_n} \quad \text{and} \quad C_2 = -\frac{1-2\zeta^2}{\sqrt{1-\zeta^2}\omega_n},$$

and the complete solution is

$$\theta_o = t - \frac{2\zeta}{\omega_n} + e^{-\zeta\omega_n t}\left(\frac{2\zeta}{\omega_n}\cos\sqrt{1-\zeta^2}\omega_n t - \frac{1-2\zeta^2}{\sqrt{1-\zeta^2}\omega_n}\sin\sqrt{1-\zeta^2}\omega_n t\right) \tag{16.23}$$

or

$$\theta_o = t - \frac{2\zeta}{\omega_n} + \frac{e^{-\zeta\omega_n t}}{\sqrt{1-\zeta^2}\omega_n}\cos\left(\sqrt{1-\zeta^2}\omega_n t - \phi\right), \tag{16.24}$$

where the phase angle is

$$\phi = \tan^{-1}\left(\frac{2\zeta^2-1}{2\zeta\sqrt{1-\zeta^2}}\right).$$

16.10 Analysis of Proportional-Error Feedback Systems

As the name implies, a proportional-error feedback control system is one that operates by applying a correction that is directly proportional to any error that might exist. Such a system must be analyzed in order to determine the following:

1. The response time, or the time required to reach steady-state operation after the application of a disturbance to the system; since a system with viscous damping never reaches a steady-state condition, this is usually defined as the time required for the transient to decay to 5% of its initial value.

Fig. **16.10** Response to a
unit-step input.

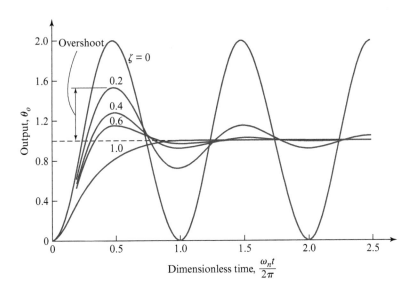

2. The natural frequency.
3. The steady-state error, that is, the difference between the output and input during steady-state operation.
4. The maximum overshoot, or the maximum deviation between output and input during transient conditions.

Output response of the system to a unit-step input is given by Eq. (16.21) and has been plotted in Fig. 16.10 for various values of the damping ratio. The abscissa is measured in dimensionless time, $\omega_n t/2\pi$, so the curves can be applied to any physical system. Thus, if the undamped natural frequency, ω_n, is known for a given system, then the response time can be calculated simply by multiplying dimensionless time by the quantity $2\pi/\omega_n$. The response for zero damping is included for its academic interest; an automatic control system with no damping at all would give completely unsatisfactory performance.

The response for critical damping, $\zeta = 1$, shows no overshoot, and the steady-state condition is reached at about $\omega_n t/2\pi = 1.0$.

It is convenient to define an *overshoot ratio* as the ratio of overshoot with damping to the overshoot that would exist if no damping were present. Figure 16.11 shows a plot of this ratio versus the damping ratio. For $\zeta = 0.40$, the overshoot ratio is approximately 0.25; this quantity can also be read from Fig. 16.10. Note that the steady-state condition is not reached for $\zeta = 0.40$ until about $\omega_n t/2\pi = 2.0$. The response time, therefore, is twice that for critical damping.

Since the response time is proportional to $\omega_n t/2\pi$, we can make this time short simply by making the undamped natural frequency, ω_n, large.

In practice, ζ is usually between 0.4 and 0.7 in a physical system. If a certain deviation from the steady state is permitted, say 5%, then a system with $\zeta = 0.7$ will reach its steady state before one having $\zeta = 1.0$ since the early portion of its response curve is steeper.

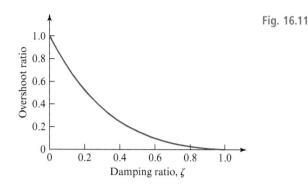

Fig. 16.11

Figure 16.10 demonstrates that the steady-state value of the output is unity. Consequently, there is no steady-state error for step-input functions when applied to proportional-error feedback systems.

Local Disturbance.   Another look at the behavior of a system can be obtained by considering that the input signal is constant or zero and that the load is subjected to a disturbance. Selecting the zero input condition, for simplicity, Eq. (16.12) then becomes

$$\ddot{\theta}_o + 2\zeta\omega_n\dot{\theta}_o + \omega_n^2\theta_o = -\frac{T}{I}, \tag{16.25}$$

where T represents a constant torque suddenly applied to the input. The steady-state component of the solution is

$$\theta_{o,s} = -\frac{T}{I\omega_n^2}, \tag{a}$$

so that the complete solution must be

$$\theta_o = e^{-\zeta\omega_n t}\left(C_1\cos\sqrt{1-\zeta^2}\,\omega_n t + C_2\sin\sqrt{1-\zeta^2}\,\omega_n t \right) - \frac{T}{I\omega_n^2}. \tag{b}$$

The initial conditions are $\theta_o = 0$ and $\dot{\theta}_o = 0$ at $t = 0$. Using these conditions, we can solve Eq. (b) for the two constants

$$C_1 = \frac{T}{I\omega_n^2} \quad \text{and} \quad C_2 = \frac{\zeta T}{\sqrt{1-\zeta^2}\,I\omega_n^2}.$$

Substituting these into Eq. (b) and transforming to a single trigonometric term, the result is

$$\theta_o = \frac{T}{I\omega_n^2}\left[\frac{e^{-\zeta\omega_n t}}{\sqrt{1-\zeta^2}}\cos\left(\sqrt{1-\zeta^2}\,\omega_n t - \phi\right) - 1 \right], \tag{16.26}$$

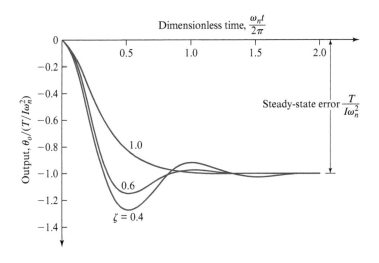

Fig. 16.12 Response of proportional-error feedback control.

where

$$\phi = \tan^{-1}\left(\frac{\zeta}{\sqrt{1-\zeta^2}}\right).$$

Equation (16.26) is plotted in Fig. 16.12 for three values of the damping ratio to show what happens. As shown, the disturbance decays at a rate that is dependent on the amount of damping. It finally reaches a steady-state condition that is not zero, but is of an amount

$$\varepsilon_s = \frac{T}{I\omega_n^2}. \tag{16.27}$$

This is the steady-state error. If we substitute $\omega_n^2 = k/I$, the equation becomes

$$\varepsilon_s = \frac{T}{k}. \tag{16.28}$$

Hence, the only method of reducing the magnitude of this error is to increase the gain (stiffness), k, of the system.

We have seen that the control system is defined by the parameters I, c, and k. Of these three, the gain, k, is usually the easiest to change. Some variation in the damping, c, is usually possible, but the employment of dashpots or friction dampers is not often a good solution. Variation in the inertia, I, is the most difficult change to make, since this is fixed by the design of the driven element, and, furthermore, improvement always requires a decrease in the inertia.

Unit-Step Velocity Input.   For the unit-step velocity input, we have $\theta_i = t$, and from Eq. (16.18),

$$\theta_{o,s} = t - \frac{2\zeta}{\omega_n}$$

after the transient decays. Therefore, the steady-state error is

$$\varepsilon_s = \frac{2\zeta}{\omega_n} = \frac{c}{k}, \tag{16.29}$$

which is obtained by substitution of the value of ζ from Eq. (16.10). This error persists only as long as the input function signals for a constant velocity. Again, we see that the magnitude can be reduced by increasing the gain, k.

The widely used automotive cruise-control system is an excellent example of an electromechanical governor. A transducer is attached to the speedometer cable, and the electrical output of this transducer is the signal, θ_o, fed to the error detector of Fig. 16.8. In some cases, magnets are mounted on the driveshaft of the car to activate the transducer. In the cruise-control system, the error detector is an electronic regulator, usually mounted under the dashboard. The regulator is turned on by an engagement switch under, or near, the steering wheel. A power unit is connected to the carburetor throttle linkage; the power unit is controlled by the regulator and gets its power from a vacuum port on the engine. Such systems have one or two brake-release switches as well as the engagement switch. The accelerator pedal can also be used to override the system.

16.11 Introduction to Gyroscopes

A gyroscope is defined as a rigid body capable of three-dimensional rotation with high angular velocity about any axis that passes through a fixed point called the center, which may or may not be its center of mass. A child's toy top fits this definition and is a form of gyroscope; its fixed point is the point of contact of the top with the floor or table on which it spins.

The usual form of a gyroscope is a mechanical device in which the essential part is a rotor having a heavy rim spinning at high speed about a fixed point on the rotor axis. The rotor is then mounted so as to turn freely about its center of mass by means of double gimbals called a Cardan suspension; an example is pictured in Fig. 16.13.

The gyroscope has fascinated students of mechanics and applied mathematics for many years. Indeed, once the rotor is set spinning, a gyroscope seems to act like a device possessing intelligence. If we attempt to move some of its parts, it seems not only to resist this motion but even to evade it. It seemingly fails to conform to the laws of static equilibrium and of gravitation.

The early history of the gyroscope is rather obscure. Probably the earliest gyroscope of the type now in use was constructed by Bohnenburger[1] in Germany in 1817 [1]. In 1852, Leon Foucault,[2] of Paris, constructed a very refined version to demonstrate the rotation of the Earth; it was Foucault who named the instrument the gyroscope from the Greek words *gyros*, circle or ring, and *skopien*, to view [4]. The mathematical foundations of gyroscopic theory were laid by Leonhard Euler in 1765 in his work on the dynamics of rigid bodies [3]. Gyroscopes were not

---

[1] Johann Gottlieb Friedrich Bohnenberger (1765–1831), University of Tübingen, Germany.
[2] Jean-Bernard-Léon Foucault (1819–1868), French physicist.

put to practical and industrial use until the beginning of the twentieth century in the United States, at which time the gyroscopic compass, the ship stabilizer, and the monorail car were all invented. Subsequent uses of the gyroscope as a turn-and-bank indicator, artificial horizon, and automatic pilot in aircraft and missiles are well known.

We may also become concerned with gyroscopic effects in the design of machines, although not always intentionally. Such effects are present in riding a motorcycle or bicycle; they are also present, owing to the rotating masses, when an airplane or automobile makes a turn. Sometimes these gyroscopic effects are desirable, but more often they are undesirable and designers must account for them in their selection of bearings and rotating parts. It is certainly true that, as machine speeds increase and as factors of safety decrease, we must stop neglecting gyroscopic forces in our design calculations, since their values will become more significant.

16.12 Motion of a Gyroscope

Although we have noted above that a gyroscope seems to have intelligence and appears to avoid compliance with the fundamental laws of mechanics, this is not truly the case. In fact, we have thoroughly covered the basic theory involved in Chapter 12, where we studied dynamic forces with spatial motion. Still, since gyroscopic forces are of increasing importance in high-speed machines, we will look at them again and try to explain the apparent paradoxes they seem to raise.

To provide a vehicle for the explanation of the simpler motions of a gyroscope, we will consider a series of experiments to be performed on the one pictured in Fig. 16.13. In the following discussion, we assume that the rotor is already spinning rapidly and that all pivot friction is negligible.

Fig. 16.13 Laboratory gyroscope.

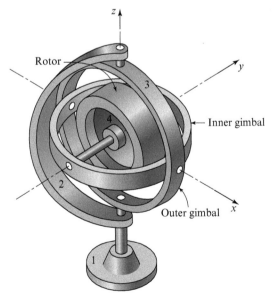

As a first experiment, we transport the entire gyroscope about the table or around the room. We find that even though we travel along a curved path, the orientation of the rotor's axis of rotation does not change as we move. This is a consequence of the law of conservation of angular momentum. If the orientation of the axis of rotation is to change, then the orientation of the angular momentum vector must also change. However, this requires an externally applied moment that, with the three sets of frictionless bearings, has not been supplied for this experiment. Therefore, the orientation of the rotation axis does not change.

As a second experiment, while the rotor is still spinning, we lift the inner gimbal out of its bearings and move it about. We again find that it can be translated anywhere but that we meet with definite resistance if we attempt to rotate the axis of spin. In other words, the rotor persists in maintaining the orientation of its plane of rotation.

As a third experiment, we replace the inner gimbal back into its bearings with the axis of *spin* horizontal, as shown in Fig. 16.13. If we now apply a steady downward force to the inner gimbal at one end of the spin axis, say by pushing on it with a pencil, we find that the end of the spin axis does not move downward as we might expect. Not only do we meet with resistance to the force of the pencil, but the outer gimbal begins to rotate about the vertical axis, causing the rotor to skew around in the horizontal plane, and it continues this rotation until we remove the force of the pencil. This skewing motion of the spin axis is called *precession*.

Although it may seem strange and unexpected, this precession motion is in strict obedience to the laws of dynamic equilibrium as expressed by the Euler equations of motion in Eqs. (12.111). Yet, it is easier to understand and explain this phenomena if we again think in terms of the angular momentum vector. The downward force applied by the pencil to the inner gimbal produces a net external moment, **M**, on the rotor shaft through a pair of equal and opposite forces at the bearings and results in a time rate of change of the angular momentum vector, **H**. As we demonstrated in Eq. (12.121), the time rate of change of the angular momentum can be written as

$$\frac{d\mathbf{H}}{dt} = \sum \mathbf{M}. \tag{16.30}$$

Since this applied external moment cannot change the rate of spin of the rotor about its own axis, it changes the angular momentum by making the rotor rotate about the vertical axis as well, thus causing the precession.

As our next set of experiments, we repeat the previous experiment, watching carefully the directions involved. We first cause the rotor to spin in the positive direction, that is, with its angular velocity vector in the positive y direction. If we next apply a positive torque (moment vector in the positive x direction) using downward force of our pencil on the negative y end of the rotor axis, the precession (rotation) of the outer gimbal is found to be in the negative z direction. Further experiments indicate that either a negative spin velocity or a negative moment caused by the pencil result in a positive direction for the precession.

For a final experiment, we apply a moment to the outer gimbal in an attempt to cause the rotor to rotate about the z axis. Such an attempt meets with definite resistance and causes the inner gimbal and the spin axis to rotate. Note again in this case that the angular momentum vector is changing as the result of the application of external torque. If the spin axis starts in the vertical direction (aligned along z), however, then the gyroscope is in stable equilibrium and the outer and inner gimbals can be turned together quite freely.

16.13 Steady or Regular Precession

Consider the heavy rotor shown in Fig. 16.14 that is spinning with a constant angular velocity ω_s about a spin axis that is tipped at a constant angle, θ, from the vertical. At the same time, assume that the spin axis has a constant angular velocity, ω_p. The state of motion just described is called *steady* or *regular* precession.

We should take careful note of several things in Fig. 16.14. First, we have shown the rotor at a posture for which the inner gimbal is rotated to quite a different angle than in Fig. 16.13; yet the xyz coordinate system is still oriented with x along the pivot axis between the inner and outer gimbals, and z is still along the true axis of precession. Second, we have designated a new coordinate system $x'y'z'$ with the x' axis coincident with x and the z' axis aligned along the axis of the rotor, the spin axis. Third, we should note that neither of these two coordinate systems is stationary and, similarly, that $x'y'z'$ is not attached to the rotor and does not experience the angular velocity ω_s. The $x'y'z'$ coordinate system remains fixed to the inner gimbal, whereas xyz remains fixed to the outer gimbal.

Nevertheless, we see that, for a disk-shaped rotor mounted as described, the $x'y'z'$ axes are the principal axes of inertia of the rotor. We designate the corresponding principal mass moments of inertia $I^{z'z'} = I^s$ and $I^{x'x'} = I^{y'y'} = I$.

Recognize that the constant angular spin velocity, ω_s, is not truly an absolute angular velocity; it is the "relative" or *apparent* angular velocity of the rotor with respect to the inner gimbal. If we designate the rotor as body 4, with the inner and outer gimbals being links 3 and 2, respectively, then

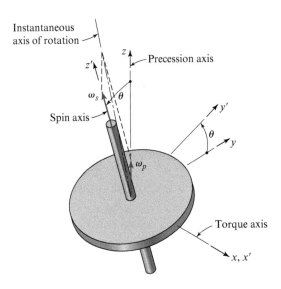

Fig. 16.14 Angular velocity components.

$$\boldsymbol{\omega}_{4/3} = \boldsymbol{\omega}_s = \omega_s \hat{\mathbf{k}}'. \tag{a}$$

Similarly, the angular velocity of precession is really the angular velocity of the outer gimbal:

$$\boldsymbol{\omega}_2 = \boldsymbol{\omega}_p = \omega_p \hat{\mathbf{k}} = (\omega_p \sin\theta)\hat{\mathbf{j}}' + (\omega_p \cos\theta)\hat{\mathbf{k}}'. \tag{b}$$

Since we are interested in steady precession here, the angle θ has been assumed constant; thus, there is no rotation between the inner and outer gimbals:

$$\boldsymbol{\omega}_{3/2} = \dot{\theta}\hat{\mathbf{i}}' = 0. \tag{c}$$

We can now use Eqs. (a)–(c) to find the absolute angular velocity of the rotor:[3]

$$\begin{aligned} \boldsymbol{\omega}_4 &= \boldsymbol{\omega}_2 + \boldsymbol{\omega}_{3/2} + \boldsymbol{\omega}_{4/3} = \boldsymbol{\omega}_p + \boldsymbol{\omega}_s \\ &= (\omega_p \sin\theta)\hat{\mathbf{j}}' + (\omega_s + \omega_p \cos\theta)\hat{\mathbf{k}}'. \end{aligned} \tag{16.31}$$

If we assume that the masses of the gimbals are negligible in comparison with that of the rotor, then the angular momentum of the system is

$$\mathbf{H} = I(\omega_p \sin\theta)\hat{\mathbf{j}}' + I^s(\omega_s + \omega_p \cos\theta)\hat{\mathbf{k}}'. \tag{16.32}$$

To find the time rate of change of the angular momentum, we must recognize that $\hat{\mathbf{j}}'$ and $\hat{\mathbf{k}}'$ are not constant; they rotate with the angular velocity of the $x'y'z'$ coordinate system. From Eqs. (b) and (c), this is

$$\begin{aligned} \boldsymbol{\omega}_3 &= \boldsymbol{\omega}_2 + \boldsymbol{\omega}_{3/2} = \boldsymbol{\omega}_p \\ &= (\omega_p \sin\theta)\hat{\mathbf{j}}' + (\omega_p \cos\theta)\hat{\mathbf{k}}'. \end{aligned} \tag{d}$$

Thus, the time rate of change of $\hat{\mathbf{j}}'$ and $\hat{\mathbf{k}}'$ are

$$\frac{d\hat{\mathbf{j}}'}{dt} = \boldsymbol{\omega}_p \times \hat{\mathbf{j}}' = -(\omega_p \cos\theta)\hat{\mathbf{i}}', \tag{e}$$

$$\frac{d\hat{\mathbf{k}}'}{dt} = \boldsymbol{\omega}_p \times \hat{\mathbf{k}}' = (\omega_p \sin\theta)\hat{\mathbf{i}}'. \tag{f}$$

Finally, using Eqs. (e) and (f) to differentiate Eq. (16.32), the net external moment that must be applied to the rotor to sustain the steady precession motion can be written as

$$\begin{aligned} \sum \mathbf{M} &= \frac{d\mathbf{H}}{dt} = \boldsymbol{\omega}_p \times \mathbf{H} \\ &= -I(\omega_p \sin\theta)(\omega_p \cos\theta)\hat{\mathbf{i}}' + I^s(\omega_s + \omega_p \cos\theta)(\omega_p \sin\theta)\hat{\mathbf{i}}', \end{aligned} \tag{16.33}$$

---

[3] We should note here that the true instantaneous axis of rotation of the rotor is not the spin axis, but accounts for the precession rotation also; this axis is shown in Fig. 16.14. It may be our tendency to picture the spin axis as the true axis of rotation of the rotor that leads us to the intuitive feeling that a gyroscope does not follow the laws of mechanics.

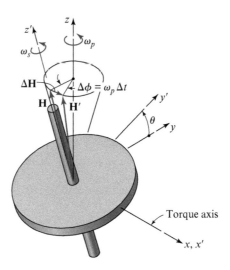

Fig. 16.15 Angular momentum vector.

or as

$$\sum \mathbf{M} = \left[I^s \omega_s + (I^s - I)\omega_p \cos\theta \right] \omega_p \sin\theta \hat{\mathbf{i}}', \tag{16.34}$$

or as

$$\sum \mathbf{M} = \left[I^s + (I^s - I)\frac{\omega_p}{\omega_s} \cos\theta \right] (\boldsymbol{\omega}_p \times \boldsymbol{\omega}_s). \tag{16.35}$$

Any one of these equations can be referred to as a gyroscopic formula.

We note that when the angular velocity of spin, ω_s, is much larger than that of precession, ω_p, which is usually the case, then the second term in the square brackets of Eqs. (16.34) and (16.35) is negligible with respect to the first. Therefore, the two equations reduce to

$$\sum \mathbf{M} = I^s \omega_p \omega_s \sin\theta \hat{\mathbf{i}}' = I^s (\boldsymbol{\omega}_p \times \boldsymbol{\omega}_s). \tag{16.36}$$

Figure 16.15 shows the same rotor in the same orientation as Fig. 16.14, but this time the angular momentum vector $\mathbf{H}$ rather than the angular velocities is shown. We have already noted that the angular momentum vector includes the effects of both the spin and the precession angular velocities. We can also see that it continually precesses, sweeping out a cone with apex angle θ about the precession axis.

In Fig. 16.15 we see the angular momentum vector $\mathbf{H}$ at some instant t and also its changed orientation $\mathbf{H}'$, after a short time interval, Δt, and we note that it has not changed magnitude, only orientation. Spanning the tips of these two vectors, we see the vector change in angular momentum $\Delta \mathbf{H}$ over this short time interval. We note that, in the limit as Δt approaches zero, the direction of the $\Delta \mathbf{H}$ vector approaches the direction of the positive x and x' axes. This is totally consistent with Eq. (16.34), which indicates that $\Delta \mathbf{H}/\Delta t$ is a vector in the $\hat{\mathbf{i}}'$ direction.

To maintain a steady precession, Eq. (16.35) demonstrates that an external moment must continually be applied to the rotor; if this moment is not applied, regular precession does not continue. Note that the axis of the applied moment must be along the x' axis, which is continually changing during the precession. Note also that the sense of the moment is the same as that which would seem required to increase the angle θ. Thus, we see that there really is no paradox at all; the externally applied moment, $\sum \mathbf{M}$, about the x axis causes a change in angular momentum in exactly the same direction as the moment is applied.

16.14 Forced Precession

We have already observed that as speeds are increased in modern machinery, the engineer must be mindful of the increased importance of gyroscopic moments. Common mechanical equipment, which, in the past, were not thought of as exhibiting gyroscopic effects, do in fact often experience such moments, and these will become more significant as speeds increase. The purpose of this section is to present a few examples that, although they may not look like the standard gyroscope, do present gyroscopic moments that should be considered during the design of the equipment.

Much was presented in the earlier sections on the phenomenon of precession. This type of motion is almost certainly accompanied by gyroscopic moments. Yet, lest we think that precession is only an unintended wobbling motion of a toy top, we will look at examples where this precession is recognized and even designed into the operation of some mechanical devices.

Let us first consider a vehicle such as a train, automobile, or racing car, moving at high speed on a straight road. The gyroscopic effect of the spinning wheels is to keep the vehicle moving straight ahead and to resist changes in direction. But when an external moment is applied that forces the wheel to change its direction, gyroscopic reaction forces immediately come into play.

To study this gyroscopic reaction in the case of a vehicle, let us consider the following example.

Example 16.2

Consider a pair of wheels connected by a straight axle, rounding a curve of constant radius as shown in Fig. 16.16. For simplicity, the roadbed is assumed not to be banked. This wheel ensemble may be considered a gyroscope. Such rounding of the curve is a *forced* precession of the wheel–axle assembly around a vertical axis through the center of curvature of the track.

Find the gyroscopic torque exerted on the wheel–axle assembly.

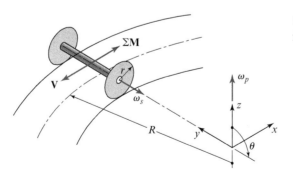

Fig. 16.16 Axle and wheels following a curved path.

Example 16.2 (cont.)

Solution

The notation used here is: r is the radius of the wheels; I^s is the combined mass moment of inertia about the axis of rotation; R is the radius of the curve; and V is the velocity of the center of the axle.

The angular velocity of precession of the assembly is

$$\omega_p = V/R. \tag{1}$$

Similarly, the average angular velocity of spin of the two wheels is

$$\omega_s = V/r. \tag{2}$$

These are shown in Fig. 16.16.

The angle of the precession cone, that is, the angle between ω_p and ω_s, is $\theta = 90°$. Substituting this value into Eq. (16.35) gives

$$\sum \mathbf{M} = I^s \omega_p \times \omega_s. \tag{3}$$

Then, substituting Eqs. (1) and (2) into Eq. (3) and performing the cross-product gives

$$\sum \mathbf{M} = \tfrac{I^s V^2}{Rr} \hat{\mathbf{i}}. \qquad\qquad Ans. \text{ (4)}$$

This is the additional external moment applied on the wheel–axle assembly caused by the gyroscopic action of the wheels while rounding the turn (forced precession) over and above the static and other steady-state dynamic loads. We note that the direction of this additional moment is such as to increase the tendency of the vehicle to roll over during the turn. This moment is applied by the tires by increasing the upward force on the outside tire and decreasing the upward force on the inside tire.

To make a quantitative comparison of the gyroscopic and centrifugal moments on the vehicle, let us consider the case of a racing car. The weight of the car including the driver is $W = 1\,800$ lb, and the height of the center of mass above the road is $h = 20$ in. We estimate that the radius of each wheel is $r = 18$ in, the radius of gyration is $k = 15$ in, and the weight of each wheel is $w = 100$ lb. From these data we can determine the centrifugal moment $\mathbf{M}^*$ tending to cause rollover of the vehicle. Taking moments about the contact point of the outer tire, the result is

$$\mathbf{M}^* = \frac{hWV^2}{gR}\hat{\mathbf{i}} = \frac{(20 \text{ in})(1\,800 \text{ lb})\,V^2}{gR}\hat{\mathbf{i}} = 36\,000\,\frac{V^2}{gR}\hat{\mathbf{i}} \text{ in} \cdot \text{lb}. \tag{5}$$

Since the vehicle has two axles and each has two wheels, the combined mass moment of inertia is

$$I^s = 2mk^2 = \frac{2(100 \text{ lb})(15 \text{ in})^2}{g} = \frac{45\,000 \text{ in}^2 \cdot \text{lb}}{g}. \tag{6}$$

Example 16.2 (cont.)

Then, using Eq. (4) for two axles, the additional external moment applied on the wheel–axle assembly caused by the gyroscopic action of the wheels while rounding the turn is

$$\sum \mathbf{M} = 2\frac{I^s V^2}{Rr}\hat{\mathbf{i}} = 2\frac{(45\,000 \text{ in}^2 \cdot \text{lb})\,V^2}{gR(18 \text{ in})}\hat{\mathbf{i}} = (5\,000 \text{ in} \cdot \text{lb})\frac{V^2}{gR}\hat{\mathbf{i}} = 0.138\,9\mathbf{M}^*. \qquad (7)$$

Thus, we see that the gyroscopic effects of the tires add almost 14% to the tendency of the vehicle to roll over in a turn, and we see that this is independent of both the radius of the turn and the speed of the vehicle.

We note from this example that the problem itself looks nothing like a gyroscope; yet it has gyroscopic effects, and these may be missed in an oversimplified analysis. We note also that the precession motion is not just an unexpected result of the motion characteristics of the system. It is deliberately caused by the driver who steers the car; it is a forced precession.

Another gyroscopic effect in automobiles is that caused by the flywheel. Since the rotation of the flywheel of a rear-wheel-drive vehicle is along the longitudinal axis and since the flywheel rotates counterclockwise as viewed from the rear, the spin vector $\boldsymbol{\omega}_s$ points toward the rear of the vehicle. When the vehicle makes a turn, the axis of the flywheel is forced to precess about a vertical axis as in the previous example. This forced precession brings into existence an applied gyroscopic moment about a horizontal axis through the center of the turn. The effect of this gyroscopic moment is to produce a bending moment on the driveshaft, tending to bend it in a vertical plane. The size of this gyroscopic moment causing bending in the driveshaft is usually of minor importance compared with the torsion loading because of the relatively small mass of the flywheel.

Lest we conclude that gyroscopic forces are always a disadvantage with which we must cope, consider the following example in which the gyroscopic effect is put to good use.

Example 16.3

The edge mill shown schematically in Fig. 16.17 is a gyroscopic grinder used for crushing ore, seeds, grain, and such. It consists of a large steel pan in which one or more heavy conical rollers, called *mullers*, roll without slipping on the bottom of the pan, and, at the same time, revolve about a vertical shaft passing through the central axis of the pan. The mullers rotate about either horizontal or inclined axes that are attached to the vertical shaft, which is rotated under power. The radius of a muller is $r = 18$ in, and the length of its axle is $l = 30$ in; its weight is $W = 2\,200$ lb, and its mass moment of inertia can be approximated by that of a cylindrical disk. Assume that the vertical shaft is driven at a constant angular velocity of $\omega_p = 40$ rev/min and that the muller rolls without slipping. Determine the optimum inclination angle θ for the muller axle, which maximizes the crushing force between the muller and the pan.

Example 16.3 (cont.)

Fig. 16.17 Schematic diagram of an edge mill.

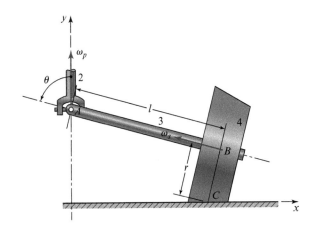

Solution

First, we use the assumption of rolling without slip to find the angular velocity of the muller. From Fig. 16.17, the angular velocity of the muller axis, link 3, is

$$\boldsymbol{\omega}_p = \boldsymbol{\omega}_3 = 40\hat{\mathbf{j}} \text{ rev/min} = 4.19\hat{\mathbf{j}} \text{ rad/s}, \tag{1}$$

whereas the angular velocity of spin of the muller about its axle is

$$\boldsymbol{\omega}_s = \boldsymbol{\omega}_{4/3} = \omega_s\left(-\sin\theta\hat{\mathbf{i}} + \cos\theta\hat{\mathbf{j}}\right). \tag{2}$$

Since the muller undergoes precession in addition to spin about its axle, the absolute angular velocity of the muller is

$$\boldsymbol{\omega}_4 = \boldsymbol{\omega}_3 + \boldsymbol{\omega}_{4/3} = (-\omega_s \sin\theta)\hat{\mathbf{i}} + (4.19 \text{ rad/s} + \omega_s \cos\theta)\hat{\mathbf{j}}. \tag{3}$$

Since point A is fixed, the velocity of point B of link 3 is

$$\begin{aligned}
\mathbf{V}_B &= \boldsymbol{\omega}_3 \times \mathbf{R}_{BA} \\
&= \left(4.19\hat{\mathbf{j}} \text{ rad/s}\right) \times \left(30\sin\theta\hat{\mathbf{i}} - 30\cos\theta\hat{\mathbf{j}} \text{ in}\right) \\
&= -125.7\sin\theta\hat{\mathbf{k}} \text{ in/s}.
\end{aligned} \tag{4}$$

Also, since there is no slip at point C, the velocity of point B of link 4 is

$$\begin{aligned}
\mathbf{V}_B &= \boldsymbol{\omega}_4 \times \mathbf{R}_{BC} \\
&= \left[(-\omega_s \sin\theta)\hat{\mathbf{i}} + (4.19 \text{ rad/s} + \omega_s \cos\theta)\hat{\mathbf{j}}\right] \times \left(18\cos\theta\hat{\mathbf{i}} + 18\sin\theta\hat{\mathbf{j}} \text{ in}\right) \\
&= -18(\omega_s + 4.19\cos\theta \text{ rad/s})\hat{\mathbf{k}}.
\end{aligned} \tag{5}$$

Example 16.3 (cont.)

Equating Eqs. (4) and (5), the angular velocity of spin of the muller about its axle is

$$\omega_s = 6.98 \sin\theta - 4.19\cos\theta \text{ rad/s.} \tag{6}$$

From Eqs. (1) and (2), the cross-product is

$$\left(\boldsymbol{\omega}_p \times \boldsymbol{\omega}_s\right) = \left(4.19\hat{\mathbf{j}} \text{ rad/s}\right) \times \omega_s\left(-\sin\theta\hat{\mathbf{i}} + \cos\theta\hat{\mathbf{j}}\right)$$
$$= 4.19\omega_s\sin\theta\hat{\mathbf{k}} \text{ rad/s.}$$

Then, substituting Eq. (6) into this result gives

$$\left(\boldsymbol{\omega}_p \times \boldsymbol{\omega}_s\right) = 4.19(6.98\sin\theta - 4.19\cos\theta)\sin\theta\hat{\mathbf{k}} \text{ rad/s}^2. \tag{7}$$

Using the formula for a round disk from Table A.5 in Appendix A, the mass moment of inertia of the muller is

$$I^s = \frac{mr^2}{2} = \frac{(2\,200 \text{ lb})(18 \text{ in})^2}{(386 \text{ in/s}^2)2} = 923.3 \text{ in}\cdot\text{lb}\cdot\text{s}^2,$$
$$I = \frac{mr^2}{4} + ml^2 = 5\,590 \text{ in}\cdot\text{lb}\cdot\text{s}^2. \tag{8}$$

Then, substituting Eqs. (7) and (8) into Eq. (16.35), the net externally applied moment on the muller is

$$\sum\mathbf{M} = \left[923.3 + \frac{-4\,667(4.19)\cos\theta}{6.98\sin\theta - 4.19\cos\theta}\right](6.98\sin\theta - 4.19\cos\theta)4.19\sin\theta\hat{\mathbf{k}} \text{ in}\cdot\text{lb.}$$

Rearranging this equation, the moment applied to the muller can be written as

$$\sum\mathbf{M} = \left(27\,001\sin^2\theta - 98\,082\sin\theta\cos\theta\right)\hat{\mathbf{k}} \text{ in}\cdot\text{lb.} \tag{9}$$

This moment must be externally applied to the muller in order to sustain the forced precession. It must come from the net effect of the weight of the muller and the crushing force between the muller and the pan. Formulating these effects for the other side of the equation, we obtain

$$\sum\mathbf{M} = \mathbf{R}_{BA} \times \mathbf{W}_4 + \mathbf{R}_{CA} \times \mathbf{F}_c$$
$$= \left(30\sin\theta\hat{\mathbf{i}} - 30\cos\theta\hat{\mathbf{j}} \text{ in}\right) \times \left(-2\,200\hat{\mathbf{j}} \text{ lb}\right)$$
$$+ \left[(30\sin\theta - 18\cos\theta)\hat{\mathbf{i}} + (18\sin\theta - 30\cos\theta)\hat{\mathbf{j}} \text{ in}\right] \times \left(F_c\hat{\mathbf{j}}\right) \tag{10}$$
$$= (-66\,000\sin\theta + 30F_c\sin\theta - 18F_c\cos\theta)\hat{\mathbf{k}} \text{ in}\cdot\text{lb.}$$

Equating Eqs. (9) and (10), the crushing force becomes

$$F_c = \frac{27\,001\sin^2\theta - 98\,082\sin\theta\cos\theta + 66\,000\sin\theta \text{ in}\cdot\text{lb}}{30\sin\theta - 18\cos\theta \text{ in}}. \tag{11}$$

Example 16.3 (cont.)

We note that the crushing force is a function of the angle θ of inclination of the muller axis; we now hope to choose this angle θ to maximize the crushing force. To do this we differentiate Eq. (11) with respect to θ and set the result equal to zero; that is

$$\left(54\,002 \sin\theta\cos\theta - 98\,082\cos^2\theta + 98\,082\sin^2\theta + 66\,000\cos\theta\right)(30\sin\theta - 18\cos\theta)$$
$$- (30\cos\theta + 18\sin\theta)\left(27\,001\sin^2\theta - 98\,082\sin\theta\cos\theta + 66\,000\sin\theta\right) = 0.$$

This equation simplifies to

$$2\,456\,442\sin^3\theta + 810\,030\sin^2\theta\cos\theta - 972\,036\sin\theta\cos^2\theta + 1\,765\,476\cos^3\theta - 1\,188\,000 = 0,$$

and can be solved numerically (the equation has multiple roots). When this is done, the root that maximizes the crushing force F_c of Eq. (11) is found to be

$$\theta = 115.9°. \hspace{3cm} \textit{Ans.}$$

This is the optimum angle of inclination of the muller axle. Substituting this value into Eq. (11), the crushing force of the mill is

$$F_c = 3437\text{ lb}. \hspace{3cm} \textit{Ans.}$$

We should note that this crushing force is more than 50% larger than the weight of the muller; the additional force is attributable to choosing the inclination angle θ for which both the gyroscopic and the centrifugal force effects contribute as much as possible to the crushing effect of the mill. There are other designs for crushing machines in which the pan rotates under a muller that has a fixed axis. In such a design there is no gyroscopic action, and the crushing action is due solely to the weight of the muller.

An airplane propeller spinning at high speed is another example of a mechanical system exhibiting gyroscopic torque effects. In the case of a single propeller airplane, a turn in compass heading, for example, is a forced precession of the propeller's spin axis about a vertical axis. If the propeller is rotating clockwise as viewed from the rear, this forced precession induces a gyroscopic moment, causing the nose of the plane to move upward or downward as the heading is changed to the left or right, respectively. Turning the nose upward rather suddenly causes the plane to turn to the right, whereas a sudden turn downward causes a turn to the left.

Note that it is not the propeller forces that cause this effect, but the spinning mass and its gyroscopic effect during a change in direction of the plane (forced precession). The very substantial spinning mass of a radial engine (Fig. 1.27b) of early aircraft markedly exaggerated this gyroscopic effect and was, at least in part, responsible for the disappearance of use of rotary engines on airplanes. Since the effect comes from the precession of the spinning mass and not from the propeller, does the same danger not exist from the spinning mass of the rotor of a turbojet engine? When an airplane is equipped with two propellers rotating at equal speeds in opposite directions (or with counter-rotating turbines), however, the gyroscopic torques of the two can annul each other and leave no net effect on the plane as a whole.

PROBLEMS

16.1 Table P16.1 lists the output torque for a single-cylinder, four-stroke engine running at
4 600 rev/min. (a) Find the mean output torque. (b) Determine the mass moment of inertia
of an appropriate flywheel using $C_s = 0.025$.

Table P16.1 Torque data for the single-cylinder, four-stroke engine

θ_i (deg)	T_i (N·m)	θ_i (deg)	T_i (N·m)	θ_i (deg)	T_i (N·m)	θ_i (deg)	T_i (N·m)
0	0	180	0	360	0	540	0
10	17	190	−344	370	−145	550	−344
20	812	200	−540	380	−150	560	−540
30	963	210	−576	390	7	570	−577
40	1 016	220	−570	400	164	580	−572
50	937	230	−638	410	235	590	−643
60	774	240	−785	420	203	600	−793
70	641	250	−879	430	490	610	−893
80	697	260	−814	440	424	620	−836
90	849	270	−571	450	571	630	−605
100	1 031	280	−324	460	814	640	−379
110	1 027	290	−190	470	879	650	−264
120	902	300	−203	480	785	660	−300
130	712	310	−235	490	638	670	−368
140	607	320	−164	500	570	680	−334
150	594	330	−7	510	576	690	−198
160	544	340	150	520	540	700	−56
170	345	350	145	530	344	710	−2

16.2 Using the data of Table 16.2, determine the moment of inertia for a flywheel for a four-
cylinder, 90°V engine having a single crank. Use $C_s = 0.0125$ and a nominal speed of
4 600 rev/min. If a cylindric or disk-type flywheel is to be used, what should be the
thickness if it is made of steel and has an outside diameter of 400 mm? Use
$\rho = 7.8$ Mg/m^3 as the density of steel.

16.3 Using the data of Table 16.1, find the mean output torque and the flywheel inertia
required for a three-cylinder, in-line engine corresponding to a nominal speed of
2 400 rev/min. Use $C_s = 0.03$.

16.4 The load torque required by a 200-ton punch press is displayed in Table P16.4 for one
revolution of the flywheel. The flywheel is to have a nominal angular velocity of 2 400 rev/
min and to be designed for a coefficient of speed fluctuation of 0.075. (a) Determine the
mean motor torque required at the flywheel shaft and the motor horsepower needed,
assuming a constant torque–speed characteristic for the motor. (b) Find the moment of
inertia needed for the flywheel.

Table P16.4 **Torque data for the 200-ton punch press**

θ_i (deg)	T_i (in·lb)	θ_i (deg)	T_i (in·lb)	θ_i (deg)	T_i (in·lb)	θ_i (deg)	T_i (in·lb)
0	857	90	7 888	180	1 801	270	857
10	857	100	8 317	190	1 629	280	857
20	857	110	8 488	200	1 458	290	857
30	857	120	8 574	210	1 372	300	857
40	857	130	8 403	220	1 115	310	857
50	1 287	140	7 717	230	1 029	320	857
60	2 572	150	3 515	240	943	330	857
70	5 144	160	2 144	250	857	340	857
80	6 859	170	1 972	260	857	350	857

16.5 Find T_m for the four-cylinder engine whose torque displacement is that of Fig. 16.4.

16.6 In a pendulum mill, shown schematically in Fig. P16.6, the grinding is by a conical muller that is free to spin about a pendulous axle that, in turn, is connected to a powered vertical shaft by a Hooke universal joint. The muller presses against the inner wall of a heavy steel pan, and it rolls around the inside of the pan without slipping. The weight of the muller is $W = 4350$ N; its principal mass moments of inertia are $I^s = 13.14$ kg·m^2 and $I = 9.56$ kg·m^2. The length of the muller axle is $l = R_{GA} = 1$ m, and the radius of the muller at its center of mass is $R_{GB} = 250$ mm. Assuming that the vertical shaft is to be inclined at $\theta = 30°$ and will be driven at a constant angular velocity of $\omega_p = 240$ rev/min, find the crushing force between the muller and the pan. Also determine the minimum angular velocity ω_p required to ensure contact between the muller and the pan.

Fig. P16.6

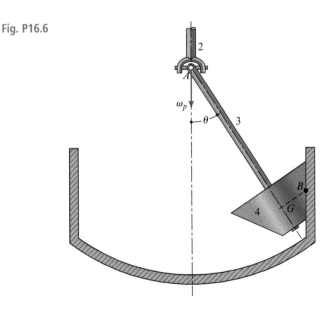

16.7 Using the gyroscopic formulae, Eqs. (16.33)–(16.35), solve the problem presented in Example 12.9 of Chapter 12.

16.8 The oscillating fan precesses sinusoidally according to the equation $\theta_p = \beta \sin 1.5t$, where $\beta = 30°$; the fan blade spins at $\boldsymbol{\omega}_s = 1\,800\hat{\mathbf{i}}$ rev/ min. The weight of the fan and motor armature is 23.3 N, and other masses can be assumed negligible; gravity acts in the $-\hat{\mathbf{j}}$ direction. The principal mass moments of inertia are $I^s = 0.007\,06$ kg·m^2 and $I = 0.002\,71$ kg·m^2; the center of mass is located at $R_{GC} = 100$ mm to the front of the precession axis. Determine the maximum moment, M^z, that must be accounted for in the clamped tilting pivot at C.

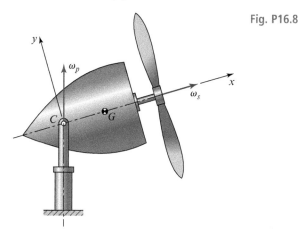

Fig. P16.8

16.9 The propeller of an outboard motorboat is spinning at high speed and is caused to precess by steering to the right or left. Do the gyroscopic effects tend to raise or lower the rear of the boat? What is the effect, and is it of notable size?

16.10 A large and very high-speed turbine is to operate at an angular velocity of $\omega = 18\,000$ rev/ min and will have a rotor with a principal mass moment of inertia of $I^s = 24.4$ kg·m^2. It has been suggested that, since this turbine will be installed at the North Pole with its axis horizontal, perhaps the rotation of the Earth will cause gyroscopic loads on its bearings. Estimate the size of these additional loads.

REFERENCES

[1] Bohnenburger, J. G. F., 1817. Beschreibung einer Maschine zur Erläuterung der Geseze der Umdrehung der Erde um ihre Axe, um der Veränderung der Lage der letzteren, *Tübinger Blätter für Naturwissenschaften und Arzneikunde*, **3**, 72–83.

[2] Bollinger, J. G., and N. A. Duffie, 1988. *Computer Control of Machines and Processes*, Reading, MA: Addison-Wesley.

[3] Euler, L., 1765. Theoria motus corporum solidorum seu rigidorum, in *Commentarii Academiae Scientiarum Imperialis Petropolitanae*, trans. R. W. Willis, 1841 in *Principles of Mechanisms*, London: John W. Parker.

[4] Scarborough, J. B., 1958. *The Gyroscope: Theory and Applications*, New York: Interscience Publishers.

APPENDIX A

Tables

Table A.1 Standard SI prefixes[1,2]

Name	Symbol	Factor
yotta	Y	$1\,000\,000\,000\,000\,000\,000\,000\,000 = 10^{24}$
zetta	Z	$1\,000\,000\,000\,000\,000\,000\,000 = 10^{21}$
exa	E	$1\,000\,000\,000\,000\,000\,000 = 10^{18}$
peta	P	$1\,000\,000\,000\,000\,000 = 10^{15}$
tera	T	$1\,000\,000\,000\,000 = 10^{12}$
giga	G	$1\,000\,000\,000 = 10^{9}$
mega	M	$1\,000\,000 = 10^{6}$
kilo	K	$1\,000 = 10^{3}$
hecto[3]	H	$100 = 10^{2}$
deka[3]	Da	$10 = 10^{1}$
deci[3]	D	$0.1 = 10^{-1}$
centi[3]	C	$0.01 = 10^{-2}$
milli	M	$0.001 = 10^{-3}$
micro	μ	$0.000\,001 = 10^{-6}$
nano	N	$0.000\,000\,001 = 10^{-9}$
pico	P	$0.000\,000\,000\,001 = 10^{-12}$
femto	F	$0.000\,000\,000\,000\,001 = 10^{-15}$
atto	A	$0.000\,000\,000\,000\,000\,001 = 10^{-18}$
zepto	Z	$0.000\,000\,000\,000\,000\,000\,000 = 10^{-21}$
yocto	Y	$0.000\,000\,000\,000\,000\,000\,000\,000 = 10^{-24}$

[1]If possible, multiple and submultiple prefixes are used in steps of 1 000. For example, lengths are specified in millimeters, meters, or kilometers. In a combination unit, prefixes are only used preceeding the numerator; for example, meganewton per cubic meter (MN/m^3) is used, but not newton per cubic centimeter (N/cm^3).
[2]Spaces are used in SI instead of commas to group numbers to avoid confusion with the practice in some countries of using commas for decimal points.
[3]Not recommended but sometimes encountered.

Table A.2 **Conversion from US customary units to SI units**

To convert from	to	Multiply by	
		Accurate*	Common
Foot (ft)	Meter (m)	0.304 800*	0.305
Foot·pound (ft·lb)	Newton·meter (N·m)	1.355 818	1.36
	Joule (J)	1.355 818	1.36
Foot·pound/second (ft·lb/s)	Watt (W)	1.355 818	1.36
Horsepower (hp)	Watt (W)	745.699 9	746
Inch (in)	Meter (m)	0.025 400*	0.025 4
Inch·pound (in·lb)	Newton·meter (N·m)	0.112 984 8	0.113
	Joule (J)	0.112 984 8	0.113
Inch·pound/second (in·lb/s)	Watt (W)	0.112 984 8	0.113
Mile, US statute (mi)	Meter (m)	1 609.344*	1 610
Pound force (lb)	Newton (N)	4.448 222	4.45
Pound mass (lb)	Kilogram (kg)	0.453 592 4	0.454
Pound/ft^2 (lb/ft^2)	Pascal (Pa)	47.880 26	47.9
Pound/in^2 (lb/in^2), (psi)	Pascal (Pa)	6 894.757	6 890
Revolutions/min (rev/min)	Radian/second (rad/s)	0.104 719 8	0.105
Ton, short (2 000 lb)	Kilogram (kg)	907.184 7	907

*An asterisk indicates that the conversion is exact.

Table A.3 **Conversion from SI units to US customary units**

To convert from	to	Multiply by	
		Accurate	Common
Joule (J)	Foot·pound (ft·lb)	0.737 562 1	0.738
	Inch·pound (in·lb)	88.507 44	8.85
Kilogram (kg)	Pound mass (lb)	2.204 622	2.20
	Ton, short (2 000 lb)	0.001 102 311	0.001 10
Meter (m)	Foot (ft)	3.280 840	3.28
	Inch (in)	39.370 08	39.4
	Mile (mi)	0.000 621 371	0.000 621
Newton (N)	Pound (lb)	0.224 808 9	0.225
Newton·meter (N·m)	Foot·pound (ft·lb)	0.737 562 1	0.738
	Inch·pound (in·lb)	8.850 744	8.85
Newton·meter/second (N·m/s)	Horsepower (hp)	0.001 341 022	0.001 34
Pascal (Pa)	Pound/foot^2 (lb/ft2)	0.020 885 43	0.020 9
	Pound/inch^2 (lb/in^2), (psi)	0.000 145 037 0	0.000 145
Radian/second (rad/s)	Revolutions/minute (rev/min)	9.549 297	9.55
Watt (W)	Horsepower (hp)	0.001 341 022	0.001 34
	Foot·pound /second (ft·lb/s)	0.737 562 1	0.738
	Inch·pound/second (in·lb/s)	8.850 744	8.85

Table A.4 **Properties of areas**

A = area
I = area moment of inertia
J = polar area moment of inertia

k = centroidal radius of gyration
$\bar{y}$ = centroidal distance from bottom

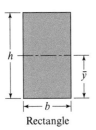

Rectangle

$A = bh$
$I = bh^3/12$

$k = 0.289h$
$\bar{y} = h/2$

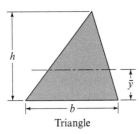

Triangle

$A = bh/2$
$I = bh^3/36$

$k = 0.236h$
$\bar{y} = h/3$

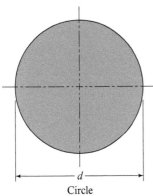

Circle

$A = \pi d^2/4$
$I = \pi d^4/64$
$J = \pi d^4/32$

$k = d/4$
$\bar{y} = d/2$

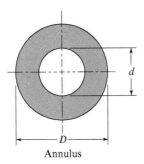

Annulus

$A = \pi(D^2 - d^2)/4$
$I = \pi(D^4 - d^4)/64$
$J = \pi(D^4 - d^4)/32$

$k = \sqrt{D^2 + d^2}/4$
$\bar{y} = D/2$

Table A.5 **Mass moments of inertia**

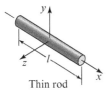

Thin rod

$$m = \rho \pi r^2 l$$
$$I^{yy} = I^{zz} = ml^2/12$$

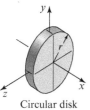

Circular disk

$$m = \rho \pi r^2 t$$
$$I^{xx} = mr^2/2$$
$$I^{yy} = I^{zz} = mr^2/4$$

Rectangular prism

$$m = \rho abc$$
$$I^{xx} = m(a^2 + b^2)/12$$
$$I^{yy} = m(a^2 + c^2)/12$$
$$I^{zz} = m(b^2 + c^2)/12$$

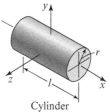

Cylinder

$$m = \rho \pi r^2 l$$
$$I^{xx} = mr^2/2$$
$$I^{yy} = I^{zz} = m(3r^2 + l^2)/12$$

Annular cylinder

$$m = \rho \pi (b^2 - a^2) l$$
$$I^{xx} = m(a^2 + b^2)/2$$
$$I^{yy} = I^{zz} = m(3a^2 + 3b^2 + l^2)/12$$

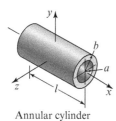

Cone

$$m = \rho \pi r^2 h/3$$
$$I^{xx} = 3mr^2/10$$
$$I^{yy} = I^{zz} = m(12r^2 + 3h^2)/80$$

Table A.5 **(cont.)**

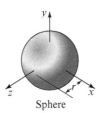
Sphere

$$m = \rho 4\pi r^3 / 3$$
$$I^{xx} = I^{yy} = I^{zz} = 2mr^2/5$$

Table A.6 **Involute function**

$\varphi(\text{deg})$	$\text{Inv}(\varphi)$	$\text{Inv}(\varphi + 0.1°)$	$\text{Inv}(\varphi + 0.2°)$	$\text{Inv}(\varphi + 0.3°)$	$\text{Inv}(\varphi + 0.4°)$
0.0	0.000000	0.000000	0.000000	0.000000	0.000000
0.5	0.000000	0.000000	0.000000	0.000000	0.000001
1.0	0.000002	0.000002	0.000003	0.000004	0.000005
1.5	0.000006	0.000007	0.000009	0.000010	0.000012
2.0	0.000014	0.000016	0.000019	0.000022	0.000025
2.5	0.000028	0.000031	0.000035	0.000039	0.000043
3.0	0.000048	0.000053	0.000058	0.000064	0.000070
3.5	0.000076	0.000083	0.000090	0.000097	0.000105
4.0	0.000114	0.000122	0.000132	0.000141	0.000151
4.5	0.000162	0.000173	0.000184	0.000197	0.000209
5.0	0.000222	0.000236	0.000250	0.000265	0.000280
5.5	0.000296	0.000312	0.000329	0.000347	0.000366
6.0	0.000384	0.000404	0.000424	0.000445	0.000467
6.5	0.000489	0.000512	0.000536	0.000560	0.000586
7.0	0.000612	0.000638	0.000666	0.000694	0.000723
7.5	0.000753	0.000783	0.000815	0.000847	0.000880
8.0	0.000914	0.000949	0.000985	0.001022	0.001059
8.5	0.001098	0.001137	0.001178	0.001219	0.001283
9.0	0.001305	0.001349	0.001394	0.001440	0.001488
9.5	0.001536	0.001586	0.001636	0.001688	0.001740
10.0	0.001794	0.001849	0.001905	0.001962	0.002020
10.5	0.002079	0.002140	0.002202	0.002265	0.002329
11.0	0.002394	0.002461	0.002528	0.002598	0.002668
11.5	0.002739	0.002812	0.002894	0.002962	0.003039
12.0	0.003117	0.003197	0.003277	0.003360	0.003443
12.5	0.003529	0.003615	0.003712	0.003792	0.003882
13.0	0.003975	0.004069	0.004164	0.004261	0.004359
13.5	0.004459	0.004561	0.004664	0.004768	0.004874
14.0	0.004982	0.005091	0.005202	0.005315	0.005429
14.5	0.005545	0.005662	0.005782	0.005903	0.006025
15.0	0.006150	0.006276	0.006404	0.006534	0.006665
15.5	0.006799	0.006934	0.007071	0.007209	0.007350
16.0	0.007493	0.007637	0.007784	0.007932	0.008082

Table A.6 **(cont.)**

φ(deg)	Inv(φ)	Inv($\varphi + 0.1°$)	Inv($\varphi + 0.2°$)	Inv($\varphi + 0.3°$)	Inv($\varphi + 0.4°$)
16.5	0.008234	0.008388	0.008544	0.008702	0.008863
17.0	0.009025	0.009189	0.009355	0.009523	0.009694
17.5	0.009866	0.010041	0.010217	0.010396	0.010577
18.0	0.010760	0.010946	0.011133	0.011323	0.011515
18.5	0.011709	0.011906	0.012105	0.012306	0.012509
19.0	0.012715	0.012923	0.013134	0.013346	0.013562
19.5	0.013779	0.013999	0.014222	0.014447	0.014674
20.0	0.014904	0.015137	0.015372	0.015609	0.015850
20.5	0.016092	0.016337	0.016585	0.016836	0.017089
21.0	0.017345	0.017603	0.017865	0.018129	0.018395
21.5	0.018665	0.018937	0.019212	0.019490	0.019770
22.0	0.020054	0.020340	0.020630	0.020921	0.021216
22.5	0.021514	0.021815	0.022119	0.022426	0.022736
23.0	0.023049	0.023365	0.023684	0.024006	0.024332
23.5	0.024660	0.024992	0.025326	0.025664	0.026005
24.0	0.026350	0.026697	0.027048	0.027402	0.027760
24.5	0.028121	0.028485	0.028852	0.029223	0.029598
25.0	0.029975	0.030357	0.030741	0.031130	0.031521
25.5	0.031917	0.032315	0.032718	0.033124	0.033534
26.0	0.033947	0.034364	0.034785	0.035209	0.035637
26.5	0.036069	0.036505	0.036945	0.037388	0.037835
27.0	0.038287	0.038696	0.039201	0.039664	0.040131
27.5	0.040602	0.041076	0.041556	0.042039	0.042526
28.0	0.043017	0.043513	0.044012	0.044516	0.045024
28.5	0.045537	0.046054	0.046575	0.047100	0.047630
29.0	0.048164	0.048702	0.049245	0.049792	0.050344
29.5	0.050901	0.051462	0.052027	0.052597	0.053172
30.0	0.053751	0.054336	0.054924	0.055519	0.056116
30.5	0.056720	0.057267	0.057940	0.058558	0.059181
31.0	0.059809	0.060441	0.061779	0.061721	0.062369
31.5	0.063022	0.063680	0.064343	0.065012	0.065685
32.0	0.066364	0.067048	0.067738	0.068432	0.069133
32.5	0.069838	0.070549	0.071266	0.071988	0.072716
33.0	0.073449	0.074188	0.074932	0.075683	0.076439
33.5	0.077200	0.077968	0.078741	0.079520	0.080305
34.0	0.081097	0.081974	0.082697	0.083506	0.084321
34.5	0.085142	0.085970	0.086804	0.087644	0.088490
35.0	0.089342	0.090201	0.091066	0.091938	0.092816
35.5	0.093701	0.094592	0.095490	0.096395	0.097306
36.0	0.098224	0.099149	0.100080	0.101019	0.101964
36.5	0.102916	0.103875	0.104841	0.105814	0.106795
37.0	0.107782	0.108777	0.109779	0.110788	0.111805
37.5	0.112828	0.113860	0.114899	0.115945	0.116999
38.0	0.118060	0.119130	0.120207	0.121291	0.122384

Table A.6 (cont.)

φ(deg)	Inv(φ)	Inv($\varphi + 0.1°$)	Inv($\varphi + 0.2°$)	Inv($\varphi + 0.3°$)	Inv($\varphi + 0.4°$)
38.5	0.123484	0.124592	0.125709	0.126833	0.127965
39.0	0.129106	0.130254	0.131411	0.132576	0.133749
39.5	0.134931	0.136122	0.137320	0.138528	0.139743
40.0	0.140968	0.142201	0.143443	0.144694	0.145954
40.5	0.147222	0.148500	0.149787	0.151082	0.152387
41.0	0.153702	0.155025	0.156358	0.157700	0.159052
41.5	0.160414	0.161785	0.163165	0.164556	0.165956
42.0	0.167366	0.168786	0.170216	0.171656	0.173106
42.5	0.174566	0.176037	0.177518	0.179009	0.180511
43.0	0.182023	0.183546	0.185080	0.186625	0.188180
43.5	0.189746	0.191324	0.192912	0.194511	0.196122
44.0	0.197744	0.199377	0.201022	0.202678	0.204346
44.5	0.206026	0.207717	0.209420	0.211135	0.212863
45.0	0.214602				

APPENDIX B

Answers to Selected Problems

1.3 $\gamma_{\min} = 53.1°$, $\gamma_{\max} = 98.1°$, at $\theta = 40.1°$; $\gamma = 59.1°$, at $\theta = 228.6°$; $\gamma = 90.9°$.

1.5   (a) $m = 1$; (b) $m = 1$; (c) $m = 0$; (d) $m = 1$.

1.7   7 variations.

1.9 $m = 1$; correct.

1.11 $m = 1$; correct.

1.13 $j_1 = 5, j_2 = 1$; no slipping at A.

1.15 $Q = 1.099$.

1.17   0.208 33 in to the left.

1.19   One solution is $r_1 = 32.92$ in, $r_2 = 10$ in, $r_3 = 28$ in, $r_4 = 20$ in.

1.21 $m = 1$; θ_2 or θ_3 as input.

1.23 $m = 1$; θ_2 or θ_5 or s_4 as input.

1.25 $m = 1$; θ_2 or θ_4 or s_7 as input.

1.27 $MA = 1.86$.

1.29 $m = 1$; θ_2 or θ_4 as input.

1.31   One solution is $r_1 = 4.10$ in, $r_2 = O_2A = 8.88$ in, $r_3 = AB = 13.62$ in.

1.33 $m = 1$; θ_2 or θ_4 as input.

1.35   Dead-centers: $\theta_2 = 114.05°$, $\theta_4 = 162.82°$, and $\theta_2''' = -114.05°$, $\theta_4''' = -162.82°$; toggles: $\theta_2' = 56.50°$, $\theta_4' = 133.14°$, and $\theta_2'' = -56.50°$, $\theta_4'' = -133.14°$.

1.37 C is a j_2 joint.

1.39 $m = 1$; θ_2 or θ_4 or θ_8 as input.

1.41 $m = 1$; θ_2 or θ_5 or s_8 as input.

1.43 $m = 1$; θ_6 or θ_7 or s_2 as input.

1.45 $MA = 0.50$.

2.1   Spiral.

2.3 $\mathbf{R}_{QP} = -7\hat{\mathbf{i}} - 14\hat{\mathbf{j}} = 15.652\angle{-116.57°}$.

2.5 $\Delta\mathbf{R}_A = -4.5a\hat{\mathbf{i}}$.

2.7   Clockwise; $\mathbf{R}(0) = 4\angle 0°$; $\mathbf{R}(20) = 404\angle 0°$; $\Delta\mathbf{R} = 400\angle 0°$.

2.9 $\Delta\mathbf{R}_{P_3} = -2.121\hat{\mathbf{i}}_1 + 3.879\hat{\mathbf{j}}_1$; $\Delta\mathbf{R}_{P_3/2} = 3\hat{\mathbf{i}}_2$.

2.11 $\Delta\mathbf{R}_Q = 47.548\hat{\mathbf{i}} + 27.452\hat{\mathbf{j}}$ mm $= 54.904$ mm$\angle 30°$.

2.13 $\overset{?\surd}{\mathbf{R}_C} = \overset{\surd\surd}{\mathbf{R}_A} + \overset{\surd I}{\mathbf{R}_{BA}} + \overset{\surd?}{\mathbf{R}_{CB}}$; $R_C = 50\cos\theta_2 + \sqrt{19\,200 - 2\,000\sin\theta_2 - 2\,500\sin^2\theta_2}$ mm.

2.15 $\overset{\surd I}{\mathbf{R}_{AO_2}} + \overset{\surd?}{\mathbf{R}_{BA}} - \overset{?\surd}{\mathbf{R}_{BC}} - \overset{\surd\surd}{\mathbf{R}_{CO_5}} - \overset{\surd\surd^y}{\mathbf{R}_{O_5O_2}} - \overset{\surd\surd^x}{\mathbf{R}_{O_5O_2}} = \mathbf{0}$; with $\rho_5\Delta\theta_5 = -\Delta R_{BC}$, (C1). θ_2 is a suitable input.

Known quantities are $R^x_{O_5O_2}$, $\theta_1 = 180°$, R_{AO_2}, R_{BA}, $\theta_4 = 90°$, ρ_5, R_{CO_5}, $\theta_6 = 0$, $R^y_{O_5O_2}$, $\theta_{15} = 90°$.

Unknown variables are, R_{BC}, θ_3, and $\Delta\theta_5$.

2.17 $\overset{\sqrt{I}}{\mathbf{R}_{BO_2}} - \overset{\sqrt{?}}{\mathbf{R}_{BA}} - \overset{\sqrt{?}}{\mathbf{R}_{AO_5}} + \overset{\sqrt{\sqrt{}}}{\mathbf{R}_{O_2O_5}} = 0$, with $R_2\Delta\theta_2 = \rho_3\Delta\theta_3$ $(C1)$.

θ_2 is a suitable input. Known quantities are $R_{O_2O_5}$, $\theta_1 = 0$, R_{BO_2}, ρ_3, R_{BA}, and R_{AO_5}.
Unknown variables are θ_3, θ_4, and θ_5.

2.19 $\overset{\sqrt{\sqrt{}}}{\mathbf{R}_{O_2O_5}} + \overset{\sqrt{?}}{\mathbf{R}_{AO_2}} + \overset{\sqrt{?}}{\mathbf{R}_{CA}} - \overset{\sqrt{I}}{\mathbf{R}_{CO_5}} = 0$, and $\overset{\sqrt{\sqrt{}}}{\mathbf{R}_{O_2O_5}} + \overset{\sqrt{C1}}{\mathbf{R}_{BO_2}} + \overset{\sqrt{?}}{\mathbf{R}_{DB}} - \overset{\sqrt{?}}{\mathbf{R}_{DO_5}} = 0$,

with $\theta_{22} = \theta_2 + \alpha$ $(C1)$.

θ_5 is a suitable input. Known quantities are $R_{O_2O_5}$, θ_1, R_2, R_3, R_4, R_5, R_{22}, and R_{35}.
Unknown variables are θ_2, θ_3, θ_4, θ_{22}, and θ_{35}.

2.21 $\overset{\sqrt{\sqrt{}}}{\mathbf{R}_1} + \overset{\sqrt{\sqrt{}}}{\mathbf{R}_{11}} + \overset{\sqrt{I}}{\mathbf{R}_2} + \overset{??}{\mathbf{R}_3} = 0$ and $\overset{\sqrt{\sqrt{}}}{\mathbf{R}_{11}} + \overset{\sqrt{I}}{\mathbf{R}_2} + \overset{\sqrt{C1}}{\mathbf{R}_{34}} + \overset{\sqrt{?}}{\mathbf{R}_4} - \overset{\sqrt{?}}{\mathbf{R}_9} = 0$ with $\theta_{34} = \theta_3 + \alpha$. $(C1)$.

Angle θ_2 is a suitable input. Known quantities are R_1, θ_1, R_2, R_4, R_9, R_{11}, θ_{11}, and R_{34}.
Unknown variables are R_3, θ_3, θ_4, θ_9, and θ_{34}.

2.25 $\theta_2 = \pm(2k+1)\pi/2 = \pm90°, \pm 270°, \ldots$

2.27 $\overset{\sqrt{\sqrt{}}}{\mathbf{R}_{11}} + \overset{\sqrt{?}}{\mathbf{R}_3} + \overset{\sqrt{?}}{\mathbf{R}_6} - \overset{\sqrt{?}}{\mathbf{R}_5} - \overset{?\sqrt{}}{\mathbf{R}_{15}} - \overset{\sqrt{\sqrt{}}}{\mathbf{R}_{19}} = 0$, $\overset{\sqrt{I}}{\mathbf{R}_2} + \overset{\sqrt{C1}}{\mathbf{R}_{22}} - \overset{\sqrt{C3}}{\mathbf{R}_{55}} - \overset{?\sqrt{}}{\mathbf{R}_{15}} - \overset{\sqrt{\sqrt{}}}{\mathbf{R}_{19}} = 0$, $\overset{\sqrt{\sqrt{}}}{\mathbf{R}_{11}} + \overset{\sqrt{C2}}{\mathbf{R}_{33}} +$

$\overset{\sqrt{?}}{\mathbf{R}_7} - \overset{?\sqrt{}}{\mathbf{R}_4} - \overset{\sqrt{\sqrt{}}}{\mathbf{R}_{14}} = \mathbf{0}$ with $\theta_{22} = \theta_2 + \gamma$ $(C1)$, $\theta_{33} = \theta_3 + \beta$ $(C2)$, and $\theta_{55} = \theta_5 + \alpha$ $(C3)$.

Angle θ_2 is a suitable input.

Known quantities are R_2, R_3, $\theta_4 = -90°$, R_5, R_6, R_7, R_{11}, θ_{11}, R_{14}, $\theta_{14} = 0$, $\theta_{15} = 180°$, R_{19}, $\theta_{19} = -90°$, R_{22}, R_{33}, R_{55}, α, β, and γ.
Unknown variables are θ_3, R_4, θ_5, θ_6, θ_7, R_{15}, θ_{22}, θ_{33}, and θ_{55}.

2.29 $\overset{\sqrt{I}}{\mathbf{R}_2} + \overset{\sqrt{?}}{\mathbf{R}_3} - \overset{?\sqrt{}}{\mathbf{R}_4} - \overset{\sqrt{\sqrt{}}}{\mathbf{R}_{14}} = 0$, $\overset{\sqrt{I}}{\mathbf{R}_2} + \overset{\sqrt{?}}{\mathbf{R}_9} - \overset{\sqrt{?}}{\mathbf{R}_5} - \overset{\sqrt{\sqrt{}}}{\mathbf{R}_{15}} = 0$, $\overset{\sqrt{?}}{\mathbf{R}_3} + \overset{\sqrt{?}}{\mathbf{R}_7} - \overset{\sqrt{?}}{\mathbf{R}_8} - \overset{\sqrt{C}}{\mathbf{R}_6} - \overset{\sqrt{?}}{\mathbf{R}_9} = 0$,

with $-\frac{\rho_6}{\rho_3} = \frac{\Delta\theta_3 - \Delta\theta_9}{\Delta\theta_6 - \Delta\theta_9}$ $(C1)$.

Angle θ_2 is a suitable input.

Known quantities are R_2, R_3, $\theta_4 = 90°$, R_5, R_6, R_7, R_8, $R_9 = \rho_6 + \rho_3$, R_{14}, $\theta_{14} = 180°$, R_{15}, and $\theta_{15} = 0$. Unknown variables are θ_3, R_4, θ_5, θ_6, θ_7, θ_8, and θ_9.

2.31 $\theta_4 = 139.94°$ and $\theta_4 = 262.96°$.

2.33 (a) $\theta_3 = 11°$, $\theta_4 = 95°$; $\theta'_3 = -71°$, $\theta'_4 = -155°$; (b) $\theta_3 = 11°$, $\theta_4 = 94.90°$; $\theta'_3 = -71°$, $\theta'_4 = -154.90°$; (c) $\theta_3 = 11.00°$, $\theta_4 = 94.90°$; $\theta'_3 = -71.00°$, $\theta'_4 = -154.90°$; (d) $\theta_3 = 11.0026°$, $\theta_4 = 94.9037°$.

2.35 $\overset{\sqrt{\sqrt{}}}{\mathbf{r}_1} + \overset{\sqrt{C1}}{\mathbf{r}_{13}} + \overset{\sqrt{?}}{\mathbf{r}_3} + \overset{?C2}{\mathbf{r}_{32}} - \overset{\sqrt{I}}{\mathbf{r}_2} - \overset{\sqrt{C3}}{\mathbf{r}_{12}} = 0$, $\overset{\sqrt{C1}}{\mathbf{r}_{13}} + \overset{?C4}{\mathbf{r}_{34}} - \overset{\sqrt{?}}{\mathbf{r}_4} - \overset{\sqrt{\sqrt{}}}{\mathbf{r}_{14}} = 0$, with $\theta_3 - \theta_{13} - \gamma + \pi = 0$

$(C1)$,

$\theta_{32} - \theta_{13} = 0$ $(C2)$, $\theta_2 - \theta_{12} - \beta + \pi = 0$ $(C3)$, and $\theta_{34} - \theta_3 = 0$ $(C4)$.

Angle θ_2 is a suitable input. Known quantities are $r_1, \theta_1, r_2, r_{12}, r_{13}, r_3, r_4, r_{14}, \theta_{14}, \beta, \gamma$.
Unknown variables are: $\theta_{12}, \theta_{13}, \theta_3, r_{32}, \theta_{32}, r_{34}, \theta_{34}, \theta_4$.

2.37
$$\left\{\begin{array}{l} \overset{\sqrt{\sqrt{}}}{\mathbf{r}^y_1} + \overset{\sqrt{\sqrt{}}}{\mathbf{r}^x_1} + \overset{\sqrt{I}}{\mathbf{r}^x_2} + \overset{\sqrt{?}}{\mathbf{r}^x_{34}} - \overset{\sqrt{?}}{\mathbf{r}^x_4} = 0, \\[6pt] \overset{\sqrt{\sqrt{}}}{\mathbf{r}^y_1} + \overset{\sqrt{\sqrt{}}}{\mathbf{r}^x_1} + \overset{\sqrt{\sqrt{}}}{\mathbf{r}^y_{15}} + \overset{?\sqrt{}}{\mathbf{r}^x_{15}} + \overset{\sqrt{?}}{\mathbf{r}^y_{67}} - \overset{\sqrt{?}}{\mathbf{r}^y_7} - \overset{\sqrt{C1}}{\mathbf{r}^y_{37}} - \overset{\sqrt{?}}{\mathbf{r}^y_4} = 0, \\[6pt] \overset{\sqrt{\sqrt{}}}{\mathbf{r}^y_1} + \overset{\sqrt{\sqrt{}}}{\mathbf{r}^x_1} + \overset{\sqrt{\sqrt{}}}{\mathbf{r}^x_{15}} + \overset{?\sqrt{}}{\mathbf{r}^y_{15}} + \overset{\sqrt{C2}}{\mathbf{r}_{63}} - \overset{?C3}{\mathbf{r}^x_{36}} - \overset{\sqrt{C4}}{\mathbf{r}^y_{36}} + \overset{\sqrt{?}}{\mathbf{r}^x_{34}} - \overset{\sqrt{?}}{\mathbf{r}_4} = 0, \end{array}\right\} \text{with } \theta^x_{34} - \theta^y_{37} - 90° = 0$$

$(C1)$, $\theta_{67} + \beta - \theta_{63} = 0$ $(C2)$, $\theta^y_{36} - \theta^x_{34} = 0$ $(C3)$, and $\theta^y_{36} + 90° - \theta^x_{34} = 0$ $(C4)$.

Angle θ_2 is a suitable input.

Known quantities are $r_1^y, \theta_1^y, r_1^x, \theta_1^x, r_2, r_{34}^x, r_4, r_{15}^x, \theta_{15}^x, \theta_{15}^y, r_{67}, r_7, r_{37}^y, r_{63}, r_{35}^y$.

Unknown variables are $\theta_{34}^x, \theta_4, r_{15}^y, \theta_{67}, \theta_7, \theta_{37}^y, \theta_{63}, r_{36}^x, \theta_{36}^x, \theta_{36}^y$.

2.39 $\overset{\sqrt{\sqrt{}}}{\mathbf{r}_1} + \overset{?\sqrt{}}{\mathbf{r}_2} - \overset{?\mathrm{I}}{\mathbf{r}_3} = \mathbf{0}$.

Angle θ_3 is a suitable input. Distances r_2 or r_3 are other possible inputs.

Known quantities are r_1, θ_1, θ_2. Unknown variables are r_2, r_3.

2.41 $\rho_3 \Delta\theta_3 = r_2 \Delta\theta_2$.

2.43 $\overset{\sqrt{?}}{\mathbf{r}_4} + \overset{\sqrt{\sqrt{}}}{\mathbf{r}_1} + \overset{?\sqrt{}}{\mathbf{r}_{43}} - \overset{\sqrt{\sqrt{}}}{\mathbf{r}_3} - \overset{\mathrm{I}\sqrt{}}{\mathbf{r}_2} = \mathbf{0}$ with $\Delta r_{43} = r_3 \Delta\theta_{13}$. Distance r_2 is a suitable input.

Known quantities are $\theta_4, r_1, \theta_1, \theta_{43}, r_3, \theta_3, \theta_2$. Unknown variables are: r_4, r_{43}, θ_{13}.

2.45 $\overset{\sqrt{\mathrm{I}}}{\mathbf{r}_2} + \overset{\sqrt{?}}{\mathbf{r}_4} - \overset{?\sqrt{}}{\mathbf{r}_5} - \overset{\sqrt{\sqrt{}}}{\mathbf{r}_1} = \mathbf{0}$, with $\rho_3 \Delta\theta_3 = r_2 \Delta\theta_2$. Angle θ_2 is a suitable input.

Known quantities are $r_1, \theta_1, r_2, r_4, \theta_5, \rho_1, \rho_3$. Unknown variables are θ_4, r_5, θ_3.

3.1 $\dot{\mathbf{R}} = 6.283$ m/s$\angle 162°$.

3.3 $\mathbf{V}_{BA} = \mathbf{V}_{B_3/2} = 82.61$ mi/h N $24.8°$E.

3.5   Dia $= 70$ in; $\mathbf{V}_{AB} = 3\,000\hat{\mathbf{j}}$ in/s; $\mathbf{V}_{BA} = -3\,000\hat{\mathbf{j}}$ in/s; $\omega_2 = 200$ rad/s cw.

3.7   (a) Straight line at N $48°$ E; (b) no change.

3.9 $\omega_3 = 1.44$ rad/s ccw; $\omega_4 = 15.40$ rad/s ccw.

3.11 $\mathbf{V}_C = 7.118$ m/s$\angle -75.8°$; $\omega_3 = 0.335$ rad/s ccw.

3.13 $\mathbf{V}_C = 16.08$ in/s$\angle 151°$; $\mathbf{V}_D = 11.60$ in/s$\angle -111°$.

3.15 $\omega_3 = \omega_4 = 22$ rad/s ccw; $\mathbf{V}_B = 191.6$ in/s$\angle 96.5°$.

3.17 $\omega_6 = 4.03$ rad/s ccw; $\mathbf{V}_B = 0.231$ m/s$\angle 180°$; $\mathbf{V}_C = 0.484$ m/s$\angle -152.4°$;

       $\mathbf{V}_D = 0.484$ m/s$\angle -153.8°$.

3.19 $\omega_3 = 3.24$ rad/s ccw; $\mathbf{V}_B = 5.070$ m/s$\angle -56.3°$.

3.21 $\mathbf{V}_C = 361$ in/s$\angle 137.8°$.

3.23 $\mathbf{V}_B = 10.65$ m/s$\angle -120°$; $\mathbf{V}_C = 12.17$ m/s$\angle -93.3°$; $\mathbf{V}_D = 9.47$ m/s$\angle -60°$.

3.25 $\mathbf{V}_B = 0.313$ m/s$\angle -23.0°$.

3.27 $\mathbf{V}_{C_3} = 1.526$ m/s$\angle -156.8°$; $\mathbf{V}_{C_6} = 0.861$ m/s$\angle -118.6°$; $\mathbf{V}_E = 1.548$ m/s$\angle -98.9°$;

       $\omega_5 = \omega_6 = 9.67$ rad/s cw; $\mathbf{V}_{D_5} = 1.290$ m/s$\angle -149.9°$.

3.29 $\omega_4 = 4.354$ rad/s ccw.

3.31 $\mathbf{V}_A = 0.026$ m/s$\angle 180°$; $\mathbf{V}_B = 0.1155$ m/s$\angle 180°$.

3.41 $\theta_4' = 2.30936$ rad/m; $r_4' = 1.15468$ m/m; $\omega_4 = -0.17321$ rad/s (cw);

       $\omega_3 = 0.51959$ rad/s ccw.

3.43 $\theta_3' = 2.3094$ rad/m; $r_4' = 1.1547$ m/m; $\omega_3 = -0.289$ rad/s (cw);

       $V_{O_3/4} = -0.1443$ m/s.

3.45 $r_3' = 1.1547$ in/in; $\theta_3' = -0.2887$ rad/in; $\theta_4' = 0.8660$ rad/in; $\omega_3 = 2.165$ rad/s ccw;

       $\omega_4 = -6.495$ rad/s (cw); $V_E = 7.502$ in/s$\angle 150.01°$; $V_{B_4/3} = -8.660$ in/s$\angle 120°$.

3.47 $\theta_3' = -12.5$ rad/m; $\theta_4' = 0$; $\omega_3 = -3.75$ rad/s (cw); $\omega_4 = 0$; $\Delta = 0$ when

       $\theta_3 = \theta_4 = 68.91°$.

3.49 $\theta_3' = -0.200$ rad/rad; $\theta_4' = 0$, $\theta_5' = -0.500$ rad/rad; $\omega_3 = -1.00$ rad/s (cw); $\omega_4 = 0$,

       $\omega_5 = -2.50$ rad/s (cw).

3.51 $\theta'_{13} = 2.778$ rad/rad; $r'_2 = -5.0$ in/rad; $\theta'_3 = -8.333$ rad/rad; $\omega_3 = -250$ rad/s (cw);
$V_{A_3/2} = -150.0$ in/s.

3.53 $r'_3 = 0.25882$ in/in; $\theta'_3 = -0.03220$ rad/in; (a) $\omega_3 = -0.386$ rad/s (cw);
(b) $V_{A_2/3} = 3.104$ in/s$\angle 45°$; (c) $V_B = 19.319$ in/s$\angle -45°$.

3.55 $\theta'_3 = 2$ rad/rad; $\theta'_4 = 4/3$ rad/rad; $\theta'_5 = 0.8$ rad/rad; $\omega_3 = -30$ rad/s (cw);
$\omega_4 = -20$ rad/s (cw); $\omega_5 = -12$ rad/s (cw).

3.57 $\theta'_3 = -23.094$ rad/m; $R'_4 = 0.57735$ m/m; $\omega_3 = -2.771$ rad/s (cw);
$V_4 = 0.06928$ m/s.

3.59 $\theta'_3 = \theta'_4 = 0.667$ rad/rad; $\omega_3 = \omega_4 = -6.667$ rad/s (cw); $V_{D_4C_3} = 0.7454$ m/s$\angle 26.57°$.

3.61 $\theta'_3 = \theta'_4 = 1$ rad/rad; $R'_C = 0.4206$ m/rad$\angle 162°$; $V_C = 6.308$ m/s$\angle 162°$.

4.1 $\ddot{\mathbf{R}} = -0.100\hat{\mathbf{i}}$ m/s^2.

4.3 $\hat{\mathbf{u}}^t = 0.30071\hat{\mathbf{i}} - 0.95372\hat{\mathbf{j}}$; $A^n = -1.747$ in/s^2; $A^t = 0.503$ in/s^2; $\rho = -16.219$ in.

4.5 $\mathbf{A}_A = 303579$ in/s$^2\angle 161.6°$.

4.7 $\mathbf{V}_B = 3.60$ m/s$\angle -90°$; $\mathbf{V}_C = 2.510$ m/s$\angle 12.1°$; $\mathbf{A}_B = 118.5$ m/s$^2\angle 165°$;
$\mathbf{A}_C = 63.06$ m/s$^2\angle -119.7°$.

4.9 $\omega_2 = 38.64$ rad/s cw; $\alpha_2 = 5571.3$ rad/s^2 cw.

4.11 $\alpha_3 = 563.3$ rad/s^2 ccw; $\alpha_4 = 123.7$ rad/s^2 ccw.

4.13 $\mathbf{A}_C = 931.4$ m/s$^2\angle 114.4°$; $\alpha_3 = 1741$ rad/s^2 ccw; $\alpha_4 = 3055$ rad/s^2 ccw.

4.15 $\mathbf{A}_C = 781.3$ m/s$^2\angle -68.9°$; $\alpha_4 = 1495$ rad/s^2 ccw.

4.17 $\mathbf{A}_B = 4.00$ m/s$^2\angle 0°$; $\alpha_3 = 17.50$ rad/s^2 ccw; $\alpha_6 = 10.82$ rad/s^2 cw.

4.19 $\alpha_2 = 4181$ rad/s^2 ccw.

4.21 $\mathbf{A}_C = 18024$ in/s$^2\angle -104.4°$; $\alpha_3 = 74.08$ rad/s^2 cw.

4.23 $\mathbf{A}_B = 732.2$ m/s$^2\angle -120°$; $\mathbf{A}_D = 1209$ in/s$^2\angle 120°$.

4.25 $\theta_3 = 171.01°$; $\theta_4 = 195.54°$; $\omega_3 = 70.45$ rad/s ccw; $\omega_4 = 47.57$ rad/s ccw;
$\alpha_3 = 3197$ rad/s^2 ccw; $\alpha_4 = 3331$ rad/s^2 ccw.

4.27 $\theta_3 = 28.32°$; $\theta_4 = 55.88°$; $\omega_3 = -0.633$ rad/s (cw); $\omega_4 = -2.156$ rad/s (cw);
$\alpha_3 = 7.822$ rad/s^2 ccw; $\alpha_4 = 6.704$ rad/s^2 ccw.

4.29 $\theta_3 = 38.42°$; $\theta_4 = 155.60°$; $\omega_3 = -6.855$ rad/s (cw); $\omega_4 = -1.235$ rad/s (cw);
$\alpha_3 = 62.50$ rad/s^2 ccw; $\alpha_4 = -96.51$ rad/s^2 (cw).

4.31 $\mathbf{V}_B = 4.600$ m/s$\angle -19.10°$; $\mathbf{A}_B = 67.57$ m/s$^2\angle -182.50°$; $\omega_4 = 6.572$ rad/s cw;
$\alpha_4 = 86.12$ rad/s^2 ccw.

4.33 $\mathbf{A}_E = 144.7$ m/s$^2\angle -107.20°$.

4.35 $\mathbf{A}_B = 0.5058$ m/s$^2\angle 77.90°$; $\alpha_3 = 1.247$ rad/s^2 cw.

4.37 $\mathbf{A}_{C_4} = 15.00$ m/s$^2\angle 91.10°$; $\alpha_3 = 6.00$ rad/s^2 ccw.

4.39 $\mathbf{A}_G = 7.028$ m/s$^2\angle -64.50°$; $\alpha_5 = 0$; $\alpha_6 = 110.0$ rad/s^2 cw.

4.41 $\theta''_3 = 0.240$ rad/rad^2; $\theta''_4 = 0.150$ rad/rad^2; $\theta''_5 = 0$; $\alpha_3 = 6.0$ rad/s^2 ccw;
$\alpha_4 = 3.75$ rad/s^2 ccw; $\alpha_5 = 0$.

4.47 (a) $\rho_C = 1750$ in; (b) $v = 2262$ in/s; (c) $A_I = 97430$ in/s^2; (d) $V_A = 1959$ in/s;
$V_C = 1764$ in/s; (e) $A_A = 30320$ in/s^2; $A_C = 15160$ in/s^2.

4.49 (a) $\gamma = 24.23°$ccw; (b) $\rho_B = 75$ mm; $\rho_E = 106$ mm; (d) $A_B = 1.400$ m/s^2;
$A_E = 4.140$ m/s^2; $A_I = 3.33$ m/s^2.

4.51 (*a*) $\hat{\mathbf{u}}_D^t = -\hat{\mathbf{j}}$; $\hat{\mathbf{u}}_D^n = \hat{\mathbf{i}}$; (*b*) $\rho_D = -4.50$ in; (*c*) $x_{CC} = -6.00$ in; $y_{CC} = 3.00$ in; $\mathbf{V}_D = 375.00$ in/s$\angle{-90°}$; $\mathbf{A}_D = 13\,343$ m/s$^2\angle{-159.52°}$.

4.53 (*a*) $x'_B = -25.856$ in/rad; $y'_B = 0$; $x''_B = 111.424$ in/rad^2; $y''_B = 36.0$ in/rad^2; (*b*) $\hat{\mathbf{u}}_B^t = \hat{\mathbf{i}}$; $\hat{\mathbf{u}}_B^n = \hat{\mathbf{j}}$; (*c*) $\rho_B = 18.571$ in; (*d*) $x_{CC} = 8$ in; $y_{CC} = 14.571$ in; $\mathbf{A}_B = 16862$ in/s$^2\angle{17.91°}$.

4.55 $\alpha_3 = -30$ rad/s^2 (cw); $\alpha_4 = -20$ rad/s^2 (cw); $\alpha_5 = -12$ rad/s^2 (cw).

4.57 $\rho_C = -4.583$ in; $x_{CC} = 0$; $y_{CC} = 3.0$ in; $\mathbf{A}_C = 24698$ in/s$^2\angle{-151.94°}$.

4.59 $\rho_B = -2.828$ in; $x_{CC} = -2$ in; $y_{CC} = -2$ in; $\mathbf{A}_C = 316.2$ in/s$^2\angle{161.57°}$.

4.61 $x'_C = -400.000$ mm/rad; $y'_C = 129.904$ mm/rad; $x''_C = -880.450$ mm/rad^2; $y''_C = -400.00$ mm/rad^2; (*a*) $\hat{\mathbf{u}}_C^t = -0.951\,10\hat{\mathbf{i}} + 0.308\,88\hat{\mathbf{j}}$; $\hat{\mathbf{u}}_C^n = -0.308\,88\hat{\mathbf{i}} - 0.951\,10\hat{\mathbf{j}}$; (*b*) $\rho_C = 271.117$ mm; (*c*) $x_{CC} = 46.175$ mm; $y_{CC} = 142.175$ mm; $\mathbf{A}_C = 217.587$ m/s$^2\angle{-155.57°}$.

5.1 $\mathbf{V}_{P_4} = 94.20$ in/s$\angle{15.62°}$; $\mathbf{A}_{P_4} = 5029.20$ in/s$^2\angle{-66.60°}$.

5.3 $\omega_3 = -43.64$ rad/s (cw); $\omega_5 = 47.20$ rad/s ccw.

5.5 $\mathbf{V}_{P_3} = 1.179$ m/s$\angle{-175.13°}$; $\mathbf{A}_{P_3} = 11.18$ m/s$^2\angle{-79.70°}$.

5.7 $\theta_4 = 101.39°$; $\theta_{5/4} = 61.45°$.

5.9 $\omega_4 = 10.066$ rad/s cw; $\omega_{5/4} = 17.178$ rad/s (ccw).

5.11 $\alpha_4 = 91.944$ rad/s^2 ccw; $\alpha_5 = 37.818$ rad/s^2 ccw.

5.13 $x'_{Px} = 1.667$ in/in; $y'_{Px} = 0$; $x'_{Py} = 0$; $y'_{Py} = 1.667$ in/in.

5.15 $\theta'_{32} = -1.000\,00$ rad/rad; $\theta'_{42} = 8.660\,26$ rad/rad; $\theta'_{35} = 0$; $\theta'_{45} = -2.166\,67$ rad/rad; $\theta''_{322} = -9.474$ rad/rad^2; $\theta''_{422} = 46.021$ rad/rad^2; $\theta''_{325} = 0$; $\theta''_{425} = 0$; $\theta''_{355} = 0$; $\theta''_{455} = 0$; $\omega_3 = 50$ rad/s cw; $\omega_4 = 487.18$ rad/s ccw; $\alpha_3 = 23\,685$ rad/s^2 cw; $\alpha_4 = 115\,009$ rad/s^2 ccw.

6.3 Face ≥ 7.8 in from pivot.

6.5 $\dot{y}\left(\frac{\beta}{2}\right) = \frac{\pi L}{2\beta}\omega$; $\dddot{y}\left(\frac{\beta}{2}\right) = -\frac{\pi^3 L}{2\beta^3}\omega^3$; $\ddot{y}(0) = \frac{\pi^2 L}{2\beta^2}\omega^2$; $\ddot{y}(\beta) = -\frac{\pi^2 L}{2\beta^2}\omega^2$.

6.7 *AB*: dwell, $L_1 = 0$, $\beta_1 = 60°$; *BC*: full-rise eighth-order polynomial motion, Eq. (6.14), $L_2 = 62.500$ mm, $\beta_2 = 62.442°$; *CD*: half-harmonic return motion, Eq. (6.20), $L_3 = 2.060$ mm, $\beta_3 = 7.750°$; *DE*: uniform motion, $L_4 = 25.000$ mm, $\beta_4 = 60°$; *EA*: half-cycloidal return motion, Eq. (6.25), $L_5 = 35.440$ mm, $\beta_5 = 169.808°$.

6.9 $\Delta t = 0.025$ s; $\dot{y}_{max} = 12.656$ m/s; $\dot{y}_{min} = -1.002$ m/s; $\ddot{y}_{max} = 486.426$ m/s^2; $\ddot{y}_{min} = -486.426$ m/s^2.

6.11 $\dot{y}_{max} = 41.888$ rad/s; $\ddot{y}_{max} = 7\,896$ rad/s^2.

6.13 Face width $= 55.000$ mm; $\rho_{min} = 75.000$ mm.

6.15 $R_0 > 227.228$ mm; face width > 334.052 mm.

6.17 $\phi_{max} = 12°$; $R_r < 362.5$ mm.

6.19 $R_0 > 2.25$ in; $\ddot{y}_{max} = 1480$ in/s^2.

6.21 $R_0 > 2.85$ in; $\ddot{y}_{max} = 1580.5$ in/s^2.

6.23 $u = (R_0 + R_c + Y)\sin\Theta + y'\cos\Theta$; $v = (R_0 + R_c + Y)\cos\Theta - y'\sin\Theta$.

6.27 (*a*) $y' = 0.866$ in/rad; $y'' = 1.000$ in/rad^2; $y''' = -3.464$ in/rad^3; (*b*) $u_{cam} = 1.462$ in; $v_{cam} = 0.350$ in; (*c*) $\rho = -3.481$ in; (*d*) $\hat{\mathbf{u}}^t = 0.746\hat{\mathbf{i}} - 0.666\hat{\mathbf{j}}$; $\hat{\mathbf{u}}^n = 0.666\hat{\mathbf{i}} + 0.746\hat{\mathbf{j}}$; (*e*) $\phi = 8.2°$.

6.31 (a) $y' = -86.26625$ mm/rad; $y'' = -168.40450$ mm/rad^2; (b) $u_{cam} = -213.575$ mm; $v_{cam} = -115.025$ mm; (c) $\rho_{cam} = -148.625$ mm; (d) $\phi = 32.19°$.

6.33 (a) $y' = 38.197$ mm/rad; $y'' = 0$; $y''' = -687.544$ mm/rad^3; (b) $\rho_{cam} = -68.082$ mm; (c) $\hat{\mathbf{u}}^t = -0.67455\hat{\mathbf{i}} - 0.73823\hat{\mathbf{j}}$; $\hat{\mathbf{u}}^n = 0.73823\hat{\mathbf{i}} - 0.67455\hat{\mathbf{j}}$; (d) $u_{cam} = 24.761$ mm; $v_{cam} = -71.949$ mm; (e) $\phi = 17.58°$.

6.35 (a) $y' = -28.648$ mm/rad; $y'' = -99.239$ mm/rad^2; $y''' = 343.772$ mm/rad^3; (b) $\rho_{cam} = -36.300$ mm; (c) $\hat{\mathbf{u}}^t = -0.30701\hat{\mathbf{i}} + 0.95170\hat{\mathbf{j}}$; $\hat{\mathbf{u}}^n = -0.95170\hat{\mathbf{i}} - 0.30701\hat{\mathbf{j}}$; (d) $u_{cam} = -64.988$ mm; $v_{cam} = -48.850$ mm; (e) $\phi = 22.12°$.

6.37 (a) $F = m\omega^2 e\cos\omega t + mg$; (b) $\omega = 19.81$ rad/s.

6.39 (a) $F_{c\ max} = 15.442$ lb; $F_{c,\ min} = 0$; (b) $\Theta = 163.07°$.

6.41 $\omega = 260.2$ rev/min.

7.1 $m = 2.5$ mm/tooth.

7.3 $P = 10$ teeth/in.

7.5 $m = 2.75$ mm/tooth; $D = 110$ mm.

7.7 $P = 1.964$ teeth/in; $D = 18.33$ in.

7.9 $D = 231.1$ mm.

7.11 $N_2 = 18$ teeth; $N_3 = 54$ teeth.

7.13 $a = 6$ mm; $d = 7.5$ mm; $c = 1.5$ mm; $p_c = 18.85$ mm/tooth; $p_b = 17.713$ mm/tooth; $t = 9.425$ mm; $R_2 = 72$ mm; $R_3 = 108$ mm; $r_2 = 67.66$ mm; $r_3 = 101.49$ mm; $CP = 15.00$ mm; $PD = 14.18$ mm; $m_c = 1.647$ teeth avg.

7.15 $\alpha_2 = 18.58°$; $\alpha_3 = 6.32°$; $\beta_2 = 16.04°$; $\beta_3 = 5.45°$; $m_c = 1.63$ teeth avg.

7.17 $CP = 0.789$ in; $PD = 0.650$ in; $m_c = 1.52$ teeth avg.

7.19 (a) $CP = 35.1$ mm; $PD = 28.7$ mm; $m_c = 1.80$ teeth avg. (b) $C'P' = 26.3$ mm; $P'D' = 28.7$ mm; $m_c' = 1.55$ teeth avg; no change in pressure angle.

7.25 $t_b = 0.6857$ in; $t_a = 0.2715$ in; $\varphi_a = 32.78°$.

7.27 $t = 28.99$ mm.

7.29 $t_b = 3.822$ mm; $t_a = 0.0406$ mm; $\varphi = 35.41°$.

7.31 (a) $d = 4.206$ mm; (b) 107.41 mm.

7.33 $m_c = 1.664$ teeth avg.

7.35 $m_c = 1.771$ teeth avg.

7.37 $\phi' = 26.64°$.

7.39 $a_3 \leq 33.45$ mm.

7.41 $a_3' = 17.81$ mm; $d_3' = 40.94$ mm; $a_2' = 32.19$ mm; $d_2' = 26.56$ mm; $e_2 = -1.56$ mm; $e_3 = -15.94$ mm; $m_c' = 1.49$ teeth avg.

7.45 $\theta_{72}' = 0.26250$; $\omega_5 = 1.759$ rad/s cw; $\omega_7 = 3.299$ rad/s cw.

7.47 One solution: $N_6 = 20$; $N_5 = 30$; $N_8 = 35$ teeth; $N_3 = 30$ teeth; $N_4 = N_7 = 25$ teeth; $N_{10} = 40$ teeth.

7.49 $\omega_3 = 222.1$ rev/min ccw.

7.51 $\omega_A = 644.8$ rev/min cw.

7.53 77.3%, opposite to input sense.

7.55 $N_5 = 84$ teeth; $R_3 = 6.5$ in; $\omega_3 = 8.08$ rev/min ccw.

7.57 $\omega_A = (9/34)\omega_2$; Levai type F.

7.59 $\Delta t_D = 60$ min/rev; $\Delta t_H = 12$ hr/rev; $\omega_F = 2$ rev/day.

8.1 $p_t = 12.57$ mm/tooth; $p_n = 8.89$ mm/tooth; $m_n = 2.83$ mm/tooth; $R_2 = 30.00$ mm;
$R_3 = 48.00$ mm; $N_{e2} = 42.43$ teeth; $N_{e3} = 67.88$ teeth.

8.3 $\psi_b = 33.34°$; $m_x = 1.69$ teeth avg.

8.5 $N_3 = 72$ teeth; $R_2 = 50.00$ mm; $R_3 = 225.00$ mm; $p_n = 18.08$ mm/tooth;
$m = 6.25$ mm/tooth; $F > 92.49$ mm.

8.7 $m_n = 1.75$ teeth avg; $m = 2.64$ teeth avg.

8.9 $\psi_3 = 15°\text{RH}$; $N_3 = 39$ teeth; $R_2 = 24.11$ mm; $R_3 = 41.99$ mm.

8.11 $\gamma_2 = 18.43°$; $\gamma_3 = 71.57°$.

8.13 $\gamma_2 = 27.00°$; $\gamma_3 = 93.00°$.

8.15 $R_2 = 25.50$ mm; $R_3 = 42.00$ mm; $\gamma_2 = 34.83°$; $\gamma_3 = 70.17°$; $a_3 = 2.048$ mm;
$a_2 = 3.952$ mm; $d_2 = 2.886$ mm; $d_3 = 4.790$ mm; $F \approx 13.40$ mm; $N_{e2} = 20.71$ teeth;
$N_{e3} = 82.54$ teeth.

8.17 $N_2 = 1$ tooth; $R_2 = 43.13$ mm; $N_3 = 60$ teeth; $R_3 = 119.37$ mm.

8.19 $N_3 = 123.7$ teeth; $R_3 = 196.80$ mm; $\psi = 20.0°$; $R_2 = 13.12$ mm;
$R_2 + R_3 = 209.92$ mm.

8.21 59.1%.

8.23 $\omega_6 = 657.77$ rev/min; $\omega_5 = 700.21$ rev/min; $\omega_3 = 678.99$ rev/min.

9.1 For six points, 0.170 37, 1.464 47, 3.705 91, 6.294 09, 8.535 53, and 9.829 63.

9.3 Typical solution: $r_2 = 176.63$ mm; $r_3 = 568.88$ mm.

9.5 Typical solution: $r_1 = 2.434$ m; $r_2 = 0.950$ m; $r_3 = 2.858$ m.

9.7 Typical solution: O_2 at $r_1 = 79.16$ in; $r_2 = 13.84$ in; $r_3 = 91.52$ in.

9.9 Typical solution: $O_2A = AB = O_4B = O_2O_4$ and spring AO_4 with chosen free length.

9.13 and **9.23** $r_2/r_1 = 1.834$, $r_3/r_1 = 2.238$, $r_4/r_1 = -0.693$.

9.15 and **9.25** $r_2/r_1 = -3.499$, $r_3/r_1 = 0.878$, $r_4/r_1 = 3.399$.

9.17 and **9.27** $r_2/r_1 = 0.625$, $r_3/r_1 = 1.309$, $r_4/r_1 = -0.401$.

9.19 and **9.29** $r_2/r_1 = -1.801$, $r_3/r_1 = 0.908$, $r_4/r_1 = 1.274$.

9.21 and **9.31** $r_2/r_1 = -0.610$, $r_3/r_1 = 0.565$, $r_4/r_1 = 0.380$.

10.1 $m = 2$ including one idle freedom. Path of B is the intersection of cylinder of radius BA
about the y axis and sphere of radius BO_3 about O_3.

10.3 $\omega_2 = -2.570\hat{j}$ rad/s; $\omega_3 = 1.158\hat{i} - 0.086\hat{j} + 0.643\hat{k}$ rad/s;
$V_B = -3.854\hat{i} - 2.000\hat{j} + 6.676\hat{k}$ in/s.

10.5 and **10.7** $\omega_3 = \omega_4 - 25.714\hat{i}$ rad/s; $V_A = 4.500\hat{j}$ m/s; $V_B = -5.785\hat{k}$ m/s;
$\alpha_3 = 1\,543\hat{j}$ rad/s²; $\alpha_4 = -343\hat{i}$ rad/s²; $A_A = 270\hat{i}$ m/s²; $A_B = -149\hat{j} - 77\hat{k}$ m/s².

10.9 $\Delta\theta_4 = 46.5°$; $Q = 1.012$.

10.11 and **10.13** $\omega_3 = 4.083\hat{\mathbf{i}} - 7.071\hat{\mathbf{j}} + 3.334\hat{\mathbf{k}}$ rad/s; $\omega_4 = 10.797\hat{\mathbf{i}}$ rad/s;
$\mathbf{V}_A = -46.76\hat{\mathbf{i}} - 27.0\hat{\mathbf{j}}$ in/s; $\mathbf{V}_B = 48.28\hat{\mathbf{j}} + 102.36\hat{\mathbf{k}}$ in/s; $\alpha_3 = 273\hat{\mathbf{i}} + 115\hat{\mathbf{j}} - 148\hat{\mathbf{k}}$ rad/s^2;
$\alpha_4 = 130\hat{\mathbf{i}}$ rad/s^2; $\mathbf{A}_A = 972\hat{\mathbf{i}} - 1\,684\hat{\mathbf{j}}$ in/s^2; $\mathbf{A}_B = -522\hat{\mathbf{j}} + 1\,756\hat{\mathbf{k}}$ in/s^2.

10.15 $\omega_3 = 7.753$ rad/s; $V_B = 345$ mm/s.

10.17 and **10.19** $\omega_2 = 12.00\hat{\mathbf{i}} - 20.78\hat{\mathbf{j}}$ rad/s, $\omega_3 = 1.333\hat{\mathbf{i}} + 6.906\hat{\mathbf{j}} - 1.823\hat{\mathbf{k}}$ rad/s. $\mathbf{V}_A = 0.520\hat{\mathbf{i}} + 0.300\hat{\mathbf{j}} + 1.039\hat{\mathbf{k}}$ m/s; $\mathbf{V}_B = 0.653\hat{\mathbf{i}}$ m/s.

10.21 and **10.23** $V_B = 2.809$ m/s; $\omega_3 = -4.190\hat{\mathbf{i}} + 19.444\hat{\mathbf{j}} - 5.471\hat{\mathbf{k}}$ rad/s; $\omega_4 = 19.444\hat{\mathbf{j}}$ rad/s,

10.25
$$T_{15} = \begin{bmatrix} \cos(\phi_1 + \phi_2 - \phi_4) & \sin(\phi_1 + \phi_2 - \phi_4) & 0 & 250\cos\phi_1 + 250\cos(\phi_1 + \phi_2)\,\text{mm} \\ \sin(\phi_1 + \phi_2 - \phi_4) & -\cos(\phi_1 + \phi_2 - \phi_4) & 0 & 250\sin\phi_1 + 250\sin(\phi_1 + \phi_2)\,\text{mm} \\ 0 & 0 & -1 & 250 - \phi_3\,\text{mm} \\ 0 & 0 & 0 & 1 \end{bmatrix};$$

$$R_1 = \begin{bmatrix} 250\cos\phi_1 + 250\cos(\phi_1 + \phi_2)\,\text{mm} \\ 250\sin\phi_1 + 250\sin(\phi_1 + \phi_2)\,\text{mm} \\ 212.50 - \phi_3\,\text{mm} \\ 1 \end{bmatrix}.$$

10.27
$$T_{15} = \begin{bmatrix} -\sin\phi_4 & -\cos\phi_4 & 0 & \phi_2 \\ 0 & 0 & -1 & -\phi_3 - 2\ \text{in} \\ \cos\phi_4 & -\sin\phi_4 & 0 & \phi_1 \\ 0 & 0 & 0 & 1 \end{bmatrix};$$

$$R_1 = \begin{bmatrix} \phi_2 \\ -\phi_3 - 3.80\ \text{in} \\ \phi_1 \\ 1 \end{bmatrix}.$$

10.29
$$\dot{R}_5 = \begin{bmatrix} 0 \\ -1.6\ \text{in/s} \\ 0 \\ 0 \end{bmatrix}; \quad \ddot{R}_5 = \begin{bmatrix} 0 \\ 0 \\ 0 \\ 0 \end{bmatrix}.$$

10.31 $\phi_1 = 3.6t + 9$ in; $\phi_2 = 4.8t + 12$ in; $\phi_3 = 4$ in; $\phi_4 = -90°$.

10.33 $\tau_1 = 0$; $\tau_2 = 4.5$ lb; $\tau_3 = -2.25t$ lb; $\tau_4 = -0.113 - 0.27t - 0.27t^2$ in $\cdot$ lb.

11.3 $P = 329.33$ lb.

11.5 $T_{12} = 18.38$ N $\cdot$ m.

11.7 $T_{12} = 42.46$ N $\cdot$ m.

11.9 $T_{12} = 28.1$ N $\cdot$ m.

11.11 $\mathbf{T}_{12} = -81.0\hat{\mathbf{k}}$ N $\cdot$ m; $\mathbf{F}_{12} = -\mathbf{F}_{32} = 974.5$ N$\angle-123.74°$;
$\mathbf{F}_{14} = 1\,374.6$ N$\angle-61.70°$; $\mathbf{F}_{34} = 811.4$ N$\angle87.09°$.

11.13 $\mathbf{F}_{12} = -\mathbf{F}_{32} = 307\text{ kN}\angle{-}129.59°$; $\mathbf{F}_{14} = -\mathbf{F}_{34} = 387\text{ kN}\angle 59.70°$;
$\mathbf{T}_{12} = -113\hat{\mathbf{k}}\text{ kN}\cdot\text{m}$.

11.15 $\mathbf{T}_{12} = 47.26\hat{\mathbf{k}}\text{ N}\cdot\text{m}$.

11.17 *(a)* $F_{13} = 11\,840\text{ N}\angle 0°$; *(b)* $F_{13} = 3\,990\text{ N}\angle{-}135°$; *(c)* $F_{13} = 10\,960\text{ N}\angle 135°$.

11.19 $\mathbf{F}_C = 1\,025\,000\text{ N}\angle{-}171.13°$; $\mathbf{F}_D = 1\,659\,000\text{ N}\angle 163.41°$.

11.21 $\mathbf{F}_C = 118\hat{\mathbf{i}} + 140\hat{\mathbf{j}} - 251\hat{\mathbf{k}}\text{ lb}$; $\mathbf{F}_D = -71\hat{\mathbf{i}} - 1551\hat{\mathbf{k}}\text{ lb}$.

11.23 $\mathbf{F}_E = 768\hat{\mathbf{i}} - 904\hat{\mathbf{j}} + 1\,678\hat{\mathbf{k}}\text{ N}$; $\mathbf{F}_F = 520\hat{\mathbf{j}} + 682\hat{\mathbf{k}}\text{ N}$.

11.25 $T_{12} = 547\text{ in}\cdot\text{lb cw}$.

11.29 $\mathbf{F}_{12}^{F} = 4977\hat{\mathbf{j}}\text{ N}$; $\mathbf{f}_{12} = 1701\hat{\mathbf{i}}\text{ N}$; $\mathbf{F}_{12}^{R} = 5577\hat{\mathbf{j}}\text{ N}$; $\mathbf{F}_{13} = 3402\text{ N}\angle 120°$;
$\mathbf{F}_{23} = 2304\text{ N}\angle 42.41°$; $\mu \geq 0.34$.

11.31 $\mathbf{T}_{12} = 1.528\hat{\mathbf{k}}\text{ in}\cdot\text{lb}$.

11.33 $T = F_B d\sin^2\theta = F_B d(R/x)^2$; $x \geq R\sqrt{1/\mu^2 + 1}$.

11.35 *(a)* $T_{12} = 8.17\text{ in}\cdot\text{lb cw}$; *(b)* $\mathbf{F}_{14} = 4.10\text{ lb}\angle{-}106.7°$;
$\mathbf{F}_{34} = \mathbf{F}_{23} = \mathbf{F}_{12} = 7.86\text{ lb}\angle 30°$; *(c)* slipping.

11.37 $\mathbf{P} = 2.54\text{ lb}\angle 45°$; $F_{12} = F_{32} = F_{23} = F_{43} = F_{34} = 1.8\text{ lb}$; $\mathbf{F}_{14}^{B} = 1.20\text{ lb}\angle 90°$;
$\mathbf{F}_{14}^{C} = 3.00\text{ lb}\angle{-}90°$; tipping.

11.39 *(a)* $S_{r_2} = 184.76$; $P_{cr_2} = 589\,300\text{ lb}$; $N_2 = 3.78$; *(b)* $D_{\min} = 2.06\text{ in}$.

11.41 $d = 1.18\text{ in}$; *(a)* true; *(b)* false; *(c)* true.

11.43 $d = 2.70\text{ in}$.

11.45 *(a)* $P = 5760\text{ lb}$; *(b)* $P_c = 4975\text{ lb}$; *(c)* $N = 3.61$; $mg \leq 14\,000\text{ lb}$.

11.47 *(a)* $(S_r)_D = 98.4$; *(b)* $P_{cr} = 624.5\text{ lb}$; $P_{cr}/A = 19\,876\text{ psi}$; *(c)* $F = 405.6\text{ lb}$;
$P_{cr} = 1\,041\text{ lb}$; $S_r = 92.3$.

11.49 *(a)* $S_{r3} = \frac{13.86\text{ in}}{t}$; *(b)* $(S_r)_{D3} = 141.23$; *(c)* $t = 0.317\text{ in}$.

12.1 $I_O = 0.003\,193\text{ kg}\cdot\text{m}^2$.

12.3 $\mathbf{F}_{23} = 752.44\text{ N}\angle 21.86°$; $\mathbf{F}_{34} = 226.45\text{ N}\angle 65.84°$; $\mathbf{T}_{12} = -21.02\hat{\mathbf{k}}\text{ N}\cdot\text{m}$.

12.5 $\mathbf{F}_{23} = \mathbf{F}_{12} = 6627\text{ N}\angle{-}177.35°$; $\mathbf{F}_{14} = 1932\text{ N}\angle{-}90°$; $\mathbf{F}_{34} = 3\,594\text{ N}\angle 147.48°$;
$\mathbf{T}_{12} = 334.85\hat{\mathbf{k}}\text{ N}\cdot\text{m}$.

12.7 $\mathbf{F}_{23} = \mathbf{F}_{12} = 3172\text{ lb}\angle{-}16.97°$; $\mathbf{F}_{14} = 3443\text{ lb}\angle{-}163.43°$; $\mathbf{F}_{34} = 3337\text{ lb}\angle 8.55°$;
$\mathbf{T}_{12} = -35\,764\hat{\mathbf{k}}\text{ in}\cdot\text{lb}$.

12.9 $\boldsymbol{\omega}_3 = 3.020\hat{\mathbf{k}}\text{ rad/s}$; $\boldsymbol{\omega}_4 = 3.607\hat{\mathbf{k}}\text{ rad/s}$; $\boldsymbol{\alpha}_3 = -0.390\hat{\mathbf{k}}\text{ rad/s}^2$; $\boldsymbol{\alpha}_4 = -40.816\hat{\mathbf{k}}\text{ rad/s}^2$;
$\mathbf{A}_{G_3} = 1\,751\text{ in/s}^2\angle{-}16.28°$; $\mathbf{A}_{G_4} = 771.05\text{ in/s}^2\angle{-}3.50°$; $\mathbf{F}_{23} = \mathbf{F}_{12} =$
$500\text{ lb}\angle{-}126.71°$; $\mathbf{F}_{14} = 1\,609\text{ lb}\angle{-}84.54°$; $\mathbf{F}_{34} = 955\text{ lb}\angle{-}166.87°$;
$\mathbf{T}_{12} = 5\,362\hat{\mathbf{k}}\text{ in}\cdot\text{lb}$.

12.11 $\mathbf{F}_{23} = \mathbf{F}_{12} = 1\,222\text{ lb}\angle 55.67°$; $\mathbf{F}_{14} = 2179\text{ lb}\angle{-}90.26°$; $\mathbf{F}_{34} = 934\text{ lb}\angle 157.34°$;
$\mathbf{T}_{12} = 8\,268\hat{\mathbf{k}}\text{ in}\cdot\text{lb}$.

12.13 $\boldsymbol{\omega}_3 = 10.554\hat{\mathbf{k}}\text{ rad/s}$; $\boldsymbol{\omega}_4 = -4.314\hat{\mathbf{k}}\text{ rad/s}$; $\boldsymbol{\alpha}_3 = -199.62\hat{\mathbf{k}}\text{ rad/s}^2$;
$\boldsymbol{\alpha}_4 = -352.36\hat{\mathbf{k}}\text{ rad/s}^2$; $\mathbf{A}_{G_3} = 6\,206.2\text{ in/s}^2\angle{-}86.28°$; $\mathbf{A}_{G_4} = 1\,764.2\text{ in/s}^2\angle 43.91°$;
$\mathbf{F}_{23} = \mathbf{F}_{12} = 169\text{ lb}\angle{-}105.82°$; $\mathbf{F}_{34} = 103\text{ lb}\angle{-}39.28°$; $\mathbf{F}_{14} = 103\text{ lb}\angle 132.31°$;
$\mathbf{T}_{12} = -82\hat{\mathbf{k}}\text{ in}\cdot\text{lb}$.

12.15 $\boldsymbol{\omega}_3 = -2.567\hat{\mathbf{k}}$ rad/s; $\mathbf{V}_B = 47.04\hat{\mathbf{i}}$ in/s; $\boldsymbol{\alpha}_3 = 77.23\hat{\mathbf{k}}$ rad/s^2; $\mathbf{A}_B = 1016.2\hat{\mathbf{i}}$ in/s^2;
$\mathbf{A}_{G_3} = 669.6$ in/s$^2\angle{-28.39°}$; $\mathbf{F}_{23} = \mathbf{F}_{12} = 733.3$ lb$\angle{-13.04°}$; $\mathbf{F}_{34} = 488.8$ lb$\angle{-18.13°}$;
$\mathbf{F}_{14} = 152.1$ lb$\angle{90.00°}$; $\mathbf{T}_{12} = -2805\hat{\mathbf{k}}$ in $\cdot$ lb.

12.17 $\mathbf{F}_{23} = 37.0$ lb$\angle{-25.23}$; $\mathbf{F}_{14} = 2.25$ lb$\angle{-90.00°}$.

12.19 $\boldsymbol{\omega}_3 = 0.610\hat{\mathbf{k}}$ rad/s; $\mathbf{V}_B = -46.48\hat{\mathbf{i}}$ in/s; $\boldsymbol{\alpha}_3 = 6.27\hat{\mathbf{k}}$ rad/s^2; $\mathbf{A}_B = 218.0\hat{\mathbf{i}}$ in/s^2;
$\mathbf{F}_{12} = \mathbf{F}_{23} = 6502.1$ lb$\angle{-67.64°}$; $\mathbf{F}_{14} = 21498.3$ lb$\angle{90.00°}$; $\mathbf{T}_{12} = 8641.4\hat{\mathbf{k}}$ in $\cdot$ lb.

12.21 $\mathbf{F}_{12} = \mathbf{F}_{23} = 91.9$ N$\angle{-120.00°}$; $\mathbf{F}_{14} = \mathbf{F}_{43} = 202.0$ N$\angle{72.75°}$; $\mathbf{T}_{14} = 5.36\hat{\mathbf{k}}$ N $\cdot$ m.

12.25 $\mathbf{T}_{12} = -5.72\hat{\mathbf{k}}$ in $\cdot$ lb.

12.27 $\boldsymbol{\alpha}_2 = -298.8\hat{\mathbf{k}}$ rad/s^2; $\mathbf{F}_{12} = 3143$ N$\angle{16.70°}$.

12.29 (a) $R'_S = 1$ in/in; $R'_C = 1.732$ in/in; (b) $m_{EQ} = 0.061$ lb$\cdot$s^2/in; (c) $P = -1139$ lb.

12.31 (c) $\mathbf{F}_{12} = -18.5\hat{\mathbf{i}} + 142.9\hat{\mathbf{j}}$ lb; $\mathbf{F}_{23} = -12.3\hat{\mathbf{i}} + 119\hat{\mathbf{j}}$ lb; $\mathbf{F}_{34} = 82.2\hat{\mathbf{j}}$ lb;
$\mathbf{F}_{14} = \pm82.2\hat{\mathbf{i}}$ lb (ccw couple). (d) $\mathbf{T}_2 = 2086.5\hat{\mathbf{k}}$ in $\cdot$ lb. (e) Contact at top right and bottom left.

12.33 (c) $\mathbf{F}_{12} = 21.98\hat{\mathbf{j}}$ lb; $\mathbf{F}_{23} = 15.38\hat{\mathbf{i}} + 15.65\hat{\mathbf{j}}$ lb; $\mathbf{F}_{13} = 7.97$ lb$\angle{-120°}$.
(d) $\mathbf{P} = 22.22\hat{\mathbf{i}}$ lb. (e) Contact at 6 in to right of G_3.

12.35 (a) $R'_S = 1$ in/in; $R'_C = -1$ in/in; (b) $x'_{G_3} = 1$ in/in; $y'_{G_3} = 0$, $x''_{G_3} = -0.0375$ in/in^2;
$y''_{G_3} = -0.06495$ in/in^2; (c) $m_{EQ} = 0.01153$ lb$\cdot$s^2/in; (d) $\mathbf{P} = 722.4\hat{\mathbf{i}}$ lb.

12.37 $w_B = w_F = 1.333$ lb.

12.39 $w_B = w_E = 0.0667$ kg; $\theta_B = \theta_E = 180°$.

12.41 $I_G = 9.691$ in $\cdot$ lb$\cdot$s^2; $k_G = 9.552$ in.

12.45 $\omega_3 = (1 - 3R/4r)\omega_2$; $\omega_3 = 0$ for $r = 3R/4$.

12.47 $\alpha_{in} = 2.23(10)^9$ rad/s^2; $F_{2A} = 1397.65$ N; $F_{11} = 2261.75$ N; unbalanced moment 904.7 N$\cdot$m during startup.

12.49 $\mathbf{F}_{12} = -2.69\hat{\mathbf{j}}$ N; $\mathbf{T}_{12} = -0.925\hat{\mathbf{j}}$ N $\cdot$ m; $\mathbf{F}_{35} = -7.35\hat{\mathbf{i}} - 9.25\hat{\mathbf{k}}$ N; $\mathbf{M}_{35} = 2.122\hat{\mathbf{k}}$ N $\cdot$ m;
$\mathbf{F}_{45} = 7.35\hat{\mathbf{i}} + 9.25\hat{\mathbf{k}}$ N; $\mathbf{M}_{45} = -2.122\hat{\mathbf{k}}$ N $\cdot$ m; $\mathbf{F}_{15} = \mathbf{0}$; $\mathbf{M}_{15} = 1.85\hat{\mathbf{j}}$ N $\cdot$ m.

12.51 $T_{12} = 22.10$ in $\cdot$ lb; $\mu \geq 0.113$.

12.53 $\mathbf{F}_{35} = 112.87$ lb$\angle{131.2°}$; $\mathbf{F}_{25} = 152.79$ lb$\angle{-55.96°}$.

12.55 (a) $R'_S = 0$; $\theta'_S = 0.020$ rad/in; $R'_C = 1.732$ in/in; (b) $T = 1562.4$ in $\cdot$ lb;
(c) $\mathbf{P} = -97.89\hat{\mathbf{j}}$ lb.

12.57 $x'_{G_2} = 0.866$ in/in; $y'_{G_2} = 0.500$ in/in; $x''_{G_2} = y''_{G_2} = 0$; $R'_S = 1.000$ in/in;
$R'_C = -1.000$ in/in; $\ddot{R}_2 = 1179.8$ in/s^2.

12.59 (a) $R'_S = 2.000$ in/rad; $\theta'_S = 0$; $R'_C = -1.50$ in/rad; $\theta'_C = -0.50$ rad/rad;
(b) $T = 634.18$ in $\cdot$ lb; (c) $\mathbf{F} = -111.57\hat{\mathbf{i}}$ lb.

13.1 (a) $m\ddot{x} + kx = ky$; $\omega_n = \sqrt{k/m}$; (b) $m\ddot{x} + c\dot{x} + kx = F$; $\omega_n = \sqrt{k/m}$;
(c) $m\ddot{x} + c\dot{x} + (k_1 + k_2)x = 0$; $\omega_n = \sqrt{(k_1 + k_2)/m}$; (d) $m\ddot{x} + (k_1 + k_2)x = k_2 y$;
$\omega_n = \sqrt{(k_1 + k_2)/m}$; (e) $m\ddot{x} + c\dot{x} + kx = c\dot{y}$; $\omega_n = \sqrt{k/m}$;

(f) $m\ddot{x} + \left[k_1 + k_2 + \frac{k_3 k_4}{k_3 + k_4}\right]x = 0$; $\omega_n = \sqrt{\dfrac{k_1 + k_2 + \frac{k_3 k_4}{k_3 + k_4}}{m}}$.

13.3 (a) $k = 36.00$ lb/in; (b) $\delta_{st} = 0.06111$ in; (c) $\tau = 0.250$ s/rev; (d) $f = 4$ Hz;
(e) $\dot{x} = 17.915$ in/s; $\ddot{x} = 229.412$ in/s^2; $F = -13.075$ lb.

13.5 $\omega_n = 3.010$ Hz.

13.7   (a) $\omega_n = \sqrt{\dfrac{g}{\ell}\left[\dfrac{(k_1+k_2)a^2}{W\ell} - 1\right]}$; (b) $\dfrac{\ell}{a} \geq (k_1 + k_2)\dfrac{a}{W}$.

13.9   (a) First era: $0 \leq t \leq 0.040$ s, $x = (0.056\,25\text{ in})(1 - \cos 171t)$; Second era: $t > 0.040$ s, $x = (0.0309\text{ in})\cos(171t - 74.04°)$;
(b) first era: $X = 0.056\,25$ in; second era: $X = 0.003\,09$ in.

13.11   (a) $\omega = 47.75$ rev/min; (b) $\omega_n = 6.458$ rad/s; (c) $\gamma_1 = 11.459°$; $\gamma_2 = 3.180°$.

13.13   5. 36 m.

13.15 $c_c = 16.95$ lb·s/in; $\omega_d = 36.420$ rad/s; $\tau = 0.177$ s/cycle; $\delta = 1.283$.

13.17   (a) $\omega_d = 35.544$ rad/s; (b) $\tau = 0.177$ s/cycle; $\delta = 1.283$.

13.21 $X = 0.097$ in; $\phi = 56.52°$.

13.23 $k = 13\,663$ lb/in; $X = 0.006$ in.

13.25   (a) Upper limit ($\omega_1 \leq 112$ rad/s) is Rayleigh–Ritz, lower limit ($\omega_1 \geq 100$ rad/s) is Dunkerley. Exact is $\omega_1 = 110$ rad/s. (b) $m = 0.002\,81$ lb·s^2/in; (c) $a_{12} = 0.010\,17$ in/lb.

13.27   (a) $\omega_1 = 665.6$ rad/s; (b) $\omega_1 = 667.3$ rad/s.

13.29 $\omega_1 = 500.02$ rad/s; $\omega_2 = 6\,138.20$ rad/s; (a) $\omega_1 = 509.97$ rad/s; (b) $\omega_1 = 500.02$ rad/s.

13.31 $a_{22} = 1.25 \times 10^{-4}$ in/lb.

13.33   (a) $m = 63.1194$ kg; (b) $\omega_2 = 288.2$ rad/s.

13.35 $\omega_1 = 146.6$ rad/s.

13.37   (a) $\omega_1 = 460.13$ rad/s; (b) $\omega_1 = 475.90$ rad/s.

13.39 $\omega_1 = 10.12$ rad/s; $\omega_2 = 126.45$ rad/s; (a) $\omega_1 = 784.08$ rad/s; (b) $\omega_1 = 10.085$ rad/s.

13.41   (a) $\omega_1 = 152$ rad/s; (b) $m = 0.000\,417$ lb·s^2/in; (c) $a_{12} = 0.021\,65$ in/lb.

14.1   At $X = 30\%$;, $p_e = 1\,512$ kPa;, $p_c = 338$ kPa..

14.3   See Fig. AP14.3.

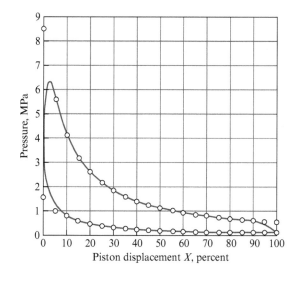

Fig. AP14.3

14.5 $\mathbf{F}_{41} = 1010\,\text{N}\angle-90°$; $\mathbf{F}_{34} = 4254\,\text{N}\angle-13.7°$; $\mathbf{F}_{32} = 4196\,\text{N}\angle163.7°$;
$\mathbf{T}_{21} = 89.59\hat{\mathbf{k}}\,\text{N}\cdot\text{m}$.

14.9 $\mathbf{F}_{41} = 575\,\text{N}\angle-90°$; $\mathbf{F}_{34} = 2999\,\text{N}\angle191.0°$; $\mathbf{F}_{32} = 9918\,\text{N}\angle-39.0°$;
$\mathbf{T}_{21} = 183\hat{\mathbf{k}}\,\text{N}\cdot\text{m}$.

14.11   At $\omega t = 60°$; $X = 40.9\%$; $P = 4\,445\,\text{N}$; $\ddot{x} = -1\,335\,\text{m/s}^2$; $F_{14} = 561\,\text{N}$;
$T_{21} = 110.6\,\text{N}\cdot\text{m}$.

15.1 $\mathbf{F}_A = 14.741\,\text{lb}\angle76.11°$; $\mathbf{F}_B = 3.685\,\text{lb}\angle76.11°$; $w_C = 3.60\,\text{lb}$; $\theta_C = 76.11°$.

15.3 $\mathbf{F}_A = 35.023\,\text{N}\angle-14.42°$; $\mathbf{F}_B = 11.678\,\text{N}\angle165.59°$; $m_C = -1.183\,\text{kg}$;
at $\theta_C = 165.58°$.

15.5 $\mathbf{F}_A = 3.000\,\text{lb}\angle90.0°$; $\mathbf{F}_B = \mathbf{0}$.

15.7 $m_L\mathbf{R}_L = 4.957\,\text{g·m}\angle-13.88°$; $m_R\mathbf{R}_R = 0.621\,\text{g·m}\angle94.48°$.

15.9   Remove $w_L\mathbf{R}_L = 1.108\,\text{oz·in}\angle179.63°$; and $w_R\mathbf{R}_R = 0.287\,\text{oz·in}\angle-58.56°$.

15.11   See Section 15.8 for answers.

15.13 $(w_1)_{\text{new}} = 23.74\,\text{lb}$; $\theta_1 = 27.43°$; $(w_3)_{\text{new}} = 2.44\,\text{lb}$; $\theta_3 = -59.98°$.

15.15 $w_{C1} = -22.58\,\text{lb}$; $\theta_{C1} = 74.96°$; $w_{C2} = -16.50\,\text{lb}$; $\theta_{C2} = -120°$.

15.17 $\mathbf{F}_A = 1.413\,\text{lb}\angle-141.21°$; $\mathbf{F}_B = 0.861\,\text{lb}\angle108.35°$; $w_{C1} = 2.229\,\text{lb}$; $w_{C2} = 1.235\,\text{lb}$.

15.19 $F_A = 467.10\,\text{N}\angle-30°$; $F_B = 119.40\,\text{N}\angle-141°$.

15.21 $\mathbf{F}_{21A} = 13.63\,\text{lb}\angle-153°$; $\mathbf{F}_{21B} = 7.50\,\text{lb}\angle85°$.

15.23   (a) $F_P^x = 8\,108.1\cos\theta + 10\,086.7\sin\theta\,\text{lb}$; $F_P^y = 724.2\cos\theta - 2\,702.7\sin\theta\,\text{lb}$;
(b) $w_{C1} = 3.30\,\text{lb}$; $\gamma_1 = 120°$; $w_{C2} = 4.67\,\text{lb}$; $\gamma_2 = 135°$; (c) $F_P^x = 12\,065.2\,\text{lb}$;
$F_P^y = -724.2\,\text{lb}$.

15.25   (a) $F_1 = 0$; $F_2 = 230.83\,\text{lb}$; (b) $w_{C1} = 0$; $\gamma_1 = \text{undefined}$; $w_{C2} = 5.5\,\text{lb}$; $\gamma_2 = 90°$.

15.27   (a) $S_P^x = -0.116P\cos\theta - 0.790P\sin\theta$; $S_P^y = -1.299P\cos\theta + 2.250P\sin\theta$;
(b) $F_1 = 3\,144\,\text{lb}$; $F_2 = 4\,522\,\text{lb}$.

15.29   (a) $S_P^x = 42\,019.4\cos\theta - 26\,887.3\sin\theta\,\text{lb}$; $S_P^y = 14\,405.4\cos\theta + 11\,797.9\sin\theta\,\text{lb}$;
(b) $w_{C1} = 104.7\,\text{lb}$; $\gamma_1 = 217.5°$; $w_{C2} = 50.5\,\text{lb}$; $\gamma_2 = 202.4°$; (c) $S_P = 29\,396\,\text{lb}$;
$\tau = 38.69°$.

16.1 $T_m = 70.88\,\text{N}\cdot\text{m}$; $I = 0.154\,\text{kg}\cdot\text{m}^2$.

16.3 $T_m = 277.7\,\text{in}\cdot\text{lb}$; $I = 1.842\,\text{in}\cdot\text{lb·s}^2$.

16.5 $T_m = 1\,111\,\text{in}\cdot\text{lb}$.

16.6 $F_c = 174\,700\,\text{N}$; $\omega_p \geq 30.85\,\text{rev/min}$.

16.8 $\mathbf{M}^z = 0.282\hat{\mathbf{k}}\,\text{N}\cdot\text{m}$.

16.10 $\sum\mathbf{M} = -0.537\hat{\mathbf{k}}\,\text{N}\cdot\text{m}$ (negligble).

Index